HANDBOOK OF RECYCLING

HANDBOOK OF RECYCLING

State-of-the-art for Practitioners, Analysts, and Scientists

SECOND EDITION

Edited by

CHRISTINA MESKERS
SINTEF Manufacturing, Trondheim, Norway;
Department of Energy and Process Engineering, NTNU—Norwegian University of Science and Technology, Trondheim, Norway

ERNST WORRELL
Utrecht University, Utrecht, The Netherlands

MARKUS A. REUTER
WASM: Minerals, Energy and Chemical Engineering, Faculty of Science and Engineering, Curtin University, Perth, Australia;
SMS Group GmbH, Düsseldorf, Germany

ELSEVIER

Elsevier
Radarweg 29, PO Box 211, 1000 AE Amsterdam, Netherlands
The Boulevard, Langford Lane, Kidlington, Oxford OX5 1GB, United Kingdom
50 Hampshire Street, 5th Floor, Cambridge, MA 02139, United States

ISBN: 978-0-323-85514-3

For information on all Elsevier publications
visit our website at https://www.elsevier.com/books-and-journals

Publisher: Matthew Deans
Acquisitions Editor: Stephen Jones
Editorial Project Manager: John Leonard
Production Project Manager: Prem Kumar Kaliamoorthi
Cover Designer: Greg Harris

Typeset by STRAIVE, India

Contents

6. Developments in collection of municipal waste

GIJS LANGEVELD AND MAARTEN GOORHUIS

7. The path to inclusive recycling: Developing countries and the informal sector

SONIA VALDIVIA AND MATHIAS SCHLUEP

2

Recycling from a product perspective

8. Physical separation

KARI HEISKANEN

9. Sensor-based sorting

XIAOZHENG CHEN, NILS KROELL, ALEXANDER FEIL, AND KATHRIN GREIFF

10. Mixed bulky waste

SANDOR KARREMAN

11. Packaging

ERNST WORRELL

12. End-of-life vehicles

PIETER KUIPER AND SUSANNE VAN BERKUM

13. Electrical and electronic equipment (WEEE)

MATHIAS SCHLUEP AND FABIAN OTTIGER

14. Photovoltaic and wind energy equipment

OLGA IGNATENKO, NEILL J. BARTIE, CHRISTINA MESKERS, AND MARKUS A. REUTER

15. Buildings

DAVID PECK

16. Construction and demolition waste

VIVIAN W.Y. TAM, MAHFOOZ SOOMRO, AND ANA CATARINA JORGE EVANGELISTA

17. Industrial by-products

JAANA SORVARI AND MARGARETA WAHLSTRÖM

18. Mine tailings

PRISCILLA P. NELSON AND D. ERIK SPILLER

3

Recycling from a material perspective

19. Steel

THILO KRESCHEL AND OLENA VOLKOVA

20. Aluminum

DON DOUTRE AND ANNE KVITHYLD

21. Copper

MARK E. SCHLESINGER

22. Lead

S. ARNOUT, B. BLANPAIN, D.R. SWINBOURNE, M. CHINTINNE, AND M. STEVENSON

23. Zinc

J. ANTREKOWITSCH, S. STEINLECHNER, G. HANKE, AND U. BRANDNER

24. Ferroalloy elements

STEFAN LUIDOLD AND EVA GEROLD

25. Precious and technology metals

CHRISTINA MESKERS AND MARKUS A. REUTER

26. Concrete and aggregates

VIVIAN W.Y. TAM, MAHFOOZ SOOMRO, AND ANA CATARINA JORGE EVANGELISTA

27. Cementitious binders incorporating residues

Y. PONTIKES AND R. SNELLINGS

28. Glass

THOMAS D. DYER

29. Lumber

AMBROSE DODOO, LEIF GUSTAVSSON, AND ROGER SATHRE

30. Paper☆

HARALD GROSSMANN, TONI HANDKE, AND TOBIAS BRENNER

31. Plastic recycling

LI SHEN AND ERNST WORRELL

32. Black rubber products

H.C. VAN ZYL AND MARKUS A. REUTER

33. Textiles

JANA M. HAWLEY

34. Carbon fibers

MAREEN ZÖLLNER, DOROTHEA HAMANN, THOMAS KRAMPITZ, SOPHIA THÜM, AND HOLGER LIEBERWIRTH

4

Recycling and the circular economy

35. From waste management to circular economy—The role of recycling in policy

ERNST WORRELL

☆ Revised by Ernst Worrell, Copernicus Institute of Sustainable Development, Utrecht University, the Netherlands.

36. Geopolitics of resources and recycling

BENJAMIN SPRECHER AND ERNST WORRELL

37. Information and communication

ERNST WORRELL

38. Recycling, law, and the circular economy

CHRIS BACKES

39. Extended producer responsibility

W.J.V. VERMEULEN AND K. CAMPBELL-JOHNSTON

40. Exploring the economics of recycling in a dynamic global context

PIETER VAN BEUKERING, ONNO KUIK, FRANS OOSTERHUIS, AND JAN BRUSSELAERS

41. Economic policy instruments

MAARTEN DUBOIS AND JOHAN EYCKMANS

42. Economic aspects of metal recycling

CHRISTIAN HAGELÜKEN AND CHRISTINA MESKERS

5

Recycling fundamentals

43. Physical separation fundamentals

KARI HEISKANEN

44. Thermodynamics

PATRICK WOLLANTS AND B. BLANPAIN

45. Exergy—Quantification of resource dissipation

ALEJANDRO ABADÍAS LLAMAS, NEILL J. BARTIE, CHRISTINA MESKERS, AND MARKUS A. REUTER

46. Process simulation—Thermodynamics and process technology to understand recycling systems

NEILL J. BARTIE, ALEJANDRO ABADÍAS LLAMAS, CHRISTINA MESKERS, AND MARKUS A. REUTER

47. Life cycle assessment (LCA)

CHRISTINA MESKERS, NEILL J. BARTIE, AND MARKUS A. REUTER

Contributors

Julian M. Allwood University of Cambridge, Cambridge, United Kingdom

J. Antrekowitsch University of Leoben, Leoben, Austria

S. Arnout InsPyro, Leuven, Belgium

Chris Backes Utrecht Centre for Water, Oceans and Sustainability Law (UCWOSL), Utrecht University, Utrecht, The Netherlands

Neill J. Bartie Helmholtz Centre Berlin for Materials and Energy (HZB), PVcomB, Berlin, Germany

Romain G. Billy Industrial Ecology Programme, Department of Energy and Process Engineering, Norwegian University of Science and Technology, Trondheim, Norway

B. Blanpain Department of Materials Engineering, KU Leuven, Leuven, Belgium

U. Brandner University of Leoben, Leoben, Austria

Tobias Brenner Papiertechnische Stiftung (PTS), Heidenau, Germany

Jan Brusselaers Institute for Environmental Studies (IVM), Vrije Universiteit Amsterdam, Amsterdam, The Netherlands

K. Campbell-Johnston Department of ICT, Strategy and Policy, TNO, The Hague, The Netherlands

Xiaozheng Chen Department of Anthropogenic Material Cycles, RWTH Aachen University, Aachen, Germany

M. Chintinne Aurubis Beerse, Beerse, Belgium

Jonathan Cullen Department of Engineering, Cambridge University, Cambridge, United Kingdom

Ambrose Dodoo Department of Building Technology, Linnaeus University, Växjö, Sweden

Don Doutre Novelis Inc., Kingston, Canada

Maarten Dubois Deloitte Consulting Belgium, Sustainability & Circular Economy, Brussels, Belgium

Thomas D. Dyer School of Science and Engineering, University of Dundee, Dundee, United Kingdom

D. Erik Spiller Colorado School of Mines, Golden, CO, United States

Ana Catarina Jorge Evangelista Engineering Institute of Technology, Perth, WA, Australia

Johan Eyckmans Center for Economics and Corporate Sustainability (CEDON), KU Leuven Brussels, Belgium

Alexander Feil Department of Anthropogenic Material Cycles, RWTH Aachen University, Aachen, Germany

Eva Gerold Department of Metallurgy, University of Leoben, Leoben, Austria

Maarten Goorhuis De Vuilnisman, Arnhem, The Netherlands

Kathrin Greiff Department of Anthropogenic Material Cycles, RWTH Aachen University, Aachen, Germany

Harald Grossmann Technische Universität Dresden, Dresden, Germany

Leif Gustavsson LG, Silogatan 1, Lund, Sweden

Christian Hagelüken Director Government Affairs (Retired), Umicore, Hanau, Germany

Dorothea Hamann Institute for Mineral Processing Machines and Recycling Systems Technology, Technical University Bergakademie Freiberg, Freiberg, Sachsen, Germany

Toni Handke Technische Universität Dresden, Dresden, Germany

G. Hanke University of Leoben, Leoben, Austria

Jana M. Hawley College of Merchandising, Hospitality and Tourism, University of North Texas, Denton, TX, United States

Kari Heiskanen Metso, Espoo, Finland

Olga Ignatenko Independent Professional, Kyiv, Ukraine

Sandor Karreman Renewi, Nieuwegein, The Netherlands

Thomas Krampitz Institute for Mineral Processing Machines and Recycling Systems Technology, Technical University Bergakademie Freiberg, Freiberg, Sachsen, Germany

Thilo Kreschel Institute of Iron and Steel Technology, Technical University Bergakademie Freiberg, Freiberg, Germany

Nils Kroell Department of Anthropogenic Material Cycles, RWTH Aachen University, Aachen, Germany

Onno Kuik Institute for Environmental Studies (IVM), Vrije Universiteit Amsterdam, Amsterdam, The Netherlands

Pieter Kuiper Auto Recycling Nederland, Breukelen, The Netherlands

Anne Kvithyld SINTEF, Trondheim, Norway

Gijs Langeveld Project Gijs B.V., Amersfoort, The Netherlands

Holger Lieberwirth Institute for Mineral Processing Machines and Recycling Systems Technology, Technical University Bergakademie Freiberg, Freiberg, Sachsen, Germany

Gang Liu College of Urban and Environmental Sciences, Peking University, Beijing, China

Alejandro Abadías Llamas Nordenham Metall GmbH, Nordenham, Germany

Stefan Luidold Department of Metallurgy, University of Leoben, Leoben, Austria

Christina Meskers Department of Industrial Ecosystems, SINTEF Manufacturing; Department of Energy and Process Engineering, Faculty of Engineering Sciences, NTNU—Norwegian University of Science and Technology, Trondheim, Norway

Daniel B. Müller Industrial Ecology Programme, Department of Energy and Process Engineering, Norwegian University of Science and Technology, Trondheim, Norway

Priscilla P. Nelson Colorado School of Mines, Golden, CO, United States

Frans Oosterhuis Institute for Environmental Studies (IVM), Vrije Universiteit Amsterdam, Amsterdam, The Netherlands

Fabian Ottiger World Resources Forum Association, St. Gallen, Switzerland

David Peck Circular Built Environment and Critical Materials, Faculty of Architecture and Built Environment, Delft University of Technology, Delft, The Netherlands

Evi Petavratzi Decarbonisation and Resource Management, British Geological Survey, Keyworth, United Kingdom

Y. Pontikes KU Leuven, Department of Materials Engineering, Leuven, Belgium

Helmut Rechberger Institute for Water Quality and Resource Management, Vienna University of Technology, Vienna, Austria

Markus A. Reuter WASM: Minerals, Energy and Chemical Engineering, Faculty of Science and Engineering, Curtin University, Perth, Australia

Roger Sathre Sustainable Built Environment Group, Department of Built Environment and Energy Technology, Linnaeus University, Växjö, Sweden

Mark E. Schlesinger Missouri University of Science and Technology, Rolla, MO, United States

Mathias Schluep World Resources Forum Association, St. Gallen, Switzerland

Li Shen Copernicus Institute of Sustainable Development, Utrecht University, Utrecht, The Netherlands

Mark U. Simoni Industrial Ecology Programme, Department of Energy and Process Engineering, Norwegian University of Science and Technology; Section Mineral Resources, Geological Survey of Norway, Trondheim, Norway

R. Snellings KU Leuven, Department of Materials Engineering; KU Leuven, Department of Earth and Environmental Sciences, Leuven, Belgium

Mahfooz Soomro School of Engineering, Design and Built Environment, Western Sydney University, Penrith, NSW, Australia

Jaana Sorvari Finnish Environment Institute, Helsinki, Finland

Benjamin Sprecher Delft University of Technology, Delft, The Netherlands

S. Steinlechner University of Leoben, Leoben, Austria

M. Stevenson Global Lead Technologies, Melbourne, VIC, Australia

D.R. Swinbourne School of Civil, Environmental and Chemical Engineering, RMIT University, Melbourne, VIC, Australia

Vivian W.Y. Tam School of Engineering, Design and Built Environment, Western Sydney University, Penrith, NSW, Australia; College of Civil Engineering, Shenzhen University, Shenzhen, China

Sophia Thüm Institute for Mineral Processing Machines and Recycling Systems Technology, Technical University Bergakademie Freiberg, Freiberg, Sachsen, Germany

Sonia Valdivia World Resources Forum Association, St. Gallen, Switzerland

Susanne van Berkum Auto Recycling Nederland, Breukelen, The Netherlands

Pieter van Beukering Institute for Environmental Studies (IVM), Vrije Universiteit Amsterdam, Amsterdam, The Netherlands

Antoinette van Schaik MARAS—Material Recycling and Sustainability, The Hague, The Netherlands

H.C. van Zyl Curtin University, Perth, Australia

W.J.V. Vermeulen Copernicus Institute of Sustainable Development, Utrecht University, Utrecht, The Netherlands

Olena Volkova Institute of Iron and Steel Technology, Technical University Bergakademie Freiberg, Freiberg, Germany

Margareta Wahlström VTT Technical Research Centre of Finland, VTT, Espoo, Finland

Patrick Wollants Department of Materials Engineering, KU Leuven, Leuven, Belgium

Ernst Worrell Copernicus Institute of Sustainable Development, Utrecht University, Utrecht, The Netherlands

Mareen Zöllner Institute for Mineral Processing Machines and Recycling Systems Technology, Technical University Bergakademie Freiberg, Freiberg, Sachsen, Germany

About the editors

Christina Meskers, Ph.D. is senior research scientist at SINTEF, in the Industrial ecosystems department. Prior to this, she was senior advisor at the Industrial Ecology programme of the Norwegian University of Science and Technology (2021-2022), and senior manager open innovation, and market intelligence & business research at Umicore, Belgium. She graduated from Delft University of Technology with degrees in resource engineering and materials science. Christina has a passion for innovation and strategy, connecting people and ideas across disciplines, industries, organisations and value chains. Her work focuses on the contribution of metals and materials industry to the transition to sustainable products and a sustainable society. She has over 15 years of experience in the (raw) materials sector, including co-authoring a United Nations' International Resource Panel report on recycling (2013), and contributing to *Principles of metal refining & recycling* (2021), as well as organizing numerous symposia. She served on the advisory boards of international master programs, innovation networks, the Board of Directors of The Minerals, Metals and Materials Society (2020-2023), and is associate editor of the Journal of Sustainable Metallurgy. Christina was a 2020 Brimacombe Medalist and awarded the TMS Young Leaders Professional Development Award (2008).

Ernst Worrell (PhD) is a professor of Energy, Resources and Technological Change at Utrecht University in the Copernicus Institute, The Netherlands. His research focuses on the technical, economic, and policy aspects of the energy-resource nexus, including energy, material, and resource efficiency improvement, as well as waste management and processing. The research includes the transition to a circular economy working on recycling and material efficiency. He led the industrial energy assessment work at Lawrence Berkeley National Laboratory until 2008 and was Director,

Energy Use and Efficiency at the sustainable energy consulting company Ecofys between 2004 and 2010. He is (co-) author of over 300 publications, of which over 200 are peer-reviewed articles. For 12 years, until 2013, he was Editor-in-Chief of the peer-reviewed journal *Resources, Conservation and Recycling*, and he also serves as Associate Editor of *Energy, the International Journal* and of *Energy Efficiency*.

Markus A. Reuter is a chief expert with SMS Group in Düsseldorf, Germany. Prior to holding this position, Reuter was director at Helmholtz Institute Freiberg for Resource Technology; chief technologist, Ausmelt Australia; and director of technology management, Outotec (now Metso) Australia and Finland. He also worked at Mintek and Anglo American Corporation, both in South Africa. In addition, he has served as a full professor at TU Delft in the Netherlands and the University of Melbourne in Australia, and has held honorary and adjunct professorships at Aalto University in Finland, at Central South University in China, and presently ongoing positions at TU Bergakademie Freiberg in Germany, as well as Curtin University Perth in Australia. He holds honorary doctorates from the University of Liège in Belgium and the University of Stellenbosch in South Africa; DEng and PhD degrees from Stellenbosch University; and a Dr. habil. From RWTH Aachen in Germany. Recent honors include receiving The Minerals, Metals & Materials Society (TMS) Extraction & Processing Division Distinguished Lecture Award in 2016; delivering the TMS Extraction & Processing Division (EPD) and Light Metals Luncheon Lectures in 2020 and 2022, respectively; receiving three TMS Science Awards in 2022 and 2014; as well as receiving an Outotec Technology award in 2014.

PART 1

Recycling in context

CHAPTER

1

Introduction

Ernst Worrell[1], *Markus A. Reuter*[2], *and Christina Meskers*[3]

[1]Copernicus Institute of Sustainable Development, Utrecht University, Utrecht, The Netherlands [2]WASM: Minerals, Energy and Chemical Engineering, Faculty of Science and Engineering, Curtin University, Perth, Australia [3]Department of Industrial Eco-systems, SINTEF Manufacturing, Trondheim, Norway

1.1 THE CHALLENGES

Materials form the fabric of our society; they are everywhere and life as we know it would be impossible without them. Terms such as the Bronze Age and Iron Age demonstrate that materials have defined our society in the past and will in the future. Materials are the enabler in the transition of our society towards sustainability. They are essential for renewable energy production, clean mobility, cities of the future, and the drive to further digitalization, to name just a few areas. The demand for all materials has been increasing and they are used in increasingly complex products.

The challenge of a sustainable society is rooted in the way that material resources are processed to make materials and products, how these are used, and the fact that these often are discarded at the end of life. Besides using material resources, energy and water are consumed and waste and other emissions generated. This *linear* economy is now running into its limits, given the large demand for materials and resources of an increasing (and increasingly affluent) global population. We are increasingly exceeding planetary boundaries (Persson et al., 2022), largely related to the way that we produce, use, and discard materials. This is reflected in the sustainable development goals (SDGs) of the United Nations, as well as in the global challenges formulated by various organizations (e.g., European Commission, World Economic Forum). Governments around the world are starting to address unsustainable production and consumption. Hence, around the world countries and corporations have set off on a pathway to a so-called *circular* economy, i.e., an economy that builds on resource efficiency, of which recycling is an essential part.

Recycling is the essential, loop-closing step in the circular economy. All materials and products—whether their life has been long or short—will have to be recycled at some point. During recycling, the materials' quality (and quantity) need to be preserved so they can stay as long as possible in the cycle at very high functionality and performance levels. In this way the unsustainable pattern of mine-make-use-discard can be broken. Although this sounds straightforward, the practical reality is much more complicated and complex. Recycling effectiveness is impacted not only by the performance of the recycling processes and

Handbook of Recycling. **https://doi.org/10.1016/B978-0-323-85514-3.00008-7**

technologies themselves, but also by what happens upstream in the value chain:

- choices during the design of products, including materials selection,
- length and wear of product and part life,
- collection at product end of life and the application lifetime extension strategies,
- quantity and composition of residues and wastes generated during any life cycle stage.

In addition, it is affected by the demands made downstream. For materials to find their way into new products, they need to meet the composition and performance specifications (quality requirements) for those products.

Thus realizing the optimized and efficient material recycling systems society needs requires on the one hand a deep understanding of the fundamentals of material recycling systems themselves. On the other hand, it requires a systemic understanding of the recycling systems' role within the broader context. The combination of both is essential to unravel the complex interplay between materials, products, processes in value chains, and the different stakeholders.

This book provides the basis, in terms of the fundamental theory of recycling, recycling technology, material collection and recovery, economics, and policy and regulatory aspects, to understand the state of the art and the challenges of recycling in the larger context. It is intended to provide students, professionals, analysts, and decision makers with a solid background in order to mitigate the challenges.

1.2 THE ROLE OF MATERIALS IN SOCIETY

Industrial society has become extremely dependent on resources, as it produces more, builds an increasingly complex society, and accumulates an incredible volume of resources. Figure 1.1 depicts the global production of the key materials used in society and shows an extremely rapid growth in the past few decades, as emerging economies like China develop. Industrialization started in Europe and North America and these countries developed important material-producing infrastructure, leading global production. Today, China produces over half of all the cement, steel, and other commodities in the world. Figure 1.1 also shows the increasing importance of metals such as nickel, which used to be called "minor metals." Now these are essential ingredients for new technologies such as rechargeable batteries and play a key role in energy storage in the (renewable) energy transition.

Humankind now dominates the global flows of many elements of the periodic table. It has been estimated that the mass of all human-made structures and products now exceeds the mass of all biomass on the planet (Elhacham et al., 2020). The global material *footprint* has increased dramatically from about 8.8 t/y in 2010 to about 12.2 t/y in 2017. It varies from as low as 4.7 t/capita in low-middle income countries to highs of 26.3 t/capita in high-income countries (United Nations Statistics Division, 2023). The global average is growing rapidly, given expected population growth and developing patterns for the majority of the population living in developing countries. Despite a weak decoupling between economic growth [measured in terms of gross domestic product (GDP)] in some countries, i.e., the (apparent) material consumption growth rate is lower than that of GDP, global material consumption is still growing.

However, the increasing global consumption makes it necessary to reevaluate the way that we use resources.

The earth's resources are not infinite, but until recently, they have seemed to be: the demands made on them by manufacturing throughout the industrialization of society appeared infinitesimal, with the rate of new discoveries outpacing the rate of consumption. This is no longer the case. Access to materials has also become an issue of national security of many nations, especially to ensure that emerging new

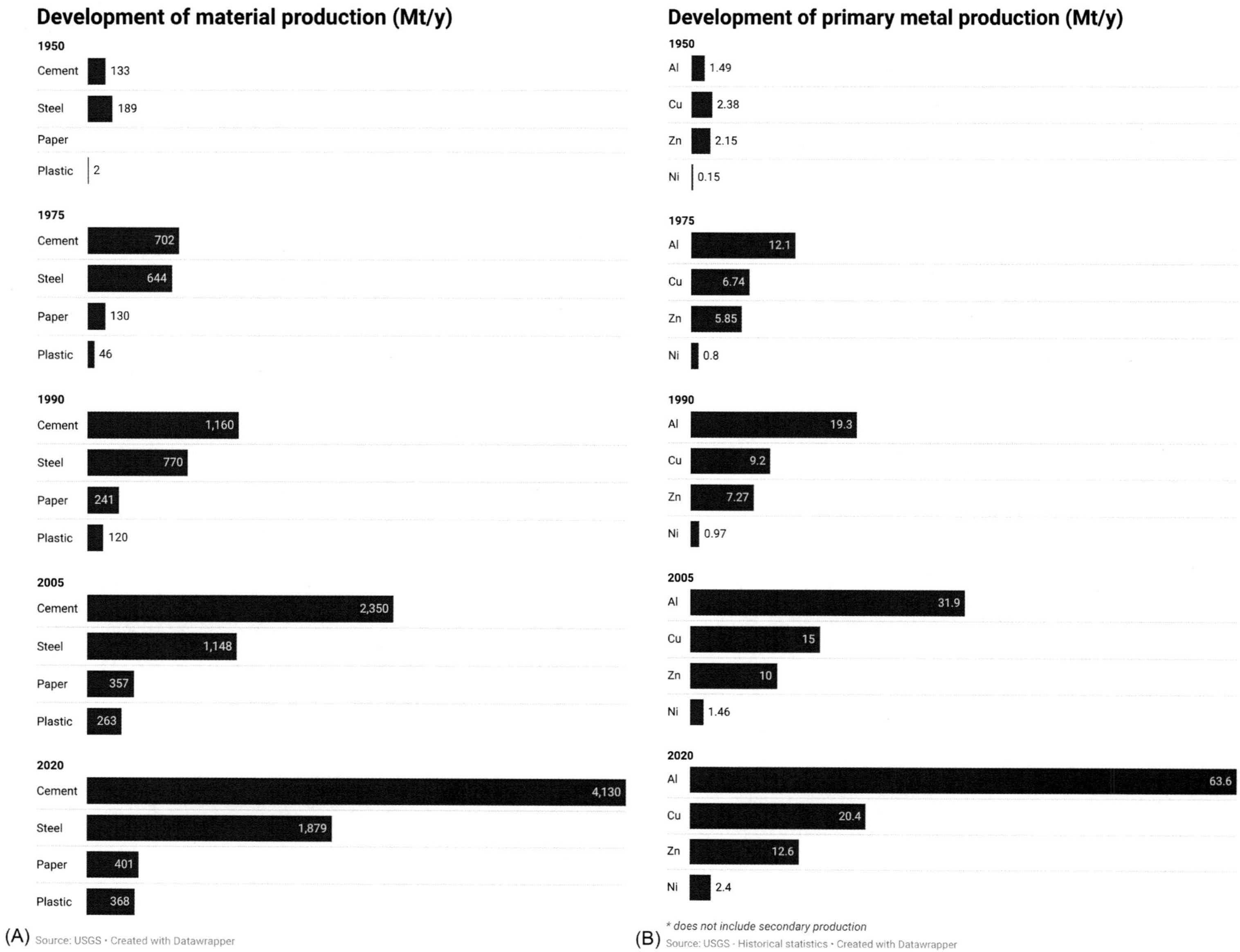

FIGURE 1.1 Development of the production of key materials. (A) Cement, steel, paper, and plastic. (B) Primary production of aluminum, copper, zinc, and nickel. *Data from USGS (2020), Worldsteel (2022), FAO, Geyer et al. (2017).*

"sustainable" technologies can be supplied with metals and materials. This has led many countries to designate specific materials as critical for their economy, while developing strategies to maintain access to these materials. The increased demand for materials has also led to price spikes, even as small fractions of global production may be affected due to political developments, strikes, mine closures, or accidents. Long term, some elements may even become geologically scarce within a few lifetimes (Henckens et al., 2014), opening up a future quest for substitutes and resource efficiency.

To maintain our level of welfare, services by resources should be provided more efficiently, using less (environmental) resources per unit of activity, i.e., improve the resource efficiency of our society.

1.3 FROM LINEAR TO CIRCULAR ECONOMY

1.3.1 The Linear Economy

Historically, industry has operated as an open, linear system, transforming resources to products that are eventually discarded to the environment. This, coupled with the massive increase in the use of resources, has led to growing impacts on the environment. Large consumption of energy, water, and materials, as well as emissions of greenhouse gases, wastewater, particulate matter, and solid wastes, for example, are directly tied to the production and use of the resources, while also affecting the use of land, water, and other environmental resources.

Moreover, our resources use results in tremendous amounts of solid waste. Waste is a large problem, as we are running out of land for landfilling, or if uncontrolled, landfilling leads to negative environmental impacts through leaching. Other end-of-life waste treatment options (e.g., incineration) also have negative environmental and health impacts. This is especially a problem for emerging economies, where material use (and hence discarding of it) is growing very rapidly, while a limited waste management infrastructure exists. It is also a problem, in both industrialized and developing economies, for new products (e.g., electronics, batteries), for which no or insufficiently effective infrastructure currently exists to recover and recycle the products and materials.

1.3.2 The Circular Economy

The transition to a resource-efficient, or circular, economy is essential to reduce the negative environmental impacts. A plethora of definitions has appeared (Kirchherr et al. 2017). Figure 1.2 shows a representation of the materials and product cycles in a circular economy paradigm. Recycling is the "loop-closing" step in the circular economy. Many other representations of the circular economy suggest that the circles are closed. In practice, completely closing all loops is virtually impossible (Chapter 2), which makes the "100% circular economy" a good rallying goal but also an enigma. Figure 1.2 depicts the dissipative losses of materials, and the material and energy resources needed to make up for the losses. The need to minimize the dissipative losses, and to maintain the quality of materials, is essential for a well-functioning circular economy.

Furthermore, the figure shows the connections between the circular economy and the energy system. Embedded in the circular economy paradigm is the use of renewable energy for all energy needs. For each step in the circular economy, energy input is needed. That energy is produced with energy infrastructure made of materials and products that are part of the circular economy (or materials cycle). In addition to energy, water is also needed. Thus the material, energy, and water cycles come together in the circular economy concept. Finally, the

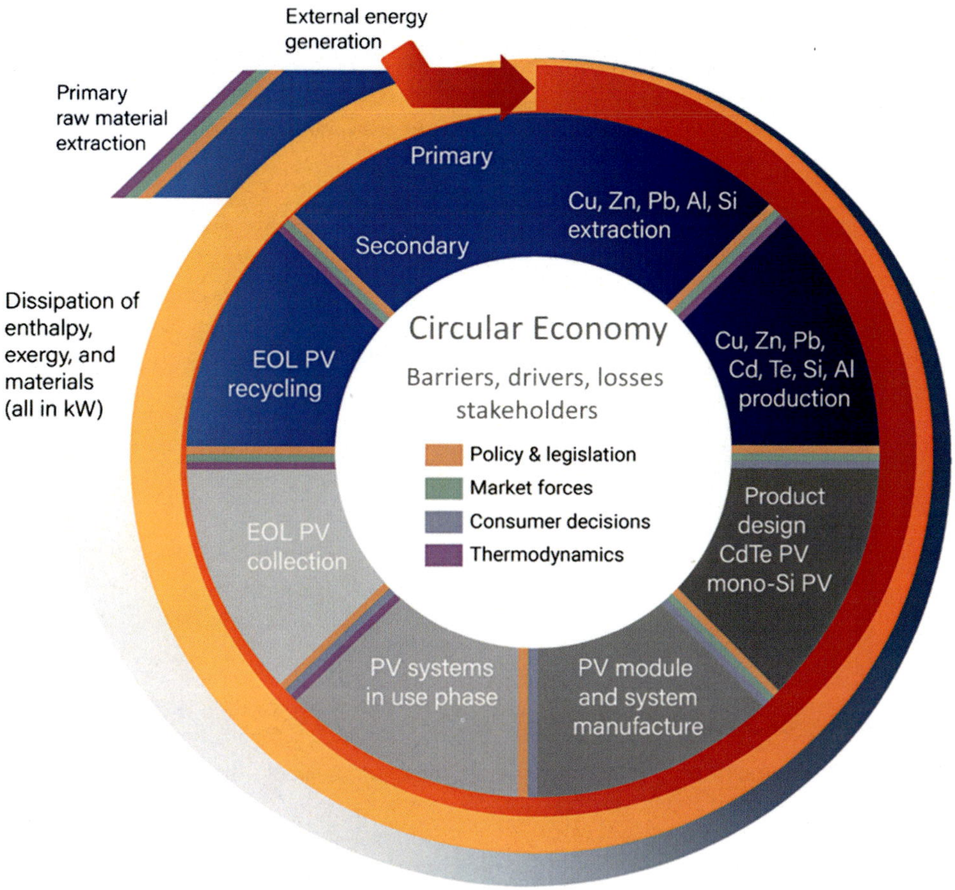

FIGURE 1.2 Material and product life cycle (circular economy) showing its interaction with the energy cycle. The key mediators between stakeholders are indicated. *Adapted from Bartie et al. (2019, 2022).*

figure highlights the key barriers and stakeholders in different parts of the circular economy.

Since the first edition of this book, society has been moving towards a circular economy. Still, there remains a huge potential to close the circles and reduce the volume and speed of the material flows through the cycles.

1.3.3 The R-Strategies

There are several ways that resource efficiency can be improved; in fact, within a circular economy, up to 10 levels (called the R-Ladder or hierarchy) are distinguished (Potting et al., 2017; Reike et al., 2018), which can be summarized as (as shown in Figure 1.3):

- Reduce the need for a (material) service, hence saving the material; by refusing materials and products, by rethinking products and processes, and reducing the materials needed and wasted;
- Use resources more efficiently in the provision of an activity or product;
- Extend the lifespan of products and parts through repair, refurbishing, or reusing and remanufacturing;

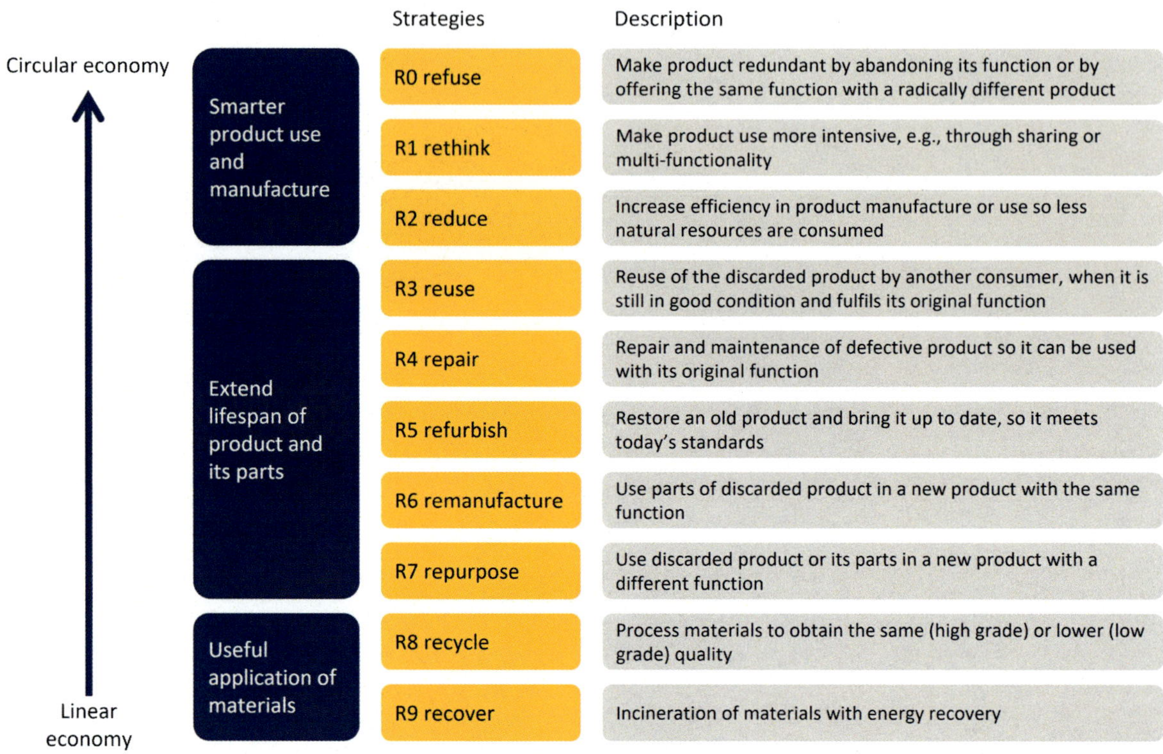

FIGURE 1.3 The 9R framework or R-ladder listing circularity strategies within the production chain, in order of priority. *Adapted from Kirchherr et al. (2017) and Potting et al. (2017).*

- Recycle the resources and materials in products (and parts).

1.4 RECYCLING IN THE CIRCULAR ECONOMY

1.4.1 Defining Recycling

Waste is only waste if it cannot be used again (i.e., loss of functionality or performance) or if its economic value, including waste management costs, is not sufficient to make its exploitation economically viable. Waste can arise at any stage in the lifecycle—during primary resource mining, processing, part and product manufacturing, use, collection, and even during recycling.

Recycling is the reprocessing of recovered materials at the end of a product's life, returning them into the supply chain. As all products (and parts) at one time will become obsolete and cannot be reused or repaired, recycling can be seen as the *foundation of the circular economy*, on which other strategies and opportunities (e.g., product reuse, refurbishing) should be built. Recycling enables waste material to become a resource again. Theoretically, recycling can be done at a rate comparable to the rate with which we discard resources.

In order to realize a circular economy, the system must be carefully designed to minimize (inevitable) losses. Recycling is part of a larger system ("the recycling chain"), starting with collection and recovery, concentration, and recycling (mechanical or chemical). In between the different steps, recovered material may be traded and transported. Each process in the system consumes resources, generates emissions and unavoidable material, and energy losses occur. Generally, the energy needed for recycling (also called secondary materials production) is substantially less compared to the energy needed to produce the material from ores (also called primary or virgin production).

A distinction between "short loop" and "long loop" recycling can be made. Short loop recycling is the recycling of waste materials arising from the production of materials and products. Typical examples are scrap materials from processing metals and manufacturing metal parts, e.g., runners, turnings, off-spec material, and rejects. Also, off-spec products that are not put on the market are recycled. The time needed for the recyclable materials to be recovered and put back in the value chain is very short, as is the recycling chain itself, since the origin and composition of the materials are usually well-known, the material is clean, and there is a strong economic driver for recycling. Long loop recycling refers to the recycling that takes place after parts and products (materials) have completed their life cycle, and is distinctly different from short loop materials. In this book, examples from both types of loops are discussed.

1.4.2 Product and Material-Centric Perspectives

To understand the challenges in recycling systems, one needs to look at the recycling systems from a product perspective as well as a material perspective.

Products, be it a phone, a car, a bottle, or a newspaper, reach end of life and enter the recycling chain. The collection infrastructure is often set up based on product groups, e.g., small electronic equipment, batteries, cars; or product-material combinations, e.g., glass bottles or plastic packaging. The collected material within a product group is very heterogeneous. It is a combination of end-of-life products from different brands, equipment types, models, lifetimes, etc., that together make up the composition of the collected material.

Before the materials from the products can be physically or chemically recycled, the different materials within the products must be separated from each other, so they can enter the appropriate recycling process. This happens during the separation and sorting step. In this step, the perspective moves from a product perspective to a materials perspective. Specific material fractions are obtained, e.g., steel, iron, aluminum, or copper, which may be further sorted based on alloy composition. During final material recovery, especially in chemical processes, deep knowledge about the behavior of the different elements is crucial to remove impurities and obtain material, e.g., alloys, with the required specifications, so it can enter the value chain again.

The fraction of a material that can reenter the life cycle will depend on the material itself, the composition and complexity of the product from which it is being recovered, and the performance of the collection, separation, and sorting, and final material recovery technologies/processes for that specific material, as (still) the quality of the recovered material determines its future applicability.

While some (bulk) materials are well recycled, others can (currently) not be recycled, especially due to their complex connections that provide the functionality in consumer products. Secondary resources are often dispersed in our society, distributed over many users (households, offices, factories, ...). And, because of the many different applications, they occur in more complex compositions when compared to natural ores. The dispersed nature of secondary resources requires sophisticated collection, recovery, and separation systems.

For example, despite all the attention given from policy and industry actors to the strategic supply problems of rare earth metals, less than 1% of the rare earth metals in waste are currently being recycled. Instead, these are lost due to the aforementioned complexity. This is also valid for other elements recognized as critical to our economy.

1.4.3 Environmental-Economic-Social Dimensions

So far, the main focus has been on the physical quantity and quality dimension of recycling (Figure 1.4). Understanding the physical dimension is essential, as the environmental, economic, and societal impacts are all using the data and insights from the physical dimension.

Although recycling has far-reaching environmental and social benefits, market forces determine if a material or complex product can ultimately be recycled and their contained metals, materials, and compounds recovered. The economics of recycling also depend on the degree to which the material has become dispersed or dissipated, as well as the matrix/product composition (e.g., complex consumer product, building, transport, packaging, etc.) within which it is recovered.

Recyclable wastes are often collected by cities and municipalities, selling it into a market of traders and secondary processors who reprocess the materials to, eventually, sell them to manufacturers. In the recycling market, prices fluctuate according to the balance of supply and demand, the prices of primary materials, as well as the behavior and organization of markets and their stakeholders (e.g., the role of increased market power concentration, and speculation with resources/resource prices).

If recycled materials approach the same quality as primary materials, the price of the recycled material is coupled to that of the primary or virgin material. However, this becomes more complex when minor and critical elements associated with specific metal ores are priced. The supply of these elements is geologically linked to the extraction of the bulk, base metals such as aluminum, copper, nickel, lead, and zinc, as few or no separate mines exist for these metals. Furthermore, material prices may be affected by price developments in other markets, e.g., the price of (virgin) plastics is directly linked to the oil price, which in turn is driven by the demand of transportation fuels and has fluctuated wildly.

A lively international trade in recycled resources has emerged, due to the difference in local costs of separating materials from products, and due to the increasing resources appetite from rapidly developing nations like China and India. Yet markets for recycled materials

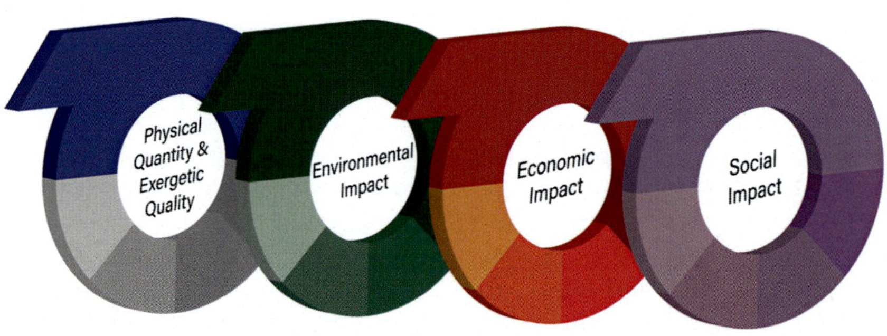

FIGURE 1.4 The various perspectives that have to be integrated to analyze the cycles of CE and progress towards achieving SDGs, i.e., the physical description, environmental, economic, and social impacts. *From Bartie et al. (2019).*

are still far from perfect, as demand for recycled materials may still be limited in important markets (e.g., packaging). The global trade has made recycling a strategic and geopolitical issue as well.

The recycling markets are also affected by economic or policy interventions. Market forces often fail to value externalities such as environmental pollution, future scarcity, or labor conditions, making for an "uneven" playing field. In various countries, these externalities and (other) market failures have provided the incentive to design policies to support resource efficiency in general, and recycling specifically. For example, in the European Union a broad portfolio of strategies, directives, and regulations has been implemented.

Legislation setting a required level of collection and/or recycling for vehicles, electronic products, and packaging wastes is now in force in most European countries, while other nations have similar programs and plans for more exist. Landfill bans and taxation of landfilling or incineration are other important policies that directly affect the recycling industry. The policies have driven the development of new infrastructure, processes, and financing and business models.

Policy is especially important to prepare the industry for future changes in product (e.g., batteries, electronics) and product compositions, as markets alone do not provide a timely (price) incentive. Ecodesign policies are examples of policies that focus on whole product chains to address the sustainability challenges in a more integrated way, including design for recycling. The European Union's Proposal for Ecodesign for Sustainable Products Regulation (European Commission, 2022) provides an example.

Policies can support value chain collaboration and support the flow of information between value chain actors and so support the performance of recycling systems. For example, information about the location of batteries and hazardous components in electronics is provided from the manufacturers to recyclers, and digital product passports are developed to facilitate information sharing across the value chain (Koppelaar et al. 2023).

Legislation, policies, and standards can also, unintentionally, create barriers for recycling. Concerns about the impact of products on the environment and human health have resulted in the banning of certain materials and compounds, e.g., mercury and per- and polyfluoroalkyl substances (PFAS), and strict limitations on others, e.g., in the European REACH legislation. This may hinder the possibility of recycled materials finding their way into new products.

In short, recent economic and global developments have put recycling in the spotlight again, necessitating a thorough understanding of the fundamentals of recycling, a critical assessment of the role of recycling in the context of a resource efficient society, and the transition to resource efficient circular economies. Moreover, these transitions in our society will affect the future of recycling. Policies guiding the transitions affect technology choice and hence the future need for types and quantities of materials. This highlights the need for a resource efficient economic system building on recycling to contribute to a sustainable world.

1.5 THE BOOK

The *Handbook of Recycling*, Second Edition, presents the current practices in recycling, the current and upcoming challenges, and the tools (and practices) to move forward to a new, more circular society. Each chapter is authored by key experts from academia, industry, and the policymaking community. This provides a strong basis from theory to practice, describing the lessons learned and the state of the art, to understand future challenges in recycling of a wide variety of products, materials, and waste flows.

Part I—Recycling in context provides the context for recycling by discussing the basics of recycling in science and society, and putting recycling in the perspective of resource efficiency and the *circular economy*. Recycling forms the foundation of a circular economy. However, there are limits to the circular economy (and recycling) that need to be addressed. Understanding the fundamentals (physics) of recycling systems (Chapter 2), product and material flows in society (Chapter 3), recycling relative to other R-strategies (Chapter 4), and the role of design for recycling (Chapter 5) are first discussed. This is followed by an overview of the developments in collection of municipal waste (Chapter 6). Lastly, insight is provided into inclusive recycling (Chapter 7) and how the informal sector in developing countries can be included in the recycling system.

Part II—Recycling from a product perspective starts with an overview of the concepts and technologies used for physical separation (Chapter 8) and sensor-based sorting (Chapter 9). It then continues with the recycling of key product and residue flows (Chapters 10–18), including mixed bulky waste (e.g., furniture) packaging, electrical and electronic equipment (WEEE), and construction and demolition waste. The part has been expanded with chapters on vehicles (ELVs), as well as products that will become more important in future: photovoltaics, wind turbines, batteries, and buildings. Lastly, industrial by-products and mine tailings are discussed. For each, the chapters provide a description of the product itself, the market situation, collection, and the sorting and separation technologies and resulting fractions. Each chapter closes with a reflection on the challenges and future developments.

Part III—Recycling from a material perspective follows logically from Part II, as it discusses the recycling of a variety of material streams from a technical perspective. Starting with metals, first the base metals are presented followed by a selection of metals, many considered critical raw materials (Chapters 19–25). Then bulk materials, such as concrete, glass, lumber, paper, and plastic are discussed (Chapters 26–31). This part finishes with black rubber, textile, and carbon fiber recycling (Chapters 32–34). For each material stream, the chapters cover the market and application areas of the material, technological state of the art, and future (technical) developments. Each chapter closes with a reflection on the key issues and challenges that need to be resolved to "close the loop."

Part IV—Recycling and the circular economy looks at recycling systems from economic and societal perspectives, which are backdrop and determine the context in which the recycling system operates. Different stakeholders and their perspective are introduced. The transition from waste management to a circular economy looks at recycling in these different paradigms (Chapter 35). The role of recycling in enabling supply chain resilience and the geopolitical perspective is covered in Chapter 36. This part continues with how key aspects of recycling are formulated in legislation (Chapters 37 and 38), and discusses extended producer responsibility in detail (Chapter 39). The last chapters focus on economics, from a policy instrument perspective (Chapters 40 and 41) and from a business perspective (Chapter 42).

Part V—Recycling fundamentals provides introductions to some of the fundamental topics needed to understand the opportunities for recycling, such as physical separation and sorting (Chapters 43), thermodynamics for extractive metallurgy (Chapter 44), exergy (Chapter 45), process simulation (Chapter 46), and life-cycle assessment (Chapter 47). The application and challenges in a recycling context are highlighted, and appropriate examples are used. The readers are guided to more advanced texts on these topics. This part is crucial for understanding the physics underlying resource efficiency. This knowledge provides the basis for innovation in the system, creating innovative new solutions and technologies, and also determining which systems and processes should be made redundant to ensure resource efficiency is maximized.

TABLE 1.1 Limits and challenges for the circular economy concept.

Limit	Description
Thermodynamic	Cyclical systems consume resources and create wastes and emissions
System boundary	– Spatial: problems are shifted along the product life cycle – Temporal: short-term nonrenewable use can build long-term renewable infrastructure
Physical scale of the economy	– Rebound effect – Jevons paradox – Boomerang effect
Path-dependency and lock-in	First technologies retain their market position despite inefficiency
Governance and management	Intraorganizational and intrasectoral management of interorganizational and intersectoral physical flows of materials and energy
Social and cultural definitions	– The concept of waste has a strong influence on its handling, management, and utilization – The concept is culturally and socially constructed – The concept of waste is always constructed in a certain cultural, social, and temporal context, and this context is dynamic and changing

From Korhonen et al. (2018).

As can be seen, the limits and challenges of the circular economy listed in Table 1.1, and by extension for recycling, are addressed throughout this book. Digitalization and the possibilities of Industry 4.0 will have a large impact on the design, use, and recycling of products and the processes used. The possibilities of process simulation and sensor-based sorting are highlighted in the book.

Combined, the *Handbook* brings together a unique and up-to-date collection of important insights on recycling, providing state-of-the-art discussions and information from a wide variety of backgrounds and experiences. This will help the reader to deeply understand the fundamentals, dynamics, complex interactions, opportunities, and challenges of recycling, within the larger picture of a circular system. We hope that this will contribute to the realization of a circular economy and efficient use of resources in our society.

References

Bartie, N.J., Abadías Llamas, A., Heibeck, M., Fröhling, M., Volkova, O., Reuter, M.A., 2019. The simulation-based analysis of the resource efficiency of the circular economy—the enabling role of metallurgical infrastructure. Miner. Process. Extract. Metall. Trans. Instit. Min. Metall. 129 (2), 229–249. https://doi.org/10.1080/25726641.2019.1685243.

Bartie, N., Cobos-Becerra, L., Fröhling, M., et al., 2022. Metallurgical infrastructure and technology criticality: the link between photovoltaics, sustainability, and the metals industry. Miner. Econ. 35, 503–519. https://doi.org/10.1007/s13563-022-00313-7.

Elhacham, E., Ben-Uri, L., Grozovski, J., Bar-On, Y.M., Milo, R., 2020. Global human-made mass exceeds all living biomass. Nature 588, 442–444. https://doi.org/10.1038/s41586-020-3010-5.

European Commission, 2022. Proposal for a Regulation Establishing a Framework for Setting Ecodesign Requirements for Sustainable Products and Repealing Directive 2009/125/EC. COM(2022) 142 Final. Directorate General for Environment. https://environment.ec.europa.eu/publications/proposal-ecodesign-sustainable-products-regulation_en. (Accessed 16 February 2023).

Geyer, R., Jambeck, J.R., Law, K.L., 2017. Production, use, and fate of all plastics ever made. Sci. Adv. 3, e1700782.

Henckens, M.L.C.M., Driessen, P.P.J., Worrell, E., 2014. Metal scarcity and sustainability, analyzing the necessity to reduce the extraction of scarce metals. Resour. Conserv. Recycl. 93, 1–8. https://doi.org/10.1016/j.resconrec.2014.09.012.

Kirchherr, J., Reike, D., Hekkert, M., 2017. Conceptualizing the circular economy: An analysis of 114 definitions. Resour. Conserv. Recycl. 127, 221–232. https://doi.org/10.1016/j.resconrec.2017.09.005.

Koppelaar, R.H.E.M., Pamidi, S., Hajósi, E., Herreras, L., Leroy, P., Jung, H.-Y., Concheso, A., Daniel, R., Francisco, F.B., Parrado, C., Dell'Ambrogio, S., Guggiari, F., Leone, D., Fontana, A., 2023. A digital product passport for critical raw materials reuse and recycling. Sustainability 15, 1405. https://doi.org/10.3390/su15021405.

Korhonen, J., Honkasalo, A., Seppälä, J., 2018. Circular economy: the concept and its limitations. Ecol. Econ. 143, 37–46. https://doi.org/10.1016/j.ecolecon.2017.06.041.

Persson, L., et al., 2022. Outside the safe operating space of the planetary boundary for novel entities. Environ. Sci. Technol. 56 (3), 1510–1521. https://doi.org/10.1021/acs.est.1c04158.

Potting, J., Hekkert, M., Worrell, E., Hanemaaijer, A., 2017. Circular Economy: Measuring Innovation in the Product Chain. https://www.pbl.nl/sites/default/files/downloads/pbl-2016-circular-economy-measuring-innovation-in-product-chains-2544.pdf. (Accessed 22 June 2022).

Reike, D., Vermeulen, W.J.V., Witjes, S., 2018. The circular economy: new or refurbished as CE 3.0?—exploring controversies in the conceptualization of the circular economy through a focus on history and resource value retention options. Resour. Conserv. Recycl. 135, 246–264. https://doi.org/10.1016/j.resconrec.2017.08.027.

U.S. Geological Survey, 2020. Historical Statistics for Mineral and Material Commodities. T.D. Kelly, G.R. Matos With Major Contributions Provided by D.A. Buckingham, C. A. DiFrancesco, K.E. Porter and USGS Mineral Commodity Specialists https://www.usgs.gov/centers/national-minerals-information-center/historical-statistics-mineral-and-material-commodities. (Accessed 16 February 2023).

United Nations Statistics Division, 2023. Report on SDG 12. https://unstats.un.org/sdgs/report/2019/goal-12/. (Accessed 16 February 2023).

Worldsteel, 2022. World Steel in Figures 2022. World Steel Association, Brussels, Belgium. www.worldsteel.org.

CHAPTER

2

The fundamental limits of circularity quantified by digital twinning

Markus A. Reuter[1], *Ernst Worrell*[2], *and Christina Meskers*[3,4]

[1]WASM: Minerals, Energy and Chemical Engineering, Faculty of Science and Engineering, Curtin University, Perth, Australia [2]Copernicus Institute of Sustainable Development, Utrecht University, Utrecht, The Netherlands [3]Department of Industrial Eco-systems, SINTEF Manufacturing, Trondheim, Norway [4]Department of Energy and Process Engineering, Faculty of Engineering Sciences, NTNU—Norwegian University of Science and Technology, Trondheim, Norway

2.1 INTRODUCTION

This chapter discusses recycling in terms of a material and, more importantly, a product perspective. This provides a basis to discuss the various intricacies, opportunities, and limits in detail within the context of the product and the circular economy (CE) paradigm (Reuter et al., 2019).

Usually, it is convenient to discuss recycling of metals on an element-by-element basis, e.g., aluminum recycling, steel recycling, gold recycling, or aluminum can recycling. This is not sufficient in a CE paradigm. Ignoring that most materials used in society are composed of a combination of materials and elements, e.g., in alloys, functional material combinations with plastics, coatings, etc., means ignoring their effect on the recycling rate, on the final material quality, and thus on the applicability in the same product. In other words, the essentials of the CE paradigm are not considered.

The inconvenient truth is that complete or 100% "closing" of the loop is not possible. The laws of thermodynamics dictate this (Abadías Llamas et al., 2020).

Figure 2.1 shows the CE of photovoltaic modules (Bartie et al., 2021, 2022). It highlights the barriers between the stakeholders for each part of the circular economy, which are discussed in the different chapters in this book. The thermodynamic barrier appears throughout the CE: between use and collection, between collection and recycling, between recycling and secondary resource extraction, with primary resource extraction, and finally between resource extraction and metal production. The background shows the many technologies that have to be integrated to produce, for example, a photovoltaic module. The outer rings (Figure 2.1) illustrate the external energy requirement and the dissipation of enthalpy, exergy, and materials (in kilowatts), respectively. The external energy requirement is highest in metal extraction and production, and then decreases. The dissipation ring is exactly the opposite. By the time the product is recycled,

Handbook of Recycling. https://doi.org/10.1016/B978-0-323-85514-3.00013-0

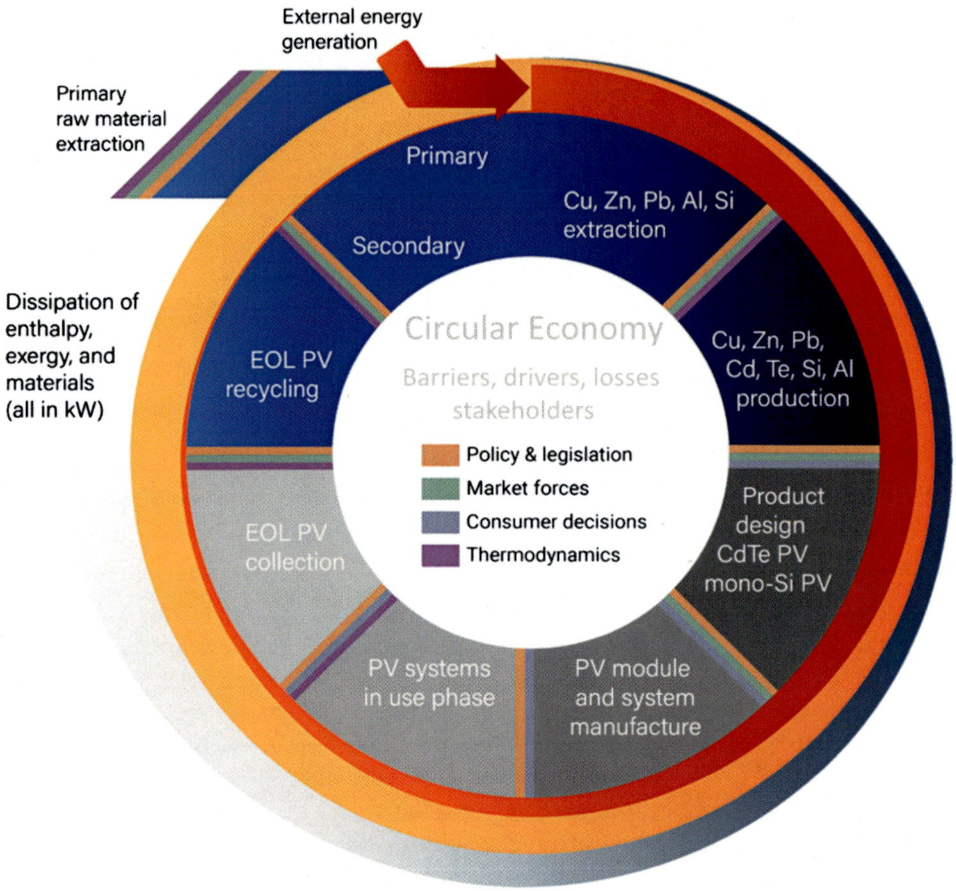

FIGURE 2.1 An overview of the CE, stakeholders, and barriers between them for photovoltaic (PV) cells as an example. *Adapted from Bartie et al. (2022).*

the dissipation has become highest, indicating that the loop is not closed. Knowing the physical limits of "closing" the loop within the CE requires details, down to the nitty-gritties, to be able to quantify all streams in terms of kilowatts. The fundamental basis is the enthalpy (energy) and entropy. Together they quantitatively capture the degradation of the quality of all materials and energy flows (including fuels) within the CE system. Counteracting this loss of quality is what drives the CE system.

The next sections dive deeper into the intricacies of the recycling of individual materials and of products composed of a multitude of tightly connected functional materials within the CE. A more detailed discussion of thermodynamics, exergy, and process simulation can be found in Chapters 44–46.

2.2 A PRODUCT AND MATERIAL FOCUS ON RECYCLING WITHIN THE CE

2.2.1 The Metal Wheel

Maintaining the quality of materials and energy within the CE system requires detailed thermochemical knowledge. The Metal Wheel, depicted in Figure 2.2, best explains the

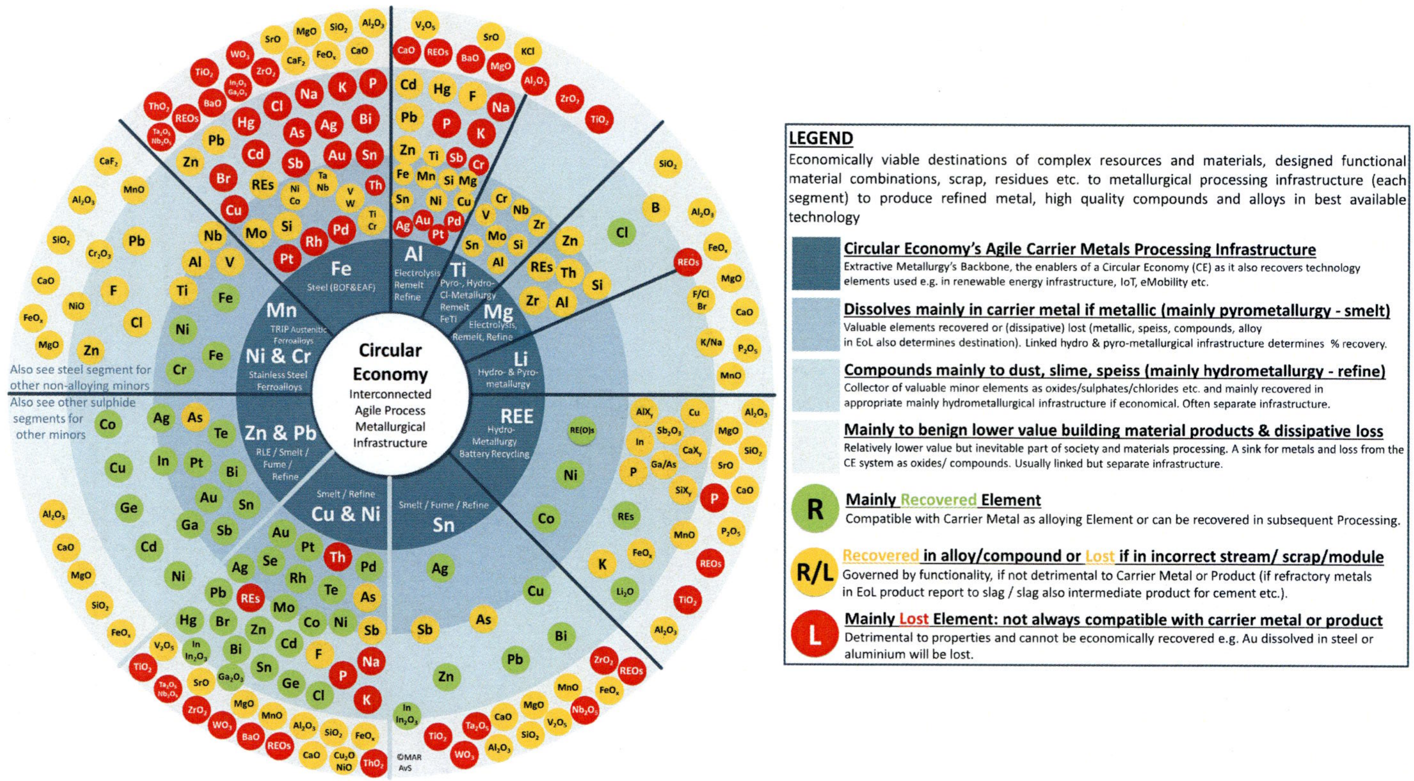

FIGURE 2.2 The Metal Wheel showing the full interlinkage of all elements in the CE system and the segments of process metallurgy where these are best recovered to minimize the overall material and exergy losses.

difference between product and material recycling. It explains how the different metallurgical processing systems (shown as segments) best unmix complex material mixtures into individual materials for new products. Think of this in terms of recycling a cup of coffee from yesterday into an aromatic cup of coffee today. First there is the unmixing of the old cup of coffee into water, coffee, milk, and sugar and then remixing of the individual ingredients into a new cup of coffee (UNEP, 2013).

The center ring reflects a material point of view, i.e., one that only considers elements in a limited manner, as it does not include the degradation of material and energy quality of all intimately linked materials in a product. It cannot describe the energy and exergetic cost required to produce high-quality metals, alloys, and materials from the end-of-life (EoL) materials or downgrade into products of lesser economic value that can at most be deposited in well-managed ponds and residue dumps.

From a CE point of view, the center ring is the "ball bearing" of the CE, which makes the CE turn. Using the metallurgical processing infrastructure (center ring of each segment) as a basis, materials and metals can be recycled into products that find application again. The outer rings indicate where the other materials (elements) that enter the process end up, and it is indicated if the element can be recovered, or not.

This center ring is in fact the metallurgy that supplies all the technology elements (Reuter et al., 2019) used in various functional combinations in products. At end of life (EoL), products return to this center ring for processing and recovery of materials, elements, and energy from them. This is product-focused recycling, i.e., considering an EoL product or scrap mixtures in their full complexity in a CE paradigm, as shown in Figure 2.1.

Rigorous process simulation of the CE system (creation of digital twins) is required to fully understand the complexity of product-focused recycling. At the heart of the recycling of products is deep knowledge; a brief explanation of what is meant by this is illustrated by simple examples.

2.2.2 Material-Centric Recycling: Aluminum Alloys

Metals and alloys are mainly mixtures of elements and can be recycled in this form through various chemical processes.

Aluminum (Al), for example, in pure form or clean scrap can be well recycled. The aluminum industry is doing much to maximize the recycling of the various aluminum-containing metal alloys and other materials recovered from various EoL products (e.g., packaging, building, and automotive products). Aluminum alloys, like any metal, can be recycled repeatedly without loss of properties, if the alloying components and impurities that are inevitably picked up and dissolved during its many cycles are maintained within the element tolerances of the different alloy types (shown in the outer rings of the aluminum segment of Figure 2.2).

The high value of aluminum scrap is a key incentive for recycling. It is well documented that recycling of aluminum saves about 95% of the energy required as compared to primary aluminum production, avoiding corresponding emissions including greenhouse gases. However, this number does not reveal quality degradation, aluminum losses due to dross (a complex mix of Al_2O_3 and other oxides) formation, and general exergetic losses from the CE system. Aluminum recycling is discussed in detail in Chapter 20, Aluminum.

If aluminum (alloys) are not physically separated from the other material parts in the EoL product, they enter, for example, steel or copper recycling processes. Then they usually end up in the slag of pyrometallurgical processing as alumina (Al_2O_3) (Figure 2.2, the outer ring of the Cu segment, for example), or as various other compounds such as Al_2SiO_5, $CaAl_2SiO_6$, $CaAl_2O_4$, $FeAl_2O_4$, and $Fe_3Al_2Si_3O_{12}$. These can precipitate as solid particles in the molten slag and sometimes create very disruptive operational issues due to viscosity, flow, foaming, and separation problems. Slag chemistry also affects the distribution of elements into various phases and therefore must be carefully managed. If aluminum enters hydrometallurgical processes, it can possibly be dissolved and precipitated as $Al(OH)_3$. Together with various coprecipitated hydroxides, it then creates a rather complex mixture, which makes economic separation into high-quality products quite challenging.

Thus care should be taken during physical separation to remove the aluminum (alloy) parts, before they are lost in the slag and in addition disrupt and/or limit operation and the recovery of elements from the end-of-life product the aluminum alloy is linked to.

2.2.3 Material-Centric Recycling: Nonferrous Metals

Metals arise in numerous compounds due to, for example, their possible valency states. Which compounds are present depends on what the conditions are within the process technology, whether high temperature (pyrometallurgy, pyrolysis, salt slag electrolysis, etc.) or aqueous (hydrometallurgy) processing. This implies that metals can be recovered in the metal phase, can be volatilized into a flue dust and collected for further processing, and can be recovered in a slag, speiss, or aqueous phase. It may be recovered or be lost from the material cycle if the dust or slag cannot be processed economically. To illustrate this, consider indium-tin-oxide (ITO), as well as As, Ce, Er, Ga, In, Nd, Sn, Ta, Tb and W containing recyclates. Processing under different operating conditions in a furnace will result in different phases such as metal, matte, speiss, gaseous species (e.g., Br, Cl, F, O, S species), or oxidic species, for example, and may thus be recovered or be lost. This is determined by, e.g., the prevailing partial oxygen pressure created by the addition of carbon or hydrogen, through waste plastics, methane, or similar reductant types.

Nonferrous metals recycling is discussed in Part 3 of this book.

2.2.4 Material-Centric Recycling: Polymer and Plastic Recycling

Depending on the way polymers and plastics are integrated into complex products, it is often best to use their energy to fuel the recycling processes (although this may result in CO_2 emissions) if the polymers and plastics enter a metal recycling process. The presence of plastics/polymers can result in metal losses due to oxide and carbide formation, e.g., in aluminum recycling, while the fillers (e.g., TiO_2), pigments, and flame retardants (e.g., Sb_2O_3, phosphates) present in the polymer/plastic can react, dissolve, or otherwise impact the metal recycling process. This impact can be quantified using exergy (Meskers et al., 2008).

Obviously, an alternative exists if the polymers can be broken down to monomers (dictated by thermodynamics) to be used to build polymers (through so-called chemical plastic recycling technologies).

2.2.5 Product-Centric Recycling

The functionality of products necessitates that materials be designed and combined to interact in a complex functional manner. By default, this creates very complex waste streams, and requires consideration of the Metal Wheel in its entirety. To recover metals and materials from these products and waste flows requires a deep understanding of the physics of separation as well as the complete system. True Design for Recycling (DfR) tools will incorporate the complex physics into the design tools to be able to show the inevitable losses that will appear due to the functional connections in products.

Note that in various cases DfR makes no sense, as the material/metal value is so low and the functionality makes it impossible to design for recycling. Instead, time and money must be spent on designing the recycling systems for optimal resource recovery and resource efficiency. Therefore design for resource efficiency should be the driver for creating a recycling system that is using best available technology (BAT) and is certified to meet this requirement in a CE paradigm. Neglecting this totally inclusive metallurgical view into the CE system will surely crash the system, as Figure 2.3 shows.

Design for resource efficiency (DfRE), of which DfR is a subset, demands knowledge of:

- liberation behavior of materials within a product or modules within the product,
- particulate quality of recyclates both in terms of materials and exergy,
- separation efficiency linked to the compatibility and thus recovery and/or loss of material in metallurgical processing (Figure 2.2).

Furthermore, it requires:

- Modeling of the CE system, as a function of design choices, connection type, and connected materials.
- A technology- and industrial process-driven basis for DfR to optimize resource efficiency and closure of material cycles for both commodity and critical and scarce elements in complex consumer products, such as cars and e-waste/WEEE products.

Not all the metals can be extracted from EoL goods. The common commodity metals like steel, aluminum, magnesium, and copper can be recovered relatively easily, as these are often used in relatively simple applications. However, the small amounts of precious/critical/valuable metals in, for example, WEEE can be harder to recover. These complex waste streams often consist of over 60 elements, combined in all kinds of functional materials.

A material compatibility matrix and the design for resource efficiency Metal Wheel have been developed as advisory DfR tools for designers (Reuter and van Schaik, 2012). These tools capture the necessary physics of recycling. The simulation models and the Metal Wheel are based on system models that link product design to the complete recycling system. Although different products may all have similar suites of functional materials loosely called "mineralogies," i.e., contained elements and the functional connections, their recovery is not the same and hence their recycling rates are different. Depending on the product and the combinations of materials, the recovery of metals may be different due to chemistry, concentration, and metallurgical processes being incompatible, as reflected by the Metal Wheel.

Therefore forcing recycling rate quotas for, especially, the minor metals would be a fallacy. Rather, a focus should be to maximize recovery of the elements in their entirety from a product, taking cognizance of the complexity of that product. The more complex the product, the more materials will interact and affect the recovery and the production of high-quality materials, alloys, and metals. Given the correct economic basis, BAT infrastructure and market-driven policy forcing, for example, a ban on landfilling will help to maximize recovery and recycling rates. It has always been the case in any metallurgical plant to recover valuable elements. If there is an economic incentive to do so, recovery will happen.

Summarizing, the physics and thus the exergy dissipation of the system will determine the economics and how best to manage the CE system. Figure 2.4 reflects various aspects that will be discussed elsewhere in this book, as follows:

- As is known from primary production, mineralogy is complex and mostly requires detailed processing to "unmix" the valuable and gangue minerals from each other. In physical separation, understanding interfacial phenomena is key to maximizing the grade and recovery of minerals. That can

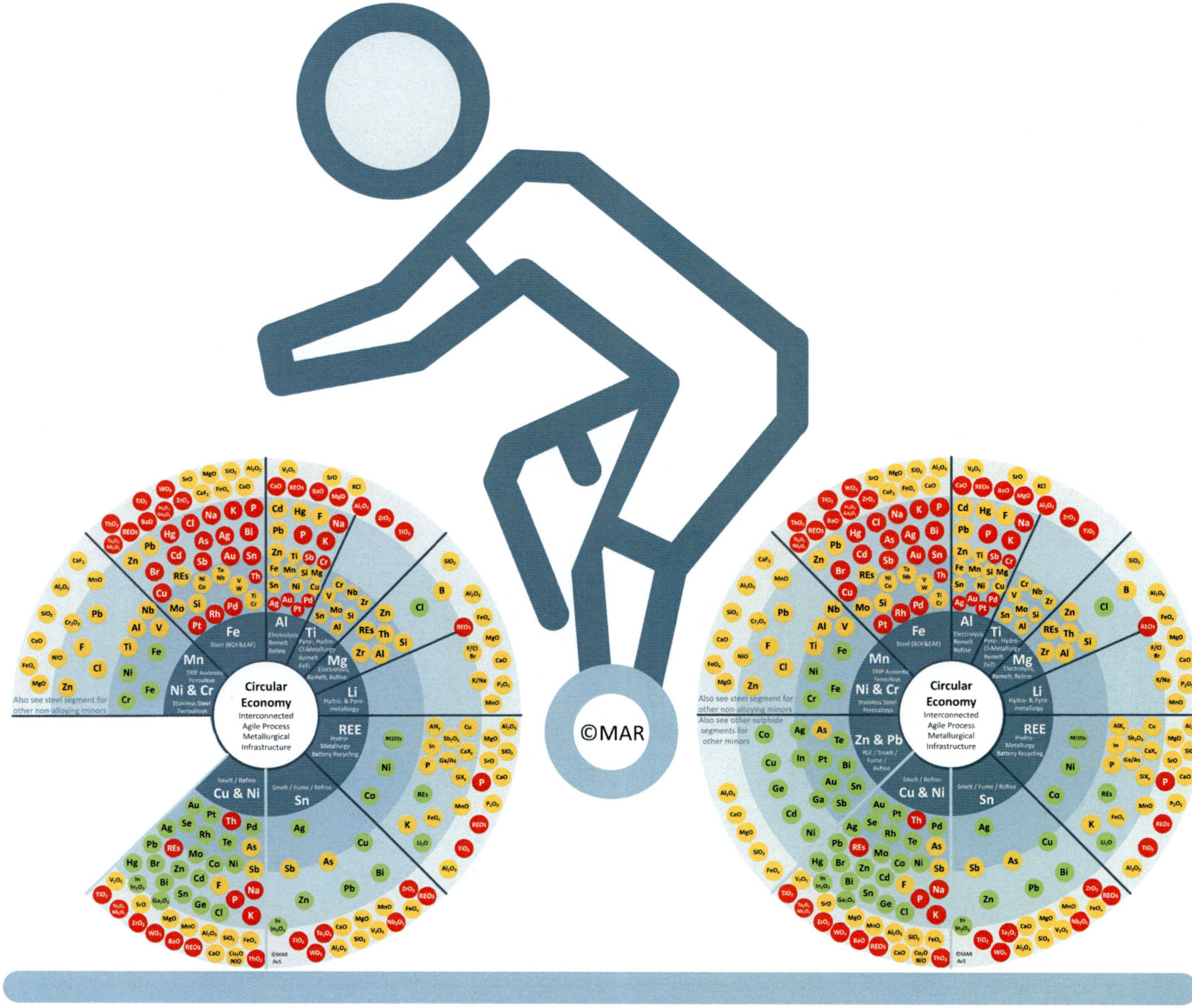

FIGURE 2.3 Unwise decisions and policies impacting the metallurgical system will have catastrophic results for the circular economy. Here is shown the consequence of removing the lead and zinc systems. *From Blanpain et al. (2019).*

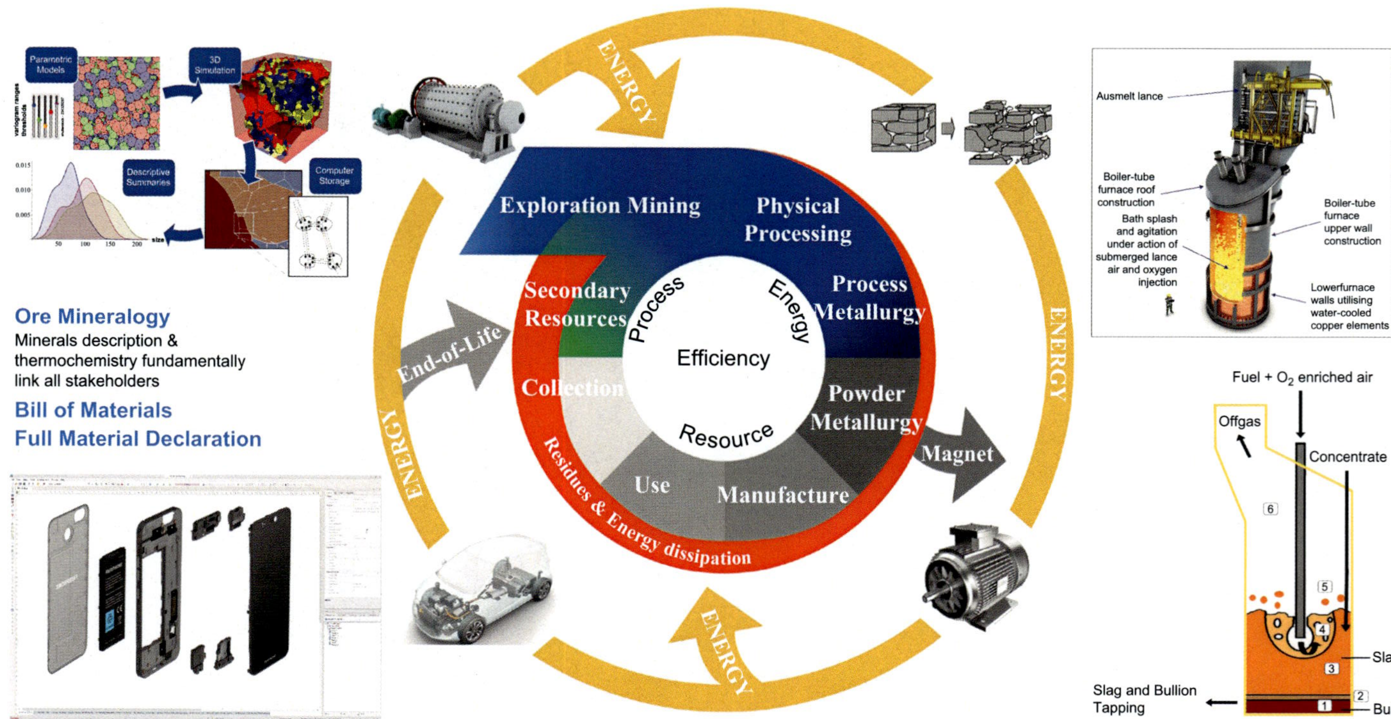

FIGURE 2.4 Digital twinning of the circular economy system, linking the physics to product design and the mineralogy of ores to understand exergy dissipation and the limits of the circular economy (Reuter et al., 2019).

never occur 100% and therefore losses occur (top left Figure 2.4).

- Modern products are even more complex than geological minerals, which makes separation even more challenging. It creates complex mixtures that are challenging to separate. Once again, the interfaces between functional materials determine the recoveries. Using the understanding of interfacial phenomena joining different materials and therefore unjoining them is of key importance and should also be understood in terms of product design (bottom left Figure 2.4).
- Interfacial mass and energy transfer between the different material phases in processing reactors, both hydro- and pyrometallurgical, define the limits of circularity. This is at the heart of not being able to "close" the loop. Therefore understanding the interfacial phenomena in products and in reactors is at the heart of understanding the CE, in which all materials at some point in time will have to be recycled (right Figure 2.4).

Analogous to these interfacial physics, the interfaces between the different stakeholders of the CE (Figures 2.1 and 2.4) matter. The improved closing of the loop will be possible if the "interfaces" between the stakeholders are well understood and harmonized to maximize transfer of information and knowledge. Digital twinning and digitalization can help with this.

Figure 2.4 suggests that enabling the CE in the broadest sense requires a deep and harmonious understanding of all interfacial and interstakeholder phenomena. Research should focus on these phenomena.

2.3 DIGITAL TWINNING OF THE CE SYSTEM: UNDERSTANDING THE OPPORTUNITIES AND LIMITS

Understanding the CE well requires a detailed understanding of the system and its interfaces, as shown in Figures 2.1 and 2.4. Digital twinning of the system is a powerful and transparent manner of understanding the complexity of the physics of the CE system (Bartie et al., 2021, 2022). This section gives a brief overview of the topic. The other chapters in the book, the references included, and their further references provide for a wide overview, and provide indicative data for, e.g., vehicles, electronic goods, batteries, lamps, and photovoltaic modules.

Maximizing resource efficiency requires a detailed understanding of the complete value chain within the CE paradigm, for recovering the maximum number of metals in hydro- and pyrometallurgical processing facilities. This is only possible through a holistic systems approach. It starts with product design, and then optimizes the technical and organizational setup of dismantling, shredding, and separation technologies. Only then can the relevant material fractions be directed to the most appropriate metallurgical end process, as explained by the Metal Wheel within a CE paradigm, while minimizing exergy dissipation. The integrated digital platforms required to do this are depicted in Figure 2.5.

Figure 2.5 summarizes the digital platforms and tools that will help to better understand the limits of circularity and that are the foundation for physics-based informed policy (Bartie et al., 2021):

- The depicted four subsystems (for example, since more can be discerned) of the CE covering a wide spectrum of disciplines can be linked by appropriate digital tools and approaches.
- Digital platforms make it possible to integrate different tools using the fundamental language of physics, engineering, and economics that combines different approaches.
- A big challenge is to link this to the sustainable development goals (SDGs) in a logical and transparent manner.

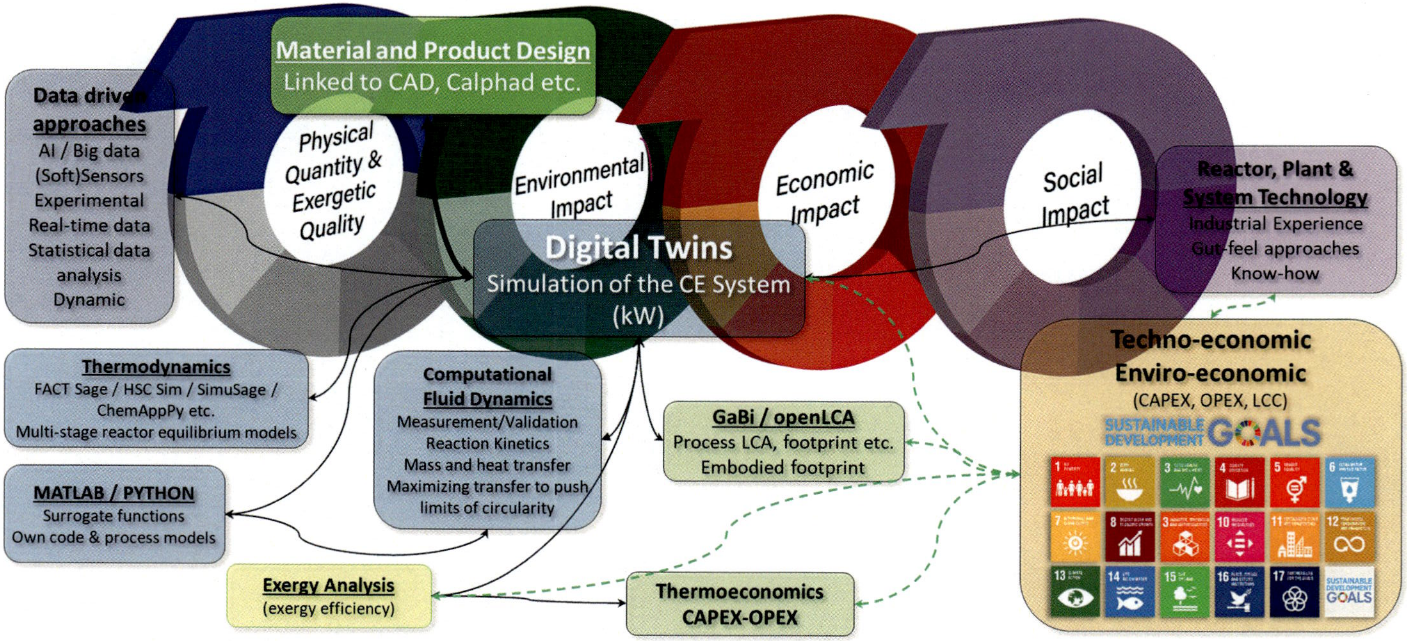

FIGURE 2.5 Integrated digital platform for twinning large-scale recycling and CE systems, and quantifying resource efficiency in kilowatts, to understand exergy dissipation and the degradation of material quality of all streams of the system. *Modified from Bartie et al. (2021).*

- With such digital twinned systems in place, complete supply chains can be mapped, ultimately also providing a basis to estimate the costs of the system well; above all such a system will be able to estimate the exergetic dissipation from the system, which in fact is the true loss from the system and quantifies the true limit of circularity.

2.4 OPPORTUNITIES AND CHALLENGES

Chapter 4, Material Efficiency: Squaring the Circular Economy, as well as Reuter et al. (2019) discuss the opportunities and challenges of the CE in more detail, articulating the limits identified by Korhonen et al. (2018) in a more fundamental way. Some of these are listed here and can help to direct future research:

- In theory, CE provides a wonderful opportunity to clean up society and nature, but then all the tools must be based on the laws of nature.
- All material and product streams must be defined in terms of thermochemical data to be able to quantify energy and material quality in terms of exergy, so that the losses can be understood qualitatively and quantitatively.
- While mass flow analysis (MFA) approaches have a role, understanding the true exergetic losses, and therefore the economics of systems, requires detailed simulation to map and link all the stakeholders shown in Figure 2.1.
- To speak of criticality of materials is not enough. It should be complemented with "criticality of materials processing infrastructure." The availability of the processing infrastructure to produce (and recycle) metals, alloys, and compounds that find functionality in future products should all be mapped. Understanding the impact of products and their (re)designs on circularity, and the infrastructure required for circularity, is even more critical.
- A great opportunity is to link the processing simulation and modeling tools depicted in Figure 2.5 and use them for assessing progress toward the SDGs.
- The discussion on recycled content in products will be much clearer when the exergy losses of all streams associated with creating the recycled content in a product are well understood.
- In the CE no stream that flows into a product can disregard any other connected stream in the system. Figure 2.1 makes it clear how complex the integrated system is to create a PV cell for producing clean energy. Exergy dissipation of the complete system can already be quantified in terms of exergy to fully understand the nexus energy, fuels, and materials for all materials that are used to create a product.

References

Abadías Llamas, A., et al., 2020. Simulation-based exergy analysis of large circular economy systems: zinc production coupled to CdTe photovoltaic module life cycle. J. Sustain. Metall. 6 (1), 34–67. https://doi.org/10.1007/s40831-019-00255-5.

Bartie, N.J., Cobos-Becerra, L., Fröhling, M., Schlatmann, R., Reuter, M.A., 2021. The resources, exergetic and environmental footprint of the silicon photovoltaic circular economy: assessment and opportunities. Resour. Conserv. Recycl. 169, 105516. https://doi.org/10.1016/j.resconrec.2021.105516.

Bartie, N., Cobos-Becerra, L., Fröhling, M., et al., 2022. Metallurgical infrastructure and technology criticality: the link between photovoltaics, sustainability, and the metals industry. Miner. Econ. 35, 503–519. https://doi.org/10.1007/s13563-022-00313-7.

Blanpain, B., Reuter, M., Malfliet, A., 2019. Lead metallurgy is fundamental to the circular economy, policy brief. https://kuleuven.sim2.be/wp-content/uploads/2019/02/SOCRATES-Policy-Brief-2019-Lead.pdf.

Korhonen, J., Honkasalo, A., Seppälä, J., 2018. Circular economy: the concept and its limitations. Ecol. Econ. 143, 37–46. https://doi.org/10.1016/j.ecolecon.2017.06.041.

Meskers, C., Reuter, M.A., Boin, U., Kvithyld, A., 2008. A fundamental metric for metal recycling applied to coated magnesium. Metall. Trans. B 39 (3), 500–517. https://doi.org/10.1007/s11663-008-9144-8.

Reuter, M.A., van Schaik, A., 2012. Opportunities and limits of recycling: a dynamic-model-based analysis. MRS Bull. 37 (4), 339–347. https://doi.org/10.1557/mrs.2012.57.

Reuter, M.A., et al., 2019. Challenges of the circular economy—a material, metallurgical and product design perspective. Annu. Rev. Mater. Res. 49, 253–274. https://doi.org/10.1146/annurev-matsci-070218-010057.

UNEP, 2013. In: Reuter, M.A., Hudson, C., van Schaik, A., Heiskanen, K., Meskers, C., Hagelüken, C. (Eds.), A Report of the Working Group on the Global Metal Flows to the International Resource Panel, Metal Recycling—Opportunities, Limits, Infrastructure. 316 p.

CHAPTER

3

Maps of the physical economy to inform sustainability strategies

Daniel B. Müller[1], *Romain G. Billy*[1], *Mark U. Simoni*[1,2], *Evi Petavratzi*[3], *Gang Liu*[4], *Helmut Rechberger*[5], *and Jonathan Cullen*[6]

[1]Industrial Ecology Programme, Department of Energy and Process Engineering, Norwegian University of Science and Technology, Trondheim, Norway [2]Section Mineral Resources, Geological Survey of Norway, Trondheim, Norway [3]Decarbonisation and Resource Management, British Geological Survey, Keyworth, United Kingdom [4]College of Urban and Environmental Sciences, Peking University, Beijing, China [5]Institute for Water Quality and Resource Management, Vienna University of Technology, Vienna, Austria [6]Department of Engineering, Cambridge University, Cambridge, United Kingdom

3.1 INTRODUCTION

The size and complexity of global material flows have increased drastically over the past decades because of population growth, industrialization, migration and urbanization, globalization, and lifestyle changes. The accelerating use of materials has been accompanied by rising energy consumption, waste generation, and emission flows, and consequently has spurred concerns for sustainability. In fact, many sustainability challenges are linked through material and energy flows (Figure 3.1). In the case of climate change mitigation, material production alone is currently responsible for 23% of global greenhouse gas emissions (Allwood et al., 2010; Fischedick et al., 2014; Hertwich, 2021). Both climate change mitigation and adaptation are linked to material needs and therefore indirectly affect greenhouse gas emissions. It is impossible to change the energy system without altering the material system and vice versa. Effective climate change mitigation strategies need to address transformation of both the energy and the material systems (Allwood et al., 2012).

Indeed, many Sustainable Development Goals (SDGs) (UN General Assembly, 2015) can be benchmarked using systems of matter and energy. For example, climate action (SDG 13) to reduce greenhouse gas emissions in the transport sector through electric mobility requires a supply of affordable and clean energy (SDG 7), and of responsibly sourced battery raw materials such as lithium (SDG 12). Lithium mining in turn can have severe impacts on ecosystems such as in the Andes, including biodiversity loss (life on land, SDG 14) and may increase competition for access to clean water and sanitation (SDG 6). Addressing individual SDGs in isolation may thus cause problem shifts and change the solution space for addressing other SDGs. In other words, sustainable development requires a targeted transformation of the entire physical economy (Figure 3.1).

Different terms have been used to describe these systems of matter and energy (Müller et al., 2013), such as anthropogenic metabolism (Baccini and Brunner, 1991), social or socioeconomic metabolism (Fischer-Kowalski, 1998; Haberl et al., 2019), or physical economy (Turner and Poldy, 2001). In this chapter, we use the term "physical economy."

Our activities as producers and consumers of goods and services collectively shape the physical economy, which in turn transforms both the built and the natural environment. Due to the systemic interconnections of the physical economy and its links to the environment, it is impossible to change the built environment without also altering the natural environment and vice versa. However, we have a very poor understanding of the domino effects of our actions. We may consciously try to "improve" our local environment, but these changes may have unintended side effects in other parts of the supply chain. And changes of the larger system, for example the climate, will inevitably affect the local system. Staying with the climate example, this could be extreme weather events or population migration, which in turn drive material demand as infrastructures

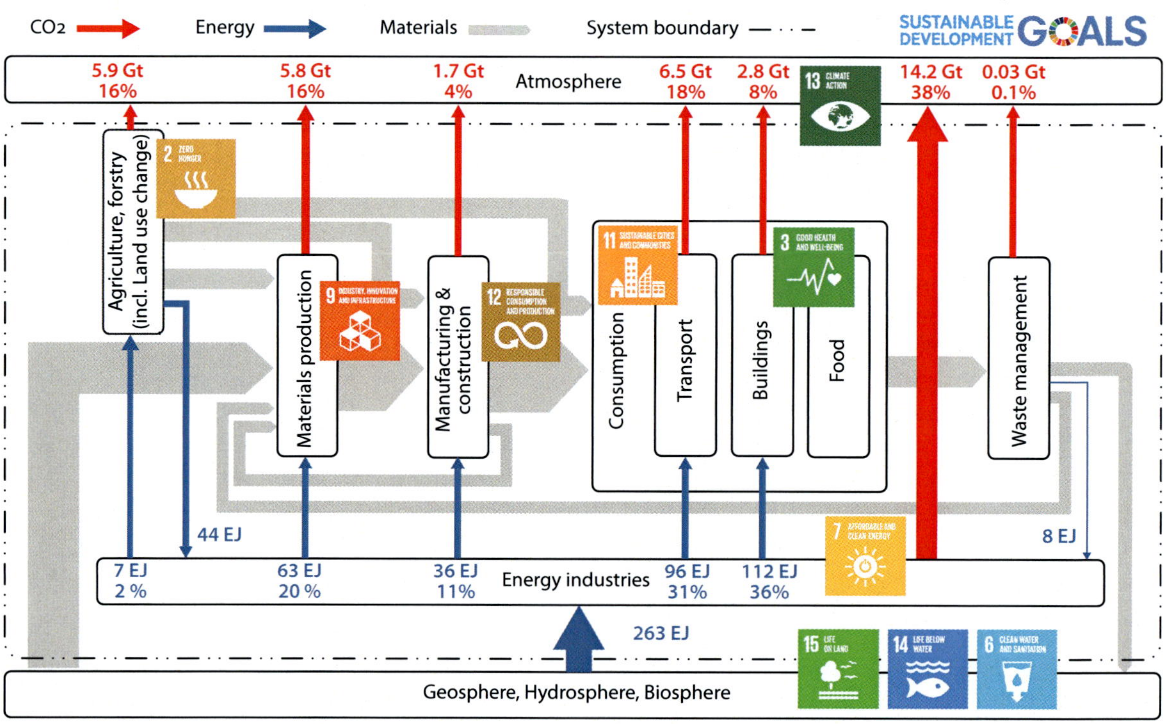

FIGURE 3.1 Global material *(gray)*, energy *(blue)*, and greenhouse gas emission flows *(red)* form the physical economy. Since SDGs are linked through material and energy flows, there is a risk of problem shifts if the linkages are not considered adequately. *Modified from Müller et al. (2013).*

need to be rebuilt and new housing will have to be provided. Therefore a better understanding and robust mapping of the physical economy is urgently needed to orientate and guide the sustainable transformation of the socioeconomic system.

Several tools are used to describe and evaluate socioeconomic systems, including traditional economic models, life cycle assessments, footprint analyses, or material flow analyses. However, these tools have different scopes that do not necessarily hinge on a robust understanding of the physical economy, such as

- the chemical and physical transformations (mass and energy balance consistency),
- the role of stocks in providing services to the consumers and society and in creating inertia, and
- the collective interactions between human activity and the natural environment.

Economic models tend to describe societal problems in monetary terms. An important principle in economic models is the assumption of the conservation of nominal money, which tends to hold only over short periods of time. Monetary analysis has the advantage of allowing for the integration of all factors of production, including land and natural resources, labor, and capital at different scales from individual business to economy. However, economic analysis is limited to the description of a small part of societal processes that involve financial transactions, excluding the natural environment as well as any societal processes that do not involve financial transactions. Economic instruments and aspects are discussed in Chapters 40–42 (Economic drivers and valuation; economic policy instruments; economic aspects of metal recycling).

Life cycle assessment (LCA) and footprint analysis are laden with moral questions of the allocation of responsibility for environmental impacts to specific functional units. Cattle, for example, produce methane emissions that contribute to global warming and climate change, but how can the responsibility or blame for these emissions be fairly attributed to the consumer products meat and milk, or to by-products such as leather, bone meal, or manure? This problem can be addressed in different ways, such as allocation according to total mass, dry matter, or financial value. However, independent of the chosen approach, the main principle is that the different fractions of the blame should add up to 100% to avoid double-counting. In other words, there is an underlying assumption of a "blame balance" consistency. LCA models further assume that parts, like individual products or individual processes, can be optimized independent of the context. This is often not a robust assumption when considering the interlinkages and complexity of profound sustainability transformations. Principles of LCA can be found in Chapter 47—Life cycle assessment (LCA).

Material flow analysis (MFA) or substance flow analysis (SFA) is typically used for mapping and modeling the physical aspects of the economy and its interactions with the environment, sometimes also termed the anthropogenic, industrial, social, or socioeconomic metabolism (Baccini and Brunner, 1991, 2012; Ayres and Simonis, 1994; Fischer-Kowalski, 1998; Brunner and Rechberger, 2004, 2017; Graedel and Allenby, 2010; Haberl et al., 2019). The basic principle of MFA is the conservation of mass and energy. Since mass and energy are exchanged between the anthroposphere and the environment, MFAs can describe the interactions between the two and thereby provide a robust basis for their integration.

The notion that understanding the physical basis of our economy is essential for humanity's survival has been growing in recent decades. Government agencies, nongovernmental organizations, research institutions, and industry associations have started to map a wide variety of aspects of the global physical economy, focusing on different countries or regions (e.g.,

UNEP, 2013; EU JRC, 2022), individual materials or substances (e.g., IAI, 2022), or specific sectors (e.g., IEA, 2021). Nevertheless, there is still a lack of understanding of how the different studies are linked together. Efforts to connect the individual parts are severely impeded by fragmented, inconsistent, nontransparent, and incomplete data collection and reporting systems. Economy-wide MFAs (Wiedenhofer et al., 2019) attempt to provide such an overview; however, they tend to lack the high spatial, temporal, material, and life cycle resolution of traditional MFAs. For example, many critical materials are not included due to their small weight compared to bulk materials such as aggregates, paper, or steel. We are still far away from a sufficiently consistent and robust understanding of the global physical economy, which is necessary to effectively inform the strategies and decision making of governments and businesses alike. Here, we present a framework for monitoring the global physical economy in four dimensions using seven steps or components. We further introduce principles and strategies for improving the robustness of these components and the monitoring of the physical economy.

3.2 DIMENSIONS OF MFA

As an approach that analyzes stocks and flows of matter and energy in systems defined in space and time (Brunner and Rechberger, 2004), MFA is deemed appropriate for monitoring the physical economy. A main characteristic of MFAs is that they explicitly cover the four dimensions of Stages, Trade, Layers, and Time (European Commission, 2020) (Figure 3.2).

3.2.1 Stages and Trade

The dimensions *Stages* (dimension 1, D1) and *Trade* (D2) define together the system observed, including the processes (balance volumes) and the goods exchanged between them (flows), either within the anthroposphere or exchanges with the environment. While the stages dimension defines the sequence of processes involved *within* a region or country, the trade dimension differentiates the exchanges *between* regions or countries that collectively form a larger system. Stages and trade can be applied at different scales; for example, an MFA of an industrial plant may comprise several manufacturing processes (stages, D1), and may purchase raw

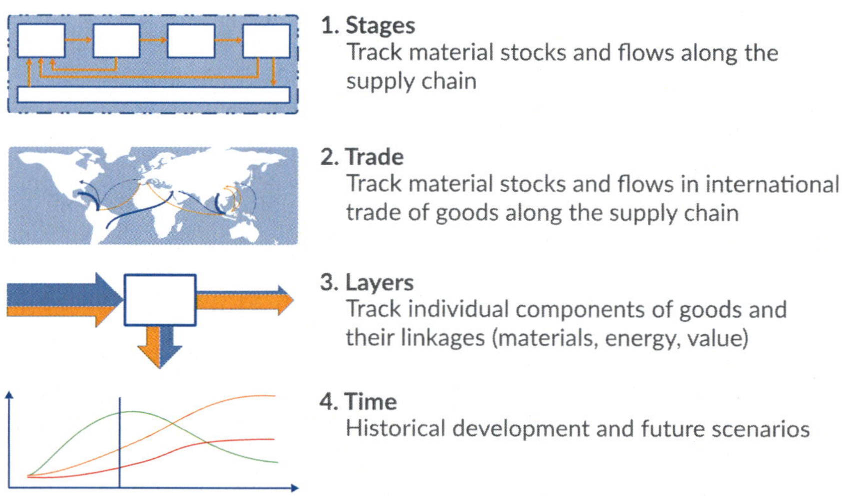

FIGURE 3.2 Four dimensions of MFAs. *Reproduced with permission from the MinFuture project (European Commission, 2020).*

materials from other companies and sell intermediate or final products to other customers (trade, D2).

3.2.2 Layers

The dimension of *Layers* (D3) describes the aspects of the system analyzed. They can either include the entire goods, measured in terms of mass or energy consistent with thermodynamic principles, or number of items or value, which may not be conserved and thus can be used only as supporting layers. Alternatively, or in addition, layers can include specific parts of the goods, such as individual chemical elements or materials such as wood or components. While the mass of chemical elements is conserved in transformation processes, it may not be conserved in transformation processes for materials such as wood or components such as batteries. The individual layers are linked with each other through concentrations (kg/kg), energy densities (MJ/kg), or prices (USD/kg). However, while the concentrations and energy densities are intrinsic properties that do not change in different contexts, prices and footprints are extrinsic or relational properties that depend on the context and are subject to value appraisal. They are allocated to the goods and are thus not equally robust as the intrinsic properties that are consistent with thermodynamics. This is particularly relevant when discussing long-term and deep transformations of the physical economy. Total mass and mass of individual elements are conserved throughout MFA systems (except for nuclear reactions), while the number of items and the value may not be conserved; products can be created and destroyed and prices can fluctuate based on market dynamics and government policy, while the matter and energy they are made of are conserved. MFA studies therefore tend to focus on intrinsic, physical, measurable properties and use extrinsic layers only as proxies for estimating the physical layers of mass and energy.

3.2.3 Time

The dimension of *Time* defines the temporal scope of the study. It can include past (historical analysis) or future time periods (scenario analysis). Historical analyses are often used to identify existing development patterns (Zoboli et al., 2016) that are subsequently used as premises for scenario analysis, such as the saturation of per-capita iron stocks (Müller et al., 2006, 2011; Pauliuk et al., 2013).

3.2.4 Dimensions in Other Tools

Different aspects and dimensions of the physical economy can also be explored using other industrial ecology tools (Graedel and Lifset, 2016) such as life cycle assessment (LCA) and input-output analysis (IOA) (Miller and Blair, 2022), as illustrated in Table 3.1. Classical product- or service-based LCAs use a functional unit to compare alternative options and differentiate stages, but tend to omit trade, layers, and time. In theory, any MFA (differentiating any dimensions) could be used as an inventory for an LCA. However, this might involve the LCA abandoning the functional unit approach: systems evolving over time have countless functions in society: steel is used in many products, and each stage in the system may provide jobs, income, or local identity. Consequently, the allocation of blame and the possibility to concisely compare alternative approaches against a given functional unit would be lost as well. While MFAs can be used to inform LCA inventories, LCAs do not depend on MFAs for the inventory stage: the blame balance can be established based on monetary flows using economic input-output tables, ignoring the mass balance principles. In theory, additional layers such as elemental balances could be added in an LCA inventory; however, this would make the development of the blame balance significantly more complicated and less transparent.

TABLE 3.1 Dimensions covered by various industrial ecology tools.

Dimension		Tool MFA/SFA	IOA	LCA*
Stages		Yes	Yes, no stocks	Yes
Trade		Yes	Yes	No
Layers	Total mass	Yes	Possible, no balance	Possible, no balance
	Elemental mass	Yes	No	No
	Energy	Yes	Possible, no balance	Possible, no balance
	$ (money)	Possible, no balance	Yes	Possible, no balance
	Blame	No	Possible, assuming B follows $	Yes
Time		Yes	Yes, usually 1 year	No real-time, only allocated time

* *Assumes attributional LCA for individual products or services.*
MFA, material flow analysis; *SFA*, substance flow analysis; *IOA*, input-output analysis; *LCA*, life cycle assessment; *M*, mass; *E*, energy; *$*, money; *B*, blame.

IOAs differentiate stages, trade, layers, and time dimensions. However, in IOAs, stages usually include only production processes, excluding environment and use phase, and end-of-life management. Trade refers mainly to international trade among countries or regions, and less at subnational levels. The layers are usually limited to a single layer of financial flows. In some cases (hybrid IOAs), IOAs are conducted using multiple layers, usually using the economic layer as the fundamental layer that carries the other physical layer and the "embodied" environmental impacts. However, the physical and the financial layers may be inconsistent because value—unlike matter or energy—can be created and destroyed and because not all physical flows have financial equivalents. For example, air, water, wood, or even excavated local geological material for construction purposes could be considered as "free" if they are consumed but do not have to be paid for. In addition, physical and financial flows tend to not align very well because physical and chemical processes act on products independent of whether financial transactions take place. The resolution of financial transactions thus cannot be accurately adjusted to the granularity of the physical layer, resulting inevitably in inconsistencies between the layers. To reduce these inconsistencies, integration needs to consider the physical layer as the fundamental starting point or layer, with a higher granularity, and the financial layer as auxiliary, not vice versa. Ironically, the fragmented data landscape often requires researchers to use financial data to estimate physical stocks and flows. A robust monitoring of the physical economy should therefore always start with a monitoring of the intrinsic properties and be balanced, as done in MFA.

3.3 COMPONENTS FOR MONITORING THE PHYSICAL ECONOMY

MFAs differ widely in their scope and approach; however, they always consist of the same seven conceptual components that are critical for monitoring the physical economy (Figure 3.3): systems, data, models and scenarios, uncertainty analysis, visualization, indicators, and strategy and decision support. Each component and how they link together are described here.

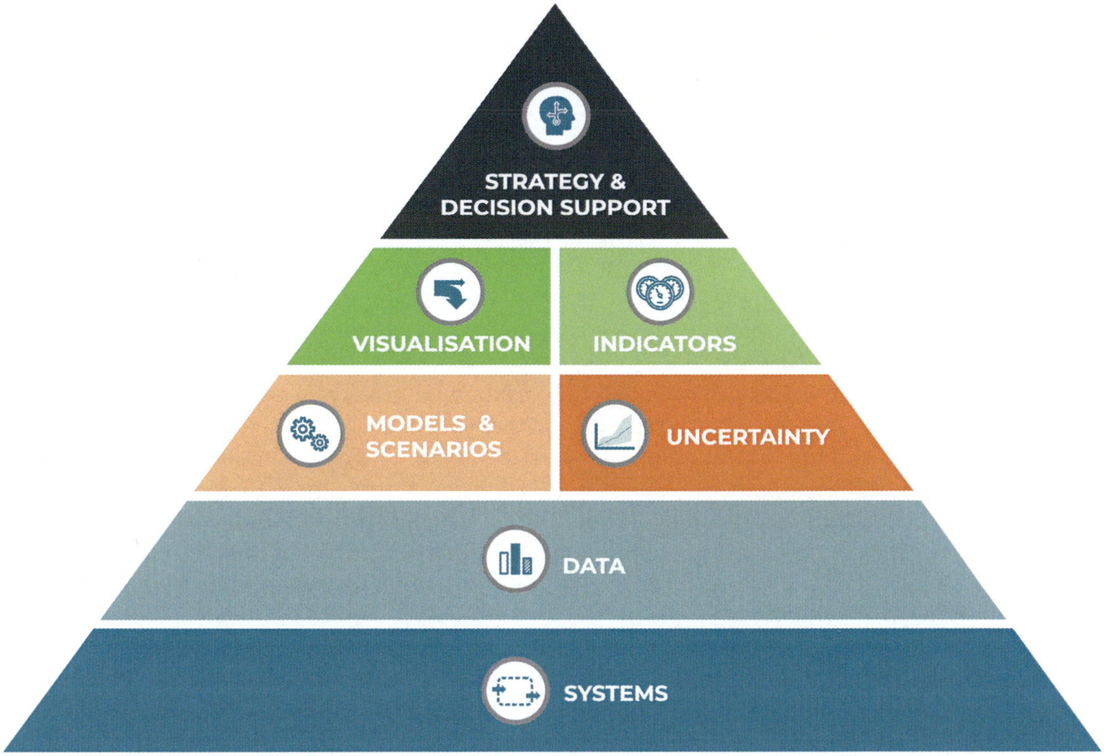

FIGURE 3.3 The seven components for monitoring the physical economy, presented in a hierarchy: the robustness of each component is limited by the robustness of the underlying components. *Reproduced with permission from the MinFuture project (European Commission, 2020).*

Systems describe where the materials in question are located in space and time (processes, stocks) and where they are going (flows) within a defined boundary. Systems form the foundation of any MFA as they describe both the reference frame ("coordinate system") and the "coordinates" of the observed stocks and flows. The granularity of the system is determined by the purpose of the MFA.

Data represent measurements or estimations of either stocks at a given point in time or of flows over a given period. The data can refer to the total goods or additional layers, or they can refer to data used to calculate the stocks and flows, such as concentrations or transfer coefficients.

Models are here defined as mathematical representations of metabolic systems. They consist of balance equations, representing the system, and model approach equations, representing the drivers and their quantitative impacts on the system. Dynamic MFA models include an additional type of equation called "intrinsic equations" that describe the link between stocks and stock changes. Models are used to simulate historical changes in metabolic systems, to identify patterns, or to develop scenarios for exploring alternative development pathways.

Uncertainty is inherent to all system descriptions and can result from errors in the system definition or uncertainty in the data. Approaches to uncertainty analysis aim at making uncertainties transparent and at reducing them (Allesch and Rechberger, 2018). Tailor-made MFA software facilitates the application of error propagation within and data reconciliation for a system (http://www.stan2web.net).

Indicators in MFA are statistical values derived from the quantified systems. They represent specific characteristics or aspects of interest within complex systems and are often used to benchmark and monitor the performance of systems, to define targets, or to measure progress toward reaching these targets.

Visualizations illustrate complex systems using graphical representations. They are used to facilitate the analysis, the interpretation, and the communication of the results. Material cycles are often illustrated using Sankey diagrams (Lupton and Allwood, 2017), often in combination with other visual representations.

Strategy and decision support represents the last step of an MFA, where the results are interpreted in terms of their policy (or business) relevance. For example, the findings of MFA could inform forthcoming national policies on security of supply, decarbonization targets, waste strategies, and so on. At the corporate level, MFA could provide invaluable information to sustainability strategies, procurement practices, and identify hotspots that may result in significant business risks, such as decision for capacity increase with potential supply constraints. In addition, the findings can be interpreted from a methodological perspective, for example related to critical improvements of the monitoring of stocks and flows.

The strategy and decision support step is both the first and the last one. It is the first because it defines the questions raised for the MFA and as such determines the system, level of detail, data needs, and all the other levels below. Once all the work is completed, it is used to inform strategies and potentially to raise new questions, starting a new cycle with different system boundaries.

The seven components form a hierarchy, illustrated with a pyramid (Figure 3.3). The hierarchical structure indicates that the robustness of each component is informed by the robustness of the underlying levels. Systems can be described qualitatively without quantification of stocks and flows; however, data without a system context are meaningless. The MFA system definition represents the reference frame for describing the location of material stock and flow measurements. The robustness of the data therefore rests upon the robustness of the system. Similarly, a system can be quantified through observations that provide the necessary data without using a mathematical model and scenarios, but mathematical models and scenarios are useless if not based on robust systems and data. Uncertainty analysis, in turn, is dealing with uncertainties of the system definition, of the data of stocks and flows, and of the model approach equations/parameters.

The higher levels tend to be regarded as more (policy) relevant and tend to attract more attention. However, the robustness of these higher levels entirely depends on the robustness of the systems and data components. Paradoxically, it is therefore strategically smart to invest most of the efforts in the enhancement of the components at the bottom of the pyramid.

3.4 APPLICATION OF THE FRAMEWORK: MAPS OF THE ALUMINUM CYCLE

We illustrate the use of MFA with the example of aluminum and show how different system definitions, including layers and scales, can be used to describe and solve systemic challenges. We start with the aluminum cycle at a global scale and show how a system combining aluminum and carbon layers can provide insights into the potential of recycling to mitigate global greenhouse gas (GHG) emissions. We then add layers for alloying elements to discuss how quality issues related to downcycling can limit recycling. Finally, we present a plant-level MFA for the layers goods, aluminum, and carbon to illustrate how a physical accounting at individual plants can be used to identify potentials for saving resources, energy, emissions, and waste, and therefore for reducing costs.

3.4.1 Global Aluminum Cycle—Energy Use and GHG Emissions

The global aluminum cycle (Figure 3.4A) starts with primary production by mining bauxite ore, which is refined to produce alumina (Al_2O_3). An electrolytic smelting process reduces alumina to aluminum metal. The secondary route produces aluminum by recycling scrap (Chapter 20—Aluminum), either from manufacturing losses (preconsumer scrap) or from products discarded after use (postconsumer scrap). Recycling provides 42.6 Mt/year of Al compared to 29.4 Mt/year from primary production (Liu et al., 2013). However, when looking at the carbon layer (Figure 3.4B), most of the GHG emissions are generated in the primary smelting process. This energy-intensive process generates emissions in the form of process emissions (e.g., from anode consumption) and direct CO_2 emissions from on-site use of fossil fuels, but first and foremost, indirect CO_2 emissions from electricity production.

As visualized in (Figure 3.4B), LCA studies have shown that recycling uses about 10–20 times less energy per kilogram of aluminum and hence also significantly reduces GHG emissions (International Aluminum Institute, 2009; IEA-ETSAP, 2012; European Aluminum, 2017). One might conclude from this functional unit perspective that an important climate change mitigation strategy would be to increase recycling. Often legislation requiring a higher recycled content is proposed, but does this make sense here? From a system perspective (Figure 3.4A), recycling is already dominant; however, most of the recycled scrap is preconsumer scrap generated in the semimanufacturing and manufacturing industry. Preconsumer scrap tends to be pure, is generated in large quantities, and has a high value, and thus is recycled efficiently. Nevertheless, preconsumer scrap generation is an inefficiency of the system: its recycling requires additional energy. Ironically, less efficient production systems result in higher recycling rates. Is energy truly saved with preconsumer scrap recycling? Aluminum first goes through primary production, then in addition through recycling, causing higher energy use than only primary production. In a system perspective, to increase material efficiency, one should focus on reducing waste generation in manufacturing. Preconsumer scrap should therefore be reduced, thereby lowering overall recycling rates, and thus recycled content. Conversely, policies that encourage a higher recycled content in products might provide incentives to reduce efficiency and therefore increase overall energy use and GHG emissions.

The time dimension is also an important factor when assessing the future potential of recycling for transforming the global aluminum cycle. Even though a large share of aluminum goes into building and construction (12 Mt/year), and the in-use stock is 70× higher than in packaging products, this sector only generates 1.2 Mt/year of postconsumer scrap, compared to 3.4 Mt/year from packaging, the main contributor to recycling flows (Figure 3.4A). Of course, a building lifetime is decades or centuries, versus weeks or months for packaging. This highlights the importance of considering the dynamic nature of stocks and the lifetime of products in different sectors. While the long lifetime delays the potential for recycling and circularity, it also reduces the energy demand for keeping the material in the loop: stocks built in the past also provide services in the future

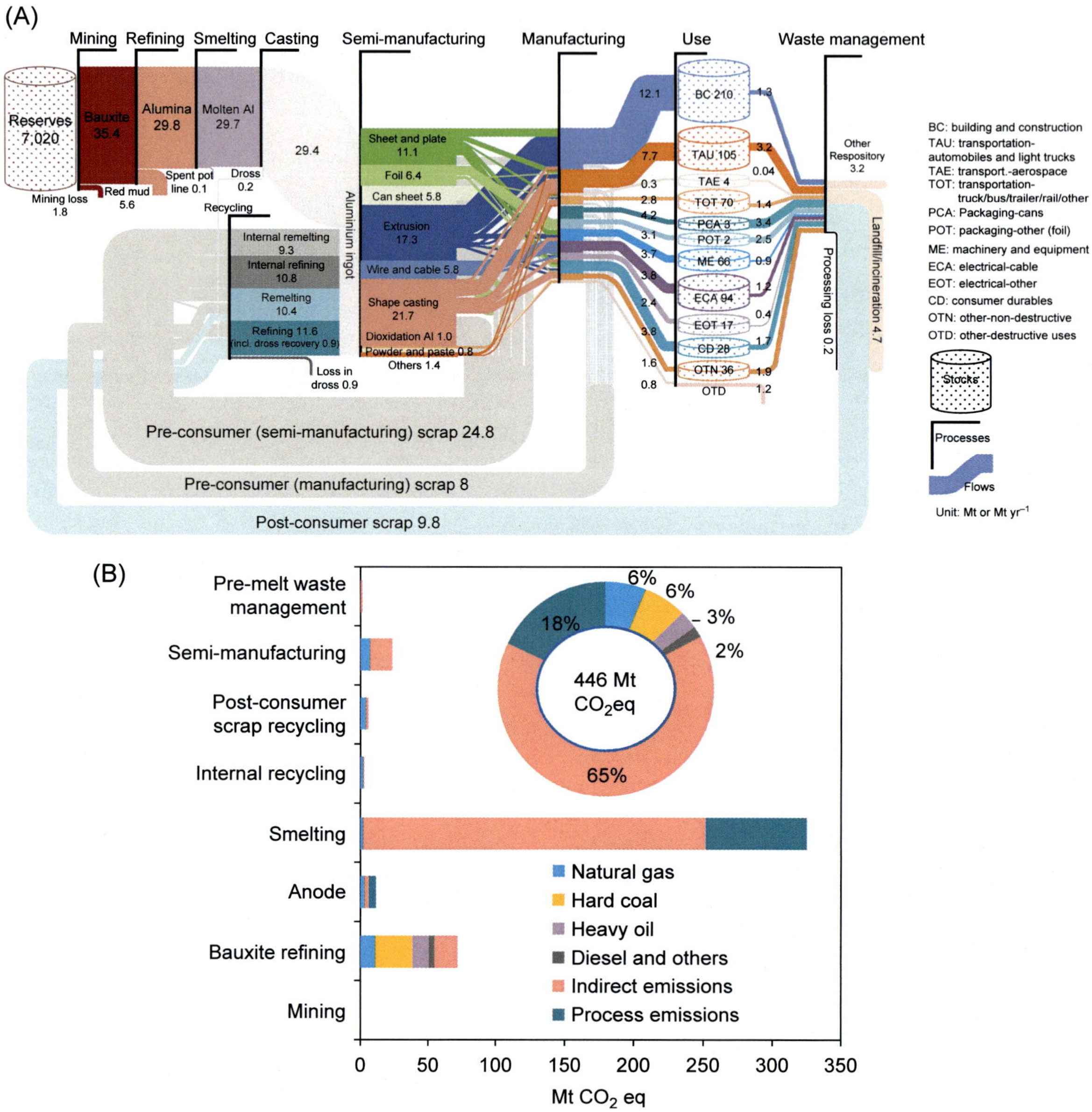

FIGURE 3.4 (A) Global anthropogenic aluminum cycle in 2009 and (B) GHG emissions in all production stages along the global aluminum cycle in 2009. *From Liu et al. (2013).*

without a need to spend energy for material production or recycling. Recycling comes with different technical and thermodynamic challenges in different parts of the system, emphasizing the need for a holistic perspective to maximize resource and energy efficiency.

3.4.2 Global Aluminum Alloy Cycle—Cascading Use and Concern for Scrap Surplus

The common perception that metals can be recycled indefinitely needs to be nuanced. Aluminum is usually not used in a pure form; instead it is alloyed with alloying elements to give it specific properties needed in the manufacturing and the use stages. The accumulation of alloying elements and impurities in the scrap is a severe hindrance to closed-loop recycling. The thermodynamic properties of aluminum make it very difficult to refine this metal and remove unwanted elements. The energy and materials needed to remove or dilute unwanted elements often offset the benefits of recycling (Løvik and Müller, 2014). Removing those alloying elements and impurities through refining is not economically feasible due to the high energy requirements. Therefore wrought aluminum scrap can be used to produce secondary castings but not vice versa. This problem is solved in the primary production by refining the alumina in the Bayer process, before it enters the smelter. At the secondary production, this problem is today largely addressed through a cascadic use or downcycling. Postconsumer aluminum scrap, for example from ELVs and WEEE, is mixed, shredded, and refined to make secondary castings alloys, characterized by their high content of alloying elements (especially silicon and copper) and impurities (Modaresi et al., 2014). Refining here means adding, not removing, alloying elements. The automotive industry is the largest user of secondary castings, mainly for engine blocks and other combustion engine drivetrain parts. These applications act as the main sink in the global aluminum system.

The cascadic use strategy has worked very well for many decades. It effectively limited the need for sorting aluminum scrap while making use of the alloying elements or even impurities in secondary casting applications. However, dynamic MFA models have shown that this strategy may not work in the future (Løvik et al., 2014; Modaresi et al., 2014; Buchner et al., 2015). The demand for secondary castings, which is closely tied to the demand for internal combustion engines, is expected to decrease with the transition to electric mobility. At the same time, wrought alloys are increasingly needed for battery parts and for lightweighting as a replacement for steel, further increasing the amount of mixed scrap in vehicles. A continuation of the cascadic use strategy would therefore result in rapidly growing amounts of mixed scrap that have a shrinking market in engine parts. A scrap surplus that cannot be recycled would come at a very high energy and greenhouse gas cost. To avoid a scrap surplus, the cascadic use strategy needs to be replaced with a circular economy strategy. The focus no longer lies on closing the aluminum cycle, but on closing the individual aluminum alloy cycles.

This transition to a circular economy in the aluminum sector requires monitoring of the global aluminum cycle and corresponding models and scenarios to anticipate the consequences of alternative strategies. Figure 3.5 illustrates such a monitoring system for guiding the transition in the vehicle part. It tracks not only aluminum, but differentiates additional layers: individual alloying elements, alloys, and parts within different types of vehicles.

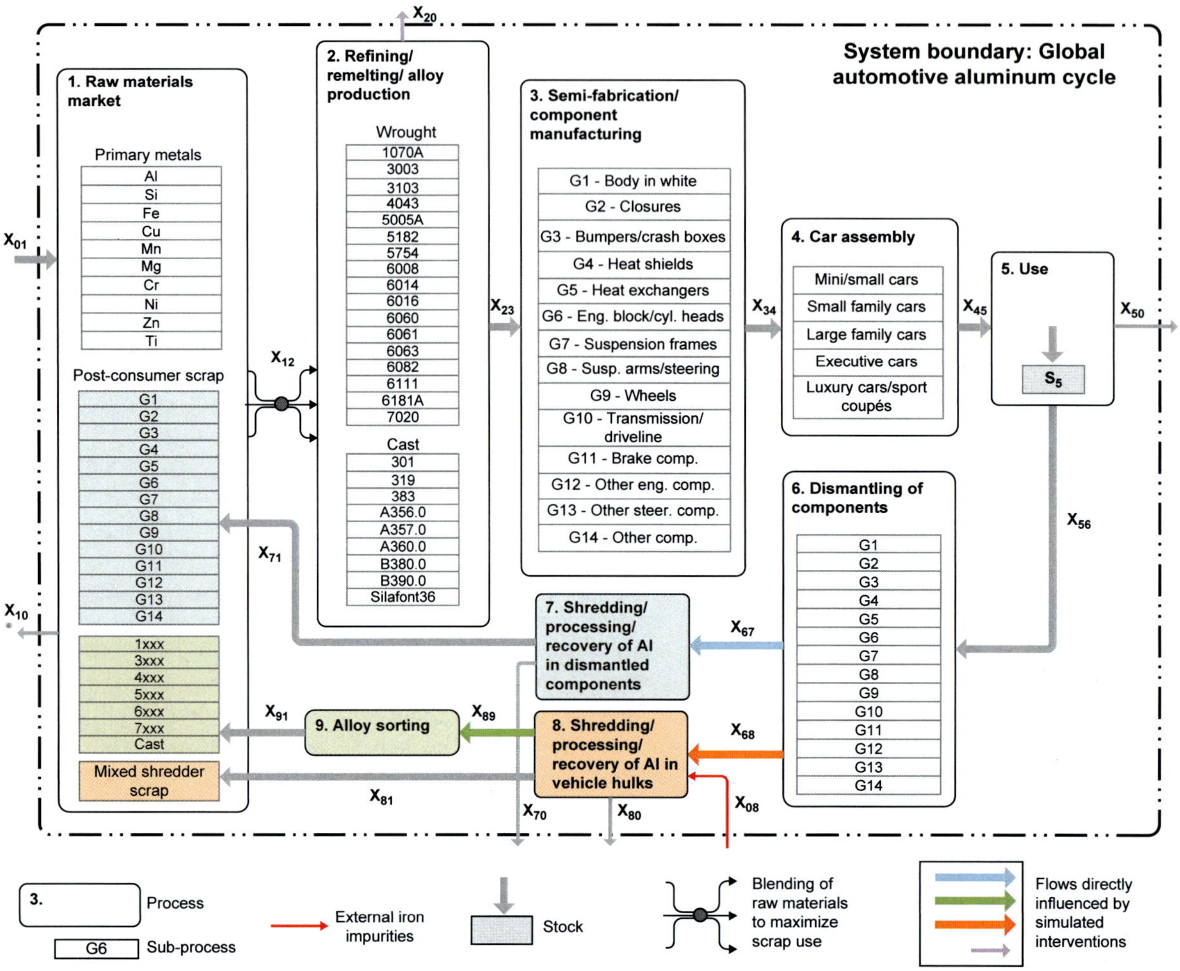

FIGURE 3.5 System definition of the global use, production, and recycling of aluminum and its alloying elements in automobiles. Different alloys and corresponding elements are tracked in the whole system in different layers. *From Løvik et al. (2014).*

It also includes new technologies for sorting, to be evaluated through scenario simulations. The simulation results based on this model (Løvik et al., 2014) showed that the implementation of a circular economy strategy for aluminum—avoiding a scrap surplus—is possible, but it requires coordinated actions of many players in the aluminum sector, including aluminum producers, automobile manufacturers, dismantlers and shredder operators, and scrap dealers. Since vehicles will no longer be able to absorb large amounts of mixed scrap from other sectors, similar models are also needed for the other aluminum applications, including in the building, appliances, and packaging sectors. For aluminum recycling technologies, see also Chapter 20—Aluminum.

3.4.3 Plant-Level Physical Maps—Illustration of the Multilayer Concept

While global and country level metal cycles are useful for understanding the main issues in the global system, they only use average values and do not capture the diversity in performance between different plants and processes at each stage of the value chain. Smaller scale models are needed to quantify the material and energy efficiency of single plants, to study potential improving options, and to follow their implementation.

Models that aim to represent material flows within a single plant have been used for several decades by the steel industry, initially using the flowsheeting approach for optimizing industrial processes and later integrating it with decision support systems (Porzio et al., 2013). This was achieved by exergy analysis, enabling the identification of trade-offs between material and energy efficiency (Gonzalez Hernandez et al., 2018). Even if these methods have also been used to some extent by the aluminum industry, it is only in the recent years that more examples have been published. Hannula et al. (2020) developed a simulation-based flowsheet for aluminum recycling and studied resource efficiency through exergy analysis and LCA. The objective of the work was to study the effects of particle properties, mechanical separation parameters, and desired alloy type on the resource consumption of an aluminum recycling process. The quantity of virgin aluminum required to dilute impurities and to meet predefined alloy specifications was estimated using process simulation.

The use of multilayer MFA for physical accounting at plant level provides detailed insights on the real metabolism of industrial plants and can help identifying internal recycling options. Billy et al. (2022) performed a three-layers (goods, aluminum, and carbon) MFA of a primary aluminum plant, comprising two smelting lines, a foundry, and anode baking and recycling facilities. Eighty flows, linking 12 processes and 14 inventories, were quantified using corporate data, laboratory analyses, and estimates from literature. Results were then tested with sensitivity analysis and mass balance conservation.

Taking a carbon layer at plant level gives more detailed insight into where in the plant and from which source streams emissions occur than conventional carbon accounting methods (Figure 3.6). At the same time, this method adds robustness thanks to mass balance consistency at process and plant level, which would not necessarily be the case with detailed flowsheeting approaches at process level. This MFA-based accounting framework allows testing of systemic material efficiency measures and can be used to optimize resource use and emissions within plants, without risking optimizing one process at the expense of another one. A set of system-based indicators can be derived from the physical accounting and used to build a consistent framework that reconciles company-level performance indicators with carbon reporting, providing insights for strategy and decision support that are consistent with the system and data layers.

Differentiating the carbon and aluminum layers provides insights about waste composition and flows (Figure 3.7). Waste flows with a high aluminum content tend to be recycled to maximize the yield of the plant. Carbon-rich flows are either recycled or, if they contain high levels of impurities, incinerated for energy recovery. Reclamation of waste flows with a mixed chemical composition is more challenging for physical and economic reasons, and they usually end up being landfilled. Such waste composition charts can be used for optimizing internal recycling. They enable plant managers to identify waste streams that are not currently reclaimed and could potentially be recycled or sent to energy recovery with increased sorting or material separation. Besides, a multilayer MFA physical accounting ensures a mass-balance consistent link between waste flows

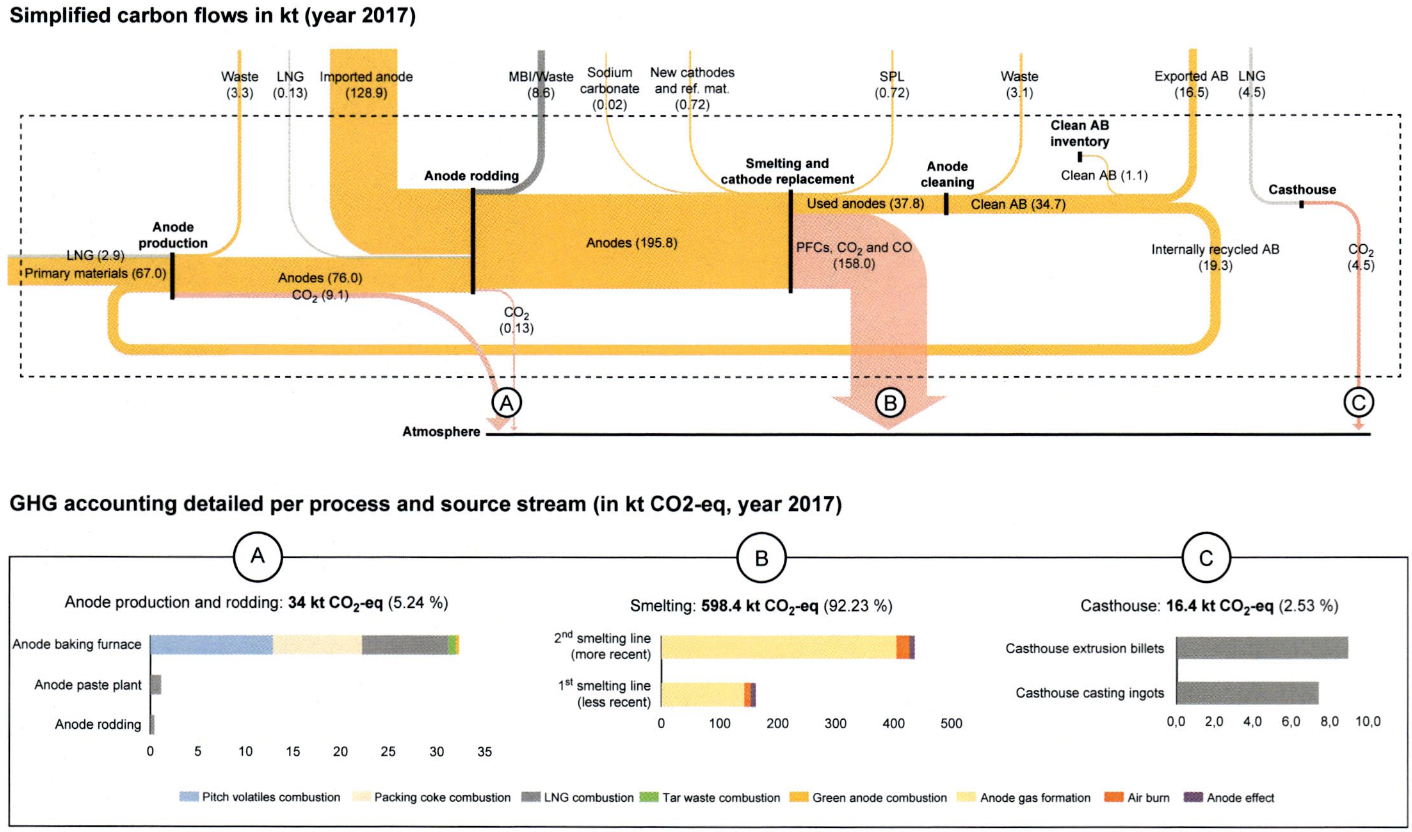

FIGURE 3.6 Simplified carbon flows (in kt of carbon) and associated GHG emissions (in kt CO_2-eq) in a primary aluminum production plant. Fuel flows in *light gray*, mass balance inconsistency in *darker gray*, GHG emissions flows in *red*, other materials in *yellow*. *MBI*, mass balance inconsistency; *AB*, anode butts; *PFCs*, perfluorinated compounds; *LNG*, liquid natural gas; *ref. mat.*, refractory materials; *SPL*, spent pot lining. *From Billy et al. (2022).*

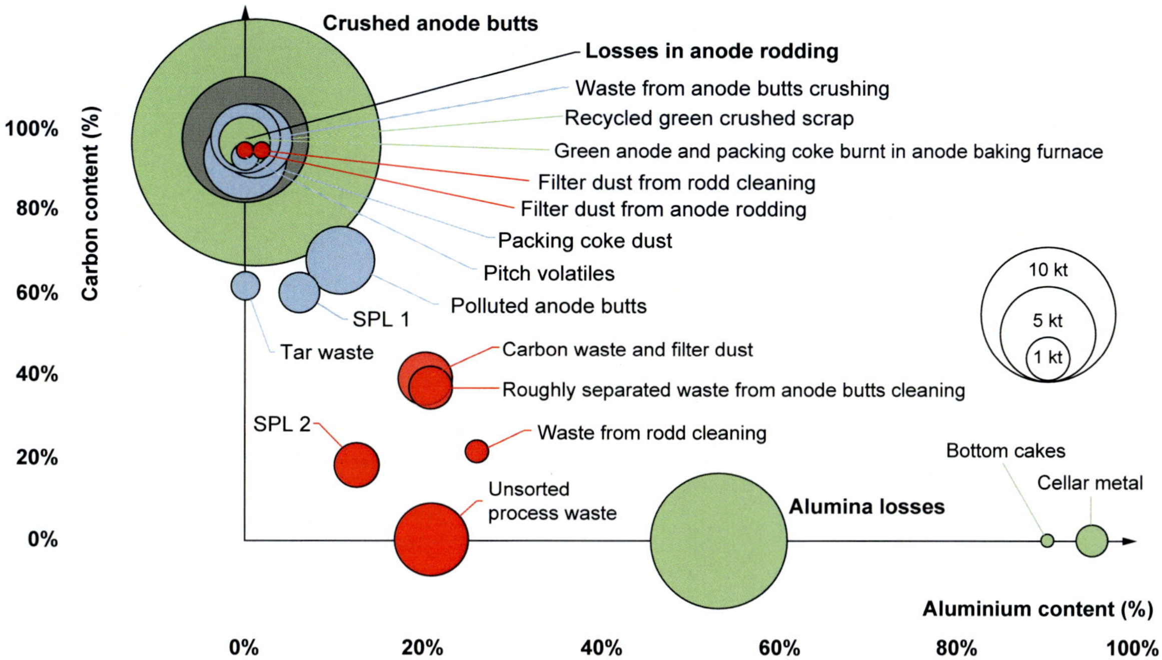

FIGURE 3.7 Waste composition chart. Waste flows generated in 2017 according to their Al and C content and their EoL management method. *Green* = recycled, *blue* = energy recovery, *red* = landfilled. *SPL1*, spent potlining first cut, i.e., carbon fraction; *SPL2*, spent potlining second cut, i.e., refractory fraction. *From Billy et al. (2022).*

and GHG emissions and facilitates the quantification of the emission reduction potential of different recycling and material efficiency options.

3.5 RECOMMENDATIONS

We conclude this chapter with some lessons learned from applications of the principles laid out in the previous sections. These lessons are often not in line with traditional assumptions and common claims about recycling and circular economy strategies. We would like to debunk in particular the myths laid out in Table 3.2. Those myths have in common that they might appear sensible when single materials, processes, plants, or countries are considered in isolation. However, they can be proven wrong when we consider the wider system: shortcomings and negative side effects become clear, showing the importance of system context.

These insights may be summarized in two sentences: Circularity should not be regarded as an end in itself; it should be seen as one among several means for a sustainable development. How far circularity strategies contribute to each SDG depends on the specific context.

Recycling and circular economy are pivotal elements for human society in the twenty-first century. But, as the road to hell is paved with good intentions, recycling can have highly desirable or undesirable, even damaging, consequences dependent on the specific circumstances and the way it is implemented. To ensure that recycling provides a benefit to society, we need to understand how it affects the larger socioeconomic metabolism at different scales and over time.

TABLE 3.2 Common myths about recycling debunked.

Myth	Debunking
Metals can be recycled indefinitely	– Alloying elements and impurities are often difficult to remove, resulting in downcycling unless scrap is sorted very well – Recycling becomes increasingly challenging the more circular an economy becomes due to the accumulation of undesired substances
Circular economy will make mining obsolete	– As long as material stocks keep growing, a circular economy cannot satisfy material needs and mining remains essential – Primary materials are required for dilution of scrap impurities – Primary materials are required to replace any losses along the cycle, which sometimes are unavoidable
More recycling (higher recycled content) always saves resources, energy, and GHGs	– High recycling rates are often due to recycling of preconsumer scrap generated in manufacturing processes. This represents an inefficiency in the system that results in higher, not lower, energy consumption – Recovering materials from waste can require more energy than producing it from ore. There is always an optimum for total energy use, which is rarely 100% recycling – Many demand-side measures (e.g., extend lifetimes, reduce material intensity) effectively save resources, energy, and GHGs, but also reduce total recycling flows
Recycling performance depends on technologies only	– Recycling performance depends on coordinated measures of different actors in the system, including product design for easy disassembly, consumer behavior, and willingness for separate collection, convenient collection infrastructure, markets for recycled materials, as well as regulatory frameworks facilitating recycling
An efficient solution at small scale is also efficient when scaled up	– The effectiveness of solutions depends on the context: e.g., upscaling can be limited by the availability of natural resources and scrap, which also varies by region
A system that is efficient today should be preserved	– Some systems, such as the cascadic use of metals, are highly efficient in saving resources and energy at their current state. However, they may become very inefficient or impossible to maintain in the future, for example when scrap amounts increase and/or the bottom reservoir of the cascade shrinks. Preserving a system's performance may require it to change

MFA maps of the physical economy can inform decision makers about this context and provide a robust basis for designing and monitoring recycling and circular economy strategies. Given the severity and urgency of many environmental problems and the enormous amount of data collected today about human activities, it is striking that our maps of the physical economy are still so rudimentary and fragmented.

Similarly, we risk wasting a lot of resources for well-intended circular economy infrastructures if the strategies are not rooted in a robust understanding of the system we are transforming.

Insights for recycling and circularity strategies are usually limited by the system and data—the bottom of the pyramid (Figure 3.3). The quantitative system understanding requires a common language between traditional disciplines analyzing individual sectors or countries. Investments made into the quantitative system understanding—including data harmonization—can help to identify smarter, cheaper, and more effective strategies that generate synergies with other goals. Due to the urgency of the global sustainability challenges, robust maps of the physical economy are no longer a luxury, but a necessity.

Acknowledgments

This chapter is an outcome of the EU-funded MinFuture project (European Commission, 2020), Grant agreement ID: 730330. We thank Maren C. Lundhaug and the other partners of the MinFuture project for their contributions.

References

Allesch, A., Rechberger, H., 2018. Compilation of Uncertainty Approaches and Recommendations for Reporting Data Uncertainty. MinFuture Deliverable D3.3. TU Wien, Vienna. 87 pp. Available from: https://minfuture.eu/downloads/D3.3_uncertainty.pdf.

Allwood, J.M., Cullen, J.M., Milford, R.L., 2010. Options for achieving a 50% cut in industrial carbon emissions by 2050. Environ. Sci. Technol. 44 (6), 1888–1894. https://doi.org/10.1021/es902909k.

Allwood, J.M., Cullen, J.M., Carruth, M.A., Cooper, D.R., McBrien, M., Milford, R.L., Moynihan, M.C., Patel, A. C., 2012. Sustainable Materials: With Both Eyes Open. UIT Cambridge, Cambridge, ISBN: 1906860076.

Ayres, R.U., Simonis, U.E., 1994. Industrial Metabolism: Restructuring for Sustainable Development. United Nations University Press, Tokyo, New York, Paris, ISBN: 9280808419, p. 376.

Baccini, P., Brunner, P.H., 1991. Metabolism of the Anthroposphere. Springer, Berlin, New York, ISBN: 9783540537786. 157 pp.

Baccini, P., Brunner, P.H., 2012. Metabolism of the Anthroposphere: Analysis, Evaluation, Design, second ed. MIT Press, ISBN: 9780262016650. 408 pp.

Billy, R.G., Monnier, L., Nybakke, E., Isaksen, M., Müller, D. B., 2022. Systemic approaches for emission reduction in industrial plants based on physical accounting: example for an aluminum smelter. Environ. Sci. Technol. 56 (3), 1973–1982. https://doi.org/10.1021/acs.est.1c05681.

Brunner, P.H., Rechberger, H., 2004. Practical Handbook of Material Flow Analysis. CRC Press/Lewis Publishers, Boca Raton, FL. 318 pp. ISBN:00-203-50720-7.

Brunner, P.H., Rechberger, H., 2017. Practical Handbook of Material Flow Analysis: For Environmental, Resource, and Waste Engineers, second ed. CRC Press, Boca Raton, ISBN: 978-1-4987-2134-9. 423 pp.

Buchner, H., Laner, D., Rechberger, H., Fellner, J., 2015. Future raw material supply: opportunities and limits of aluminium recycling in Austria. J. Sustain. Metall. 1 (4), 253–262. https://doi.org/10.1007/s40831-015-0027-3.

EU JRC, 2022. Material System Analysis (MSA) Reports. European Commission Joint Research Centre (JRC). Available from: https://rmis.jrc.ec.europa.eu/?page=msa-reports-b922fd. (Accessed 31 March 2022).

European Aluminium, 2017. Sustainable Development Indicators 2015. Brussels, Belgium, 14 pp. Available from: https://european-aluminium.eu/media/3258/pro_sustainable-devolopment-indicators-2015_20170614.pdf.

European Commission, 2020. MinFuture—global material flows and demand-supply forecasting for mineral strategies. EU H2020 Project Grant Agreement ID: 730330 https://www.minfuture.eu/. Available from: https://cordis.europa.eu/project/id/730330.

Fischedick, M., Roy, J., Acquaye, A., Allwood, J.M., Ceron, J. P., Geng, Y., Kheshgi, H.S., Lanza, A., Perczyk, D., Price, L.K., Santalla, E.M., Sheinbaum, C., Tanaka, K., 2014. Industry. In: Edenhofer, O., Pichs-Madruga, R., Sokona, Y., Farahani, E., Kadner, S., Seyboth, K., Minx, J.C. (Eds.), Climate Change 2014: Mitigation of Climate Change. Contribution of Working Group III to the Fifth Assessment Report of the Intergovernmental Panel on Climate Change. Cambridge University Press, Cambridge, United Kingdom, pp. 743–793 (Chapter 10).

Fischer-Kowalski, M., 1998. Society's metabolism—the intellectual history of materials flow analysis, part I, 1860-1970. J. Ind. Ecol. 2 (1), 61–78. https://doi.org/10.1162/jiec.1998.2.1.61.

Gonzalez Hernandez, A., Lupton, R.C., Williams, C., Cullen, J.M., 2018. Control data, Sankey diagrams, and exergy: assessing the resource efficiency of industrial plants. Appl. Energy 218, 232–245. https://doi.org/10.1016/j.apenergy.2018.02.181.

Graedel, T.E., Allenby, B.R., 2010. Industrial Ecology and Sustainable Engineering, International ed. Pearson Prentice Hall, Upper Saddle River, NJ. 352 pp. ISBN: ISBN10: 0138140340; ISBN13: 9780138140342.

Graedel, T.E., Lifset, R.J., 2016. Industrial ecology's first decade. In: Clift, R., Druckman, A. (Eds.), Taking Stock of Industrial Ecology. Springer International Publishing, Cham, ISBN: 978-3-319-20571-7, pp. 3–20.

Haberl, H., Wiedenhofer, D., Pauliuk, S., Krausmann, F., Müller, D.B., Fischer-Kowalski, M., 2019. Contributions of sociometabolic research to sustainability science. Nat. Sustain. 2 (3), 173–184. https://doi.org/10.1038/s41893-019-0225-2.

Hannula, J., Godinho, J.R.A., Llamas, A.A., Luukkanen, S., Reuter, M.A., 2020. Simulation-based exergy and LCA analysis of aluminum recycling: linking predictive physical separation and re-melting process models with specific alloy production. J. Sustain. Metall. 6 (1), 174–189. https://doi.org/10.1007/s40831-020-00267-6.

Hertwich, E.G., 2021. Increased carbon footprint of materials production driven by rise in investments. Nat. Geosci. 14 (3), 151–155. https://doi.org/10.1038/s41561-021-00690-8.

IAI, 2022. The Global Aluminium Cycle—Aluminium Stocks and Flows Visualization. International Aluminium Institute. Available from: https://alucycle.international-aluminium.org/. (Accessed 31 March 2022).

IEA, 2021. World Energy Outlook 2021. International Energy Agency, Paris. 386 pp. Available from: https://www.iea.org/reports/world-energy-outlook-2021.

IEA-ETSAP, 2012. Aluminium production. In: Technology Brief I10. International Energy Agency (IEA) Energy Technology Systems Analysis Programme (ETSAP). 5 pp. Available from: https://iea-etsap.org/E-TechDS/PDF/I10_AlProduction_ER_March2012_Final%20GSOK.pdf.

International Aluminium Institute, 2009. Global Aluminium Recycling: A Cornerstone of Sustainable Development. International Aluminium Institute, London, United Kingdom. 36 pp. Available from: https://international-aluminium.org/wp-content/uploads/2021/09/Global-Aluminium-Recycling.pdf.

Liu, G., Bangs, C.E., Müller, D.B., 2013. Stock dynamics and emission pathways of the global aluminium cycle. Nat. Clim. Chang. 3 (4), 338–342. https://doi.org/10.1038/nclimate1698.

Løvik, A.N., Müller, D.B., 2014. A material flow model for impurity accumulation in beverage can recycling systems. In: Light Metals 2014. TMS, The Minerals, Metals & Materials Society, pp. 907–911.

Løvik, A.N., Modaresi, R., Müller, D.B., 2014. Long-term strategies for increased recycling of automotive aluminum and its alloying elements. Environ. Sci. Technol. 48 (8), 4257–4265. https://doi.org/10.1021/es405604g.

Lupton, R.C., Allwood, J.M., 2017. Hybrid Sankey diagrams: visual analysis of multidimensional data for understanding resource use. Resour. Conserv. Recycl. 124, 141–151. https://doi.org/10.1016/j.resconrec.2017.05.002.

Miller, R.E., Blair, P.D., 2022. Input-Output Analysis: Foundations and Extensions, third ed. Cambridge University Press, Cambridge, ISBN: 9781108484763.

Modaresi, R., Løvik, A.N., Müller, D.B., 2014. Component- and alloy-specific modeling for evaluating aluminum recycling strategies for vehicles. JOM 66 (11), 2262–2271. https://doi.org/10.1007/s11837-014-0900-8.

Müller, D.B., Wang, T., Duval, B., Graedel, T.E., 2006. Exploring the engine of anthropogenic iron cycles. Proc. Natl. Acad. Sci. 103 (44), 16111–16116. https://doi.org/10.1073/pnas.0603375103.

Müller, D.B., Wang, T., Duval, B., 2011. Patterns of iron use in societal evolution. Environ. Sci. Technol. 45 (1), 182–188. https://doi.org/10.1021/es102273t.

Müller, D.B., Liu, G., Løvik, A.N., Modaresi, R., Pauliuk, S., Steinhoff, F.S., Brattebø, H., 2013. Carbon emissions of infrastructure development. Environ. Sci. Technol. 47 (20), 11739–11746. https://doi.org/10.1021/es402618m.

Pauliuk, S., Wang, T., Müller, D.B., 2013. Steel all over the world: estimating in-use stocks of iron for 200 countries. Resour. Conserv. Recycl. 71, 22–30. https://doi.org/10.1016/j.resconrec.2012.11.008.

Porzio, G.F., Fornai, B., Amato, A., Matarese, N., Vannucci, M., Chiappelli, L., Colla, V., 2013. Reducing the energy consumption and CO2 emissions of energy intensive industries through decision support systems—an example of application to the steel industry. Appl. Energy 112, 818–833.

Turner, G.M., Poldy, F., 2001. Let's get physical: creating a stocks and flows view of the Australian Economy. MODSIM 2001 International Congress on Modelling and Simulation, Canberra, Australia, 10–13.12.2001, Perth, 1637–1642 pp.

UN General Assembly, 2015. Transforming our world: the 2030 Agenda for Sustainable Development. United Nations, New York, USA. 35 pp. Available from: http://www.un.org/ga/search/view_doc.asp?symbol=A/RES/70/1&Lang=E.

UNEP, 2013. In: Reuter, M., Hudson, C., Van Schaik, A., Heiskanen, K., Meskers, C., Hagelüken, C. (Eds.), Metal recycling: opportunities, limits, infrastructure. United Nations Environmental Programme UNEP, Nairobi, ISBN: 978-92-807-3267-2. A report of the working group on the global metal flows to the international resource panel. 320 pp. Available from: https://wedocs.unep.org/20.500.11822/8423.

Wiedenhofer, D., Fishman, T., Lauk, C., Haas, W., Krausmann, F., 2019. Integrating material stock dynamics into economy-wide material flow accounting: concepts, modelling, and global application for 1900–2050. Ecol. Econ. 156, 121–133. https://doi.org/10.1016/j.ecolecon.2018.09.010.

Zoboli, O., Laner, D., Zessner, M., Rechberger, H., 2016. Added values of time series in material flow analysis: the Austrian phosphorus budget from 1990 to 2011. J. Ind. Ecol. 20 (6), 1334–1348. https://doi.org/10.1111/jiec.12381.

CHAPTER

4

Material efficiency—Squaring the circular economy: Recycling within a hierarchy of material management strategies

Julian M. Allwood

University of Cambridge, Cambridge, United Kingdom

4.1 IS A CIRCULAR ECONOMY POSSIBLE OR DESIRABLE?

Kenneth Boulding's famous 1966 essay, "The Economics of the Coming Spaceship Earth," raised awareness of the contrast between an "open economy" with unlimited input resources and output sinks, and a "closed economy" in which resources and sinks are bounded and remain forever a part of the concerns of the economy (Boulding, 1966). Boulding's essay is often cited as the origin of the phrase "circular economy," which is now widely used as an axiom in environmentally concerned discussions. At present, in parallel with the (failing) search for the miracle of unlimited renewable energy supplies and the (forlorn) creative approaches to hiding (sequestering) undesirable outputs underground, the dogma of today's proenvironmental discussions in politics and mass media reporting assumes that aspiring to a "circular economy" is one of the key technical fixes that will solve our environmental problems and allow the economy to keep on growing.

Governments, celebrities, NGOs, and academics compete to be leading champions of the "circular economy," but is that really correct? If global demand for steel has quadrupled in under 50 years (USGS, 2012a), largely to make high-rise buildings that are replaced on average every 40 years, is the pursuit of recycling really the most important option for reducing the impacts of your material production? If you are concerned about the "criticality" of elements used in modern electronics, is the best option to try to separate out all those elements from the complex mixed streams of waste electronics? Assuming he owned a car, should Kenneth Boulding's descendants today be driving a car made entirely from the atoms of his car built in the 1950s?

The key image created by the "circular economy" is of a fixed number of atoms currently formed into today's products that should be repeatedly reorganized into future products without requiring any further injection of new atoms. Is that really desirable? These rhetorical questions challenge the attractiveness of this target: if demand is growing, the circle cannot remain closed, and it may be a much more important priority to reduce the rate at which new material is required; if today's products require precise and complex mixing of atoms to create high-performance properties, the energy required to separate them from these products may be very much greater than the energy required to extract new material from ore or tailings; if technology is advancing so that the design requirements for today's products are unrelated to those of the past, it may not be possible to create tomorrow's products from today's stock of atoms in use.

Behind these challenges lies one primary question: how much energy is required to operate a circular economy? Almost all recycling processes operate by breaking down a solid waste stream into a liquid, which is then purified by some means. With today's technologies, some wastes cannot be broken down, and some liquids cannot be purified, but in principle there is no atomic bond that could not in future be separated under human control and with sufficient energy. However, the energy required to achieve this may be very much greater than that required to liberate an equivalent stream of atoms from other sources (e.g., ores, or even mining tailings), and the process of doing so may lead to other environmentally harmful consequences such as the release of toxins or greenhouse gases.

The idea of a circular economy might—if global demand for both the volume and compositions of products stabilized—be technically feasible, but for many materials would require significantly greater energy inputs than creating products by other means. If energy is produced from fossil fuels, as seems inevitable in the short-to-medium term future, the pursuit of this circular economy would therefore accelerate the release of greenhouse gases and consequent warming of the climate. Is this what its proponents intend?

Rather than having circularity as a goal, a more pragmatic vision for a material future would be to aim to meet human needs while minimizing the environmental impact of doing so (Allwood et al., 2011). Because of the urgency of the challenge to curb global warming, the primary impact of concern is the emission of greenhouse gases, and in turn therefore the primary challenge is to deliver products with minimum energy. For some materials in some contexts, recycling—reducing old products to liquid, and then reforming the liquid to new materials—may be the least energy-intensive solution, hence the importance of this book overall. However, a wider set of strategies with a greater potential impact across all material classes can be found under the umbrella term "material efficiency"—describing the aim to deliver material services with less input of material. Successful recycling can reduce demand for new ore or biomass, but successful material efficiency reduces the total demand for material processing. This chapter aims to explore the means by which material efficiency can be achieved and to discuss how it might be applied and brought about for different major material groups.

But before setting out on that exploration, we must return to Boulding's famous essay. The statement earlier suggested that a circular economy could be achieved

> if global demand for both the volume and composition of products stabilised.

That 12-word condition describes an environmental nirvana that virtually defies all imagination in current growth-driven economies, but prior to the Industrial Revolution was largely inevitable. Wrigley (2013) describes the operation of the United Kingdom prior to the exploitation of fossil fuel resources as an "organic economy," which was self-limiting: a householder with access to a certain amount of land could use the land for growing crops for food, for animal husbandry, for fiber (for clothing or structure), or for fuel (for comfort, cooking, or material production). However, with a finite amount of land per householder these options must be traded, so the population could not grow in number or in material demand without sacrificing comfort or security. The availability of fossil fuels eliminated this constraint, so has allowed both forms of growth—without limit to date. Biologists and anthropologists report countless examples of species failing to self-regulate, and thus creating their own extinction by exhausting their support systems, and yet nearly 100 years ago, John Maynard Keynes anticipated that the goal of economic development was simply to reach a new equilibrium with a higher quality of life:

> For many ages to come the old Adam will be so strong in us that everybody will need to do some work if he is to be contented. We shall do more things for ourselves than is usual with the rich today, only too glad to have small duties and tasks and routines. But beyond this, we shall endeavour to spread the bread thin on the butter to make what work there is still to be done to be as widely shared as possible. Three-hour shifts or a fifteen-hour week may put off the problem for a great while. For three hours a day is quite enough to satisfy the old Adam in most of us! … Thus for the first time since his creation man will be faced with his real, his permanent problem—how to use his freedom from pressing economic cares, how to occupy the leisure, which science and compound interest will have won for him, to live wisely and agreeably and well. … The strenuous purposeful money-makers may carry all of us along with them into the lap of economic abundance. But it will be those peoples, who can keep alive, and cultivate into a fuller perfection, the art of life itself and do not sell themselves for the meansof life, who will be able to enjoy the abundance when it comes.
> *Keynes (1930)*

Unfortunately—at least when assessing human impacts on the environment—Keynes was overoptimistic, and apart from specific (often religious) communities such as monastic communities or the Shakers, human demand for energy and material has proven insatiable. Figure 4.1 gives evidence of this, showing per capita demand for energy, electronics, cars growing apparently with no sign of saturation.

Growth—measured in particular by national GDPs—is now the primary goal of economic policy in virtually all nations, and this both nurtures and feeds off the transient human desires for novelty and change that can be satisfied by material purchasing. A growing literature suggests that the pursuit of economic growth in this form does not lead to human well-being (Kasser (2002, 2017) and Layard (2006), for example, provide extensive datasets to support this thesis), yet the reductionist view of mainstream economics that well-being (or "welfare") arises from purchasing is both pervasive and supported by observations of purchasing behavior. Figure 4.2 gives a compelling demonstration that, unless energy prices are very high, car purchasers in the United States prefer larger cars with greater acceleration: "more" (acceleration, volume, weight) with "less" (money) is an economic success arising from tremendous progress in techniques of production, but apparently does not lead to increased human satisfaction, despite its consequence of increased material and energy production. This, and much other, evidence of individual demand for consumption, supported by government goals of GDP growth, creates the exact reverse of the 12-word condition required to allow circularity.

Boulding's proposed solution to this apparently insatiable human habit of demand for more and more goods was to move from economic measures based on annual flows (in particular, the GDP) to measures based on stocks. In contrast to the promotion of recycling inherent to the vision of a circular economy, such stock-based measures would promote conservation, maintenance, and renewal as the keys to delivering material well-being. An economy measured by its stocks would count increased material

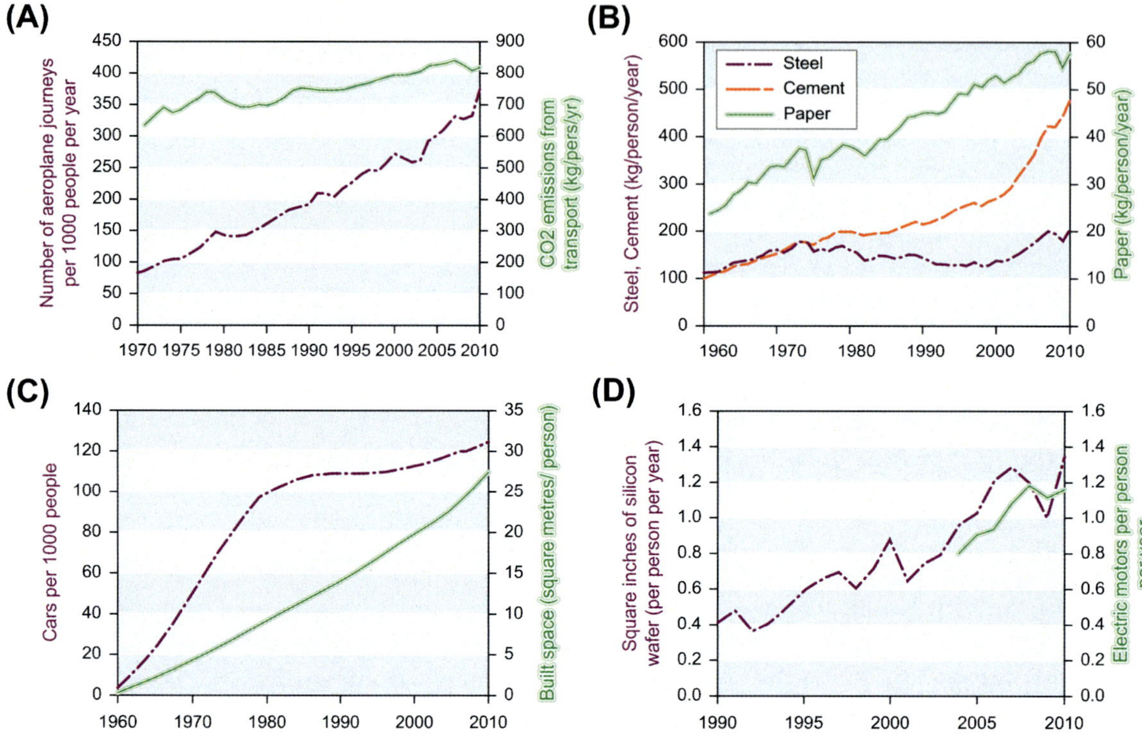

FIGURE 4.1 Trends of growth in global demand for material. The graphs have been normalized by population data from the World Bank (2013), which correlates closely with UNDESA (2011): (A) Airplane passengers carried and transport CO_2 emissions from World Bank (2013). (B) Steel data from USGS (2012a), cement data from USGS (2012b), and paper data from FAO (2012). (C) Global car production data from USDoT (2012, Tables 1–23), converted to in-use stocks, assuming a 20-year car life with zero stock in 1960; Built space calculated from cement production USGS (2012b), assuming all cement used as concrete to create building floors lasting for 40 years. Cement usage estimated at 300 kg/m^2 based on bottom-up estimates for 200 mm floor slab in Goodchild (1993) and van Oss and Padovani (2008), and top-down estimates of Chinese construction in Fu et al. (2013). (D) Silicon wafer production data from Winegarner (2011) and electric motor data from Zhou (2011). *Adapted from Gutowski et al. (2013).*

flows as a failure rather than an economic success. Demand for new cement within this vision would reflect a failure of conservation rather than a success in creating economic growth, where success would arise from the maintenance of lightweight, low-energy products over a long duration, with individual measures of success derived from quality of life, leisure, creativity, and other constructive values rather than from income alone.

Our vision of a future sustainable material economy is therefore not prescribed by the ambition to create a circular economy, but aims to minimize its total environmental impact. The key measures to promote this are frequently reported as "hierarchies" of options, and one such hierarchy is used to structure a wider search for material conservation options in the next section.

4.2 HIERARCHIES OF MATERIAL CONSERVATION

"Reduce, reuse, recycle" is a critical and intelligent mantra for the future of material management, but as Figure 4.1 demonstrated, our

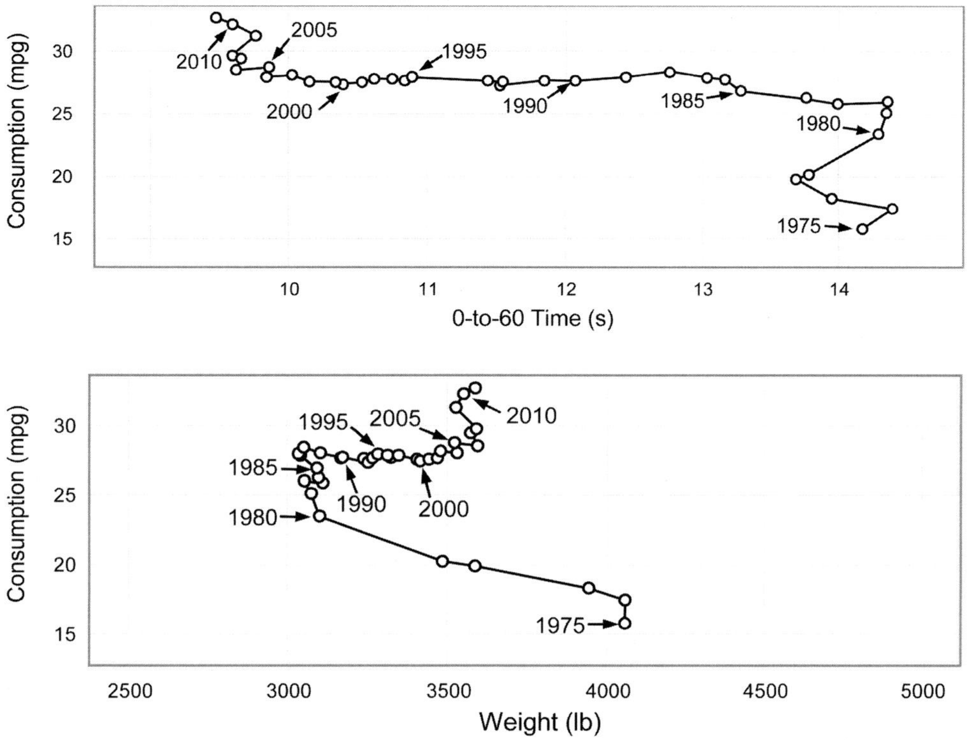

FIGURE 4.2 Evolution of car performance, consumption, and weight in the United States 1975–2010. Consumption is measured in a laboratory test, reflecting vehicle design parameters rather than driver behavior. The figures show that the energy crisis in the late 1970s drove regulation leading to improved consumption and reduced weight; once the crisis resolved, regulation was not tightened further, so manufacturers deployed technology developments to improve vehicle acceleration rates (dramatically) while also steadily increasing weights; this trend has shifted only recently, as the financial crisis of the late 2000s drove tighter regulation, leading to slightly improved consumption, and no further increase in weight. *From EPA (2012).*

preference is in reality "Redouble, replace, recycle-a-bit-if-it's-easy, reject." The British public is absolutely convinced of the benefits of recycling green glass bottles, and having done our bit for the environment, we can fly off to Spain for another short break because we must have our dose of winter sunshine or our human rights would be violated.

But "reduce, reuse, recycle" (the "3 Rs") as promoted in the United States, United Kingdom, and many other countries remains the correct ambition for reducing environmental impact, and is one statement among many versions (the Dutch prefer Lansink's ladder as described by Parto et al. (2007)) of a hierarchy of material management options. (In some cases, this is referred to as a hierarchy of "waste management." However, we prefer the term "material management hierarchy," as the label "waste" potentially disguises many other options for better managing materials to reduce the total impact of their production and processing.) Today, a number of hierarchies have been introduced for material management within a circular economy (see, e.g., Reike et al., 2018), yet all build on the 3 Rs. The fact that recycling—the topic of this book—is third choice is significant and correct: recycling is neither "green" nor free of impacts, so is a less attractive option than reducing demand for material or reusing materials without energy-intensive processing. However, the challenge of such hierarchies is not to capture them in a short phrase, but to turn the phrase into practice. Therefore we'll explore in this section the means by which such a hierarchy of material management could be brought about, and in the next section, we will explore its applications to the major material classes.

4.2.1 Reduce

Demand for material production would be reduced if the overall demand for services created by materials was reduced, and such reduction in service demand remains the "gold medal" of environmentally motivated materials management. Alternatively, but also with great potential for reducing the impacts of material production, the amount of material required to deliver each unit of service could be reduced by changes in design or production of goods. The strategy of "reduce" is at the top of any materials management hierarchy, because it leads to a reduction in demand for primary materials production, which for bulk materials is generally the most energy-intensive process of material conversion.

4.2.1.1 Simplicity, Austerity, Poverty: Reducing Demand for Material Services

Our social status, as humans, is determined by an infinitely subtle comparison of ourselves with our peers. Among animals, this status appears to be determined rather more simply. Bulls: I'm stronger than you are. Sharks: I eat more than you do. Budgerigars: I sing better than you, and look at my lovely tail feathers. But among humans, the competition is on many dimensions: I'm more influential than you, I'm more of an individual than you are, and always among the metrics, I've got more stuff than you. The gravestones in UK churches record families' key memories of their departed, with "much loved father/mother" the most common memory, and yet during their lives—which may have been shortened as a result—those fathers (in particular) competed fiercely to get more stuff. Kasser (2002) surveyed a class of final year students about to graduate and move to their first employment, and asked two questions. Firstly: "Next year, would you rather earn $100,000 knowing that your classmates were on average earning $200,000, or would you rather $50,000 knowing the others averaged $25,000." The answer was unequivocally the second—we measure our wealth relative to our peer group, not as an absolute. Secondly: "Next year, would you rather have four weeks holiday knowing your peers had eight, or would you rather have two weeks holiday knowing your peers had just one." In this case, the answer was clearly the first—we measure our leisure in absolute not relative terms. Yet despite an absolute preference for increased leisure hours, government policy and business reward schemes focus on our instinctive competition for wealth which—as Layard (2006) clearly explains—is one with no end-point: if we measure our wealth relative to our peers, then the (brief) happiness created for an individual whose wealth increases is exactly balanced by a relative unhappiness in the peer group, while meanwhile the newly wealthy individual soon compares themselves against a new, slightly wealthier, peer group. Changes in ranking—our relative assessment of our earnings—can only be a zero-sum game.

The competition for status linked to wealth is an axiom of economic policy in most countries, is the driving force for developing an economy fueled by money lending, and is a widely assumed norm of recruitment and reward among employers. Yet there is significant evidence that this does not actually tie up with individual preferences. When extra public holidays are announced, for example for the most recent Royal Wedding in the United Kingdom, most people did not use the extra day to take on extra work, but enjoyed an expanded weekend.

More broadly, even though working hours in the United Kingdom are higher than for most European countries (for the employed) with a current average of 43 h per week (ONS, 2011), personal aspirations for a better lifestyle say that the top priority of most employees is to have more leisure hours. (Humphrey et al. (2011), in an extensive UK survey of attitudes, of those who had thought about what they would want to do in later life, by far the most common aspiration (81% of responses) was to have more leisure.) With unemployment rates around 8% in the United Kingdom, we could apparently recalibrate our economy to eliminate unemployment if we all accepted an 8% cut in pay and working hours. Let's assume that a typical employed person in the United Kingdom spends 40 h at work, 10 h commuting, 80 h sleeping, washing, and eating, and 20 h managing their domestic life, in which case they currently have 18 h for chosen leisure. Therefore the potential reward for a 10% cut in pay and working hours is at least a 25% increase in chosen leisure time—which is what 81% of us aspire to "in later life." Surely this is a tremendous payback? But apparently it is not one we wish to pursue. Instead, popular lifestyle magazines about houses and celebrities appear to demonstrate to us that it's only *after* you've earned more money that you can afford to have more leisure time. However, a conundrum in the glossy magazines of house decoration is that some of the most desirable internal decoration requires having very little stuff in the room: although it's difficult to accept that real people live in this way, the most upmarket properties have space around each beautiful item of furnishing or decoration so it can be seen well, just as in traditional Japanese tea ceremonies, very few but beautiful objects are used.

"Simplicity" seems then to be a viable aspiration for the rich—with "decluttering" of a lifestyle seen as a sign of health. In parallel with the requirement that to be thin/healthy in the United States you must be rich, the lifestyle magazines suggest that if you are rich, you can afford to own fewer and more beautiful objects. This is clearly not borne out in practice—Druckman and Jackson (2008) provide clear evidence that income is the only predictor of personal carbon footprints in the United Kingdom—but to balance the apparently automatic urge to earn and spend more, we have a long history of cultural questioning about whether a simper lifestyle is in fact healthier or more liberated. This is strongly supported by the major religions—whether through Christ's suggestion to the rich man that to he give away all he owns, through the voluntary simplicity of Buddhist monastics, or through the disciplines of Lent or Ramadan. Valuing the stock of what we already own—as Boulding proposed—as opposed to valuing the rate at which we increase or replace that stock, has a long cultural history despite being an opposite to standard economic policy aiming at growth in flows.

However, despite these hints, few people make the voluntary choice to earn less, and the idea of pursuing the health of simplicity is untouchable for politicians, for fear that they might be promoting "austerity." In contrast to the chosen values of simplicity, "austerity" implies an enforced constraint on the freedom to earn or spend and has no place in any current political discussion. Only in times of war or crisis is austerity a politically possible choice: during the Second World War, the population of the United Kingdom accepted rationing of many goods, and voluntarily contributed excess material—such as iron railings—to support the cause of the war; following the 2011 tsunami in Japan, which led to the shutdown of many of Japan's nuclear power stations, the population of Tokyo voluntarily reduced their summer electricity consumption by 25%. Without such a crisis, it is difficult to divert spending from high to low intensity activities—to avoid rebound effects—because the recipient of our apparently low intensity spending (e.g., in a service industry) may simply use the additional income to purchase high-intensity goods. In effect the voluntary pursuit of demand reduction requires either that the population choose to pay more for less service, or to earn less—and therefore to reduce GDP. Thus Boulding's alternative measures of economic health are central—until we have metrics that demonstrate growth in quality of life while GDP declines, politicians will be unable to pursue policies other than those that aim to increase GDP.

And behind the enforced constraints of austerity lies the undesirable state of "poverty" where basic needs for security, food, warmth, and shelter are not met. Clearly any discussion of strategies to reduce overall demand for the services provided by materials must recognize that in parallel with this overall demand reduction is a rebalancing between current rich and poor.

We seem to be swerving far from the theme of this book, and must be careful not to stretch the editors' trust too far—but if the goal of recycling is to reduce the environmental impact of delivering benefits from materials, we have little choice but to discuss the challenging fact that a greater reduction in impact will occur if in rich countries we can choose to want less new material to provide those benefits, and without that choice there is a danger that any other action will be only transient. Indeed there is a broad literature on "rebound effects" (e.g., Sorrell et al., 2009) that might be used to call into question whether any action other than reducing demand will truly be effective: if recycling saves energy and cost, where will the "released" energy and money be used instead? If material efficiency strategies lead to a reduction in demand for steel, how will the steel companies react in order to stimulate future demand for their products?

The preceding discussion has revealed hints that, despite a strong urge to consume more (to purchase more material goods), there may be real and important motivations to choose a different path with less material requirements in return for an improved quality of living. However, unfortunately we have not found significant evidence of individuals making that choice—and instead the evidence of personal choices about diet makes depressing reading: even though it is widely understood that obesity leads to greater chances of serious illness and reduced life expectancy, rates of obesity in rich countries are increasing rapidly in part driven by preferences for highly processed prepared meals, rather than healthier simpler ingredients prepared at home and at lower cost. If, as a population, we understand that obesity worsens our own life and yet we choose it (or at least choose not to take the actions that would avoid or counter it) what chance do we have that the population will choose to reduce their material consumption for the benefit of others, unseen, unknown, and even unborn?

Sadly, despite suggestions that we (in richer countries) may in some senses prefer a simpler life, current evidence of human preferences suggests that voluntary choice will be insufficient to lead to reduced demand for the services provided by materials. It appears that, as humans, individually, even when presented with conclusive evidence of the self-harm created by our choices, we are incapable of constraining ourselves. However, collectively, it seems that we can develop sufficient shared trust in the evidence to agree to regulations that then constrain our individual choices—and that is ultimately the way that we must aim to reduce the impacts of our material production. Regulation that, for example, restricts cars to a mass under 300 kg per vehicle, guarantees that all buildings must be used for 500 years before being replaced, or prices the retirement of electronic goods above the cost of replacing them, is the key to reducing the environmental impacts of material provision. And, this far-reaching discussion is therefore central to a book on recycling, because without regulation that leads to a reduction in material output, we are greatly constraining the degree to which we can reduce the harmful impacts that are the cause of our interest in recycling.

4.2.1.2 *Intensifying Use: Reducing Demand for Excess Capacity*

The 60 million of us living in the United Kingdom own 28 million ~5-seat cars, which we use on average for 4 h per week with 1.6 people in each. If we could organize ourselves to be in groups of five at the right time, and weren't too fixed on where we wanted to start or end up, we could provide the same amount of car-time with just 213,000 cars and so reduce the environmental impacts of materials production for cars by 99.2%.

Our bedrooms are unoccupied for two-thirds of each day, as are our nightclubs and restaurants, and our offices are used for 45 h out of 168, so if we slept, danced, and ate in our offices, we could reduce our need for built space in the United Kingdom to under a third of present levels. And in our streets at home, if we shared our lawnmowers, power-drills, washing machines, hair dryers, and coffee machines, we could comfortably save 95% of our material purchasing.

But we don't. And we don't want to—because we own goods not just to use them, but to use them where and when we want them, without having to consult anyone else, and to show them to each other. I want to go to the shops now, and not wait 20 min for the bus, and I want the freedom to decide on my own whether to mow the lawn at 11:00 or 15:15. "Convenience" is a watchword of growing prosperity, and perish the thought that I should "inconvenience" myself through shared use of the pillars of my liberty such as my lawnmower and washing machine.

The opportunity to reduce material demand through more intense use of a reduced stock of products has had very little attention to date, although if a real constraint were applied to material supply (for example in war), shared use would instantly become normal practice. However, over the past 20 years, several attempts have been made to share car ownership and increase the intensity of car use through carpools and sharing schemes. To date, take-up of these schemes has been low, and several reasons have been proposed for this: Prettenthaler and Steininger (1999), from studying a car-sharing scheme in Austria, found that while considering just the journeys completed, 70% of the households would save money by car sharing, in reality the car had other functions—as a meeting place, for storage, as a symbol—and when these were taken into account, 9% of households would still benefit.

The rise of "budget airlines" has occurred because the capacity utilization of conventional airlines was relatively low, and new operators were able to reduce the capital costs per flight by increased use of their planes. In other examples (such as in high-performance computing, extreme scientific equipment, production equipment in continuous lines, and elements of road infrastructure) high-value equipment is used near to capacity to allow spreading of the capital costs. However, the reverse remains the case for domestic scale equipment and goods, and in particular for cars, and there is a rich opportunity to unlock more value from these largely idle assets. One of the ambitions of the research groups examining "product service systems" has been to identify the conditions under which companies might benefit from selling the service of using a product rather than selling the product itself. But as yet no clear strategy to persuade customers to prefer this option has emerged.

Shared use of goods appears to be a powerful weapon to reduce material demand, yet runs counter to the norms of recent Westernized development. Potentially our increasing preference for urbanization might allow a reexploration of shared ownership—rates of car ownership in large cities with good public transport infrastructure are lower than those in rural areas, and with increasing land costs leading to smaller dwelling spaces, other shared services may become attractive to future city-dwellers.

4.2.1.3 *Life Extension: Reducing Demand for Replacement*

Approximately half of the world's emissions of CO_2 from industry arise from producing materials for construction, and as we've seen already, most of the resulting built space is unoccupied for most of the time. But on top of this, buildings designed to last for 100 years or more are knocked down and replaced on average every 40 years in the United Kingdom or nearer to every 20 years in China. Not only is the "embodied energy" of construction underused through excess capacity, it is also underexploited and could deliver its intended service over a much longer period.

Figure 4.3 provides a summary of reasons why owners choose to replace goods, based on whether their needs have changed, or whether alternative offerings on the market today are more attractive than what's already owned. Exploration of the application of this simple grid to a range of steel-intensive products suggests that relatively few products—mainly those related to infrastructure (or in China, buildings that were constructed with poor quality control)—are replaced because they have worn out. Much more common is replacement because users' needs have changed or replacement because a new offering is more attractive. For nonresidential buildings, decisions to replace may arise because of changes in regulations (allowing a taller building or one for a different purpose, on the same site, or requiring different safety features), changes in fashion (from multiple offices to open plan space) or technology developments (in heating ventilation and air conditioning (HVAC), or communications technology, for example). As the costs of construction in developed economies are dominated by labor, and therefore the costs of a complete building refurbishment are comparable to the costs of replacing the building, owners will often prefer replacement as a means to an uncompromised development.

What can be done to counter this preference for replacement over refurbishment? A striking contrast to the example of commercial buildings in the United Kingdom given earlier is found in the history of the world's steel rolling mills, as virtually every rolling mill ever made continues to operate today. Figure 4.4 illustrates how the cost and the embodied energy (the energy required to make the materials) in both products builds up. Both for the commercial building and the rolling mill, the bulk of their embodied energy is in their structural frame, and this forms a relatively low fraction of their total cost. However, the highest value part of the rolling mill is the control system, which includes only a very small fraction of the embodied energy, is easily removed and replaced, and which due to technology development is most likely to lead to a need for product replacement. Therefore, both for the owner of the mill, and for the supplier of mill equipment, it makes commercial sense to upgrade the control system to allow the mill to last longer. In contrast, for the office building, the highest fraction of cost is the HVAC and communication services, which must be replaced long before the main structure, but are difficult to access without also removing

	Degraded	**Inferior**
The performance of the product has declined ...	... relative to when it was bought	... relative to what is currently available
	Unsuitable	**Worthless**
The desire for the product has changed ...	... in the eyes of its current user	... in the eyes of all users

FIGURE 4.3 A framework for understanding product replacement decisions (from Skelton, 2013). The two rows of the framework distinguish between failure that arises from a change in the state of the product, and failure that arises from a change in the desires of the user. The columns distinguish between changes that affect only the current individual product and user, and more systemic changes that come about through developments elsewhere in the market.

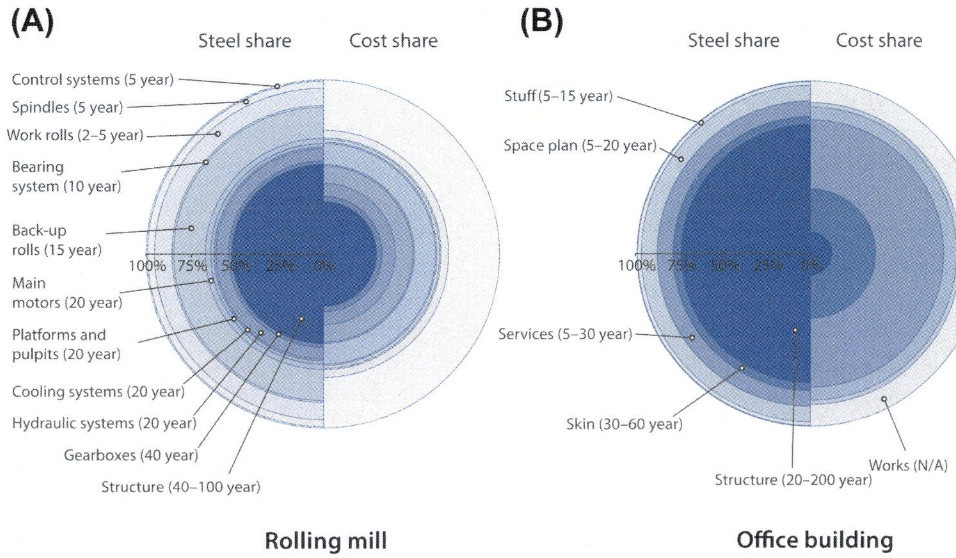

FIGURE 4.4 "Onion skin" models to illustrate the composition of steel and cost in (A) a rolling mill and (B) a representative office building (developed with data from industrial case studies, and described in Allwood et al. (2012) and reproduced with permission).

all other interior fittings in the building. It is the "failure" of these services (combined with changed requirements for building layout) that therefore currently drives most replacement decisions for commercial buildings. Countering this preference would require that buildings be designed with easily removed service systems, and excess capacity to allow the installation of novel service technology and layout adjustments in future.

Most products are a combination of subassemblies and components, and as the onion skins of Figure 4.4 illustrate, it is rare for all subassemblies to fail after the same lifespan. A rapidly developing challenge to the ambition to create longer-lasting products—particularly to maintain for longer their structural core, which dominates embodied energy—is that products are increasingly dependent on electronics. (For example, Braess and Seiffert (2005) describe how the average number of microelectronic control units in VW cars rose from around 10 in 1999 to over 40 by 2005.) The number of electric motors and CPUs in most goods is increasing, but the technical lifespan of these components remains short: most IT related devices have been completely superseded within 3 years. The insight created by the onion skin diagrams in Figure 4.4 is of vital importance to ensure that the embodied energy of the structural core of products is allowed to continue in use: unless the motors and CPUs and their associated wiring, sensors, and interfaces can be replaced repeatedly, the embodied energy of structural components will be written off as rapidly as the electronics, leading to a rapid increase of demand for material production. This is a particular concern for car manufacturing—where the use of electronics is rapidly expanding and becoming more and more integrated through the whole car—but also for "active" buildings, with sensors or small actuators built into structural components. Design for long life requires that the different layers of the onion skin model for the product can easily be separated and replaced, in order to exploit the structural core for as long as possible. Without such design being integrated into the next generation of cars, we will see car lifetimes dropping from around 15 years at present to nearer to 5.

4.2.1.4 *Lightweight Design: Reducing Excess Material Use*

Buying material costs money, so surely no one would use more material than they needed to make their product—it wouldn't make any sense, would it? It seems extremely unlikely that any designer or business would want to use more material than necessary in their product, but this is only the case if considering material costs in isolation. In reality, two other features of cost drive different decisions: the economies of scale favor the production of standardized components that can be used in many applications over those that are optimized for a particular product; labor costs, particularly in developed economies, are high relative to the costs of bulk materials, which are produced with extraordinary efficiency, so if it's possible to reduce labor by increasing material use, that is often the choice taken to minimize costs. This section will look at the effect of these two cost drivers.

A small number of industries, particularly in aerospace, aim to produce the lightest possible product that will meet their customer's needs: every extra kilogram of material used to make a plane or rocket is a kilogram less of cargo that could be lifted and hence could generate revenue. Reducing the weight of planes and rockets is so important that the industry has always been ready to incur extra production costs to save material. However, this is not the case in most applications: the average car in the United Kingdom is already 20 times heavier than the average passenger, and as we saw in Figure 4.2 customers prefer larger, heavier cars with more features; separately, the performance of buildings is unrelated to their weight. Buildings must be built with sufficient material to meet local safety regulations, but customers are largely unaware of how much material is used beyond this. Customers for cars and buildings are therefore largely indifferent to the mass of the product they purchase, while remaining highly sensitive to the cost of making their building.

Cars and buildings could be made lighter if their components were further optimized: the panels and plates that form the structure of the car could be made from ribbed sheets and have an internal frame of tailored lightweight members rather than heavy floorpans; the steel or reinforced-concrete beams that form the horizontal structure of commercial buildings could be given the "fish-belly" shape shown in Figure 4.5, with more material placed where the bending loads on the beam are greatest. Optimizing individual components for their specific load conditions typically saves around one-third of the material currently used (this is confirmed in a set of case studies by Carruth et al. (2011))—so, potentially, we could reduce demand for material production and therefore its harmful impacts by one-third by efficient design. Surely that's rational? It's clearly an important strategy, but there is no commercial incentive for this approach at present, because the savings would require a much more tailored approach to producing each component. If every beam in a steel framed structure were optimized, each beam would be made individually and the economies of scale denied, as tooling and labor was used specifically in each case. Furthermore, downstream on the construction site, effort would be required to manage and control a large supply of individual components, rather than having a stock of identical parts that can be swapped between each other. As a result, for many products, at least one-third more material is used than required, and this leads to cheaper products than if they were lighter.

However, even without this optimization, we still prefer excess material use if it allows us to save labor. A study of 24 commercial building projects in London has shown that the final

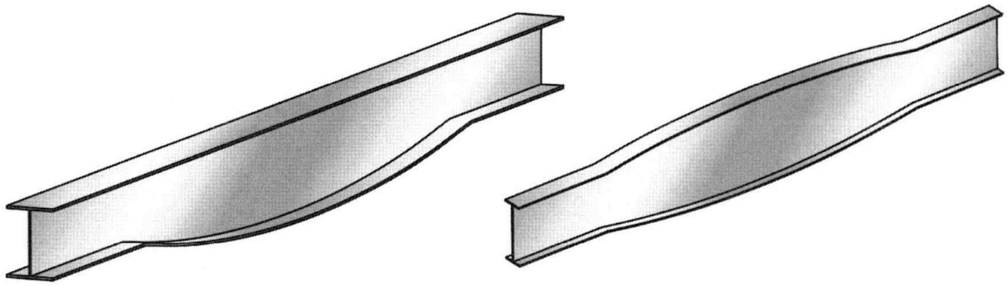

FIGURE 4.5 Schematic illustrations of optimized "fish-belly" beams with more material at the location of maximum bending moment, allowing an overall reduction in beam mass. *From Carruth and Allwood (2012).*

buildings, as designed by the leading structural engineers in the United Kingdom, used on average double the amount of material that would have been required to meet the Eurocode safety standards. Interviews with the engineers behind these choices revealed that this is a consequence of cost-saving: the cost of a day of engineering design time is roughly equivalent to the cost of a tonne of structural steel, so it is not economically sensible to design the building to meet the code, when less time can be used to create a design that definitely exceeds it (Moynihan and Allwood, 2014).

So, counter to expectations, it makes good business sense to overspecify materials when doing so allows a greater savings in labor costs, and this is a difficult issue to overcome. Product certification may be helpful to raise customer awareness of the inefficiencies of design, and in due course, to create added brand value for products that are designed to avoid excess. In transport, lighter vehicles have the benefit of reduced fuel consumption, and rising fuel prices and new regulation may stimulate customers to pursue more efficient vehicles. (While the world-record performance is 15,000 mpg and the UK fleet average is 35 mpg, there's certainly a lot of space for improvement—and the key to creating fuel-efficient cars is to make them smaller and lighter.) Lightweight design of some components—particularly in static products such as buildings—may reduce the loads on others, so create a virtuous circle of benefit. Finally, novel and more flexible production technologies—such as flexible rolling processes, or computer-based layout systems for reinforcement bars—may reduce the cost penalty of component optimization, and equipment manufacturers may be able to find new market opportunities in delivering such equipment.

4.2.1.5 *Improving Yield: Reducing Scrap Rates in Production*

The energy-intensive materials industries produce products that no one actually wants. No one has a coil of strip steel, an aluminum bar, or a bag of cement powder at home to show their friends. These products are instead intermediate goods designed to trade off the benefits of the economies of scale against the particular requirements of a family of customers. However, because they are "averaged" geometries, these intermediate products actually don't meet any specific requirements, so converting them to final components in products creates scrap. Scrap is the automatic consequence of the economies of scale in material production, and just as the materials industries have been extremely efficient in creating intermediate products, so downstream manufacturing (supported by its tutors in the computer-controlled machining industry) have become experts in the efficient generation of scrap. The construction sector has some skill in scrap creation. The car industry is extremely good at it, and confidently scraps about 30% of what it purchases, but the real

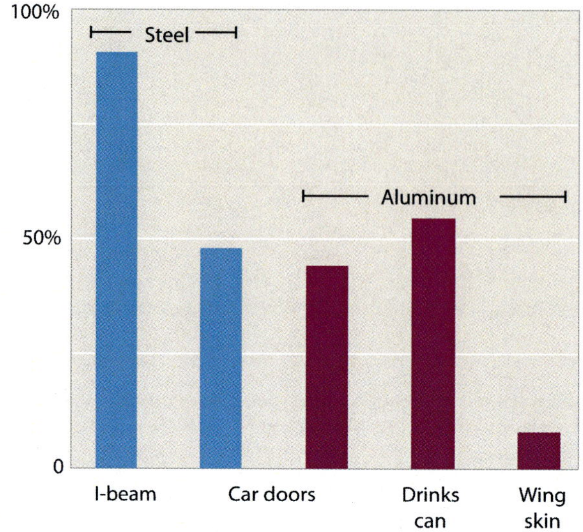

FIGURE 4.6 Yield losses in production of five representative components, illustrated as a percentage ratio of the mass of the final component to the mass of metal originally cast. Data from industrial case studies as described in Milford et al. (2011) and image adapted from Allwood et al. (2012) with permission.

experts at scrap generation are the aerospace industry. Typically the manufacture of an aeroplane involves the conversion of 90–95% of all purchased materials into scrap, with the aeroplane left over as a minor by-product. That's efficiency!

Figure 4.6 shows the ratio of the mass of liquid metal produced to the mass of various final components, and Figure 4.7 shows the effect of this scrap production on the embodied energy of a final component—a car door panel in this case. Losing around half of the liquid metal, particularly due to the blanking and stamping processes required to form the door panel, causes the embodied energy in the part to double.

Once again, the choice to create all this scrap is based on good business logic—it minimizes costs—but at the cost of causing a significant excess of material production. Around 25% of all steel made in the world each year is converted into scrap (Cullen et al., 2012) and never features in a product, and for aluminum the figure is around 50% (Cullen and Allwood, 2013).

Can anything be done?

Two broad approaches can be used to reduce yield losses. Firstly, collaboration between designers of products and processes along complete supply chains is likely to reveal significant opportunities for material saving, for example by adjusting final component designs so that they can be made from tessellating blanks. Secondly, new technology developments may allow more efficient use of intermediate stock materials without generating so much scrap. For example, the clothing and textiles industry is more efficient than the sheet metal industry at cutting blanks from fixed width strips, because it uses laser cutting that allows clever tessellation of different components in the same strip. As a separate example, Figure 4.8 shows alternative routes to producing metal washers: a conventional route creating scrap between the outer circumference of the washer, and scrap from the punched out hole; a novel route invented by the Sakamura Press Company in Japan and now licensed to the San Shing Fastech Company in Taiwan, in which washers are

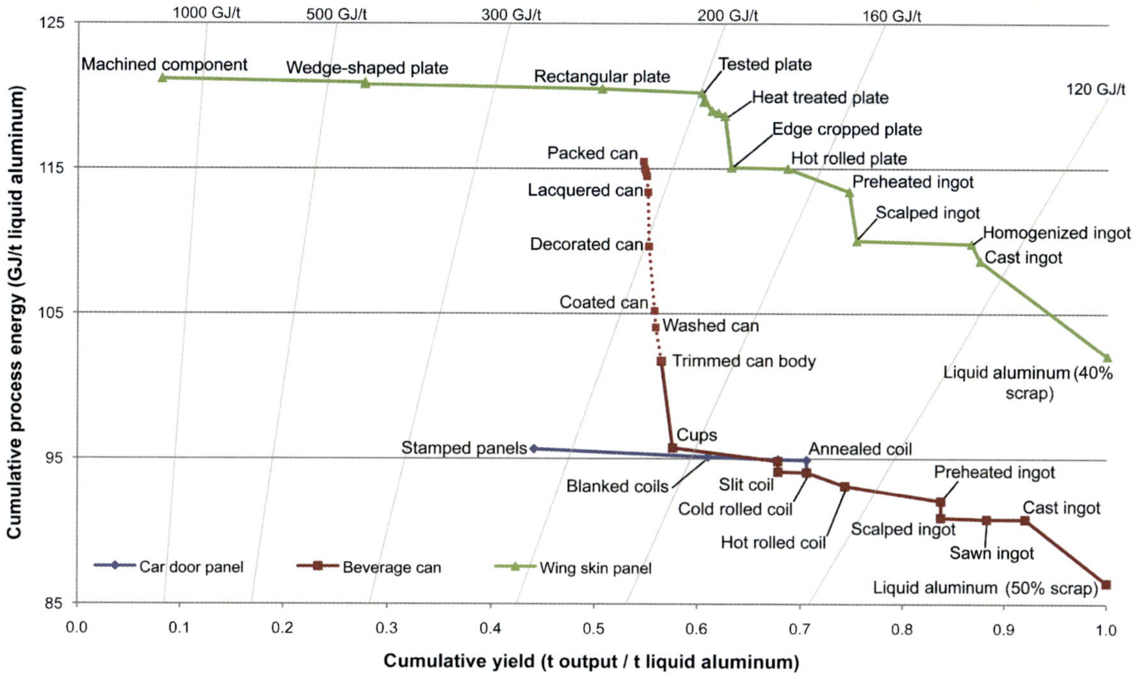

FIGURE 4.7 The process causes of yield losses for three aluminum components, with contours showing the effect of yield loss on the "embodied energy" of the resulting component. *From Milford et al. (2011).*

made by cold-forging and then piercing a bar, with no scrap at all. This is a wonderful example of material efficiency in action, and can operate at a sufficiently high production rate to compete effectively with the conventional process in the open market for metal washers.

The key to reducing yield losses is to design production processes that minimize cutting and trimming, and following years of technology development focused on the economies of scale producing standardized parts, there is great potential for invention and innovation in this area.

4.2.2 Reuse

The strategies of Section 4.2.1 under the banner "reduce" occur at the top of the hierarchy of material management options, because reducing demand for the production of materials eliminates all of its associated impacts. They also apply to all material goods—whether they are "consumable" (used up while providing a service) or "durable" (lasting to provide service more than once). Strategies leading to the reuse of materials follow in this section, applicable only to durable goods, and where possible are preferable to recycling, because the impacts of production are primarily associated with the creation of materials, rather than their conversion into products. Reusing material implies additional deconstruction and manufacturing steps, avoiding the energy intensity and long supply chains of recycling.

In developed economics today the idea of "reuse" carries overtones of poverty and substandard goods, and yet in other countries and former times, reuse has been entirely normal practice and in some cases even adds to the value of goods: in the "organic economy"

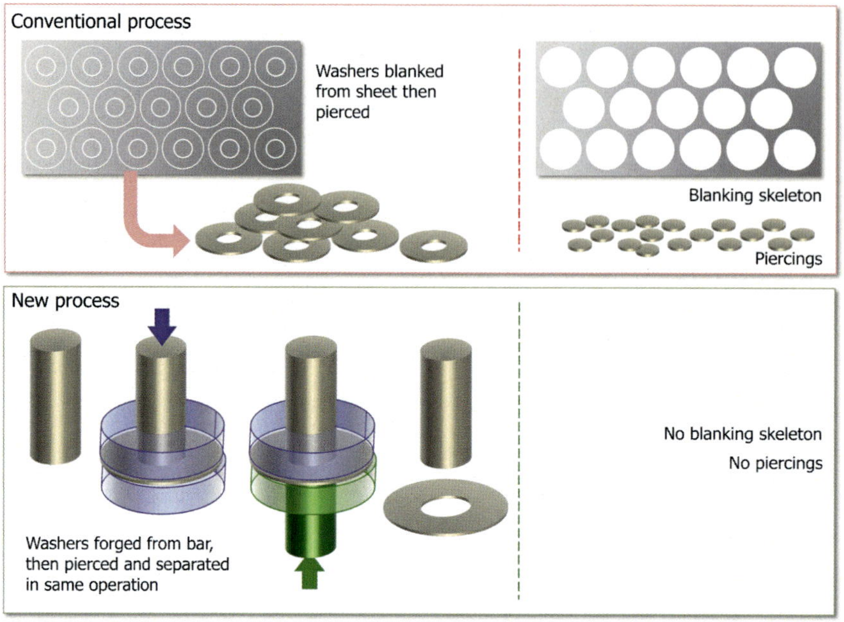

FIGURE 4.8 A novel process for producing metal washers from a solid bar, with no yield losses: the bar with diameter equal to the pierced hole in the washer is forged through a die onto a solid anvil, so the base of the rod spreads out to the thickness and outer diameter of the washer; a second tool set allows a punch of the same diameter as the bar to separate the newly spread washer from the bar.

discussed earlier, where iron could be produced only by redirecting biomass for fuel from other competing uses, all iron goods would be reused because the material had so much intrinsic value; Figure 4.9, from the Rongo Town in Kenya, shows reuse in action today, including both production from scrap metal and component resale.

Four forms of reuse are considered here—reuse of products, components, and materials in turn, and the diversion of manufacturing scrap prior to recycling.

4.2.2.1 *Reusing Products: Upgrade and Second-Hand Sales*

In most economies, second-hand markets are well established for many goods. Despite Akerlof's famous concern about "lemons" (Akerlof, 1970)—the buyer cannot easily learn as much as the seller knows about the condition of the goods—such markets emerge easily when goods are discarded not because they have "worn out" but because the current owner's needs have changed.

The challenge of creating an effective second-hand market is to find a way for buyers and sellers to share information, and to reduce the transaction costs of doing so to the point that it is attractive to trade. For housing, second-hand sale is the most common, and is well regulated to deal with Akerlof's lemon. For cars, second-hand markets are also well developed, albeit not regulated—the transaction costs are low, a large market exists, and when values become too low for further trades within one country, cars can be shipped to other countries where their relative value is higher. For low value goods in developed economies, such as most clothing, books, and crockery for example, the transaction costs are generally too high for

FIGURE 4.9 Examples of material reuse in Rongo Town, Kenya (courtesy of Chris Cleaver): (A) a scrap metal hawker; (B) small hoe blades formed from scrap metal; (C) cooking stoves and metal boxes made from scrap metal.

individuals to trade, so charity shops exist to provide an alternative to disposal. However, the relatively low value of used clothing is problematic even when supplied for free, as the costs of sorting may exceed the price potential buyers may offer—so donated clothing in the United Kingdom is largely baled and shipped to Eastern European countries where labor costs are sufficiently low to justify sorting used clothes by type and quality. For goods with values between cars and clothes, such as white goods, furniture, and leisure equipment, the internet has created a means to reduce transaction costs sufficiently to create new second-hand markets, and similar services in business-to-business trades create markets for industrial equipment and commercial furniture and fittings.

However, second-hand markets cannot prosper for goods that are rapidly superseded (such as electronics) or where regulation has changed so that a previous generation of goods cannot be operated legally (e.g., single-skin oil tankers have been outlawed by the International Maritime Organization so must be scrapped while new double-skinned tankers are produced to take their place). In these cases, it is not possible to resell the entire product; however, it may be possible to disassemble it into modules, components, or individual materials that retain value greater than their scrap price.

4.2.2.2 Reusing Components: Modularity

As discussed in Section 4.2.1.3, the end of a product's life is generally determined by one subassembly, module, or component—while the remainder of the product has residual life. If this life cannot be extended for the whole of the original product, it may be possible to exploit the residual life of the modules through partial disassembly of the original product into modules, some of which can then be reused in other assemblies. Famous examples of this strategy include the much discussed reuse of modules in Xerox copying machines (Ayres et al., 1997), reuse of modules from car-breakers (Subramoniam et al., 2009), and reuse of "disposable" cameras (Bogue, 2007): strategies that are now used by multiple manufacturers around the world (e.g., Caterpillar, Canon) for a variety of products or services.

The costs of disassembly are generally greater than those for assembly due to the loss of economies of scale: assembly plants have been optimized since the days of Henry Ford to simplify repetitive tasks, as the same operation is repeated many times—where in an equivalent disassembly line, each succeeding operation would be different, due to the variety of goods and their different condition at end of life. However, many strategies of design for disassembly have been developed, to create joints that can be reversed with ease, and to plan for the separation of component materials.

The real challenge of component reuse is not the logistics and management of the operation, but the design of long-lasting interface protocols that allow subassemblies to last through several generations of products. Famously, Michael Dell "won" the battle for the market for personal computers in the 1980s by developing a modular design strategy, in which the various key components of a computer (motherboard, disk drive, monitor, etc.) could evolve at different rates, allowing customers to choose the most up-to-date configurations. This was possible because the physical and software connections between the components evolved much more slowly than the components themselves. The Lego toy series has the same property—the geometry of the connectors has remained constant for many years, allowing a continuous development of new components, which can integrate into existing kits. However, for many goods such standardized interfaces either don't exist, or would be difficult to maintain, and this inhibits a much broader application of component reuse. For example, steel-framed commercial buildings are generally bolted together—so could be disassembled relatively easily at end of life. However, each building is designed to maximize the floor-space created on each individual plot of land, so without standardization of the spacing of columns in commercial buildings, it is difficult to develop a market for used structural beams. Similarly, the technology of cars develops sufficiently rapidly—for example, with the design of different fuel injection, air filtering, or electronic systems—that the volume and geometry of shape required to accommodate each new generation of engines is different and has little leeway—so it is very difficult to reuse structural parts of the car, even when they can be disassembled and have a high residual life when the original car is discarded.

The development of standardized open architecture and interfaces as a basis for new product development would greatly enhance the possibility that components and subassemblies could be reused to exploit their full lifespans, and potentially creates new business opportunities based on nurturing the value of subassemblies over longer periods. A few emerging examples of this approach—with the design of flat-pack buildings, and modular hotel structures—show that some of this commercial potential is immediate. However, counter-examples—such as the regular redesign of power-supply connectors for Apple computer products or Hewlett Packard printer cartridges—also demonstrate how module reuse can be prevented to maintain revenue streams based on replacement sales rather than component maintenance.

4.2.2.3 *Reusing Material: Cutting Bits Out*

If modules cannot be disassembled and reused, one last strategy for avoiding new material production is to reuse single-material components, following further manufacturing steps. The intermediate goods created by the materials processing industry are typically of a large geometry, to maximize the number of components that can be cut from the stock downstream, and most manufacturing therefore involves cutting smaller pieces out of bigger pieces of material. Although old material can retain perfect properties to allow further use, and metals often retain sufficient ductility to allow some reshaping, the habit of manufacturing to cut small pieces from large ones creates a difficulty for material reuse: unless the component can be reused directly, as discussed in the previous section, old material can only be used to make smaller components in future—so material reuse generally occurs only for large pieces of single material. Such reuse is common for wood, particularly for valuable hardwoods that may be difficult to source from new forests under more cautious management than previously. However, in developed economies it is rare for other materials, except at a domestic scale.

In developing economies, as illustrated in Figure 4.9, material reuse is common and a large scale example of this occurs with ship breaking in India. Asolekar (2006) has documented the flow of steel from the coast at Alang, where old ships are driven onto the beach, via manual disassembly, to hot rolling mills where the old ship plates are rolled into reinforcing bars for new construction. This example of material reuse is practiced at significant scale—potentially providing up to one-sixth of India's current steel demand—but has limited potential, and even in this case, the reinforcing bar made from old ship plate is less strong than new bar, and recycling by melting and with careful control of alloying, might allow sufficient material saving in construction to justify the additional energy cost of melting.

4.2.2.4 *Diverting Scrap*

A final option for reusing material without reducing it to liquid form in a recycling process occurs upstream with the opportunity to divert manufacturing scrap to an alternative use rather than sending it for recycling. Allwood et al. (2012) describe how Abbey Steel in Kettering in the United Kingdom has for 30 years bought blanking scrap from car body panel makers in the United Kingdom, and cut it into regular blanks for manufacturers of smaller components. This strategy could be applied for the scrap material of all sheet materials—and reflects the material inefficiency of current practice where each manufacturer cuts out their own blanks, creating the 20–30% scrap mentioned earlier. If suitable cutting technology exists (such as laser cutting of textiles), yield losses would be minimized by tessellating the largest possible number of geometries from the same sheet—so a parallel to diverting scrap would be to coordinate blanking requirements over a much wider range of customers.

A specific opportunity for diverting aluminum scrap from machining operations has received attention in research—albeit not yet in practice: aluminum is a highly reactive metal that can be welded in solid state at temperatures well below the melting temperature. Thus solid-state welding, for example by extruding machining chips (see, e.g., Tekkaya et al., 2009), could in future be used to divert aluminum scrap into new productive uses without melting.

4.2.3 Recycle

The rest of this volume provides the most up-to-date assessment of recycling, so it is included here to complete the hierarchy of waste management options, and allow a brief introduction to the limits to recycling that will guide the discussion on options by material in Section 4.3. Recycling is usually practiced when it is technically possible and when it saves energy (and emissions) compared to primary production. It may also be important if a resource is scarce, either due to an absolute shortage (concern about "critical elements," most recently the rare earths, raises the specter of exhausting the world's total stock, but usually this is a transient problem while new mines are opened), or due to competing uses (e.g., the growing competition for biomass for food and fuel may in future compete with the use of biomass for paper and board).

However, recycling is not a universal panacea:

- For some materials, notably cement, ceramics, and composites, there is as yet no large-scale recycling route by which the material can be returned to its original structure and quality.
- For some materials, recycling generally involves a loss of quality: for many bulk metals it is currently impossible to control the alloy content of recycled material with the same precision applied to virgin metal; for paper, the processes of pulping used paper lead to shortening of fibers and a reduction in quality; the great variety of compositions used in plastics creates great difficulty in generating new high-quality material from mixed waste streams. As a result, in many cases recycled material must be mixed with virgin material to produce acceptable products, thus reducing the net benefits of recycling.
- For many "critical" metals used in compounds (as alloys, or in electronics applications) the energy required to separate them as part of a recycling process may be significantly greater than the energy needed for virgin production.
- For some materials the energy required for recycling may be comparable to that required for virgin production. For other materials, such as paper, the emissions benefit of recycling may be different than the energy savings, if primary papermaking is powered by used biomass (see Laurijssen et al. (2010) for a discussion).

Notwithstanding these caveats, recycling remains an attractive option within the hierarchy of material management options for reducing the environmental impacts of some material use. However, the intention of this chapter is to set recycling in context: if the primary motivation for interest in recycling is to reduce the environmental impacts of materials production, it is only one of a portfolio of options, and not a universal goal to be sought and celebrated.

4.2.4 Downcycle, Decompose, Dispose

Even if the previous strategies to reduce material demand are applied vigorously, and recycling applied wherever possible, human activities will continue to lead to streams of waste material. The limiting option for this waste is to bury it underground—a widely practiced management technique for difficult problems (see also nuclear waste, carbon sequestration, red mud, toxic dumping, and psychotherapy). This can be done with more or less care, but such waste streams may also have value as substitutes for lower value material, or some benefit may be had from their decomposition.

"Downcycling" refers to the practice of using unwanted material for an application of less value than its original purpose. Common examples include the use of mixed streams of high-purity aluminum alloys to create less pure casting alloys, and the use of crushed construction and demolition waste (and in the United Kingdom, crushed green glass) as a substitute for aggregates in road and other construction. This practice is fundamentally wasteful—it would have been better to maintain the original higher value of the material—but as an option of "last resort" at least may save some of the energy and resource consumption of the lower value of material that is replaced.

Two forms of decomposition in common use in waste management are composting biomass waste and incinerating municipal waste for energy. Anaerobic digestion is increasingly used to convert biodegradable waste and sewage sludge to methane and carbon dioxide, which are then combusted for heat and power or upgraded to biomethane. This option is near the bottom of the materials management hierarchy because the energy recovered is substantially less than that required to produce the biobased products being decomposed—and, for example, it would have been a better strategy to avoid wasting food rather than to have it available for anaerobic digestion. However, as landfill gases are released from biomaterials left to rot in the ground, and for wood fibers methane releases may have a higher global warming potential than the CO_2 absorbed during tree growth, so such digestion is a better option than straight disposal.

"Energy recovery" refers to the process of combusting waste materials for heat and eventually power generation. It includes the substitution of municipal waste for fossil fuels in cement kilns, and the incineration of wood, plastics, and other waste for power generation.

If downcycling or decomposition are not possible, the remaining—"last-ditch"—materials management option is to store the unwanted material, usually underground in a landfill. This process is much studied, and research over decades has driven regulation to better management of landfill sites to avoid soil and water contamination from leaching.

4.3 WHEN IS RECYCLING NOT THE ANSWER?

The opening section of this chapter questioned whether the "circular economy" is really a desirable goal, and the second section has shown how recycling is in fact one among many strategies for conserving the value of materials and reducing the total impact of their production and disposal. This section now examines how these strategies play out in practice for the major material groups discussed elsewhere in this handbook. For each material, we first explore whether recycling is possible now, or will be in future, and evaluate the potential for recycling as a long-term material management strategy. This evaluation is then set against the other options for material management raised in Section 4.2, to provide practical guidance on the strategies that should be deployed for each material in order to have the most effect in reducing the environmental impact of delivering the services they provide. The sections are ordered according to the GHG emissions impact of current world production of each material group.

4.3.1 Iron and Steel

Global demand for steel production quadrupled between 1960 and 2020 and has now reached 1800 million tonnes per year—equivalent to over 200 kg/year for every person alive on the planet. Because it is ferromagnetic, steel is easily separated from other materials in mixed waste streams, so has one of the highest recycling rates, estimated to be up to 90% of arising waste (Graedel et al., 2011). The main "losses" of steel from this return loop occur with reinforcing bars inside reinforced concrete and steel used for subsurface applications, where extraction is too costly. Other steel that is not separated for recycling but dumped in landfills rapidly oxidizes into rust, which can be processed back into steel by conventional production routes. Manufacturing steel goods from scrap rather than ore leads to around half the total GHG emissions per product, and the cost benefit of this energy savings provides sufficient commercial motivation for recycling to need no further legislative stimulus. However, despite the high collection rate, recycled steel contributes only around one-third of current demand because of the rate of demand growth.

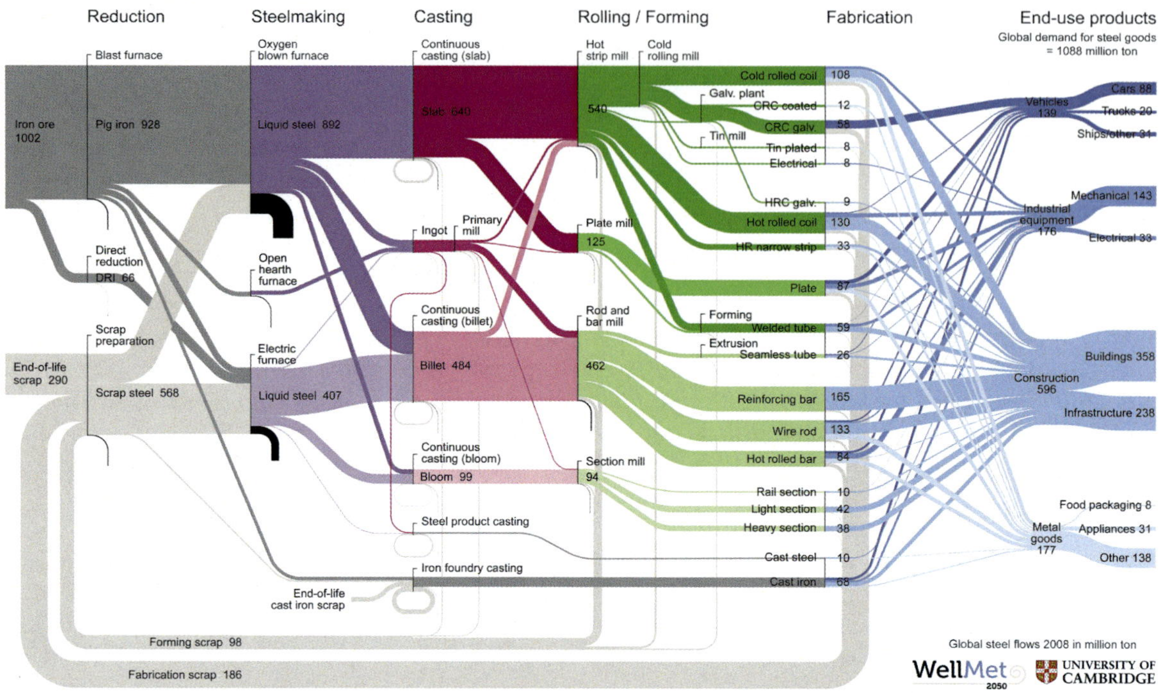

FIGURE 4.10 The annual global flow of steel in 2008 from production (from ore or scrap) to end use, including scrap generation and return. *From Cullen et al. (2012).*

Figure 4.10 illustrates global flows of steel in 2008 and demonstrates that steel recycling is closely coupled to primary production: scrap steel is used in primary production (from ore) to assist in cooling the melt, and virgin steel is used in secondary production (from scrap) to control metal composition. The flow of metal from secondary production into products demonstrates that secondary steel is widely used for producing reinforcing bars and other bars and sections. However, the higher grade applications of steel—as rolled sheet or plate for subsequent forming into panel components—are entirely made from virgin metal: developments of steel metallurgy have allowed significant increases in strength through precisely controlled alloying with more diverse elements, but these have not been matched by equivalent developments in refining unwanted elements out of the molten metal created from scrap. Removing copper from liquid iron is particularly challenging, so manual sorting is used even in developed economies to remove possible sources of copper from scrap feeds. More generally, due to the economics of current steelmaking, the main approach used at present to purify liquid steel is to bubble oxygen through the melt, to remove those impurities that preferentially bond with oxygen. More advanced extraction methods such as electrolysis, used in processing metals from waste electronics, could be developed further to improve steel refining, but these are not yet applied.

The achievement of a circular economy for steel—the most recycled metal—is thus a distant dream for two key reasons: demand growth limits the relative availability of scrap and processes for refining molten scrap have not been developed to allow composition control comparable to what is achieved with virgin production.

However, the strategies of material efficiency could be applied to allow delivery of greater value from less steel, and the high rates of steel use at present are in part due to the extraordinary efficiency of its production, and hence its low price: a tonne of steel in the UK costs around £800, which is comparable to the cost of a day's work by a structural engineer. Around half of all steel is used in construction, as reinforcing bars, sections, and cladding, and total requirements could be reduced by avoiding overspecification, selection of ideal compositions (particularly for reinforcing bars in China and India, which are currently much less strong than those used in the EU and the United States), and by maintaining buildings or building components for much greater periods. In the United Kingdom, designing buildings that met rather than exceeded the requirements of the EuroCode safety standards would halve constructional steel demand, and if in addition buildings were maintained for 200 years rather than the current 40 before replacement, we could maintain the same building stock with one-tenth of the new steel.

Steel use in cars—typically 70% of the mass of a finished car is steel, but with yield losses of around 50%, the average 1.3 tonne car in the United Kingdom has required production of 2 tonnes of steel—would be cut immediately by a move toward lightweight vehicles. Fuel consumption in cars is strongly correlated with vehicle mass (see Figure 4.11) so the key to developing a future low emitting future car fleet is not to create electric cars (the current target of EU policy, despite there being no excess supply of "low carbon" electricity available) but to create light cars. Small light cars (say 300 kg steel per car, comparable to Volkswagen's L1 concept car, and in line with the 500 kg aluminum Lotus Caterham 7, and the 600 kg steel Tata Nano) made with low yield losses (say 10% losses) could provide the same level of car service with around one-sixth of today's steel requirement. This would be halved again if cars were maintained for twice as long.

These two examples, from construction and cars, underline the potential for demand reduction as the key strategy to reduce the environmental impacts of steel production. Steel recycling is already virtually universal, but cannot at present create the same quality of material as production from virgin ore, unless a different scrap stream is created and maintained with perfect cleanliness for each distinct composition. Reducing the number of compositions in use would simplify future recycling loops, and some opportunities to control steel properties by processing rather than alloying are under development, but there is strong commercial pressure against this.

4.3.2 Cement

Cement is not currently recycled, because doing so would require energy inputs comparable to making new cement, and there is no

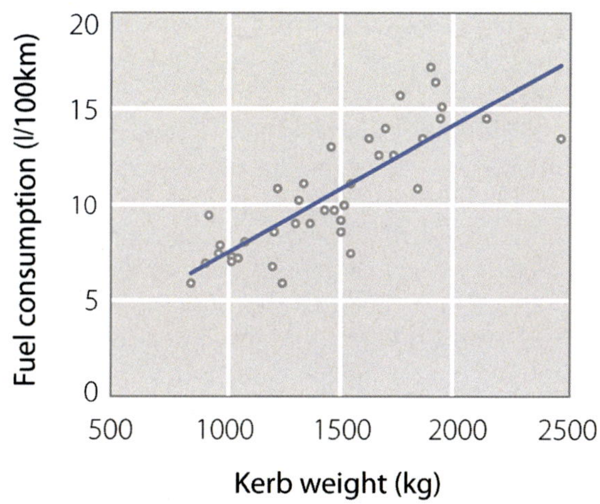

FIGURE 4.11 The effect of vehicle mass on fuel consumption for a representative range of cars on the road in the UK today, clearly demonstrating that fuel efficiency arises from mass reduction. *From Allwood et al. (2012), with permission.*

shortage of resources to make new cement. So the recycled product would require as much energy and be of lower quality than new cement, so there are no commercial processes for cement recycling. However, significant confusion has been created by efforts within the industry to promote the (rather questionable) benefits of "concrete recycling." Any discussion of a future circular economy for cement must therefore begin by clarifying the distinction between cement and concrete. Cement is a binder and is never used on its own, but provides the adhesion of a composite material—generally either mortar (cement plus sand plus water) or concrete (cement plus sand plus water plus small stones, known as aggregate).

Most cement (Portland cement in particular) is supplied as a dry powder, to which water is added to trigger a hydration reaction. This reaction cannot easily be reversed, and a circular cement economy is not currently a possibility. The one possibility being explored in this area is that there is some evidence that following use, the hydration reaction may not be complete, so finely ground used cement can be used in small amounts as a partial substitute for new cement. In contrast to the ideal of a "circular economy," cement is therefore a one-way material.

Faced with this knowledge, the cement industry has wisely looked for alternative strategies to reduce the impact of cement, and given the public popularity of the word "recycling" has attached it not to cement, but to concrete: old concrete, which would otherwise be dumped in the ground (along with any embedded steel reinforcing bars, which are too difficult to extract) may instead be crushed, and the resulting granular material used as a substitute for the aggregate required by new concrete. Estimates of the value of this approach vary, but a concern is that the range of particle sizes in crushed concrete is much broader than the range in new aggregate, and in turn this may drive up requirements for cement to make new concrete, and hence worsen the overall impacts. There is no evidence that recycling concrete in this form provides a substantial reduction in emissions.

In contrast to this problem with recycling, there is significant potential to reduce the requirement for new cement when making concrete—through avoiding overuse and through partial substitution by other materials, such as ground granulated blast furnace slag or pulverized fly ash, which are produced with less additional emissions than new cement. As already described, finely ground used cement can also be used as a substitute.

Demand for new cement can therefore be reduced by dilution of cement with other powders. But in addition, as with the use of steel in construction, demand for concrete can be reduced by maintaining buildings and infrastructure over longer lifespans and by intelligent design, as illustrated in Figure 4.12.

As yet, the idea of modular design of reusable concrete components has not developed, but interest in off-site fabrication of modules is growing rapidly, as clients aim at shorter construction times. At present, such modules are typically joined "wet" on site—so that at end of life, the building cannot be disassembled, but if new techniques allow "dry" mechanical joints between prefabricated modules, then potentially those joints can at end of life be reversed and the modules reused. This potentially offers an interesting new business model to cement companies—rather than selling a low-margin powder in the greatest volume possible, selling or leasing structural modules, whose value can be maintained over several generations of building.

Without any option for recycling, there can never be a circular economy for cement. The opportunity to create concrete blocks and components that are themselves reused has had little if any attention, but the fact that bricks (particularly from older houses) are commonly reused suggests that this may have some potential. However, without a credible return loop, and given its very high environmental impact, the

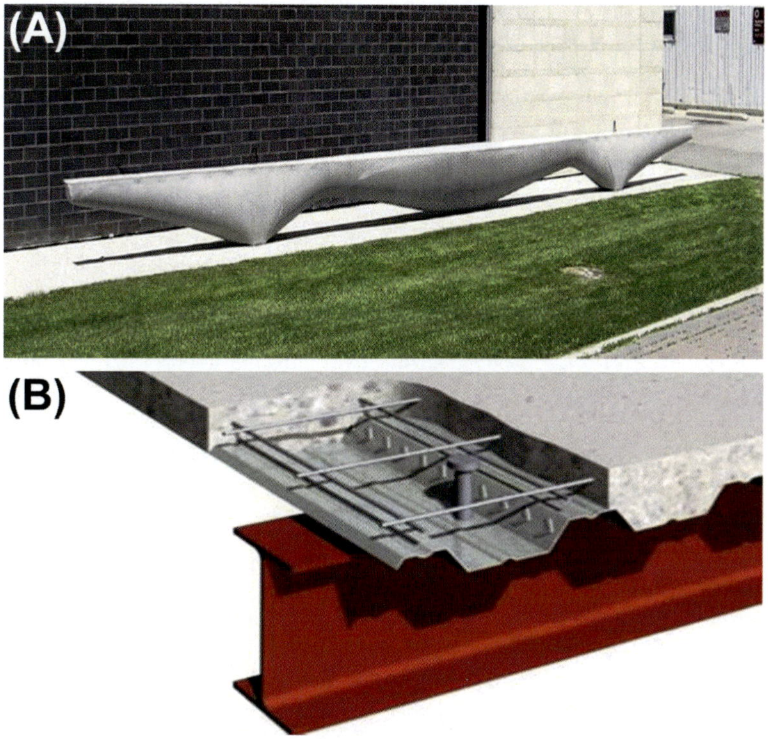

FIGURE 4.12 Lightweight design applied to concrete (A) through the use of fabric formwork to shape the beam to a "fishbelly" as seen above for steel and (B) through composite design with shear connectors between steel and concrete allowing full use of concrete's compressive strength while avoiding tension. *Panel (A) from http://www.rpschmitz.com, figure 13. Panel (B) from Tata Steel ComFlor 210 brochure: https://www.tatasteeleurope.com.*

role of cement in the world's material economy requires reexamination: cement is abundant, cheap, strong, and convenient but harmful. An environmentally constrained economy would reframe cement as an exotic material to be used in niche applications. With no possibility of creating a circular economy for cement, reducing its environmental impacts depends entirely on reducing its production.

4.3.3 Plastics

Plastics are an extraordinarily versatile family of materials with two particularly attractive properties that have caused an explosion in their use (Geyer et al., 2017): their composition can be adjusted easily to create an infinite variety of colors, textures, and mechanical properties that can be tailored to each application; final parts are produced directly from molten feedstocks, allowing precise optimization of every design with cheap molds in standard processes. These two properties, and plastics' low cost, have led to enormous diversity in the uses of plastics, and unfortunately this mitigates very strongly against the creation of a circular plastic economy: thermoplastics can be recycled—production scrap is often fed directly back into the machine that created it—but only if the material being recycled is of a consistent composition. As one of the key attractions of plastics is their variety in composition, this inhibits plastics recycling, and despite increasing sophistication in labeling different plastics, rates of recycling remain relatively low.

An interesting beacon of hope for a circular economy for plastics arose with the development of several processes claiming to turn "anything to oil"—by thermal depolymerization. However, such processes have yet to be demonstrated at scale, as it has run into practical problems (e.g., due to the presence of various chemical compounds and elements like chlorine and sulfur). The energy demand of these thermal recycling processes is high, resulting in low recycling rates as plastic is consumed for energy in these processes.

The key to reducing the impacts of plastic production lies at the top of the hierarchy of options, in reducing overall demand or switching to, e.g., reusable packaging, as packaging represents 40% of plastics used (Coelho et al., 2020). Several studies, such as Hekkert et al. (2000a,b), have shown how plastic packaging requirements can be cut. There may be fewer opportunities for reusing plastics in other applications—because they are used to make small complex components within larger assemblies, and are difficult to separate.

Achieving a circular economy for plastics is technically easy if their composition is constrained, and there is an energy benefit in achieving this. The two key alternatives appear to be to use waste plastic as a fuel for "energy recovery" resulting in CO_2 emissions, or to reduce overall demand for the type of products in which plastics are used.

4.3.4 Paper

Paper is widely recycled, and in the UK paper recycling is now widely practiced and accepted through curbside separation of domestic and commercial waste. Interestingly, its benefits are less clear than for the metals—the energy requirements for recycling paper are lower than those for primary pulpmaking, but recycling degrades the fibers in the paper and the energy for primary processes is often supplied at least partially from the waste biomass of tree harvesting. So, the GHG emissions of paper recycling may be greater than those of primary paper making, although recycling carries a cobenefit of reducing the demand for new biomass, which will face increasing competition in future as demand for food and fuels grows (Laurijssen et al., 2010). Recycling (or combustion) also avoids the complications of paper decomposition in landfill, which leads to methane releases. A closed circular economy for paper is not technically possible, due to fiber shortening during the process (limiting recycling to seven to eight loops), and because some paper applications (e.g., for sanitary use) have no return loop. However, public interest in paper recycling is high, so collection rates are good.

The services created by paper use could be delivered with less paper: office paper could be thinner, and print-on-demand services could avoid overproduction of books, magazines, and newspapers. There has been considerable discussion about whether "eReaders" will cause a reduction in paper use—or whether, as with claims dating back to the 1970s about the "paperless office," the advent of computers will continue to increase paper consumption due to the wide availability of printers. This discussion remains unresolved—until future figures on paper demand can be correlated against eReader use—and the use of paper for archiving will remain attractive long into the future, due to the rapid turnover of computer technologies.

A form of circular economy for paper could be developed if, instead of recycling, the paper were cleaned for reuse without destroying its structure. This was normal practice in the use of ancient palimpsests, or indeed of chalkboards in schools, and Leal-Ayala et al. (2012) have demonstrated that toner print on office papers could be removed by carefully calibrated lasers, to allow reuse of paper. However, as with the steel, cement, and plastics, the astonishing efficiency of current production leads to very low prices for paper, so there is little motivation for either user or supplier to develop alternative material loops.

4.3.5 Aluminum

Aluminum production from ore is extremely energy intensive due to the electrolysis step required to extract alumina from bauxite. Because this process uses 20 times more energy than that required to melt an equivalent mass of solid aluminum, the aluminum industry has expended significant marketing effort making the claim that aluminum is a "green material" and that recycling uses only 5% of the energy of primary production. In reality, this statement is rarely true—creating a product from recycled material normally involves a full supply chain with many processes, so without question recycling uses less energy than primary production, but usually the reduction is around 30% rather than the much vaunted 5% figure.

Furthermore, aluminum recycling is inhibited much more than for steel by the difficulty of purifying the liquid metal, as a multitude of aluminum alloys are produced and used. Figure 4.13 illustrates the consequences of this problem: nearly half of all aluminum produced each year becomes internal scrap, and is recycled within the factories in which it was created. End-of-life recycling makes only a small contribution to the overall flow, and apart from the contribution of closed-loop can recycling, all end-of-life recycling for aluminum involves downcycling from the high purity wrought alloy systems to the lower purity cast alloys.

A circular economy for aluminum is therefore not possible at present, because it would require perfect separation and cleaning of all alloys.

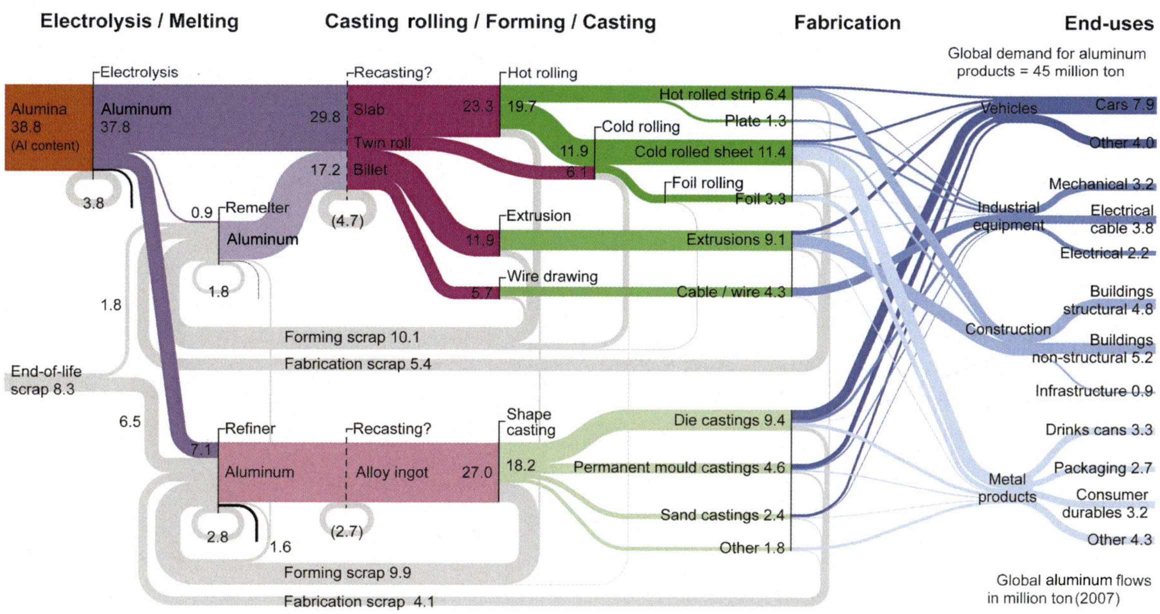

FIGURE 4.13 The global flow of aluminum in 2007, showing that production from ore dominates, in total only one-half of the liquid aluminum made each year reaches final products, and that the (higher purity) wrought and (lower purity) casting alloys have some separate recycling loops, but recycling leads on average to a downgrading of ore purity. *From Cullen and Allwood (2013).*

Within individual sites this occurs, and much effort is spent on managing internal scrap by alloy type. However, it cannot in general be achieved for postuse waste streams, and the proliferation of alloys developed by the industry to meet specific user requirements further inhibits the possibility of future alloy separation. The one example where a circular economy is approached is in the recycling of drink cans—where melting the three components of the can (the body, the head, and the ring-pull, each made from different alloys) leads to a composition that, when diluted by around 5% virgin aluminum, can be used for making new can body stock. The energy cost of this route is around 30% that of making cans entirely from virgin aluminum, and rightly the industry would like to expand it, if collection rates could be increased. Unfortunately the convenience of cans is that they can be used and discarded easily, so despite the relatively high value of the material in a used can (around 3–4 UK pence per can) collection rates are estimated to be around 50%, with the remainder being lost in landfill dumps.

The most striking feature of Figure 4.13 is the high yield losses of the aluminum supply chain, and better integration between companies combined with new process development could significantly reduce these. In general, this would reduce the size of the recycling loop, but would not reduce requirements for primary aluminum—unless recycling leads to degrading wrought alloys to casting alloys, which must then be replaced by new virgin production. This happens in the aerospace industry which, while it uses only 1–2% of the world's aluminum production, turns 90–95% of what it purchases into scrap (machining chips), which are not separated, so are downcycled.

Aluminum has properties that make it attractive in many applications—including its low density, high corrosion resistance, high conductivity, and high ductility. Using such a valuable material for disposable packaging, much of which is lost through disposal seems extravagant, and the properties of the material would be better exploited in more advanced applications. However, as with plastics, aluminum is generally used in rather small components where its properties have particular value, and these are already optimized and difficult to reuse.

Two significant opportunities for reuse to reduce overall demand for aluminum occur in construction and power cables. Aluminum extrusions and roll-formed sheets are used as the purlins and cladding of light industrial buildings, which typically have short lifespans, and in which the components are not damaged. They could therefore be reused, and already at least one UK supplier offers 20-year leases on such components, to ensure they are reused at end of life. Aluminum extrusions are also widely used in curtain walls and window frames, and while developments in Europe are aiming to ensure that these materials are separated from old buildings prior to demolition to ensure the metal is recycled, they could also potentially be reused in future. Similarly, aluminum in electrical cables could potentially be reused at end of life.

Development of a circular economy for aluminum is currently not technically feasible, due to the difficulty of alloy control, but given the energy benefit of recycling, a priority in the short term is to increase the rate of separation and collection of aluminum at end of life of mixed material products, with construction as the highest priority.

4.3.6 Clothing and Textiles

Clothing in developed economies is generally discarded before it is worn out, and most of the population own far more clothing than needed for comfort and function. The rise of "fast fashion" with short fashion seasons and new outlets in supermarkets led to an increase of clothing. As a result, several studies have attempted to estimate the amount of unused clothing in the nations' wardrobes, for example claiming that only one-third of all clothes in the United Kingdom are worn in each year.

Clothing in this sense is an icon of the problem of reducing the impacts of material production, and for clothing in particular, recycling offers little benefit. There are no recycling routes for natural fibers, and although most of the growth in clothing purchased in the United Kingdom has been through man-made fibers, such as nylon and rayon, they are increasingly used in mixed-material composites, so cannot easily be separated for end-of-life recycling. In fact, it isn't clear that recycling the materials used to make clothing would offer much benefit even if widely practiced: the dominant environmental impacts of clothing occur in use (washing and drying are energy intensive), in coloring and texturing fibers (requiring intense use of potentially toxic chemicals) and in (fertilizer and insecticide intensive) cotton agriculture, and only the last of these would be avoided by fiber recycling. The most widely quoted application of textiles recycling occurs in man-made carpets, with the Interface company apparently a leading practitioner.

Instead, the key to reducing the impacts of clothing production would be to reduce the rate of purchasing and hence production—but in developed economies, proposing this is rather akin to suggesting that plants shouldn't flower! Leasing, which is widely used for working uniforms and hotel bed-linen, could in principle be used more widely and a few well-motivated groups have explored clothes sharing. Charity shops create some hierarchies of cascading, but these have limited effect on demand in developed economies, and may not be entirely beneficial to developing economies where early economic growth that might occur through domestic clothing manufacturing is stifled by a free supply of discards.

We could live well with smaller stocks and flows of clothing, and indeed with repair and upgrade could largely have a circular clothing economy, but prefer not to do so. Acting to change this preference voluntarily (it would be accepted without question in a crisis) requires a sociological shift of extraordinary scale, and as yet the mechanisms required to create this shift are unknown. However, the eco-fashion designer Kate Fletcher (Fletcher, 2008) has created a positive first hint of possibility by observing that fast fashion creates disposable clothing: easily purchased and easily discarded after one outing. In contrast, a shirt embroidered or repaired by a friend or relative for personal reasons converts the commodity to a part of a life story, which takes on a significance far greater than its appearance.

4.3.7 Glass

The British public love putting their bottles in recycling bins, and according to several studies, consider this to be the most important action they should take to conserve the environment for future generations. The fact that we largely discard green glass, because we like drinks made in other countries, whereas we make mainly clear or brown glass because that's what we need for our own famous drinks, somehow doesn't register. Thank goodness we're busy making new roads and can use the crushed old glass within them!

Glass recycling is restricted by color and has some energy benefit: the most energy-intensive step of both primary and recycled glass production is melting, which is the same for both routes, so the benefit of recycling is to avoid the heat of reaction as the raw materials combine (Sardeshpande et al., 2007); see also Chapter 28—Glass. This leads to energy savings of around 25–30% for the recycling route—although this is partially offset by the additional energy costs of collection.

Glass is mainly used for flat applications such as windows (made by the float glass process) and containers (made by blowing). Glass bottles have become lighter with improved process control, but there is limited potential for further savings. Plate glass made by the Pilkington float glass process is already so well controlled that there is no opportunity for reducing thicknesses.

Glass degrades in use only very slowly—glass is a supercooled liquid, so the 2000-year-old Roman glass in the British Museum has become cloudy, but the 500-year-old Elizabethan glass remains clear—so could in principle be reused many times. However such reuse is rare: windows from old buildings are smashed as part of the demolition route, as the cost of deconstruction, storage, and resale would not be met by low second-hand prices; glass bottles, which used to be the standard for drinks purchasing, have been replaced in many cases by plastic bottles, and even when still used, the costs of collection, cleaning, and refilling are too great to allow commercially viable reuse.

A circular economy for glass would save only a limited amount of energy compared to the current largely one-way system, so the key to reducing impacts is to make less. This would occur if buildings lasted longer (so less window production were required) or if we reduced our total requirement for glass containers. New approaches to cleaning and returning glass containers—possibly based on refilling the same containers within supermarkets, rather than returning old containers for recycling and buying new ones—might be possible.

4.4 DISCUSSION

The phrase "circular economy" creates an attractive image of a beneficial loop of continuous material recycling without the harmful environmental impacts of new production. In fact, this chapter has shown that recycling is only one among the lower orders of a hierarchy of options for reducing those impacts, and the "circular economy" is rarely if ever the key to doing so. Reducing demand, and reusing products, components and materials are strategies with greater potential to reduce unwanted impacts, and can be applied across all material groups. These strategies have had relatively little attention to date, and possibly therefore have the most potential for development. In contrast, recycling is already well developed when it is commercially attractive, is impossible for some materials, and for most recycling systems it is only possible to maintain material quality by mixing recycled material with virgin material.

The survey of options applied to key material groups in the previous section has raised several themes that span across materials:

- Economists will often respond to engineer's reports of technical limits by denouncing their failure to think of substitutes: surely if it's difficult to recycle cement, some clever chaps in the future will find a way to substitute it with something else? Substitution is obviously a warm comfort blanket for economists, but sadly has much less potential than they realize: we use such vast quantities of steel and cement that the only two materials available in comparable quantity are wood and stone. We used to use wood and stone, and moved to steel and cement because they were stronger and more convenient, and the (at the time) free availability of fossil fuels made them readily available. Stone is in fact considerably less versatile than cement, because it can't be poured, and apart from the problem of competing demands for biomass, wood production at present requires sufficient energy (for kiln drying) that its benefit as a substitute for steel is marginal. Ashby's book on ecomaterials selection provides a thorough basis for exploring substitution (Ashby, 2009), but demonstrates few if any "easy" options to reduce impacts.

- "What about innovation then?" cry the economists—"you technical people never have the imagination to see what might happen in future." Well, those same technical people have been thoroughly interested in looking for innovative materials for decades, and have been searching the outer reaches of the periodic table to find them. As a result, innovative materials of recent discovery tend to have exotic properties (conductive polymers, super-strong graphene, carbon nanotubes), but these will be useful in highly technical applications, not as substitutes for the few bulk materials that dominate the environmental impacts of production. Steel has become stronger with new exotic alloying elements, but as that comes at the cost of a loss of ductility, the value of further strength improvements is unclear: we don't make reinforcement bars from ceramics, because we want them to cope with some aspects of building failure. At present, a range of new formulations for cement are being promoted heavily (IEA ETSAP, 2010, provides a survey), but as yet without clear evidence of product properties or total process impacts. It is much more likely that innovation in materials will deliver new electrical properties than deliver a replacement for steel and cement.
- Can we reduce the total number of compositions in use for each class of material? For every recycling loop in current operation, a reduction in the number of compositions would simplify and improve the quality of recyclate, yet this is absolutely counter to commercial developments, and would be extremely difficult to enforce. Creating simpler waste streams with less variety is more likely to occur when materials have sufficient value that companies will collaborate to simplify their return.

In effect, all of the challenges to a circular economy raised in this chapter arise fundamentally from the low cost of bulk materials, which in turn arises from the extraordinary efficiency with which they are made, and the relatively low cost of energy—particularly to the energy-intensive industries. Platinum is recycled perfectly because everyone involved with it understands the value of retaining it and using it again, so a massive increase in energy pricing that raised steel and cement to the value of platinum would bring about the instantaneous adoption of every strategy in the materials management hierarchy described earlier!

However, such an increase is unlikely to happen. Twenty years of international negotiation aiming at a cross-border carbon price have so far failed, and there is not even any political will to impose a fuel tax on aviation fuel. Climate change presents too many urgent challenges to wait longer for such agreement, so instead we need to pursue other implementation mechanisms, of which the key appears to be regulation. At a simple level, glass bottles are reused in some European countries because of regulation enforcing a deposit to be paid against their return. More broadly, smog in cities, river water quality, and acid rain have been controlled in Europe by regulation, and globally, CFCs were eliminated by regulation. The primary goal of work around materials management and its environmental impact is therefore to nurture the conditions under which the public will be ready to vote in politicians to pursue regulation. This happens easily in war or crisis (Fukushima), and also where a broad idea has become familiar. Therefore a primary goal of material activism today is both to reduce material demand, and to prepare a mindset for overall demand regulation.

Regulation is a negative end-point for this chapter, so instead we'll end with a celebration of the cultural value of materials. Tanya Harrod's wonderful survey of the symbolism of materials in different cultures (Harrod, 2013) reminds us that materials create objects of beauty, that working with them allows us to express a part of our humanity, and our shared appreciation of them is part of our cultural identity. Ready access to fossil fuels and a narrow set of measures of economic success have in less than 100 years reduced our perception of materials as being the building blocks of our legacy to commodities of convenience. If we can take back some of the wonder of creativity and association offered by the best of material goods, maybe the regulation we need could be at our common request, and not a constraining imposition.

References

Akerlof, G.A., 1970. The market for 'lemons': quality uncertainty and the market mechanism. Q. J. Econ. 84 (3), 488–500. https://doi.org/10.2307/1879431.

Allwood, J.M., Ashby, M.F., Gutowski, T.G., Worrell, E., 2011. Material efficiency: a white paper. Resour. Conserv. Recycl. 55, 362–381. https://doi.org/10.1016/j.resconrec.2010.11.002.

Allwood, J.M., Cullen, J.M., Carruth, M.A., Cooper, D.R., McBrien, M., Milford, R.L., Moynihan, M.C., Patel, A.C. H., 2012. Sustainable Materials: With Both Eyes Open. UIT Cambridge, Cambridge, UK.

Ashby, M.F., 2009. Materials and the Environment: Eco-informed Material Choice. Butterworth-Heinemann, Burlington, MA, USA.

Asolekar, S.R., 2006. Status of management of solid hazardous wastes generated during dismantling of obsolete ships in India. In: Proceedings of in the International Conference on "Dismantling of Obsolete Vessels", 2006.

Ayres, R.U., Ferrer, G., van Leynseele, T., 1997. Eco-efficiency, asset recovery and remanufacturing. Eur. Manage. J. 15 (5), 557–574. https://doi.org/10.1016/S0263-2373(97)00035-2.

Bogue, R., 2007. Design for disassembly: a critical twenty-first century discipline. Assem. Autom. 27 (4), 285–289. https://doi.org/10.1108/01445150710827069.

Boulding, K.E., 1966. The economics of the coming spaceship earth. In: Jarrett, H. (Ed.), Environmental Quality in a Growing Economy. Resources for the Future/Johns Hopkins University Press, Baltimore, MD, pp. 3–14.

Braess, H.H., Seiffert, U. (Hrsg.), 2005. Handbuch Kraftfahrzeugtechnik, 4. Aufl. Vieweg, Wiesbaden.

Carruth, M.A., Allwood, J.M., 2012. The development of a hot rolling process for variable cross-section I-beams. J. Mater. Process. Technol. 212 (8), 1640–1653. https://doi.org/10.1016/j.jmatprotec.2012.03.006.

Carruth, M.A., Allwood, J.M., Moynihan, M.C., 2011. The potential for reducing metal demand through lightweight product design. Resour. Conserv. Recycl. 57, 48–60. https://doi.org/10.1016/j.resconrec.2011.09.018.

Coelho, P.M., Corona, B., ten Klooster, R., Worrell, E., 2020. Sustainability of reusable packaging – current situation and trends. Resour. Conserv. Recycl.: X 6, 100037. https://doi.org/10.1016/j.rcrx.2020.100037.

Cullen, J.M., Allwood, J.M., 2013. Mapping the global flow of aluminium: from liquid aluminium to end-use goods. Environ. Sci. Technol. 47, 3057–3064. https://doi.org/10.1021/es304256s.

Cullen, J.M., Allwood, J.M., Bambach, M., 2012. Mapping the global flow of steel: from steelmaking to end-use goods. Environ. Sci. Technol. 46 (24), 13048–13055. https://doi.org/10.1021/es302433p.

Druckman, A., Jackson, T., 2008. Household energy consumption in the UK: a highly geographically and socio-economically disaggregated model. Energy Policy 36 (8), 3177–3192. https://doi.org/10.1016/j.enpol.2008.03.021.

EPA, 2012. Light-duty automotive technology, carbon dioxide emissions, and fuel economy trends: 1975 through 2011. U.S. EPA-420-R-12-001, Office of Transportation and Air Quality, March 2012. https://nepis.epa.gov/Exe/ZyPURL.cgi?Dockey=P100DYUC.txt (Accessed 26 June 2023).

FAO, 2012. FAOSTAT. Food and Agriculture Organization of the United Nations. https://www.fao.org/faostat (Accessed 28 June 2023).

Fletcher, K., 2008. Sustainable Fashion and Textiles: Design Journeys. Earthscan, London.

Fu, F., Pan, L., Ma, L., Li, Z., 2013. A simplified method to estimate the energy-saving potentials of frequent construction and demolition process in China. Energy 49, 316–322. https://doi.org/10.1016/j.energy.2012.10.021.

Geyer, R., Jambeck, J.R., Law, K.R., 2017. Production, use, and fate of all plastics ever made. Sci. Adv. 3, e1700782. https://doi.org/10.1126/sciadv.1700782.

Goodchild, C.H., 1993. A Report on the Comparative Costs of Concrete & Steel Framed Office Buildings. British Cement Association, Crowthorne, UK.

Graedel, T.E., Allwood, J.M., Birat, J.-P., Buchert, M., Hagelüken, C., Reck, B.K., Sibley, S.F., Sonnemann, G., 2011. What do we know about metal recycling rates? J. Ind. Ecol. 15 (3), 355–366. https://doi.org/10.1111/j.1530-9290.2011.00342.x.

Gutowski, T.G., Allwood, J.M., Herrmann, C., Sahni, S., 2013. A global assessment of manufacturing: economic development, energy use, carbon emissions, and the potential for energy efficiency and materials recycling. Annu. Rev. Environ. Resour. 38, 12.1–12.26. https://

www.annualreviews.org/doi/10.1146/annurev-environ-041112-110510.

Harrod, T., 2013. 'Visionary rather than practical': craft, art and material efficiency. Phil. Trans. R. Soc. A 371, 20110569. https://doi.org/10.1098/rsta.2011.0569.

Hekkert, M.P., Joosten, L.A.J., Worrell, E., Turkenburg, W.C., 2000a. Reduction of CO_2 emissions by improved management of material and product use: the case of primary packaging. Resour. Conserv. Recycl. 29, 33–64. https://doi.org/10.1016/S0921-3449(99)00056-7.

Hekkert, M.P., Joosten, L.A.J., Worrell, E., 2000b. Reduction of CO_2 emissions by improved management of material and product use: the case of transport packaging. Resour. Conserv. Recycl. 30, 1–27. https://doi.org/10.1016/S0921-3449(00)00046-X.

Humphrey, A., Lee, L., Green, R., 2011. Aspirations for later life, UK Government Department of Work and Pensions, Research Report no. 737. Accessed online (September 2013) at www.gov.uk/government/uploads/system/uploads/attachment_data/file/214511/rrep737.pdf.

IEA ETSAP, 2010. Cement production, Technology Brief I03, June 2010. www.etsap.org.

Kasser, T., 2002. The High Price of Materialism. MIT Press, Cambridge, MA.

Kasser, T., 2017. Living both well and sustainably: a review of the literature, with some reflections on future research, interventions and policy. Phil. Trans. R. Soc. A 375, 20160369. https://doi.org/10.1098/rsta.2016.0369.

Keynes, J.M., 1930. Economic possibilities for our grandchildren. In: Keynes, J.M. (Ed.), Essays in Persuasion. 1963. W.W.Norton & Co., New York, pp. 358–373.

Laurijssen, J., Marsidi, M., Westenbroek, A., Worrell, E., Faaij, A., 2010. Paper and biomass for energy? The impact of paper recycling on biomass availability, energy and CO_2 emissions. Resour. Conserv. Recycl. 54, 1208–1218. https://doi.org/10.1016/j.resconrec.2010.03.016.

Layard, R., 2006. Happiness: Lessons From a New Science. Penguin.

Leal-Ayala, D.R., Allwood, J.M., Schmidt, M., Alexeev, I., 2012. Toner-print removal from paper by long and ultrashort pulsed lasers. Proc. Royal Soc. A 468 (2144), 2272–2293. https://doi.org/10.1098/rspa.2011.0601.

Milford, R.L., Allwood, J.M., Cullen, J.M., 2011. Assessing the potential of yield improvements, through process scrap reduction, for energy and CO_2 abatement in the steel and aluminium sectors. Resour. Conserv. Recycl. 55 (12), 1185–1195. https://doi.org/10.1016/j.resconrec.2011.05.021.

Moynihan, M.C., Allwood, J.M., 2014. Utilization of structural steel in buildings. Proc. R. Soc. A. 470, 20140170. https://doi.org/10.1098/rspa.2014.0170.

ONS, 2011. Hours Worked in the Labour Market. UK Office for National Statistics. Accessed online (September 2013) at www.ons.gov.uk/ons/dcp171776_247259.pdf.

Parto, S., Loorbach, D., Lansink, A., Kemp, R., 2007. Transitions and institutional change: the case of the Dutch waste subsystem. In: Parto, S., Herbert-Copley, B. (Eds.), Industrial Innovation and Environmental Regulation: Developing Workable Solutions. United Nations University Press, Tokyo.

Prettenthaler, F.E., Steininger, K.W., 1999. From ownership to service use lifestyle: the potential of car sharing. Ecol. Econ. 28, 443–453. https://doi.org/10.1016/S0921-8009(98)00109-8.

Reike, D., Vermeulen, W.J.V., Witjes, S., 2018. The circular economy: new or refurbished as CE 3.0?—exploring controversies in the conceptualization of the circular economy through a focus on history and resource value retention options. Resour. Conserv. Recycl. 135, 246–264. https://doi.org/10.1016/j.resconrec.2017.08.027.

Sardeshpande, V., Gaitonde, U.N., Banerjee, R., 2007. Model based energy benchmarking for glass furnace. Energy Convers. Manage. 48, 2718–2738. https://doi.org/10.1016/j.enconman.2007.04.013.

Skelton, A.C.H., 2013. The Motivations for Material Efficiency: Incentives and Trade-Offs Along the Steel Sector Supply Chain (PhD thesis). University of Cambridge.

Sorrell, S., Dimitropoulos, J., Sommerville, M., 2009. Empirical estimates of the direct rebound effect: a review. Energy Policy 37, 1356–1371. https://doi.org/10.1016/j.enpol.2008.11.026.

Subramonium, R., Huisingh, D., Chinnam, R.B., 2009. Remanufacturing for the automotive aftermarket—strategic factors: literature review and future research needs. J. Clean. Prod. 17 (13), 1163–1174. https://doi.org/10.1016/j.jclepro.2009.03.004.

Tekkaya, A.E., Schikorra, M., Becker, D., Biermann, D., Hammer, N., Pantke, K., 2009. Hot profile extrusion of AA-6060 aluminium chips. J. Mater. Process. Technol. 209, 3343–3350. https://doi.org/10.1016/j.jmatprotec.2008.07.047.

UNDESA, 2011. World Population Prospects, the 2010 Revision (updated 28 June 2011). United Nations, Department of Economic and Social Affairs.

USDoT, 2012. Research and Innovative Technology Administration Bureau of Transportation Statistics National Transportation Statistics. U.S. Department of Transportation.

USGS, 2012a. Iron and Steel Statistics and Information. United States Geological Survey online (Accessed 8 January 2013).

USGS, 2012b. Cement Statistics and Information. United States Geological Survey (Accessed 8 January 2013).

van Oss, H.G., Padovani, A.C., 2008. Cement manufacture and the environment Part II: environmental challenges and opportunities. J. Ind. Ecol. 7, 93–126. https://doi.org/10.1162/108819803766729212.

Winegarner, R.M., 2011. Silicon Industry 2011, Sage Concepts. Available from: http://www.sageconceptsonline.com/report1.htm. (Accessed 8 January 2013).

World Bank, 2013. The World Bank Data. Available from: http://data.worldbank.org. (Accessed 8 January 2013).

Wrigley, E.A., 2013. Energy and the English industrial revolution. Phil. Trans. R. Soc. A 371, 20110568. https://doi.org/10.1098/rsta.2011.0568.

Zhou, V., 2011. Global Motor Market Study. International Copper Association Ltd. Online at industrial-energy.lbl.gov (Accessed 8 January 2013).

CHAPTER

5

Material and product-centric recycling: design for recycling rules and digital methods

Markus A. Reuter[1] and Antoinette van Schaik[2]

[1]WASM: Minerals, Energy and Chemical Engineering, Faculty of Science and Engineering, Curtin University, Perth, Australia [2]MARAS—Material Recycling and Sustainability, The Hague, The Netherlands

5.1 INTRODUCTION

Design for recycling (DfR) is an ecodesign strategy. It should result in products for which functional groupings in the product, when dismantled, produce the maximum recovery when recycled to a specific segment in the Metal Wheel (Figure 2.2), all within the circular economy (CE) (Figure 2.1–2.5). Obviously, each functional group may have metals and materials that are incompatible, but these will then be lost from the system. Thus DfR should be guided by the possible metallurgical infrastructure that can recover metals economically, while using processing simulation tools; therefore digital twinning is the basis. If product design can keep thermodynamically compatible materials close together, then metallurgical technology can deal with them. For example, the recycling of printed circuit boards is often carried out with copper metallurgy. This means that silver and gold can be easily recycled, but aluminum will be lost, as it becomes aluminum oxide instead (see Figure 2.2). To reduce this loss, designers could avoid the presence of aluminum with copper, or could plan a design that facilitates the removal of aluminum during preprocessing. For example, heat dissipaters are often made of aluminum and could be designed for easy manual extraction (called design for disassembly). For other parts containing precious materials, the recovery ratio during dismantling or sorting could be increased by visually identifying such components.

It is not easy to design a complex product, such as an LED or fluorescent lamp, photovoltaics (PVs), cars, electronic products, etc., to be simpler, as the metals in close proximity create the product's specific functionality. Functionality will always be at the heart of product design. Nevertheless, designers often have some leeway in the choice or arrangement of materials. Continuous innovation in materials can aim at delivering similar functionality with other, thermodynamically compatible, materials. In short, it is a truly complex challenge to design products for optimal recycling and what is good for DfR could, and most likely will, defeat functionality, especially that of a complex product with numerous tightly connected functional materials.

Applying first principles, knowledge of material combinations in design linked to compatibility in metallurgical processing, and hence recovery and losses of materials (governed by thermodynamics), is critical in ensuring the supply and recovery of commodities. In particular it is essential for the recovery and supply of "critical raw materials" from end-of-life (EoL) products (Hayes, 2021). DfR thus needs a supportive infrastructure to become reality.

This starts in the design phase where functional, material, and material connection choices are made. Fortunately, the tools, e.g., Metal Wheel, simulation-based digital twins, as discussed earlier, now exist that provide designers with the required information on the recycling outcomes for different designs. These are based on the increasing and limiting factors for various combined materials in electric and electronic equipment (EEE) on the basis of liberation, sorting, and second law of thermodynamics as shown by the Metal Wheel, and they allow the assessing of different designs. Where these are widely known and used, recycling outcomes can be improved.

Besides tools, the creation of an economic or policy framework that motivates designers to consider recycling is necessary. Consumers may reward good design by preferring products that are known to have high standards. It takes time to have effects on a large scale. More commonly, public policy can require, or influence, product manufacturers to take recycling into account, thus encouraging material innovation without constraining functionality. Ecolabeling, the topic in the next section, can support this and provide insight for consumers on the resource and environmental performance of products on a physics and industrial-reality basis. At the same time, such ecolabeling can help producers and original equipment manufacturers (OEMs) to distinguish resource-efficient design from lesser products, hence providing a driver for design for resource efficiency (DfRE). Another approach involves requirements on ecodesign for products, as expressed by the European Commission in the Ecodesign Directive (Directive 2009/125/EC) (European Commission, 2009) and the recent proposal on Ecodesign for Sustainable Products Regulation (European Commission, 2022).

Another aspect of DfR relates to supporting markets for recycled metal products. When recycling produces metals, or other compounds, that have the same purity (or grade) as primary metal, they are easily sold at equivalent prices. This applies in most cases where state-of-the-art processes are used. When recycling produces metal with impurities, this might fail to find markets, even when it has the physical or chemical properties needed for use in specific products. Manufacturers may not choose to use such recycled metals, either because they do not know the precise mix of elements and compounds on offer, or because they are unsure of the quality or consistency of the recycled metal on offer. These are not technical but marketing issues, which nevertheless influence the success of recycling. Designers can help to create markets by being willing to take up suitable recycled material.

Lastly, all DfR will only become reality when products are actually recycled. Well-working collection and recycling systems and infrastructures need to be present and used, so that the product *is* recycled and the expected recycling performance is obtained.

Within a CE paradigm this would mean materials from a product must return with the same quality to the same product, and simulation detail is the only way to ensure this. This means product design is linked via simulation tools to metallurgical processing to understand if the metals, alloys, and materials in the product can be returned to the same quality to be used in the same product with a minimal exergetic dissipation.

5.2 RECYCLABILITY INDEX AND ECOLABELING OF PRODUCTS

The interactions in the recycling system are two-directional along the chain: from design, disposal, collection, sorting, and (metallurgical) recycling and back. Each of the stages from design downwards impacts final recycling through the changes it creates in the physical properties of the material going for recycling. In the other direction, the realization of economic value from final recycled metals is the natural driver for collection activities, incentives for disposal and, potentially, DfRE. If products are produced out of a single metal, the system would be relatively simple, and the interactions linear, up and down the recycling "chain." In practice, as each product contains many metals as well as other advanced materials imparting functionality, there are many recycling chains (from product to recycled metal) for each metal. These chains interact during design, collection, and recycling within the CE. This creates a complex system, with a level of complexity that needs to be understood in a workable way by actors in the chain and policymakers.

Figure 5.1 gives a general overview of various technologies applied during recycling and the processing of waste materials. It shows the top level of the simulations of the interlinked system of technologies and activities, developed in the HSC Sim 10 (2023) software (www.mogroup.com). Rigorous simulation will also provide a thermodynamic basis for the environmental impact analysis. The generated information can be linked to GaBi (2022) (www.sphera.com) or openLCA (www.openlca.org) software, for example. Through this link, it is possible to evaluate the environmental impact of different designs and scenarios based on actual environmental impact, linked to the mass and material flows and the detailed compositions of each stream. This includes the design and recycling route-dependent recoveries, losses, and the environmental footprint of any created residue. This is not possible with the often applied LCA/environmental performance calculations based on general databases. These are incapable of capturing the essential detail of recycling technology and design-dependent recyclability in view of recycling optimization and DfR, especially if novel new flowsheets are required. Policymakers and OEMs should take care when applying these simple linear calculations for recycling indexes, since these will lead to false conclusions with respect to resource efficiency improvement. As a consequence, this might lead to harmful decisions for the industry. Furthermore, using physics-based models as the basis can provide a rigorously based estimation of design rules, which are in the next section. However, it must be noted that these are only loose estimations (much favored by the product designers who have no thermodynamics grounding). Best practice would be to estimate recyclability of products with tools such as shown here run by experts in the recycling field who have knowledge of metallurgy and process technological experience.

A result of this physics-based modeling approach of the system, from a design using computer-aided design (CAD) for recycling (Heibeck et al., 2021), is also a rigorously based estimation of the recyclability of products, also estimating liberation and particle size distribution by finite element methods (FEMs). This can be used to determine an **ecolabel** for a product (Figure 5.2). The proposed ecolabel combines the recyclability and exergy dissipation from the CE system for a specific product, as these can be directly linked to design tools and the product's bill of materials, full material declaration, and/or chemical content analysis. Figures 2.4 and 2.5 in Chapter 2 show the basis

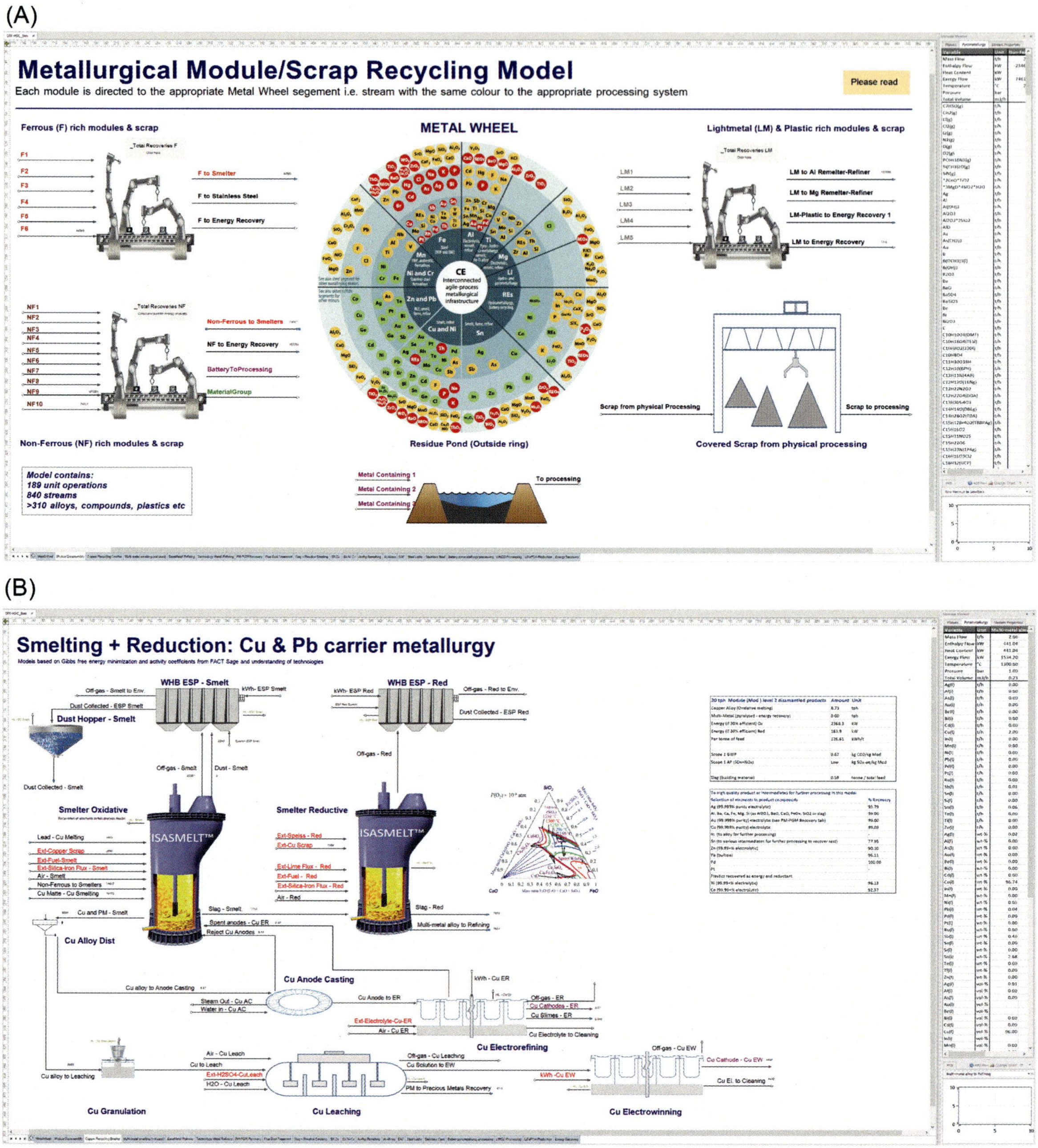

FIGURE 5.1 A complete model for the recycling of a mobile phone made in process simulation software. The tabs below show all the involved technologies to realize the CE recycling. The robotic separation tab (A) and copper processing tab (B) are shown. The many linked tabs can be observed at the bottom of the windows. In total, there are over 200 metallurgical reactors, 1000 species, and 1000 streams linking them in this model.

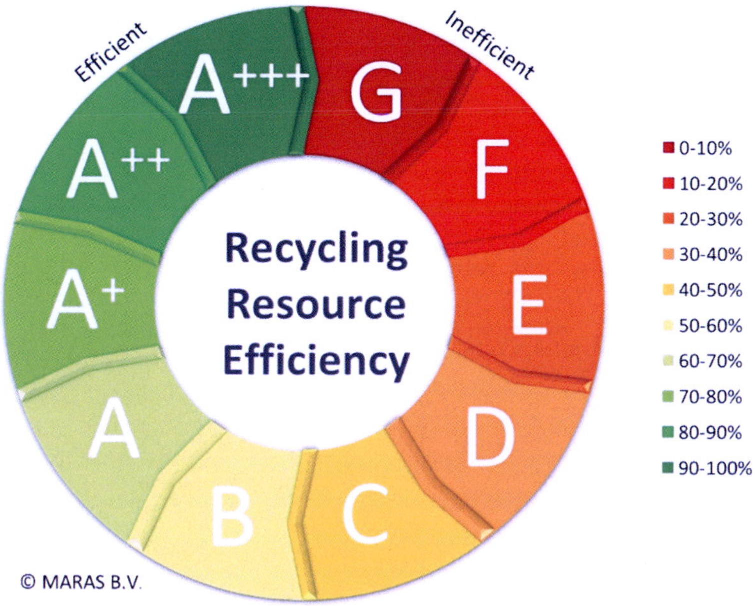

FIGURE 5.2 Recycling label depicting the total recycling of materials back into the same product, calculated by a model such as depicted in Figure 5.1 (van Schaik and Reuter, 2016; Reuter et al., 2018, 2019).

for the creation of a rigorous recyclability index based on BAT technology. Moving away from the simplistic indexes still being developed and applied thoughtlessly by nonrecycling experts (e.g., OEMs and component designers), which do not represent the complexities and depth of recycling technology, its physics, chemistry, and economics, is a prerequisite for achieving DfRE and a CE. A recycling knowledge-based **ecolabel** (Figure 5.2) has the depth to distinguish differences in product designs and to distinguish the more resource-efficient design from the other. Implementation of this ecolabel might provide a driver for OEMs to DfRE.

This approach differs fundamentally from many general DfR studies and applications as well as recycling indices as performed nowadays, which rely on simplified one-dimensional DfR rules and recycling calculations. These rules do not represent and reflect the complex realities of recycling, recyclate quality as a function of design, liberation and sorting efficiency, and metallurgical process recoveries of different metals/materials as a function of recyclate quality and concentrations of materials in it, as captured by the models as discussed here.

The general approaches are not based on a fundamental process simulation basis that can include this required depth of nonlinear design into recycling (including slag chemistry and metallurgical processing efficiencies) to capture the real issues of resource efficiency. The discussed approach of Figure 5.1, applying a product-centric, physics-based understanding of recycling, allows by its depth for the evaluation of changes in design (for recycling) by including all effects that one particular change or substitution of material might have on the entire recycling system.

Chapter 2 discusses and shows, especially in Figure 2.4, the link between product design and the fundamental mass and energy transfer processes in reactors and the CE system.

5.3 DFR RULES AND GUIDELINES

The following 10 DfR rules (five fundamental and five derived) have been developed (Reuter and van Schaik, 2015). They address the technological and economic possibilities and limits in the entire recycling system—from design to metallurgy—in relation to material interactions, recovery, losses and emissions, and resource efficiency.

Rule 1 DfR rules are product-specific and recycling system-specific;

- Oversimplification of recycling by defining general DfR rules will not produce the intended goal of resource efficiency.
- Due to its functional and unique mix of materials, each product has a unique recyclability profile. This implies that every product has a unique set of DfR guidelines, which are product and recycling system specific.
- These DfR guidelines are separately derived, and iteratively refined for each product group. This is realized by the application of recycling simulation models that can map any (BAT) recycling system and its opportunities and limits.

Rule 2 DfR needs model- and simulation-based quantification.

- DfR demands a tool (process simulation model) that is capable of quantifying the product's recycling profile and performance to:
  - pinpoint DfR issues of importance,
  - prioritize which design adjustments are to be implemented, and
  - give insight in the effect of improved design on resource efficiency (recycling rate, toxicity, scarce material recovery/losses, environmental impact, etc.).
- The detail must include, in addition to the flow of all elements, compounds, etc., the entropy and enthalpy of the system. See additional reading for the details, e.g., various flow sheets for recovery, the extraction metallurgy details, the technology, etc.

Rule 3 Design data should be accessible and available in a consistent format that is compatible with the detail required to optimize and quantify recycling performance of products for all metals, materials, and compounds present.

- Detail and format of the product data on product material composition (including chemical compounds) and construction should have the resolution to quantify, identify, and localize the commodity, critical, and disturbing materials.

Rule 4 Economically viable technology infrastructure and rigorous tools must be in place for realizing industrial DfR rules and methodology.

- Design must be based on a robust physical separation and sorting infrastructure that is at least capable of producing economically valuable recyclates.
- A robust metallurgical infrastructure and system must be in place to ensure maximum recovery of all "critical" materials from complex recyclates and dismantled functional subunits of a product.
- Environmental footprint and ecodesign tools should include the whole chain of processing to ensure that all materials, residues, and fugitive emissions are tracked. This requires suitable global policy to be in place.

Rule 5 CAD and process and system design tools that are linked to recycling system process simulation tools are needed to realize technology based, realistic, and economically viable DfR (Heibeck et al., 2021).

- Linking of existing and industry applied process simulation tools to CAD/design tools is a necessary step forward to realize realistic and economically viable DfR. Using FEMs to estimate module/part/material liberation will make this possible, as shown by Heibeck et al. (2021).

- This is a rigorous basis for industrially useful DfR rules and methodology. The example in Figure 5.1 shows that, based on rigorous recycling process simulation, detailed product compositional data can be derived from CAD/design tools. Recycling performance indicators and recycling indices including environmental analyses can be derived and provide the basis for DfR.
- Ecolabels for a product can be derived on this basis, which provide the depth to rigorously distinguish differences in product designs and identify the most resource-efficient designs.

Various DfR guidelines have been derived by applying the previously listed fundamental DfR rules and principles (Reuter, 2016). Recycling process simulation tools are applied to define, validate, and quantify the set of guidelines per product (of which the following list shows some possible guidelines). This physics-based approach sets priorities between the different guidelines and quantifies the necessity and potential result of DfR. The examples that follow the DfR guidelines given in the following paragraphs are embraced by general DfR approaches. However, the implementation of the fundamental rules given earlier, with simulation as a basis, makes it possible to derive unique sets of guidelines. These derived guidelines are product-specific as a function of material mix and (BAT) recycling systems, including a mindful consideration of product functional demands. Note that a fixed set of all possible guidelines will leave no room for the designer to design/construct a product. Input from recyclers was also used in defining examples for the certain guidelines. Some examples of these DfR guidelines are:

Rule 6 Identify and minimize the use of materials that will cause losses and contaminations in recycling due to material characteristics and behavior in sorting.

- Examples of materials that cause loss and contamination are: the use and construction of concrete in washing machines; glass plates in refrigerators; application of isolation materials; and use of colored (other than brown or green) printed wire boards.

Rule 7 Identify components/clusters in a product that will cause problems and losses in recycling due to combined and applied materials.

- A compatibility matrix (see Reuter and van Schaik, 2012) is derived per product based on the knowledge behind the Metal Wheel (Figure 2.2) and is useful for a quick first screening.
- It is important to realize that compatibility tables are not a DfR tool on their own, since they give *no* indication of resource efficiency, material recovery, losses, and fugitive emissions.

Rule 8 Design clusters or subunits in products that can be easily removed and that match with the final treatment recycling options (i.e., Metal Wheel), e.g.:

- Examples are the use or removal of Al applied as heat sinks on a printed wire board; separate PWBs for different functions, e.g., power boards (high Fe content, which will be lost in PWB processing route); control boards, removal of Ta capacitors, etc.

Rule 9 Label (including carefully considered standardization) products/components based on recovery and/or incompatibility so that they can be easily identified from recyclates and waste streams.

- Design for waste stream sorting or design for (automated) dismantling/sorting are important.
- Examples are:
  - the use of color- or identification-based labeling and easy to break connections for Ta capacitors (of crucial importance is metallurgical knowledge as captured in process simulation tools to understand what the quality requirements are for the

existing industrial Ta production/recycling infrastructure);
- labeling and identification-based waste stream sorting of CFL from LED lamps;
- standardization of marking and identification (e.g., type of marking and position) of cooling liquid and gas and marking of type of liquid and gas applied in fluid system and foam (these might be different); and
- marking of the tapping point on compressors.

Rule 10 Be mindful of liberation of materials in design (design for liberation).

- Simulation and knowledge of liberation behavior in relation to design, particulate and recyclate quality, and recycling (metallurgical) efficiency are crucial.
- Examples are: avoiding bolts/rivets of dissimilar materials (e.g., Fe bolts to Al, PWB, plastics, etc.), as these generally produce a liberation problem and therefore create cross-contamination of the different recyclate fractions; and minimizing the use of nonreversible adhesives for incompatible and undesired material combinations, such as gluing of glass on the steel mask of a CRT TV, shrink films on tube lamps, sealed batteries, PUR foam glued to steel/aluminum/plastic, wood glued to plastic, etc.
- The most resource-efficient designs.

These rules are the foundation of developing the recycling label depicted in Figure 5.2.

5.4 PRODUCT-CENTRIC RECYCLING

This section gives a brief overview of product-centric recycling of various products, product types, etc. Also consult Chapter 2 and Part 2 of this book, the references, and their references for a wide overview, which provides some indicative data for EoL automotive, electronic goods, batteries, lamps, etc.

5.4.1 Dynamics of the CE and the Urban Mine

The availability, weight, number, and type of products [end-of-life vehicles (ELVs), WEEE, batteries, catalysts, etc.] and the material composition from these products constitute the urban mine that is a source for resource recovery (van Schaik and Reuter, 2004, 2007, 2010). The technological and economic potential to recover the resources from this modern mine, consisting of a wide variety of different and time-changing products and materials, is hence determined by the EoL product developments and their weight, composition, and component content (e.g., printed wire boards containing critical resources, batteries, catalysts, etc.) for different products (models) within these product groups. These will all change over time as a function of factors such as:

- distribution of EoL products over time as a function of lifetime (usage) distributions of products;
- changing product design and product weight;
- product composition in terms of chemical compounds, linkages, etc. (e.g., as a function of legislative restrictions such as the ban on the use of chlorofluorocarbon coolants in refrigerators, lead-free soldering, changing technology, consumer demands, lightweight designs, energy demands in design, new vehicle concepts such as the electric vehicle, etc.);
- consumer behavior;
- policy; and
- disposal behavior and stocks, among others.

Predicting the material recovery from EoL developments demands insight into all of the previously listed points and requires a dynamic modeling approach capable of predicting EoL dynamic distributions of product quantities, weights, and composition (and construction) in relation to the dynamic (time-varying) material recovery (van Schaik and Reuter, 2004;

Reuter et al., 2019). Such mathematical and recycling dynamic modeling is at the core of driving changes and controlling and optimizing resource efficiency over time.

Predicting and understanding the potential of resource efficiency from the urban mine not only requires detailed compositional knowledge on the total flow of materials and recyclates but also requires these data as (statistical) standard-deviation and average-distribution values for each flow. The shape of this distribution (normal, Weibull, or even multimodal) must be known as well and how these change over time (see Chapter 43: Physical Separation—The Basics). In addition, as the materials/metals contained in recyclates are the economic target, their average and standard-deviation distributions in the input recyclates should be known as well in compositional detail. A rigorous mathematical basis for how to combine these different distribution functions of quantities and weights of products with changing composition distribution function is required to estimate the urban mine and its material content over time (Reuter and van Schaik 2012), to capture EoL product flow and its content within the CE. Within a context of physical and metallurgical recycling technology and its economics, this detail must permit the calibration of detailed mass-balance models to estimate recycling rates as a function of time-changing EoL distribution. This requires data to have a format compatible with simulation tools such as HSC Sim 10 (2023) (www.mogroup.com). It also facilitates the construction of detailed DfRE models for maximizing resource efficiency. Such theoretical knowledge is common practice in mineral processing and extractive metallurgy, used daily for optimizing metal recovery at all levels of the business. A brief overview of the state of the art of the modeling of recycling within the CE paradigm follows. Also, consult the work using discrete element modeling to predict particle sizes and liberation (Heibeck et al., 2021).

5.4.2 Some First Principles of Recycling

Recycling of complex, multimaterial consumer products such as cars, electronics/WEEE, PV modules, LEDs, and complex components such as batteries and catalysts demands an extended product-centric network of different types of processes. These separate and produce many different (intermediate) recyclates (e.g., ferrous recyclate, copper recyclate, aluminum recyclate, PWB recyclate, plastic recyclate, etc.) and recycling product flows (such as metal, speiss, slag, matte, flue dust) to recover the wide range of materials present in the product and components (van Schaik and Reuter, 2004; Hannula et al., 2020).

The possibilities and limits of recycling are strongly influenced by the performance of the liberation and separation processes (Castro et al., 2004, 2005). Recycling of complex, multimaterial consumer products uses a network of different types of processes to first liberate and separate the different interconnected materials into a range of recyclates (steel recyclate, copper recyclate, aluminum recyclate, plastic recyclate, etc.), and then takes those as the input to the final treatment processes that work on the thermodynamics, such as metallurgical processing. The selection and arrangement of processes, together with a thoughtful trade-off between collection, sorting, and/or separate or combined processing of products for recycling, determine the ultimate quality of intermediates and recyclates and hence the material/energy recovery from them. This also implies that government policy should stress the building and maintenance of a suitable global metallurgical processing infrastructure and of well-channeled preliminary pathways, to ensure a maximum recovery of metals, minimize theft, and optimize manual sorting in the informal sector (Xinwen et al., 2011). The following aspects can be consulted in the cited literature and the large body of literature cited in these papers as well as in the different parts of this book.

- *Product design, liberation, recyclate quality, and losses* (van Schaik and Reuter, 2004, 2007, 2010)

  The physics of comminution liberates material connections as a function of the material properties. This is directly related to particulate characteristics of recyclate flows. This is linked to sorting physics, chemistry, and thermodynamics of high-temperature processing and resource recovery from recyclate streams. Imperfection in separation and contamination of recyclates and material losses will also inevitably occur during mechanical sorting (sorting efficiency will never be 100%) due to the nature of sorting, physics, and statistics in these types of processes (the classical grade/recovery relationship, where 100% recovery can never be achieved in combination with 100% grade). Therefore metals enter metallurgical processing as functionally mixed, rather than as single metal streams. The materials/compounds all interact in metallurgical processing, so in turn could increase the treatment charges incurred by custom smelters and processors of the recyclates. The losses are a function of physical recycling and/or dismantling and result in unwanted losses of materials/metals and creation of residue streams in metallurgical processing due to prevailing chemistry and thermodynamics.
- *Nonlinear interconnected nature of design and recycling of complex multimaterial products* (Bartie et al., 2020; Fernandes et al., 2020)

  Metals are lost because the mix of metals brought together in design cannot be sorted from each other (by liberation and mechanical sorting). Therefore when entering metallurgical processes, they have physical and thermodynamic properties that do not allow the processes used to separate all of the elements from each other. The loss of metals when they cannot be separated during processing into thermodynamic and economic treatable fractions is often more common than generally assumed.

  Therefore a thermodynamic and systemic understanding of the recycling chain and the CE is so important, inclusive of the interactions between different materials/elements/compounds in different ratios in order to avoid unnecessary loss of metals.
- *Metallurgical processing infrastructures, process thermodynamics, and recycling of complex products* (Ignatenko et al., 2007; Reuter et al., 2019)

  The Metal Wheel (Figure 2.2) illustrates what generally happens to different metals in an EoL product entering the iron (Fe), aluminum (Al), copper (Cu), zinc (Zn), or lead (Pb) processes. Each circle indicates the destination of a metal found in mixed, EoL (electronic) product (waste), be it metal, slag, intermediate, flue dust, etc. The ideal situation for recycling would be a system that connects scrap and residues to all carrier (base) metals processes. This is a system that could be conceptualized as the center of the Metal Wheel so that scrap and residues then flow, as determined by thermodynamics and economics, into the appropriate carrier metal technologies (or Metal Wheel segments), with further links between each technology (or segment) for processing the residues from each technology. This is obviously utopian and cannot be totally achieved with existing technology. Nevertheless, the conceptualization can help in understanding the intricate system and knowledge required for recycling and waste/residue processing, what is implied by a systemic view to recycling, and what deep thermodynamic and process knowledge is required to "close" the loop. This suggests clearly that cradle-to-cradle thinking fails for even simple products due to the (limits imposed by the) laws of nature (thermodynamically) and illustrates that product resource efficiency has its limits.

- *Thermodynamic detail is required to maximize resource efficiency* (Abadías Llamas et al., 2019, 2020)

  If the CE is understood well, the thermodynamic properties of each metal and their various compounds/alloys are particularly important. If metals and their compounds in EoL goods and recyclates have thermodynamic and physical properties that are compatible with a particular base metal metallurgical infrastructure, the metallurgical processing technologies used by metals producers and refiners can usually separate and recover them economically from the various streams. These are recovered into, e.g., metal, matte, alloy, speiss, sludges, precipitates, slimes, flue dusts, fumes, slags, etc., minimizing the losses to streams that have a dumping/ponding cost attached to them. This is also the basis for DfR within the constraints of product functionality and performance demands (Figure 2.2).

  The economics of recycling are affected by the degree to which metals can be separated, recovered, and transformed into high-quality materials that can be applied, for example, in sustainability-enabling products. Where metals do not separate, they can either reduce the quality of the primary recycled metal product, because they cannot be recovered for use as separate metals themselves; and/or they tend to increase the energy needed during the recycling process. Elements within complex products, recyclate, or sludges are not recycled individually. Instead, they pass through one of a wide range of combinations of processes. The choice of process is an economic and physics-based technological optimization puzzle for the recycling operation, one that is driven by the changing values of the metal and high-end alloy products. Secondary materials can be fed into the overall metallurgical flowsheet. This leads to the conclusion that a metal production infrastructure must exist that allows flows of metals between carrier metal processes. This needs to be developed and nurtured over a long time, not least due to investment lead times, significant capital costs, the need for cross-carrier metal expertise, and because for each product type a different Metal Wheel exists.

  Generally, it has been shown that economically it is wise first to treat EoL products, after suitable sorting and concentration, through pyrometallurgy (a first rough separation at higher temperature that also permits energy recovery from plastics), followed by hydrometallurgy, producing the refined metals and materials of sufficiently high purity to return to the EEE and automotive products. For residues, sludges, etc., treatment at high temperature is often preferred to capture elements such as iron in benign slags, as hydrometallurgy would then create further residues such as jarosite and goethite. In some cases, hydrometallurgy is preferred if materials are well sorted, especially for some high-value materials such as rare earths, but then contamination must be limited.

  Figure 2.2 shows that to recover metals from complex mixtures of recyclates, very careful control and understanding of the theory of metallurgical reactor technology is required to obtain metals that report to fractions from which these can be recovered economically.

- *Developments in recycling and metallurgical technology for current and future products* (Reuter, 2011, 2016)

  Due to rapidly changing material compositions and EoL products, it will be rather wise to invest much more time and effort in innovating, optimizing, and maintaining the metallurgical processing infrastructure rather than (re)inventing new processes, as the basic structure in operation can handle and recover many elements already. Also, the CAPEX (Capital Expenditure) of metallurgical infrastructure is high, hence detailed understanding of this system is crucially important rather than simply (re-) inventing new process and wasting taxpayer R&D money on projects that are doomed to fail due to economy of scale, technological issues, as well as lack of know-how. Of crucial importance is the existence of a system that can take care of all these mixes of elements while maximizing metal recovery and obviously profit, and minimizing ecological damage.

- *Process and recycling system optimization: optimization of resource efficiency and environmental performance of complex products* (Bartie et al., 2021)

Optimization and understanding of recycling require deep systemic knowledge and understanding of the physics-based link between product design, material applications, and combinations with the final treatment process technology (metallurgical processes) in the entire recycling system within the CE. Deep metallurgical process knowledge is paramount as well as a thorough understanding of the process technology and its economic drivers to maximize resource efficiency from complex products by a mindful balance of product mixing/sorting for processing, dismantling in combination with shredding/sorting, and (pyro- and hydro-) metallurgical processing of recyclates from these products. The outcomes from recycling, i.e., the range, selection, and purity of metals produced, are in turn a key determinant of the economic profitability and viability of recycling in its entirety. Economy of scale recycling technologies and processes often already exist that can economically separate many metals into high-quality products, but whether the metals entering recycling are separated depends on the choice of processes used and also product functional design. Rapidly changing product design and product replacement (e.g., LCD display technology replaced by LED/OLED technology) poses challenges to physical recyclers to have recycling technology keep pace with rapidly changing product properties and material usage, which poses different demands on sorting technology development and plant configuration in order to most optimally recover materials. It should be realized that a critical mass (over a long enough timespan) of EoL products with a more or less similar character is required to economically invest and build on technology development and plant design for shredding and sorting.

- *The simulation tools for analyzing the CE system* (Reuter, 2016; Bartie et al., 2021)

Rigorous economic and physics-based understanding of technology and systems for recycling make it possible to make informed decisions over the CE system boundary. Computer-based modeling and process simulation tools of the recycling technology and system performance of products as shown by Figure 5.1 have been developed and applied for the recycling of complex products by the authors to understand and capture the factors that improve design for resource and exergy efficiency. These models include expert rule-based modeling of the relation between design and liberation behavior, particulate quality, physics of sorting efficiency, recyclate quality, distribution of materials/compounds over all produced recyclates, and recycling products and metallurgical process efficiency as a function of design and recyclate composition. This implies that sophisticated tools exist that enable the evaluation of resource recovery in metal and material processing systems (Reuter et al., 2019). These produce consistent mass and energy balances for all compounds and materials in a processing system. This can be for complete copper plants from rock to refined metal or a complete recycling system for each liberated and unliberated mineral (physical material) and chemical element as a function of design, as shown in using the HSC Sim tool (www.mogroup.com), which is linked to environmental software such as GaBi (2022) (www.sphera.com).

5.4.3 Digital Twinning of Systems

In summary, recycling is clearly a complex thermodynamic and economic puzzle to solve, obviously with no one answer or set of DfR rules, and not solvable with beautiful credos such as cradle-to-cradle. It would also be self-evident that mining the urban mine, which has the sound of hype to it, will be rather a complex and even (economically) impossible task, as reflected by the nonlinear effects and the complexity of the urban ore body and its mineralogy. It would also be clear that for each mixture and each condition this separation will be different, implying that general material flow analysis (MFA), life cycle assessment (LCA), and DfR methodologies that do not address this depth of mineralogy (i.e., compounds, alloys, transformations, thermochemistry, etc.) will obviously and inevitably lead to false conclusions. Furthermore, uneconomic and unrealistic technological and policy recommendations may result.

Hopefully these methods will be adopted in the mainstream and link the recycling industry and the sophistication of product design, and its CAD and other tools, as also shown in Figure 2.4. Understanding the CE demands very rigorous modeling to reveal the opportunities and limits of recycling and true dissipation of exergy from the system, as shown in Figure 5.2. Design for resource efficiency captures this detail, as discussed by Reuter et al. (2019) and Castro et al. (2004, 2005).

5.5 EXAMPLES OF RECYCLING SYSTEM SIMULATION

Resource efficiency, environmental impact quantification of recycling, as well as ecolabeling and DfR and resource efficiency are briefly discussed based on this physics- and knowledge-based background by providing examples and discussions for the different products. Consult the references, and the references in these papers, for more details of these applications.

The examples show that the simulation-based approach permits the analysis of the system in terms of:

- Mass flows of produced recyclates/recycling products and distribution/dispersion of (critical) materials over different recyclate streams in terms of full composition and details of each stream.
- Calculation of quality and composition of recyclate (grade) streams and recycling products and quantify this in terms of thermochemistry, i.e., enthalpy and entropy/exergy as a function of any input to the system.
- Predicting the impact of different operating modes of technology and the system architecture (i.e., the complete technology infrastructure behind the Metal Wheel—Figure 5.1) on the total recycling of minor and commodity metals.
- Identification of the dispersion, occurrence, and appearance (chemical phases/compounds) of possible toxic/harmful elements in all effluent streams that interface with nature.
- Quantifying the flows of all streams in terms of kilowatts, to define both enthalpy and entropy/exergy, making it possible to understand the fundamental resource-efficient performance of the CE system and the product.

A key value of this is to understand the exergy dissipation from the system, as shown in Figure 5.1.
- Linking the process simulation to environmental footprint software, design tools, sustainability development goals (SDGs), etc. (see Figure 2.4 in Chapter 2).
- These physics-based simulation models are a requirement for the application of DfR tools and rules, which can link to predictive product design.

5.5.1 SuperLightCar

The body-in-white (BIW) for the EU's 6th framework SuperLightCar project is a novel construction of steel, light metals, and polymers, reducing the weight by 35% for a substantially lower manufacturing cost per kilogram (Hirsch et al., 2009; Krinke et al., 2009). For this project the developed models have been applied to provide detailed and physics-based calculations on recycling/recovery of ELVs, including various new design body in white concepts. From these models the optimal recycling concepts (combination and arrangement of processes) for a car body could be established, while considering and including the following aspects (Hirsch et al., 2009; Krinke et al., 2009):

- product design and liberation,
- separation physics of automated sorting,
- chemistry and thermodynamics of metal production and recycling systems,
- recyclate and recycling product quality (physical and chemical) as a function of product design choices and calorific values of the (intermediate) recycling streams,
- losses and emissions,
- optimization and selection of plant/flowsheet architecture with changing product design for different lightweight multimaterial car body concepts.

For this project van Schaik and Reuter (2007) uniquely developed and applied fuzzy recycling models to provide real-time design-related recycling calculations. This approach is well suited for creating a link to product design through CAD software, as shown by Heibeck et al. (2021) using finite element modeling.

5.5.2 Recycling of Waste Electrical and Electronic Equipment

Dynamic recycling simulation models have been developed and applied by van Schaik and Reuter (2010, 2012) for the recycling of waste electrical and electronic equipment (WEEE). This simulation approach has been applied to determine the recyclability of the Fairphone (www.fairphone.com) as depicted in Figure 5.1 and is documented in Reuter et al., 2018. For the first time, the recycling label for a product (Figure 5.2) has been published.

This analysis was made possible due to the availability of full compositional design detail of the complete Fairphone, i.e., information on the "layout" (i.e., construction, connections, etc.) of the design, all linked materials and their composition, quantities, major modules, etc. This detailed analysis made it possible to improve the recyclability of the Fairphone 3 compared to Fairphone 2 (Reuter et al., 2018).

5.5.3 Recycling of Photovoltaic (PV) Cells

The detailed analysis of large-scale systems has been discussed for silicon and CdTe PV cells by Abadías Llamas et al. (2020) and Bartie et al. (2020, 2021). Uniquely, these system analyses integrate various aspects illustrated in Figure 5.1, i.e., product design, process simulation, exergy analysis, and environmental impact assessment, and permit the detailed optimization of the CE system for these examples in terms of exergy and environmental impact. In addition, the effect of different

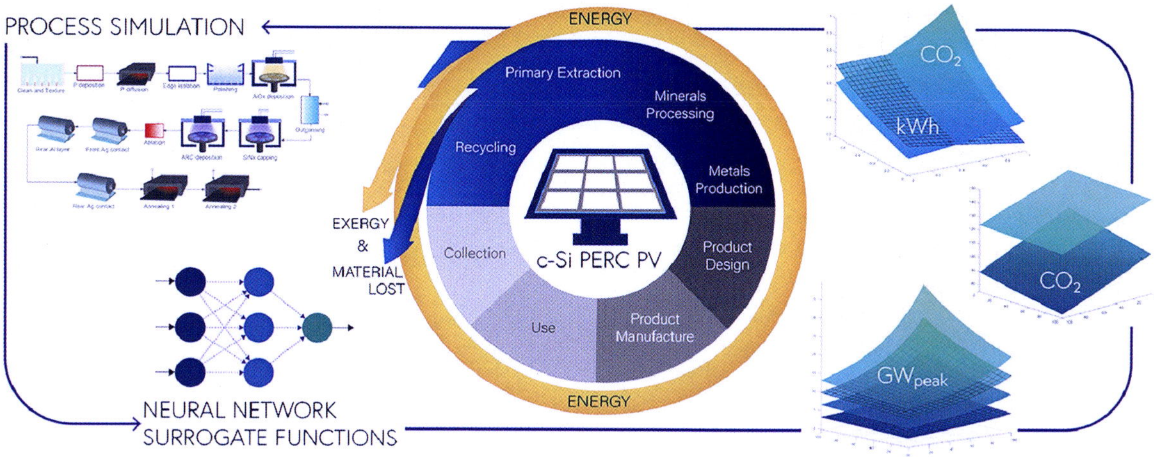

FIGURE 5.3 Optimizing the CE PV system, from mineral to metal, to product, recycling, and back into PV, showing the exergy dissipation as well, as for example, to the carbon footprint of the complete system. *From Bartie et al. (2021).*

process routes can be evaluated. Figure 5.3 shows some results and shows how a digital twin can be used to describe the complete system to derive useful information to minimize the footprint as well as exergy dissipation.

5.5.4 Recycling of Rare Earth Magnets

Similar to the PV examples, Fernandes et al. (2020) linked mining, minerals processing, and extractive metallurgy to the manufacture and recycling of rare earth FeNdB magnets. Uniquely the exergy dissipation of the system could be highlighted along with the environmental footprint of the complete FeNdB magnet CE system. This unique example shows how primary and secondary resources are linked via material production, manufacture, and EoL in the CE by a simulation model to fully quantify recyclability as well as exergy dissipation in addition to the system footprint.

5.6 SUMMARY

To understand the recycling of products and the inherent losses produced due to the design complexity, a deep understanding of the physics and thermodynamics as well as all mass and heat transfer processes of the CE system are required. This chapter refers to work that discusses this in detail to obtain a correct value of the material recycling to once again produce the quality of materials that can flow back into the same product, while minimizing the exergy dissipation from the system that defines the circularity of each individual product. The recycling label has been developed to show the true material recycling back into the same product. A similar label can be defined that shows the dissipation of exergy and therefore the exergetic performance, which will always be less than 100%, mirroring also the loss of quality in materials due to dilution and dissipation. Therefore material recycling will always be less than 100%. The degree less is defined by the complexity of the product; the more complex, the more the losses, mostly defined by complex nonlinear thermodynamics, as well as mass and heat transfer during processing and material production.

5.7 FUTURE CHALLENGES

UNEP (2013) and Reuter et al. (2019) discuss the various challenges for recycling. Considering the detail of a product such as a PV cell, as discussed previously, the challenges are various (nonexhaustively):

- Functionality of products and modules necessitate that materials are functionally connected and therefore by default not easy to separate,
- Levels of the elements are often low in the products, which from a second law of thermodynamics point of view and therefore from economics makes these difficult to recover,
- Product design must enhance, e.g., the recovery of magnets from products, so that these can easily be disassembled and sorted into bins to minimize cross contamination, and
- The unavailability of the complex interlinked metallurgical infrastructure that can on an economy of scale recover the elements, which are often in low concentrations, within the chemistry, thermodynamics, and economics as reflected by the Metal Wheel.

While DfR is often proposed to help "close the loop" of the CE, this is not easy due to the closely linked elements in the functional materials of the products being considered in this chapter (and also elsewhere in this book).

Using digital simulation tools, the simulation-based approach that links product design to metallurgical processing infrastructure, as discussed in Chapter 2, has shown the technological and economical limits of recycling, e.g., see Reuter et al. (2019) for the Fairphone example. Furthermore, the examples by Bartie et al. (2021, 2022) discuss the metallurgical infrastructure and technology criticality in terms of the link between PVs, sustainability, and the metals industry.

What is missing is the detail to especially understand the liberation of materials when passing through the physical separation, i.e., cutting and shredding processes. This begs the question as to what can be separated and what must be processed as a module in its totality, as recently shown for complex car parts (Iglesias-Émbil et al., 2023), using exergetic approaches to optimize the product/module design. Recently innovative work has been done by Heibeck et al. (2021) in cutting-edge work that shows how FEMs can fundamentally simulate shredding of complex multimaterial products to understand their liberation as well as energy consumption.

Acknowledgements

Chapter 5.2 is based on the following publications:
Bartie et al. (2019, 2020, 2021, 2022), Reuter and van Schaik (2015), and Reuter et al. (2015, 2018, 2019).

References

Abadías Llamas, A., et al., 2019. Simulation-based exergy, thermo-economic and environmental footprint analysis of primary copper production. Miner. Eng. 131, 51–65.

Abadías Llamas, A., et al., 2020. Simulation-based exergy analysis of large circular economy systems: zinc production coupled to CdTe photovoltaic module life cycle. J. Sustain. Metall. 6 (1), 34–67.

Bartie, N., et al., 2020. The simulation-based analysis of the resource efficiency of the circular economy—the enabling role of metallurgical infrastructure. Miner. Proces. Extract. Metall. (TIMM C) 129 (2), 229–249.

Bartie, N.J., et al., 2021. The resources, exergetic and environmental footprint of the silicon photovoltaic circular economy: assessment and opportunities. Resour. Conserv. Recycl. 169, 105516.

Bartie, N., Cobos-Becerra, L., Fröhling, M., Schlatmann, R., Reuter, M., 2022. Metallurgical infrastructure and technology criticality: the link between photovoltaics, sustainability, and the metals industry. Miner. Econ. 35 (3–4), 503–519.

Castro, B., Remmerswal, H., Boin, U., Reuter, M.A., 2004. A thermodynamic approach to the compatibility of materials combinations for recycling. Resour. Conserv. Recycl. 43 (1), 1–19.

Castro, B., Remmerswal, H., Brezet, H., van Schaik, A., Reuter, M.A., 2005. A simulation model of the comminution-liberation of recycling streams: relationship between product design and the liberation of materials during recycling. Int. J. Miner. Process. 75 (3-4), 255–281.

European Commission, 2009. Directive 2009/125/EC of the European Parliament and of the Council of 21 October 2009 Establishing a Framework for the Setting of Ecodesign Requirements for Energy-Related Products. Official Journal. http://data.europa.eu/eli/dir/2009/125/2012-12-04.

European Commission, 2022. Proposal for a Regulation Establishing a Framework for Setting Ecodesign Requirements for Sustainable Products and Repealing Directive

2009/125/EC.COM(2022) 142 Final. Directorate General for Environment. https://environment.ec.europa.eu/publications/proposal-ecodesign-sustainable-products-regulation_en.

Fernandes, I.B., Abadías Llamas, A., Reuter, M.A., 2020. A simulation-based exergetic analysis of NdFeB permanent magnet production to understand large systems. J. Met. 72 (7), 2754–2769.

GaBi, 2022. Sphera. www.sphera.com.

Hannula, J., et al., 2020. Simulation-based exergy analysis of physical separation and re-melting of aluminum scrap to specific alloys. J. Sustain. Metall. 6 (1), 174–189.

Hayes, P., 2021. Process Principles in Minerals and Materials Production, fourth ed., ISBN: 978-0-6450274-5-7. 755 pp.

Heibeck, M., Rudolph, M., Modler, N., Reuter, M.A., Filippatos, A., 2021. Characterizing material liberation of multi-material lightweight structures from shredding experiments and finite element simulations. Miner. Eng. 172, 107142.

Hirsch, J.R., Bassan, D., Lahaije, C., Goede, M., 2009. Aluminium International Today DMG World Media. ISSN: 1475-455X News Article ID: 15697(Volume 21 No. 5—Sept.–Oct. 2009), pp. 46–49.

HSC 10, 2023. Metso Outotec Research 1974–2022. www.mogroup.com.

Iglesias-Émbil, M., Abadías, A., Valero, A., Calvo, G., Reuter, M.A., Ortego, A., 2023. Criticality and recyclability assessment of car parts—a thermodynamic simulation-based. Approach Sustain. 15 (1), 91. https://doi.org/10.3390/su15010091.

Ignatenko, O., van Schaik, A., Reuter, M.A., 2007. Recycling system flexibility: the fundamental solution to achieve high recycling rates. J. Clean. Prod. 16 (4), 432–449.

Krinke, S., et al., 2009. Recycling and DfR of multi-material vehicles (as part of 'Life cycle assessment and recycling of innovative multi-material applications'). In: Proceedings of the International Conference 'Innovative Developments for Lightweight Vehicle Structures', May, 26–27th 2009, Wolfsburg, Germany (Volkswagen Head Office)., ISBN: 978-3-00-027891-4, pp. 196–208.

Reuter, M.A., 2011. Limits of design for recycling and "sustainability": a review. Waste Biomass Valor. 2, 183–208.

Reuter, M.A., 2016. Digitalizing the circular economy—circular economy engineering defined by the metallurgical Internet of Things-, 2016 TMS EPD Distinguished Lecture, USA. Metall. Trans. B 47 (6), 3194–3220.

Reuter, M.A., van Schaik, A., 2012. Opportunities and limits of recycling: a dynamic-model-based analysis. MRS Bull. 37 (4), 339–347.

Reuter, M.A., van Schaik, A., 2015. Product-centric simulation-based design for recycling: greenprinting of LED lamp recycling. J. Sustain. Metall. 1 (1), 4–28.

Reuter, M.A., van Schaik, A., Gediga, J., 2015. Simulation-based design for resource efficiency of metal production and recycling systems, Cases: Copper production and recycling, eWaste (LED Lamps), Nickel pig iron. Int. J. Life Cycle Assess. 20 (5), 671–693.

Reuter, M.A., et al., 2018. Limits of the circular economy: fairphone modular design pushing the limits. World Metall.—ERZMETALL 71 (2), 68–79.

Reuter, M.A., et al., 2019. Challenges of the circular economy—a material, metallurgical and product design perspective. Annu. Rev. Mater. Res. 49, 253–274.

UNEP, 2013. In: Reuter, M.A., Hudson, C., van Schaik, A., Heiskanen, K., Meskers, C.M., Hagelüken C. (Eds.), United Nations Environmental Protection (UNEP) Report "Metal Recycling: Opportunities Limits Infrastructure" Report. http://www.resourcepanel.org/reports/metal-recycling.

van Schaik, A., Reuter, M.A., 2004. The time-varying factors influencing the recycling rate of products. Resour. Conserv. Recycl. 40 (4), 301–328.

van Schaik, A., Reuter, M.A., 2007. The use of fuzzy rule models to link automotive design to recycling rate calculation. Miner. Eng. 20, 875–890.

van Schaik, A., Reuter, M.A., 2010. Dynamic modelling of e-waste recycling system performance based on product design. Miner. Eng. 23, 192–210.

van Schaik, A., Reuter, M.A., 2016. Recycling indices visualizing the performance of the circular economy. World Metall.—ERZMETALL. 69 (4), 201–216.

Xinwen, C., et al., 2011. Informal electronic waste recycling: a sector review with special focus on China. Waste Manag. 31 (4), 731–742.

Additional Reading

Adams, M.D., Gold Ore Processing, second ed., pp. 937–958. Elsevier, doi:https://doi.org/10.1016/C2015-0-00699-2.

Crundwell, F., Moats, M.S., Ramachandran, V., Robinson, T., Davenport, W.G., 2011. Extractive Metallurgy of Nickel, Cobalt and Platinum Group Metals. Elsevier, https://doi.org/10.1016/C2009-0-63541-8.

Seetharaman, S. (Ed.), 2014. Treatise on Process Metallurgy. Elsevier Ltd., https://doi.org/10.1016/C2010-0-67121-5.

Schlesinger, M.E., Sole, K.C., Davenport, W.G., Alvear, G., 2021. Extractive Metallurgy of Copper, sixth ed. Elsevier, Amsterdam, ISBN: 9780128218754.

Ullmann, Ullmann's Encyclopedia of Industrial Chemistry, Wiley-VCH Verlag GmbH & Co. KgaA, Weinheim, Germany. Doi:https://doi.org/10.1002/14356007.

USGS, 2020. Mineral Commodity Summaries 2020. US Geological Survey, Reston VA, USA.

CHAPTER

6

Developments in collection of municipal waste

Gijs Langeveld[1] *and Maarten Goorhuis*[2]

[1]Project Gijs B.V., Amersfoort, The Netherlands [2]De Vuilnisman, Arnhem, The Netherlands

6.1 INTRODUCTION

When someone asks me what I am doing for a living, I explain that I am a garbageman who works in an office. In almost any case, my answer is the start of a conversation: how should they manage their waste, questions about the fundamental importance of doing things right, and how to improve waste management. Because everybody consumes products, waste decision making is part of our daily experience. Waste is a universal issue affecting every person on the globe.

There are three reasons why municipal solid waste (MSW) management is essential:

- To prevent the spread of pests and diseases;
- To reduce the impact of waste on our environment and climate;
- To recover materials from waste.

Initially, the main driver to manage MSW was the need to provide for sanitation. In the previous eras uncontrolled dumping of MSW led to the spread of vermin and pest diseases. To improve sanitation, it became necessary to collect the waste and take it out of the cities and villages. Until the Industrial Revolution, MSW consisted predominantly of organic waste, and was spread on the land as natural compost. With increased welfare and increased industrialization, the composition of the waste slowly shifted and became more inorganic. Due to the increases in consumption and population, the amounts of MSW grew. Instead of spreading MSW on the land, waste was increasingly dumped in landfills. At the end of the nineteenth century, the first waste incineration plants were erected, mostly in or near large cities where the amounts of MSW were more significant. However, landfills remained the dominant disposal method.

In the second half of the twentieth century, it became apparent that many landfills contained hazardous materials that were leaking out of the landfill and were causing environmental pollution. Also, waste incineration and flue gas cleaning techniques appeared insufficient to prevent the creation and spread of dioxins. MSW management developed further to provide, next to sanitation, also environmental protection. Consequently, the concept of a sanitary landfill, where waste is contained and monitored, was

Handbook of Recycling. https://doi.org/10.1016/B978-0-323-85514-3.00041-5

developed. Additionally, improved waste incineration and flue gas cleaning techniques were developed.

Waste directly impacts climate change. "If food waste were a country, it would be the third-largest greenhouse gas emitter in the world" (FAO, 2013). And this is food waste only. Other waste streams such as plastics and textiles have an enormous greenhouse gas impact if not reused or recycled. "Reducing human-caused methane emissions is one of the most cost-effective strategies to rapidly reduce the rate of warming and contribute significantly to global efforts to limit temperature rise to 1.5°C. (..) Existing targeted measures could reduce methane emissions from the waste sector by 29–36 Mt/yr by 2030. The greatest potential is in the improved treatment and disposal of solid waste. As much as 60 percent of waste-sector targeted measures have either negative or low cost" (United Nations Environment Programme and Climate and Clean Air Coalition, 2021).

Waste directly impacts our environment when materials that do not belong there enter our environment due to improper waste management. If waste is not collected correctly, it will be dumped, littered, or burned. "As waste has no value, it will always flow to the lowest treatment option (..), it is like water, it flows to the lowest point. Therefore, waste is everywhere: in our bodies, our food, our water, even our air, oceans, rivers and of course on land" (Newman, 2020). Our failure to organize a good collection and recycling system can be seen all around us. Recently, an international team of researchers assessed the impact on the stability of the earth's system of the cocktail of synthetic chemicals and other "novel entities" flooding the environment. Humanity has already exceeded a planetary boundary related to environmental pollutants, including plastics (Persson et al., 2022).

In the last decades a third reason has become more prominent: the recovery of resources from our waste to reduce our material usage. "The global material footprint rose from 43 billion metric tons in 1990 to 54 billion in 2000, and 92 billion in 2017—an increase of 70 percent since 2000, and 113 percent since 1990. (..) What's more, the global material footprint is increasing at a faster rate than both population and economic output" (UNSD, 2021).

The waste hierarchy has become a widely accepted guideline for worldwide waste management to improve sanitation, protect the environment and climate, and to provide for resource recovery. The waste hierarchy stipulates a priority order in waste prevention and management legislation policy. Based on the European Waste Framework Directive, Figure 6.1 shows the waste hierarchy (directive 2008/98/EC)[1] (European Union, 2018).

Although the waste hierarchy seems to be quite clear, in practice several uncertainties exist. On the one hand, the term recycling stands for a great variety of processes and methods that significantly differ in terms of environmental and economical benefit. The recycling of mixed plastic, as opposed to material separated plastic, for instance, has a very different environmental and economic impact. Generally speaking, recycling operations that lead to materials with the same quality as the original have a better environmental performance than recycling operations that lead to materials with a lower quality. Take plastics recycling as an example: Various recycling techniques differ in terms of environmental and economic impact. The phrase "Shit in is shit out" is regularly used in the sector, which means the better you can sort the plastic fractions into mono materials, the better the quality of the produced recyclate. Life cycle assessments (LCAs) can support decision

[1] The precise interpretation may vary from one place to another.

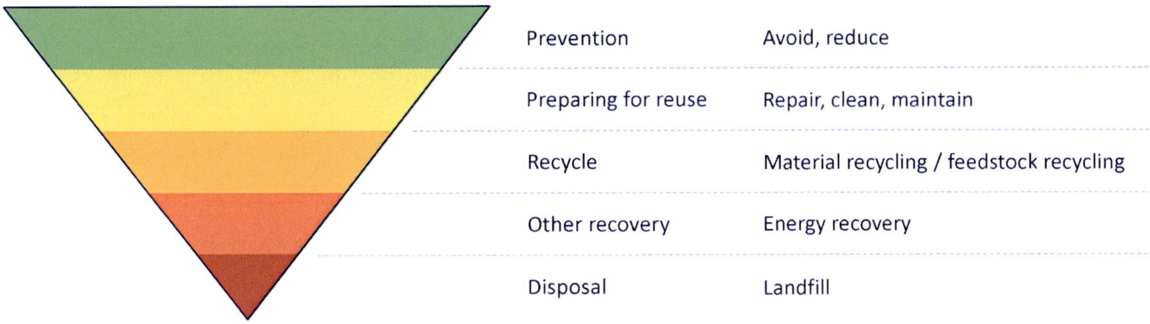

FIGURE 6.1 Waste hierarchy according to EU directive 2008/98/EC. *Modified from Goorhuis and Bartl (2011).*

making. Still, we should be aware of the limitations of this technique: it does not include what is not measurable, like value chain innovation or products ending up in the environment.

This chapter aims to deepen our expertise on municipal solid waste (MSW) by giving an overview of the global developments in the collection. Afterwards, the reader has an overview of the importance of municipal waste management, the introduction of definition and model, an overview of the current state in the different regions, how products are collected and other essentials for good collection, relevant trends and developments, and where gaps and challenges are.

6.2 DEFINITIONS AND MODELS

6.2.1 Definitions

The definition of municipal solid waste varies from country to country. Municipal solid waste is also often referred to as municipal waste or urban (solid) waste.

The Organization for Economic Cooperation and Development (OECD) defines MSW as "waste collected and treated by or for municipalities. It covers waste from households, including bulky waste, similar waste from commerce and trade, office buildings, institutions and small businesses, yard and garden waste, street sweepings, the contents of litter containers, and market cleansing waste. The definition excludes waste from municipal sewage networks and treatment, as well as waste from construction and demolition activities" (OECD, 2011).

The European Landfill Directive defines MSW as "Waste from households, as well as other waste which, because of its nature or composition, is similar to waste from households" (European Union, 1999).

Eurostat, which gathers the European data on waste management, uses a slightly different definition: "waste collected by or on behalf of municipal authorities and disposed of through waste management systems. The municipal waste consists mainly of waste generated by households, although it also includes similar waste from sources such as shops, offices and public institutions" (Eurostat, 2013).

In the United States, the US Code defines MSW (United States Code, 2006), where it says: "municipal solid waste" means waste material:

(1) generated by a household (including a single or multifamily residence); and
(2) generated by a commercial, industrial, or institutional entity, to the extent that the waste material:
  (a) is essentially the same as waste normally generated by a household;
  (b) is collected and disposed of with other municipal solid waste as part of normal municipal solid waste collection services; and

(c) contains a relative quantity of hazardous substances no greater than the relative quantity of hazardous substances contained in waste material generated by a typical single-family household.

The Pan American Health Organization (PAHO) defines MSW as: "Solid or semi-solid waste produced through the general activities of a population centre. Includes waste from households, commercial businesses, services, and institutions, as well as common (non-hazardous) hospital waste, waste from industrial offices, waste collected through street sweeping, and the trimmings of plants and trees along streets and in plazas and public green spaces" (Espinoza et al., 2011).

Although the definition of MSW varies around the world, there are some common characteristics:

- MSW is waste from private households;
- MSW includes similar waste from small enterprises, offices, and other institutions;
- MSW is collected through the municipal waste collection system;
- MSW includes both organic fractions, such as food and garden waste, and inorganic materials.

6.2.2 Integrated Sustainable Waste Management

Cities can use the Integrated Sustainable Waste Model to identify and tackle solid waste issues. It is an integrated model that has been developed and improved since the 1990s. Waste is embedded in the political, economic, sociocultural, and environmental context (Wilson et al., 2013) (Figure 6.2).

The left triangle comprises the three critical physical elements that we have introduced previously: public health, environment and climate, and resource recovery. The right triangle focuses on ISWM "software"—the governance strategies that all need to be addressed to deliver a

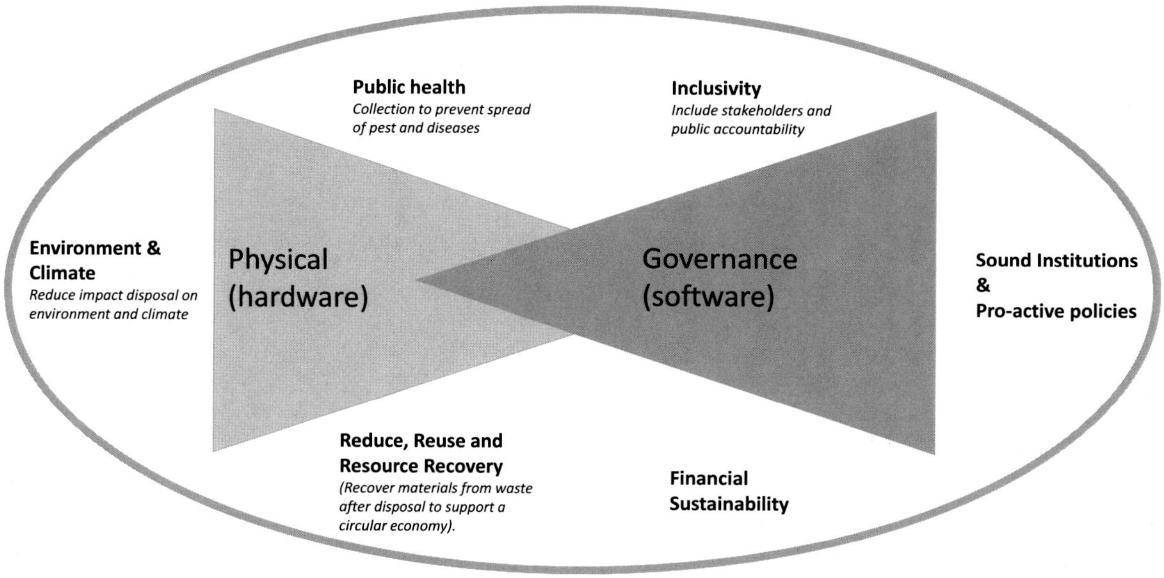

FIGURE 6.2 The "two triangles" representation of the Integrated Solid Waste Management (ISWM) framework. *Based on Wilson et al. (2013).*

well-functioning system. The system as a whole needs to:

- rest on a base of sound institutions and proactive policies
- be financially sustainable, which means cost-effective and affordable
- be inclusive, allowing stakeholders to contribute as users, providers, and enablers (Wilson et al., 2013)

6.2.2.1 Institutions and Policies

Usually, national law defines the responsibilities for the management of MSW and the performance criteria regarding what the result of waste management should be. In Europe, national legislation is based upon the EU Waste Directive. How it is achieved is then up to the municipalities. Cities are either fully accountable for collecting and treating all municipal solid waste, or in some cases, the responsibility has been limited to managing household waste and waste from public spaces. In the latter case, the management of similar waste from small enterprises, offices, and other institutions is the responsibility of the waste holder.

6.2.2.2 Financial Sustainability

Waste management is a service like education or health services: it provides a public service for which the costs need to be covered. Assuming cost-effectiveness and affordability is always a target, we should be aware that the more service we demand, the higher the costs will likely be. In Figure 6.3 we have mapped the reasons for waste management versus the costs involved.

In a financially sustainable model, the producer or the consumer covers the costs structurally. For example, via fees and taxes, supplemented by extended producer responsibility (EPR) systems for specific waste streams.

6.2.2.3 Inclusivity

Generally speaking, the city government is responsible for waste management. However, they will need to interact with other stakeholders such as users, providers, and external agencies, to ensure inclusivity and accessibility to waste services for all citizens or users. Also, raising public awareness is required to improve results.

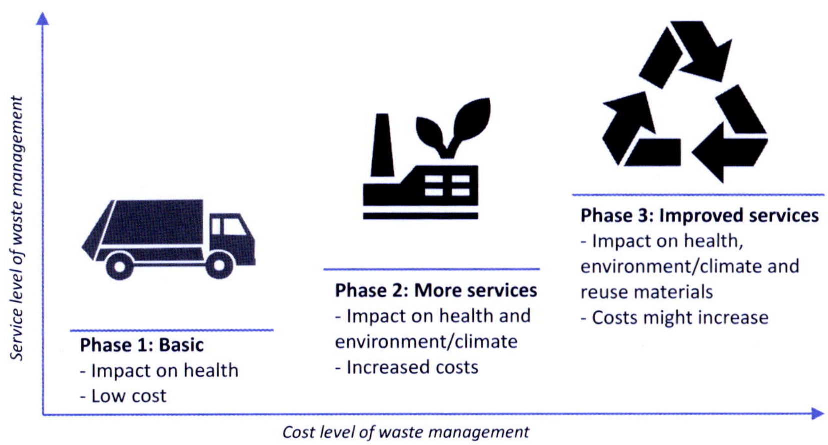

FIGURE 6.3 Stages of waste management versus the costs.

6.3 A GLOBAL PICTURE OF SWM

This section is based on research from the World Bank, in particular World Bank, 2018. This is an adaptation of an original work by the World Bank. Views and opinions expressed in the adaptation are the sole responsibility of the author or authors of the adaptation and are not endorsed by the World Bank.

6.3.1 Volumes

The world generated 2.01 billion tonnes of municipal solid waste in 2016 (Figure 6.4). Waste generated per person per day averages 0.74 kg but ranges widely, from 0.11 to 4.54 kg. Waste generation volumes are correlated with income levels and urbanization rates. The East Asia and Pacific region generates most of the world's waste, at 23%, and the Middle East and North Africa region produce the least in absolute terms, at 6%.

The World Bank expects the world will generate 3.40 billion tonnes of MSW by 2050 under a business-as-usual scenario. Waste is growing the fastest in sub-Saharan Africa, South Asia, and the Middle East North Africa regions. By 2050, total waste generated is expected to approximately triple, double, and double, respectively. Daily per capita waste generation in high-income countries is projected to increase by 19% by 2050 (from 1.58 kg/capita per day in 2016 to 1.87 kg/capita per day in 2050), compared to low- and middle-income countries where it is anticipated to increase by approximately 40% or more (from 0.53 kg/capita per day in 2016 to 0.79 kg/capita per day in 2050).

Municipal solid waste is one of several waste streams that countries and cities manage. Other typical waste streams include industrial waste, agricultural waste, construction and demolition waste, hazardous waste, medical waste, electronic waste, or e-waste. Some waste streams, such as industrial waste and construction and demolition waste are generated in much higher quantities than municipal solid waste: 12.73 and 3.35 kg/capita per day, respectively (Kaza et al., 2018).

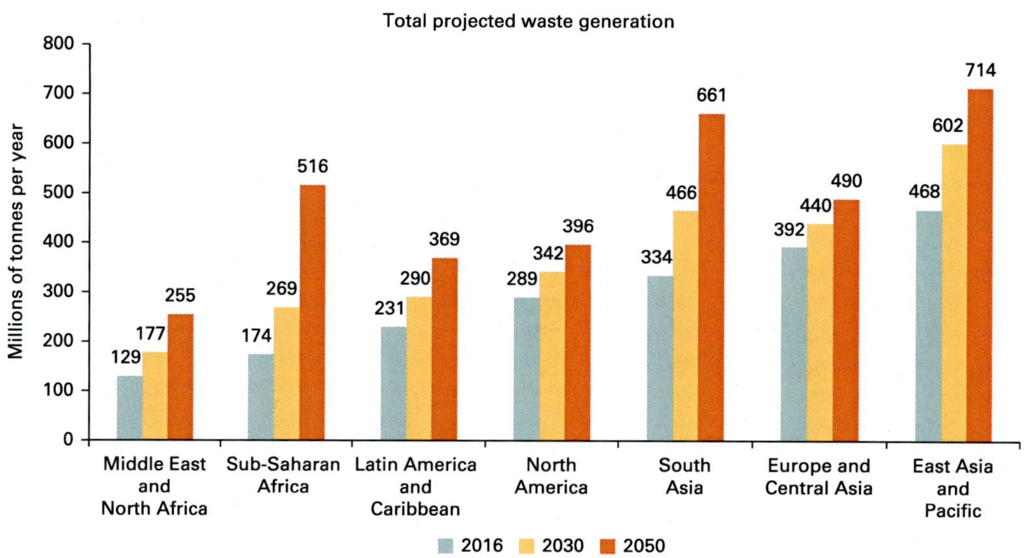

FIGURE 6.4 Projected waste generation, by region (millions of tonnes/year). *From Kaza et al. (2018).*

6.3.2 Waste Composition

Food and green waste is the largest waste category (Figure 6.5), globally making up 44%, and in low- and middle-income countries, its share is more than 50%. In high-income countries, the amount of organic waste is comparable in absolute terms but, because of more significant amounts of packaging waste and other nonorganic waste, the fraction of organics is about 32%. High-income countries purchase and dispose of more packaging and products, and this is directly reflected in their waste composition.

6.3.3 Waste Collection

In high-income countries, almost all waste is collected (Figure 6.6). In lower-middle-income countries, collection rates are about 51%, and in low-income countries, about 39%. Low-income countries collect about 48% of waste in cities, but this proportion drops drastically to 26% outside of urban areas. Uncollected waste is often managed independently by households that burn their waste on site or by the informal sector such as rag pickers, scrap collectors.

6.3.4 Waste Disposal

Around the world, almost 40% of waste is disposed in sanitary landfills. A sanitary landfill is a modern engineering landfill where waste is allowed to decompose into biologically and chemically inert materials in a setting isolated from the environment (Waste Management, 2011). About 19% undergoes materials recovery through recycling and composting, and 11% is treated through modern incineration. Globally 33% of waste is still openly dumped (Figure 6.7).

Waste disposal practices vary significantly by income level and region (Figure 6.8). Open dumping is prevalent in lower-income countries, where sanitary landfills are not yet widely available. As nations prosper economically, waste management uses more sustainable methods. In a sense waste is a luxury issue: both the volume of waste is growing with increased

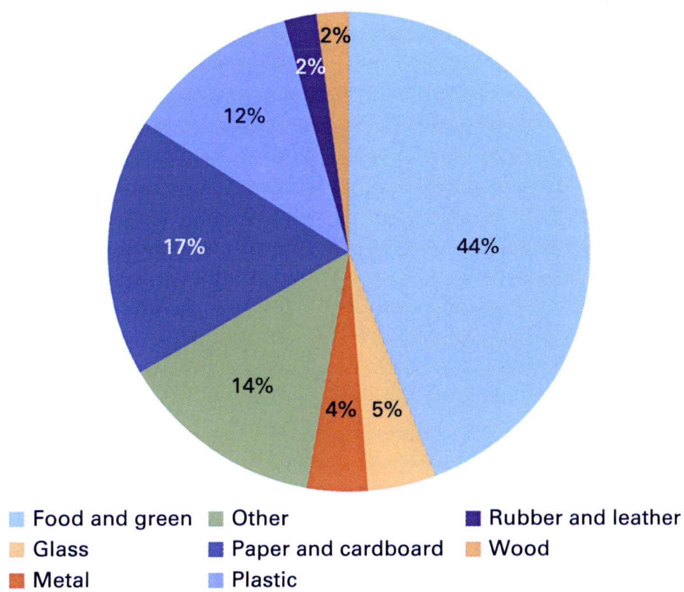

FIGURE 6.5 Global waste composition in 2016 (percent). *From Kaza et al. (2018).*

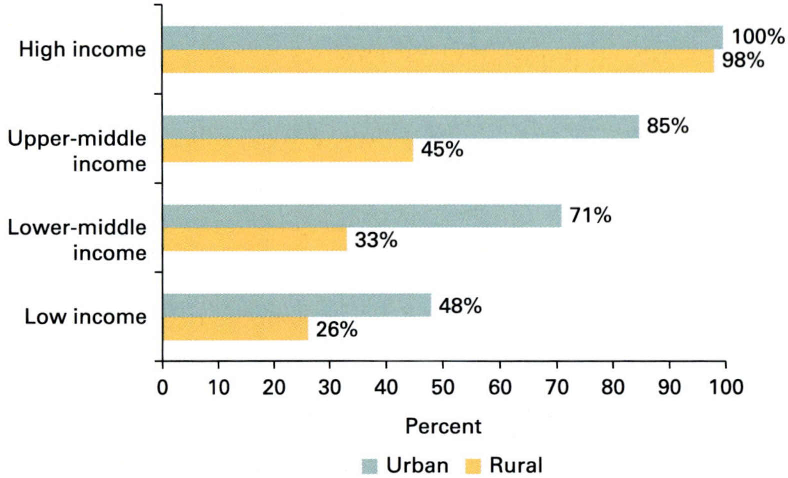

FIGURE 6.6 Urban and rural waste collection rates, by income level (percent). *From Kaza et al. (2018).*

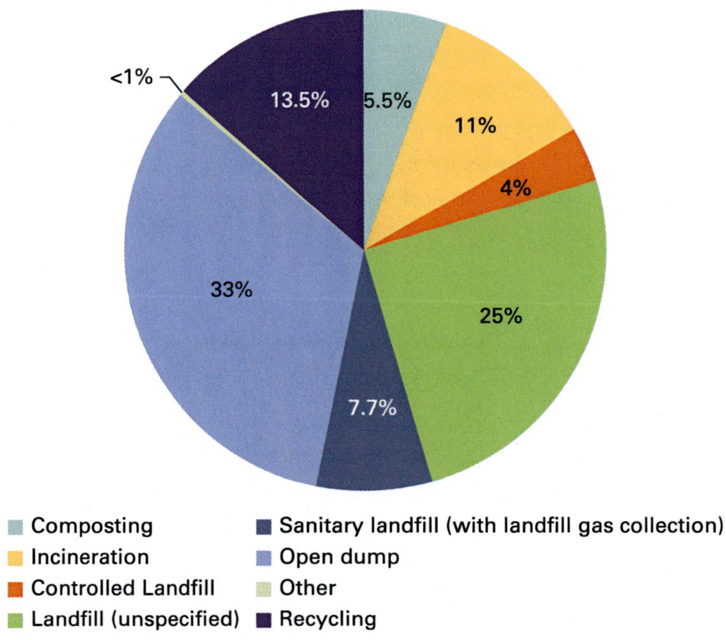

FIGURE 6.7 Global treatment and disposal of waste (percent). *From Kaza et al. (2018).*

income as the awareness is increased on the negative spillover effects on society. Construction and use of sanitary landfills is commonly the first step toward sustainable waste management. In high-income countries, incineration is also more common.

6.3.5 Administration, Operations, Financing, and Cost Recovery

In most countries, solid waste management operations are typically a local responsibility. Nearly 70% of countries have established

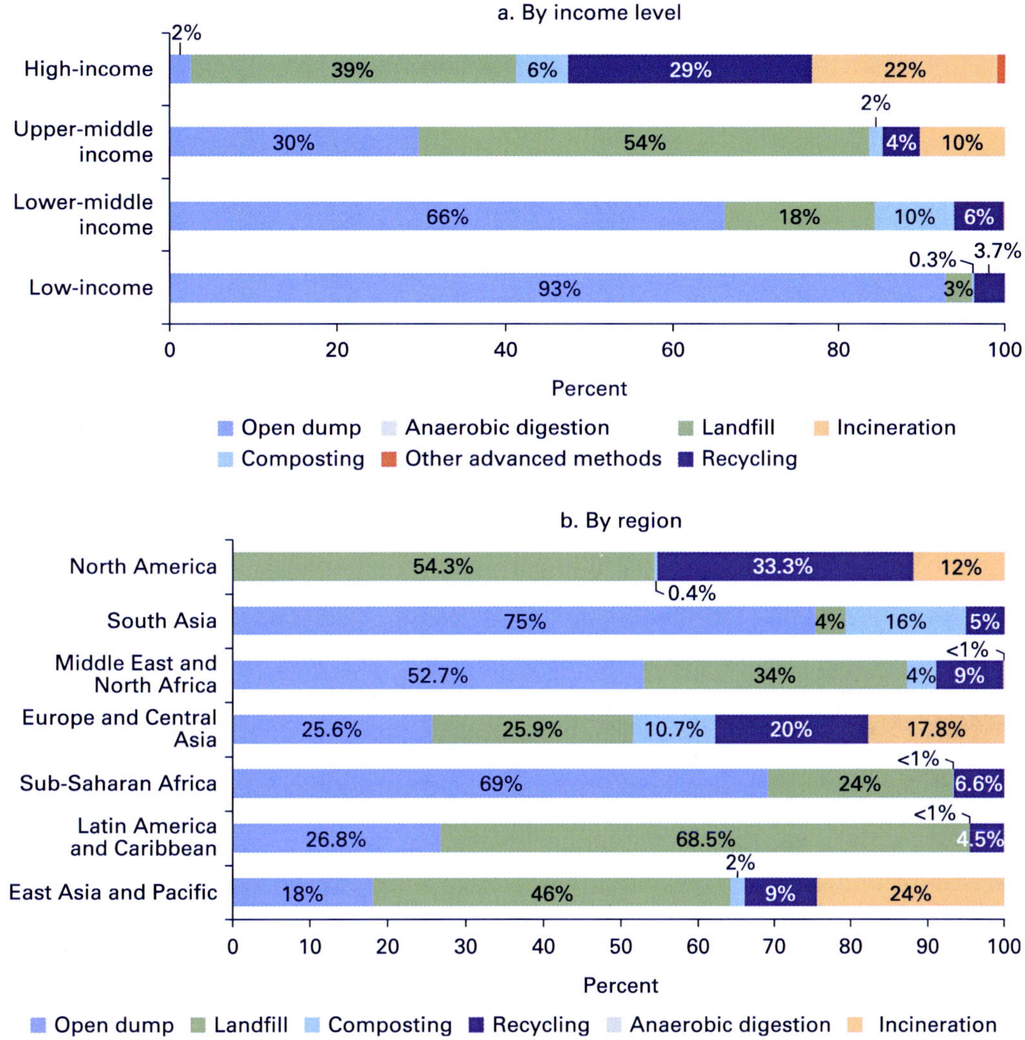

FIGURE 6.8 Disposal methods (A) by income and (B) by region. *From Kaza et al. (2018).*

institutions responsible for policy development and regulatory oversight in the waste sector (Table 6.1). About two-thirds of countries have created targeted legislation and regulations for solid waste management, though the degree of enforcement varies drastically.

The private sector is typically engaged through management or concession contracts for the collection, treatment, and disposal, and contracts generally last fewer than 10 years. Successful partnerships with the private sector for financing and operations tend to succeed only under conditions with appropriate incentive structures and enforcement mechanisms. Therefore they are not always the ideal solution.

Public reporting on waste policies and waste data is primarily limited to high-income countries and some middle-income countries.

TABLE 6.1 Existence of national waste management regulation across the world.

Income group	Total number of countries	Number of countries with defined solid waste management laws or guidelines	Number of countries without defined solid waste regulations or guidelines	Number of countries with no information	Shares of countries with defined solid waste management laws or guidelines (%)
High income	78	75	2	1	96
Upper-middle income	56	47	4	5	84
Lower-middle Income	53	47	1	5	89
Low income	30	18	1	11	60
All	217	187	8	22	86

Adapted from Kaza et al. (2018).

In high-income countries, operating costs for integrated waste management, including collection, transport, treatment, and disposal, generally exceed US$100 per tonne. Lower-income countries spend less on waste operations in absolute terms, with costs of about US$35 per tonne and sometimes higher, but these countries experience much more difficulty recovering costs. In low-income countries, nearly 20% of municipal budgets are spent on solid waste management, more than 10% in middle-income countries, and 4% in high-income countries.

The most considerable financial challenge for cities is usually covering operational expenditures for labor, fuel, and equipment servicing. Across collection and disposal operations, waste collection typically accounts for 60–70% of total costs (Table 6.2).

User fees range from an average of US$35 per year in low-income countries to US$170 per year in high-income countries, with full or nearly complete cost recovery being limited mainly to high-income countries. User fee models may be fixed or variable based on the type of user being billed. Typically, local governments cover about 50% of investment costs for waste systems, and the remainder comes mainly from national government subsidies and the private sector.

TABLE 6.2 Waste management costs by disposal type in US$/tonne.

	Low-income countries	Lower-middle-income countries	Upper-middle-income countries	High-income countries
Collection and transfer	20–50	30–75	50–100	90–200
Controlled landfill to sanitary landfill	10–20	15–40	20–65	40–100
Open dumping	2–8	3–10	–	–
Recycling	0–25	5–30	5–50	30–80
Composting	5–30	10–40	20–75	35–90

Based on Kaza et al. (2018).

6.4 COLLECTION AND RECOVERY SYSTEMS

6.4.1 Collection Systems

If waste is collected, there are three general types of collection systems:

- Mixed waste collection
- Mixed waste collection for recycling
- Segregation at source

The most straightforward collection system is to collect all waste materials in a single bin for mixed waste (Figure 6.9). Collected waste is transported directly to a facility for final disposal, such as a landfill or incineration plant. The main advantage of a mixed waste collection system is the simplicity for the users, the simplicity of logistics, and the improved health conditions compared to a situation without a collection system: waste no longer poses a risk to inhabitants. Also, waste no longer is burned or disposed of in an uncontrolled way. It is either landfilled, or used as input for a waste-to-energy installation (incineration). Energy recovery is possible, but there is no official resource recovery of wasted materials. In low-income countries, informal waste pickers will recover some valuable recyclables from the waste disposal site.

The second type of collection system is to collect all waste materials in a single bin, so that the waste can be partly recycled (Figure 6.10). After collection, waste is sorted for recycling. In a material recovery facility (MRF), recyclable materials and sometimes organics are sorted out. The sorting can be done either by hand or automatically. Low-technology MRFs can be very effective in realizing good quality output for recycling or energy production. Organic waste can produce energy in an anaerobic digester, and combustible components can be used as input for a refuse derived fuel (RDF) facility. In low-income countries, the output is sometimes also used for composting. In high-income countries, this route is usually not feasible because it does not meet quality standards for compost that can be used in agriculture. Instead, this compost can be used to cover landfills. The advantage of this type of collection and recovery system is again the simplicity for users and logistics. The main disadvantage is the low recovery rate compared to segregation at source.

Recently, some high-income countries are also increasingly using postcollection sorting plants to sort out plastics from mixed waste. The advantage is the convenience for users and easy logistics. The disadvantage is potentially lower qualities of materials due to contamination.

The third collection system is households separating their waste at home or segregation at source (Figure 6.11). In this case, citizens separate their waste into different mono streams. Examples of separately collected waste streams are organic waste, paper and cardboard, (plastic) packaging waste, glass, textiles, and e-waste. Some cities choose a system to collect a mix of these mono streams and sort their recyclables in an MRF.

The advantage of source separation systems is that they support better quality recycling, as it reduces contamination. In Europe, organic

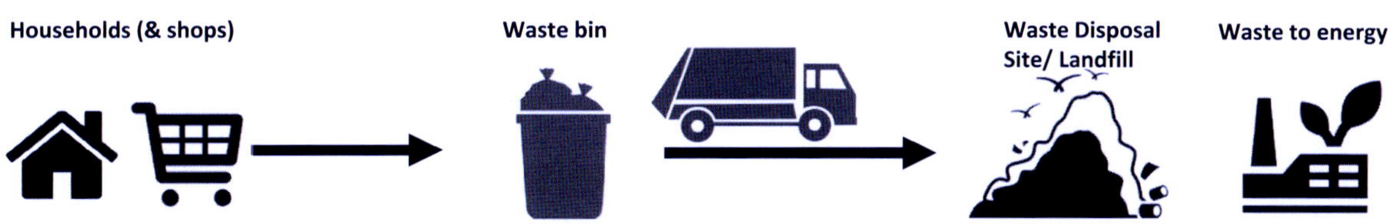

FIGURE 6.9 Mixed waste collection.

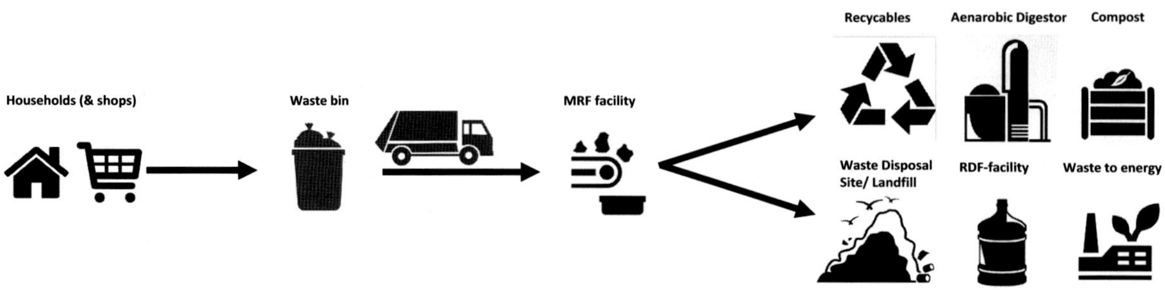

FIGURE 6.10 Mixed waste collection with sorting for recycling. *MRF*, material recycling facility; *RDF*, refuse derived fuel.

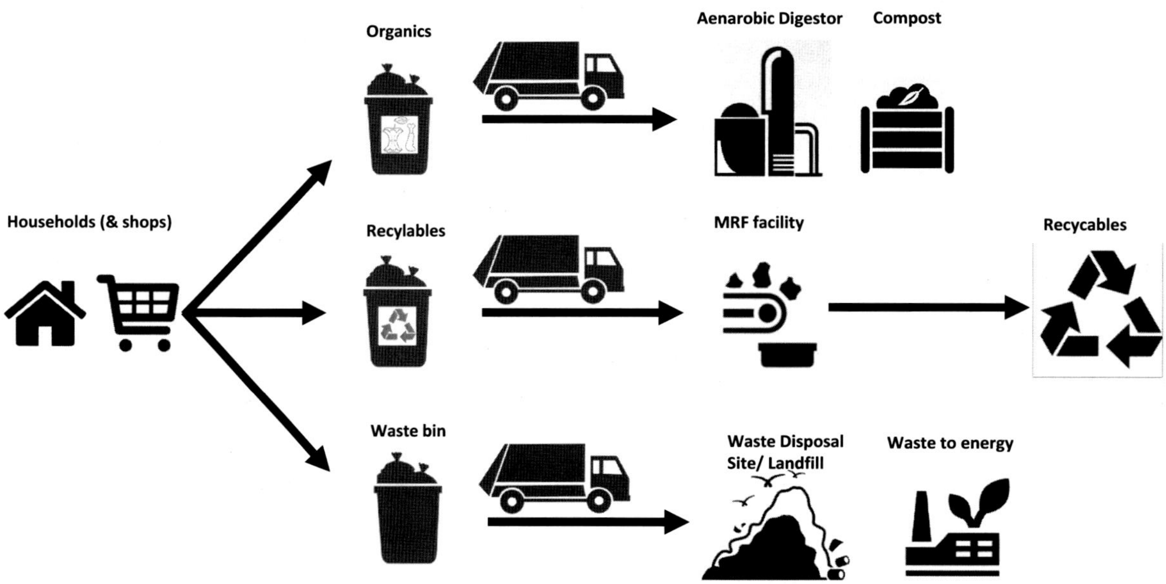

FIGURE 6.11 Segregation at source.

waste, paper and cardboard, and textiles can only go into a recycling process if they have been collected separately. Also, users become more aware of how much waste they generate, which could affect their purchases. A disadvantage is that users need more space to store their waste, which is challenging in highly dense urban areas. And if users do not correctly dispose of their waste, there is no quality improvement, but collection costs increase. It requires effort for increased participation rates, because users need to be(come) motivated to segregate at the source. National and local governments play an important role in raising awareness, education of the public, and creating the right incentives. This also includes systems that incentivize residents to increase their waste separation, like pay-as-you-throw (charging per unit) or reversed waste collection (collection of only recyclables at home).

In many countries, packaging waste is collected as a mix of dry recyclables, including plastics, metal packaging, beverage cartons, and sometimes glass. In Germany, this concept has been further developed into the so-called "Wertstofftonne" (value-bin). The collection of mixed dry recyclables is not exclusive for packaging but for all plastic and metal, sometimes combined with other items such as small electrical appliances and compact disks. A collection system for mixed dry recyclables requires a sorting facility where the mixed dry recyclables are separated into different mono-streams. Such a sorting plant consists of varying separation techniques such as sieves, wind shifters, magnetic separation, eddy-current separation, and near-infrared. Depending on the precise composition of the input material and the desired quality of the output, handpicking can be used in the last phase of the sorting process.

The performance of each system on the three aims of MSW—health, environment, and resource recovery—is shown in Table 6.3.

All three collection systems can perform well to improve health conditions. From an environmental perspective, collection for recyclables is preferred. Source segregation systems have the potential to increase resource recovery, at least for the organic streams, because there is less contamination. But these systems do need more effort on communication to realize better results. Segregation at source and collection for recycling thus performs better on the waste hierarchy (Figure 6.1).

TABLE 6.3 Performance of each collection system on the aims of MSW management.

	Health	Environment	Resource recovery
1. Mixed collection	++	+	−−
2. Mixed collection for recycling	++	++	+
3. Segregation at source	++	++	++

6.4.2 Collection Methods

In each collection system, the collection method can vary between door-to-door curbside pickup and bring systems, such as drop-off sites and recycling centers.

Curbside collection systems usually lead to a higher participation rate and better quality of recyclables compared to bring systems (Dahlén, 2009). Curbside collection systems are advantageous in areas with enough space to store waste, such as rural areas or cities' outskirts. In these cases, people can be provided with bins to store and offer separated waste streams for collection. Collection frequencies may vary depending on climate conditions and waste production. For citizens, a curbside collection system provides the highest service.

A bring system using drop-off sites is more suitable in densely populated areas with little room to store waste in or near the house. In this case, more effort is needed to offer separated waste for collection, which usually leads to a lower participation rate and lower yields than curbside collection systems.

Recycling centers are the backbone of any collection system, enabling citizens to offer separated and unseparated waste at their convenience and to offer garbage in larger quantities.

6.4.3 Treatment of MSW

Processes of MSW treatment can be classified into the categories of landfill (disposal), incineration (thermal recovery), and recycling (including composting/digestion). Bartl and Goorhuis (Unpublished date) classified the level of development of the waste management system based on the advancement of treatment methods according to the waste hierarchy (Table 6.4). Six different classes are distinguished.

Class I (i.e., landfilling >80%) is more or less the starting point of waste management. Even if sanitary landfills have replaced open dumping at this stage, many resources and energy are lost.

TABLE 6.4 Classification of advancement in waste treatment.

Class	Criteria	Remark
I	Landfill: > 80%	Large room for improvement
II	Landfill: 40–80%	Considerable advances achieved; further improvements necessary
III	Landfill: 20–40%	On average
IV	Landfill: <20% Recycling: ≤65%	Far advanced in avoiding landfills, the potential to increase recycling
V	Landfill: <20% Recycling: >65%	Current state-of-art with room for improvement
VI	Recycling: >100%	Towards theoretical desired state (100% recycling not possible)

From Bartl and Goorhuis (Unpublished date).

In class II the fraction of waste going to landfills is below 80% but still above 40%. The countries within this area have already achieved considerable advances in avoiding landfills. However, more than 40% of the waste ends up in landfills, and thus these countries have much room for improvement.

In case the rate of landfill is further decreased and lies within the range of 20% and 40%, class III is reached. Class IV is designated for countries that are more advanced in waste management and already have a landfill rate below 20%. However, it seems necessary to define an additional region, class V. On the one hand, it is characterized by a very low rate of landfill (<5%; i.e., near to "zero"). On the other hand, incineration has to be avoided where possible (<40%).

Theoretically, there is also a class VI, which represents optimal waste management where all waste is recycled. However, this situation is not achievable in practice since every recycling operation requires energy and input streams and results in output streams (waste). Every recycling cycle means an inevitable decrease in quality (i.e., downcycling) and waste generation (Bartl, 2014). The degree of landfill and incineration is very low. However, a recycling rate of 100% is, therefore, a theoretical idea.

Figure 6.12a shows the current state for nine selected European countries and the EU average. The advancement in waste treatment varies in Europe between Class II and Class V. Countries in Western and Northern Europe tend to have relatively low landfill rates and high recycling rates. Figure 6.12b sketches the significant improvements of European waste management from the past (1990) to the present (until 2020). In 1990 European waste management depended heavily on (74%) landfill. It had just crossed the border between class I and class II. Over the subsequent 15 years the landfill rate was drastically reduced, thus moving from class II into class III by about 2007. However, the speed of improvement dramatically decreased in the next years and the progress since 2015, when the recycling rate was 47%, it has only marginally improved (49% in 2020). It is obvious that Europe will not reach area IV within a midterm range and there is no indication that Europe will ever reach area V.

6.5 FUTURE DEVELOPMENTS

Some of the most important global trends and developments that impact the collection of municipal waste are discussed in this section.

6.5.1 Societal Development and Pressure

When looking forward, global waste is expected to grow to 3.40 billion tonnes by 2050, more than double the population growth over the same period. Overall, there is a positive correlation between waste generation and income level. Daily per capita waste generation in high-income countries is projected to increase by 19% by 2050, compared to low- and middle-income countries, where it is expected to

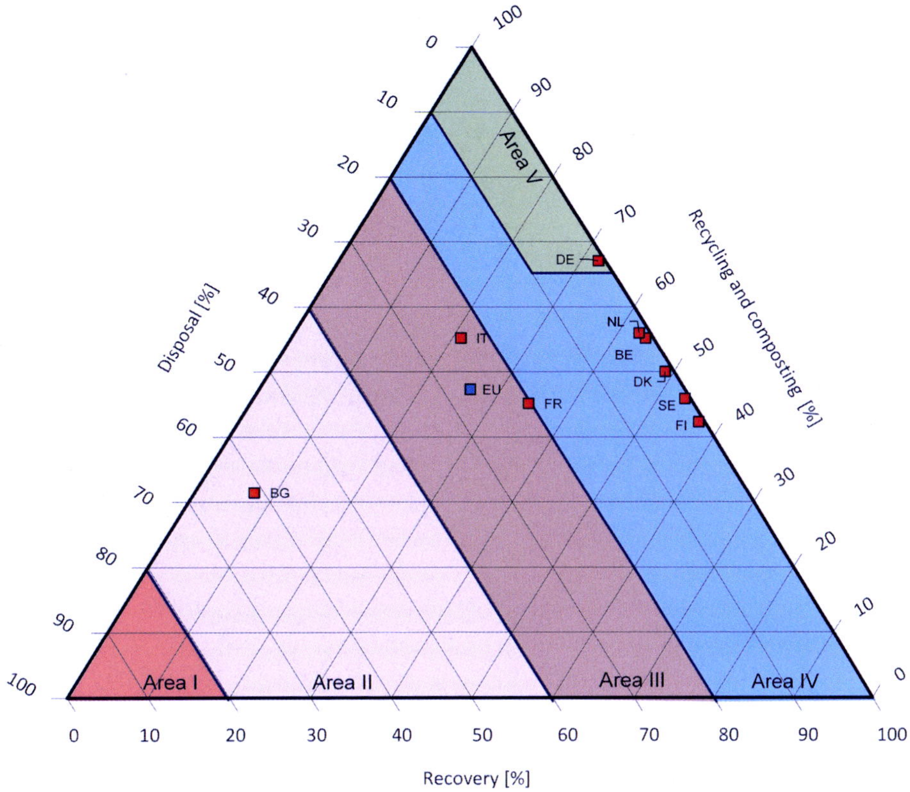

FIGURE 6.12A Proportions of recycling, incineration, and landfilling in European countries (circles) in 2018. *Figure produced by Andreas Bartl based upon Eurostat (2022).*

increase by approximately 40% or more (Kaza et al., 2018). This will increase the need for good collection and proper treatment facilities. And this is true specifically for uncontrolled waste disposal sites and hazardous waste streams like e-waste or hospital waste, which already cause direct deaths and injuries.

As waste becomes more apparent and public awareness increases, the societal pressure to do something about our current waste problems is rising. NGOs are calling for action on issues like plastics in the ocean, litter, and proper waste management. Governments are increasingly focusing on climate change and other environmental problems. Governments pose bans on single-use plastics, and Extended Producer Responsibility schemes are extended into other product categories and new countries. Deposit schemes will become more and more the standard.

Producers are increasingly pledging for circular value chains as part of their increased corporate social responsibility (CSR). And as consumers become more aware of greenwashing, they need the other value chain stakeholders. Therefore, global and regional platforms to collaborate on a circular economy, like the New Plastic Initiative (https://ellenmacarthurfoundation.org/topics/plastics/overview) or CEFLEX (www.ceflex.eu), are getting tailwinds.

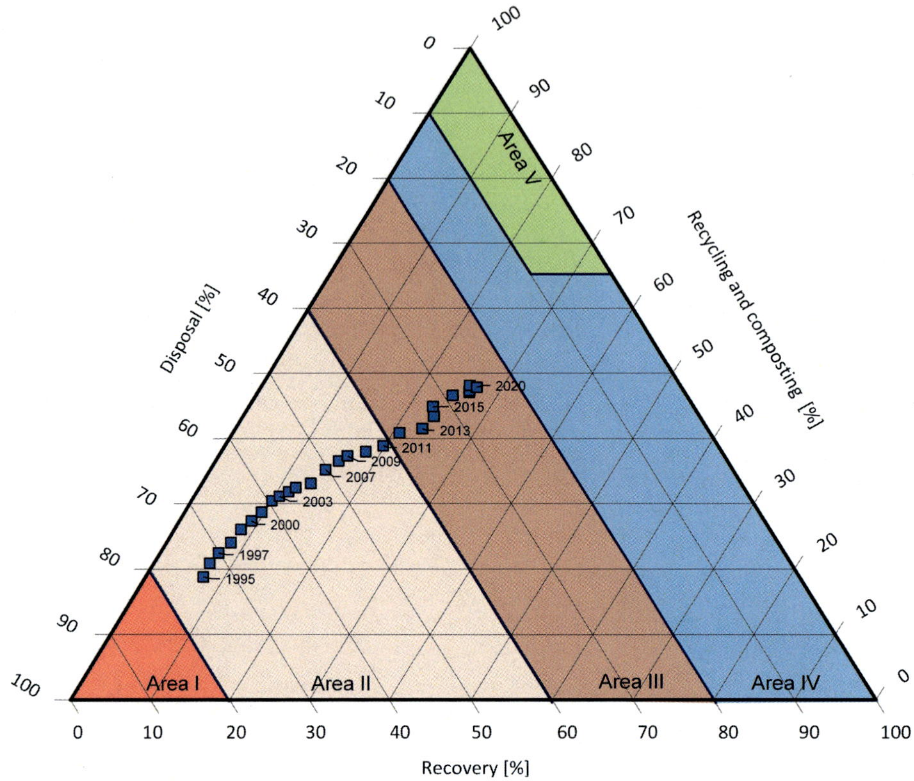

FIGURE 6.12B Waste management paths in the EU-28 (EU-27 up to 2007) from 1990 to 2020. *Figure produced by Andreas Bartl based upon Eurostat (2022).*

6.5.2 Sustainable Impact

The necessity to develop into a circular economy will lead to stronger product regulation, a growing demand for high-quality recyclables, and new business models. In Europe the traditional waste legislation seems to develop into more holistic product policies, which take the life cycle of products into account. This type of legislation may include requirements for the expected liftetime of products, the amount of recyclate in new products, the reusability and repairability of products, as well as requirements for the separate collection and recycling.

The value of organic waste becomes more and more apparent: It reduces methane gas emissions, improves degraded soils, and can be an essential source of clean energy. In low- and middle-income countries, waste-to-energy is increasingly considered a viable alternative for their current waste disposal sites.

There is an increased need for reliable and smart data. Governments need to show their performance based on the Sustainability Goals, and companies are interested in real-time environmental impact data. For example, the reduction of CO_2 emissions in waste management contributes to national targets.

6.5.3 Value Chain Integration

Companies like IKEA and Lidl are vertically integrating to ensure supply of recyclates. Reusable products like Loop (https://loopstore.com/brand/loop-update) have become more and more accepted. On the other hand, companies have become increasingly aware that low-quality recyclables are not valuable. We expect more countries to take a route like China, banning the import of (low-grade) recyclables. Recycling will need good sorting to scale up production of high-quality output.

The role of technology is and will be increasing. The sector will need technologies like connected devices, integrated hardware systems, cleaner vehicles, digitalization, robotization, artificial intelligence, blockchain, and watermarking to improve the sector's effectiveness, moving towards a circular economy at reasonable costs. The trend can be both productive and nonproductive: digitalization, for example, reduces the need for paper. However, it increases the demand for energy and cloud service providing infrastructure. An example of how Industry 4.0 can increase recycling rates is sorting of waste streams with new sorting techniques like watermarking (the holy grail) or artificial intelligence. These techniques can potentially lead toward sorting more material into higher quality fractions.

Waste companies have become increasingly professional in delivering environmental services instead of being logistics providers. As procurement shifts towards performance-based contracts, sustainable impact indicators are becoming more critical to winning contracts. As an example, emission-free trucks are increasingly becoming a condition for winning a contract. Also, more products and areas will use closed-loop systems such as deposit systems for drinking bottles. If designed correctly, deposit systems can achieve over 90% return rates, showing that these types of systems could pave the way to achieve ambitious targets.

6.6 CONCLUSION AND OUTLOOK

Because everybody consumes products, waste decision making is part of our daily experience. Everybody understands, consciously or unconsciously, waste management is a service to improve the quality of our lives. Everybody wants to get rid of their waste, but nobody likes to dump it in their backyard.

However, our failure can be seen all around us: in the oceans, water, air, and even our bodies. Three-fourths of our waste is dumped, and globally one-third of our waste is not collected at all. And increased population and income will increase our consumption and thus our waste. If we don't address the problem, the impact of waste on our planetary boundaries will move beyond those. The first and most crucial step moving towards this system is actually to collect waste. All countries should realize 100% collection to prevent litter in our environment via air pollution (after burning) or waste entering our environment. Proper collection is needed for good sorting, recycling, and end-of-life, leading to an effective value chain.

Technically, there are no significant barriers to a good collection. However, as James Carville said (in 1992): "it's the economy, stupid": As it is a public service, costs are involved. "(..) The whole issue is financial. This is the clearest example of 'who benefits and who pays?' that I know of" (Newman, 2020, p. 151). Or, in other words, we need good institutions for waste management to prevent a global tragedy of the commons by wasting our health, environment, climate, and materials. It doesn't make sense to see waste as a source of profit. Although that is true for specific value chain stakeholders, it is a service to keep our globe livable.

References

Bartl, A., 2014. Moving from recycling to waste prevention: a review of barriers and enablers. Waste Manage. Res. 32 (9). https://doi.org/10.1177/0734242X14541986.

Dahlén, L.E., 2009. Evaluation of recycling programmes in household waste collection systems. Waste Manage. Res. 28 (7), 577–586. https://journals.sagepub.com/doi/10.1177/0734242X09341193.

Espinoza, P.T., et al., 2011. Regional Evaluation on Urban Solid Waste Management in Latin America and the Caribbean. Pan American Health Organization.

European Union, 1999. Council Directive 1999/31/EC on the Landfill of Waste.

European Union, 2018. Directive 2008/98/EC of the European Parliament and of the Council of 19 November 2008 on Waste and Repealing Certain Directives.

Eurostat, 2013. Eurostat. Retrieved June 23, 2013, from Glossary: Municipal Waste: http://epp.eurostat.ec.europa.eu/statistics_explained/index.php/Glossary:Municipal_waste.

Eurostat, 2022. Municipal Waste by Waste Management Operations. Online available from: https://appsso.eurostat.ec.europa.eu/nui/show.do?dataset=env_wasmun&lang=en.

FAO, 2013. Food Wastage Footprint & Climate Change. Food and Agriculture Organisation of the United Nations. http://www.fao.org/3/bb144e/bb144e.pdf.

Goorhuis, M., Bartl, A., 2011. ISWA key issue paper on waste prevention, waste minimization and resource management. In: International Solid Waste Association (ISWA) Working Group Recycling and Waste Minimization, Vienna, Austria. Online available from: https://ec.europa.eu/environment/resource_efficiency/pdf/ISWA%20International%20Solid%20Waste%20Association.pdf.

Kaza, S., Yao, L., Bhada-Tata, P., Van Woerden, F., 2018. What a Waste 2.0: A Global Snapshot of Solid Waste Management to 2050. Urban Development Series, World Bank, Washington, DC, https://doi.org/10.1596/978-1-4648-1329-0. Online available from: https://openknowledge.worldbank.org/bitstream/handle/10986/30317/9781464813290.pdf and https://datatopics.worldbank.org/what-a-waste/trends_in_solid_waste_management.html.

Newman, D., 2020. Everything Is Connected: Understanding a Complicated World. Waterside Productions, pp. 47–151. https://www.amazon.nl/Everything-Connected-Understanding-Complicated-English-ebook/dp/B08GGCSHHL.

OECD, 2011. OECD Factbook 2011–2012: Economic, Environmental and Social Statistics. OECD Publishing.

Persson, L., Carney Almroth, B.M., Collins, C.D., Cornell, S., de Wit, C.A., Diamond, M.L., Fantke, P., Hassellöv, M., MacLeod, M., Ryberg, M.W., Jørgensen, P.S., Villarrubia-Gómez, P., Wang, Z., Hauschild, M.Z., 2022. Environ. Sci. Technol. https://doi.org/10.1021/acs.est.1c04158.

United Nations Environment Programme and Climate and Clean Air Coalition, 2021. Global Methane Assessment: Benefits and Costs of Mitigating Methane Emissions. United Nations Environment Programme, Nairobi. ISBN: 978-92-807-3854-4. https://www.unep.org/resources/report/global-methane-assessment-benefits-and-costs-mitigating-methane-emissions.

United States Code, 2006. United States Code. Supplement 4, Title 42 – The public Health and Welfare. United States of America.

UNSD, 2021. United Nations Statistics Division: "Shrinking our material footprint is a global imperative" from report on SDG 12. https://unstats.un.org/sdgs/report/2019/goal-12/.

Waste Management, 2011. (Chen et al., 2003; Pruss et al., 1999). Sanitary Landfill. 31(4). https://www.sciencedirect.com/topics/earth-and-planetary-sciences/sanitary-landfill.

Wilson, D.C., Velis, C.A., Rodic, L., 2013. Integrated sustainable waste management in developing countries. Proc. Inst. Civil Eng. Waste Resour. Manage. 166 (2), 52–68. ISSN 1747-6526. https://doi.org/10.1680/warm.12.00005.

CHAPTER

7

The path to inclusive recycling: Developing countries and the informal sector

Sonia Valdivia and Mathias Schluep

World Resources Forum Association, St. Gallen, Switzerland

7.1 INTRODUCTION

Urban waste management services organized by local governments are typically based on integrated and articulated collection systems and infrastructure that are either private or municipal. The infrastructure consists of recycling technologies for recovering valuable products, components and materials recovery, composting, incineration (burning), and landfilling (Chapter 6—Developments in Collection of Municipal Waste). In developing countries, due to the lack of capacities, resources, and enabling conditions for legislation enforcement towards proper waste management, such systems are mostly executed by a large urban informal workforce. These are usually referred to as the "informal sector." This informal workforce makes a living by collecting, sorting, recycling, and selling valuable products, components, and materials recovered from waste (ILO, 2013). Estimates indicate that the proportion of materials recycled or recovered from waste (recycling rate) by the informal sector in selected cities may vary greatly: from 30% in India up to 50% in Pakistan and the Philippines (Figure 7.1). General national statistics consider as average the following waste groups: paper and cardboard, plastics, glass, metals (ferrous and nonferrous), textiles, among others. According to Allen (2019) informal recyclers collect 90% of what is recycled in Brazil and 80–90% of post-consumer packaging and paper recovered in South Africa. The high recycling rates by the informal sector reveal their relevance in waste management in developing countries and their contribution to the circular economy.

According to Elgin et al. (2021), globally the informal economy accounted for 32–33% of GDP and 31% of employment over the period 1990–2018. They infer that a higher level of economic development, as measured by per capita income, is associated with lower informality. Thus informality tends to be considerably more pervasive in the emerging and developing economies than in advanced economies:

https://doi.org/10.1016/B978-0-323-85514-3.00006-3

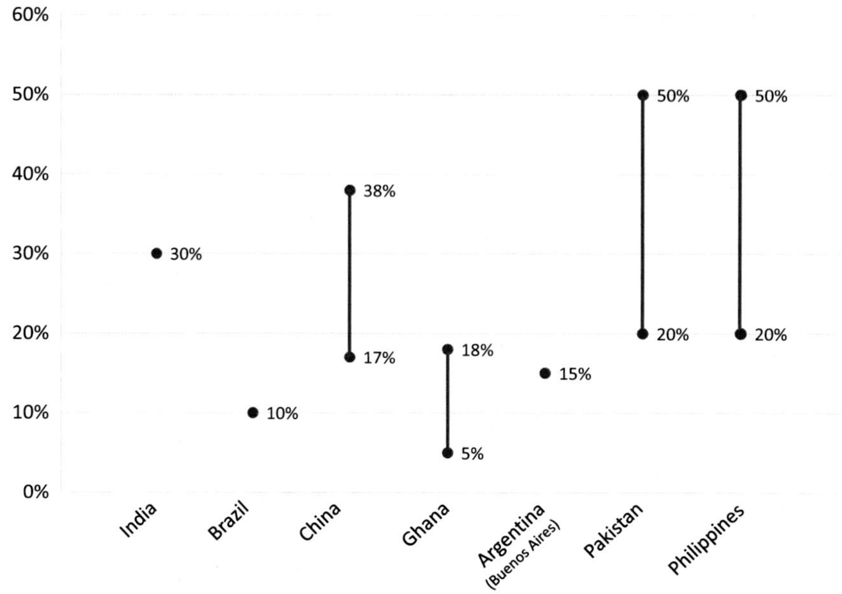

FIGURE 7.1 Recycling rates by the informal sector in different countries. *Data from Di Maria et al. (2019), Luz (2021), Linzner and Salhofer (2014), Oduro-Appiah et al. (2019), Van Eijk et al. (2021), Allen (2019).*

- in the emerging and developing economies it accounts for 36–37% of GDP and 39% of employment,
- in advanced economies, it accounts for about 19% of GDP and 16% of employment, on average.

On one hand, the informal sector contributes to poverty reduction in developing countries by providing employment, livelihoods, and income for millions of workers and business owners, including women. ILO's estimations indicate that between 0.3% and 1% of the population in developing countries survives by recovering materials from waste (ILO, 2013).

On the other hand, the poor working conditions created within the informal economy are questioned. There is ample evidence of difficult working conditions, child labor, lack of legal protection, numerous legal, health, and physical risks, often low social standing, and low and fluctuating incomes. Women tend to be more severely impacted among the informal workers. Furthermore, the identified worst practices applied in informal waste management, such as open burning, chemical leaching, and uncontrolled depollution of wastes, among others, cause severe damage to the environment. At a societal level, tax evasion, illegal trade, and corruption associated with waste trading affect the stable and healthy growth of the local economies (Karcher et al., 2018).

Formalizing the informal sector and introducing more sustainable practices are pivotal to:

1. Reach the objectives of the Agenda 2030 (UN, 2015) and its Sustainable Development Goals (SDGs), notably SDG 1—No Poverty, SDG 5—Gender Equality, SDG 8—Decent Work and Economic Growth, SDG 10—Reduced Inequalities, and SDG 16—Peace, Justice, and Strong Institutions.
2. Transit towards a more circular economy by increasing the availability of materials, especially those with growing scarcity such as critical raw materials.

However, in some developing countries formalization processes may represent excessive costs, especially for the lowest-paid workers with poor benefits such as low quality of public health-care programs and pension schemes. According to The Economist (2016), in 5 of 18 Latin American countries covered in a research study, the share of taxes that the poorest tenth would have to pay if becoming formal workers would be higher than their incomes (Figure 7.2). This includes a country like Mexico. In Colombia the tax share would amount to about 90% of the earnings of the poorest tenth. This explains why workers engaged in subsistence activities prefer—in some countries more than in others—to remain in the shadow rather than to become formalized.

This chapter discusses the "informal sector" definition and interdependencies with the formal sector. It further aims to impart an understanding of how informal waste management is functioning, and what the most severe environmental and socioeconomic effects are. These aspects are illustrated based on the case of waste tire management, with focus on worst practices in the Accra, Tema, and Koforidua regions in Ghana. Furthermore, in this chapter business model, standardization, and policy-related approaches on how to link informal and formal waste management activities are pointed out.

7.2 DEFINITION AND LINKS WITH THE FORMAL SECTOR

Despite having been the subject of political and scientific discussions for decades, no international consensus has been reached on the definition of the informal sector. Reasons lay in differences of the nature of the informal sector in a specific geographical setup, depending on the legal and economic framework, cultural background, and other socioeconomic factors.

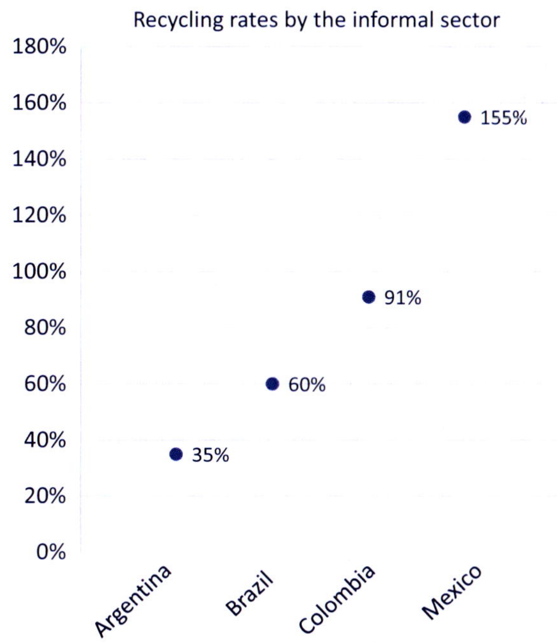

FIGURE 7.2 Taxes owed by the tenth poorest population. *Data from The Economist (2016).*

It is worth noting that in the literature very often terms such as "informal economy," "informal activities," "shadow economy," and "informal workforce" are used interchangeably when referring to the "informal sector."

Some convergence exists around the ILO (2013) definition of the informal sector, which covers any economic activity or source of income that is not subject to government regulation, taxation, or observation. This includes all economically active persons who do not receive social protection through their work. These workers are excluded from safety-net programs such as pension plans, unemployment insurance, and some public health services. This means that the informal sector covers a wide range of activities that are not included in national accounts, such as "subsistence activities," e.g., in farming and waste management, small-scale enterprises, and illegal activity.

Depending on the local regulations, individuals or groups of them engaged in "subsistence activities" may fall under a properly regulated activity. This is the case of e-waste collectors in Ghana, which according to the Ghana hazardous and electronic waste control and management act 2016 (ACT 917) (Ghana EPA, 2016) are considered as Tier 1 workers. Hence, they are allowed to collect under given conditions (determined routes, with uniform and protection) and sell their items to formal collection sites or recyclers. Similarly, in Peru, thanks to the Ley del Reciclador N.° 29419 (MINAM, 2010), after their registration at corresponding municipalities and their training and vaccination, informal collectors are allowed to collect wastes separately. In most cases, however, "subsistence activities" are not well defined in local regulations. They are systematically categorized as informal activities and are not tolerated in formal economy transactions, even if registered, e.g., in local government offices. In countries lacking proper guidance on whether and how to engage with "subsistence activities," two groups of individuals coexist. Both are operating in the same way: while one group is considered to be informal sector workers with no permission to trade formally, the other workers group is accepted as formal suppliers of collected waste.

There is a need for enhanced consensus on what we understand under "subsistence activities" for easing their inclusion in local waste management systems. An ISO definition explains that "subsistence activities (SA)" cover workers operating individually, in families or groups (even constituted as associations), that earn the minimum to subsist and below the minimum tax threshold required per national laws and regulations to pay taxes (ISO IWA 19, 2017).

It also has been acknowledged that there are many interdependencies between the informal and the formal sector (ILO, 2018; Dasgupta, 2016). Connections between the informal and formal sectors occur through individual transactions, subsector networks of commercial relations, or subcontracted relationships. The terms and conditions for trading are typically set by the formal sector. The individual transactions cover the trade of goods, materials, tools and equipment, and acquisition of skills and know-how.

In the waste management field, recoverable secondary materials traded by the informal sector represent an important input to the formal facilities. This is the case in steel and aluminum production, which contain more than 50% scrap, which in developing countries is notably supplied by the informal workforce.

7.3 INFORMAL WASTE TIRE RECYCLING: CHALLENGES AND OPPORTUNITIES

According to the World Resources Forum Association (WRFA) (2018), each year some 1.5 billion used car and truck tires are discarded globally. The current state of the art and challenges in developed countries are discussed in

Chapter 32—Black Rubber Products and Chapter 12—Vehicles (ELV). In developing countries with no extended producer responsibility programs, weak enforcement of policies, and the lack of authorized facilities, a huge portion of waste tires is absorbed by the informal sector, which applies worst practices (such as open burning) or dumps these wastes in open yards, into the rivers or seas, or next to the roadside.

The motivation for unsavory and unsafe open burning practices is the surprising high amount of metal in tires: in particular, steel that is twisted or braided into strong cables and embedded into tires to give them strength (Chapter 32—Black Rubber Products).

Workers subsisting from rudimentary forms of recycling in an environment largely devoid of any safe or sound economic opportunities, burn waste tires typically at low, uncontrolled temperatures for recovering the steel wire, which is then sold as steel scrap. This burning creates harmful (toxic) emissions, mostly evident as black, thick, acrid smoke clouds that threaten both human health and the environment. Moreover, impact on climate change is also a key issue resulting from this practice: to retrieve around 1.36 kg scrap steel per tire, 23.2 kg CO_2 emissions are generated, equivalent to driving a typical passenger diesel car for about 76 km (WRFA, 2018).

Legislation typifies open burning of any kind of waste as an illegal activity. So is the case of open burning of waste tires in Ghana where in hotspot areas, such as in Accra, Koforidua, and Tema, these are burnt by informal workers, typically young men, for recovering metals such as steel (Arthur et al., 2020). Some informal workers are aware of the negative implications of such worst practices. Nevertheless they continue, motivated by the attractive scrap iron/steel prices in the market and the lack of alternatives.

According to Arthur et al. (2020), with a scrap steel price of around 0.15 €/kg, about 0.20 €/tire can be earned by selling the steel retrieved from the burning of one passenger car tire. In contrast, 0.25 €/tire can be obtained if the steel tread obtained from a tire is obtained through controlled practices such as shredding (and not through worst ones like open burning) and sold at commercial prices in the formal market. This confirms that open "*...burners earn less than collectors of waste tyres who supply recycling and even upcycling facilities... Nevertheless, it is striking that processes with so massive pollution effects are conducted for only very little economic return.*"

In addition to open burning, other worst practices conducted in some informal activities include, for example:

- the uncontrolled manual reconditioning of lead-acid batteries releasing toxic substances (ILA, 2021),
- the uncontrolled manual dismantling of refrigerators releasing ozone depleting gases (WRFA, 2018), and
- the chemical leaching of printed circuit boards to facilitate the recovery of precious metals contained in them (Karcher et al. 2018).

7.4 APPROACHES TOWARDS INCLUSIVE RECYCLING

Strengthened and creative partnerships with the private sector prove to be successful measures for the inclusion of informal sector workers in value chains and increased acceptance of their recovered materials or components. This is illustrated with successful approaches from South Africa and India, highlighting the key enabling factors.

7.4.1 Cooperation With South African Informal E-Waste Sector Workers

Founded in 1992, DESCO is the largest South African e-waste recycler and pioneer in incorporating informal workers as subcontractors into its business model and as part of its Corporate Social Responsibility Policy. Potential subcontractors are identified among (a) former employees running their own enterprises or willing to start one, or (b) suppliers from whom DESCO has been procuring waste fractions for 1 year or longer (Ottiger and Schluep, 2020).

Support with the formalization process. Potential subcontractors that are not yet formalized are supported by DESCO during the formalization process with:

- on-site training, e.g., on health and safety standards,
- administrative assistance, e.g., for registering their companies as formalized enterprises and making them compliant with corporation and tax laws,
- the provision of free personal protective equipment (PPE), and the use of dismantling tools at DESCO's premises.

Cooperation with suppliers. Once potential partners become formalized subcontractors, DESCO provides start-up support in the form of preprocessed stock and starting capital. DESCO provides free of charge use of vehicles for transporting heavy WEEE.

Business model. The subcontractors have two income sources:

1. Subcontractors buy e-waste products, components, or materials from DESCO to recover valuable components or materials after their dismantling off-site. Recovered items are then sold back to DESCO or to any other potential clients such as traders or other recyclers. This applies to all WEEE, components, and materials streams except for printed circuit boards, which must be sold back to DESCO.
2. Subcontractors are hired to support the dismantling and processing of e-waste products, components, and materials. This work can be done on-site or at the premises of other clients.

As of 2020, 10 subcontractors work with DESCO, five of which remain at the premises and five operate off-premises. It is worth noting that founded companies usually have lower operating costs than DESCO has. These companies can therefore afford collecting e-waste in a larger regional radius, which ultimately also benefits DESCO.

Benefits. This approach not only benefits DESCO with securing a technical labor force for collection and manual dismantling, and sufficient e-waste volumes for its core operations, but also benefits subcontractors by strengthening their capacities, providing access to formal jobs, PPE, tools, and infrastructure, thus facilitating their inclusion in formal value chains.

7.4.2 Formal Coworking Space in India for Informal Workers

E[co]work (2022) (www.ecowork.international) is a start-up business founded in 2022 that aims to empower informal microentrepreneurs in the Indian waste sector by providing solutions for safe and inclusive e-waste management. This includes coworking space with efficient machinery and tools, storage possibilities, training, and PPE.

The coworking space at E[co]work is located in the northeast part of New Delhi, in an area of more than 1000 m^2.

Business model. The operational and administrative costs of the coworking space are expected to be fully covered with the fees paid by the workers for using the working areas and tools. The operations are environmentally sound and in line with the Indian legal requirements concerning health and safety risks. Additional paid services are offered by E[co]work to the broader public such as workshops and events organization and a marketplace, e.g., through an online app.

Incomes for the microentrepreneurs result from the selling of preprocessed products, components, and materials to any potential recycler or trader interested in them. The ownership of generated valuable items remains with the workers.

Benefits. Informal microentrepreneurs using the coworking space benefit from a stable and legal working environment with no risk of forced closure. They also benefit from improved and flexible working conditions, which can be based on short-term agreements. Informal microentrepreneurs also get support in their transition to formal entrepreneurship.

7.5 POLICIES AND STANDARDIZATION DEVELOPMENTS FOR INCLUSIVE RECYCLING

Enabling conditions for inclusive recycling include national regulations and recognized standards initiatives. Policies, regulation, and procedures provide organizations with a roadmap while ensuring compliance with specific requirements, giving guidance for decision-making, and streamlining internal processes. Noteworthy policies and standardization developments for inclusive recycling are described here.

7.5.1 ISO Standards

Due to the robust internationally multistakeholder-based consensus processes followed for their elaboration, ISO standards provide a solid base on which to create public policy that helps further SDG goals such as human rights, water and energy efficiency, public health, and more. There can be no sustainable and inclusive recycling if the activities of informal workers are not properly recognized and managed, and if the negative impacts associated with this work, e.g., through worst practices, are not adequately addressed. Two ISO standards address informal sector aspects and their inclusion in recycling systems:

- ISO IWA 19 (2017) standard on Guidance Principles for the Sustainable Management of Secondary Metals, and
- ISO/CD 59014 (2023) standard on Environmental Management and Circular Economy—Sustainability and Traceability of Secondary Materials—Principles and Requirements.

The latter is still under development.

7.5.1.1 ISO IWA 19:2017

The ISO IWA 19 (2017) standard on Guidance Principles for the Sustainable Management of Secondary Metals provides a set of sustainability requirements, traceability mechanisms, and a roadmap for economic operators, including workers engaged in subsistence activities, to reach conformity with the requirements articulated along five principles, with objectives (Table 7.1):

Principle 1 aims to improve the working conditions of workers involved in the collection and recycling processes.

Principle 2 aims to ensure that collection and recycling processes uphold local communities' rights and help to build social inclusion and resilience.

Principle 3 aims to minimize and mitigate the impact of collection and recycling processes on the environment and natural resources.

Principle 4 aims to improve the recovery of secondary metals by recommending informal workers and all economic operators to move away from worst practices and implement technologies that increase the amount and quality of secondary metals recovered from processes along the value chain.

Principle 5 is an overarching principle that recommends developing and implementing a sustainable management plan, appropriate to the size and nature of operations, following the results of the baseline environmental and socioeconomic conditions.

TABLE 7.1 Principles and objectives for sustainable and inclusive recycling of metals.

PRINCIPLE 1—Enabling safe, healthy and equitable working conditions

- **Objective 1.1**—Enable safe and healthy workplaces.
- **Objective 1.2**—Establish working terms and conditions that are decent and equitable.
- **Objective 1.3**—Eliminate child labor, forced labor, harassment, and all forms of discrimination.
- **Objective 1.4**—Ensure freedom of association and the right to collective bargaining.
- **Objective 1.5**—Provide clear channels for communication, transparency, and dialogue with workers.

PRINCIPLE 2—Building local community relations and resilience

- **Objective 2.1**—Respect and foster local communities' rights.
- **Objective 2.2**—Enable the social inclusion of workers in the community.
- **Objective 2.3**—Establish clear channels for communication, transparency, and dialogue with local communities and affected stakeholders.

PRINCIPLE 3—Conserving and protecting the environment and natural resources

- **Objective 3.1**—Conserve and protect water, air, and soil resources.
- **Objective 3.2**—Restore severely damaged areas from metals recovery operations.
- **Objective 3.3**—Conserve and protect biodiversity, ecosystems, and ecosystem services.

PRINCIPLE 4—Improving recovery of secondary metals

- **Objective 4.1**—Develop, implement, and promote technologies and strategies to increase secondary metals recovery related to quantity and quality.

PRINCIPLE 5—Implementing a sustainable management approach

- **Objective 5.1**—Document and evaluate the existing baseline conditions of secondary metals operations in the areas addressed by the Principles and Objectives.
- **Objective 5.2**—Mitigate negative impacts and strengthen positive impacts of secondary metal operations through the development, implementation, and continuous improvement of a Management Plan.
- **Objective 5.3**—Strengthen the organizational capacity of Economic Operators involved in secondary metals operations.
- **Objective 5.4**—Ensure compliance with local and/or national laws and regulations.
- **Objective 5.5**—Eliminate bribery, money laundering, and corruption.

From Valdivia et al. (2016).

This standard is also useful for due diligence purposes by any economic operator interested in evaluating the compliance of suppliers with the sustainability requirements. For this purpose, the due diligence is conducted through second-party or third-party audits. The ultimate goal of the ISO IWA 19 (2017) standard is the formalization of the activities of informal workers engaged in the waste management value chains.

Due to limited resources and lack of capacities in subsistence activities, a gradual implementation taking up to about 4–5 years is a reasonable approach, which was a key recommendation from an ISO IWA 19 stakeholder consultation. Specific measures that would be taken by the informal workforce during this period (Table 7.2) range from identification of worst practices in place, through their transition until their formalization and compliance with local regulations. It is worth noting that the suggested time frame for the examples activities in Table 7.2 is flexible and may need to be adjusted according to the context.

TABLE 7.2 Example of suggested time frame for subsistence activities in their journey for compliance with the ISO IWA 19 (2017) requirements until their formalization.

Time frame	Summary of steps
Short-term	Prioritize training on collection of waste and physical segregation while avoiding worst practices and health and safety in consultation with affected workers Identify the formalization requirements for becoming an association Focus on improving working conditions to reduce negative impacts on health **Target by the end of year 1**: Partially comply (~10%)* with specific requirements. Activities moved away from worst practices
Mid-term	Continue improving working conditions to reduce impacts on health Start with the formalization process Implementation of a self-assessment for timely identification of potential severe impacts **Target by the end of year 2**: Partially comply (~40%)* with specific requirements. Formalization process started
Long-term**	Become a legal entity (an association) Implement second-party-based assessment supported by the formal recycler (main client) **Target by the end of years 4–5**: Reach full compliance (100%) with the requirements. A legal form is achieved

**Percentages are proxy values. The Guidance Principles do not recommend a unique method but require the proposal of a well-founded and sound estimation method in the Management Plan.*
***Up to 4–5 years.*
Modified from ISO IWA 19 (2017).

7.5.1.2 ISO 59014

As per ISO procedures, after 3 years ISO IWA 19 (2017) needed to evolve towards the next step. This implied either a revision, a withdrawal if no interest by the community was raised, or a conversion into an International ISO standard. With the endorsement of the ISO committees 323 on Circular Economy and ISO 207/SC5 on Environmental Management/Life Cycle Assessment, this standard was positively voted by the ISO community to be converted into an International ISO standard. ISO 59014 on Secondary Materials: Principles, Sustainability and Traceability Requirements is expected to be published in 2024. This standard has a broader materials scope. It aims to provide a framework for and guidance on implementing operational, environmental, social, managerial, traceability, and circular economy related requirements in the context of activities and processes for generating secondary materials. Activities and processes covered are within both the formal and informal sectors, so that the transition from informal to formal operations is facilitated.

Requirements are developed under the following principles:

1. Respect for interested parties' concerns
2. Value chain responsibility
3. Responsibility towards interested parties engaged in subsistence activities
4. Safe, healthy, and equitable working conditions
5. Protection of the environment
6. Resource conservation
7. Enabling of circular resource flow using secondary materials
8. Life cycle perspective

Both ISO standards are voluntary and are the first in the ISO family that acknowledge and address the issues of the informal sector in value chains. Based on a feasibility assessment of the ISO IWA 19 (2017) conducted by WRFA (Hagedorn and Valdivia, 2018), the application of ISO IWA 19 (2017) is generally feasible by formal recycling facilities with existing health, safety, and environmental management systems. This is not the case for microrecyclers or groups of individuals operating in the informality that need additional resources and support from larger economic operators upstream, as well as from the local governments, to facilitate their inclusion in value chains. Supporting mechanisms for these informal workers or microrecyclers need to be put in place. These include:

- Provision of PPE and tools, e.g., under lending or pay-as-you-use schemes;
- Training and capacity building;
- Legal assistance on regulations and laws, as part of formalization;

The DESCO and E[co]work cases illustrate how tailored supporting mechanisms can be applied successfully.

7.5.2 Ghana Technical Guidelines

An innovative policy approach for engaging individuals from the informal sector is the Ghana hazardous and electronic waste control and management Act 2016 (ACT 917). ACT 917 (Ghana EPA, 2016) sets out the framework for collecting a fee (called an eco-levy) on electronics imports, which shall be used to finance sustainable e-waste management. Based on ACT 917 (Ghana EPA, 2016), the Guidelines for environmentally sound e-waste management for collectors, collection centers, transporters, treatment facilities, and final disposal (Lambert et al., 2018) were issued. These Guidelines open the possibility in Ghana to include individual workers, also from the informal sector, in local e-waste management activities. This is a major step towards inclusive recycling, as roughly 95% of the e-waste in this country is collected and later recycled by the informal sector.

A five-tier approach is established in the Guidelines and in each tier the guiding principles, roles, and responsibilities of actors in partnerships are clarified (Figure 7.3). The Guidelines foresee an incremental approach, with lowest requirements in Tier 1 for collectors registered through a simplified process. This registration process proves to be feasible and reasonable for individual collectors, mainly individuals previously working informally. These collectors (Tier 1) must deliver e-waste (whole equipment or components) to a collection center (Tier 2) authorized by the (Ghana) EPA. Here e-waste is temporarily stored and dismantled before its transportation (Tier 3) to further authorized treatment facilities (Tier 4) or final disposal (Tier 5).

Based on a recent assessment (Hinchlife et al., 2020), collection in Ghana is notably conducted in a formal-informal partnership modus, where the collection center is operated by formal actors while supplied by informal ones. The Guidelines, supported with a longer-term finance scheme for enabling payments to collectors, have allowed deviating e-waste streams from the worst-practices pathway to formal recycling channels. These guidelines have also contributed to reducing worst practices such as open burning practices, e.g., of cable for recovering copper. Thus it is demonstrated that bridging the gap between the informal and formal sector is feasible.

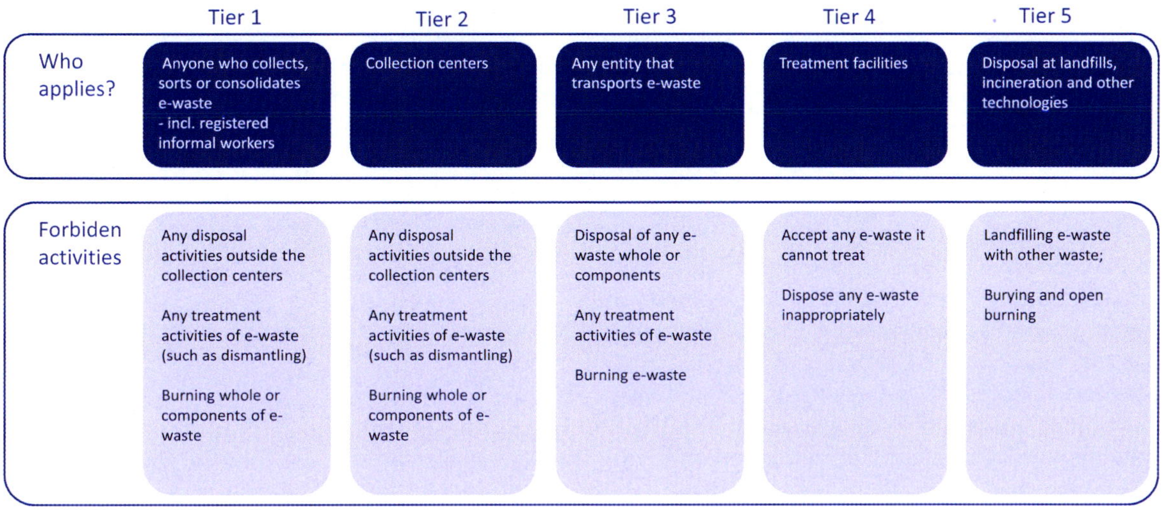

FIGURE 7.3 Tier approach for inclusive recycling towards collectors in Ghana—A simplified version. *Modified from Lambert et al. (2018).*

7.6 CONCLUSION AND OUTLOOK

The informal sector plays an important role in waste management value chains and the transition to a circular economy, especially in developing economies. Its importance transcends the socioeconomic spheres due to the generation of jobs, contribution to the economy, the concerning risks to health and the environmental spheres because of severe impacts on ecosystems, and its great potential in the recovery of secondary materials, some of which are increasingly necessary (take critical metals as an example) in consumer products worldwide.

Worst practices, with severe negative impacts on health and the environment, are associated with certain individuals or groups of individuals involved in the informal sector. Such worst practices vary from open burning to recovering material of value (e.g., burning cables to recover copper) through uncontrolled dismantling of used products while spilling and emitting hazardous substances (e.g., breakage of lead acid batteries), and child labor (e.g., in the collection and separation of waste), among others.

Despite its strategic role in national economies and international value chains, the informal sector has not been sufficiently regarded from a holistic, strategic, or an integrated perspective, which is why many potential and economic opportunities are lost and negative impacts are exacerbated. Very concretely, there is much talk about the need for their inclusion in national economies, but the necessary processes by the public institutions are not facilitated. In the absence of supportive national frameworks, local governments are the closest organizations for developing inclusion pathways.

There are some successful initiatives and programs in emerging economies that prove that informal sector inclusion in an orderly and gradual manner is possible, with benefits for the key actors. On the government side, the entities with the most direct relationship with the informal sector are the local governments, who have the capacity to issue specific municipal ordinances and supervise the application of the guidelines promulgated by law. The Ghana Guidelines (ACT 917) provide an example of how the principle of "tier-approach application" allows for the necessary flexibility and graduality towards an effective inclusion of the informal sector in waste management. Partnerships with the private sector have also proven to be a mechanism with high success chances, for example, through shared working spaces, training, and the provision of tools. International or local technical standards represent a powerful tool complementing or guiding the private-sector-led initiatives, especially in cases where the law does not yet cover the possibility of including the informal sector.

The inclusion of the informal sector in value chains is possible with the intervention of key economic operators in waste management (for example, formal recyclers requiring collected input) and of local governments as process catalysts. Responsible sourcing or ethical sourcing can contribute to gaining scale and speed, provided this is supported by legally binding mechanisms at national or local government levels guiding the inclusion of the informal sector. The cases cited are intended to provide sources of inspiration to initiate these much-needed processes.

References

Allen, C., 2019. An Inclusive Recovery. The Social, Environmental & Economic Benefits of Partnering with Informal Recyclers. GAIA. www.doi.org/10.46556/VPKH5682.

Arthur, E., Prosper, D., Hilbert, I., Nartey Kyere, V., Kwesi Ocran, P., 2020. Waste Tyre Management. Sustainable Recycling Industries (SRI). World Resources Forum and Empa, ISBN: 978-3-906177-27-4. www.sustainable-recycling.org/wp-content/uploads/2020/10/SRI_2020_WasteTyres-Ghana.pdf. (Accessed 10 March 2022).

Dasgupta, M., 2016. Moving From Informal to Formal Sector and What It Means for Policymakers. World Bank Blogs. 20 September 2016 https://blogs.worldbank.org/jobs/moving-informal-formal-sector-and-what-it-means-policymakers. (Accessed 10 March 2022).

Di Maria, F., Mersky, R.L., Daska, S., Ayalon, O., Ghosh, S., 2019. Preliminary comparison among recycling rates for developed and developing countries: the case of India, Israel, Italy and USA. In: Ghosh, S. (Ed.), Sustainable Waste Management: Policies and Case Studies. Springer, Singapore, https://doi.org/10.1007/978-981-13-7071-7_1.

E[co]work, 2022. E[co]work is a Co-Working Space for E-Waste Dismantlers. https://www.ecowork.international/ecoworkspace. (Accessed 10 March 2022).

Elgin, C., Ayhan Kose, M., Ohnsorge, F., Shu, Y., 2021. Understanding the informal economy: concepts and trends. In: Ohnsorge, F., Yu, S. (Eds.), The Long Shadow of Informality. World Bank Group. https://www.Thedocs.worldbank.org/en/doc/37511318c092e6fd4ca3c60f0af0bea3-0350012021/related/Informal-economy-Chapter-2.pdf.

Ghana EPA, 2016. Hazardous and Electronic Waste Control and Management Act 2016 (ACT 917). http://www.epa.gov.gh/epa/sites/default/files/downloads/publications/Hazardous%20and%20Electronic%20Waste%20Control%20and%20Mgt%20Act%20917.pdf. (Accessed 10 March 2022).

Hagedorn, A., Valdivia, S., 2018. Pilot Testing of the ISO IWA 19:2017: Feasibility Report. WRFA Publication. https://www.sustainable-recycling.org/pilot_testing/.

Hinchlife, D., Gunsilius, E., Wagner, M., Hemkhaus, M., Batteiger, A., Rabbow, E., Cheng, C., de Fautereau, B., Ott, D., Kumar, A., Smith, E., 2020. Partnerships Between the Informal and the Formal Sector for Sustainable E-Waste Management. StEP Solving the E-waste Problem. https://step-initiative.org/files/_documents/publications/Partnerships-between-the-informal-and-the-formal-sector-for-sustainable-e-waste-management.pdf. (Accessed 10 March 2022).

International Labour Organization (ILO), 2013. R204—Transition From the Informal to the Formal Economy. Recommendation, 2015 (No. 204) https://www.ilo.org/dyn/normlex/en/f?p=NORMLEXPUB:12100:0::NO::P12100_ILO_CODE:R204. (Accessed 10 March 2022).

International Labour Organization (ILO), 2018. Women and Men in the Informal Economy: A Statistical Picture, third ed. https://www.ilo.org/global/publications/books/WCMS_626831/. (Accessed 10 March 2022).

International Lead Association (ILA), 2021. Recycling in the Informal Sector. http://members.ila-lead.org/UserFiles/File/Recycling%20in%20the%20Informal%20Sector.pdf.

ISO IWA 19, 2017. Guidance Principles for the Sustainable Management of Secondary Metals. ISO. https://www.iso.org/standard/69354.html. (Accessed 10 March 2022).

ISO/DIS 59014, 2023. Environmental Management and Circular Economy—Sustainability and Traceability of Secondary Materials—Principles and Requirements. https://www.iso.org/standard/80694.html. (Accessed 10 March 2022).

Karcher, S., Valdivia, S., Schluep, M., 2018. From Worst to Good Practices in Secondary Metals Recovery. SRI—Sustainable Recycling Industries, ISBN: 978-3-906177-19-9. www.sustainable-recycling.org/wp-content/uploads/2018/07/worst-practices-web-25.7.18.pdf. (Accessed 10 March 2022).

Lambert, F., Atiemo, S., Tuekpe, L., Appiah, D., Manhart, A., Schleicher, T., 2018. Technical Guidelines on Environmentally Sound E-Waste Management for Collectors, Collection Centers, Transporters, Treatment Facilities and Final Disposal in Ghana. https://www.sustainable-recycling.org/wp-content/uploads/2018/03/eWaste-Guidelines-Ghana_2018_EPA-SRI.pdf.

Linzner, R., Salhofer, S., 2014. Municipal solid waste recycling and the significance of informal sector in urban China. Waste Manag. Res. 32 (9), 896–907. https://doi.org/10.1177/0734242X14543555.

Luz, F.G., De Castro, M.C.A.A., Rozenský, L., Hájek, M., 2021. Recycling capacity through formal and informal activities in six cities in São Paulo State, Brazil. Int. J. Environ. Waste Manag. 27 (2), 183–200. https://doi.org/10.1504/IJEWM.2021.112950.

Ministerio de Medio Ambiente de Peru (MINAM), 2010. Ley del Reciclador N.° 29419. Ley que regula la actividad de los recicladores. https://sinia.minam.gob.pe/normas/reglamento-ley-no-29419-ley-que-regula-actividad-recicladores. (Accessed 10 March 2022).

Oduro-Appiah, K., Abraham, A., Neequaye Kotey, V., de Vries, N., 2019. Working with the informal service chain as a locally appropriate strategy for sustainable modernization of municipal solid waste management systems in lower-middle income cities: lessons from Accra, Ghana. Resources 8, 12. https://doi.org/10.3390/resources8010012.

Ottiger, F., Schluep, M., 2020. Formal-Informal Sector Interface Models. GIZ Publication. https://www.sustainable-recycling.org/reports/formal-informal-sector-interface-models/. (Accessed 24 March 2022).

The Economist, 2016. Informality in Latin America. Casual Mondays to Fridays. The High Cost of Joining the Formal Sector. Edition of Oct. 22, 2016 https://www.economist.com/the-americas/2016/10/22/casual-mondays-to-fridays. (Accessed 10 March 2022).

United Nations (UN), 2015. The 2030 Agenda for Sustainable Development.

Valdivia, S., Sureda, M., Schluep, M., Widmer, R., 2016. ISO guidance principles for the sustainable management of secondary metals. In: Proceedings of 2016 Electronics Goes Green.

Van Eijk, F., Huisman, H., Keesman, B., Breukers, L., 2021. Waste Management in the LATAM Region. Netherlands Enterprise Agency. Report nr. RVO-036-2021/RP-INT www.rvo.nl/sites/default/files/2021/02/Report_LATAM_Waste_Management_feb_2021.pdf. (Accessed 10 March 2022).

World Resources Forum Association (WRFA), 2018. Recovering Iron From Used Tyres. Sustainable Recycling Industries (SRI). Series on Worst Practices No. 3 https://www.sustainable-recycling.org/reports/recovering-iron-from-used-tyres-a-fact-sheet/. (Accessed 24 March 2022).

PART 2

Recycling from a product perspective

CHAPTER

8

Physical separation

Kari Heiskanen

Metso, Espoo, Finland

8.1 INTRODUCTION

Recycling materials may consist of end-of-life product streams, by-products and waste streams from original equipment manufacturing, components production, and finally rejects, by-products, and waste streams from raw material producers. A common feature is that all do consist of compounds. These compounds can either be dominant in the product or form varying parts of them. The elements of the compounds can be recycled only by chemical or metallurgical means.

Each piece of material to be recycled has a set of properties, physical or chemical in nature, like compound or element concentration, density, magnetic susceptibility, and color (and more). The selection of physical separation is based on the distribution of the selected property.

Each particle/unit has its unique discrete property value selected as the basis for separation. A collection of pieces forming the lot to be treated thus has a distribution of property values, described as the property space. The variation arises from the particle's mass/area distribution of the compounds.

Metal alloys and polymers have a constant property value and cannot be changed by breakage or other physical means. In practice, mixtures of very fine matter in a continuous matrix, like pigments or flame retardants in plastic, show similar behavior to miniaturized electronic devices. The required particle size distribution for property variation is too small for practical purposes for such materials.

For "simple" products to be recycled, the property is a function of particle size. Additional breakage will move the property distribution to have distinct peaks. A freight railroad car is a good example of very few materials and easily breakable joints between the compounds (alloys).

Most of the products manufactured are characterized by a variable degree of the scale, showing some particle size dependency. Some of them are very complex and often contain components that fall into the first category where complexity cannot be reduced. Good examples of this are end-of-life electronics. The physical separation of such devices is restricted.

Many of the waste streams, like metal production slags, have voluminous matrix components, which cannot be reduced in complexity. These can be treated by removing the matrix into streams where the valuable components are concentrated.

Compounds in a recyclable stream are distributed and connected in many different ways

https://doi.org/10.1016/B978-0-323-85514-3.00045-2

affecting the size dependency. Particles may consist of a single compound or mixtures of several compounds. The metric of single compound content is termed *liberation*. If the particle size is made finer by breakage, the liberation will change in a way typical of the recyclable material and method of particle reduction. This characteristic is termed a *liberation curve*, the naming coming from similar phenomena with natural minerals. Particles in a recycled set will contain different mass fractions of other compounds. These particles will also exhibit different physical and chemical properties that react differently to physical forces and chemical environments. The chemical interactions may become very complex and negatively lead to unwanted reactive results, affecting recycling rates.

This essential feature can be captured in a product-based approach to recycling rather than a material-based approach. The former implicitly takes the whole cycle from the EOL phase to the production of renewed raw material into account. It also allows modeling the interactions caused by the stream complexities to improve recycling rates through technology and systems development. The latter is adequate for bookkeeping but does not address the effects of complexity at all. For example, an error often made in material-based recycling discussions is mixing steel recycling with iron recycling. Steels form a vast family of iron-based alloys. Many of the over 6000 alloys can be recycled together. However, to produce recycled steel, there are limits on the content of several other metals included in the scrap, which either need to be diluted with primary material or separated away from the feed scrap stream or lost into production slags, fumes, and dusts. One can mention, as an example, copper and tin.

In addition to the complexity of a single product, the practical recycling streams have another level of complexity arising from the collection phase, where different products are collected into a combined stream. The increased complexity may dilute the economic value of a stream substantially by changing the mass ratio of high-value compounds relative to the low-value compounds. Different products have different liberation curve characteristics, so combining different kinds of streams will also increase the complexity of the liberation characteristics. The collection of recyclable materials should be designed to avoid an unnecessary increase in stream complexity. Much can be done at the origin of the recycling process. In industry, cuttings, turnings, rejects, etc. should be sorted carefully. The same applies to household EOL goods, a much more difficult task. Even collecting the same type of appliances produced at different times causes additional complexity.

A complex set of properties makes recycling difficult. For example, freight railroad car recycling does not pose difficulties, while mobile phone recycling does.

The optimal degree of presorting is dictated by the collection system costs and structures, location and process capabilities of treatment facilities, and the economic incentives available for different actors.

However, the streams tend to be too complex and unsuitable for direct final processing after collection. Thus the usual step following collection is a physical separation stage comprised essentially of two unit processes: breakage to create particles with the required characteristics and separating the obtained particles into a concentrated stream and a reject stream. These unit processes can be repeated and combined in different ways to obtain the required final product. Physical treatments range from manual sorting to sophisticated automated and controlled systems. The properties of the particles, determined by the material (combinations), form the basis for selecting separation methods. The property chosen for separation can be any physical or chemical property of interest that correlates with the technical and economic goals set. Common properties exploited in recycling are particle size, shape, density, magnetic susceptibility, conductivity, chemical composition, color, and surface free energy. The chosen property and its distribution in the material to be separated often correlates with particle size and size distribution. The principle of separation is to design and use a set-up where the sum of forces acting on particles with the property are different from those not having the property. In all these steps, individual pieces' particle size, shape, and density affect the recycling process outcome.

The fundamental methods and metrics used to evaluate the recycling processes, such as sorting and separation, their efficiency, and variation are discussed in Chapter 43—Physical separation.

The separated streams are either purified physically further or sent for further processing by chemical or metallurgical means. Some material streams may already reach a salable commercial quality after a physical treatment.

8.2 PROPERTIES AND PROPERTY SPACES

Any particle can be characterized by its properties averaged over its volume or surface. This property can simply be its chemical composition, but more useful are the composition of compounds and the set of consequent physical properties. Any set of particles will have a distribution of the property values (Figure 8.1). This distribution can be treated as any statistical distribution having a mean and variance.

Let us denote the particle size class as **i** and the properties as **j, k,** …

The size classes can be single ISO 565 fractions, or any combination deemed applicable or any other sizes. It can, for example, consist of only three categories: "too fine for processing" −4 mm, "optimal for processing" 4–64 mm, and "too coarse and un-liberated for processing" +64 mm.

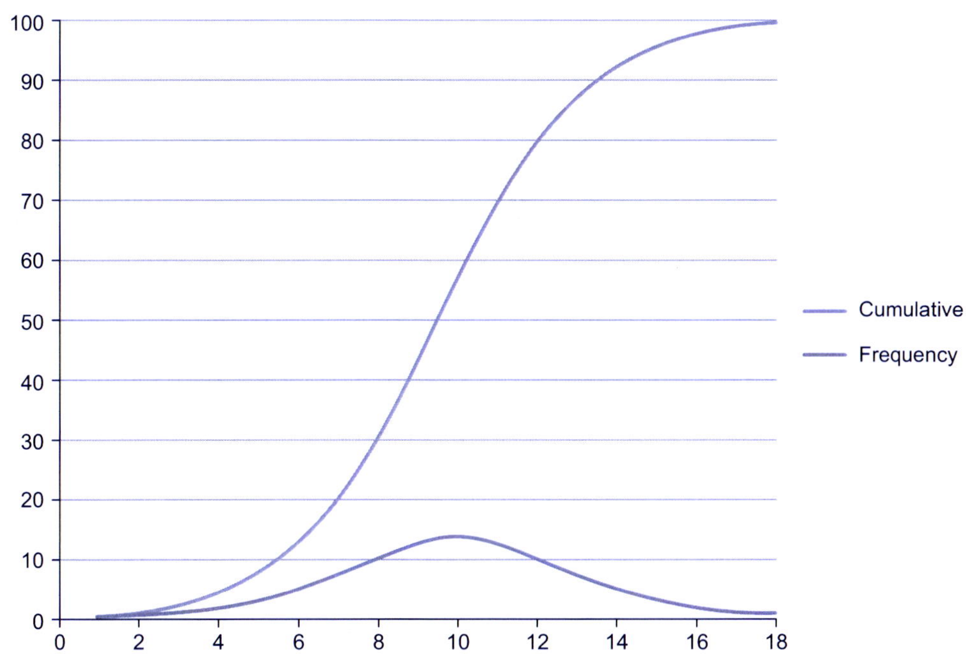

FIGURE 8.1 Property distribution.

Properties and their number of classes (**j, k, …**) must be considered by their utility. Only important properties for the separation stage and further processing requirements must be considered. There should be as few property classes as possible.

When a low purity steel scrap is treated, the most straightforward technique is magnetic separation. As most carbon steels exhibit ferromagnetism, magnetic susceptibility is the property (**j**) that will describe the separation adequately. We can simplify it to consist only of two classes: with ($\mathbf{j_1}$) or without this property ($\mathbf{j_2}$). However, if the steel scrap contains elements detrimental to the steelmaking process, we need to add composition properties like the fractional content of copper.

Models that track the changes in numbers of particles between the different volumes of property space are called population models (Reuter et al., 2005; King, 2001). The simplest space is a one-dimensional binomial space of similar-sized particles, either having the property or not. The general model is an N-dimensional space (Figure 8.2).

Particles can move to and from a property class volume by several mechanisms. Particles can move through the boundary by mechanical abrasion, chemical reactions, or melting. They can also move by finite steps due to breakage or agglomeration. There can also be physical additions and withdrawals of particles into and from a given property volume.

8.3 BREAKAGE

Breakage is the most important process in the physical treatment to move particles in the property space **R**. Breakage outcome is a composite outcome of brittle, ductile, or viscous behavior under the forces imposed on the particles.

Brittle materials are characterized by a small elastic deformation and ideally no plastic deformation (Hooke's law) under stress. In reality, there also is a small plastic deformation

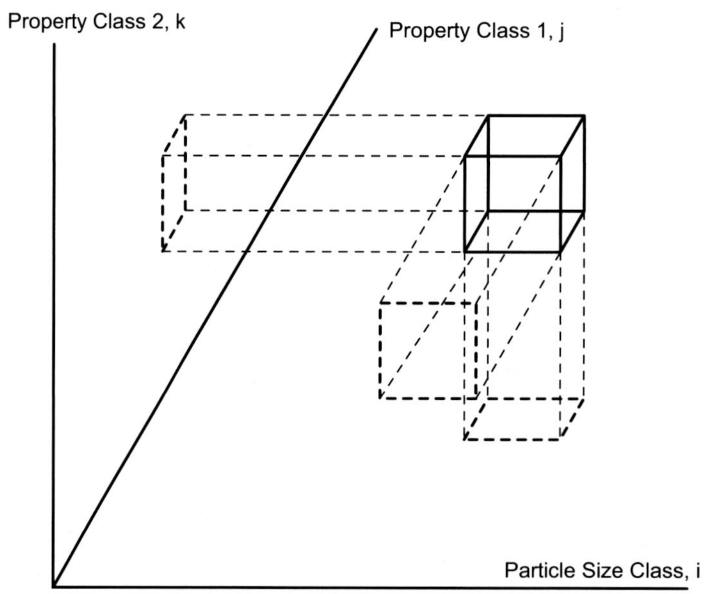

FIGURE 8.2 A 3-D property space R with a volume of particles belonging to a class $R_{i,j,k}$ (with a surface area of S_c).

component. Typically, crystalline materials are brittle, while amorphous materials tend to be ductile. Ductile materials show a large plastic deformation at stresses producing fracture. Viscous behavior shows a time- and temperature-dependent plastic deformation. Many viscous materials show a brittle (amorphous glass) behavior at ambient temperatures, while others (plastics) are ductile. The three features mentioned here are present in different degrees in all materials subjected to breakage.

Recyclable minerals, rocks, ceramics, and slags are brittle and can be broken by tensile stress produced by compression or impact. The tensile stress initiating the fracture is perpendicular to the direction of the compression. At some applied compressive stress, the resulting tensile stress causes the microcracks in the material to become critical, e.g., when the energy release rate becomes larger than the crack resistance, a material property. For brittle materials, the crack resistance is 1–10 J/m^2. A small fraction of the energy creates a new surface area. Most energy input is released as heat, typically 96–99%. Technical alternatives are to use guided physical tools, hammers, or compression surfaces, or unguided tools like grinding balls, beads, or the material itself. Producing large progeny particles with slow compression is energy efficient; the energy input is below 1–2 kWh/t. Fine particles, smaller than 50 μm, are relatively easy to produce but require high energy inputs, in the order of tens of kWh/t. Even submicron particles can be produced, but the energy expenditure is considerable (Jankovic, 2008). Particle size distributions follow roughly simple distributions (log-normal or Weibull).

Elastomers have a high elastic deformation compared to crystalline materials but a similar behavior at stresses not causing breakage. Due to the high elasticity, the breakage phenomenon is closer to ductile than brittle behavior.

Metals form a particular group of room-temperature ductile materials due to their unique electron configurations. The crack resistance varies between 10^2–10^5 J/m^2. For plastics, another large group of recycled ductile materials, the crack resistance varies between 10–10^3 J/m^2.

Breaking ductile materials requires a high shear force to be applied. Different types of shears are used. Impact crushers can also create large enough shear forces (Figure 8.3). Metal particles typically become bent and folded depending on their dimensions and plasticity. Particles break at the time of pressure release. Fine particles are difficult to produce.

Plastic particles that start to show viscous behavior at lower temperatures than metals can form long streamers due to the heat released during comminution. At ambient temperatures, plastics will break only at high impact velocities. At cryogenic temperatures and high impact velocities, viscous particles will behave in a way that resembles brittle behavior. However, unlike brittle materials, no fines will form at a high compression zone close to the impact point. At temperatures higher than ambient, the behavior is different. An impact causes substantial deformation but

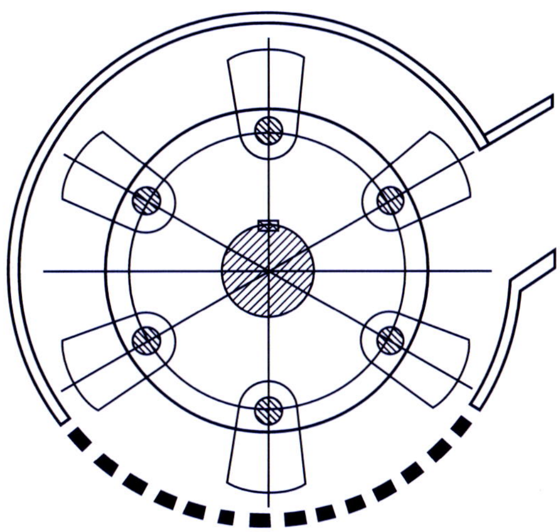

FIGURE 8.3 Impact crusher.

may not produce breakage. If breakage occurs, long streamers are common. Fine particle production is not possible.

Random breakage models, useful for brittle breakage, are not very useful for modeling ductile breakage. When recycled products consist of materials having various behaviors, the breakage becomes problematic. Especially in ductile materials, the character of the joints between material types will dictate the outcome (Van Schaik and Reuter, 2008). Joints may even prevent good liberation. Thin layers joined on other materials are problematic, and such layers can only be removed by abrasive breakage or a chemical leaching treatment.

A warning: The heat released during breakage may cause exothermic chemical reactions to start. As an example, materials containing alkali metals must be tested before application.

8.4 PARTICLE SIZE CLASSIFICATION

Size classification is a prerequisite for many mechanical separations for recycled materials. For coarser materials, classification is commonly done by physical screens. Screens can be applied either dry or wet. Intermediate moistures cause clogging of screen openings. The finer end (typically below a millimeter) can be performed by wet or dry dynamic separators or down to about 50 μm also with screens.

8.4.1 Screens

Screens are suitable separators for sizes often encountered in recycling. Screens can be applied from several hundreds of millimeters to about 50 μm. The most common ones are vibrating deck screens, roller, gyratory, and trommel screens. Deck materials can be made of steel wire, rubber, or polyurethane. Vibration is most commonly created by one or two (seldom three) shafts with unbalanced weights. Electromagnetic exciters can also induce vibration. The vibration pattern is often elliptical to move material on the screen deck. When the material enters the screen surface, it stratifies, and the finest material passes through rapidly as a zero-order process $dC/dt = k$. After the finest particles, finer than about half of the screen opening D_a, have passed, the process becomes a probability process of first-order $dC/dt = gC$. Each particle has a probability of passing the screen at each vibrational cycle. The probability goes to zero when two of the main dimensions are larger than square openings.

8.4.2 Static Size Separators

Static separators depend on the different settling velocities of particles in a gravitational field. The simplest is a vertical draft tube where the medium has a defined upward flow. Coarse particles with a higher settling velocity than the upward flow will settle to the bottom. A common and more efficient method is the "zig-zag" separator, a modified draft tube where angled corners create flow rotations to disperse the particle stream, enhancing the separation efficiency. As the settling velocity is a function of particle volume and density, these separators also separate based on density. They are typically used to remove plastics from heavy shredded material. There are also horizontal separators, where particles settle out from a horizontal flow and are collected into bins.

8.4.3 Dynamic Size Separators

For wet operations, the most commonly used device is the hydrocyclone, where the flow is fed tangentially to the outer perimeter wall of a cylindrical body. The cylindrical body is connected to a conical bottom. Flows enter from the cyclone on the centerline. The vortex finder is placed on the top of the cylindrical body, extending into the cyclone deeper than the feed

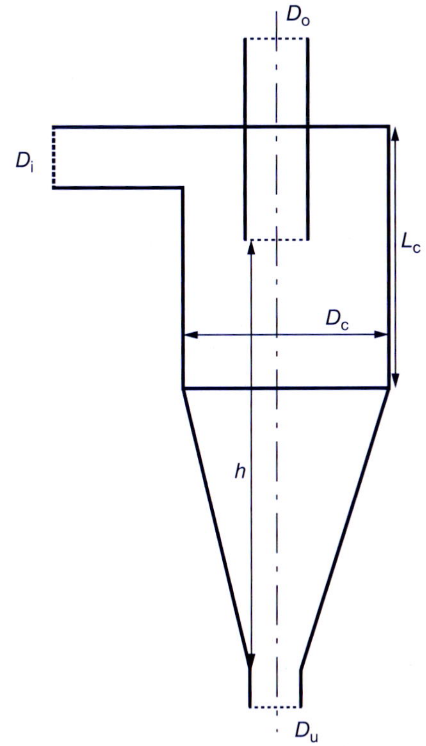

FIGURE 8.4 Hydrocyclone.

point at the wall. The apex is at the bottom of the conical part (Figure 8.4). The induced tangential flow creates a swirling flow. The flow needs to go from the perimeter to the centerline, causing a radial component of flow directed to the centerline. The forming flow pattern combines tangential, radial, and vertical flows.

The tangential movement of flow causes a centrifugal force. The radial component transports particles toward the centerline. As the centrifugal force is related to the mass of a particle and radial flow drag force to the area of the particle, large particles are more affected by the centrifugal force and the fine particles by the drag force. As a result, the coarse particles will remain at the perimeter, and the fine particles move toward the centerline. The vertical and tangential movement combination will create two helices rotating in the same lateral but different vertical direction. As a result, we have a downward spiraling flow at the perimeter and an upward flow at the cyclone's center. The helices transport the particles either up to the vortex finder or down the cone and the apex. Fines will report to the vortex finder, and the coarse come out from the apex. By changing the orifice sizes, feed volume and cyclone curvature (diameter), we can obtain the required performance (Nageswararao et al., 2004).

Dry materials are separated with a similar principle, wherein a radial flow, a surface force (drag force), is opposed by a body force (centrifugal force). A rotating vane cage produces the tangential velocity. The surface force is related to the second, and the body force to the third power of the particle dimension, causing fines to pass through the rotating cage. Dry cyclones can also be used. A challenge in dry separation is dispersion.

8.5 GRAVITY SEPARATION

Gravity separations are feasible if the concentration criterion *Cr* between the particles to be separated is

$$Cr = \left| \frac{\rho_{\text{heavy}} - \rho_{\text{fluid}}}{\rho_{\text{light}} - \rho_{\text{fluid}}} \right| > 2.5 \qquad (8.1)$$

Higher *Cr* values make the separation efficient even with broad particle size ranges, while a low *Cr* separation requires a controlled particle size distribution for efficient operation.

8.5.1 Jigs

Jigs can be used to separate closely sized coarse material. The smaller the concentration criterion, the narrower the size fraction that must be utilized. The largest sizes usable depend on density but are typically less than 10 cm. The jigging action sinusoidally pulsates water through a particle bed lying on a screen. At the upward pulsing, the bed loosens, and an initial acceleration occurs. The early acceleration is almost relative to density as fluid drag forces are negligible due to the slow velocity at the beginning of the sinusoidal cycle. When the fluid movement increases, the particles start settling down in a hindered way in the upward stream. Hindered settling velocity is a function of both size and density. The described cycle causes fine light particles to come to the top of the loose bed. As the flow velocity starts to slow, large coarse grains settle further faster toward the bottom of the bed. At a given point, the bed becomes consolidated, and trickling of fine particles takes place. As this pulsation goes on, the bed stratifies to light particles at the top and heavy particles at the bottom. For a proper operation of a jig, ragging (coarse heavy particles) needs to be introduced on the screen before the feed stream is introduced (Figure 8.5).

8.5.2 Shaking Tables

Another common gravity separator in recycling is the shaking table, where a thin film of suspension at about 25% solids by weight is introduced with dressing water to an inclined surface. Due to the friction, the flow on the surface becomes a Couette flow with a velocity profile increasing from zero at the table surface to a higher value at the top (free liquid-air surface).

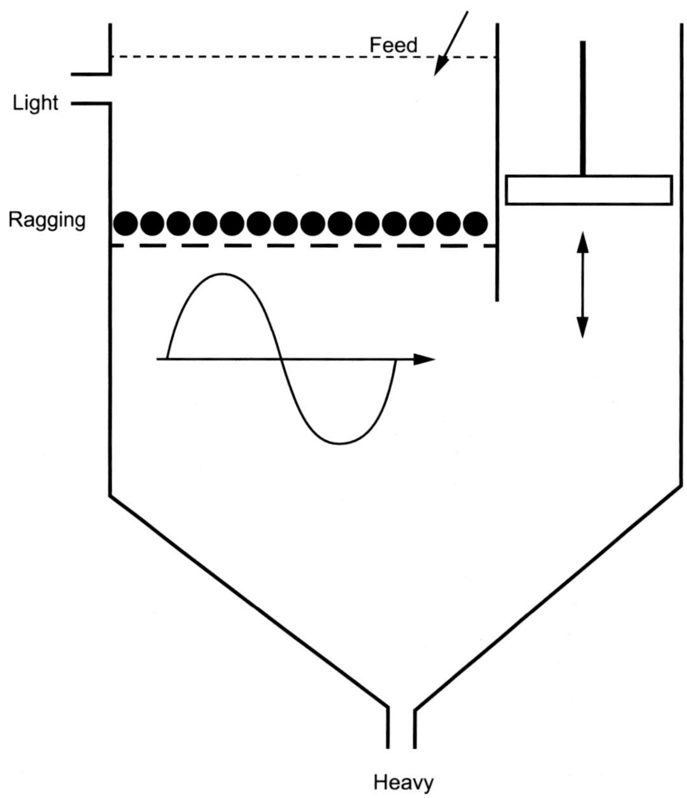

FIGURE 8.5 A jig.

The flow pattern causes stratification by density and size. The heavy and fine particles are found at the bottom and large and light on the top. The stratified bed is moved along by longitudinal strokes with a slow forward and a rapid return movement. The table is equipped with riffles, which become progressively lower in the flow direction. As the stratified particle bed moves slowly forward, particle layers are exposed to the dressing water, which sweeps first coarse light particles away from the riffles. Only fine heavy particles remain at the point where riffles taper off (Figure 8.6). Shaking table feed sizes can vary from 100 μm to 15 mm. Depending on particle size, the capacity varies for a standard 72″ table from 0.5 to 11 t/h. Tables can also be operated by air. Then the bottom of the table is aerated to provide fluidization. The upward airflow causes a drag force which is counteracted by gravity. The dense particles will segregate to the bottom of the riffles.

Other gravity separators include spirals and Mozley multigravity separators, and centrifugal gravity separators like the Knelson and the Falcon separators. These types are not common in recycling.

8.5.3 Dense Media Separation

Particles are introduced into a media with an (apparent) density between the material densities to be recovered and rejected. Typical separations made using dense media separators are plastics, rubber and textiles from metals, and aluminum and magnesium from other metals. Industrial dense media separations (sink float) are usually performed in ferrosilicon ($\rho = 6.7$–6.9) slurries, magnetite slurries, with the heavy material itself or, in the case of plastics and textiles, in salt solutions. Atomized ferrosilicon (16–17% Si) can be used at densities up to 3700–3800 kg/m^3. After separation, both the sink and the float streams must be washed to minimize medium losses. The separation is quite sensitive to the presence of fines, which increases the medium apparent viscosity too much.

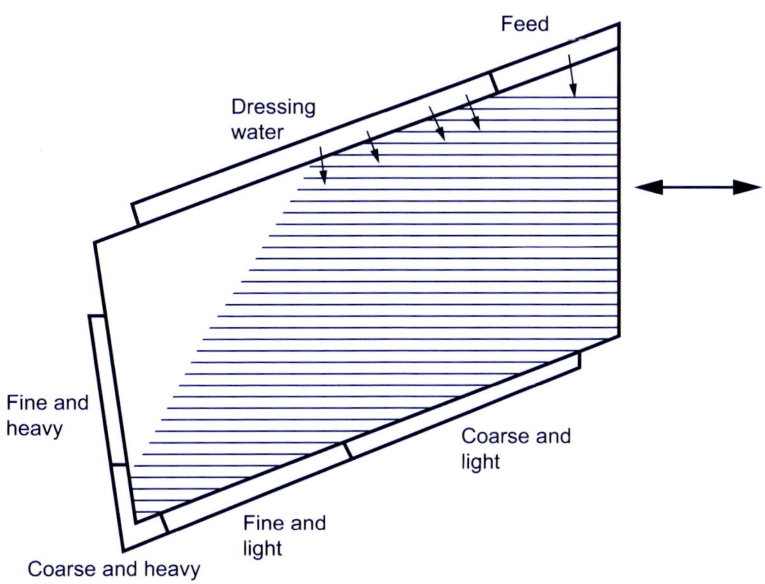

FIGURE 8.6 A shaking table.

8.6 FLOTATION

Flotation in recycling has limited use in separating shredded materials but has more use in separating valuable materials from granular recycling feeds like slags and recently spent battery recyclates. The typical particle size for flotation is light solids from a few millimeters and heavy solids ($\rho > 2700\,\text{kg/m}^3$) from 300 μm to about 10 μm. Flotation is widely used in deinking. Flotation is a physicochemical process where gas bubbles are attached to the particles in a fluid mixture and lifted by the buoyancy to the top of the flotation cell, forming a froth, which is then collected (Figure 8.7).

Flotation is based on the idea of manipulating the apparent free surface energy with reagents and choosing the operational conditions in a way that the difference between the particles to be separated and the rejected particles becomes as large as possible. The difference must be adjusted relative to the surface tension of the liquid used in flotation (most typically water, but not necessarily). If cohesion in the fluid is larger than the adhesion between the surface and the fluid, the fluid tends not to wet the surface and gas bubbles will adhere to the surface. Such surfaces are hydrophobic. Plastics have free surface energy in the range of 30–45 mJ/m^2, which renders them all hydrophobic and makes separating them by flotation very difficult. Most minerals tend to have free surface energy in the range of 300–500 mJ/m^2, causing them to become easily wetted or hydrophilic. The surface tensions (γ_{xy}) of the three interfaces of the solid-gas-liquid system obey the Young equation

$$\gamma_{SG} - \gamma_{SL} = \gamma_{LG}\cos(\theta) \tag{8.2}$$

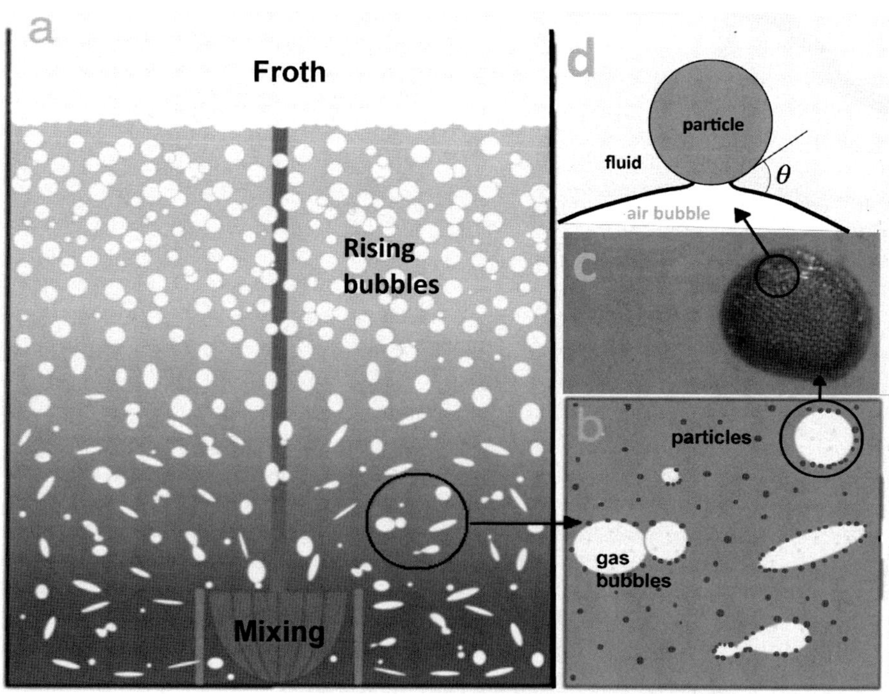

FIGURE 8.7 Flotation.

The contact angle θ[1] for flotation must typically be greater than 60 degrees (see also Figure 8.7).

Collectors are used to modifying the surface free energy. They contain a polar part that contacts the surface and a nonpolar end that causes the change in the free surface energy. The most common ones are anionic thiols and fatty acids, and cationic amines. Thiols are used for sulfides and metals, where electrochemical reactions occur. For systems with electrochemical collector reactions, pH is the primary controlling variable. Fatty acids are used for oxides, carbonates, and phosphates. Cationic amines are used for silicate materials. Due to the similarity of surfaces with silica tetrahedrons, selectivity is difficult to obtain for the silicates.

Further control of the chemistry can be achieved by controlling reagents like O_2, CN^- and HS^- ions. Flotation machines can be mechanical, pneumatic, or dissolved air devices. In mechanical flotation cells, the air is introduced via a mechanical stirrer (rotor), mixing the suspension and dispersing air into fine bubbles. The turbulent mixing causes particles and bubbles to collide and attach if the contact angle is adequate. Mechanical cells come in sizes from 1 L to 630 m^3/axle. Pneumatic cells, often used in deinking of paper, have separate mixing chambers where bubbles dispersed by either a sparger or a porous plate are contacted with the suspension. Dissolved air flotation dissolves air into water at a higher pressure and mixes the saturated water into a suspension with particles. The pressure release creates a large number of fine bubbles at material surfaces, enabling flotation.

8.7 MAGNETIC SEPARATION

Magnetic separation is based on the difference in magnetic susceptibility (*S*). For recycling purposes, we can distinguish three regimes, depending on how they react to a magnetic field.

$$S = \frac{M}{H} \tag{8.3}$$

where *M* is the intensity of magnetization and *H* is field intensity. Diamagnetic materials have a negative susceptibility, meaning that the field inside a particle is less than outside, and thus they are not amenable for magnetic separation. Such materials include rubber, plastics, glass, most metals (e.g., $S_{Cu} \approx -1.0 \cdot 10^{-7}$), slags, and dusts. Paramagnetic materials have a small positive susceptibility and can be separated from diamagnetic matrix by a high magnetic field intensity. Field intensities can be up to 2 tesla (T). Some metals like Ti, V, Cr, Mn, and platinum group metals are paramagnetic. Many Fe-containing compounds exhibit a paramagnetic property. The third group is ferromagnetism, where the susceptibility is high. Of pure metals, Fe, Co, and Ni are ferromagnetic. Most steels (not austenitic stainless steels) are ferromagnetic. The magnetic field intensity is much higher inside a particle and thus gives rise to high magnetic force. Used separator field intensities are typically 0.1–0.2 T. Magnetic force F_M is related to particle volume, its susceptibility, magnetic field intensity *H*, and its local spatial gradient. This is important for separating paramagnetic materials, where the force is weak without the gradient effect.

8.7.1 Low-Intensity Separation

Low-intensity separators are suitable for a large particle size range. Dry cobbers (belt separators) can separate large particles up to a few hundred millimeters. Permanent magnets are installed inside a belt pulley. It is a typical pre-separation device. Wet separators are usually meant for fine ferromagnetic materials down to micrometer range. They consist of a rotating

[1] The convention is to always measure the contact angle via the fluid phase.

drum with stationary permanent magnets inside. In concurrent separators, the flow of slurry is concurrent with the separator. A counter-rotation device has the nonmagnetic flow direction counter-current, but the flow of magnetic material is concurrent for a long pickup zone and thus a high recovery. The counter-current separator is almost similar but with a feed point closer to the final magnetic material discharge. The concurrent devices can tolerate up to 8 mm particles. A typical diameter for these separators is 1200 mm, with varying lengths up to 3.6 m.

8.7.2 High-Intensity Separation

High-intensity separators have a field strength of more than 1 T. Such separators introduce the feed flow into a volume containing a three-dimensional steel matrix placed in the magnetic field. The matrix creates high field gradients attracting the paramagnetic material. The matrix is water flushed under the magnetic field to remove nonattached particles. In the last phase, the matrix is moved outside the magnetic field and the paramagnetic material is flushed. Any ferromagnetic material entering the matrices will be magnetized and remain in the matrix.

8.8 EDDY CURRENT SEPARATION

An alternating magnetic field (change of magnetic flux (dH/dt)) induces a circulating (eddy) current into conducting particles (Figure 8.8). This in turn creates a force (Lorenz force) that is perpendicular to the flux and the direction of the eddy current. The created force is proportional to the ratio between conductivity and density of the material. Aluminum and magnesium have high ratios and can be separated easily; copper and silver have medium-high ratios and they can be separated with high relative magnet speeds (flux changes); zinc and gold have quite low and lead and stainless very

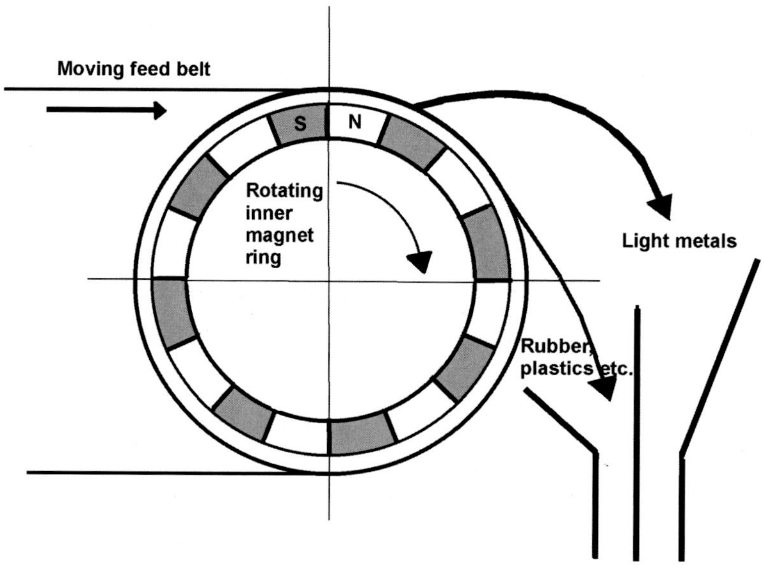

FIGURE 8.8 Eddy current separator.

low ratios, causing their separation to be almost impossible; glass, rubber, and plastics have no significant conductivity and are not affected by the eddy current.

8.9 ELECTROSTATIC SEPARATION

Electrostatic separation uses electric conductivity for separation. Particles can be charged by high-voltage corona charging with a high electron flow. When the feed passes to an earthed roll, conductors will lose their charge and fly off the roll by centripetal force, while nonconduction particles are attached to the roll and are brushed off to a separate bin. Particles can also be charged by letting them pass by charged plates (anode). This will change the trajectories of conductors to those of the nonconductor particles. Feed needs to be dry and dust-free. As a single pass is not very efficient, separators are often arranged in cascades. They are suitable for millimeter size and finer feeds.

8.10 SORTING

Several particle properties can be identified using sensors and then project an external force to remove from the stream particles with the identified property. The sorting decision can be based on a measured value of every particle or the average of a lot.

The simplest way of doing sorting is manual sorting, not discussed here. For mechanical sorting, the whole electromagnetic spectrum from X-rays to infrared can be used for detection. Visible light frequencies (color) are the simplest. Also, spectral analysis is commonly used to identify materials after they have been excited by, e.g., a laser pulse (LIBS). Methods like prompt gamma neutron activation (PGNAA) or nuclear magnetic resonance (NMR) can also be used for detection. Sensor fusion combining several measurements is nowadays a common practice to enhance sorting. Detailed information on sensor-based sorting can be found in Chapter 8—Sensor-based sorting.

Methods using electromagnetic spectra for identification are measuring surface properties. PGNAA and NMR measure volume averages.

Analysis time and particle size are crucial for the operation. A 200 ms identification time allows treating 18,000 particles per hour/sensor. If every tenth particle needs to be removed and if they are, say, square particles of 100×100 mm with a thickness of 5 mm having a density of 2.7, the capacity is about 240 kg/h for the removed stream. An 80×80 mm size reduces the capacity to 150 kg/h/sensor.

When a particle needs to be removed from the stream, it can be diverted by a moving gate, blown by a directed air blast/blasts from the stream or be picked from the stream by a robotic arm. Sorting technologies are limited at the upper limit by the weight of particles and at the lower limit by particle size. If particles are large, the air blasts do not have enough energy to remove unwanted particles from the streams. If gates or moving flaps are used, their slow movement limits capacity. At the finer end, they are limited by CAPEX (capital expenditure) costs becoming high compared to the capacity.

8.11 CONCLUSION

After collection, the first step to enhance the circular economy is to direct the collected materials as fractions to the optimal extraction route. Most efficiently, both economically and resource efficiency-wise, this can be done by applying the unit processes of physical separation to the collected materials. The complexity of the streams for extraction can be controlled and adjusted. This chapter has given a short overview of physics and procedures in physical separation. These basics allow developing physical separation circuits further. Advanced flowsheets with multiple operation nodes, targeted breakage steps, and online sensor technologies combined with real-time knowledge of mass balances will be needed in the future to treat efficiently the ever more complex collected feeds.

References

Jankovic, A., 2008. A review of regrinding and fine grinding technology – the facts and myths. In: Metso Miner Process Technol. Asia-Pac.

King, R.P., 2001. Modeling and Simulation of Mineral Processing Systems. Butterworth-Heinemann Publications, Oxford, Great Britain. ISBN 0750648848.

Nageswararao, K., Wiseman, D.M., Napier-Munn, T.J., 2004. Two empirical hydrocyclone models revisited. Miner. Eng. 17 (5), 671–687. https://doi.org/10.1016/j.mineng.2004.01.017.

Reuter, M.A., Heiskanen, K., Boin, U., van Schaik, A., Verhoef, E., Yang, Y., 2005. The Metrics of Material and Metal Ecology, Harmonizing the Resource, Technology and Environmental Cycles. Elsevier BV, Amsterdam. ISBN: 13 978-0-444-51137-9, ISBN: 10: 0-444-51137-7, ISSN: 0167-4528.

Van Schaik, A., Reuter, M., 2008. The use of fuzzy rule models to link automotive design to recycling rate calculation. Miner. Eng. 20, 875–890. https://doi.org/10.1016/j.mineng.2007.03.016.

CHAPTER

9

Sensor-based sorting

Xiaozheng Chen, Nils Kroell, Alexander Feil, and Kathrin Greiff

Department of Anthropogenic Material Cycles, RWTH Aachen University, Aachen, Germany

Sensor-based sorting (SBS) has been widely applied in the recycling industry for various waste types for more than two decades. Compared to manual sorting, SBS achieves higher throughputs, reproducible sorting results, and higher automation degrees. While the sortable particle features of traditional sorting methods (e.g., wind sifting or magnetic separation) are limited by the underlying separation forces (e.g., magnetic force), SBS methods are characterized by the independence of detected particle features (e.g., color and chemical structure) and the applied separation forces (e.g., air valves). Thus, a variety of different particle features can be used and combined for corresponding sorting tasks, and the sorting process can be easily adjusted by the applied sorting program. Therefore, sorting tasks such as sorting according to different colors or polymer types that were difficult or impossible to achieve with traditional sorting equipment can be achieved by SBS. Currently, sensor-based sorters are applied to various material flows, and SBS (cascades) represent a central element of many modern sorting processes.

9.1 MECHANICAL TREATMENT OF WASTE

Mechanical recycling chains generally consist of two treatment levels: *pretreatment,* the enrichment of valuable product fractions through preconditioning and sorting, and *recycling*, the production of secondary raw materials (Pretz et al., 2020). Taking lightweight packaging (LWP) waste as an example, it is first mechanically treated to enrich plastics according to their polymer type in sorting plants. Subsequently, sorted fractions are further refined and processed in recycling plants to produce high-quality recyclates. These are fed back into the anthropogenic material cycle to substitute primary raw materials in new good production.

In both treatment levels, sensor-based sorters equipped with application-specific sensors and in different construction types (e.g., belt and chute sorters) are applied. Due to distinct material compositions and characteristics, sensor-based sorters are located at different positions in the process chain. In most cases, SBS unit is positioned after necessary preconditioning processes (shredding, sieving, and sifting) as SBS

Handbook of Recycling. https://doi.org/10.1016/B978-0-323-85514-3.00028-2

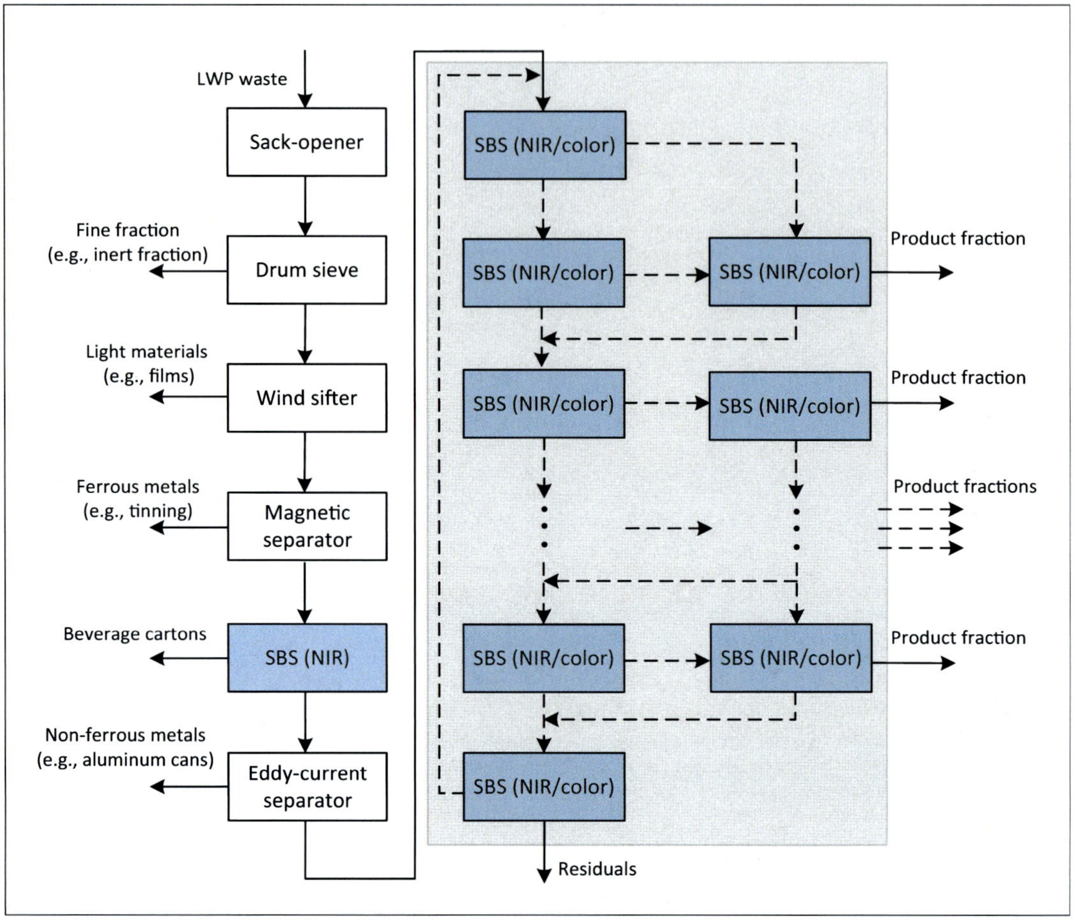

FIGURE 9.1 Exemplary and simplified mechanical treatment of lightweight packaging material in Germany. *Gray area*, SBS cascade.

processes require preconditioned materials. Figure 9.1 shows a typical sequence of mechanical treatment processes for LWP in sorting plants in Germany.

In LWP sorting plants, SBS is an important step through which valuable product fractions are sorted according to polymer types and colors. As shown in Figure 9.1, multiple sensor-based sorters with different sensors are placed at the end of the mechanical treatment process and generate the desired preconcentrates. Modern LWP sorting plants contain sorting cascades with typically more than 20 sensor-based sorters to meet the high requirements on material recovery and purity of the generated preconcentrates.

9.2 PRINCIPLE OF SENSOR-BASED SORTING

Despite the variety of covered sorting applications and construction types, all sensor-based sorters follow a similar principle of consecutive steps, as shown in Figure 9.2:

(1) feeding,
(2) detection, and
(3) separation.

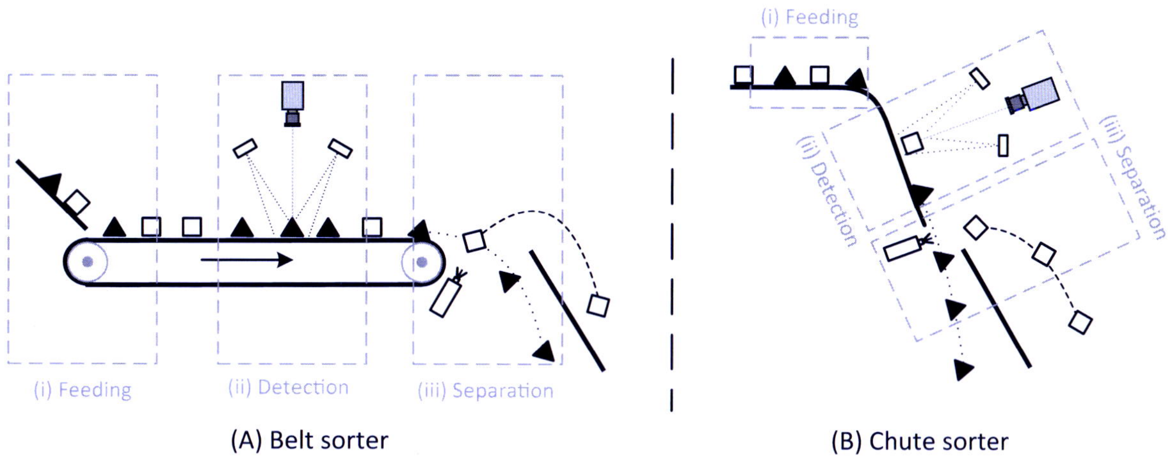

FIGURE 9.2 Construction types of sensor-based sorters. The three steps in sorting are indicated. The *black triangles* are the drop fraction; the *white squares* are the eject fraction.

In the feeding step, the material flow is uniformly distributed over the conveyor surface. The goal is to present the material flow as a *singled monolayer*, i.e., avoid overlapping or touching of particles and guarantee sufficient spacing between different particles.

In the detection step, the material flow passes through a detection area where one or more sensor units measure the (surface) properties of each object. Depending on the measured characteristics, the objects are typically irradiated with light from emitters in a particular wavelength area, and the sensor captures the reflected or transmitted radiation. The captured sensor data is then transferred to a calculation unit. In the calculation unit, (machine learning) algorithms classify the objects into different material classes or target fractions. Based on the classification result and a user-defined sorting recipe, a sorting decision is made for each particle.

Based on the sorting decision, a separation unit sorts the input material flow into two or more target material flows in the last step: separation. According to the sorting strategies, the target fraction, in the case of *positive sorting*, or nontarget fraction, in the case of *negative sorting*, are blown to one fraction called *eject*. The other materials are collected in the other fraction called *drop*.

Different mechanical separation forces can be applied to sort each particle in the desired target material flow. The most common separation systems by far are compressed air nozzle bars. With compressed air nozzles, particles with the sorting decision "drop" follow a parabolic trajectory at the end of the conveyor belt and end into the drop fraction. In contrast, particles with the sorting decision "eject" are shot with a few milliseconds of compressed air blast from coordinate-specific air nozzles into the eject fraction. In specific cases, the material flow can be separated into more than two fractions with more separation plates to different chambers through suitable compressed air nozzle settings or applying multiple nozzle bars. Other separation systems use mechanical paddles to change the trajectory of falling particles or mechanical grippers or vacuum cups to sort individual particles from the conveyor belt.

9.2.1 Construction Types

Two types of sensor-based sorters have become established in the state of the art: *belt sorters* and *chute sorters*. Belt sorters (Figure 9.2A) are typically applied in sorting plants for pretreatment of objects in relatively large sizes (≥10 mm, e.g., plastic bottles), while chute sorters (Figure 9.2B) are mainly applied in recycling plants for smaller particles (<10 mm, e.g., plastic flakes).

In SBS processes with belt sorters, the usually heterogeneous input materials are unloaded by a vibration chute or delivered from other conveyor belts with a lower speed to the acceleration belt with much higher belt speed (about 3 m/s) to ensure that the material flow is presented as a singled monolayer to the SBS unit. The throughput of a belt sorter depends mainly on the width (0.7–2.8 m depending on the application) and speed (up to about 4.5 m/s) of the belt conveyor.

Chute sorters differ from belt sorters in the transport of material flow: instead of being transported with a conveyer belt, objects go through the detection and separation area by sliding down an inclined surface; see Figure 9.2B. Compared to belt sorters, objects sorted with chute sorters have less relative movement, which is advantageous, e.g., for rolling objects. In addition, a transmission measurement with color cameras is possible, as the background of chute sorters is transparent, which leads to the wide application of chute sorters in glass sorting. Moreover, the object extraction and detection are—owing to the transparency of background—easier in the case of similar characteristics of objects and background, e.g., in sorting black plastics on black conveyor belts. However, due to the limited sliding speed, a chute sorter typically achieves relatively lower throughputs than a belt sorter.

9.2.2 Working Modules of SBS

Depending on the target of a single sorting step (high yield or high purity grade), a sensor-based sorter can work in different modules: rougher, cleaner, or scavenger; see Figure 9.3. The module *rougher* aims to sort a material flow with maximum yield, which results in a product fraction with a high amount of impurities. The valuable materials are then fed to a *cleaner* stage, the target of which is to sort the material with high purity, and the valuable material fraction is the end-product fraction. The residues of both rougher and cleaner can be re-sorted with a *scavenger* stage to recover valuable materials fed back to the rougher. Only a multistage sorting process can achieve both high purity of the output streams and high yield of valuable materials (Feil et al., 2021).

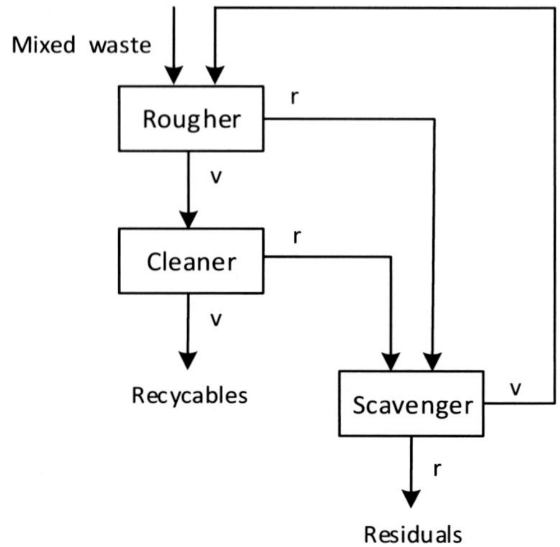

FIGURE 9.3 Different working modules of sensor-based sorters: rougher, cleaner, and scavenger. v, valuable materials; r, residuals.

9.3 REQUIREMENTS FOR OPTIMAL SORTING RESULTS

In SBS, adequate material flow presentation and preconditioning of the material flow is of utmost importance. Optimal conditions require adherence to the following requirements:

(1) Material flow presentation as a singled monolayer:

Overlapping or touching particles hamper the correct detection and sorting of particles. As most sensors measure the particle surfaces or have a limited penetration depth, only the material layers on top are detected, based on which the sorting decision is made. Depending on the combination of overlaying and underlaying materials, overlapping leads to false discharges and lower product purity (non-target material is overlapped by target material) or lower yield (target material is overlapped by non-target material). Furthermore, touching or overlapping particles may be recognized and sorted as one object, resulting in lower purity or yield depending on the applied classification algorithms. For air-nozzle-based sorting, touching particles or limited particle distances may also result in particle entrainment of lighter particles close to target materials and thus lower product purities. In addition, for transmissive sensor measurements, e.g., XRT or RGB, or sensors with limited penetration depth, e.g., NIR, overlapping particles can lead to misclassification (e.g., mixed NIR spectra of overlapping particles) and false discharges.

(2) Limitation of grain size ratio:

As the compressed air settings and optimal air nozzle distances depend on the particle sizes, a limited grain size ratio of the material flow is necessary. Furthermore, the reflected radiation differs between different particle heights. Optimally, the maximum to minimum grain size ratio for sensor-based sorters should be lower than 3 to 4.

(3) Removal of flyable 2D fractions:

To guarantee a particle discharge at the right time and position, the relative movement of objects on the conveyor surface must be minimized. Consequently, flyable 2D fractions such as plastic films should be separated before SBS, or adapted airflow guidance should be used to minimize their relative movements on the conveyor surface.

(4) Removal of fines:

The available sensor resolution and air nozzle distance determine the minimum particle size to be sorted. Fine particles are more likely to be swirled up on acceleration belts. They can deposit on conveying surfaces or components of the sorting unit due to air turbulence during particle discharge with air nozzles. To avoid such process disturbances and a decrease in the sorting performance, fine fractions should be removed during preconditioning before the SBS process.

(5) Minimization of composite materials:

Composite particles and particle surfaces with contaminations (e.g., adherent organics) negatively affect the SBS process. In some applications, composite particles may be liberated by adequate comminution processes. Further preconditioning may be applied to remove contaminations from particle surfaces before SBS, such as hot washing of plastic flakes. For waste collected in bags, effective bag opening is of vital importance for sufficient sorting results.

9.4 AVAILABLE SENSORS

The choice of appropriate sensors for the intended applications is crucial to capture distinct particle and material characteristics for optimal sorting results. In the last two decades, various sensor technologies have been further developed and applied to SBS. Some sensors are able to detect similar characteristics but are applied in different material flows or positions in a process chain due to their specific measuring principle and the economic aspects (conflict between cost-benefit and cost-expenditure).

9.4.1 Color Detection

Sensors for color and particle shape detection are widely used in SBS. Image sensors measure the intensity of visible light (VIS), covering the wavelength range of 380–780 nm. An image sensor consists of a large number of light-sensitive semiconductor elements, which convert the incident electromagnetic radiation into charge units. The charge units are accumulated over the exposure time and converted into a digital number that describes the intensity of the radiation. The radiation incident on a semiconductor element is transferred into brightness information (pixels). For each spectral region, the radiation intensity of that spectral region is measured.

Traditional and widely applied VIS sensors are red, green, and blue (RGB) sensors, which measure the intensity in three spectral regions. In the last decades, hyperspectral imaging in the VIS area was introduced and is increasingly applied for more accurate color detection and classification.

9.4.1.1 RGB Sensors

With RGB sensors, the radiation intensity of the three basic colors RGB is measured, through which their color components can be determined. The original color is identified with the three color components by additive color mixing. In this way, the color is represented approximately, and the number of possible colors depends on the depth of each spectral channel. Widely used RGB cameras have 8-bit or 12-bit spectral channels. For an 8-bit channel, each color component (R, G, and B) can have 256 (= 2^8) different brightness levels; thus 256^3 different colors can be identified by mixing the three primary color brightnesses. The detection of the RGB components can be technically implemented in various ways, presented in Figure 9.4.

Three-chip color cameras (Figure 9.4A) have a separate sensor for each primary color. The incident light is split into the three primary colors RGB via upstream prisms and directed to the respective sensor. Each sensor thus receives the brightness of a color channel.

In color detection via a color filter mask (Figure 9.4B), a color filter that only allows specific spectral ranges to pass through is applied directly to a sensor. A pixel thereby only receives the brightness of a specific color channel. There are various color filters, with Bayer filters (Figure 9.4B) being the most commonly used ones due to their low cost.

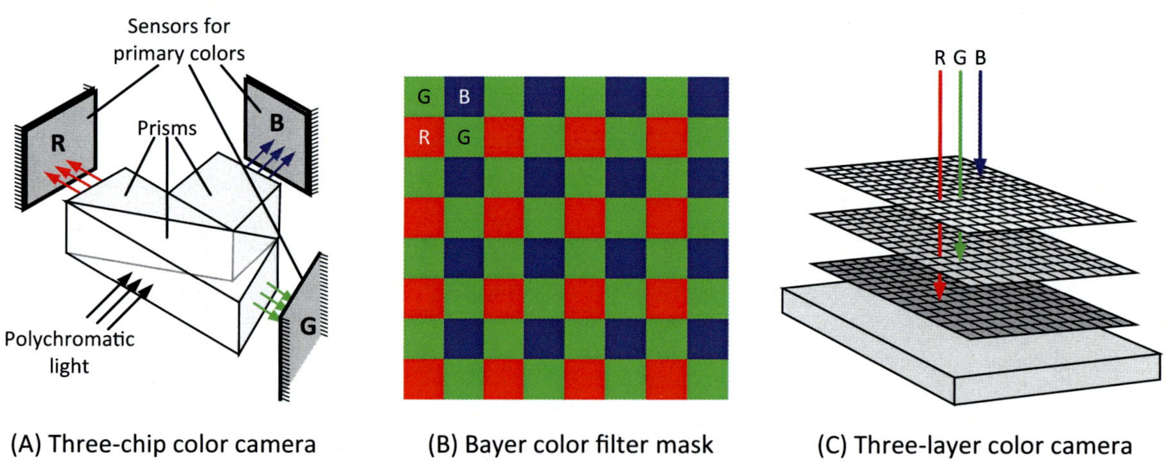

FIGURE 9.4 Technical implementation of RGB cameras, showing the camera setup (A), the color filter mask (B), and a three-layer color sensor (C).

In a three-layer color sensor (Figure 9.4C), three pixel layers are arranged one above the other. In each pixel layer, the light of a different wavelength range is absorbed and converted into charges, which enable the determination of the RGB color components.

RGB cameras can be applied in almost all applications where color and shape play an important role. For example, in glass and plastic recycling, RGB cameras are applied to sort flakes into different color fractions. Furthermore, they can be applied in separating glass, ceramics, stones, porcelain, and metals.

9.4.1.2 *Hyperspectral Imaging*

While RGB sensors only measure the intensities of the three RGB spectral ranges, it is possible to capture the intensities of more than 100 spectral ranges using hyperspectral color imaging. This image information can be displayed in a hyperspectral cube (Figure 9.5A). In the hyperspectral cube, the spatial image information of a sensor (x,y) is represented in two dimensions and the spectral image information λ in a third dimension. The number of elements in the (x,y)-plane corresponds to the spatial resolution in pixels. Depending on the spectral resolution, there are k elements in the λ-direction: $k = 3$ corresponds to RGB cameras, and for hyperspectral imaging, k has the value of ≥ 100.

There are several ways to acquire hyperspectral image data. A similar basic principle applies to all acquisition modes: a maximum of two dimensions can be read from the hyperspectral data cube using a two-dimensional area sensor. The third dimension is acquired by multiple sequential measurements. Since point area scanning does not allow the sample to move, a line scanning model (Figure 9.5B) is used in SBS.

In line scanning, the incident light of the object to be measured enters the camera through an entrance aperture (slit). A two-dimensional dispersing element (diffraction grating or prism) generates a spectrum for each point of the line and distributes it over the second dimension of the detector. Therefore, the wavelength range can be arranged according to the position of the detector. By moving the object to be captured or the sensor unit forward, the t-direction in Figure 9.5B, objects can be detected line by line. Hence, this approach is particularly suitable for conveyor belt systems and real-time applications.

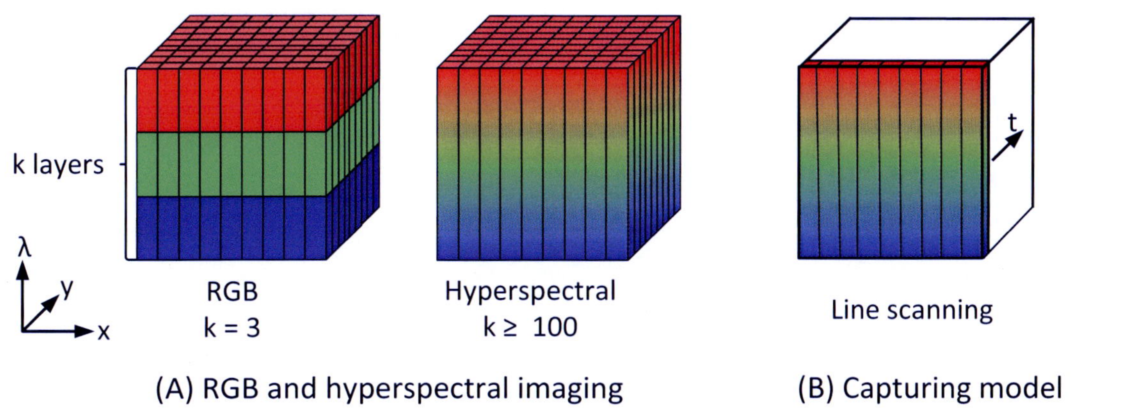

FIGURE 9.5 RGB and hyperspectral imaging showing in the hyperspectral cube and the capturing model in sensor-based sorting.

9.4.2 Near-infrared Sensor

Near-infrared (NIR) spectroscopy is based on the interaction between electromagnetic radiation in the NIR region and molecular structures of the material. NIR-sensitive chemical structures of specific materials absorb the energy of irradiation in specific wavelengths, which results in different levels of decreases in the reflected spectrum. The observation of absorption bands makes the analysis of material physical properties and chemical structures possible and thus enables their classification. The irradiation of NIR covers the wavelength interval between 780 and 2500 nm. As most NIR sensors have more than 100 spectral ranges, the capturing of NIR spectra also belongs to hyperspectral imaging.

NIR spectroscopy is dominated by overtones and combination bands, especially H-containing functionalities, e.g., CH, OH, NH, due to the low mass of hydrogen (H). Most polymer backbones, for example, contain NIR-sensitive functionalities such as CH, OH, NH, C=O and, frequently, the combination of them. These specific chemical structures lead to a characteristic spectrum of each material. Figure 9.6 shows the general raw spectra of five widely used polymer types: polyethylene terephthalate (PET), polyethylene (PE), polypropylene (PP), polystyrene (PS), and polyvinyl chloride (PVC).

In state-of-the-art sorting plants, the applied NIR sensors cover the irradiation range of about 900–1800 nm (*gray marked area* in Figure 9.6). Different manufacturers of SBS equipment may

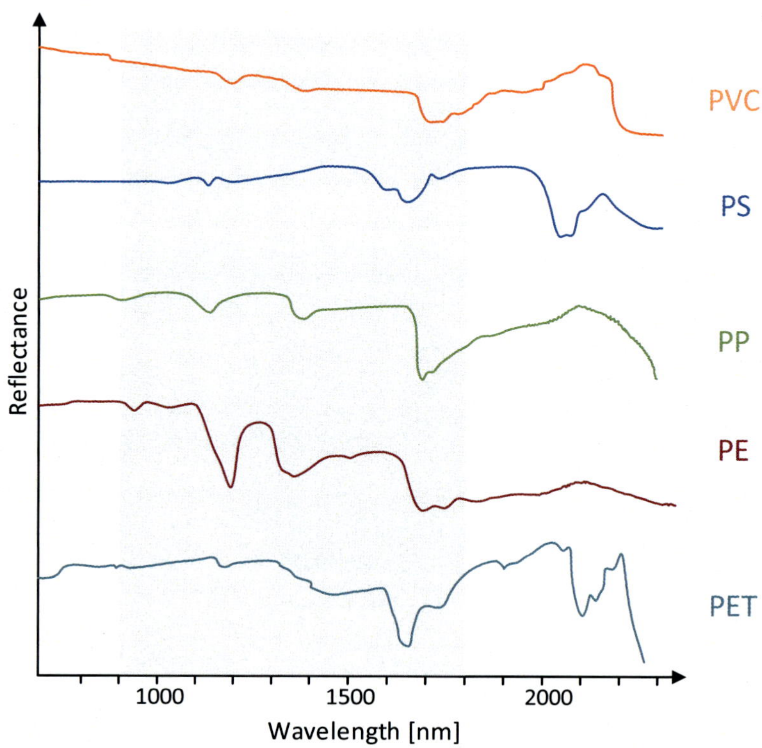

FIGURE 9.6 NIR spectra of widely applied polymers. *Gray marked area* is the typical range of sensors in sorting plants.

choose different wavelength ranges, depending on the specific target applications.

Due to the NIR-active chemical structures of polymers, wood, paper, and minerals, NIR sensors are widely used in, e.g., plastic, wood, paper waste, and construction and demolition (C&D) waste recycling. However, NIR spectroscopy has a limitation in detecting materials containing carbon black as a coloring agent, as they do not reflect sufficient radiation for spectral analysis.

9.4.3 3D-Laser Triangulation

3D sensors used in waste recycling are noncontact measurement methods for determining the three-dimensional shape of objects. For material flows, the 3D information is mainly used for particle detection and extraction and can be used for volume flow calculation. The most often used measurement is based on 3D-laser triangulation (3DLT).

In the 3DLT light-sectioning method, a laser line is irradiated onto the surface to be measured. An area camera then captures the laser beams reflected from the surface at a specific triangulation angle α (see Figure 9.7). This triangulation angle between the laser and the camera results in a lateral deflection of the laser beam as a function of the respective object height. With trigonometric relations, the object height and the one-dimensional profile of the measured object can be calculated from the lateral deflection of the laser line. Several one-dimensional profiles can be captured and assembled into a three-dimensional shape through relative movements between the object and the laser line. In SBS applications, objects are moved through the laser line and camera capturing area by a conveyor belt. Figure 9.7 shows the principle of laser triangulation using the light-sectioning method as an example.

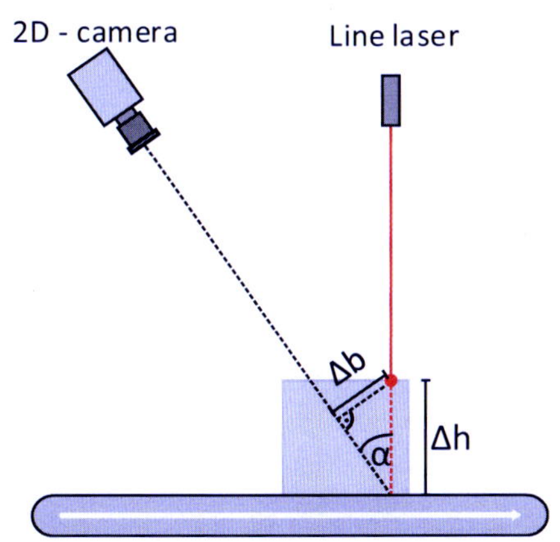

FIGURE 9.7 Working principle of 3D-laser triangulation (3DLT).

The size and location of the camera determine the limits of the measurable height range. Objects with surfaces containing large slopes or steps can shadow or cause multiple reflections of both the projected laser light and the light radiating to the camera. Shadowing caused by inclines, steps, or low depressions can be avoided by dual-camera 3DLT. In dual-camera 3DLT, a second camera is placed at the other side of the laser, with which both the front and back of objects can be captured.

Due to the high spectral power density of the laser radiation, the detection of the objects is not influenced by ambient light. Thus, surface detection is mostly independent of the object's color and material design. Exceptions are absorbing as well as reflecting surfaces. These cause too little diffuse reflection of the laser light, so that these surfaces cannot be adequately detected. In the case of transparent or partially transparent materials (e.g., transparent PET bottles in lightweight packaging waste), internal scattering of the laser light occurs, which widens the measurement spot and causes misdetection. The light can also be reflected twice, as it forms a light spot on the front and back sides, leading to detection error in object height. As hollow spaces can often not be illuminated by the applied laser, they cannot be detected and are thus included in calculating volume, leading to a higher measured volume than the actual volume.

9.4.4 Laser-Induced Breakdown Spectroscopy

Laser-induced breakdown spectroscopy (LIBS) is widely used in aluminum recycling for alloy detection. While other sensors used in recycling processes are line or area sensors, LIBS measures only one point at the same time. LIBS is an automatic emission spectroscopy through plasma ignition (at about 10,000–20,000°C) with a high-energy laser pulse as an excitation source. The constituent elements and molecules of the target material are decomposed, and the electrons are lifted to a higher energetic level from their ground state. During the plasma cooling process, the emitted electrons revert to their original energy state and emit characteristic optical radiation.

In practice, the laser energy concentration is increased by a focusing lens to form a plasma on the sample surface; see Figure 9.8. The plasma emission is collected with a collection lens and transferred with fiber optics to a spectrometer to disperse light and then measured with a detector. The detector usually consists of color sensors with a spectral range of 200–800 nm. The results of LIBS are shown as spectra containing peaks at certain wavelength areas. Depending on the spectral range and resolution of the detector, specific elements can be determined. A quantitative elemental composition determination can be achieved with an element-specific calibration, as the detected peak intensities reflect the quantitative composition.

As all elements emit light during plasma, LIBS can be used to determine the elemental composition of any object, regardless of its physical state (solid, liquid, or gas). Besides, as plasma is generated, no or only marginal sample preparation is necessary for most applications. In specific applications, coordinate-specific laser shots can clean the surface and remove impurities before applying LIBS. However, as LIBS forms plasma and is thus a destructive process, it damages the surface of samples. To avoid damage to the conveyer belt, LIBS is often combined with other sensors, e.g., 3DLT, to determine the object's position. As LIBS measures only characteristics of one specific point of an object, determining the elemental composition of inhomogeneous objects can be problematic.

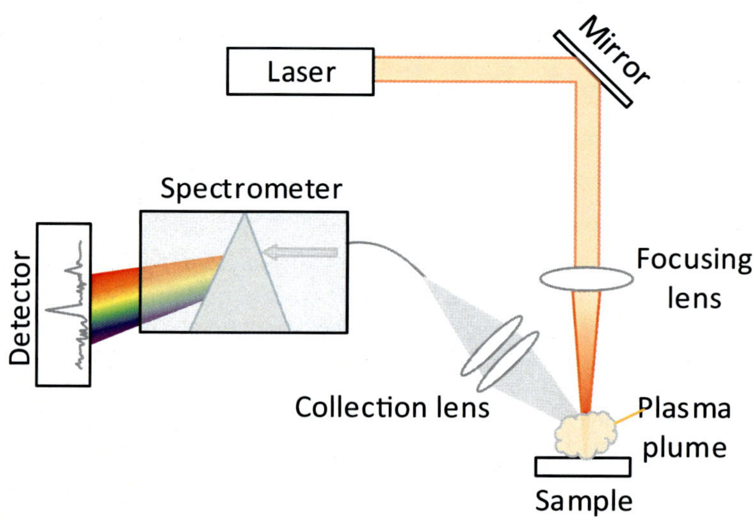

FIGURE 9.8 Setup for LIBS capturing and analysis.

9.4.5 X-Ray Sensors

X-ray covers the wavelength range of 0.001–10 nm. In waste-sorting applications, X-ray is mainly used in two different ways: fluorescence and transmission. With both technologies, the X-ray radiation is generated artificially, e.g., by an X-ray tube. As X-rays ionize matter and may damage the DNA of tissue, its illumination area must be protected.

9.4.5.1 X-Ray Fluorescence

X-ray fluorescence (XRF) analysis is based on observing secondary fluorescence caused by the excitation of atoms and ions using X-ray radiation. The sufficient energy of X-ray dislodges the electrons from the atom's inner shell, making the atom unstable. Electrons from the outer shell replace the released electrons, during which characteristic energy (fluorescence) is emitted (Figure 9.9). This energy corresponds to the binding energy of the inner and outer shell, which is characteristic for a specific element and thus enables a determination of the element type.

The fluorescence is captured through wavelength or energy dispersion. While *wavelength dispersive XRF* requires collimators and a crystal to parallelize and scatter radiation and uses a scintillation counter as a detector, *energy dispersive XRF* requires no specific optics and uses a semiconductor detector to differentiate between different wavelengths. This leads to a much higher spectral resolution of wavelength dispersive XRF and a much shorter measuring time of energy dispersive XRF.

XRF analysis is suitable for detecting heavy elements, for example, in sorting precious metals, metal scrap, and lead glass. Because of the high requirement on measuring time in sorting applications, only a limited number of elements can be detected instead of the complete composition of each particle.

9.4.5.2 X-Ray Transmission

X-ray transmission (XRT) sorting is based on the characteristic of X-ray that it can penetrate objects and thereby is attenuated. The attenuation of X-ray correlates with the material type (atomic densities) and the material thickness.

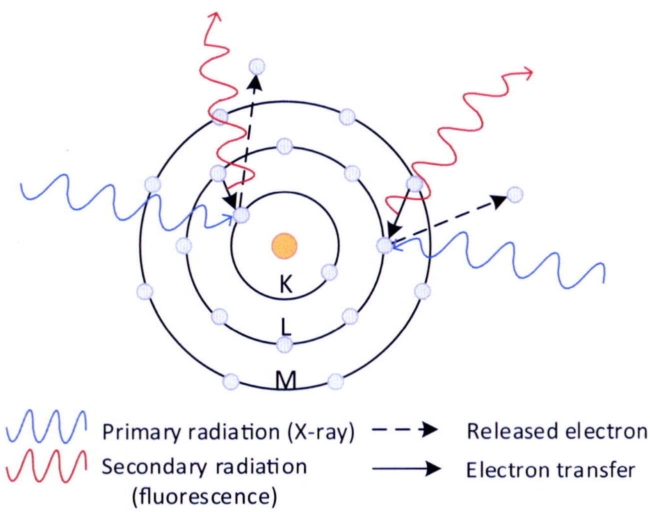

FIGURE 9.9 Working principle of X-ray fluorescence.

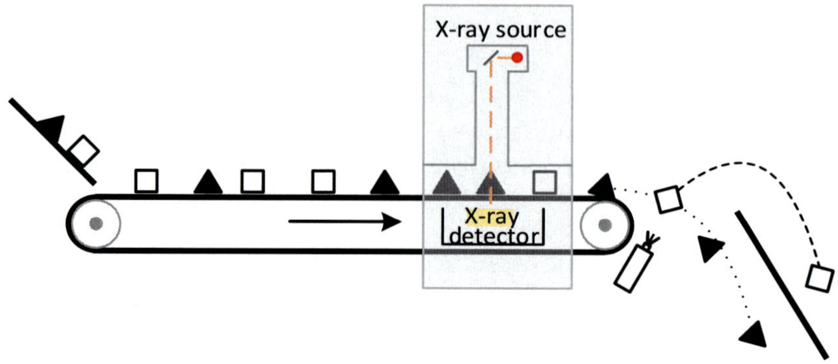

FIGURE 9.10 XRT-based sorter construction. *Gray area*, encapsulated area.

By capturing the remaining transmitted radiation with an X-ray detector, the characteristic intensities make the material determination possible. As the transmission is measured, the X-ray detector and X-ray generating source are placed on different sides of the measured objects and the conveyer belt (Figure 9.10).

As both thickness and the absorption coefficient influence the absorption/transmission intensities of X-ray, interpretation problems may arise in the case that objects with different thickness and different atomic densities are measured and analyzed. For example, a thicker object with low atomic density and a thinner object with high density may have similar intensities, which makes a determination of the material type impossible. To address this problem, dual-energy XRT with two single energy detectors is applied through two simultaneous measurements in different wavelengths: a filter is applied to ensure that only high-energy radiation can reach the second detector, through which the two detectors work at different energy levels. Combining intensity data from both detectors makes eliminating the influence of object thickness possible: compared to reference materials, objects can be classified as elements in high density and low density.

In sorting applications, materials with different atomic densities are principally sortable with XRT technology, and it is widely applied in battery and metal recycling, for example, sorting copper-containing particles from shredder scrap. However, the measurable particle thickness is limited by radiation source power, and compared to other light sources, X-ray technology requires great effort for shielding and radiation protection.

9.4.6 Induction Sensors

In sorting plants, metallic objects are often separated from other materials with conventional processes such as magnetic separators and eddy-current separators (Figure 9.1). However, some metallic components, e.g., stainless steel and composite materials, cannot be sorted with these processes. In this case, an induction sensor based on the measurement of material electrical conductivity is applied. One sensor generates an electromagnetic field at the active surface by an oscillator, and the approach of conducting materials to the active area causes a change in current in the oscillator. By detecting the current change, conducting materials can be detected.

Unlike other sensors, no radiation sources are necessary for induction sensors. To detect conducting materials over the entire width, a sensor bar consisting of several small sensors emitting electromagnetic waves is positioned under the conveyer belt or chute. The position of sensors corresponds to the position of measured objects, which leads to limited spatial resolution (>10 mm) due to the minimum sensor size. In addition, the signal intensity depends on the object's position on the belt or chute.

Induction sensors are mainly used to recover residual metals, as all kinds of conducting materials can be detected even if they are relatively small. For example, all kinds of metals can be detected for particle size greater than 1 mm. Induction sensors are applied in sorting different kinds of wastes, for example, shredder scrap and electronic scrap.

9.4.7 Other Sensors

Other sensors that can be used in sorting applications are, for example, mid-infrared (MIR) spectroscopy, radar/lidar, Raman spectroscopy, laser-induced fluorescence, and acoustic sensors. However, these technologies are under investigation and not widely used in existing sorting or recycling plants yet, or mainly used in the mineral industry. MIR spectroscopy, for example, can be applied in plastic classification and overcomes NIR's limitations in black material classification and additive detection in plastics (Becker et al., 2017; Signoret et al., 2020) but shows lower spectral resolution and currently has higher cost. Radar/lidar is increasingly used in research for volume determination to overcome the limitations of 3DLT, e.g., the limited measurement of transparent objects.

9.5 APPLICATION OF DIFFERENT SENSORS IN RECYCLING

Due to the different measuring principles, different sensors are used for different applications, and in the same plant, several sensors are applied for different purposes. For example, in LWP sorting plants, NIR sensors are used for sorting according to plastics type, and RGB sensors are applied in color sorting (e.g., sorting PET into blue, green, and transparent; see Figure 9.1). Table 9.1 shows different sensor characteristics and application examples of each sensor in the recycling industry.

TABLE 9.1 Characteristic and application examples of sensors in recycling.

Sensor	Radiation wavelength (nm)	Active materials	Measured characteristic	Application in recycling
RGB/ VIS-HSI	380–780	All visible materials	Color, 2D shape	Plastic, glass, paper, ceramics, mineral
NIR	780–2500	E.g., plastic, beverage cartons, wood, paper, minerals	Material type	LWP, paper, C&D waste
3DLT	- (Laser)	All nontransparent materials	Particle 3D shape, volume	Volume calculation; position determination (e.g., with LIBS)
LIBS	- (Laser)	Principally all elements	Material type	Aluminum, metal alloys
XRF	0.001–10	Heavy elements	Material type	Metal, lead glass, aluminum, zinc
XRT	0.001–10	Materials with a difference in atomic density	Atomic density	Battery, aluminum, wood, compost
Induction	- (electromagnetic field)	Metals	Electrical conductivity	Shredder scrap, electronic scrap

Recent research has broadened possible applications for different sensors by extracting specific characteristics. For example, in plastic recycling, XRT can detect flame retardant in plastics (Vrancken et al., 2017), and LIBS, in addition to MIR, can detect and classify black plastics (Huber et al., 2014). Also, NIR can be applied in determining calorific value, moisture, and chlorine content in refuse-derived fuel (Vrancken et al., 2017).

9.6 RECENT DEVELOPMENTS

In the last several years, many investigations in SBS and recycling have been conducted, through which the detection and classification accuracy, the sorting efficiency, and yield were increased (Gundupalli et al., 2017). The sensor resolution, frame-rate, and sensitivity were improved, which led to a more precise detection in both spatial and spectral direction.

The separation unit was upgraded with robotic arms in specific applications, which pick up valuable materials through grabbing or pneumatic force (Sarc et al., 2019). Compared to air nozzles, robotic arms enable the separation of multiple target fractions in one sorting step and result in fewer false-negative discharges. However, they are characterized with a lower throughput as the robotics arms can only sort one object after another. Thus, they currently achieve much lower throughputs and yields compared to air-nozzle-based SBS.

To detect and extract the material characteristics more precisely, multisensors are applied in specific applications. For example, in sorting C&D waste with robotic arms, color sensors, 3DLT, and NIR sensors are applied. Through sensor fusion, a product fraction in right color and right material type with high purity can be generated.

In data analysis, machine learning models were improved both in prediction accuracy and computational time. More and more machine learning models were tested and applied to reach optimal sorting results. Recently, deep learning techniques, e.g., convolutional neural networks, were applied to sort materials into more detailed classes (Gruber et al., 2019; Yu et al., 2020). As a result, PE silicon cartridges, for example, can now be separated from PE product fractions based on form and texture.

9.7 OUTLOOK

With these improvements, more and more characteristics of material flows can be extracted. Besides applying sensor technology in recycling processes for sorting, applications of sensor-based material flow characterization (SBMC) are an upcoming research topic (Kroell et al., 2021).

In SBMC, the goal is to characterize material flows based on acquired sensor data. Based on the extracted material flow characteristics, new applications and optimization potentials in sorting and recycling plants are envisioned. Possible new applications include real-time characterization of input material flows, sensor-based monitoring and early detection of process disturbances, predictive maintenance, real-time product quality monitoring, implementing adaptive process control algorithms, and obtaining a better process understanding (Kroell et al., 2022).

For example, through volume determination and a comparison to the capacity of different aggregates, the feeding of the plant or specific machines can be adjusted. Without adding new sensors to the sorting plant, the data of existing sensors could be used for SBMC, and machine parameters could be adapted online.

Sensor technologies have provided possibilities in sorting and recycling different waste material flows through rapid detection and classification and accurate separation of product fractions. Today, sensor technologies face challenges such as more complex material compositions, higher requirements on production fraction purities, and higher recycling rates to enable enhanced material recirculation in the context of a circular economy. Improvements in sensors, peripherals, and data analysis are further investigated to meet increasing requirements and demands. Sensor technologies still offer considerable potential in contributing to performance increases of sorting plants, recycling plants, and the value chain through SBMC in upcoming years.

References

Becker, W., Sachsenheimer, K., Klemenz, M., 2017. Detection of black plastics in the middle infrared spectrum (MIR) using photon up-conversion technique for polymer recycling purposes. Polymers 9. https://doi.org/10.3390/polym9090435.

Feil, A., Kroell, N., Pretz, T., Greiff, K., 2021. Anforderungen an eine effiziente technologische Behandlung von Post-Consumer Verpackungsmaterialien in Sortieranlagen (Requirements for an efficient technological treatment of post-consumer packaging materials in sorting plants). Müll und Abfall. https://doi.org/10.37307/j.1863-9763.2021.07.04.

Gruber, F., Grählert, W., Wollmann, P., Kaskel, S., 2019. Classification of black plastics waste using fluorescence imaging and machine learning. Recycling 4, 40. https://doi.org/10.3390/recycling4040040.

Gundupalli, S.P., Hait, S., Thakur, A., 2017. A review on automated sorting of source-separated municipal solid waste for recycling. Waste Manage., 56–74. https://doi.org/10.1016/j.wasman.2016.09.015.

Huber, N., Eschlböck-Fuchs, S., Scherndl, H., Freimund, A., Heitz, J., Pedarnig, J.D., 2014. In-line measurements of chlorine containing polymers in an industrial waste sorting plant by laser-induced breakdown spectroscopy. Appl. Surf. Sci. 302, 280–285. https://doi.org/10.1016/j.apsusc.2013.10.070.

Kroell, N., Chen, X., Maghmoumi, A., Koenig, M., Feil, A., Greiff, K., 2021. Sensor-based particle mass prediction of lightweight packaging waste using machine learning algorithms. Waste Manage. 136, 253–265. https://doi.org/10.1016/j.wasman.2021.10.017.

Kroell, N., Chen, X., Greiff, K., Feil, A., 2022. Optical sensors and machine learning algorithms in sensor-based material flow characterization for mechanical recycling processes: a systematic literature review. Waste Manage. 149, 259–290. https://doi.org/10.1016/j.wasman.2022.05.015.

Pretz, T., Raulf, K., Quicker, P., 2020. Waste, 4. Recycling. In: Ley, C. (Ed.), Ullmann's Encyclopedia of Industrial Chemistry. Wiley, Chichester, pp. 1–39.

Sarc, R., Curtis, A., Kandlbauer, L., Khodier, K., Lorber, K.E., Pomberger, R., 2019. Digitalisation and intelligent robotics in value chain of circular economy oriented waste management – a review. Waste Manage. 95, 476–492. https://doi.org/10.1016/j.wasman.2019.06.035.

Signoret, C., Caro-Bretelle, A.-S., Lopez-Cuesta, J.-M., Ienny, P., Perrin, D., 2020. Alterations of plastics spectra in MIR and the potential impacts on identification towards recycling. Resour. Conserv. Recycl. 161, 104980. https://doi.org/10.1016/j.resconrec.2020.104980.

Vrancken, C., Longhurst, P.J., Wagland, S.T., 2017. Critical review of real-time methods for solid waste characterisation: informing material recovery and fuel production. Waste Manage. 61, 40–57. https://doi.org/10.1016/j.wasman.2017.01.019.

Yu, Y., Zou, S., Yin, K., 2020. A novel detection fusion network for solid waste sorting. Int. J. Adv. Robot. Syst. 17. https://doi.org/10.1177/1729881420941779. 172988142094177.

CHAPTER

10

Mixed bulky waste

Sandor Karreman

Renewi, Nieuwegein, The Netherlands

10.1 INTRODUCTION

Large municipal solid waste (LMSW) originates from households. This stream is generally rarely collected door to door because it is too large or too heavy to collect as normal household waste. This chapter discusses how mixed bulky waste can be separated and recycled into reusable materials, taking the Netherlands as a case study.

In the Netherlands, LMSW is collected and treated separately from regular household waste. Table 10.1 shows which waste streams are part of LMSW, along with an estimate of the processing method.

Approximately 25% of LMSW consists of mixed bulky waste in the Netherlands. Mixed bulky waste streams often contain products with different materials or sizes (e.g., sofas, furniture, mattresses, or skis). Mixed bulky waste poses many challenges for the sorting and recycling processes. Usually, households generate mixed bulky waste during sporadic events, e.g., attic or basement cleaning, forcing waste collectors to take it along with the regular household waste. Mixed bulky waste also has a fire risk due to its composition or its collection process (e.g., it may include small gas tanks, batteries, or waste that can overheat or ignite).

There are two main systems for collecting mixed bulky waste in the Netherlands:

- The first system asks households to separate and deliver the waste themselves to waste depots. All waste that is not, or cannot be, separated is collected as a generic waste stream.
- The second system asks municipal waste services to collect mixed bulky waste at the source, i.e., the household. The mixed bulky waste is collected in bulk and then sorted at sorting centers or municipal waste depots. Waste collected at the source has a better quality than the waste delivered to waste depots due to better separation.

10.2 THE CIRCULAR PROCESS FOR MIXED BULKY WASTE

Discarded products should ideally be reusable for the same purpose. Mixed bulky waste can achieve this status, but only if an extra step in the value chain is added. In the Netherlands, this extra step partially exists through second-hand stores selling reusable discarded products to their customers, or recycling agencies offering a separate collection system to municipalities, or through nongovernmental organizations.

TABLE 10.1 Composition and volume of large municipal solid waste collected in the Netherlands in 2018.

Waste fractions	Volume (kt)	Processing method
Mixed bulky waste	525	60% Energy recovery 40% Recycling
Carpets and floor covering	9	Energy recovery
Electronics waste	81	Recycling
Bulky garden wastes	492	Recycling
Furniture	47	Recycling
Glass (windows)	13	Recycling
Metals	76	Recycling
Wood	425	Energy recovery
Clean debris	440	Recycling
Asbestos-containing wastes	12	Landfill
Tires	4	Recycling
Clean soil/dirt	93	Recycling
Bituminous roofing	12	Energy recovery
Impregnated wood	48	Energy recovery
Other plastics	27	Recycling
Others	1	Energy recovery
Total LMSW	2305	

Data from Statistics Netherlands—Statline (2021).

Nevertheless, a large share of the discarded products cannot be reused any longer once they suffer critical damage, or due to wear or tear. Disassembling these products into smaller components for reuse or recycling is the next step in their value chain to lower their CO_2 footprint. Although initiatives exist, they are mostly confined to government-supported initiatives. The relatively high labor cost results in a lack of economically viable dismantling processes in the Netherlands.

If reusing the product components is impossible, sorting and separating the reusable fractions can lower the CO_2 emissions during their end-of-life cycle. The equipment quality of the municipal recycling centers has increased in recent years, allowing a reduction of unprocessed mixed bulky waste from ~780 kton in 2000 to ~505 kton in 2018, i.e., a decrease of 35% (Figure 10.1).

10.3 CONDITIONS FOR ECONOMICALLY VIABLE SORTING

To sort mixed bulky waste with economically viable solutions, the following base conditions must be met:

- Creating a separate collection system.
- Having appropriate sorting technologies, installations, and knowledge.

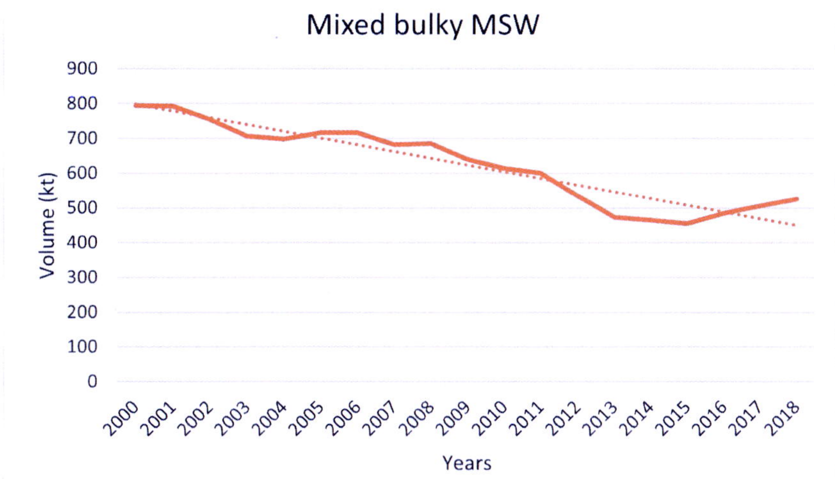

FIGURE 10.1 Amount of mixed bulky waste over time in the Netherlands. *Data from Statistics Netherlands—Statline (2021).*

- The revenues from collecting and sorting must outweigh the costs.
- The government must stimulate sorting of mixed bulky waste. This is especially important in a situation where other less sustainable treatment options, e.g., incineration or landfilling, are more economically appealing to disposers.

10.4 SORTING OF MIXED BULKY WASTE

Renewi, an international waste-to-product company, is highly experienced in sorting and processing mixed bulky waste. Its state-of-the-art separation technologies will serve as a case study for this chapter.

They consist of:

- a sorting crane to remove parts of the waste stream that cannot be properly reduced with the existing shredder (e.g., rolls of carpet, iron bed spiral);
- a shredder to shred the presorted waste into smaller pieces;
- a sorting line to sort the shredded waste with different technologies based on mechanical (e.g., density or weight) and optical properties.

Renewi has three sorting installations (Amsterdam, Nieuwegein, and Wateringen) capable of processing mixed bulky waste. Each of the lines began sorting waste in early 2011 and have been modernized several times ever since to improve the quality of the sorting process. The latest additions to the sorting line in Nieuwegein are two optical separators used to separate wood that can be used to produce chipboard from a mixed wood stream (e.g., MDF, chipboard, and wood) that is used as fuel. A scheme of the sorting installation and process in Nieuwegein is shown in Figure 10.2. An impression of the sorting line is given in Figure 10.3. The sorting line covers an area of approximately 75 by 40 m.

Nieuwegein's sorting installation uses several standard sorting techniques like sieves, magnets, wind shifters, and optical separators to sort and separate waste. These techniques vary in sorting by size (small/large), weight

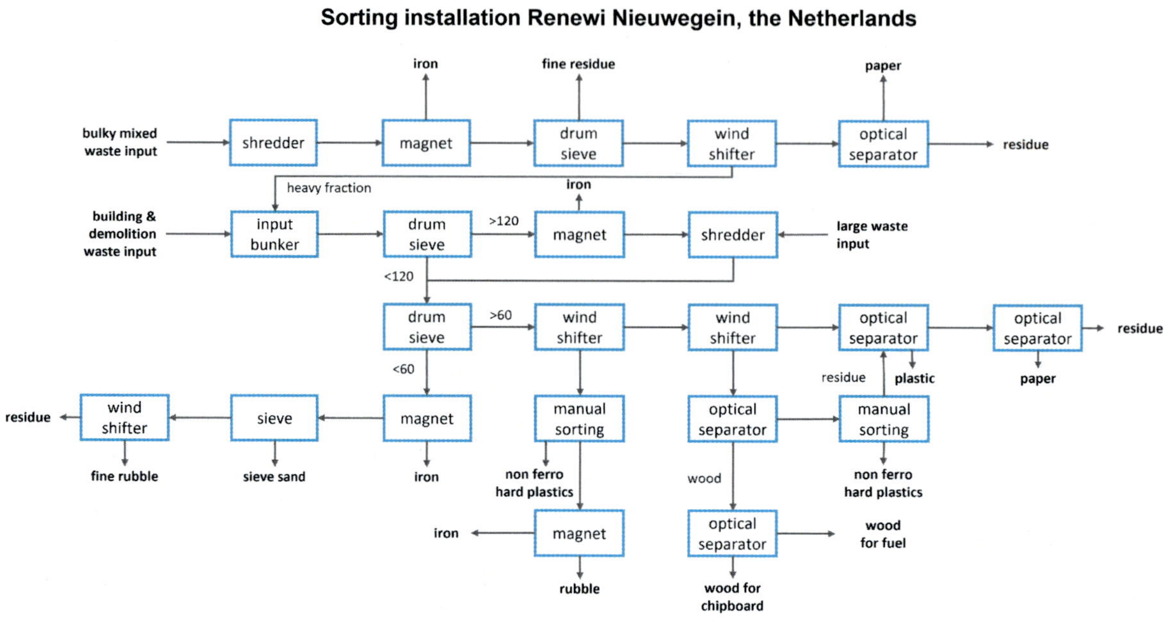

FIGURE 10.2 Sorting installation Renewi Nieuwegein.

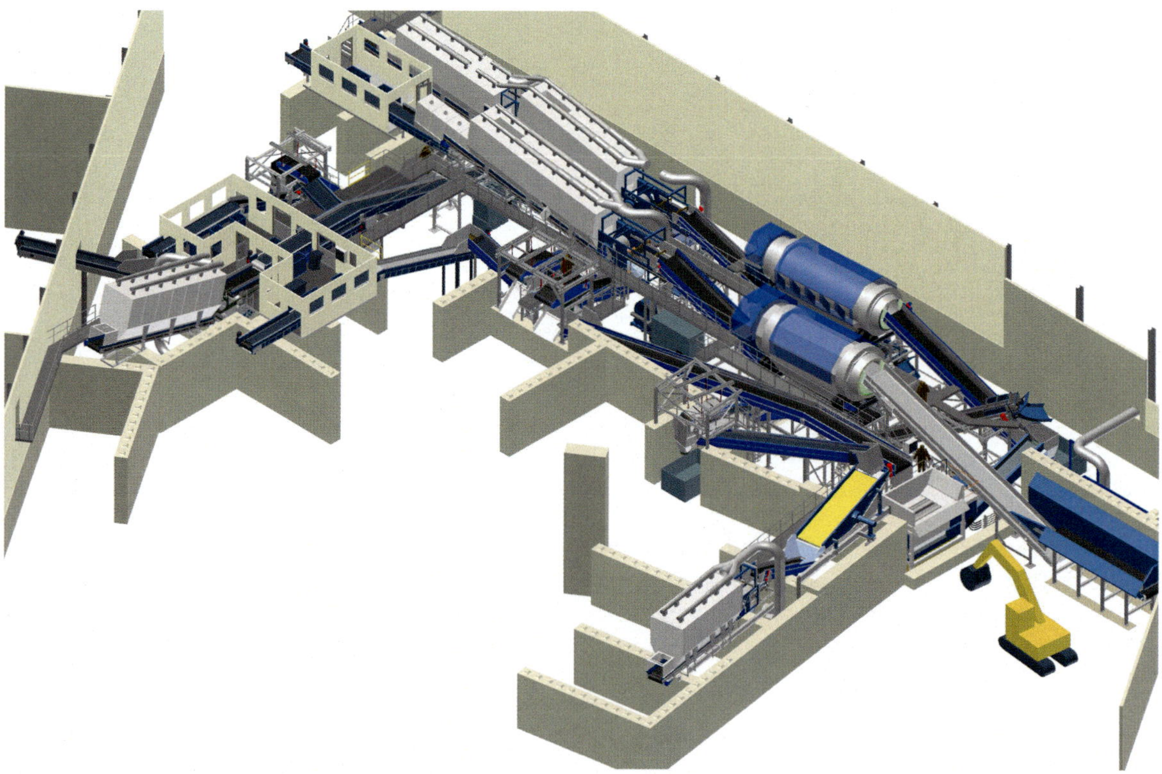

FIGURE 10.3 An impression of the sorting line of Renewi Nieuwegein.

(light/heavy), or composition. The order in which these are used is key for the effectiveness and efficiency of the sorting process. After years of experimentation, Renewi's sorting installations can now handle 25–30 t/h of mixed waste per hour and more than 40 t/h of building and construction waste.

10.5 SORTING PROCESS

The sorting process consists of three main steps that are described in the following paragraphs.

10.5.1 Large Fraction Separation

At the start of the mixed bulky waste sorting, a crane separates large fractions like mattresses, bed spirals, and large chunks of carpet that cannot fit in the sorting line. Clean mattresses go to a company called Retourmatras, a company specializing in recycling waste mattresses. If the mattresses are too dirty to be recycled, they are shredded for incineration.

10.5.2 Light Fraction Processing

At this stage, the presorted bulky waste (Figure 10.2) is fed into a shredder to further reduce its size. A magnet then removes iron from the reduced bulky waste, to be sent next to a drum sieve (Figure 10.4). The sieve separates the fine fraction to deliver it to incineration plants. The other fraction ends up in a wind sifter to separate the light fraction from the heavy fraction. An optical separator ultimately extracts paper (Figure 10.5C) from the light fraction and sends the remaining part of the light fraction to incineration plants (Figure 10.5D). This phase processes about 60% of the input.

10.5.3 Heavy Fraction Processing

The remaining heavy fraction is mixed with building and demolition waste. This creates a mixture that is more constant in composition. The result of this is that the sorting yields are better than if bulky waste were processed separately. This blend is first fed into a drum sieve that shakes it to break it apart. The smaller fraction (<120 mm) falls through the sieve holes and is sent to a second sieve. The larger fraction (>120 mm) leaves the sieve where a magnet removes the iron. This is the first mono stream to leave the process. The remaining larger fraction is crushed and fed into the second sieve. This sieve has smaller holes to let the smaller fraction (<60 mm) fall out and pass through:

- a magnet to sort and remove metals,
- a sieve to sort and remove sand,
- a wind shifter to sort and remove small rubble.

The remaining residue is transported to a waste incinerator to be converted into energy.

The larger fraction (>60 mm) in the second sieve leaves the sieve and passes through two wind shifters that blow out the lighter fractions, i.e., a mix of materials like plastics. The resulting fraction is then fed into two consecutive optical dividers to separate plastics and paper. The remaining residue is transported to either our energy pellet factory to make refuse derived fuel or an incinerator.

The larger fraction (>60 mm) that is not blown away by the wind shifter enters the sorting cabin. A manual sorting takes place to separate and recover recyclable fractions (e.g., plastics and nonferrous materials). The remaining fraction then passes under magnets to recover any remaining iron, resulting in a wood and rubble only fraction. The wood fraction passes a double optical divider installation to separate the wood into a so-called a-wood fraction and a b-wood fraction. A-wood is used for the production of chipboard (Figure 10.5A). B-wood is used for the production of chipboard or as biomass for energy production (Figure 10.5B).

During the process, waste may fall out, e.g., because it is too heavy to be separated by the wind shifter. In such a case, the waste is fed back

FIGURE 10.4 Mixed bulky waste.

into the sorting line and processed again, to increase the recycling efficiency.

10.6 RECYCLING EFFICIENCY

For all waste flows that fall under the LMSW category (Table 10.1) the recycling efficiency is about 64%, 35% goes to energy recovery, and the remaining 1% ends up in landfill. These percentages strongly depend on how the mixed bulky waste is collected and how citizens offer their LMSW to available collection systems.

Processing the mixed bulky waste fraction of LMSW leads to the following sorting results within Renewi's sorting installations (Table 10.2).

10.7 CONCLUSION AND OUTLOOK

Recycling will play a key role in advancing the circular economy. Currently, more and more technologies can sort mixed bulky waste, and Renewi sorting installations prove that this is an economic reality. Future developments are necessary to continue improving the percentage

FIGURE 10.5 Output fractions from sorting mixed bulky waste. (A) a-wood fraction, (B) b-wood fraction, (C) paper, (D) residue fraction.

TABLE 10.2 Recycling efficiency mixed bulky waste.

Fraction	Share of mixed bulky waste (%)	Destination
Clean debris	5	Recycling
Wood	22	Recycling
Impregnated wood	0.5	Energy recovery
Metals	4	Recycling
Paper and cardboard	2	Recycling
Plastic foils	1	Recycling
Other plastics	2	Recycling
Other	3	Recycling
Residue	59.5	Energy recovery
To recycling	40	
To energy recovery	60	

of secondary raw materials reinputted into the market.

These improvements include:

- better use of existing sorting techniques,
- developing new sorting techniques to further improve the efficiency of the sorting process,
- a product design that improves its end-of-life reuse,
- legislation requiring the use of recycled raw materials in the production process.

To make this practically and economically feasible, better cooperation between all stakeholders is critical.

Reference

Statistics Netherlands, 2021. https://opendata.cbs.nl/statline/.

CHAPTER

11

Packaging

Ernst Worrell

Copernicus Institute of Sustainable Development, Utrecht University, Utrecht, The Netherlands

11.1 INTRODUCTION

To manage the environmental impact and increasing scarcity of resources, there is a strong need to change the way society produces and consumes materials. Reducing primary material production and consumption will not only save resources, but also the energy and resources required for production and waste processing. In our present society, the most important material flows consist of material chains in the build environment, transportation, and packaging. Packaging materials have a relatively high environmental impact. In the European Union, packaging waste has grown steadily to 78 Mt in 2018, equivalent to 174 kg/capita. The European packaging waste consisted of 40% paper and board, 19% plastic, 19% glass, 16% wood, and 5% metals. Around 35% of the municipal solid waste in Western Europe can be ascribed to packaging materials, and at least 33% of all solid waste in the United States. As income levels increase and lifestyles change, packaging waste volumes are likely to increase in transitioning and developing countries as well. For example, 2018 annual packaging waste generation in Bulgaria was estimated at around 70 kg/capita in 2018 compared to 43 kg/capita in 2010. In contrast, in Germany 228 kg/capita compared to 196 kg/capita was generated. Similarly, the composition of waste will vary between countries. As society is becoming more aware of the problems associated with packaging, there is an increasing focus on sustainability in the development of future packaging (Bruijnes et al., 2020) next to the other functions that packaging supplies (e.g., food safety, communication).

In this chapter, the developments in packaging waste are discussed, focusing on the volume generated as well as the composition. This will be followed by a discussion of different options to collect and recover packaging waste. As there is a wide variety of collection schemes, we will discuss typical characteristics and a few examples. The chapter finishes with some general conclusions.

11.2 PACKAGING WASTE

The increasing use of packaging and the subsequent increasing generation of packaging waste can be illustrated based on developments in the European Union, as Eurostat reports packaging waste figures since 2000. Figure 11.1 depicts the total volume, showing a steady growth of the volume, interrupted only by the financial crisis of 2009. Packaging use varies

Handbook of Recycling. **https://doi.org/10.1016/B978-0-323-85514-3.00032-4**

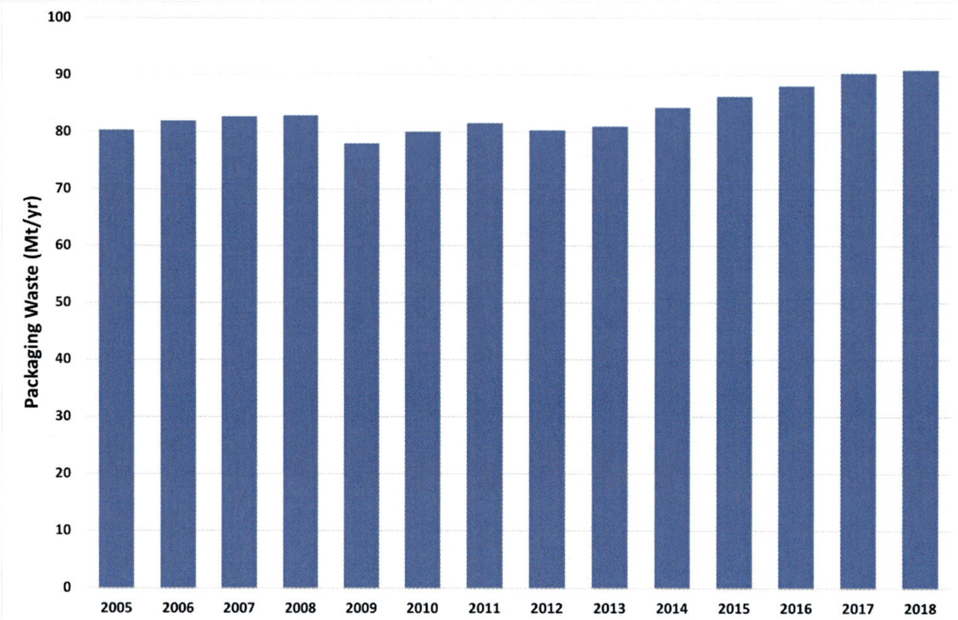

FIGURE 11.1 Development of packaging waste in the European Union (EU-28, including UK and Norway) from 2005 until 2018 (expressed in kt/year). *Based on data from Eurostat.*

by country depending on various factors, e.g., population, economic development, distribution of household sizes, and national traditions in packaging. Policies will affect the total volume of packaging waste. A large variety of policy instruments have been introduced in many places to manage packaging waste, varying from voluntary approaches to taxation and bans of specific packaging applications or materials. An example of the latter is the ban on single-use plastic bags in many countries, as these contributed to littering and subsequent environmental and public health problems. This particular example is a good illustration of the trade-offs faced, as alternative single-use bags may lead to other environmental impacts (López Gómez and Serna Escobar, 2022), while it also shows the limits of assessment methods as the environmental impact of littering are not (yet) included in life cycle assessment (LCA), the primary assessment tool used. Generally (and as this example shows), the effectiveness of policy may vary between instruments and over time (see, e.g., Rouw and Worrell, 2011; Dikgang et al., 2012), depending on a range of context-specific factors (e.g., culture, demographics, wider economic and policy environment).

To understand the underlying trends in packaging material use and waste generation, the concept of the Environmental Kuznets Curve (EKC) can be used. The EKC concept has been applied to several material studies that examined the relationship between material use and economic development. The studies show that there is no linear growth of the intensity of material use over time and outline the factors that influence material consumption development over time. Intensity of use is expressed as the demand for material in kilograms per unit of income (measured as gross domestic product, or GDP) (see Eq. 11.1).

$$IU = X_i/GDP \tag{11.1}$$

where IU, intensity of material use; X_i, the consumption of a specific material during year i; GDP, gross domestic product.

The EKC concept argues that material intensity first increases and starts decreasing after a certain level of development has been reached, while income may continue to grow, resulting in a reduced material intensity. This process is called dematerialization or decoupling. Dematerialization refers to the "absolute or relative reduction in the quantity of materials used and/or the quantity of waste generated in the production of a unit of economic output." Decoupling means that economic growth and material use is unlocked. The observed trend of the dematerialization process may be "distorted" by material substitution. Due to (technological) innovations and changing demands, materials replace one another over time. In structural applications, for example, the use of wood is successively replaced by iron, steel, aluminum, and plastic. Although the applied EKC depicts the intensity of material use (kg/GDP) over time, it does not necessarily explain the development of absolute material use over time. In fact, when the intensity of use declines, absolute material use can still increase, albeit to a smaller extent than GDP growth. For this reason, De Bruyn and Opschoor (1997) distinguish "weak dematerialization," which only implies a decline in the intensity of use, and "strong dematerialization,' which also includes a decline in absolute material use.

Income shows a strong correlation with packaging consumption in Europe, but as Figures 11.2 and 11.3 depict, population is statistically a slightly more robust variable, given the comparable income levels of the original EU-15 member states. The relationship between population and packaging consumption is linear for the EU-15 member states over the studied period. The relationship between GDP and packaging consumption shows leveling off over time, or in other words demonstrates a certain

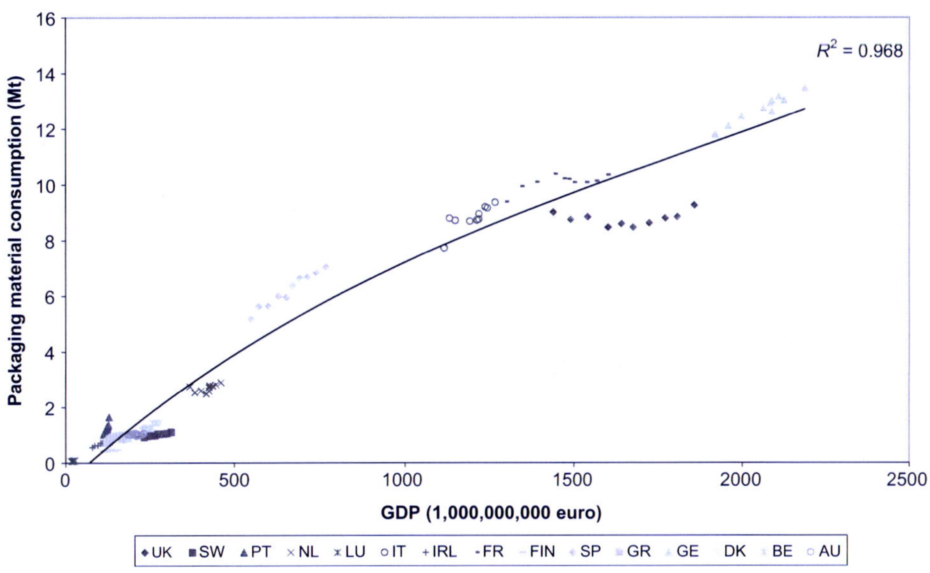

FIGURE 11.2 The relationship of packaging material consumption and national income for EU-15 countries in the period 1997–2006. *From Rouw and Worrell (2011).*

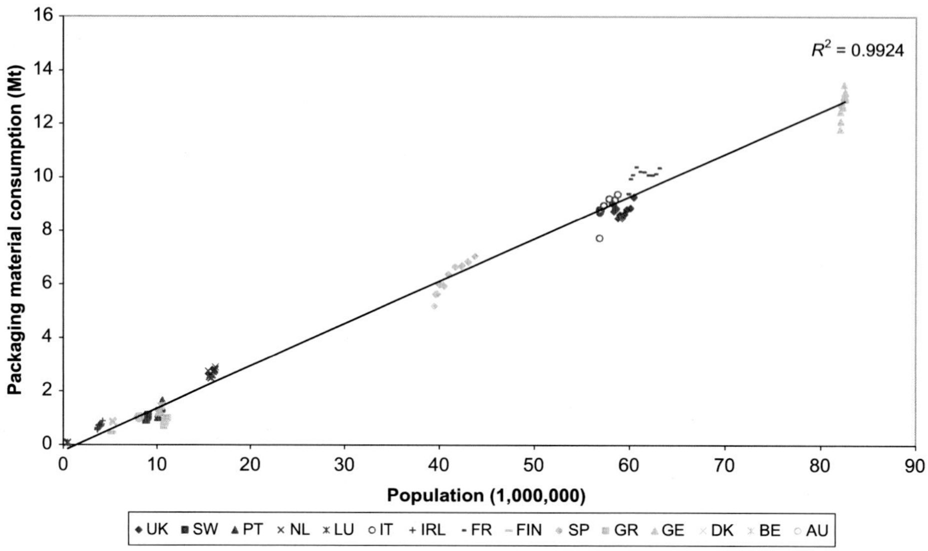

FIGURE 11.3 The relationship of packaging material consumption and population for EU-15 countries in the period 1997–2006. *From Rouw and Worrell (2011).*

degree of decoupling. This is most likely classified as "weak dematerialization," as the total volume of packaging material is still increasing, even in the EU-15 countries. Other analyzed variables, such as GDP/capita, households, number of single households as well as distribution of household sizes, show no statistically significant relationship with packaging consumption or packaging consumption per capita. This means that, given available data from the EU, there is no indication that single households consume more packaging materials per person than households with multiple people.

Despite the continuous growth in the use of material for packaging, large reductions are potentially feasible in the volume of packaging material (see, e.g., Hekkert et al., 2000; Wever, 2009). These may even result in further cost savings along the total supply chain (Wever, 2009). However, typically only relatively minor opportunities for material reduction are realized due to a variety of factors (van Sluisveld & Worrell, 2013).

11.3 COMPOSITION

As consumption patterns change over time, the composition of packaging waste will also change. In the past decades, plastics have taken an increasingly larger share of the packaging market, due to its versatility. In industrialized countries over the past years, these shares show only minor changes (Table 11.1), as various trends, e.g., e-retail, as well as policies affect the choice of packaging material. The rise of plastics is because they are relatively easy to shape and can be more or less tailored to various packaging functions using barrier layers, composites, and additives. For example, more and more liquids are packed in plastic PET and HDPE bottles. In industrialized countries and developing countries alike, plastics are forming an increasing part of packaging waste, despite its low weight. Moreover, national traditions in packaging and technology changes will affect the material composition of packaging and packaging waste. An analysis of the use of

TABLE 11.1 Approximate composition of packaging waste in the United States and the European Union. Note that differences in the definition of packaging waste (e.g., inclusion of transport packaging like pallets) may affect the distribution.

	United States		European Union	
Material/Year	**2010**	**2018**	**2010**	**2018**
Wood	12%	14%	15%	16%
Paper & Board	50%	51%	40%	41%
Plastics	18%	18%	19%	19%
Glass	12%	12%	20%	19%
Metals	6%	5%	6%	5%

Data from US EPA and Eurostat.

beverage containers in the Netherlands over time by Chappin et al. (2005) demonstrates the impact of changes in packaging on the material choice, while statistical data on packaging waste composition collected in Europe demonstrate the overall changes in the volume and composition of total packaging waste. Traditions in packaging affect the composition as well. In contrast to most other countries, beverage cans in the Netherlands were until recently mainly made of steel and not aluminum. This also explains why there is yet no separate collection system for aluminum cans in this country, which is found in many other countries to retrieve the valuable metal. Also, national legislation affects the composition of packaging materials through the ban of certain materials due to environmental or health concerns. Table 11.1 provides a breakdown of the typical composition of packaging waste as found in the EU and the United States. There is little data on the composition of packaging waste in developing countries, despite the growing use of packaging in these countries.

Plastics in packaging consist of multiple types of plastics, which will need to be separated before recycling. The key plastics used in packaging are:

- PET: mainly for bottles, trays, and containers,
- HDPE: e.g., nonfood bottles,
- LDPE/LLDPE: e.g., plastic bags, films,
- PP: e.g., dairy and frozen foods packaging, and
- PS: e.g., drinks, foams.

PVC is still used in packaging, but its share is declining. The specific characteristics of PVC are only needed in niche applications, so the current use of PVC in packaging is generally driven by economics. Other plastics, including biobased plastics, are also used in lesser quantities. Note that even when the main constituent is a particular plastic type, the use of additives may make them different for recycling purposes. For example, PET containers (e.g., for fruit) should not be mixed with bottle-grade PET due to the use of additives in containers. Moreover, color may also necessitate the further separation of materials before recycling, such as with plastics and glass. This has resulted in more and more places in recovery systems that distinguish glass by color.

11.4 RECOVERY AND RECYCLING

Packaging waste is increasingly recovered for recycling. It is important to distinguish recovery and recycling when comparing the current achievements. Recovery is defined as the amount of material retrieved or recovered from the waste stream, both through separate collection and in waste separation plants, to replace (other) materials. Recycling is determined by the actual volume of material that is recycled to replace virgin material. This excludes the volume of material that is of too low quality to be recycled and is converted to, e.g., refuse-derived fuel (RDF), incinerated, or landfilled. The point of measurement is also important in the latter, as it can be measured at the entrance of the recycling facility, or at the exit. As incoming material may not meet quality demands or quality standards, it will be rejected in the sorting and recycling process. Depending on the quality of the incoming material mix, the point of metering may hence considerably affect the reported recycling rate. Definitions may vary from one country to another, making comparisons sometimes difficult. Therefore the European Commission has recently (2020) issued monitoring guidelines to harmonize recycling rates across member states (Eurostat, 2022). Moreover, monitoring of the volume of packaging brought to market is based on estimates (e.g., waste generation), resulting in uncertainties in any recovery or recycling rates reported.

In the European Union, the recycling rate of packaging has increased from about 42% in 1997 to close to 66% in 2018 measured by total weight, with the trend slowing in recent years. The estimated recycling rate in the United States is about 54%. The remainder of the packaging waste is incinerated and landfilled. The European Union is considered one of the leaders in packaging recycling. Note that within the EU there is a large variation between the different member states, and between materials. Historically, metals, paper, and glass have achieved high recycling rates, building on long-term experience in many countries. In many Western European countries, the recovery and recycling rates of paper, glass, and steel are typically in the range of 70–80%, or even higher in selected cases. Aluminum recycling in a few selected countries, and states in the United States, approximates these high recycling rates, but is typically 50% or less in most countries. In contrast, recovery and recycling of plastics is in most countries still very limited due to, e.g., the variety of plastics. The large variety of plastics, the use of additives to shape material properties, and the use of multilayer or composites has made (high-quality) recycling, particularly of plastics, difficult. Hence, the advantage of plastics (i.e., the tailoring of material properties to a specific application) generates a problem for plastics recycling. A few countries have a longer experience with or more effective policies and collection systems for plastics recycling and they typically achieve higher recovery and recycling rates (of around 50%-60%), compared to an average recycling rate of 41% in the EU-27 (all figures are for 2019). Note that only a small part of recycled packaging plastic is recycled into (new) packaging, as the vast majority is used for other purposes. This, so-called open loop recycling, may lead to downcycling of material quality and does not contribute sufficiently to a transition to circular packaging. Policies in several European countries now require a specified degree of recycled content for specific types of packaging, and this has led to increased use of recycled PET in bottles. In the United States the Environmental Protection Agency estimates the plastic recycling rate at only 14% (mainly PET, HDPE, and PP), against 80% for paper and board and 74% for steel (2018). Figure 11.4 depicts the recycling rates reported for the EU-27 and a few selected member states, to illustrate the differences between countries, materials, and over time. It clearly demonstrates the differences between countries for the recycling of different materials due to differences in history, markets, and policy success.

While in many developing countries there are few or no formal recycling schemes for packaging, there is a large informal sector of waste scavengers that collect, reuse, recover, and recycle packaging and packaging materials (Chapter 7—The path to inclusive recycling). The informal waste recovery sector often provides an important part of income for low-income households. The large role of the informal sector makes it difficult to establish formal recovery and recycling for developing countries. Developing countries have taken initiatives to ban certain packaging to reduce the amount of waste in cities, as the lack of waste management leads to excessive littering with consequences for, e.g., public health. For example, a growing number of developing countries now ban free plastic shopping bags to reduce plastic litter. In fact, several developing countries (e.g., Kenya, Rwanda) have been

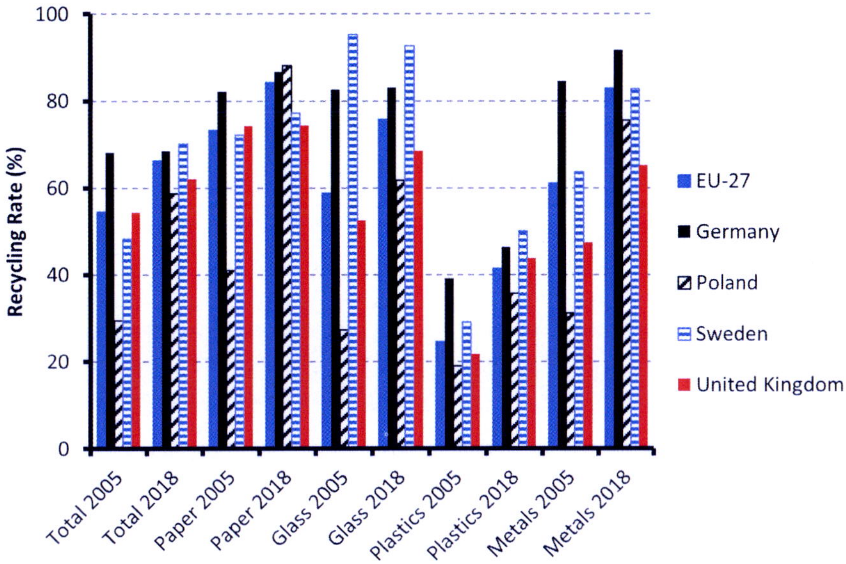

FIGURE 11.4 Recycling rate of packaging materials (total), plastic packaging, paper and board packaging, glass packaging, and metals packaging in the EU-27 and selected member states in 2005 and 2018. The countries are selected to illustrate the differences between countries. *Data from Eurostat.*

global leaders in banning single-use plastic bags, and leapfrog over some of the waste management problems found in industrialized countries, as plastic bags are typically hard to recycle or are downcycled when recycled.

11.5 COLLECTION AND RECOVERY SCHEMES

Due to the large share of packaging waste in MSW, packaging has always received a lot of attention in attempts to reduce waste. Policies and instruments have been introduced at all levels of government, varying from local initiatives by cities and towns, to national governments, and the organizations like the European Union. In contrast, the United States has no federal policy on packaging waste. Consequently, large differences in policies are found throughout the various states. However, while governments have actively pursued options to reduce packaging waste, the packaging industry introduces thousands of new packages a year, of which the majority are produced without a lot of attention paid to the environmental consequences of the packaging in the waste stage (e.g., van Sluisveld and Worrell, 2013). A difficulty in realizing opportunities for waste reduction and material efficiency is the relationship between the packaging and the packed product. If a change in packaging might result in damage to the product, the environmental impact may be higher. However, little empirical research has studied the role of packaging, product damage, and user and consumer behavior in depth.

To decrease packaging waste, many governments in industrialized countries have reverted to extended producer responsibility (EPR) schemes, albeit with varying success (European Environment Agency Report, 2005). Packaging waste policy typically approaches the problem from a waste management perspective. This is slowly changing towards a more holistic system that includes a waste hierarchy. The hierarchy starts with prevention or material efficiency

(Chapter 6—Developments in Municipal Waste), followed by product reuse, recycling, incineration with energy recovery, and incineration/landfilling. In contrast to the hierarchy, we see that in many EPR schemes, prevention and reuse receive little or no attention. Most EPR schemes today seem to be focused on cost reduction in waste management through collective collection and recovery schemes. Table 11.2 summarizes selected examples of recovery and recycling schemes as found in many industrialized countries.

11.5.1 Reduce and Refuse

As the awareness on the negative impact of the (excessive) use of packaging materials on the environment has increased, many people and, increasingly, stores and outlets, are enabling or rewarding bring your own (reusable) packaging. Moreover, policies (e.g., the Single-Use Plastics Directive in the European Union (European Commission, 2019), support this transition. The use of reusable bags to replace banned single-use plastic bags has been discussed earlier, but has also been supported by discount systems in selected stores in many countries. This is part of a growing trend, and various schemes are used for, e.g., to-go drinks and other containers in specialized stores while recent European legislation increases the number of single-use packaging types that are prohibited or for which reusable alternatives are required.

11.5.2 Reuse

Reuse of the packaging itself may result in higher environmental benefits, as the packaging does not need to be produced, and instead is cleaned, rinsed, or refurbished to be used again. Collection cost, product design requirements for reuse, and cleaning (energy and water) costs may partially offset these gains. Typically, a refund system is used to encourage recovery. The best-known example in many countries is the deposit system for bottles and crates. The bottles are collected at the point of sale, transported to the manufacturer (or industry), cleaned, and refilled. There are small variations in the system. For example, in the Netherlands most common beer bottles are of the same design, allowing for the bottles to be reused by most breweries. The recovery rates of deposit systems are generally very high (over 90%) and limit the contamination with other packaging or materials. An analysis of beverage

TABLE 11.2 Selected packaging recovery and recycling schemes.

Type	Mechanism	Examples
Reuse	Deposit fee	Bottles, e.g., Germany, the Netherlands Pallets, e.g., Europe Boxes and crates in various countries
Separate collection for recycling	Refund/deposit fee	Bottles and cans, e.g., California and other states in the United States PET bottles, the Netherlands
	Voluntary	Glass, paper
		PET bottles, France Plastic bottles, Belgium
Co-mingled/mixed collection for recycling	Fee-based	DSD system, Germany
	Voluntary	UK (varies by city)

containers in the Netherlands over the period 1984–1999 has shown that reuse, next to material recycling, had the most impact on reducing waste (Chappin et al., 2005).

A less visible system is the pooling of pallets by transport companies in Europe. The wooden pallets have a uniform design and are pooled by all participants in the system. They can be refurbished when damaged. Pallet and transportation companies are part of the pooling system for storage, refurbishing, and replacement. In other sectors (e.g., moving companies, B2B) crates and boxes are also reused, sometimes combined with a deposit system. The pressure on single-use plastics and growing interest in the circular economy around the world has renewed interest in reusable packaging, leading to a variety of new packaging systems in many different markets and products (Megale Coelho et al., 2020).

11.5.3 Separate Collection

Of the recovery systems for recycling, separate collection can guarantee a higher quality of the collected material, limiting contamination with other materials. Historically, separate collection has been used for paper and glass for decades in industrialized countries, which has resulted in high recovery and recycling rates. Separate collection systems can be classified in different ways. Systems can be based on a refund or deposit fee system or based on voluntary action of the user. Voluntary systems can be separated into curbside collection and "drop off" systems, where the user brings it to a given location for collection. Even when established, these systems require regular information and education support (see also Chapter 37) to inform new users to correctly separate and collect packaging wastes.

In packaging, *refund or deposit fees* have been used for a long time for reuse systems but also for collecting recyclables. For years, many states in the United States have charged a small fee per bottle or can (typically between 3 and 5 US cents). The fee is returned when the container is returned at specific collection points. The refund system in the state of Michigan has been monitored for over 20 years and has shown on average collection rates of over 93%. The fees collected for remaining containers that were not returned covered the administrative costs of the system. Similar systems are found in other states of the United States, as well as in European countries, e.g., Scandinavia. The Duales System Deutschland (DSD) system in Germany and some other systems charge fees based on the packaging material of manufacturers, but this fee is not refundable.

Voluntary systems build on the behavior of users and consumers to separate the recyclable materials from other waste. *Curbside systems* do collect the materials or packages at home on a regular schedule. Households have special containers or bags to separate and store the materials until collection. The containers can separate different materials or packages or combine various materials in one. The latter is co-mingled or mixed packaging waste and is discussed in the following text. In some countries and communities part of the packaging waste is collected this way. For example, in the Netherlands, typically paper and paperboard packaging is collected separately with other paper, while other forms of packaging may be combined in collection systems (Brouwer et al., 2019). The alternative for a curbside system is a *"drop off" system*. For example, selected packaging, including glass bottles and plastics (and in some communities metal cans), is collected in containers, which are distributed through neighborhoods and near shopping centers. In the Netherlands, glass is separated by color (i.e., clear, amber, and green), to allow for optimal recycling. Similar combinations of systems can be found in other countries.

Note that the choice of system may affect the collection rate, as well as the contamination of packaging materials (see, e.g., Brouwer et al., 2019). Contamination may result in higher reject rates or downcycling. As evidenced by the high recovery rates of some of these materials, separate collection systems can result in effective and efficient recovery and recycling systems, generating relatively pure and clean recycled material streams. In the case of glass and plastics, separation of the various types can help to improve the recyclability of the collected material.

11.5.4 Mixed Collection

In some countries and/or urban communities, packaging waste is collected as a mix. The different systems vary, depending on the types of packaging and materials collected. In Germany, all packaging material consumption is part of the EPR-based DSD system. For the different types of materials and the volumes used, the packagers and producers pay a fee to the DSD system (recognizable as the "Green Dot") for packaging sold in Germany. In the German system, paper is collected separately, but other packaging is collected as a mix in a separate container. The mix is treated in postcollection recovery centers to separate the different materials (including key plastics). In the UK, several communities collect all recyclables combined, including paper. While this potentially reduces the costs of collection and recovery, it can result in reduced quality of the recovered material. Especially, recovered paper is sensitive to contamination from food wastes, resulting in lower quality recyclable paper and fibers (Miranda et al., 2013). This will result in a lower recycling rate, despite a high recovery rate. Hence, a mixed collection system needs to be carefully designed to maximize recovery *and* recycling.

The economics of packaging recycling are not different from those of recycling in general. Due to the importance of the volume of packaging in waste, there have been a lot of systems in place in different countries to collect and recycle packaging materials. The economics are influenced by the costs and the benefits. Note that these may vary between different stakeholders, dependent on the financial organization of a recycling scheme (see, e.g., Ferreira da Cruz et al. (2012) for an analysis of the packaging waste collection system in Portugal). The overall costs are also affected by the volume and quality of the recyclables collected. The costs are affected by the operation of the collection system, transport, separation (if necessary), and depreciation of capital equipment for each of the steps. The benefits are the results of reduced collection costs of MSW, avoided waste treatment (i.e., incineration, landfilling), and the value of the recyclable material. The latter is affected by markets and quality of the recyclable material.

11.6 CONCLUSION AND OUTLOOK

Packaging is the key factor in the volume and composition of municipal solid waste in most industrialized countries. Packaging material consumption is still increasing in industrialized countries, and even rapidly increasing in developing countries, despite policy attempts to limit the volume of packaging material, or ban specific materials or specific types of packaging. Packaging policy is still driven by traditional waste management policy concerns, and we see still only limited attempts to achieve a more holistic approach to manage packaging material use, recycling, and waste, e.g., as a result of the problems associated with plastic (packaging). Policy in most countries has focused on increasing recycling of packaging material, which has been successful for various types of packaging and materials. The overall environmental and economic impacts of this focus are generally positive and may vary due to a variety of factors. In selected cases, however, the focus may have shifted away from more environmentally friendly packaging or product concepts.

Future developments will also affect the volume, composition, and recyclability of packaging, as they have done in the past. An important development that may have an impact on paper and cardboard use is the emergence and increasing importance of e-retail, and policies around the world targeting single-use plastics. Simultaneously, we witness emerging initiatives for more holistic packaging policies in selected countries with a growing interest in reusable packaging concepts, and increased attempts to reduce the volume of packaging waste and littering, in industrialized and developing countries alike.

As the world becomes more aware of the environmental trade-offs of packaging, we see a worldwide interest towards increased sustainability of product delivery and the role and design of packaging (Bruijnes et al., 2020). This is part of the rapidly changing dynamics in retailing, product development, and the policy environment. This is already resulting in a rapid increase in interest in reusable packaging systems, while recycling targets are growing in many countries. Moreover, climate change is becoming a major driver for the transition to a circular economy by reducing the use of those materials that have a high climate footprint (e.g., plastics). The development of more material-efficient packaging systems or the switch to more climate-friendly materials (including biobased materials) is accelerated by the need for quick reductions in greenhouse gas emissions. Combined, these forces drive a growing interest to come to inherently sustainable packaging systems with a much smaller and different footprint.

References

Brouwer, M., Picuno, C., Thoden van Velzen, E.O., Kuchta, K., De Meester, S., Ragaert, K., 2019. The impact of collection portfolio expansion on key performance indicators of the Dutch recycling system for Post-Consumer Plastic Packaging Waste, a comparison between 2014 and 2017. Waste Manag. 100, 112–121.

Bruijnes, C., Diepenmaat, H., ten Klooster, K., van Soest, J.P., Langeveld, G., Balk, V., 2020. The State of Sustainable Packaging. Netherlands Institute for Sustainable Packaging, The Hague, The Netherlands. https://kidv.nl/the-state-of-sustainable-packaging-1. (Accessed 10 March 2022).

Chappin, M.M.H., Hekkert, M.P., van Duin, R., 2005. Decomposition analysis of Dutch beverage packaging waste: an analysis of material efficient innovations. Resour. Conserv. Recycl. 43, 209–229.

De Bruyn, S.M., Opschoor, J.B., 1997. Developments in the throughput-income relationship: theoretical and empirical observations. Ecol. Econ. 20, 255–268.

Dikgang, J., Leiman, A., Visser, M., 2012. Analysis of the plastic-bag levy in South Africa. Resour. Conserv. Recycl. 66, 59–65.

European Commission, 2019. Directive 2019/904 on the Reduction of the Impact of Certain Plastic Products on the Environment. Official Journal of the European Union. 25 June 2019. https://eur-lex.europa.eu/eli/dir/2019/904/oj. (Accessed 10 March 2022).

European Environment Agency Report, 2005. Effectiveness of Packaging Waste Management Systems in Selected Countries: An EEA Pilot Study, Copenhagen, Denmark. https://www.eea.europa.eu/publications/eea_report_2005_3. (Accessed 10 March 2022).

Eurostat, 2022. https://ec.europa.eu/eurostat/web/circular-economy/indicators/monitoring-framework. (Accessed 17 March 2022).

Ferreira da Cruz, N., Simões, P., Marques, R.C., 2012. Economic cost recovery in the recycling of packaging waste: the case of Portugal. J. Clean. Prod. 37, 8–18.

Hekkert, M.P., Joosten, L.A.J., Worrell, E., Turkenburg, W.C., 2000. Reduction of CO_2 emissions by improved management of material and product use: the case of primary packaging. Resour. Conserv. Recycl. 29, 33–64.

López Gómez, I.D., Serna Escobar, A., 2022. The dilemma of plastic bags and their substitutes: a review on LCA studies. Sustain. Prod. Consump. 30, 107–116.

Megale Coelho, P., Corona, B., ten Klooster, R., Worrell, E., 2020. Sustainability of reusable packaging—current situation and trends. Resour. Conserv. Recycl. X 6, 100037.

Miranda, R., Monte, M.C., Blanco, A., 2013. Analysis of the quality of the recovered paper from commingled collection systems. Resour. Conserv. Recycl. 72, 60–66.

Rouw, M., Worrell, E., 2011. Evaluating the impacts of packaging policy in The Netherlands. Resour. Conserv. Recycl. 55, 483–492.

van Sluisveld, M.A.E., Worrell, E., 2013. The paradox of packaging optimization—a characterization of packaging source reduction in the Netherlands. Resour. Conserv. Recycl. 73, 133–142.

Wever, R., 2009. Thinking About the Box: A Holistic Design Engineering Approach of Packaging for Durable Consumer Goods. PhD Thesis, Delft University of Technology, Delft, The Netherlands.

CHAPTER

12

End-of-life vehicles

Pieter Kuiper and Susanne van Berkum
Auto Recycling Nederland, Breukelen, The Netherlands

12.1 INTRODUCTION

By the end of the 19th century and early 20th century several developments enabled the modern car: the internal combustion engine (ICE) had evolved over time into an efficient gasoline engine, industrial scale production of steel, copper, and other metals and materials was built up, and essential car parts such as tires and transmissions were developed (Bellis, 2019). Additionally, Henry Ford's factories with assembly lines allowed for cheap, mass production of vehicles. Within a couple of decades vehicle production went from a thousand a year to millions a year.

Today about 85 million light duty vehicles are sold globally every year, of which about 3 million are electric vehicles, in addition to heavy duty vehicles such as trucks and busses (IEA, 2021). They represent a massive material resource. After its useful life, when the car cannot be repaired or reused anymore, it becomes an end-of-life vehicle (ELV). In the EU, about 6.1 million ELVs were recycled in 2019 (Eurostat, 2021).

Although vehicle dismantling has been around since the first car was put on the road, industrial processing of ELVs dates from the 1960s. Then car ownership and thus the number of ELVs to be dismantled increased dramatically so that hammer mills were first used to chop cars into pieces. The ELV composition has changed from only steel and some wood to many different materials, some of which are difficult to recycle.

Geographical differences result in diverse methods of handling ELVs: when distances are large, the value from parts and materials is too low for transport to a location specializing in ELV treatment. Therefore, in sparsely populated countries ELVs are often left in deserted areas, which seldom happens in densely populated countries. The main environmental risk from rusting ELVs is soil- and groundwater contamination by the fluids present in the car.

Worldwide, environmental risks and recycling opportunities have led to the introduction of legislation. In Europe, the current ELV Directive is the base for ELV legislation in the member states (European Union, 2000). Since the introduction of the ELV-Directive, the use of hazardous heavy metals like lead, mercury, cadmium, and hexavalent chromium has been nearly phased out. The ELV Directive also focuses on the classification of the new purpose of ELV waste streams. There is not just a minimal recycling rate required, but a distinction is made in the Directive between "reuse," "recycling,"

Handbook of Recycling. **https://doi.org/10.1016/B978-0-323-85514-3.00025-7**

and "(energy) recovery." Member States must ensure that reuse and recycling is minimally 85% and reuse and recovery is minimally 95% of the average vehicle weight per year. These targets have been tightened compared to the 2006–2015 targets, which were 80% for reuse and recycling and 85% for reuse and recovery. With over 85% of ELV weight being recycled or reused, ELV recycling has a major impact on resource supply.

An increasingly important detail of the European recycling legislation is that the weight of a traction battery, which is removed for recycling, counts for 100% in the "reuse and recycling" category of the ELV-directive although its true recycling rate, as governed by the Battery Directive, is lower. The ELV Directive will be reviewed and updated in 2022 (European Commission, 2021). The relatively high ELV-recycling rates in Europe do not apply to all deregistered cars, meaning dismantled plus exported, only to dismantled cars.

Compliance with the ELV Directive is organized in some member states by the government, in some by individual brands or other organizations. In the Netherlands, all brands cooperate in the foundation Auto Recycling Nederland (ARN).

This chapter is a case study of ELV recycling in the Netherlands, organized by ARN, to illustrate the challenges and possibilities of vehicle recycling. First, developments in ELV composition and volumes are discussed, and then the steps of the recycling chain are introduced, followed by the recycling technologies for specific car parts and materials. The chapter concludes with future developments and outlook.

12.2 VEHICLE COMPOSITION

Over time, material usage in vehicles has changed considerably. First, fuel efficiency and light weighting was a driver for increased use of aluminum alloys, plastics, and composites. Second, the need for more comfort and automation resulted in the use of more electric motors and car electronics, e.g., for navigation and car management. Finally, the introduction of battery-powered electric vehicles (BEVs) led to cobalt, nickel, and lithium becoming part of the vehicle. By now the entire periodic table of elements is present in the car, in the form of a wide range of alloys and compounds, and the car has become a highly complex product (Reuter et al., 2006).

In the Netherlands, the average age of a vehicle arriving at a dismantler's yard was 18.9 years in 2020 (ARN, 2021a). This means that the average ELV today was manufactured almost 20 years ago and consists of a mixture of the materials used then. When taking into account the lifetime of cars, it takes a long time before the evolutions and developments in material usage in the car, start to impact the average ELV composition. van Schaik et al. (2002) show that the age of ELV vehicles dismantled in a specific year can be described by a Weibull distribution curve. That year's recycling rate will be strongly influenced by the average vehicle composition and weight of each ELV's production year. This means that the composition and weight will be distributed as well, and the ELV recycling system has to be able to deal with the diversity of cars and thus compositions entering the system (van Schaik et al., 2002).

The average ELV-weight in the Netherlands has been fairly constant for the last 10 years at 1070 kg in 2020, as Figure 12.1 shows. As the average vehicle weight is a result of many variables, it is difficult to pinpoint specific causes and effects.

The composition of the entire vehicle population on the road or of the dismantled population is difficult to determine. ADEME in France determined the share of each material in 2015 (Monier et al., 2017). Categorizing these materials and calculating with the ELV weight mentioned earlier (1070 kg) yields the following ELV breakdown (Table 12.1).

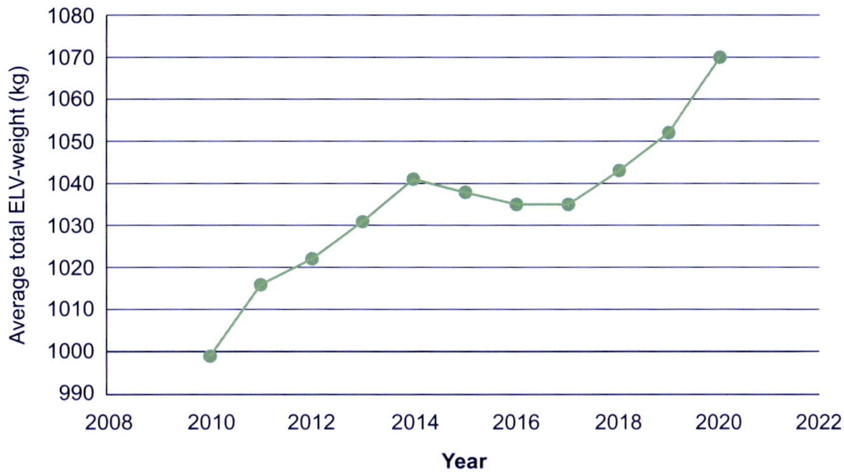

FIGURE 12.1 Development of average total ELV-weight in the Netherlands. *Courtesy: ARN.*

TABLE 12.1 Average composition of an ELV in 2015 in percent, and in kilograms using average ELV weight of 1070 kg.

Material category	Share of category in car (%)	Material	Share of material in car (%)	Material in car (kg)
Ferrous metals	70.0	Ferrous metals	70.0	748
Plastics	10.0	Polypropylene (PP)—other parts	4.4	107
		ABS, PVC, PC, PMMA, PS, etc.	2.2	
		Polypropylene (PP)—bumpers	1.1	
		Polyamides (PA)	1.0	
		Polyethylene (PE)—fuel tanks	0.8	
		Polyethylene (PE)—other parts	0.5	
Nonferrous metals	5.0	Nonferrous metals (excluding wiring harnesses)	4.0	53
		Wiring harnesses	1.0	
Foam and textiles	3.7	Polyurethane foam	2.0	40
		Textiles, other	1.7	
Tires	3.4	Tires	3.4	36
Glass	3.0	Glass	3.0	32
Lead-acid starter battery	1.4	Lead-acid starter battery	1.4	15

Continued

TABLE 12.1 Average composition of an ELV in 2015 in percent, and in kilograms using average ELV weight of 1070 kg—cont'd

Material category	Share of category in car (%)	Material	Share of material in car (%)	Material in car (kg)
Fluids, gases, oil filters	1.2	Spent oil and filters	0.7	13
		Cooling or brake fluids	0.4	
		Air conditioning fluids	0.1	
Rubber compounds other than tires	1.1	Other rubber compounds	1.1	12
Paint	0.8	Paint	0.8	9
Catalytic converters	0.5	Catalytic converters	0.5	5
Total*	100.0		100.0	1070

**Total of individuals does not add to 100 due to rounding.*
Data from Monier et al. (2017) and ARN (2021a).

The calculation of the properties of "average ELVs" requires exact information about the population that is dismantled. The extensive number of possible combinations of brand, model and type means that this population is changing continuously and that almost every ELV has unique properties. The ELV breakdown in Table 12.1 is therefore indicative. For example, the weight of the catalytic converter: Table 12.1 indicates 5 kg, whereas Umicore data indicates 4 kg (Umicore, 2021). Similarly for glass: 32 kg for an average ELV, whereas the literature reports 35 kg (Sato et al., 2018).

Significant differences between ICE vehicles, having a drivetrain with combustion engine and gearbox, and BEVs, with electric motors and traction battery, mean that the frequency of appearance of certain parts in the population is as important for their average weight in the population as their true average weight. In the reference years for Table 12.1, the number of BEVs in the total ELV population was too small to have a significant influence.

12.3 RECYCLING CHAIN

ELV recycling started with the typical car salvage junk yard that sold parts and occasionally sold its hulks to a scrap buyer. It has developed into a professional recycling chain with specialized segments, focused on both the prevention of the loss of hazardous materials and environmental damage as well as the recovery of valuable materials.

In Europe, the consecutive segments in the recycling chain are (Figure 12.2):

- Deregistration,
- Dismantling and processing the materials from dismantling and depollution,
- Shredding, followed by mechanical and physical processing of the (metal-rich) heavy fraction, and
- Postshredder treatment (PST) of the light and heavy residue fractions.

Each step, actor, and segment will be discussed in detail next.

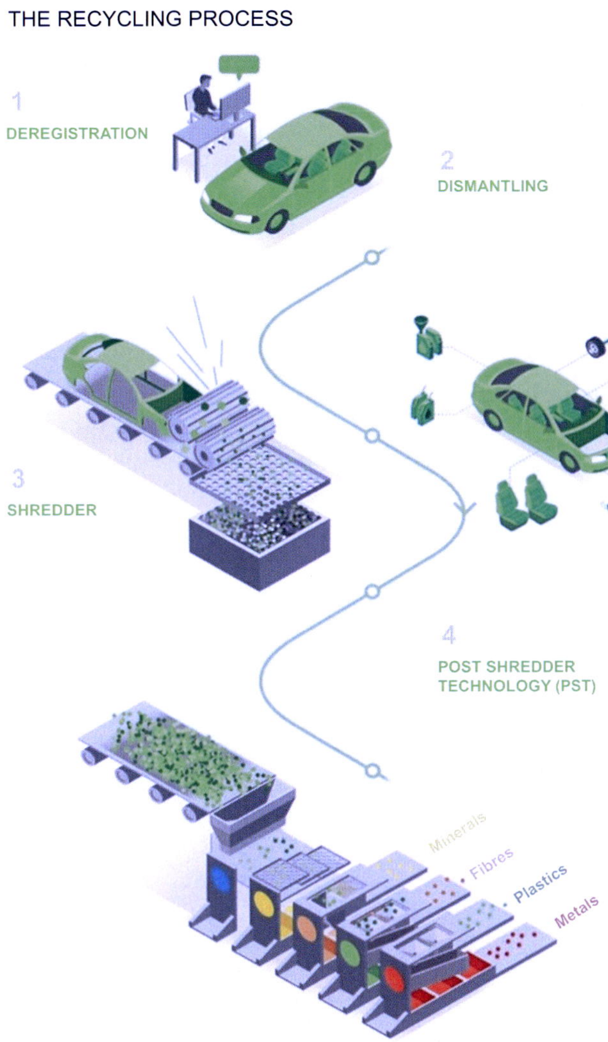

FIGURE 12.2 Recycling chain of end-of-life vehicles. *Courtesy: ARN.*

12.3.1 Deregistration and Dismantling

The dismantlers site is the place where the last owner hands over his old vehicle. In Europe, operating a dismantling site is well regulated and an authorization is necessary. Official dismantling locations are therefore called "authorized treatment facilities" or ATFs.

Depending on the local situation, the last owner of the vehicle can either sell his old car to the dismantler or just hand it over. After that, the last owner receives a Certificate of Destruction (CoD) and the vehicle is deregistered from the national register of vehicles, if the national government keeps such a register.

After accepting the ELV from the last owner, the dismantler removes valuable parts for reuse as spare parts. For this, state-of-the-art sales software is available as an aid to identify much requested parts from ELVs. In addition, national parts sales platforms are operated, which also indicate which parts are searched for. Valuable parts are either dismantled directly or the complete vehicle is placed in the yard for dismantling later. Parts like starter engines and alternators can be sold to remanufacturing companies, depending on the brand.

After receiving the vehicle (now called an ELV) on his yard, according to Dutch law the dismantler must depollute the ELV and perform obligatory dismantling within 10 days. There are hazardous materials that during the shredding process cannot be separated from other materials, can cause disturbances, e.g., explosion or combustion, or would be dispersed, contaminating all other materials. These materials must be removed prior to shredding. This removal process is called depollution. In practice this means that all fluids, battery, and tires are removed from the vehicle.

In the Netherlands, use of specialized fluid removal equipment has become the industry standard in the last 15 years. These so-called drainage stations are equipped with a hydraulic lift to elevate the vehicle to enable ergonomic labor conditions (Figure 12.3). Furthermore, the stations are fitted with hollow tank drills, as shown in Figure 12.4, equipped with air-powered diaphragm pumps and they use suction hoses, pumps, and containers to drain residual fuel, windshield washer fluid, coolant and brake fluids. This ensures both high quality of the collected liquids and a drainage process without spills. Drainage systems enable car dismantlers to drain an ELV in 8–12 min. Fluids like Eolys and AdBlue, fuel borne catalysts, can be recovered using separate pumps, where the water-based AdBlue can be stored separately and the oil-based Eolys can be mixed with the waste engine and gearbox oil.

Dismantling of the following parts is also obligatory within 10 days in the Netherlands (VROM, 2007):

- LPG-tanks
- lead-acid starter batteries
- traction battery
- oil filters
- polychlorinated biphenyl (PCB)- and polychlorinated terphenyl (PCT) containing capacitors
- small batteries, e.g., alkaline
- all explosive parts, unless neutralized, e.g., air bag deployment units and seatbelt tensioners.

Furthermore, catalytic converters, parts containing lead, mercury, or hexavalent chromium need to be removed, however, but not within 10 days since these parts are less hazardous.

Tires, glass, and large plastic parts may be left on the hulk when these can be recycled in or after the shredding process. When this is not possible, these parts must be dismantled. In the Netherlands, glass can be recovered for recycling as a material using PST. However, some dismantlers prefer to dismantle the glass to keep a clean yard and obtain a higher scrap price from the shredding company.

All liquids, traction batteries of electric vehicles, lead-acid starter batteries, glass, and oil filters are collected separately and transported to a dedicated recycling company. Because of their high value, catalytic converters are sold to catalyst collectors, often small traders, who sell to precious metals refiners where final metal recovery takes place (Chapter 25—Precious and technology metals and Chapter 42—Economic aspects of metal recycling). An average catalytic converter consists of a 3 kg metal can, containing a 1 kg ceramic monolith containing thin layers of the precious metal acting as catalyst (Umicore, 2021). Batteries are valuable as well: the lead acid battery is brought to lead recyclers (Chapter 22—Lead), and the traction battery is sent for Ni, Co, Mn, and Li recovery.

FIGURE 12.3 Integrated drainage system with liquid-specific tank drilling, pumps, and containers. *Courtesy: ARN.*

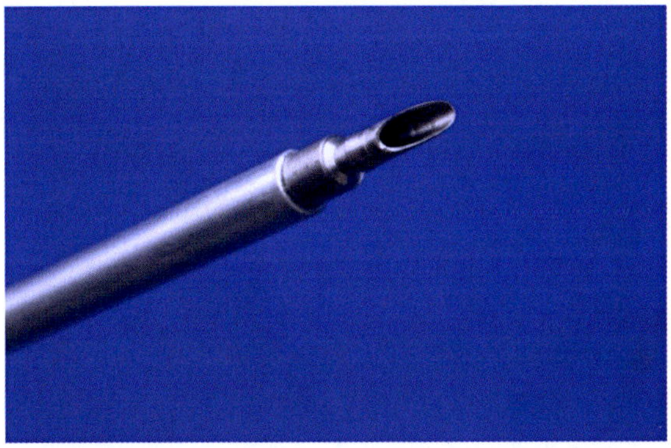

FIGURE 12.4 Hollow drill for fluid removal. *Courtesy: ARN.*

12.3.2 Shredding

By Dutch law, an ELV may be stored at the ATF's yard for a maximum of 3 years. Then, at the latest, the dismantler sells the hulk for its scrap metal value to a metal recycling company. The best way to recover a clean ferrous scrap from car hulks is to use a shredder. In contrast to, e.g., a scrap shear, a shredder is equipped with an air stream that removes light materials and can create a clean ferrous scrap fraction, without residual nonmetals. This prevents nonmetallic materials, e.g., plastics and rubber, from being included in the ferrous fraction, entering an electric arc furnace (Chapter 19—Steel), and incinerating without good flue gas cleaning. Hence, the Dutch government has made hulk processing using a shredder obligatory (VROM, 2007). The large air stream of the shredder separates the shredder output in a light and a heavy fraction. The light fraction, also called light automotive shredder residue or light ASR, contains most of the foam and textile particles, but also copper wires mingled with these and most of the small particles. The light ASR is often sized, producing a fine shredder light fraction (SLF-fine) and a coarse shredder light fraction (SLF-coarse) (Khodier et al., 2018). The heavy fraction is subjected to magnetic separation to remove ferrous metals, the main product of the shredder. This ferrous fraction is further cleaned from copper-containing starter engines and alternators by hand-sorting. The remaining heavy material, called heavy fraction, is rich in nonferrous metals, but also contains heavier plastic shreds and rubber.

12.3.3 Postshredder Treatment

The heavy fraction is treated in special installations for the recovery of nonferrous metals. These installations are equipped with, among others, sensor-based sorting machines and sink-float separators (Chapters 8 and 9—Physical Separation and Sensor-Based Sorting), yielding nonferrous metal fractions ready for recycling at smelters and refiners. The residue of the installation is heavy ASR. This together with the light ASR from the shredder forms the nonmetallic shredder output and is called ASR. Khodier et al. indicate that ASR is about 24% of the ELV input weight processed by the shredder (Khodier et al., 2018). Processing of ASR is described in more detail in Section 12.6.

12.4 RECYCLING OF AUTOMOTIVE PARTS

12.4.1 Traction Battery

The introduction of lithium-ion batteries (LIBs) as a traction battery for automotive applications has caused a shift from ICEs towards BEVs. Moreover, the Paris Agreement in 2015 and more strict EU regulations on the CO_2 emission of vehicles have strongly promoted electric vehicles sales (European Union, 2019): in the Netherlands the number of EVs on the road has increased from 30,000 in 2014 to 139,000 in 2019 (Statistics Netherlands (CBS), 2019). Globally, the number of EVs on the road is predicted to increase from 12 million vehicles in 2021 to 54 million in 2025 (McKerracher, 2021).

An obvious route for LIBs entering the recycling chain is as part of ELVs, corresponding to 145 kt in 2020 globally. However, most end-of-life batteries are currently collected due to recalls. The lifespan of a lithium-ion traction battery is approximately 10 years and is expected to create a waste stream of 1680 kt globally in 2030 (Pilllot, 2021; Yoon, 2021).

Batteries are discarded for various reasons: the battery has a reduced usable capacity, and does no longer meets the criteria, or it is damaged or defective, due to a car accident for example. In the first case, although the LIB is no longer suitable for automotive applications, it can still be used for alternative applications, also called second use or second life. In the latter case, damaged or defective LIBs can be dangerous and may ignite spontaneously. In such cases, the LIB is recycled via a metallurgy route.

An automotive battery pack requires sufficient capacity; therefore, for practical and safety reasons, the pack consists of several modules (Figure 12.5). Each module is separately controlled and monitored by the battery management system (BMS) and can be replaced separately to enable enhanced longevity. The modules consist of separate cells, varying in number of cells, as well as type of cells: the most common cells are pouch, cylindrical and prismatic (Faessler, 2021).

The first hybrid EVs put on the market contained NiMH batteries, such as the Toyota Prius. However, due to their lower energy density, the NiMH chemistry is not suitable for BEVs (Sabihuddin et al., 2015). LIBs are now employed in automotive applications, where nickel-manganese-cobalt (NMC) and nickel-cobalt-aluminum (NCA) chemistries dominate because of their high performance and high specific energy (Harper et al., 2019).

The rise in demand for batteries for the automotive industry, as associated with ethical concerns about mining of cobalt, has pushed development of new cathode materials with reduced cobalt content or complete replacement by other metals. However, such developments have several disadvantages, such as a lower

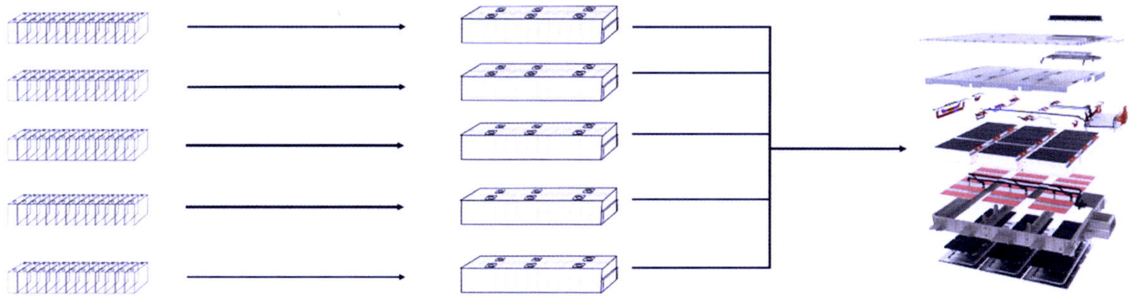

FIGURE 12.5 Schematic figure of the composition of a battery pack, from left to right: multiple cells form a module, and a battery pack comprises several modules with additional electronics such as the BMS. *Modified from Chen et al. (2021).*

thermal and air stability or decreased cycle stability (Manthiram, 2020) Cobalt-free cathode materials such as lithium-sulfur and lithium-iron phosphate still have a lower energy density and remain inferior for automotive applications compared to the current NMC chemistry. Thus, they are not expected to be employed in vehicles in the short term.

12.4.1.1 *Pretreatment of LIBs*

Pretreatment of LIBs consists of two parts: deep discharging and dismantling. To enable safe handling, LIBs are subjected to deep discharging. Most often executed at large scale is electrical deep discharging down to a maximum 2.5 V per cell. This process cannot be automated yet, due to the large variety of LIBs. In addition, one must have access to the BMS in order to execute the discharging process safely (Nembhard, 2019). The regained energy can be stored in other batteries or directly reused on site.

Dismantling LIBs is becoming common practice to increase the recovery rate and to promote high quality recycling. Dismantling consists of removing the casing, wiring, and the BMS (Duesenfeld, 2021; Sortbat, 2021). These components consist of plastics, stainless steel, aluminum, and copper, representing approximately 40% of the total weight of a battery pack (Elwert et al., 2016). Subsequently, the battery pack is dismantled to take out the modules. The modules can go directly into pyrometallurgical treatment or are further physically processed followed by hydrometallurgical processes.

12.4.1.2 *Second Use of LIBs*

LIBs from electric vehicles are mostly discarded due to decreased battery capacity, indicated by the relative battery capacity, the state of health (SoH). A SoH of approximately 70% is regarded as insufficient to meet customer expectations as the driving range has decreased significantly. Ageing of a LIB is strongly dependent on the use of the vehicle: frequently driving short distances and fast charging enhance the ageing process, as well as subjecting the LIB to extreme temperatures (Faessler, 2021).

Second use is defined as the use of LIBs in an application other than the original application, whereas reuse is defined as the use of LIBs in an equivalent application. Due to the reduced SoH, reuse of LIBs is less favored. However, for battery energy storage system (BESS) applications, the SoH of these batteries is still sufficient. BESS applications do not require the highest available energy density since the dimensions of the BESS are more flexible. Most BESSs are used for energy storage from photovoltaic systems to enable increased self-consumption or to shift energy purchases for economic reasons (Faessler, 2021).

When the LIB shows no external damaging or the use history is unknown, the LIB will be put in quarantine and the temperature is monitored (Sortbat, 2021). When the LIB shows no temperature deviations, the LIB is dismantled and the modules and, if possible, cells are extensively evaluated visually and electrically (EcarACCU, 2021; Kotak et al., 2021). Remanufacturing on the level of cells is currently impossible due to the use of irreversible joints on several locations of the module, for example electrical joints and the connection to the outer housing (Schäfer et al., 2020).

Waste materials arising during preparation for second use are similar to dismantling: plastics, stainless steel, aluminum, and copper, and defective battery modules. These materials are sent for material recycling. The recycling efficiency of the second use route is based on modules only and can increase up to 98% (EcarACCU, 2021). Evidently, when battery modules in second use applications reach end-of-life, they will be sent for material recycling.

12.4.2 Tires

Tires are high-tech and complex products, generally constructed as indicated in Figure 12.6. Belts are made of steel, with a nylon cap ply, and the inner liner is made of halobutyl rubber and the bead of steel. The rubber part usually consists of 10–15 components, each made separately and each made from two or three different rubber blends, with each blend consisting of 10–15 different raw materials. The complexity of the product makes a recycling process that breaks it down to its individual components virtually impossible.

In the Netherlands, collection and recycling of end-of-life tires (ELTs) from passenger cars and light commercial vehicles (LCVs) are organized by ARN for the tires from ELVs and by RecyBEM for tires from the replacement market. Tires from ELVs are removed by a dismantler from the wheel rim for efficient loading of the container. If the tire or the wheel rim is sufficiently valuable, the tire is removed from the rim using a tire changing machine. However, removal is performed more efficiently using a crusher, leaving both tire and rim as scrap.

Reusable tires are separated from scrap tires by hand at a collection facility. Valuable tires are sorted, in terms of quality and size, and have their lifespan extended by retreading or sale as a second-hand tire. Both re-use as second-hand tires and retreading are examples of product recycling. Recyclers process the scrap tires into high-grade rubber granulate, steel and textile, as discussed in Chapter 32—Black rubber products. Secondary raw materials from ELTs are important resources for industries, such as construction, automotive and cement production.

Yearly approximately 10 million car tires are collected in the Netherlands: in 2020 RecyBEM collected 9.5 million replacement car tires, equivalent to 77,000 t material (van Oostenrijk, 2021). ARN processed approximately 800,000 ELV tires in 2020, which equals about 5400 t, resulting in a total of 82,400 t in the Netherlands (ARN, 2021b).

As the EU landfill directive makes no distinction between tires from passenger cars and LCVs on the one hand and truck and bus tires on the other hand, these are counted together in European statistics. The European Tire and Rubber Manufacturers Association (ETRMA) indicates that on average 94% of 3.2 million tons of collected ELTs in the EU were treated in 2018. The majority of the treated ELTs, that is 62%, was recovered through material recovery or re-use and retreading (ETRMA, 2021).

Dutch tire recycling legislation differentiates between tires from trucks and buses and tires from passenger cars and LCVs. The recycling classification for the latter is shown in Table 12.2. On average, a Dutch ELV has 36 kg tires, corresponding to 3% of the average ELV-weight (Table 12.1, ARN, 2021a). Approximately 20% of the tires collected

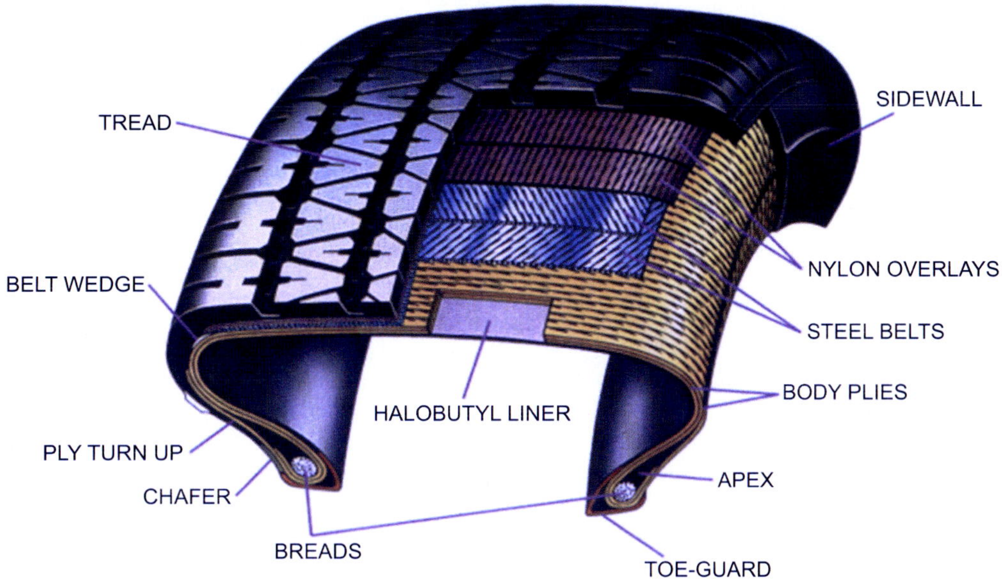

FIGURE 12.6 Tire construction. *From Taylor (2021).*

TABLE 12.2 Breakdown of classification of replacement and ELV-tire recycling in the Netherlands in 2020.

Classification	Tires from replacement		Tires from ELV	
Material recovery	**6.37 million**	**67%**	**2.8 kton**	**52%**
Re-use/retreading	2.85 million	30%	1.2 kton	22%
Alternative use/energy recovery	0.29 million	3%	1.4 kton	26%

Data from ARN (2021a) and van Oostenrijk (2021).

from ELVs and aftermarket are re-used as their profile is still sufficient, and 2% are re-used after retreading (ARN, 2021b). ELTs with insufficient profile have a negative value. The driving force behind higher-end applications is often legislation. In Europe, the Landfill Directive and the Directive on ELVs have restricted the landfilling and have promoted recycling (European Union, 1999, 2000). The EU prohibited landfilling of complete and shredded tires in 2003 and 2006, respectively. The Directive only distinguished between "allowed to be landfilled" and "not allowed." Landfilling of bicycle tires and tires with an outer diameter larger than 1400 mm continues to be allowed, while landfilling of all other tires of any type of vehicle is prohibited.

12.5 RECYCLING OF AUTOMOTIVE FLUIDS

Automotive fluids removed from the ELV during depollution are typically a mixture of oils or alcohols with water and solid contaminations. Generally, three steps take place during recycling of the fluids (Pinheiro et al., 2020):

- removal of solids by filtration
- removal of water by distillation.
- separation and purification of the hydrocarbons by processes such as flocculation, centrifugation, and distillation.

A description of each fluid is given here.

12.5.1 Oils

The oils are recovered in the depollution stage. Here, engine oil and gearbox oil are mixed, together with the Eolys fluid. At the processing plant, first contaminants such as oxidations products and additives are removed by filtration, and water (about 8%) is extracted. Next, distillation and centrifugation separate the base oil from the remaining phase, containing oil used for heating purposes (23%), the distillation column bottom residue (17%), a small trace of gasoline (2%), residual water, and minor contaminants. The valuable base oil can be used to produce new engine oil and corresponds to approximately 50% of the original waste stream.

12.5.2 Brake Fluid

Different brake fluid compositions are used; however, they mainly consist of glycol ether and mono ethylene glycol (MEG) (TotalEnergies, 2021). An important additive is borate, to handle water that is absorbed by the brake fluid. When brake fluid contains water, this could form bubbles whose compressible nature would have a disastrous effect on the braking power of the vehicle. Corrosion inhibitors and antifoam additives have an important function as well.

The recycling process starts with removal of the small amount of low-boiling point oil. Subsequently, water and MEG are removed by distillation. A complete evaporation step separates the polyglycols from the solid residues, e.g., the borates. The polyglycols are then purified by a second distillation and can be readily used in new brake fluid.

12.5.3 Engine Coolant

Engine coolant is a mixture of water, MEG, and additives, such as corrosion inhibitors and lubricants. The recycling process of engine coolant starts with dewatering by partial distillation with subsequently a complete distillation to produce three fractions at different heights of the distillation column: residual water, MEG, and oils. MEG is the valuable fraction and is purified by active carbon before reuse as new engine coolant.

12.5.4 Windscreen Wash Fluid

Mainly consisting of water, windscreen wash fluids also contain alcohols and glycols and low concentrations of additives. Recycling starts with the removal of any oil contamination by decantation and removal of other contaminants by a flotation and flocculation step. Biological treatment is then used to convert alcohols and glycols into water and carbon dioxide. The remaining material is clean water and is disposed of through the sewer.

12.5.5 Fuel Mixture

Any clean diesel or gasoline recovered at the ATF will be used by the dismantler for his own vehicles. The fluid disposed of is a contaminated fuel, consisting of diesel, gasoline, water, and liquid and solid contaminants. Upon receipt at the plant, the water is separated by decanting. After that, the flash point of the residual fuel is determined. Low flash point fuel mixtures require specific fire- and explosion prevention measures and are therefore processed separately. The water is cleaned by a floatation and flocculation process, removing residual fuel, oils, and other organic contaminants. After additional cleaning, the water is disposed of. Using emulsion breakers, both fuel mixtures, high- and low flash point mixtures, are then cleaned from water bound to the fuel that could not be removed by decanting.

The final application of the fuel mixtures is as a heat source in industrial processes.

12.5.6 Air Conditioning Refrigerant

Removal of the air conditioning refrigerant requires specific apparatus. Air conditioning refrigerants have a high global warming potential (GWP) and must be carefully removed from the vehicle to prevent dissipation of the harmful gasses. The apparatus is attached to a cylinder for appropriate collection of the waste stream. Before treatment of the material, it is analyzed to determine the composition (mixtures of refrigerants), purity, and the presence of water and oils. When the gas is of sufficiently high purity, it is reclaimed and recycled into new products. When the material contains too a high concentration of impurities, the material is destroyed by incineration (Castro et al., 2021; Westfalen Gassen, 2021).

12.6 AUTOMOTIVE SHREDDER RESIDUE

When processing ELVs, both the shredding process and the separation of the "heavy residue" fraction yield ASR, up to 24% of total ELV-input weight processed by the shredder, or 257 kg assuming a 1070 kg ELV weight (Khodier et al., 2018). In Europe, the ELV-Directive indicated that no later than 1 January 2015, the average vehicle weight re-used and recycled should be 85% (European Union, 2000). This means that also part of the ASR also needs to be recycled, as opposed to energy recovery, to meet this requirement. Assuming 24% of the ELV is ASR, and the rest of the ELV materials (76%) already being re-used or recycled, this means that the additional 9% must come from treating or processing the ASR fraction, corresponding to 37.5% of the ASR weight.

12.6.1 Composition

ASR consists of all nonmetals present in the shredded hulks, together with the metal particles that were not liberated sufficiently to be recovered in either the shredding process, or the processing of the "heavy residue" fraction. The SLF-coarse fraction has a particle size of 30–150 mm (Figure 12.7), and the SLF-fine fraction typically contains particles $\leq$30 mm. As the heavy fraction is sieved during processing as well, its residue normally consists of a coarse heavy fraction with a typical

FIGURE 12.7 Coarse ASR fractions: shredder light fraction (SLF-coarse, 30–150 mm, left) and shredder heavy fraction (SHF-coarse, 30–100 mm, right). *Courtesy: ARN.*

particle size range of 30–100 mm (SHF-coarse, Figure 12.7) and a fine heavy fraction (SHF-fine, typically <30 mm).

The process parameters of the shredder and the separation equipment for the heavy fraction influence whether specific materials end up in a "fine" or "coarse" fraction and a "light" or "heavy" fraction. Typical compositions of the ASR, its subfractions and their typical ratios are listed in Table 12.3.

12.6.2 Recycling Technologies

Until the 1990s, landfilling of shredder residues was the industry standard. However, recycling technologies for ASR gained attention because of environmental awareness. Moreover, the bulk density of ASR is typically 300–400 kg/m^3, thus requiring large landfill volumes.

In Japan, Toyota has developed a sorting/thermal ASR recycling process that uses various sorting methods to separate the metals from the foam and fiber materials in the ASR. The foam and fiber-mixture is applied as soundproofing for buildings. The remaining plastic and rubber mixture is turned into so-called melt-bricks and is used as a heat source for electric arc furnaces (Toyota, 2017).

This process would need improvements in recycling of the plastics and mineral fractions to reach the European ELV Directive target for recycling.

Thermal ASR-processing routes operational before 2015, such as the process of Citron in Le Havre, could not guarantee meeting the 85% recycling requirement (Cossu and Lai, 2015). Thus European companies started to develop and build large sorting installations (Ignatenko et al., 2008).

A typical flowsheet for the sorting of shredder light residue fraction and the metal-free part of shredder heavy residue fraction may look as shown in Figure 12.8.

In ASR sorting, the basic components (metals, plastics, fibers, and minerals) are separated as much as possible using, among others, differences in magnetic behavior, density, drop speed and electrical conductivity (Chapter 8—Physical separation and Chapter 43—Physical separation). Additional sieving and comminution are necessary to perform the separation of particle populations that are similar in size and shape and produce products that are more homogeneous in composition to meet the product specifications. Specific separating equipment applied in ASR-processing is listed in the following text.

TABLE 12.3 Typical compositions and ratios of ASR and ASR-subfractions.

	Amount in subfraction ASR				
Material	**SLF-fine**	**SLF-coarse**	**SHF-fine**	**SHF-coarse**	**Amount in ASR**
Ferrous (impure)	0.7%	0.7%	0.1%	0.3%	0.5%
Nonferrous	0.8%	0.3%	1.1%	0.9%	0.7%
Plastics	21.2%	15.9%	29.8%	72.0%	36.4%
Foam/textiles (fibers)	10.9%	53.3%	2.6%	9.2%	26.6%
Minerals (impure)	59.3%	17.0%	59.9%	2.0%	23.7%
Other (e.g., elastomers, wood)	7.0%	12.8%	6.5%	15.7%	12.1%
Total	**17%**	**40%**	**11%**	**32%**	**100%**

Data from ARN.

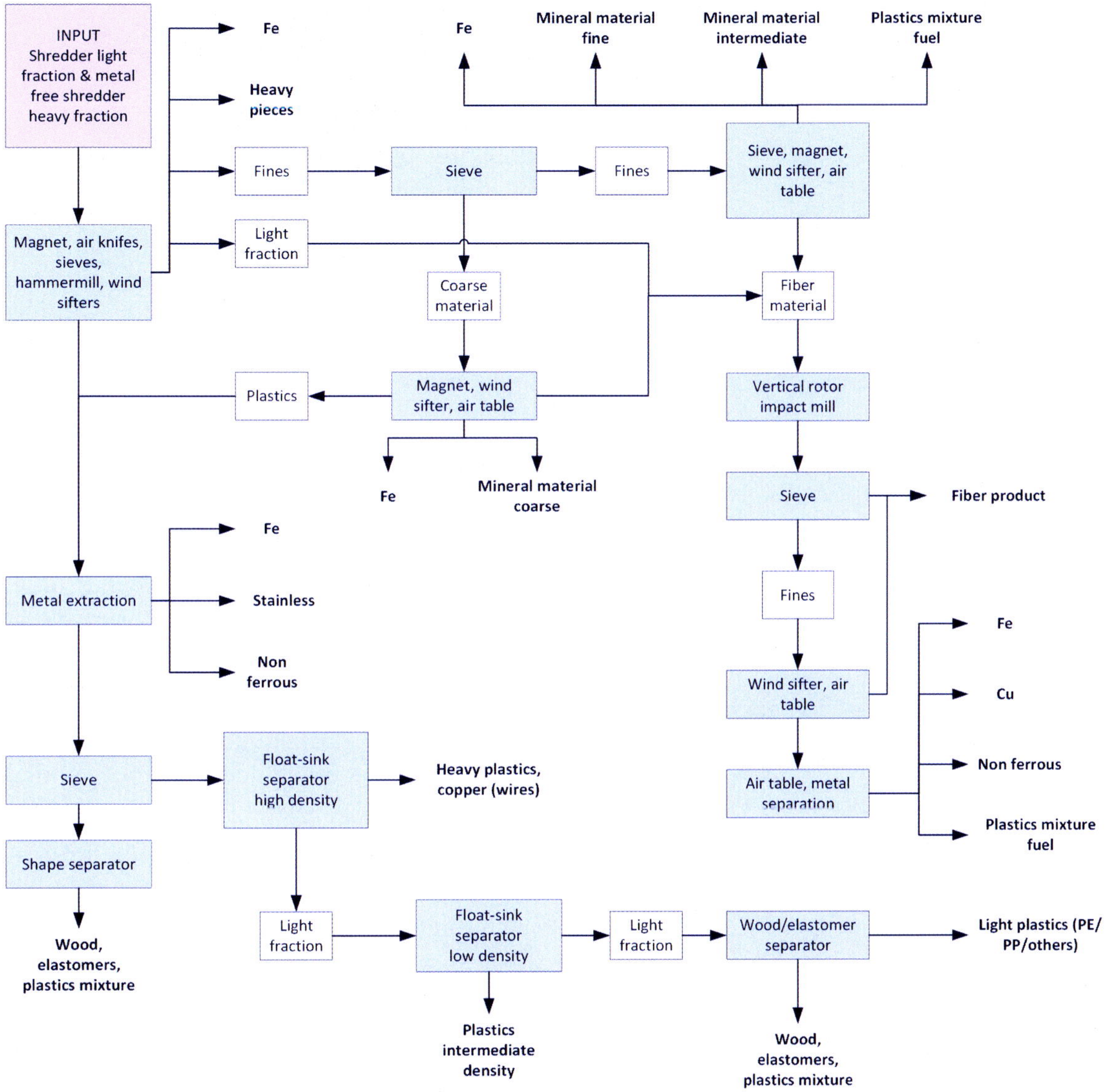

FIGURE 12.8 Shredder residue sorting flowsheet. *Reproduced with permission from ARN.*

Flip-flop screens (Figure 12.9) are used in several steps of the ASR separation process. The elastic screen panels are tensioned and stretched by the machine. During tensioning, the holes are widened, freeing particles that are stuck, and a trampoline movement forces particles from the surface, making these screens self-cleaning. The mixture of wires and textile present in ASR makes processing using a stationary sieve impossible.

Separating tables consist of a screen with an air flow passing through it. Specific screen motion

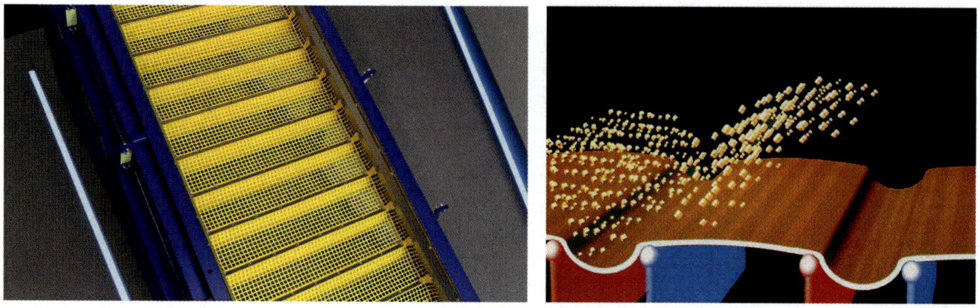

FIGURE 12.9 Self-cleaning "flip-flop" screens (left) top view and (right) operation of the screens: red and blue parts of the screen frame move relative to each other, stretching half of the screen deck in turn. *Reproduced with permission from Hein Lehmann B.V.*

moves heavier particles upwards. Air blown from the bottom fluidizes the lighter particles and they move downwards as a result of the screen inclination (Figure 12.10A). As the amount of air, the screen type, and the screen motion can be varied, these tables offer many separating options. In ASR-processing, they are used to separate plastics from a mixture of coarse glass, metals and plastics.

A sink-float separator contains a suspension with relatively high density. In this way, plastics with density higher than the density of the suspension sink, while plastics with lower density float (Figure 12.10B). The density can be varied using, for example, a lime suspension. As the bath density is altered by entrained dirt washed from the plastics, the density needs to be adjusted continuously.

Vertical rotor impact mills are important in recovering value from ASR. Although ASR is a residue stream with negative value, it has the benefit that it contains many of the small copper wires applied in vehicles. The sharp ends of these are stuck in the foam and textiles of the ASR. Recovery of this copper is important for ASR processing economics. However, pieces of wire that are separated from foam or textile particles tend to get stuck in the next foam or textile piece immediately. This is prevented by the use of a vertical rotor impact mill, which has the unique feature that it can reshape the wire-shaped copper into small copper balls. These copper balls can be easily separated from foam and textile particles (Figure 12.10C).

12.6.3 Final Products

Examples of final applications of the typical main output fractions of the flowsheet (Figure 12.8) are listed in Table 12.4 and shown in Figure 12.11.

The need for a circular economy and the minimum of 85% re-use and recycling in the European ELV-Directive are drivers for product development and innovation. An example of this is the production of planks and beams from a mixture of fiber pellets and plastics, thus creating the first recycling of the ASR fibers as a material instead of recovery as an energy source.

12.7 FUTURE DEVELOPMENTS AND OUTLOOK

Over the past two decades, technological improvements and the urgency to reduce CO_2 emissions have stimulated significant developments in the automotive industry, resulting in changing waste streams in the recycling industry. To decrease the CO_2 emissions during the life span of a vehicle, automotive producers are shifting from regular steel to lower

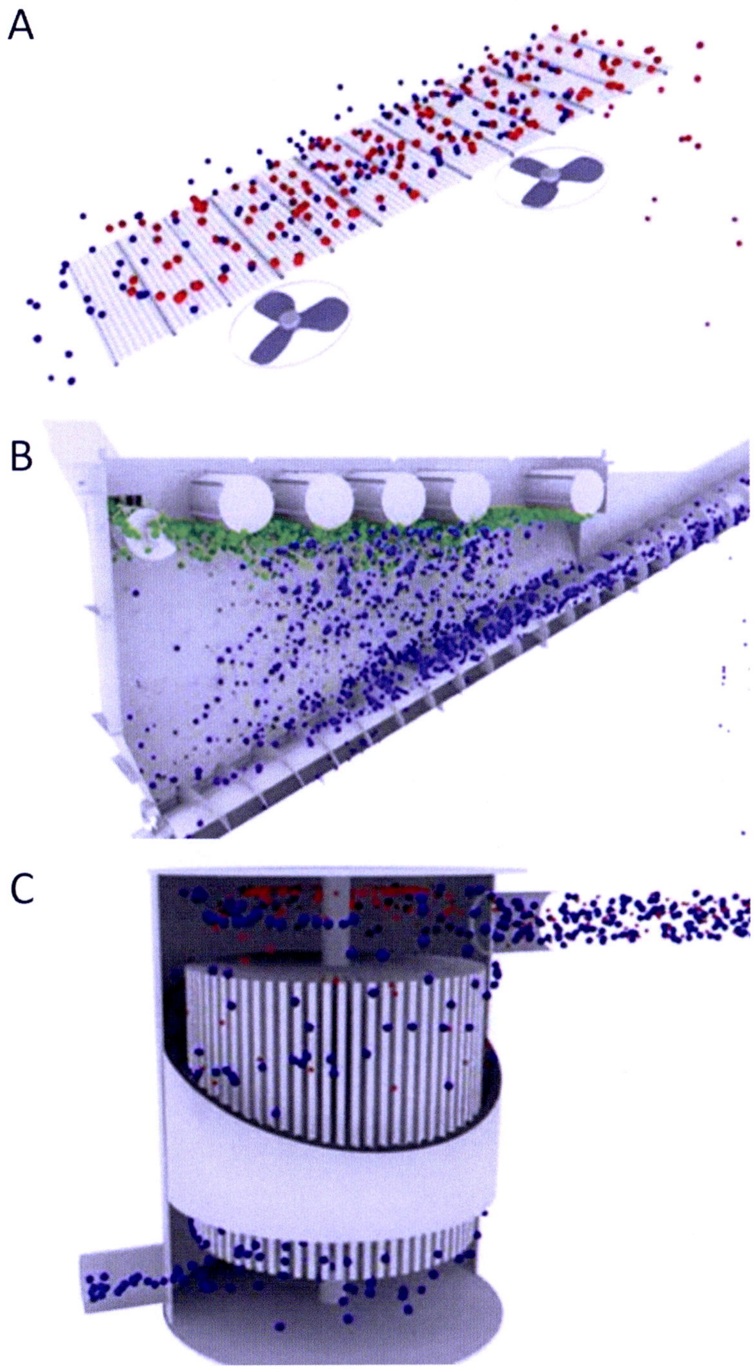

FIGURE 12.10 Specific equipment for separation of ASR: (A) Separating table, (B) sink-float bath, and (C) vertical rotor impact mill. *Courtesy: ARN.*

TABLE 12.4 Typical output fractions of ASR sorting installations and their applications (MSWI, municipal solid waste incinerator).

Material	Final application
Ferrous fractions of several grain sizes	Re-use as metal
Nonferrous fractions of several grain sizes	Re-use as metal
Fine minerals (consisting of glass, sand, paint and other materials)	Construction
Coarse minerals (also consisting of different materials)	Construction
Mixture of light plastics (PE, PP, and others)	Automotive (after regranulation)
Mixture of plastics with intermediate density (e.g., polycarbonate and talcum-filled PP)	Reducing agent (blast furnaces)
Mixture of plastics with high density (e.g., PVC and polyamides (nylon))	Energy recovery (MSWI)
Fibers (pelletized/nonpelletized)	Extrusion planks (pellets)/energy recovery (cement kiln/MSWI)
Residual streams of plastics with wood or elastomers or plastics sludges	Energy recovery (MSWI)
Dust	Energy recovery (MSWI)

FIGURE 12.11 Typical output fractions of an ASR sorting installation (from left to right): fine minerals 0-1 mm, plastics medium density and fibers. *Courtesy: ARN.*

density materials such as aluminum and carbon fiber reinforced plastics or are able to make parts lighter using high-strength steel (Dai et al., 2016). The increased variety in materials, and more importantly composite materials, complicates recycling. As more materials with similar physical properties, e.g., low density, are shredded and mixed, it becomes very complex to separate the materials and mixed waste streams only suitable only for low-quality recycling purposes are created, thus moving the circular economy further away instead of closer.

Technological improvements have led to a significant increase of embedded electronics in vehicles. A typical ELV in 2014 contained eight electric parts on average, whereas a new vehicle today holds 21 electric parts (Restrepo et al., 2017). With the current shift towards battery electric vehicles, the average number of embedded electronics will increase dramatically. In addition, several electronic parts are powered by small separate batteries, such as the alarm, GPS, or tire pressure sensors (IDIS, 2021). These electric and electronic parts, copper wires, and corresponding batteries are not dismantled since they are not easily accessible and finally end up in the ASR.

The metal sorting and the residue sorting, such as ASR sorting, will benefit from the decreasing cost of advanced sorting machines. Throughput levels of detector and camera-based sorting systems will increase further because of faster and cheaper image processing.

The change in vehicle drivetrains has large consequences for the composition of the car: a combustion engine mainly consisting of steel is replaced by an electric motor, which contains copper and a variety of rare earth metals such as neodymium. In addition, the amount of copper in a vehicle increases from 39 kg for a hybrid electric vehicle to 80 kg of copper for a battery electric vehicle (International Copper Association, 2017). However, the energy source transforms the composition of the car: the fuel tank made from a single plastic is replaced by a heavy battery containing several metals, such as aluminum for the casing and cobalt and nickel for the cathode. The battery weight impacts the recycling rate of the vehicle and thus compliance with the ELV Directive. The battery weighs 100 kg for an average plug-in hybrid vehicle (PHEV) and up to 700 kg for high-end BEVs, whereas a plastic fuel tank weighs about 14 kg (American Chemistry Council, 2021; ARN, 2021a).

In summary, current technical developments in the automotive industry complicate recycling significantly. However, as demand remains high for critical metals, such as gold for electronics, cobalt for batteries and neodymium for electromotors, and bulk metals like copper, and the production of low CO_2 EAF steel requires ever more steel scrap, recycling of these materials becomes more profitable.

12.8 CONCLUSIONS

Environmental legislation has been a big driver for professionalizing ELV recycling. Presently, developments in the automotive industry are mainly driven by the urge to reduce CO_2 emissions. This is achieved by shifting from ICEs towards electric vehicles as well as by reduction of the weight of the vehicle. Such trends require new materials, often composite materials, as well as innovative construction methods, leading to more complex and often more costly recycling techniques. The major recycling issues for ICE driven vehicles have been resolved; however, new recycling challenges are already expected as a result of continuous technological developments in vehicle design.

References

American Chemistry Council, 2021. Plastic Fuel Tanks can Decrease Vehicle Weight and Incidence of Corrosion. https://www.automotiveplastics.com/wp-content/uploads/FuelTank.pdf. (Accessed 22 January 2022).

ARN, 2021a. Personal Communication With Auto Recycling Nederland. Breukelen, the Netherlands.

ARN, 2021b. Infographic banden circulair. In: ARN duurzaamheidsverslag 2020. Available at: https://duurzaamheidsverslag2020.arn.nl/en/autorecycling-2/infographic-banden-circulair/. (Accessed 22 January 2022).

Bellis, M., 2019. History of the Automobile. ThoughtCo. https://www.thoughtco.com/automobile-history-1991458. (Accessed 22 January 2022).

Castro, P., Aráujo, J., Martinho, G., Pereiro, A., 2021. Waste management strategies to mitigate the effects of fluorinated greenhouse gases on climate change. Appl. Sci. 4367. https://doi.org/10.3390/app11104367.

Chen, Y., Kang, Y., Zhao, Y., Wang, L., Liu, J., Li, Y., Liang, Z., et al., 2021. A review of lithium-ion battery safety concerns: the issues, strategies, and testing standards. J. Energy Chem., 83–99. https://doi.org/10.1016/j.jechem.2020.10.017.

Cossu, R., Lai, T., 2015. Automotive shredder residue (ASR) management: an overview. Waste Manag., 143–151. https://doi.org/10.1016/j.wasman.2015.07.042.

Dai, Q., Kelly, J., Elgowainy, A., 2016. Vehicle Materials: Material Composition of U.S. Light-Duty Vehicles. Energy Systems Division, Argonne National Laboratory. https://greet.es.anl.gov/files/light-duty-vehicle-2016. (Accessed 22 January 2022).

Duesenfeld, 2021. Ecofriendly Recycling of Lithium-Ion Batteries. https://www.duesenfeld.com/recycling_en.html. (Accessed 22 January 2022).

EcarACCU, 2021. The Journey of a Refurbished Battery Cell. https://ecaraccu.nl/refurbished/. (Accessed 22 January 2022).

Elwert, T., Goldmann, D., Buchert, M., Römer, F., Merz, C., Schueler, D., Sutter, J., 2016. Current developments and challenges in the recycling of key components of (hybrid) electric vehicles. Recycling, 25–60. https://doi.org/10.3390/recycling1010025.

ETRMA, 2021. End of Life Tyres Management—Europe 2018 Status. European Tyre & Rubber Manufacturers' Association (ETRMA). https://www.etrma.org/wp-content/uploads/2020/09/Copy-of-ELT-Data-2018-002.pdf. (Accessed 22 January 2022).

European Commission, 2021. End-of-Life Vehicles. European Commission—DG Environment. https://ec.europa.eu/environment/topics/waste-and-recycling/end-life-vehicles_en. (Accessed 22 January 2022).

European Union, 1999. Council Directive 1999/31/EC of 26 April 1999 on the Landfill of Waste. European Union, pp. 0001–0019. Official Journal L 182, 16/07/1999 http://data.europa.eu/eli/dir/1999/31/oj.

European Union, 2000. Directive 2000/53/EC of the European Parliament and of the Council of 18 September 2000 on End-Of Life Vehicles. European Union, pp. 0034–0043. Official Journal L 269, 21/10/2000 http://data.europa.eu/eli/dir/2000/53/oj.

European Union, 2019. Regulation (EU) 2019/631 of the European Parliament and of the Council of 17 April 2019 Setting CO_2 Emission Performance Standards for New Passenger Cars and for New Light Commercial Vehicles. European Union, pp. 13–53. Official Journal L 111, 25.4.2019 http://data.europa.eu/eli/reg/2019/631/oj.

Eurostat, 2021. End-Of-Life Vehicle Statistics. https://ec.europa.eu/eurostat/statistics-explained/index.php?title=End-of-life_vehicle_statistics. (Accessed 22 January 2022).

Faessler, B., 2021. Stationary, second use battery energy storage systems and their applications: a research review. Energies, 2335. https://doi.org/10.3390/en14082335.

Harper, G., Sommerville, R., Kendrick, E., Driscoll, L., Slater, P., Stolkin, R., Walton, A., et al., 2019. Recycling litium-ion batteries from electric vehicles. Nature, 75–86. https://doi.org/10.1038/s41586-019-1682-5.

IDIS, 2021. The International Dismantling Information System. www.idis2.com. (Accessed 22 January 2022).

IEA, 2021. Global EV Outlook 2021. International Energy Agency (IEA), Paris. https://www.iea.org/reports/global-ev-outlook-2021.

Ignatenko, O., van Schaik, A., Reuter, M.A., 2008. Recycling system flexibility: the fundamental solution to achieve high energy and material recovery quotas. J. Clean. Prod., 432–449. https://doi.org/10.1016/j.jclepro.2006.07.048.

International Copper Association, 2017. Copper Intensity in the Electrification of Transport and the Integration of Energy Storage. Geopend. January 22, 2022 https://copperalliance.org/resource/copper-intensity-in-the-electrification-of-transport-and-the-integration-of-energy-storage/. (Accessed 22 January 2022).

Khodier, A., Williams, K., Dallison, N., 2018. Challenges around automotive shredder residue production and disposal. Waste Manag., 566–573. https://doi.org/10.1016/j.wasman.2017.05.008.

Kotak, Y., Marchante Fernández, C., Canals Casals, L., Satishbhai Kotak, B., Koch, D., Geisbauer, C., Trilla, L., Gómez-Núñez, A., Schweiger, H., 2021. End of electric vehicle batteries: reuse vs. recycle. Energies 14 (8), 2217. https://doi.org/10.3390/en14082217.

Manthiram, A., 2020. A reflection on lithium-ion battery cathode chemistry. Nat. Commun., 1550. https://doi.org/10.1038/s41467-020-15355-0.

McKerracher, C., 2021. Electric Vehicle Outlook 2021. Bloomberg NEF.

Monier, V., Salès, K., Lucet, L., Benhallam, R., 2017. Annual Report of the End-Of-Life Vehicle Sector Observatory—2015. ADEME. https://librairie.ademe.fr/dechets-economie-circulaire/807-automotives-end-of-life-vehicles-sector-observatory-2017.htm. (Accessed 22 January 2022).

Nembhard, N., 2019. Safe, Sustainable Discharge of Electric Vehicle Batteries as a Pre-Treatment Step to Crushing in the Recycling Process. KTH School of Industrial Engineering and Management, Stockholm. http://kth.diva-portal.org/smash/record.jsf?pid=diva2%3A1339943&dswid=8769. (Accessed 22 January 2022).

Pilllot, C., 2021. The rechargeable battery market and main trends 2020-2030. In: 26th International Congress for Battery Recycling, Geneva.

Pinheiro, C., Quina, M., Gando-Ferreira, L., 2020. Management of waste lubricant oil in Europe: a circular economy approach. Crit. Rev. Environ. Sci. Technol., 2015–2050. https://doi.org/10.1080/10643389.2020.1771887.

Restrepo, E., Lovik, A., Wäger, P., Widmer, R., Lonka, R., Müller, D., 2017. Stocks, flows, and distribution of critical metals in embedded electronics in passenger vehicles. Environ. Sci. Technol., 1129–1139. https://doi.org/10.1021/acs.est.6b05743.

Reuter, M., van Schaik, A., Ignatenko, O., de Haan, G., 2006. Fundamental limits for the recycling of end-of-life vehicles. Miner. Eng., 433–449. https://doi.org/10.1016/j.mineng.2005.08.014.

Sabihuddin, S., Kiprakis, A., Mueller, M., 2015. A numerical and graphical review of energy storage technologies. Energies, 172–216. https://doi.org/10.3390/en8010172.

Sato, F., Furubayashi, T., Nakata, T., 2018. Energy and CO_2 benefit assessment of reused vehicle parts through a material flow approach. Int. J. Autom. Eng. 9 (2), 91–98. https://www.jstage.jst.go.jp/article/jsaeijae/9/2/9_20184094/_pdf/-char/en.

Schäfer, J., Singer, R., Hofmann, J., Fleischer, J., 2020. Challenges and solutions of automated disassembly and condition-based remanufacturing of lithium-ion battery modules for a circular economy. Procedia Manuf., 614–619. https://doi.org/10.1016/j.promfg.2020.02.145.

Sortbat, 2021. Dismantling EV Batteries. Geopend. January 22, 2022 https://www.sortbat.be/expertise/dismantling. (Accessed 22 January 2022).

Statistics Netherlands (CBS), 2019. Number of All-Electric Cars has Doubled. https://www.cbs.nl/en-gb/news/2019/19/number-of-all-electric-cars-has-doubled. (Accessed 22 January 2022).

Taylor, P., 2021. Tires and the tire market. In: Tire Waste and Recycling. Elsevier, pp. 15–25, https://doi.org/10.1016/B978-0-12-820685-0.00022-3.

TotalEnergies, 2021. Brake Fluid Composition. 15 November https://lubricants.totalenergies.com/brake-fluid-composition?field_topics_tid=18. (Accessed 22 January 2022).

Toyota, 2017. Vehicle Recycling. Toyota Motor Corporation. https://global.toyota/en/sustainability/report/vehicle-recycling/. (Accessed 22 January 2022).

Umicore, 2021. Personal Communication.

van Oostenrijk, K., 2021. AVV-Monitor Autobanden 2020. RecyBEM, Den Haag. https://www.recybem.nl/sites/recybem.nl/files/user/avv_monitor_autobanden_2020.pdf. (Accessed 22 January 2022).

van Schaik, A., Reuter, M., Boin, U., Dalmijn, W., 2002. Dynamic modelling and optimisation of the resource cycle of passenger vehicles. Miner. Eng., 1001–1016. https://doi.org/10.1016/S0892-6875(02)00080-8.

VROM, 2007. Activiteitenregeling milieubeheer. Ministerie van Volkshuisvesting, Ruimtelijke Ordening en Milieubeheer, The Netherlands. https://wetten.overheid.nl/BWBR0022830/2021-10-16. (Accessed 7 December 2021).

Westfalen Gassen, 2021. Return Process for Used Refrigerants. Westfalen Gassen Nederland BV. https://westfalen.com/fileadmin/user_upload/Dateien_NL/PDF/recycling_flyer_compressed.pdf. (Accessed 22 January 2022).

Yoon, J., 2021. Battery Recalls and Supply Crunch Challenge Electric Vehicle Revolution. Financial Times. 7 September www.ft.com. (Accessed 22 January 2022).

Further Reading

US EPA, 2021. Environment Protection Agency. 13 December https://www.epa.ie/our-services/monitoring–assessment/waste/national-waste-statistics/tyres/. (Accessed 22 January 2022).

CHAPTER

13

Electrical and electronic equipment (WEEE)

Mathias Schluep and Fabian Ottiger
World Resources Forum Association, St. Gallen, Switzerland

13.1 INTRODUCTION

The management of waste electrical and electronic equipment (WEEE), or e-waste for short, has seen greater development of professionalization over the last decade. Experiences around the globe have shown that specialized takeback schemes are required to adequately address this complex waste stream. Two paradigms have spurred these developments: a *circular economy* and the *extended producer responsibility (EPR).*

13.1.1 Policy Development

Some countries, mainly in Europe, started to experiment with new approaches in managing this high-tech waste stream more than 30 years ago. The European WEEE Directive (European Union, 2012), based on the concept of extended producer responsibility (EPR) (see Chapter 39—Extended producer responsibility) as an environmental policy, has set the global pace and standard in regulating e-waste management. As a result, producers accepted the end-of-life responsibility for their products and initiated producer responsible organizations (PROs) to manage the material flows and the financing of unprofitable processing steps. The recycling industry went through a rapid evolution where specialists emerged among others for manual dismantling, mechanical processing, or final material recovery of secondary raw materials. The legislators carefully developed regulations defining responsibilities and promoting a constant improvement of the system efficiency. And last, but not least, the consumer, from the big corporate to the small household, increasingly demanded a convenient and sustainable option to dispose of e-waste.

According to a recent e-waste policy study (GSMA, 2020), 78 countries have either a policy, legislation, or regulation governing e-waste in place. This covers 71% of the world's population and marks an improvement of roughly doubling the coverage of regulative e-waste management in one decade. Although this sounds promising, in many countries, policies are nonlegally binding strategies, and must be understood as recommendations mainly (Forti et al., 2020). In many countries in the Global South, professional e-waste management is still not available. Across Africa and Asia, for example, there are

 https://doi.org/10.1016/B978-0-323-85514-3.00020-8

only 19 countries with legally binding legislation on e-waste, five countries with an e-waste policy but nonlegally binding legislation, and 31 countries with policies in development (GSMA, 2020).

13.1.2 Objectives of E-Waste Management

E-waste is usually regarded as a waste problem, which can cause environmental damage if not dealt with in an appropriate way. However, the enormous resource impact of EEE is widely overlooked. EEE is a major driver for the development of demand and prices for a range of metals. Whereas the global copper consumption was at approximately 25 Mt in 2020 (ICSG, 2021), it is estimated that EEE accounts for roughly 30% of the total consumption (Deetman et al., 2018). From a world production of 8.5 Mt of silicon (USGS, 2022), approximately 21% is refined to semiconductor purity and subsequently used in electronics (Schnebele, 2021). Around 10% of the world's gold production was used in electronics in 2017 (George, 2020). Consequently, inappropriate disposal of e-waste not only leads to significant environmental problems but also to a systematic loss of secondary materials (Hagelüken and Meskers, 2008). Hence the appropriate handling of e-waste can both prevent serious environmental damage and recover valuable materials.

Besides the positive impact on resources, state-of-the-art recycling operations also contribute to reducing greenhouse gas emissions (Hischier et al., 2005). Primary production, i.e., mining, concentrating, smelting, and refining, especially of precious and special metals is energy intensive and hence has a significant CO_2 impact. "Mining" our old computers to recover the contained metals—if done in an environmentally sound manner—needs only a fraction of this energy input (Hagelüken and Meskers, 2008).

Furthermore, the environmentally sound management of end-of-life refrigerators, air-conditioners, and similar equipment is significant in mitigating the climate change impact (Menikpura et al., 2014). The ozone-depleting substances in these devices, such as chlorofluorocarbon (CFC), hydrochlorofluorocarbons (HCFCs), fluorocarbons (FCs), and hydrofluorocarbons (HFCs), have a very high global-warming potential (Swico, SENS, SLRS, 2021) and should not escape into the environment.

Hence, the main services a comprehensive WEEE or e-waste management system has to deliver in order to ensure sustainability are:

- collection of all e-waste,
- recovery of valuables such as secondary raw materials using state-of-the-art methods and technologies and
- segregation and safe disposal of hazardous waste.

Costs for unprofitable processes and proper disposal as well as for administration, monitoring, and control to ensure quality have to be associated with all of these activities (Huisman and Magalini, 2018).

13.2 WASTE CHARACTERIZATION

The most widely used classification of WEEE refers to the EU WEEE Directive (European Union, 2012). Originally the WEEE directive listed 10 categories, which however were subject to a transitional period until 2018. From August 2018 all Electrical and Electronic Equipment (EEE) are classified within the six categories set out in Annex III of the EU WEEE directive (Table 13.1). It is estimated that in Europe an average person owns 44 EEE products plus another 45 (energy saving) lamps and light fittings (Huisman and Leroy, 2017). Also, EEE products from commercial, industrial, or institutional organizations that are similar to household products are considered WEEE.

TABLE 13.1 Categories of EEE covered by the EU WEEE Directive after the transitional period (after August 2018) and the estimated average amount and corresponding weight an average European person owned in 2015.

	Category	Amount per person	Weight per person (kg)
1.	Temperature exchange equipment	1.1	52
2.	Screens, monitors, and equipment containing screens having a surface greater than 100 cm^2	3.2	21
3.	Lamp	45	15
4.	Large equipment (any external dimension more than 50 cm) including, but not limited to: Household appliances; IT and telecommunication equipment; consumer equipment; luminaires; equipment reproducing sound or images, musical equipment; electrical and electronic tools; toys, leisure and sports equipment; medical devices; monitoring and control instruments; automatic dispensers; equipment for the generation of electric currents*. This category does not include equipment included in categories 1 to 3	3.8	99
5.	Small equipment (no external dimension more than 50 cm) including, but not limited to: Household appliances; consumer equipment; luminaires; equipment reproducing sound or images, musical equipment; electrical and electronic tools; toys, leisure and sports equipment; medical devices; monitoring and control instruments; automatic dispensers; equipment for the generation of electric currents*. This category does not include equipment included in categories 1 to 3 and 6	27	49
6.	Small IT and telecommunication equipment (no external dimension more than 50 cm)	9.3	12

**E.g., photovoltaic panels or small equipment with integrated photovoltaic panels.*
Adapted from Huisman and Leroy (2017).

Electronics have gone through a rapid evolution both in terms of functionality and in the use of raw materials. While in the 1980s a typical analog phone required an average of 12 chemical elements, this number increased to over 60 elements nowadays for modern smartphone technology (Zepf et al., 2014). In addition to a complex material mix, mechanical properties, the polymer composition, and the chemical composition of WEEE vary not only between equipment types with different functions, but also between single appliances within one equipment type (Chancerel and Rotter, 2009). As a result, a particular WEEE category is very heterogeneous and subject to changes over time. The increased use of many compounds, bound together in a complex manner for functional reasons, dictates the physical and chemical properties of the materials available for recycling. Therefore effective recycling requires a product-centric view rather than a material-centric view (Reuter et al., 2013).

Metals make up the largest fraction of recyclable materials in WEEE (62%), followed by plastic/metal mixtures (17%), and plastics (9%). The particularly valuable printed circuit boards account for roughly 1.4% of the total quantity (Swico, SENS, SLRS, 2021). It has to be noted that these ratios represent averages over all WEEE categories and greatly vary between individual categories. E.g., the share of plastics in small equipment can be as high as 30% (Wäger et al., 2009). E-waste is categorized as hazardous waste due to the presence of hazardous substances according to the Basel Convention (UN Environment, 1989). The share of hazardous substances is approximately 1%, whereas more than half of the pollutants stem from batteries, followed by capacitors and components containing mercury.

These numbers do not include other problematic substances found as additives in plastics. E-waste consists of plastics in the form of various polymers, mainly ABS, PP, PS, and PC-ABS. Such plastics, however, contain a wide range of additives such as flame retardants, fillers, pigments, and stabilizers that collectively impact the recycling of WEEE plastics (Haarman et al., 2020). Some brominated flame retardants (BFRs) are known as persistent organic pollutants (POPs) and therefore are restricted for their use in newly marketed EEE (European Union, 2003). These substances are still found in WEEE (mainly octaBDE and decaBDE used in external housings and HBCD used in foams) and need to be removed during the treatment process. It must be noted that the problematic BFRs are mainly found in older plastic casings of screens and small household appliances for high-temperature applications (Wäger et al., 2012). Recent studies indicate that BFRs in WEEE plastics have decreased significantly over the last 10 years (Haarman et al., 2020): a decrease of Br levels in CRT screens by 50% and by 40% in small appliances between 2010 and 2014–15.

According to the Global E-waste Monitor 2020 (Forti et al., 2020) the world generated approximately 53.6 Mt of e-waste in 2019, which is equal to 7.3 kg/capita. The global e-Waste Monitor applies a combined methodology to estimate e-waste generation, based on empirical data from trade statistics as well as a sales lifespan modeling for individual appliances. Although the numbers might miss local characteristics, the Global E-waste Monitor provides the only reliable e-waste statistics on a global scale and gives an excellent indication of the size and evolution of this waste stream in different parts of the world. For more detailed data on country level, selected regional and national waste monitors are available (UNU & UNITAR, Sustainable Cycles (SCYCLE) Programme, 2021). National reports, inventories, and illegal trade assessments can be accessed through the LibraWEEE website (WEEE Forum, 2021).

Consumption and waste generation volumes vary greatly between the world regions, Europe producing the highest amount of WEEE (16.2 kg/capita) and Africa being at the lower end (2.5 kg/capita). Related to a varying purchasing power in different regions, the type and age of appliances consumed and the subsequent composition of WEEE generated can also vary (Forti et al., 2020). In addition, the product lifespan and the subsequent storage time before an appliance is disposed of are varying factors for the estimation of WEEE flows (Thiébaud et al., 2018). Cultural and economic aspects—such as hoarding, reuse, and repair—influence it. Lower income countries often have a different view on what is still usable or not. As a result, appliances found in collected WEEE tend to be older than, e.g., in OECD countries (Secretariat of the Basel Convention, 2011).

13.3 RECYCLING CHAIN AND TECHNOLOGIES

The recycling chain for WEEE consists of three steps (Figure 13.1):

- Collection of end-of-life appliances,
- preprocessing of appliances into material fractions, and
- end processing (including final material recovery and disposal) into secondary raw materials as a marketable product.

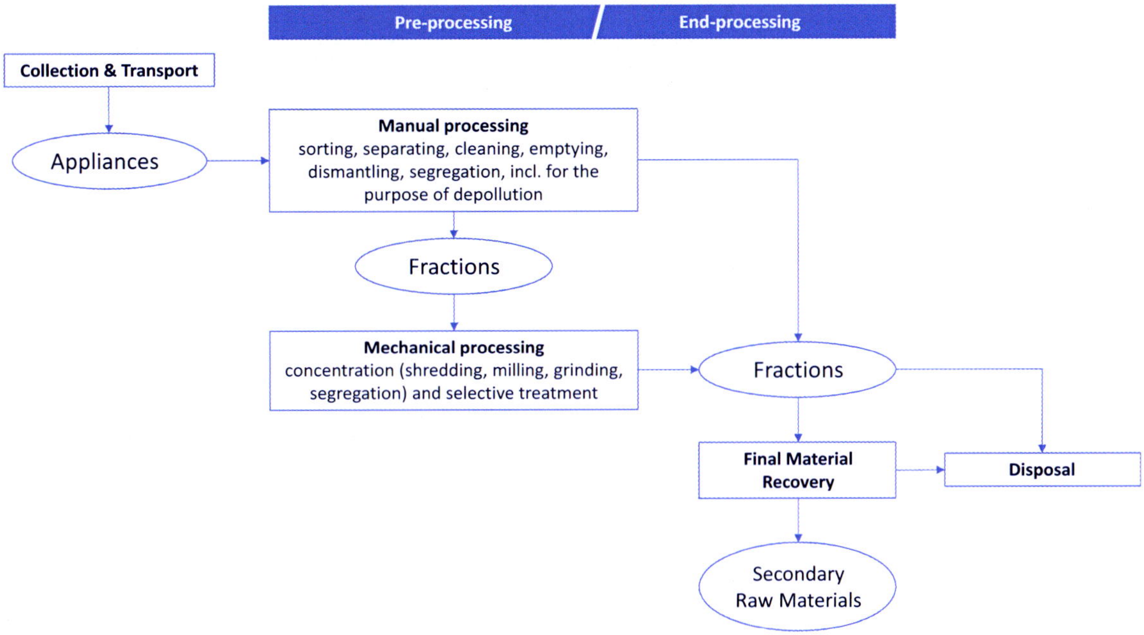

FIGURE 13.1 Schematic overview of the key processes in the recycling chain.

For each of these steps and WEEE category, specialized operators and facilities exist. This infrastructure is also used to process (industrial) equipment or fractions thereof that are similar to WEEE, such as large IT and telecom infrastructure and photovoltaic energy generation. The material recovery efficiency of the entire recycling chain depends on the efficiency of each step and on how well the interfaces between these interdependent steps are managed. If, for example, for a material the efficiency of collection is 50%, the combined preprocessing efficiency is 70% and the end-processing (materials recovery) efficiency is 95%, the resulting net material yield along the chain would be only 33%. High efficiencies in each step, starting from the beginning of the chain, are thus essential. One can have very efficient preprocessing and end-processing steps, but if collection is low then the overall recycling efficiency will remain low despite improvements in preprocessing and end-processing technology.

The preprocessing step involves receiving whole appliances from collection and transport activities and includes both manual and mechanical processing. During the manual processing step (manual dismantling), appliances are taken apart into fractions. Some can be sent directly to end-processing, such as printed circuit boards, metals, or plastic housings. Fractions containing a mix of materials, which cannot be separated further manually, require further mechanical processing to obtain fractions suitable for end-processing. Manual processing is crucial for the purpose of depollution of appliances and fractions, separating hazardous from valuable materials, respectively.

Steps following preprocessing are usually considered as processes leading to secondary raw materials as marketable products. This includes metal refining, which usually entails technologies with high investment requirements, such as for large integrated smelters, and is subject to an international market.

13.3.1 Collection and Transport

The first step in the recycling chain is the collection and subsequent transport to processing locations. Collection is of fundamental importance for all WEEE collection groups, as it determines the amount of WEEE that enters the recycling chain and thus the overall recycling efficiency as well as the necessary downstream processing capacity. When discussing collection systems and recycling technologies, it is crucial to include social and societal factors, as they are key to a high collection rate. Building awareness and environmental responsibility at the consumer level is crucial for a successful collection system (see also Chapter 37—Information and communication). Other factors influencing the collection rate are legal conditions. For example EPR, if well implemented, contributes to a higher collection rate.

Despite clear legal conditions, achieving high collection rates is still a challenge for takeback schemes. In Europe the average collection of WEEE was 9.5 kg/capita in 2018, revealing a large difference between countries, spanning 6.2–13.5 kg/capita (Balde et al., 2020). A combination of interlinked factors is affecting these collection rates and can be grouped into:

- other WEEE flows outside of the formal WEEE system,
- how the WEEE legislation has been implemented at the national level, and
- behavioral and economic factors.

Balde et al. (2020) estimated that the other WEEE flows, outside of the formal WEEE system, are equal to the formally collected waste volume. The distribution of formal collection and other WEEE flows is depicted in Figure 13.2.

Generally, investment and technology requirements are low in this step of the recycling chain. Typical means for collection include specific collection points at convenient locations, and designated companies specializing in collection. Especially in the Global South, a key factor

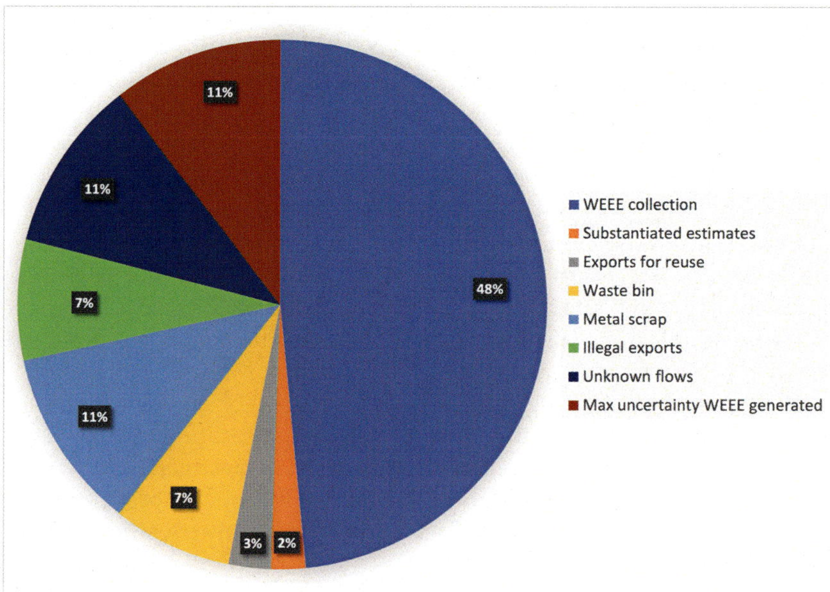

FIGURE 13.2 Average distribution of accounted formal WEEE collection and estimated other WEEE flows in Europe. *Adapted from Balde et al. (2020).*

in collection is the informal sector engaged in waste picking (for more details refer to Chapter 7—The path to inclusive recycling). Informal waste sectors are often organized in a network of individuals and small businesses of collectors. As many poor people rely on small incomes generated in this chain, impressive collection rates of up to 95% of waste generated are achieved (Secretariat of the Basel Convention, 2011), which is far above what can be achieved by today's formalized takeback schemes, which are typically found between 30% and 60% (Balde et al., 2020).

The type of transport depends on the type of appliances. For example, lamps, CRT screens, and monitors or refrigerators must be transported with more care than small telecommunication and IT equipment, due to the risk of leakage of hazardous substances when the appliances are damaged.

13.3.2 Manual Preprocessing

The goal of the manual preprocessing step in the recycling chain is to separate materials from WEEE and to prepare them for adequate treatment in the final steps of the recycling chain, thus separate appliances into fractions that fit in the end-processing step available. This also means that hazardous substances must be removed and either treated or stored safely. This step is done to avoid the dilution of and/or contamination of other fractions with toxic substances during the downstream processes. Critical components include, e.g., lead glass from CRT screens, light bulbs, and batteries.

At the same time, valuable components can either be reused directly or directed to efficient recovery processes or adequate subsequent final treatment processes.

Preprocessing requires less capital and technology investments compared to end-processing, as it often relies on manual labor or requires simpler mechanical size reduction and separation technologies at lower costs (Wang et al., 2012). To maximize the recovery yield of certain materials, manual processing is often the preferred option over a pure mechanical process. E.g., the manual removal of printed wiring boards (PWBs) from electronics will always be most effective from a precious metal point of view (Meskers et al., 2009), as intensive size reduction of PWB-containing equipment spreads the precious metals over all fractions, including fractions from which they cannot be recovered (see Metal Wheel, Figure 2.2 in Chapter 2 in this book) (Chancerel et al., 2009). Groups of components best removed by hand for producing the cleanest recyclates include PWBs, wires, parts made out of steel, wood and plastic housings, massive aluminum parts (e.g., heat sinks, housing), and copper coils and getters from CRT screens (Reuter et al., 2013). However, manual dismantling has its limits; at some point in the process, semimechanical treatment of the remaining fractions becomes the better option, from both an economical and an eco-efficiency perspective (Gmünder, 2007).

13.3.3 Mechanical Preprocessing

Mechanical preprocessing, the next step in e-waste treatment, is normally an industrial and rather large-scale operation to concentrate recyclable materials in a dedicated fraction suitable for end-processing, as well as to further separate hazardous materials. For a complex waste stream, such as WEEE, this means physically separating the materials and sorting them into its constituent parts and into clean fractions. This is also called "liberation," i.e., liberating materials from each other by physical force (Reuter et al., 2013). Typical components of a mechanical processing plant are crushing units, shredders, and magnetic, eddy-current, and air separators. The gas emissions are filtered, and effluents are treated to minimize environmental impact. Separation technologies are discussed in detail in Chapters 8 and 9 (Physical separation and Sensor-based sorting) and Chapter 43—Physical separation.

Although a variety of separation technologies exist, none of them will ultimately lead to pure streams of materials but will only increase purity, as not all materials can be fully liberated. There will always be particles consisting of two or more materials. Limitations of separation technologies to liberate materials are given by the complexity of the material composition, large differences in physical and chemical characteristics, and connected materials that affect the physical properties and therefore separation (Reuter et al., 2013). More information about the impact on sorting performance can be retrieved from Chapters 5, 8, and 43.

As is often the case for WEEE, the recyclers deal with a complex mix of materials. This is problematic since the physical property distributions used in separation may not be compatible with the (often not well known) chemical property distributions. This is important for end-processing, as the degree of liberation of the different materials determines the quality and recyclability of materials in different recycling streams (Van Schaik et al., 2004, Chapter 5—Material and product-centric recycling and design for recycling rules and digital methods).

Mechanical processing typically includes the following processing types:

- Concentration processes: crushing (shredding, milling), sizing (sieving and screening, separation based on particle dimension), sorting (magnetic, electric field, sensor based, by density, by force separator, or other physical characteristics), removal of dust.
- Selective treatment: For selective treatment of WEEE, the European Committee for Electrotechnical Standardization (CENELEC, 2014) offers a comprehensive overview of requirements for the removal and separate treatment of substances, mixtures, and components with hazardous characteristics.

13.3.4 End-Processing and Disposal

Most of the fractions recovered in the preprocessing steps need to be refined, in the case of metals, or conditioned to be sold as secondary raw materials. Some residues may require special treatment to be disposed of in a final disposal site.

During the end-processing step, three flows of materials are paid special attention: metals, plastics, and glass. For metal refining, from metal or printed circuit board fractions (Chapter 25—Precious and technology metals), the obtained metals have the same quality as those from primary production. The same is valid for glass. For plastics the picture is different, as discussed in Chapter 31—Plastic. The recovery yield in refining processes depends strongly on the quality of the input stream, and the performance of the process itself. The limitation of recycling is given by how well the interfaces of the three main steps in the recycling chain (as described previously) are managed. An optimized collection system and liberation process through manual and mechanical separation techniques will lead to a higher quality (i.e., with lower harmful impurity content) or optimized material stream entering the refining process.

Nonrecoverable fractions or hazardous waste fractions need to be disposed of in hazardous waste treatment facilities, in special landfills, or sent for incineration.

End-processing of the different material fractions is discussed in more detail in Part 3—Recycling from a materials perspective of this book.

13.4 FUTURE DEVELOPMENTS

13.4.1 The Miniaturization Paradox

It is likely that the quantities of discarded EEE will increase substantially in the foreseeable future because of fast innovation cycles and increased market penetration of cheap appliances, the latter being the main driver of e-waste volumes in developing countries and countries in transition (Schluep et al., 2013).

An important long-term trend that affects the waste flows is the ever-higher integration and miniaturization of digital electronics and, related to this, an increasing complexity of the material composition (Bangs et al., 2016). The physical mass needed to provide capacities for storing, processing, and transmitting data is decreasing at a rapid pace (roughly along the lines of Moore's Law), which leads to a decrease of the average physical mass per device in use, despite increasing functionality. It has been observed using the example of mobile phones that this trend does not lead to a decrease in total mass flow, because at the same time the number of devices is increasing faster (Hilty et al., 2005). Historically, this "miniaturization paradox" can be explained by the general trend that processing capacity "is getting cheaper faster than it is getting smaller" (Hilty, 2008). The miniaturization paradox has three effects on e-waste streams:

1. Total mass flow increases despite smaller and more lightweight devices.
2. More devices enter other waste streams because they are small and unremarkable; this trend increases with embedded electronics in textiles, smart buildings, Internet of Things, their increasing usage in (electrical) mobility, etc.
3. Informal recycling becomes more difficult because of the higher integration density of the devices.

E-waste management, the formal as well as the informal recycling industry, and even the definition of e-waste will have to adapt to this general trend if the dissipation of valuable materials is to be slowed down in the long term.

13.4.2 Critical Raw Materials

A strong focus has been set on critical raw materials (CRMs) contained in EEE in the last decade. Among others, the European Union has created a list of critical raw materials (CRMs), combining raw materials of high importance to the EU economy and of high risk associated with their supply (European Commission, 2020). Relevant concentrations of CRMs can be found in the following types of equipment and components (Manoochehri et al., 2021) (Figure 13.3):

1. Printed circuit boards from IT equipment, hard disc drives, and optical disc drives.
2. Batteries from WEEE and end-of-life vehicles.
3. Neodymium iron boron magnets (NdFeB magnets) from hard disc drives and electrical engines of end-of-life vehicles (ELVs).
4. Fluorescent powders from cathode ray tubes (CRTs; in TVs and monitors) and fluorescent lamps.

CRMs are currently not adequately addressed in e-waste management strategies and related standards. Together with comparable low primary raw material prices for CRMs and missing incentives for the recycling industry, individual CRMs are currently not recovered, or recovered at a very low rate. New developments, however, are addressing the issue through, e.g., new standards for CRM recycling. Closely related to the existing CENELEC WEEE standard (CENELEC, 2014), a new certification scheme for collection, transport, and treatment facilities of key types of waste containing significant amounts of valuable and critical raw materials was developed in 2021 under the umbrella of the Horizon 2020 CEWASTE project (Manoochehri et al., 2021). As standardization measures only will not trigger higher recovery rates for CRMs, the project suggested that legislation should require recovery of specific CRMs in future recasts. In addition, financial and fiscal incentives should spur the economic viability of recovering CRMs. This should include the general promotion of using secondary CRMs in new products and market platforms where demand for recycled components, materials, and CRMs meets supply.

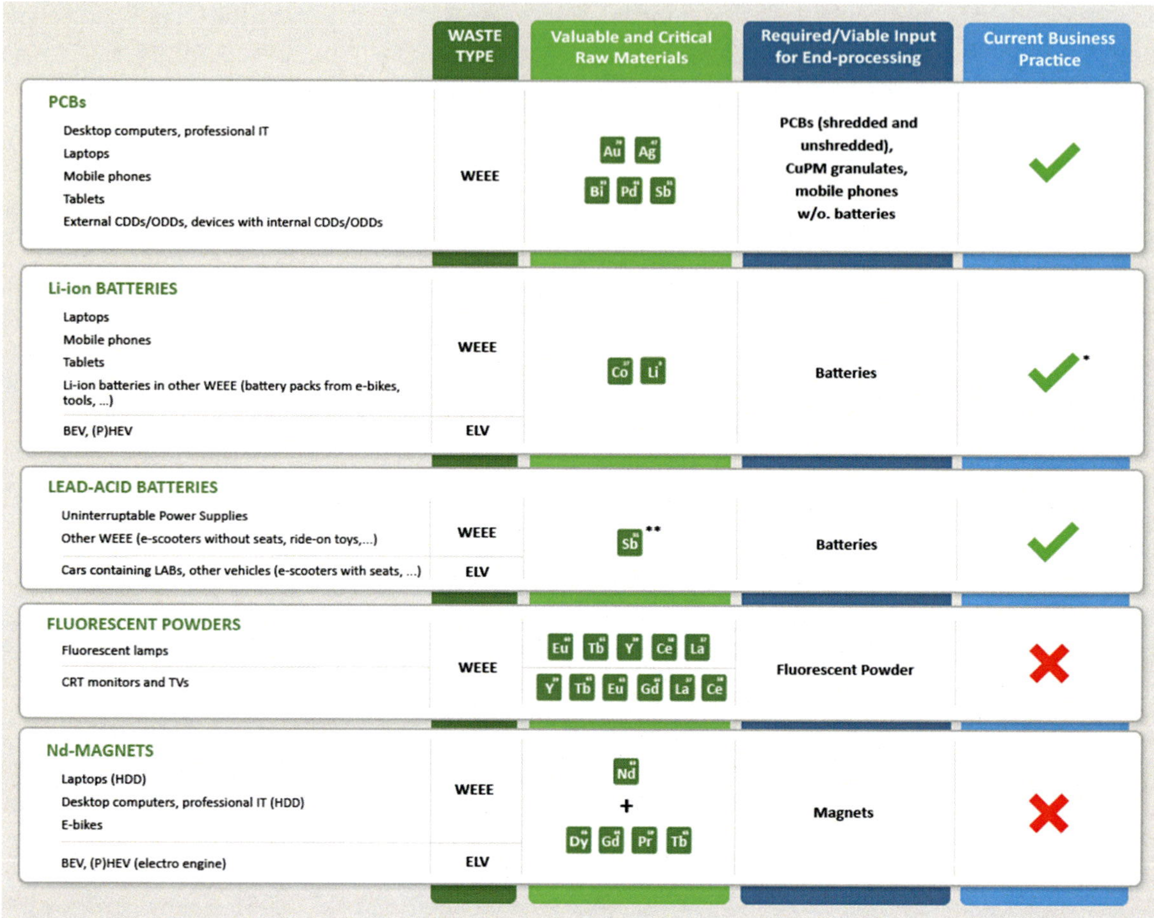

FIGURE 13.3 Key fractions and equipment from WEEE containing relevant amounts of critical raw materials. *Reproduced with permission from Manoochehri et al. (2021).*

Actors involved in the collection of e-waste should raise awareness of the importance of recycling of CRMs, while producer responsibility organizations (PROs) should consider consolidating fractions of CRM-rich products in adequate quantities ("clustering") to make recycling attractive. It is also thought that actors in the value chain should be in the position to access information on the CRM rich components, thereby making monitoring of actual recycling of CRMs easier.

13.4.3 Basel Convention vs Circular Economy

The Basel Convention (UN Environment, 1989) was adopted more than three decades ago and remains a bulwark in combating the environmental justice of "the toxic trade" of wastes. Of special concern have been the highly adverse environmental and social effects of crude and unsound treatment. The question of whether, and under what conditions, exports of nonfunctional hazardous electronic equipment should be allowed, has been hotly debated. With the current intention to move from a linear to a circular economy on a global level the Basel Convention has come under scrutiny by some producers of electronics as well as some in the recycling industry. The main purpose of the Convention, to combat toxic trade and promote minimization of waste and its cross-border movement, is indisputable. However, concerns are raised about whether the Convention is hindering a transition to a circular economy. Some stakeholders from the recycling industry voice strong doubts as to whether the current rules and notification procedure is compatible with the globalized nature of supply chains and does not facilitate the need to give used products a second life and efficiently recover secondary raw materials from complex waste streams.

Stakeholders share the view that the concepts of the Basel Convention and Circular Economy are in principle very compatible, but improvements to the implementation process of the Basel rules need to be made, especially in terms of traceability, digitalization, transparency, and efficiency (World Resources Forum Association, 2021). The Basel Convention was installed to protect poor countries from the negative externalities of linear thinking. The recycling industry, however, especially points to persistent problems with implementing the Basel Convention properly in developing countries (PREVENT, 2022). To increase capacity, the process should be digitalized and be made more transparent, becoming easier and quicker. Additional capacity building, expanding recycling companies around the world, and establishing bilateral cooperation between countries could help to create jobs on both sides and support the implementation of the Basel Convention. In terms of repair and secondhand goods, there are still loopholes in the Convention that need to be looked at to ensure that nonfunctional, nonrepairable items are not ending up in developing countries.

13.4.4 Towards a Global Harmonization of Extended Producer Responsibility

Extended producer responsibility (EPR) schemes (Chapter 39—Extended producer responsibility) have been successfully introduced to tackle the issue of postconsumer products over the last two decades, such as for WEEE. Typically, EPR is implemented at the national level with diverse levels of implementation and enforcement on one hand, and with many countries not having such policies in place at all on the other hand. This creates a major burden for producers to comply with a variety of approaches. At the same time, there is a need for locally adapted approaches, since there is no one solution that fits every political and cultural context. That's also why responsibility is not only mandated to the producer of consumer products, but this is often distributed across different actors in the value chain. In addition, EPR was mainstreamed at a time when the mindset was focusing on "solutions for a waste problem" rather than on "strategies to improve the circularity of products."

Recent debates (World Resources Forum Association, 2021) are approaching this issue and suggesting that EPR needs a harmonization process on a global level. More discussions will go into the feasibility of a global EPR scheme. Related targets need to be feasible, reachable, and scalable across locations and defined in the format of global treaties, which sets some specific targets on waste reduction and infrastructure development throughout the entire supply chain of wastes. Strong incentives will be required for companies in different markets, such as not to settle on the lowest common denominator and ensure a level playing field and genuine progress.

13.4.5 Responsible Sourcing of Secondary Raw Materials

Secondary raw materials are increasingly sourced from a global market, including from countries in the Global South. Driven by the value of secondary raw materials, developing countries have seen a steep growth in recycling activities over the past decades. Individuals and businesses in the recycling sector are providing economic subsistence to themselves and their communities and are contributing positively to the circular economy. In the absence of appropriate rules, control mechanisms, and technical capacities, however, their activities often have negative impacts on humans and the environment. Collection and crude recycling techniques of e-waste mainly happens in the so-called informal sector, causing various issues, such as exploitation of the most vulnerable individuals (including child labor), tax avoidance, and illegal practices, as well as unfair competitive advantages over formal players through the externalization of environmental and social costs (for more details see also Chapter 7—The path to inclusive recycling).

Such issues point towards responsible sourcing also becoming a requirement for secondary raw materials. With the absence of transparent supply chains and control mechanisms, a notion of unsound and unfair business practices will be increasingly associated with secondary raw materials sourced from the Global South, similar to the issues of conflict minerals in primary raw materials supply chains. Though sustainability standards related to the responsible sourcing of secondary raw materials are in their infancy, an early framework addressing the concept of inclusive recycling, allowing the fair trade of secondary raw materials while including and developing the prevailing informal sector in developing countries, was published in 2017 as an International Workshop Agreement under ISO (ISO, 2017). Based on this framework, a corresponding ISO standard addressing "Principles, sustainability, and traceability requirements for the management of secondary materials" is currently being developed (ISO, 2023). The private industry is also slowly increasing their efforts to address responsible sourcing for secondary resources, such as Apple, who supports responsible sourcing for primary and secondary materials through their conflict minerals approach, which includes recycling facilities (Apple, 2020).

13.5 CONCLUSIONS

As pointed out in this chapter, WEEE represents a complex waste stream, not only in relation to its changing composition, but also to its requirements for an efficient recovery of valuable and critical raw materials. This presents a constant challenge to all actors in the value chain, who need to apply a product-centric view rather than a material-centric view in the design of effective collection systems as well as the development and application of recycling technologies. In addition, a sustainable management of WEEE requires enabling framework conditions, providing a level playing field as well as economically viable and transparent value chains.

References

Apple, 2020. Conflict Minerals Report 2020. Apple Inc., Cupertino, California. https://www.apple.com/supplier-responsibility/pdf/Apple-Conflict-Minerals-Report.pdf.

Balde, C.P., Wagner, M., Iattoni, G., Kuehr, R., 2020. In-depth Review of the WEEE Collection Rates and Targets in the EU-28, Norway, Switzerland, and Iceland. United Nations University (UNU)/United Nations Institute for Training and Research (UNITAR) – co-hosting the SCYCLE Programme, Bonn. https://weee-forum.org/wp-content/uploads/2020/11/In-depth-review_WEEE-Collection-Targets-and-Rates_UNITAR_2020_Final.pdf.

Bangs, C., Meskers, C., Van Kerckhoven, T., 2016. Trends in electronic products – the canary in the urban mine? In: Electronic Goes Green 2016+, Belin, Germany.

CENELEC, 2014. EN 50625-1, Collection, Logistics & Treatment Requirements for WEEE – Part 1: General Treatment Requirements. European Committee for Electrotechnical Standardization.

Chancerel, P., Rotter, S., 2009. Recycling-oriented characterization of small waste electrical and electronic equipment. Waste Manage. 29 (8), 2336–2352.

Chancerel, P., Meskers, C.E.M., Hagelüken, C., Rotter, V.S., 2009. Assessment of precious metal flows during preprocessing of waste electrical and electronic equipment. J. Ind. Ecol. 13 (5), 791–810.

Deetman, S., Pauliuk, S., van Vuuren, D.P., van der Voet, E., Tukker, A., 2018. Scenarios for demand growth of metals in electricity generation technologies, cars, and electronic appliances. Environ. Sci. Technol. 52 (8), 4950–4959.

European Commission, 2020. Critical Raw Materials Resilience: Charting a Path Towards Greater Security and Sustainability. Brussels https://eur-lex.europa.eu/legal-content/EN/TXT/?uri=CELEX:52020DC0474.

European Union, 2003. EU directive 2002/95/EC of the European Parliament and of the Council of 27 January 2003 on the Restriction of the use of certain Hazardous Substances in Electrical and Electronic Equipment (RoHS). https://eur-lex.europa.eu/legal-content/EN/LSU/?uri=CELEX:32002L0096.

European Union, 2012. Directive 2012/19/EU of the European Parliament and of the Council of 4 July 2012 on Waste Electrical and Electronic Equipment (WEEE) (Recast). EC, Brussels. https://eur-lex.europa.eu/legal-content/EN/LSU/?uri=CELEX:32012L0019.

Forti, V., Baldé, C.P., Kuehr, R., Bel, G., 2020. The Global E-Waste Monitor 2020, Quantities, Flows, and the Circular Economy Potential. United Nations University (UNU)/United Nations Institute for Training and Research (UNITAR) – co-hosted SCYCLE Programme, International Telecommunication Union (ITU) & International Solid Waste Association (ISWA), Bonn/Geneva/Rotterdam. https://ewastemonitor.info/gem-2020.

George, M.W., 2020. 2017 Minerals Yearbook, Gold (Advanced Release). U.S. Geological Survey, Reston, Virginia. https://www.usgs.gov/centers/national-minerals-information-center/gold-statistics-and-information.

Gmünder, S., 2007. Recycling – From Waste to Resource: Assessment of Optimal Manual Dismantling Depth of a Desktop PC in China Based on Eco-efficiency Calculations. Master thesis, Swiss Federal Institute of Technology (ETH) and Swiss Federal Laboratories for Materials Science and Technology, St. Gallen, Switzerland.

GSMA, 2020. GSMA CleanTech e-Waste Policy Study. https://www.gsma.com/mobilefordevelopment/cleantech/e-waste/.

Haarman, A., Magalini, F., Courtois, J., 2020. Study on the Impacts of Brominated Flame Retardants on the Recycling of WEEE plastics in Europe. Sofies SA, Geneva. https://www.bsef.com/wp-content/uploads/2020/11/Study-on-the-impact-of-Brominated-Flame-Retardants-BFRs-on-WEEE-plastics-recycling-by-Sofies-Nov-2020-1.pdf.

Hagelüken, C., Meskers, C., 2008. Mining our computers – opportunities and challenges to recover scarce and valuable metals from end-of-life electronic devices. In: Reichl, H., Nidden, N.F. (Eds.), Electronic Goes Green 2008+. Frauenhofer IRB Verlag, Stuttgart, pp. 623–628.

Hilty, L., 2008. Information Technology and Sustainability: Essays on the Relationship Between ICT and Sustainable Development. Books on Demand, Norderstedt. ISBN13: 9783837019704.

Hilty, L., Behrendt, S., Binswanger, M., Erdmann, L., 2005. The Precautionary Principle in the Information Society – Effects of Pervasive Computing on Health and Environment. Vols. TA46e/2005, STOA 125 EN, Swiss Center for Technology Assessment (TA-SWISS), Scientific Technology Options Assessment at the European Parliament, Bern.

Hischier, R., Wäger, P., Gauglhofer, J., 2005. Does WEEE recycling make sense from an environmental perspective?: the environmental impacts of the Swiss take-back and recycling systems for waste electrical and electronic equipment (WEEE). Environ. Impact Assess. Rev. 25 (5), 525–539.

Huisman, J., Leroy, P., 2017. Prospecting Secondary Raw Materials in the Urban Mine and mining wastes (ProSUM) – Final Report. In: ProSUM Project, Brussels.

Huisman, J., Magalini, F., 2018. WEEE Recycling Economics, The Shortcomings of the Current Business Model. European Electronics Recycler Association (EERA), United Nations University (UNU), Amsterdam. https://www.eera-recyclers.com/files/unu-eera-brochure-online-v5-002.pdf.

ICSG, 2021. The World Copper Factbook 2021. International Copper Study Group, Lisbon, Portugal.

ISO, 2017. Guidance principles for the sustainable management of secondary metals. In: International Workshop Agreement (IWA) 19:2017. International Standardization Organisation, Geneva, Switzerland.

ISO, 2023. Management and Circular Economy – Sustainability and traceability of secondary materials recovery – Principles and Requirements. In: ISO/DIS 59014:2023, under development. International Standardization Organisation, Geneva, Switzerland.

Manoochehri, S., Schluep, M., Valdivia, S., 2021. A Contribution to Future Critical Raw Materials Recycling CEWASTE Project Final Report. World Resources Forum Secretariat, St. Gallen. https://cewaste.eu/wp-content/uploads/2021/04/CEWASTE-Final-Public-Raport.pdf.

Menikpura, S.N.M., Santo, A., Hottaa, Y., 2014. Assessing the climate co-benefits from Waste Electrical and Electronic Equipment (WEEE) recycling in Japan. J. Clean. Prod. 74, 183–190.

Meskers, C., Hagelüken, C., Salhofer, S., Spitzbart, M., 2009. Impact of pre-processing routes on precious metal recovery from PCs. In: European Metallurgical Conference. GDMB, Innsbruck, Austria, pp. 527–540.

PREVENT, 2022. Practical Experiences with the Basel Convention: Challenges, Good Practice and Ways to Improve Transboundary Shipments of WEEE in low and middle income countries. Discussion paper, PREVENT Waste Alliance, Solving the e-Waste Problem (StEP).

Reuter, M.A., Hudson, C., van Schaik, A., Heiskanen, K., Meskers, C., Hagelüken, C., 2013. Metal Recycling: Opportunities, Limits, Infrastructure, A Report of the Working Group on the Global Metal Flows to the International Resource Panel. IRP, Paris. https://www.resourcepanel.org/reports/metal-recycling.

Schluep, M., Müller, E., Hilty, L., Ott, D., Widmer, R., Böni, H., 2013. Insights from a decade of development cooperation in e-waste management. In: Proceedings of the First International Conference on Information and Communication Technologies for Sustainability, ICT4S, Zurich.

Schnebele, E.K., 2021. 8 Minerals Yearbook, Silicon (Advanced Release). U.S. Geological Survey, Reston, Virginia. https://www.usgs.gov/centers/nationalminerals-information-center/silicon-statistics-andinformation.

Secretariat of the Basel Convention, 2011. Where Are WEee in Africa? Findings From the Basel Convention e-Waste Africa Programme. Geneva/Switzerland http://www.basel.int/Portals/4/download.aspx?d=UNEP-CHW-EWASTE-PUB-WeeAfricaReport.English.pdf.

Swico, SENS, SLRS, 2021. Technical Report 2021. Swico Recycling, Zurich.

Thiébaud, E., Hilty, L.M., Schluep, M., Widmer, R., Faulstich, M., 2018. Service lifetime, storage time, and disposal pathways of electronic equipment: a Swiss case study. J. Ind. Ecol. 22, 196–208.

UN Environment, 1989. Basel Convention on the Control of Transboundary Movements of Hazardous Wastes and Their Disposal. Secretariat of the Basel Convention, Geneva, Switzerland. www.basel.int.

UNU & UNITAR, Sustainable Cycles (SCYCLE) Programme, 2021. E-Waste Monitor. https://ewastemonitor.info. (Accessed 29 December 2021).

USGS, 2022. Mineral Commodity Summary 2022. U.S. Geological Survey, Reston, Virginia.

Van Schaik, A., Reuter, M.A., Heiskanen, K., 2004. The influence of particle size reduction and liberation on the recycling rate of end-of-life vehicles. Miner. Eng. 17, 331–347.

Wäger, P.A., Böni, H., Buser, A.M., Morf, L., Schluep, M., Streicher, M., 2009. Recycling of Plastics from Waste Electrical and Electronic Equipment (WEEE) – Tentative Results of a Swiss Study. Empa, Technology and Society Laboratory, St. Gallen, Switzerland.

Wäger, P., Schluep, M., Müller, E., Gloor, R., 2012. RoHS regulated substances in mixed plastics from waste electrical and electronic equipment. Environ. Sci. Technol. 46, 628–635.

Wang, F., Huisman, J., Meskers, C., Schluep, M., Stevels, A., Hagelüken, C., 2012. The Best-of-2-Worlds philosophy: developing local dismantling and global infrastructure network for sustainable e-waste treatment in emerging economies. Waste Manage. 32 (11), 2134–2146.

WEEE Forum, 2021. LibraWEEE. https://libraweee.com. (Accessed 29 December 2021).

World Resources Forum Association, 2021. WRF2021 Conference Report. WRFA, St. Gallen/Switzerland.

Zepf, V., Reller, A., Rennie, C., Asthfield, M., Simmons, J., 2014. Materials Critical to the Energy Industry. An Introduction. BP International, London.

CHAPTER

14

Photovoltaic and wind energy equipment

Olga Ignatenko[1], Neill J. Bartie[2], Christina Meskers[3], and Markus A. Reuter[4]

[1]Independent Professional, Kyiv, Ukraine [2]Helmholtz Centre Berlin for Materials and Energy (HZB), PVcomB, Berlin, Germany [3]Department of Energy and Process Engineering, Faculty of Engineering Sciences, NTNU—Norwegian University of Science and Technology, Trondheim, Norway [4]WASM: Minerals, Energy and Chemical Engineering, Faculty of Science and Engineering, Curtin University, Perth, Australia

14.1 INTRODUCTION

Wind turbines and photovoltaic (PV) modules are vital elements for transitioning to the carbon neutrality urgently needed under the current global climate crisis and providing sustainable energy use for the global economy. The International Renewable Energy Agency (IRENA) states that to meet the 1.5°C scenario by 2050, about 90% of the energy generation should be from renewables, compared to 26.4% in 2022. To be on track this should be 65% by 2030, where wind and solar together provide 42% of the total energy generated. This corresponds to an installed capacity of over 5200 GW solar, nearly 3000 GW onshore wind, and about 380 GW offshore wind (IRENA, 2022). Sustainable Development Goal 7—affordable and clean energy—further supports this goal. It also brings in a different dimension, that of universal and equitable access to energy. Both wind and solar energy can be used connected to the grid and off-grid, so people that don't have access to the grid can have electricity. This is especially important for emerging and developing economies.

In 2021 the cumulative installed PV capacity was about 843 GW globally with about 57% installed in Asia. PVs generate about 3% of global electricity. The PV sector has grown at over 35% annually during the last decade, with the crystalline silicon (c-Si) technology as the dominant technology, while the CdTe and CIGS technologies have a much smaller market share (IRENA, 2022). Wind power also has experienced significant growth, with wind installations increasing fourfold between 2010 and 2021. In 2021, the cumulative installed capacity of onshore wind power reached about 769 GW globally. The offshore wind capacity is much smaller: 28 GW (IRENA, 2019a).

Wind and solar energy use a wide range of elements in various combinations. This includes

https://doi.org/10.1016/B978-0-323-85514-3.00031-2

Ag, Al, Au, B, Cd, Cr, Cu, Fe, Ga, Ge, In, Mg, Mo, Ni, P, Pb, Se, Si, Sn, Te, Ti, Zn, and rare earth elements. Many of these are considered critical raw materials.

Figure 14.1 shows the materials required to produce 500 MW of power using solar and wind energy. The figure has a material-centric perspective, showing that the materials providing the key functionality of solar or wind, Si (plus Ag, Cu, and dopants) and NdFeB, respectively, make up only a small portion of the total materials demand, while the majority consists of base materials. Considering the global metal usage, PV and wind applications require roughly 40% of tellurium production (U.S. Geological Survey, 2021), 11.3% of silver production (The Silver Institute, 2021), 8% of indium production, and a minor share of cadmium production (U.S. Geological Survey, 2021). About 21% of silicon production is used in semiconductor and PV applications combined (Schnebele, 2021). Permanent magnets account for 29% of the rare earth production. About 9% of the permanent magnets are used for wind turbines (European Commission, 2017).

The additional solar and wind capacity to be built by 2030, about 4× today's capacity (IRENA, 2022), requires a tremendous material and (fossil) energy investment and will have considerable impact on the material markets.

The significant increase in wind energy harvesting together with the turbine's estimated

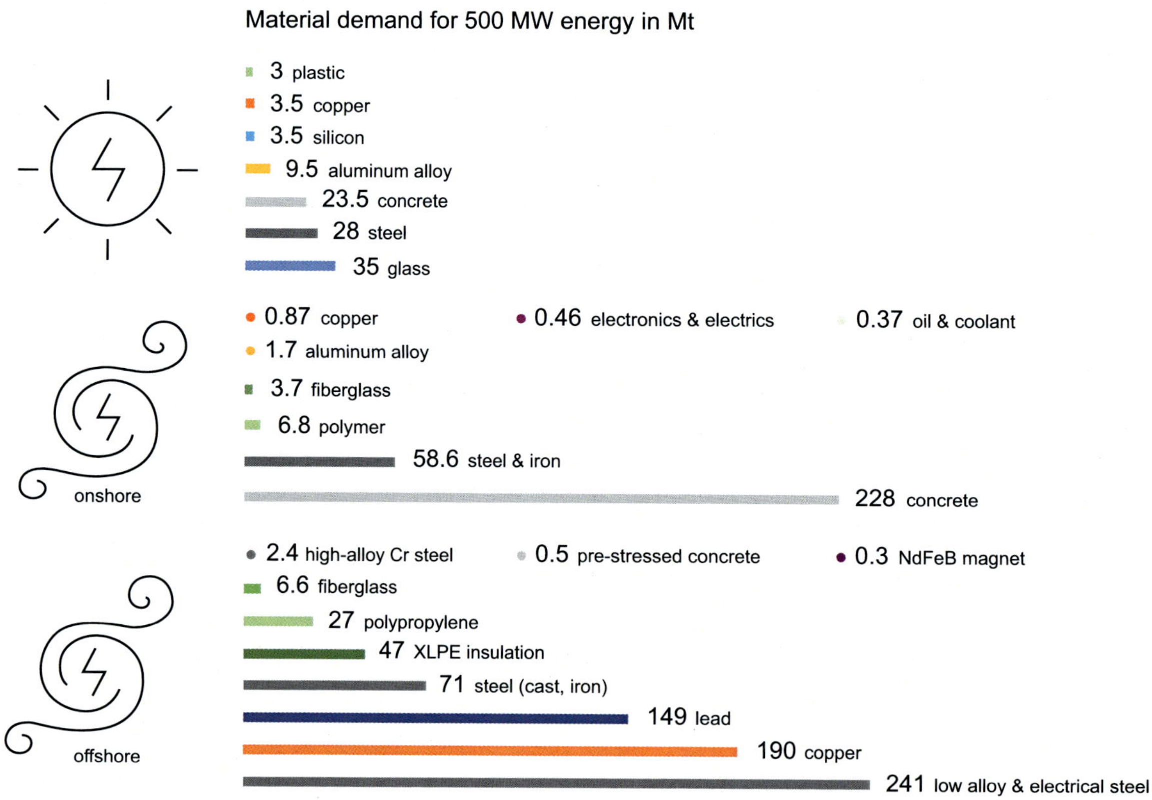

FIGURE 14.1 Materials required (in Mt) to produce 500 MW solar, onshore wind, and offshore wind power. *Data from IRENA (2019a, b).*

average lifetime of about 20 years will produce a large stream of end-of-life wind turbines in the near future. Already the end-of-life wind turbines constituted over 50 kt in 2022 in the EU alone and that number increases annually. Worldwide, the end-of-life wind turbine rotor blades stream is estimated to reach over 2 Mt by 2050 (Beauson et al., 2022).

The lifetime of a typical PV module is usually considered to be the time it takes until its performance is about 80% of the original performance specified by the manufacturer. This corresponds to 20–30 years, but the first PV installations and damaged modules are coming back today. Global cumulative waste volumes were about 0.2 Mt in 2021, according to IRENA, and were projected to increase to 4 Mt in 2030, while between 2030 and 2040 another 46 Mt will arise in a 1.5°C scenario (IRENA, 2022).

Therefore recycling is of vital importance to ensure the future of green energy supply and global sustainable development within the circular economy (CE) paradigm (SDG 12). This includes CE strategies such as adapted design, lifetime extension, reuse, and recycling of wastes arising during material and equipment production, and at end of life.

However, to do so one needs to understand and harmonize the material and energy flows, e.g., as discussed in Chapters 44–46. Thus the energy-materials nexus within the CE must be addressed in detail to fully understand the material and exergy efficiency of the systems.

In this chapter the key aspects of the CE of wind turbines and PV modules are discussed, including product complexity, how it impacts recycling today, and what steps can be taken to increase circularity.

14.2 WIND TURBINES

Depending on the manufacturer and model, wind turbines' material composition varies significantly, but usually includes steel, ferrous and nonferrous metals, carbon and glass fibers, and polymers. By mass it can be estimated as 66–79% steel, 11–16% fiberglass, resin, or plastic, 5–17% iron or cast iron, 1% copper, and up to 2% aluminum (Mone et al., 2017). Additionally, concrete is also used, either as some structural part or for the foundation of the wind turbines. For instance, the estimated amount of concrete for a 500 MW onshore wind turbine installation is 228 Mt (IRENA, 2019a). There are three main parts of wind turbines, each with its own specific composition: the tower, the nacelle, and the rotor blades. The tower is used for structural support and as a necessary elevation stand for the nacelle and rotor blades. It is made of steel, mainly low alloy steel and iron, and concrete.

14.2.1 Nacelle

The nacelle, mounted on the tower, houses the electrical generator and mechanical parts for the turbine's operation (Figure 14.2). Its composition is very complex as it includes, among others, polymers, wires, cast iron or aluminum, rare earth magnets, etc., which are combined and connected in joints of complex products inside the nacelle. These are the generator, electric and mechanical control units, and sensors. The content of the nacelle is, therefore, comparable to end-of-life electric vehicle drive trains or to waste from electrical and electronic equipment (WEEE).

There are also small quantities of rare elements present. Neodymium, praseodymium, and dysprosium are used in the permanent magnets of the electrical generators in most direct-drive turbines (Farina and Anctil, 2022). This is the case for nearly all offshore wind turbines in Europe and in approximately 76% of offshore wind turbines worldwide (Alves Dias et al., 2020). However, the use of a permanent magnet can be avoided, at least for onshore applications, by using different generator types, like multipolar synchronous generators, squirrel cage induction

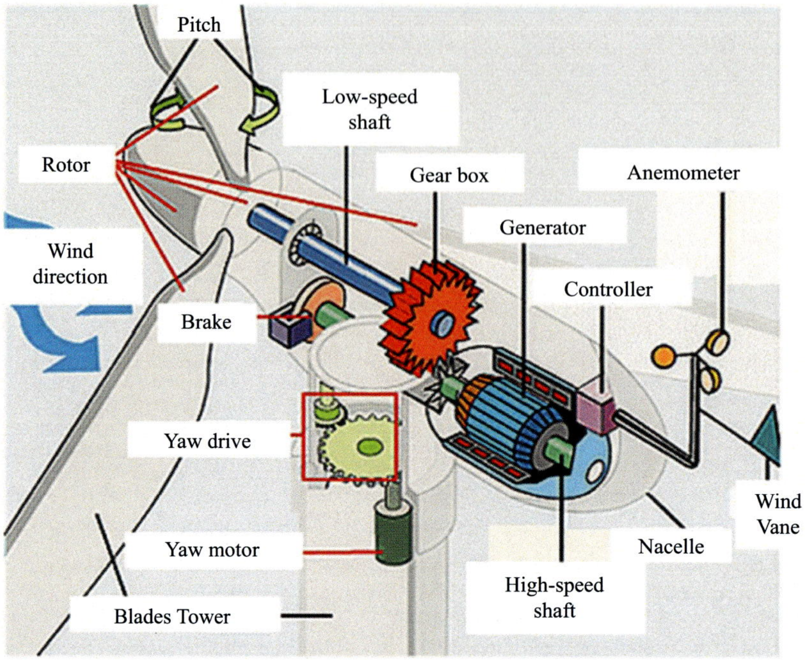

FIGURE 14.2 Structure of double fed wind turbine nacelle. *From Mahmoud and Xia (2012).*

generators, or superconductor-based generators. Alternatively, it is possible to reduce the need for rare earth elements by two-thirds by using hybrid drive generators that require smaller-sized permanent magnets (Alves Dias et al., 2020).

14.2.2 Rotor Blades

The rotor blades, mounted on the nacelle, are usually made of composite materials including fiberglass, coating polymers, and metals. In some cases, especially for small installations, natural materials like wood are used instead, or lightweight metal alloys. The structure of the blade consists of several elements, i.e., shear webs, load-carrying beam, leading and trailing edge, and aerodynamic shell, each consisting of a specific type of composite, while the blade is manufactured as a one-piece component (Figure 14.3). Depending on the manufacturer, the design and the arrangement of these elements differ. Generally, they are composed of:

- reinforcement fibers, e.g., glass, carbon, aramid, or basalt;
- thermosets polymer matrix, e.g., epoxies, polyesters, vinyl esters, polyurethane, or thermoplastics;
- sandwich core of wood or foams like polyethylene terephthalate;
- coatings materials like polyethylene or polyurethane;
- copper wiring;
- steel bolts.

The majority of the blades are manufactured from glass fiber reinforced polymer composites, which consist of 60–70% reinforcing fibers and 30–40% resin (Jensen and Skelton, 2018). It is clear that "unmixing" these complex materials into virgin materials is a challenge that needs attention.

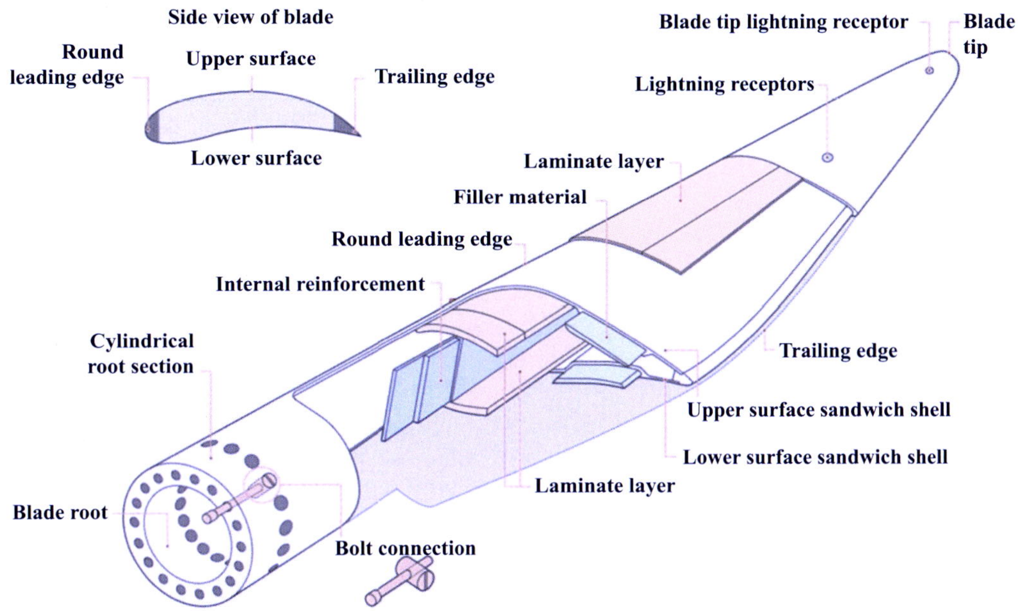

FIGURE 14.3 Material composition of a wind turbine blade cross section. *From Katsaprakakis et al. (2021).*

14.3 PHOTOVOLTAIC MODULES

PV modules started to be deployed on a larger scale in the early 2010s with large installations using monocrystalline silicon, cadmium telluride (CdTe), or copper-indium-(gallium)-diselenide (CI(G)S) technology. Today crystalline silicon (c-Si) is the dominant technology, with over 90% of the market (VDMA, 2020–2022), while CdTe is the main other technology. Most important for PVs is the efficiency with which sunlight is converted into electrical energy. The quest for increased efficiency and less material use drives the development of new PV materials.

Several other technologies are used for niche applications. Concentrator photovoltaics (CPV) and high-concentration photovoltaics (HCPV) use lenses or curved mirrors to focus the sunlight on a small, very efficient multijunction cell, which layers for example InGaP (top), InGaAs (middle), and Ge (bottom) cells on top of each other. These are used in areas with high amounts of sun such as the southern part of the United States, Spain, and Saudi Arabia. Spacecraft also use multijunction cells. Discussion of these uses is outside the scope of this chapter.

14.3.1 Si-Based PV

Crystalline silicon modules come in different variations depending on the manufacturer, though they share the same basic construction. Modules have to perform under widely varying conditions: blistering sun, frost and snow, storm, hail, and rain, and are subjected to sand, dust, and salt depending on the location. This puts high demands on the construction. A typical module (Figure 14.4) thus has a sealed, layered structure where the solar cells are sandwiched between two polymeric encapsulation layers, all of which are laminated between a protective polymeric (or sometimes glass) backsheet, and a sun-facing front glass cover.

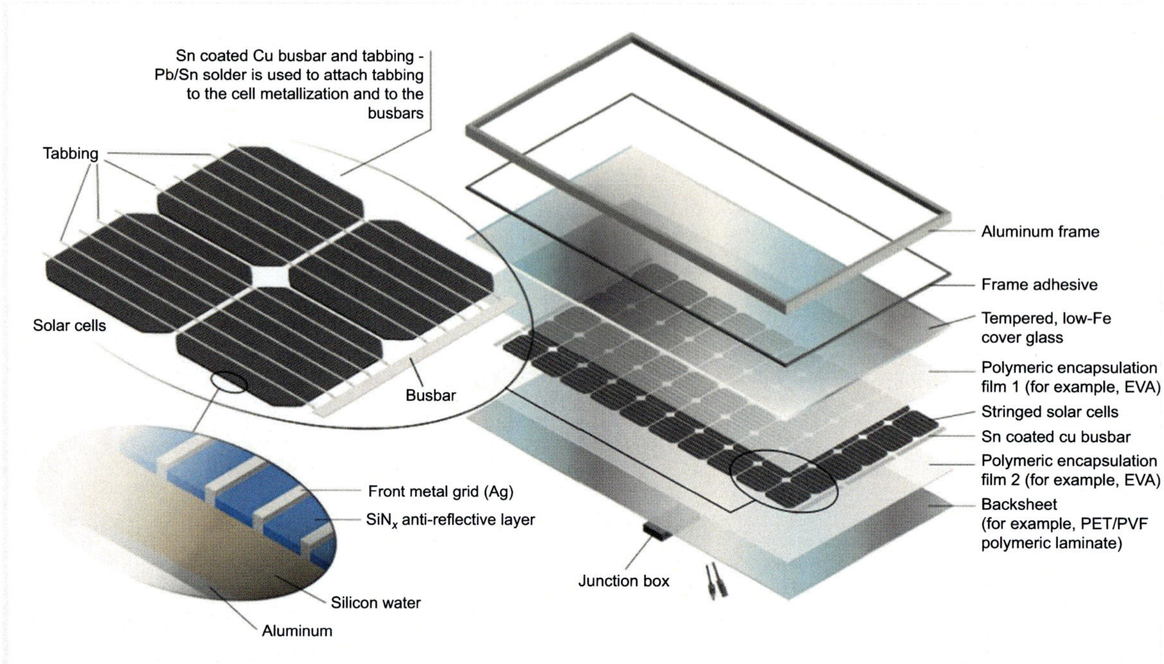

FIGURE 14.4 Conventional crystalline silicon (c-Si) module design. Detailed structures such as passivation, doping layers, etc. omitted. EVA: ethyl vinyl acetate. PVF: polyvinyl fluoride. PET: polyethylene terephthalate. *From Heath et al. (2020).*

Everything is sealed together, making separation of the materials for recycling particularly difficult.

The life cycle starts with the production of metallurgical and solar-grade silicon, followed by, in the case of monocrystalline wafers, the manufacturing of single crystal silicon ingots, which are cut into wafers. The cells are then manufactured and, finally, the module is put together. Figure 14.5 shows the life cycle, main steps, and materials involved. Detailed descriptions can be found in Bartie et al. (2021); a summary is given here.

Metallurgical grade (MG) Si is produced by reducing silica (SiO_2) with carbon in an electric arc furnace followed by ladle refining to remove impurities such as Al, Ca, Mg, Fe, B, and P. Further refining is commonly done using the Siemens process to obtain at least 6N (99.9999%) pure solar-grade (SG) Si. The Siemens process is energy intensive, resulting in a polysilicon production power consumption of about 71 kWh/kg Si, with 40 kWh/kg Si attributed to the Siemens reactor itself (Masson and Kaizuka, 2019). The alternative silane fluidized bed reactor (FBR) process uses around 10 kWh/kg Si. It had a market share of 5% in 2021, which is expected to increase to about 12% by 2031 (VDMA, 2020–2022). In this process MG-Si is converted to silane, purified by fractional distillation, and then fed to an FBR to produce SG-Si granules. Widespread adoption of the FBR process would be important for lowering the energy consumption and CO_2 emissions associated with c-Si manufacturing.

Monocrystalline Si ingots are pulled from a molten Si bath in a Czochralski furnace, and then cut into wafers, typically 170 μm thick (VDMA, 2020–2022) using diamond wire cutting. The cutting process results in a cutting loss, referred to as kerf, of about 26% of the total amount of ingot consumed per wafer (VDMA, 2020–2022), and is the largest and most expensive material inefficiency in the life cycle. Kerf (recycling) remains a challenge for the PV industry.

The cells are manufactured through several steps of material and metallization layer deposition on the Si wafers, using techniques such as sputtering chemical vapor deposition and screen printing. In the module assembly step, the Si value chain and other material value chains come together (Figure 14.5). To conduct current from the cell a silver metal grid is printed on the front of the wafer, and the back is covered with aluminum with silver contact pads. The cells are strung together with tin-coated copper ribbons and soldered. The module is put together top to bottom. First, a sheet of low-iron glass is covered with an encapsulation layer, typically EVA. Next, the cells are placed, and then another encapsulation layer followed by the protective backsheet (e.g., PET sandwiched between two layers of PVF) is placed on top (Figure 14.4). After lamination the module edges are sealed and framed with an $AlMg_3$ alloy and the junction box is connected. The module is then tested before shipping. Broken cells and rejected modules arising during production and manufacturing are recycled.

Summarizing, a typical module composition is: 2.73 kg Al (excl. frame), 0.33 kg Ag, 10.86 kg Cu, 0.078 kg Pb, 6.05 kg SnO_2, 30.98 kg Si, 661 kg glass, 111 kg polymers, and 1.24 kg other, per ton modules (Bartie et al., 2021).

Many other module configurations exist. For example, some modules do not use a backsheet

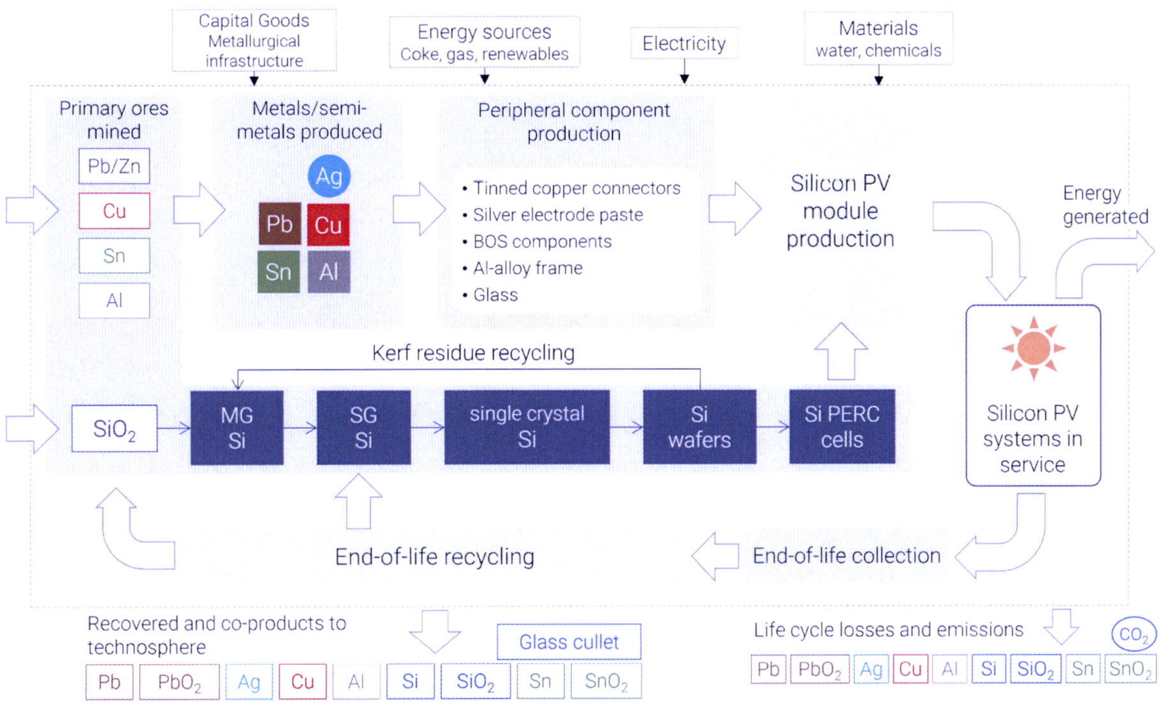

FIGURE 14.5 Life cycle of mono-Si PV module with passivated emitter and rear cell (PERC) cell configuration, showing the physical flows associated with the key components. *From Bartie et al. (2022).*

but rather a second layer of glass, and some are not framed at all. Different encapsulation and backsheet materials are also used.

14.3.2 CdTe-Based PV

The raw materials to produce CdTe modules are by-products from zinc, lead, and copper primary production (Figure 14.6). These production routes are interconnected, as there is an exchange of by-products between the copper and lead-zinc production routes (Bartie et al., 2020; Abadías Llamas et al., 2020). Furthermore, to produce the CdTe necessary for one module and considering the other applications of Cd and Te (267 and 554 g, respectively), the system depicted in Figure 14.6 would automatically produce 104 kg Cu, 74 kg Zn, and 35 kg Pb (Bartie et al., 2022).

The basic structure of a CdTe module is similar to a c-Si module. The main difference is that the CdTe cell is not in the form of a wafer. Instead, it is a second-generation PV technology in which a thin film absorber and other layers are deposited directly onto a glass substrate through vapor transport deposition of (traditionally) CdTe and CdS (Bonnet, 2012; Fthenakis, 2004). More recently, CdSe and/or $CdTe_{(1-x)}Se_{(x)}$ are also used, and MgZnO instead of CdS, among others (Lingg et al., 2019). Like c-Si, the module is built up by placing layers on the front glass sheet made of heat-treated borosilicate or soda-lime glass. First, a transparent conductive oxide layer consisting of indium

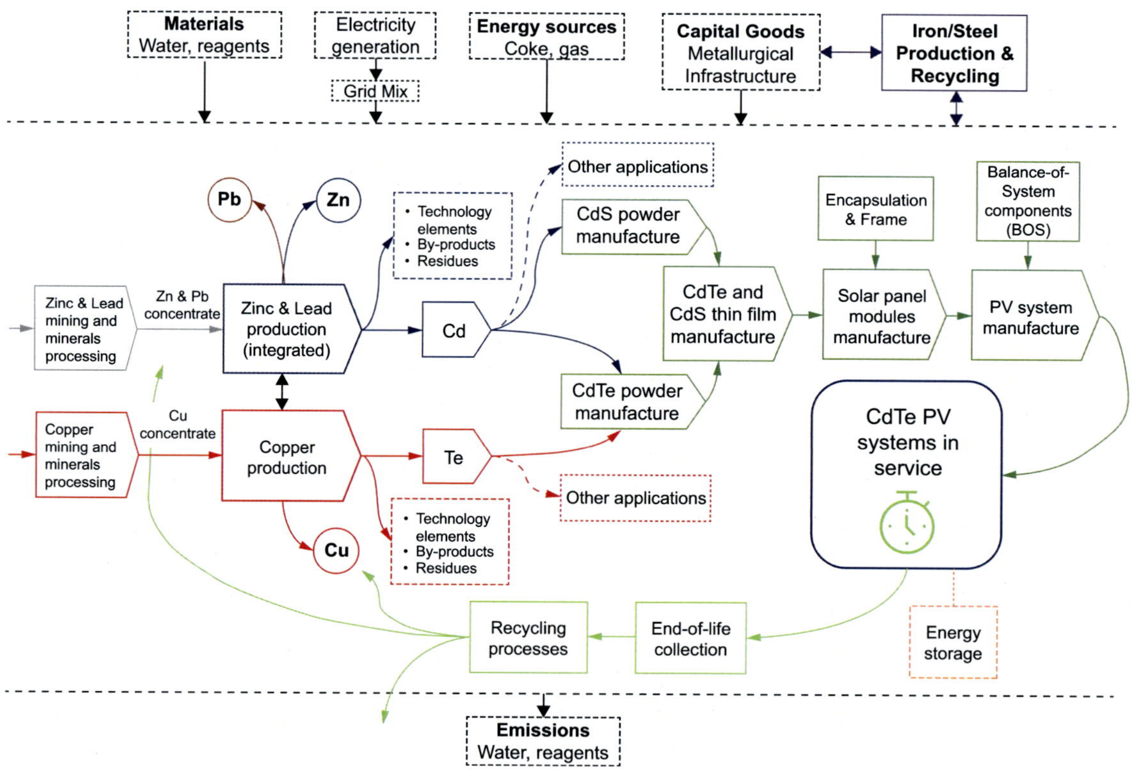

FIGURE 14.6 Life cycle of CdTe PV modules, showing the physical flows associated with the key components. *From Bartie et al. (2020).*

tin oxide (ITO), SnO_2, or Cd_2SnO_4 is deposited. This is followed by layers of CdS and CdTe for the traditional case, and a conducting back contact of carbon paste with copper or other metals (NREL, 2022). The entire layer is about 3 μm thick (Fthenakis et al., 2008), which is much thinner than a silicon wafer. Module edges are sealed, and an encapsulation layer added, followed by a protective barrier of tempered glass. The module frame is added (if required) and the junction box connected.

This manufacturing process is highly material efficient. The same functionality (PV power output) is achieved with 3 μm of material instead of a 170 μm Si wafer in c-Si. The wastes arising during deposition of the thin films and manufacturing of the modules, such as spent sputtering or vapor deposition targets, are usually recycled.

14.4 WIND TURBINE RECYCLING

The recycling of wind turbines starts with dismantling various parts, usually at the location of the wind energy installation. At this stage, all the materials that can be easily separated, for instance steel and concrete from the tower, are collected and sent for further processing; see Chapter 19, Steel, and Chapter 26, Concrete and Aggregates, for detailed information. Parts that remain too complex for further processing undergo further mechanical, thermal, or chemical treatment to separate materials streams and reuse them within the manufacturing cycle.

14.4.1 Fiberglass and Polymers Composites

Currently, nearly 2.5 Mt of composite materials are in use in the wind energy sector (IRENA, 2019a). Recycling blades is a challenge as there is a wide variation in materials in the structure of the blade elements, and the blade is manufactured as one piece (Figure 14.3). There is thus no possibility for dismantling for easy materials separation. In addition, the blades will be found in various conditions, as the decommissioning of wind turbines can be not only after their operational term has expired, but also earlier due to replacement with a newer and more efficient model or due to sustaining damage. Also, the amount of material coming from decommissioned blades fluctuates over time: very small quantities from a single turbine and very large quantities from an entire windfarm. Therefore there is a large variation in the amount, composition, and quality of materials, making the development of a sustainable recycling solution very challenging. Most of the blades worldwide are currently either landfilled or incinerated with energy recovery (Cooperman et al., 2021).

Options for further processing of the carbon and glass fiber composites include material recovery by mechanical crushing, high-voltage fragmentation, or chemical solvent dissolution, and thermal treatment with material recovery by pyrolysis or in cement production (Cooperman et al., 2021). Slag-based nonferrous smelters can use these rather unreactive carbon sources as reductants and energy sources. Chapter 34, Carbon Fiber, provides detailed information about available technologies and challenges.

14.4.2 Electric and Electronic Components of the Nacelle

The parts of the complex design used in the nacelle, like the generators and control boxes, are rarely designed for easy dismantling for materials separation. Therefore further shredding is required to liberate various materials connected/joined together and is followed by mechanical separation and sorting into different material streams.

The mechanical separation includes removal of the light fraction, which contains small lightweight pieces of polymers, wires, and mineral dust, magnetic separation for ferrous metal, and eddy current separation for nonferrous metals. The main outcomes here are a ferrous fraction (steel and cast iron), nonferrous fractions of aluminum and copper, and polymer fractions. Metal fractions can be sent for further processing in specialized plants, while polymer and light fractions, depending on the mixture, are more challenging for material recycling and often directed for energy recovery. If the generator is shredded, the magnet particles end up in the magnetic fraction going to steel recycling and are lost. The reader is referred to Part 3, Recycling from a material perspective, for detailed information on recycling of the different fractions.

14.4.3 Rare Earth Magnets From the Nacelle

The electric generators often contain rare earth magnets. For example, direct-drive wind turbines use 700–1200 kg/MW NdFeB magnets or 175–420 kg/MW of pure neodymium (Binnemans et al., 2018). As the magnets have a longer life span than wind turbines, they can be removed and reused directly in new generators. However, removal of the magnets requires manual dismantling and, to be safely extracted, they should be demagnetized, which may reduce the quality of the magnets to be reused (Alves Dias et al., 2020). Another challenge is that currently there is neither infrastructure for the collection of rare earth-containing products, nor mature technology for their recycling. Only research projects, start-ups, and pilots are available.

14.5 PV RECYCLING

14.5.1 Collection

PV modules are used in commercial settings, i.e., PV power plants and on commercial buildings, and in the consumer market, i.e., on roofs of houses. The collection mechanisms are different for each setting and vary between countries. In Europe an extended producer responsibility system, PV CYCLE (www.pvcycle.org), was established by the industry in 2007, well before legislation came into place. The leading producer of CdTe modules established the first voluntary global collection and recycling program in 2005 (Sinha et al., 2012). This is important, as in most places there is not a policy in place.

Korea, Australia, India, and Japan are developing requirements, while Washington State (United States) requires recycling (Heath et al., 2020). The European Union has had regulation in place since 2012, when PV was added to the WEEE directive (Directive 2012/19/EU; European Union, 2012). It requires at least 85% of the modules put on the market to be collected, and at least 80% prepared for reuse and recycled. The latter can be easily achieved by recycling the frame, glass, and junction box that make up most of the weight of a module, while the critical semiconductor materials that make PV work are lost. Collection rates are well above 85%, with about 95% collected. The requirements for collection, logistics, and treatment are formalized in the European standard EN 50625-2-4 and the technical specifications CLC/TS 50625-3-5:2017 (CENELEC, 2017a, b). After collection, the modules are sorted so the modules with thin-film technology can go to a dedicated recycling process.

14.5.2 c-Si Module Recycling

Today the modules are usually recycled through existing metal or (laminated) glass recycling processes, which recover the bulk materials such as the Al frame, glass sheet, and junction box, which enter the existing recycling processes. The EVA layer is often incinerated. The junction box can be processed through WEEE recycling processes (Duflou et al., 2018). The materials with the most potential economic value and environmental impact such as Si, Ag, Cd, Te, Cu, Sn, and Pb, which make up less than 1% of the weight, are not recovered (Heath et al., 2020).

Improved recycling processes are under development to address the challenges. First is overcoming the encapsulation so the sandwich structure can be opened, and the glass and laminated polymer layers can be separated from the wafers. Second is separation and purification of the wafers and the other metals. Heath et al. (2020) gave an overview of all the processes explored. Only a few have made it to pilot or industrial scale.

The hot-knife process uses a heated cutter to cut the EVA layer, yielding a glass fraction and a wafer fraction for further separate processing (Ito, 2016). The Veolia process uses only mechanical methods including, after frame and junction box removal, cutting the module into pieces followed by grinding, sifting, and eddy current separation. Optical sorting is used to get a clean glass fraction. Nine other fractions are obtained, including silicon, plastic, and metals (Veolia, 2018). The FRELP process (Latunussa et al., 2016; Ardente et al., 2019) and the ASU process (Huang et al., 2017) integrate both steps in one process and include recovery of Si, Ag, and Cu, among others. The ASU process is described in more detail in Chapter 46, Process simulation and is summarized here. After dismantling of the frame and junction box, the rest of the module is thermally treated to remove the EVA layer by combustion, and the glass is separated from the Si wafers, ribbons, and contacts. These are leached first with HNO_3 to dissolve Ag, Cu, and Pb, while Sn precipitates. The dissolved metals are recovered using electrowinning. Next, the wafers are treated with HF to remove the Si-nitride layer and Al electrode, followed by treatment with NaOH to remove the positively and negatively doped regions on either side of the Si wafer to obtain an SG-Si grade wafer core that can be recycled to the Czochralski process, avoiding the energy-intensive Siemens process (Huang et al., 2017; Bartie et al., 2021).

Using process simulation (Chapter 46), Bartie et al. (2021) compared the impact of end-of-life module recycling using the ASU process to that of no recycling, looking at resource efficiency and the environmental footprint of the entire life cycle. The evaluation shows that recycling lowers the power consumption for module production from 116 to 75 kWh/m^2 and lowers the primary SiO_2 need from 9.64 to 5.12 kt/GW$_p$ for modules with a 23% nominal power conversion efficiency. Material recoveries obtained with the recycling process are: 70.3% Ag, 78.9% Cu, 94.0% Pb, 94.1% Sn as SnO_2, and 86.9% SG-Si (Bartie et al., 2021).

14.5.3 CdTe Module Recycling

The recycling process largely follows the same steps as for c-Si modules and is described in detail in Fthenakis, 2004; Sinha et al., 2012; Wade, 2013. Dismantling removes the frame, if present, and the junction box before shredding. Subsequently, the shredded module is leached with an H_2SO_4 and H_2O_2 solution to dissolve the Cd and Te. The glass is separated from the solution, and then Cd and Te are precipitated into a sludge and dewatered. The unrefined semiconductor material is further processed in a hydrometallurgical flowsheet, recovering metals that are returned to semiconductor manufacturing. Recovery of the semiconductor material is claimed to be over 90%, and that of glass about 90% (Sinha et al., 2012).

In a pyrometallurgical scenario the unrefined semiconductor material can be further processed in a lead smelting flowsheet, as indicated in Figure 14.6. Cd reports mainly to the off-gas and the Te reports to the bullion, after which it is recovered through the lead-refining processes. In this case, about 40% of Te put on the market consists of recycled Te (Bartie et al., 2020). One needs to keep in mind that a high concentration of tellurium in lead affects the production and refining of lead. This highlights that the quantity and quality of material fed to any smelter must be controlled to ensure maximal resource efficiency.

14.6 FUTURE DEVELOPMENTS

14.6.1 Wind Energy

The key improvements in wind turbine technologies are increasing the rotor diameter and the hub height to obtain more power. The largest onshore wind turbine, with a 170 m rotor diameter, was presented by Siemens-Gamesa. Several research projects or prototypes focus on innovations in design, material requirements, and manufacturing techniques for a wide range of applications. The research on rotor blade design focuses on improving the blades' aerodynamics and materials used to maximize energy production, reduce maintenance costs, and improve performance in hard conditions. The innovations sought for the power inverters are improved reliability and dimensions, while the digital technology for smart turbines aims to improve turbine monitoring and controls to maximize the overall energy output and reduce damaging, which ultimately expands wind farm lifetimes.

Improving the recyclability of end-of-life turbines is another key area.

For rotor blades, innovation in blade and material design as well as recycling technology is pursued. The Dreamwind project (www.dreamwind.dk) works on development of new materials incorporating biobased resources for high-strength composite, a new design for the blades that allows for disassembly after use, and a recycling process to separate the glass from the plastic fibers. In this way the large and expensive fiberglass components from wind turbines can be recycled in the future. The ZEBRA project (https://irt-jules-verne.fr/en/projets/zero-waste-blade-research-project/) is another project focusing on 100% recycling for wind blades. It has established a cross-sector consortium that represents the full value chain: from development of materials to blade manufacturing, to wind turbine operation and decommissioning, and finally recycling of the decommissioned blade material. It particularly addresses design for recycling in both the materials used and construction of the blades, employing a well-recyclable thermoplastic resin type. It also uses CANOE's innovative recycling process for fiber-reinforced acrylic composite material, which uses a dissolution method to recover recycled methyl methacrylate monomer and recycled fiber (Nehls, 2022; LM Wind Power, 2022).

Development of an efficient and cost-effective method for the recycling of rare earth elements from permanent magnets improves availability of the critical rare earth element as well as the recyclability of wind turbines. For example, the REE4EU project (www.ree4eu.eu) has developed the integrated high-temperature electrolysis (HTE) and ion liquid extraction (ILE) method of extraction for rare earth alloys from end-of-life permanent magnets (European Commission, 2020). In the pilot, several tons of waste have already been treated, generating an alloy that can be reused further in the manufacturing of permanent magnet products.

Also, the EU has included the identification and reuse of rare earth in permanent magnet motors as one of the key areas to be addressed in the updated regulation on eco-design requirements for electric motors and variable speed drives. The update, expected in 2023, will bring changes into the design of wind turbine generators in future (Gauss et al., 2021).

14.6.2 PV

In PVs, the key developments focus on increasing cell and module efficiency in existing technologies and with new cell compositions, reducing material and energy consumption (CO_2 footprint) per GW solar energy produced, recycling processes focused on recovering Si, as well as circular business models, including Design for X and lifetime extension.

Reduction in material consumption, often in combination with increased cell efficiency, has taken place over the past decade and is expected to continue. Around 2008 the c-Si wafer thickness was 270–300 μm (Fthenakis et al., 2008); currently it is 170–175 μm and is expected to decrease to 150–160 μm by 2031 (VDMA, 2020–2022). This reduces the amount of Si per module considerably. Still, the growth in PV (material) demand is outpacing the material savings. This material efficiency improvement achieved by decreasing the wafer thickness from 175 to 150 μm results merely in a 5% reduction in CO_2-equivalent emissions (Bartie et al., 2022). Thinner wafers are more fragile, making the recovery of intact cells during recycling less likely, reducing possibilities for wafer reuse.

In parallel, the reduction and substitution of silver is ongoing. Silver is expensive and has future resource constraints (Norgren et al., 2020). Its use has declined by 70% since 2010 (Heath et al., 2020) and silver usage per cell is expected to halve in the next decade (VDMA, 2020–2022). On the one hand, the grid lines printed on the front of the cell become thinner, and on the other hand, silver is replaced by low-cost copper and copper/nickel. Silver is an important source of revenue in recycling processes, and its reduction will impact recycling economics.

The module design is changing too. One trend is frameless modules. This would reduce the use of aluminum, removing a source of recycling revenue and creating trade-offs in the use phase as well. Lower weight and volume of frameless modules reduce transport cost and impact but may increase module breakage (Norgren et al., 2020). Another trend is the replacement of the backsheet with a glass sheet, a glass/glass design, combined with reduction of the glass thickness to 2–3 mm (VDMA, 2020–2022). This eliminates the fluorine-containing polymer backsheet, and the need for fluorine capture during recycling. It also changes the operation and economics of recycling processes. Ensuring that the glass can be used again in PV modules will be essential (Norgren et al., 2020). Contrary to the removal of frames, however, replacing the polymer-based backsheet with glass significantly increases module weight and the associated transport cost and environmental impacts.

New PV cell compositions are being developed. The perovskites, which use lead halide-based perovskites as the absorber (Lal et al., 2017; Mohammad Bagher, 2015), have seen tremendous growth in power conversion efficiency and, like thin-film technologies, use very little material. Si/perovskite tandem modules use c-Si with a perovskite layer on top (DOE, 2022). The tandem modules combine the life cycles of two different technologies, increasing resource complexity and material losses in the production and recycling of these modules. However, with the potentially much higher power conversion efficiencies achievable with tandem technology, fewer modules would be needed to generate a given amount of energy compared to current commercial PV types.

14.7 KEY ISSUES AND CHALLENGES

14.7.1 Wind Turbines and Their Recycling

The recycling of wind turbines remains challenging and creates environmental concerns. Challenges include logistics for large parts, difficulties of material recovery from a complex multimaterial design like nacelle and rotor blades, and the low recyclability of composite materials of the rotor blades. These challenges result in low recycling rates of wind turbines, low grade of the materials recovered, and limited applicability of the recovered materials within the manufacturing cycle. Another challenge lies in the lack of commercial infrastructure for the recycling of wind turbines and mature solutions and technology able to deal with the challenges that the end-of-life wind turbines represent. Bringing the wind turbine production and recycling up to the requirements of the CE requires further research into and development of solutions that strike an environmentally and economically sound balance in materials recovery and that achieve high recycling rates.

Currently, recycling of wind turbines already represents an environmental problem that undermines the shift toward a CE. Many wind turbines come to the end of life today, being replaced by new, more efficient models, and there is a constant stream of decommissioned parts after damage repair or maintenance, representing a significant materials stream that increases annually. However, the solution for wind turbine recycling has to be commercially available before 2040, when the wind turbines start being decommissioned in mass, in order to ensure the continuous function of the global circular and carbon-neutral economy.

As many countries have pledged the transition of their economies to carbon neutrality by the mid-century, the global reliance on wind power will grow and use of wind turbines will increase. The future recyclability of the wind turbines must be ensured through an improved design, in which the recyclability of materials and their combinations and the possibility for easy dismantling and separation of various materials have been considered. The details are described in Chapter 5—Material and product-centric recycling and design for recycling rules and digital methods, DfR, providing insight into what is to be considered in the design stage in order to ensure not only the maximal recycling rates of the products but also the maximum recovery of the exergy embodied within.

Specific to wind turbine recycling are the large sizes—rotor blades alone currently can reach over 80 m, which brings in the transportation issue. Currently, many wind turbines are located in remote areas, and the large size increases the cost for transport to the recycling facility and therefore the costs of recycling (Jensen and Skelton, 2018). This can be overcome by smart design solutions for easy on-site disassembling of the large parts.

Another issue that can be addressed by smart design solutions is the recyclability of the nacelle, which contains electrical and mechanical components with complex product designs and a range of materials. For instance, the recovery of the critical element of rare earth elements embedded into the permanent magnets of the electric generators used in wind turbines is a big concern. As was mentioned earlier, the magnets themselves can be removed and reused directly. However, it is currently a difficult task since the existing generators used in the wind turbines are not designed for easy extraction of the magnets. Furthermore, there may be a different geometry requirement for the magnets in the new models of generators and therefore reused magnets may not always be an option. Also, when the current technology for nacelle recycling is used, it results in secondary material streams with low quality due to the contamination with other materials, including critical materials. Their extraction can be very difficult and therefore expensive, so they are lost within other material streams. A design for a reuse/remanufacturing/recycling approach is required for the removal of permanent magnets before the standard processing of end-of-life generators. This will help to maintain both high recycling rates for metals recovery and recovery of the earth elements.

Also, the low recyclability of the rotor blades represents a significant challenge: the design and combination of materials make it very difficult to separate these materials. Moreover, the polymers and reinforcing fibers used are not always recyclable even in pure streams, and even if materials recovery is possible, the recovered secondary materials have reduced properties compared with virgin materials, making recovered secondary streams very limited in further application. This is another issue to be addressed in the design stage via the choice of specific materials that are easily recyclable.

Major challenges remain in how to strike a reasonable balance between the quality of the recycled components and achieving high recycling rates for the overall end-of-life wind turbine, which is a common issue for most complex consumer goods. Obviously, this must all happen economically. A key fundamental aspect that underpins this is the exergy dissipation of the recycling system and therefore downcycling material qualities. For instance, for metals recycling the grade of the resulting secondary stream and further applicability for secondary metal alloys for new product manufacturing need to be addressed. Low quality of the metal recyclates often results in the need for diluting the received secondary metals stream with primary metals to achieve an economically attractive product. More details on the recycling of various metals and materials and their specific challenges are discussed in the other chapters of this book.

14.7.2 PV Modules and Their Recycling

A key challenge in the recycling of PV is the delamination process. The encapsulation layer is essential for the service life of the module, but complicates recycling. It causes imperfect liberation and cross-contamination of the fractions. Combustion of the encapsulation results in CO_2 emissions. Finding alternative encapsulation materials and different module designs that can be delaminated easily, or improved deencapsulation processes, will be important for handling the expected future recycling volumes.

The generation of large volumes of kerf, which are hard to recycle today, is a major issue in the c-Si life cycle. Although kerf loss per wafer is expected to further reduce by 2030 (VDMA, 2020–2022), the increase in c-Si PV production will likely result in increasing the total kerf volume. While no-kerf wafering methods exist, they are not expected to gain significant market share due to the maturity of the diamond wire cutting process (VDMA, 2020–2022). The challenges with kerf recycling are, first, that the fine Si particles combined with a water-based cutting fluid can result in hydrogen formation and dust explosions. Second, the kerf is insufficiently pure to be recycled as SG-Si, and thus must go through the energy-intensive Siemens process. Improving the recycling processes so the kerf can enter the life cycle again at MG quality or SG quality will be key. Stakeholders have been working on finding solutions for years. Some recent examples are the CABRISS project (www.spire2030.eu/cabriss) in which ReSiTec has achieved kerf recycling up to 4N purity; a kerf processing plant by REC (Verdu, 2020) and ROSI Solar has been one of the first to be awarded funding for their work in this field under the EU Green Deal (ROSI Solar, 2020). The nonlinear effects of kerf recycling and end-of-life recycling on the c-Si life cycles are discussed in detail in Bartie et al. (2021, 2022).

The challenges with kerf recycling and with the current module recycling processes illustrate the importance of developing high-value recycling processes in which material quality is maintained and that are able to handle the future volumes. Recycling the silicon wafer from end-of-life modules back into the PV cycle at SG quality will require further improvement of the recycling processes, and a good understanding of the impurities in the materials from recycling and how these can be removed or impact the manufacturing processes (Heath et al., 2020). Although SG-Si has 6N purity, often material of 11N purity is used (Heath et al., 2020). Every increase in purity comes with additional use of energy and material resources to produce the material from primary or secondary sources. This may make it even more difficult to use recycled Si materials in the PV life cycle and lower the life cycle's footprint at the same time.

14.8 CONCLUSIONS AND OUTLOOK

As with most modern products situated in the CE, PV and windmills have complex material linkages to impart their specific functionalities and lower carbon footprint energy. Therefore "unmixing" the material in these products into quality materials that find an application in the same renewable energy infrastructure is challenging.

Large numbers of PV and wind energy installations will be put on the market in the next decades, reaching end of life 20–30 years later. Thus, Design for Recycling and Design for X principles (see Chapter 5, Material and product-centric recycling and design for recycling rules and digital methods) urgently need to be applied and recycling technologies need to be developed to have a CE. The recycling infrastructure to be built up will have to be flexible enough to deal with the variations in design and technology developments.

While the focus has been largely on recycling, the possibility of lifetime extension, another key CE aspect, in the form of remanufacturing and reuse has been unexplored (Tsanakas et al., 2020). To realize this, circular business models need to be developed, so that a high-quality CE can be financed.

In renewable energy equipment the material and energy cycles literally come together. For a sustainable renewable energy future, a deep understanding of the materials-energy nexus across its entire life cycle, and the complex interactions in it, is required. This must be done on the rigorous basis discussed in this book to ensure realistic options are being developed, which adapt agilely with future demands. This means Design for X should be embedded in the CE and linked to the metallurgical processing infrastructure, as discussed by Bartie et al. (2022).

References

Abadías Llamas, A., Bartie, N.J., Heibeck, M., Stelter, M., Reuter, M.A., 2020. Simulation-based exergy analysis of large circular economy systems: zinc production coupled to CdTe photovoltaic module life cycle. J. Sustain. Metall. 6 (1), 34–67. https://doi.org/10.1007/s40831-019-00255-5.

Alves Dias, P., Bobba, S., Carrara, S., Plazzotta, B., 2020. The Role of Rare Earth Elements in Wind Energy and Electric Mobility: An Analysis of Future Supply/Demand Balances. European Commission, Joint Research Centre, Publications Office, https://doi.org/10.2760/303258.

Ardente, F., Latunussa, C.E.L., Blengini, G.A., 2019. Resource efficient recovery of critical and precious metals from waste silicon PV panel recycling. Waste Manag. 91, 156–167. https://doi.org/10.1016/j.wasman.2019.04.059.

Bartie, N.J., Abadías Llamas, A., Heibeck, M., Fröhling, M., Volkova, O., Reuter, M.A., 2020. The simulation-based analysis of the resource efficiency of the circular economy—the enabling role of metallurgical infrastructure. Miner. Process. Ext. Metall. 129 (2), 229–249. https://doi.org/10.1080/25726641.2019.1685243.

Bartie, N.J., Cobos-Becerra, Y.L., Fröhling, M., Schlatmann, R., Reuter, M.A., 2021. The resources, exergetic and environmental footprint of the silicon photovoltaic circular economy: assessment and opportunities. Resour. Conserv. Recycl. 169 (2021), 105561. https://doi.org/10.1016/j.resconrec.2021.105516.

Bartie, N., Cobos-Becerra, L., Fröhling, M., Schlatmann, R., Reuter, M., 2022. Metallurgical infrastructure and technology criticality: the link between photovoltaics, sustainability, and the metals industry. Miner. Econ. 35, 503–519. https://doi.org/10.1007/s13563-022-00313-7.

Beauson, J., Laurent, A., Rudolph, D.P., Pagh Jensen, J., 2022. The complex end-of-life of wind turbine blades: a review of the European context. Renew. Sust. Energ. Rev. 155 (2022), 111847. https://doi.org/10.1016/j.rser.2021.111847.

Binnemans, K., Jones, P.T., Müller, T., Yurramendi, L., 2018. 2018: Rare Earths and the Balance Problem: How to Deal with Changing Markets? J. Sustain. Metall. 4, 126–146. https://doi.org/10.1007/s40831-018-0162-8.

Bonnet, D., 2012. CdTe thin-film PV modules. In: McEvoy, A., Markvart, T., Castañer, L. (Eds.), Practical Handbook of Photovoltaics, second ed. Elsevier, pp. 283–322, https://doi.org/10.1016/B978-0-12-385934-1.00009-X.

CENELEC, 2017a. CLC/TS 50625-3-5:2017 Collection, logistics & Treatment Requirements for WEEE. Part 3–5. Technical Specification for De-Pollution—Photovoltaic Panels. CENELEC, Comite Europeen de Normalisation Electrotechnique, Brussels.

CENELEC, 2017b. EN 50625-2-4:2017 Collection, Logistics & Treatment Requirements for WEEE. Part 2–4. Treatment Requirements for Photovoltaic Panels. CENELEC, Comite Europeen de Normalisation Electrotechnique, Brussels.

Cooperman, A., Eberle, A., Lantz, E., 2021. Wind turbine blade material in the United States: quantities, costs, and end-of-life options. Resour. Conserv. Recycl. 168 (2021), 105439. https://doi.org/10.1016/j.resconrec.2021.105439.

DOE, 2022. Perovskite Solar Cells. U.S. Department of Energy. Office of Energy Efficiency & Renewable Energy.

https://www.energy.gov/eere/solar/perovskite-solar-cells. (Accessed 16 April 2022).

Duflou, J.R., Peeters, J.R., Altamirano, D., Bracquene, E., Dewulf, W., 2018. Demanufacturing photovoltaic panels: comparison of end-of-life treatment strategies for improved resource recovery. CIRP Ann. 67 (1), 29–32. https://doi.org/10.1016/j.cirp.2018.04.053.

European Commission, 2017. Study on the Review of the List of Critical Raw Materials: Critical Raw Materials Factsheets. European Commission, Directorate-General for Internal Market, Industry, Entrepreneurship and SMEs. https://data.europa.eu/doi/10.2873/398823.

European Commission. (2020). Recycled Permanent Magnets Provide a Source for Rare Earth Elements. CORDIS EU Research Results. cordis.europa.eu/article/id/415387-recycled-permanent-magnets-provide-a-source-for-rare-earth-elements (Accessed 16 April 2022).

European Union, 2012. Directive 2012/19/EUon Waste Electrical and Electronic Equipment. https://eur-lex.europa.eu/legal-content/EN/LSU/?uri¼CELEX:32012L0019. (Accessed 16 April 2022).

Farina, A., Anctil, A., 2022. Material consumption and environmental impact of wind turbines in the USA and globally. Resour. Conserv. Recycl. 176 (2022), 105938. https://doi.org/10.1016/j.resconrec.2021.105938.

Fthenakis, V.M., 2004. Life cycle impact analysis of cadmium in CdTe PV production. Renew. Sust. Energ. Rev. 8 (4), 303–334. https://doi.org/10.1016/j.rser.2003.12.001.

Fthenakis, V.M., Kim, H.C., Alsema, E., 2008. Emissions from photovoltaic life cycles. Environ. Sci. Technol. 42 (6), 2168–2174. https://doi.org/10.1021/es071763q.

Gauss, R., Burkhardt, C., Carencotte, F., Gasparon, M., Gutfleisch, O., Higgins, I., Karakjic, M., Klossek, A., Mäkinen, M., Schäfer, B., Schindler, R., Veluri, B., 2021. Rare Earth Magnets and Motors: A European Call for Action. Rare Earth Magnets and Motors Cluster of the European Raw Materials Alliance. https://eit.europa.eu/sites/default/files/2021_09-24_ree_cluster_report2.pdf.

Heath, G.A., Silverman, T.J., Kempe, M., Deceglie, M., Ravikumar, D., Remo, T., Cui, H., Sinha, P., Libby, C., Shaw, S., Komoto, K., Wambach, K., Butler, E., Barnes, T., Wade, A., 2020. Research and development priorities for silicon photovoltaic module recycling to support a circular economy. Nat. Energy 5 (7), 502–510. https://doi.org/10.1038/s41560-020-0645-2.

Huang, W.H., Shin, W.J., Wang, L., Sun, W.C., Tao, M., 2017. Strategy and technology to recycle wafer-silicon solar modules. Sol. Energy 144, 22–31. https://doi.org/10.1016/j.solener.2017.01.001.

IRENA, 2019a. Future of Wind. International Renewable Energy Agency, Abu Dhabi. https://www.irena.org/publications/2019/Oct/Future-of-wind. (Accessed 16 April 2022).

IRENA, 2019b. Future of Solar Photovoltaic. International Renewable Energy Agency, Abu Dhabi. https://www.irena.org/publications/2019/Nov/Future-of-Solar-Photovoltaic. (Accessed 16 April 2022).

IRENA, 2022. World Energy Transitions Outlook 2022: 1.5°C Pathway. International Renewable Energy Agency, Abu Dhabi. https://www.irena.org/publications/2022/Mar/World-Energy-Transitions-Outlook-2022. (Accessed 16 April 2022).

Ito, M., 2016. Development of Recycling Technology of Glass and Metals From Photovoltaic Panels by Separation with a Heated Cutter. NPC Incorporated, Japan. https://www.nedo.go.jp/content/100806683.pdf.

Jensen, J.P., Skelton, A., 2018. Wind turbine blade recycling: experiences, challenges and possibilities in a circular economy. Renew. Sust. Energ. Rev. 97 (2018), 165–176. https://doi.org/10.1016/j.rser.2018.08.041.

Katsaprakakis, D.A., Papadakis, N., Ntintakis, I., 2021. A comprehensive analysis of wind turbine blade damage. Energies 14, 5974. https://doi.org/10.3390/en14185974.

Lal, N.N., Dkhissi, Y., Li, W., Hou, Q., Cheng, Y.-B., Bach, U., 2017. Perovskite tandem solar cells. Adv. Energy Mater. 7 (18), 1602761. https://doi.org/10.1002/aenm.201602761.

Latunussa, C.E.L., Ardente, F., Blengini, G.A., Mancini, L., 2016. Life cycle assessment of an innovative recycling process for crystalline silicon photovoltaic panels. Sol. Energy Mater. Sol. Cells 156, 101–111. https://doi.org/10.1016/j.solmat.2016.03.020.

Lingg, M., Buecheler, S., Tiwari, A.N., 2019. Review of CdTe1−xSex thin films in solar cell applications. Coatings 9 (8), 520. https://doi.org/10.3390/coatings9080520.

LM Wind Power, 2022. Our Blade Recycling Initiatives. LM Wind Power. www.lmwindpower.com/en/sustainability/zero-waste-blades. (Accessed 16 April 2022).

Mahmoud, M.S., Xia, Y., 2012. Some industrial systems. In: Applied Control Systems Design. Springer, London, https://doi.org/10.1007/978-1-4471-2879-3_2.

Masson, G., Kaizuka, I., 2019. Trends in Photovoltaic Applications 2019. IEA PVPS. https://iea-pvps.org/trends_reports/trends-in-pv-applications-2019/. (Accessed 16 April 2022).

Mohammad Bagher, A., 2015. Types of solar cells and application. Am. J. Opt. Photon. 3 (5), 94. https://doi.org/10.11648/j.ajop.20150305.17.

Mone, C., Hand, M., Bolinger, M., Rand, J., Heimiller, D., Ho, J., 2017. Cost of Wind Energy Review. National Renewable Energy Laboratory (NREL). Reportnr: NREL/TP-6A20-66861, revised May 2017 https://www.nrel.gov/docs/fy17osti/66861.pdf. (Accessed 16 April 2022).

Nehls, G., 2022. ZEBRA Project Launched to Develop First 100% Recyclable Wind Turbine Blades. Composites World. https://www.compositesworld.com/news/

zebra-project-launched-to-develop-first-100-recyclable-wind-turbine-blades. (Accessed 16 April 2022).

Norgren, A., Carpenter, A., Heath, G., 2020. Design for recycling principles applicable to selected clean energy technologies: crystalline-silicon photovoltaic modules, electric vehicle batteries, and wind turbine blades. J. Sustain. Metall. 6 (4), 761–774. https://doi.org/10.1007/s40831-020-00313-3.

NREL, 2022. Cadmium Telluride Solar Cells. National Renewable Energy Laboratory (NREL). https://www.nrel.gov/pv/cadmium-telluride-solar-cells.html. (Accessed 16 April 2022).

ROSI Solar, 2020. ROSI Is Awarded First "Green Deal" Funding. 29 July 2020. https://www.rosi-solar.com/rosi-is-awarded-first-green-deal-funding. (Accessed 16 April 2022).

Schnebele, E., 2021. Silicon. In: USGS Mineral Yearbook 2018. U.S. Geological Survey. https://www.usgs.gov/centers/national-minerals-information-center/minerals-yearbook-metals-and-minerals. (Accessed 16 April 2022).

Sinha, P., Cossette, M., Ménard, J.-F., 2012. End-of-life CdTe PV recycling with semiconductor refining. In: 27th European Photovoltaic Solar Energy Conference and Exhibition, pp. 4653–4656, https://doi.org/10.4229/27thEUPVSEC2012-6CV.4.9.

The Silver Institute, 2021. World Silver Survey. The silver institute and Metal Focus, Washington, D.C., USA. www.silverinstitute.org/all-world-silver-surveys/. (Accessed 16 April 2022).

Tsanakas, J.A., Heide, A., Radavičius, T., Denafas, J., Lemaire, E., Wang, K., Poortmans, J., Voroshazi, E., 2020. Towards a circular supply chain for PV modules: review of today's challenges in PV recycling, refurbishment and re-certification. Prog. Photovolt. Res. Appl. 8 (6), 454–464. https://doi.org/10.1002/pip.3193.

U.S. Geological Survey, 2021. Mineral Commodity Summaries 2021. U.S. Geological Survey, https://doi.org/10.3133/mcs2021. 200 pp.

VDMA, 2020–2022. In: Fischer, M., Woodhouse, M., Herritsch, S., Trube, J. (Eds.), International Technology Roadmap for Photovoltaic (ITRPV), 2022 ed. VDMA. https://www.vdma.org/international-technology-roadmap-photovoltaic. (Accessed 16 April 2022).

Veolia, 2018. Veolia Opens the First European Plant Entirely Dedicated to Recycling Photovoltaic Panels. Veolia, France. https://www.veolia.com/en/newsroom/news/recycling-photovoltaic-panels-circular-economy-france. (Accessed 16 April 2022).

Verdu, D., 2020. Taking a Big Step in Making Solar Industry Circular. REC, Norway. https://prosin.no/media/905959/david-verdurec-solar-taking-a-big-step-in-making-solar-industry-circular-1108-prosin-2020.pdf. (Accessed 23 April 2022).

Wade, A., 2013. Evolution of First Solar's Module Recycling Technology. First Solar Inc.

CHAPTER

15

Buildings

David Peck

Circular Built Environment and Critical Materials, Faculty of Architecture and Built Environment, Delft University of Technology, Delft, The Netherlands

15.1 THE WHY: BUILDINGS AND CIRCULARITY

Buildings are an essential part of living for everyone in the world. They provide at the most basic level shelter and safety. Buildings are where people live, work, and play.

The construction sector in the European Union is, however, the highest producer of waste when compared to other economic sectors, accounting for 35% of the total waste generated (Eurostat, 2018). At the same time the creation, running, maintenance, and refurbishment of buildings are highly resource and energy intensive. Up to 50% of globally extracted resources are used to construct buildings and associated infrastructures (UNEP, 2020). Greenhouse gas (GHG) emissions from material extraction, manufacturing of construction products, construction, and renovation of buildings are estimated at 5–12% of total national GHG emissions. Construction and demolition material (CDM) efficiency and construction and demolition waste (CDW) prevention could significantly reduce those emissions.

As the world population grows and becomes wealthier, the challenge for all the stakeholders working in the building and construction sector is to respond to the global demand for building and infrastructure needs, while significantly reducing environmental impacts. The EU has committed to reducing emissions by 55% by 2030 and this can only be achieved by close collaboration between all stakeholders, including the whole value chain, from mining to design, through multiple life cycles to recycling. The building sector has difficulties in reintroducing CDW into new construction cycles, as the level of reuse and recycling of these materials is still very low (see Chapter 26—Concrete and aggregates and Chapter 16—Construction and demolition for details). The reasons for this low reuse and recycling rate are, according to the building stakeholders, a lack of training and knowledge, the loss of construction material quality, and the limited availability of final products that can replace traditional primary material-based ones. Another very important aspect is that in most new construction and refurbishment projects there is a lack of consideration of design for deconstruction principles, which are discussed in Durmisevic (2010) and US EPA (2015), for example.

Much of the material demand for the construction sector is from primary mined material sources. This demand and primary supply have

 https://doi.org/10.1016/B978-0-323-85514-3.00034-8

seen an exponential rise in the twenty-first century. This is driven by developments such as growing world population, aging populations, growing wealth in many economies, increasing expectations of places to live and work, and changing family and relationship patterns. The world is becoming increasingly urban as people are increasingly living in cities, also leading to the rise in the number of megacities, which are cities with a population of over 10 million people. The embodied carbon in building materials and components also increases as the primary material demand increases.

Importantly, there is the urgent need to lower the environmental and societal impacts of running buildings. Energy consumption comes from cooking, lighting, heating, cooling, and ventilating buildings as well as providing digital services. Buildings have become "smarter" and demand more sophisticated technologies (UNEP, 2020). Besides smarter, buildings will also be more connected with other systems: for example, through interaction with the electricity grid (using, storing, and providing energy), with mobility systems by providing charging infrastructure, and with district heating systems, as shown in Figure 15.1. This in turn drives the demand for a wider palette of technology materials. At the same time, the demand for the more conventional bulk materials such as concrete, bricks, steel, aluminum alloy, wood, and glass, etc. continues to increase too. In mass terms, the bulk materials form the majority, while the rise in technology material use poses new challenges, such as supply chain and criticality aspects, discussed in Chapter 36—Geopolitics of resources and recycling.

The current predominant pattern for building construction and refurbishment tends to follow the linear economic model of "take-make-waste," where primary resources are "taken" or extracted, via primary mining and harvesting. These raw materials are then processed into usable materials, which are used on the construction site directly. It is often the case that materials are used to manufacture components and products in factories, which are then directly installed and used in the building. At the point of refurbishment or complete building demolition, materials can be recycled; however many materials are then downcycled, and some critical materials are lost, being placed beyond economically realistic recovery.

This chapter aims to explore the importance of buildings as a secondary resource for materials while exploring the material complexity of buildings. The focus is on not just the bulk building materials, but also the technology and critical materials that are essential for the future functionality.

The chapter looks at current state-of-the-art of building waste recycling, to show what approaches work and what leads to missed opportunity. An overarching approach is taken, where the how, who, when, and what are considered (Hassi et al., 2009). In this, reuse, repair, refurbishing, remanufacturing, and recycling strategies are used to extend the life of buildings and enhance flows into eventual recycling activities.

15.2 THE HOW AND WHO: A FRAMEWORK

Researchers at Delft University of Technology, Faculty Architecture and Built Environment, "Circular Built Environment Hub" collaborated to develop a framework to help explain and better understand materials in the Circular Built Environment (CBE) context.

The method used was cocreation workshops by experts. A cross-disciplinary group of experts from the Delft CBE hub came together to develop solution pathways in on-line workshops. Their backgrounds and expertise covered domains such as materials, architectural design, building engineering and technology, product design, management in the built environment, urbanism, education and learning, societal, food and water, business, policy and climate, and environment.

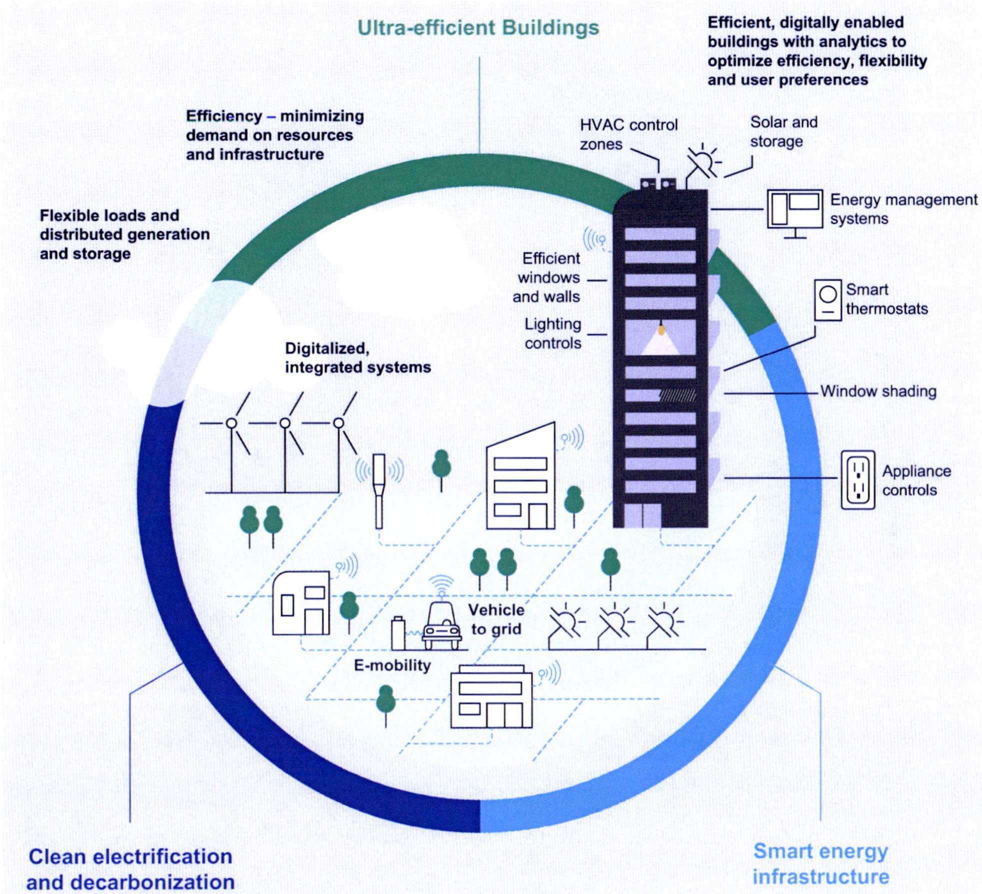

FIGURE 15.1 Buildings as part of integrated energy systems in cities. *From World Economic Forum (2021).*

The group was typically 10–12 people, meeting on-line over 3 months, a total of five times, and using a video call platform and an online collaborative whiteboard platform that allowed visual and interactive cocreation workshops. The collaborative whiteboard platform became the information storage repository. This section is partly derived from some of the information on the whiteboard platform.

The group of experts first developed a framework upon which they could work. The full framework (Figure 15.2) covers different scales in the inner circle of the figure, so that it scales from the smallest part of the built environment (materials) to the largest, systemic part (regions). The latter goes beyond the scope of this work and is therefore excluded. On the outer ring the six interrelated domains (or aspects) are listed.

Together they make up the (draft) definition of CBE (TU Delft, 2022):

> The Circular Built Environment (CBE) is a system designed for closing resource loops at different spatial-temporal levels by transitioning cultural, environmental, economic & social values towards a sustainable way of living (thus enabling society to live within the planetary boundaries).

FIGURE 15.2 The Delft "Circular Built Environment Hub" framework used by the Circular Built Environment hub. The scales (materials, components, buildings, neighborhoods, cities, and regions) are represented by the inner circle. These interact with the interrelated domains (technology, design, resource flows, stakeholders, economy, and management) shown in the outer circle. *Reproduced with permission from TU Delft (2022).*

The interaction between each scale and the domains are explained next.

15.2.1 Materials

First the scale of materials is discussed across the domains of technologies, management, design, stakeholders, resources, and economic aspects.

Technology: A CBE transition requires technologies, which require materials. The new technologies offer solution pathways, be they data, energy, new designs, etc. It is challenging for primary supply alone to meet the CBE demand for these materials. The fundamental approach of a CBE requires that primary materials use be minimized.

Management: Materials are managed from either a primary or secondary perspective. Circular management strategies generally fall into either material (product) life extension (PLE) or recycling. PLE strategies often require additional materials.

Design: This domain has two aspects at the materials scale:

- the design of new materials *and*
- the design of products/components and systems to facilitate PLE and/or recycling.

Both these design activities can facilitate a CBE. Design plays a key and transformative role in addressing materials challenges and opportunities. Both material design and product/component design are closely interlinked.

Stakeholders can include citizens, policy makers, companies, not-for-profit organizations, and NGOs. Organizations outside the company—for profit loci—tend to be less engaged and understanding is typically lower. Citizens, covered under the term of "wider society," tend to be less involved and included in the stakeholder dialogue on materials, although their involvement is higher on secondary materials. Different (time and geographical) scales apply across the materials value chain, both primary and secondary. The aims of stakeholder engagement can be many and varied, including across the Global South and Global North. The societal aspects of complex material chains are not well addressed across the spectrum of stakeholders in relation to primary material supply from the Global South.

CBE material **resource flows** break down into a variety of material resources, with material types such as: plastics, biobased materials, e.g., wood, metals, and minerals. The analysis of the materials can differ, but common tools can be used. Current systems of resource flows are, in the context of sustainability, broken. The demand and supply of materials are, over the near-term period up to 2030, unsustainable at present rates of increase and at the present balance of primary and secondary supply.

"Circular" **economy** is strongly influenced by linear economic approaches, such as financial pricing, markets, economic growth, macro- and microeconomics, etc. Actual circular economic approaches require a radically new approach. Merely enhancing end-of-life recycling, in a CBE context, in the short to medium term, cannot address the challenges societies face. Nevertheless, it remains highly important and must be rapidly enhanced.

Summarizing the materials aspects, a CBE transition demands a wide range of technologies, which require materials supply, both primary and secondary. The CBE demand for materials is challenging to manage and supply, and thus requires a more systemic perspective.

The deployment of new technologies, circular business models, policy shift, and societal engagement in a CBE offers solution pathways when the design of materials, *and* of products and systems, is done at the same time. Although materials differ, common tools can be used for analysis.

Circular economic approaches require a radically new approach, leaving linear economic approaches behind, and include engagement of the material stakeholders across material value chains, geographies, and perspectives.

Materials do have to be applied and it is in the components and building scale (Figure 15.2) where this application happens. These aspects are addressed in the following sections.

15.2.2 Components

The CBE Hub workshops undertook a review at the scale of components across the domains of technologies, management, design, stakeholders, resources, and economic aspects. This section looks at each of those domains in turn.

Technology on the component scale concerns production technologies and processes and component hardware, but also digital technologies. R-strategies (Figure 15.3) such as reuse, repair, refurbish, remanufacturing, and recycling (Kirchherr et al., 2017), are typically proposed when it comes to component technologies.

Management of the component lifecycle and the reverse supply chain, but also (inter)organizational company operations, e.g., knowledge and expertise, fabrication facilities, outsourcing, sales, and marketing, are important factors to be understood.

Design by application of frameworks that allow for:

- regeneration,
- narrowing, e.g., through material choice, design optimization,

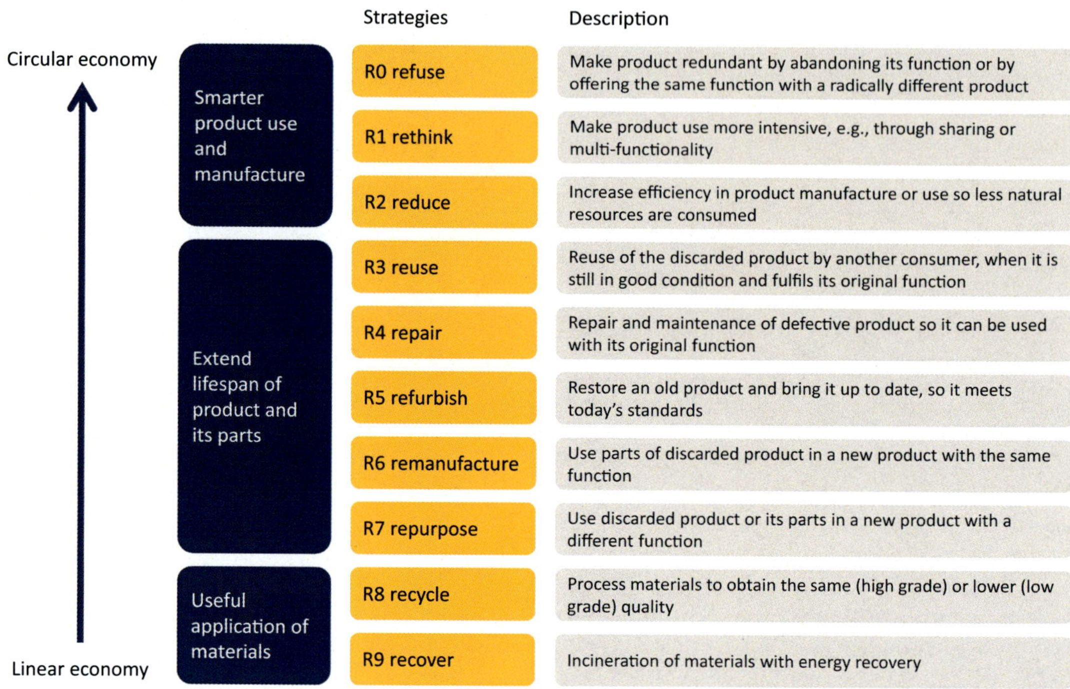

FIGURE 15.3 The 9R framework or R-ladder listing circularity strategies within the production chain, in order of priority. *Adapted from Kirchherr et al. (2017) and Potting et al. (2017).*

- slowing, through product life extension using, e.g., Design for Disassembly, reversible design,
- closing loops, e.g., design for recycling.

This is discussed in more detail in Chapter 2—The fundamental limits of circularity and Chapter 4—Material efficiency of this book.

Stakeholders are mostly related to industry and design such as product suppliers, builders, architects, engineers, and facility managers, but also standardization and norming bodies.

Complex **resource** and value chains must be programmed into components, considering a potential building's use, before the building design has actually started. Flows need to be tracked (i.e., through material flow analysis) and organized to a certain level of granularity (see Chapter 3—Maps of the physical economy to inform sustainability strategies, and Chapter 46—Process simulation).

The circular value chain requires new **economic** and business models that promote the circular use of components. Complex economic and relational structures emerge, as local supply chains are preferred over today's global supply chains.

15.2.3 Buildings

At this scale, components come together to form buildings through physical connections that should enable potential future disconnections, to facilitate a CBE. This can be enabled through the use of building information modeling (BIM) **technologies** at an early stage. It also includes digital technologies to support, for example, system monitoring, building maintenance, and system performance (Meyer, 2018).

The **management** of building actors and processes, building company operation, and procurement is an important factor. On the level of the construction system this requires laws and norms with building specifications, as well as fiscal and financial policy stimuli.

In a Circular Built Environment, buildings are **designed** as carbon neutral, with temporary configurations of component assemblies, creating flexible spaces for multiple and varied future uses and at the same time stimulating the well-being of its users. The buildings can educate and inform their users through their design and appearance. The temporality of configurations is designed using frameworks, which can focus on slowing and closing of loops across the lifecycle of the building, its components, and materials.

Stakeholders are currently investors and clients, product and component suppliers, builders, architects, engineers, and facility managers, but also the building users that aim to realize circular ambitions on a project-to-project basis. Building stakeholders providing essential, but "transient" resources, like energy, water, and food, are also involved.

Circular buildings, from an **economic** perspective, require an approach based on constraints and sufficiency, rather than excess. This needs business models that include the pricing of circular material use, and revolve around true costs, residual value, and social functions. True costs will include the cost of environmental and societal impacts.

15.3 THE WHEN: SHEARING LAYERS

It is important to understand the movement of materials and components in buildings. Much of the current thinking on buildings and materials is focused on building demolition and the recovery of materials via recycling of the bulk materials. This section shows that there is not a single point in time when buildings release products, components, and materials for circular activities. Rather, buildings continuously release a variety of products, components, and materials that change over time as a function of the lifetime and composition of each (see Chapter 2—The fundamental limits of circularity and Chapter 5—Material and product-centric recycling

The concept of "shearing layers" is useful to understand the time aspects of buildings. The concept was developed by Stewart Brand in his book *How Buildings Learn: What Happens After They Are Built* (Brand, 1995), which expands the concept of Duffy (1992). As shown in Figure 15.4, the layers start with the site, which is normally the ground upon and into which the building is located. The structure is then created. A skin of a building is then attached to the structure. Services and a space plan are then put into and onto the building. Finally, users of the building bring stuff into the building.

The materials that make up the different layers can cover a wide spectrum of material types and combinations such as plastics, bio-based materials (textile, wood), metals, minerals, etc. in a wide range of components and such as a products. In terms of elements, buildings, including the "stuff" occupants bring in, cover the entire spectrum of elements in industrial use, with even elements like uranium used in buildings such as nuclear power stations or research labs.

In any circular built environment, a range of R-strategies takes place. The 9R framework (Figure 15.3) provides an overview of the type of circular activity that can be applied when a material or component becomes available. Businesses cannot operate with any circular activity unless they can be certain of when a material or component will "supply" their operations. A combination of the building layers' expected

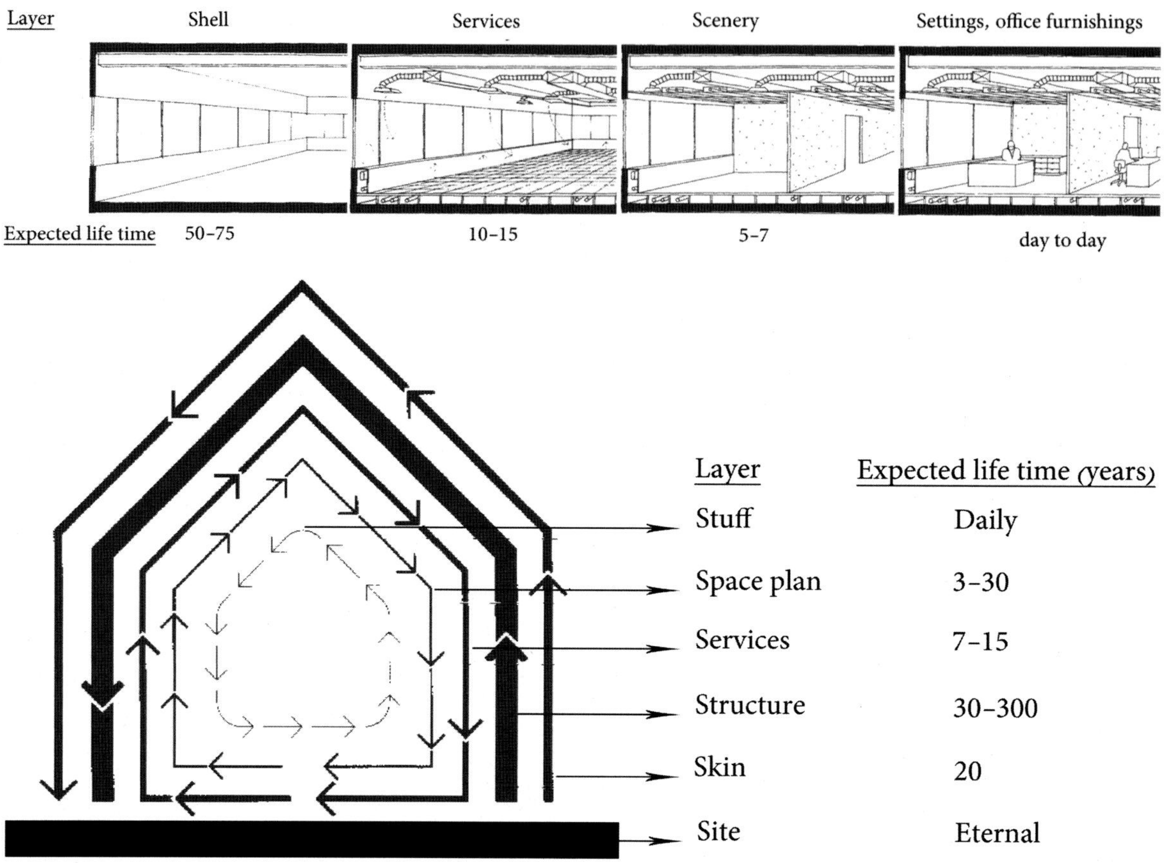

FIGURE 15.4 Building layers and their expected lifetime according to Duffy (1992) and Brand (1995), showing how components and materials flow through buildings over time. *From Salama (2017), who took the (original) figure from Durmisevic (2010).*

lifetime and R-strategies and activities provides a starting point for planning circular activities for buildings.

15.4 THE WHAT: MATERIALS IN BUILDINGS

A building structure can be concrete and steel, which will contain a limited number of chemical elements, while the services, such as heating, ventilation, and air conditioning (HVAC), can contain 60–80 elements. This is even more the case when "controllers" and other digital technologies in HVAC systems are considered. In terms of mass, the rare earth elements in electronics, for example, will be very low.

It is interesting how buildings and the construction sector are viewed by different stakeholders. Figure 15.5 shows the critical materials used in Europe's industrial ecosystems.

This shows how the EU suggests construction uses a range of critical raw materials (CRMs), but the construction of buildings also requires the installation and use of the other ecosystems:

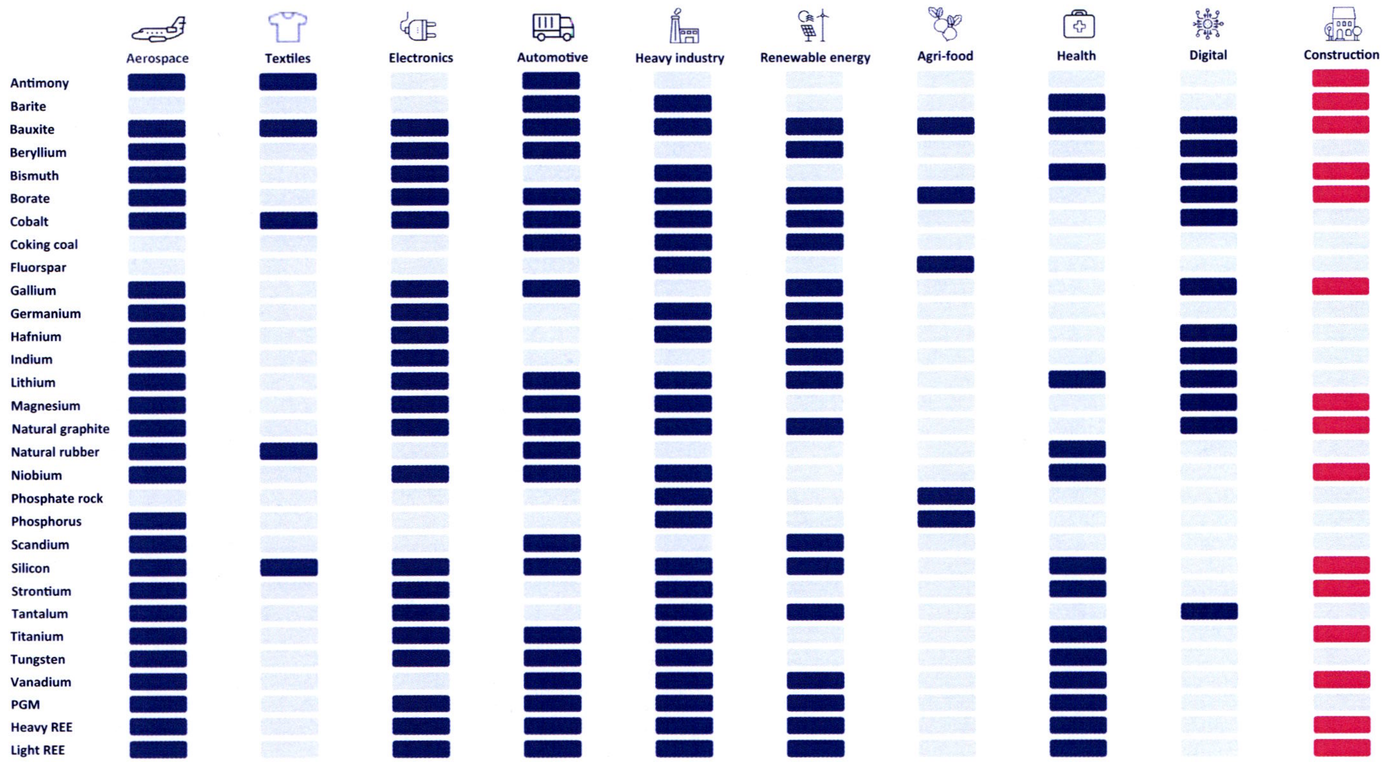

FIGURE 15.5 The critical materials used in Europe's industrial ecosystems. *Modified from European Commission (2020).*

digital, electronics, and textiles, even renewable energy technologies. Indeed, when the Brand model perspective is taken, all sectors, and therefore all materials, are covered at some point by buildings. The scope as to which product/component/building/infrastructure any material is in is poorly defined and, it can be argued, is hard to define.

For many stakeholders buildings are made of concrete, bricks, steel, wood, and glass. At that the differentiation ends. While those materials do make up the greatest mass of a building, they may not be the materials where the widest range of circular opportunities lies.

15.5 IMPROVING DATA ON MATERIALS

Digital technologies are essential for the transition to a Circular Built Environment and many other sectors (Pagoropoulos et al., 2017). Digital technologies can enhance and support various circular strategies, as shown in Figure 15.3, in particular understanding resource flows (Pagoropoulos et al., 2017) and the tracking and tracing of products and components for product life extension strategies.

To date much of the focus has been on building information modeling (BIM) and digital twins to support CBE actions. The geographical information system (GIS) is, however, used at an urban scale in the decision-making process (Wandl et al., 2019). BIM platforms have been identified to show how they can be utilized for even challenging materials such as critical materials, and this is explored in more detail next (Meyer, 2018).

Despite the acknowledged opportunities that these digital technologies offer, there are three main challenges in the application to the CBE.

First, from a digital perspective, not a single integrated digital approach has been developed supporting CBE throughout the building layers and their lifetimes so far; therefore providing a comprehensive overview of materials movement through the entire supply chain is not possible. In a building construction process, the management of materials can be considered a vital part of managing the project. Failure to do so will cause losses and increase waste generation. It requires supervision at the entry and exit of material in a construction process that calls for monitoring material circulation aiming to prevent material loss.

Secondly, from a practical perspective, to develop the systemic change a CBE needs, it is also crucial to change behaviors. The construction industry in Europe currently has:

- A difficulty in finding and developing enough skilled workers,
- a lack of both awareness and understanding of digital technologies,
- a lack of knowledge on how to shift toward circular business models while maintaining competitiveness, and
- a lack of incentives such as the use of recycled materials via public sector procurement.

Thirdly, from a technical perspective, one of the biggest challenges in materials management in a CBE is the variety in size and properties of materials, products, and components, which makes it difficult to sort them automatically with standard equipment.

As highlighted earlier, a positive development in the circular built environment transition is the increasing development and use of building information modeling (BIM). A BIM model makes it easy to collect data of the whole life cycle of a building and share it among all involved stakeholders in the process. This data can be easily exported to other formats for further use.

In 2018 Charley Meyer conducted research to test if BIM is compatible to facilitate knowledge of, and solutions for, critical materials in buildings (Meyer, 2018). BIM was assessed to see if there was an approach that could process CRM information into standardized datasets to make it easily accessible and comparable.

As a case study, Meyer selected an electronic component, a sensor, to assess the current use and composition of objects containing CRMs. Meyer demonstrated that the recycling rate of critical materials in such electronic components in buildings is low. More efficient use of the materials can be achieved through technical changes, in designing for a longer lifetime, and new business models summarized as product life extension strategies. Then enhanced recycling at the end of the final use cycle could be achieved.

Meyer went on to show how documentation of all the required information to lengthen the lifetime of CRMs, via circular strategies, in BIM appears to be an appropriate solution through the use of so-called IFC files. Herein standardized datasets could be created and exported into many formats for further use. Based on specified information in IFCs, objects in a BIM model can be found by searching for instance for "critical materials," "gallium" or "Ga," and schemes can be exported outlining this information.

Meyer concluded by showing a framework that provides information on critical materials, material selection processes, optimized used of the materials, and finally converts this data into the so-called "property sets" suitable for IFC. The sensor example used by Meyer does highlight the complexity of materials in buildings and shows the technology convergence, where digital, electronic, advanced HVAC, and low carbon energy generation technologies converge to produce a very low carbon or net zero building. At the same time, the embodied carbon in the structural materials of buildings, in particular of buildings using mainly steel and concrete in the structure, need enhanced recycling when they are beyond any further safe use.

The challenge of data availability and data accuracy on CRMs, however, presented the work with challenges. If the framework is implemented with data on both "bulk" materials and CRMs, interested stakeholders could have a holistic overview of not only what is in a building, but also the optimal circular product life extension strategy and eventual optimal circular recycling strategy.

Platforms are being developed that draw on the idea to integrate different existing monitoring tools for materials in a CBE, including:

- Building Information Modeling (BIM),
- Material Flow Assessment and Material Passport,
- GIS-based material stream, and
- Industrial Symbiosis platform.

These approaches can be optimized by utilizing real-time on-site progress data and AI sensor technologies, where a system could automatically capture and trace on-time and on-site material movements. Consequently, by automating registration processes and reintroducing materials management systems, this is beneficial for both circular product life extension strategies and for circular recycling.

15.6 THE HOW, WHO, WHEN, AND WHAT

As buildings in a low carbon, sustainable, circular built environment become power stations, farms, biocooling constructions, and much more, there is an increasing diversity in the material choice, construction methods, and types of buildings. Buildings using wood as the main material is nothing new, but recent innovations on the design and joining of timber is producing new opportunities.

At the same time, modular buildings that are built in factories are reemerging as a sustainable alternative. These modules can be new build and used as renovation or refurbishing components. These designs can revolutionize product life extension and, at eventual end of final use cycle, recycling.

One important opportunity for closing the loop on building components and even modular building sections, is, as shown in Figure 15.3, remanufacturing (Boorsma et al., 2019). Even though there is no single definitive definition of remanufacturing, the elements used to construct the available definitions have commonalities. The definitions are generally constructed of three elements, which are:

- the use of a previously used product as an input for the remanufacturing activity.
- an industrial production process where varying emphasis is given to different operational activities, such as inspection, cleaning, disassembly, inspection, testing, coating, parts replaced, reassembly, testing, and dispatch.
- the intended outcome, referred to as at least equal to original product grade, even "better than" the original.

Remanufacturing can take a product or component and return it back to a factory where it will be processed to a condition that is as good as, or better than, new, in a box, with a warranty. The new user cannot tell the difference between a newly manufactured product and a remanufactured one. The "better than" new can be realized by the incorporation of modifications and upgrades during the remanufacturing process.

Remanufacturing is not repairing, reconditioning, or refurbishing. This is important, as a remanufactured product can meet all the applicable regulatory and safety requirements in a building, because it is as good as (or better than) new. This ensures it meets all the requirements.

The suitability of a product for remanufacturing depends on the physical product design, as well as the success of its interactions with the system it operates in. Typical criteria suggest that a remanufacturable product should be long-lasting, failing functionally, constructed of highly standardized parts with stable technology, and containing remaining useful life after use.

It is important that the cost to obtain the return product, known as cores, should be economically viable and the target market should be aware of what is on offer. Logistics plays a part too. The timing of incoming cores and the distance and cost of transportation, for example, are influenced by core management activities. Cores need to return when needed. A service-based business model may facilitate remanufacturing.

As a product is disassembled, inspected, and parts replaced, the old worn/defective parts are placed into the recycling stream. A recycler has a cleaned and monomaterially separated stream of "waste." Even better, it will arrive on a schedule. When the core can no longer be remanufactured, it can be fully disassembled, and all the parts go into recycling. It is everything a modern recycler wishes for.

15.7 OUTLOOK

This is the future of the built environment: to yield the greatest positive impact in closing material loops. It requires a combination of understanding the timings when things come out of buildings, their exact material composition, their value in terms of life extension, a steady stream of separated materials for recycling, and business models that make sense for all stakeholders. The result will be securing *both* critical and bulk materials in appropriate quality and lowering climate emissions by a significant factor, far beyond the current goals set for 2030.

The how, who, when, and what can help stakeholders develop circular built environment pathways to address the resource and climate challenges we face. The when, the time challenge, is key. The demand and supply of critical and other materials presents the built environment and its stakeholders with unprecedented challenges in the short term. The climate challenge will not wait. The key is a reduction in demand via circular strategies, as outlined in this chapter.

References

Boorsma, N., Tsui, T., Peck, D., 2019. Circular building products, a case study of soft barriers in design for remanufacturing. In: Ijomah, W. (Ed.), Proceedings of the International Conference of Remanufacturing 2019, Amsterdam, The Netherlands, pp. 1–17.

Brand, S., 1995. How Buildings Learn: What Happens After They're Built. Penguin Books. Paperback, October 1, 1995.

Duffy, F., 1992. The Changing Workplace. Phaidon Press Ltd. ISBN: 978-1854541208.

Durmisevic, E., 2010. Green Design and Assembly of Buildings and Systems, Design for Disassembly a Key to Life Cycle Design of Buildings and Building Products. VDM Verlag Dr. Müller, Saarbrucken, Germany. ISBN 978-3-639-11262-7.

European Commission, 2020. COM (2020) 474 Final Communication From the Commission to the European Parliament. The Council, The European Economic and Social Committee and The Committee of the Regions; Critical Raw Materials Resilience: Charting a Path towards greater Security and Sustainability. https://eur-lex.europa.eu/legal-content/EN/TXT/?uri=CELEX:52020DC0474.

Eurostat, 2018. Eurostat Statistics for Waste Flow Generation 2016. European Commission. ec.europa.eu/Eurostat/web/waste. (Accessed January 2022).

Hassi, L., Peck, D., Dewulf, K., Wever, R., 2009. Sustainable innovation – organization and goal finding. In: Andersen, R.D., Lehmann, M. (Eds.), Joint Actions on Climate Change. Aalborg University, Denmark. ISBN: 978-87-91830-30-3.

Kirchherr, J., Reike, D., Hekkert, M., 2017. Conceptualizing the circular economy: an analysis of 114 definitions. Resour. Conserv. Recycl. 127 (2017), 221–232. https://doi.org/10.1016/j.resconrec.2017.09.005.

Meyer, C., 2018. Optimizing the Use of Critical Materials in the Built Environment Using Building Information Modelling (BIM), TU Delft Architecture and the Built Environment; Contributors Peck, D (mentor) & Bähre, B (mentor), Delft University of Technology Repository. http://resolver.tudelft.nl/uuid:7bbca8cd-473e-44e2-b787-044a31df7831. (Accessed 22 June 2022).

Pagoropoulos, A., Pigosso, D.C.A., McAloone, T.C., 2017. The emergent role of digital technologies in the Circular Economy: a review. Procedia CIRP 64, 19–24. https://doi.org/10.1016/j.procir.2017.02.047.

Potting, J., Hekkert, M., Worrell, E., Hanemaaijer, A., 2017. Circular Economy: Measuring Innovation in the Product Chain. https://www.pbl.nl/sites/default/files/downloads/pbl-2016-circular-economy-measuring-innovation-in-product-chains-2544.pdf. (Accessed 22 June 2022).

Salama, W., 2017. Design of concrete buildings for disassembly: an explorative review. Int. J. Sustain. Built Environ. 6 (2), 617–635. https://doi.org/10.1016/j.ijsbe.2017.03.005.

TU Delft, 2022. Circular Built Environment Hub. Faculty of Architecture and the Built Environment, Delft University of Technology, Delft, the Netherlands. https://www.tudelft.nl/en/architecture-and-the-built-environment/research/research-themes/circular-built-environment/. (Accessed 22 June 2022).

United Nations Environment Programme (UNEP), 2020. 2020 Global Status Report for Buildings and Construction: Towards a Zero-Emissions, Efficient and Resilient Buildings and Construction Sector. Nairobi https://globalabc.org/resources/publications/2020-global-status-report-buildings-and-construction. (Accessed 22 June 2022).

US EPA, 2015. Design for Deconstruction Manual. United States Environmental Protection Agency. https://www.epa.gov/sites/default/files/2015-11/documents/designfordeconstrmanual.pdf. (Accessed 30 June 2022).

Wandl, A., Balz, V., Qu, L., Furlan, C., Arciniegas, G., Hackauf, U., 2019. The circular economy concept in design education: enhancing understanding and innovation by means of situated learning. Urban Plan. 4 (3), 63–75. https://doi.org/10.17645/up.v4i3.2147.

World Economic Forum, 2021. Net Zero Carbon Cities: An Integrated Approach. World Economic Forum, Geneva. https://www.weforum.org/reports/net-zero-carbon-cities-an-integrated-approach.

CHAPTER

16

Construction and demolition waste

Vivian W.Y. Tam[1,2], *Mahfooz Soomro*[1], *and Ana Catarina Jorge Evangelista*[3]

[1]School of Engineering, Design and Built Environment, Western Sydney University, Penrith, NSW, Australia [2]College of Civil Engineering, Shenzhen University, Shenzhen, China [3]Engineering Institute of Technology, Perth, WA, Australia

16.1 INTRODUCTION

Population growth, continuous industrial development, construction of infrastructure, and house building activities create huge amounts of construction and demolition (C&D) waste. C&D debris generally consists of asphalt, concrete, brick, dirt, wood, metal, wall-board, roofing and insulation materials, plastics, cardboard, glass, and miscellaneous trash (Jiménez, 2013). There is hence a dire need for waste recycling. The construction industry has been a major consumer of natural resources and, hence, the global natural aggregate production increased 58% from 21,000 Mt in 2007 to 50,000 Mt in 2017 and is expected to rise to 60,000 Mt in 2030. Countries with the highest consumption of natural aggregate, going hand in hand with their economic development, are China (40%), India (10%), other Asian countries such as Indonesia, Malaysia, Thailand (16%), and Africa (8%) (O'Brien, 2019), as illustrated in Figure 16.1.

Countries such as China, India, Indonesia, Malaysia, Thailand, the Gulf States, Turkey, Russia, Brazil, and Mexico have recorded some of the strongest increases in the demand for waste recycling (Tam et al., 2018). Hence, vast depletion of natural resources and the growing awareness of sustainable waste management by the developed and emerging economies have given ever-increasing relevance to recycle and reuse of C&D waste and the use of recycle aggregate in the infrastructure and civil engineering projects.

Although a much higher portion of construction material could be replaced by reprocessed and upgraded C&D waste, these options are not yet considered and applied in many of the developing economies, due to insufficient regulatory frameworks and lack of knowledge.

Meanwhile, in the developed countries, the process of stimulus of the utilization of C&D waste varies in scale from country to country. It is hoped and expected that the utilization of recycled aggregate will increase and become a substantial part of the aggregate market in the near future.

The global aggregate production of a colossal 50,000 Mt annually is an indication of the vast

Handbook of Recycling. **https://doi.org/10.1016/B978-0-323-85514-3.00036-1**

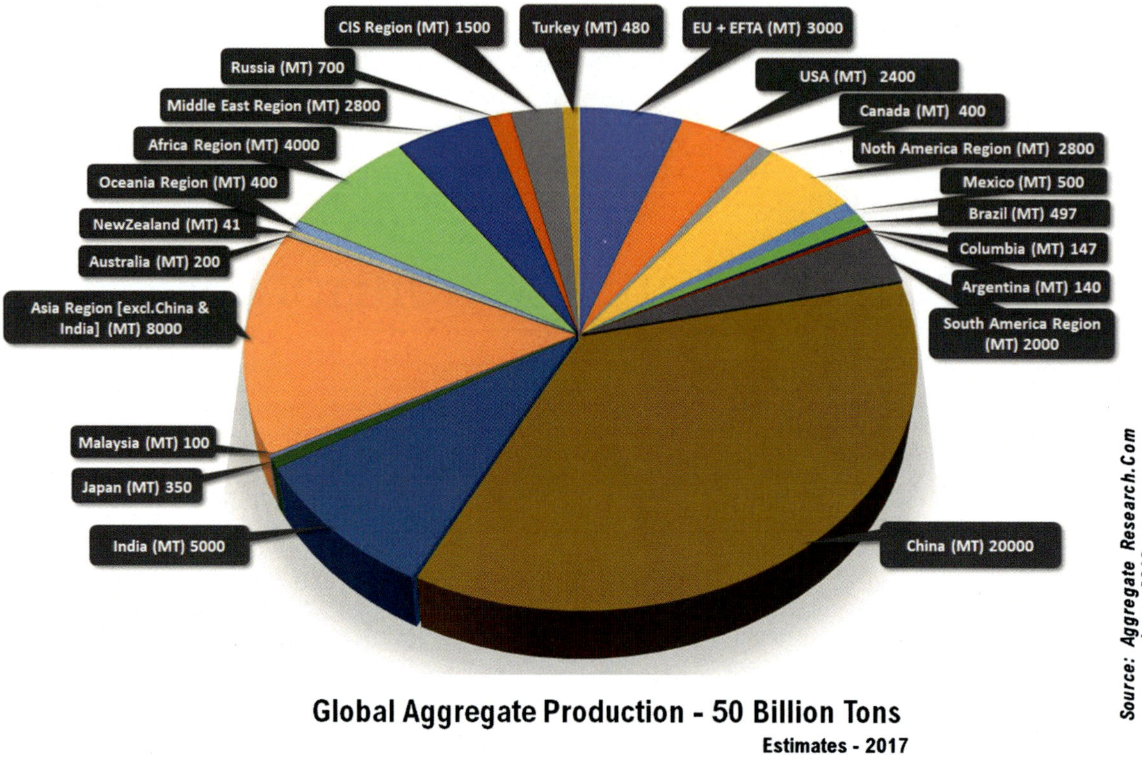

FIGURE 16.1 Estimated global aggregate production in 2017. *Data from O'Brien (2019).*

development projects that at present are materializing around the world. As the land available for landfills becomes scarcer, the ways to use C&D waste are gaining in importance due to regulatory legislation and its cheaper cost and availability. However, research and development to increase recycling and utilization of C&D waste are constantly needed to sustainably utilize alternative materials in the production of concrete, incorporating recycled aggregate. The current tendency in several developed economies is to view waste as a resource or a by-product that can be utilized in a variety of infrastructure and building construction projects.

Depending on their quality, the recycled aggregate produced from C&D waste can be employed in various civil engineering projects. The significance and aim of this chapter are to review the literature on the production and utilization of recycled aggregate in roadway construction, concrete pavements, load-bearing concrete structures, and other civil engineering works and include some discussion on the reduction of CO_2 emissions in the atmosphere. Additionally, this chapter approaches to the circular economy concept that replaces the end-of-life concept, discussing the alternatives to recycle, recover, and reuse of construction and demolition materials.

16.2 C&D WASTE USE

Since the 1980s, considerable progress has been seen in C&D waste management systems in the developed economies, particularly in

Australia, Western Europe, and North America. In the current decade, the Asia/Pacific, Russia, and South America regions have been the largest producers of aggregate as well as its sales, because of their rapidly rising construction activity, especially in China, India, Indonesia, Malaysia, Thailand, Gulf States, Turkey, Russia, Brazil, and Mexico. China alone accounted for nearly half of all the new aggregate demand worldwide during the 2010–15 period (The Freedonia Group, 2012). According to Persistence Market Research (2021), the demand for construction aggregates increased at 3.8% per year (CAGR) in 2016–20. The trend continued until 2020, as the unprecedented outbreak of COVID-19 caused supply-chain disruption. However, construction aggregate sales recovered in 2021, reaching a valuation of US$ 344.4 Bn. Furthermore, Future Market Insights (2021) has projected the demand to grow with a 6.8% CAGR between 2021 and 2031 based on the recovery of construction and infrastructure activities, especially in emerging economies. The global recycled aggregate consumption estimates (by regions) are shown in Figure 16.2.

The tendency of environmentally conscious communities and enterprises is not only to recycle a large percentage of C&D waste, but to aim for zero waste, which means ensuring that all products are made to be reused, repaired, or recycled back into the marketplace or nature, eliminating all discharges to land, water, and air. While this concept first emerged in the state of California in the United States in 1975, zero waste plans have been adopted around the world, especially by local governments in Australia and New Zealand. As a plentitude of research work shows, there has been a notable increase not only in the methods of recovery from C&D waste into recycled aggregate, but also of the means and techniques of its utilization in the construction industry (Marzouk and Azab, 2014).

16.3 RECYCLING

According to da Rocha and Sattler (2009), the C&D waste generation occurs during all the main phases of the building life cycle, i.e., construction, renovation, and demolition. The

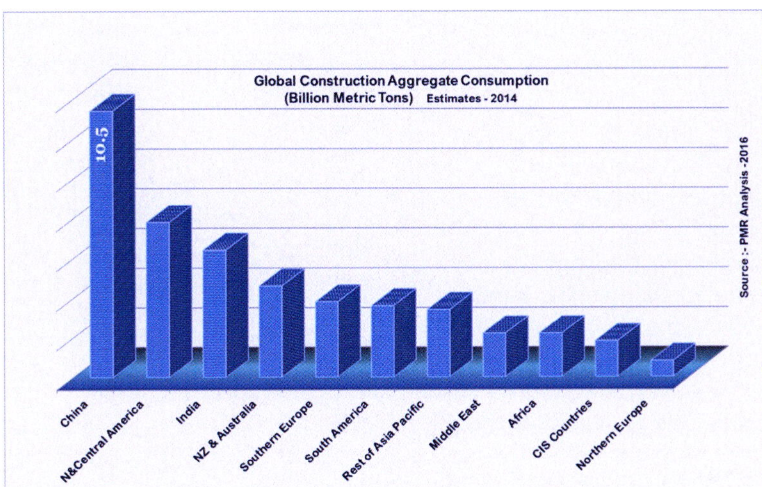

FIGURE 16.2 Worldwide consumption of recycled aggregate (by region) in 2014. *Data from PMR Analysis (2016). Figure from Tam et al. (2018).*

demolition phase therefore seems to be the specific key to be considered for the adoption of more sustainable practices, to contribute to higher percentages of the C&D waste generated.

Many countries have recycling schemes to recover recycled aggregate from C&D waste, which can be reused as aggregate, while in many other countries concrete recycling is a well-established industry and most of the C&D waste is being crushed and reused as aggregate mainly in infrastructure and nonstructural concrete. Recovery rates of recycled aggregate vary from almost 100% in some parts of the world to nearly zero in countries where most construction waste ends up in the landfill. For example, Japan is a leading country in recycling concrete waste and has been implementing 98% recycling and using it for structural and nonstructural concrete applications (Tam et al., 2010).

In the European Union, the revised Waste Framework (Directive (2008/98/EC), 2008) stated that the member states set an ambitious goal of achieving a 70% level of recovery for the recycling and reuse of nonhazardous C&D waste generated at buildings and public works construction sites by the year 2020. However, the 2011 statistics for the countries in the European Union showed that the level of recycling and recovery of material from C&D waste fluctuated considerably, i.e., between less than 10% and over 40% across the European Union (Oliveira Neto et al., 2017). In Denmark, the main focus has been put on obtaining a better quality of recycling C&D waste, together with maintaining a high recycling rate, but on the contrary, in France C&D waste management is an emerging issue and lacks political will and consumer interest (Oliveira Neto et al., 2017).

The Union Européenne des Producteurs de Granulats (UEPG) (European Aggregates Association) was formed in 1987 to work for the promotion of the interests of the European Aggregate Industry and continues to aim for a sustainable supply of aggregates and towards its competitiveness and its growth. In 2014, production from recycled and reused aggregate increased to 228 Mt, representing 8.6% of the total output of 2650 Mt (UEPG, 2016).

In Australia, the management of C&D waste is not legislated by the Central Government but the environmental issues, including all waste streams, are primarily the responsibility of the Australian state and territory governments. In 2006–2007 in Australia, 43.78 Mt of waste was generated, 38% of which was from the C&D streams (DEA, 2012).

16.4 RECYCLING TECHNOLOGIES AND PRACTICE

C&D waste could be demolished concrete structures, broken waste concrete, rejected concrete products on production line, broken pavements, and bricks from buildings. Thus recycled aggregate could come from demolition of concrete structures, airport runways, bridge supports, concrete roadbeds, or rejected concrete products on the production line, etc. Concrete made using such aggregate is referred to as recycled aggregate concrete (RAC).

Existing technology for producing recycled aggregate from C&D waste by means of mechanical crushing is relatively inexpensive and readily available; therefore the process of converting C&D waste to recycled aggregate can be done in both the developed and developing countries (Malešev et al., 2010). There are two categories of plants available for processing C&D waste: stationary and mobile (Thomas and Bremner, 2012). The same processes take place in both type of plants to separate the contaminants from bulk material and to obtain a useful grading (Pellegrino and Faleschini, 2016):

- separation
- crushing
- separation of ferrous elements
- screening, and
- decontamination and removal of impurities (i.e., wood, paper, plastics)

Hansen cited in De Brito (de Brito and Saikia, 2013) reported that recycled aggregate is generally used as bulk backfill, in subbase, base, or surface material in road construction, lean concrete bases, hydraulically bound materials, and in the manufacture of new concrete. Depending on its quality, recycled aggregate can be used in hydraulically bound material or in the production of new concrete only if the recycled aggregate coming from waste materials does not have hazardous contaminants. The contaminants in most cases may be found only on the surface layers of old concrete, so that no more than traces of potentially harmful contaminants will be found in the bulk of recycled aggregate. Under other circumstances, potentially harmful contaminants may not be water soluble, which in reality makes them harmless in recycled aggregate concrete.

In Australia, recycled aggregate (both as coarse and as fine aggregate) has been the most common C&D waste used in (nonstructural) concrete production. About 5.0 Mt of recycled concrete and masonry are available, mainly in the Sydney and Melbourne areas, out of which 0.5 Mt is recycled aggregate (Cement Concrete & Aggregates Australia, 2008). In Europe, recycled aggregate from C&D waste accounts for 6%–8% of the total aggregate in Europe, with a significant difference between the EU countries (UEPG, 2014). The big users are the United Kingdom, Germany, the Netherlands, Belgium, and Switzerland (WBCSD, 2009). In the United Kingdom, the quantities of recycled aggregate produced have increased gradually over the last 30 years and are being used in a wide range of construction applications. In 1980, the estimated use of recycled aggregate was 20 Mt/year, in 2007 the usage increased to 71 Mt/year, but decreased to 50 Mt/year in 2009, which accounts for 26% of the total aggregate used in the United Kingdom (MPA, 2009).

A study by the National Ready-mix Concrete Association (NRMCA) in the United States concluded that up to 10% of recycled aggregate is suitable as a substitute for natural (virgin) aggregate for most concrete applications, including structural concrete (Obla et al., 2007). United Kingdom research indicated that up to 20% of recycled aggregate can be used for most applications, including structural (WRAP and NIEA, 2013). Australian guidelines state that up to 30% recycled aggregate can be used in concrete, including structural concrete (WBCSD, 2009). German guidelines state that, under certain circumstances, recycled aggregate can be used for up to 45% of the total aggregate, depending on the exposure class of the concrete (DAfStb, 2010).

Significant potential remains for increasing the use of coarse recycled aggregate in concrete. In some countries, notably Germany, Switzerland, and Australia, premix concrete containing recycled aggregate is now being marketed. Boral's "Green" concrete is premixed concrete using recycled aggregate that has been used in a number of building projects in Australia (WBCSD, 2009).

From a recent review study, it was found that there are well-established recycling technologies for C&D waste in developed countries. Recycling technologies for 10 typical C&D waste types have been investigated; the results are summarized in Table 16.1.

16.5 FUTURE DEVELOPMENTS

Significant potential remains for increasing the use of recycled aggregate in concrete. In some countries, notably Germany, Switzerland, and Australia, concrete containing recycled aggregate is now being marketed. In the United States, the state of Virginia has found a novel way to utilize recycled aggregate. An artificial reef is created using recycled concrete aggregate, which is then covered with crushed oyster shells, creating an oyster bed. Since the recycled material is being placed in the marine environment, concrete particularly with high chloride content is acceptable.

TABLE 16.1 Summary of the experience on recycling technology and practice.

Construction and demolition waste	Recycling technology	Recycled product
Asphalt	– Cold recycling – Heat generation – Minnesota process – Parallel drum process – Elongated drum – Microwave asphalt recycling system – Finfalt – Surface regeneration	– Recycled asphalt – Asphalt aggregate
Brick	– Burn to ash – Crush into aggregate	– Slime burnt ash – Filling material – Hardcore
Concrete	– Crush into aggregate	– Recycled aggregate – Cement replacement – Protection of levee – Backfilling – Filler
Ferrous metal	– Melt – Reuse directly	– Recycled steel scrap
Glass	– Reuse directly – Grind to powder – Polishing – Crush into aggregate – Burn to ash	– Recycled window unit – Glass fiber – Filling material – Tile – Paving block – Asphalt – Recycled aggregate – Cement replacement – Man-made soil
Masonry	– Crush into aggregate – Heat to about 900°C to ash	– Thermal insulating concrete – Traditional clay brick – Sodium silicate brick
Nonferrous metal	– Melt	– Recycled metal
Paper and cardboard	– Purification	– Recycled paper
Plastic	– Convert to powder by cryogenic milling – Clipping – Crush into aggregate – Burn to ash	– Panel – Recycled plastic – Plastic lumber – Recycled aggregate – Landfill drainage – Asphalt – Man-made soil

TABLE 16.1 Summary of the experience on recycling technology and practice—cont'd

Construction and demolition waste	Recycling technology	Recycled product
Timber	– Reuse directly – Cut into aggregate – Blast furnace deoxidization – Gasification or pyrolysis – Chipping – Molding by pressurizing timber chip under steam and water	– Whole timber – Furniture and kitchen utensils – Lightweight recycled aggregate – Source of energy – Chemical production – Wood-based panel – Plastic lumber – Geofiber – Insulation board

In the European Union, the processing of construction waste (concrete, brick, asphalt, and stone) into aggregate complying with European Standards for aggregate as well as the development of material specifications enabling this aggregate to be used in construction applications is essential for meeting the Waste Framework Directive of European Union of 70% recycle and recovery target for 2020 (Barritt, 2015).

The Northern Ireland Environment Agency (NIEA) and Waste & Resources Action Programme (WRAP) in consultation with industry and other regulatory stakeholders developed a "Quality Protocol" applicable in England, Wales, and Northern Ireland. The process of converting inert waste material into a product is classified as a waste recovery operation and is subject to the waste management controls set out in the Waste Framework Directive and domestic legislation; whereas the aggregate must be produced in compliance with the criteria outlined in Sections 2.3–2.5 considering: acceptable input materials, BS EN-12620: "Aggregates for Concrete" standard, and the aggregate must require no further processing (WRAP and NIEA, 2013). Ghaffar et al. (2020) mentioned that, despite the gained recognition and market of CDW being used into new construction applications, there are barriers such as logistics (41%), cost (29%), regulations (12%), and others (6%). Additionally, other researchers (Chinda, 2016; Tam et al., 2007) have identified the lack of standards for C&D recycling and reuse, certification, and effective sorting. Also, the lack of balance between the demand and supply in the recycling and reuse market as other factors affecting the progress of circular construction concept.

16.6 CONCLUSION AND OUTLOOK

The acceptability of recycled aggregate is hampered due to a poor image associated with recycling activity and lack of confidence by the constructors, builders, and general consumers in the finished product made from recycled material. Even though recycled aggregate is being utilized in roadway and pavement construction in substantial quantities, it is still difficult to overcome the barriers that prevent the wider use of recycled aggregate in structural and high-rise building construction.

In several countries, the majority of the aggregate output of recycling plants is currently non-certified aggregate, but due to stricter demands from consumers, who would prefer to use recycled aggregate conforming to specifications and with guaranteed quality, certification has become of prime importance. A lack of government support and commitment towards development of the recycling industry is also an impediment. Developing appropriate policy supported by a proper regulatory framework can provide the necessary impetus. For example, in 2020, the Australian Government launched the Sustainable Procurement Guide aiming to encourage the transformation of waste into a resource, where most products and services can be continually used, reused, recycled, and reprocessed as part of a circular economy (Australian Government, 2020). This will also help in data compilation, documentation, and education of consumers on its use and control over disposal of waste material.

Acknowledgments

This chapter is an adaptation of:

Tam, W. Y. V., Soomro, M. and Evangelista, A. (2018), "A Review of Recycled Aggregate in Concrete Applications (2000–2017)", Construction and Building Materials 172: 272–92.

Tam, W. Y. V. (2011), "Cost effectiveness of using low cost housing technologies in construction", Procedia Engineering, The Twelfth East Asia-Pacific Conference on Structural Engineering and Construction, 14, 156–160.

Tam, W. Y. V. and Tam, C. M. (2008), Re-use of construction and demolition waste in housing development, Nova Science Publishers, Inc., United States.

Tam, W. Y. V. and Tam, C. M. (2006), "A review on the viable technology for construction waste recycling", Resources, Conservation and Recycling, 47(3), 209–22.

References

Australian Government, 2020. Sustainable Procurement Guide: A Practical Guide for Commonwealth Entities. Retrieved from: http://www.environment.gov.au/protection/.

Barritt, J., 2015. An overview on recycling and waste in construction. Proc. ICE Constr. Mater. 169 (3), 1–5. https://doi.org/10.1680/coma.15.00006.

Cement Concrete & Aggregates Australia, 2008. Use of Recycled Aggregates in Construction. https://www.ccaa.com.au/documents/Library%20Documents/CCAA%20Reports/RecycledAggregates.pdf.

Chinda, T., 2016. Investigation of factors affecting a construction waste recycling decision. Civ. Eng. Environ. Syst. 33 (3), 214–226. https://doi.org/10.1080/10286608.2016.1161030.

DAfStb, 2010. Concrete in Accordance With DIN EN 206-1 and DIN 1045-2 With Recycled Aggregates in Accordance With DIN EN 12620. Deutscher Ausschuss für Stahlbeton e. V., Berlin, Germany.

da Rocha, C.G., Sattler, M.A., 2009. A discussion on the reuse of building components in Brazil: an analysis of major social, economical and legal factors. Resour. Conserv. Recycl. 54 (2), 104–112. https://doi.org/10.1016/j.resconrec.2009.07.004.

de Brito, J., Saikia, N., 2013. Concrete with recycled aggregates in international codes. In: Recycled Aggregates in Concrete. Elsevier, https://doi.org/10.1007/978-1-4471-4540-0.

DEA, 2012. National Waste Information Baseline. Department of Environment Affairs, South Africa. http://sawic.environment.gov.za/documents/1880.pdf.

Directive (2008/98/EC), 2008. Directive 2008/98/EC of the European Parliament. Official Journal of the European Union. Retrieved from: http://eur-lex.europa.eu/LexUriServ/LexUriServ.do?uri=CELEX:32008L0098:EN:NOT.

Future Market Insights, 2021. Construction Aggregates Market by Product Type, Application Type & Region for 2021–2031 | Global Sales Analysis and Opportunity 2021–2031. Retrieved January 4, 2022, from: https://www.futuremarketinsights.com/reports/construction-aggregates-market.

Ghaffar, S.H., Burman, M., Braimah, N., 2020. Pathways to circular construction: an integrated management of construction and demolition waste for resource recovery. J. Clean. Prod. 244, 118710. https://doi.org/10.1016/j.jclepro.2019.118710.

Jiménez, J.R., 2013. Handbook of Recycled Concrete and Demolition Waste. Elsevier, https://doi.org/10.1533/9780857096906.3.351.

Malešev, M., Radonjanin, V., Marinković, S., 2010. Recycled concrete as aggregate for structural concrete production. Sustainability 2 (5), 1204–1225. https://doi.org/10.3390/su2051204.

Marzouk, M., Azab, S., 2014. Environmental and economic impact assessment of construction and demolition waste disposal using system dynamics. Resour. Conserv. Recycl. 82, 41–49. https://doi.org/10.1016/j.resconrec.2013.10.015.

MPA, 2009. MPA Aggregates Information Sheet 3.

O'Brien, J., 2019. Global Aggregate Growth Examined. Aggregateresearch.com, pp. 1–5. https://www.aggbusiness.com/feature/global-aggregates-growth-examined-gain-convenor-jim-obrien.

Obla, K., Kim, H., Lobo, C., 2007. Crushed Returned Concrete as Aggregates for New Concrete. RMC Research & Education Foundation, p. 51.

Oliveira Neto, R., Gastineau, P., Cazacliu, B.G., Le Guen, L., Paranhos, R.S., Petter, C.O., 2017. An economic analysis of the processing technologies in CDW recycling platforms. Waste Manage. 60, 277–289. https://doi.org/10.1016/j.wasman.2016.08.011.

Pellegrino, C., Faleschini, F., 2016. Sustainability and the Concrete Industry. Elsevier, https://doi.org/10.1007/978-3-319-28540-5.

Persistence Market Research, 2021. Global Market Study on Construction Aggregate. New York, USA https://www.persistencemarketresearch.com/market-research/global-construction-aggregates-market.asp.

Tam, V.W.Y., Shen, L.Y., Fung, I.W.H., Wang, J.Y., 2007. Controlling construction waste by implementing governmental ordinances in Hong Kong. Constr. Innov. 7 (2), 149–166. https://doi.org/10.1108/14714170710738522/FULL/XML.

Tam, V.W.Y., Tam, L., Le, K.N., 2010. Cross-cultural comparison of concrete recycling decision-making and implementation in construction industry. Waste Manage. 30 (2), 291–297. https://doi.org/10.1016/j.wasman.2009.09.044.

Tam, V.W.Y., Soomro, M., Evangelista, A.C.J., 2018. A review of recycled aggregate in concrete applications (2000–2017). Constr. Build. Mater. 172, 272–292. https://doi.org/10.1016/j.conbuildmat.2018.03.240.

The Freedonia Group, 2012. Global demand for construction aggregates to exceed 48 billion metric tons in 2015. Concrete Construction Magazine. https://www.concreteconstruction.net/business/global-demand-for-construction-aggregates-to-exceed-48-billion-metric-tons-in-2015_o.

Thomas, M., Bremner, T., 2012. Performance of lightweight aggregate concrete containing slag after 25 years in a harsh marine environment. Cem. Concr. Res. 42 (2), 358–364. https://doi.org/10.1016/j.cemconres.2011.10.009.

UEPG, 2014. European Aggregates Association – Annual Review 2013–2014.

UEPG, 2016. European Aggregates Association – Annual Review 2015–2016. Union Européenne des Producteurs de Granulats, https://doi.org/10.1016/0038-092X(81)90058-X.

World Business Council for Sustainable Development (WBCSD), 2009. Recycling Concrete. The Cement Sustainability Initiative at the Worlds Business Council for Sustainable Development. Geneva, Switzerland https://docs.wbcsd.org/2009/07/CSI-RecyclingConcrete-FullReport.pdf.

WRAP, NIEA, 2013. Aggregates From Inert Waste: End of Waste Criteria for the Production of Aggregates From Inert Waste. pp. 1–24. https://www.gov.uk/government/publications/quality-protocol-production-of-aggregates-from-inert-waste.

CHAPTER

17

Industrial by-products

Jaana Sorvari[1] *and Margareta Wahlström*[2]

[1]Finnish Environment Institute, Helsinki, Finland [2]VTT Technical Research Centre of Finland, VTT, Espoo, Finland

17.1 WASTE, BY-PRODUCT, OR PRODUCT?

Besides the desired products, industrial processes generate unintentional residues that have value and could substitute for some virgin materials as such, or after processing. Such residues are generally called by-products. While process conditions are mainly adjusted to attain the highest quality of the product, usually no attention is paid to the quality of the residues to maximize their utilization.

Whether a residue is classified as a waste, a by-product, or a product has implications for its recycling and reuse because any professional treatment, disposal, or reuse of waste is regulated by waste legislation. Waste legislation imposes on each respective owner of waste a liability of its due and safe management. As for a product, producer responsibility extends to the lifetime of the product if it is used and discarded in accordance with the instructions, and even beyond the use phase in the case of extended producer responsibility.

Until 2008, European legislation did not recognize the by-product concept. Hence, the outcomes of an industrial process were either products, i.e., outcomes that the process intended to produce, or waste. Since then, the by-product concept was adopted in the European Waste Framework Directive (WFD) (EU, 2018). This Directive defines a by-product as "a substance or object, resulting from a production process, the primary aim of which is not the production of that item." It is worth noting that, according to the WFD, a by-product is a material that has never been waste. The WFD also includes the generic End of Waste (EoW) criteria, which define when a residue ceases to be waste and turns into a product. As soon as the residue has become a product, it is potentially regulated by the Registration, Evaluation, Authorisation and Restriction of Chemicals (REACH) directive (EC, 2006) and existing specific legislation, for example the regulation for construction products or fertilizers. The generic EoW criteria require that:

- the residue has undergone a recovery operation (covers recycling);
- it is commonly used for specific purposes;
- a market or demand exists for it;
- it fulfills the technical requirements for the specific purposes and meets the existing legislation and standards applicable to equivalent products;

Handbook of Recycling. https://doi.org/10.1016/B978-0-323-85514-3.00044-0

- and it causes no overall adverse impacts to the environment or human health.

These criteria and the methodology developed by the Joint Research Centre (JRC) (Delgado Sancho et al., 2009) serve as the basis for defining waste-specific EoW criteria. So far, at the EU level, specific EoW criteria exist for only a few materials, e.g., iron, steel, aluminum and copper scrap, and glass cullet (as of October 2021). Thus, from the European regulatory perspective, most residues generally referred to as by-products are in fact considered wastes. The decision on exempting a residue from waste status per the EOW criteria can also be made case-specifically by the competent authority at the local, national, or regional level. In fact, to promote the use of industrial residues, many member states have adopted specific national EoW criteria as well as regulations for waste-derived aggregates (Velzeboer and van Zomeren, 2017), and have applied such criteria to revoke the waste status of some residues. The lack of specific European EoW criteria has probably contributed to the differences of the EoW approaches and the approaches to recognize the by-product status in the member states (Umweltbundesamt GmbH and ARCADIS Belgium, 2020). The European Joint Research Centre (JRC) has provided a list of waste streams for which it is necessary to develop EU-wide EoW criteria (Orveillon et al., 2022). JRC has also identified the most potential waste streams such as, mineral fraction of CDW, for the EOW criteria development.

In this chapter, no stand is taken regarding whether a specific residue is to be classified as a by-product from a regulatory viewpoint. Therefore an industrial by-product is understood to be a solid inorganic residue that is formed in the metal and mining industry, energy production, forest industry, chemical industry, building industry, or industrial scale thermochemical treatment of waste (particularly municipal waste) and has potential for reuse, recycling, or recovery of valuable substances or materials. Thus wastes or residues that are not formed in industrial activities, such as municipal solid waste (MSW) and its different source separated fractions (e.g., glass and plastics), residues composed mainly of organic material (e.g., wastewater treatment sludge, tires) and liquid by-products from the chemical industry are excluded, as well as the residues that are generally classified as hazardous wastes (e.g., jarosite). The lists of by-products and their applications presented here are not exhaustive since the primary aim is to provide a brief overview of the most common and large volume by-products generated around the world, and their traditional as well as some more novel applications. Recycling of construction and demolition waste, recycling and recovery techniques, and recovery of specific metals are not addressed in detail, since these topics are discussed in other chapters of this book.

17.2 MAJOR BY-PRODUCTS

Various industries generate recyclable or reusable by-products (Table 17.1). The process whereby a specific by-product is formed determines its generic properties and composition. However, process conditions, equipment, and raw materials eventually dictate its specific technical and environmental properties and, hence, suitability for recycling. Generally speaking, waste concrete from construction and demolition, reclaimed asphalt from road maintenance, and slag from crude iron and steel production are recycled to a large extent all over the world. During the past few years, the recycling of coal combustion fly ash (CCFA) has also increased considerably worldwide, and in some countries practically all CCFA is currently recycled (Marinina et al., 2021). The volume of CCFA is expected to decrease, however, due to ongoing replacement of coal with biomass as an energy

TABLE 17.1 By-products generated in different industries and their generic properties, relevant from the viewpoint of recycling.

Industrial sector	By-product	Formation	Main characteristics and remarks
Construction and demolition (including manufacture of building materials)	Concrete	1. Dismantling of roads, runways and structures, e.g., bridges, buildings; any hazardous materials need to be removed (⇒ reclaimed concrete) 2. Generated continuously in the manufacture of hollow-core slabs 3. Washing of tank trucks that transport concrete, and concrete mixers in concrete plants. Forms in the settling of washing liquid	1. Can include subbase soil material and bitumen (asphalt-concrete pavement in roads) or sealants (buildings). Upgrading can be done on-site or off-site. The quality depends on the type and use of the structure, and the accuracy of the selective demolition system. Potential contaminants include PCBs (from sealants used in concrete buildings), asbestos, dioxins, heavy metals (cadmium, chromium, mercury, lead), molybdenum, CFC, oil, PAH, phenols 2. Typically almost pure concrete containing only minor residues of mineral and plant oil used in molding 3. Granularity and water content varies. Can contain some hardened concrete. Can be recycled back to the manufacturing process
	Bricks	Dismantling of structures, building	Consists of clay and mortar. Can include some impurities originating from the raw material or from the construct (e.g., chimneys). Generated and used often together with waste concrete from demolition, difficult to clean due to formation of dust that is harmful to human health (barrier to reuse as such without processing)
	Mineral wool	1. Dismantling of buildings 2. Manufacture of mineral wool tiles when the raw material is melted and defibrated and further processed to shaped products	Comprises amorphous synthetic, vitreous fibers produced from mineral raw materials. Covers glass wool, rock wool, or slag wool, depending on the starting material. Includes 18% of sodium, potassium, calcium, magnesium, and barium oxides
	Reclaimed asphalt	Road maintenance and construction: 1. Cutting of pavement 2. Crushing of asphalt tiles	Properties resemble those of virgin stone materials. Includes mainly stone material and few percentages of bitumen. Good technical properties owing to low fines content. Since the fines are bound into bitumen, the sensitivity to water is lower. In general, recycling does not pose significant risks to the environment or human health. Deterioration of quality is a problem when asphalt is recycled several times

Continued

TABLE 17.1 By-products generated in different industries and their generic properties, relevant from the viewpoint of recycling—cont'd

Industrial sector	By-product	Formation	Main characteristics and remarks
	Glass	Dismantling of buildings	Main components include silicon oxide, calcium carbonate, and sodium carbonate. Amorphous and inert material, and lighter than equivalent mineral soil materials. A part can be reused as such
	Plasterboard and wallboards	Dismantling of buildings	Consists of gypsum (see gypsum below). Contains impurities, such as nails, screws, wallpaper, and other wall coverings
	Excess soil	Construction in general	Physically similar to virgin soil material. May contain impurities emitted from previous activities at the construction site, such as heavy metals or organic contaminants, which determine the environmental acceptability. Poor technical properties can hinder recycling in earth construction
Metal industry	Blast furnace slag (BFS)[1]	Manufacture of crude iron. Forms in the blast furnace 1. When the slag is rapidly cooled with water at a rate sufficient to ensure that no crystallization occurs = granulated BFS 2. When the slag is allowed to cool in air = air-cooled BFS	Contains mainly natural rock minerals, with the oxides of silicon, calcium, magnesium, and aluminum being the main constituents. Porous material that has low density, low thermal conductivity, high friction, and low wear resistance. Water-cooled BFS is vitreous material and its good hydraulicity enables its use as a binder agent. Sulfate (particularly in air-cooled BFS) and vanadium are potential most soluble contaminants
	Steel slag	1. Primary steelmaking,[2] i.e., manufacture of crude steel by combusting crude iron in a converter in the presence of quicklime, dolomite, and oxidizing agents of the ore 2. Secondary steelmaking (see EAFS below)	Consists mainly of calcium, silicon, magnesium, and aluminum oxides. Particle size distribution equivalent to gravel, but coarser with higher bulk density, mechanical properties depend on the porosity. Unreacted lime results in heterogeneous and unstable material. Contains usually various additional elements as potential contaminants that can hinder recycling, such as cadmium, chromium, molybdenum, nickel, selenium
	Ferrochromium (FeCr) slag[3]	Manufacture of ferrochrome (used in the production of stainless steel) from chromite. Forms in the cooling of molten slag	Consists of silicon, magnesium, and aluminum oxides. From the viewpoint of technical properties, can be used as a substitute of sand, gravel, and crushed rock. Particles are sharper compared with sand. Chromium content is high (up to ca. 10%) which prevents recycling in some applications, e.g., in applications where dusting and spreading via air is possible
	EAFS	Manufacture of crude steel from steel scrap by the electric arc furnace (EAF) process. Forms when the	Porous material with a risk of volume expansion. Wear-resistant material with similar or better technical

		liquid slag is first solidified and further cooled with water	properties than natural materials. Can have a high content of chromium
	Zinc slag	Galvanization of steel to avoid corrosion. Forms when the slag is skimmed from the molten zinc bath	Contains more than 50% zinc, which can be recovered
	Nonferrous slag	Recovery and processing of nonferrous metal from natural ores. Forms in smelting and subsequent cooling of molten slag	Rocklike or granular materials, such as copper (Cu), nickel (Ni), phosphorus, lead (Pb), and zinc (Zn) slag. Characteristics vary depending on the ore, process, and slag cooling method and rate. Some slag are classified as *hazardous wastes* (e.g., Cu slag). Processed air-cooled and granulated Cu, Ni, and phosphorus slag are mechanically sound, resistant to abrasion, and stable. Particularly granulated slag generated by rapid quenching tends to be vitreous which reduces friction and skid resistance and can be a problem in some applications, e.g., pavements
	Flue dust	Manufacture of ferrous and nonferrous metals. Flue dust forms in the separation of particles from flue gas	Composition depends on the raw material and process. In those cases where the dust contains several metals as impurities, it is often classified as hazardous waste
	Foundry sands	Casting of molds. Forms as a residue in both ferrous and nonferrous metal casting process	Composition varies depending on the sand type (e.g., green sand, furan sand, alkaline phenolic sand). The basic constituent is natural silica sand, which can contain traces of metals originating from the metal molded. Contains generally also traces of organic binders. Most of the foundry sands are recycled back to casting process, but some need to be removed after several rounds. Some spent foundry sands are characteristically hazardous, e.g., those from brass and bronze foundries
Mining and enrichment of ores	Red mud (ilmenite)	Production of bauxite. Forms in the refinement of bauxite to produce aluminum	Consists of titanium-iron oxide. Finely crushed, alkaline material (pH >12). Contains several heavy metals and metals, e.g., arsenic, chromium, lead, nickel, selenium, zinc, as impurities with varying concentrations
	Waste rock and fines	Mining and quarrying: part of the ore body, which is separated at the excavation front in order to get access to the ore	Contains low grades of value minerals, which cannot be mined and processed profitably. Differing characteristics depending on the ore. Waste rock from open-pit mines and quarries are mostly inert natural

Continued

TABLE 17.1 By-products generated in different industries and their generic properties, relevant from the viewpoint of recycling—cont'd

Industrial sector	By-product	Formation	Main characteristics and remarks
			materials if they do not contain sulfides, which might generate acidic drainage
	China clay (kaolin) waste	Mining. Forms in the extraction of kaolin mineral when the sand and mica residues are separated from the mineral slurry	Properties equivalent to primary aggregates used in earth construction. Consists of overburden, waste rock, coarse sand, and micaceous residue, coarse sand being the main component. Mica can cause changes in the density. Almost 90% of the coarse sand is quartz. Production of 1 t of china clay results in 9 t of waste
	Final tailings	Enrichment of metal ores. Forms when the ore is treated chemically to separate the desired metal or mineral	Differing characteristics depending on the raw material (ore) and process. Includes mineral material, metal residues, and chemicals used in the enrichment
	Lime (calcium oxide), e.g., lime kiln dust (LKD)	Refining of limestone (= calcium carbonate). Forms during filtering of the exhaust gases from heating limestone in a kiln	Very fine, white powdery material of uniform size; principal constituents are calcium and magnesium carbonates. Chemical and physical properties vary depending on the type of lime being manufactured, feedstock and fuel type, process, type of kiln, and dust collection method. Based on reactivity determined by the amount of free lime and magnesia, distinction is made between two types: calcitic (chemical lime, quicklime, etc.) and dolomitic
Energy production	Fly ash[4]	Combustion of coal, biomass, or mixture of these[5]. Forms when a portion of light, noncombustible mineral matter is emitted from the furnace along with flue gases and volatilized minerals and separated as fly ash in mechanical or electric particulate control devices, e.g., a dust precipitator	Fly ashes have generally better thermal insulation compared to natural materials. Being fine-textured can absorb humidity and silt. Frost susceptibility depends on particle size distribution. The composition of fly ashes varies significantly depending on boiler type and used fuels. Main constituents of coal combustion fly ash (CCFA) include silicon, aluminum, and iron compounds with smaller amounts of oxidized alkali metals and some heavy metals as impurities. CFFA is fine powdery material with a grain size of 0.002–0.1 mm corresponding to loam. CCFA is a consolidating material. Consolidation can be accelerated by activators (e.g., concrete, lime, FGD slag). The properties of co-combustion with biomass (peat, wood, etc.) fly ashes are in principle similar to coal fly ashes, but the quality variations are usually more severe. The main components in the fly ash from

		peat combustion (PCFA) are the oxides of iron, magnesium, potassium, and organic matter. Concentration of heavy metals is lower compared to ash from wood combustion. Corrosiveness can be a problem in some applications. The high concentration and/or potential leaching of (heavy) metals, such as antimony, arsenic, boron, barium, cadmium, lead, selenium, molybdenum, and zinc, can also prevent recycling. Some ashes, e.g., PCFA, can contain radioactive elements
Bottom ash[4] and boiler slag from the combustion of coal, peat, wood, or mixture	Combustion of coal, peat, wood, or mixture of these[5]. Forms when a portion of noncombustible mineral matter is retained in the bottom of the furnace. Slag is generated in wet-bottom boilers where bottom ash is kept in a molten state and then put in contact with water	The amount of bottom ash (vs fly ash) depends on the combustion technique. Fixed bed combustion generates more bottom ash than fluidized bed combustion. Granularity of bottom ash is equivalent to coarse or gravelly sand, some particles are glassy. It is non pozzolanic (unlike fly ash); therefore it does not consolidate as easily Boiler slag is coarse and granular, it can be crushed into different sizes
FGD (flue gas desulfurization) slag	Combustion of coal. Forms in the treatment of flue gases using lime for removing sulfur, based on 1. Wet-scrubbing process (or wet process) which results in a dry or wet end product, generates also a filter cake (waste water treatment) 2. Dry additive injection process: waste collected with a particulate collection device 3. Wet-dry process (or spray-absorption/semidry process): waste collected with a particulate collection device	1. Consists mainly of contaminated gypsum (see phosphogypsum below), content of impurities reported to be up to 6% 2. Consists of fly ash and reaction products of additive input fed into the furnace (e.g., limestone, lime). The amount of ash in the waste varies between 30% and 80% 3. A dry mixture of fly ash and desulfurization waste, consisting mainly of calcium sulfite and calcium sulfate and some unreacted lime and calcium carbonate. The fly-ash content depends on the separation degree of fly ash before desulfurization
Biochar	Forms in the pyrolysis, hydrothermal carbonization, gasification, flash carbonization or torrefaction of biomass (e.g., wood, peat[5]). Pyrolysis is the most common technique	Consists mainly of carbon, but includes also nitrogen, oxygen, hydrogen, potassium, metals, among others. The physical properties and chemical composition vary depending on the feedstock, production technique, and process conditions (e.g., temperature, residence time, heating rate). Biochar is porous and has a high surface area (which makes it an efficient

Continued

TABLE 17.1 By-products generated in different industries and their generic properties, relevant from the viewpoint of recycling—cont'd

Industrial sector	By-product	Formation	Main characteristics and remarks
			adsorbent) and high cation exchange capacity (capacity to supply nutrient cations for plant uptake)
Forest industry	Paper mill sludge (i.e., fiber sludge)	Manufacture of paper from wood pulp. Forms in the separation of the primary sludge generated in the mechanical wastewater treatment	Quality and composition vary according to the manufacturing process and wastewater treatment method. Sometimes primary sludge is mixed with sludges from other parts of the process. In general, fiber sludge consists of organic material and caoline (around 50/50%). Caoline is very stable material while the organic matter mainly consists of cellulose and wood fibers. Impurities include heavy metals that come from wood, chemicals used in the process, tubes and fillers
	Deinking sludge	Manufacture of recycled pulp from waste paper. Forms in the removal of impurities by filtering and further removal of ink by flotation	Includes mostly short fibers, i.e., organic matter and clay (caoline, soapstone, bentonite) with impurities such as ink, heavy metals, and plastics; the final waste has been usually incinerated on site after thickening and drying. Although the organic matter content is high, deinking sludge has proved suitable for some earth construction applications due to low biodegradability
	Green liquid precipitate	Manufacture of wood pulp for papermaking. Forms in the soda recovery plant belonging to the chemicals recovery line when the precipitate is removed from green liquid	Consists of calcium carbonate and different insoluble impurities of green liquid, such as soot, metal oxides, sodium compounds, silicates and impurities of wood. Generally considered as inert waste due to low organic matter content. Granularity equivalent to loam with high water content. Compression strength is too low to be used in earth construction without additives, such as fiber sludge and fly ash
Chemical industry	Iron calcinate/hematite filler	Manufacture of sulfuric acid. Forms when pyrite is combusted in calcining plant	Consists of iron (III) oxide, which is chemically stable, durable, and resistant to corrosion
	Kiln dust	Manufacture of cement. Forms in cement kiln, recovered from the dust collection system for the control of air emissions	Covers two different dusts: cement kiln dust (CKD) and lime kiln dust (LKD). CKD is a fine powdery material equivalent to Portland cement. The principal constituents are compounds of lime, iron, silica, and alumina. The free lime tends to be in coarser particles closest to the kiln. Its content in LKD can be significantly higher than in CKD. Calcium and magnesium carbonates are the principal mineral

			constituents of LKD, the composition varies depending on the product (chemical, hydrated or dolomitic lime or quicklime) manufactured. Part of the dust can be separated and returned to the kiln
	Gypsum	Forms in various chemical processes, e.g., in the manufacture of titanium pigment, in the reaction between phosphorus mineral and sulfuric acid in the manufacture of phosphorus fertilizer (phosphogypsum), in flue-gas desulfurization (FGD, see above)	Composed of calcium-sulfate-dihydrate ($CaSO_4{\cdot}2H_2O$). Large quantities are formed in the phosphoric acid process of the phosphorus fertilizer production. Gypsum is moderately water-soluble and it is used particularly in soil amendment and in building industry. Depending on the origin of the raw material (phosphate rock), phosphogypsum can include impurities, particularly fluoride and radioactive elements that can prevent its recycling
Waste treatment	Bottom ash and slag	Combustion of municipal waste (solid MSW, sewage sludge) Formation: see above "Energy production"	Bottom ash is the main solid residue from MSW combustion. Composition varies depending on the feed, i.e., waste material, but silica, calcium, and iron are the most abundant elements. Heavy metal concentrations can limit recycling if not removed, See also above: "Energy production"
	Biochar	Pyrolysis[6] of different waste materials, e.g., MSW, sewage sludge, car tires Formation: see above "Energy production"	See above: "Energy production"

The list aims to present the most important materials with recycling potential and it is not exhaustive.

[1] *Two additional types of BFS, i.e., expanded or foamed BFS and pelletized BFS, are also known; the former forms when the slag is cooled and solidified by controlled quantity of water, air, or steam and the latter when the slag is cooled and solidified with water and air quenched in a spinning drum.*

[2] *Primary steelmaking slag covers slag from Linz-Donawitz process (LD slag), 5converter slag, slag from continuous casting (CC slag), and slag from basic oxygen furnace (BOF slag).*

[3] *In Finland, FeCr slag used in construction is considered a product per the decision of the Supreme Administrative Court (KHO200590).*

[4] *Ashes often come from mixed combustion of different raw materials, e.g., wood, peat, coal, and ashes with similar generic physical properties are formed also in the combustion of municipal solid waste; the chemicals involved then vary depending on the waste material.*

[5] *Covering wood-based and other plant-based residues, e.g., from agriculture and forestry.*

[6] *Thermochemical process where an organic material is treated in a high temperature (>300°C) in the absence (or near absence) of oxygen, includes several techniques with different process temperature, heating rate, and residence time.*

Data from various sources, e.g., Recycled Materials Resource Centre (RMRC) (n.d.) FHWA (2016), Mroueh et al. (2000), Mäkelä and Höynälä (2000), Jha et al. (2001), Rao (2006), US EPA (2007), Meyer and Tran (2019), Yao et al. (2015), Pancharathi et al. (2020), Shukla et al. (2020).

source (Hjelmar et al., 2022). Moreover, the gradual abandonment of landfill disposal as a means of treating MSW, as per the waste policy, has increased the number of MSW combustion and pyrolysis plants and, thus, the generation of residues such as boiler slag, ash, and biochar. This made the development of recycling options for these residues necessary.

The recycling rate of different by-products varies in different countries, however. The inadequacy in documenting the volumes and final destination of by-products as well as variation in classifying waste and its management practices (e.g., whether filling mines is considered recycling) has aggravated the availability of exact information on recycling rates.

17.3 WHERE AND HOW TO USE BY-PRODUCTS

The prevailing applications of by-products are found in the construction/building industry and, to a lesser extent, in agriculture and horticulture. In some cases, recovery of valuables or use as a raw material of new products is an option, for example from zinc slags. Table 17.2 summarizes potential applications for the major by-products presented in Table 17.1. Practically, regardless of application, the use of a by-product requires knowledge of its composition, physical properties, and variations in quality, as well as of the specifications of the virgin material for which it would substitute. These factors also

TABLE 17.2 Main recycling options and examples of by-products and techniques for processing them suitable for the specific application.

Application	By-product	Examples of processing
Construction (civil engineering) • As aggregate in different structures: roads[1], landfills, buildings, noise barriers, embankments, fills • Pavement • Groundwater protection layers, i.e., landfill liners • Roofing • Masonry aggregate • Soil stabilizer or solidifying agent in soft or wet soils	Crushed concrete (both coarse and fine fraction), all ashes, BFS[2], steel slag[3], Ni slag, Cu slag, foundry sand[4] waste rock and quarry fines, concrete sludge, china clay waste, final tailings from limestone mines, red mud, excess soil, green liquor precipitate (landfills[5]). All ashes, reclaimed asphalt, red mud, steel slag, mineral wool Fiber sludges from forest industry (paper mill sludge, deinking sludge), green sand (foundry sand), Ni and Cu slag, Phosphorus slag, Ashes, lime (LKD), cement and lime kiln dusts, mixture of fiber sludge and ash	Ash: stabilization using cement, FGD, or equivalent; Ash: modification of technical properties using fiber sludges; Air-cooled BFS, steel slag, mineral wool: crushing or grinding; Air-cooled BFS: sorting to required fractions; Steel slag: recovery of metallic particles, conditioning and weathering; Reclaimed concrete: removal of metals and impurities, advanced dry recovery (ADR) technology to separate fine materials; Fiber sludge: refinement[6] (e.g., with bentonite); Concrete sludge: homogenization, drying, and crushing; China clay sand: grading and washing only (equivalent to primary aggregates)
Agriculture and horticulture: as fertilizer or soil amendment	Granulated BFS, FGD gypsum, red mud, steel slag, cement kiln dust, ashes from the combustion of wood[7] and peat, biochar, foundry sand[4], some fiber sludges, dusts from steel making, foundries and smelters, residues from galvanizing	Fine materials, e.g., ash: granulation (to avoid dusting) followed by sieving Ashes: addition of boron (if needed)

TABLE 17.2 Main recycling options and examples of by-products and techniques for processing them suitable for the specific application—cont'd

Application	By-product	Examples of processing
Reuse[8]	Glass, concrete, bricks and tiles from construction and demolition	Manual dismantling, sorting and cleaning (i.e., separation from mortar, grout, and adhesives), crushing, sieving, and magnetic separation
	Phenolic sand (to replace new sand in green sand molding system in gray iron foundry)	Mechanical reclamation
Building industry • As additive and binder in construction materials (e.g., source of fines): In (Portland) cement and concrete, asphalt concrete • In concrete bricks and tiles • As constituent or fine aggregate/binder or filler in asphalt/bitumen and bitumen bound materials • As constituent in roofing felt • As insulator in structures	• Fly ash,[9] BFS, red mud, FGD gypsum, flue dust form BFS and steel making, Ni slag, Cu slag, foundry sand, tailings, fiber sludge • Red mud • Foundry sand, red mud, ashes, cement kiln dust, steelmaking slag, Ni slag, Cu slag, Zn slag, phosphorus slag • Ni slag • Ferro-chromium slag	Flue dust: sintering Foundry sands: removal of very fine particles Fiber sludge: drying Granulation or crushing (air-cooled FeCr slag)
Manufacture of new products • Portland cement • Plasterboards • Ceramic tiles • Sandblasting sand • Mineral wool • Lightweight aggregate • Glass aggregate for floor tiles, abrasives, roofing shingles, etc.	• BFS, foundry sand, phosphorus slag • FGD gypsum • Zn slag • Ni slag • Glass, foundry sand • Fiber sludge mixed with fly ash • Fiber sludge	Foundry sand: briquetting, reduction of silica content Pelletization followed by heating and cooling Melting in high temperature, followed by rapid cooling with water
Metallurgical industry • As auxiliary material to form silicates (slag) in secondary refinement process of copper	Foundry sands	
Mining industry: backfilling of mines	Waste rock, tailings	No processing needed
Waste management, wastewater treatment and soil remediation	Neutralizing (alkaline) materials, stabilizing or solidifying materials, e.g., ashes, FGD gypsum, red mud, biochar	

[1] *Freezing can be a problem.*
[2] *Used particularly in road construction.*
[3] *Due to strength and good abrasion and impact resistance, particularly well-suited for road constructs used by heavy vehicles.*
[4] *Particularly sand from iron, steel, and aluminum foundries (US EPA, 2007).*
[5] *In the water-retaining layer (surface structures), often mixed with ash or fiber sludge in order to increase structural strength and to ensure water-retaining capacity.*
[6] *The need of processing depends on the properties of the material and the requirements set by the application; use without processing is also possible.*
[7] *Use in forestry has been questioned because of the potential long-term risks associated with cadmium content (Pitman, 2006).*
[8] *Has to be free of dangerous substances and undamaged structures, etc.*
[9] *Ash from coal combustion (CCFA) is used more frequently than other ash.*
Source: Data compiled from various sources (shown in the main text and in Table 17.1).

determine the need for and type of pretreatment. In most cases, pretreatment, processing, or both are required to render a by-product usable. Treatment and processing options include:

- sorting to improve the quality of the material, e.g., removal of impurities or unwanted fractions through selective demolition;
- conditioning and weathering to stabilize the material—for example, to enhance the hydration of the free lime and magnesia of steel slag used in earth construction;
- crushing or grinding to remove porous particles and to convert granular monolithic materials;
- drying;
- sieving to separate different particle size fractions;
- filtration;
- sintering to reduce porosity and enhance the strength or electrical and thermal conductivity of fine materials;
- pelletizing, granulation, and briquetting to homogenize heterogeneous materials, to avoid turning fine materials to dust, or to ease storage and use;
- modification of chemical properties, e.g., to increase the solubility of micronutrients or to improve fertilizing properties by adding nutrients;
- magnetic separation or thermal treatment using a rotary kiln or plasma to recover metals; and
- vitrification to immobilize soluble constituents.

Processing of residues to be suitable for substitution of a virgin material incurs costs due to the need of energy, equipment, labor, and potential auxiliary materials. The total costs of processing depend on the specific characteristics of the residue and the requirements regarding the application. Transportation and potential storage add to the costs. On the other hand, the market value of the final by-product is generally low, especially if virgin material is easily available with moderate costs. Hence, the price of virgin materials can significantly affect the demand of by-products. The end-user value of a by-product ultimately depends also on its quality, including homogeneity, and availability (volume-wise, time-wise).

17.3.1 Civil Engineering

In civil engineering, by-products are widely used as aggregates in different structures and fills in road and landfill construction and maintenance, and in the improving of soil properties by stabilization (Figure 17.1). The basic components of earthworks are pavement structures, load-dissipating structures (base course and subbase), and foundation structures. The type of structure and quality of subsoil determine which components are in fact necessary in a particular construction. Different layers call for different material properties (Table 17.3), the major determinants being stability, hydraulic conductivity, bearing capacity, and freeze-thaw resistance (in northern regions).

In road building, by-products can be used as bulk fills, aggregates, or binders. As aggregates, they are used either as unbound, i.e., granular, materials (primarily to substitute gravel) or bound. Bitumen, concrete, lime, and their combinations are common binders, which are replaceable by some by-products, such as blast furnace slag (BFS) and fly ash. In addition, by-products can be used for thermal insulation and as a drainage layer in filtering layers, in noise barriers, as fillers in landscaping and pavement, as relief materials, and as sealing materials. The properties of metal slag residues, e.g., BFS, are favorable since their thermal resistance is generally higher compared to virgin materials and their bulk density is lower. Therefore, the volume of material needed to build the desired structure is lower, allowing thinner structure layers compared to natural materials. This is a relevant feature considering freezing. Mineral wool residue does not freeze as easily

FIGURE 17.1 Road building using reclaimed concrete.

as soil material, which enables its use during winter in cold regions.

By-products with suitable geotechnical properties for use in landfill structures come from various industries. Specific applications include top covers, and particularly the impermeable mineral layer. Fly ash mixed with paper sludge or sewage sludge, among others, has been used in top cover layers. Use of by-products in the bottom layers of landfills is less frequent, most probably due to the technical standards, which require adequate hydraulic conductivity and have requirements for environmental quality. Moreover, the bottom layer rarely can be removed if the by-product turns out to be unsuitable.

The geotechnical properties of a single by-product are generally improved by mixing it with other materials, which can be either by-products or virgin materials. In the case of the mineral layer, variations in quality and the influence of weather conditions during mixing and compaction at the site are key issues impacting the geotechnical properties. Furthermore, estimation of the long-term stability and degradation of any organic material present in by-products often involves high uncertainty.

17.3.2 Agriculture and Horticulture

Use of a by-product in agriculture or horticulture presumes that it has properties favorable to plant growth. The positive effect can be related to the supply of nutrients; improving of soil texture, aeration, percolation, and water retention; or conditioning soil by altering its chemical, physical, or biological composition. To function as a fertilizer, i.e., a substance that provides nutrients, the nutrients need to be soluble or transformable to a form available to plants. Also, in the EU the latest regulation on fertilizers (EU, 2019a) needs to be followed. This regulation stipulates the acceptable contents of heavy metals in inorganic fertilizers, liming agents, and soil improvers.

By-products that have liming properties and can thus correct soil acidity and thereby increase crop yield have been used for ground improvement. BFS, steelmaking slag, and ash from wood combustion are the most common by-products used for this purpose. Steelmaking slag and sludge also contain some plant nutrients: they have been found to increase the content of calcium and magnesium in

TABLE 17.3 Technical requirements for by-products used in earth construction, per the Technical Guidelines issued in Finland (Mäkelä and Höynälä, 2000) and their updates in accordance with the Eurocode 7 (Transport Agency, 2017; Ministry of the Environment, 2018) Applications: roads, parking lots, courtyards.

Application	Purpose and technical requirements	Examples of suitable by-products
Pavement	Should serve as an even and wear resistant foundation for traffic and prevent penetration of water to lower layers. The requirements therefore include: strength, robustness, resistance to cracking, deformation, and wear	Metallurgical slag, reclaimed asphalt, ashes from coal and peat combustion, waste rock, quarry fines
Base course	Should sustain tensions and deformations arising from traffic, distribute the load to wider area of lower layers, form an even base for pavement and serve as drying material for the water penetrating though pavement. Should have: sufficient carrying capacity and water permeability, freeze-thaw resistance, frost resistance, suitable granularity	Crushed metallurgical slag, granulated BFS, crushed concrete, waste rock, stabilized fly ash
Subbase	Serves as a dissipating layer and increases carrying capacity and frost resistance, leads away the water penetrating from upper layers and prevents the capillary rise of water. Should possess: sufficient carrying capacity particularly in melting phase, sufficient frost resistance, freeze-thaw resistance, suitable granularity, thermal insulation capacity or good freezing resistance, adequate hydraulic conductivity	Like base course, also stabilized bottom ash and concrete sludge
Protective course	Should prevent capillary rise of water, separate pavement from ground base and prevent its mixing with upper layers, dissipate water from the structure, prevent or slow down frost penetration, and secure sufficient carrying capacity during melting	Crushed metallurgical slag, e.g., FeCr slag, crushed concrete, waste rock, consolidating or stabilized ash, ash from coal and peat combustion, crushed rock wool, crushed bricks, and concrete sludge
Roadbed (embankment)	Serves as homogenic material that ensures even frost heaves and dislocations during melting. Requirements therefore include load and frost resistance, and sufficient strength during melting	Metallurgical slag, waste rock, nonconsolidating ash (to limited extent)
Embankment relief	Should reduce loads falling on soft roadbed. Used in berms of road and street structures and courtyards and parking areas. Requirements include load resistance, sufficient strength during melting and lightness	Granulated BFS

herbage. Kiln dust, too, is suitable for agricultural purposes. In the United States zinc-bearing hazardous dust from electric arc furnaces has been widely used in the production of zinc micronutrient fertilizers (US EPA, 2001). The international Basel Convention, which aims at protecting human health and the environment against the adverse effects of hazardous wastes, also acknowledges such use (UNEP, 2010).

CCFA has several merits as a fertilizer and soil amendment (e.g., Yao et al., 2015). It contains micro- and macronutrients, which are essential for plants and crops and, due to its alkalinity, it neutralizes soil acidity and consequently reduces the toxicity of heavy metals, among others. It also serves as a pesticide; its activity as an insecticide arises from the presence of abrasive silica (Vitekari et al., 2012). Granulated ash fertilizers are generally less soluble than the equivalent powdered fertilizers and products stabilized by self-hardening.

Energy production from biomass, such as wood, has increased because of the replacement of fossil raw materials with renewable ones. The European bioeconomy strategy (EC, 2018) also supports this trend. Particularly peat forests benefit from fertilization with the ash generated in wood-based energy plants, since they are generally limited in the mineral nutrients that are abundant in wood ash, but typically sufficient in nitrogen, which is not present in ash.

The use of red mud as a soil amendment has been studied extensively already during the twentieth century. In Australia, red mud was found to have the potential to reduce eutrophication of rivers and waterways by retaining nutrients, particularly phosphorus, in infertile sandy soils (Summer et al., 1996) and to increase pasture production. The potential release of harmful substances, e.g., lead, cadmium, nickel, and zinc, to the environment needs to be considered, however. Seawater-neutralized red mud has shown good capacity to immobilize soluble acid and metals from acid sulfate soil solutions, making it a good alternative for lime if leaching is an issue.

The various beneficial effects of biochar (from pyrolysis of different biomass or MSW) on soil quality and fertility have been the main reason for its use in soil amendment. However, biochar alone does not provide soil with enough nutrients (Varalta and Sorvari, 2020). Biochar is recognized also as a potential long-term carbon sink and therefore it serves to mitigate climate change.

Low bioavailability of essential nutrients as well as high concentrations of toxic heavy metals, which hinder soil biological activity, can be a problem when a by-product, e.g., ash, is applied directly to soil. Mixing with compost material, manure, or other biomass can solve this problem by fostering microbial activity and the consequent mineralization, and alleviating the toxicity. Phytotoxicity can also arise from an unsuitable pH value of the by-product. For example, high alkalinity causes the phytotoxicity of red mud, but mixing with organic material makes it suitable for soil amendment (Berta et al., 2021).

17.3.3 Recovery of Valuable Elements

Slag from metallurgical processes often contains useful components that can be recovered. Manganese, silica, magnesia, alumina, and niobium from steelmaking slag, and nickel and cobalt from copper slag, are examples of recoverable metals in by-products (Rao, 2006). Steelmaking slag could be used as a secondary source of iron. Dust from the steelmaking process also contains valuable metals, e.g., silver, that can be separated by leaching and subsequently recovered.

Construction and demolition waste contains significant amounts of metals that are generally sold as scrap to be used as raw materials. The metals are first shredded, and then iron is removed by magnetic separation. Finally, nonferrous metals such as copper, aluminum, and stainless steel are removed by density separation.

Silica present in ash can be separated and used, for example, in coatings and the manufacture of glass. It is also possible to recover metals from ashes, and their high iron fraction can be utilized for iron ore feed and cement. Development of recovery techniques has in fact enabled the increased separation of the valuable metals from different ashes and slags, e.g., from combustion of MSW (e.g., Šyc et al., 2020). The mineral fraction remaining after the recovery can be further used as an aggregate in different structures, for example, in road construction, providing that the leaching of any remaining contaminants does not pose a risk to the environment, and in the manufacture of concrete. Whether the aim is to maximize the recovery of metals without using the remaining mineral fraction, or to utilize both, determines the needed pretreatment methods.

17.3.4 Reuse or Recycling On-site

Reuse or recycling on-site is the preferred option from the viewpoint of the environmental impact arising from storing, transporting, and processing or refining residues. On-site recycling is suitable for by-products that are equivalent to manufactured products such as concrete residues generated in the manufacture of hollow core slabs and sludges of mixing plants, and reclaimed asphalt generated in the renewal of pavement.

On-site recycling in the manufacturing process has been presented as an alternative means of recycling steelmaking slag (e.g., Rao, 2006). In this option, the phosphorus content of the slag needs to be reduced before it can be returned to the process. Spent foundry sands are typically circulated in the process if their technical properties remain suitable. Here, a distinction needs to be made between different foundry sands. Only silica-based foundry sands, i.e., sands from iron, steel, and aluminum foundries, are generally considered safe for reuse and recycling.

17.3.5 Manufacture of New Products

Some by-products have proven to be suitable raw materials for new products. For example, BFS and steel slag have been widely used in the manufacture of concrete due to having properties similar to Portland cement, whereas fiber sludge has proven a suitable raw material in the manufacture of cement. Several by-products can serve as constituents in cement clinker production (see Chapter 27—Cementitious binders incorporating residues of this book). Fiber sludge has also been used as the base raw material in some industrial sorbent and animal bedding products (Meyer and Tran, 2019). BFS can be combined with silica or alumina and converted to fibers for rock wool. Spent foundry sand can be used as a source of silica in this process.

Fly ash can be mixed with the main raw material to produce ceramics, bricks, floor and wall tiles, sound insulation panels, fillers in polymers and rubbers, zeolites, and inorganic fibers (Kumar et al., 2007; Akhtar and Tarannum, 2018), among others. Polymer composites made of fly ash (filler) and plant-based fabric (reinforcement) can replace wood in different inner constructs, e.g., in flooring tiles, ceiling, wall paneling. Fly ash can also replace white cement used in the manufacture of distemper, which is used in the interior surface in buildings. The large surface area of biochar makes it suitable as a supporting material for catalysis (e.g., Yaashikaa et al., 2020).

Besides ashes, particularly foundry sands are used as additives to provide fines in cement and concrete. Some applications, such as the manufacture of cement, may require mechanical activation that enables the use of higher proportions of ash and the attainment of improved quality for the new product. The potential effect on the durability of concrete structures has raised some discussion about the suitability of ash for this purpose, however. Both the EU and ASTM International have issued standards for the quality of fly ash that can be used in concrete (DIN EN 450, ASTM International C618—12a).

The US EPA (Environmental Protection Agency) supports the use of foundry sand from iron, steel, and aluminum foundries as an ingredient in several building materials (US EPA, 2007). Use as a manufactured soil, among others, can also be accepted provided that the material fulfills the state regulations.

Iron ore tailings have been used in cement-based composites and as a geopolymer. They could also serve as a pigment in the production of paints for buildings (Galvão et al., 2018).

17.3.6 Environmental Engineering

Various by-products have proved their usefulness in reducing the adverse impacts caused by waste, wastewater, flue gas emissions, and contaminated soil. Ash and cement kiln dust, among others, have been used to stabilize and solidify waste. Several studies have shown the benefits of using wood ash in composting plants of biowastes. These benefits include acceleration of the composting process; enhanced hygienization due to the reduced number of pathogens and temperature increase; increased mineralization; and higher nutrient content of the final compost (e.g., Fernández-Delgado et al., 2015). Mixing fly ash with organic wastes, e.g., manure, paper mill sludge, sewage sludge, or crop residues, improves the quality of these wastes in soil amendment by reducing the availability of heavy metal and sludge pathogens and by increasing the content of fine-grained minerals along with the improvement of soil properties and biological activity.

Biochar, too, has been proven to be beneficial in composting of biowastes. It accelerates the degradation process and reduces the availability of heavy metals, among others (e.g., Varalta and Sorvari, 2020). Besides the various benefits of biochar to the composting process and the quality of the compost product, reduction of the emissions of greenhouse gases, i.e., gaseous NOx, during composting is a notable extra benefit. Hence, integration of composting and pyrolysis could serve as a truly sustainable and circular solution for the management of organic wastes.

In the remediation of contaminated environments, ash from combustion generated in the forest industry has been used to stabilize soil contaminated by heavy metals such as lead, while biochar, being an efficient adsorbent of contaminants, can be used also for the treatment of contaminated water (e.g., landfill leachate) (e.g., Yaashikaa et al., 2020). Red mud, too, is suitable for both soil and water treatment (Elbaz et al., 2019). In addition, some studies show the suitability of red mud for the treatment of flue gases. Besides heavy metals, red mud can remove nonmetallic ions of fluorine, arsenic, and phosphorus as well as phenolic compounds and dyes from wastewater (Wang et al., 2019).

Green liquid precipitate formed in the manufacture of wood pulp has proven suitable for neutralizing wastewater, although its phosphorus content somewhat restricts its use.

According to the studies of Asokan et al. (2010), a mixture of coal combustion residues can immobilize the toxic elements of jarosite formed in the refinement of zinc ore. The solidified end-product could be used in construction as bricks and blocks. Ferrous slag has also been used together with jarosite to make building materials. It is worth noting that jarosite is generally considered hazardous waste, and thus such end products may not be environmentally acceptable.

Cement kiln dust is used as a chemical conditioner and stabilizer of sewage sludge instead of lime (Elbaz et al., 2019). The purpose of stabilization is to eliminate health hazards and noxious smells from the sludge besides the immobilization of the heavy metals. Cement kiln dust has also been demonstrated to be an efficient material for the removal of sulfur dioxide from cement kiln flue gas, with a removal efficiency of 90% and above (US DOE, 2001). Additional applications where kiln dust can substitute for virgin materials include soil remediation and wastewater treatment.

The ability of some by-products, such as recycled concrete aggregate, ash, kiln dust, and steel slag, to sequestrate CO_2 through carbonation serves to reduce the emissions of this greenhouse gas (e.g., Kaliyavaradhan and Ling, 2017). Carbonation has been originally used for improving the properties of a building material (i.e., curing concrete): it makes the material harder and more durable. Carbonation is a natural process where atmospheric CO_2 reacts with calcium compounds and forms stable calcium carbonate ($CaCO_3$). In weathering carbonation, the CO_2 sequestration occurs naturally (natural carbonation) while active carbonation (accelerated carbonation) occurs under controlled conditions. The several contributing environmental and by-product specific factors aggravate the generation of reliable quantitative, universal data on the natural carbonation of by-products when used in construction applications.

17.4 TECHNICAL AND ENVIRONMENTAL REQUIREMENTS

Uniform, constant quality is perhaps the most important prerequisite that any by-product must fulfill to be used on a large scale. By-products are also generally required to be technically at least as good as the virgin materials for which they substitute. Particularly by-products used in earth construction should have a service life comparable to virgin materials. The producer is responsible for ensuring that the by-product complies with the defined quality criteria and that quality control is carried out according to the requirements. Then, the user must use the material in defined applications according to the producer's instructions.

Any waste to be recycled generally needs to go through an extensive basic characterization that integrates information about its generic physicochemical properties and long-term behavior, as well as the concentration of harmful substances and their environmental fate. A by-product that is used continuously usually does not need to go repeatedly through characterization unless its quality changes significantly, for example, because of changes in the production process or raw materials. However, it is important to regularly control the by-product's quality to ensure that its properties meet specified requirements. A quality control system to ensure conformity with environmental and technical specifications should therefore be established in connection with the characterization. Particular attention should be paid to sampling, to produce representative samples for both characterization and quality control. Here, the sampling method, number of samples, and time of sampling are important factors. Any sampling plan should be based on the characteristics of the specific by-product as well as on the objectives of the sampling.

17.4.1 Quality Control

The quality control system and the testing needs included are material- and application-specific. An example of potential parameters to measure is shown in Table 17.4. The technical characteristics relevant to the application and the most critical environmental characteristics are selected for periodic checking. Homogeneous materials with well-known characteristics require less frequent monitoring than heterogeneous and less thoroughly investigated materials. The quality control program should include a description of the properties to be monitored, monitoring and sampling methods, sampling and control frequencies, limit values for quality and acceptable errors, as well as reporting methods.

The quality of a by-product can be proved through certification. The certificate can contain a detailed description of the properties that need to be controlled. In addition, use of a by-product as a product requires information about its long-term behavior and suitable applications. For example, in the case of earth construction, this generally requires building and monitoring pilot constructions and preparing design manuals and user guidelines. Therefore several countries have manuals for the use of by-products in landfill construction, among others. These manuals usually contain a description of geotechnical and environmental requirements to be met and

TABLE 17.4 Relevant technical and environmental features and their measurement in the case of a by-product planned to be recycled in earthworks (Mroueh et al., 2000; Arm, 2003; Transport Agency, 2017; Ministry of the Environment, 2018).

Material property	Parameter to be measured	Measurement/basis of determination
Technical features		
Chemical aggressiveness, corrosiveness	Several	pH, salt concentration, electrical conductivity
Wear resistance	Ball mill-value	Ability to withstand degradation in a ball mill machine (EN 1097-9)
	Los Angeles-figure	Degradation of a material sample in a rotating drum with steel spheres, several standards (e.g., EN 1097-2, ASTM C131A–D)
Carrying capacity	Modulus Compression	See below
	CBR value	Force needed to achieve the defined penetration (2.5/5.0 mm) when a steel plunger is pressed into the surface, several standards (e.g., EN 13286-47, ASTM D1883)
Firmness	Friction angle	Maximum angle before which one of the items will begin sliding, describes the grain-to-grain frictional resistance
	Cohesion	Force that holds the material together
Transformation	Modulus	Several measurements available
Compression	Modulus	Deformation of material when compressed
Packing	Maximum dry volume weight Optimum water content	Density of dry material Determined from the moisture content/dry density curve
Frost susceptibility	Frost heave	Granularity (particle size and distribution) Water permeability
	Segregation potential	Capillarity: extent of the material to elevate or depress the surface of a liquid contact with it
Thermal properties	Heat conductivity	Heat energy transferred by the material per unit time and per unit surface area, divided by the temperature difference
Stability	Change in granularity	Granularity
Erosion sensitivity	Speed of water current	Granularity
Granularity	Granularity curve Granule size	Separation of different particle size fractions by sieving
Water permeability	Permeability coefficient	Volume of water flowing in unit time through a unit volume of a porous material (unit pressure difference maintained across)
Capillarity	Height of capillary action	Granularity
Specific gravity	Volume weight of the solid material	Ratio of the weight of a volume of material to the weight of an equal volume of water

Continued

TABLE 17.4 Relevant technical and environmental features and their measurement in the case of a by-product planned to be recycled in earthworks (Mroueh et al, 2000; Arm, 2003; Transport Agency, 2017; Ministry of the Environment, 2018)—cont'd

Material property	Parameter to be measured	Measurement/basis of determination
Environmental characteristics		
Dry matter and water content	Concentration of dry matter	Residue after direct drying at 105°C or water content determined by titration or distillation (EN 15934:2012)
Composition	Total concentration of substances included in the material	Method is substance-dependent, several standardized options, e.g., digestion using acids (EN 13656:2020, EN 13657:2002) for metals, gravimetric and gas chromatographic method for hydrocarbons
Organic matter content	Total concentration of organics	Total organic carbon (TOC) by dry combustion (EN 15936:2012)
Biodegradability	Amount of matter that can be consumed (degraded) by microorganisms	Consumption of O_2 or generation of CO_2 (aerobic cond.) or generation of CH_4 (anaerobic cond.) when inoculated a specified time with microorganisms, several standardized tests (e.g., Ultimate Aerobic Biodegradation, ISO 17556)
Leaching of substances in time from a granular material	Concentration of substances in eluate	Cumulative concentration after an upflow during a specified time through a column filled with the material (characterization test, EN 16627-3)
Leaching of substances from a monolithic or stabilized material	Concentration of substances in eluate	Concentration after a specified time when the material is immersed in water (characterization test); Horizontal dynamic surface leaching test (EN 16637-2)
Solubility of substances in different pH conditions, acid/base neutralization capacity	Concentration of substances in eluate	Concentration in the eluates of parallel extractions at different pH (characterization test, EN 14429:2015, EN 14997:2017, CEN/TS 15364:2006)
Leaching of substances from a granular material	Concentration of substances in eluate	Concentration in a one- or two-stage batch leaching test (quality control test, EN12457)
Leaching of substances from the surface of nonpermeable materials or material with low permeability	Concentration of substances in solute	Diffusion measured using a surface solubility test, e.g., Dynamic Monolithic Leaching Test (DSLT) (EN 15863:2015, NVN 7347)
Ecotoxicity (solid material)	Several alternatives, e.g., LD_{50}/LC_{50}, NOEC	Toxicity response of soil organisms (e.g., Enchytraeids), several standardized tests available
Ecotoxicity of leachate	Several alternatives, e.g., LD_{50}/LC_{50}, NOEC	Toxicity response of aquatic organisms (e.g., *Daphnia magna*), several standardized tests available

CBR, California bearing ratio; *LD_{50}/LC_{50}*, lethal dose/concentration, concentration where 50% of test animals are killed; *NOEC*, no-observed-effect concentration, highest concentration that does not cause adverse effects in test organism.

cover the various work stages and monitoring. In Europe, by-products used in construction are also regulated by the Construction Products Regulation (EU, 2011), which includes requirements regarding the technical performance and safety. Construction products marketed inside EU must have a CE marking as per this regulation. A CE mark is a producer's declaration that the product fulfills EU directives or national requirements. Harmonized test methods must be used to prove compliance in case testing is required.

17.4.2 Environmental Compliance

In the recycling of any residue, environmental suitability is a key issue. "Zero pollution" for a toxic-free environment is the ambition mentioned in the European Green Deal (EC, 2021), which supports actions for clean material loops. Special attention should be paid to persistent organic pollutants (POPs), as the law bans the use of wastes and products containing POPs (EU, 2019b).

Potential hazards to the environment and human health depend on the properties and use of the by-product, e.g., the existence of harmful substances, the type of application, environmental conditions in the place where it is managed, and the existence and characteristics of receptors that may be exposed or might receive emissions from it. Environmental and health risks are therefore also different in different stages of recycling. During storing, processing, building or construction, maintenance, and demolition of a by-product structure, or application of soil amendments and fertilizers, the focus is often on occupational hazards. Inhalation of dust and skin contact are generally the major routes of human exposure at these stages. As for materials used in buildings, e.g., by-product–based gypsum board, the potential release of volatiles and radiation into indoor air during the use phase is an important issue to acknowledge. Here, quality standards for indoor air can be used as a reference in assessing the acceptability of the by-product. The release of harmful substances also poses environmental risks. Leaching of harmful substances is a major issue in the case of by-products used in earth construction; therefore case-specific information on the long-term leaching behavior of harmful substances is required. Leaching of contaminants should also be considered, particularly in the case of soil amendments and fertilizers.

Several internationally standardized laboratory tests are available to test the leaching behavior of by-products (Table 17.4). Benchmarks have been issued to compare against the test results and to assess environmental suitability. In testing by-products to be used in earth construction, the leaching mechanism, which depends on the material properties and environmental conditions, determines the test methods. Typically, there is a difference between granular and monolithic materials. In the case of the former type of material, it is assumed that water percolates through the material layer and the solubility of the compounds determines their leaching. The so-called percolation test is then used as a study method. Unlike granular materials, monolithic (or stabilized/solidified) materials have low hydraulic conductivity and maintain their shape when in contact with water. Water is assumed to be in contact only with the surface and not to percolate through it the material. Leaching behavior is thus surface dependent, often dictated by diffusion, and therefore studied using the so-called diffusion test.

Several standardized toxicity tests are available for further investigation into the toxicity of the leachate beyond chemical analysis. According to the European Committee for Standardization (CEN) dealing with the assessment of the release of dangerous substances from construction products (CEN, 2017), in the case of inorganic construction materials, data regarding environmental impact attained by ecotoxicity tests is limited. Hence, chemical analysis of eluates would be sufficient. If the by-product is used as a soil amendment or fertilizer, ecotoxicity testing can provide important additional information on their environmental impacts, however. In some cases, a site-specific risk assessment is required to consider the specific application and conditions at the location where the by-product is used.

The existence and concentration of harmful and toxic substances in a by-product depend on the raw materials and process conditions during its generation. In addition, any weathering, pretreatment, or processing can alter a by-product's physicochemical properties, or the speciation of its elements, and consequently change their availability. In the case of fertilizers and soil amendments, the type of application and volume or rate and frequency of application finally determine the probability of the adverse impact.

The environmental and health risks caused by by-products can be reduced by processing, i.e., by eliminating or transforming toxic substances into nontoxic substances or by decreasing their availability, e.g., by stabilization. Hazardous substances can also be removed by extraction based on different solvents or use of living organisms (bioleaching). In earthworks, emissions to the environment can also be limited in situ by coating or paving. However, in the long term, cover materials require regular maintenance because of wearing and breakage. Another approach to minimizing environmental and health risks is to limit the use of by-products in sensitive areas and applications. This can mean that by-products are not allowed to be in contact with groundwater, acknowledging the seasonal and weather-related alterations of the water table. In addition, the use can be limited to outside the important groundwater areas and soils of high hydraulic conductivity, such as gravel and sand. Application of by-products in the environment can also require the establishment of safety zones. For example, some studies have suggested buffer strip zones between wood ash–treated areas and watercourses and lakes, to avoid potential adverse effects that could arise from the increase in pH and nutrient content and potential mobilization of toxic compounds (Aronsson and Ekelund, 2004).

17.5 SUSTAINABILITY ASPECTS

Sustainable use of natural resources, which means using environmentally compatible raw and secondary materials, is one of the basic requirements set in the European regulation on construction products (EU, 2011) and (Caldeira et al., 2022). An update of this regulation is ongoing and expected to be adopted in early 2025 (EC 2022). The suggested revision sets some obligations to manufacturers such as, mandatory reporting of the global warming potential. Ensuring sustainable circularity is one of the objectives of the Safe and Sustainable by Design (SSbD) framework presented by the European Commission's Joint Research Centre (Caldeira et al., 2022). This framework covers a step-wise process to assess the safety and sustainability of chemicals and materials.

In the determination of overall environmental sustainability of by-products vs virgin materials, life cycle assessment (LCA) is the most common technique. However, since LCA focuses on the assessment of products, services, or processes, it usually relies on generic rather than on site-specific data. It is therefore poorly suited for assessing the potential adverse environmental impacts at a specific site. Here other methods, such as laboratory tests, e.g., leaching tests (Table 17.4) and/or modeling, are useful when assessing the risks. Also, any information gained from the monitoring of pilot or large-scale applications is valuable for the verification of environmental safety.

Due to the growing concern of climate change, the potential of some by-products used in earthworks, in soil amendment, or in building materials to act as carbon sinks has emerged as an additional benefit to be considered in LCA.

Particularly waste concrete and biochar (Gupta and Kua, 2017) have been studied from this viewpoint. Recycling of by-products also serves in the fight against biodiversity loss caused by the extraction of natural resources. The methods to measure biodiversity loss and the extent of carbon sequestration attainable by the recycling of by-products are still under development, however.

Consideration of sustainability assumes that all its elements, i.e., environmental, societal, and economic, are taken into account. LCA and risk assessment provide information only on the environmental aspects, while particularly economic aspects are also crucial when considering the use of by-products instead of virgin materials. Other LCA-based methods, such as life cycle costing (LCC), can be adopted to assess the economic impact. In some cases, societal aspects, such as public acceptability, can also be an issue: these can be studied in a social-LCA. The information generated by the different methods that focus on different elements of sustainability should finally be integrated. Therefore some studies have developed sustainability indicators and adopted multicriteria decision-making methods (MCDM) for selecting the most sustainable building materials.

17.6 CONCLUSIONS, CHALLENGES, AND OUTLOOK

The current need to tackle the global climate change crisis and the depletion of natural resources, which ultimately lead to biodiversity loss and adverse effects on human health, have resulted in the adoption of circular economy as a key objective at the policy level (EC, 2020a) The New Circular Economy Action Plan (CEAP) is one of the main constituents of the European Green Deal (EC, 2019). In CEAP, "making sustainable products as a norm in Europe" is a key measure and it also aims to ensure that resources are kept within the EU economy for as long as possible. The CEAP is an overarching policy that covers waste legislation. Hence, policies related only to waste legislation or circular economy cannot be distinguished. It is evident that reuse and recycling of industrial residues (by-products), which can substitute for virgin materials in different applications, have a crucial role in the needed transition to a circular society. Such utilization needs to be implemented without compromising the quality of the environment, human health, and safety to all living organisms, however, as per the EU's Zero Pollution Action Plan (EC, 2021) supported by the Chemicals Strategy for Sustainability (EC, 2020b). How much of the reuse and recycling potential of the by-products is used in practice finally depends on many factors and thus can vary considerably in different countries and regions.

First and foremost, whether any residue is considered a waste, a by-product, or a product is a crucial issue. Professional and large-scale management, treatment, or recycling of waste generally requires an environmental permit if no separate statute exists that exempts it from this obligation. A lack of clear environmental criteria can result in time-consuming permit processes, because the authorities need to conduct a case-by-case acceptability evaluation—and time is money. Productization would solve this problem and guarantee the end user the technical and environmental compliance of the residue. In Finland, the considerable time required to attain a permit was a driver to decree a statute (Ministry of the Environment, 2017), which exempts some residues used in earth construction from permit obligation when they fulfill the specific environmental criteria set in it (Sorvari, 2008). Moreover, in 2022 Finland adopted an End-of-Waste legislation for reclaimed concrete (Government Decree 466/2022). These regulations should foster the utilization of the residues involved.

Particularly road construction generally assumes fast deliveries, and the material to be used in the structures is not necessarily selected well beforehand. In colder regions, by-products from energy production, such as ash, are mainly generated during the cold season, whereas the construction works that could use them are run in the summertime. Also, the energy plants that generate ash can be small, and hence the volume of ash from a single plant is low. These situations assume the existence of a storage system or some logistic center where separate batches of ash can be safely stored and mixed in such a way that the technical and environmental properties of the mixture remain acceptable for substituting virgin materials. Storage and controlling any release to the environment creates extra costs and increases the price of by-products. Added to these are the analysis costs. Availability and logistics are an issue because the road transport of by-products over long distances is generally not profitable and, due to the consequent emission of greenhouse gases, it can be seen as environmentally disadvantageous. Moreover, if the prices of virgin materials are very low, there is no economic incentive to replace them with by-products. Furthermore, the heterogeneous quality of some by-products can reduce their desirability compared with virgin materials. A lack of information about the long-term environmental fate of by-products and of their harmful substances has also prevented recycling. Experience gained from experimental and large-scale applications with monitoring has increased the knowledge, but comparable data are still needed, particularly on the overall sustainability through the whole life cycle of by-products.

A major challenge in the utilization of by-products is the need for balancing between the environmental risks caused by any remaining contaminants in them and benefits gained through saving of natural resources. Finding the balance requires a holistic approach. For example, in Finland, strict environmental criteria or a lack of them have/has in many cases been a barrier to get the required approval from local authorities. At the same time, circular economy should be promoted. Controversies regarding acceptability in different regulatory context adds its own flavor to this complex: the environmental criteria might not be logically connected with each other and hence the level of protection varies. For example, in Finland, by-products suggested for earth construction need to go through a characterization procedure that usually includes testing of leaching of contaminants (e.g., test CEN/TS 16637-3:2016, Table 17.4) to define the risk of their migration in the environment. At the same time, only "total" concentrations (dissolving by a strong acid, e.g., aqua regia) need to be analyzed if the same by-product, e.g., ash, is proposed for soil amendment or fertilizing. Consequently, also the environmental criteria in these two applications are not in line with each other. Hence, there is clearly a need to also harmonize the risk assessment approaches used in the different regulatory contexts and consider potential mixture effects of hazardous substances when assessing risks. This is one of the main objectives of the EU's recent strategy regarding chemicals (EC, 2020b).

Using all the recycling potential of by-products requires removal of the previously mentioned barriers. Some policy instruments, for example, economic incentives, could be a solution. In addition, we need systems that help the demand meet the supply. In fact, material exchange databanks have been established in some countries for this purpose. Process technology, too, could provide some new and more economical solutions for transforming by-products into products of a high and stable technical and environmental quality to render them desirable for different applications. New feasible applications would also help maximize the volume of by-products ending up in recycling. When aiming at a high recycling rate, the goal should be value creation and retention and avoidance of downcycling. Hence, when possible, by-products should be recycled in high-value applications instead of "easy solutions," i.e., low-value applications, such as backfilling. Here, the possibilities to improve the quality by some feasible refinement methods should be acknowledged. This implies careful consideration of the alternative applications where the by-product could substitute for valuable virgin materials.

Whether by-products are or will be recycled in the future on a large scale ultimately depends on the overall benefits attained by recycling. Such benefits can be environmental (e.g., saving of natural resources), economic (e.g., cost savings), and societal (e.g., increased employment, new technical innovations, and business opportunities). These elements are the building blocks of sustainability. Identification of the key factors affecting sustainability and the best solutions calls for methods to measure it. Here, clear, measurable sustainability indicators and multicriteria decision-making tools can be useful, especially since they also enable the involvement of and communication with different stakeholders to find out their priorities and the attainment of consensus between them.

References

Akhtar, M.N., Tarannum, N., 2018. Flyash as a resource material in construction industry: a clean approach to environment management. In: Hemeda, S. (Ed.), Sustainable Construction and Building Materials. IntechOpen, https://doi.org/10.5772/intechopen.82078.

Arm, M., 2003. Mechanical Properties of Residues as Unbound Road Materials – Experimental Tests on MSWI Bottom Ash, Crushed Concrete and Blast Furnace Slag. (Doctoral thesis). KTH Land and Water Resources Engineering, Stockholm. Report No. 64. Swedish Geotechnical Institute. Available from: https://www.sgi.se/globalassets/publikationer/rapporter/pdf/sgi-r64.pdf.

Aronsson, K.A., Ekelund, N.G.A., 2004. Biological effects of wood ash application to forest and aquatic ecosystems. J. Environ. Qual. 33, 1595–1605.

Asokan, P., Saxena, M., Asolekar, S.R., 2010. Recycling hazardous jarosite waste using coal combustion residues. Mater. Charact. 61 (12), 1342–1355.

Berta, K.M., Kurdi, R., Lukács, P., Penk, M., Somogyi, V., 2021. Red mud with other waste materials as artificial soil substitute and its effect on Sinapis alba. J. Environ. Manage. 287, 112311.

Caldeira, C., Farcal, L., Garmendia Aguirre, I., Mancini, L., Tosches, D., Amelio, A., Rasmussen, K., Rauscher, H., Riego Sintes, J., Sala, S., 2022. Safe and Sustainable by Design Chemicals and Materials—Framework for the Definition of Criteria and Evaluation Procedure for Chemicals and Materials. Publications Office of the European Union, Luxembourg. JRC128591. https://doi.org/10.2760/487955.

CEN, 2017. Construction products – assessment of release of dangerous substances – guidance on the use of ecotoxicity tests applied to construction products. CEN/TR 17105, p. 2017.

Delgado Sancho, L., Catarino, A., Eder, P., Litten, D., Luo, Z., Villanueva Krzyzaniak, A., 2009. End-of-Waste Criteria. EUR 23990 EN, European Commission, Luxembourg, https://doi.org/10.2791/28650. JRC53238.

EC (European Commission), 2006. Regulation (EC) No 1907/2006 of the European Parliament and of the Council of 18 December 2006 concerning the Registration, Evaluation, Authorisation and Restriction of Chemicals (REACH), establishing a European Chemicals Agency, amending Directive 1999/45/EC and repealing Council Regulation (EEC) No 793/93 and Commission Regulation (EC) No 1488/94 as well as Council Directive 76/769/EEC and Commission Directives 91/155/EEC, 93/67/EEC, 93/105/EC and 2000/21/EC.

EC (European Commission), 2018. A Sustainable Bioeconomy for Europe: Strengthening the Connection Between Economy, Society and the Environment. Updated Bioeconomy Strategy, https://doi.org/10.2777/792130.

EC (European Commission), 2019. The European Green Deal. COM(2019) 640 final. Brussels, 11.12.2019.

EC (European Commission), 2020a. A New Circular Economy Action Plan – For a Cleaner and More Competitive Europe. COM(2020) 98 final. Brussels, 11.3.2020.

EC (European Commission), 2020b. Chemicals Strategy for Sustainability Towards a Toxic-Free Environment. COM(2020) 667 final. Brussels, 14.10.2020.

EC (European Commission), 2021. Pathway to a Healthy Planet for All EU Action Plan: Towards Zero Pollution for Air, Water and Soil. COM(2021) 400 final. Brussels, 12.5.2021.

EC (European Commission), 2022. Proposal for a Regulation Laying Down Harmonised Conditions for the Marketing of Construction Products, Amending Regulation (EU) 2019/1020 and Repealing Regulation (EU) 305/2011. COM(2022) 144 final. Brussels 30.3.2022. Available from: https://ec.europa.eu/docsroom/documents/49315.

Elbaz, A.A., Aboulfotoh, A.M., Dohdoh, A.M., Wahba, A.M., 2019. Review of beneficial uses of cement kiln dust (CKD), fly ash (FA) and their mixture. J. Mater. Environ. Sci. 10 (11), 1062–1073.

EU (European Union), 2011. Regulation (EU) No 305/2011 of the European Parliament and of the Council of 9 March 2011 laying down harmonised conditions for the marketing of construction products and repealing Council Directive 89/106/EEC.

EU (European Union), 2018. Directive (EU) 2018/851 of the European Parliament and of the Council of 30 May 2018 amending Directive 2008/98/EC on waste.

EU (European Union), 2019a. Regulation (EU) 2019/1009 of the European Parliament and of the Council of 5 June 2019 laying down rules on the making available on the market of EU fertilising products and amending Regulations (EC) No 1069/2009 and (EC) No 1107/2009 and repealing Regulation (EC) No 2003/2003.

EU (European Union), 2019b. Regulation (EU) 2019/1021 of the European Parliament and of the Council of 20 June 2019 on persistent organic pollutants.

Fernández-Delgado, J.M., Prähauser, B., Walter, A., Insam, H., Franke-Whittle, I.H., 2015. Co-composting of bio-waste and wood ash, influence on a microbially driven-process. Waste Manage. 46, 155–164.

FHWA (Federal Highway Administration), 2016. User Guidelines for Waste and Byproduct Materials in Pavement Construction. FHWA-RD-97-148. Available from: https://www.fhwa.dot.gov/publications/.

Galvão, J.L.B., Andrade, H.D.G., Brigolini, J., Peixoto, R.A.F., Mendes, J.C., 2018. Reuse of iron ore tailings from tailings dams as pigment for sustainable paints. J. Clean. Prod. 200, 412–422.

Gupta, S., Kua, H.W., 2017. Factors determining the potential of biochar as a carbon capturing and sequestering construction material: critical review. J. Mater. Civ. Eng. 29 (9), 04017086.

Hjelmar, O., Hyks, J., Korpisjärvi, K., Wahlström, M., Grönholm, R., 2022. BAT (Best Available Techniques) for Combustion and Incineration Residues in a Circular Economy, Nordic Council of Ministers. TemaNord 542/2022. Available from: https://pub.norden.org/temanord2022-542.

Jha, M.K., Kumar, V., Singh, R.J., 2001. Review of hydrometallurgical recovery of zinc from industrial wastes. Resour. Conserv. Recycl. 33, 1–22.

Kaliyavaradhan, S.K., Ling, T.-C., 2017. Potential of CO_2 sequestration through construction and demolition (C&D) waste—an overview. J. CO2 Util. 20, 234–242.

Kumar, R., Kumar, S., Mehrotra, S.P., 2007. Towards sustainable solutions for fly ash through mechanical activation. Resour. Conserv. Recycl. 52, 157–179.

Mäkelä, H., Höynälä, H., 2000. Sivutuotteet ja uusiomateriaalit maarakenteissa – Materiaalit ja käyttökohteet (By-products and Recycled Products in Earth Construction – Materials and Applications). Teknologiakatsaus 91/2000 Tekes – the Finnish Funding Agency for Technology and Innovation: Helsinki.

Marinina, O., Nevskaya, M., Jonek-Kowalska, I., Wolniak, R., Marinin, M., 2021. Recycling of coal fly ash as an example of an efficient circular economy: a stakeholder approach. Energies 14, 3597.

Meyer, T., Tran, H., 2019. Utilization of Kraft Pulp Mill Waste. In: Proceedings of the 2019 PEERS Conference, St. Louis, USA, October 27–30, 2019. Available from: https://www.tappi.org/content/Events/19PEERS/19PEE33.pdf.

Ministry of the Environment, 2017. Government Decree on the Recovery of Certain Wastes in Earth Construction (843/2017). Available from: https://www.finlex.fi/en/laki/kaannokset/2017/en20170843.

Ministry of the Environment, 2018. Strength and Stability of Structures. Design of Foundations. National Building Code of Finland.

Mroueh, U.-M., Wahlström, M., Mäkelä, E., et al., 2000. By-products in Earth Construction – Assessment of Acceptability. Technology Review 96/2000 Tekes – the Finnish Funding Agency for Technology and Innovation: Helsinki.

Orveillon, G., Pierri, E., Egle, L., Gerbendahl, A., Wessman, P., Garcia John, E. and Saveyn, H., 2022. Scoping Possible Further EU-Wide End-of-Waste and By-Product Criteria. EUR 31007 EN. Publications Office of the European Union, Luxembourg. JRC128647. https://doi.org/10.2760/067213.

Pancharathi, R.K., Sangoju, B., Chaudhary, S. (Eds.), 2020. Advances in Sustainable Construction Materials. Select Proceedings of ASCM 2019. Lecture Notes in Civil Engineering. Springer Nature, Singapore.

Pitman, R.M., 2006. Wood ash use in forestry – a review of the environmental impacts. Forestry 79 (5), 563–588.

Rao, S.R., 2006. Resource Recovery and Recycling From Metallurgical Wastes. Waste Management Series. vol. 7 Elsevier B.V.

Recycled Materials Resource Centre (RMRC), n.d. Materials. College of Engineering, Madison. Available from: http://rmrc.wisc.edu/.

Shukla, S.K., Barai, S.V., Mehta, A. (Eds.), 2020. Advances in Sustainable Construction Materials and Geotechnical Engineering. Select Proceedings of TRACE 2018. Lecture Notes in Civil Engineering. Singapore, Springer Nature.

Sorvari, J., 2008. Developing environmental legislation to promote recycling of industrial by-products – an endless story? Waste Manage. 28, 489–501.

Summer, R.N., Smirk, D.D., Karafilisc, D., 1996. Phosphorus retention and leachates from sandy soil amended with bauxite residue (red mud). Aust. J. Soil Res. 34, 555–567.

Šyc, M., Simon, F.G., Hykš, J., Braga, R., Biganzoli, L., Costa, G., Funari, V., Grosso, M., 2020. Metal recovery from incineration bottom ash: state-of-the-art and recent developments. J. Hazard. Mater. 393, 122433.

Transport Agency, 2017. Eurokoodin soveltamisohje – Geotekninen suunnittelu – NCCI 7 (Guideline for application of Eurocode – Geotechnical design – NCCI 7). Siltojen ja pohjarakenteiden suunnitteluohjeet 21.4.2017 (Guidelines for the design of bridges and foundations). Liikenneviraston ohjeita 13/2017. In Finnish.

Umweltbundesamt GmbH and ARCADIS Belgium, 2020. Study to Assess Members States (MS) Practices on By-product (BP) and End-of Waste (EoW). Final version 30.4.2020, European Commission, Brussels, https://doi.org/10.2779/279259. Available from: https://op.europa.eu/s/uXH8.

UNEP (United Nations Environment Programme), 2010. Technical Guidelines on the Environmentally Sound Recycling/Reclamation of Metals and Metal Compounds (R4). Basel Convention Technical Guidelines. UNEP/SBC/2010/21. United Nations, Geneva. Available at: https://digitallibrary.un.org/record/699747.

US DOE (US Department of Energy), 2001. Cement Kiln Flue Gas Recovery Scrubber Project: A DOE Assessment. DOE/NETL-2002/1163, https://doi.org/10.2172/789334.

US EPA, 2001. The Micronutrient Fertilizer Industry: From Industrial Byproduct to Beneficial Use. EPA-305-B-01–006. Available from: https://www.epa.gov/nscep.

US EPA, 2007. Foundry Sands Recycling. EPA530-F-07-018. 2 p. Available from: https://nepis.epa.gov.

Varalta, F., Sorvari, J., 2020. Biochar-compost mixtures as a promising solution to organic waste management within a circular holistic approach. In: Hettiarachchi, H., Caucci, S., Schwärzel, K. (Eds.), Organic Waste Composting Through Nexus Thinking: Practices, Policies, and Trends. Springer, pp. 213–232, https://doi.org/10.1007/978-3-030-36283-6_10.

Velzeboer, I., van Zomeren, A., 2017. End of Waste Criteria for Inert Aggregates in Member States. ECN-E-17-010, ECN, The Netherlands.

Vitekari, H.N., Talele, A.P., Mane, R.G., Gaikwad, V.S., Shah, J.V., 2012. Fly ash based biopesticides: a comprehensive review. Int. J. Pharm. Biol. Sci. 2 (1), 76–82.

Wang, L., Hu, G., Lyu, F., Yue, T., Tang, H., Han, H., Yang, Y., Liu, R., Sun, W., 2019. Application of red mud in wastewater treatment. Minerals 9, 281.

Yaashikaa, P.R., Kumar, P.S., Varjani, S., Saravanan, A., 2020. A critical review on the biochar production techniques, characterization, stability and applications for circular bioeconomy. Biotechnol. Rep. 28, e00570.

Yao, Z.T., Ji, X.S., Sarker, P.K., Tang, J.H., Ge, L.Q., Xia, M.S., Xi, Y.Q., 2015. A comprehensive review on the applications of coal fly ash. Earth-Sci. Rev. 141, 105–121.

CHAPTER

18

Mine tailings

Priscilla P. Nelson and D. Erik Spiller

Colorado School of Mines, Golden, CO, United States

18.1 INTRODUCTION

The world's mining industry is rapidly changing and faces challenges that must be addressed with a sociotechnical and environmental approach, which integrates across the entire materials flow from mountain to tailings and beyond, embracing the value chain and engaging the circular economy.

Modern society around the world is notably based on mined rock materials that provide metals such as iron and copper, building materials such as aggregate and cement, minerals for chemical-based products such as fertilizers, energy minerals such as uranium and coal, and more. While creating these values through mining and processing, the industry also produces immense amounts of waste products commonly called tailings. These tailings contain rock of no apparent commercial value, remnant material/minerals of value which were not recovered in processing typically because of the recovery cost, and other unrecognized values or contaminants (often toxic). These wastes or tailing materials are characteristically both coarse and fine, are benign to environmentally toxic, and exist as stored piles of dry solids to sloppy slurries contained behind constructed dams. The relative abundance of tailing material as a function of mineral commodity is shown in Figure 18.1.

If these tailing challenges are managed sustainably, the mining industry will be recognized as an ethical, ecological, and diverse industry that can offer long-term social and environmental benefits and attract future generations with jobs that address aspirational goals. This last point is incredibly important for the mining industry and its workforce of the future. The industry needs more workers, more skills, and more diversity. The incoming generations will need to identify the industry with a noble purpose and meaningful careers, rather than providing jobs that do not engage and generate career alienation. To paraphrase the comments of Blauner (1964), alienation develops when workers are unable to control their immediate work processes, when workers cannot find a sense of purpose and function that connects their jobs to the overall organization, and when workers cannot relate their assigned responsibilities to a personal expression of noble purpose and professional achievement. To address the future labor supply needs, the mining industry must change the image of mining work and increase the attractiveness of working in the sector, especially for young women and men—a vision for mining is to become a high-technology industry that speaks to today's young people.

Handbook of Recycling. **https://doi.org/10.1016/B978-0-323-85514-3.00037-3**

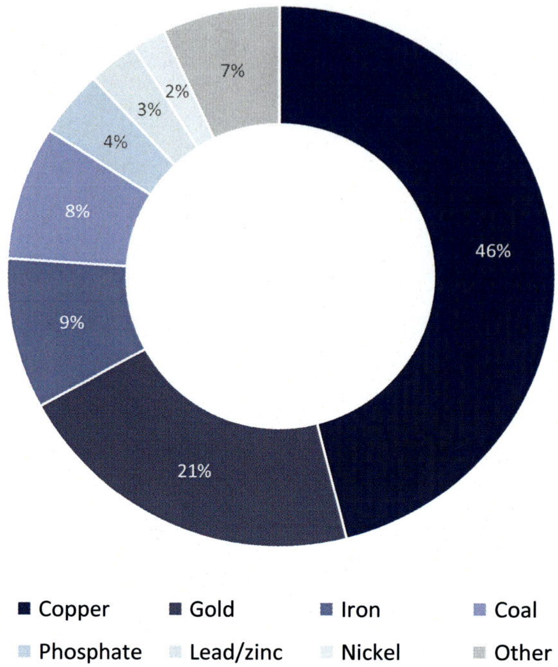

FIGURE 18.1 Share of each commodity in the global tailings volume. *Modified from: Visual Capitalist. Original source of data: Global Tailings Review, ICMM, UNEP and PRI.*

Many limitations to increasingly sustainable operations of the mining and the workforce environment can be traced back to initial physical planning and design. Since mining is characterized by huge investments and long-term operations, it is particularly important to have a well-designed physical production system. If the initial plan is falsely or overconstrained, there may be limitations and negative consequences throughout the life of the mine and beyond. Retaining options and flexibility in the initial design phases of every major development project is therefore critical to establish the possibilities for mining to incorporate new technologies that may approach multidimensional sustainability. Furthermore, the mines of the future need to be designed to promote cooperation and creative problem-solving in multiskilled teams, all while minimizing its legacy of tailings. Reuse and recycling of tailings is a critical part of tailings management in the context of sustainability.

Society has recognized the need to address the sustainability of mining, i.e., the tailings both legacy and future, through the United Nation's Sustainability Development Goals (SDGs) (United Nations, 2015; World Economic Forum, 2016; Columbia Center on Sustainable Investment, 2020) and the evolving Environmental, Social and Governance (ESG) mandate that demands an integrated and responsible mining industry (Barclay, 2020).

There is a need for a new vision for the whole industry based on a social/political/environmental mandate that demands an integrated mining industry. The mine planners shape the work environment, and if the planners design an inflexible process and redesign becomes necessary, it is likely to be very expensive. The industry must take opportunities in mining system design to optimize all along the value chain rather than suboptimizing parts or individual processes. Mining leadership, therefore, has the heavy responsibility to rethink mine planning through to its tailings legacy. There is a need to generate new conceptual designs that will promote high productivity and a good economy, safety and a healthy work environment, and social and environmental acceptability.

The minerals required to meet the escalating needs for the increasing global population and expectations for quality of life (e.g., more electric vehicles, battery storage, photovoltaic solar, and other alternative energy sources) rely on earth resource extraction, processing, and refining; this means workforce growth, more water for processing, and more storage for mine tailings. This leads to the focus on minimizing tailings and the use of water. One opportunity is to develop mining in situ, where as much as possible of the production and processing takes place underground. Such technology is not without environmental risks and risks for health and safety.

A modern mine operation can be described in four generalized and sequential steps. Figure 18.2 visually describes the four-step process that results in both a value commodity and tailings for storage/disposition. Most ferrous, nonferrous, and industrial minerals mines follow this operational sequence.

Mines often use water, especially during processing, and water usage is hugely important in the ESG mandate. Water is often the single most important environmental issue for the mining sector and includes both its quality and quantity. Reuse and recycling of mine tailings in whatever form is key to tailings minimization, which in turn is an important foundation to a global sustainable mining industry. Modern mines strive to recycle water and recent advances in filtered and stacked tailings technology (often referred to as dry stacking) are becoming more accepted and applied, due to new higher capacity filtration equipment.

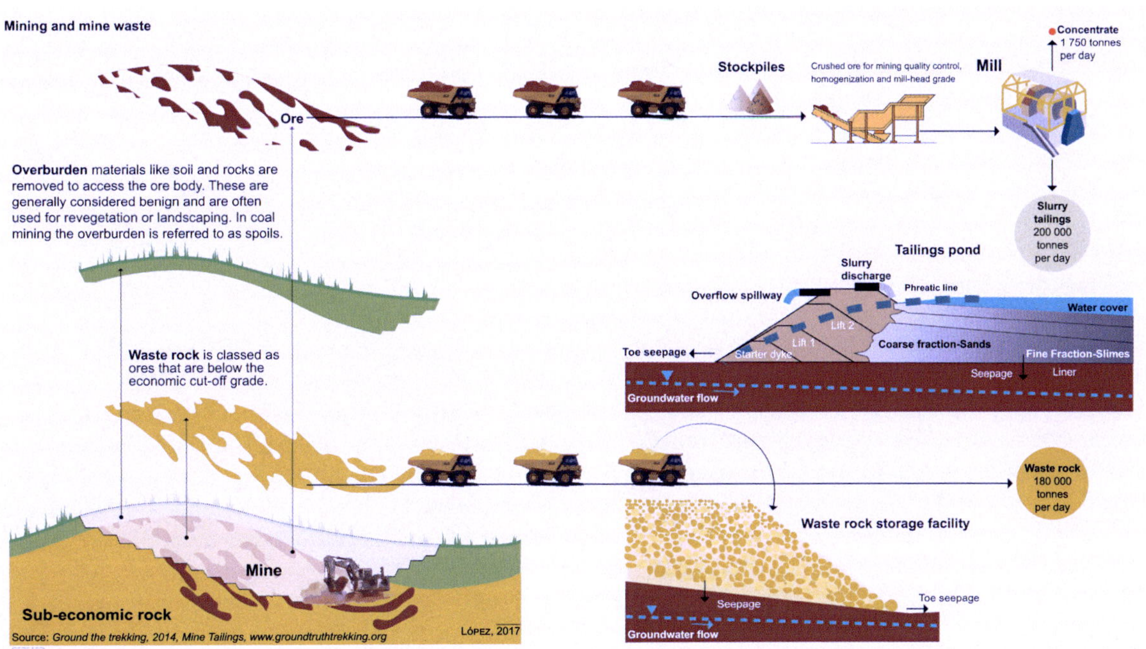

FIGURE 18.2 From mine to tailings dam. *Reproduced with permission from Kristina Thygesen (https://www.grida.no/resources/11415).*

18.2 FUTURE OPPORTUNITIES FOR TAILINGS MANAGEMENT

The estimated worldwide generation of solid wastes from the primary production of mineral and metal commodities is over 100 Gt/year (Tayebi-Khorami et al., 2019), which is expected to increase in the future due to the higher utilization of low-grade ores to meet societal demand/growth. The split between coarse-grained wastes from mining and fine-grained wastes or tailings from mineral processing can be 50:50, which is typical for base and precious metal mines, with lower proportions of tailings generated in the production of bulk commodities such as iron ore and coal. With this unprecedented rate of mining and tailings production and the recent fatal tailings dam failures, there is an increasing concern worldwide about tailings management and the potential consequences from tailings dam seepage and failure. With lower grade ores and the possibility of more extreme weather events associated with global climate change, the industry must address the rising concerns associated with mine tailings management.

Engineered and holistic tailings management involves interventions throughout the value chain. The value chain includes the ore body itself (characterization and modification of ore and gangue), excavation, transport, mineral processing, and tailings management, including disposal and storage. This requires a staff that can work through geometallurgical and materials flow considerations, including an integrated selection of physical, geometallurgical, chemical, and biological processes to achieve maximum extraction (including remining, reprocessing, and recycling) and minimized tailings volumes, with acceptable social and environmental impacts (minimized negatives, and maximized positives). Intervention can be designed to occur at any step in the value chain.

For the future and as envisioned by the International Council for Mining and Metals (ICMM) (2016), the value chain must be extended to engage the circular economy through downstream production of value-added products that minimize the need for tailings disposal and take advantage of the embodied energy and mining work done in tailings production. In tailings the rock has already been size comminuted through high energy consumptive crushing and grinding. As downstream products and markets develop, it may come to be that the desired tailings properties (engineered tailings) may drive changes in the upstream value chain to yield enhanced downstream values, with integrated decision making considering combinations of economics, social, and environmental drivers and with a whole-life-of-mine perspective that will lead to sustainable outcomes.

Much has been written about the potential use of mine waste for a variety of purposes, as is evidenced by Almeida et al. (2020), whose report on mine waste use includes technical papers published in the past 60 years. The uses vary from aggregate and fill material to structural additives for civil engineering applications. Waste materials can also have certain chemical properties or trace elements that make them attractive as a Portland cement additive or for use in the ceramics industry. As pressure to lower carbon emissions increases and carbon trading schemes develop, the emerging technology of using appropriate mine waste for the permanent storage of anthropogenic carbon emissions could offer significant economic opportunities, as well as a practical solution to lowering carbon emissions. It can even be argued that there are potentially more applications for the use and reuse of mine waste rock than perhaps for the ore commodity(ies) being mined initially.

18.3 MAIN DRIVERS FOR CHANGE

There is a need for a new vision for the whole mining industry based on the SDGs and the evolving ESG driven goals for industry. Responsible investment may well be the driving force that will transform the mining industry at a pace never seen before.

Concern over liability for failure of a tailings storage facility (TSF) is certainly a driver for change. However, other driving trends include deeper, higher temperature and more remote mines, lower grades resulting in increased tailings volumes, escalating concern for water quality and quantity, complex ore mineralogy leading to finer grinds for liberation, and significant clay mineral contents and complicated surface chemistries with difficult solids and slurry characteristics. All these challenge reagent and tailings dewatering selections, as well as tailings transport and placement.

In deciding about changes, the industry must think holistically about responding to the following needs:

- Demonstrate contributions to ESG and SDG goals.
- Minimize land use for TSFs.
- Minimize and reuse tailings.
- Minimize costs and risks (investors and the public are increasingly risk-averse).
- Facilitate and encourage responsible product design, use, reuse, recycling, and disposal.
- Incentivize innovative methods and technologies for resource recovery and reuse.
- Design and monitor TSFs in anticipation of the impacts of global climate change including increasing frequency of extreme weather events.
- Successful closure to accommodate any climate/weather events in perpetuity.

In addition, aspects that influence safe tailings facility operation and management include:

- Technology: good engineering design and plan for TSF monitoring is required.
- Operations: the best design can be operated poorly and, conversely, a poorly engineered facility can be understood and operated safely by a good operator.
- Corporate governance: the company must be committed to doing a good job, including allocation of resources for tailings management.
- Regulations and guidelines: guidelines are generated in response to failures, they do not prevent them. Funding must be made available in the long term for effective management of long-term liabilities.

The world's 29,000–35,000 existing active, inactive, and abandoned TSFs contain approximately 223 Gt (534 billion m^3) of tailings, and an additional 40 to 50 Gt will be generated over the next 5 years (World Mine Tailings Futures, 2021). Most exist at legacy sites or are under current construction, and these are designed to previous standards and may not meet current standards. Values contained in these TSFs vary widely and only a few successful attempts at remining and recovery of values have occurred, as it is a difficult undertaking. Only a few new TSFs are constructed each year, so emerging technologies for new facilities are needed to address both new and existing facilities. Most TSFs are long-term liabilities that must be managed as defined in the Global Industry Standard on Tailings Management (Global Tailings Review, 2020).

18.4 EMERGING TECHNOLOGIES

The mining industry has a great opportunity to incorporate new technologies into tailings management. New capabilities in real-time sensing with unprecedented accuracy mean that material characterization can occur continuously along the full material flow system from mountain to tailings to ensure that mineralogical knowledge is used to inform mine waste management at all stages of the mining life cycle (Figure 18.2). The strategic use of digital technologies will allow mine operators to build complex data systems that can provide continuous remote surveillance using mobile, Internet of Things (IoT)-enabled devices, satellite data, and video feed from drones or fixed points.

Instantaneous access to such data systems and artificial intelligence (AI) can support improved decision making that can reduce both risks and costs.

In addition, new technologies are available to incorporate into mine operational design, such as mass in-situ and preprocessing comminution with coarse ore sortation and pretreatment methods, including thermal (via oven, microwave, or radiofrequency), chemical additive, electric, magnetic, ultrasonic, and biomilling. Although not new, there are new applications for electrokinetics to control water and chemistry, and automation and distributed sensing to improve operational safety and reliability. Furthermore, evolving biological processes offer the possibility for selectivity and reaction rate enhancements in many applications.

18.4.1 Opportunities Along the Materials Flow

As mining companies begin to contemplate a wider range of technical options for the challenges and opportunities of tailings management, they will inevitably participate in a larger ecosystem of equipment suppliers, service providers, downstream by-product manufacturers, and industrial symbiosis partners. They should consider carefully who is best equipped to solve the problem at hand, and if not the mining company, then what the commercial and contractual relationships should be.

Consequences of technical failure in the mining waste field have proven to be very high, so traditional procurement methods (including efforts to maximize risk transfer to partners) may not be appropriate except where accountabilities can be very clearly defined. A further complicating factor is reputational risk. Symbiotic relationships with service providers and by-product companies, particularly if they are local enterprises, can be very beneficial for a mining company's reputation by creating shared value in host communities and boosting ESG scores. On the other hand, if there is a failure (for example, in a contracted TSF) then the mining company may incur a severe reputational penalty even if it believes it has successfully transferred risk ownership in the contract. Given this spread of both positive and negative potential impacts, companies may consider collaborative business relationships (CBRs) and relational contracting methods to build partnerships with companies in waste and by-product value chains. The growth of a circular economy will increase interaction between all industrial sectors, and CBRs will become an increasingly valuable corporate competency.

18.4.2 Upstream Opportunities

As the number and variety of test data become available, geometallurgical approaches to understanding mineral characteristics and the nature of the ore formation will allow the spatial and temporal distributions of the materials flow to be better predicted. Microwave technologies are beginning to be applied to selective and preprocessing comminution that may reduce overall energy consumption and change the character of the mill feed to allow lower-cost extraction (Monti et al., 2016).

Other technologies can be applied to reduce the quantity of tailings that need to be managed. These include:

- In situ recovery (ISR) extraction operations that avoid the production of tailings.
- Bio-enhanced processing to improve extraction and reduce comminution and transportation costs.
- Hydro-electrokinetic processes that can operate in situ mining and leaching processes to control the flow of water and increase water recovery.
- Enhanced in situ fracture and fragmentation using microwave and ultrasonic technology applications.

- Development of new excavation equipment to allow selective mining that reduces the volume of waste rock and may increase the grade delivered to the processing facilities.
- In-pit or onsite sorting/classification (see Chapter 8, Physical separation) of ore that can decrease transportation costs and increase ore grade to the processing facilities, while producing a coarser, easier to handle waste.

18.4.3 Downstream Opportunities

Both the management and reuse of tailings materials require knowledge of the temporal and spatial variability of the materials flow. This includes understanding the rheology of tailings slurries, tailings transport, and design optimization. The opportunity for real-time characterization of materials flow (including particle size distribution, mineralogy, water content, and chemistry), perhaps accomplished during dry stacking, can support waste stream segregation or partitioning by relatively simple gravity and particle size classification. The reuse of heap leach waste materials should also be considered.

Tailings reuse, remining, and recycling represent major opportunities for tailings management in the future. In most usages, it is imperative that mine operations include management of temporal and spatial variability of tailings characteristics. In fact, for most potential usages more knowledge about the tailings materials will be required than is currently and typically being assessed. Four major areas for downstream opportunities are identified herein.

In some cases, tailings have been used directly as agricultural supplements. Given the global sand shortage (UNEP, 2019), the simple separation of inert sand from a tailings supply can make economic sense. The major research focus for tailings reuse has generally been on substitutions for or in conventional construction materials.

Traditional construction is commonly cement-based, such as concrete and some mortars. Concrete is the second most-consumed substance on Earth, after water, and the manufacture of concrete involves billions of tons of feedstock, with a use rate of an estimated four tons of concrete per person per year globally. By 2050, concrete production is expected to be four times higher than in 1990.

The manufacture of cement is a major cause of greenhouse gas emissions (Balaji et al., 2017), and about 10% of the global emissions of CO_2 are due to construction materials production (Kappel et al., 2017). By 2025, around 3.5 Gt of carbon dioxide is anticipated to be released into the atmosphere during cement production. Cement contents and raw materials could be fully or partially replaced by eco-friendly secondary resources with lower embodied energy. Experience indicates that such substitutions should be taken with caution, as mining wastes may contain harmful compounds, such as heavy metals that may leach (Candeias et al., 2013). Strategies to remove or neutralize these elements are also important to optimize the use of mining residues with different compositions (see Chapter 17, Industrial by-products).

Silica and alumina are relevant elements to produce construction products, and pozzolanic activity of the natural pozzolans is closely interrelated with their content of reactive silica. When submitted to a relatively low thermal treatment, the silica and alumina state may change to an amorphous form and acquire pozzolanic reactivity, enabling the development of different construction products by partial conventional binder replacement. Secondary resources such as engineered tailings may have latent pozzolanic reactivity, and if present, the pozzolanicity of tailings can enhance mortars and concrete durability. More information can be found in Chapter 27—Cementitious binders incorporating residues. Examples of the produced materials and technologies in use or under development include the following:

- Geopolymers are a class of inorganic polymers that have an amorphous structure of [SiO_4] and [AlO_4], generally produced by mixing a raw aluminosilicate source in the form of a powder with an alkaline silicate solution followed by curing. Geopolymerization may be a key factor to stabilize hazardous compounds in mixtures with mining residues, avoiding leaching problems.
- Nanoadditive materials applied for mine tailings stabilization and toxic leachate containment (Brammer, 2021).
- Mortars: tailings have been used in hydraulic lime conventional mortars, mortars with complete replacement of natural aggregate by tailings, and mortars with partial replacement of lime with tails.
- Concrete/cement substitution/filler: tailings may be substituted for cement if they can be activated to act as a pozzolan.
- Aggregates: e.g., sintering can produce lightweight aggregate, and hot mix asphalt (HMA) has been developed containing mining waste as aggregates for making medium to low traffic roads.
- Sintering/ceramics (e.g., tiles, panels): e.g., iron ore tailings with a high percentage of silica have been used as part of raw materials for ceramic floor and wall tiles production, and tungsten tailings after magnetic separation.
- Bricks (fired or nonfired): tailings have been used, but not if of very low plasticity. Incorporation of tailings tends to reduce the plasticity of the feed before molding, so that additives such as mineral binders may be needed to improve plasticity.
- Polymer-based construction materials: composite products can be constituted by several sizes of particles/aggregates linked by a polymeric resin matrix. This aggregate/polyester matrix is optimized for mechanical and durability properties. Such advanced composite materials are compacted under intense vibration, vacuum, and pressure, resulting in dense and nonporous panels.

A new technique of nanosynthesis has also been applied to tailings. Zhang et al. (2020) report on the use of iron ore tailings (IOTs) as the silica source with potential for the preparation of zeolites with a well-defined microporous and mesoporous structure with a large surface area and high acidic strength. Such zeolites with elaborate micropore structures have uses in a broad range of applications as catalysts and adsorbents in the chemical industry.

Kim and Park (2020) report on an advanced synthetic process that produces wear-resistant glass ceramics from mining wastes and smelting byproducts (mixing gold tailings, red mud, waste limestone, and ferronickel slag). As the total generation of red mud in the world is estimated to be around 120 Gt/year, and the disposed red mud exceeded 3 Gt between 2010 and 2015, any reuse of this material is of economic and environmental consequence.

Naturally formed geologic melts abound, in their cooled form as igneous rocks such as basalt. Remelting of rocks, and subsequent controlled cooling for forming a solid with desired properties, is a relatively recent concept. For example, the recorded history of melting basalt rock and manufacturing glass fiber from the melt dates back to 1922 (US Patent US1438428), and additional developments occurred during and after World War II. Much of the activity during the Cold War period occurred in the Soviet Union, with fibers pulled from basalt melts being investigated for aerospace and military purposes, including insulation and textile applications (Acar et al., 2017). After the dissolution of the Soviet Union, basalt fibers began to be produced and used on a commercial scale as the research was declassified, with the primary locus of fiber production being in Ukraine.

Basalt fiber applications include many different composite applications (potential substitutes for glass and carbon fibers for strength and stiffness) and as ballistic materials. Due to its resistance to high temperature, basalt fiber is often used for high-temperature applications such as flame-retardant materials, disk brakes, and thermal insulation applications. Basalt fibers have excellent chemical resistance to alkaline attack and, due to their good electric insulating properties, basalt fibers have been used in printed circuit boards. Overall, basalt fibers exhibit better physical and mechanical properties than glass fiber, and the cost of basalt fiber is less than carbon fiber and comparable to glass fiber. Composites made with basalt fibers are of higher high-tensile strength than steel or E-glass, and are competitive in many applications with carbon-fiber composites.

In most cases of metals mining, the minerals that make up tailings are predominantly silicates—plagioclase, biotite, amphibole, and pyroxene—similar to the mineralogy of basalt. Current research at the Colorado School of Mines (Nelson et al., 2020) is focused on determining the chemical and mineralogical characteristics of mine and mineral processing tailings, which will allow the determination of melt composition and variability possible to achieve desired melt temperatures and other process-pertinent properties. In multicomponent earth-resource mixes, the relationships between chemistry and physical properties (viscosity, interfacial/surface tension, crystallization temperature) and thermodynamic properties (T_L, T_G, and chemical activities) constitute a complex network of chemistry-temperature-structure-property relationships. With the mineralogy and geochemistry of a tailings material known, it is possible to predict its melting point, glass formation temperature, and viscosity through thermodynamic and thermophysical property modeling (Verein Deutscher EisenhuttenLeute, 1995). This is important to better understand which of the tailings samples are most suitable for fiber production. The goal is to start producing fiber and composites by 2024.

18.4.4 Environmental Applications

Tailings and mine waste have been the subject of extensive research related to proactive management of environmental impacts (e.g., acid mine drainage (AMD), heavy metals, radioactive minerals) including CO_2 sequestration to remediate the effects of climate change. Interestingly, Harrison et al. (2012) reported on accelerated sequestration of CO_2 by tailings, and McCutcheon et al. (2019) reported later on chemical and biological reactions for CO_2 sequestration. Biogenic magnesium carbonate mineral precipitation from magnesium-rich tailings generated by many mining operations was investigated to offset net mining greenhouse gas emissions, while simultaneously giving value to mine waste products. These precipitates occurred as micrometer-scale mineral grains and microcrystalline carbonate coatings that entombed filamentous cyanobacteria.

18.5 CONCLUSIONS AND OUTLOOK

Catastrophic failures of tailings facilities in recent years have resulted in step-change developments and new industry standards of care and regulatory requirements. These enhanced programs require a significant increase in the number of and training available for engineers and scientists to meet the requirements of improved safe design, construction operation, and closure of tailings facilities. The release of the industry standard on tailings management in 2020 (Global Tailings Review, 2020) represents a major turning point in the regulation of tailings management and transparency, as does the evolving ESG mandate that demands an integrated and responsible mining industry that may well be the driving force for mining industry transformation at a pace not seen before.

Change is needed regarding tailings management. The current focus on ESG represents a huge opportunity, and the mining industry of the future can contribute much more to economic and social development in communities. The material demands of the world cannot be met without mining, but the industry's success depends on effective community engagement and adaptability, flexibility, and agility so that innovations can be adopted and integrated as they become available.

References

Acar, V., et al., 2017. Basalt fibers. In: Fiber Technology for Fiber-Reinforced Composites. Elsevier, https://doi.org/10.1016/B978-0-08-101871-2.00008-4.

Almeida, J., et al., 2020. Overview of mining residues incorporation in construction materials and barriers for full-scale application. J. Build. Eng. https://doi.org/10.1016/j.jobe.2020.101215.

Balaji, M., et al., 2017. Experimental study on the behavior of self compacting self cured concrete using chemical admixtures and metakaolin. In: International Conference on Emerging Trends in Engineering, Science and Sustainable Technology (ICETSST-2017), 2017.

Barclay, D., 2020. Rebooting the Mining ESG Narrative. Published November 30, 2020 https://www.linkedin.com/pulse/rebooting-mining-esg-narrative-dan-barclay/.

Blauner, R., 1964. Alienation and Freedom: The Factory Worker and His Industry. Chicago University Press.

Brammer, T.J., 2021. Comparison of nano meso inorganic polymers to Portland cement and geopolymers for mine tailings stability and containment of toxic RCRA 8 metals and other leachates. In: Proceedings of the 2021 Annual Meeting of the Society for Mining, Metallurgy and Exploration.

Candeias, C., et al., 2013. Heavy metal pollution in mine-soil-plant system in S. Francisco de Assis—Panasqueira mine (Portugal). Appl. Geochem. 44 (2013), 12–26. https://doi.org/10.1016/j. apgeochem.2013.07.009.

Columbia Center on Sustainable Investment (CCSI), 2020. Mining and the SDGs: a 2020 Status Update. CCSI and the Responsible Mining Foundation. https://www.responsibleminingfoundation.org/mining-and-the-sdgs/.

Global Tailings Review, 2020. The Global Industry Standard on Tailings Management. Global Tailings Review. https://globaltailingsreview.org/global-industry-standard/.

Harrison, A.L., et al., 2012. Accelerated carbonation of Brucite in mine tailings for carbon sequestration. Environ. Sci. Technol. 2013 (47), 126–134. https://doi.org/10.1021/es3012854.

International Council on Mining and Metals, 2016. Mining and Metals in the Circular Economy. International Council on Mining and Metals (ICMM). https://www.icmm.com/website/publications/pdfs/mining-minerals/2016/research_circulareconomy-2016.pdf.

Kappel, A., et al., 2017. Colour, compressive strength and workability of mortars with an iron rich sewage sludge ash. Constr. Build. Mater. 157 (2017), 1199–1205. https://doi.org/10.1016/j.conbuildmat.2017.09.157.

Kim, Y., Park, H., 2020. A value-added synthetic process utilizing mining wastes and industrial byproducts for wear-resistant glass ceramics. ACS Sustain. Chem. Eng. 2020 (8), 2196–2204. https://doi.org/10.1021/acssuschemeng.9b05884.

McCutcheon, J., et al., 2019. Carbon sequestration in biogenic magnesite and other magnesium carbonate minerals. Environ. Sci. Technol. 2019 (53), 3225–3237. https://doi.org/10.1021/acs.est.8b07055.

Monti, T., et al., 2016. High-resolution dielectric characterization of minerals: a step towards understanding the basic interactions between microwaves and rocks. Int. J. Miner. Process. 151 (10), 8–21. https://doi.org/10.1016/j. apgeochem.2013.07.009.

Nelson, P.P., et al., 2020. Glass from tailings. In: Proceedings of the Tailings and Mine Waste Conference. Colorado State University.

Tayebi-Khorami, A., et al., 2019. Re-thinking mining waste through an integrative approach led by circular economy aspirations. Fortschr. Mineral. 2019 (9), 286. https://doi.org/10.3390/min9050286.

UNEP, 2019. Sand and Sustainability: Finding New Solutions for Environmental Governance of Global Sand Resources. GRID-Geneva, United Nations Environment Programme, Geneva, Switzerland. ISBN:978-92-807 https://wedocs.unep.org/bitstream/handle/20.500.11822/28163/SandSust.pdf.

United Nations, 2015. Transforming our world: the 2030 agenda for sustainable development. https://sustainabledevelopment.un.org/post2015/transformingourworld.

Verein Deutscher Eisenhuttenleute (VDEh) (Ed.), 1995. Slag Atlas. Verlag Stahleisen GmbH, Dusseldorf.

World Economic Forum (WEF), 2016. Mapping Mining to the Sustainable Goals: An Atlas. Published with Columbia Center on Sustainable Investment (CCSI), the United Nations Development Program (UNDP), and the UN Sustainable Development Solutions Network (UNDSN). https://www.undp.org/content/undp/en/home/librarypage/poverty-reduction/mapping-mining-to-the-sdgs–an-atlas.html.

World Mine Tailings Futures, 2021. Estimate of World Tailings Portfolio 2020. World Mine Tailings Futures (WMTF). https://worldminetailingsfailures.org.

Zhang, Y., et al., 2020. Characterization of mining-related aromatic contaminants in active and abandoned metal(loid) tailings ponds. Environ. Sci. Technol. 2020 (54), 15097–15107. https://doi.org/10.1021/acs.est.0c03368.

Further Reading

Roche, C., Thygesen, K., Baker, E. (Eds.), 2017. Mine Tailings Storage: Safety Is No Accident. A UNEP Rapid Response Assessment. United Nations Environment Programme and GRID-Arendal, Nairobi and Arendal. www.grida.no. https://www.grida.no/publications/383. ISBN:978-82-7701-170-7.

PART 3

Recycling from a material perspective

CHAPTER

19

Steel

Thilo Kreschel and Olena Volkova

Institute of Iron and Steel Technology, Technical University Bergakademie Freiberg, Freiberg, Germany

19.1 INTRODUCTION

In a circular economy, materials are reduced, reused, remanufactured, and recycled as much as possible. For the Fe and steel industry, this concept means that the secondary resource steel scrap is to be preferred over the primary resource Fe ore whenever possible (Figure 19.1). Although theoretically every product made from steel is 100% recyclable, the maximum practical recycling rate is around 70% because of losses due to corrosion and dissipation (Jahrbuch, 1992).

Steel is a very common and popular material. In 2018, 240 kg of crude steel per person was used worldwide (Jahrbuch, 2020). That is several times more than the consumption of other materials such as Al, Mg, plastic, C fiber, and composites. According to the statistics in 2018, 1093 kg of crude steel was used per person in South Korea, 903 kg in Taiwan, and 906 kg in the United Arab Emirates (Jahrbuch, 2020). The lowest consumption of steel in 2018 was in Venezuela, with an estimated consumption of 18 kg of crude steel per person (Jahrbuch, 2020).

Iron and steelmaking processes are very energy intensive and emit considerable amounts of CO_2. In the primary route, CO_2 gas is emitted in the amount of 1880 kg/t crude steel directly in the coking plant, sintering plant, blast furnace, basic oxygen furnace (BOF), and subsequent casting and forming processes (Luengen, 2020). The blast furnace process accounts for the largest share of these CO_2 emissions. H_2 is the only alternative reducing agent to carbon in Fe ore reduction. In the direct reduction of Fe ores, hydrogen-rich natural gas has been used as a reducing agent in MIDREX, HYL processes since the early 1970s.

With the scrap-based secondary route via the electric arc furnace (EAF), only part of the CO_2 emissions are generated in the production processes. Most of the CO_2 emissions come from the CO_2 load of the externally drawn electrical energy utilized in the processes, since the electric furnace route itself does not generate any process gases that can be used for energy. The CO_2 emissions of this route with a CO_2 load of the electrical energy of 300 g/kWH are 410 kg/t crude steel (Luengen, 2020).

Unfortunately, it is not technologically possible to produce steel in a completely C-free manner. C is used in the EAF mixing process for CO bubble formation, EAF slag foaming with the aim of reducing energy consumption, as binder for refractory, for crude steel deoxidation, and as an alloying element. However, not all steel grades can be produced in EAFs.

 https://doi.org/10.1016/B978-0-323-85514-3.00023-3

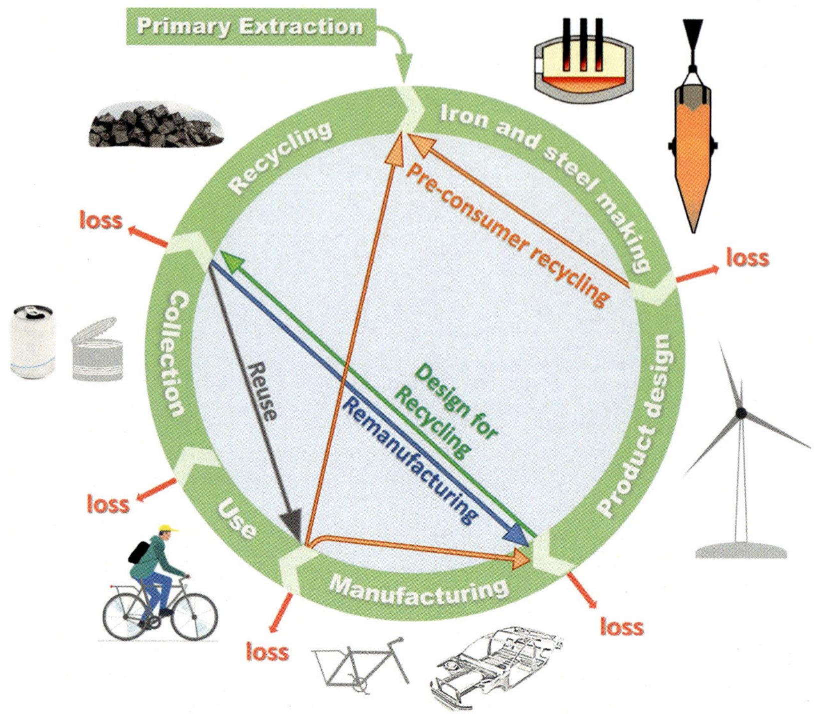

FIGURE 19.1 Circular economy loop of Fe-containing products.

For example, the production of flat steel products in EAFs still lags behind today (Grummes, 2019).

Steel is not the only useful product obtained from the steelmaking process. By-products include slag, dust, exhaust gas, steam, and waste heat. The use of Fe-containing by-products has thus led the steel industry to save natural resources and to reduce its environmental impact. Integrated steel plants (BOF, primary route) generate large amounts of solid by-products, which could comprise about 400–500 kg/t steel (Sista et al., 2019). In one scrap-based steel plant (EAF, secondary route), this amount is only 200 kg/t steel (Branca et al., 2020). Major solid by-products such as slags, dusts, sludge, and scales contain Fe and Fe oxide. For example, mill scale consists of about 70% Fe in the form of Fe_2O_3, Fe_3O_4, FeO, and metallic Fe. Therefore separated Fe from slags, briquetted dust, sludge, and scale are used as Fe carriers in steel making. In 2019, 12.3 Mt of ironmaking slag was produced in Germany (Merkel, 2020), and 11.4 Mt of this slag was used as a building material in the cement and construction industry and as fertilizers. This value corresponds to a continued high usage rate of 93% of ironmaking slag. Slag and dust contain valuable elements such as Cr, V, Mo, and rare earth elements, which are often lost. The combustible exhaust gas from blast furnaces, BOFs, and coking plants is generally used in plants to heat pig iron or cast ladles and slabs before rolling, among other useful functions. The useful waste heat potential in German integrated iron and steel plants combined was estimated to amount to 0.322 GJ/t solid crude steel (Sprecher et al., 2019), offering further opportunities for maximizing sustainability.

19.2 USE PHASE AND RECYCLING EXAMPLES

Due to the technical developments of the last decades, very few products are manufactured today that consist exclusively of steel. Instead, products today are often very complex in terms of their applications and properties and therefore consist of diverse mixes of materials. The palette ranges from metals to glass, plastics, and natural materials. The steel-containing products have diverse use phases. Depending on the function, regular use phases of a few months (for packaging), several years (for household appliances and road vehicles), or up to many decades (for buildings and structures) can be assumed.

The life span of a rail vehicle, for example, can be divided into purchase, commissioning, use, maintenance, and decommissioning phases. As a rule, several repair and deployment phases occur. These serve to restore the so-called wear stock, for example by replacing the wheel tires or drive motor, with economic and ecological considerations as the basis. The use phase can thus extend over several decades. During the entire use phase, there are changes in the direct material flow (due to corrosion or abrasion, among other factors) and induced material flow (for example, through the conversion of electrical energy into kinetic energy) (Main, 2017).

At the end of the product life, the material resources included in completely different products must be made available for subsequent use. This characteristic is a global social requirement due to economic and ecological demands in terms of sustainable use of natural resources and protection of the environment. Appropriate processing is necessary for this purpose. In principle, this approach enables infinite reuse (Püchert, 1996; KrWG, 2020). Following are three examples.

Ships are largely composed of steel, including the hull, engines, gearboxes, supply lines, and on-board lifting gear. Other metals, glass, plastics, and natural materials such as wood are also found on ships. When scrapping ships, it is important to separate these materials from each other for recycling purposes. However, it must be noted that there is a great shortage of ship-dismantling facilities worldwide that ensure ecological and social sustainability. Consequently, to this day, many merchant ships end up as scrap metal on beaches in South Asia.

Reinforced concrete is a frequently used building material worldwide. Inside the composite formed by concrete and reinforcing steel, the concrete is particularly suitable for absorbing compressive stresses and the reinforcing steel for absorbing tensile stresses (Bleck and Moeller, 2018). At the end of the service life of a reinforced concrete structure, separation of the two materials is mandatory for recycling. Mechanical crushing technologies are used for this purpose, followed by magnetic and screening systems to separate the metallic fraction from the concrete.

During their use, rails are subject to system-related mechanical wear, among other types of wear, by abrasion. This wear essentially determines the service life of the track. After a rail has been removed, it is recycled. The dimensions of rails make it necessary to shred them for later reuse in metallurgical smelting units. Due to the mechanical properties of modern rail materials, using rail breakers is an effective alternative to flame cutting.

19.3 CLASSIFICATION OF STEEL SCRAP

Steel scrap is a metallic resource that is an essential raw material in metallurgical smelting processes. Such scraps can be subdivided according to their material composition, physical properties, or source (BDSV et al., 1998). An example of such a subdivision is the European Steel Scrap Specification (BDSV, 1995).

Regarding the material composition, the exact chemical composition (alloyed or unalloyed) and degree of contamination by other materials (nonferrous metals, cast Fe, and other substances) are of interest. Regarding the physical properties, the dimensions, shape, piece size, and bulk density are important criteria for the use of scrap.

When considering scrap sources in the steel industry, a distinction is commonly made between own scrap arisings and purchased scrap. Own scrap arisings include the scrap produced in steelworks or rolling mills in the course of melting, casting, and rolling. The material compositions and physical properties of these scraps are very easy to calculate and sort. Purchased scrap is scrap that the steel industry buys in processed form from the steel recycling industry. It is composed of new and old scrap (Figure 19.2). New scrap accumulates in the steel processing industry, for example as punching waste or chips, and therefore its material composition and physical properties are also very easy to calculate and sort. In the case of end-of-life scrap, steel-containing products are the source at the end of their use phase.

19.4 REQUIREMENTS FOR SCRAP

The requirements that the steel industry applies to scrap relate to transport, handling, and use in the melting unit (BDSV et al., 1998).

19.4.1 Physical Requirements

The important physical requirements are the individual piece size, individual piece mass, scrap density, and bulk mass.

There are company-specific restrictions on the individual piece size due to the melting unit opening size, in the cases of both EAFs and BOFs. Furthermore, the relationship between the piece size and melting behavior must be taken into account in order not to disturb the technological process. Particularly in the case of EAFs, it must be considered that overly long or bulky scrap pieces can lead to such problems as impairment of arc ignition, impairment of current conduction due to cavities, and risk of electrode fractures due to scrap collapsing.

The individual piece mass is usually limited to less than 1 t for use in an EAF and to less than 3 t for use in an oxygen converter. Otherwise,

FIGURE 19.2 Scrap types. (A) Old scrap at a scrap yard of a steelmaking plant. (B) New scrap (left side) and own scrap (right side) at a scrap yard of a steelmaking plant.

incomplete dissolution of individual scrap pieces within the usual tap-to-tap or blowing times is to be expected. Regarding scrap density or bulk mass, a large volume with a low bulk mass entails an increased effort for transport, handling, and loading during charging. The optimum bulk density is above $1\,t/m^3$.

In addition, the use of packaged sheet scrap is subject to such requirements as a sufficient Fe content of at least 90% and a maximum of 7% nonmetallic content. For the use of steel chips as a significant portion of new scrap, a dosed application is needed. In an EAF, an upper layer of chips promotes arc ignition when the electrodes are applied.

19.4.2 Chemical Requirements

Steel scrap contains scale, moisture, corrosion products, and impurities from coatings of Zn, Sn, Cu, paint, enamel, plastics, and other materials. Old scrap is primarily affected here, but new scrap is as well. These impurities can contaminate the chemical composition of steel and affect the metallurgical processes in different ways, such as by changing the melting losses. Nonmetallic fractions increase the slag quantity, which increases the energy consumption and cost.

19.4.3 Safety Requirements

For safety, old scrap in particular must be monitored. Details on this issue are provided in accident prevention regulations issued by the liability insurance associations of employers.

The safety requirements include that no hollow or explosive bodies may be contained in the scrap. Hollow bodies that are not sufficiently open harbor the risk of accumulating liquids, especially precipitation water. In contact with liquid metal, sudden evaporation (i.e., explosion) in the melting unit is to be expected.

Unopened gas cylinders with flammable or explosive contents, such as liquid gas residues, represent a particular hazard. As a matter of principle, explosive devices and ammunition residues must not be present in scrap because of the risk of explosion.

To prevent possible contamination and hazards, incoming purchased scrap is generally subjected to radioactivity measurement. For this purpose, measuring facilities are available in the steel industry, such as in the form of gate passages for road and rail transports.

19.5 TREATMENT PROCESS

19.5.1 Dividing

A distinction can be made between disassembly and thermal and mechanical comminution methods. Disassembly is mainly used for complex steel-containing products such as road vehicles at the end of their service lives. In this way, still-usable assemblies or individual parts can be obtained as replacement parts. Here, manual activities predominate. These activities also include the draining of all liquid operating materials and their professional disposal. Depending on the requirements, disassembly can be followed by crushing, sorting, or compacting of the residual material.

Thermal comminution is performed by flame cutting. It is mainly used for very large units, vehicles, or bulky individual parts such as long girders. A typical application example is the shredding of rail vehicles or ships at the end of their service life.

The tools for mechanical comminution are more diverse than those for thermal comminution and include mechanical shears, skull crackers, chip crushers, rail crushers, and shredding plants. These are all operated with different sizes. Mechanical shears used are mainly alligator, guillotine, and excavator shears. Mobile plants may also be available from the manufacturers, depending on the size. In recent years, steel grades with very high strength values have increasingly been launched on the market. As a result, the requirements placed on cutting tools in terms of strength and wear resistance are already increasing. Skull crackers are suitable for large-volume scrap pieces, especially castings. Size reduction is performed with the aid of a drop pear or drop ball, which is mechanically lifted off and then dropped on to the scrap from a great height. The considerable noise and dust generation is problematic. Splinter protection is indispensable. Chip crushers can be used to reduce the volume of chips from mechanical processing significantly. They have hardened shear blades and a grinding mechanism in combination with an appropriately wear-resistant lining.

Rail crushers are capable of crushing railroad, tramway, and crane rails that are much too long in their original state for use in the melting unit. They can also be used, for example, to crush hardened bar stock. Rail crushers are available as stationary equipment or mobile attachments, for example for hydraulic excavators.

Shredding plants can be used, for example, to shred entire cars on a large scale after the operating fluids have been drained off. First, they are torn into pieces about the size of a fist. The individual materials are then sorted in several stages. Consequently, fractions such as steel, nonferrous metals, plastics, and glass are separated.

19.5.2 Sorting

Sorting processes are necessary depending on the origin of the scrap and its subsequent use. The main objectives of these processes are separation of nonferrous metals, separation of nonmetallic materials, separation of alloyed steel scrap from unalloyed steel scrap, and fulfilment of safety requirements.

For the separation of nonferrous metals and nonmetallic materials, for example after shredding, magnetic and air-flow separators are used. Magnetic separators use ferromagnetism to separate materials. Air-flow separators use the density differences between materials to separate them.

The main goal of separating alloyed steel scrap from unalloyed scrap is to use alloyed scrap specifically to produce alloyed steels. This approach results in savings on cost-intensive alloying agents and creates the prerequisites for meeting the quality requirements for unalloyed and low-alloy steels.

19.5.3 Compacting

Compacting can be considered as a processing method whenever the scrap density is to be increased or the handling of small scrap pieces is to be improved.

Scrap metal presses are the best-known devices. They can press old scrap such as wrecked cars or tin cans, as well as new scrap from metal processing, into packages of various dimensions. Problems can always arise if nonferrous metals or nonmetallic fractions have not previously been separated to a large extent. Briquetting of steel scrap is used much less frequently.

19.6 STEEL SCRAP SMELTING PROCESS

Crude steel is essentially produced by two different methods: the path from Fe iron ore to steel using a blast furnace or smelting reduction together with the BOF, or the path from scrap to steel using an EAF (Figure 19.3). The iron ore-based steelmaking can be called the primary/integrated route or oxygen steelmaking, and scrap-based steelmaking can be called the secondary route or electric steelmaking.

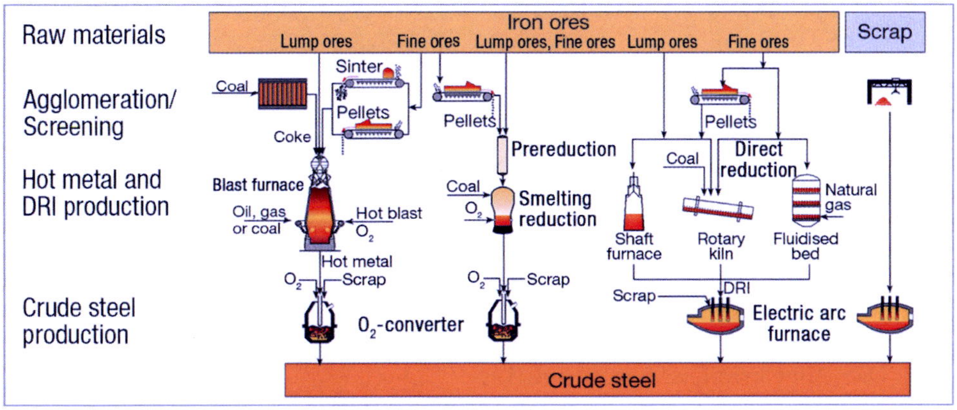

FIGURE 19.3 Schematic process scheme of iron and steelmaking based on ores or scrap. *Reproduced with permission from Steel Institute VDEh.*

19.6.1 BOF Steelmaking

A typical BOF is a vertical vessel lined with basic refractory, typically with C-bonded magnesia bricks (MgO-C). The BOF vessel consists of a spherical bottom, a cylindrical shell, and an upper cone. The specific working volume (defined as converter volume (m^3) per charge weight (t)) is typically in the range of 0.65–0.75. Thus around 20% of the converter volume is filled with hot metal and scrap (Jalkanen and Holappa, 2014). The maximum BOF capacity today is 400 t crude steel. The BOF can rotate around its horizontal axis. Tilting is necessary to charge the raw materials, sample the steel and slag, and tap the steel melt and slag. Oxygen is blown using a water-cooled top lance. The oxygen flow rate is normally in the range 6–10 m^3/min per t crude steel with an oxygen pressure of 1–1.5 MPa (Jalkanen and Holappa, 2014). To improve the mixing of the metal bath, accelerate the decarburization, decrease the amount of dissolved oxygen, and decrease the tap-to-tap time, the melt bath during the oxygen blowing is stirred using the bottom tuyeres. Therefore N_2 and Ar are used as stirring gases, with the switching point from N_2 to Ar depending upon the requirements for the nitrogen content in the steel. Bottom-blown BOFs are also available, where with several tuyeres in the bottom of the vessel, the oxygen, with or without lime powder, is injected into the metal melt.

Liquid hot metal is used in BOF as the main metal raw material. The hot metal tapped from the blast furnace is always saturated with C and typically contains about 4.5 wt.% C. In addition, depending on the iron ores used, the hot metal may contain Mn, Si, P, Ti, Cr, or other elements. C, Mn, Si, P, or other elements react with blowing oxygen with heat evolution. Therefore, large quantities of coolant must be added to the BOF. The coolant in the BOF could be steel scrap, iron ore, beach iron, briquetted dust, or briquetted sludge. Moreover, the scrap ratio in the current BOF process is about 15–30%. However, the industrial use of scrap in BOFs is known to be up to 50% (Fritz and Gebert, 2005). To produce 1 t crude steel in a BOF, 1050–1150 t metallic entry is required. Part of the Fe from hot metal and scrap is oxidized and passes into slag, although part is lost with the flue gases and deposited as dust. The Fe oxide content in the BOF slag thus depends upon the production conditions and has generally been reported to be 14–30 wt.%. This amount of FeO_t in the BOF slag is necessary for the successful dissolution of lime added as slag former and the dephosphorization process. The other slag formers in BOF are lime, gangue in scrap, and dissolved MgO from the refractory.

19.6.2 EAF Steelmaking

In the secondary steelmaking route, crude steel is produced from steel scrap using an EAF. As a raw/charge material, 100% steel scrap, or scrap with the addition of direct reduced iron (DRI), hot briquetted iron (HBI), or even liquid hot metal can be used. However, EAFs typically use 100% steel scrap as the metallic charge. As the main energy source, electricity is used in EAF. A distinction is made between EAFs using alternating current (AC) and direct current (DC), although most EAFs today use AC. In contrast to an AC EAF with three artificial graphite electrodes, a DC EAF has one graphite electrode and one mostly steel anode in the furnace bottom. The development and implementation of slag foaming through injection of a C source and oxygen, bottom tapping, bottom stirring, scrap preheating, and oxy-fuel burners have led to substantial decreases in the electric energy consumption, tap-to-tap time, and electrode consumption of the current EAFs. The electricity consumption of modern EAFs based on 100% scrap is currently about 345 kWh/t crude steel. For comparison, in 1965, the electricity consumption was 630 kWh/t crude steel. The tap-to-tap time decreased from 180 to 40 min between 1965 and 2010 (Madias, 2014). Just like in the BOF process, some Fe in the form of iron oxide passes to the slag. The iron oxide content in EAF slag during the production of C or construction steels is higher than that in BOF slag, which is estimated to be about 25–30 wt.%. An EAF is a smelting unit with a high efficiency but is not a very efficient reactor type for performing refining reactions. Refining in an EAF is limited to dephosphorization, decarburization, and temperature adjustment.

Austenitic and ferritic stainless steels have high Cr and Ni contents. They are typically produced via electric steelmaking using an EAF and are based on stainless steel scrap containing Cr and Ni. However, the lack of stainless steel scrap makes it necessary to use ordinary steel scrap and to form alloys with high-C ferrochromium, ferronickel, or pure Ni. For decarburization without incurring Cr oxidation, the secondary metallurgy of stainless steel production includes additionally treatment of molten steel in Ar oxygen decarburization (AOD) or vacuum oxygen decarburization (VOD) units.

19.7 STEEL

With more than 2500 specified steel grades, various property combinations are available for different areas of application. Many new types of steel or steel ceramic composites, especially those with tailored properties, have been developed in recent years.

According to DIN EN 10020, steels are divided into three classes according to their chemical composition:

- Unalloyed steels. Unalloyed steels are those whose content in relation to any alloying element does not reach the respective limit value, according to Table 19.1.

TABLE 19.1 Limit values between unalloyed and alloyed steels, wt.%.

Al	B	Bi	Co	Cr	Cu	La	Mn	Mo	Nb	Ni	Pb	Se	Si	Te	Ti	V	W	Zr	*
0.3	0.0008	0.1	0.3	0.3	0.4	0.1	1.65	0.08	0.06	0.3	0.4	0.1	0.6	0.1	0.05	0.1	0.3	0.05	0.1

*Others (except C, N, P, and S) (each).
Data from DIN (2000).

- Stainless steels. Stainless steels are those with at least 10.5% Cr content and at most 1.2% C content.
- Other alloyed steels. Other alloy steels are steels that are not stainless steels and exceed at least one of the limit values in Table 19.1. Chemical compositions of various existing steels and their properties are presented in Stahlschluessel (2019).

Stainless steels are further classified according to their Ni content as Ni content <2.5 wt.% and Ni content ≥2.5 wt.%; according to properties such as corrosion, heat, and scale resistance; and according to their microstructure, which can be ferritic, austenitic, martensitic, or austenitic-ferritic (duplex steels).

The other alloyed steels are classified in accordance with the main grades as weldable fine grain steels; alloyed steels for rails, sheet piles and pit extensions; alloyed steels for hot or cold rolled flat products; alloyed steels in which Cu is the only specified alloying element; and alloyed electrical sheets.

19.8 ALLOY OR TRAMP ELEMENTS?

The most common alloying elements are C, Si, Mn, Cr, and Ni. Alloying with C increases the strength and enables steel to be hardened by quenching. Unfortunately, as the C content in the steel increases, the toughness decreases. The maximum C solubility in steel is 2 wt.%.

Si increases the yield and tensile strengths needed without a significant reduction in total elongation. The spring steels, tool steels, heat and scale-resistant steels, and electrical steels are alloyed with Si. Modern electrical steel can be alloyed with Si up to 6.5 wt.%.

Alloying with Mn increases hardenability, strength, toughness, and hot workability. The most important austenitic steel containing Mn is Hadfield steel with 12–14 wt.% Mn and 1.2–1.4 wt.% C (Mn:C = 10:1) (Bleck and Moeller, 2018). In structural steel, a Mn content up to 1.7 wt.% sets S and forms MnS, preventing the formation of low-melting FeS phases. In addition, quenching and partitioning (Q&P) processing has been applied to metastable austenitic CrMnNi steel containing nitrogen and carbon to generate outstanding mechanical properties required for applications in third-generation advanced high-strength steels (AHSS) (Wendler et al., 2017).

Alloying with Cr increases the tensile strength, and the elongation is only slightly impaired. Cr forms carbides, which further increase the hardness. A high Cr content leads to improved heat and temper resistance. Steels with more than 12 wt.% Cr form a passive layer and are thus protected against corrosion in aqueous media.

Ni is the most important alloying element for austenitic corrosion-resistant steels. Ni increases hardenability and toughness and lowers overheating sensitivity.

Mo in the range 0.2–0.4 wt.% increases the hardenability and heat resistance and reduces the tempering embrittlement of Cr and Mn-containing quenching and tempering (Q&T) steels. Alloying with Mo in the 2–5 wt.% range improves the pitting corrosion resistance of high-alloy steels.

Cu improves weather and corrosion resistance, especially in the presence of P and Mo. In addition, Cu additives increase the yield strength, tensile strength, and hardenability.

Alloying with Ti is performed to achieve hardening by grain refinement due to prevention of recrystallization and austenite grain growth as well as precipitation hardening through Ti carbides.

Al contents of up to 8 wt.% are used to reduce weight. Besides density reduction, Al addition to austenitic steels produces desirable combinations of strength and ductility, enhances the wear resistance, and increases the resistance against hydrogen embrittlement (Rahimi et al., 2016).

In the presence of Cu, P improves the resistance of structural steels to atmospheric corrosion. The known examples for P alloying are the hot-rolled high-strength sheet steel with 0.04 wt.% P and P-alloyed interstitial free (IF) steel.

Free-cutting steels are alloyed up to 0.15–0.3 wt.% with S and Pb. A short-breaking chip is obtained during machining and the surface finish is improved.

Dissolved oxygen may also be desirable in some steels. Deep-drawing steels are cast with dissolved residual O up to 90 ppm.

Figure 19.4 provides an overview of the typical alloy and tramp elements in steels. Unfortunately, elements that are useful alloying elements for some steels may not be desirable for other steels. For example, the C content of modern IF steels can be as low as 30 ppm. Mn is undesirable in some electrical steels. P causes segregation during solidification, which can further lead to hot cracks. In the cases of some other steel grades, P can also serve as alloying element. P tends to accumulate at grain boundaries (segregation) in steel, which can lead to brittle failure through grain boundary cracking. S generally worsens the mechanical properties. Cu, Sn, Mo, Cr, and Ni are tramp elements in most alloyed and low alloyed steels. Cu increases the risk of hot shortness and cold work hardening in low-alloyed steels, which is even more pronounced if Sn is also present in the alloy (Sandig et al., 2019). These negative influences can be considered as loss of ductility in steel in the temperature range of 1100–1300°C. Moreover, the excessive amount of Cu leads to a radical decrease in surface quality in heat treatment processes such as hot rolling. Cu contamination of more than 0.2 wt.% has a snowballing impact on crack depths of hot rolling mill products (Leroy et al., 1995). Sn exists in the form of segregation in solid steel (Zhang et al., 2019), which affects the thermoplasticity (Matsuoka et al., 1997; Nagasaki and Kihara, 1997), temper brittleness (Zhao et al., 1995; Zhou et al., 1998), and thermal processing properties (Zhang et al., 2019) of steel. Thereby, Sn leads to solidification segregation, grain boundary segregation, and surface segregation (Zhang et al., 2019). Table 19.2 shows the maximum

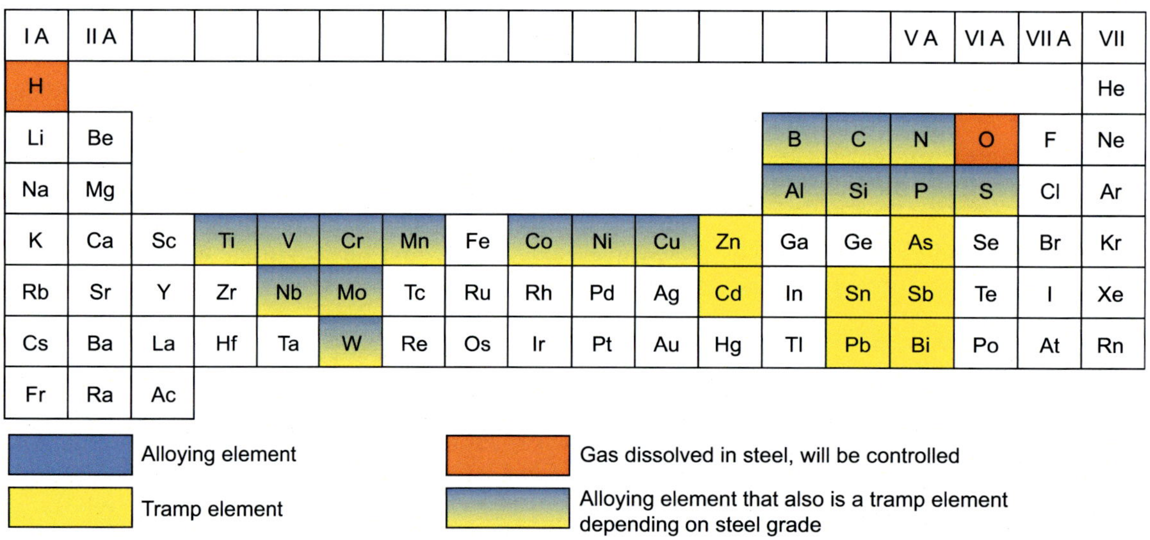

FIGURE 19.4 Common elements in steel.

TABLE 19.2 Maximum allowed contents of Cu, Sn, and As in low-C steel for Compact Strip Production (CSP) plants in China, Japan, and the United States, wt.%.

CSP plain for low-C steels	Sn	Cu	As	Cu+8Sn+5As
American Riverdale	0.003	0.04	0.018	0.154
Japanese Kawasaki	0.002	0.03	0.015	0.125
Lianyuan Steel	0.012	0.07	0.043	0.381
Vf Zhujiang Steel	0.014	0.10	0.04	0.412

Based on Zhang (2009) and Zhang et al. (2019).

allowed contents of Cu, Sn, and As in low-C steel in China, Japan, and the United States.

19.9 PURIFICATION OF SCRAP

Purification of steel solutions must be understood in the context of solution chemistry of molten steel. Texts such as Turkdogan (1980) and Guthrie (1992) provide the necessary thermodynamic and kinetic background.

19.9.1 Decopperization

The Cu content of steel melts in an EAF can currently be up to 0.2 wt.%. Due to the increased use of scrap from the electronics and automotive sectors, this Cu share will continue to increase in the future. In contrast, there are requirements for modern steel grades that require Cu content of $\leq$0.1 wt.%. Within the Sn plate grades, the Cu content must not exceed 0.02 wt.%. Because the oxygen affinity of Cu is lower than that of Fe, selective oxidization of Cu from steel melts is not possible.

Reviews of decopperization techniques have been given by Savov et al. (2003), Jimbo et al. (1988), Noro et al. (1997), and Sandig et al. (2019). In principle, the methods of Cu removal can be divided into:

- Scrap pretreatment, sorting, and physical separation;
- Dilution by addition of Cu-free material (hot metal, sponge Fe) or balancing out of harmful Cu effects by addition of other alloying elements;
- Evaporation/gasification;
- Decopperization using sulfate or sulfide-containing slags (Na_2SO_4, FeS, BaS, K_2S, SrS, Na_2S, Li_2S) or other reactants;
- Filtration.

Cu can be separated from steel scrap by a nonferrous metal bath. The method is based on the solubility of Cu in a metal bath of Pb or Al, which is higher than that of steel. Wang et al. (1991b) found that for the reduction of the Cu content in 1 t of scrap from 0.3% to 0.1%, 1000 kg Pb is necessary. The major disadvantage of the proposed process is that molten metal adheres to the solid scrap after it is removed from the bath, leading to the contamination of scrap with Al, Pb, and Cu.

Dilution by addition of Cu-free material (hot metal or sponge Fe) cannot be the long-term metallurgical solution to the Cu problem. The harmful effects of Cu can be balanced out by the addition of alloying elements such as Ni and Al. The addition of Ni in a proportion of 2:1 or 1:1 neutralizes the harmful properties of Cu (Cochrane, 2000).

The evaporation/gasification methods fall into two categories: vacuum evaporation, where Cu is vaporized as a metal (Savov, 1999; Savov et al., 2000; Zaitsev et al., 2004a, b; Łabaj, 2012;

Jung, 2014), and gasification that involves chemical reactions to bind Cu to a more volatile chemical compound, such as CuCl (selective chlorination), which may be performed under reduced or atmospheric pressure (Hu et al., 2013). Technically, the evaporation of Cu is difficult to achieve in steelworks and is therefore not economically viable. In addition, the chlorination of steel melts at higher temperatures in steel plants is not easy.

Decopperization with sulfate or sulfide-containing slags proceeds exothermically with a change in the Gibbs energy of dG_0 = –19.814 kJ/mol according to the following reaction:

$$2[Cu] + (FeS) \rightarrow [Fe] + (Cu_2S)$$

The disadvantages of Cu removal via slag treatment include a low degree of Cu removal and the sulfurization of the steel melt. Studies by Wang et al. (2009), Li et al. (1999), and Liu and Jeffes (1989) with different slag compositions showed that 4%–37% of the Cu content could be removed within 5–40 min. In addition, part of the Cu entry into the melt occurs shortly before it is cast by adding the cooling scrap to set the temperature and thus cannot be removed by slag treatment.

Treatment with elements such as Se to bind Cu in Fe alloys as Cu_2Se for accumulation and then transfer it into the slag phase is effective on the laboratory scale (Savov, 1999). However, very high amounts of relatively expensive Se would be required for decopperization on an industrial scale. Furthermore, toxic Se vapor, which poses risks to human health and the environment, is released during treatment (Savov et al., 2000; Glazov et al., 2000; Bernardini et al., 1982).

Filtration studies (Li et al., 1998; Wieliczko et al., 1998) used filters composed of spinel materials based on Al, Zn, and Ti, such as Al_2O_3-MgO, Al_2O_3-ZnO, Al_2O_3-Zn-C, Al_2O_3-Zn-Al, TiC, and TiO_2-C. Zigalo et al. reported that 33% of the initial Cu content may be removed from steel utilizing a granular Al_2O_3 ZnO filter packaging in the tundish over 30 min, and decopperization of nearly 70% may be achieved by Al_2O_3-ZnO-C powder blowing (Zigalo et al., 1991). Li reported 19% decopperization in 45 s using a Al_2O_3-ZnO-C filter material for steel (Li et al., 1998).

19.9.2 Detinning

The next critical tramp element in steel is Sn, whose content in scrap increases with the recycling rate. It comes from the tinplate of beverages, food cans, and other packages. Tinplate is a type of cold-rolled steel plate coated with a thin layer of Sn, where 2–8 kg Sn is used to coat a 1 t steel plate depending on the thickness of the steel plate and Sn layer. Thus the steel melt can have 0.2–0.8% Sn after the smelting of the tinplate scrap.

The following methods for Sn removal are known:

- Electrolytic detinning.
- Sulfation.
- Chlorination.
- Vacuum distillation.
- Treatment with Ca or Ca-based slags.
- Treatment with rare earth elements (La, Se, and Ce).

In electrolytic detinning, tinplate scrap is pressed into bundles and immersed in an alkaline aqueous solution bath (NaOH, Na_2SnO_3, or Na_2CO_3) (Savov, 1999; Zhang et al., 2019). The bundles serve as anodes, whereas the Fe/steel plate is used as a cathode. Sn is deposited on the Fe/steel cathode as a sponge material. When the treated scrap is rinsed, a residual Sn content of $\leq$0.02% is achieved (Savov, 1999).

Detinning can also be achieved by the treatment of tinplate scrap with gases of a given S potential (Tu and Janke, 1994; Savov, 1999). Sn reacts with S gas and forms a solid SnS layer, which is easily removable from the steel surface. If $H_2S_{(g)}$ is used, then H_2 is formed as well.

The application of the sulphation method has a few important negative effects. First, the use of S-containing gases requires additional measures to protect the environment and health of the personnel. Secondly, a fraction of Fe is sulfidized in addition to Sn. Subsequently, this approach leads to the loss of Fe with the easily removable Sn-containing sulfide layer.

Detinning by chlorinating the tin coating is based on the volatility of $SnCl_4$ (Savov, 1999, Zhang et al., 2019). $SnCl_4$ has a low boiling point (113°C) and a large vapor pressure at room temperature. However, the chlorination process requires that the Sn-coated steel sheet scrap does not contain moisture or organic matter. The principle of chlorination is relatively simple, but the practice is more difficult.

The vacuum distillation of melt is based on the difference between the vapor pressure of the Sn and Fe in a vacuum state. For example, at 1600°C, the vapor pressure of Sn is 118.9 Pa, whereas that of Fe is only 5.176 Pa (Zhang et al., 2019).

The Ca treatment process can partially remove residual elemental Sn and has good application prospects, as studies by Kitamura et al. (1985), Ochifuji et al. (1995), Tsukihashi (2001), and Ghosh (2009) show. The formation of Ca_2Sn can proceed only under highly reducing conditions due to the great affinity of Ca for oxygen. Ca also features a strong affinity toward S. If S and Sn are present in the melt, Ca will tend to react with S at first. For successful detinning, the steel melt must be deeply deoxidized and desulfurized before detinning with Ca. Ca-containing slag, such as CaC_2 or CaF_2, is typically used rather than Ca metal (Tsukihashi, 2001).

Detinning Fe melts by Se addition assumes that the affinity of the solute element Sn toward Se is greater than that of Fe (Savov, 1999). The use of Se in steelmaking is narrow due to its toxic nature.

Rare earth elements (La and Ce) are used in ferrous metallurgy as strong deoxidizing and desulfurizing reagents. A thermodynamic calculation demonstrated that at temperatures lower than 1140°C, Sn and La dissolved in a high-C iron melt can form the stable compound $LaSn_3$ (Savov, 1999). Nonetheless, even if detinning proves to be possible, the high price of rare earth elements and their losses by oxidation will hinder the development of this method.

19.9.3 Dezincing

Coating steel with Zn is used for corrosion protection and is widespread. The number of applications of galvanized steel is increasing as new applications of steel are being discovered. The list of these applications is quite vast, ranging from home appliances to building materials like roofing and cladding, bridges, cars, etc. During steelmaking, Zn evaporates (the boiling temperature of Zn is 906°C) and is collected in the exhaust filters together with other metallic and oxidic materials as dust. A typical melting operation of the EAF generates about 10–15 kg dust/ton of crude steel, and it contains 15–25% Zn (Janjua, 2008).

The scrap dezincing techniques can be classified as:

- Leaching with electrolysis,
- Formation of Fe_3Zn_{10} phase or Fe-Zn alloy, and
- Evaporation/gasification of Zn.

Hudson and Brandiff, 1975 disclosed a process for the recovery of metallic Zn from galvanized steel scrap by leaching with concentrated sulfuric acid. Hissel (1985) described a dezincing process incorporating electrolytic cells using Hg and an alkaline solution. Leeker and Neidringhaus (1992) and Niedringhaus et al. (1992) showed the possibility of using a solution of caustic soda as the electrolyte for dezincing galvanized scrap. An advantage of this approach is that the Fe layer is stable in a caustic solution, and as a result, the Zn/Fe separation after treatment is not a major problem. LeRoy and Houlachi (1994) and LeRoy and Janjua (1994) described another dezincing method, where galvanized steel scrap is immersed in a caustic electrolyte and electrically connected to a cathodic material from Ni- or Co-based materials. Wijenberg et al. (1997) and Alonzo et al. (2002) investigated the dissolution of Zn from steel scrap in alkaline solutions. Morgan et al. (1998) and Morgan (1999, 2001) presented different variations of a dezincing process in which galvanized steel is immersed in a hot electrolytic solution of Na or K hydroxide and Zn is galvanically corroded.

Formation of a Fe_3Zn_{10} phase occurred when shredded scrap was baked in a heat treatment furnace to convert most of the Zn into a brittle Fe_3Zn_{10} phase that could be removed by shot-blasting (Fujio and Kazuhiro, 1992). They reported a Zn removal rate of 85% after baking at 750°C and subsequent shot blasting for 5 min. Franzen and Pluschkell (1999) performed similar experiments by annealing Zn-coated samples between 560–620°C for a holding time of 300–1000 s. The delta phase Fe-Zn alloy layer was created and removed by mechanical bending.

Okada and Fujio, 1994 offered a method of recycling plated metal—such as Zn or Al—from shredded scrap having nonmetallic contaminants such as paint, oil, and plastics from car shredders. Thereby, the scrap is purified in two steps: first evaporating the organic material at 200–500°C under reduced pressure and subsequently evaporating the metallic coating at 500–950°C under reduced pressure. Saotome et al., 1996 published similar experiments using a vacuum for the removal of Zn from automotive process scraps. In their study, evaporation was conducted between 700°C and 1200°C in a vacuum, and the reported purity of Zn was 99.86%. Ozturk and Fruehan, 1996 conducted experiments on Zn vaporization from galvanized scrap in N, air, CO, and CO_2 atmospheres. Tee and Fray, 1999 performed lab-scale tests for Zn removal from electrogalvanized and Fe-Zn galvanized scrap using air and chlorine mixtures. Franzen and Pluschkell, 2001 investigated Zn evaporation from scrap in Ar, N_2, and He and found that Zn could be evaporated effectively by holding the galvanized samples at 600°C for 5 min.

Numerous methods for removing Zn from steel scrap have been investigated over the years, but very few have been commercialized. To incorporate new methods commercially, they must be cost-competitive with the conventional means of dealing with Zn in steel scrap. The low content of Zn in the dust generated from the primary ore-based steelmaking route makes it very costly for Zn recovery. Consequently, the strategy used by many steel producers has been to purchase only rather clean scrap as Fe carrier and coolant. The market scarcity of clean scrap in the future could make the integrated steel industry concerned about scrap rinsed from Zn, if the price is comparable to the cost of using ores as coolant.

19.10 OUTLOOK

Steel scrap recycling is an important example of recycling, and steel scrap processing is a global industry currently. Losses are inevitable within processing and handling of any material. For steel, the recycling is limited due to by-product streams and waste in steelmaking (slag, dust, sludge), corrosion, and dissipation/waste by using steel products and handling steel scrap. These losses in production and handling can and will be decreased as technology develops and awareness of the recycling necessity grows among all the stakeholders involved. Corrosion losses can be reduced through the development of new steels with tailor-made properties and utilization of various protective coatings. Dissipation/waste losses occurring during use can be reduced through a responsible attitude of consumers towards resources and the environment.

The long-term sustainability and recycling in the steel and scrap processing industries primarily depend on scrap quality. The ideal steel scrap of steelmakers must be "black," meaning an absence of any metallic and organic coverings/coatings, attachments, and tramp elements. The steel scrap must have a bulk weight of not less than $1\,t/m^3$ and be sorted according to the alloy content. In reality, the steel scrap contains different metallic impurities such as tramp and alloying elements, surface layer impurities, and sand and dirt (gangue). The current gangue content of the steel scrap is 1.5–5 wt.%. Further increase in the share of electric steelmaking will promote an increase in scrap utilization and life cycle, leading to the accumulation of tramp elements in the scrap, especially metals with low affinity to oxygen, such as Cu, Sn, Ni, and Mo. Several strategies can be involved to minimize the detrimental effects of tramp elements. This includes scrap dilution with ore-based Fe materials, such as DRI, HBI, or "pure" scrap grades; processing improvement at steel plants; sorting improvement at shredder plants; design for recycling; and tolerance to impurities based on the development of new alloys.

The dilution of steel melt with "pure" Fe carriers is the most common method currently applied in steelmaking, though that often means that high purity scrap and primary materials are used to produce an alloy of lesser purity. Steelmakers carefully grade and regularly check the quality of the steel scrap. The number of grades of steel scrap at a modern steel plant can reach several hundred. Thanks to digitalization and automation (Level 2), the operating models can calculate the optimal mix of different scrap grades in a fraction of a second. Herewith, the impurities in the steel scrap can serve as alloying elements. The negative effects of tramp elements can also be reduced through process improvement. For example, a decreased secondary cooling rate in continuous casting of Cu-containing construction steels appeared to decrease the crack depth (Utaise et al., 1996); whereas thin strip casting units (CSP, ISP, and others) can maintain low levels of Cu and Sn segregation due to rapid crystallization.

Steel is one of the most used and popular materials due to its wide range of properties, low cost, and almost complete recyclability. To cover the needs of society for advanced industries, steels with tailor-made properties have been and are being further developed. This leads to a continued appearance of steels with completely different and incompatible chemical compositions. Chemical elements that are useful in some steels for certain properties can be detrimental in other steels. Thousands of different low-alloyed, alloyed, and high-alloyed grades of steel are available and can be combined in different applications (e.g., in the construction, automotive, and transportation industries). This existing and further expanding chemical diversity already complicates steel scrap recycling, and will do so even more in the future. Thus improved understanding of steel flows based on the steel grade level is necessary to model and control the possible accumulation of tramp and impurity elements, which is essential for the sustainable economic and technological development.

References

Alonzo, V., Darchen, A., Hauchard, D., Paofai, S., 2002. Acceleration of zinc corrosion in alkaline suspensions containing iron oxides or iron hydroxides. Electrochim. Acta 48, 951–955.

Bernardini, G., Corsini, F., Mazzetti, G., Trosti-Ferroni, R., 1982. Phase relations in the CuFeSe system at 300°C. Mater. Res. Bull. 17 (8), 981–991.

Bleck, W., Moeller, E., 2018. Handbuch Stahl. Auswahl, Verarbeitung, Anwendung. Carl Hanser Verlag München, ISBN: 978-3-446-44961-9.

Branca, T.A., Colla, V., Algermissen, D., Granbom, H., Martini, U., Morillon, A., Pietruck, R., Rsendahl, S., 2020. Reuse and recycling of by-products in the steel sector: recent achievements paving the way to circular economy and industrial symbiosis in Europe. Metals 10, 345. https://doi.org/10.3390/met10030345.

Bundesvereinigung Deutscher Stahlrecycling- und Entsorgungsunternehmen e.V, June 1995. European Steel Scrap Specification.

Bundesvereinigung Deutscher Stahlrecycling- und Entsorgungsunternehmen e.V, Willeke, R., Stahlrecycling, F., 1998. Vom Rohstoff Schrott zum Stahl (German). Reed Elsevier Deutschland.

Cochrane, B., 2000. Alloyed success for steel recycling. Mater. World 8 (7), 17–19.

DIN EN 10020 (2000-07-00), 2000. Definition and Classification of Grades of Steel.

Franzen, A., Pluschkell, W., 1999. Removal of zinc layers from coated steel strip by thermal and mechanical processing. Steel Res. 70 (4-5), 141–146.

Franzen, A., Pluschkell, W., 2001. Low Temperature Investigations on the Reaction in the System Fe/Zn. European Commission, Technical Steel Research. Primary Steelmaking, Report EUR 19468 EN.

Fritz, E., Gebert, W., 2005. Milestones and challenges in oxygen steelmaking. Can. Metall. Q. 44 (2), 249–260.

Fujio, S., Kazuhiro, S., 1992. Removal of zinc by baking of automotive body scrap. In: Proceedings of the 75th Steelmaking Conference, Toronto, pp. 749–756.

Ghosh, D., 2009. Removal of tin and copper from liquid iron by Al_2O_3-saturated Ca-$CaCl_2$ slags at 1448 to 1648 K. Metall. Mater. Trans. B Process Metall. Mater. Process. Sci. 40, 508–522.

Glazov, M., Pashinkin, A.S., Fedorov, V.A., 2000. Phase equilibria in the Cu-Se system. Inorg. Mater. 36 (7), 641–652.

Grummes, F., 2019. Rapider Technologiewechsel in der Volksrepublik? Stahl Eisen 139 (7–8), 66.

Guthrie, R.I.L., 1992. Engineering in Process Metallurgy. Oxford Science Publications. 548 pp.

Hissel, J.A., 1985. Process and apparatus for the recuperation of zinc from industrial wastes containing zinc. CA Patent No. 1193996.

Hu, X., Yan, Z., Jiang, P., Zhu, L., Chou, K., Matsuura, H., 2013. Removal of copper from molten steel using FeO-SiO_2-$CaCl_2$ flux. ISIJ Int. 53, 920–922.

Hudson, H.G., Brandiff, R., 1975. Electrolytic zinc salvaging method. US Pat. No. 3905882.

Jahrbuch, 1992. Jahrbuch Stahl 1992, German. Verlag Stahleisen, Düsseldorf, Germany. Nov. 1991.

Jahrbuch, 2020. Statistisches Jahrbuch der Stahlindustrie 2019/2020. Wirtschaftsvereinigung Stahl.

Jalkanen, H., Holappa, L., 2014. Converter steelmaking. In: Seetharaman, S. (Ed.), Treatise on Process Metallurgy, Industrial Processes, Part A. vol. 3. Elsevier, ISBN: 978-0-08-096988-6, pp. 223–270.

Janjua, R., 2008. Optimisation of Electric Arc Furnace Dust Recycling and Zinc Recovery by Scrap De-zincing. Doctor Thesis, TU Bergakademie Freiberg.

Jimbo, I., Sulksy, M.S., Fruehan, R.J., 1988. The refining of copper from ferrous scrap. I&SM 15, 20–23.

Jung, S.H., 2014. Reaction Mechanism and Kinetics of Evaporation of Cu and Sn from Liquid Steel. PhD Thesis, Pohang University of Science and Technology (POSTECH), KR.

Kitamura, K., Takenouchi, T., Iwanami, Y., 1985. Removal of impurities from molten steel by CaC_2. Tetsu-to-Hagané 71, 220–227.

KrWG, October 2020. Gesetz zur Förderung der Kreislaufwirtschaft und Sicherung der umweltverträglichen Bewirtschaftung von Abfällen – Kreislaufwirtschaftsgesetz (German).

Łabaj, J., 2012. Kinetics of coper evaporation from the Fe-Cu alloys under reduced pressure. Arch. Metall. Mater. 57, 165–172.

Leeker, J.W., Neidringhaus, J., 1992. Alkaline leaching of galvanized steel scrap. US Pat. No. 5106467.

LeRoy, R.L., Houlachi, G., 1994. Power assisted dezincing of galvanized steel. US Pat. No. 5302261.

LeRoy, R.L., Janjua, M.B., 1994. Galvanic dezincing of galvanized steel. US Pat. No. 5302260.

Leroy, V., Haeyer, R.D., Defourny, J., Hoogendoorn, T., 1995. Effects of Tramp Elements in Flat and Long Products. ECSC, EUR 16672EN, ISBN: 928275636-X.

Li, L., Xiang, X., Zhao, P., Wang, C., Tian, H., Li, S., 1998. Decopperization in steel melt through filtration. J. Iron Steel Res. 10 (3), 5–7.

Li, L., Li, S., Xiang, C., Cao, J., Liu, K., Li, S., 1999. Copper removal from molten steel by gasification. J. Iron Steel Res. Int. 6 (2), 10–13.

Liu, X., Jeffes, J.H.E., 1989. Decopperisation of molten steel by various slags. Ironmak. Steelmak. 16 (5), 331–334.

Luengen, H.B., 2020. Wege zur Minderung von CO2-Emissionen in der Stahl- und Eisenindustrie in Europa. Stahlinstitut VDEh, Duesseldorf.

Madias, J., 2014. Electric furnace steelmaking. In: Seetharaman, S. (Ed.), Treatise on Process Metallurgy, Industrial Processes, Part A. vol. 3. Elsevier, ISBN: 978-0-08-096988-6, pp. 271–301.

Main, 2017. EN 13306. Maintenance-Maintenance Terminology.

Matsuoka, H., Osawa, K., Ono, M., Ohmura, M., 1997. Influence of Cu and Sn on hot ductility of steels with various C content. ISIJ Int. 37, 255–262.

Merkel, T., 2020. Produktion und Nutzung von Eisenhüttenschlacke in Deutschland im Jahr 2019. Report FEhS Inst. 27 (1), 29–30.

Morgan, W.A., 1999. Process for dezincing galvanized steel. US Pat. No. 5855765.

Morgan, W.A., 2001. Process for dezincing galvanized steel using an electrically isolated conveyor. US Pat. No. 6258248.

Morgan, W.A., Dudek, F.J., Daniels, E.J., 1998. Process for dezincing galvanized steel. US Pat. No. 5779878.

Nagasaki, C., Kihara, J., 1997. Effect of copper and tin on hot ductility of ultra-low and 0.2% carbon steels. ISIJ Int. 37, 523–530.

Niedringhaus, J.C., Rodabaugh, R.D., Leeker, J.W., Streibick, A.E., 1992. A technical evaluation of de-zincification of galvanized steel scrap. In: Proceedings of the 75th Steelmaking Conference, Toronto, pp. 725–741.

Noro, K., Takeuchi, M., Mizukami, Y., 1997. Effects of Cu and other tramp elements on steel properties. Necessity of scrap reclamation technologies and present conditions of technical development. ISIJ Int. 37, 198–206.

Ochifuji, Y., Tsukihashi, F., Sano, N., 1995. The ACTIVITY OF CALCIUM IN CALCIUM-METAL-fIUORIDE FLUXES. Metall. Mater. Trans. B Process Metall. Mater. Process. Sci. 26, 789–794.

Okada, Y., Fujio, S., 1994. Method and apparatus for removing plated metal from steel sheet scraps. US Pat. No. 5350438.

Ozturk, B., Fruehan, R.J., 1996. Vaporization of zinc from steel scrap. ISIJ Int. 36, S239.

Püchert, H., 1996. Ein Ansatz zur strategischen Planung von Kreislaufwirtschaftssystemen. Deutscher Universitätsverlag, ISBN: 978-3-322-99540-7.

Rahimi, R., Ullrich, C., Klemm, V., Rafaja, D., De Cooman, B. C., Biermann, H., Mola, J., 2016. Influence of Al on the temperature dependence of strain hardening behavior and glide planarity in Fe-Cr-Ni-Mn-C austenitic stainless steels. Mater. Sci. Eng. A 649, 301–312.

Sandig, E.F., Chebykin, D., Prutchykova, V.V., Fabrychnaya, O., Volkova, O., 2019. Review: possibilities of steel scrap decopperization. MSF 959, 145–160.

Saotome, Y., Nakazawa, Y., Yokoyama, Y., 1996. Vacuum aided recycling systems technology (VARS Tech) as a restoration systems technology of Ecofactory. Vacuum 47 (6–8), 833–836.

Savov, L., 1999. Fundamental Study on the Evaporation of the solute Elements Copper and Tin from iron-based Melts at reduced Pressure. PhD Thesis, TU Bergakademie Freiberg.

Savov, L., Tu, S., Janke, D., 2000. Methods of increasing the rate of tin evaporation from iron-based melts. ISIJ Int. 40, 654–663.

Savov, L., Volkova, O., Janke, D., 2003. Copper and tin in steel scrap. RMZ-Mater. Geoenviron. 50, 627–640.

Sista, K.S., Dwarapudi, S., Nerune, V.P., 2019. Direct reduction recycling of mill scale through iron powder synthesis. ISIJ Int. 59, 787–794.

Sprecher, M., Luengen, H.B., Stranziger, B., Rosenmann, H., Wolfgang, A., 2019. Abwärmenutzungspotentiale in Anlagen integrierter Hüttenwerke der Stahlindustrie. Stahl Eisen 139, 27–33.

Stahlschluessel, 2019. Key to Steel 2019. Stahlschluessel Wegst GmbH, ISBN: 392-259-93-54.

Tee, J.K.S., Fray, D.J., 1999. Removing impurities from steel scrap using air and chlorine mixtures. J. Miner. Met. Mater. Soc. 51 (8), 24–27.

Tsukihashi, F., 2001. Physical chemistry of removal of tramp elements from steel. In: Proceedings of the 9th China-Japan Symposium on Science and Technology of Iron and Steel, Xian, China, 8–9 November 2001, pp. 59–66.

Tu, S.W., Janke, D., 1994. Fundamental metallurgical investigations on the recyclability of Sn- and Zn-coated steel scrap. In: Warlimont, H. (Ed.), Environmental Aspects in Materials Research. DGN Informationsgesellschaft mbH Verlag, ISBN: 3-88355-205-4, pp. 103–107.

Turkdogan, E.T., 1980. Physical Chemistry of High Temperature Technology. Academic Press, New York. 446 pp.

Utaise, K., Yotsuya, S., Tanahashi, A., 1996. Scrap situation and its effective application in Japanese EAF-steel industry. ISIJ Int. 36 (Supplement), S18–S21.

Wang, C., Nagasaka, T., Hino, M., Ban-Ya, S., 1991b. Copper distribution between FeS-alkaline or -alkaline earth metal sulfide fluxes and carbon saturated iron melt. ISIJ Int. 31, 1309–1315.

Wang, J., Guo, S., Zhou, L., Li, Q., 2009. Slag for decopperization and sulphur control in molten steel. J. Iron Steel Res. Int. 16 (2), 17–21.

Wendler, M., Ullrich, C., Hausera, M., Krüger, L., Volkova, O., Weißa, A., Mola, J., 2017. Quenching and partitioning (Q&P) processing of fully austenitic stainless steels. Acta Mater. 2017, 346–355.

Wieliczko, A.G., Griszczenko, J.N., Krucinski, M., 1998. Technological efficiency of liquid steel decopperizing. Hutnik Wiadomosci. Hutnicze 65 (7), 256–259.

Wijenberg, J.H.O., Stevels, J.T., De Wit, J.H.W., 1997. Galvanic coupling of zinc to steel in alkaline solutions. Electrochim. Acta 43, 649–657.

Zaitsev, A.I., Zaitseva, N.E., Shakhpazov, E.K., Mogutnov, B. M., 2004a. Evaporation of copper from iron melts. ISIJ Int. 44, 639–646.

Zaitsev, A.I., Zaitseva, N.E., Shakhpazov, E.K., Mogutnov, B. M., 2004b. Potentialities of simultaneous removal of tin and copper from molten iron through evaporation. ISIJ Int. 44, 957–964.

Zhang, L., 2009. Influence of residual elements in steel on the quality of CSP products. Met. Mater. Metall. Eng. 37, 12–15.

Zhang, X., Ma, G., Liu, M., Zhi, L., 2019. Removal of residual tin in the ferrous metallurgy process: a review. Metals 9, 834. https://doi.org/10.3390/nn9080834.

Zhao, B., Wang, J., Du, Y., Li, J., 1995. Effect of As, Sn and Sb on temper brittleness of 30CrMnSiA steel. Trans. Mater. Heat Treat. 16, 13–17.

Zhou, S., Guo, Y., Shen, G., 1998. Control critical range of P and Sn of Cr-Mo-V series gun steel. Ordnance Mater. Sci. Eng., 13–19.

Zigalo, I., Baptizmanskij, Y., Vyatkin, Y., Velichko, A., Shakhpazov, E., Grishchenko, Y., 1991. Copper in steel and problems of removing it. Steel USSR 21 (7), 299–302.

Further Reading

Matsuo, T., 2000. Acceleration of copper and tin removal from molten steel by decarburization under reduced pressure. Tetsu-to-Hagane 86 (11), 33–39.

Preßlinger, H., Hibler, H., 1984. Entkupferung von Roheisenschmelzen mit Sulfatschlacken. BHM 129 (9), 333–338.

Savov, L., Janke, D., 2000. Evaporation of Cu and Sn from induction-stirred iron-based melts treated at reduced pressure. ISIJ Int. 40, 95–104.

Uchida, Y., Matsui, A., Kishimoto, Y., Miki, A., 2015. Fundamental investigation on removal of copper from molten iron with Na_2CO_3-FeS fluxes. ISIJ Int. 55, 1549–1557.

Wang, C., Nagasaka, T., Hino, M., Ban-Ya, S., 1991a. Copper distribution between molten FeS-$NaS_{0.5}$ flux and carbon saturated iron melt. ISIJ Int. 31, 1300–1308.

Xiang, C., Li, L., Wang, C., Tian, H., Li, S., 1997. Decopperization in steel melt with compound decopperizing agents of oxides. J. Sci. Technol. Beijing 19, 538–541.

Yamaguchi, K., Ono, H., Usui, T., 2010. Oxidation removal of Cu from carbon saturated iron via Ag phase. Tetsu-to-Hagane 96, 531–535.

CHAPTER

20

Aluminum

Don Doutre[1,*] *and Anne Kvithyld*[2]

[1]**Novelis Inc., Kingston, Canada** [2]**SINTEF, Trondheim, Norway**

20.1 INTRODUCTION

Aluminum is lightweight, strong, formable, corrosion resistant, and an excellent conductor of both heat and electricity. During the twentieth century these properties contributed to its widespread application in a range of industries including aeronautic, construction, electrification, transport, and packaging (Figure 20.1). Most aluminum is used in relatively long-term applications, and it is estimated that 75% of the aluminum ever produced is still in use in one form or another (Bertram et al., 2017). Since 1970, world production has increased exponentially at a rate of 4.0% p.a., a trend that is expected to continue. The production of secondary (recycled) aluminum has grown even faster at 5.3% p.a. It now represents roughly one-third of the total annual production.

The recycling of aluminum is beneficial in that it avoids most of the energy, emissions, and solid wastes associated with primary production from bauxite ores. Aluminum scrap is also valuable, helping to justify collection and processing from a variety of waste streams.

The in-use inventory of aluminum by application area (Figure 20.1B) represents the resource base for the secondary aluminum industry. The origin of the collected end-of-life scrap inputs to the secondary aluminum industry (Figure 20.1C) is different from the in-use inventory. Although packaging represents only ~1% of the resource base, it is the second largest secondary input due to its short lifetime. Cans have a turnaround time of 2–3 months, compared to architectural facades, which may last 50 or more years.

This chapter discusses the major issues associated with end-of-life aluminum scrap recycling, starting with a description of the limited intercompatibility of the main alloy families and blending and batching strategies. Oxidative losses to dross, melting systems, and dross are outlined. Two examples of product recycling are described: used beverage cans (UBCs) and automotive wheels. Possible purification/refining techniques are discussed. The chapter closes with some observations on current challenges and future trends.

* Senior Principal Scientist (retired), Novelis Inc.

 https://doi.org/10.1016/B978-0-323-85514-3.00003-8

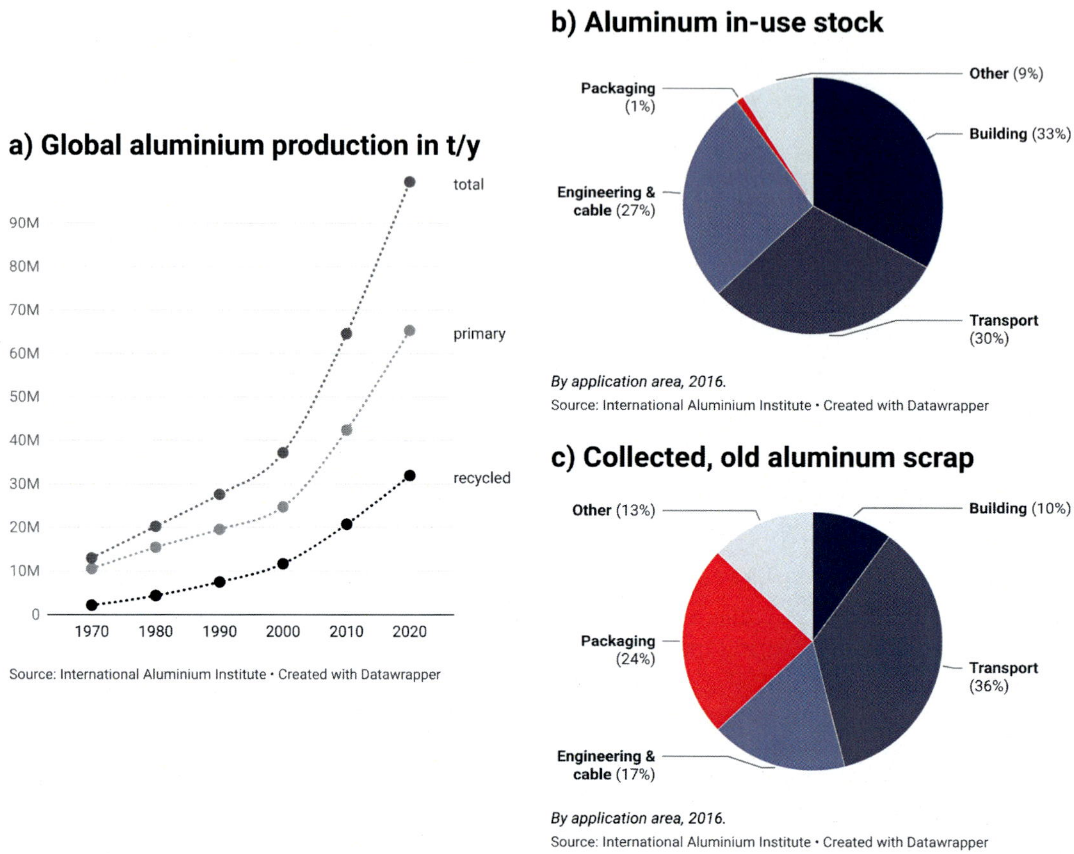

FIGURE 20.1 (A) Global aluminum production from primary and recycled material. (B) In-use stock by application in 2016, and (C) collected old scrap by application area in 2016. *From International Aluminium Institute (2016, 2022).*

20.2 ALLOYS AND THEIR RECYCLING

Many of the scrap inputs are available in sizable quantities and relatively consistent size and shape. Consequently, the value of old scrap from these sources is relatively high, which drives recycling. This contributes to collection rates of over 90% for aluminum scrap from automotive and construction sources (Natural Resources Canada, 2022). In contrast, less than 50% of aluminum from packaging applications is reclaimed as there are often other materials associated with the aluminum either by design (e.g., laminates) or co-mingled during collection, necessitating pretreatment. The Institute of Scrap Recycling Industries, for example, maintains a list of specifications for different aluminum scrap grades covering composition, degree of compaction, and permissible levels of contaminants such as water and lubricants (ISRI, 2021). Details and examples can be found in Aluminum Recycling (Schlesinger, 2014).

Except for electrical applications, aluminum is seldom used in pure form. The Aluminum Association, representing the industry, categorizes alloys into two families: cast and wrought (Table 20.1), which are further subdivided according to their major alloying elements. Each

TABLE 20.1 Wrought alloy families, major alloying element(s), and typical applications.

AA alloy designation	Alloying element(s)	Application
1xxx		Electrical cable, reflectors
2xxx	Cu	Aerospace and defense
3xxx	Mn (+/− Mg)	Building and transportation (e.g., soffits, truck siding) +1% Mg: beverage can bodies
4xxx	Si	Brazed heat exchangers, welding wire
5xxx	Mg	Medium-strength sheet, can ends, automotive inners. Transportation (marine applications, rail cars, and truck tanks)
6xxx	Mg and Si	Extrusion profiles and automotive sheet
7xxx	Zn- (Mg/Cu/ Zr)	High-strength alloy for aerospace, bumpers
8xxx	Others: e.g., Fe, Li, ...	Foil (Fe), aerospace (Al-Li)

has its own structure, mechanical properties, etc., making it suitable for a specific application (Davis, 2001).

Cast alloys are identified using a different 3-digit system. The 3xx series casting alloys are by far the most common, as they can be used in a myriad of applications. Silicon is present to improve castability and is generally present at levels of 5% or more. In contrast, silicon levels seldom exceed 1% in any of the wrought alloy systems. Iron contamination represents a major impediment to recycling, as even at low levels it can seriously reduce ductility. Iron dissolves readily in molten aluminum and is virtually impossible to remove once present.

The Aluminum Association lists over 400 unique compositional specifications for wrought alloys and about 250 for casting alloys (Kammer, 1999). At first sight this diversity appears to be a great enabler for recycling. In practice this is not so straightforward. The majority of the casting alloys contain high levels of silicon, an element that is generally present at much lower levels in wrought alloys. Consequently, wrought alloys can be recycled back into wrought alloys or downgraded and serve as an input to casting alloys, whereas casting alloys are not commonly recycled into wrought products. Figure 20.2 illustrates this general progression of alloys through their life cycles, cascading from wrought to cast alloys to their final destination, material that does not meet any specification: deoxidation of steel (Chapter 19—Steel).

This cascade is driven largely by the accumulation of impurities, principally iron and silicon, which originate from imperfect scrap liberation and separation (both aluminum alloys from other metals and different aluminum alloys from each other), adhering sand and rust particles, and reaction/dissolution of the iron tools and equipment used during handling and processing of scrap.

The final useful cast alloy, before becoming deoxidation material, is often one of the 380 series die casting alloys. These have very wide compositional limits. Alloy C380 is an aluminum—silicon (7.5–9.5%)—copper (3.0–4.0%) alloy that finds a myriad of applications ranging from escalator treads to barbecue covers,

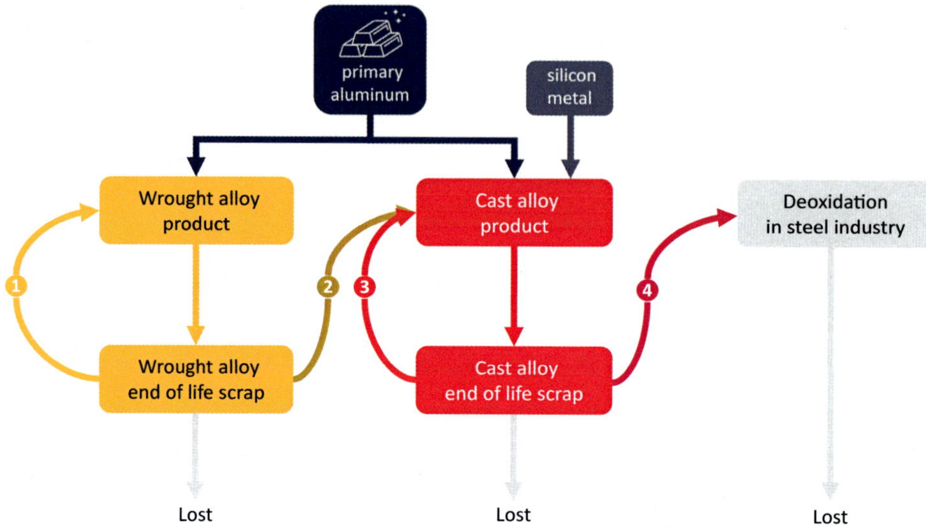

FIGURE 20.2 Progression of aluminum alloys in recycling. ① Recycling to wrought alloy; ② wrought alloy recycled into cast alloy; ③ recycling to cast alloy; ④ final use as deoxidant. *Modified from Tabereaux and Peterson (2014).*

automotive transmission casings, and other components of gasoline engines. The other alloying elements have simply maximum values: ≤1% Fe, ≤0.5% Mn, ≤0.5% Ni, ≤2.9% Zn, ≤0.35% Sn, ≤0.5% others (total). Interestingly, alloy 380, along with alloys 319 and 333 with their wide tolerances for Cu, Zn, and Mn, were specifically developed to absorb the large amount of scrap from aluminum aircraft during and immediately after the Second World War (Rooy, 2003).

In order to meet an alloy specification, several strategies are available:

1. *Avoiding* contamination and/or removing contaminants during the collection and preparation of the scrap. Basically, applying shredding/sorting and separation techniques, including decoating.
2. *Blending* of scrap materials from different sources. This can be challenging but can be done, particularly if the composition of available materials is uniform and compatible. Wrought alloys can, in general, be used as inputs for casting alloys, while opportunities for blending, even within the wrought alloys, are generally limited (Peterson, 2015).
3. *Dilution* by adding primary metal. If suitable scraps for blending are unavailable, dilution provides the only practical means of absorbing some forms of scrap. Although inconsistent with the theme of recycling, dilution with primary metal can make sense, for example diluting the excess magnesium rather than removing it using chlorination.
4. *Refining*. Most elements, once dissolved in aluminum, are very difficult and costly to remove.

Usually, a combination of strategies is deployed. Hannula et al. illustrate the interconnectedness between scrap characteristics (particle size and form, alloy composition), sorting and separation pathways, and target alloy compositions (Hannula et al., 2020). The impacts of the avoiding, blending, diluting strategies are quantified.

Two on-line videos demonstrate how the avoidance recycling strategy is performed in practice (Norsk Hydro, 2015; Discovery, 2018).

20.3 MELT LOSS

Molten aluminum has a very high affinity for oxygen and reacts instantaneously to form aluminum oxide films on all exposed surfaces. The thickness of the oxide layer also increases with both time and temperature. Dross that forms during melting usually entraps metallic aluminum that is lost from the process when the furnace is skimmed. Rossel's experiments clearly demonstrated that dross formation increases with magnesium concentration and for thinner gauge materials (Figure 20.3) (Rossel, 1990).

Melt losses for 2 mm thick sheet were between 1% and 4% and increased to between 5% and 15% for 0.2 mm thick sheet. As a point of reference, the thickness of an aluminum can varies between 0.1 and 0.3 mm, whereas a vehicle automotive closure sheet is on the order of 2 mm in thickness. Melt losses for the thicker gauge materials (20 and 200 mm) were well below 1%. This behavior by and large dictates the selection of furnaces for different types of scrap inputs. Other key considerations in the choice of melting technology are the level and nature of contaminants and coatings, the moisture level, and the bulk density of the scrap (e.g., compressed vs loose turnings).

20.3.1 Direct Fired Dry Hearth Furnace

Also simply called "reverb" furnaces, these are widely used to melt thicker gauge scrap and large pieces such as sows/ingots weighing hundreds to, in some cases, thousands of kilograms. Depending on the scale of the operation, they may be top charged, side charged, or both. In all cases the scraps must be free of moisture to avoid the possibility of steam explosions. This furnace design is not well suited for coated or otherwise contaminated scrap.

Hot gases from the burners pass over the charge and transfer heat by both convection and radiation. Heat recovery systems are very common and use the residual heat in the combustion products to preheat the incoming combustion air. Oxygen enrichment is sometimes used to increase flame temperatures and melt

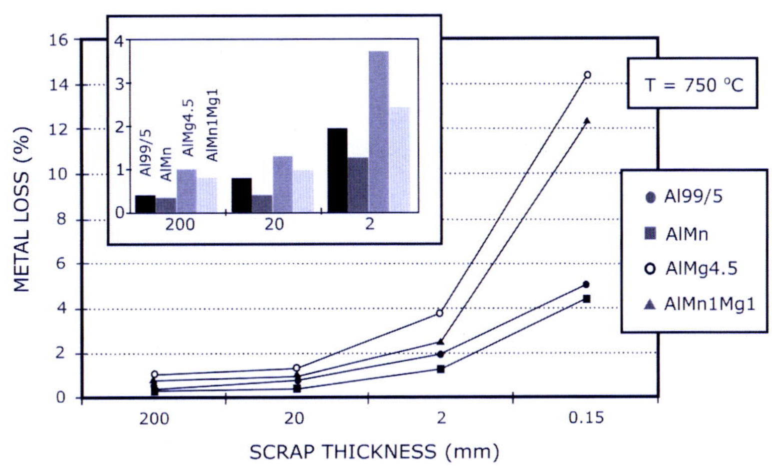

FIGURE 20.3 Metal loss (dross formation) with respect to scrap thickness. *From Rossel (1990).*

rate. After melting, the furnace is skimmed to remove the dross, which is typically 2–4% of the charge weight and typically contains between 50% and 90% metal.

20.3.2 Multichamber Furnace

As the name suggests, these furnaces have two or more interconnected chambers (Figure 20.4a). Scrap is placed on a sloping hearth and heated by drawing a portion of the hot combustion gases from the main hearth over and through the charge. As the charge heats, the organic contaminants are broken down and burned in the main hearth to provide some of the energy required for melting. Once the charge has been sufficiently preheated, it is pushed into the molten metal, which is moved from the main hearth using an electromagnetic pump. These furnaces are more accurately described as integrated melting systems. They require dedicated charging equipment and very sophisticated control systems, and consequently can handle a variety of scrap inputs.

20.3.3 Sidewell Melter

A sidewell melter is simply a conventional reverb furnace equipped with a large open well that connects with the main hearth through a submerged weir or arches. During operation hot molten metal is drawn from the main hearth into the sidewell where scrap is being added and the now cooled metal returned to the main hearth. Separating the charging and melting areas in this way has several advantages, including:

- No exposure of the charge to hot combustion gases. This improves recovery when melting thin gauge material.
- Salt fluxes can be used to remove oxides and other contaminants.

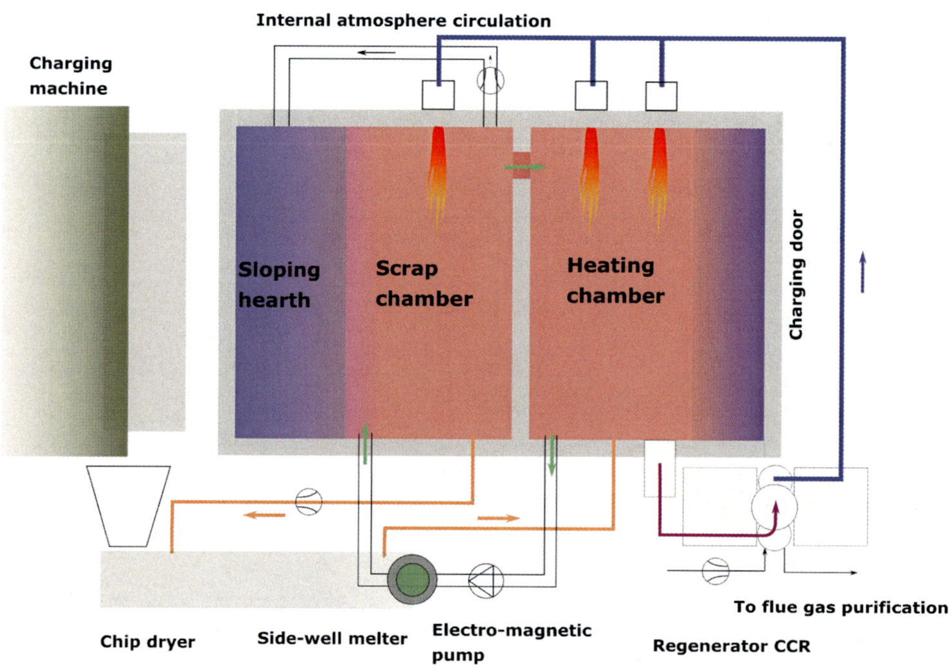

FIGURE 20.4A Schematic cross section of a multichamber furnace as viewed from the top.

- Low density materials such as decoated UBCs can be conveyed and fed continuously to the sidewell.
- Burners can operate continuously as there is no need to open doors for charging.

These furnaces tend to be large and, to operate efficiently, the arches have to remain submerged. Generally, only a portion of the total melt is transferred to a holding furnace during each cycle.

20.3.4 Tilting Rotary Furnace

The tilting rotary furnace (Figure 20.4b) has emerged as the technology of choice for aluminum dross processing and is also widely used for scrap melting. These furnaces typically are equipped with door-mounted air or oxy-fuel burners that swing open to allow access for charging and emptying. During operation the furnace rotates slowly (3–5 rpm), gently agitating the charge material and continuously bringing the hot furnace refractory beneath the charge to enhance heating. Scrap may be charged along with dross to increase the metal recovery from the dross. When processing furnace drosses, a salt mixture (typically NaCl/KCl) is added to promote metal coalescence while collecting the oxides as a crumbly solid known as salt cake. Prompt metal recoveries are typically 90% with about 5% of the metal lost through oxide formation and 5% entrained in the salt cake. This salt cake is in some cases further processed by grinding to recover entrapped metal. Zhou et al. (2005) and Boin et al. (2004) have described and modeled the influence of operating parameters on mass and energy balances in an industrial rotary furnace (Figure 20.5).

20.3.5 Molten Metal Treatment

Regardless of the melting technology used, molten metal generally undergoes further treatment prior to use in order to remove/control impurities such as nonmetallic inclusions, dissolved hydrogen gas, and traces of alkali metals. Note that other metallic elements are not removed. In many plants this involves transferring the molten metal to a separate holding furnace or to ladles for distribution (Engh et al., 2021).

Hydrogen gas is much more soluble in liquid than solid aluminum and will form gas bubbles during solidification. Although the solubility of hydrogen is only 0.9 ppm by weight, this represents about 2.5% by volume. Hydrogen is

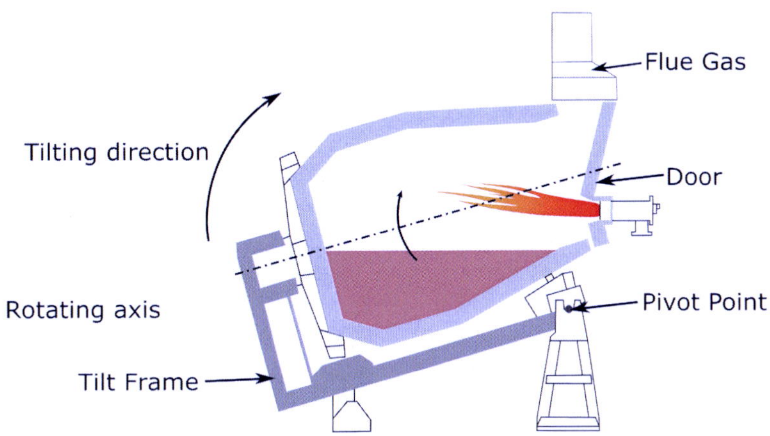

FIGURE 20.4B Schematic cross section of a tiltable rotary furnace. *From Schmitz (2014).*

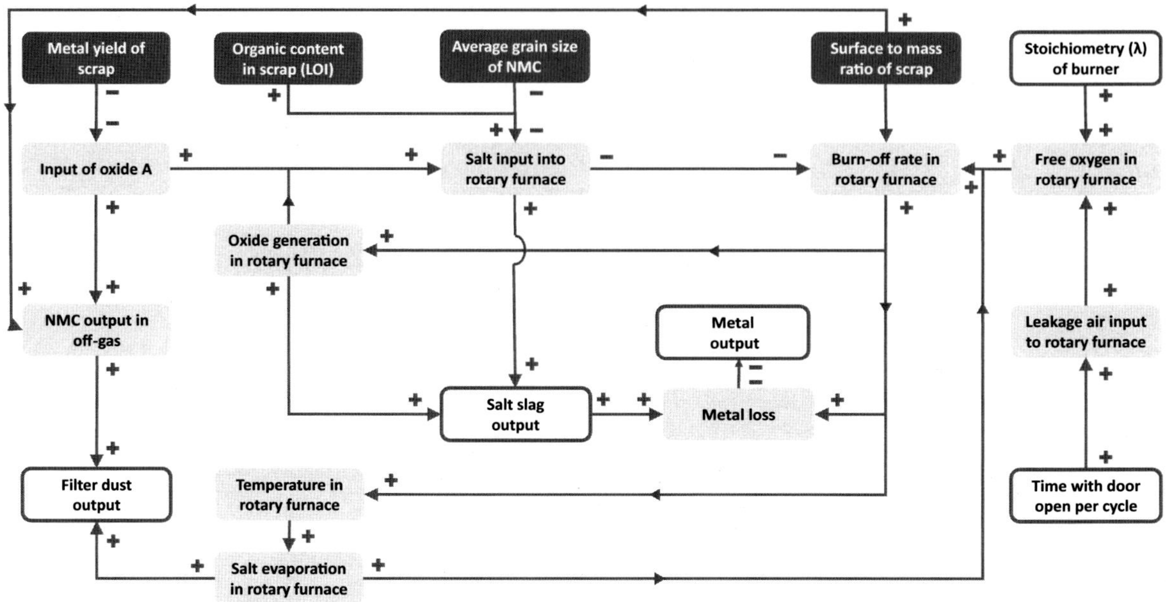

FIGURE 20.5 Influence of operating parameters on burn-off rate and resulting metal yield. *From Boin et al. (2004).*

removed by bubbling an inert gas (Ar or N_2) through the molten metal (Sigworth and Began, 2012). This can be done either as a batch process carried out in a furnace or transfer ladle, or continuously during DC (direct chill casting) of large ingots. In many instances a small addition of either chlorine gas (1–2%) or a fluxing salt ($MgCl_2$) is added to also promote inclusion removal and to reduce the amount of dross generated during the treatment. In the case of DC casting, metal may also be passed through a filter in order to further reduce the inclusion content (Sigworth, 2014). A wide variety of metal treatment technologies have been developed such as the spinning nozzle inert flotation (SNIF), aluminum compact degasser (ACD), rotary flux injector (RFI), siphon inert reactor (SIR), ceramic foam filters (CFFs), and deep bed filters (DBFs) (Neff et al., 2018). A description and discussion of the principles of these technologies can be found in Engh et al. (2021).

Arc-spark optical emission spectroscopy (OES) is used to measure/monitor chemical composition off-line and LIBS-based systems have been developed for on-line/real-time measurements (Leosson et al., 2022). A wide variety of techniques, both on and off-line, have been developed for measuring nonmetallic inclusions and dissolved hydrogen: porous disk filtration apparatus (PoDFA), Prefil, liquid metal cleanliness analyzer (LiMCA), Batscan, Alscan, Hycal, and reduced pressure test (RPT) (Le Brun, 2008).

20.4 USED BEVERAGE CAN (UBC) RECYCLING

20.4.1 Can Design/Alloy Selection

Since its first appearance in the 1960s, the drawn and ironed, two-part beverage can has become one of the dominant applications for rolled aluminum products and one of the most familiar and identifiable aluminum objects used by consumers.

Recycling of UBCs involves almost all of the sorting and cleaning operations commonly used with any aluminum scrap stream. Recycling UBCs presents some unique challenges because the material is relatively thin, is painted/coated, and consists of two mechanically intertwined wrought alloys. The can body is a manganese iron alloy containing about 1% magnesium while the end is a relatively pure aluminum with high magnesium content (~4.5%) (Table 20.2). Note that the can body itself is "scrap friendly" as the permissible levels of iron, copper, and silicon are relatively high. The allowable limits for these elements in the can end are considerably lower. In effect, the can end provides "sweetener" to help keep these elements within (body alloy) specifications by compensating for the inevitable iron and silicon pickup that occurs due to the presence of contaminants in shredded UBCs.

Note the very large difference (0.6%) between the actual and predicted magnesium levels after remelting (Table 20.2). This is the result of preferential attrition and oxidation of the can ends during the decoating process and to Mg losses upon melting and holding. In contrast iron and copper levels increased due to dissolution of ferrous contaminants (e.g., banding wire) and reduction of silica (from glass bottles).

TABLE 20.2 Composition of the beverage can body stock (CBS) and can end stock (CES), the calculated composition of a scrapped can (80 wt% CBS + 20 wt% CES), and the actual alloy composition of remelted shredded UBCs.

	Body	End	0.8 CBS + 0.2 CES	Remelted shreds
Mg	1.20	4.55	1.87	1.29
Mn	0.87	0.40	0.78	0.84
Fe	0.44	0.22	0.40	0.52
Si	0.24	0.08	0.21	0.22
Cu	0.14	0.04	0.12	0.16

20.4.2 Collection, Shredding, and Sorting

Preparation and melting of UBCs involves several steps (Figure 20.6). Incoming bales of UBCs arrive by truck and are inspected/tested for excess moisture or contaminants. The bales are then broken apart and conveyed past a magnetic separator to remove any steel cans, (steel) banding and other contaminants. Air knives are used at most belt transfer points to separate loose plastics as well as "heavies" such as red metals and lead. Trommel screens remove loose dirt and other fine contaminants. The loose UBCs are then shredded using a knife or a hammer shedder and passed over vibrating screens, separating "cold fines" from the now shredded cans. Optical (near IR) sorting is sometimes used to remove plastics from heavily contaminated waste streams. The goal is to reduce the organic levels as much as possible prior to the decoating step.

20.4.3 Decoating

Prior to melting, UBCs undergo a thermal treatment in order to remove the paints and lacquers used to decorate the outside and protect the inside of the can from contact with the beverage. These coatings represent about 3% of the total can weight. If remelted directly, the coatings would release large volumes of pyrolysis products, leading to both reduced metal recovery and emissions to the atmosphere.

Decoating is typically achieved by contacting the loose shred with preheated air, most commonly in a kiln, at 550–570°C. Decoating is a complex process where a balance needs to be found between removal of the organic coatings and avoiding oxidation of the aluminum and magnesium present in the alloy (Kvithyld et al., 2008; Meskers et al., 2008). Fines, consisting primarily of silica glass, aluminum dust, carbon, and inorganic pigments, are removed from the hot material as it exits the kiln and the hot shreds are then directly conveyed to the melting furnaces.

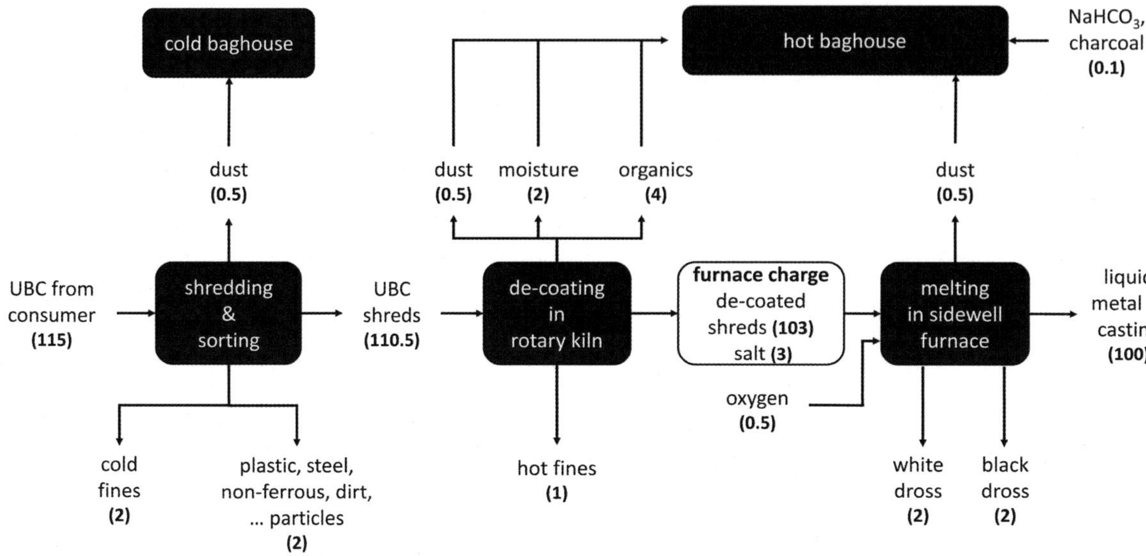

FIGURE 20.6 Idealized mass balance of UBC processing.

20.4.4 Remelting

During melting, salt is added to remove any remaining contaminants and promote metal coalescence, generating a salt-rich black dross (Van Linden and Stewart, 1989; Peterson, 1990; Roy and Sahai, 1997; Utigard et al., 1998). White dross forms in the main hearth because of flame exposure and melting of other solid inputs such as RSI (remelt secondary ingot) or alternate scraps.

From a typical mass balance (Figure 20.6), it can be seen that reclaiming aluminum from UBCs generates considerable quantities of by-products and wastes, approximately 15% of the incoming weight. Some of these contain enough metallic aluminum to justify further processing (e.g., furnace drosses and cold fines). Others are more problematic. The hot fines contain large quantities of soda-lime glass along with finely divided Mg-rich powder from the preferential high-temperature attrition of the can ends. Lead particles that make it past the cold air knife separators melt and oxidize during decoating and report to the hot fines as well. Overall recovery of the metallic aluminum present in the initial UBC is in excess of 95%.

20.4.5 Challenges

Challenges in UBC recycling include:

- Compositional control. The availability and nature of the inputs vary continuously. This can be accomplished by operating multiple, large capacity melter/holder furnaces and adjusting inputs to each, to be able to blend molten metal of differing compositions upon transfer to the casting furnace(s).
- Combustible dust and emissions control. While cleaning the UBC scrap, the capture and control of combustible dust accumulation at both the cold (shredding and sorting) and hot (decoating) end is an ongoing activity. Emission controls systems can make up 20% or more of the plant footprint.
- Equipment maintenance and material handling: The process involves a large number

of unit operations arranged in series that operate continuously. Shredded UBCs tend to "set" in silos, making it difficult to get the material out of the silos, and since they have a low bulk density (~150 kg/m^3) it is impractical to provide intermediate storage in stockpiles. Thus everything has to run with minimal interruption for the overall process to function smoothly.

Despite all this, the process in one form or another has been widely adopted and worldwide approximately 70% of all cans are currently recycled. Thus to improve the recycling of UBCs, their collection needs to be improved.

20.5 WHEEL RECYCLING

In contrast to UBC recycling, the recycling of aluminum wheels from cars appears to be relatively straightforward. Nonetheless, it provides an example of some of the compositional and alloy quality issues that can arise with even well-pedigreed postconsumer scrap.

20.5.1 Design/Alloy Selection

Aluminum wheels are produced primarily from low iron (largely primary-based) A356.2 alloy. This alloy contains 6.5% Si, 0.4% Mg, and a typical limit of 0.12% Fe. The limit on Fe content is to ensure that the wheels are highly ductile, so they resist both impact events and failure by fatigue. The alloy also contains a minor addition, strontium, to modify the aluminum-silicon eutectic and increase ductility.

In North America approximately 70% of the vehicles are equipped with aluminum wheels with an average weight of 8 kg (DuckerFrontier, 2020). At 14 million cars per year and 4 wheels per car, this represents a postconsumer scrap source of 320,000 tons annually.

20.5.2 Collection, Sorting, and Pretreatment

Wheels are one of the first items removed from end-of-life vehicles (Chapter 12—Automotive (ELV)). After the tires are stripped, the raw wheels are collected and shipped to recyclers for further processing. At this point the wheels still are contaminated with paint, traces of bead sealant, valve stems, and balance weights. The latter are removed manually and the wheels are separated into three categories based on visual appearance: forged, chrome plated, and painted. Forged wheels are made from a different alloy, 6061, and redirected to extrusion billet producers. Chromed wheels are redirected to piston alloy manufacturers, largely because of the nickel underlay or strike coat applied prior to chrome plating. This step is very important to avoid contamination of the eventual 356 alloy with nickel upon remelting. Even at trace levels nickel promotes what is known as filiform corrosion where unsightly "wormy-like" streaks grow on the surface of the wheel face (Courval et al., 1997). Chrome wheels contain a surprising amount of Ni, up to 1%. Even at this level the nickel value alone justifies the additional handling and shipping costs. Finally, the remaining wheels can be coarsely shredded and the adhering paints/pigments mechanically removed by grit blasting using iron pellets. The end result can then be packaged for sale as a substitute for primary based A356.2 alloy (Morin et al., 2019).

20.5.3 Challenges

Wheel recycling in Europe is complicated by a few technical and legislative conditions. In addition to using alloy 356, a significant fraction of wheels is produced using the Al-Si11 alloy, and these are visually indistinguishable from A356-based wheels. This would necessitate either dilution of the melt with low-Si material if the mixed product is directed to A356 consumers, or the addition of Si if sent to Al-11Si users (Das et al., 2007). In addition, several European foundries use antimony (Sb) as a eutectic refiner, which interferes with the more common eutectic modifiers, sodium (Na) and strontium (Sr). To date, attempts to separate the alloys using LIBS or XRF have not proven commercially successful. Legislative regulations relating to vehicle recertification have also led to the majority of European vehicles being reexported to Africa or Eastern Europe, reducing the number of wheels available for recycling locally (Morin et al., 2019).

20.6 DROSS PROCESSING

20.6.1 White Dross, Black Dross, and Salt Cake

In addition to the dross generated during melting scrap, dross is also generated during filling, alloying, furnace fluxing, holding, and during metal transfer. These collectively are called white dross (Figure 20.7A). They are very rich in metallic aluminum (15–80%) and aluminum oxide and contain little if any salt or cryolite. Depending on the alloy being processed and the temperature in the furnace, white dross may also contain significant amounts of aluminum nitride. Black dross (Figure 20.7B) refers to the dross resulting when salt is added to the remelter and contains about 30–50% salt flux (generally a near eutectic mixture of NaCl and KCl) along with oxides and other contaminants that may have accompanied the charge (bottle glass, rock dust, metallic particles (Fe), etc.). Metal levels can vary widely but are generally below 10% with good feed preparation, decoating, and melting practices. Salt cake (35–10% metal, 20–60% oxide, and 20–80% salt) (Figure 20.7C) is very similar to black dross, but the term is generally used for the nonmetallic waste generated from rotary salt furnaces.

Dross and salt cake are considered a hazardous waste in many jurisdictions (Jafari et al., 2014). Virtually all drosses will react to some degree with moisture due to the presence of metal nitrides, phosphides, carbides, and selenides, generating hazardous or toxic fumes, and thus require treatment prior to disposal or reuse in most jurisdictions (Tolaymat and Huang, 2015; Xiao et al., 2005). Both white dross and black dross are commonly treated in rotary furnaces, alone or in combination with other available/compatible scrap, to recover the contained metal in the dross.

The tilting rotary salt furnace has emerged as the most appropriate furnace/reactor design for treating black and white drosses (Peterson, 2011). Currently, there are two distinct approaches to dross treatment: salt-based and salt-free.

FIGURE 20.7 (A) White dross prior to skimming, (B) black dross, and (C) salt cake after skimming.

20.6.2 Salt-Based Rotary Furnace Dross Treatment

The process steps involved in the salt-based approach are shown in Figure 20.8.

The rotary salt furnace (RSF) (Figure 20.4b) rotates slowly, providing a gentle rolling agitation that promotes flux distribution and metal coalescence. Oxy-fuel burners with a slight excess of oxygen to provide a compact flame and reduce nitride formation are preferred. Flue gas temperature and drive motor torque are monitored and a sudden drop in the latter indicates when the charge is fully molten. Furnace sizes can vary from less than 1 to 25 t or more. At the end of a cycle the rotation is stopped, the furnace tilted, and a weir is swung into position to hold back the salt cake as the metal is emptied from the furnace. With proficient operation, over 90% yield of the metal available in most scraps and drosses can be attained. Salt additions vary widely, depending on the nature of the charge. The aim is to end up with a granular product that is easier to handle in subsequent salt cake recycling steps.

20.6.3 Salt Cake Processing

Since the direct disposal of salt cake is banned in most jurisdictions, it is treated prior to disposal of the oxide residues (Mahinroosta and Allahverdi, 2018).

The first step in this operation (Figure 20.8) is crushing and sieving to recover the remaining large ~1–3 mm metal particles that failed to coalesce in the RSF. Crushing also speeds up the extraction of salt during the dissolution process. A range of flammable/toxic gases including hydrogen, ammonia, hydrogen sulfide, and phosphine evolve during the leaching step as the carbides, sulfides, nitrides, and phosphides react with water. Following leaching, the nonmetallic product (NMP), consisting mostly of aluminum-magnesium oxides, silica, aluminosilicates, and titania, is separated by filtration, washed, and dried. The salt is recovered by evaporation and

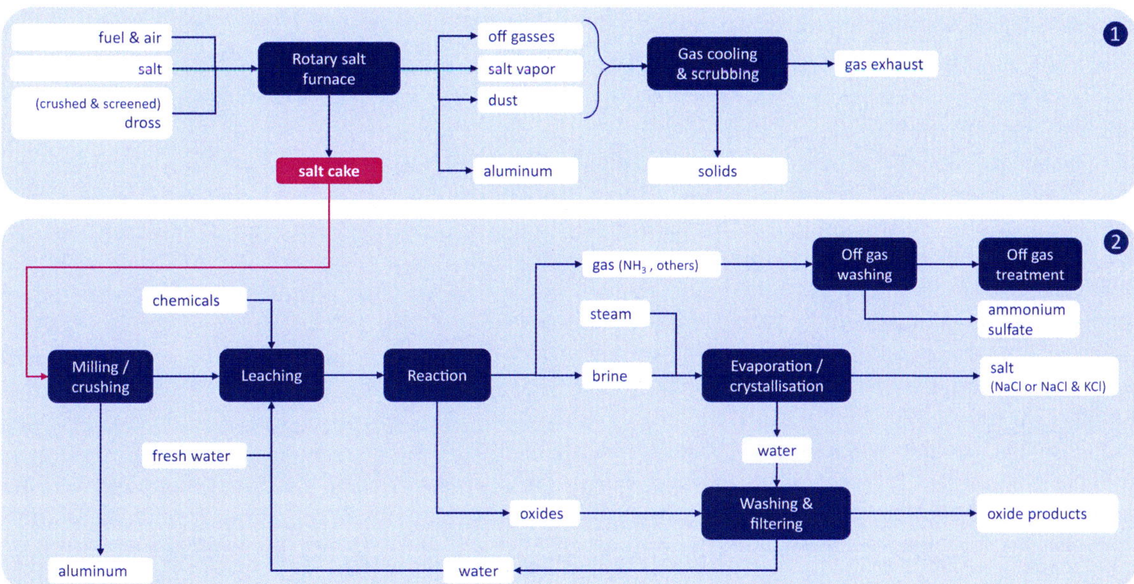

FIGURE 20.8 Flowsheets of the salt-based rotary furnace dross treatment (1) and salt cake processing (2). *Based on Lavoie et al. (1991) and European Commission et al. (2017).*

sold for use in aluminum smelting and recycling operations. Water from the evaporation stage is recycled back to the dissolution and washing stages, to reduce the intake of water. Finding a "home" for NMP represents a challenge, as the material still may contain residual nitrides and very small aluminum metal particles. Due to its variable composition and mineralogical form, it is generally unsuited as a feed material for the Bayer or Hall-Héroult processes, although it has been used as an input for the production of calcium aluminate for steelmaking slags and as a high alumina input for cement production.

20.6.4 Salt-Free Dross Processing

To overcome the issues associated with processing by-product salt cake, several salt-free alternatives have been proposed (Ünlü and Drouet, 2002). Although some of these have been operated commercially, they have not made significant inroads in supplanting the RSF. Most of these processes have been applied to the immediate treatment of hot drosses, where the entrained metal is still in the liquid state.

The Alcan plasma dross (APD) process uses a plasma torch instead of the usual gas burner in the RSF to provide very intense heating. This, in combination with the tumbling action of the furnace and the presence of minor amounts of cryolite, allows the metal to coalesce. A major advantage of the APD process is the greatly reduced gas volumes and of course the lack of salt in both the metal and NMP. The major drawback of the process is the presence of high levels of aluminum nitride in the NMP, resulting in the release of ammonia when the material comes in contact with moist air.

The DROSCAR process uses a DC arc struck between two graphite electrodes to heat the charge as the vessel slowly rotates. Again, the combination of intense heating and the tumbling are said to promote metal coalescence. Since the vessel can be sealed and an inert atmosphere (Ar) can be used, nitride formation is avoided.

In the DROSSRITE process hot dross is charged to a preheated vessel, which is then sealed and rotated under an argon atmosphere. After sufficient time for the metal in the dross to coalesce, the vessel is opened and the metal poured out. At this point the vessel is resealed and oxygen is injected to generate heat by reaction with the residual aluminum present in the remaining oxides. After the aluminum is consumed, the oxides are poured out and the now hot vessel is ready to receive the next charge.

20.7 PURIFICATION AND REFINING

Several proposals for technologies to remove specific unwanted alloying elements have been made over the years (Gaustad et al., 2012). In comparison to dilution and blending, none of the proposed methods are feasible from a business viewpoint even if they are technically sound.

20.7.1 Electrorefining

In principle, electrorefining can be used to remove impurities from molten aluminum (Lu et al., 2022) and indeed the Hoopes cell (Hoopes, 1925) has been used for many years to produce high purity aluminum. In the Hoopes cell a high-density Al-Cu alloy forms at the anode and aluminum ions are preferentially transferred through the electrolyte, depositing pure metal on the surface, and leaving the impurities (principally Fe and Si) behind in the anodic metal pool. The process is energy intensive and best suited to upgrade relatively pure metal (>99.9% Al). To date electrorefining has not been applied for scrap purification on a commercial scale due to the high energy consumption (comparable to primary aluminum smelting) and build-up of intermetallics in the anode layer.

20.7.2 Chemical Techniques

The technical literature abounds with reports describing the selective removal of various elements from aluminum. Silicon can be reduced by adding excess calcium, iron can be precipitated by excess manganese, or its solubility driven down by adding large concentrations of magnesium. All these approaches replace one element with another and require a solid/liquid separation. None has been applied commercially. An exception to this is the removal of trace amounts of Ti and V from smelter-grade metal by the addition of boron, forming Ti/VB_2. This is done to improve the electrical conductivity of material for electrical transmission cables (Khaliq et al., 2014; Kieft, 2006).

20.7.3 Chlorination for Magnesium Removal

Magnesium is the only major alloying element that can be removed from molten alloys on a commercial scale. Magnesium dissolved in aluminum will react preferentially with chlorine gas according to Eq. (20.1):

$$Mg + Cl_2 = MgCl_2 \tag{20.1}$$

This reaction is virtually instantaneous, and Mg levels can be reduced to well below 0.1% with 100% yield. As the stoichiometry indicates, 2.96 kg chlorine is required per kg of Mg removed. For a 100 t furnace, the removal of 0.1% Mg thus requires the addition of 300 kg chlorine gas and generates just under 400 kg of molten magnesium chloride. Upon contact with moist air $MgCl_2$ will hydrolyze according to Eq. (20.2):

$$MgCl_2 + H_2O = MgO + 2HCl_{(g)} \tag{20.2}$$

releasing gaseous hydrochloric acid.

Chlorine gas injection is also used to reduce trace levels of sodium and lithium in wrought alloys.

Other methods of Mg removal have been proposed, including:

- reaction with fine silica particles to form a magnesium-aluminum oxide (Barrera-Méndez et al., 2010),
- reaction with $KAlF_4$ (fluoride salts) (Hess, 1971),
- vacuum distillation, utilizing the low vapor pressure of magnesium (Sigworth and Engh, 1982),
- electrolysis (Gesing et al., 2016).

None of these approaches has proven commercially successful.

20.7.4 Fractional Crystallization for Iron Dilution

Fractional crystallization takes advantage of the difference in the solubility of impurities between the solid and the liquid state. Hence, the initial solid that forms when an alloy begins to freeze will contain a lower concentration of dissolved impurities. This approach has been applied to pretreat inputs to the Hoopes cell, reducing the rate at which intermetallic impurities accumulate in the anode. It is also potentially useful to produce aluminum with very low iron content, so that the amount of diluent (i.e., primary aluminum) required to lower iron levels in molten metal can be reduced. For example, consider an alloy with an Fe limit of 0.2% where the batch analysis comes back at 0.21%. If primary aluminum (0.1% Fe) was added, then an addition of 10% w/w would be required. In addition to the time and energy required to melt and reheat, the other elements of the alloy would be diluted and likely require further alloy element additions to meet the alloy specifications. By using a 0.01% Fe primary aluminum input instead, the addition required drops to 5.3% w/w, saving time, energy, alloying element additions, and thus money.

20.7.5 Other Purification Techniques

Very significant development efforts went into trying to develop alternate methods of producing primary aluminum using carbon as a reductant. These approaches resulted in a crude aluminum containing significant levels of silicon and other elements. In order to produce pure aluminum from these alloys, a number of other purification techniques were developed. Two of the most prominent were the Gross process (monochloride process) and liquid metal solvent extraction. Both were demonstrated at the pilot scale, but neither were ultimately commercially successful in displacing the Hall-Héroult process. Nonetheless, it was recognized at the time that these approaches could be used to upgrade impure scrap.

The monochloride process, originally described by P. Gross in 1945 as a means of producing or purifying aluminum (Gross, 1949), is based on the temperature-dependent equilibrium between molten aluminum trichloride ($AlCl_3$) and aluminum monochloride ($AlCl$). Above 1000°C $AlCl_3$ reacts with aluminum to form AlCl (Eq. 20.3):

$$AlCl_3 + 2Al = 3AlCl\ (> 1000°C) \qquad (20.3)$$

When the temperature is lowered the AlCl will decompose to reform $AlCl_3$ and Al (Eq. 20.4):

$$3AlCl = AlCl_3 + 2Al\ (700°C) \qquad (20.4)$$

The net effect is the transfer of molten aluminum from a high-temperature reservoir to a low-temperature one. The impurities remain behind, and a purer aluminum metal is obtained.

Liquid metal solvent extraction uses molten zinc to preferentially dissolve aluminum, leaving behind impurities such as silicon and iron. After extraction, the zinc-aluminum alloy is filtered, and the zinc removed by distillation. This process was investigated by the US Bureau of Mines as a means of recovering aluminum from aluminum-silicon alloys produced by carbothermic reduction (St.Clair and Blue, 1949). Energy requirements were estimated at 2–6 kWh/kg, compared to 10 Wh/kg for direct electrolysis (Saevarsdottir et al., 2020).

20.8 FUTURE TRENDS AND CHALLENGES

The aluminum industry is facing an ongoing challenge to reduce CO_2 emissions associated with global warming. While this challenge is particularly acute for primary (smelter) producers, there is also a need for the secondary aluminum industry to reduce fossil fuel use in all areas of the fabrication process, as well as to minimize primary inputs and their embedded CO_2 costs.

The often-quoted figure for recycled aluminum requiring just 5% of the energy needed to produce primary aluminum only applies to melting; there can be significant energy costs associated with scrap pretreatment and downstream operations following casting. When these additional operations are included, producing a finished product from recycled aluminum requires close to 25% of the energy needed compared to using a primary aluminum (Allwood and Cullen, 2012).

In order to improve melting efficiency, electromagnetic subhearth stirring has now become commonplace and high-efficiency regenerative burners are widely used in large scale melting operations. Oxy-fuel burners are also being used to reduce direct gas consumption and CO_2 emissions, although if the oxygen is generated using fossil fuels, the overall CO_2 reduction is compromised.

"Near net" casting processes, such as twin roll or twin belt casting, are used in order to reduce energy and material losses (scalping, reheating, trimming) associated with ingot casting and fabrication. This lowers the internal scrap volumes to be recycled significantly, and thus the losses of Al to dross/oxide formation during recycling of the clean scrap.

The recycling of secondary aluminum depends critically on the efficient sorting, separation, and handling of the scrap inputs prior to melting. Except for Mg removal, the opportunities for chemical purification in the liquid state are economically nonexistent. Blending and dilution remain the only options to reduce the levels of dissolved elements. With continued development, particularly in terms of throughput, analysis-based separation technologies such as LIBS and XRT combined with further digitalization have the potential to disrupt the current "downcycle" system of the recycling loop.

One unknown is how the adoption of electric vehicles, and the consequent displacement of the cast aluminum engine, will impact the demand for "downcycled" aluminum. In this regard, sorting technologies such as LIBS and XRF may become more attractive as the need to identify and separate wrought from cast alloys increases. On the other hand, developments such as the "giga-casting" of large sections of the car body may replace the demand for the scrap and impurity tolerant secondary aluminum alloys currently used to produce internal combustion engine components.

References

Allwood, J.M., Cullen, J.M., 2012. Sustainable Materials With Both Eyes Open. UIT Cambridge Ltd., Cambridge, UK. https://www.uselessgroup.org/publications/book.

Barrera-Méndez, F., Escobedo-Bocardo, J.C., Cortés-Hernández, D.A., Almanza-Robles, J.M., Muñoz-Arroyo, R., Hernández-Garcia, H.M., 2010. Magnesium removal from molten Al-Si alloys using zeolite. Can. Metall. Q. 49 (2), 163–170. https://doi.org/10.1179/cmq.2010.49.2.163.

Bertram, M., Ramkumar, S., Rechberger, H., Rombach, G., Bayliss, C., Martchek, K.J., Müller, D.B., Liu, G., 2017. A regionally-linked, dynamic material flow modelling tool for rolled, extruded and cast aluminium products. Resour. Conserv. Recycl. 125, 48–69. https://doi.org/10.1016/j.resconrec.2017.05.014.

Boin, U., Reuter, M.A., Probst, T., 2004. Measuring – modelling: understanding the Al scrap melting process inside a rotary furnace. Erzmetall 57 (5), 264–269.

Courval, G.J., Chamberlain, B., Pattemore, D., 1997. Effects of alloy composition and condition on filiform corrosion performance of cast aluminum wheels. SAE Technical Paper 970021, https://doi.org/10.4271/970021.

Das, S.K., Green, J.A.S., Kaufman, J.G., 2007. The development of recycle-friendly automotive aluminum alloys. JOM 59 (11), 47–51. https://doi.org/10.1007/s11837-007-0140-2.

Davis, J.R., 2001. Light metals and alloys. In: Davis, J.R. (Ed.), Alloying: Understanding the Basics. ASM International, https://doi.org/10.31399/asm.tb.aub.9781627082976.

Discovery, 2018. How Are Aluminium Cans Recycled? How Do They Do It? YouTube video https://www.youtube.com/watch?v=KmMP67eC2tg.

DuckerFrontier, 2020. North America light vehicle aluminum content and outlook. Final report summary. Prepared for the Aluminum-Association https://drivealuminum.org/wp-content/uploads/2022/05/DuckerFrontier-Aluminum-Association-2020-Content-Study-Summary-Report-FINAL.pdf.

Engh, T.A., Sigworth, G.K., Kvithyld, A., 2021. Principles of Metal Refining and Recycling. Oxford University Press, Oxford, UK, https://doi.org/10.1093/oso/9780198811923.001.0001. Oxford Scholarship Online.

European Commission, Joint Research Centre, Delgado Sancho, L., Roudier, S., Farrell, F., et al., 2017. Best available techniques (BAT) reference document for the non-ferrous metals industries: *Industrial Emissions Directive 2010/75/EU (integrated pollution prevention and control)*. Publications Office https://data.europa.eu/doi/10.2760/8224.

Gaustad, G., Olivetti, E., Kirchain, R., 2012. Improving aluminum recycling: a survey of sorting and impurity removal technologies. Resour. Conserv. Recycl. 58, 79–87. https://doi.org/10.1016/j.resconrec.2011.10.010.

Gesing, A.J., Das, S.K., Loutfy, R.O., 2016. Production of magnesium and aluminum-magnesium alloys from recycled secondary aluminum scrap melts. JOM 68, 585–592. https://doi.org/10.1007/s11837-015-1720-1.

Gross, P., 1949. Process for the Production and Refining of Aluminum. US Patent 2470305,.

Hannula, J., Godinho, J.R.A., Abadias Llamas, A., Luukkanen, S., Reuter, M.A., 2020. Simulation-based exergy and LCA analysis of aluminum recycling: linking predictive physical separation and re-melting process models with specific alloy production. J. Sustain. Metall. 6, 174–189. https://doi.org/10.1007/s40831-020-00267-6.

Hess, P.D., 1971. Magnesium Removal From Aluminum Alloy Scrap. US Patent 3620716,.

Hoopes, W., 1925. Electrolytically Refined Aluminum and Articles Made Thereof. US Patent 1534315,.

International Aluminium Institute (Ed.), 2016. Global Aluminium Flow Model, 2016 ed. International Aluminium Institute.

International Aluminium Institute, 2022. Global Aluminium Cycle. https://alucycle.international-aluminium.org/public-access/. (Accessed 22 June 2022).

ISRI, 2021. ISRI. Scrap Specifications Circular 2021. Institute of Scrap Recycling Industries. http://www.scrap2.org/specs/ and www.isri.org (Accessed 5 January 2022).

Jafari, N.H., Stark, T.D., Roper, R., 2014. Classification and reactivity of secondary aluminum production waste. J. Hazard. Toxic Radioact. Waste 18 (4). https://doi.org/10.1061/(ASCE)HZ.2153-5515.0000223.

Kammer, C., 1999. Aluminium Handbook. Vol. 1: Fundamentals and Materials, first ed. Aluminium-Verlag, Düsseldorf. ISBN: 9783870172619.

Khaliq, A., Rhamdhani, M.A., Brooks, G.A., Grandfield, J., 2014. Removal of vanadium from molten aluminum – Part III. Analysis of industrial boron treatment practice. Metall. Mater. Trans. 45B, 784–794. https://doi.org/10.1007/s11663-013-0017-4.

Kieft, R., 2006. Molten aluminium purification – purification by formation and removal of inter-metallics. Final Technical report, European Community project GRD1-2001-40715 (MAP),.

Kvithyld, A., Meskers, C.E.M., Gaal, S., Reuter, M., Engh, T. A., 2008. Recycling light metals: optimal thermal de-coating. JOM 60 (8), 47–51. https://doi.org/10.1007/s11837-008-0107-y.

Lavoie, S., Dubé, C., Dubé, G., 1991. The Alcan plasma dross treatment process. In: Rooy, E.L. (Ed.), Light Metals 1991. Minerals, Metals, & Materials Society, pp. 981–985.

Le Brun, P., 2008. Melt treatment—evolution and perspectives. In: DeYoung, D.H. (Ed.), Light Metals 2008. John Wiley & Sons, pp. 621–626.

Leosson, K., Meirbekova, R., Gudmundsson, S.H., Salloum-Abou-Jaoude, G., 2022. Continuous chemical analysis of molten aluminum. In: Eskin, D. (Ed.), Light Metals 2022. Springer, Cham, pp. 723–727, https://doi.org/10.1007/978-3-030-92529-1_95.

Lu, X., Zhang, Z., Hiraki, T., et al., 2022. A solid-state electrolysis process for upcycling aluminium scrap. Nature 606, 511–515. https://doi.org/10.1038/s41586-022-04748-4.

Mahinroosta, M., Allahverdi, A., 2018. Hazardous aluminum dross characterization and recycling strategies: a critical review. J. Environ. Manage. 223, 452–468. https://doi.org/10.1016/j.jenvman.2018.06.068.

Meskers, C.E.M., Reuter, M.A., Boin, U., Kvithyld, A., 2008. A fundamental metric for metal recycling applied to coated magnesium. Metall. Mater. Trans. 39B, 500–517. https://doi.org/10.1007/s11663-008-9144-8.

Morin, G., Hartlieb, M., Chiesa, F., Marin, G., 2019. eccomelt356.2 As a substitute or addition to Primary A356.2 to increase Competitiveness and Reduce Carbon Footprint of Castings. In: Presentation at the Castexpo & Metalcasting Congress. American Foundry Society, Atlanta, GA.

Natural Resources Canada, 2022. Aluminum Facts. Government of Canada. https://www.nrcan.gc.ca/our-natural-resources/minerals-mining/minerals-metals-facts/aluminum-facts/20510. (Accessed May 2022).

Neff, D., Sigworth, G.K., Gallo, R., 2018. Melting and melt treatment of aluminum alloys. In: Anderson, K., Weritz, J., Kaufman, G.J. (Eds.), ASM Handbook. vol. 2A. Aluminum Science and Technology. ASM International, pp. 143–164, https://doi.org/10.31399/asm.hb.v02a.a0006535.

Norsk Hydro, 2015. Aluminium Recycling – How Is Aluminium Recycled. YouTube video https://www.youtube.com/watch?v=chIFt2A9MRI.

Peterson, R.D., 1990. Effect of salt flux additives on aluminum droplet coalescence. In: Van Linden, J.H.L., Stewart, D.L., Sahai, Y. (Eds.), Second International Symposium – Recycling of engineered Materials. The Minerals Metals and Materials Society, pp. 69–84.

Peterson, R.D., 2011. A historical perspective on dross processing. Mater. Sci. Forum 693, 13–23. https://doi.org/10.4028/www.scientific.net/MSF.693.13.

Peterson, R.D., 2015. Recycling of automotive wrought alloys. In: Hyland, M. (Ed.), Light Metals 2015. John Wiley & Sons, Inc, pp. 1023–1028, https://doi.org/10.1002/9781119093435.ch172.

Rooy, E., 2003. A history of aluminum foundry alloy development. In: Proceedings from the AFS International Conference on Structural Aluminum Casting. American Foundry Society, USA, pp. 87–98.

Rossel, H., 1990. Fundamental investigations about metal loss during remelting of extrusion and rolling fabrication scrap. In: Bickert, C. (Ed.), Light Metals 1990: Proceedings of the Technical Sessions Presented by the TMS Light Metals Committee at the 119th TMS Annual Meeting, Anaheim, California, February 18–22, 1990. TMS, Warrendale, USA, pp. 721–729. ISBN: 9780873391146.

Roy, R.R., Sahai, Y., 1997. Coalescence behavior of aluminum alloy drops in molten salts. Mater. Trans. JIM 38 (11), 995–1003. https://doi.org/10.2320/matertrans1989.38.995.

Saevarsdottir, G., Kvande, H., Welch, B.J., 2020. Aluminum production in the times of climate change: the global challenge to reduce the carbon footprint and prevent carbon leakage. JOM 72 (1), 296–308. https://doi.org/10.1007/s11837-019-03918-6.

Schlesinger, M.E., 2014. Aluminum Recycling, second ed. CRC Press by Taylor & Francis Group.

Schmitz, C., 2014. Handbook of Aluminium Recycling, second ed. Vulkan-Verlag, Essen, Germany. ISBN: 9783802730061.

Sigworth, G.K., 2014. Best Practices in Aluminum Metalcasting. American Foundry Society, Schaumburt, IL. Part 2, Chapter 5. p. 73. Inclusions and inclusion removal https://www.asminternational.org/web/ims/store/-/

journal_content/56/10192/20130822/PUBLICATION. ISBN: 978-0-87433-416-6.

Sigworth, G.K., Began, B., 2012. Control and measurement of hydrogen in aluminum. AFS Transactions. 116 paper 12-002.

Sigworth, G.K., Engh, T.A., 1982. Refining of liquid aluminium – a review of important chemical factors. Scand. J. Met. 11, 143–149.

St.Clair, H.W., Blue, D.D., 1949. Recovery of aluminum from crude aluminum-silicon alloy by extraction with molten zinc. US Dept. of the interior – Bureau of mines. Report R.A 4535 https://catalog.hathitrust.org/Record/005951623.

Tabereaux, A.T., Peterson, R.D., 2014. Aluminum production. In: Seetharaman, S. (Ed.), Treatise on Process Metallurgy, Volume 3: Industrial Processes. Elsevier, pp. 839–917, https://doi.org/10.1016/B978-0-08-096988-6.00023-7. Chapter 2.5.

Tolaymat, T., Huang, X., 2015. Secondary Aluminum Processing Waste: Salt Cake Characterization and Reactivity. U.S. Environmental Protection Agency, Washington, DC. EPA/600/R-15/109 https://nepis.epa.gov/Exe/ZyPDF.cgi?Dockey=P100NRFR.pdf.

Ünlü, N., Drouet, M.G., 2002. Comparison of salt-free aluminum dross treatment processes. Resour. Conserv. Recycl. 36 (1), 61–72. https://doi.org/10.1016/S0921-3449(02)00010-1.

Utigard, T.A., Friesen, K., Roy, R.R., Lim, J., Silny, A., Dupuis, C., 1998. The properties and uses of fluxes in molten aluminum processing. JOM 50 (11), 38–43. https://doi.org/10.1007/s11837-998-0285-7.

Van Linden, J.H.L., Stewart, D.L., 1989. Molten salt flux composition effects in alumium scrap remelting. In: Campbell, P.G. (Ed.), Light Metals 1989. The Minerals Metals and Materials Society, pp. 173–180.

Xiao, Y., Reuter, M.A., Boin, U., 2005. Aluminium recycling and environmental issues of salt slag treatment. J. Environ. Sci. Health. A Tox. Hazard. Subst. Environ. Eng. 40 (10), 1861–1875. https://doi.org/10.1080/10934520500183824.

Zhou, B., Yang, Y., Reuter, M.A., Boin, U.M.J., 2005. Modelling of aluminium scrap melting in a rotary furnace. Miner. Eng. 19 (3), 299–308. https://doi.org/10.1016/j.mineng.2005.07.017.

CHAPTER

21

Copper

Mark E. Schlesinger

Missouri University of Science and Technology, Rolla, MO, United States

In 2018, global production of copper totaled approximately 29.060 Mt (ICSG, 2019). Of this, approximately 8.72 Mt (~30%) were obtained from secondary materials—scrap and other copper-containing wastes. Copper was one of the first materials to be recycled, and the use of copper scrap is significant in the production of primary copper from ores in addition to its processing as a separate industrial operation. This chapter provides an introduction to copper recycling, starting with the sources of secondary copper, its collection and processing, and ultimate smelting and refining. As the usage pattern of copper changes, the type of material available for copper recovery changes as well, and the impact of this on the processing strategy for secondary copper is explained as well.

21.1 SOURCES OF COPPER SCRAP

Figure 21.1 illustrates the global flow of copper from production through fabrication, use, and end-of-life disposition. About two-thirds of fabricated copper production occurs in Asia (ICSG, 2019), two-thirds of that occurring in China; nearly all of the rest is produced in the Americas or in Europe. About 62% of fabricated copper is produced as pure-Cu wire rod; most of the rest is evenly divided between rods, bars, and sections (RBS), plates, sheets, and strips (PSS), tubing, and alloy wire.

Copper use is divided among five sectors of the global economy (Ciacci et al., 2017; Gordon et al., 2006; ICSG, 2019):

- Equipment (31%), which is largely appliances, consumer electronics, military and commercial ordnance, and coinage. The average life of copper products in this sector is 5 years (Gómez et al., 2007).
- Building and construction (28%), which consists primarily of interior wiring; plumbing, heating, and architectural uses; and air conditioning and commercial refrigeration. Copper used in this sector stays in service an average of 30 years.
- Infrastructure (16%), which includes power-generating utilities, telecommunications, lighting, and business electronics. The average service life of copper in this sector is also 30 years.
- Transport (13%), which is divided into motor vehicles (automobiles, trucks, and buses) and other transportation equipment, e.g., aircraft, ships, railroad, and aerospace equipment. An average 17-year life span in this sector is expected.

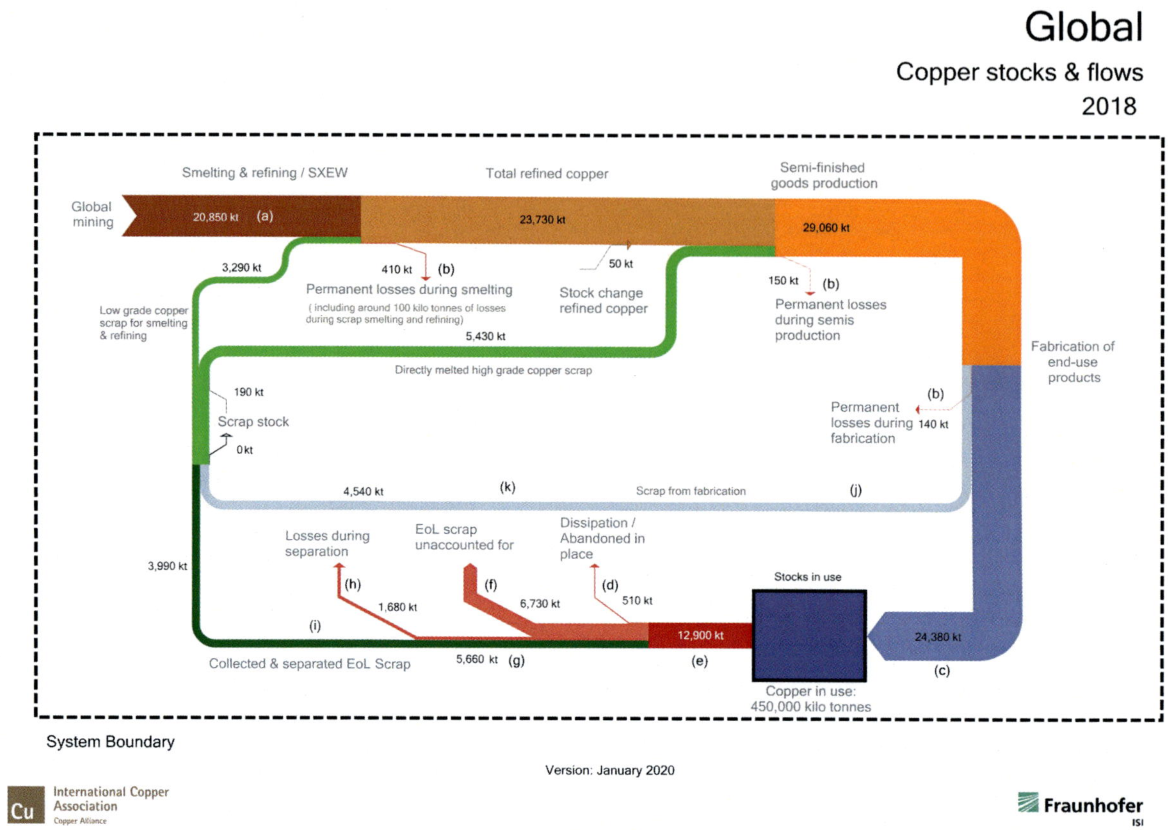

FIGURE 21.1 Global stocks and flows of copper (2018), in kt. *From Copper Alliance, https://copperalliance.org/policy-focus/society-economy/circular-economy/stocks-flows/.*

- Industrial (12%), which includes in-plant equipment, industrial valves and fittings, nonelectrical instruments, and off-highway vehicles. Life spans of 20 years are expected in this sector.

The vast majority of copper put in service is "pure" copper, with little or no alloying elements. Copper is not commonly added as an alloying element to other metals, so relatively little is removed from the product stream in other base metal alloys. Given the increasing demand for copper and its lengthy service life, the reservoir of copper in use has grown considerably with time; it was an estimated 433 Mt in 2018 (ICSG, 2019). The annual growth of the reservoir is slightly more than half of the copper put into use.

Copper recovered for reuse can be divided into two broad classifications: *new* and *old* scrap (Schlesinger et al., 2021). New scrap (also called prompt industrial, prime, and internal arising scrap; "scrap from fabrication" in Figure 21.1) is generated by manufacturing operations, and comprises about 15% of all copper produced (ICSG, 2019). Most new scrap can be recycled simply by putting it back into the fabrication process (Wood et al., 2011), since it contains no attachments and requires no upgrading; a small amount requires processing to remove coatings or attachments. Old (or obsolete, postconsumer) scrap is copper recovered from end-of-life (EoL) objects that have been taken out of service. It is more expensive to process, due to its unpredictability, scattered location, and greater levels of attachment and coatings. Because of that, most copper from EoL items is not recovered, as Figure 21.1 shows.

The recovery of copper from old scrap depends on the source of the EoL, as well as its location and condition. EoL goods are classified into six categories (Ruhrberg, 2006; Schlesinger et al., 2021):

- End-of-life vehicles (ELVs) include cars, trucks, and buses, but not planes, ships, or trains. The copper in ELVs consists primarily of wiring, in particular electric motors and the wiring "harness" that connects the lights and motors (McGlothlin, 2018). The overall copper content of ELVs is about 0.8%; this will increase as electric vehicles (>10% copper) become part of the ELV stream.
- Construction and demolition (C&D) waste includes copper from the wiring of buildings, pipes and fittings from the plumbing, and brass from door handles and frames, among other items (Wildnauer, 2019). C&D debris as a whole contains about 0.30% copper, or 0.05–0.24 kg/m^3 (Koutamanis et al., 2018). The current high price of copper encourages copper recovery from buildings; unfortunately, this often occurs as theft from job sites or vandalism.
- Waste from electrical and electronic equipment (WEEE) is the largest source of old copper scrap (Glöser et al., 2013). WEEE has three subcategories: white goods, domestic electrical appliances such as washing machines, refrigerators, dishwashers, etc.; brown goods that includes audiovisual appliances such as televisions and DVD players; and gray goods, computers and telecommunication appliances. The estimated copper content of WEEE ranges from 3% to 21% (Barnwal et al., 2018).
- Industrial electrical equipment waste (IEW) includes power cables (underground and surface), transformers, and other electrical equipment. The relative amount of this material is small, but its copper content is high: 5–80%.
- Industrial nonelectrical equipment waste (INEW), sometimes grouped with IEW, includes large transportation equipment (planes, ships, and trains), spent ammunition and ordnance, and other machinery. Although no formal estimates of its copper content exist, the percentage is likely to be small (Ruhrberg, 2006).
- Municipal solid waste (MSW) is the hardest waste stream to process and contains very little copper (0.05–0.20%). Loose coins in MSW contain much of its copper, and small appliances contain much of the rest.

Recovery of old copper scrap from waste streams depends on the type of waste, the location of the waste stream, and the price of copper (Vexler et al., 2004). High-grade streams such as IEW are profitable to process and recycle, so most of the copper from this waste is recovered (Glöser et al., 2013). Although few statistics are kept, the same is likely true for INEW. The presence of infrastructure for dismantling and processing ELVs means that recovery of copper from this stream is higher in the developed world than in less-developed areas. Recovery of copper from MSW is always low. Recovery of copper from WEEE has also been low but is increasing as a result of government mandates to process it.

Schlesinger et al. (2021) describe the processes for upgrading old copper scrap. The most significant of these technologies are those for processing spent wire and cable. The process described by Sullivan (1985) is still the basis for most cable processing flowsheets (see Figure 21.2). Granulation is the process of chopping wire and cable

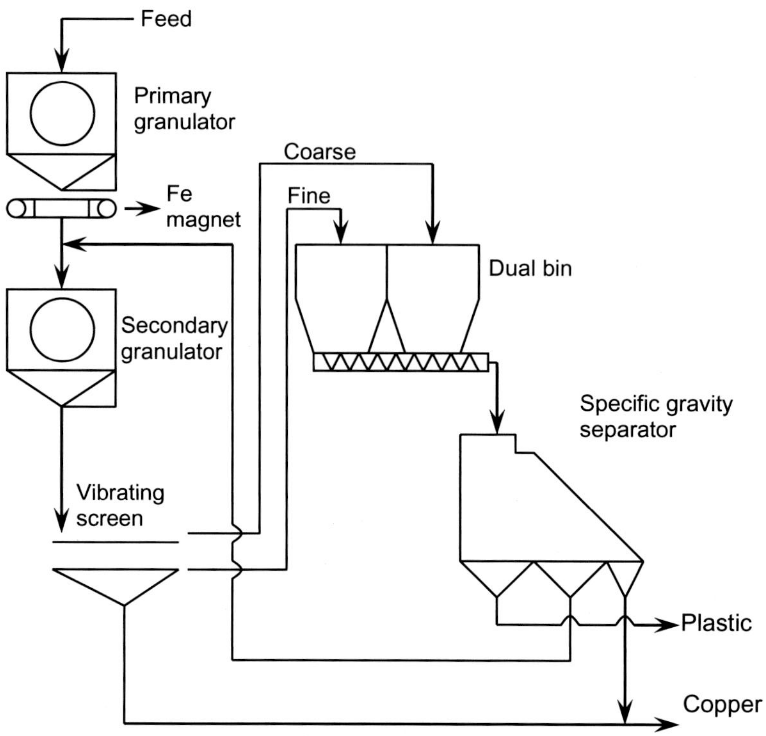

FIGURE 21.2 Process flowsheet for recovery of copper from wire scrap. *After Sullivan (1985).*

into progressively smaller sections, until the wire inside separates from the rubber or plastic insulation. The metal is then recovered based on its higher specific gravity.

As the volume of electronic scrap grows, methods for processing it to upgrade its metal value are of increasing importance (Kaya, 2016; Schlesinger et al., 2021). The wide variety of e-waste types makes development of a common processing strategy difficult. For white goods and brown goods in particular, the strategy often entails:

1. disassembly to recover large items
2. shredding to reduce the size of the remaining material and liberate the metal
3. separation of metals based on differences in density and electrical conductivity.

Shredding is typically done using low-speed, high-torque knife shredders. A primary shredding step generates relatively large particles (4 cm × 4 cm), from which copper is recovered by air tabling, screening, or hand-picking. Secondary shredding generates a finer material with better liberation. Copper recovery from this stream is accomplished using ECS or electrostatic separation. For the printed circuit boards in gray goods, this approach is less effective. Instead, the PCBs are often recovered whole and sent directly for smelting (Kaya, 2016).

The increasing amount of copper in end-of-life vehicles (ELVs) makes recovery an important process. About 90% of the copper in an ELV is found in the various wire harnesses (McGlothlin, 2018); lesser amounts are found in the entertainment system, motors, alternator, sensors, and transmission. As a result, the dismantling of ELVs produces very little copper to recycle. Most of the copper from ELVs is in the nonferrous metal scrap stream remaining after the vehicle has been shredded and its iron and steel have been magnetically removed. Three metals dominate this stream: aluminum, copper, and zinc. Several techniques are used to separate the metals in this fraction. Eddy-current separation (ECS) uses the different electrical conductivity of the metallic and nonmetallic particles in the shredded material to remove most of the nonmetallics. Heavy-media separation removes the lower-density aluminum and magnesium from the metallic fraction. What remains is Zebra, a mixture of mostly zinc, brass, and copper.

Two methods can be used to upgrade Zebra, both based on the difference in color of the copper, brass, zinc, and other metals in it (Pretz and Mutz, 2006). The first is hand-sorting, which is widely practiced in China and India but may become less viable as labor costs there increase over time. The second is automated sorting, which has been commercially practiced for over 10 years (Steinert, 2020).

Because copper scrap is not usually recycled where it was recovered, a global market exists for both new and old scrap. The American Institute for Scrap Recycling Industries (ISRI, 2021) lists over 50 copper-containing commercially traded grades of scrap. The most common grades are:

- № 1 copper wire (Barley): Bare, uncoated, unalloyed copper wire, not smaller than No. 16 gauge, free of burnt wire that is brittle.
- № 2 copper wire (Birch): Miscellaneous, unalloyed copper wire with a nominal 96% copper content (minimum 94%). Free of excessively leaded, tinned, soldered copper wire; brass and bronze wire; excessive oil content, iron, and nonmetallics; copper wire from burning, containing insulation; hair wire; burnt wire that is brittle; and free of ash.
- № 3 copper (Cliff): Miscellaneous, unalloyed copper scrap with a nominal 96% copper content (minimum 94%). Free of excessively leaded, tinned, soldered copper scrap; brasses and bronzes; excessive oil content, iron, and nonmetallics; copper tubing with other than copper connections or with sediment; copper wire from burning, containing insulation; hair wire; burnt wire that is brittle; and free of ash.
- № 4 copper wire nodules (Clove): Bare, uncoated, unalloyed copper wire scrap nodules, chopped or shredded, free of tin, lead, zinc, aluminum, iron, other metallic impurities, insulation, and other foreign contamination. Minimum copper 99%.
- № 5 copper wire nodules (Cocoa): Unalloyed copper wire scrap nodules, chopped or shredded, minimum 99% copper. Maximum metal impurities not to exceed 0.50% aluminum and 1% each of other metals or insulation.
- № 6 heavy copper (Candy): Clean, unalloyed, uncoated copper clippings, punchings, bus bars, commutator segments, and wire not less than 1/16 of an inch thick, free of burnt wire that is brittle; but may include clean copper tubing.
- № 7 light copper (Dream): Miscellaneous, unalloyed copper scrap with nominal 92% copper content (minimum 88%), consisting of sheet copper, gutters, downspouts, kettles, boilers, and similar scrap. Free of burnt hair wire; copper clad; plating racks; grindings; copper wire from burning, containing insulation; radiators and fire extinguishers; refrigerator units; electrotype shells; screening; excessively leaded, tinned, soldered scrap; brasses and bronzes; excessive oil, iron, and nonmetallics; and free of ash.
- № 8 insulated copper wire scrap (Druid): Copper wire scrap with various types of insulation.

Pictures of several copper scrap grades are shown as Figure 21.3. The prices of scrap copper

FIGURE 21.3 Grades of copper scrap: (A) No. 1 copper wire (Barley), (B) No. 2 copper wire (Birch), (C) No. 3 copper (Cliff), (D) copper wire nodules (Clove), (E) copper wire nodules (Cocoa), (F) heavy copper (Candy), (G) light copper scrap (Dream), (H) insulated wire scrap (Druid). *Courtesy: ISRI International.*

grades track the price of primary copper; they also depend on purity, the types of impurity, and particle size. Copper-containing wastes from industrial processes are also added to recycling streams; these include plating sludges, foundry slags, and the residue from leached anode slimes.

21.2 SMELTING AND REFINING OF COPPER SCRAP

21.2.1 Characteristics of Secondary Copper

Several factors determine what happens to copper-containing scrap and waste. These include:

- The condition of the scrap. The different types of № 1 scrap contain very small levels of oxidation or impurities, and need only be remelted in a nonoxidizing environment. As a result, this scrap commands a higher price, and is more likely to be charged to furnaces requiring a purer grade of input material. № 2 scrap is more oxidized and will require refining. Heavily oxidized waste materials like plating sludges require remelting in a reducing environment to avoid copper loss. Material recovered from e-waste (see Chapter 13—Electronic and electronic equipment (WEEE)), such as printed circuit boards (PCBs), is generally sent to a limited number of smelters specifically configured to handle it. These include Noranda in Canada, Dowa in Japan, Aurubis in Germany, LS Nikko in South Korea, Umicore in Belgium, Sims Metal in Australia, Nerin in China, and Boliden in Sweden (Ye et al., 2021).
- Location. Copper scrap is often collected at locations that are very far from primary smelters. The cost of transporting it determines where it winds up.
- Alloying elements. Impurity elements in secondary copper often have sufficient value to be recovered in their own right. This includes the gold and silver content of electronic scrap, the lead in copper drosses, and the lead, tin, and zinc in brass and bronze scrap. Recovering these elements often dictates recycling strategy.
- Government regulations. Some governments place export tariffs on copper-containing scrap to encourage recycling at home. Restrictions on the exporting of some electronic scrap have also encouraged local reprocessing efforts. Restrictions on air emissions from processing plants have had an impact on processing strategy (Tang, 2016). The recent decision by the Chinese government to restrict imports of outside scrap material (including copper) has created ongoing turmoil that impacts processing decisions (Home, 2019).

21.2.2 Scrap Processing in Primary Copper Smelters

Secondary copper can be added at three locations in the primary coppermaking process. The most common is the converting furnace, but additions are also made in the smelter and anode furnace.

Because secondary copper has no sulfur in it, adding it to a furnace is a net energy consumer. This makes it difficult to add to matte smelting furnaces, which already require some fuel use. Making the size of scrap or waste particles small enough to use in flash-furnace concentrate burners is also difficult. However, small-size shredded scrap can be added in limited quantities to flash furnaces (Maeda et al., 2000). In fact, the smelting furnace is preferred to the converter for feeding electronic scrap, due to its plastic content. The plastic has fuel value that provides heat for smelting (Tang, 2016). In addition, when burned intermittently, plastic often gives off smoke and other particulates that might escape through the mouth of a Peirce-Smith converting furnace, adversely affecting workplace hygiene (Yamashita et al., 2016). Burned in a sealed flash furnace, these particulates are efficiently captured by dust collection devices.

Feeding scrap and waste to other types of primary smelting furnace is easier, and both the Noranda and Mitsubishi smelting processes have been adapted to include scrap in the feed (Huitu, 2003). Figure 21.4 illustrates the flowsheet for scrap usage in the Mitsubishi process at Naoshima (Ariizumi et al., 2016; Oshima et al., 1998). Particulate scrap is mixed with concentrate and blown into the smelting furnace through its rotating lances. Larger scrap pieces are charged to the smelting and converting furnaces through roof and wall chutes. Mitsubishi converting is particularly exothermic, allowing large amounts of scrap to be melted in the converting furnace. The Onahama smelter in Japan has for some time used reverberatory furnaces to add ASR to sulfide matte. The enthalpy of combustion of the organic matter in the residue helps offset the extra fuel required for this (Kikumoto et al., 2000).

In contrast to smelting furnaces, scrap is usually needed in copper converters, to use the excess heat created by the converting reaction. Depending on matte grades, oxygen enrichment, and scrap quality, the fraction of scrap used in converter charges can range from 0% to 35% or higher.

A higher quality of scrap is needed for primary converters: low-alloy scrap, № 1 and № 2 scrap if available, along with compressed turnings and anode scrap. Low-grade material and plant reverts may also be fed if their plastic content is not too large (Oshima et al., 1998).

Scrap can also be added to anode furnaces, although the most common addition is directly recycled scrap anodes, rather than externally purchased scrap. Because anode furnaces were

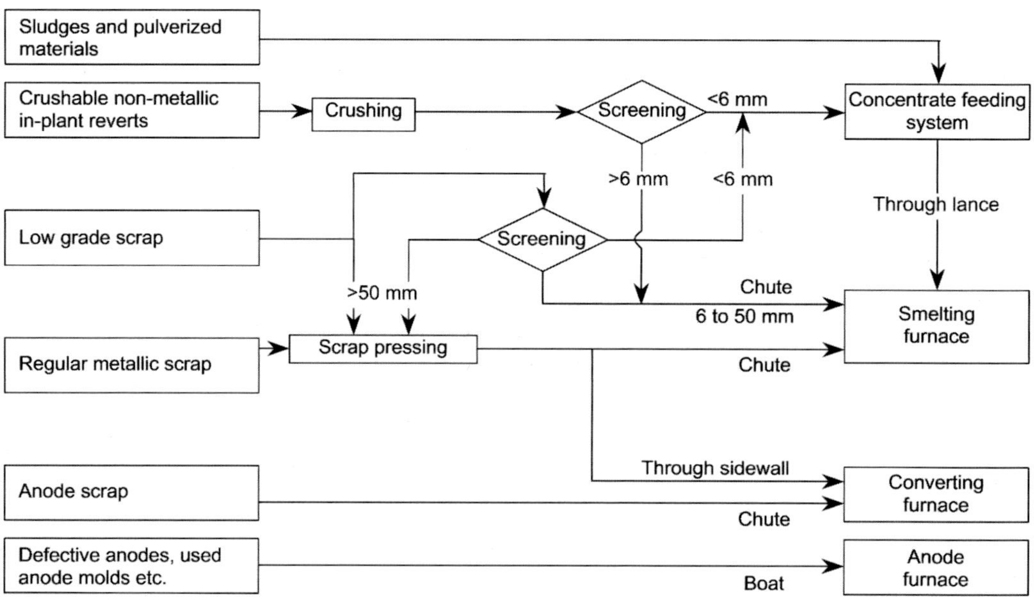

FIGURE 21.4 Scrap smelting in the Mitsubishi smelter, Naoshima. Low Cu-grade particulate scrap is fed into the smelting furnace through its rotating lances. Larger scrap is fed through roof and sidewall chutes. Large high-grade scrap is fed into the anode furnace. *After Oshima et al. (1998).*

designed to handle molten blister copper, scrap melting can be slow (Potesser et al., 2007; Suping and Qinghua, 2013). New furnace and burner designs can make anode furnaces better for melting scrap (Deneys and Enriquez, 2009).

21.2.3 The Secondary Copper Smelter

Figure 21.5 illustrates the possible routes copper scrap can take when processed in secondary copper smelters. Although each secondary copper smelter is unique, they can be divided into two groups. The first type is the *metal* smelter, which treats only higher-grade metallic scrap. Many of these smelters are located in China (Risopatron, 2007; Tang, 2016). The second is the *black copper* smelter, which processes low-grade scrap and waste along with higher-value scrap (Shuva et al., 2017). Hoboken in Belgium and Brixlegg in Austria are examples of this type of smelter (Hagelüken, 2006; Messner et al., 2005).

Rinnhofer and Zulehner (2005) described the typical process for scrap remelting in tiltable reverberatory furnaces. The furnaces typically have a small capacity (5–30 tonnes) and a fairly long cycle time (up to 24h tap-to-tap; Risopatron, 2007). They have the advantage of

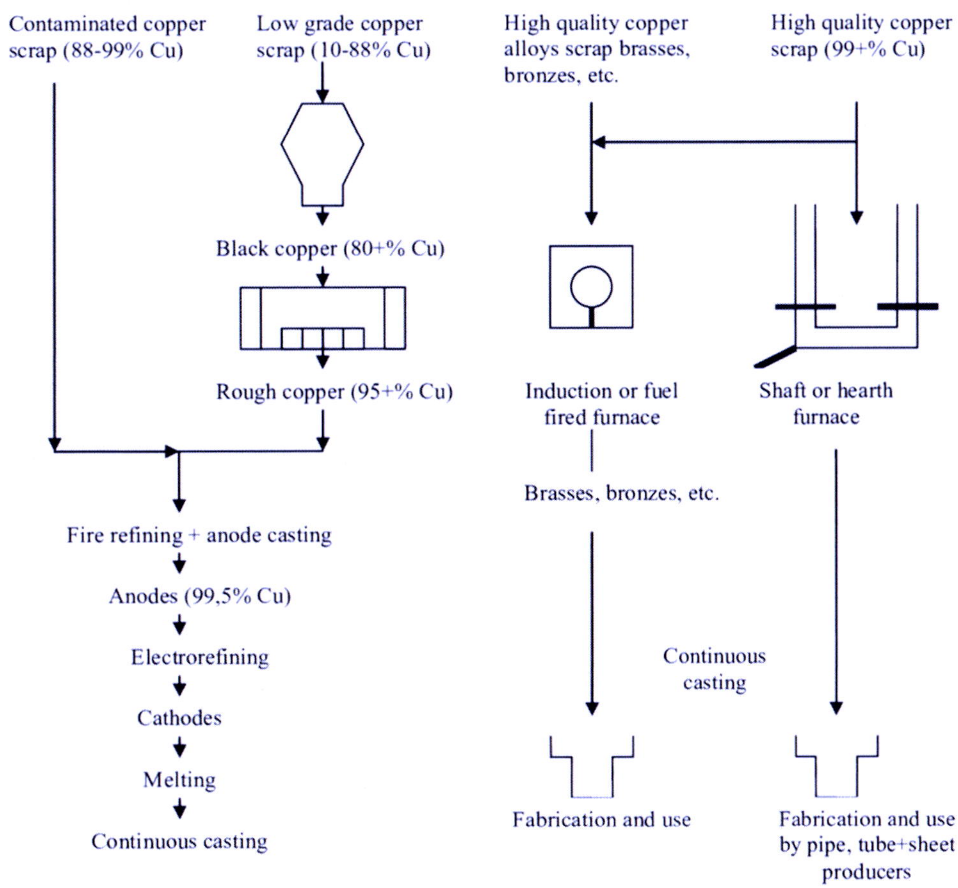

FIGURE 21.5 Flowsheet of processes for recovering copper and copper alloys from scrap. *From Schlesinger et al. (2021).*

flexibility and can be used for both oxidation of impurities and deoxidation afterward. Side tuyeres are used both for injection of air during oxidation and natural gas during reduction (Tang, 2016).

While reverbs are effective furnaces for refining operations, their efficiency as scrap melters is low (Potesser et al., 2007), and this has created incentive for more effective furnace design. Recent designs have added a reverb to a shaft furnace, which delivers molten copper to the reverb at a much quicker rate (Air Products, 2007; Tang, 2016). The refining capability of the reverb allows the use of more scrap in casting facilities than normal.

Figure 21.6 is a flowsheet for pyrometallurgical processing of low-grade scrap in a secondary black-copper smelter.

The "melting/reducing furnace" at the top accepts copper-bearing scrap, including:

- automobile shredder product from which the copper and iron cannot be separated (Shelmo), along with motors, switches, and relays (Elmo)
- dross from decopperizing lead bullion
- dusts from copper melting and alloying facilities
- sludges from copper electroplating operations
- auto shredder residue (ASR)

The feed to these furnaces is low grade and highly oxidized. It requires reduction to metallic copper. Major metallic impurities are lead and tin (from bronze scrap, solder, and decopperizing dross), zinc (from scrap brass), iron (from automotive scrap), and nickel (from scrap monel and other alloys). These elements are often present as mixtures of metal and oxide.

Heat and CO reductant are supplied to this type of furnace by combusting metallurgical

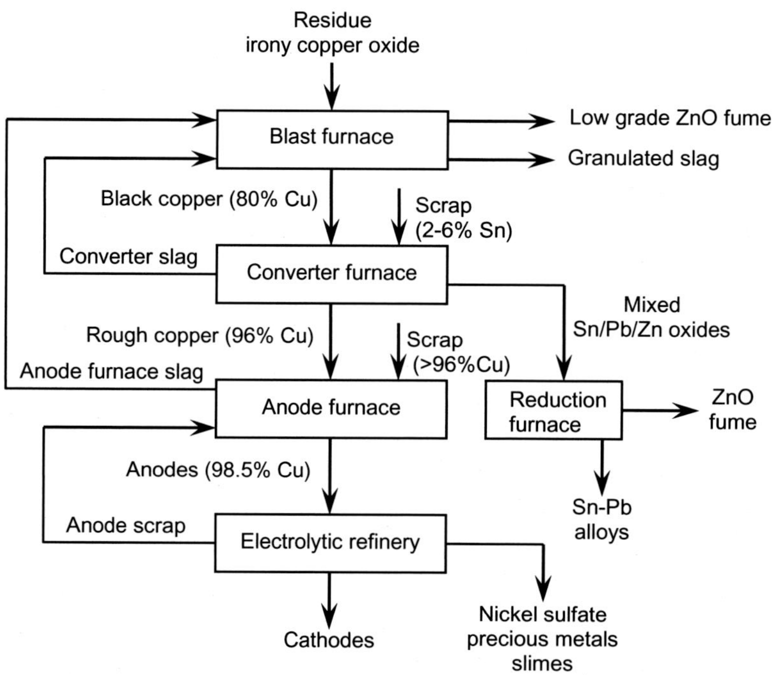

FIGURE 21.6 Flowsheet for low-grade copper scrap blast-furnace treatment. *From Schlesinger et al. (2021).*

coke or natural gas. The carbon monoxide in turn reduces the oxides of the feed to metal or a lower oxide. Metallic iron in the scrap also performs some reduction, especially of easily reduced oxides like Cu_2O. As a result of these reactions, secondary copper reducing furnaces generate three products. They are:

- molten *black copper*, 74–80 wt% Cu, 6–8% Sn, 5–6% Pb, 1–3% Zn, 1–3% Ni, and 5–8% Fe (Ayhan, 2000)
- molten slag containing FeO, CaO, Al_2O_3, SiO_2 along with 0.6–1.0% Cu (as Cu_2O), 0.5–0.8% Sn (as SnO), 3.5–4.5% Zn (as ZnO) and small amounts of PbO and NiO
- offgas containing CO, CO_2, H_2O, and N_2, plus metal and metal oxide vapors.

Cooling and filtering of offgas recovers oxide dust containing 1–2% Cu, 1–3% Sn, 20–30% Pb, and 30–45% Zn (Suping and Qinghua, 2013). The dust also contains chlorine from chlorinated plastics in the feed. It is always reprocessed to recover its metal content (Hanusch and Bussmann, 1995).

Traditionally a blast furnace was used as the vessel for initial melting and reducing of secondary copper. However, this furnace has all but disappeared in recent years (Tang, 2016), due to its thermal inefficiency and the need for expensive metallurgical coke. Two other types of furnace have become more prominent for this purpose.

The first is a Kaldo furnace (Outotec, 2018; Tang, 2016). This is an upgraded version of the well-known top-blown rotary converter (TBRC). The Kaldo furnace (Figure 21.7) is a batch processor, which burns natural gas or oil. The use of oxygen or enriched air reduces heat losses, and the oxygen/fuel ratio can be adjusted to produce a reducing or oxidizing atmosphere. The furnace rotates during operation, which improves kinetics, and the overall fuel usage is less than that of a blast furnace. Because they rotate, they are generally preferred for smaller

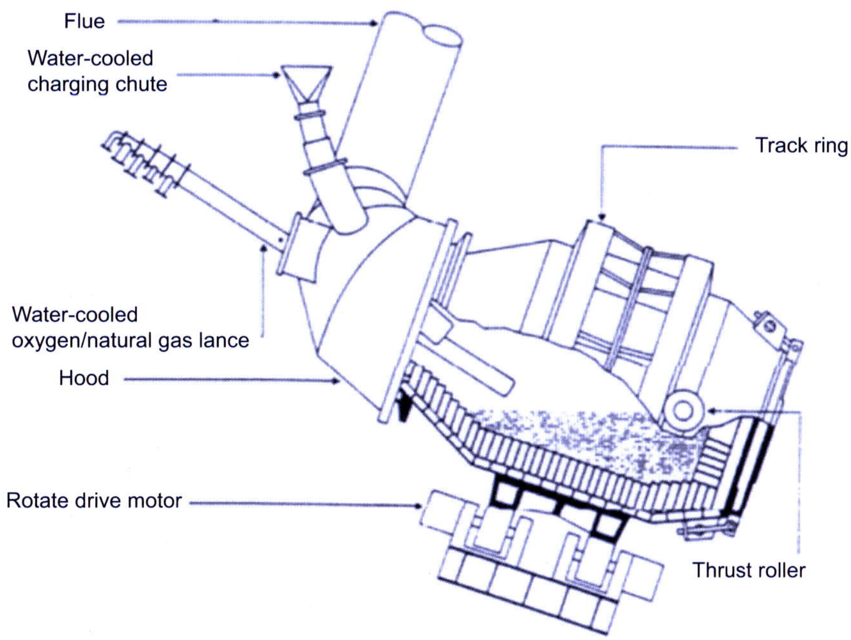

FIGURE 21.7 Kaldo top-blown rotary furnace used for melting and reducing copper. *From Samuelsson and Björkman (2014).*

operations. Kaldo furnaces are currently in use for secondary copper processing in Belgium, China, India, and Sweden (Pesl, 2010; Tang, 2016; Zhu and Liu, 2000). A Chinese variation of the technology has also been put into service.

The second type of commonly used furnace for secondary copper melting and reduction is the top submerged lance furnace, which is increasingly prevalent for sulfide matte smelting and converting (Hughes et al., 2019; Wood et al., 2011). Like the Kaldo furnace, Ausmelt and Isasmelt furnaces accept a wide variety of feeds and can process almost anything with copper in it (Alvear Flores et al., 2014; Ayhan, 2000). As stationary furnaces, they can burn coal or coke fines as a fuel, along with fuel oil and natural gas. They can also use the plastic in electronic scrap and auto shredder residue as a fuel (Brusselaers et al., 2006). They can change from a reducing to an oxidizing environment like a Kaldo furnace, which allows them to remove the impurities from black copper without needing another furnace. Isasmelt furnaces are being used for secondary copper production in Belgium and Germany, and three Ausmelt furnaces are in service in Asia.

Tang (2016) lists other types of furnaces developed in recent years in China. These include a horizontal rotary furnace (100 kt/year), which operates using a pure O_2 environment (Suping and Qinghua, 2013); a tilting stationary furnace (200 kt/year); and a top-blown furnace (200 kt/year), used at one Chinese smelter (Jinchuan).

The impurities in black copper can be divided into two groups—those that are more easily oxidized than copper (Fe, Pb, Sn, Zn) and those that are difficult or impossible to remove by oxidation (Ni, Ag, Au, platinum group metals). These impurities are removed sequentially by a strategy similar to that for purifying primary copper.

The first step in refining black copper is oxidation. The black copper produced by blast furnaces or Kaldo units is typically oxidized in a Peirce-Smith converter (Lennartsson et al., 2018). Black copper produced in a TSL furnace is refined as a second stage of operation in that furnace (Wood et al., 2011). In either case, air is blown into the molten black copper, oxidizing Fe, Pb, Sn, and Zn along with some Ni and Cu. Alloyed copper scrap is also added to the converter. Most of its impurities are also oxidized.

This oxidation generates slag containing 10–30% Cu, 5–15% Sn, 5–15% Pb, 3–6% Zn, and 1–5% Ni, depending on the composition of the converter feed (Ayhan, 2000). The slag can be returned to the blast furnace to recover the copper and nickel (Messner et al., 2005; Pesl, 2010), or processed first to recover the lead and tin (Hagelüken, 2006). Lead, tin, and zinc will ultimately be recovered either from slag or dust generated by one of the furnaces, but several trips through the circuit may be required before these elements wind up in a separate phase.

An offgas is also generated which, when cooled and filtered, yields dust analyzing 0.5–1.5% Cu (as Cu_2O), 0.5–1.5% Sn (as SnO), 10–15% Pb (as Pb and PbO), and 45–55% Zn (as ZnO). This dust is usually reduced to recover its Pb and Sn as solder (Ayhan, 2000).

Oxidation of black copper provides little heat to the converter. Heat for the converting process must, therefore, be provided by burning hydrocarbon fuel.

The main product of the converter is molten *rough copper*, 95–97% Cu. It is added to an anode furnace for final, controlled oxidation before being casted as anodes. № 1 scrap is also often added to the anode furnace for melting (Schlesinger et al., 2021).

Rough copper usually contains nickel or tin, which is not completely removed by oxidation converting. It may also contain appreciable amounts of gold, silver, and platinum group metals from the original scrap. Recovery of these metals is important to the profitability of a recycling facility. As a result, the anode furnace product is almost always cast as anodes for electrorefining. The impurity level in secondary anodes is higher than that in most anodes from primary smelting operations. As a result, electrolyte purification facilities need to be larger. Otherwise, plant practice is similar to that described by Schlesinger et al. (2021). The principal electrorefining products are high purity cathode copper, nickel sulfate from electrolyte purification, and anode slimes.

21.3 CONCLUSIONS AND OUTLOOK

High prices for copper ensure that recycled material will continue to be a significant fraction of world copper production. However, improving copper recovery from waste streams will be a focus of continued effort. Copper recovery from e-waste is a particular problem, because typical mineral processing techniques are not well-suited to upgrading the copper concentration, and other elements in e-waste are penalty elements that reduce its value as smelter feed (Mälhammar and Berglund, 2001). Udayakumar et al. (2022) have discussed the suitability of hydrometallurgical techniques for recovering copper and other metals from scrap PCB; the need to recover gold and other precious metals as well as copper makes developing a commercial process difficult. Copper recovery from brown and white goods is a difficult problem, because the recycling of these goods is primarily driven by the process of recovering the aluminum and iron with which they are built, rather than the small amounts of copper wire. As the number of electric vehicles grows, improving the recovery of the greater amounts of copper wire in these vehicles will be a growing concern (Sprovieri, 2014).

References

Air Products, 2007. Melting and Refining Technology for Copper Produced From Scrap Charge. Air Products and Chemicals Inc. Pub. No. 334-07-004-US http://www.airproducts.net.br/~/media/Files/PDF/industries/metals-melting-refining-technology-copper-scrap-charge.pdf. (Accessed 5 January 2022).

Alvear Flores, G.R.F., Nikolic, S., Mackey, P.J., 2014. ISASMELT™ for the recycling of E-scrap and copper in the U.S. case study of a new compact recycling plant. JOM 66 (5), 823–832.

Ariizumi, M., Takagi, M., Inoue, O., Oguma, N., 2016. Integrated processing of E-scrap at Naoshima Smelter and Refinery. In: Proceedings of Copper 2016. MMIJ, Tokyo, pp. 2587–2597.

Ayhan, M., 2000. Das neue HK–verfahren für die verarbeitung von kupfer–sekundärmateralien. Intensivierung Metallurgische Prozesse 27, 197–207.

Barnwal, A., Vishvakarma, S., Dhawan, N., 2018. Comparison of different routes for recovery of metals from electronic scrap. Mater. Today: Proc. 5, 17046–17054.

Brusselaers, J., Mark, F.E., Tange, L., 2006. Using Metal-Rich WEEE Plastics as Feedstock/Fuel Substitute for an Integrated Metals Smelter. PlasticsEurope. https://www.glencoretechnology.com/.rest/api/v1/documents/b1a9cdf54f69f9f63555a2e7b1fab37e/Using_metal-rich_WEEE_plastics_as_feed.pdf. (Accessed 5 January 2022).

Ciacci, L., Vassura, I., Passarini, F., 2017. Urban mines of copper: size and potential for recycling in the EU. Resources 6. https://doi.org/10.3390/resources6010006.

Deneys, A., Enriquez, A., 2009. Scrap melting in the anode furnace and the development of coherent jet technology in copper refining. In: Kapusta, J., Warner, T. (Eds.), International Peirce–Smith Converting Centennial. TMS, Warrendale, PA, pp. 321–338.

Glöser, G., Soulier, M., Tercero Espinoza, L.A., 2013. Dynamic analysis of global copper flows. Global stocks, postconsumer material flows, recycling indicators, and uncertainty evaluation. Environ. Sci. Technol. 476, 6564–6572.

Gómez, F., Guzman, J.I., Tilton, J.E., 2007. Copper recycling and scrap availability. Resour. Policy 32, 183–190.

Gordon, R.B., Bertram, M., Graedel, T.E., 2006. Metal stocks and sustainability. PNAS 103, 1209–1214.

Hagelüken, C., 2006. Recycling of electronic scrap at Umicore's integrated metals smelter and refinery. World Metall.—Erzmetall 59, 152–161.

Hanusch, K., Bussmann, H., 1995. Behavior and removal of associated metals in the secondary metallurgy of copper. In: Queneau, P.B., Peterson, R.D. (Eds.), Third Int. Symp. Recycling of Metals and Engineered Materials. TMS, Warrendale, PA, pp. 171–188.

Home, A., 2019. China Is Redrawing the Global Copper Scrap Map. Reuters https://www.reuters.com/article/metals-copper-ahome/column-china-is-redrawing-the-global-copper-scrap-map-andy-home-idUSL8N28L4GB. (Accessed 5 January 2022).

Hughes, S., Jåfs, M., Johto, H., Stål, J., Karonen, J., 2019. Outotec solutions for processing E-scrap. In: Gaustad, G., et al. (Eds.), REWAS 2019. TMS, Warrendale, PA, pp. 283–294.

Huitu, K., 2003. Using of secondary raw materials in copper processing. In: Helsinki Univ. Tech. Publ. Mater. Sci. Metall, pp. 43–51. TKK-MK-152.

ICSG, 2019. The World Copper Factbook 2019. International Copper Study Group. www.icsg.org. (Accessed 5 January 2022).

ISRI, 2021. ISRI. Scrap Specifications Circular 2021. Institute of Scrap Recycling Industries. http://www.scrap2.org/specs/ and www.isri.org (Accessed 5 January 2022).

Kaya, M., 2016. Recovery of metals and nonmetals from electronic waste by physical and chemical recycling processes. Waste Manage. 57, 64–90.

Kikumoto, N., Abe, K., Nishiwaki, M., Sato, T., 2000. Treatment of industrial waste material in reverberatory furnace at Onahama Smelter. In: Taylor, P.R. (Ed.), EPD Congress 2000. TMS, Warrendale, PA, pp. 19–27.

Koutamanis, A., van Reijn, B., van Bueren, E., 2018. Urban mining and buildings: a review of possibilities and limitations. Resour. Conserv. Recycl. 138, 32–39.

Lennartsson, A., Engström, F., Samuelsson, C., Björkman, B., Pettersson, J., 2018. Large-scale WEEE recycling integrated in an ore-based Cu-extraction system. J. Sustain. Metall. 4, 222–232.

Maeda, Y., Inoue, H., Kawamura, S., Ohike, H., 2000. Metal recycling at Kosaka Smelter. In: Stewart, D.L., Stephens, R., Daley, J.C. (Eds.), Fourth Int. Symp. Recycling of Metals and Engineered Materials. TMS, Warrendale, PA, pp. 691–700.

Mälhammar, G., Berglund, C., 2001. The influence of copper smelter requirements on design rules for electronics. In: Proc. Second Int. Symp. Environmentally Conscious Design and Inverse Manufacturing. IEEE, New York, pp. 128–131.

McGlothlin, S., 2018. Copper Recycling Process Technology in End of Life (ELV) Shredder Plants. www.metalbulletin.com/events/download.ashx/document/speaker/6539/a0ID000000X0jUWMAZ/Presentation. (Accessed 21 January 2020).

Messner, T., Antrekowitsch, H., Pesl, J., Hofer, M., 2005. Prozessoptimierung durch stoffstromanalyse am schachtofen der Montanwerke Brixlegg AG. BHM 150, 251–257.

Oshima, E., Igarashi, T., Hasegawa, N., Kumada, H., 1998. Recent operation for treatment of secondary materials at Mitsubishi Process. In: Asteljöki, J.A., Stephens, R.L. (Eds.), Sulfide Smelting '98. TMS, Warrendale, PA, pp. 597–606.

Outotec, 2018. Outotec® PM Kaldo Furnace. https://www.mogroup.com/portfolio/pm-kaldo-process/. (Accessed 27 December 2021).

Pesl, J., 2010. State of slag cleaning in secondary copper smelting. In: Presented at Short Course on Slag Cleaning at Copper 2010, Hamburg, Germany, 05.06.10.

Potesser, M., Antrekowitsch, H., Hollets, B., 2007. The elliptical anode furnace—a metallurgical comparison—part one. In: Warner, A.E.M., Newman, C.J., Vahed, A., George, D.B., Mackey, P.J., Warczok, A. (Eds.), Cu2007, Vol. III (Book 1): The Carlos Díaz Symposium on Pyrometallurgy. CIM, Montreal, pp. 667–680.

Pretz, T., Mutz, S., 2006. Metal recovery in automobile recycling. Acta Metall. Slovica 12, 456–462.

Rinnhofer, H., Zulehner, U., 2005. Gas-fired furnaces for copper and copper alloys. In: von Stark, A., Mühlbauer, A., Kramer, C. (Eds.), Handbook of Thermoprocessing Technologies: Fundamentals, Processes, Components, Safety. Vulkan-Verlag, Essen, pp. 358–364.

Risopatron, C., 2007. Copper Scrap Recycling Project. http://www.economia-dgm.gob.mx/dgpm/doctos/cedoc/Tres.pdf. (Accessed 16 November 2007).

Ruhrberg, M., 2006. Assessing the recycling efficiency of copper from end-of-life products in Western Europe. Resour. Conserv. Recyc. 48, 141–165.

Samuelsson, C., Björkman, B., 2014. Copper recycling. In: Handbook of Recycling: State-of-the-Art for Practitioners, Analysts, and Scientists, first ed. Elsevier, Amsterdam, pp. 85–94.

Schlesinger, M.E., Sole, K.C., Davenport, W.G., Alvear, G., 2021. Extractive Metallurgy of Copper, sixth ed. Elsevier, Amsterdam. ISBN: 9780128218754.

Shuva, M.A.H., Rhamdhani, M.A., Brooks, G.A., Masood, S., Reuter, M.A., Firdaus, M., 2017. Analysis for optimum conditions for recovery of valuable metals from E-waste through black copper smelting. In: Hwang, J.Y., et al. (Eds.), 8th International Symposium on High–Temperature Metallurgical Processing. TMS, Warrendale, PA, pp. 419–427.

Sprovieri, J., 2014. Wire harness recycling. Assembly Magazine https://www.assemblymag.com/articles/92263-wire-harness-recycling. (Accessed 5 January 2022).

Steinert, 2020. Non-Ferrous Metals. https://steinertglobal.com/us/metal-recycling/non-ferrous-metal-recycling/#. (Accessed 28 January 2020).

Sullivan, J.F., 1985. Recycling scrap wire and cable: the state of the art. Wire J. Int. 18 (11), 36–50.

Suping, Y., Qinghua, L., 2013. Process features and operation practices of NGL furnace in secondary copper smelter. In: Proceedings of Copper 2013. IIMCH, Santiago, pp. 323–335.

Tang, X., 2016. The Technological Developments of Scrap Copper Smelters in China. http://www.metalbulletin.com/events/download.ashx/document/speaker/8955/a0ID000000ZP0tpMAD/Presentation. (Accessed 21 January 2020).

Udayakumar, S., Razak, M., Ismail, S., 2022. Recovering valuable metals from Waste Printed Circuit Boards (WPCB): a short review. Mater. Today Proc. 55, 3062–3070.

Vexler, D., Bertram, M., Kapur, A., Spatari, S., Graedel, T.E., 2004. The contemporary Latin American and Caribbean copper cycle: 1 year stocks and flows. Resour. Conserv. Recycl. 41, 23–46.

Wildnauer, M., 2019. Life Cycle Assessment of Copper Tube and Sheet. Copper Development Association Inc. www.copper.org/environment/sustainability/pdfs/copper_life_cycle_assessment_tube_and_sheet.pdf. (Accessed 5 January 2022).

Wood, J., Creedy, S., Matusewicz, R., Reuter, M., 2011. Secondary copper processing using Outotec Ausmelt TSL technology. In: Proceedings of MetPlant 2011. AusIMM, Melbourne, pp. 460–467.

Yamashita, M., Hoshi, M., Suzuki, Y., 2016. JX Nippon Mining & Metals Group's comprehensive copper recycling process. In: Proceedings of Copper 2016. MMIJ, Tokyo, pp. 2575–2586.

Ye, F., Liu, Z., Xia, L., 2021. Material and energy balance of E–waste smelting – an industrial case study in China. Metals 11. https://doi.org/10.3390/met11111814.

Zhu, Y., Liu, W., 2000. Kaldo technology for copper scrap treatment. Youse Jinshu 52, 72–74.

Further Reading

Zhang, S., Forssberg, E., 1999. Intelligent liberation and classification of electronic scrap. Powder Technol. 105, 295–301.

CHAPTER

22

Lead

S. Arnout[1], B. Blanpain[2], D.R. Swinbourne[3], M. Chintinne[4], and M. Stevenson[5]

[1]InsPyro, Leuven, Belgium [2]Department of Materials Engineering, KU Leuven, Leuven, Belgium [3]School of Civil, Environmental and Chemical Engineering, RMIT University, Melbourne, VIC, Australia [4]Aurubis Beerse, Beerse, Belgium [5]Global Lead Technologies, Melbourne, VIC, Australia

22.1 INTRODUCTION

Lead is a soft, malleable, ductile, bluish-white, dense metallic element, extracted chiefly from galena (PbS) and found in ores together with zinc, silver, and copper. Lead is one of the first metals produced by humans, with beads of it dating from 6500 BCE having been found in Anatolia, Turkey (De Keersmaecker et al., 2018).

Today about 80% of lead is used in lead-acid batteries. A further 6% of lead is used in the form of lead sheet by the building industry. There are several other smaller volume applications for metallic lead, such as radiation shielding, cable sheathing, and various specialized applications, such as earthquake dampers. Recycling rates are high, and three-quarters of the lead production now originates from secondary sources.

Processing of spent lead-acid batteries occurs mainly in dedicated smelters. Their challenge is to capture sulfur, either from the paste before smelting, in a slag or matte during smelting, or from the smelting off-gases. As a result, many different processes are in use. The battery breaker-rotary furnace combination is the dominant route. Other wastes are processed in primary smelters, for which the furnace technology, even after drastic innovations in the last decades, is still evolving at a quick pace. In both segments, Chinese-developed technology plays a major role, increasingly also outside China.

Further technology development, focused on hydrometallurgy rather than furnaces, aims at closing the loop to battery materials without the emissions often associated with smelting.

22.2 MATERIAL USE

The mine and metal production, and the metal use of the last 5 years, are given in Table 22.1. It is notable that in spite of the gradually more stringent ban of lead from a number of its traditional applications (such as pigment, antiknocking agent, solder alloy, plumbing, gun shot), the level of lead usage has doubled in the last 30 years (ILZSG, 2021; US Geological Survey, 2012). About 90% of all lead is used in

TABLE 22.1 World refined lead production and usage 2016–20 (in '000 tons) (ILZSG, 2021).

Year	2016	2017	2018	2019	2020
Mine production	4704	4602	4572	4684	4501
Metal production	11,584	11,961	12,244	12,273	11,894
Metal use	11,541	12,104	12,290	12,225	11,754

readily recyclable products, conserving precious ore reserves for future generations. Lead enjoys one of the highest recycling rates of all materials in common use today. CHR Metals estimates secondary lead to account for 75% of total refined lead metal output globally (Roberts, 2020). In a decade, secondary production, without secondary production at primary smelters, rose from about 5 million t/y (Mt/y) in 2008 to over 8 Mt/y in 2018. Additionally, secondary production at primary smelters rose from 0.9 to 1.5 Mt/y. This growth can be largely attributed to China, where combined capacity increased to over 4 Mt/y. Because of the relatively unreactive nature of lead and the resulting ease of refining lead, the purity of the recycled lead is identical to that of primary metal from mining.

Generally, there are three major sources of secondary lead scrap feed:

- Lead-acid batteries and battery-derived streams such as lead parts, plates, and paste from spent battery dismantling or production waste.
- Metallic lead (alloy) from plumbing, weights, and cable sheathing.
- Intermediate products from nonferrous smelting and refining processes such as drosses and slags.

It is estimated that over 80% of the secondary lead originates from used lead-acid batteries (White, 2013). Most recycling processes combine the first two sources, although there are some using the third, leading to more complex bullion and refining treatments afterwards. Today there is a large range of smelting technologies in use for secondary smelting. We can distinguish between dedicated recycling technologies such as rotary furnaces, reverberatory furnaces, and the Varta blast furnace (secondary smelters), and flexible technologies such as QSL, Ausmelt, Isasmelt, SKS, Kivcet, and the traditional lead blast furnace, which primarily treat primary lead concentrates (primary smelters). Apart from primary lead smelters, copper smelters treating complex concentrates and recycling, e.g., printed circuit boards, produce small amounts of lead metal.

In 2018, of a world total production of lead from secondary sources of approximately 9.54 Mt, about 8.12 Mt came from dedicated lead-acid battery smelters, 0.85 Mt from battery scrap processed through primary smelting, and the remaining 0.57 Mt from other lead-bearing secondaries. The proportion of secondary lead recovered through primary smelters worldwide is therefore about 15% and the percentage recovered from battery scrap is about 9% (Roberts, 2020).

The use of a particular technology is as much a result of the expectations of the complexity of the feed at the time of plant design as it is of the inertia of capital investments in changing primary and secondary resource markets.

Lead production is one of the pivotal metallurgical flow sheets in primary and secondary metal production (UNEP, 2013), as through lead production associated metals such as Cu, Sn, Sb, Bi, and Ag are coproduced. It should also be noted that nonlead plants generate substantial amounts of lead and depend on it for separating metals and critical elements. Interesting examples of such secondary smelting operations can be found for precious metals recycling and Cu/Sn recycling (Campforts et al., 2013; Goris, 2013). As a result of its metallurgical carrier metal function in several flow sheets, lead is a crucial component in the circular economy, enabling recycling options and allowing for high recycle rates of several other metals (Blanpain et al., 2019).

22.3 THE LEAD-ACID BATTERY

Representing over 80% of the use of lead, batteries are the dominant application of lead as well as the dominant secondary resource for lead metal production. The lead acid battery was invented in 1859 by Gaston Planté and is now widely used as a stationary battery and SLI-type battery (Starting-Lighting-Ignition) (see, for instance, Crompton, 2000). In hybrid vehicles, it is widely used for energy storage and in electric vehicles as a back-up for safety features. Especially in China, it is used as the single energy source in electric bikes and trikes (Roberts, 2020).

The negative plate, or anode, is composed of a lead grid; the positive plate, or cathode, is also a metallic grid pasted with PbO_2. The electrolyte solution is an aqueous sulfuric acid solution (Figure 22.1). Due to the discharge reaction

$$PbO_2 + 2H_2SO_4 + Pb + 2H_2O \rightarrow 2PbSO_4 \quad (22.1)$$

the two electrodes are converted into $PbSO_4$. The half-reaction during discharge at the anode is:

$$Pb + SO_4^{2-} \rightarrow PbSO_4 + 2e^- \quad (22.2)$$

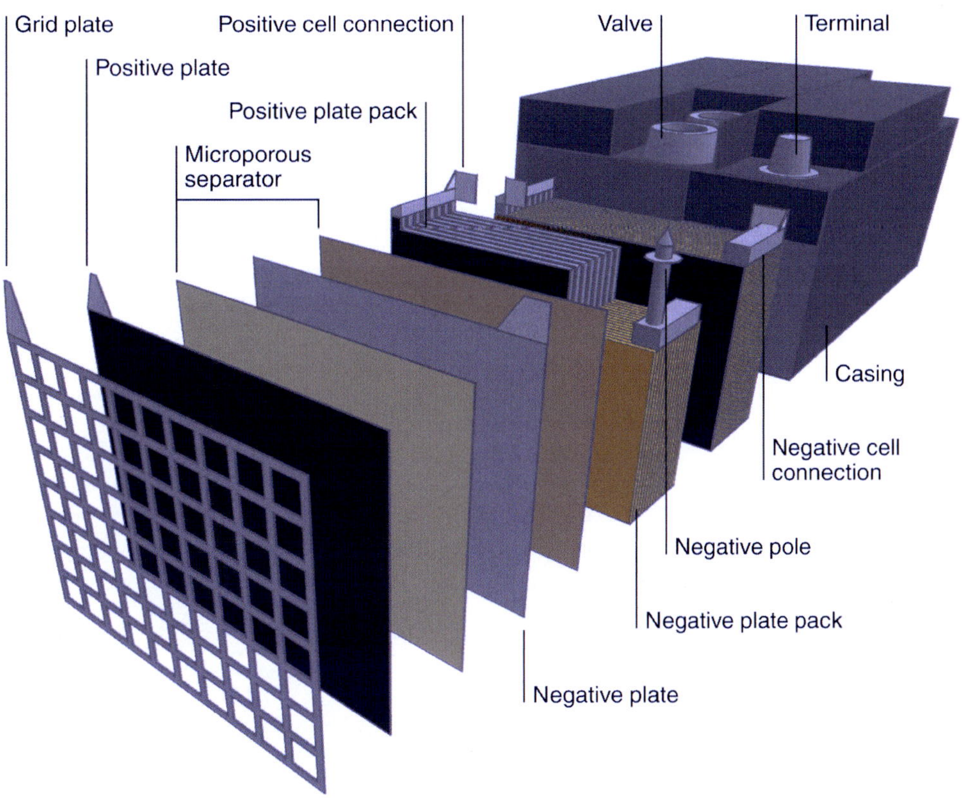

FIGURE 22.1 Schematic of an automotive lead-acid battery. *After InsPyro.*

and the half-reaction during discharge at the cathode is:

$$PbO_2 + 2H_2SO_4 + 2e^- \rightarrow Pb^{2+} + 2H_2O \quad (22.3)$$

During recharging these reactions proceed from right to left. The cell voltage is 2.0 V and the power capacity of these batteries can now be up to 35–40 Wh/kg. The weight continues to be the most important disadvantage of these batteries, notwithstanding the remarkable electrical performance achieved. The grid plates are produced from lead alloys to improve the mechanical strength, creep resistance, and castability, while minimizing the hydrogen evolution reaction. Commonly used materials are an Sb alloy with traces of Sn, Se, Cu, and As, and a Ca alloy containing Sn, Al, and often Ag (Siegmund and Prengaman, 2001).

The lead-acid battery is now a complex consumer product, made of several materials and additives. The composition of a lead-acid battery is shown in Table 22.2. The main components are lead, either as a metal, oxide, or sulfate, sulfuric acid, and polypropylene. The latter is a valued product as well and can be recycled (Jolly and Rhin, 1994).

According to research by the Battery Council International in the United States, more than 99% of all lead-acid batteries are recycled today, the highest recycle rate of all consumer products (Battery Council International, 2021).

22.4 RECYCLING TECHNOLOGIES

22.4.1 Battery Preprocessing

Lead-acid batteries are composed of several materials, as indicated previously. An important step in their recycling is therefore the separation of the distinct material fractions.

22.4.1.1 Acid Drainage

Upon reception at the recycling plant, a first step common to all sound recycling approaches is the removal of the sulfuric acid. Various simple but effective strategies are in place to crack the battery casing and drain the battery of its acid. Dropping the batteries from a height of a few meters into a storage bunker or driving over them with a front loader are common. The acid is collected into a basin for settling. In some cases, a press filter is used to recover additional solids and deliver cleaner acid. Depending on the local situation, it can be either transported for use in other applications (e.g., alkali waste neutralization (Stevenson, 2004)), or neutralized using sodium or calcium hydroxides. In the case of lime neutralization, the resulting gypsum

TABLE 22.2 Composition of typical lead-acid battery scrap (Gravita, 2013a).

Component	(wt%)
Lead (alloy) components (grid, poles, ...)	25–30
Electrode paste (fine particles of lead dioxide and lead sulfate)	35–45
Sulfuric acid (10–20% H_2SO_4)	10–15
Polypropylene	5–8
Other plastics (PVC, PE, etc.)	4–7
Ebonite	1–3
Other materials (glass, ...)	<0.5

needs to be disposed of, or may be mixed with the battery paste for smelting.

22.4.1.2 Breaking and Sorting

Breaking or dismantling batteries further, after drainage, separates the lead-bearing materials from the plastics or other fractions. Automated breakers are used in the most advanced plants, but manual or semiautomated dismantling lines are also in use. Large stationary batteries with a metal casing cannot be treated in automated breakers. They must be dismantled in less automated ways, often with custom-built auxiliary equipment.

Modern automated breaking and sorting lines are available from dedicated engineering companies and can treat over 50 t/h of used batteries. However, more often plants are smaller (Engitec, 2013a; Wirtz, 2023), e.g., treating 15–25 t/h, because the high transport costs for heavy battery scrap limit recyclers to the use of locally available source materials. The unit consists of a hammer mill or roller crusher, followed by hydrodynamic or sink/float separators to remove and sort plastics, and a further separation of the metals from the paste, based on their size (Stevenson, 2004). A flow sheet of a typical plant with a breaker and separators is given in Figure 22.2.

Separating the different materials has the advantage that the polypropylene casing can be recovered and sold for recycling into new plastic products, therefore increasing the recycling efficiency (European Commission, 2020). Further, besides the lead-rich fractions, a mixed heavy plastic fraction is also formed, which contains combined plastic-metal particles and plastics other than polypropylene. This fraction contains the polyethylene, silica, as well as decreasing amounts of PVC, from separators, with other materials such as fillers, ebonite, or glass. It can be added to a lead smelting furnace or used for energy recovery in a hazardous waste incinerator (Behrendt and Fisher, 2013).

The fractions used for lead recycling are the metallic fraction and the battery paste. The metallic fraction, containing the grids and connectors, is generally treated in a rotary furnace in the same plant, without the need for an extensive smelting cycle, as the amount of oxides/

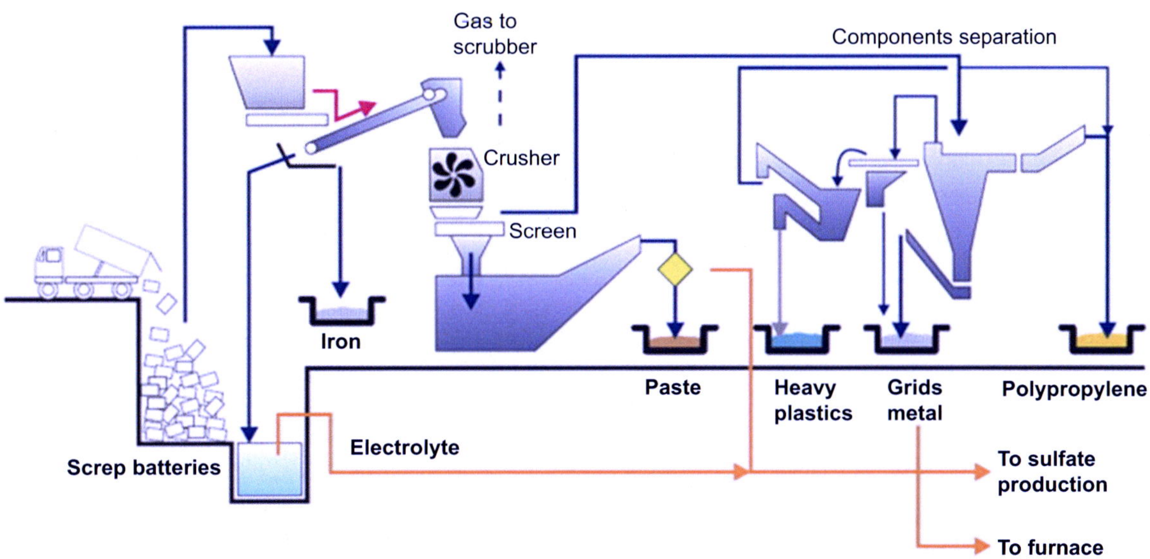

FIGURE 22.2 Typical flow sheet for a preprocessing plant: acid drainage, crusher, and sorting. *From Engitec (2013b).*

sulfates is minimal. The paste is also locally smelted, possibly after desulfurization, but can also be shipped to a primary smelter.

22.4.1.3 *Desulfurization of the Paste*

To facilitate local smelting, sulfur is often removed from the batteries' paste fraction. Paste desulfurization relies on the reaction of lead sulfate to lead carbonates, by the addition of a reagent to the slurry. The most used reagent is sodium carbonate, with the reaction being:

$$PbSO_4\ (s) + Na_2CO_3\ (aq) \geq PbCO_3\ (s) + Na_2SO_4\ (aq)$$

An excess of carbonate is needed, as it is also consumed to neutralize any remaining acid. The desulfurized paste contains about 0.5% insoluble sulfur and 1.5–2.5% sodium, which can be further reduced to <0.2% Na and < 0.2%S using more advanced superdesulfurization (Olper and Maccagni, 2009). This process uses CO_2 to fully convert Na-Pb carbonate-hydrates to $PbCO_3$ in the paste.

Other reagents that can be used are ammonium carbonate or urea, which lead to ammonium sulfate. The reactor design is more complicated, but in some countries the market for ammonium sulfate, used for fertilizer production, is larger than for sodium sulfate, used to produce detergents and glass (Olper and Maccagni, 2009). A process based on urea was recently installed in Nigeria (Scura et al., 2021). Instead of sodium carbonate, sodium hydroxide may also be used, leading to lead oxide instead of carbonate (Stevenson, 2004).

The resulting sodium or ammonium sulfate needs to be removed from the solution and this is achieved by evaporation and crystallization. The water is then recycled to the breaker. A flow sheet of a typical plant for desulfurization is given in Figure 22.3.

22.4.2 Smelting

22.4.2.1 *Secondary Smelting*

Here we discuss dedicated secondary smelting technologies. The choice of technology is dependent on a number of items, including the battery pretreatment method, size of the operation, source, market and availability of scrap feed, and the technical capability of the

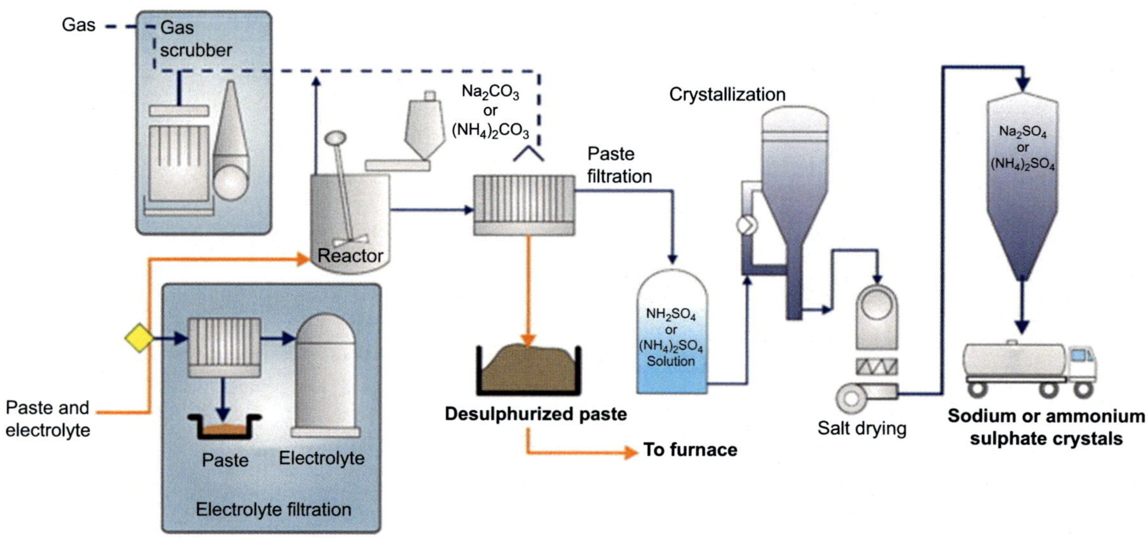

FIGURE 22.3 Typical flow sheet for a preprocessing plant: hydrometallurgical desulfurization. *From Engitec (2013b).*

country/region. There are three major pretreatment options:

- no separation,
- separation without desulfurization, and
- separation with desulfurization.

In all processes, reduction of the lead compounds (sulfate, oxide, or carbonate) is the critical chemical reaction. The most-used process with separation is the short rotary furnace, whereas without separation, it is the shaft furnace.

Smelters specializing in battery scrap typically handle small amounts of other waste fractions, such as cable sheathing or lead sheet from building applications. Clean scrap fractions can also be recycled in remelting facilities. For example, a lead sheet producer may collect nearly uncontaminated lead sheet and combine it with lead bullion from other plants to produce new alloys.

Short Rotary Furnace

By far the most important smelting technology consists of the numerous variants of the rotary furnace. The ease and simplistic nature of their operation, although hiding a metallurgical complexity, has allowed many new entrants into the market, with turnkey plants available and "up and running" within 6 months. Coupled with the soda-iron slag, the system has a wide tolerance for error, allowing for the operator to smelt lead without needing full control over the metallurgy and the production rate.

If the various battery fractions have been well separated (i.e., plastics, metals, oxides/sulfates) they can receive specific treatment. The oxide/sulfate fractions need reductive smelting to produce metallic lead bullion. Desulfurized paste, although low in sulfur, also needs reduction. The metal fraction may move straight to the refinery, but as it is contaminated with sulfate due to corrosion (Olper and Maccagni, 2009), and may contain some plastics, it is mostly treated in the same reductive smelting process to avoid emissions.

The sorting of battery fractions allows full use of the batch nature of the short rotary process. If operations allow for different batch cycles, paste can be smelted separately from metal fractions. The paste leads to a very pure soft lead, whereas the metallics contain the antimonial lead parts and other alloying elements. The antimony content, although limited, can then be concentrated into bullion for use at the refinery when producing antimonial lead.

The short rotary furnace is operated as a batch process (Figure 22.4). The units are relatively small, and it is common to operate two to four furnaces for one breaker. Heating is provided by a fuel or gas burner, with air or increasingly with oxygen, at the front or back of the furnace. The furnace is fed through a front opening, generally with a dedicated feeder. This feeder consists of a hopper, wide enough to be filled using a front loader, and an Archimedes screw on the bottom pushing the material through a tube into the furnace (e.g., Gravita, 2013b). Conveyor belts may also be used (Forrest and Wilson, 1990). The feed consists of lead containing fractions (paste, metal, and recirculated drosses), together with iron, and one or more fluxes.

The purpose of the iron is to capture the sulfur as FeS. To this extent, scrap is added as a flux, and reacts with $PbSO_4$, according to the global reaction:

$$Fe + PbSO_4 + CO \rightarrow FeS + Pb + 2CO_2$$

As FeS melts at relatively high temperatures for the rotary furnace (around 1200°C), fluxes are needed to decrease the melting point of the matte. Two strategies are employed: creating a sodium-rich slag/matte, commonly termed the soda-iron slag, or creating a calcium ferrosilicate slag. The soda-iron method is more widely adopted throughout the world.

The first strategy consists of the addition of soda ash (Na_2CO_3), which leads to a sulfur capture in sulfate as well as in sulfide form. In less

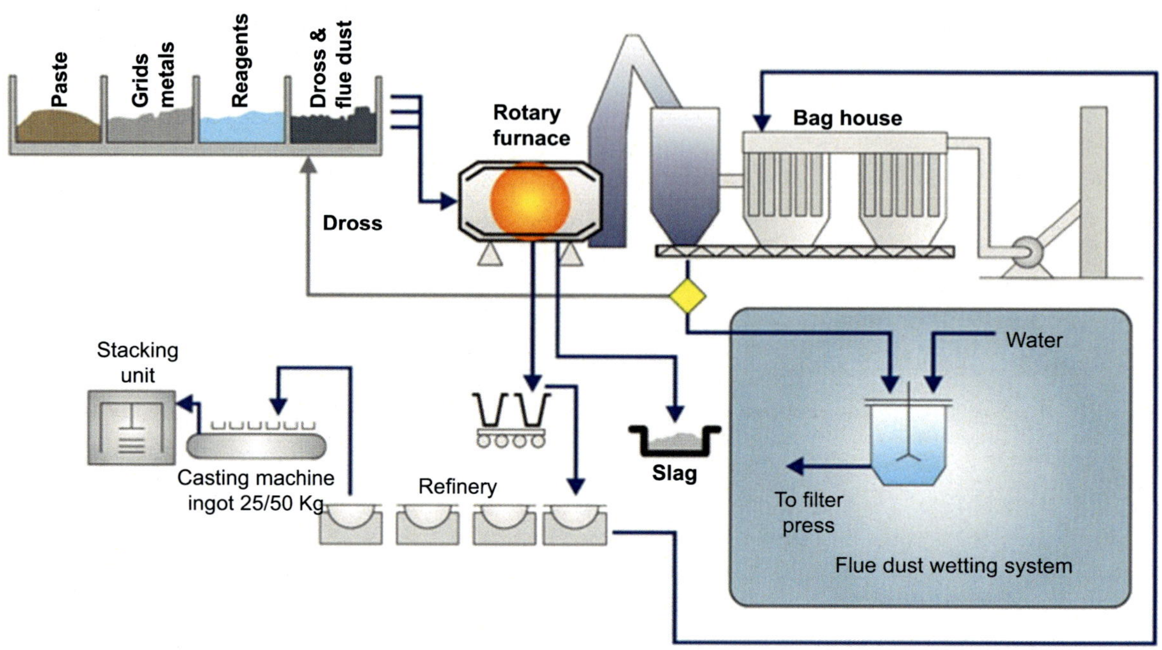

FIGURE 22.4 Rotary furnace secondary lead production flow sheet. *From Engitec (2013b).*

reducing conditions, the following global reaction can take place:

$$PbSO_4 + Na_2CO_3 + CO \rightarrow Pb + Na_2SO_4 + 2CO_2$$

Both Na_2CO_3 (Na_2O) and Na_2SO_4 have melting points below 900°C, which facilitates the reaction. At higher reduction, sodium sulfide is formed:

$$PbS + Na_2CO_3 + CO \rightarrow Pb + Na_2S + 2CO_2$$

Sodium sulfide also has a relatively high melting point (1180°C), but when FeS and Na_2S are combined in comparable quantities (25–75% FeS), they form a sulfide matte with melting points below 800°C (Kopylov and Novoselov, 1964). Similar to pure FeS, dissolution of other compounds such as iron oxide and metallic iron is expected (Arnout et al., 2011), but is not well understood. The low melting points lead to a simple and robust process that can operate below 1000°C. The environmental stability of the slag is a source of concern. The slag is hygroscopic and pulverizes easily during the reaction to hydrates (such as erdite) and further on to compounds such as iron oxides, sodium sulfate, and sodium carbonate. Further, the solubility of sodium compounds in water may allow for easy release of heavy metals to the environment (Stevenson, 2009). Research to study strategies for stabilization and reuse is ongoing (e.g., Knežević et al., 2010). Still, soda-iron slags are considered nonhazardous in some countries (Stevenson, 2020).

If soda-iron slag production is not allowed or leads to too-high landfill or treatment costs, a second strategy is used. As before, iron is added to capture sulfur as the sulfide. In this approach, lime and silica fluxes are added to create a calcium ferrosilicate slag (Stevenson, 2004). Depending on the scrap, the slag may already contain considerable amounts of iron oxide, or silica from separators. The lead alloys may lead to calcium and aluminum oxide formation in the slag. The silicate slag has considerable affinity and solubility for calcium and iron sulfide. This approach may require operating at higher temperatures. Foam depressants or slag modifiers may be used to prevent slag foaming or to decrease slag viscosity (Stevenson, 2004). The silicate structure allegedly may provide a more stable matrix for the heavy metals, resulting in lower leaching rates in landfills. Leaching can be further reduced by transforming the slag into a geopolymer material (Onisei et al., 2012), and by full reduction of heavy metals (e.g. fuming), potentially allowing reuse.

In the case of prior paste desulfurization, the smelting operation in the Short Rotary furnace is simplified, because of lower quantities of sulfur in the charge. Similar mattes or slags are created, but in much smaller quantities, down to 5–10% of lead produced according to Maccagni and Nielsen, 2015.

Shaft Furnace

Batteries are treated in a one-step continuous process, using a shaft furnace or cupola with afterburner, as shown in Figure 22.5. The Varta process (Martensson, 2013), which is a common shaft furnace design, can use whole batteries with casing; only drainage is required.

The smelting energy is provided by (possibly preheated or oxygen enriched) air firing up the coke in the hot burden. The burden of the furnace is composed of coke, iron scrap, drained batteries or battery fractions, and other recyclables such as cable lead or refinery drosses. In the shaft, as the burden moves down, it is gradually heated by the hot gases formed at the tuyere level and flowing upward. The plastics pyrolyze or disintegrate with slowly rising temperatures, and a large fraction evaporates in the form of volatile and complex organic compounds. Therefore afterburning is required. Next, the metallic fractions melt and collect in the hearth. At higher temperatures, deeper down in the shaft, decomposition and reduction of $PbSO_4$ and PbO take place by Fe and CO gas. As in the short rotary process, metallic iron scrap is added to capture sulfur as FeS.

This reaction produces a matte, which is treated in other metallurgical processes. Lead losses are limited, since with increasing iron content, the solubility for lead in the matte decreases to

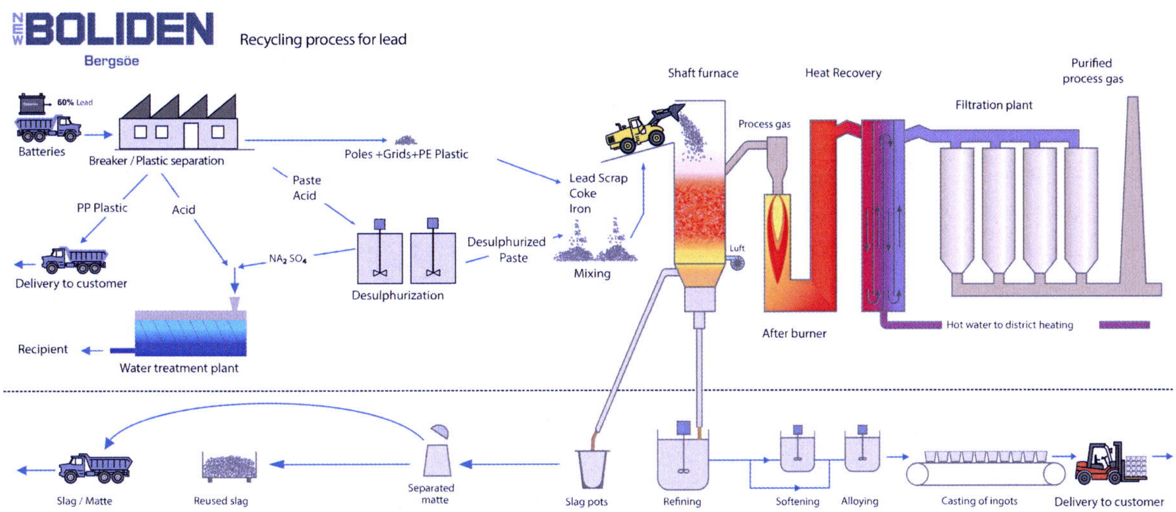

FIGURE 22.5 Shaft furnace secondary lead production. *After Martensson (2013).*

a few percent; 95% of the total sulfur input is captured in the matte. The remaining sulfur input leaves the furnace as SO_2 gas and is scrubbed in the gas cleaning section.

In the hearth, the slag melts at temperatures around 1100°C. The slag and matte are tapped at regular intervals at slightly higher temperatures. The slag is of the calcium ferrosilicate type, and a large fraction of it is recycled to the furnace feed. The lead is collected through a syphon.

The process gases leaving the furnace are afterburnt, preferably with heat recovery, to eliminate CO and volatile organic compounds. Injection of lime at high temperature reduces the SO_2 emissions. After cooling and dedusting, the gases are released through a stack.

The filter dust needs further treatment. The lead fraction in the filter dust justifies recycling to the furnace, but due to the chlorine content it is often first hydrometallurgically treated (Kunicky, 2013).

Reverberatory Furnace

After phasing out older style blast furnaces in the 1980s and 1990s, the United States (and United Kingdom) installed two-stage smelting processes, with the first part using a single reverberatory furnace and the second a slag/dross treatment furnace. In oxidizing conditions, a soft bullion is produced in the reverberatory, and the dross/slag contains the other elements such as antimony and tin. In this case, any sulfur not removed before smelting is largely captured by scrubbing SO_2 from the gas phase (Stevenson, 2020). The second furnace (set) can be rotary, electric, or another type, and recovers a highly alloyed bullion.

More recently, plants in China have used round reverberatory furnaces, with the ability to inject oxygen, for lead recycling. The permit application for a similar plant in Australia seems to indicate that these plants do not desulfurize paste nor capture sulfur in the slag, but rather turn sulfur into SO_2 during smelting and rely on gas scrubbers to capture it (Stevenson, 2020).

22.4.2.2 Primary Smelting

Primary smelting processes principally treat a lead-bearing concentrate from mined ores, but they can accept some recycled lead materials. These are often termed "secondaries" and include materials generated within the plant such as fume, drosses, and zinc-processing residues, as well as recycled lead-acid batteries. In times of restricted concentrate supplies, secondary materials provide a way for smelters to maintain output. Availability of secondary materials has also increased as a result of environmental legislation, so many smelter groups have found it cost effective to direct lead-acid battery waste to primary smelters (Hassall and Roberts, 2000).

Lead ores containing "galena" (PbS) are upgraded to a very fine-grained concentrate. The traditional approach to extracting lead from galena concentrates utilizes the blast furnace, fed from a sintering process, in which galena is oxidized and partially fused into large, strong, and porous lumps. The sinter/blast furnace route cannot accept significant amounts of battery paste (Sinclair, 2009), although a few traditional smelters have moved to accept 10–25% of their feed as secondaries, especially in Japan (Siegmund, 2000).

Since the 1980s lead smelters have been under pressure to improve their practices. The result was the introduction of novel "direct smelting processes" that directly feed concentrates, fluxes, and oxygen into one or two closed reactors. Such reactors achieve much greater levels of SO_2 capture, much more efficient use of the heat of oxidation of sulfur, and reduced manpower through automation. They utilize oxygen or oxygen-enriched air and so greatly lower the volume of off-gases produced and therefore the quantity of emissions released (Siegmund, 2000).

Their significance to lead recycling is that they open the possibility of feeding battery paste and metallics alongside concentrates into the reactors. Feeding secondary materials to primary smelters also has the advantage that expensive desulfurization of battery paste is unnecessary. However, the endothermic oxidation of $PbSO_4$ causes difficulties with the heat balance in the reactor, the presence of chloride in PVC separators in batteries affects the recovery of minor elements, and the very low concentrations of silver in battery scrap robs the smelters of a significant revenue stream (Stevenson, 2004).

Siegmund (2003), Stephens (2005), and Hayes et al. (2010) have reviewed the direct smelting technologies, which fall into two groups according to their process chemistry: three that produce most lead directly from oxidation of PbS, which are all bath smelting processes: QSL, OBBF, and Ausmelt Outotec/Isasmelt TSL; and the two processes that produce all lead from reduction of slag produced by the complete oxidation of PbS to PbO in a flash reactor: Kivcet and LOFS.

As mentioned before, the fraction of lead produced from secondary materials by primary smelters is small. However, in China in 2018 approximately 25% of the lead produced from secondary sources was produced through primary smelters (Roberts, 2020), mostly by OBBF (oxygen bottom-blown furnace) technology for the oxidation stage.

Li and Suo (2010) and Wu et al. (2020) have reviewed the evolution of the OBBF, a cylindrical furnace developed by SKS, and state that battery paste can comprise up to 50% of the feed.

A recent development is termed the OSBF (oxygen side-blown furnace), described by Chen et al. (2014, 2015). It utilizes the principles of the Russian Vanyukov furnace, in which a reactor similar in shape to the lead blast furnace is charged with molten slag and lump coal and in which the bath is violently agitated with oxygen or air injected through the tuyeres.

The QSL furnace has been described by Siegmund (2003). A German QSL smelter has operated with more than 50 wt% of secondaries in the feed (Meurer et al., 2005), while "secondaries" typically comprise approximately 17 wt% of the feed in this type of smelter. Note that this includes recycled fume, zinc plant residues, etc. (Deininger et al., 1994).

Top submerged lance (TSL) smelting was developed in Australia and is marketed by Outotec/Ausmelt (Sofra and Hughes, 2005; Hughes et al., 2008) and Isasmelt (Burrows et al., 2015). Two TSL smelters are reported to have a significant proportion of battery scrap in their feed: Weser-Metall GmbH (WMG) at Nordenham in Germany and KCM 2000 Group at Plovdiv in Bulgaria. The German TSL reactor accepts a feed having about 65 wt% as recycled materials, including battery paste (Pullenberg, 2010; Kerney, 2010). The slag contains approximately 50 wt% Pb. In 2018 they commissioned an OSBF, to reduce the slag to a lead content of less than 2 wt% (Weber et al., 2020; Recylex, 2019; Kerney, 2020). KCM 2000 Group in Bulgaria commissioned a new TSL lead smelting plant in mid-2014. Approximately 13% of the lead bearing feed comprises battery paste (Starev and Doganov, 2020).

The Kivcet process has been described by Siegmund (2003). Being a flash process, the Kivcet reactor can accept battery paste in the feed but faces problems handling solid battery scrap (Rich, 1994). All feed materials are fed through burners so must be dried and screened to remove coarse components, e.g., grids, poles, and plastics (Siegmund, 2003). Kivcet furnaces in Italy and Canada were reported to treat battery waste (Perillo et al., 1989; Siegmund, 2003).

In China, a process is in use called lead oxygen-enriched flash smelting (LOFS), which is very similar to the Kivcet process. In 2009 it was demonstrated that it could process lead-acid battery waste (Gao et al., 2012).

22.4.3 Refining

The lead bullion may contain numerous impurities, depending on the original lead alloys in the battery. Liquid lead is an excellent solvent for a wide range of elements and the bullion can contain a large number of elements such as Cu, Ni, S, As, Ag, Sn, Sb, and Bi. Secondary lead, especially from lead-acid batteries, is typically less complex in composition than primary bullions from lead concentrate smelting. It tends to contain higher amounts of Sb and considerably lower amounts of precious metals.

In refining, we can distinguish between primary and complex secondary bullions, secondary bullion, and oxidized bullion.

1. For primary and complex secondary bullion, the full range of refining is required, starting with Cu liquation, followed by Sn, Sb, As oxidation, precious metals refining, and Bi removal. Electrolytic refining is also a possibility.
2. For secondary bullion, Cu, Sn, and Sb are the main impurities to be removed. As the input of As, Ag, and Bi is limited, no specific refining is carried out for these elements.
3. Some smelters operate with highly oxidizing conditions. Impurities such as Sn, As, and Sb report to the Pb-rich slag, and the bullion only contains Cu and Ag. The subsequent refining process is then limited to the removal of Cu and Ag.

Pyrorefining is the most widely used refining method (Hayes et al., 2010). It consists of a series of treatments, all of which aim at the elimination of one, or a family, of similar impurities, as a dross from the lead bullion surface. It typically occurs in a sequence of batch processes:

- Decoppering. By decreasing the temperature, Cu precipitates, and by adding sulfur or pyrite, CuS and Cu_2S are formed with low solubilities.
- Removal of Sn, As, Sb (softening). The liquid lead is heated and oxidized. Oxidation can be done by air/oxygen injection or by sodium salts (Harris process, decreasing use).
- Separation of the precious metals. In primary plants, Ag and Bi impurities are removed with the Parkes and Kroll-Betterton process, respectively. The low levels of these impurities in recycled lead do not necessitate their removal in secondary plants, nor allow for an economic recovery of these metals.
- Final cleaning and alloying. Residues of previous steps are removed and, in many cases, alloying additions are made to produce alloys to the specification of battery producers.

22.5 FUTURE DEVELOPMENTS

Research and development of battery recycling processes aim at solutions with lower waste and lower emissions. Existing plants are upgraded with breakers, paste desulfurization, enhanced ventilation, gas scrubbers, and regenerative thermal oxidizers to reduce exposure to the environment and personnel. Personal protective equipment and procedures are continuously optimized to reduce blood lead levels in personnel.

Paste desulfurization is an important step to avoid SO_2 emissions, and recent developments have led to alternative chemistries to the sodium carbonate scheme, such as sodium hydroxide or urea, as discussed earlier.

Full hydrometallurgical treatment would avoid any smelting and therefore drastically reduce any concentrated and diffuse emissions of lead and sulfur. For example, further hydrometallurgical treatment of the desulfurized paste is possible using a chloride-based flow cell technology for electrolytic reduction to lead metal (Pizzocri and Maccagni, 2013). The system consists of ammonia chloride and hydrogen peroxide leaching. Pilot and demonstration plants have been built. Other processes with decreased interest are based on fluorosilic acid (Prengaman and McDonald, 1990) and chloride (Andrews et al., 2000) leaching and electrowinning.

An alternative to electrowinning metal may be recovering lead oxide, which could theoretically be a source for battery production, but rarely meets the requirements to date. Also, apart from recovering lead, many challenges remain to manage and recover the other elements in the battery.

The following hydrometallurgical processes are currently investigated in the pilot stage. The company Aqua Metals developed a process of recovery of lead by the dissolution of desulfurized paste in methane sulfonic acid (MSA) and plating the pure lead onto rotating anodes. A full plant was built but did not operate to capacity and, since a fire, has been pulled back to pilot plant stage. Residues from the process need processing in a furnace (Stevenson, 2020).

Based on research at Cambridge University, the company Aurelius developed a method involving sodium hydroxide and citric acid to produce PbO, suitable for batteries. A pilot plant has been constructed (Fox, 2019).

Clearly, even if hydrometallurgy comes with challenges, the approach opens a pathway to even lower lead exposure in the environment or for workers in the industry. Additionally, by enabling recycle loops directly from end-of-life products to technological materials, further economic and efficiency gains may be realized.

22.6 KEY ISSUES AND CHALLENGES

Lead production and use have continued their growth over the last 15 years, mainly due to the fact that lead-acid batteries have remained the battery of choice for standard SLI applications and stationary systems. It is expected that this will not change in the near future (Prengaman, 2009; Wilson, 2011), but developments in Li-based batteries may render them competitive in terms of cost and environmental impact. At present, the climate impact at the production stage of lead-acid batteries is considerably lower than for Li-based batteries (Usbeck et al., 2020). Mainly in China, electric motorbikes have adopted the lead-acid battery, although the outlook for this application may change (Roberts, 2020). At the same time, specific lead-acid battery designs are widely used in hybrid electric vehicles (Battery Council International, 2013).

Due to lead's physicochemical properties as well as its use in products that are conducive to closing its materials life cycle, lead recycling has been widely practiced and may be considered exemplary. Still, a small but unacceptable portion of lead recycling occurs in an informal sector without environmental, health, and safety policies, especially in developing countries. Using simple extraction methods, the readily recoverable metal is collected, leaving high Pb waste streams, which can contaminate villages long after the operations are halted. Many countries are increasingly enforcing standards, but as many people's income depends on the informal sector, the transition to a fully clean recycling sector requires efforts from both governments as well as the formal sector to adapt regulations, social policies, trading schemes, and technologies (ILA, 2021). Despite health concerns, the practice of lead metallurgy will need to be continued, as lead plays a vital role in the extraction and recycling schemes of several critical elements (Blanpain et al., 2019).

A wide range of lead production technologies is in use. Each technology has its advantages and drawbacks, and choices are made based on feed types as well as local situation and preference. The adaptability of most technologies to treat secondary materials, at least partially, is remarkable, so much so that it is in some cases difficult to make a clear distinction between primary and secondary lead producers. Focused secondary producers efficiently recover lead as well as PP casings. The challenge remains to avoid Pb and SO_2 emissions, as well as to deal with the trace elements and separator materials. Additionally, the production of environmentally stable slags with low heavy metal content, which allow for reuse, remains challenging. In view of climate change, carbon-free or climate-neutral recycling technologies will need to be developed. Hence, it may be expected that the drive toward the development and use of flexible, clean, and energy-efficient technologies will continue.

References

Andrews, D., Raychaudhuri, A., Frias, C., 2000. Environmentally sound technologies for recycling secondary lead. J. Power Sources 88, 124–129. https://doi.org/10.1016/S0378-7753(99)00520-0.

Arnout, S., Nagels, E., Blanpain, B., 2011. Thermodynamics of lead recycling. In: Proceedings of EMC 2011, p. 363.

Battery Council International, 2013. The Advanced Lead-Acid Battery, the Greatest Unknown Recycling Story. The Advanced Lead-Acid Battery Consortium (02.11.13). http://www.alabc.org/.

Battery Council International, 2021. National Recycling Rate Study. https://batterycouncil.org/page/Recycling Study.

Behrendt, H.-P., Fisher, M., 2013. Lead recycling and hazardous waste incinerator with tradition and future. In: Proceedings of EMC 2009, GDMB, p. 411.

Blanpain, B., Reuter, M., Malfliet, A., 2019. Lead Metallurgy is Fundamental to the Circular Economy, Policy Brief. https://kuleuven.sim2.be/wp-content/uploads/2019/02/SOCRATES-Policy-Brief-2019-Lead.pdf.

Burrows, A.S., Azakenov, T., Zatayev, R., 2015. Lead ISASMELT™ operations at Ust Kamenogorsk. In: Proceedings of Lead Zinc 2015, GDMB, pp. 245–256.

Campforts, M., Foerier, S., Vasseur, K., Meskers, C., 2013. Valorisation of critical metals from high temperature residues at Umicore. In: 3rd Slag Valorisation Symposium, pp. 235–247.

Chen, L., Yang, T., Bin, S., Liu, W., Zhang, D., Bin, W., Zhang, L., 2014. An efficient reactor for high-lead slag reduction process: oxygen-rich side blow furnace. J. Met. 66 (9), 1664–1669.

Chen, L., Hao, Z., Yang, T., Liu, W., Zhang, D., Zhang, L., Bin, S., Bin, W., 2015. A comparison study of the oxygen-rich side blow furnace and the oxygen-rich bottom blow furnace for liquid high lead slag reduction. J. Met. 67 (5), 1123–1129.

Crompton, T.R., 2000. Battery Reference Book, third ed. Butterworth-Heinemann, Oxford, UK.

De Keersmaecker, M., Dowsett, M., Adriaens, M., 2018. A short historical overview on the use of lead. In: Adriaens, M., Bioletti, S., Rabin, I. (Eds.), Chemical Interactions Between Cultural Artefacts and Indoor Environment. ACCO, Leuven, Belgium; The Hague, The Netherlands, pp. 193–214.

Deininger, L., Choi, K.C., Siegmund, A.H.-J., 1994. Operating experience with the QSL-plants in Germany and Korea. In: Warren, G.W. (Ed.), Proc. EPD Congress 1994. TMS, pp. 477–501.

Engitec, 2013a. Lead Recycling References. http://www.engitec.com/PDF/LEADREF.pdf. November 2013.

Engitec, 2013b. CX® Breaker, Separation and Filtration Unit. http://www.engitec.com/PDF/CXSYSTEM.pdf. November 2013.

European Commission, 2020. Guidelines on the Application of Commission Regulation EU 493/2012 Laying Down Detailed Rules Regarding the Calculation of Recycling Efficiencies of the Recycling Processes of Waste Batteries and Accumulators. https://ec.europa.eu/environment/waste/batteries/pdf/Guidelines%20on%20RE.pdf. Retrieved January 2021.

Forrest, H., Wilson, J.D., 1990. Lead recycling utilising short rotary furnaces. In: Lead-Zinc '90, TMS.

Fox, A., 2019. Out of the furnace and into the leaching pot. In: 6th International Secondary Lead Conference, September.

Gao, W., Wang, C., Yin, F., Chen, Y., Yang, W., 2012. Situation and technology progress of lead smelting in China. Adv. Mater. Res. 581–582, 904–911.

Goris, D., 2013. Do you know metallo chimique? In: Recycling Metals From Industrial Waste, Workshop, Golden, CO, June.

Gravita, 2013a. Lead Battery Scrap. http://www.gravitatechnomech.com/Lead-scrap/Lead-battery-scrap.html#. 2 November 2013.

Gravita, 2013b. Rotary Furnace. http://www.gravitatechnomech.com/equipments/rotary-furnace.html. November 2013.

Hassall, C., Roberts, H., 2000. Market fundamentals of lead and zinc supplies. In: Dutrizac, J.E., Gonzalez, J.A., Henke, D.M., James, S.E., Siegmund, A.H.-J. (Eds.), Proc. Lead-Zinc 2000. TMS, pp. 1–16.

Hayes, P.C., Schlesinger, M.E., Steil, H.-U., Siegmund, A., 2010. Lead smelter survey. In: Siegmund, A., Centomo, L., Geenen, C., Piret, N., Richards, G., Stephens, R.I. (Eds.), Lead Zinc 2010, 5th Decennial Symposium. TMS, pp. 343–414.

Hughes, S., Reuter, M.A., Baxter, R., Kaye, A., 2008. Ausmelt technology for lead and zinc processing. In: International

Symposium on Lead and Zinc Processing; Lead and Zinc 2008. SAIMM, pp. 147–161.

ILA, 2021. Recycling in the Informal Sector. ILA—International Lead Association. http://members.ila-lead.org/UserFiles/File/Recycling%20in%20the%20Informal%20Sector.pdf.

ILZSG, 2021. Lead and Zinc Statistics. ILZSG—International Lead and Zinc Study Group. https://www.ilzsg.org/static/statistics.aspx. January 2021.

Jolly, R., Rhin, C., 1994. The recycling of lead-acid batteries: production of lead and polypropylene. Resour. Conserv. Recycl. 10, 137–143.

Kerney, U., 2010. Recyclex PB production in Nordenham with bath smelting technology—an update. In: Siegmund, A., Centomo, L., Geenen, C., Piret, N., Richards, G., Stephens, R.I. (Eds.), Lead Zinc 2010, 5th Decennial Symposium. TMS, pp. 415–427.

Kerney, U., 2020. Experiences with the new reduction furnace in Nordenham, Presented at the Meeting of the GDMB Lead Experts Committee, 2020 (Virtual), October 29–30.

Knežević, M., Korać, M., Kamberović, Ž., Ristić, M., 2010. Possibility of secondary lead slag stabilization in concrete with presence of selected additives. Metall. Mater. Eng. (MJoM) 16 (3), 195–204.

Kopylov, N.I., Novoselov, S.S., 1964. The Cu2S-FeS-Na2S system. Russ. J. Inorg. Chem. 9 (8), 1038.

Kunicky, Z., 2013. Hydrometallurgical treatment of flue dust from lead blast furnace, especially for chlorine and thallium removal. Presented at the Meeting of the GDMB Lead Experts Committee, Aachen, Germany, May 15–17.

Li, D., Suo, Y., 2010. Oxygen bottom-blowing smelting lead smelting technology by bottom blowing electrothermal reduction. In: Siegmund, A., Centomo, L., Geenen, C., Piret, N., Richards, G., Stephens, R.I. (Eds.), Lead Zinc 2010, 5th Decennial Symposium, TMS, pp. 483–486.

Maccagni, M., Nielsen, J., 2015. Improvements in secondary lead smelting. In: Proceedings of Pb-Zn 2015, p. 307.

Martensson, C., 2013. New Boliden Bergsöe, Landskrona, Sweden: safety and environmental work at Bergsöe. In: GDMB Lead Experts Meeting.

Meurer, U., Pűllenberg, R., Griesel, J., 2005. Developing QSL technology to an economically superior process. In: Fujisawa, T., Dutrizac, J.E., Fuwa, A., Piret, N.L., Siegmund, A.H.-J. (Eds.), Proc. International Symposium on Lead and Zinc Processing; Lead & Zinc '05. MMIJ (Mining and Materials Processing Institute of Japan), pp. 547–558.

Olper, M., Maccagni, M., 2009. The secondary Pb challenge—the Engitec approach. In: Proceedings of EMC, p. 441.

Onisei, S., Pontikes, Y., Van Gerven, T., Angelopoulos, G.N., Velea, T., Predica, V., Moldovan, P., 2012. Synthesis of inorganic polymers using fly ash and primary lead slag. J. Hazard. Mater. 205–206, 101–110.

Perillo, A., Carlini, G., Ibba, R., 1989. Performance of the KSS lead smelter in Portovesme, Italy. In: Jaeck, M.L. (Ed.), Primary and Secondary Lead Processing, Halifax, Nova Scotia, Canada. The Metallurgical Society of CIM, pp. 15–26.

Pizzocri, M., Maccagni, M., 2013. FAST Pb process: its impact on the new secondary lead facility. In: Pizzocri, M., Maccagni, M. (Eds.), Proceedings of EMC 2013, p. 549.

Prengaman, R.D., 2009. Lead industry—transition, evolution, and growth (abstract). In: Proceedings of EMC 2009, p. XVII.

Prengaman, R.D., McDonald, H., 1990. Lead-Zinc '90. The Minerals, Metals and Materials Society, pp. 1045–1076.

Pullenberg, R., 2010. Integrated battery recycling at Berzelius Metall GmbH Germany—a closed loop! In: Siegmund, A., Centomo, L., Geenen, C., Piret, N., Richards, G., Stephens, R.I. (Eds.), Lead Zinc 2010, 5th Decennial Symposium. TMS, pp. 703–708.

Recylex, 2019. The New Reduction Furnace of Weser-Metall. Website https://recylex.eu/en/the-new-reduction-furnace-of-weser-metall-gmbh/. Accessed: 6 January 2022.

Rich, V., 1994. The International Lead Trade, first ed. Woodhead Publishing, Cambridge, UK, p. 48.

Roberts, H., 2020. Evolution of global secondary lead production. In: Siegmund, A.H.-J., Alam, S., Grogan, J., Kerney, U., Shibata, E. (Eds.), 9th International Symposium on Lead and Zinc Processing, PbZn 2020. TMS, pp. 3–15.

Scura, F., Bergamaschini, A., La Sala, G., Urgese, S., 2021. New route for the lead paste desulphurization U4Lead process. World Metall. ERZMETALL 74, 33.

Siegmund, A.H.-J., 2000. Primary lead production—a survey of existing smelters and refineries. In: Dutrizac, J.E., Gonzalez, J.A., Henke, D.M., James, S.E., Siegmund, A.H.-J. (Eds.), Proc. Lead-Zinc 2000. TMS, pp. 55–116.

Siegmund, A.H.-J., 2003. Modern applied technologies for primary lead smelting at the beginning of the 21st century. In: Kongoli, F., Itagaki, K., Yamauchi, C., Sohn, H. Y. (Eds.), Proc. Yazawa International Symposium Metallurgical and Materials Processing: Principles and Technologies, High-Temperature Metal Production. vol. 2. TMS, pp. 43–62.

Siegmund, R., Prengaman, R.D., 2001. Grid alloys for automobile batteries in the new millennium. JOM 53 (1), 38–39.

Sinclair, R., 2009. The Extractive Metallurgy of Lead (CD). Australasian Institute of Mining and Metallurgy, ISBN: 978 1 921522 02 4.

Sofra, J., Hughes, R., 2005. Ausmelt technology operation at commercial lead smelters. In: Fujisawa, T., Dutrizac, J.E., Fuwa, A., Piret, N.L., Siegmund, A.H.-J. (Eds.), Proc. International Symposium on Lead and Zinc Processing; Lead & Zinc '05. MMIJ (Mining and Materials Processing Institute of Japan), pp. 511–528.

Starev, N., Doganov, G., 2020. KCM–innovator in the Pb metal production through Ausmelt technology and variable SO_2 concentration off-gas utilization. In: Siegmund, A.H.-J., Alam, S., Grogan, J., Kerney, U., Shibata, E. (Eds.), 9th International Symposium on Lead and Zinc Processing, PbZn 2020. TMS, pp. 321–326.

Stephens, R.L., 2005. Advances in primary lead smelting. In: Fujisawa, T., Dutrizac, J.E., Fuwa, A., Piret, N.L., Siegmund, A.H.-J. (Eds.), Proc. International Symposium on Lead and Zinc Processing; Lead & Zinc '05. MMIJ (Mining and Materials Processing Institute of Japan), pp. 45–72.

Stevenson, M.W., 2004. Recovery and recycling of lead-acid batteries. In: Mosely, P.T., Garche, J., Parker, C.D., Rand, D.A.J. (Eds.), Valve-Regulated Lead-Acid Batteries. Elsevier, pp. 491–512 (Chapter 15).

Stevenson, M., 2009. Presentation at the International Secondary Lead Conference, Macau.

Stevenson, M., 2020. Technical review of the proposed secondary lead smelter at Hazelwood North, Victoria. Report to EPA Australia, August.

US Geological Survey, 2012. Lead; world refinery production, by country. Matos, G.R. (comp.), In: Historical Global Statistics for Mineral and Material Commodities (2015 Version). U.S. Geological Survey Data Series, https://doi.org/10.3133/ds896. 896, x p.

UNEP, 2013. In: Reuter, M.A., Hudson, C., van Schaik, A., Heiskanen, K., Meskers, C., Hagelüken, C. (Eds.), Metal recycling: opportunities, limits, infrastructure. A report of the working group on the global metal flows to the international resource panel.

Usbeck, V.C., Bonell, M., Gediga, J., 2020. Comparative Life Cycle Assessment of Batteries for Automotive Applications. Sphera. August. Executive Summary at https://www.acea.auto/uploads/publications/ES-CLCA-V5.pdf.

Weber, P., Behrmann, D., Breuer, T., Margraf, R., 2020. Installation of a brown field slag reduction furnace: state of the art off-gas treatment with dry gas cleaning for SO_2 capture. In: Siegmund, A.H.-J., Alam, S., Grogan, J., Kerney, U., Shibata, E. (Eds.), 9th International Symposium on Lead and Zinc Processing, PbZn. TMS, pp. 187–200.

White, P., 2013. Private communication, 12 November 2013; ILZSG Press release, 4 October 2013.

Wilson, D., 2011. Lead—a bright future for the grey metal? World of Metallurgy Erzmetall 4, 196.

Wirtz, 2023. https://www.wirtzusa.com/product/brs-battery-breaking-and-separation-systems/.

Wu, W., Xin, P., Wang, J., 2020. The latest development of oxygen bottom blowing lead smelting technology. In: Siegmund, A.H.-J., Alam, S., Grogan, J., Kerney, U., Shibata, E. (Eds.), 9th International Symposium on Lead and Zinc Processing, PbZn 2020. TMS, pp. 327–336.

CHAPTER

23

Zinc

J. Antrekowitsch, S. Steinlechner, G. Hanke, and U. Brandner

University of Leoben, Leoben, Austria

23.1 INTRODUCTION

The annual global production of zinc is 13.8 Mt (2020). More than 60% of this is used for galvanizing, while the rest is mainly split into brass production (17%), zinc-based alloys (17%), semimanufacturers (6%), and zinc compounds such as zinc oxide and zinc sulfate (6%). The estimated global capacity of zinc oxide production was 2 Mt in 2015. The actual consumption was 1.5 Mt worldwide, split into the usage of 55% in rubber and tire, 20% in the field of ceramics, 8% in the chemical industry, as well as 7% agricultural usage, 3% in paintings and coatings, and 7% in other utilization areas (Schlag, 2016).

Zinc is recycled at all stages of production and use—for example, from scrap that arises during the production of galvanized steel sheet, from scrap generated during manufacturing and installation processes, and from end-of-life products. The presence of a zinc coating on steel does not restrict steel's recyclability and all types of zinc-coated products are recyclable. Zinc-coated steel is recycled along with other steel scrap during the steel production process—the zinc volatilizes and is then recovered. For the zinc and steel industries, recycling of zinc-coated steel provides an important source of raw material. Historically, the generation of zinc-rich dusts from steel recycling was a source of loss from the life cycle (landfill); however, technologies exist today that provide an incentive for steel recyclers to minimize waste. Thus the recycling loop is endless—both zinc and steel can be recycled again and again without losing any of their physical or chemical properties (World Bureau of Metal Statistics, 2018; International Zinc Association, 2015, 2016).

Depending on the composition of the scrap being recycled, it can either be remelted or returned to the refining process.

The overall recycling rate is difficult to define due to the wide lifetime range of the various zinc products. Therefore, typical values can range between 8% and 30% depending on which types of scraps or residues are considered as recycled materials. The recycled content for refined zinc is currently about 8% (scrap combined with concentrate as feedstock at the smelter). When considering the sum of all material used to produce zinc-bearing goods, the recycled content jumps to approximately 25–30%. However, the recycling rate of zinc is increasing, underlining the considerable importance of this industry in the world's zinc (World Bureau of Metal Statistics, 2018; International Zinc Association, 2015, 2016).

23.2 RECYCLING TECHNOLOGIES

In the following sections, the main processes for zinc recycling from different scraps and residues are described. An overview of the various residues as well as the related recycling processes and products is provided in Figure 23.1.

The remelting of alloys is a major field in zinc recycling. Most of the furnaces typically used in foundries allow a remelting of various scrap types. Furnaces are heated either electrically or by gas. Also, induction furnaces are in use. Depending on the product range and the purpose (premelting, melting, holding), various furnace designs are available.

Remelting of zinc scrap consumes less than 10% of the energy required to produce primary zinc. To save transport and storage costs, flexible in-house recycling seems to have become more popular than external remelting (Tan and Khoo, 2005).

The other scraps and residues are, depending on their purity, converted to zinc dust, high-quality zinc oxide, zinc oxide concentrate or impure concentrate, or zinc compounds.

23.2.1 Zinc Oxide Production From Dross

One method still used for zinc oxide production today was developed by Le Claire in France in 1840, by burning metallic zinc in air—the so-called "French process" or "indirect" process. At the same time in America, a different production route was developed—the so-called "American" or "direct" process. The name originated from the possibility of making direct use of oxidic ore and other oxidic secondary materials such as ashes, clinker, and top dross from hot dip galvanizing (Steinlechner, 2013). The commonly utilized raw material for the American process is zinc dross. It is generated during the hot dip galvanizing and is mainly the solid material floating on top of the zinc bath, accumulating impurities as well as oxidized zinc.

As shown in Figure 23.2, the following distinction can be made for the main processes used for high-purity ZnO production, including their raw materials.

Both the French and American processes make use of the volatility of metallic zinc. Metallic zinc is melted, or oxides are reduced to a metal and vaporized at furnace temperatures above 907°C, typically around 1000°C. Due to an excess amount of air, the gaseous metallic zinc immediately reacts to solid ZnO particles in an exothermic reaction with a bright luminescence. Attributable to the volatilization of zinc within the processing of materials, purification can be achieved by exploiting the different partial pressures of compounds present within the

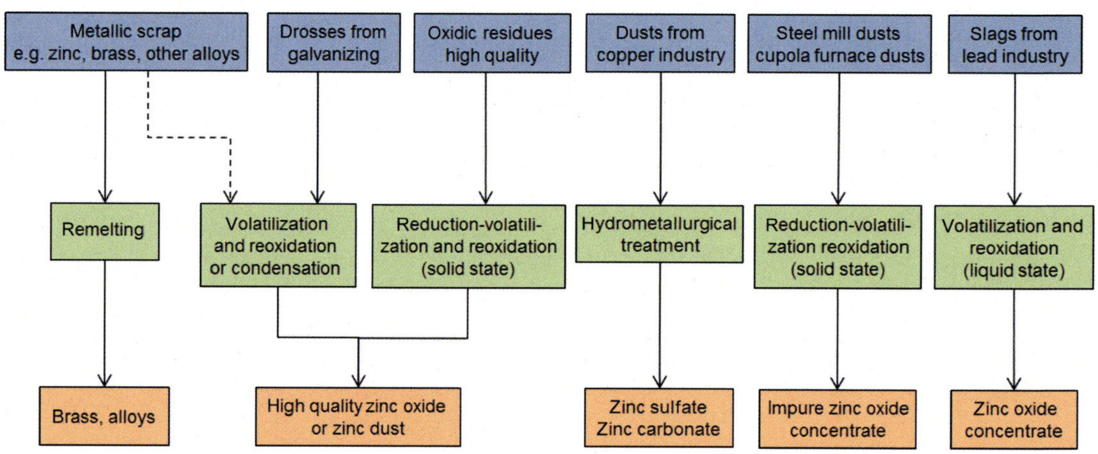

FIGURE 23.1 Main processes for zinc recycling from residues.

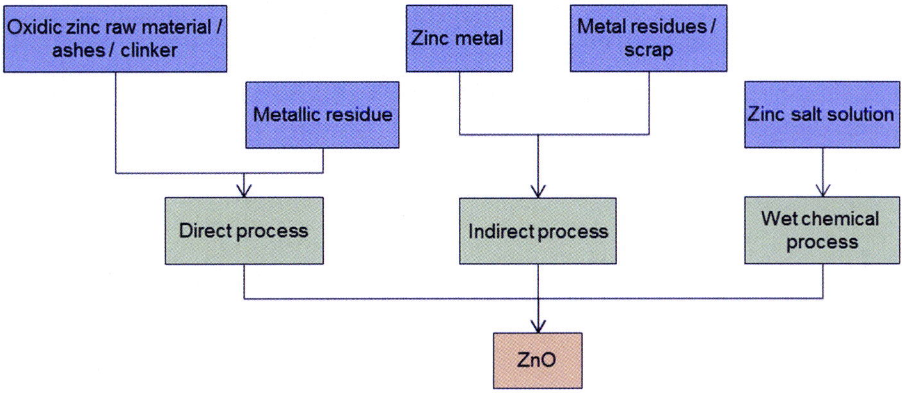

FIGURE 23.2 Different methods for ZnO production and utilized raw materials. The direct process is also called the American process and the indirect process is also called the French process.

production process. Critical elements or compounds are those with similarly high or higher vapor pressure as metallic zinc, such as cadmium, bismuth, and magnesium or manganese compounds. In both the French and American processes, air is used as a cooling and oxidation agent for the vaporized zinc. Furthermore, the excess air works as a transportation medium for the formed zinc oxide particles to the product filter house (Tan and Khoo, 2005; Steinlechner, 2013).

The third method used for ZnO production is a hydrometallurgical one based on zinc hydroxide or zinc carbonate production as a semifinished product. The advantage of this method is the possibility of several cleaning steps before the product is obtained. Due to that, a wide variety of possible raw materials can be utilized, but this goes hand in hand with increasing process complexity. The semifinished product is then processed to ZnO by way of thermal treatment. Depending on the thermal conditions, the specific surface area is up to 100 m^2/g (Tan and Khoo, 2005; Steinlechner, 2013).

Alternatively to the three processes for high-purity zinc oxide production, a fourth option is possible, namely the Larvik furnace. The Larvik furnace can be split into three zones: melting, vaporization, and separation. The input material is similar to what is used in the French process. The difference is that this furnace type has a significantly higher acceptance of impurities, like lead, iron, etc. As an example, recovery as a lead and iron alloy can be mentioned; the lead and iron are casted together and separated by density in the casting mold. Possible materials are again bottom dross, crude skimming, or grinded skimming. The required melting heat is provided by the enthalpy of the off gas in the vaporization zone. Nonmeltable components are removed from the liquid metal surface in the melting zone. Special attention is given to aluminum in the input material, due to the possible skin formation on the zinc metal surface by aluminum oxide, preventing continuous zinc vaporization and therefore possibly causing an explosion like zinc vaporization and oxidation. Another special aspect of this process is the addition of phosphorus to form a eutectic alloy with iron, with the aim of a low liquidus temperature (lower than 1538°C). This offers the opportunity for a continuous separation of the iron as a liquid phase. The iron alloy sinks to the bottom in the same way as liquid lead, because of the higher density, and is separated during the casting. The same occurs with the lead, forming a third liquid metal layer in the furnace, next to zinc and the already mentioned iron alloy (phosphor) (Steinlechner, 2013).

Commercial grades of zinc oxide can be found—depending on their quality—in a wide range of utilization areas. This includes specialities, such as the semiconductor or tire industries, aside from daily-life products such as suncreams or pharmaceuticals, utilizing products of high purity containing typically 99.99% ZnO produced via the French process. Products with slightly lower quality are the ceramics industry or as food additives for animals, which is typically produced via the American process, generating a product with up to 99% ZnO. Aside from the chemical purity, properties such as specific surface area, coatings, or individual customer requirements influence the decision for the chosen/required ZnO production process (Steinlechner, 2013).

Compared by weight, most of the high-purity ZnO produced is manufactured by the French process, showing a market share of around 60%. In general, high quality means any zinc oxide that does not have to be returned to primary zinc production, as an ore concentrate substitute. American quality zinc oxide follows with slightly less than 40%, and only a minor amount, typically utilized in special fields, is manufactured by the wet route (Tan and Khoo, 2005).

23.2.2 Electric Arc Furnace Dust

In 2019, 517 Mt of steel were produced via the electric arc furnace (EAF) (World Steel Association, 2021). The melting of steel scrap in an EAF results in the generation of 15–25 kg electric arc furnace dust (EAFD) per ton steel (Guézennec et al., 2005). Therefore approximately 8 Mt EAFD are currently generated every year. The forecasted increase in steel production via the EAF will result in a rise in EAFD formation, leading to a predicted generation of up to 18 Mt/y by 2050 (Xylia et al., 2016). EAFD is classified as hazardous waste by most developed countries and contains significant amounts of valuable metals (Table 23.1), which is why

TABLE 23.1 Chemical composition range of EAFD from carbon steel plants.

Element	(wt.%)	Element	(wt.%)
Fe	10–45	S	0.1–3.0
Zn	21–43	Si	0.3–2.4
Ca	2.0–12	Na	0.2–2.2
Pb	0.4–10	K	0.4–2.0
Cl	0.8–5.0	Al	0.2–1.6
Mn	0.8–4.5	Cr	0.1–1.4
Mg	0.3–3.6	Cu	0.1–0.5
C	0.4–3.3	Cd	0.0–0.2

Modified from Remus et al. (2013).

processing EAFD is not only mandatory for ecological reasons, but also highly attractive from an economic point of view.

In general, the recycling of EAFD is conducted in pyrometallurgical processes based on carbothermal reduction. The Waelz process still dominates the EAFD recycling market because only a few alternative process concepts have been implemented industrially so far (Lin et al., 2017). In this section, the Waelz process and the most important alternative pyrometallurgical process concepts are described.

Waelz Process

The Waelz process comprises three steps, which are depicted in Figure 23.3.

The first step is the agglomeration of the raw materials with reducing agents (e.g., petroleum coke) and slag-forming additives (e.g., lime and quartz).

In the second step, the agglomerated raw material is charged into the kiln. The material moves through the furnace slowly due to the slightly decreasing slope and is heated up by the counter-current flowing gas (Mager et al., 2000). Figure 23.4 illustrates the cross section of the slowly rotating Waelz kiln and the occurring chemical reactions.

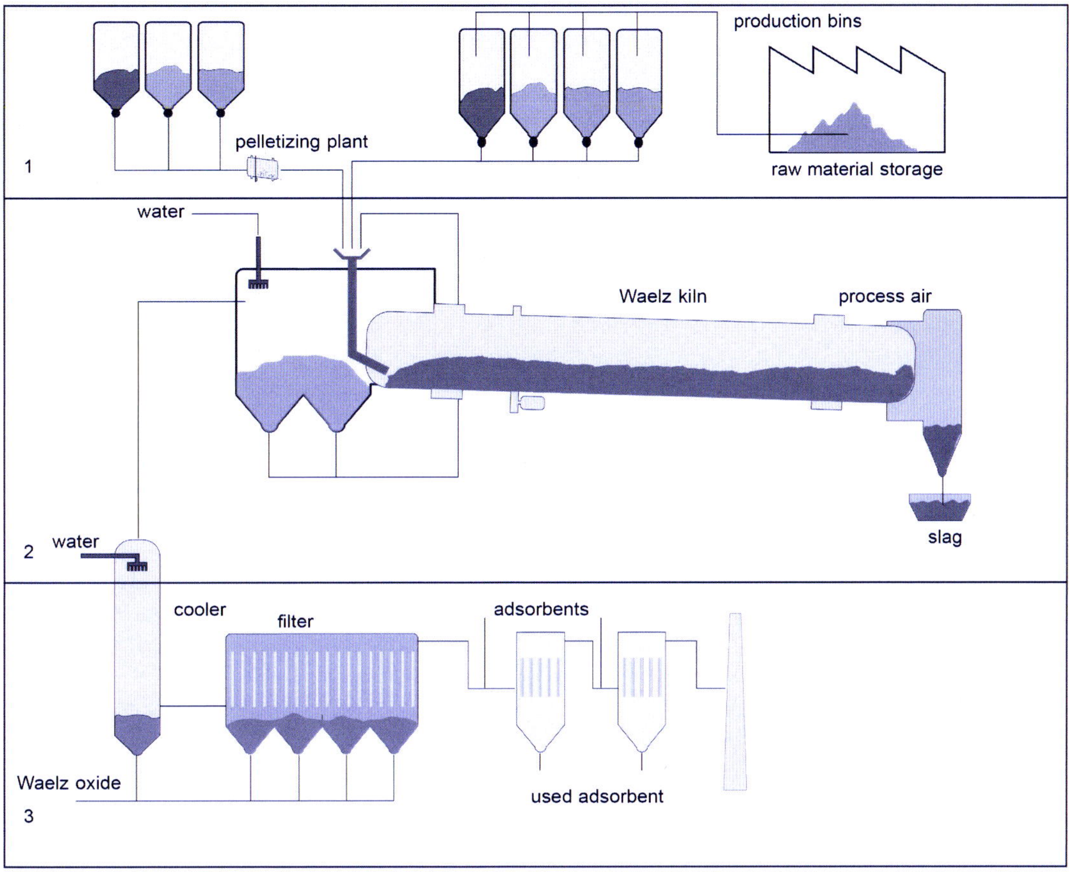

FIGURE 23.3 Schematic overview of a Waelz kiln plant consisting of three steps: 1. Material preparation and mixing; 2. Pyrometallurgical treatment in the Waelz kiln; 3. Exhaust gas treatment and Waelz oxide separation. *Modified from Mager et al. (2003).*

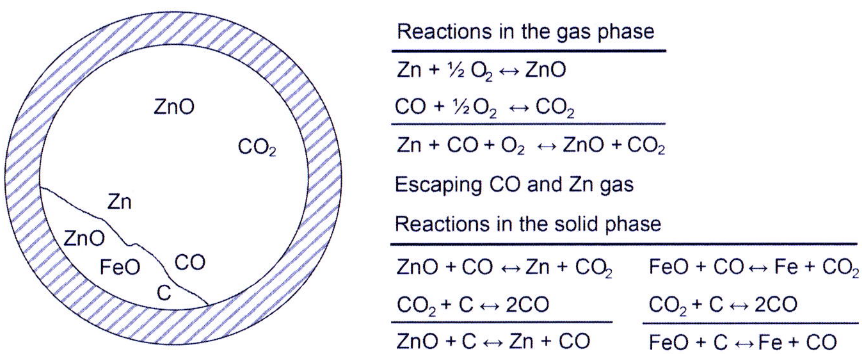

FIGURE 23.4 Waelz kiln cross section and chemical reactions of the Waelz process. *Based on Mager et al. (2000).*

After the drying of the solid material, carbon starts to reduce the iron oxide and zinc oxide. The reduced zinc evaporates due to its high vapor pressure and is transferred from the solid to the gas phase. In the oxidizing gas atmosphere, zinc combusts to zinc oxide. The zinc oxide particles leave the kiln as part of the exhaust gas. The reoxidation of zinc generates heat, which is an important part of the heat balance of the kiln (Mager et al., 2000).

In the end section of the kiln, iron is reoxidized by the addition of air via a lance. Due to the heat generated by the reoxidation of the iron phase, no additional energy supply is necessary for the endothermic reduction reactions in the kiln. About 600–650 kg of Waelz slag are generated per ton of treated EAF dust. The Waelz slag is discharged at a temperature of approximately 800°C and cooled in the granulation plant by the addition of water before it is mainly landfilled (Antrekowitsch and Offenthaler, 2010).

In the third step, zinc oxide is separated from the exhaust gas (N_2, CO_2, H_2O, CO, and O_2) and collected in the filter house (Figure 23.3). Similarly, other volatile metals and respective compounds like Pb, Cl, and F are also collected in the flue dust. These impurities contaminate the zinc oxide; thus the term "crude Waelz oxide" is used (Mager et al., 2000).

Before crude Waelz oxide can be processed in primary zinc production, unwanted compounds need to be removed, so the crude Waelz oxide needs to be pretreated. Soda leaching lowers the chlorine contamination to enable its usability in the primary zinc metallurgy without harming the electrolysis process too much. However, fluorine removal is often less effective. Despite the major amount of the halogens being removed during washing, it is generally not possible to substitute more than 15% of the primary concentrate with washed Waelz oxide (Antrekowitsch and Offenthaler, 2010).

The Waelz process is still the best available technology for EAFD processing. In particular, the reoxidation of the former reduced iron in the kiln reduces the carbon consumption and increases the throughput capacity, respectively (Mager et al., 2000).

Rotary Hearth Furnace (RHF)

The rotary hearth furnace (RHF) was originally developed for dusts from the primary iron and steel production (sinter dust, blast furnace dust, and basic-oxygen furnace dust), but EAF dust with a limited zinc content is also treatable. The exact maximum zinc content is not described explicitly (Oda et al., 2006). To ensure the reducibility and the required physical properties of the feed material, the processed dusts are mixed with carbon and slag-forming additives (e.g., lime or quartz). Before the prepared mixture can be charged into the RHF, briquetting into so-called "self-reducing pellets" is necessary. Figure 23.5 shows the schematic overview of an RHF plant, including the material preparation and flue gas routing, from which the zinc present in the EAFD can be discharged in the form of zinc oxide in the filter house (Piret, 2012).

The high process temperature of about 1300°C and the relatively high carbon content compared to the Waelz process result in the generation of direct reduced iron (DRI) at the solid discharge end of the furnace (Oda et al., 2006).

Electric Smelting Reduction Furnace (ESRF)

The ESRF technology uses an electrically heated furnace to process EAFD. For this purpose, EAFD is first agglomerated with a reducing agent and slag-forming additives and then charged into the furnace. The electrical energy supplies the heat to melt and reduce the charged material. The added carbon carrier acts as a reducing agent for iron and zinc. Zinc evaporates and reoxidizes in the exhaust gas stream. Other volatile elements also leave the furnace as part of the off gas and contaminate the zinc oxide, which is collected in the bag filters as crude zinc oxide. The reduced iron is tapped as molten pig iron and separated from the molten slag (Nakayama, 2012).

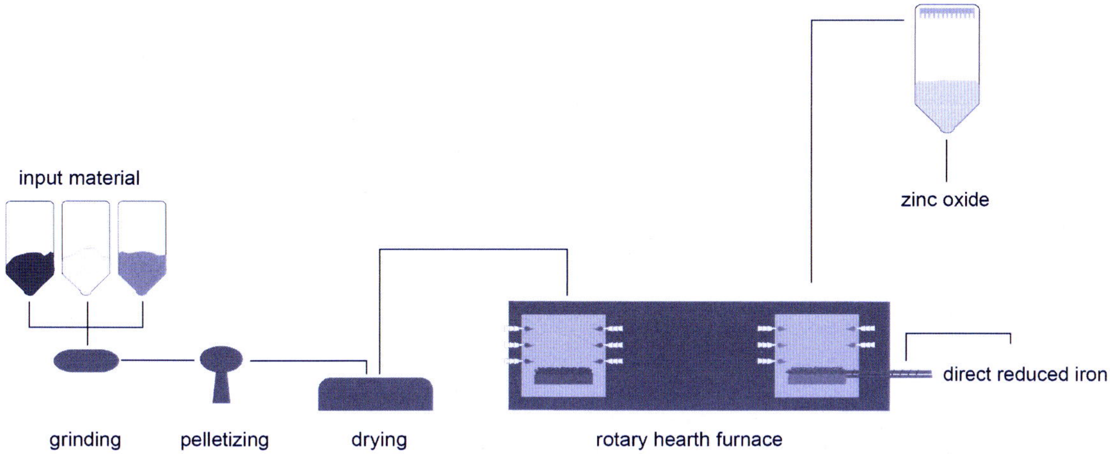

FIGURE 23.5 Schematic overview of the rotary hearth furnace (RHF). *Adapted from Oda et al. (2006).*

PRIMUS Technology

Similar to the RHF, PRIMUS technology uses the rotary hearth mechanism described earlier to process EAFD (Roth et al., 2001). However, the main difference compared to the RHF is that the furnace used for the first step in the PRIMUS concept is a multiple hearth furnace (MHF) that should realize at least a part of the reduction. The DRI discharged from the MHF is fed into a directly connected EAF. In this furnace, the DRI is reduced to an iron alloy and, in parallel, a slag is generated. Separating the zinc oxide from the other volatile compounds such as lead, potassium, and sodium chloride in the MHF, which would be an advantage compared to the Waelz process, is difficult to achieve. In addition, the complex mechanical construction, with many moving and thus tribologically stressed parts, results in high maintenance costs (Roth et al., 2001).

23.2.3 Copper Industry Dust

The primary copper production can be divided into smelting, converting, pyrometallurgical, and finally hydrometallurgical refining. Secondary materials are processed similarly. Only the smelting step and the converting conditions vary, based on the raw material input. Recycling of copper is discussed in detail in Chapter 21, Copper.

Processed primary and especially secondary raw materials in copper smelting contain significant amounts of zinc. Table 23.2 gives an overview of the copper sources used and their zinc content.

As can be seen, the secondary sources show quite high zinc concentrations, while primary

TABLE 23.2 Secondary copper raw materials.

Type of material	Cu (wt.%)	Zn (wt.%)
Copper concentrate	~30	0–2
Copper alloy scrap	36–98	0–43
Slags	10–50	2–10
Drosses	10–50	2–20
Sludges	0–40	0–20
Dusts	1–30	5–40
Nonferrous shredder	10–60	0–20
Electronic scrap	7–20	1–6

Data from Edelstein (2013), Rentz et al. (1999), Muchová et al. (2011), Ayhan (2000), Yamane et al. (2011), Barroso (2010), Balladares (2011), and Rumpold and Antrekowitsch (2013).

concentrates are rather low in zinc. Nevertheless, the necessity of adding cooling scrap in the converting and refining step leads to zinc inputs, also in the primary production route. Because of the high vapor pressure and affinity to oxygen, impurities can be removed easily via the slag and dust phase. The slags from copper metallurgy are treated by a reduction step to recover entrained copper. Consequently, enclosed zinc oxide is also reduced and volatilized. That means the major portion of zinc is concentrated in the flue dusts, which represent potential zinc secondary raw materials. Table 23.3 shows the different primary (production from ore) and secondary (recycling) process steps as well as compositions of the formed flue dusts.

Besides the given valuable metals, such materials contain harmful contaminants like halogens from secondary sources. Primary zinc production, in particular the electrowinning process, is very sensitive to chlorine and fluorine. Consequently, flue dusts from the copper industry are not suitable for the primary zinc industry, despite the high zinc contents (Antrekowitsch and Offenthaler, 2010).

Based on the complexity of these materials, the recycling requires specific procedures including hydrometallurgical or pyrometallurgical processes.

The hydrometallurgical methods offer numerous options to recover the valuable metals from flue dust. In principle they are leached completely or partly, followed by a selective extraction. Very often a halogen removal is carried out first, as moderate to low values can be achieved by simple water or soda washing (Piret, 1995).

The conventional technology for the recycling of secondary copper dusts is leaching with sulfuric acid followed by cementation and crystallization of zinc sulfate. Lead and tin are not soluble at these conditions and can be separated by filtration. The lead and tin-rich filter cake can be processed in the lead industry. In the next step, copper is removed from the solution by cementation. For this, iron scrap or zinc represent potential cementation agents. The advantage of zinc is that no iron is added in the process. When iron is already present in the solution, scrap can be used as a cementation agent as it is cheaper, and iron has to be removed anyway. Finally, zinc is recovered by the crystallization of zinc sulfate (Piret, 1995; Jha et al., 2001).

A similar approach is to separate the zinc selectively via solvent extraction from the solution and to recover it by electrolysis. This technology is known as the Zincex process, but due to the risk of a chlorine and organic contamination of the electrolyte, it is not widely applied (Piret, 1995; Jha et al., 2001; Martín et al., 2002).

Besides the sulfuric acid-based processes, ammonia can also be used to recover valuable metals. The EZINEX process represents the most common method. For this, ammonia chloride is used to dissolve the heavy metal fraction (except

TABLE 23.3 Flue dusts from different process steps in the copper industry. Composition given in wt.%.

(wt.%)	Primary smelting	Primary converting	Slag treatment	Refining	Secondary smelting	Secondary converting
Zn	0.1–10	5–70	25–60	5–40	20–60	25–70
Pb	0.1–5	2–25	2–15	2–20	5–50	5–30
Sn	0.1–1	0.1–4	–	–	0.2–5	1–20
Cu	5–30	10–25	0.5–2.5	15–25	2–12	2–15

Data from Barroso (2010), Balladares (2011), Rumpold and Antrekowitsch (2013), Cusano et al. (2017), Hanusch and Bussmann (1995), Litz (1999), and Piret (1995).

iron) followed by solution purification via cementation with zinc. In the end, the pure zinc is recovered by a novel electrolysis process, which represents the core of the EZINEX technology. Instead of the sulfate system, which is usually used in the zinc industry, a chloride-based electrolysis is applied (Piret, 1995; Jha et al., 2001; Olper and Maccagni, 2008).

The leaching of metal oxides with NH_3-$(NH_4)_2CO_3$ is an alternative ammonium treatment, but due to the interference of chlorides, a halogen removal in advance is necessary. After the solution purification by cementation, zinc carbonate can be precipitated (Jha et al., 2001; Schlumberger and Bühler, 2012).

Pyrometallurgical processes are rarely used for copper dust recycling. The available process reduces the metal compounds contained. Furthermore, soda is added to fix sulfur in the slag. High temperatures lead to a zinc concentration in the flue dust. Remaining lead and tin form a marketable alloy. In fact, only high tin and lead contents in combination with low halogen concentrations justify such a processing. Therefore this process is not suitable for modern secondary copper dusts (Piret, 1995).

23.2.4 Fuming of Slags From Lead Metallurgy

Lead concentrates are usually accompanied by zinc in the form of zinc sulfide. During roasting, the zinc sulfide turns into zinc oxide. Especially in the shaft furnace process for lead production, the reduction rate is set in such a way that lead is reduced while zinc stays as oxide and mainly ends up in the slag. Typical zinc values allowing an economic recovery are in the range of 8–12%, depending on the current zinc price (Ettler and Johan, 2014; Ullmann, 2007; Püllenberg and Höhn, 1999).

The slag fuming process is based on the reduction with carbon monoxide according to:

$$ZnO + CO \rightarrow Zn + CO_2$$

Zinc evaporates and reoxidizes in the off-gas system, generating a high zinc-containing dust that is collected in the baghouse.

A typical furnace for this process is a rectangular shaped hearth furnace. The processing temperature of the slag is above 1200°C. The char particles mixed with air are blown into the molten slag, generating reducing gas bubbles. Figure 23.6 shows the gas bubble with an enclosed char particle and explains the reactions between gas, char, and slag (Püllenberg and Höhn, 1999).

The reduced zinc accumulates as zinc vapor in the gas bubbles. Triggered by the Boudouard reaction, the generated CO_2 reacts with carbon from the char to CO. This regeneration of CO leads to constant high reducing conditions in the gas bubble, which results in a high zinc recovery yield. When a zinc-containing gas bubble reaches the bath level, the zinc vapor goes into the furnace atmosphere and reacts via a postcombustion with O_2 or CO_2 to ZnO. Typical zinc recovery yields of about 87.5% are observed. Depending on the process conditions, lead oxide is also recovered from the slag (Koch and Janke, 1984). In Table 23.4, typical data for the slag fuming process are presented.

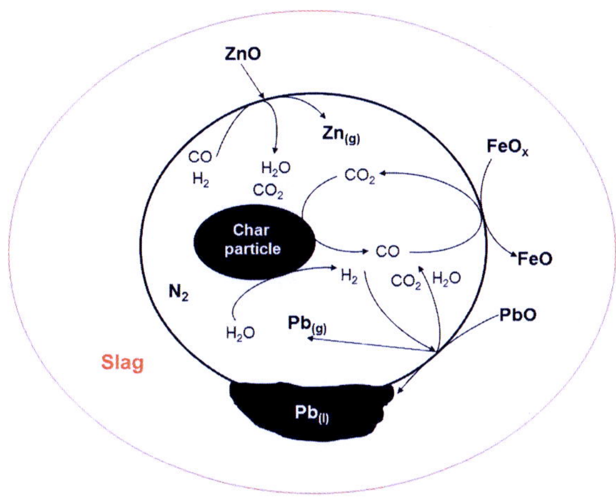

FIGURE 23.6 Reactions between gas, char, and slag during the slag fuming process. *From Püllenberg and Höhn (1999).*

TABLE 23.4 Typical data for slag fuming of lead blast furnace slag.

	Capacity (t/d)	Cu (wt.%)	Pb (wt.%)	Zn (wt.%)	Fe (wt.%)	CaO (wt.%)	SiO_2 (wt.%)
Charged slag	655	0.8	0.1–2.6	10.7–18.3	25.6–30.5	1.9–15.2	21.0–31.7
Treated slag	600	0.4	0.01–0.03	2.6–0.8	26.4–34.2	5.3–17.7	30.0–38.3
Dust	119	0.1	12.5–13.9	58.6–66.0	0.2	–	0.3
Matte	16	22.6	0.2	0.7	41.8	–	–

Data from Ettler and Johan (2014) and Jak and Hayes (2002).

23.3 KEY ISSUES AND CHALLENGES

23.3.1 Residues From Primary Zinc Production

About 90% of the world's primary production of zinc takes place by means of hydrometallurgical processes. The ore is roasted and then leached in several steps in sulfuric acid with decreasing pH values and increasing temperatures. While the first leaching step leaches only zinc oxide, some of the zinc stays in the leaching residue, mainly bound to iron as zinc ferrite, which is stable under these conditions ("neutral leaching," pH 4.0–5.5, ~60°C) (Sinclair, 2005). Zinc ferrite is a product of roasting when iron is present in the ore. Due to increasing iron content in the primary ores, compared to historical materials, the amount of zinc lost to the residue has also increased. As a result, modern zinc plants use a series of leaching steps with decreasing pH values and increasing temperatures to leach all the available zinc. An unwanted side effect is that iron is also leached. As iron would cause major problems in the following electrolysis, it must be removed from the solution. To achieve this, two different iron removal processes are used: the jarosite process ($Na_2Fe_6(SO_4)_4(OH)_{12}$) and the goethite process ($FeO{\cdot}OH$). In both cases, an iron phase is precipitated (jarosite, goethite) and removed by settling and filtration, leaving an iron-free solution suitable for subsequent purification and electrolysis (Sinclair, 2005; Hoeber and Steinlechner, 2021; Sohn and Olivas-Martinez, 2014). In total, about 0.5–1 t jarosite/goethite residues per ton zinc occur, depending on the composition of the initial concentrate (especially the iron content).

With zinc values from 1.5 to 8.4 wt.%, these residues are comparable to the corresponding ores. Additionally, lead and silver are present in most of these materials. Other elements such as copper, gold, indium, and germanium can appear in considerable amounts in specific cases. Nevertheless, these residues are, with some exceptions, not used as a metal source nowadays; instead the material is stored in controlled, industrial landfilling sites due to the material being categorized as hazardous waste (Antrekowitsch and Hanke, 2020). Table 23.5 shows the average chemical composition of

TABLE 23.5 Average chemical composition of different iron precipitation residues.

Pb (wt.%)	Zn (wt.%)	Cu (wt.%)	Ag (ppm)	In (ppm)	Fe (wt.%)	Si (wt.%)	Ca (wt.%)
0.5–7.8	1.5–8.4	0.1–0.8	<500	<300	15.2–32.4	0.8–7.2	3.2–10.3

Data from Antrekowitsch and Hanke (2020).

jarosite residues. The big range originates from the different input materials (ores) and details in the applied zinc production process scheme.

Due to the properties of jarosite and goethite residues, especially grain size, phase composition, and element distribution, existing methods of mineral processing and metallurgy do not allow an economic treatment of these materials. One rare example where zinc and lead are recovered from jarosite residues can be found in South Korea. In a large and highly flexible metallurgical plant offering multiple steps based on top submerged lance technology, jarosite residues can be treated together with other materials (Hoeber and Steinlechner, 2021; Sohn and Olivas-Martinez, 2014; Abadías et al., 2020).

Different promising pyro- and hydrometallurgical processes are currently under development. Some were originally designed for treating other zinc-bearing residues but were later adapted for their application to jarosite and goethite residues. Several processes for hydrometallurgical treatment are based on a sequence of leaching and separation steps, extracting mainly zinc but in some cases also lead, silver, gold, indium, gallium, and germanium in several fractions (Kangas et al., 2017; Jha et al., 2001; Antrekowitsch and Hanke, 2020; Vahidi et al., 2009). One pyrometallurgical approach is a roast-reduction process that also allows a multimetal winning from jarosite residues, while others recover lead and zinc by vacuum reduction and distillation (Hoeber and Steinlechner, 2021; Li and Ma, 2018). However, considerable complexity and energy consumption are major problems and are the main reasons for neither of them being realized up to now.

23.3.2 Circular Economy and Carbon Footprint

Zinc recycling will face crucial changes in the future. For instance, in steel mill dust recycling via the Waelz kiln route, drawbacks like the CO_2 footprint due to pet-coke utilization, resulting in increasing costs for emission credits as well as a more difficult dumping situation for the generated Waelz slag, will overrule the main advantage of Waelz kilns being an energy-efficient facility. Utilizing biomass-based reducing agents and recovery of iron from the Waelz slag could help to overcome these challenges. Alternative CO_2-neutral technologies, which completely replace Waelz kilns, could also be seen as another scenario, but that must be considered as a mid-to-long-term transformation.

Even though steel mill dust is one of the most investigated materials for zinc recycling, there are many other sources, as shown in this chapter. Thus a wide range of processes is present in zinc recycling, starting with the simple remelting for casted products and ending up at complex multimetal recovery concepts for sources like dust from the copper industry or jarosite. The dispersive use of zinc in various products in combination with different lifetimes does not always allow contributing to a circular economy. While a utilization, e.g., in die casting allows an easy return into the production circuit, products like zinc oxide in the ceramics industry or zinc sulfate for agricultural usage can no longer be used as zinc resources. In this context, the low zinc-containing steel mill dust, hardly recycled nowadays, would offer an important source of zinc for the future, if an economic process concept becomes available.

Nevertheless, the main challenge for the future will be the handling of the energy consumption and the CO_2 footprint, since most recycling processes are based on carbothermal reduction. Thus an overall transformation to alternative innovative process technologies is expected within 20–30 years.

References

Abadías, L.A., et al., 2020. Resource efficiency evaluation of pyrometallurgical solutions to minimize iron-rich residues in the roast-leach-electrowinning process. In: PbZn

2020: 9th International Symposium on Lead and Zinc Processing. Springer, pp. 351–364.

Antrekowitsch, J., Hanke, G., 2020. Characterization and processing of residues from hydrometallurgical zinc smelters. In: Siegmund, A., Alam, S., Grogan, J., Kerney, U., Shibata, E. (Eds.), PbZn 2020: 9th International Symposium on Lead and Zinc Processing. The Minerals, Metals & Materials Series. Springer, Cham, https://doi.org/10.1007/978-3-030-37070-1_38.

Antrekowitsch, J., Offenthaler, D., 2010. Die Halogenproblematik in der Aufarbeitung zinkhältiger Reststoffe. vol. 155 BHM, pp. 31–39.

Ayhan, M., 2000. Das neue HK-Verfahren für die Verarbeitung von Kupfer-Sekundär-materialien. Heft 87 der Schriftenreihe GDMB, pp. 197–207.

Balladares, E., 2011. Procesamiento Pirometalurgico De Minerals De Cobre En Chile: Technologias Actuales Y Emergentes. In: Presentation at XVI. Congreso Colombiano de Minera, Columbia.

Barroso, A.R.F., 2010. Copper Recycling, ENGR5187—Solid Waste Processing and Recycling. Laurentian University.

Cusano, G., et al., 2017. Best Available Techniques Reference Document for the Non-Ferrous Metal Industries. European IPPC Bureau.

Edelstein, D.L., 2013. Copper, 2011 Minerals Yearbook, US Geological Survey. pp. 20.1–20.25. https://s3-us-west-2.amazonaws.com/prd-wret/assets/palladium/production/mineral-pubs/copper/myb1-2011-coppe.pdf.

Ettler, V., Johan, Z., 2014. 12 years of leaching of contaminants from Pb smelter slags: geochemical/mineralogical controls and slag recycling potential. Appl. Geochem. 40, 97–103. https://doi.org/10.1016/j.apgeochem.2013.11.001.

Guézennec, A.-G., Huber, J.-C., Patisson, F., Siessecq, P., Birat, J.-P., Ablitzer, D., 2005. Dust formation in electric arc furnace: birth of the particles. Powder Technol. 157, 2–11. https://doi.org/10.1016/j.powtec.2005.05.006.

Hanusch, K., Bussmann, H., 1995. Behavior and removal of associated metals in the secondary metallurgy of copper. In: Third International Symposium Recycling of Metals and Engineered Materials, Point Clear, Alabama, pp. 171–788.

Hoeber, L., Steinlechner, S., 2021. A comprehensive review of processing strategies for iron precipitation residues from zinc hydrometallurgy. Clean. Eng. Technol. 4, 100214. https://doi.org/10.1016/j.clet.2021.100214.

International Zinc Association, 2015. Zinc Recycling, Closing the Loop. https://www.zinc.org/wp-content/uploads/sites/24/2015/04/Closing_the_Loop_July2015_Final.pdf. (30.06.2021).

International Zinc Association, 2016. Zinc Environmental Profile, 2016 Update, Life Cycle Assessment. https://www.zinc.org/wp-content/uploads/sites/24/2016/04/Zinc_Env_Profile_rev_April_2016.pdf. (30.06.2021).

Jak, E., Hayes, P., 2002. Phase equilibrium and thermodynamics of zinc fuming slag. Can. Metall. Quart. 41 (2), 163–174. https://doi.org/10.1179/cmq.2002.41.2.163.

Jha, M.K., Kumar, V., Singh, R.J., 2001. Review of hydrometallurgical recovery of zinc from industrial wastes. Resour. Conserv. Recycl. 33, 1–22. https://doi.org/10.1016/S0921-3449(00)00095-1.

Kangas, P., Nyström, M., Orko, I., Koukkari, P., Saikkonen, P., Rastas, J., 2017. The Jarogain Process for Metals Recovery from Jarosite and Electric Arc Furnace Dust: Process Design and Economics. https://publications.vtt.fi/pdf/technology/2017/T317.pdf.

Koch, K., Janke, D., 1984. Schlacken in der Metallurgie 1. Verlag Stahleisen mbH Düsseldorf.

Li, J., Ma, H., 2018. Zinc and lead recovery from jarosite residues produced in zinc hydrometallurgy by vacuum reduction and distillation. Green Process. Synth. 7 (6), 552–557. https://doi.org/10.1515/gps-2017-0079.

Lin, X., Peng, Z., Yan, J., Li, Z., Hwang, J.-Y., Zhang, Y., Li, G., Jiang, T., 2017. Pyrometallurgical recycling of electric arc furnace dust. J. Clean. Prod. 149, 1079–1100. https://doi.org/10.1016/j.jclepro.2017.02.128.

Litz, J.E., 1999. Flue Dusts: An Ideal Feed for Resource Recovery, Residues and Effluents—Processing and Environmental Considerations. TMS, pp. 223–239.

Mager, K., et al., 2000. Recovery of zinc oxide from secondary raw materials: new developments of the Waelz process (Hg.). In: Stewart, D.L., Daley, J.C., Stephens, R.L. (Eds.), Recycling of Metals and Engineered Materials. John Wiley & Sons, Inc, Hoboken, NJ, USA, pp. 329–344, https://doi.org/10.1002/9781118788073.ch29.

Mager, K., Meurer, U., Wirling, J., 2003. Minimizing dioxin and furan emissions during zinc dust recycle by the waelz process. JOM 55, 20–25. https://doi.org/10.1007/s11837-003-0099-6.

Martín, D., et al., 2002. Extending zinc production possibilities through solvent extraction. J. South Afr. Inst. Min. Metall. 2002, 463–467. https://www.downloadmaghaleh.com/wp-content/uploads/edd/maghaleh/1398/tinosh-gostaresh.pdf.

Muchová, L., Eder, P., Villanueva, K.A., 2011. End-of-waste Criteria for Copper and Copper Alloy Scrap: Technical Proposals. EUR 24786 EN. Publications Office of the European Union, Luxembourg (Luxembourg). https://publications.jrc.ec.europa.eu/repository/bitstream/JRC64207/reqno_jrc64207_jrc64207.pdf.pdf.

Nakayama, M., 2012. New EAF dust treatment process: ESRF. SEAISI Quart. (South East Asia Iron Steel Inst.) 41, 22–26.

Oda, H., et al., 2006. Dust Recycling System by the Rotary Hearth Furnace. Nippon Steel Technical Report No. 94, Chiyoda,.

Olper, M., Maccagni, M., 2008. From C.Z.O. to zinc cathode without any preatreatment. In: The EZINEX Process, Lead and Zinc 2008, Durban, South Africa, pp. 85–98.

Piret, N.L., 1995. Criteria for optimization of recycling processes of primary and secondary copper smelter dusts. In: Third International Symposium on Recycling of Metals and Engineered Materials, Point Clear, Alabama, pp. 189–214.

Piret, N.L., 2012. EAF dust processing: rotary heart a potential substitute for the Waelz Kiln. World Metall. ERZMETALL 65 (5), 306–316.

Püllenberg, R., Höhn, R., 1999. Schlacken in der Bleimetallurgie. vol. 83.1 GDMB-Verlag, pp. 81–94.

Remus R., Aguado Monsonet, M.A., Roudier, S. and Delgado Sancho, L. Best Available Techniques (BAT) Reference Document for Iron and Steel Production. European Commission Joint Research Centre. Publications Office, Luxembourg (2013). https://eippcb.jrc.ec.europa.eu/sites/default/files/2019-11/IS_Adopted_03_2012.pdf. DOI: https://doi.org/10.2791/97469.

Rentz, O., Krippner, M., Hähre, S., 1999. Report on BAT in German Copper Production. University Karlsruhe.

Roth, J.L., et al., 2001. PRIMUS, a new process for recycling by-products and producing virgin iron. Rev. Métall. 98, 987–996.

Rumpold, R., Antrekowitsch, J., 2013. Zinc in the copper secondary production and potential recycling strategies. In: Proceedings of Copper 2013, Santiago de Chile, pp. 265–280.

Schlag, S., 2016. Presentation: ZnO and zinc chemicals—market update. In: International Zinc and Zinc Oxide Conference, Scottsdale, Arizona, February.

Schlumberger, S., Bühler, J., 2012. Urban Mining: Metal Recovery From Fly and Filter Ash in Waste to Energy Plants. Ash Utilization, Stockholm, Sweden.

Sinclair, R.J., 2005. The Extractive Metallurgy of Zinc. Australasian Institute of Mining and Metallurgy, Victoria.

Sohn, H.Y., Olivas-Martinez, M., 2014. Chapter 2.3—Lead and zinc production. In: Seetharaman, S. (Ed.), Treatise on Process Metallurgy. Elsevier, pp. 671–700, https://doi.org/10.1016/B978-0-08-096988-6.00025-0.

Steinlechner, S., 2013. Amelioration and Market Strategies for Zinc Oxide With Focus on Secondary Sources. PhD Thesis, Leoben Austria,.

Tan, R., Khoo, H., 2005. Zinc casting and recycling. Int. J. Life Cycle Assess. https://doi.org/10.1065/lca2004.08.172.

Ullmann, 2007. Ullmann's Encyclopedia of Industrial Chemistry: Lead. John Willey and Sons.

Vahidi, E., Rashchi, F., Moradkhani, D., 2009. Recovery of zinc from an industrial zinc leach residue by solvent extraction using D2EHPA. Miner. Eng. 22 (2), 204–206. https://doi.org/10.1016/j.mineng.2008.05.002.

World Bureau of Metal Statistics, 2018. World Bureau of Metal Statistics (106th), Ware England.

World Steel Association, 2021. World Steel in Figures 2020.

Xylia, M., et al., 2016. Worldwide resource efficient steel production. In: ECEEE Industrial Summer Study Proceedings, European Council for an Energy Efficient Economy, pp. s.321–s.333. http://urn.kb.se/resolve?urn=urn:nbn:se:kth:diva-227898.

Yamane, L.H., et al., 2011. Recycling of WEEE: Characterization of spent printed circuit boards from mobile phones and computers. Waste Manage. 31, 2553–2558. https://doi.org/10.1016/j.wasman.2011.07.006.

CHAPTER

24

Ferroalloy elements

Stefan Luidold and Eva Gerold

Department of Metallurgy, University of Leoben, Leoben, Austria

24.1 INTRODUCTION

Raw materials are essential for the economy and consequently for modern society because they are required for industrial activities as well as for infrastructure and products used in everyday life. The importance of oil, gas, and coal has often been emphasized since the oil crises of the 1970s, which led to sharp price increases. However, the importance of nonenergy materials such as minerals and metals was neglected in the industrialized world due to a long period of stable and low raw material costs. This led to a considerable concentration of the mining production of individual raw materials in a small number of countries. Therefore the European Union (EU) became increasingly dependent on imports of many raw materials. The resulting high vulnerability has been demonstrated in the past by the rush for tantalum in 2000 due to the boom in mobile phones, and more recently by the extreme price increase for rare earths in 2010–11. To manage the complex challenge of ensuring the availability of raw materials, the EU started the EU Raw Materials Initiative as an integrated strategy in 2008, which is based on multiple pillars. Its first action was the identification of critical nonenergy raw materials at the EU level by an ad hoc working group after the development of a methodology for the rating of criticality (Luidold, 2013).

However, the low recycling rates are caused not only by economic factors and limited metallurgical infrastructure, but also by several additional factors. At the base are limitations set by nature, such as physics, chemistry, metallurgy, and thermodynamics. These dictate the outcome of a given input into a recycling process, the constraints of nature tying together the use of energy and metal recovery, as well as quality and quantity of the inevitable losses. The relationship between quality and recovery of a given metal or metals represents an important aspect at this stage. Another major issue is found in the way (consumer) goods are produced, and whether the product design includes recyclability or not. The structure of the waste collection system depicts another important influence, as the collection of postconsumer waste is very much a logistics challenge. There are typically several stakeholders in the logistics chain with different aims and motivations. Often, the minimization of collection costs is treated as more important than the maximization of material efficiency. Another obstacle to recycling is the lack of consumer awareness concerning collection and recycling possibilities. The leakage of waste into the “grey market” of informal, or illegal, low-technology recyclers forms an additional factor affecting the volume and availability of material for recycling (Reuter et al., 2013).

24.2 USE AND RECYCLING

Ferroalloys are master alloys of iron with one or more other elements, such as chromium, manganese, or silicon, and mainly serve for the production of steels and cast iron. Further applications of these materials include their use as reduction agents (especially ferrosilicon) and the alloying of nonferrous metals. The utilization of ferroalloys rests upon their more cost-effective production and their largely better dissolution behavior in iron compared to the pure metals. Table 24.1 summarizes the world production of most types of ferroalloys.

It is already common practice to process secondary raw materials besides primary ores, known as residual materials and by-products of different operations, for the extraction of pure metals and ferroalloys. Hence, local shortages of raw materials and mining delivery problems can be surmounted by purchasing high-quality materials and by the acceptance of residues. When using secondary raw materials, energy and investment costs for producing concentrates are often lower than mining the ores and first enrichment steps. Disadvantages of this consideration are extensive processes for the recovery of valuable metals because of the presence of complex compounds and high-melting oxides. The aim is a complete conversion of the residues into commercially salable products such as pure metals, ferroalloys, or inert material for landfilling and reuse as building materials.

TABLE 24.1 Global production of various ferroalloys in 2017 (metric tons, gross weight).

Alloy type	Token	Quantity
Silicomanganese	SiMn	11,200,000
Ferrochromium	FeCr	12,600,000
Ferrosilicon	FeSi	5,960,000
Ferromanganese	FeMn	4,620,000
Ferronickel	FeNi	4,000,000
Ferrosilicon-chromium	FeSiCr	185,000
Ferromolybdenum	FeMo	149,000
Ferrovanadium	FeV	52,400
Ferroniobium	FeNb	92,200
Ferrotitanium	FeTi	10,400
Other, unspecified		5,922,580

Modified from Schulte (2017).

Nevertheless, secondary resources did not play a significant role in the extraction of the ferroalloys with largest production volumes, such as SiMn, FeCr, FeSi, and FeMn. However, in 2020, recycled chromium contained in reported stainless steel scraps accounted for 25% of apparent chromium consumption and therefore substituted for ferrochromium in some metallurgical uses. A similar situation arose for ferromanganese, because manganese was recycled incidentally as a constituent of ferrous and nonferrous scrap, whereas scrap recovery specifically for manganese remained negligible. Additionally, the recovery of manganese occurs along with iron from steel slags. Beyond that, the recycling of silicon, specifically the production of ferrosilicon from secondary resources, remained completely insignificant (USGS, 2021).

Contrariwise, secondary resources exhibit a significant share of the recovery of nickel in alloyed form, as this accounted for approximately 50% of apparent nickel consumption in 2020. The processed nickel-containing wastes included flue dusts, grinding swarf, mill scales, and shot blasts generated during the manufacturing of stainless steels as well as filter cakes, plating solutions, spent catalysts, spent pickle liquors, sludges, and all types of spent nickel-containing batteries. Furthermore, nickel-containing alloys and stainless-steel scraps were also melted and directly used to produce new alloys and stainless steels (USGS, 2021).

Generally, catalysts, ferrous scrap, and superalloy scrap serve as feedstocks for the recycling of molybdenum. In doing so, ferrous scrap includes remnants originated in the steelmaking process (revert scrap), new scrap generated by steel mill customers and recycled by scrap collectors and processors, as well as molybdenum-bearing alloys recycled after serving their useful life (old scrap). The quantity of recycled molybdenum may be as much as 30% of the apparent Mo supply. No processes exist for the separate recovery and refining of secondary molybdenum from its alloys, steels, and superalloys, but the Mo content of the reclaimed scraps is significant and reused. Consequently, the recycling of Mo-bearing scrap will continue to depend on the markets for the principal metals alloyed with molybdenum, such as iron, nickel, and chromium (USGS, 2021).

Regarding vanadium, a significant amount was recycled from spent process catalysts that may compose as much as 40% of total vanadium catalysts, whereas the scrap recovery specifically for niobium content was negligible. The recycling of niobium only takes place via the recirculation of Nb-bearing steels and superalloys and its amount is not available, but may be as much as 20% of apparent consumption (USGS, 2021).

Finally, about 45,000 t of titanium scrap metal was consumed in 2020, of which 35,000 t was from the titanium industry, 8000 t from the steel industry, and less than 500 t from the superalloy industry, whereas the estimated quantity of secondary tungsten produced and the amount consumed from secondary resources by processors and end users in 2020 were withheld to avoid disclosing company proprietary data (USGS, 2021).

An extensive assortment of secondary raw materials generally exists for the production of ferroalloys, but only a couple of residues can be used. Reasons for this are low amounts of material or minor contents of contained valuable metals such as vanadium, nickel, and molybdenum. In addition, it proved to be complex to develop a recycling process for discontinued production of waste materials with an alternating chemical composition.

In general, the possible residues for ferroalloy production, which usually contain a combination of different elements, originate from different areas, such as:

- Residues of metallurgical processes
- Catalysts from various chemical manufacturing and refining processes
- Combustion residues from heating plants and power stations
- Other industrial by-products, in particular from metallurgy and metal processing

To give an overview of all raw material possibilities, Figure 24.1 shows the main areas, together with some typical representatives.

24.3 RECYCLING OF RESIDUES

24.3.1 Spent Catalysts

Molybdenum supported on an alumina carrier with promotors such as Ni or Co (Ni, Mo/Al_2O_3, NiO/Al_2O_3) often serves as a hydrotreating and hydrocracking catalyst used in petroleum refining to remove undesirable impurities such as sulfur, nitrogen, and metals (V and Ni) by hydrodesulfurization (HDS), denitrogenation (HDN), and demetallization (HDM) reactions. In 2007, the worldwide generation of 150–170 kt/a of waste hydroprocessing catalysts was reported, which might have further increased due to the processing of heavier feedstocks of higher impurities and enlarged production over the last several years. Further common Ni-bearing catalysts include hydrogenation catalysts for fat hardening and the Raney nickel catalysts applied in drugs manufacture, besides those for the methanation of

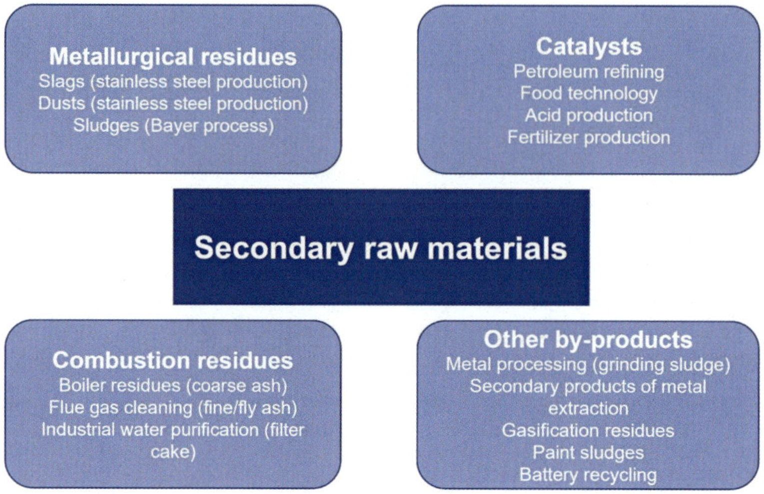

FIGURE 24.1 Overview and classification of secondary raw materials for ferroalloy production.

carbon dioxide from hydrogen and ammonia synthesis gas (NiO/Al_2O_3, Ni/SiO_2) (Meshram et al., 2019).

The initial composition of nickel catalysts depends on their operation purpose, but spent HDS catalysts generally comprise about 10–30% Mo, 1–12% V, 0.5–6% Ni, 1–6% Co, and 8–12% S, as well as 10–12% C, besides alumina. Although pyrometallurgical processes served earlier for the recycling of such materials, recent developments in their recycling mostly rest upon hydrometallurgical methods (Meshram et al., 2019).

Valdi produces ferroalloys, ultimately to be used by stainless steel manufacturers, from HDS catalysts and other secondary raw materials, such as batteries as well as wastes from the metallurgy of stainless steel, by treating the catalysts in a roasting furnace to remove sulfur and coke and then smelting the feedstock in an immersed electrode furnace. Similarly, Treibacher Industrie AG recycles catalysts together with residues containing vanadium, molybdenum, and nickel in up-to-date recycling plants at the highest environmental standards. Contrariwise, Moxba-Metrex treats HDT spent catalysts after a conditioning by mechanical and thermal processes via hydrometallurgy to extract pure metal oxides and concentrates. The product range includes not only molybdenum oxides, molybdenum concentrates, vanadates, cobalt oxides, and nickel products, but also high-grade ferromolybdenum, ferrovanadium, and ferronickel (Akcil et al., 2015).

24.3.2 Superalloys

Regarding the world's total superalloy production, 28% are classified as Ni-based and 5% as Ni-Fe-based superalloys. The first group exhibits the generic chemical composition of 42–74% Ni, 0–17% Co, 0–18% Fe, 0–15% Mo, 0–12% W and Ta, 0–6.5% Al, 0–5% Ti, 0–6% Re, and contains some other trace elements. Consequently, they constitute a good source of Ni and other metals after the service life and can be treated for resource recovery by both pyrometallurgy and hydrometallurgy (Meshram et al., 2019).

Currently, a wide range of superalloys is available, and their tough nature makes it difficult to recycle them after the lifetime of the products made from them. Additionally, the superalloy industry generates a large amount of scrap. Significant quantities accumulate, for instance, during forging and machining. Only about 15% of the superalloy may end up as finished goods for turbine parts, while other components can have a final product yield of <10%. A few processes are available for their recycling and some of them are in use. However, only sparse information about the recycling of superalloys can be found, because superalloy manufacturers frequently perform the recycling themselves and the recyclers do not want to produce business for the competitors. Commonly, the recycling of superalloys takes place by remelting after cleaning the scrap and charging them together with fresh material, for instance into an electric arc furnace. However, it has been found that recycling by pyrometallurgical techniques results in about 20% loss of alloying elements. As an alternative, many pyro-hydrometallurgical routes have been developed that are sound from a recovery and scientific point of view. Nevertheless, it seems difficult to implement them on a commercial scale due to operational complexity and lack of cost efficiency (Srivastava et al., 2014).

So, no ferroalloys are produced that serve as a raw material for the production of superalloys, but the utilization of scrap as a secondary raw material saves significant amounts of ferroalloys extracted from ores and concentrates.

24.3.3 Vanadium-Containing Slags

The production of vanadium makes use of vanadium slags on a large scale, which originate from the processing of titanomagnetite ores in the ferrous metallurgy (e.g., Highveld Steel and Vanadium Corp.). At first, a prereduction of the ore containing about 1.5% V_2O_5 by coal takes place in directly heated rotary furnaces. Further reduction proceeds in electric arc furnaces. The oxidation of the pig iron obtained from it by oxygen lances transfers vanadium into the slag. The latter contains this element in the form of an iron spinel, $FeO{\cdot}V_2O_3$, with a vanadium content corresponding to about 25% V_2O_5. This material constitutes the most important source for the production of vanadium (Bauer et al., 2005).

Today, mainly alkaline roasting is used in the extraction of vanadium oxides from such vanadium slags. This process route starts with a milling for the removal of incorporated iron grains. An oxidizing roasting of the material in multiple hearth or rotary furnaces occurs at 700–850°C after the addition of alkali salts to form water-soluble sodium vanadate. It should be noted here that Ca, Mg, and Al constitute disturbing accompanying elements, which build insoluble vanadates and therefore lower the yield of the leaching step. After the leaching of the roasted material in water, the precipitation of ammonium vanadates occurs by addition of sulfuric acid or hydrochloric acid as well as ammonium salts to the solution. At the end, a calcination of these compounds converts them into high-purity alkali free V_2O_5 (Bauer et al., 2005).

Finally, aluminothermic processing of vanadium oxides into high-percentage ferrovanadium grades occurs, because this technique is well controllable on a large scale and allows the assurance of consistent quality. The reaction of vanadium pentoxide with aluminum exhibits a self-sustaining process that even releases an excess of heat. A small amount of a mixture comprising V_2O_5 flakes, Al powder or granules, fine steel shots, and an ignition mixture (e.g., BaO_2 and Al powder) is ignited and then the charging of the further mixture takes place. Because the reaction proceeds very rapidly, the aluminum in the metal does not reach equilibrium with the vanadium oxide in the slag, so its vanadium content remains too high. To avoid this, an electric arc serves at the end of the aluminothermic reaction to keep the metal and slag liquid until the completion of the reaction. The excess aluminum in the metal can then be freshened out to a few tenths of a percent with oxides, such as iron oxide or vanadium oxide.

The small amount of thereby generated slag goes to the next batch (Antrekowitsch and Hochenhofer, 2006).

24.3.4 Fly Ashes

Coal fly ash as a by-product of coal combustion in thermal power plants accounts for 5–20% of feed coal and typically appears up to 70–85% in the form of coarse bottom ash and 15–30% as fine fly ash. The amount of annually accumulating fly ash is estimated at $750 \cdot 10^6$ t in 2015 and is expected to increase further for many years due to the world's increasing reliance on coal-fired power generation. As such ashes often contain significant vanadium and nickel content, they represent an interesting secondary raw material with high economic potential (Akcil et al., 2015).

Alkaline roasting with subsequent hydrometallurgical processing serves for the extraction of vanadium oxides not only from slags, but also from fly ashes and other V-bearing residues and spent catalysts, with a few minor differences. Precautions are often required because silicon can provoke filtration challenges and phosphorus a deterioration of the vanadium yield on the precipitation, in addition to a decrease of the leaching efficiency by the elements Ca, Mg, and Al (Bauer et al., 2005).

24.3.5 Spent Batteries

According to their composition, lithium-ion batteries (LIBs) and nickel metal hydride batteries (NiMHs) with nickel contents in the range of 10–20% and 25–49%, respectively, comprise an interesting source of nickel, apart from other valuable elements, such as Co, Li, Mn, Zn, Cu, Al and the rare earth elements (REEs). The estimation of a global market of more than 10^7 electric vehicles per year in 2025 will generate huge additional primary nickel requirements and, after the lifetime of the batteries, also a respective secondary resource for the contained metals (Meshram et al., 2019).

For instance, Pradhan et al. (2021) noted that an amount of 705,000 t/a of end-of-life LIBs are projected by 2025 and about 6,000,000 t/a by 2040. For handling the supply challenges associated with some metals contained in these batteries, new approaches have been adopted to recover them from secondary resources, in which the scraps are either subjected to pyrometallurgical or hydrometallurgical process steps to obtain alloys or metal-containing solutions. Although pyrometallurgical processes are often applied for industrial recycling of spent LIBs to extract transition metals like Ni, Co, and Cu, a literature survey indicated that hydrometallurgy represents the most researched and most commonly applied technique because it prevents the loss of lithium by slagging as well as high energy consumption and the requirement of high-temperature equipment. Figure 24.2 provides an overview about the basic recycling routes for LIBs.

Concerning the commercial recycling of batteries, the waste treatment facility of INMETCO represents an example for the production of ferronickel and other metals from secondary resources. This includes spent batteries and catalysts as well as residues from specialty steel, superalloy, plating, and surface finishing industries such as cakes, dusts, filters, grindings, mill-scale sludges and solutions, plus swarf. The battery types comprise LIB, NiCd, and NiMH; a Cd-recovery furnace first removes cadmium from the NiCd cells. The solid wastes are subsequently blended with carbon and pelletized using either liquid wastes or water prior to their reduction in a rotary hearth furnace together with spent catalysts and shredded batteries. A submerged electric arc furnace smelts the reduced material to a remelt alloy of iron with 9–19% Cr, 8–16% Ni, and about 0.8% Co, a valuable feed for stainless steel production. However, the REEs are completely transferred to the slag and therefore lost due to its use as roadbed aggregate in place of gravel. To avoid this drawback, several companies started activities

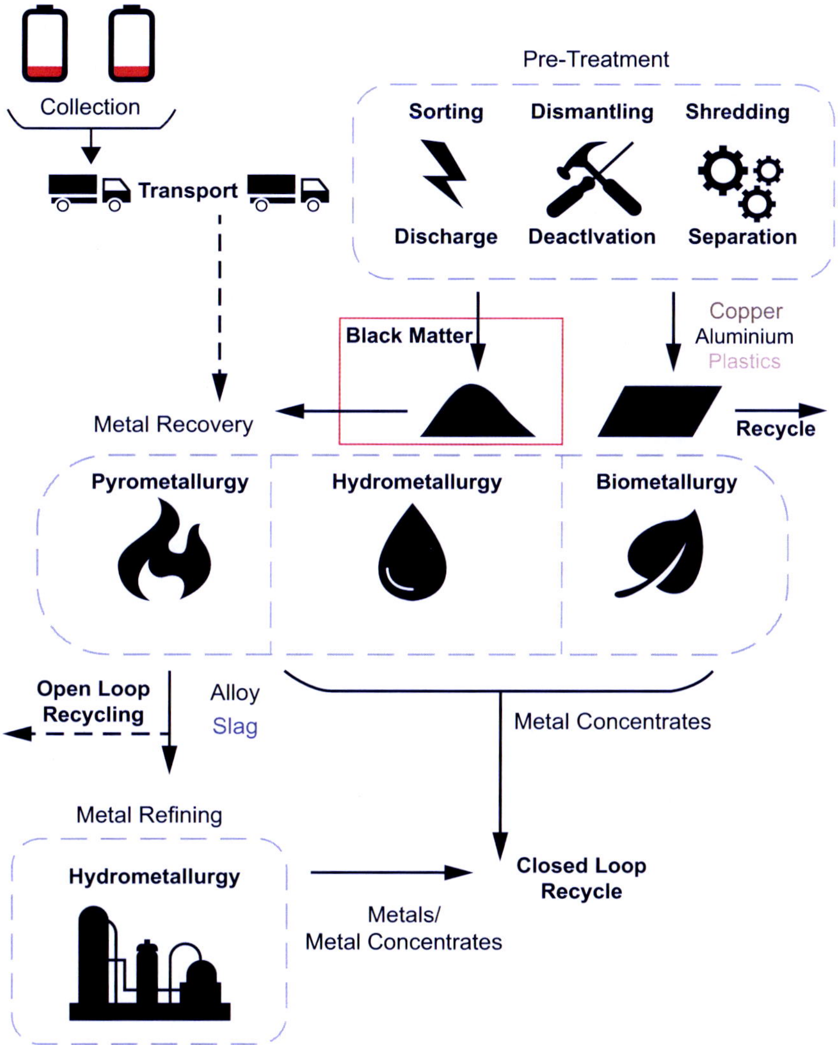

FIGURE 24.2 Possible process routes in the recycling of lithium-ion batteries in view of an open or closed recycling loop. *From Gerold (2021).*

to extract REE from NiMH, such as Umicore/Solvay, Retriev Technologies (formerly Toxco), and Honda/Japan Metals and Chemicals (Meshram et al., 2019).

Whereas all pyrometallurgical processes for the recycling of spent batteries result not only in ferroalloys but also in slagging of the REEs, the hydrometallurgy basically enables a complete recovery of high purity metals with minimal air emissions and energy demands. Their methods exhibit high selectivity but often require some pretreatment for homogenization and do not serve for the production of ferroalloys. Contrariwise, Liu et al. (2019) mention in their paper that, in general, a conventional pyrometallurgical recycling path for LIBs is short and easy to scale up (Liu et al., 2019). Furthermore, it can also deal with different types and variants of LIBs and prevents safety risks from the leakage of electrolytes.

Pinegar and Smith (2019) provided an overview of current industrial recycling processes for lithium-ion batteries (Table 24.2). Whereas European countries have developed strict regulations for battery recycling (collection rate of 45 and 50 wt.% recycling efficiency and further increase), only a few policy regulations exist in North America. Consequently, the LIB collection and recycling rate in this region could be even less than 5%, although some of the major LIB recycling companies set up their own waste-collecting network.

It has to be mentioned that no information is given as to whether the plant capacities consider only LIB or also include other types of spent batteries as well as mineral ores and scraps. The advantages of pyrometallurgical processes include the fact that they can effectively treat different types of batteries with ores and industrial wastes simultaneously. Furthermore, some processes enable a direct feed of LIB packs and charged batteries in the furnace without any pretreatment. However, the obtained alloy still requires subsequent process steps to refine cobalt, nickel, and copper separately. The economic feasibility of such pyrometallurgical operations strongly depends on the recovery yield and the metal prices. So, the recent development of LIB chemistry towards the decrease of cobalt and nickel content might jeopardize pyrometallurgy as a primary LIB recycling technology. Nevertheless, LIB can be effectively recycled in this way as a secondary feed along with other materials (Pinegar and Smith, 2019).

Due to the steadily growing sales of lithium-ion batteries, interest in raw material recycling of the materials employed is also increasing. Financial incentives as well as legal requirements represent the driving forces in this

TABLE 24.2 LIB recycling companies.

Company	Country	Capacity (t/a)	Process
Umicore	Belgium	7000	Pyro.
Umicore	China	5000	Pyro.
Retrieve	Canada/USA	4500	Hydro.
Glencore (Xstrata Nickel)	Canada/Norway	7000	Pyro.
GEM High-Tech. Co.	China	25,000–30,000	Hydro.
Brunp Co.	China	10,000	Hydro.
Batrec	Switzerland	1000	Pyro.
Accurec	Germany	6000	Pyro.
Recupyl	France	110	Hydro.
Valdi	France	20,000	Pyro.
Akkuser Ltd.	Finland	4000	Mech.
Inmetco	USA	6000	Pyro.
JX Nippon Mining and Metals	Japan	5000	Pyro.

Data from Pinegar and Smith (2019).

industry. The following legally binding regulations play a significant role in this regard:

- Battery directive: requires a minimum collection rate of 45% and recycling rate of 50% of other spent batteries since September 2016.
- End-of-life vehicles directive: since January 2015, requires the use or recycling of at least 95% of the average vehicle weight.

The aim of these directives is to optimize existing as well as develop new industrial recycling processes to enable an environmentally friendly and profitable recycling of the materials into the respective manufacturing processes.

The different stages of recycling can be divided into preprocessing, mechanical, hydrometallurgical, and pyrometallurgical processes, as shown in Figure 24.3 (Gerold, 2021). Preprocessing is considered to be any process that does not change the structure of the lithium-ion battery (e.g., sorting by battery type from mixed waste). Mechanical processing uses various techniques to expose, classify, and concentrate materials without modifying their chemistry. These techniques are based on differences in the physical properties of the materials, such as density, shape, or size, and generally occur before stages involving chemical reactions. After mechanical processing, materials are recycled through hydrometallurgical, pyrometallurgical, or combined steps. Pyrometallurgy refers to operations at elevated temperatures in which redox reactions occur to melt or reduce and purify metals. Meanwhile, hydrometallurgy involves the leaching of valuable elements from a solid matrix and subsequent modification by precipitation or crystallization. Since these techniques appear to be extremely resource intensive, there is a strong influence of economic constraints (Clark et al., 2015; Hanisch et al., 2015; Harper et al., 2019).

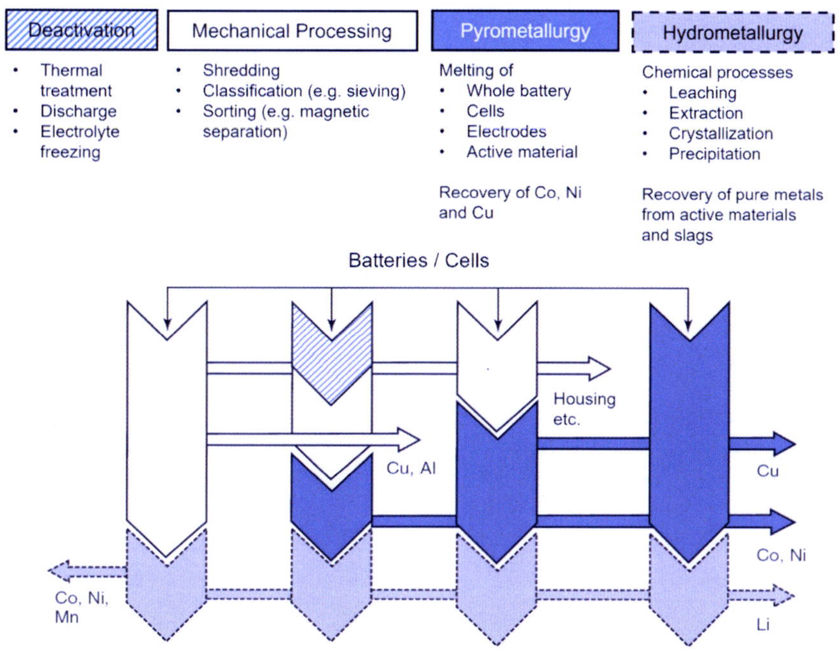

FIGURE 24.3 Unit operations in battery recycling and their possible combinations to develop efficient recycling processes. *From Hanisch et al. (2015).*

The active material or black mass is the main target of most established and emerging recycling processes due to its high economic value, especially in compositions containing cobalt. In fact, Co-containing cathode materials (e.g., $LiCoO_2$ or $LiNi_xMn_yCo_zO_2$) have higher costs than other commercial substances (e.g., $LiMn_2O_4$, $LiNiO_2$, or $LiFePO_4$) (Windisch-Kern et al., 2021). However, the recovery of battery-grade cathode material is a complex process. A large number of preprocessing and enrichment steps are required before final hydrometallurgical refining can take place. Since minor impurities affect the quality of the final product, adequate purification of the spent material represents a key task due to its high chemical sensitivity. In addition, the chemical composition of the active material significantly influences the hydrometallurgical steps (Korthauer, 2013; Gratz et al., 2014).

Due to the relatively low commercial value, efforts to recycle $LiFePO_4$ and $LiMn_2O_4$ cathode chemistries are limited. However, technologies from Battery Resources, LithoRec, and OnTo are capable of processing not only Co-containing cathode materials, but also other chemistries. Other established companies in the field of lithium-ion battery recycling include Umicore, Sumitomo-Sony, Retriev Technologies, Recupyl, Akkuser, and Accurec, and these operations use both pyro- and hydrometallurgical processes to recover valuable metals from spent LIBs (Gratz et al., 2014).

In recent decades, great efforts have been made to develop end-of-life battery recycling using processes with high recovery efficiency, simple operation, low cost, high selectivity, and environmentally friendly characteristics. In 2019, the worldwide capacity for the recycling of lithium-ion batteries worldwide consisted of less than 100,000 t/y and it is necessary to expand this drastically in the near future. The systematic evaluation of a defined recycling technology should take into account the factors of efficiency, cost-effectiveness, and environmental compatibility. For the hydrometallurgical form of metal recycling from spent lithium-ion batteries, the potential is already high and will increase steadily in the coming years. Through the combination of mechanical treatment and hydrometallurgical processes, the aspects of high selectivity as well as product purity can be equally guaranteed with a simultaneously low energy input.

24.3.6 Bayer's Sludge (Alumina Production)

During alumina production via the Bayer process, approximately 10,000–50,000 t/y vanadium are extracted, because its content in bauxite amounts to 0.05–0.25%. About 30–35% of vanadium present in bauxite is extracted and accumulated as vanadium sludge (10–18% V_2O_5) during the digestion of bauxite with caustic soda solution. Its removal from the leach solution takes place in the evaporation unit by precipitation of vanadium salts, because V adversely affects the electrical properties of aluminum (Meshram et al., 2016).

However, low V content in combination with impurities such as fluorides, phosphates, and arsenic render the sludge less attractive for vanadium extraction.

To overcome these problems, Gladyshev et al. (2015) demonstrated the treatment of vanadium sludge as well as electrostatic filter dust from an alumina calcination plant. Water leaching with subsequent neutralization by CO_2-enriched air and addition of alumina solution effectively removed phosphate (>98.5%) from the vanadium sludge prior to the precipitation of ammonium metavanadate from the solution at pH 7.2–7.6. A repulping of the obtained solid in hot distilled water followed by the addition of sulfuric acid to achieve a pH of 2 converts it into polyvanadates, which are calcined to vanadium pentoxide at 560°C.

24.4 CONCLUSION

In addition to the application of ferroalloys as an additive in the production of a wide variety of iron and steel grades and nonferrous metal alloys, they are also used as a starting material for the production of chemicals and pure metals, as well as a reducing agent.

The use of ferroalloys instead of the pure elements for alloying is due on the one hand to economic reasons, and on the other hand to process technical reasons. They can be produced more easily and thus more cheaply, but they can also be introduced into the molten steel and homogenized more quickly due to their lower melting temperatures and lower density difference from the molten steel. In general, ferroalloys are produced not only from primary raw materials such as ores and concentrates, but also from secondary raw materials, such as residues and by-products from various industrial processes. Their processing is aimed either at complete conversion into marketable products (ferroalloy, building material, etc.) or inertization of the residual materials that have no further application.

References

Akcil, A., Veglio, F., Ferella, F., Demet Okudan, M., Tuncuk, A., 2015. A review of metal recovery from spent petroleum catalysts and ash. Waste Manag. 45, 420–433. https://doi.org/10.1016/j.wasman.2015.07.007.

Antrekowitsch, H., Hochenhofer, M., 2006. Stahlveredler. Chemische Technik—Prozesse und Produkte, 6a. WILEY-VCH Verlag GmbH & Co KGaA.

Bauer, G., et al., 2005. Vanadium and vanadium compounds. In: Ullmann's Encyclopedia of Industria Chemistry, A27, seventh ed. WILEY-VCH Verlag GmbH & Co KGaA, Weinheim, Germany. https://onlinelibrary.wiley.com/doi/epdf/10.1002/14356007.a27_367.pub2. (2017 edition).

Clark, J.H., Famer, T.J., Herrero-Davila, L., Sherwood, J., 2015. Circular economy design considerations for research and process development in the chemical sciences. Green Chem. 18 (2016), 3914–3934. https://doi.org/10.1039/C6GC00501B.

Gerold, E., 2021. Entwicklung eines hydrometallurgischen Recyclingkonzeptes für Lithium-Ionen-Batterien. Dissertation (2021), Montanuniversität Leoben.

Gladyshev, S.V., Adulvaliyev, R.A., Tastanov, E.A., Beisembekova, K.O., Temirova, S.S., Deveci, H.J., 2015. Recovery of vanadium and gallium from solid waste by-products of Bayer process. Miner. Eng. 74, 91–98. https://doi.org/10.1016/j.mineng.2015.01.011.

Gratz, E., Sa, Q., Apelian, D., Wang, Y., 2014. A closed loop process for recycling spent lithium ion batteries. J. Power Sources 262, 255–262. https://doi.org/10.1016/j.jpowsour.2014.03.126.

Hanisch, C., Diekmann, J., Stieger, A., Haselrieder, W., Kwade, A., 2015. Recycling of lithium-ion batteries. In: Handbook of Clean Energy Systems, Volume 5—Energy Storage. vol. 130, pp. 1–24, https://doi.org/10.1002/9781118991978.hces221.

Harper, G., Sommerville, R., Kendrick, E., Driscoll, L., Slater, P., Stolkin, R., Walton, A., Christensen, P., Heidrich, O., Lambert, S., Abbott, A., Ryder, K., Gaines, L., Anderson, P., 2019. Recycling lithium-ion batteries from electric vehicles. Nature 575, 75–86. https://doi.org/10.1038/s41586-019-1682-5.

Korthauer, R., 2013. Handbuch Lithium-Ionen-Batterien. Springer Berlin Heidelberg, https://doi.org/10.1007/978-3-642-30653-2.

Liu, C., Lin, J., Coa, H., Zhang, Y., Sun, Z., 2019. Recycling of spent lithium-ion batteries in view of lithium recovery: a critical review. J. Clean. Prod. 228, 801–813. https://doi.org/10.1016/j.jclepro.2019.04.304.

Luidold, S., 2013. The situation regarding critical raw materials for the high technological sector in Austria. In: Raw Materials are the Future—Volume 2: EUMICON—Contributions to a Resilient Raw Materials Policy, pp. 605–627.

Meshram, P., Abhilash, Kumari, J., Pandey, B.D., 2016. Extraction of vanadium and synthesis of vanadium pentoxide from Bayer's sludge. Russ. J. Non-Ferrous Met. 57, 338–346. https://doi.org/10.3103/S1067821216040064.

Meshram, P., Abhilash, Pandey, B.D., 2019. Advanced review on extraction of nickel from primary and secondary sources. Miner. Process. Extr. Metall. Rev. 40, 157–193. https://doi.org/10.1080/08827508.2018.1514300.

Pinegar, H., Smith, Y.R., 2019. Recycling of end-of-life lithium-ion batteries. Part I. Commercial processes. J. Sustain. Metall. 5, 402–416. https://doi.org/10.1007/s40831-019-00235-9.

Pradhan, S., R. Nayak and S. Mishra 2021: A review on the recovery of metal values from spent nickel metal hydride and lithium-ion batteries. Int. J. Environ. Sci. Technol. (2021) (Accessed 30 June 2021).

Reuter, M.A., Hudson, C., van Schaik, A., Haiskanen, K., Meskers, C., Hagelüken, C., 2013. Metal Recycling:

Opportunities, Limits, Infrastructure, A Report of the Working Group on the Global Metal Flows to the International Resource Panel. UNEP.

Schulte, R.F., 2017. Ferroalloys [Advance Release], 2017 Minerals Yearbook. U.S. Geological Survey (USGS). https://prd-wret.s3.us-west-2.amazonaws.com/assets/palladium/production/atoms/files/myb1-2017-feall.pdf.

Srivastava, R.R., Kim, M.S., Lee, J.C., et al., 2014. Resource recycling of superalloys and hydrometallurgical challenges. J. Mater. Sci. 49, 4671–4686. https://doi.org/10.1007/s10853-014-8219-y.

USGS. Mineral Commodity Summaries 2021, U.S. Geological Survey (USGS) (2021). (Accessed 29 June 2021).

Windisch-Kern, S., Gerold, E., Nigl, T., Jandric, A., Altendorfer, M., Rutrecht, B., Scherhaufer, S., Raupenstrauch, H., Pomberger, R., Antrekowitsch, H., 2021. Recycling chains for lithium-ion batteries: A critical examination of current challenges, opportunities and process dependencies. Waste Manag. 138 (2022), 125–139. https://doi.org/10.1016/j.wasman.2021.11.038.

CHAPTER

25

Precious and technology metals

Christina Meskers[1,2] and Markus A. Reuter[3]

[1]Department of Energy and Process Engineering, Faculty of Engineering Sciences, NTNU—Norwegian University of Science and Technology, Trondheim, Norway [2]Department of Industrial Eco-systems, SINTEF Manufacturing, Trondheim, Norway [3]WASM: Minerals, Energy and Chemical Engineering, Faculty of Science and Engineering, Curtin University, Perth, Australia

25.1 INTRODUCTION

This chapter focuses on the precious metals (Ag, Au), platinum group metals (PGMs) (Pt, Pd, Rh, Ru), and the metals Sb, Li, Re, Sc, Ta, and Ti. These metals are essential for sustainable development and the transitions taking place in society today. This includes renewable energy, mobility, emission reduction, digitalization, healthcare, hydrogen economy, and many more. Used in metallic and compound form and in some applications in minute quantities, their unique chemical and physical properties make them difficult to replace in highly complex products (printed circuit boards, catalysts, batteries) and alloys (superalloys, light metal alloys) due to the functionality they provide. Because of this, their demand is expected to increase strongly in the coming years and decades. Compared to the base materials such as paper, concrete, glass, and steel, these metals are produced in much smaller quantities, ranging from tens of tons for rhodium, for example, to hundred thousands of tons for titanium.

Except for gold, silver, and rhenium, the other metals covered in this chapter are listed as critical raw materials/minerals by both the European Union and the United States (European Commission, 2020; U.S. Geological Survey, 2022) due to their strategic importance. Resource-efficient use and recycling are part of the strategies pursued. Furthermore, tantalum and gold are considered conflict minerals (OECD, 2013; United Nations, 2011; SEC, 2011), so their origin needs to be transparent and traceable. The traceability of preconsumer scrap and its processing in facilities that are part of the conflict-free smelter program (CFSP), originally established by EICC and GeSI, is now operated by the Responsible Minerals Initiative (www.responsiblemineralsinitiative.org). Along that same line of thought, responsible recycling is just as important.

Many of the metals discussed in this chapter are coproduced with other metals or are by-products, as shown in the Metal Wheel (Figure 2.2 in Chapter 2—The fundamental limits of circularity quantified by digital twinning) and their recycling often uses entirely or partially the same infrastructure. The precious metals and PGMs are produced from own mines and as by-products from each other as well as from copper, lead, nickel, zinc, and antimony (complex) ores. For example, primary

https://doi.org/10.1016/B978-0-323-85514-3.00019-1

silver production is about 30% by-product of Pb and Pb-Zn ores, about 20% byproduct of Cu, Cu-Ni and Cu-Ni-Co ores, 15% by-product of Au production, and only about 30% is from silver mines (The Silver Institute and Metals Focus, 2021). Tantalum is exclusively produced as a by-product of niobium (from coltan) and tin production, while bauxite residue from aluminum production is the starting material for scandium production. Rhenium is produced from the flue gases of copper concentrate smelting, which are first processed for molybdenum recovery. Even lithium is partly produced from lithium (spodumene) mines, where the Li content is the economic driver, and partly coproduced from salt brines with sodium and potassium chlorides, among others. This interlinkage of metal production chains means that production cannot be increased fast if demand suddenly increases (Hagelüken and Meskers, 2010; Chapter 36—Geopolitics of resources and recycling).

Although the value of the metals is quite high, this does not guarantee recycling. Recycling has its challenges. Dissipative uses, such as titanium dioxide as pigment and antimony trioxide as flame retardant, render these materials lost for recycling. A lack of comprehensive collection of end-of-life consumer products (e.g., WEEE) and obtaining concentrated fractions suitable for recycling, while minimizing the material losses along the way, is another challenge. For some metals and applications, the recycling infrastructure or technology can be lacking because it is not developed (i.e., the chemistry is not right), too costly (low metal content and/or production volumes), or both, so that actual recycling does not take place. Recycling of scandium in aluminum alloys or tantalum capacitors in e-waste are examples of this.

This chapter provides a condensed overview of the recycling of these metals by first looking at the applications, then the collection and dominant recycling processes, and lastly what the key challenges and opportunities are in the circular economy of technology metals. As the field is in continuous development, the authors do not claim completeness.

25.2 APPLICATIONS

25.2.1 Precious Metals

The precious metals are a group of metals that include silver (Ag), gold (Au), and the platinum group metals platinum (Pt), palladium (Pd), rhodium (Rh), ruthenium (Ru), iridium (Ir), and osmium (Os). These metals have a high economic value, and their physical and chemical properties make them suitable for a wide range of applications.

Compared to base materials such as paper, plastic, steel, and aluminum, their global supply and demand is minute, and typically expressed in troy ounces (tr.oz., 31.10 g) instead of Mt. Global gross demand in 2021, met by primary production plus recycling, was: 32,130 t Ag, 4021 t Au, 210 t Pt, 314 t Pd, 32.2 t Rh, 31.6 t Ru, and 7.2 t Ir (The Silver Institute and Metals Focus, 2021; World Gold Council, 2022; Cowley, 2022).

Because of their similar properties, the PGMs can be exchanged among each other in certain applications, depending on the price of each. Platinum and palladium are regularly replaced by each other in automotive catalysts. Furthermore, due to their high price, there is strong interest in reducing the amount used in an application while still achieving the same performance (i.e., thrifting). This is done in automotive and chemical catalysts and electronic applications, among others, so metal content has been decreasing over time (Bangs et al., 2016; Rasmussen et al., 2019).

Jewelry, silverware, coins, and investments have been the main use areas of, especially, gold and silver, for millennia. Gold use includes about 55% as jewelry, nearly 30% as bars, coins, and ETFs for investment, about 11% to central banks, in total over 90% of gold use (World Gold Council, 2022). For silver, this percentage is much lower, about 45%, divided into 18% jewelry, 24% investment, and 4% silverware (The Silver Institute and Metals Focus, 2021). Jewelry applications used about 22% of Pt and about 9% of Pd (Cowley, 2022). Pure metals are rarely used in jewelry and coins. The most important silver alloys are Ag-Cu and Ag-Au-Cu (-Pt,Pd), e.g., sterling silver Ag 925 Cu 75. Typical gold alloys are yellow gold, white gold (Au-Ag or Au-Pd (Pt)), and rose gold (Au-Cu or Au-Ag-Cu).

Catalytic applications (Lloyd, 2011) are a key use area of, especially, PGMs and include automotive exhaust emission catalytic converters (autocats), stationary emission catalytic converters, chemical and petrochemical catalysts, and catalysts for hydrogen production (e.g., fuel cells (Rasmussen et al., 2019)). Autocats accounted for about 35% of Pt, 83% of Pd, and 92% of Rh use (Cowley, 2022). Stationary emission catalysts (mainly Pd) are used to treat emissions of diesel generators, gas turbines, and manufacturing processes. Increasingly stringent emission regulations drive the demand for PGMs (Gunn, 2014; Cowley, 2022). A distinction between homogeneous catalysts and heterogeneous catalysts is made. The latter consists of a solid inert carrier, such as cordierite (alumina), on which the PGMs are deposited and the gas or liquid flows through the catalysts. Homogeneous catalysts are generally a solution and are recovered during the process by distillation and reused until the catalytic activity is too low.

Besides PGMs, Ag and Au are also used in the chemical, pharmaceutical, and petrochemical industries. Silver is used as Ag-α-Al_2O_3 catalyst in, for example, the production of ethylene oxide and as silver wire mesh or crystals in formaldehyde production (Brumby et al., 2008), while gold is used in catalyst nanoparticles to produce vinyl acetate and pharmaceuticals.

The electrical and electronic applications make up a large part of the industrial applications. Although the quantities of precious metal per unit are very low, the quantity of products and breadth of applications is so large that the precious metals are highly dispersed and omnipresent in society. They are used in electric contacts and conductors in IT and telecom equipment, home appliances, electric vehicles and their charging infrastructure, and photovoltaic modules, to name a few. The metal demand for many of these application areas is growing, driven by the transition to a low-carbon society.

The electronics industry (plating contacts and printed circuit boards) represents over 80% of the industrial applications for gold and accounts for about 6% of the total gold demand, while about 10% of the global silver demand goes to PV (The Silver Institute and Metals Focus, 2021). Pt is used in hard disk coatings, Pd is used in multilayer ceramic chip capacitors (MLCCs), and both are used for plating of integrated circuits (Gunn, 2014), while Ru is essential in hard disks for data storage. Furthermore, Pt and Pt-Rh alloys are used for industrial tools in glass and fiberglass manufacturing, crucibles, and thermocouples due to their high temperature and chemical resistance.

25.2.2 Sb, Li, Re, Sc, Ta, Ti

Antimony (Sb), lithium (Li) rhenium (Re), scandium (Sc), tantalum (Ta), and titanium (Ti) are usually not discussed together because of their differences in properties, production methods, usage, and price, for example. When grouped according to density, lithium, scandium, and titanium are clearly light metals ($<3\,kg/dm^3$) while rhenium and tantalum are some of the heaviest ones ($>15\,kg/dm^3$). Considering melting point, lithium and antimony melt below 800°C, scandium and titanium around 1500–1600°C, while tantalum and rhenium melt well above 3000°C. Titanium, lithium, and antimony are predominantly used in their oxide form. In contrast, Re, Sc, and Ta are mainly used in their metallic form (Figure 25.1). When considering their use and its consequences for recycling, these metals illustrate specific cases.

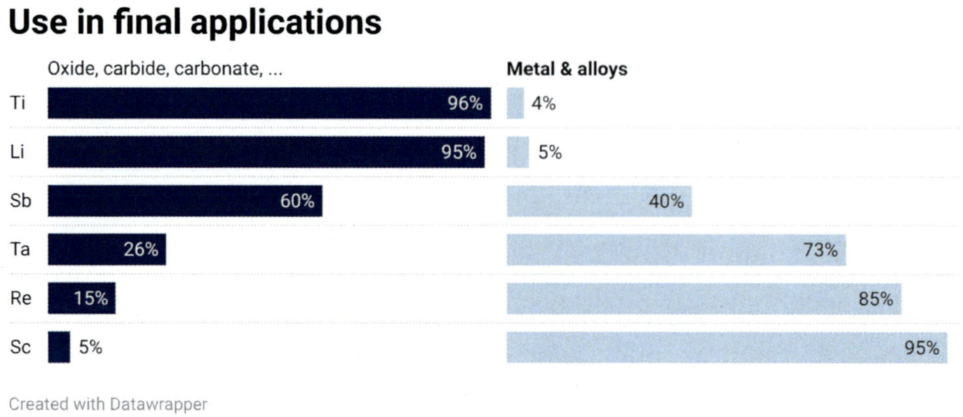

FIGURE 25.1 Share of the use of titanium, lithium, antimony, tantalum, rhenium, and scandium as compound or metal in their final applications. *Data from USGS (2020), European Commission (2020), Gunn (2014), Damm (2018), Mancheri et al. (2018).*

Oxides are typically used in pigment filler and catalytic applications. Roughly 6 Mt/y titanium, in the form of oxides, is mined and used as TiO_2 pigments and fillers in a wide range of applications, including plastics, ink and paint, pharmaceuticals, and many others (USGS, 2020). This accounts for over 95% of the usage. For antimony the situation is similar, though less extreme: about 60% of the annually mined 160 kt Sb is used as Sb_2O_3 as a flame-retardant synergist (52%), in plastics and printed circuit boards, for example, and as a catalyst (6%) in PET and polyester production and in fluid catalytic cracking (FCC) catalysts. Furthermore, Re, as oxide, combined with PGMs is used in catalysts in the petrochemical industry (Gunn, 2014) for the production of high-octane lead-free gasoline, among others (Shen et al., 2021).

Rechargeable batteries are a major use for lithium (>50%). It is used in the cathode of lithium-ion batteries as oxide (Li(Ni,Co,Mn)O_2) in NMC chemistries and as Li(NiCoAl)O_2 in NCA chemistries, as well as in LPF batteries as phosphate ($LiFePO_4$). Li salts ($LiPF_6$) dissolved in an organic solvent are used as an electrolyte. The quantity of lithium per battery varies with the battery chemistry and can range from 1.6 kg to almost 10 kg (Acatech, 2021). Other uses of Li compounds are in grease, glass ceramics, and ceramics. The metallic use represents about 5% (expressed as Li-carbonate equivalents) and includes use as electrodes in primary (nonrechargeable) lithium batteries used in car keys, watches, cameras, and medical devices. Future uses of Li metal are in solid-state and next-generation batteries. Lead acid battery alloys require antimony as an alloying element. This accounts for about 40% of antimony usage and is the main use in its metallic form. Recycling of lead-acid batteries is discussed in Chapter 22—Lead. About 5% of scandium is used in solid oxide fuel cells as Sc_2O_3 to increase life span and efficiency (European Commission, 2020).

The metal-based applications of Ti and Li make up less than 5% of their overall use. Ti alloys are used in aerospace (~50%), automotive (~20%), handheld equipment (~10%), medical implants (~10%), and metallurgical applications (~10%) (European Commission, 2020). A Boeing 787, for example, uses about 12 t of titanium alloys for the airframe, or about 14% of the frame weight (Takeda et al., 2020).

The $\alpha + \beta$ type alloys are most used, e.g., Ti-6Al-4V, and contain 2–6% α-phase stabilizers (Al, Sn, Zr) combined with 6–10% β-phase stabilizers (Mo, V, Nb, Cu, Si). They have low density, and fracture toughness, creep strength, and ductility are better than α alloys; tensile strength and fatigue resistance are better than β alloys (Sibum et al., 2017). Many alloys are biocompatible and can be used as implants. γ-TiAl alloys are the latest addition to the alloy family.

A small share (<5%) of global lithium use is as an alloying element in aluminum alloys of the Al-Cu, Al-Mg-Li, and Al-Li groups, which are used in aerospace applications (Prasad et al., 2014). Scandium is almost exclusively used as an alloying element to create very light aluminum (Al-Mg-Sc) alloys for automotive, aerospace, and sports equipment applications (Wang et al., 2011).

Rhenium is mainly an alloying element in nickel-based superalloys for high pressure gas turbines (HPTs) in aerospace and land-based power generation (European Commission, 2020). About 3–6% Re is added to nickel-based superalloys, which are casted into single-crystal turbine blades.

Tantalum is used in metallic form in capacitors (41%). The majority of capacitors are used in computers, electrical and telecom infrastructure, and automotive applications. Furthermore, Ta is used as an alloying element in superalloys for turbines (23%), which contain 3–11 wt% Ta and in sputtering targets (17%), among others. The remainder is used as carbides in tungsten carbide cutting tools (Gunn, 2014; Nassar, 2017; Damm, 2018; Mancheri et al., 2018).

25.3 SCRAP TYPES AND QUANTITIES

The recycling of, in particular, precious metal-containing scrap is driven by the economic value of the metals. Their value is often also an enabler for the recovery of additional metals that are present in scrap. Especially for the metals discussed in this chapter, a product perspective is important, as the technology metals are embedded in complex material mixtures, which can make it costly to recover them.

A distinction needs to be made between preconsumer or new scrap and postconsumer or old scrap when discussing the scrap types and their recycling. New scrap is often recycled, as it is clean, well-separated, and easily collected. Collection and separation into a concentrated fraction, in an economically feasible way, is essential for successful recycling of end-of-life scrap. Sampling and assaying to correctly determine the metal content, and hence the economic value, is essential. Especially for complex materials such as printed circuit boards or catalysts, for example, this requires specialist methods. For detailed discussion of recycling economics, please see Chapter 42—Economic aspects of metal recycling.

When considering the recycling system and the contribution of recycling, several aspects need to be taken into account:

1. The supply-demand balance. The demand is still growing, hence primary materials are necessary to compensate for market growth and inefficiencies and losses in the collection and recycling chain.
2. The lifetime of many applications is long, so there is a considerable time delay between the time when products enter the use phase and when they become end of life. E.g., for automotive catalysts the lifetime is similar to the life of the car, so 10–15 years. That means that today (2022) the catalysts from cars sold between 2007 and 2012 are recycled. When the amount of PGM recycled from catalysts is compared to the PGM demand for automotive catalysts back then, the recycled percentage is even higher (Hagelüken and Meskers, 2010; Rasmussen et al., 2019; Hagelüken and Goldmann, 2022).
3. The collection of end-of-life products and getting materials in the right recycling stream for recovery can be very difficult.

Recyclers receive devices with different lifetimes and compositions, resulting in a very heterogeneous stream, especially for materials from complex (consumer) goods. This is discussed in detail in Part 1 of this book (Chapter 2—The fundamental limits of circularity quantified by digital twinning and Chapter 5—Material and product-centric recycling and design for recycling rules and digital methods).

4. The metal purity and quality. To achieve true circularity, the metals obtained from end-of-life products need to have the appropriate quality to be used again in the same product, so product design, recycling, and metal production are linked. Unmixing the different materials connected to each other and in small quantities is difficult, as discussed in Parts 1 and 2 in this book. Exergy can be used as a quantitative metric for quality loss (Meskers et al., 2008 and Chapter 45—Exergy—Quantification of resource dissipation and Chapter 46—Process simulation—Thermodynamics and process technology to understand recycling systems).

25.3.1 Precious Metals and PGM

Recycling activities for the precious metals are well established, driven by their economic value and their noble (electro)chemical properties. The degree of recycling varies with applications and scrap type. Preconsumer scrap, business-to-business use, and nondispersive consumer applications achieve high rates, while applications for which collection is not working well have much lower rates.

The contribution of recycled materials to the supply of precious metals is considerable: about a quarter of silver, gold, and platinum gross demand comes from recycling. For palladium and rhodium this is about one-third (The Silver Institute and Metals Focus, 2021; World Gold Council, 2022; Cowley, 2022).

For silver, the main source is industrial use (catalysts) (about 55%), while photography, jewelry, and silverware make up the remainder. For gold this is reversed: the main source is jewelry (90%) and use in electronics is merely 10%. The main source of recycled PGMs is automotive catalytic converters. Spent light-duty vehicle catalysts contain 1–3 g PGM, while heavy-duty catalysts contain 12–15 g PGM per catalyst (Yakoumis et al., 2021). These represent nearly 75% of the recycled Pt, about 85% of the recycled Pd, and all of the recycled Rh. The other sources for recycled Pt are jewelry (23%) and for recycled Pd are electrical and electronics (14%) (Cowley, 2022). Materials from electronics (e.g., printed circuit boards or cellphones) have a fairly open recycling loop, where collection is less effective, and metal losses occur during size reduction (dismantling, shredding, grinding), sorting and separation (Chapter 13—Electric and electronic equipment (WEEE)). The losses vary for each metal depending on the type of product, the chosen processing route, and the operating parameters (Chancerel et al., 2009; Reuter et al., 2018, 2019).

Furthermore, chemical and petrochemical PM and PGM-containing catalysts are recycled. Their lifetime can be decades. This is done via a highly efficient, closed-loop business model, so these volumes do not appear in the statistics. Only the minimal amount of primary material used to make up for losses is included. In contrast, automotive catalysts have an open recycling loop (Rasmussen et al., 2019; Hagelüken and Goldmann, 2022).

The different scrap qualities are collected and processed separately. The high purity, high metal content (metallic) jewelry, silverware, and coin scrap go to precious metals refineries. Catalyst scrap can be treated in dedicated processes or sent to integrated metal refineries using a copper, or copper-lead, smelter, where complex materials with low precious metals content (e.g., circuit boards, cell phones) are also treated (Schlesinger et al., 2021).

25.3.2 Antimony

Sb is recycled from lead alloys, especially those used in lead-acid batteries, in the form of antimonial lead. This is the major source of recycled metal, discussed in Chapter 22—Lead. Antimony from the Sb-containing flame retardants in printed circuit boards and from plastics that are part of copper-bearing WEEE fractions are recovered via copper and copper-lead smelters. Furthermore, antimony is recovered from Sb-containing by-products from lead refineries (Dupont et al., 2016). This brings the European recycling rate to about 28% (European Commission, 2020).

Sb_2O_3 is usually not recovered from the products, i.e., plastics it is used in. For example, ABS and HIPS in Waste Electric and Electronic equipment contain 5–15 wt% Sb as Sb_2O_3 (Dupont et al., 2016). Its final destination depends on what happens to the plastic (or other material). During the sorting practices of WEEE it is the aim to separate the plastics with brominated flame retardants from the others. During this process, incorrect sorting results in unintended inclusion in the Br-flame-retardant free fraction, and recycling as part of that plastic fraction. The Br-flame-retardant plastic fraction is incinerated, as no recovery processes for Sb from this fraction exist (Filella et al., 2020). Alternative processes have been developed, e.g., using selective dissolution of the plastics to recover Sb_2O_3 (Schlummer et al., 2016), though none of them is used on an industrial scale (Dupont et al., 2016).

During incineration of municipal solid waste, which includes plastics, wrongly sorted WEEE, etc., or uncontrolled burning of plastics or printed circuit boards, the Sb compounds distribute over the bottom ash and fly ash fractions that are generated. Because of its high volatility, it is expected that all Sb ends up in the fly ash fraction. However, up to 50% can end up in bottom ash, depending on the type of waste incinerated, the combustion temperature, and presence of other elements, especially chlorine (Filella et al., 2020). Collected fly ash is often landfilled, and Sb can leach from the ash in the landfill. Therefore capturing and treating the leachate is important to reduce impact on the environment. Also, Sb can leach from bottom ash, reducing the use of bottom ash in other applications (Filella et al., 2020).

25.3.3 Lithium

Recycled lithium mainly comes from batteries used for e-mobility and portable electronics. Collection of these is essential to get the batteries into a recycling chain (Neumann et al., 2022). Recycling is driven by the value of the metals contained and supporting legislation, for example in the European Union. Two approaches are possible (An, 2019). The universal process, e.g., Umicore, treats all battery chemistries together and focuses on raw material recovery, throughput, and minimal pretreatment. The pyrometallurgical process gives an alloy with Co, Ni, Cu, Mn, and a Li-rich slag. Both go into specific hydrometallurgical processes: the alloy is treated for recovery of Co, Ni, and other metals, while the slag is treated for the recovery of lithium salts. The battery chemistry-specific processes use dismantling, shredding, and sorting to create different fractions, e.g., aluminum foils and black mass, suitable for further (hydrometallurgical) processing and refining. This approach has high requirements for collection and sorting based on battery knowledge, and not all materials are recovered (Velázquez-Martínez et al., 2019; Meskers, 2021).

25.3.4 Rhenium

Global Re metal production was 49 t in 2019 as a by-product of primary copper production. Recycling contributes about 20 t/y, to meet a global demand of 60–65 t/y. Recycling of Re-containing scrap and products is common practice, driven by the value of the material.

Superalloy scraps and products, mostly nickel-based, are recycled in a business-to-business closed loop. Spent Pt-Re catalysts are recycled mainly for the platinum value, and Re is a valuable by-product. This happens in a closed loop, business-to-business setting (Gunn, 2014; Shen et al., 2021).

25.3.5 Scandium

Scandium is obtained from tailings and by-products from the mining and processing of REEs. Furthermore, processes are known for the recovery from nickel laterites, uranium, titanium, and zirconium ores, tungsten and tin ores, and tantalum and niobium ores (Wang et al., 2011). Red mud in particular, the residue remaining after processing of bauxite ore to extract alumina, is considered an important potential source. Scandium contents vary depending on the bauxite ore and can be 60–120 g/t red mud (Akcil et al., 2018). Typically, after cementation of the red mud, extraction in acid medium, and separation with ion exchange methods, scandium extraction rates of 70–85% have been obtained (Akcil et al., 2018). Up to now the flow sheets that have been proposed have not been commercialized (Akcil et al., 2018; Botelho et al., 2021; Wang et al., 2011) and research is ongoing. Thus little infrastructure is available that could be used for refining of fractions from recycling operations.

Scandium metal is currently not recycled on a large scale. Its reported recycling rate is close to zero. During Al-Sc and Mg-Sc alloy production, scandium is partly lost to dross because of its high reactivity with O, Cl, and F. Dross from Mg-Sc alloy production can contain 12–23% Sc. This can be recovered by leaching the dross in hydrochloric acid followed by solvent extraction, precipitation, and calcining. This process is not applied industrially, as the quantity of scandium-containing dross is too small for an economically feasible process (Wang et al., 2011). Any new scrap generated during casting and manufacturing could be recycled as internal scrap.

At end of life, recycling of scandium as part of the aluminum alloys it is used in might be possible if these alloys are separately collected and recycled. This is currently not the case. The losses to dross would still occur, so dross processing to recover Sc would be necessary. As most of the scandium is used in these alloys, this could be the first area to explore to increase scandium recycling.

E-waste, in particular printed circuit boards and OLEDs, have been mentioned as a potential source for the recovery of scandium by Botelho et al. (2021), as scandium is present in minute quantities on the circuit boards. In the grand scheme of things, e-waste is an irrelevant source for Sc. The quantity of scandium available from printed circuit boards is very low, compared to the ~95% of the scandium production used in aluminum alloys. Furthermore, concentrating scandium in a fraction from which it can be recovered is technically and economically extremely challenging.

25.3.6 Tantalum

Throughout the life cycle of tantalum, different wastes and scraps are generated (Figure 25.2). Most of the recycled material originates from the processing of tantalum, followed by scraps from the manufacturing of Ta products. Considering the entire cycle and all applications together, about 30% of the supply is from recycling, and roughly 30% of end-of-life materials are recycled (Nassar, 2017).

The residues and scraps generated during primary production (internal scrap in Figure 25.2) include residues generated during the leaching of Ta_2O_5; during the reduction of K_2TaF_7 to Ta metal; and lastly metal scraps from the walls of the electron beam melting furnace (Nassar, 2017). These residues are internally recycled. Primary processing is described by

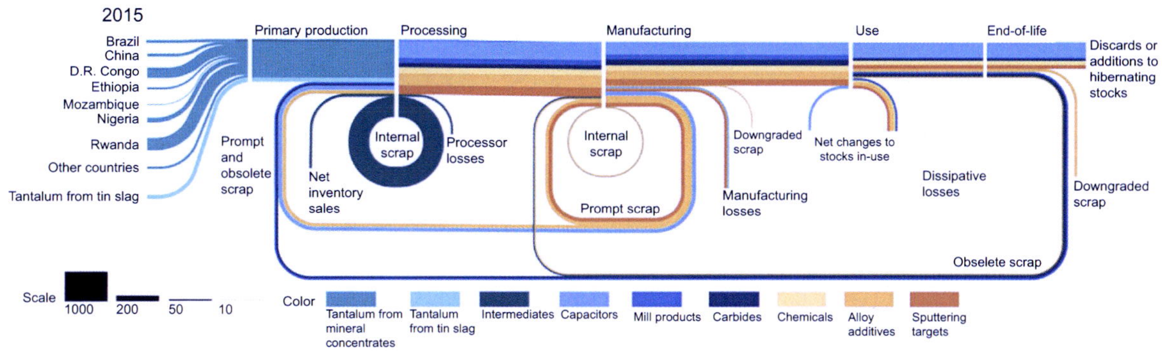

FIGURE 25.2 Global tantalum cycle of 2015. Flow width is proportional to the tantalum content; quantities are in metric tons of tantalum. Flow directions are from left to right, except for scrap recycling for which the flows are reversed. *From Nassar (2017).*

Shikika et al. (2020), while Taube et al. (2020) investigated recovery from residues and discusses the associated challenges.

Typical manufacturing scraps are from capacitor manufacturing, from the manufacturing and use of sputtering targets, Ta-containing superalloy scrap, and metallic scraps (Nassar, 2017). Most is recycled internally, except for capacitor scrap, which is returned to primary processing. Sputtering targets are considered spent when about 20–50% of the material remains. Furthermore, during the sputtering process material is deposited on the product to be coated, and on the sputtering chamber walls. The scraps from the wall and the spent targets are recycled as prompt scrap.

End-of-life scrap consists of more than half of tantalum carbides from cutting tools, which have a recycling rate of about 60% (Nassar, 2017). This is a fast cycle, as the lifetime of the cutting tools is about a year. Closed-loop business models are often used. The other half of the end-of-life scrap consists of tantalum metal used in installations in the chemical industry, which have a lifetime of 20 years or more, and from superalloys that are typically in use for 10–12 years (Nassar, 2017). Capacitors used in electronics, the major use of tantalum, are not recycled from end-of-life products. This is due to the low concentration of tantalum on the printed circuit boards, the technical and economic difficulty to liberate and concentrate the capacitors in a separate stream, and the technical difficulty of obtaining a tantalum concentrate from the capacitors with sufficient purity. The printed circuit boards are recycled via the copper-precious metals route, where all tantalum ends up in the slag because of its high affinity for oxygen and extremely low solubility in copper (Shuva et al., 2017).

25.3.7 Titanium

Growing titanium production has also increased the availability of titanium secondary raw materials (Figure 25.3). Recycling is driven by the economic value of titanium.

Primary production scrap from the Kroll process has a lifetime of less than a year. It consists of titanium metal sponge that is in direct contact with the crucible that is heavily contaminated with iron (Takeda et al., 2020; Zhang et al., 2011; van Vuuren, 2019). These parts are removed from the clean sponge and sent to the steel industry to be used as feedstock for ferrotitanium or deoxidation material. In contrast, final products have a lifetime of 10–30 years. Hence there is a tremendous time delay between the titanium part entering the use phase and leaving it.

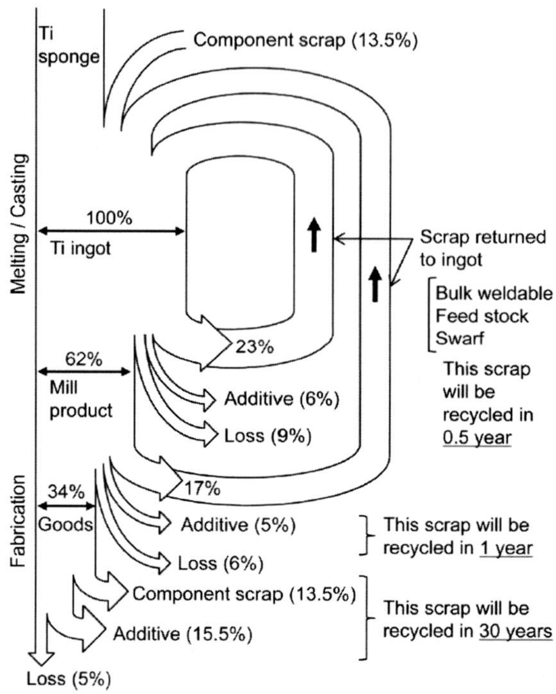

FIGURE 25.3 Material flow of titanium scrap in Japan (2007) showing both high-grade and cascade recycling and the time delay between material production and scrap arising. *From Takeda and Okabe (2020).*

Within the aviation sector, there is a high level of scrap generation, as the parts are machined from ingots, so ~85% of the ingot is converted to swarf, contaminated with oil and tool fragments (Takeda et al., 2020).

The Ti-6Al-4V alloy used has a fair tolerance for iron (max. 0.25%) and oxygen (max. 0.2%) impurities. Good scrap management can avoid high contamination with iron; oxygen contamination cannot be avoided and is difficult to remove during remelting. The recycling rate of swarf and parts is over 90% (Takeda et al., 2020).

Additive manufacturing techniques are used for part production (e.g., Tshephe et al., 2022). The (near) net shape production method reduces the need for machining and thus swarf waste generation considerably, though powder scrap is generated, which could be reused and recycled (Moghimian et al., 2021).

25.4 RECYCLING TECHNOLOGIES

25.4.1 Metallic Scrap With Precious Metals

PM winning and recycling processes are common and deployed around the world. They exploit the chemical properties of these metals (e.g., reactivity and oxidation) and use a variety of separation techniques. The recycling process consists of four main steps:

- pretreatment of feedstock, e.g., removal of organic impurities from sweeps,
- sampling and assaying,
- concentration, extraction, and separation of precious metals,
- refining to pure metals.

Typically there is much more silver and gold than PGMs in the material. Many of the processes use very reactive reagents or produce toxic cyanide residues; however, new lixiviants such as glycine (Eksteen et al., 2018) are charting new avenues with more benign reagents and processing routes. Nevertheless, containment, fail-safe systems and sealed drainage areas are used to minimize losses and impact on the environment. This is further driven by the high value of the metals. Hydrometallurgical refining is mostly used for refining high gold (>75%), non-doré scraps, as it is fast, simple, and robust (Fritz and Schmidt, 2022). This consists of dissolution of gold in aqua regia, while silver precipitates as $AgCl_2$, followed by precipitation of gold with SO_2. The remaining PGM solution can be sent to PGM refining. Detailed process and flow sheet descriptions can be found in (Adams, 2016) for gold, and Nose and Okabe (2014) and Crundwell et al. (2011) for PGM.

25.4.2 Printed Circuit Boards From WEEE

Printed circuit boards, server boards, cellphones, and other small IT devices, as well as shredded printed circuit board fractions (containing varying amounts of copper wires and plastics) are recycled for their copper and precious metal value in well-established commercial pyrometallurgical processes using copper and lead as collector metals for precious and other metals (Figure 25.4), putting the Metal Wheel (Figure 2.2) into practice.

Before further processing, the materials are sampled and assayed. The samples need to be representative of, and thus tailored to, the entire scrap shipment and the assays need to be accurate, so that metal contents ranging from percent (for copper) to parts per million (for precious metals) can be determined reliably.

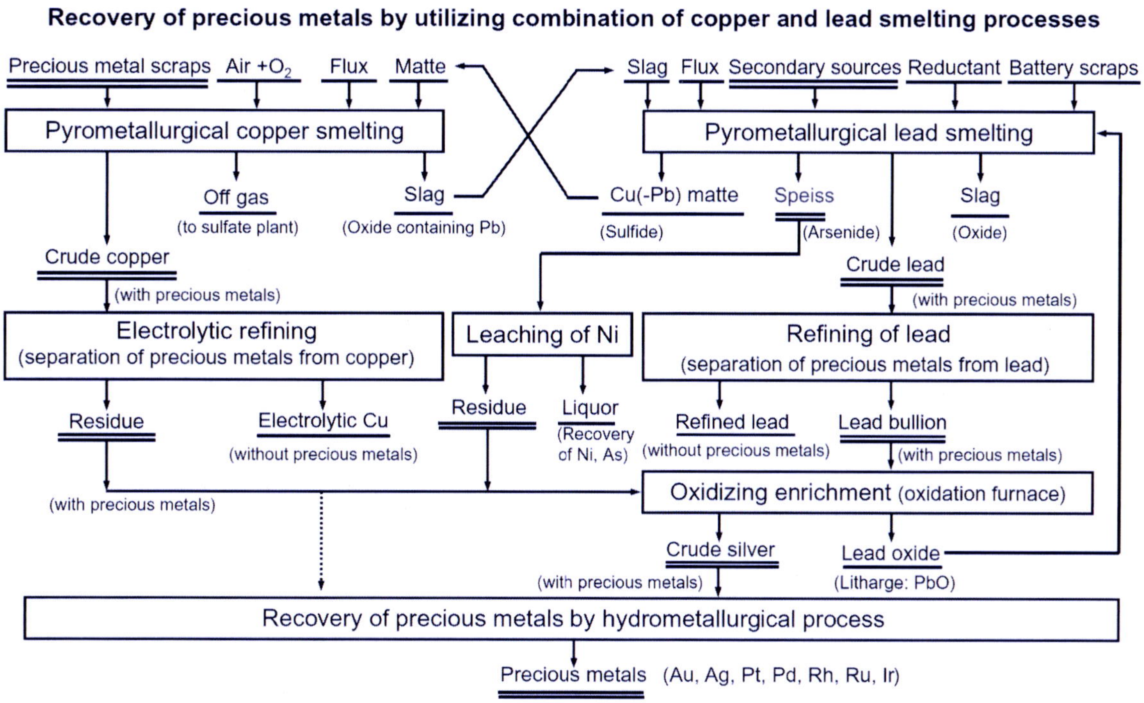

FIGURE 25.4 Example of a smelting-refining process for precious and other metals from printed circuit board using the combination of copper and lead metallurgy. *From Nose and Okabe (2014).*

The scrap is treated together with other suitable materials typically in the copper-based flow sheet with top-blown submerged lance furnaces, producing a crude copper with precious metals and an oxide slag containing elements such as Pb, Sn, Bi, Sb, In, as well as Fe, Al, Ta, and part of the silver (Lux et al., 2022; Kandalam et al., 2023a,b). The organic compounds of the feed material are used as reducing agents and converted to energy (Hagelüken and Meskers, 2010). The crude copper is separated from the precious metals using electrowinning or electrorefining to create a precious metal and PGM rich residue that can be processed in a precious metals refinery using processes as described in Nose and Okabe (2014) and Crundwell et al. (2011). The oxide slag is further processed in the lead-based flow sheet, where lead, antimony, tin, indium, and bismuth can be recovered. Silver is removed from the lead bullion and returned to the precious metals refining flow sheet. Note that residues such as copper matte and speiss from the lead flow sheet are returned to the copper and nickel processing flow sheets, respectively. An example of this type of operation is Umicore in Belgium. Elements that oxidize easily, such as iron, aluminum, rare earths, and tantalum, end up in the final slag and are not recovered (Shuva et al., 2016, 2017).

25.4.3 Catalysts

Specific processes have been developed for recycling of different catalysts, which include pyrometallurgical, hydrometallurgical, and biometallurgical processes. The latter are not yet commercially implemented (Yakoumis et al., 2021). Lloyd (2011) describes in detail the composition of different catalyst types, illustrating the material complexity. The challenges in catalyst recycling are:

- separation of the carrier material from the PGMs, as the carrier is either combustible or flammable (carbon/organic carriers) or has an extremely high melting point (cordierite carrier),
- removal of (organic) contaminants,
- separation of the individual PGMs,
- recovery or valorization of other elements such as Zr, Ti, Ce, and oxide-based carriers.

Carbon-based or organic-based catalysts are incinerated to remove the carbon, creating ashes that are then treated hydrometallurgically to dissolve and recover the PGMs. Careful material handling is key to avoid loss of PGM-containing materials.

Cordierite or alumina-based catalysts can be treated in pyro- or hydrometallurgical processes. Reforming or hydrogenation catalysts with an Al_2O_3 bearing zeolite with M: alkaline (earth) metal can be treated by dissolution of the ceramic base in sodium hydroxide or sulfuric acid. Prior to leaching, the excess carbon and hydrocarbons are burned off.

Automotive catalysts consist of a monolith/honeycomb, usually cordierite ($2MgO \cdot 2Al_2O_3 \cdot 5SiO_2$), though steel metal is also possible, covered with the washcoat materials (e.g., γ-Al_2O_3, $CeZrO_2$, activated carbon) and the PGMs, which are combined with base metals (e.g., Cu, Ni, V, W) (Yakoumis et al., 2021). This is a complex material mixture, making purification challenging. The monolith is encased in a steel canister. After removal of the can, the monolith is crushed and ground into a fine powder, followed by metal recovery in pyrometallurgical, hydrometallurgical, or biometallurgical processes; the latter is not yet commercialized. Pyrometallurgical processes use Cu, Fe, Pb, Ni, or mattes to collect the PGMs, then separate the PGMs and refine them further. The oxides from the monolith and washcoat report to the slag and are not further extracted. Hydrometallurgical processes leach the PGMs using aqua regia, cyanides, chlorides, for example, followed by separation and recovery of the PGMs. The monolith and washcoat materials could be valorized too. Emissions and waste management, energy, and reagent consumption and costs, as well as obtaining high recoveries, are attention areas for both types of processes. The process technologies are described in detail in Nose and Okabe (2014), Crundwell et al. (2011), and Yakoumis et al. (2021). Louw and Reuter (2022) present a dynamic model for a catalyst recycling plant.

Rhenium-containing catalysts (~0.3% Re) are treated for the recovery of Re and Pt. The exact processes used in industry are not disclosed. Two different approaches are described in the literature. Hydrometallurgical processing first dissolves the ceramic carrier and rhenium in sodium hydroxide or sulfuric acid. Platinum remains in the residue. Rhenium is then recovered by ion exchange. Alternatively, a pyrometallurgical process is used that first calcines the material at 1150°C with soda ash to lower the solubility of alumina and then selectively dissolves Re and Pt using sulfuric acid, which are subsequently recovered (Shen et al., 2021).

The final rhenium product obtained is catalyst-grade ammonium perrhenate, which has a high purity, to avoid contamination (poisoning) of the catalyst. Thus the combined content of As, Ca, Fe, K, Mg, Na, Pb, and Sn cannot be above 200 ppm, and the combined content of As, Co, Cr, Cu, Mn, Ni, Sb, Si, and Sn cannot be above 200 ppm (Gunn, 2014).

25.4.4 Superalloys and Metallic Scrap Containing Re, Ta

Rhenium-containing superalloy scrap is recycled back into superalloys when the scrap is new and very clean. Less clean scrap is recycled by removing Re from the alloy using oxidative roasting to obtain potassium perrhenate ($KReO_4$), which then enters the primary process for rhenium metal production. Alternative processes are the alkaline fusion process using alkali nitrates, fluoride treatment, or electrochemical dissolution (Shen et al., 2021).

Tantalum-containing superalloy scrap (3–11 wt% Ta) is recycled as part of appropriately sorted superalloy scrap. Manufacturing scrap is recycled back to superalloys, while end-of-life scrap can also be diverted to the steel industry due to concerns about contamination of the superalloy with other materials. In the steel industry it is used as alloying material (Nassar, 2017). Clean, unoxidized metallic scrap, such as sputtering targets, and wires from capacitors that contain max 1.5–3% oxygen can be recycled by melting in an electron beam furnace. Alternatively, the material can be converted to Ta hydride, which is brittle, followed by grinding and dehydrogenation at 600°C to obtain metallic powder. Instead of dehydrogenation, carburization can be done to obtain Ta carbides. If the scrap contains too much oxygen, deoxidation with magnesium can be done first (Albrecht et al., 2011). If other metal impurities are present, these can be removed by treatment with nitric and hydrochloric acid. For other complex manufacturing scraps, e.g., capacitors, after roasting the obtained material can be treated the same as a primary concentrate.

25.4.5 Cutting Tool Scrap (TaC)

Cutting tool scrap consists of cemented carbides of tungsten (WC) with 2–15 wt% tantalum carbide (TaC) (Albrecht et al., 2011). Its recycling is driven by the recovery of the tungsten carbide, while the recovery of cobalt and tantalum carbides provides additional economic value. The end-of-life recycling rate is about 60%, close to the practical limit (Nassar, 2017). Two processes are used. The direct recycling process dissolves the binder that cements the carbides together, and the obtained powder is directly used to make new cutting tool alloys. The indirect process uses chemical processing, such as roasting and treatment with caustic soda (Albrecht et al., 2011), to separate the scrap into the pure materials. Both processes are used, about 40% via direct recycling and 60% via indirect recycling (Nassar, 2017).

25.4.6 Titanium Metallic Scrap

Clean and sorted scrap is usually introduced into the remelting step of the primary route to produce titanium ingots, while loose scrap can be used for cold hearth melting without further preparation; vacuum arc remelting in specially designed furnaces mixes the scrap with titanium sponge and compresses it to electrodes. The electrodes are then remelted.

The swarf from aviation part production is first cleaned to remove oily and tool part contamination. Then it is melted using electron beam melting (EBM), plasma arc melting (PAM), or vacuum arc melting (VAR), as these technologies minimize oxygen contamination during melting. Virgin, sponge titanium is added to meet the alloy specifications. Note that if titanium demand increases dramatically, oversupply of scrap and shortage of dilution metal can occur. Oxygen contamination is the main challenge, as the oxide layer on the titanium scrap dissolves in the melt, and removal possibilities are limited. On an industrial scale, deoxidation with calcium metal and electrorefining in molten salts is used. Other processes are in the research stage. Detailed information can be found in Takeda and Okabe (2019, 2020).

Metal alloy powders from additive manufacturing can be reused (directly) in a new manufacturing cycle or need to be recycled, depending on whether the powder's composition, contamination level, and physical characteristics are still meeting the specifications for direct reuse. Out of specification powder is typically contaminated due to contact with tools, oil, reactive gases, moisture, etc. and is recycled using direct remelting or it goes back to the alloy manufacturing process together with scrap metal and built parts. Traceability is key, especially for aerospace applications (Moghimian et al., 2021).

For the recycling of γ-TiAl alloy scrap from, for example, aero engines, Reitz et al. (2011) developed a process using vacuum induction melting (VIM) or aluminothermic reduction (ATR) followed by pressure electroslag remelting (PESR) and/or vacuum arc remelting (VAR).

25.5 FUTURE CHALLENGES

The broad range of elements, their applications, and recycling discussed in this chapter have shown that the challenges are broad and diverse. Nevertheless, the commonalities can be identified as:

- materials are intricately, often on the atomic or molecular level, and functionally connected, and thus not easy to separate.
- quantities in the products or components are low (even minute), which hinders recovery from a thermodynamics point of view (Second Law of Thermodynamics) and results in unfavorable economics.
- metallurgical infrastructure necessary to recover the elements at scale is unavailable, or key linkages in the infrastructure are missing.

Advances in product design (Chapter 5—Material and product-centric recycling and design for recycling rules and digital methods) to direct components into the appropriate recycling process, along with supporting government policy, can help to overcome the challenges. This is illustrated by the case on tantalum-containing capacitors.

25.5.1 Ta-Containing Capacitors

Recycling of postconsumer capacitors located on printed circuit boards in waste electrical and electronic equipment (WEEE) and telecom infrastructure is not done commercially. There are several challenges to overcome:

- Very low Ta content per device, distributed over several capacitors
- Identification, liberation, and separation of the capacitors into a separate fraction from the product, and the printed circuit board in the product
- Separation of the tantalum-containing material (metal and oxide) from the rest of the capacitor

- Generation of a tantalum concentrate with sufficient purity to serve as feedstock for capacitor manufacturing.

The number of capacitors in electronics varies per device type, and the tantalum content and capacitor size decreased over time. The Ta content per capacitor (3–7 mm long and 1.5–3 mm wide) has been declining from several hundred mg to 10–80 mg today. This ranges from 0.02 g Ta (22 capacitors)/smartphone, to 1 g Ta (17 capacitors)/laptop, up to 2–3 g Ta (89 capacitors)/server circuit board (Nassar, 2017; Ramon et al., 2020; Agrawal et al., 2021).

Ta capacitors are small and complex products, with a layered structure, including from inside to outside: Ta metal, Ta oxide, MnO_2 electrolyte, graphite and silver paste, resin with flame retardants. The housing consists of resins, usually epoxy and phenolic novolac, silica powder, and flame retardants and seals the capacitor from the outside environment. The terminal is made of Fe-Ni (Agrawal et al., 2021; Ueberschaar et al., 2017) (Figure 25.5).

Typically, the printed circuit boards have both black capacitors and orange capacitors in different sizes on both sides of the board. The composition varies: 26–36.8 wt% Ta, 0.1–4.5 wt % Nb, 1.5–3.6 wt% Ag; the balance is the terminal and housing material (Ramon et al., 2020).

Liberation of the capacitors in a recycling process especially for capacitor recovery requires first to have a separate circuit board fraction from the devices that still have capacitors on them. Too intense shredding removes the capacitors from the boards but does not collect them in a separate capacitor fraction, hence the capacitors are "lost" and not recycled. (Chancerel et al., 2009). Several technologies have been proven, including crushing followed by sieving and mass and magnetic separation, or a combination of eddy current and gravity separation (Ueberschaar et al., 2017) to obtain a capacitor fraction. Automated or robotic dismantling has also been investigated, using RGB color and a line-depth camera combined with a neural network for identification and mechanical removal (Ramon et al., 2020). Although technically

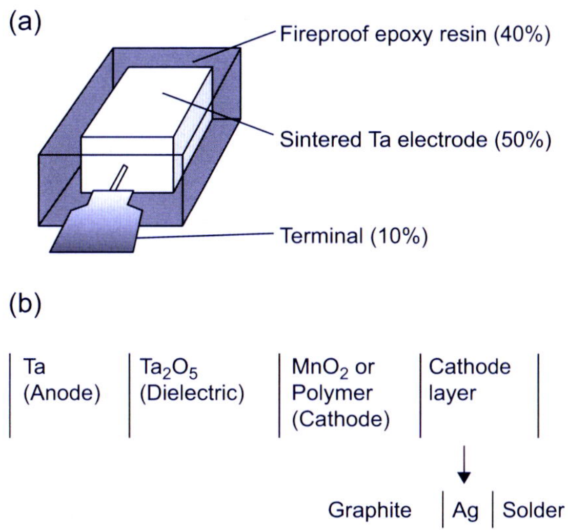

FIGURE 25.5 Schematic of tantalum capacitor reflecting its material complexity and the interlinkages between materials. (A) Key parts and (B) layered structure of electrode from center to outside. *From Mineta and Okabe (2005).*

feasible, a large volume of boards needs to be processed to obtain a sufficient capacitor tonnage to justify the investment. The tantalum-containing material needs to be separated from the rest of the capacitor. The encapsulating resin that protects the capacitor during use is very difficult to remove, providing a major challenge. Several technologies have been proposed and proven, though none is industrialized. Pyrolysis combined with sorting technologies seems to have the most potential (Agrawal et al., 2021). Besides the technical challenges that need to be overcome, the environmental performance and economic feasibility need to be considered as well.

25.6 CONCLUSIONS AND OUTLOOK

The recovery of technology elements is challenging, as explained in this chapter. The specific challenges are highly dependent on the element itself, the other elements it is combined with, the application it is used in, the lifetime of the component or product, and the recycling processes, to name a few factors. Some applications are simply dissipative, while others have recycling potential. The following aspects need particular attention.

Chemistry and technology for recovery of technology elements. Although this is often known, the complexity added by the functional material combinations in products, the difficulties in separation, and thermodynamic limitations hinder recovery in metallurgical processes (Iglesias-Émbil et al., 2023). Trade-offs need to be made, as recovering one section of the Metal Wheel means incompatible sections will be lost (Chapter 2—The fundamental limits of circularity quantified by digital twinning). New knowledge and developments are needed to push recycling processes closer to the thermodynamic limits, and meet the requirements of the sustainability and circular economy paradigms at the same time.

Collection and liberation into appropriate fractions. Enhancing this would increase the volume of material entering (metallurgical) recycling processes. Changes in the collection and recycling systems for end-of-life products or in ownership models are essential. Design for recycling, e.g., through modular design, is another important lever, though not easy. The Fairphone design approach (Reuter et al., 2018, 2019) is an example. Modular design combined with the use of digital simulation tools to link product design to physical processing (e.g., shredding (Heibeck et al., 2021, 2023)) and to metallurgical processing shows the impact of different design choices. The use of other separation methods including disassembly robots, e.g., the tantalum capacitor case in this chapter and Apple's Daisy and Dave robots (Apple, 2022), can be used to obtain specific fractions containing critical and technology metals. The performance of robots can be supported through clever design choices.

Economic feasibility. Increasing the available volumes for recycling supports the economies of scale of recycling processes. Thrifting, i.e., lowering the material content while maintaining functionality, makes recycling economics more challenging. As the material content per component decreases, thermodynamics become more unfavorable and more components are needed to obtain the same tonnage (and thus value) of material. When evaluating the economic feasibility, both the feasibility of the individual process and the economic benefits at the system level, e.g., supply chain resilience, access to raw materials, etc., need to be considered so that appropriate circular business models can be found.

Policy and legislation. Both can positively impact, among others, the economics of recycling, enable the development and implementation of recycling and metallurgical infrastructure, and support priority setting, for example in ecodesign. Often a portfolio of policies that work together is necessary because of the breadth of issues to be addressed. The Critical Raw Materials Act of the European Union is one example (European Commission, 2023).

Only by working on all of these aspects simultaneously can the circular economy for technology metals be enhanced.

FURTHER READING

Treatise on Process Metallurgy. Seetharaman, S. (ed.) (2014): Elsevier. https://doi.org/10.1016/C2010-0-67121-5.

Ullmann's Encyclopedia of Industrial Chemistry, Wiley. https://doi.org/10.1002/14356007.

References

Acatech, 2021. In: Acatech, Circular Economy Initiative Deutschland (CEID), and SYSTEMIQ Ltd. (Eds.), Resource-Efficient Battery Life Cycles – Driving Electric Mobility With the Circular Economy. Circular Economy Initiative Deutschland. https://www.circular-economy-initiative.de/publications.

Adams, M.D. (Ed.), 2016. Gold Ore Processing, second ed. Elsevier, pp. 937–958. https://doi.org/10.1016/C2015-0-00699-2.

Agrawal, M., Singh, R., Ranitović, M., Kamberovic, Z., Ekberg, C., Singh, K.K., 2021. Global market trends of tantalum and recycling methods from Waste Tantalum Capacitors: a review. In: Sustainable Materials and Technologies. vol. 29. Elsevier B.V. https://doi.org/10.1016/j.susmat.2021.e00323.

Akcil, A., Akhmadiyeva, N., Abdulvaliyev, R., Abhilash, Meshram, P., 2018. Overview on extraction and separation of rare earth elements from red mud: focus on scandium. Miner. Process. Extr. Metall. Rev. 39 (3), 145–151.

Albrecht, S., Cymorek, C., Andersson, K., Reichert, K., Wolf, R., 2011. Tantalum and tantalum compounds. In: Ullmann's Encyclopedia of Industrial Chemistry. vol. 35. Wiley-VCH Verlag GmbH & Co. KGaA, pp. 597–610, https://doi.org/10.1002/14356007.a26_071.pub2.

An, L., 2019. Recycling of Spent Lithium-Ion Batteries Processing Methods and Environmental Impacts. Springer Nature Switzerland AG, 217 p. ISBN 978-3-030-31833-8. https://doi.org/10.1007/978-3-030-31834-5.

Apple, 2022. Apple Expands the Use of Recycled Materials Across Its Products. Press release: 19 April 2022. https://www.apple.com/newsroom/2022/04/apple-expands-the-use-of-recycled-materials-across-its-products/.

Bangs, C.E., Meskers, C.E.M., Van Kerckhoven, T., 2016. Trends in electronic products – the canary in the urban mine? In: Electronics Goes Green 2016+, 7–9 September 2016, Berlin, Germany. Fraunhofer Institut, Germany. ISBN 978-3-00-053763-9.

Botelho Jr., A.B., Espinosa, D.C.R., Vaughan, J., Tenorio, J.A. S., 2021. Recovery of scandium from various sources: a critical review of the state of the art and future prospects. Miner. Eng. 172, 107148. https://doi.org/10.1016/j.mineng.2021.107148.

Brumby, A., et al., 2008. Silver, silver compounds, and silver alloys. In: Ullmann's Encyclopedia of Industrial Chemistry. Wiley-VCH Verlag GmbH & Co. KgaA, Weinheim, Germany. https://doi.org/10.1002/14356007.a24_107.pub2. http://doi.wiley.com/10.1002/14356007.a24_107.pub2.

Chancerel, P., Meskers, C.E.M., Hagelüken, C., Rotter, V.S., 2009. Assessment of precious metal flows during preprocessing of waste electrical and electronic equipment. J. Ind. Ecol. 13 (5), 791–810. https://doi.org/10.1111/j.1530-9290.2009.00171.x.

Cowley, A., 2022. PGM market report – May 2022. Johnson Matthey. https://platinum.matthey.com/documents/40646/41236/PGM-market-report-May-2022.pdf.

Crundwell, F., Moats, M.S., Ramachandran, V., Robinson, T., Davenport, W.G., 2011. Extractive Metallurgy of Nickel, Cobalt and Platinum Group Metals. Elsevier. https://doi.org/10.1016/C2009-0-63541-8.

Damm, S., 2018. In: Deutsche Rohstoffagentur (DERA) (Ed.), Rohstoffrisikobewertung – Tantal. DERA, Berlin. ISBN: 978-3-943566-47-5. https://www.bgr.bund.de/DE/Gemeinsames/Produkte/Downloads/DERA_Rohstoffinformationen/rohstoffinformationen-31.pdf.

Dupont, D., Arnout, S., Jones, P.T., Binnemans, K., 2016. Antimony recovery from end-of-life products and industrial process residues: a critical review. J. Sustain. Metall. 2 (1), 79–103. https://doi.org/10.1007/s40831-016-0043-y.

Eksteen, J.J., Oraby, E.A., Tanda, B.C., Tauetsile, P.J., Bezuidenhout, G.A., Newton, T., Trask, F., Bryan, I., 2018. Towards industrial implementation of glycine-based leach and adsorption technologies for gold-copper ores, Canadian Metallurgical Quarterly. Can. J. Metall. Mater. Sci. 57 (4), 390–398. https://doi.org/10.1080/00084433.2017.1391736.

European Commission, 2020. Study on the EU's List of Critical Raw Materials (2020), Factsheets on Critical Raw Materials. https://doi.org/10.2873/92480.

European Commission, 2023. European Critical Raw Materials Act – Proposal for a Regulation (16 March 2023). https://single-market-economy.ec.europa.eu/publications/european-critical-raw-materials-act_en.

Filella, M., Hennebert, P., Okkenhaug, G., Turner, A., 2020. Occurrence and fate of antimony in plastics. J. Hazard. Mater. 390, 121764. https://doi.org/10.1016/j.jhazmat.2019.121764.

Fritz, B., Schmidt, M., 2022. An ecological analysis of the state-of-the-art refinery of high-value gold scraps. Erzmetall 75 (2), 84–93.

Gunn, G. (Ed.), 2014. Critical Metals Handbook. John Wiley & Sons. pp, 70–98, 284–311, 361–84, 340–60.

Hagelüken, C., Goldmann, D., 2022. Recycling and circular economy – towards a closed loop for metals in emerging clean technologies. Miner. Econ. 35, 539–562. https://doi.org/10.1007/s13563-022-00319-1.

Hagelüken, C., Meskers, C.E., 2010. Complex life cycles of precious and special metals. In: Graedel, T.E., et al. (Eds.), Linkages of Sustainability – Strüngmann Forum Report. MIT Press, pp. 163–197. ISBN-10: 0-262-01358-4. https://doi.org/10.7551/mitpress/9780262013581.001.0001.

Heibeck, M., Rudolph, M., Modler, N., Reuter, M., Filippatos, A., 2021. Characterizing material liberation of multi-material lightweight structures from shredding experiments and finite element simulations. Miner. Eng. 172, 107142. https://doi.org/10.1016/j.mineng.2021.107142.

Heibeck, M., Richter, J., Hornig, A., Mütze, T., Rudolph, M., Reuter, M., Modler, N., Filippatos, A., 2023. Simulating the shredding process of multi-material structures for recyclability assessment. Materials & Design, vol. 232, 112167. https://doi.org/10.1016/j.matdes.2023.112167.

Iglesias-Émbil, M., Abadías, A., Valero, A., Calvo, G., Reuter, M.A., Ortego, A., 2023. Criticality and recyclability assessment of car parts—a thermodynamic simulation-based approach. Sustainability 15 (1), 91. https://doi.org/10.3390/su15010091.

Kandalam, A., Reuter, M.A., Stelter, M., Reinmöller, M., Gräbner, M., Richter, A., Charitos, A., 2023a. A review on top submerged lance (TSL) processing—part I: plant and reactor engineering. Metals (in press).

Kandalam, A., Reuter, M.A., Stelter, M., Reinmöller, M., Gräbner, M., Richter, A., Charitos, A., 2023b. A review on top submerged lance (TSL) processing—part II: thermodynamics, slag chemistry and plant flowsheets. Metals (in press).

Louw, S., Reuter, M.A., 2022. Integrated dynamic flowsheet simulation of Platinum Group Metals recovery from spent auto catalyst. In: Proceedings Conference: 8th International Platinum Group Metal Conference 2022 At: Sun City North West, Rustenburg, South Africa.

Lloyd, L., 2011. Handbook of Industrial Catalysts. Springer US. https://doi.org/10.1007/978-0-387-49962-8.

Lux, T., Reuter, M., Degel, R., Kaussen, F., Borowski, N., 2022. BlueMetals technology—experience from commissioning E-scrap recycling plants. In: Lazou, A., Daehn, K., Fleuriault, C., Gökelma, M., Olivetti, E., Meskers, C. (Eds.), REWAS 2022: Developing Tomorrow's Technical Cycles. The Minerals, Metals & Materials Series, vol. I. Springer, Cham. https://doi.org/10.1007/978-3-030-92563-5_6.

Mancheri, N.A., Sprecher, B., Deetman, S., Young, S.B., Bleischwitz, R., Dong, L., Kleijn, R., Tukker, A., 2018. Resilience in the tantalum supply chain. Resour. Conserv. Recycl. 129, 56–69. https://doi.org/10.1016/j.resconrec.2017.10.018.

Meskers, C., 2021. Metal processes and applications – an overview. In: Engh, T.A., Sigworth, G.K., Kvithyld, A. (Eds.), Principles of Metal Refining, second ed. Oxford University Press. https://doi.org/10.1093/oso/9780198811923.001.0001. ISBN: 9780198811923.

Meskers, C.E.M., Reuter, M.A., Boin, U., Kvithyld, A., 2008. A fundamental metric for metal recycling applied to coated magnesium. Metall. Mater. Trans. B 39B (June), 500–517. https://doi.org/10.1007/s11663-008-9144-8.

Mineta, K., Okabe, T.H., 2005. Development of a recycling process for tantalum from capacitor scraps. J. Phys. Chem. Solids 66 (2–4), 318–321. https://doi.org/10.1016/j.jpcs.2004.06.060.

Moghimian, P., Poirié, T., Habibnejad-Korayem, M., Zavala, J.A., Kroeger, J., Marion, F., Larouche, F., 2021. Metal powders in additive manufacturing: a review on reusability and recyclability of common titanium, nickel and aluminum alloys. Addit. Manuf. 43, 102017. https://doi.org/10.1016/j.addma.2021.102017.

Nassar, N.T., 2017. Shifts and trends in the global anthropogenic stocks and flows of tantalum. Resour. Conserv. Recycl. 125, 233–250. https://doi.org/10.1016/j.resconrec.2017.06.002.

Neumann, J., Petranikova, M., Meeus, M., Gamarra, J.D., Younesi, R., Winter, M., Nowak, S., 2022. Review: recycling of lithium-ion batteries—current state of the art, circular economy, and next generation recycling. Adv. Energy Mater. 2022, 2102917. https://doi.org/10.1002/aenm.202102917.

Nose, K., Okabe, T.H., 2014. Platinum group metals production. In: Seetharaman, S. (Ed.), Treatise on Process Metallurgy. vol. 3. Elsevier Ltd, pp. 1071–1097, https://doi.org/10.1016/B978-0-08-096988-6.00018-3.

OECD (Ed.), 2013. Supply Chains of Minerals From Conflict-Affected and High-Risk Areas, second ed. OECD Publishing, https://doi.org/10.1787/9789264185050-en.

Prasad, N.E., Gokhale, A.A., Wanhill, R.J.H. (Eds.), 2014. Aluminum-Lithium Alloys. Elsevier. https://doi.org/10.1016/C2012-0-00394-8.

Ramon, H., Peeters, J.R., Sterkens, W., Duflou, J.R., Kellens, K., Dewulf, W., 2020. Techno-economic potential of recycling tantalum containing capacitors by automated selective dismantling. Procedia CIRP 90, 421–425. https://doi.org/10.1016/j.procir.2020.01.110.

Rasmussen, K.D., Wenzel, H., Bangs, C., Petavratzi, E., Liu, G., 2019. Platinum demand and potential bottlenecks in the global green transition: a dynamic material flow analysis. Environ. Sci. Technol. 53 (19), 11541–11551. https://doi.org/10.1021/acs.est.9b01912.

Reitz, J., Lochbichler, C., Friedrich, B., 2011. Recycling of gamma titanium aluminide scrap from investment casting operations. Intermetallics 19 (6), 762–768. https://doi.org/10.1016/j.intermet.2010.11.015.

Reuter, M.A., van Schaik, A., Ballester, M., 2018. Limits of the circular economy: fairphone modular design pushing the limits. World Metall. – ERZMETALL 71 (2), 68–79.

Reuter, M.A., van Schaik, A., Gutzmer, J., Bartie, N., Abadías Llamas, A., 2019. Challenges of the circular economy – a material, metallurgical and product design perspective. Annu. Rev. Mater. Res. 49, 253–274. https://doi.org/10.1146/annurev-matsci-070218-010057.

Schlesinger, M.E., Sole, K.C., Davenport, W.G., Alvear, G., 2021. Extractive Metallurgy of Copper, sixth ed. Elsevier, Amsterdam. ISBN: 9780128218754.

Schlummer, M., Popp, L., Trautmann, F., Zimmermann, D., Maurer, A., 2017. Recovery of bromine and antimony from WEEE plastics. In: 2016 Electronics Goes Green 2016+, EGG 2016. https://doi.org/10.1109/EGG.2016.7829830.

SEC, 2011. Dodd-Frank Wall Street Reform and Consumer Protection Act [Public Law 111–203] SEC, 2011. Section 1502 conflict minerals, H. R. 4173. One Hundred Eleventh Congress of the United States of America. https://www.sec.gov/about/laws/wallstreetreformcpa.pdf.

Shen, L., Tesfaye, F., Li, X., Lindberg, D., Taskinen, P., 2021. Review of rhenium extraction and recycling technologies from primary and secondary resources. Miner. Eng. 161, 106719. https://doi.org/10.1016/j.mineng.2020.106719.

Shikika, A., Sethurajan, M., Muvundja, F., Mugumaoderha, M.C., Gaydardzhiev, S., 2020. A review on extractive metallurgy of tantalum and niobium. Hydrometallurgy. 198, 105496. https://doi.org/10.1016/j.hydromet.2020.105496.

Shuva, M.A.H., Rhamdhani, M.A., Brooks, G., Masood, S., Reuter, M.A., 2016. Thermodynamics data of valuable elements relevant to e-waste processing through primary and secondary copper production – a review. J. Clean. Product. 131, 795–809. https://doi.org/10.1016/j.jclepro.2016.04.061.

Shuva, M.A.H., Rhamdhani, M.A., Brooks, G., Masood, S., Reuter, M.A., 2017. Thermodynamics behaviour of palladium and tantalum during equilibrium reactions between FeO_x-CaO-SiO_2-MgO slag and molten copper. Metall. Trans. B 48 (1), 317–327. https://doi.org/10.1007/s11663-016-0839-y.

Sibum, H., Güther, V., Roidl, O., et al., 2017. Titanium, titanium alloys, and titanium compounds. In: Ullmann's Encyclopedia of Industrial Chemistry. Wiley-VCH Verlag GmbH & Co. KGaA, Weinheim, Germany, pp. 1–35. https://doi.org/10.1002/14356007.a27_095.pub2.

Takeda, O., Okabe, T.H., 2019. Current status of titanium recycling and related technologies. JOM 71 (6), 1981–1990. Minerals, Metals and Materials Society. https://doi.org/10.1007/s11837-018-3278-1.

Takeda, O., Okabe, T.H., 2020. Recycling of Ti. In: Fang, Z.Z., Froes, F.H., Zhang, Y. (Eds.), Extractive Metallurgy of Titanium. Elsevier, pp. 363–387. https://doi.org/10.1016/B978-0-12-817200-1.00016-8.

Takeda, O., Ouchi, T., Okabe, T.H., 2020. Recent progress in titanium extraction and recycling. Metall. Mater. Trans. B 51 (4), 1315–1328. https://doi.org/10.1007/s11663-020-01898-6.

Taube, M.C., Adamczyk, B., Adam, C., Feldmann, I., Ostermann, M., Reuter, M.A., Stelter, M., 2020. Tantal recycling aus pyrometallurgischen Rückständen. World Metall. – ERTZMETALL 73 (4), 196–205.

The Silver Institute and Metals Focus, 2021. World Silver Survey 2021. ISBN 978-1-9162526, ISSN 2372-2312. https://www.silverinstitute.org/all-world-silver-surveys/.

Tshephe, T.S., Akinwamide, S.O., Olevsky, E., Olubambi, P. A., 2022. Additive manufacturing of titanium-based alloys – a review of methods, properties, challenges, and prospects. Heliyon 8 (3), e09041. https://doi.org/10.1016/j.heliyon.2022.e09041.

U.S. Geological Survey, 2022. 2022 Final List of Critical Minerals. USGS, Department of the Interior. https://www.usgs.gov/news/national-news-release/us-geological-survey-releases-2022-list-critical-minerals.

Ueberschaar, M., Dariusch Jalalpoor, D., Korf, N., Rotter, V.S., 2017. Potentials and barriers for tantalum recovery from waste electric and electronic equipment. J. Ind. Ecol. 21 (3), 700–714. https://doi.org/10.1111/jiec.12577.

United Nations, 2011. Guiding Principles on Business and Human Rights. UN Human Rights Office of the High Commissioner, New York and Geneva. http://www.ohchr.org/Documents/Publications/GuidingPrinciplesBusinessHR_EN.pdf.

USGS, 2020. Mineral Commodity Summaries 2020. US Geological Survey, Reston, VA, USA.

Velázquez-Martínez, O., Valio, J., Santasalo-Aarnio, A., Reuter, M.A., Serna-Guerrero, R., 2019. A critical review of lithium-ion battery recycling processes from a circular economy perspective. Batteries 5 (4), 68. https://doi.org/10.3390/batteries5040068.

van Vuuren, D.S., 2019. A critical evaluation of processes to produce primary titanium. J. South. Afr. Inst. Min. Metall. 109, 455–461. https://www.saimm.co.za/Journal/v109n08p455.pdf.

Wang, W., Pranolo, Y., Cheng, C.Y., 2011. Metallurgical processes for scandium recovery from various resources: a review. Hydrometallurgy 108 (1–2), 100–108. https://doi.org/10.1016/j.hydromet.2011.03.001.

World Gold Council, 2022. Gold Demand Trends Full Year 2021. https://www.gold.org/goldhub/research/gold-demand-trends/gold-demand-trends-full-year-2021.

Yakoumis, I., Panou, M., Moschovi, A.M., Panias, D., 2021. Recovery of platinum group metals from spent automotive catalysts: a review. Clean. Eng. Technol. 3 (2021), 100112. https://doi.org/10.1016/j.clet.2021.100112.

Zhang, W., Zhu, Z., Cheng, C.Y., 2011. A literature review of titanium metallurgical processes. Hydrometallurgy 108 (3–4), 177–188. https://doi.org/10.1016/j.hydromet.2011.04.005.

CHAPTER

26

Concrete and aggregates

Vivian W.Y. Tam[1,2], Mahfooz Soomro[3], and Ana Catarina Jorge Evangelista[4]

[1]School of Engineering, Design and Built Environment, Western Sydney University, Penrith, NSW, Australia [2]College of Civil Engineering, Shenzhen University, Shenzhen, China [3]School of Engineering, Design and Built Environment, Western Sydney University, Penrith, NSW, Australia [4]Engineering Institute of Technology, Perth, WA, Australia

26.1 INTRODUCTION

Waste is defined as a by-product material of human and industrial activities that has no residual value (Serpell and Alarcón, 1998). Construction and demolition (C&D) waste constitutes a major portion of total solid waste production in the world; some studies have estimated that up to 90% of demolition waste going to the landfills can be recycled and reused. However, such studies need to be tested in a more comprehensive manner with indicative analysis of the actual practice. During the last 20 years the recycling of C&D waste has emerged as a socioeconomic priority mainly in the developed countries, and in the present decade, developing countries are also gradually joining in.

Construction waste occurs from building construction and building renovations and results from surplus material (excess supplies), damaged or broken material (thus unusable), cut-off pieces, processing waste (sawdust, metal spoils), dismantled shuttering, used-up tools and accessories, packaging, and garbage generated by the people on construction sites.

Demolition waste results from demolition of dilapidated concrete structures, bridges, roads, etc., from either their complete removal or renovation. It also includes demolition debris from natural disasters (earthquakes, hurricanes, and tsunamis), civil conflicts, vandalism, explosions, fires, collapse of decrepit and ramshackle structures, etc.

C&D waste is divided into several main fractions: metal, concrete and mineral, wood, plastic, electronic and electrical, and miscellaneous and unsorted mixed fractions. More precisely, it may contain:

- concrete
- bricks, tiles, and ceramics
- wood
- glass
- plastic
- bituminous mixtures and tars
- metals (ferrous and nonferrous)
- soils (contaminated) and stones
- insulation materials (including asbestos)
- gypsum-based materials (including plasterboard)

- waste electronic and electrical equipment
- chemicals (including solvents)
- packaging materials
- hazardous substances

Some materials on the list, if not managed responsibly, can pollute the environment and pose a public health risk. Hazardous substances are generally present in building material because they are used, together with concrete, for completing the structure and for realizing the finishes.

This chapter addresses the waste flows, recovery ratios, and existing applications for recycled aggregate concrete. The concrete waste recycling technologies are also discussed. The chapter concludes with a look at environmental impact and market aspects.

26.2 WASTE FLOWS

About 32.4 Mt of solid waste is generated in Australia, of which about 42% is from the construction and demolition sectors (Productivity Commission-Australian Government, 2006). From that, about 7.8 Mt of material corresponding to about 57% of C&D waste is recycled. In Australia, the management of C&D waste is not legislated by the central government; instead environmental matters, including all waste streams, are primarily the responsibility of Australian state and territory governments. According to the Department of the Environment (2020), Australia generated 27 Mt of waste (44% of all waste) from the C&D sector in 2018–19. That is a 61% increase since 2006–07. This waste stream is the largest source of managed waste in Australia and 76% of it is recycled. Among different types of C&D waste, concrete waste constitutes a major proportion, about 81.8%, of the total waste. From that, about 54% of the concrete waste is recycled. Moreover, metal's recycling rate is the highest, at about 82%. The industry is highly motivated to recycle metal waste because it is profitable (Productivity Commission-Australian Government, 2006).

Yang et al. (2017) reported that with China's rapid urbanization, an increasing amount of C&D waste is being generated, especially in the major cities. In total, China produces around 30% of the world's municipal solid waste and of this about 40% is C&D waste. The construction of new buildings generates around 100 Mt/year while demolition of old properties generates 200 Mt/year. There are regulatory policies in China for handling waste, but there is lack of explicit national legislation governing C&D waste management.

In Hong Kong, according to the Waste Statistics 2015 (Energy Protection Department, 2016), the overall construction waste was about 1.533 Mt, which was 28% of the total disposal and includes waste arising from construction activities, e.g., site clearance, renovation, demolition, refurbishment, road works, etc. It also includes waste concrete generated from concrete batching plants and cement mortar/plaster plants. The overall construction waste is sorted into inert materials (rubble, debris, concrete, and earth) that are reused in construction sites (public fill) and noninert C&D waste that is disposed of in landfills (Environmental Protection Department (EPD), 2017). However, the materials delivered to the public fills are not being recycled in Hong Kong but are shipped to mainland China for dumping as reclamation fills. The actual recycling rate of C&D waste is very low in Hong Kong.

Some demolition projects may even take a risk in estimating the amount of metal waste collected on-site and lowering the contract sum during tendering to improve the competitiveness.

In the European Union, about 850 Mt/year of C&D waste are generated across Europe (Fischer and Werge, 2009; Staunton et al., 2015). Table 26.1 presents the waste generated in different countries.

TABLE 26.1 Recovery of construction and demolition (C&D) waste.

Countries	C&D waste generation (Mt)	C&D waste recovery (Mt)	C&D waste recovery (%)	References
Oceania				
Australia	19.3	12	62.2	Pickin and Randell (2017)
Asia				
China	300	120	40	Yang et al. (2017)
Hong Kong	24.3	6.8	28	Energy Protection Department (2016)
Japan	77	62	80.5	UNEP (2015)
Taiwan	63	58	91	EPA (2017)
Thailand	10	3.2	32	National Statistical Office (2012)
Europe				
Belgium	40.2	34.57	86	European Union (2017)
Denmark	21.7	20.4	94	European Commission (DG ENV) (2011)
Finland	20.8	5.4	26	European Commission (DG ENV) (2011)
France	342.6	212.4	62	European Commission (DG ENV) (2011)
Germany	192.3	165.4	86	European Commission (DG ENV) (2011)
Ireland	16.6	13.3	80	European Commission (DG ENV) (2011)
Italy	46.3	n.a.	n.a.	European Commission (DG ENV) (2011)
Netherlands	25.8	25.28	98	European Commission (DG ENV) (2011)
Norway	1.3	0.87	67.3	European Commission (DG ENV) (2011)
Portugal	11.4	5.52	48.4	European Commission (DG ENV) (2011)
Spain	38.5	5.39	14	European Commission (DG ENV) (2011)
Sweden	10.2	n.a.	n.a.	European Commission (DG ENV) (2011)
Switzerland	7	2	28	European Commission (DG ENV) (2011)

Continued

TABLE 26.1 Recovery of construction and demolition (C&D) waste—cont'd

Countries	C&D waste generation (Mt)	C&D waste recovery (Mt)	C&D waste recovery (%)	References
United Kingdom	114.2	74.23	65	European Commission (DG ENV) (2011)
Americas				
Brazil	101	6.2	6.14	Abrecon (2015)
Canada	0.66	0.2	30	Giroux Environmental Consulting (2014)
United States	534	256.3	48	US EPA (2016)
Africa				
South Africa	4.7	0.76	16	DEA (2012)

Modified from CSI (2009).

26.3 RECOVERY RATES

There is a high potential for recycling and reuse of C&D waste, since some of its components have a high resource value. In particular, there is a reuse market for aggregates derived from C&D waste as subbase in roads, as embedment for sewer and drainage pipes, and other construction applications. Technologies for the separation and recovery of C&D waste are well established, readily accessible, and in general inexpensive. Despite its potential, the level of recycling and material recovery varies greatly (between less than 10% and over 90%) in countries across the globe.

For most parts, C&D waste is recoverable and is recycled and reused for both economic gain and environmental benefit. Thus in Germany the resource recovery rate was 91% in 2011, and in France the recycled rate was 50% of the total generated C&D waste in 2014. In the United Kingdom, approximately 62% of C&D waste was recycled in 2011, while in the United States the resource recovery rate was 48% in 2011. In Australia, approximately 64% of C&D waste was recycled in 2014, and in Hong Kong about 38% of solid waste is contributed from the construction industry (Tam and Tam, 2008). A summary of the current concrete recycling recovery data around the world is shown in Table 26.1. Many developing countries as well as emerging economies have yet to collect the recovery data of their countries; the less populated countries and the large countries are expected to have lower achievable recovery rates.

However, this percentage of concrete recovery varies from one region of the world to the other, due to vast differences in construction traditions, landfill legislation, and the perception and acceptance level of the constructors/contractors and builders. Nowadays, waste management operations are being practiced effectively in Australia, the European Union, Hong Kong, and the United Kingdom.

Stated that there has been a clear decrease in the amount of generated waste as a result of increasing efforts by the governments and councils through their legislation, and by giving positive incentives to the constructors and builders to move toward a "greener" construction industry.

For example, Article 11.2(b) of the Waste Framework Directive (2008/98/EC) states, "Member states shall take the necessary measures designed to achieve by 2020 a minimum of 70% (by weight) of non-hazardous C&D waste, excluding naturally occurring material defined in category 17 05 04 (Soils and stones, other than those mentioned in 17 05 03) in the list of wastes shall be prepared for reuse, recycled or undergo other material recovery processes. This also includes backfilling operations using waste to substitute other materials" (European Union, 2008).

Réunion Internationale des Laboratoires et Experts des Matériaux, systèmes de construction et ouvrages (RILEM—www.rilem.net) in the late 1970s, and later many universities and research centers in Europe and in many other parts of the world, for example Brazil, started research projects on recycling technology concerning material coming from building demolition. In 1981 RILEM designated the Technical Committee 37-DRC (Demolition and reuse of concrete) which was succeeded by Technical Committee 121-DRG, tasked with drafting demolition and reuse guidelines, which were published in 1993. Since 2001, a RILEM Technical Committee is operating about the "Use of recycled materials in construction" (Corinaldesi and Moriconi, 2004).

26.4 RECYCLED AGGREGATE CONCRETE APPLICATIONS

Until recently, almost all demolished concrete has been dumped. As concrete (consisting of cement paste and aggregates) is an essential, mass-produced material like steel in the construction industry, much effort has been made to recycle and conserve it. Completed and repeated recycling can be theoretically suitable for concrete, as is the case for steel and aluminum (Noguchi and Tamura, 2001). Since concrete is composed only of cementitious materials, and the powders generated during the production of recycled aggregate can be reprocessed as a cement raw material resource, this permits repeated recycling in a fully closed system. Recycling of concrete can be accomplished by reuse of concrete products, processing into secondary raw materials for uses as fill, road bases, and subbases, or aggregate for the production of new concrete (Howard Humphreys and Partners, 1994; Torring, 1999). The most common way to recycle "concrete rubble" is to use it as recycled aggregate. Table 26.2 shows examples of reusable concrete waste.

TABLE 26.2 Reuse of demolished concrete waste.

Type of concrete waste	Application
Demolished concrete member	Man-made reef, paving stone
Broken into 20–40 cm blocks	Protection of levee
Crushed (<50 mm)	Subbase, backfilling, foundation materials
Crushed and worn (<40 mm)	Concrete and asphalt concrete aggregate subbase material, backfilling material
Powder (by-product of crushing process)	Filler for asphalt concrete, soil stabilization materials

From Kawano (1996).

Deformation properties of concrete made with secondary aggregate are less favorable than those of concrete made with natural gravel. There are two potential solutions to this problem:

- substitute 100% gravel with secondary aggregate and increase the dimensions of the structure by about 10%;
- substitute about 20% natural aggregate with mixed recycled aggregate, which does not reduce concrete quality, with strength up to 65 MPa.

26.5 CONCRETE RECYCLING TECHNOLOGIES

Basic equipment and process steps used to process virgin aggregate are similar to those used for processing recycled aggregate, such as crushing, sizing, and stockpiling. A recycling plant usually comprises crushers incorporating sieves, sorting devices, and screens. The main recycling processes are crushing, sorting, and screening to produce aggregate for use in civil engineering work, landscaping, and as a substitute for gravel in concrete products (Hansen, 1986).

Although this concrete recycling method is used in many countries to produce recycled aggregate, the quality of the produced recycled aggregate is low, which limits their applications for low-grade activities such as road work, pavement, and drainage (CSIRO, 2006; Jia et al., 1986; Poon et al., 2003; Tam et al., 2005; Tam, 2005).

Japan, among other countries, is now using advanced technologies to improve the quality of recycled aggregate, so that it can be used for high-grade concrete applications (Kawano, 2002; Tanaka et al., 2002; Environmental Council of Concrete Organizations, 2006).

26.5.1 Heating and Grinding

The heating and grinding method makes the hardened cement paste, which adheres to the waste concrete aggregate, soften by heating it to about 300°C (Shima et al., 2005). After heating, parts of the hardened cement paste can be separated by a grinding process, resulting in a clean original aggregate from the concrete waste.

26.5.2 Screw Grinding

The screw grinding method uses a shaft screw consisting of an intermediate part and an outlet part with a warping cone to remove mortar (consisting of cement and sand) adhered to the aggregate's surface (Matumura, 2005). Figure 26.1 shows the procedures of the screw grinding method. In this mechanism the concrete lumps are loaded into a cylinder with twin cones cleaned by each other. This process is automatically repeated, enabling the required quality of the recycled aggregate.

26.5.3 Mechanical Grinding Method

The mechanical grinding method uses a drum body that finely separates partition boards with same-sized holes. The steel balls can move horizontally and vertically by rolling the drum. The quality of aggregate can then be improved

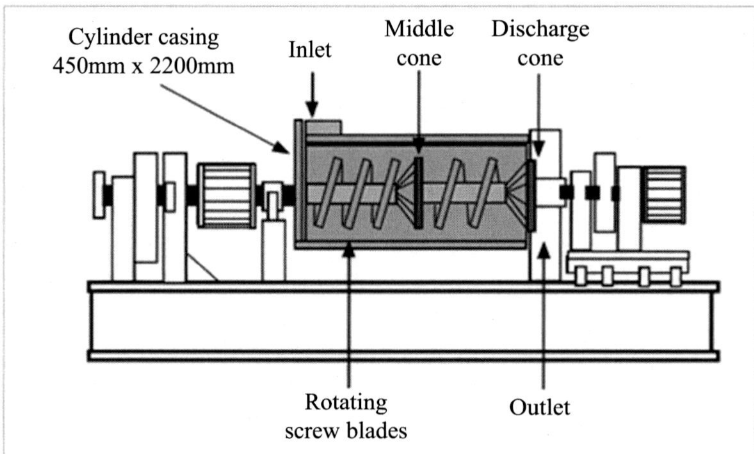

FIGURE 26.1 Screw grinding and abrading mill. *From Tomosawa et al. (2005).*

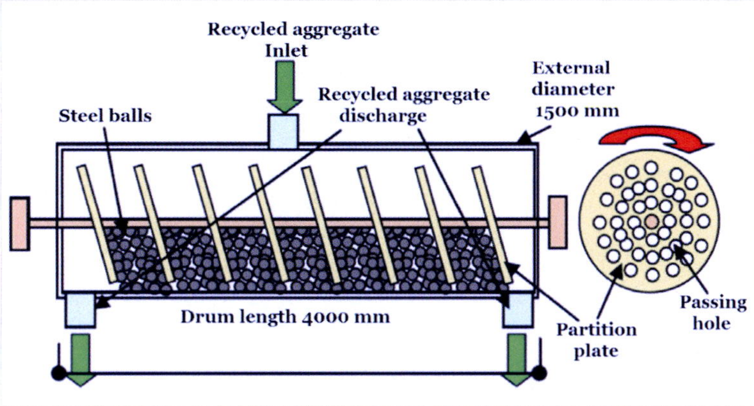

FIGURE 26.2 Mechanical mill grinding. *From Sakai (2010).*

via narrowing the inside space by using the partition boards (Dhir et al., 2011). Figure 26.2 shows the mechanical grinding equipment. The partition board holes separate the aggregates, and the steel balls separate/remove the adhered mortar.

26.5.4 Gravity Concentration

After processing with a jaw crusher, an impact crusher, and an improvement rod mill, aggregate of over 8 mm is divided into recycled coarse aggregate and mortar particles. Aggregate with sizes under 8 mm are divided into two sizes: recycled fine aggregate <5 mm size and coarse aggregate 5–8 mm size. The wet gravity concentration machine is used to move: (i) lightweight materials such as mortar particle and wood waste upward; and (ii) heavy-weight materials such as aggregate grain downward.

26.6 FUTURE DEVELOPMENTS

26.6.1 Application Areas

Worldwide, numerous studies have been conducted to evaluate the possibility of using recycled C&D waste, specifically recycled aggregate concrete (RAC) and recycled crushed bricks (RCB) in the construction of the subbase of roadways. However, more demanding applications have not been investigated as much.

Nassar and Soroushian (2016) investigated the field performance of RAC in pavement construction subjected to heavy traffic loads under aggressive weather conditions. Test results of concrete cores drilled from pavement sections after 270 days showed that RAC demonstrated performance at par or even better than that of corresponding control concrete. Later age-enhanced strength and durability attributes of RAC concrete suggest its suitability for use in concrete-based infrastructure such as pavement construction.

26.6.2 Environmental Benefits

Studies have shown that reprocessing of C&D waste into recycled aggregate brings considerable environmental benefits. These environmental gains are dependent on an efficient recycle aggregate collection and reuse supply chain. The reuse of recycled aggregate brings substantial gains in the following areas:

- Resource consumption. Substitution of quarried virgin aggregate with recycled material

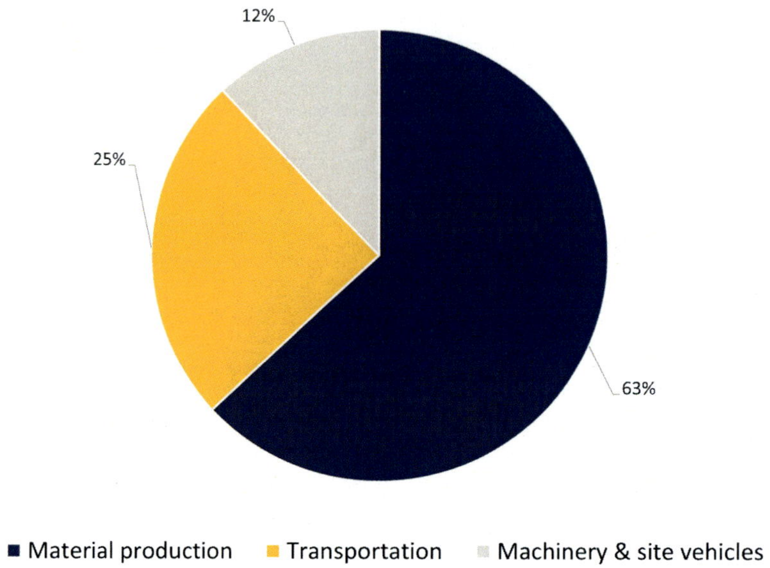

FIGURE 26.3 Typical energy consumption breakdown for a roadway construction project. *Data from Pavement Interactive (2012).*

means conserving primary, virgin aggregate for future generations.

- Quarrying. By reducing the quarry of natural aggregate, the costs for landscape recovery and impacts on biodiversity will decrease.
- Greenhouse gas (GHG) emissions: less energy is used for the production of recycled aggregates compared to virgin materials.

A typical energy consumption breakdown for a roadway construction project using primary materials is depicted in Figure 26.3. The material production accounts for about two-thirds of the energy consumption. Any reduction in the energy footprint will have a large impact.

Recycled aggregate can have lower embodied energy in addition to reduced transport emissions, especially where recycled materials are reused near the site of reprocessing.

The energy consumption and the resulting GHG emissions from recycling of aggregate has been calculated to be around 4.0 kg CO_2 per ton, which is 22–46% fewer emissions than an equivalent conventional quarry product, based on Australian data (ARRB Group, 2010).

Hossain et al. (2016) presented the LCA (life cycle assessment) analyses of recycled aggregates production from C&D waste in Hong Kong, and their results revealed that, compared with natural coarse aggregate, recycled coarse aggregate leads to a reduction of up to 65% greenhouse gas (GHG) emissions with savings of up to 58% nonrenewable energy consumption and the same was observed for producing recycled fine aggregates from C&D waste.

In India, Nayana and Kavitha (2017) evaluated that for the production of 1 ton natural aggregate, 46 kg CO_2 is emitted, compared to 2.4 kg CO_2 for 1 ton recycled aggregates. When compared to natural aggregates, recycled aggregates reduce carbon emissions by 23–28%. In the United Kingdom, the value found was 2.42 kg CO_2-eq./t, considering the distance of source to recycling site, which was fixed at 10 km and it was assumed that the material was transported using 20 ton trucks consuming diesel at 0.4 L/t per km (Korre and Durucan, 2009).

TABLE 26.3 Comparisons between quarried and recycled road base.

	CO_2 emissions (t/km)	Embodied energy (GJ/km)	Operational energy (GJ/km)	Total energy (GJ/km)
Quarried roadbase	72	762	191	953
Recycled roadbase	24	165	186	351

Data from McRobert (2008).

Similarly, US studies have found recycled aggregate to have around 30% less embodied carbon emissions than primary aggregates.

However, it is important to note that in comparing GHG emission figures, differences originate from the methodologies and system boundaries used, as well as local electricity generation factors between jurisdictions, both across Australia and worldwide.

Taking a per kilometer road perspective, McRobert (2008) compared the CO_2 emissions and the embodied energy of recycled aggregate vs quarried virgin aggregate. This was based on the construction of 1 km of roadway made entirely of recycled or quarried road base material, keeping all the other conditions constant (Table 26.3).

The embodied energy of recycled aggregate was evaluated to be approximately 30% of that of the quarried aggregate. Similarly, the CO_2 emissions of recycled aggregate were approximately 60% less than the quarried aggregate, based on the production of 150 tons per month (McRobert, 2008).

Even partial replacement with recycled aggregates makes a considerable difference. Replacement of 50% of the quarried virgin aggregate with recycled aggregate in a road construction project could reduce the embodied energy and resulting GHG emissions of the material component of the road construction by around 23% (ARRB Group, 2010).

The lower CO_2 footprint and embodied energy make clear that recycled aggregates do have very positive advantages when utilized either in roadway construction or in concrete production.

26.6.3 Recycled Aggregate Market

Different factors influence the recycled aggregate market, the primary factors being:

- Taxation in mining activity of virgin aggregate
- Taxation on landfills
- Availability and cost of recycled and virgin aggregate
- Misconception and bias against the performance of recycled materials

26.6.3.1 Certification

The marketable recycled aggregate currently produced can be divided into two categories, noncertified and certified aggregate. The majority of the output of recycling plants is currently noncertified aggregate; however, due to stricter demands from consumers, who would prefer to use specifiable and guaranteed quality, the certification of recycled aggregate is of prime importance (Trevorrow and Lyne, 1998).

All recycled product must meet the quality requirements before its use can be encouraged. However, low-quality recycled aggregate can be considered for low-grade applications. Due to traditions and acceptance barriers in constructors, builders, and the public at large, the general attitude toward recycling in the building and construction industry is largely inhibitive to the utilization of recycled material. Therefore it is of great importance that recycled materials are officially certified and accepted by the building and construction industry. Considerable emphasis must be placed on specifying the areas of utilization and quality standards for recycled materials. These must be in accordance with the local demand in order to improve confidence in the recycled material and solve problems regarding the responsibility of using recycled material (Lauritzen, 1998).

Silva et al. (2015) developed a performance-based classification system, based on the predictable relationship of basic physical properties of recycled aggregate, regardless of their size and composition, which allows accurate prediction of the material's performance and is easily understandable by all professionals in the industry. This classification system shows high reliability and reproducibility of results and can help facilitate certification of the final product.

26.6.3.2 Government Support

Lauritzen (1998) recommended that long-term strategies, e.g., for one decade or two decades, with respect to achieving goals for recycling of C&D waste, should be adopted. These must then be continuously revised in accordance with the political situation and followed up by adequate legislation and regulations at all levels, national, regional, and local.

In order to enhance the long-term application of recycled materials, it should be considered to increase the cost of sending construction and demolition waste materials to landfill and focus on the demand for recycled materials through government procurement. As a consequence, this would promote a virtuous circle that leads to a circular economy (Australian Government, 2020).

26.7 CONCLUSION

This chapter has focused on investigating construction and demolition waste generation, examining the existing applications for recycled aggregate concrete and exploring mechanisms of concrete recycling technologies. Concrete is found to be one of the most common types of construction and demolition waste. The use of concrete waste as recycled aggregate is limited to low-grade applications, which include road subbase and pavement. Four concrete recycling technologies were explored that help in improving recycled aggregate quality and thus in recycled aggregate concrete use in high-grade applications.

China and India at present are the major consumers of construction aggregate and hence have high potential for recycling and reuse of C&D waste; however, despite its potential, there is huge variation in the level of recycling and material recovery in various countries around the world, e.g., Brazil (6.14%), Denmark (94%), Netherland (98%). This variation is due to vast differences in construction traditions, legislation on landfills, and the perception and acceptance level of constructors and builders. Worldwide, there is a common sense that the reducing and recycling of waste is beneficial to the climate crisis by limiting the amount of raw/natural materials being used and reducing the amount of waste disposal in landfills.

Acknowledgment

This chapter is an adaptation of:

VWY Tam, M Soomro, ACJ Evangelista (2018), A review of recycled aggregate in concrete applications (2000–2017), Construction and Building Materials 172, 272–292. Doi: 10.1016/j.conbuildmat.2018.03.240.

Tam, W. Y. Vivian (2009), "Review on concrete waste recycling technologies", *Concrete Materials: Properties, Performance and Applications,* Nova Science Publisher, Inc., United States, ISBN: 978-1-60741-250-2, 567–573.

Tam, W. Y. Vivian and Tam, C. M. (2008), *Re-use of construction and demolition waste in housing development*, Nova Science Publishers, Inc., United States.

Tam, W. Y. Vivian (2008), "Economic comparison of concrete recycling: a case study approach", Resources, Conservation and Recycling, 52(5), 821–828.

Tam, W. Y. Vivian (2008), "Cost effectiveness of concrete recycling", Progress in Waste Management Research, Nova Science Publisher, Inc., United States, ISBN: 978-1-60456-235-4, 1–10. (Invited Article)

References

Abrecon, 2015. Abrecon-Relatorio Pesquisa Setorial (2014–2015). Associacao Brasileira Para Reciclagem de Residuos da Construcao Civil e Demolicoes, Brazil. Retrieved from: https://abrecon.org.br/reciclagem-de-entulho-residuos-da-construcao-e-demolicao-rcd/mercado/.

ARRB Group, 2010. Sustainable Aggregates-CO_2 Emission Factor Study. http://www.sustainableaggregates.com.au/docs/SASA_C02_Emission_Factor_Study.pdf.

Australian Government, 2020. Sustainable Procurement Guide: A Practical Guide for Commonwealth Entities. Retrieved from: http://www.environment.gov.au/protection/.

Corinaldesi, V., Moriconi, V., 2004. Reusing and recycling C&D waste in Europe. In: Sustainable Waste Management and Recycling: Construction Demolition Waste. Thomas Telford Publishing, USA. https://www.icevirtuallibrary.com/doi/full/10.1680/cdw.32859.0006.

CSI, 2009. Recycling Concrete. The Cement Sustainability Initiative at the Worlds Business Council for Sustainable Development, Geneva, Switzerland. https://docs.wbcsd.org/2009/07/CSI-RecyclingConcrete-FullReport.pdf.

CSIRO, 2006. Recycled Aggregate Applications as Subgrade and Pavement. Commonwealth Scientific and Industrial Research Organisation, Australia.

DEA, 2012. National Waste Information Baseline. Department of Environment Affairs, South Africa. https://docslib.org/national-waste-information-baseline-report.

Department of the Environment and Energy, Blue Environment Pty Ltd, 2020. National Waste Report 2020. Blue Environment Pty Ltd, pp. 1–126. (November). Retrieved from: https://www.environment.gov.au/system/files/resources/7381c1de-31d0-429b-912c-91a6dbc83af7/files/national-waste-report-2018.pdf.

Dhir, R., Paine, K., Brito, J., Etxeberria, M., Yong, H., Poon, C. S., Tam, V., 2011. Use of recycled and secondary aggregates in concrete: an overview. In: Dhir, R.K., Newlands, M.D. (Eds.), New Developments in Concrete Construction. Shroff Publishers, pp. 157–182. 978-93-5023-324-5.

Energy Protection Department, 2016. Monitoring of solid waste in Hong Kong, waste statistics for 2015.

Environmental Council of Concrete Organizations, 2006. Recycled Concrete and Masonry. Japan.

Environmental Protection Department (EPD), 2017. Monitoring of Solid Waste in Hong Kong-Waste Statistics for 2016. Environmental Protection Department.

EPA, 2017. Zero Waste and Resource Recycling Promotion (Taiwan). pp. 1–2.

European Commission (DG ENV), 2011. Service contract on management of construction and demolition waste – SR1. Final Report Task 2. Bio Intelligence Service, 33, pp. 1–240. https://op.europa.eu/s/vV3W.

European Union, 2008. Directive 2008/98/EC of the European Parliament and of the Council of 19 November 2008 on waste and repealing certain directives. Official Journal of the European Union. pp. 3–30. https://doi.org/2008/98/EC. 32008L0098.

European Union, 2017. Waste Statistics. Eurostat.

Fischer, C., Werge, M., 2009. EU as a Recycling Society. Present Recycling Levels of Municipal Waste and Construction & Demolition Waste in the EU. European Topic Centre on Resource and Waste Management. Retrieved from: http://scp.eionet.europa.eu/publications/wp2009_2/wp/WP2009_2.

Giroux Environmental Consulting, 2014. State of Waste Management in Canada. Canadian Council of Ministers of Environment, p. 155.

Hansen, T.C., 1986. Recycled aggregates and recycled aggregate concrete second state-of-the-art report developments 1945–1985. Mater. Struct. 19 (3), 201–246. https://doi.org/10.1007/bf02472036.

Hossain, M.U., Poon, C.S., Lo, I.M.C., Cheng, J.C.P., 2016. Comparative environmental evaluation of aggregate production from recycled waste materials and virgin sources by LCA. Resour. Conserv. Recycl. 109, 67–77. https://doi.org/10.1016/J.RESCONREC.2016.02.009.

Howard Humphreys & Partners, 1994. Managing demolition and construction wastes. In: Report of the Study on the Recycling of Demolition and Construction Wastes in the UK, for the Department of the Environment. HMSO, London.

Jia, W., Baoyuan, L., Songshan, X., Zhongwei, W., 1986. Improvement of paste-aggregate interface by adding silica fume. In: Proceedings of the 8th International Congress on the Chemistry of Cement, Rio de Janeiro, Brazil, pp. 460–465.

Kawano, H., 1996. Integrated Design and Environmental Issues in Concrete Technology. Taylor and Francis. ISBN: 9780419221807.

Kawano, H., 2002. The state of using by-products in concrete in Japan and outline of JIS/TR on recycled concrete using recycle aggregate. In: Proceedings of the Ls, Fib Congress, Session, 10.

Korre, A., Durucan, S., 2009. Life cycle assessment of aggregates – EVA025 final report. Waste and Resources Action Program (WRAP), Banbury, Oxon, UK (August), pp. 1–41.

Lauritzen, E.K., 1998. The global challenge of recycled concrete. In: Sustainable Construction: Use of Recycled Concrete Aggregate – Proceedings of the International Symposium Held at Department of Trade and Industry Conference Centre, London, UK, 11–12 November 1998.

Matumura, U., 2005. Concrete recycling technology. In: Proceeding of Annual Conference of Consultants, Hokkaido, Japan, pp. 5–10.

McRobert, J., 2008. Recycled Aggregates – Environmental Considerations. Retrieved from: https://www.wastenet.

net.au/assets/documents/content/information/perth_presentation_jencie.pdf.

Nassar, R., Soroushian, P., 2016. Use of recycled aggregate concrete in pavement construction. J. Solid Waste Technol. Manage. 42 (2), 137–144. https://doi.org/10.5276/JSWTM.2016.137.

National Statistical Office, 2012. Solid waste and Hazardous Waste in Thailand.

Nayana, A.Y., Kavitha, S., 2017. Evaluation of CO_2 emissions for green concrete with high volume slag, recycled aggregate, recycled water to build eco environment. Int. J. Civil Eng. Technol. 8 (5), 703–708. Retrieved from: http://www.iaeme.com/IJCIET/issues.asp?JType=IJCIET&VType=8&IType=5.

Noguchi, T., Tamura, M., 2001. Concrete design towards complete recycling. Struct. Concr. 2 (3), 155–167. https://doi.org/10.1680/stco.2001.2.3.155.

Pavement Interactive, 2012. Energy and Road Construction—What's the Mileage of Roadway? Pavement Interactive. Retrieved 7 January 2022, from: https://pavementinteractive.org/energy-and-road-construction-whats-the-mileage-of-roadway/.

Pickin, J., Randell, P., 2017. Australian National Waste Report 2016. A Report Prepared for the Department of Energy and Environment. Retrieved from: https://www.awe.gov.au/sites/default/files/documents/national-waste-report-2016.pdf.

Poon, C.S., Azhar, S., Kou, S.C., 2003. Recycled aggregates for concrete applications. In: Proceeding of the Materials Science and Technology in Engineering Conference—Now, New and Next, 16. Hong Kong, China.

Productivity Commission-Australian Government, 2006. Waste management. Environ. Eng. Manage. J. 38. Retrieved from: http://www.eemj.icpm.tuiasi.ro/pdfs/vol4/no2/5_Vasile Oros.pdf.

Sakai, K., 2010. Recycling concrete—the present state and future perspective. In: TCG-JSCE Joint Seminar, November, 20, Athens, Greece, pp. 1–56.

Serpell, A., Alarcón, L.F., 1998. Construction process improvement methodology for construction projects. Int. J. Proj. Manage. 16 (4), 215–221. https://doi.org/10.1016/S0263-7863(97)00052-5.

Shima, H., Tateyashiki, H., Matsuhashi, R., Yoshida, Y., 2005. An advanced concrete recycling technology and its applicability assessment through input-output analysis. J. Adv. Concr. Technol. 3 (1), 53–67. https://doi.org/10.3151/jact.3.53.

Silva, R.V., de Brito, J., Dhir, R.K., 2015. The influence of the use of recycled aggregates on the compressive strength of concrete: a review. Eur. J. Environ. Civ. Eng. 19 (7), 825–849. https://doi.org/10.1080/19648189.2014.974831.

Staunton, J., Williams, C.D., Morrison, L., Henry, T., Fleming, G.T.A., Gormally, M.J., 2015. Spatio-temporal distribution of construction and demolition (C&D) waste disposal on wetlands: a case study. Land Use Policy 49, 43–52. https://doi.org/10.1016/j.landusepol.2015.06.023.

Tam, V.W.Y., 2005. Recycled Aggregate From Concrete Waste for Higher Grades of Concrete Construction. City University of Hong Kong.

Tam, V.M., Tam, C.M., 2008. Re-Use of Construction and Demolition Waste in Housing Developments. Nova Publishers, p. 247. ISBN: 9781604563627.

Tam, V.W.Y., Gao, X.F., Tam, C.M., 2005. Microstructural analysis of recycled aggregate concrete produced from two-stage mixing approach. Cem. Concr. Res. 35 (6), 1195–1203. https://doi.org/10.1016/j.cemconres.2004.10.025.

Tanaka, R., Miura, S., Ohaga, Y., 2002. Experimental study on the possibility of using permanently recycled concrete for reinforced concrete structures. J. Soc. Mater. Sci. Jpn 51, 948–954.

Tomosawa, F., Noguchi, T., Tamura, M., 2005. The way concrete recycling should be. J. Adv. Concr. Technol. 3 (1), 3–16. https://doi.org/10.3151/jact.3.3.

Torring, M., 1999. Management of concrete demolition waste. In: Concrete Technology for a Sustainable Development in the 21st Century. Taylor and Francis Group, pp. 331–341, https://doi.org/10.1201/9781482272215-33.

Trevorrow, A., Lyne, T.J., 1998. Recycled aggregate markets: where to start. In: Telford, T. (Ed.), Sustainable Construction: Use of Recycled Concrete Aggregate. Retrieved from: https://www.icevirtuallibrary.com/doi/abs/10.1680/scuorca.27268.

UNEP, 2015. Global Waste Management Outlook. International Environmental Technology Centre, Osaka, Japan, https://doi.org/10.1177/0734242X15616055.

US EPA, 2016. Advancing Sustainable Materials Management: 2014 Fact Sheet. United States Environmental Protection Agency, Office of Land and Emergency Management, Washington, DC.

Yang, H., Xia, J., Thompson, J.R., Flower, R.J., 2017. Urban construction and demolition waste and landfill failure in Shenzhen, China. Waste Manage. 63, 393–396. https://doi.org/10.1016/j.wasman.2017.01.026.

CHAPTER

27

Cementitious binders incorporating residues

Y. Pontikes[1] *and R. Snellings*[1,2]

[1]KU Leuven, Department of Materials Engineering, Leuven, Belgium [2]KU Leuven, Department of Earth and Environmental Sciences, Leuven, Belgium

27.1 INTRODUCTION

Cements or cementitious binders are materials that can set and harden to a rock-like solid material from an initial paste and are typically used to bind other materials together. In concrete, cement acts as the binder phase between the sand and gravel aggregates. Ordinary Portland cement (OPC) is the most used binder today. The success of OPC-based construction materials (mortars, concrete) is based on their comparatively low cost, ease of use, and flexibility in terms of transport and shaping of structural elements. In addition to that, these materials are also quite resilient to errors throughout the process, and demonstrate relatively good durability, depending on the particular application.

Portland cement is a commodity product primarily composed of only five common elements (Ca, Si, Al, Fe, and O) and it comes as no surprise that cement production is geographically widespread. China leads the production of OPC, delivering approximately 54% of the total cement, whereas India, the second-ranked producer at 8%, is expected to become a major player in the years to come (IEA and CSI of WBCSD, 2018; U.S. Geological Survey, 2021). Over the past decades, cement production has been increasing steadily and reached a total of 4.1 billion tons in 2020 (U.S. Geological Survey, 2021). Considering a cement content in concrete of 10–15 wt%, this results in an estimate of 27–41 Gt of concrete production, or about 1.8 m^3 of concrete produced per person annually. The massive scale of production positions concrete as the second most-used material by humankind after water.

OPC is composed of finely ground clinker with a small amount of calcium sulfate, such as gypsum. Clinker is produced at high temperatures, typically around 1450°C in a rotary kiln, from blends of limestone with small quantities of other materials such as clays. In terms of phases, clinker is predominantly composed of (in wt%) alite (Ca_3SiO_5; 50–70), and smaller amounts of belite (β-Ca_2SiO_4; 5–25), tricalcium aluminate ($Ca_3Al_2O_6$; 5–10), and tetracalcium

Handbook of Recycling. **https://doi.org/10.1016/B978-0-323-85514-3.00029-4**

aluminoferrite ($Ca_4(Al,Fe)_4O_{10}$; 5–15); a typical OPC chemical composition is (in wt%): CaO: 65 ± 3, SiO_2: 21 ± 2, Al_2O_3: 5 ± 1.5, Fe_2O_3: 3 ± 1, MgO <5. Blending additional constituents, such as coal fly ash, limestone, natural pozzolans (e.g., volcanic ashes and tuffs), and ground granulated blast furnace slag, with the clinker creates blended cements with properties depending on the materials added. More information can be retrieved in the national standards, e.g., ASTM C150, ASTM C595, and ASTM C1157 for the United States, and EN 197-1 for the EU.

The need for high-temperature processing and the use of limestone as a main raw material contribute to the environmental footprint of OPC production. Although values vary depending on kiln technology, one ton of cement clinker production currently requires about 3800 MJ and emits approximately 0.87–0.92 tons of CO_2 (WBCSD, 2009; Habert et al., 2010). About 60% of the CO_2 emissions are derived from the limestone raw material and are hard to abate by switching to renewable energy sources. Considering the massive volumes produced, the total footprint of the cement sector is rather substantial. Data suggest that 8% of the total anthropogenic CO_2 emissions derive from the cement sector (Andrew, 2018). To mitigate its emissions, the cement industry roadmap relies on product (clinker) efficiency measures and coprocessing of residues as fuel, raw material, or cement constituent. Nevertheless, end-of-pipe carbon capture storage and/or use will still be needed to reach climate neutrality by 2050 (Cembureau and The European Cement Association, 2021). In addition to this, a range of novel binders has been proposed, offering potential alternative paths to OPC. In all these paths, the use of residues and by-products is integrated and can contribute to a less resource-intensive, "carbon neutral" circular economy. It is the scope of this chapter to describe all of these aspects, emphasizing the use of residues and by-products.

27.2 CLINKER PRODUCTION: PROCESS, AND ALTERNATIVE FUELS AND RAW MATERIALS

The use of alternative fuels and raw materials (AFRs) in cement production has occurred from the early 1970s, since it suppresses the production cost of cement while promoting a more efficient waste management. Indeed, seen in a wider framework of energy and material flows, the cement sector contributes by using a variety of household and industrial residues, preventing them from being land-filled. The latter is a dimension often neglected when the cement sector is evaluated only based on the CO_2 footprint. Still, as the notions of "coprocessing" and "industrial ecology" become more acknowledged and endorsed, cement plants are more widely regarded and operate as "coprocessing industries of alternative fuels and raw materials for sustainable cement production" (The GTZ-Holcim Public Private Partnership, 2006). For the guiding principles concerning coprocessing of AFR, a comprehensive list of principles has been published (The GTZ-Holcim Public Private Partnership, 2006). For instance, it is expected that coprocessing will take place by respecting the waste hierarchy, without affecting the environment, humans, or cement quality, by companies that are qualified and in-line with country-specific conditions. This is already an integrated part of the alternative fuel preparation in many cases (see Chatterjee and Sui, 2019 and references therein). In the future it is expected that the substitution of fuel and raw materials will increase even further, with biofuels becoming more present in the mix.

27.2.1 Alternative Fuels

The use of alternative fuels is common practice in many countries, with the thermal substitution levels of alternative fuels for conventional fossil fuels able to exceed 80% (WBCSD, 2005).

TABLE 27.1 Alternative fuels typically used in the cement industry.

Category	Fuels
Gaseous fuels	Refinery waste gas, landfill gas, pyrolysis gas, natural gas
Liquid fuels	Tar, chemical wastes, coal slurries/distillation residues, waste solvents, used oils, used oil + oiled water, wax suspensions, petrochemical waste, asphalt slurry, paint waste, oil sludge, paper fiber sludge
Solid fuels	Petroleum coke (petcoke), fine/anodes/chemical cokes, oil-bearing soils, shale/oil shales, paper/cardboard waste, rubber residues, pulp sludge, sewage sludge, used tires, battery cases, plastics residues, wood waste, impregnated sawdust, domestic refuse, meat and bone animal meal, animal fat, rice husks, refuse-derived fuel, agricultural waste, automobile shredder residue

Modified from Chinyama, M.P.M., 2011. Alternative Fuels in Cement Manufacturing. Alternative Fuel. InTech.

In general, the trend of using alternative fuels is increasing and nowadays more than 64% of the cement plants in Europe use residues as alternative fuels (Rootzén, 2012). The list of candidate alternative fuels is long, but a taxonomy can be based on the origin of the source, i.e., from biogenic or nonbiogenic processes (fossil origin materials). Table 27.1 summarizes the alternative fuel options; note that substantial differences are documented in the profile of the alternative fuels used by country (Aranda Usón et al., 2012; Chatterjee and Sui, 2019). It is obvious that other alternative fuels can be used in each case; however, it is also obvious that limitations do exist. The choice of alternative fuels has to meet different levels of considerations, ranging from business-oriented aspects to more technical ones. Certain preconditions exist, for instance, the cost of the alternative fuels is expected to be lower than that of traditional fuels, with a calorific value higher than 14 MJ/kg, low water content, and physical properties that permit efficient handling and processing. In terms of chemistry, there are strict quality controls and attention is placed on Cl (<0.2 wt%), S (<2.5 wt%), PCB (<50 ppm), and heavy metals (<2500 ppm, out of which Hg <10 ppm and Cd + Tl <90 ppm); threshold values are indicative, after Mokrzycki et al. (2003). The volatile metals, like alkalis and Hg, as well as P, Zn, and F, are also closely monitored and controlled. In general, the chemistry of the fuel should be well controlled, because it may have an impact on kiln operation, clinker and cement quality, and emissions (Chinyama, 2011; Aranda Usón et al., 2012). Moreover, there are clear limitations when it comes to hazardous wastes, e.g., radioactive or infectious medical waste cannot (should not) be processed in a cement kiln. Still, on a global level the nontechnical barriers are probably the main limiting factor in increasing the content of alternative fuels in the mix. More on this can be found elsewhere (de Beer et al., 2017; Chatterjee and Sui, 2019), with the different EU member states providing very interesting case studies.

27.2.2 Alternative Raw Materials

When it comes to alternative raw materials, again many possibilities exist, with the chemistry of OPC clinker defining the boundaries of what and at what levels they can be introduced. Since OPC clinker is primarily composed of CaO, Ca-rich residues can be introduced up to higher levels. Despite this constraint, the high-temperature clinkering process offers an important degree of freedom since it permits the use of materials with little consideration of their actual phase composition. Indeed, at the processing temperatures the system approaches thermodynamic equilibrium and the raw materials disintegrate readily to form the stable clinker phase assemblage (Noirfontaine et al., 2012).

Therefore a grouping of alternative raw materials can be made based on their major constituent; see Table 27.2. Additional residues, such as construction and demolition debris, refinery-spent catalysts, and spent aluminum pot-liners should also be considered; they are not listed in Table 27.2, as they may also have a heating value. Besides major oxides, minor oxides and heavy metals in particular (e.g., Cr) are controlled, although this is not standard practice globally (Huang et al., 2012). Depending on the type of material, different entry points in the process are selected. As with the alternative fuels, the list is not exhaustive and in principle other materials can be used, provided they meet business, logistic, handling, and chemical quality criteria. Nonetheless, unlike the alternative fuels, the percent substitution does not typically exceed 5 wt%.

The previously mentioned small substitution rate has been challenged by Eco-Cement, a family of cements developed in Japan in the 1990s that has gone to full industrial implementation since 2001, complying with national standards (Shimoda and Yokoyama, 1999; Hanehara, 2005). These cements use approximately 50 wt % incinerator bottom and fly ash from the combustion of municipal urban solid waste (MUSW) and sewage sludge as raw materials. The final cements are comparable to Portland cement but small differences in the phase assemblage and properties do exist. The key element in this process is to address the chlorides and heavy metals typically found in incinerator ashes and to that extent, the high-temperature processing during clinkering is crucial. In an alternative path for ash utilization, washing systems are installed in existing cement facilities for

TABLE 27.2 Alternative raw materials that are/can be used in OPC clinker production.

Major substitution for	Examples
Al_2O_3	Coating residues, aluminum recycling sludge, bauxite residue "red-mud," residues after secondary aluminum smelting
$CaCO_3$	Industrial lime from neutralization processes, water purification sludge, marble sludge, other sludges from cutting and polishing processes
SiO_2	Foundry sand, contaminated soil from soil remediation, sludge after sewage treatment (also high Ca possible), ceramic wastes
Fe_2O_3	Roasted pyrite from metal surface treatment, mechanical sludge from metal industry, red sludge from industrial wastewater treatment, bauxite residue "red mud," nonferrous metallurgy slags, mill scales from iron and steel production
Ca-Si-Fe-(Mg)-oxides	Steel metallurgy slags
Ca-Si-Al-oxides	Blast furnace slag
Si-Al-Ca-Fe-oxides	Fly ashes from coal combustion, bottom and fly ashes from incineration plants after treatment

Adapted from The GTZ-Holcim Public Private Partnership, 2006. Guidelines on Co-processing Waste Materials in Cement Production. Retrieved 5 of January 2022 from: https://star-www.giz.de/dokumente/bib/06-0476.pdf, with data from Verein Deutscher Zementwerke (Association of German Cement Manufacturers), U.S. EPA, 2008. Cement Sector Trends in Beneficial Use of Alternative Fuels and Raw Materials. Retrieved 5 of January 2022 from: https://archive.epa.gov/sectors/web/pdf/cement-sector-report.pdf., and private communications.

an efficient ash pretreatment. Additional information on Eco-cement, also with historical data, can be found here (Taiheiyo Cement Corporation, 2021), whereas more information on the Japanese waste management approach can be retrieved here (Ministry of the Environment (Japan), 2014).

27.3 FROM CLINKER TO CEMENT: RESIDUES IN BLENDED CEMENTS

One of the most effective demonstrated means to reduce CO_2 emissions and boost resource efficiency is the partial substitution of Portland clinker by supplementary cementitious materials (SCMs) in so-called blended cements (Juenger et al., 2019). SCMs are defined as materials that, in themselves, possess little of cementitious value but which will, in finely divided form and in the presence of moisture, react chemically with calcium hydroxide at ambient temperature to form compounds possessing cementitious properties (Snellings et al., 2012). Materials conventionally used as SCMs include industrial by-products such as blast furnace slags generated by pig iron production from iron ore, fly ashes obtained in coal-fired energy production, and silica fume collected from the stack gases of (ferro-)silicon production (Juenger et al., 2019). In addition, natural resources such as pozzolana or limestone are eligible for use as SCMs as well. The level of substitution generally depends on the CaO content of the SCMs. High-CaO blast-furnace slag, Figure 27.1, can be blended into cement up to levels of 95 wt%; see Figure 27.2. The use of SCMs has substantial benefits in terms of reducing economic and environmental costs of cement manufacture, and can provide longer service life of structures by

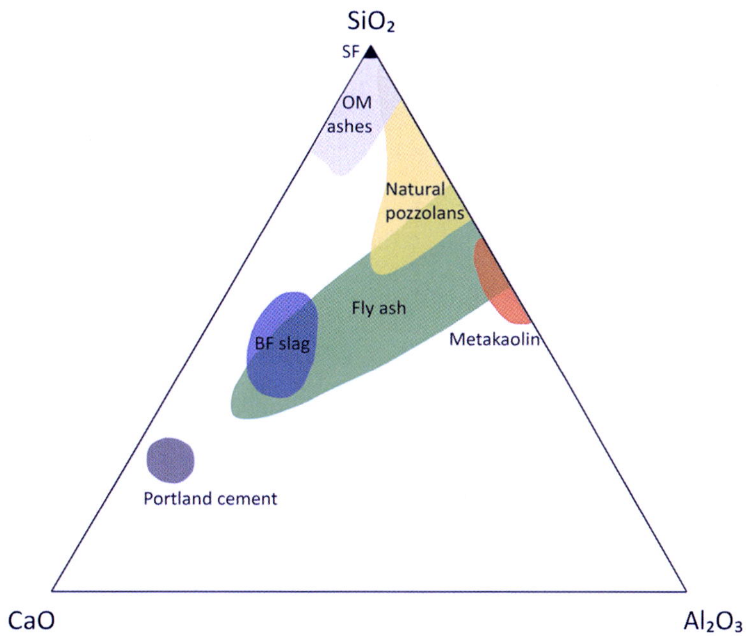

FIGURE 27.1 Chemical composition of ordinary Portland cement (CEM I) compared to the compositional ranges of conventional SCMs such as blast-furnace slags (BF slags), fly ashes, natural pozzolans, kaolinite-rich thermally activated clays (metakaolin), silica fume (SF), and organic material (OM) ashes. *Adapted from Snellings, R., Mertens, G., Elsen, J., 2012. Supplementary cementitious materials. Rev. Miner. Geochem. 74, 211–278.*

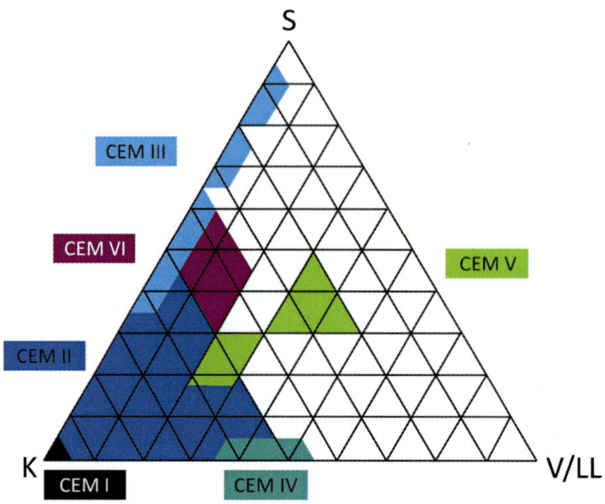

FIGURE 27.2 Cement type specifications according to EN 197-1:2011 (European Committee for Standardization, 2000) and EN 197-5:2021 (European Committee for Standardization, 2021). Cement types are defined based on the substitution levels of clinker (K) by blast-furnace slag (S), fly ash (V), and limestone (LL). The limestone proportion is limited to 20 wt%.

enhancing the resistance of concrete to deterioration (De Belie et al., 2018). In Europe, SCMs now constitute on average 24 wt% of cements sold (IEA and CSI of WBCSD, 2018). This fraction has not significantly changed over recent years, reflecting supply-side limitations. By 2050 the phase-out of carbon-intensive energy and steel production processes will reduce the supply of the most used SCMs dramatically. Further reducing or even maintaining cement CO_2 emissions at the present level will therefore require alternative SCMs to be introduced, possibly in combination with other actions.

27.3.1 Alternative SCMs

Industrial symbiosis advantages for prospective SCM suppliers and cement producers provides ample incentive to develop new SCM materials from residues. Materials of interest should meet expectations on three main levels. First, at a technical level candidate SCMs should improve or at least maintain key cement performance criteria such as compressive strength, dimensional stability and durability, or service life. Second, from an economic perspective, the SCM should be available at acceptable cost and volume to provide return on investment and competitive economic activity in the long term. Third, from an environmental point of view, significant reductions in impact are expected and compliance with legislation on emissions (contaminant leaching) is primary. Alternative SCMs include slags and sludges from metallurgical processes such as steel, nonferrous metals and alumina production, biomass and waste incineration ashes, and construction and demolition waste. Reclaimed coal combustion fly ash from historic monodeposits are of interest as well, as recent research has shown little alteration of the ashes during storage, thus preserving their potential as SCM. Finally, widely available natural resources such as pumices, tuffs, limestone, or clays are being considered as well (Juenger et al., 2019). Most candidate materials cannot be used directly as SCM and require further processing to upgrade their performance or compatibility. Common processing treatments are physical treatments such as comminution or size classification and thermal treatments such as calcination or melting followed by quenching to produce a reactive slag.

Thermal treatments, often at high temperature before slag solidification, may involve recovery or removal of metals of interest, while the remaining slag phase can be upgraded in terms of performance as SCM and environmental compatibility (Lopez Gonzalez et al., 2019).

Among new SCM sources, calcined clays provide a promising pathway that has received much attention recently (Skibsted and Snellings, 2019). Clay minerals, when calcined at temperatures between 600°C and 900°C lose their structural water, which leads to a (partially) disintegrated clay crystal structure. This "meta"-clay is an unstable material that reacts with $Ca(OH)_2$ and water (Sabir et al., 2001). In particular, kaolinite-group clay minerals have been recognized to be very reactive as pozzolans when calcined. The high reactivity of metakaolin enables Portland clinker to be substituted at high levels, with minimal losses in cement performance at early ages and improved durability parameters at late ages, as exploited in limestone calcined clay cements (LC^3). In these cements clinker levels are reduced down to 50% and substituted with a combination of calcined kaolinitic clay and ground limestone (Scrivener et al., 2018). The LC^3 (cements) have also been covered in cement standards, such as the recent EN 197-5:2021. Of interest is that kaolinitic clay resources are suitable for use as SCM from kaolinite grades as low as 30–40%, which is much lower than for other applications, and thus opens up opportunities for low-grade kaolinitic clays otherwise discarded as waste and nonkaolinitic clays of lower reactivity. Examples of clay-bearing residues of interest are mine tailings, for instance from coal or bauxite mining (Rodríguez et al., 2021), but also excavated earth from large civil works or waterway dredging sediments may be eligible secondary clay resources (Snellings et al., 2016).

Within the cement and concrete materials sector, significant progress is being made in using concrete recycling fines as SCM. The fine fraction of crushed concrete is challenging to recycle back into concrete. The fine fraction is enriched in hardened cement paste, presenting a high specific surface area. In turn, the incorporation of the fine fraction may negatively affect the water demand and the resulting performance and durability of new concrete (Khatib, 2005). Water-reducing concrete additives can provide an effective solution to these issues (de Brito and Rui, 2016). Interestingly, carbonation treatment, i.e., accelerated reaction of the hydrated cement paste with CO_2, produced calcium carbonate and pozzolanically reactive silica and alumina. As such, part of the CO_2 originally emitted during cement production is sequestered during the recycling process (Zajac et al., 2021).

27.3.2 SCM Test Methods and Standards

Current test methods for evaluating SCM reactivity combine compositional requirements with measurements of the consumption of portlandite ($Ca(OH)_2$) or lime in binary mixes, the measurement of the so-called reactive silica content (silica solubilized by an acid-base treatment), or by simply specifying minimum compressive strength levels for blends with $Ca(OH)_2$ or Portland cement. A recent review and robustness test demonstrated that these conventional test methods fail to provide reliable results for a wider range of SCMs (Snellings et al., 2016). It was found that different combinations of test methods and parameters are used in the industrial specifications, depending on the country, the SCM type, and the application (cement or concrete constituent). In response, a RILEM technical committee (RILEM TC 238 TRM) was set up to evaluate alternative reactivity test methods and develop recommendations and prestandard documents. Among all evaluated test methods, the R3 test performed best in terms of robustness and correlation to strength development (taken as a material performance benchmark) (Li et al., 2018). Instead of $Ca(OH)_2$ consumption or reactive silica, the

R3 test uses heat release and bound water as response parameters. This extends the scope of the test to materials that themselves contain significant amounts of CaO, such as blast furnace slags. In 2020 the test method was adopted as ASTM C-1897:20.

In terms of cement industry standards, a clear trend can be discerned towards the development of ternary or multiple constituent blended cements with lower clinker contents that combine various residues or, more commonly, one residue and limestone. Limestone has been shown to play a double role: on the one hand it can act as an inexpensive filler that accelerates the hydration of the clinker and that can reduce the water demand of the cement, and on the other hand, limestone can react in synergy with Al-bearing SCMs to form more space-filling hydration products (De Weerdt et al., 2011). Binary and ternary cements containing significant levels of limestone have been adopted recently into cement standards around the world, e.g., Figure 27.2 for Europe, and it can be expected that they will constitute an important share of future cement production. It is expected that further developments of cement standards will unlock new combinations, proportions, and types of constituents such as dolomitic limestone, concrete recycling fines, and others.

27.4 ALTERNATIVE CEMENTS WITH LOWER ENVIRONMENTAL FOOTPRINT

Next to OPC-based binders, a range of alternative cements has been developed or is being investigated (Gartner and Macphee, 2011). Many of the already existing alternatives have found niche applications, e.g., calcium aluminate cements for refractory uses, but due to technical (performance or durability) or economic reasons (raw materials of low cost and economy of scale for OPC) they have not replaced OPC on a wider scale. The discussion herein cannot be exhaustive. Three alternative cement families are presented, as they appear to be promising in terms of (extended) industrial implementation, partially driven by their reduced CO_2 footprint and the substantial amounts of residues incorporated:

- calcium sulfoaluminate (CSA) cements,
- alkali activated cements, and
- mineral carbonation cements.

Magnesia-based cements will not be covered herein, but the reader can find more information in Walling and Provis (2016). These cements can find applications from large-scale construction to nuclear waste immobilization and have in some cases a substantially lower CO_2 footprint compared to conventional Portland cement. There is also revived interest in calcium aluminate cements, including new blended formulations (Astoveza et al., 2021) as well as supersulfated cements, the latter being particularly interesting from a residue recycling perspective (Juenger et al., 2011).

27.4.1 Calcium Sulfoaluminate (Ferrite) Cements

Calcium sulfoaluminate-based cements contain ye'elimite ($Ca_4Al_6O_{12}(SO_4)$) as a major component next to other major phases, such as belite and ferrite. The hydration of ye'elimite in combination with added calcium sulfate is responsible for the early strength development, while the hydration of belite and ferrite contributes largely to later age strength gain. Compared to Portland cement, CSA cements can be produced at lower temperatures of about 1250°C and involve much lower CO_2 emissions from raw materials. CSA cements were initially largely produced in China as specialty cements for high-early strength or self-stressing applications and contained high levels of ye'elimite (60–70%). Bauxite is needed as raw material to achieve high ye'elimite contents and as a result these cements are relatively expensive; secondary alumina-rich resources could be explored. In general, three different groups can be distinguished, taking into account the main phases present: (a) belite-ye'elimite-ferrite binders, (b) belite-alite-ye'elimite binders, and (c) belite-ye'elimite-ternesite binders. Currently, the use of residues such as fly ashes and slags both as raw materials and as SCMs is being tested to further reduce both cost and CO_2 footprint. In addition, bauxite residue ("red mud") and gypsum from other processes can be used (Arjunan et al., 1999; Pontikes et al., 2013), considering the higher levels of Fe_2O_3 and $CaSO_4$ in the raw meal. Calciumsulfoaluminate-ferrite clinkers, incorporating 35 wt% of bauxite residue, lead to a phase assemblage of ye'elimite, belite, and ferrite phases, next to calcium aluminum ferrotitanate. Using this clinker with 10 wt% anhydrite, a mortar compressive strength of 43 MPa at 2 d and 54 MPa at 28 d was obtained (Hertel et al., 2021). Also, a novel approach was suggested regarding the use of elemental sulfur, a residue from oil and gas refining, in reducing fuel emissions. In this process, close control of the raw meal composition and kiln conditions is required to create proper conditions for ye'elimite formation and to avoid SO_x emissions (Hanein et al., 2016).

27.4.2 Alkali-Activated Cements

Unlike the cements presented so far, alkali-activated cements usually do not necessitate a high-temperature processing step for the synthesis of the solid reactants or precursors. In most cases, alkali-activated cements are developed with by-products and residues from other industrial processes, and in fact, these materials constitute the actual reactive components.

Conceptually, these cements are formed after mixing two components, a solid precursor and an alkaline activating solution. The resulting paste can set and harden, usually at room temperature. The most often used solid precursors are fly ash from coal combustion, metakaolin, a product of thermal treatment of kaolinite clays, as well as ground granulated blast furnace slag. Other residues explored are ferrous metallurgy slags, nonferrous metallurgy slags, phosphorus slag, as well as a range of other materials (e.g., Shi et al., 2006). These components can be used alone or in binary/ternary mixtures. With respect to the alkaline activators, these are usually concentrated solutions of (Na,K)-hydroxides and (Na,K)-silicates with -carbonates and -sulfates also possible to be employed (Shi et al., 2006). There are many recent works where add-water-only formulations are developed and an overview can be found herein (Luukkonen et al., 2018). That approach is expected to make the use of these binders easier, resembling a "traditional" cement, and ultimately facilitate adoption by the market.

The final binding phase depends on the starting composition, in particular the level of Ca in the Na_2O-CaO-Al_2O_3-SiO_2 system. For high-calcium compositions, an amorphous to partially crystalline C-(A)-S-H material typically forms, relatively cross-linked, with a moderate degree of Al substitution and a low C/S ratio (Wang and Scrivener, 1995). On the other hand, for low-Ca compositions, an amorphous N-A-S-(H) material develops, resembling zeolites (Bell et al., 2008), and highly cross-linked. This material contains low levels of chemically bound water (Rahier et al., 1996). For intermediate levels of Ca in the precursor blend, recent work in the area demonstrated the decisive role of Ca (Garcia-Lodeiro et al., 2011) and revealed that an interaction between C-(A)-S-H and N-A-S-(H) phases takes place over time (García-Lodeiro et al., 2013; Ismail et al., 2014) leading to a hybrid N-(C)-A-S-H phase. Apart from amorphous phases, crystalline phases can also precipitate, depending on the activator and the processing. For instance, hydroxide activators can lead to zeolite formation, which may be intensified if hydrothermal curing is applied.

Besides compositions within the Na_2O-CaO-Al_2O_3-SiO_2 diagram, additional chemistries are also explored. Fe-Al-rich amorphous silicates are interesting precursors (Pontikes et al., 2013), in addition to other works where authors tried to synthesize magnesium-containing analogs of aluminosilicate binders (MacKenzie et al., 2013) and to substitute boron for aluminum in the inorganic polymer structure (Williams and van Riessen, 2011). The role of Fe is also a topic of interest (Lemougna et al., 2013), with materials originating from nonferrous metallurgical slags typically demonstrating competitive physical and mechanical properties, e.g., Maragkos et al., 2009. Several recent works have attempted to go deeper and understand the role of Fe during the whole process, i.e., from the chemistry of the precursor all the way to curing (Peys et al., 2019; Siakati et al., 2021). In parallel with this, a process has been suggested where Fe-rich residues, e.g., tailings or hydrometallurgical residues such as bauxite residue, undergo vitrification, aiming to create a partially amorphous precursor that resembles Fe-rich slags (Hertel et al., 2016). The added-value in this approach is that a wealth of residues can be potentially employed for the production of these new binders, provided the end-chemistry of the slag (which is also the precursor downstream) is right (Van De Sande et al., 2020).

As is evident, there are many options already explored with respect to raw materials and processes when it comes to alkali-activated binders. In many cases, this enables a high content of residues to be employed in the formulation and quite often the end material has tailor-made microstructures with specific functionalities, because it is per se designed to interact with residues. Zeolite-containing, alkali-activated binders are such an example. Acknowledging there are a number of factors to be considered in developing the process (Rożek et al., 2019), the end materials could be employed in applications where adsorption or high specific surface area is critical (e.g., catalysis). Next to zeolites, refractory formulations have been developed able to withstand fires but also the casting of liquid metals (e.g., Lahoti et al., 2019). In fact, interaction with high temperature is quite distinct compared to OPC, and the alkali-activated materials can be designed to withstand thermal shock, to sinter, or not. From that perspective, these materials can also be seen as "green ceramics," precursors to be subjected to heat and become a ceramic (e.g., Marangoni et al., 2016). That posttreatment is optional and only justified if the requirements of a particular application ask for it. In most cases, elaborated microstructures can be simply obtained via room temperature processes and curing or room temperature processes and hydrothermal curing at higher temperature and pressure, usually in an autoclave. An overview can be seen in the work of Bai and Colombo (2018) on porous microstructures. One last area that appears a promising application field for alkali-activated materials is that of waste stabilization/solidification. The topic is very wide and covers most wastes imaginable, i.e., inorganic materials with heavy metals, organic wastes, and an array of radioactive wastes from nuclear power plants, e.g., Vandevenne et al. (2018).

In most of the works about alkali activation, cement is not employed in the mix design, as this would be seemingly canceling the motivation for sustainability and the development of the low environmental impact binders. This is not necessarily correct; the environmental impact and cost of the activating solutions are far from negligible, whereas controlling the shrinkage remains a nontrivial issue (Beersaerts et al., 2021). Additives such as (super) plasticizers, or shrinkage reducers, would likely help but are currently (in 2021) not available for alkali-activated materials. This limits the flexibility of these systems and makes the upscaling more challenging. A different approach is the development of alkali-activated materials with higher

CaO contents, usually described in the literature as "hybrid" alkaline cements (Garcia-Lodeiro et al., 2015). Compared to inorganic polymers and geopolymers, the amount of alkalis used here is minor. As raw materials, OPC but also calcareous fly ashes or ferrous metallurgy slags can be used, and two groups can be distinguished accordingly, one with OPC in the blend, and one without. Hybrid binders could be seen as a pragmatic step forward, towards market penetration of alkali-activated materials.

In hybrid systems, the reaction product, and consequently the performance, depend on both the precursor and the activator, and their interaction, in which the pH is a decisive parameter. In these blends, Na could replace Ca, and if Al is present, it could convert the traditional poorly crystalline C-S-H into a (N,C)-A-S-H type of gel (Fernández-Jiménez et al., 2013). Prior studies have shown that the coprecipitation of the aforementioned gels in hybrid cements is possible, e.g., Garcia-Lodeiro et al. (2011), with more recent studies revealing that the gels interact, undergoing structural and compositional changes in the process, e.g., Sánchez-Herrero et al. (2019). All these become more complex, and perhaps even more interesting, when the bulk fraction is an Fe-rich slag (Arnout et al., 2021) and when on top of that, the alkali used for the activation is part of the residue itself (Arnout et al., 2017).

The future of the alkali-activated cements remains open. These binders have a history (Provis and van Deventer, 2014) and it appears that also major cement industries are now offering such products on the market (Cemex, 2021). There are undeniably a number of challenges and what was written some time ago (Provis and van Deventer, 2014) remains most relevant and contemporary: sourcing of raw materials, costs, quality control and quality assurance, long-term performance, standardization, and acceptance from the customers are all key aspects that will have an impact on the future.

27.4.3 Mineral Carbonation Cements

The idea of using mineral carbonation products as in-situ cement was first described by Lackner et al. (1995). Based on this seminal work, CO_2 negative cements have been proposed that so far have mostly relied on calcium (and magnesium)-rich materials prone to carbonate (Fernández Bertos et al., 2004). Since even for those precursor materials natural carbonation is a slow process, all currently proposed processes require curing at elevated CO_2 concentrations (20–100% CO_2), often at increased temperatures (50–120°C) or pressures (<100 bar) (Baciocchi et al., 2010) to accelerate the carbonation and solidification reactions to economically relevant timescales of several hours. Research on mineral carbonation with the purpose of producing construction materials has mainly focused on:

- Accelerated curing of concrete. One of the early research lines into the use of accelerated carbonation is to enhance the hydration of Portland cement and reduce curing times of concrete (Young et al., 1974). This research has led to commercialized developments such as injection of CO_2 during concrete mixing, enabling to boost cement hydration and lower cement contents of concrete by 3–5% without compromising performance, e.g., CarbonCure technology (Monkman et al., 2018).
- Carbonation of brines. Desalination of seawater produces a concentrated brine rich in Ca and Mg. Introduction of CO_2 in such concentrated brines results in the formation of (Mg, Ca)-carbonates. Under controlled conditions, nesquehonite, a Mg hydroxy-carbonate hydrate, is formed that displays cementing properties when thermally treated or when heat cured. Compressive strengths obtained thus far are around 8 MPa, sufficient for making insulation panels or nonstructural blocks

(Glasser et al., 2016). The commercialization of this research is still in its early stages.

- Carbonation of hydrated lime. Traditional lime mortars harden by slow natural carbonation during long periods of exposure to the atmosphere. This process is too inefficient for modern construction, but recent research showed that curing using pressurized CO_2 can reduce carbonation times of hydrated lime to a few hours at room temperature (Vance et al., 2015). As a product, a calcium carbonate cement is produced, the properties of which can be highly variable depending on process conditions and the presence of impurities. Simultaneous marked changes in calcium carbonate polymorphs (calcite, aragonite, and vaterite) and crystal habits need to be further understood (Cizer et al., 2012), yet can lead to new binder technologies such as the ReCarb process that uses the recrystallization of vaterite into aragonite to develop compressive strengths of about 40 MPa (Hargis et al., 2021).
- Carbonation of calcium silicates. Calcium silicates and calcium silicate hydrates present or formed in cement are prone to carbonate; similarly also less reactive anhydrous calcium and magnesium silicates can be carbonated when subjected to pressurized CO_2 (Ashraf and Olek, 2016). Such calcium/magnesium silicates can be mined (e.g., wollastonite or serpentinites) (Power et al., 2013; Kelemen et al., 2020), found in industrial residues such as steel slags (Huijgen et al., 2005), or manufactured by thermal processing (Villani et al., 2014). While the former two materials may be limited by locally available resources, thermal processing introduces an additional CO_2 burden for manufactured calcium silicates. A first demonstration by Johnson et al. (2003) managed to produce blocks of 9 MPa compressive strength. More recent research achieved compressive strengths up to 72 MPa (Mo et al., 2016).

The latter research line proved to be most promising in terms of carbonated product performance and availability of precursor materials. Research into an accelerated carbonation process for stainless steel slag has led to the development of the Carbstone process (Quaghebeur et al., 2015). Moreover, the production of manufactured calcium silicates as a carbonation precursor has been piloted in industrial cement clinkering kilns by Solidia (Sahu and DeCristofaro, 2013). All of these processes mentioned depart from different precursor materials and can be used to produce a range of very different construction materials; however, they all share mineral carbonation as a core process.

27.5 CONCLUSIONS AND OUTLOOK

Cement plants today can operate as recipients of materials for energy and product enabling the "coprocessing" of large volumes of residues. Depending on economic, logistic, and legislation aspects, "ad hoc" solutions emerge. Factors such as variability, homogeneity, and volumes of these secondary streams, in conjunction with maximum acceptance levels for certain elements, pose limitations on what can be used eventually. Next to the materials used for clinker production, impressive volumes of residues are already being incorporated downstream into a wide range of blended cements. Ground granulated blast furnace slags as well as fly ashes from coal combustion are used to a significant extent. Thus, despite constraints, cement plants can be seen as interwoven, interacting elements in the socioindustrial symbiotic matrix. As a result of this, and of high-level commitments to reach climate neutrality by the mid-21st century, cement production will need to abate its CO_2 emissions and find alternatives to the resources it now takes in from carbon-intensive industrial processes, such as coal-fired energy production.

To reach carbon neutrality based on proven OPC production processes, it would be crucial to embrace circular economy measures, such as increasing resource and product efficiency, including low-clinker cement and concrete, increased recycling, modularity, etc., next to electrification and carbon capture storage and usage.

Still, new ideas emerge, aspiring to reduce the CO_2 footprint of cement today, and often "problem driven" by other industrial sectors or societal needs. In many cases, these ideas do not reach the level of industrial implementation. Yet, a certain culture is developing that is embracing and considering these alternatives, often for local or niche applications. The beneficiation of these residues, securing a proper and environmentally sound chemistry, a desirable microstructure, and an overall quality that fluctuates only in a minor, reasonable way, will facilitate their use. This would also necessitate a mentality shift, where residue-producing industries become actively involved in the valorization of their residues, without allocating the liability to end users. Industrial ecology provides the conceptual framework for these actions to emerge and to be supported by legislation, regulations, and society. For the binders in particular, a challenge would be to move from prescriptive to performance-based standards, with no compromise on their long-term durability and no illusions regarding the fact that these binders will be used in real-life conditions where quality control and quality assurance are not always meeting the most stringent requirements.

In view of this, it is apparent that a transformation is taking place where the industrial processes of today become more sustainable and where new processes and materials emerge as alternatives. Still, it is fair to say that this transition requires the engagement of each individual and that there are still objective technical and nontechnical challenges. In practice, although goals are set, quite often as a response to societal and political pressure, the path to reach these goals is to a great extent based on assumptions that necessitate investments, commitment, and ingenuity in making them a reality. Acknowledging this is merely the realization that there is still work to be done, an invitation for "more and better"; we remain optimistic.

References

Andrew, R.M., 2018. Global CO_2 emissions from cement production. Earth Syst. Sci. Data 10 (1), 195–217.

Aranda Usón, A., López-Sabirón, A.M., Ferreira, G., Llera Sastresa, E., 2012. Uses of alternative fuels and raw materials in the cement industry as sustainable waste management options. Renew. Sust. Energ. Rev. 23, 242–260.

Arjunan, P., Silsbee, M.R., Roy, D.M., 1999. Sulfoaluminate-belite cement from low-calcium fly ash and sulfur-rich and other industrial by-products. Cem. Concr. Res. 29 (8), 1305–1311.

Arnout, L., Hertel, T., Liard, M., Lootens, D., Pontikes, Y., 2017. Use of bauxite residue slurry as single activator in a hybrid binder system. In: 35th International ICSOBA Conference, pp. 519–528.

Arnout, L., Beersaerts, G., Liard, M., Lootens, D., Pontikes, Y., 2021. Valorising slags from non-ferrous metallurgy into hybrid cementitious binders: mix design and performance. Waste Biomass Valoriz. 12 (8), 4679–4694.

Ashraf, W., Olek, J., 2016. Carbonation behavior of hydraulic and non-hydraulic calcium silicates: potential of utilizing low-lime calcium silicates in cement-based materials. J. Mater. Sci. 51 (13), 6173–6191.

Astoveza, J., Trauchessec, R., Soth, R., Pontikes, Y., 2021. Properties of calcium aluminate blended cement incorporating iron-rich slag: evolution over a curing period of 1 year. Constr. Build. Mater. 282, 122569.

Baciocchi, R., Costa, G., Di Bartolomeo, E., Polettini, A., Pomi, R., 2010. Carbonation of stainless steel slag as a process for CO_2 storage and slag valorization. Waste Biomass Valoriz. 1 (4), 467–477.

Bai, C., Colombo, P., 2018. Processing, properties and applications of highly porous geopolymers: a review. Ceram. Int. 44 (14), 16103–16118.

Beersaerts, G., Ascensão, G., Pontikes, Y., 2021. Modifying the pore size distribution in Fe-rich inorganic polymer mortars: an effective shrinkage mitigation strategy. Cem. Concr. Res. 141, 106330.

Bell, J.L., Sarin, P., Driemeyer, P.E., Haggerty, R.P., Chupas, P.J., Kriven, W.M., 2008. X-ray pair distribution function analysis of a metakaolin-based, $KAlSi_2O_6 \cdot 5.5H_2O$ inorganic polymer (geopolymer). J. Mater. Chem. 18 (48), 5974–5981.

Cembureau and The European Cement Association, 2021. Carbon Neutrality. Retrieved 5 of January from: https://lowcarboneconomy.cembureau.eu/carbon-neutrality/.

Cemex, 2021. Introducing Vertua—Low Carbon Concrete. Retrieved 5 of January 2022 from https://www.cemex.co.uk/vertua-low-carbon-concrete.

Chatterjee, A., Sui, T., 2019. Alternative fuels—effects on clinker process and properties. Cem. Concr. Res. 123, 105777.

Chinyama, M.P.M. (Ed.), 2011. Alternative Fuels in Cement Manufacturing. Alternative Fuel. InTech.

Cizer, Ö., Rodriguez-Navarro, C., Ruiz-Agudo, E., Elsen, J., Van Gemert, D., Van Balen, K., 2012. Phase and morphology evolution of calcium carbonate precipitated by carbonation of hydrated lime. J. Mater. Sci. 47 (16), 6151–6165.

de Beer, J., Cihlar, J., Hensing, I., 2017. Status and prospects of co-processing of waste in EU cement plants. Case studies. Ecofys. Retrieved 15 of December 2021 from https://cembureau.eu/media/ldfdotk0/12950-ecofys-co-processing-waste-cement-kilns-case-studies-2017-05.pdf.

De Belie, N., Soutsos, M., Gruyaert, E. (Eds.), 2018. Properties of Fresh and Hardened Concrete Containing Supplementary Cementitious Materials. Springer International Publishing. RILEM State-of-the-Art Reports.

de Brito, J., Rui, S., 2016. Current status on the use of recycled aggregates in concrete: Where do we go from here? RILEM Tech. Lett. 1.

De Weerdt, K., Haha, M.B., Le Saout, G., Kjellsen, K.O., Justnes, H., Lothenbach, B., 2011. Hydration mechanisms of ternary Portland cements containing limestone powder and fly ash. Cem. Concr. Res. 41, 279–291.

European Committee for Standardization, 2000. EN 197-1, Cement. Part 1. Composition, Specifications and Conformity Criteria for Common Cements.

European Committee for Standardization, 2021. Cement. Part 5. Portland-Composite Cement CEM II/C-M and Composite Cement CEM VI.

Fernández Bertos, M., Simons, S.J.R., Hills, C.D., Carey, P.J., 2004. A review of accelerated carbonation technology in the treatment of cement-based materials and sequestration of CO_2. J. Hazard. Mater. 112 (3), 193–205.

Fernández-Jiménez, A., Zibouche, F., Boudissa, N., García-Lodeiro, I., Abadlia, M.T., Palomo, A., 2013. 'Metakaolin-Slag-Clinker Blends.' The role of Na^+ or K^+ as alkaline activators of theses ternary blends. J. Am. Ceram. Soc. 96 (6), 1991–1998.

Garcia-Lodeiro, I., Palomo, Á., Fernández-Jiménez, A., Macphee, D.E., 2011. Compatibility studies between N-A-S-H and C-A-S-H gels. Study in the ternary diagram Na_2O--CaO-Al_2O_3-SiO_2-H_2O. Cem. Concr. Res. 41 (9), 923–931.

García-Lodeiro, I., Fernández-Jiménez, A., Palomo, A., 2013. Variation in hybrid cements over time. Alkaline activation of fly ash-portland cement blends. Cem. Concr. Res. 52, 112–122.

Garcia-Lodeiro, I., Palomo, A., Fernández-Jiménez, A., Pacheco-Torgal, F., Labrincha, J.A., Leonelli, C., Chindaprasirt, P., 2015. 2—An overview of the chemistry of alkali-activated cement-based binders. In: Handbook of Alkali-Activated Cements, Mortars and Concretes. Woodhead Publishing, Oxford, pp. 19–47.

Gartner, E.M., Macphee, D.E., 2011. A physico-chemical basis for novel cementitious binders. Cem. Concr. Res. 41 (7), 736–749.

Glasser, F.P., Jauffret, G., Morrison, J., Galvez-Martos, J.-L., Patterson, N., Imbabi, M.S.-E., 2016. Sequestering CO_2 by mineralization into useful nesquehonite-based products. Front. Energy Res. 4 (3).

Habert, G., Billard, C., Rossi, P., Chen, C., Roussel, N., 2010. Cement production technology improvement compared to factor 4 objectives. Cem. Concr. Res. 40, 820–826.

Hanehara, S., 2005. Eco-cement and eco-concrete. environmentally compatible cement and concrete technology. In: COE Workshop on "Material Science in the 21st Century for the Construction Industry—Durability, Repair and Recycling of Concrete Structures", Hokkaido University Conference Hall, Sapporo, Japan.

Hanein, T., Galan, I., Elhoweris, A., Khare, S., Skalamprinos, S., Jen, G., Whittaker, M., Imbabi, M.S., Glasser, F.P., Bannerman, M.N., 2016. Production of belite calcium sulfoaluminate cement using sulfur as a fuel and as a source of clinker sulfur trioxide: pilot kiln trial. Adv. Cem. Res. 28 (10), 643–653.

Hargis, C.W., Chen, I.A., Devenney, M., Fernandez, M.J., Gilliam, R.J., Thatcher, R.P., 2021. Calcium carbonate cement: a carbon capture, utilization, and storage (CCUS) technique. Materials 14 (11), 2709.

Hertel, T., Blanpain, B., Pontikes, Y., 2016. A proposal for a 100 % use of bauxite residue towards inorganic polymer mortar. J. Sustain. Metall. 2 (4), 394–404.

Hertel, T., Van den Bulck, A., Onisei, S., Sivakumar, P.P., Pontikes, Y., 2021. Boosting the use of bauxite residue (red mud) in cement—production of an Fe-rich calcium sulfoaluminate-ferrite clinker and characterisation of the hydration. Cem. Concr. Res. 145, 106463.

Huang, Q., Yang, Y., Wang, Q., 2012. Potential for serious environmental threats from uncontrolled co-processing of wastes in cement kilns. Environ. Sci. Technol. 46 (24), 13031–13032.

Huijgen, W.J.J., Witkamp, G.-J., Comans, R.N.J., 2005. Mineral CO_2 sequestration by steel slag carbonation. Environ. Sci. Technol. 39 (24), 9676–9682.

IEA and CSI of WBCSD, 2018. Technology Roadmap: Low-carbon Transition in the Cement Industry. Retrieved 3 of December 2021 from IEA, Paris https://www.iea.

org/reports/technology-roadmap-low-carbon-transition-in-the-cement-industry.

Ismail, I., Bernal, S.A., Provis, J.L., San Nicolas, R., Hamdan, S., Van Deventer, J.S.J., 2014. Modification of phase evolution in alkali-activated blast furnace slag by the incorporation of fly ash. Cem. Concr. Compos. 45, 125–135.

Johnson, D.C., Macleod, C.L., Carey, P.J., Hills, C.D., 2003. Solidification of stainless steel slag by accelerated carbonation. Environ. Technol. 24 (6), 671–678.

Juenger, M.C.G., Winnefeld, F., Provis, J.L., Ideker, J.H., 2011. Advances in alternative cementitious binders. Cem. Concr. Res. 41 (12), 1232–1243.

Juenger, M.C.G., Snellings, R., Bernal, S.A., 2019. Supplementary cementitious materials: new sources, characterization, and performance insights. Cem. Concr. Res. 122, 257–273.

Kelemen, P.B., McQueen, N., Wilcox, J., Renforth, P., Dipple, G., Vankeuren, A.P., 2020. Engineered carbon mineralization in ultramafic rocks for CO_2 removal from air: review and new insights. Chem. Geol. 550, 119628.

Khatib, J.M., 2005. Properties of concrete incorporating fine recycled aggregate. Cem. Concr. Res. 35 (4), 763–769.

Lackner, K.S., Wendt, C.H., Butt, D.P., Joyce, E.L., Sharp, D. H., 1995. Carbon dioxide disposal in carbonate minerals. Energy 20 (11), 1153–1170.

Lahoti, M., Tan, K.H., Yang, E.-H., 2019. A critical review of geopolymer properties for structural fire-resistance applications. Constr. Build. Mater. 221, 514–526.

Lemougna, P.N., MacKenzie, K.J.D., Jameson, G.N.L., Rahier, H., Chinje Melo, U.F., 2013. The role of iron in the formation of inorganic polymers (geopolymers) from volcanic ash: a ^{57}Fe Mössbauer spectroscopy study. J. Mater. Sci. 48 (15), 5280–5286.

Li, X., Snellings, R., Antoni, M., Alderete, N.M., Ben Haha, M., Bishnoi, S., Cizer, Ö., Cyr, M., De Weerdt, K., Dhandapani, Y., Duchesne, J., Haufe, J., Hooton, D., Juenger, M., Kamali-Bernard, S., Kramar, S., Marroccoli, M., Joseph, A.M., Parashar, A., Patapy, C., Provis, J.L., Sabio, S., Santhanam, M., Steger, L., Sui, T., Telesca, A., Vollpracht, A., Vargas, F., Walkley, B., Winnefeld, F., Ye, G., Zajac, M., Zhang, S., Scrivener, K.L., 2018. Reactivity tests for supplementary cementitious materials: RILEM TC 267-TRM phase 1. Mater. Struct. 51 (6), 151.

Lopez Gonzalez, P.L., Novais, R.M., Labrincha, J.A., Blanpain, B., Pontikes, Y., 2019. Modifications of basic-oxygen-furnace slag microstructure and their effect on the rheology and the strength of alkali-activated binders. Cem. Concr. Compos. 97, 143–153.

Luukkonen, T., Abdollahnejad, Z., Yliniemi, J., Kinnunen, P., Illikainen, M., 2018. One-part alkali-activated materials: a review. Cem. Concr. Res. 103, 21–34.

MacKenzie, K.J.D., Bradley, S., Hanna, J.V., Smith, M.E., 2013. Magnesium analogues of aluminosilicate inorganic polymers (geopolymers) from magnesium minerals. J. Mater. Sci. 48 (4), 1787–1793.

Maragkos, I., Giannopoulou, I.P., Panias, D., 2009. Synthesis of ferronickel slag-based geopolymers. Miner. Eng. 22 (2), 196–203.

Marangoni, M., Arnout, L., Machiels, L., Pandelaers, L., Bernardo, E., Colombo, P., Pontikes, Y., 2016. Porous, sintered glass-ceramics from inorganic polymers based on fayalite slag. J. Am. Ceram. Soc. 99 (6), 1985–1991.

Ministry of the Environment (Japan), 2014. History and Current State of Waste Management in Japan. Retrieved 5 of August, 2021 from: https://www.env.go.jp/en/recycle/smcs/attach/hcswm.pdf.

Mo, L., Zhang, F., Deng, M., 2016. Mechanical performance and microstructure of the calcium carbonate binders produced by carbonating steel slag paste under CO_2 curing. Cem. Concr. Res. 88, 217–226.

Mokrzycki, E., Uliasz-Bocheńczyk, A., Sarna, M., 2003. Use of alternative fuels in the Polish cement industry. Appl. Energy 74 (1-2), 101–111.

Monkman, S., Kenward, P.A., Dipple, G., MacDonald, M., Raudsepp, M., 2018. Activation of cement hydration with carbon dioxide. J. Sustain. Cement-Based Mater. 7 (3), 160–181.

Noirfontaine, M.N., Tusseau-Nenez, S., Girod-Labianca, C., Pontikis, V., 2012. CALPHAD formalism for Portland clinker: thermodynamic models and databases. J. Mater. Sci. 47 (3), 1471–1479.

Peys, A., White, C.E., Rahier, H., Blanpain, B., Pontikes, Y., 2019. Alkali-activation of CaO-FeOx-SiO_2 slag: formation mechanism from in-situ X-ray total scattering. Cem. Concr. Res. 122, 179–188.

Pontikes, Y., Machiels, L., Onisei, S., Pandelaers, L., Geysen, D., Jones, P.T., Blanpain, B., 2013. Slags with a high Al and Fe content as precursors for inorganic polymers. Appl. Clay Sci. 73, 102–193.

Power, I.M., Wilson, S.A., Dipple, G.M., 2013. Serpentinite carbonation for CO_2 sequestration. Elements 9 (2), 115–121.

Provis, J.L., van Deventer, J.S.J. (Eds.), 2014. Alkali Activated Materials. RILEM State-of-the-Art Reports, Springer, Dordrecht.

Quaghebeur, M., Nielsen, P., Horckmans, L., Van Mechelen, D., 2015. Accelerated carbonation of steel slag compacts: development of high-strength construction materials. Front. Energy Res. 3 (52).

Rahier, H., Van Mele, B., Biesemans, M., Wastiels, J., Wu, X., 1996. Low-temperature synthesized aluminosilicate glasses. Part I. Low-temperature reaction stoichiometry and structure of a model compound. J. Mater. Sci. 31 (1), 71–79.

Rodríguez, J., Frías, M., Tobón, J.I., 2021. Eco-efficient cement based on activated coal washing rejects with low content of kaolinite. Constr. Build. Mater. 274, 122118.

Rootzén, J., 2012. Reducing Carbon Dioxide Emissions From the EU Power and Industry Sectors. An assessment of Key Technologies and Measures. Department of Energy and Environment, Chalmers University of Technology, Gothenburg, Sweden.

Rożek, P., Król, M., Mozgawa, W., 2019. Geopolymer-zeolite composites: a review. J. Clean. Prod. 230, 557–579.

Sabir, B.B., Wild, S., Bai, J., 2001. Metakaolin and calcined clays as pozzolans for concrete: a review. Cem. Concr. Compos. 23 (6), 441–454.

Sahu, S., DeCristofaro, N., 2013. Exploring the Chemical Properties and Performance Results of Sustainable Solidia Cement™ and Solidia Concrete™. Retrieved 7 of September 2021 from: http://solidiatech.com/wp-content/uploads/2014/02/Solidia-Cement-White-Paper-12-17-13-FINAL.pdf.

Sánchez-Herrero, M.J., Fernández-Jiménez, A., Palomo, A., 2019. Studies about the hydration of hybrid "alkaline-belite" cement. Front. Mater. 6 (66).

Scrivener, K., Martirena, F., Bishnoi, S., Maity, S., 2018. Calcined clay limestone cements (LC3). Cem. Concr. Res. 114, 49–56.

Shi, C., Krivenko, P.K., Roy, D., 2006. Alkali-Activated Cements and Concretes. Taylor & Francis, NY, USA.

Shimoda, T., Yokoyama, S., 1999. Eco-cement: a new Portland cement to solve municipal and industrial waste problems. In: Modern Concrete Materials: Binders, Additions and Admixtures, pp. 17–30.

Siakati, C., Douvalis, A.P., Hallet, V., Peys, A., Pontikes, Y., 2021. Influence of CaO/FeO ratio on the formation mechanism and properties of alkali-activated Fe-rich slags. Cem. Concr. Res. 146, 106466.

Skibsted, J., Snellings, R., 2019. Reactivity of supplementary cementitious materials (SCMs) in cement blends. Cem. Concr. Res. 124, 105799.

Snellings, R., Mertens, G., Elsen, J., 2012. Supplementary cementitious materials. Rev. Miner. Geochem. 74, 211–278.

Snellings, R., Cizer, Ö., Horckmans, L., Durdziński, P.T., Dierckx, P., Nielsen, P., Van Balen, K., Vandewalle, L., 2016. Properties and pozzolanic reactivity of flash calcined dredging sediments. Appl. Clay Sci. 129, 35–39.

Taiheiyo Cement Corporation, 2021. Ecocement System. Retrieved 5 of August 2021 from https://www.taiheiyo-cement.co.jp/english/service_product/recycle_mw/eco/index.html.

The GTZ-Holcim Public Private Partnership, 2006. Guidelines on Co-processing Waste Materials in Cement Production. Retrieved 5 of January 2022 from: https://star-www.giz.de/dokumente/bib/06-0476.pdf.

U.S. Geological Survey, 2021. Mineral Commodity Summaries. Retrieved 30 of July 2021 from: https://pubs.usgs.gov/periodicals/mcs2021/mcs2021.pdf.

Van De Sande, J., Peys, A., Hertel, T., Rahier, H., Pontikes, Y., 2020. Upcycling of non-ferrous metallurgy slags: identifying the most reactive slag for inorganic polymer construction materials. Resour. Conserv. Recycl. 154, 104627.

Vance, K., Falzone, G., Pignatelli, I., Bauchy, M., Balonis, M., Sant, G., 2015. Direct carbonation of $Ca(OH)_2$ using liquid and supercritical CO_2: implications for carbon-neutral cementation. Ind. Eng. Chem. Res. 54 (36), 8908–8918.

Vandevenne, N., Iacobescu, R.I., Carleer, R., Samyn, P., D'Haen, J., Pontikes, Y., Schreurs, S., Schroeyers, W., 2018. Alkali-activated materials for radionuclide immobilisation and the effect of precursor composition on Cs/Sr retention. J. Nucl. Mater. 510, 575–584.

Villani, C., Spragg, R., Tokpatayeva, R., Olek, J., Weiss, W.J., 2014. Characterizing the pore structure of carbonated natural wollastonite. In: 4th International Conference on the Durability of Concrete Structures, ICDCS, pp. 262–269.

Walling, S.A., Provis, J.L., 2016. Magnesia-based cements: a journey of 150 years, and cements for the future? Chem. Rev. 116 (7), 4170–4204.

Wang, S.-D., Scrivener, K.L., 1995. Hydration products of alkali activated slag cement. Cem. Concr. Res. 25, 561–571.

WBCSD, 2005. Guidelines for the Selection and Use of Fuels and Raw Materials in the Cement Manufacturing Process. Retrieved 5 of January 2022 from: https://www.wbcsd.org/contentwbc/download/2429/30121.

WBCSD, 2009. Cement Industry Energy and CO_2 Performance. "Getting the Numbers Right". Retrieved 5 of January 2022 from:, ISBN: 978-3-940388-48-3. http://docs.wbcsd.org/2009/06/CementIndustryEnergyAndCO2Performance.pdf.

Williams, R.P., van Riessen, A., 2011. Development of alkali activated borosilicate inorganic polymers (AABSIP). J. Eur. Ceram. Soc. 31 (8), 1513–1516.

Young, J.F., Berger, R.L., Breese, J., 1974. Accelerated curing of compacted calcium silicate mortars on exposure to CO_2. J. Am. Ceram. Soc. 57 (9), 394–397.

Zajac, M., Skocek, J., Skibsted, J., Haha, M.B., 2021. CO_2 mineralization of demolished concrete wastes into a supplementary cementitious material—a new CCU approach for the cement industry. RILEM Tech. Lett. 6.

CHAPTER

28

Glass

Thomas D. Dyer

School of Science and Engineering, University of Dundee, Dundee, United Kingdom

28.1 INTRODUCTION

Glass is one of the oldest synthetic materials, with evidence of its production in ancient Egypt dating back to at least 3000 BC. Its optical clarity, hardness, and strength mean that it has made its way into every aspect of human activity, with significant roles in glazing, packaging, vehicles, housewares, electrical equipment, and fibers. Moreover, the material has played an instrumental role in many advances in science and technology.

Recycling of glass has been conducted almost since its discovery, and *reuse* of glass containers has been carried out throughout the 19th and 20th centuries, both as an organized endeavor and less formally—bottles and jars naturally lend themselves to reuse. However, large-scale recycling started only in the 1970s. This has partly been driven by legislation aimed at diverting waste away from landfill. Glass is a readily recycled material, in that it can be remelted and reformed into articles with the same characteristics as the original material—"closed-loop" recycling. However, to maximize recycling levels, it has been found that alternative "open-loop" routes must also play a role. This chapter examines the issues relating to the practicability and viability of closed-loop glass recycling and examines open-loop options.

28.2 TYPES OF GLASS

Each application requires a different chemical composition. The largest glass product output is soda-lime-silica container glass, which is based on a consistent composition of about 15% sodium oxide (Na_2O), 9% calcium oxide (CaO), and slightly more than 70% silicon dioxide (SiO_2). The only significant variation in composition results from the different colors that are produced. These are most commonly clear ("flint"), brown ("amber"), and green, although other colored products are manufactured in smaller quantities. Green glass is achieved through the addition of small quantities of chromium, while amber glass contains both chromium and iron. In the case of flint glass, there is a requirement for very low concentrations—in the order of hundredths of a wt%—of metals such as iron in the raw materials (BGS, 2020).

Like flint container glass, flat glass for glazing is required to be free from metallic impurities to achieve optical clarity. However, the composition is usually similar to container glass. There is typically a proportion of magnesium oxide (MgO), largely absent from container glass.

Glass used in housewares is again dependent on the application. One important group of products is thermal shock-resistant glass cookware. These articles are made from borosilicate glass, which is typically 80% silica, 13% boron trioxide (B_2O_3), 4% sodium oxide, and 2% alumina (Al_2O_3) (Edgar et al., 2008). Borosilicate glasses are also used in scientific and medical glassware.

Glass fiber can take two forms. Continuous-filament glass fiber can be used as reinforcement in composite materials as well as, in woven form, in protective clothing and electrical insulation (Edgar et al., 2008). Glass wool is finer and tends to be used in applications such as building insulation. Glass wool is usually based on soda-lime-silica formulations. However, continuous-filament glass fiber can have a range of compositions depending on the application. High stiffness fibers ("*E*-glass") are normally composed of borosilicate glass containing an addition of aluminum. A high alumina content increases strength ("S-glass"), while a lower content leads to enhanced corrosion resistance.

The type of glass used in vehicles depends on the application, although it is usually soda-lime-silica. Windshields are composed of laminated glass, whereas side and rear windows are normally tempered glass.

The composition of glass components used in electronic equipment also varies. Conventional light bulbs are made from soda-lime-silica glass, although fluorescent lighting tubes and low-energy bulbs are usually made from borosilicate glass. Cathode ray tubes (CRTs), although becoming less common, are of a similar composition, but contain metals such as barium, strontium, zirconium and, most significantly, lead (Edgar et al., 2008). CRTs are progressively being replaced by alternative screen technologies such as liquid crystal display (LCD) screens, which usually contain borosilicate glass substrates.

28.3 MANUFACTURING

Glass is manufactured by melting a suitable combination of raw materials in a furnace at temperatures between 1400°C and 1600°C, and then forming it into the required shape before it is cooled. Furnaces can be heated by direct combustion of fossil fuels or via electricity, with outputs varying from tens of tons to more than 1000 tons per day, in the case of large float glass plants.

A wide variety of raw materials are used, but for soda-lime-silica glass the most commonly encountered materials are silica sand (SiO_2), soda ash (sodium carbonate, Na_2CO_3), and calcium carbonate ($CaCO_3$). In borosilicate glasses, boron trioxide is obtained from borax ($Na_2B_4O_7{\cdot}10H_2O$). Al_2O_3 is obtained from feldspar minerals or alumina from bauxite refinement.

Many techniques are available for manufacturing glass articles. These include:

- automated blowing of glass containers in split molds;
- float glass manufacture of flat glass: the molten glass is floated over a bath of molten tin;
- rotary wool forming for the manufacture of fibers: centrifugal force is used to push molten glass through fine holes in a rotating drum;
- mechanical drawing, where molten glass is allowed to flow under the action of gravity through holes and the resulting filaments wound into strands.

28.4 RECOVERY FOR REUSE AND RECYCLING

Glass is a material that, in many of its forms, is reusable, and savings, in terms of energy and resources, can be achieved through its recovery and reuse or recycling.

One of the earliest examples of initiatives devised to recover glass were container deposit schemes. Soft drink advertising in Ireland and England provides clear evidence of manufacturers' efforts to recover containers through such means from as early as 1800.

In many parts of the world, container-deposit legislation requires consumers to pay a small deposit on containers at the point of sale, refunded when the container is returned at a redemption point. Redemption points can take several forms, including establishments selling deposited products and automated "reverse vending machines" where a user disposes of a container and is presented with a receipt that can be exchanged for cash.

Container-deposit legislation has been put into effect in the Netherlands, Norway, Denmark, Sweden, and Germany, as well as parts of the United States, Canada, and Australia. The materials need not be glass, but glass has, in most cases, been the starting point of most of these initiatives.

While the first deposit schemes were based around refill of glass articles, and refilling still continues, there has been a shift toward recycling glass back into manufacture. Whether this is a favorable development is debatable: the benefits of recycling glass are considerable, but it is unlikely that they exceed those of reuse.

Financial incentives to return glass containers are not always necessary. Bottle banks began to appear in the 1970s, and numbers have grown: the first UK bottle bank was installed in 1977, and growth in numbers has been almost exponential—there are now more than 50,000.

In some instances, sorting glass containers in terms of color presents practical problems. In the UK, this problem has been addressed by permitting collection of "comingled" glass containers. Curbside glass collection is carried out by a number of local authorities in the UK. These schemes include both color-sorted and mixed collection.

Much of the recent growth in the recovery of container glass has been driven by legislation. In Europe this has been the Packaging and Packaging Waste Directive (1994). While efforts to recycle glass have historically been focused on container glass, due to the brief life cycle of these products, it is now recognized that glass from longer service life applications must also be recycled. The European Waste Electrical and Electronic Equipment Directive (2003) and the End-of-Life Vehicles Directive (1997) both set targets for recycling of materials from these sources.

Recycling of glass from electrical and electronic equipment has made considerable progress in recent years. In the past, because glass from these products was often contaminated with scarce elements that were both hazardous and of high market value, this tended to focus efforts on the retrieval of these elements. The interior of fluorescent tubes is coated in a layer of phosphor, which itself will also have become contaminated with toxic mercury. This rendered the quality of glass unsuitable for recycling, with emphasis consequently placed on recovering valuable mercury and rare earth phosphors. Technologies allow the distillation of mercury and the removal of phosphor, with a mercury removal efficiency of up to 99.95%, leaving the glass in a condition suitable for recycling (Kadam et al., 2019). Waste glass from CRTs from televisions and monitors presents a particular challenge to recycling. While newer technologies have now replaced CRTs, substantial stockpiles of these wastes still exist. One of the major barriers to recycling, besides the cost of recycling, is the relatively high lead content of the glass. Possible outlets that have been researched with some success include remelting of CRTs to manufacture radiation shielding, with enhanced shielding efficiency achieved through the addition of barium, and use as fine aggregate in concrete. The leaching of lead was substantially reduced by being incorporated in a cement matrix, and the concrete has potential applications as radiation shielding in structures.

Tempered glass from vehicles is readily recovered for recycling. Laminated windscreens require separation of the glass from the polymer layer. This usually involves breaking up the glass using a roller or similar mechanical action, followed by removal of the glass.

The recovery of flat glass should ideally be conducted in a manner that prevents it from being mixed with other materials from the structure it derives from. The shift toward "deconstruction"—systematic dismantling—at the end of a building's life, as opposed to demolition, means that this is likely to become more feasible in future. Moreover, there have been major advances in sensor technology in recent years, meaning that automated sorting of demolition waste allows glass to be separated (Osmani et al., 2018). However, such technology is currently not widely used.

Archeological evidence of the recycling of glass has been found in sites deriving from the earlier days of the Byzantine Empire and the later Roman Empire. However, in contemporary times, recycling restarted on a major scale in the 1970s, driven partly by concerns about energy security. Levels of recycling have steadily increased since then in most of the more developed world. In many European countries, levels of recycling have reached high levels—77% by 1994 in the Netherlands—although performance was varied. The UK is a useful example, since it has moved from a relatively low glass recycling rate to a currently more respectable level. This has in part been driven by legislation, namely the Producer Responsibility Obligations (Packaging Waste) Regulations (1997), which implemented the EU Packaging and Packaging Waste Directive (1994) in the UK. The regulations placed responsibilities on the producers and handlers of packaging to achieve progressively higher recycling targets.

The Directive initially set targets to recover 50–65% of packaging waste overall by 2001, with 25–45% being used for recycling. The Regulations have undergone periodic amendments and revisions, and targets for 2021 and 2022 were set at 81% and 82%, respectively. The overall effect of these developments is shown in Figure 28.1, which plots levels of UK glass recycling. The influence of the Directive on levels of recycling after 1997 is evident.

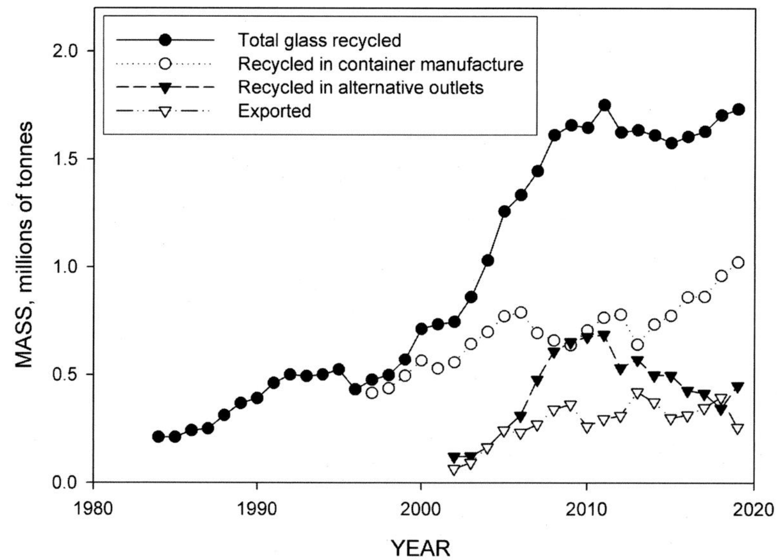

FIGURE 28.1 Recycling routes for glass in the UK (British Glass, written communication).

Not all of this recycling is closed-loop recycling. The major reason for this in the UK is the significant differences between container glass products made in the UK and those on sale to consumers. This disparity led to concerns that a green cullet surplus would arise as recycling rates increased. This did not occur, and the reason is also evident from Figure 28.1, which shows the emergence of two outlets for the surplus material: export and recycling in alternative applications—open-loop recycling. Open-loop recycling is also often necessary for the recycling of borosilicate glass products and soda-lime-silica glass with more exotic chemical additions.

The development of alternative outlets for glass has involved considerable research and development work to match the physical and chemical characteristics of the material with the needs of a wide range of different sectors.

28.5 REUSE

It is worth discussing the alternative option of reuse. Reuse of returnable glass containers is a practice that has become well established in many parts of the world. This approach is often a highly sustainable one, since the material undergoes no reprocessing prior to return into the system. A review of life-cycle assessment (LCA) studies found that reuse of glass bottles outperformed single use glass, PET, and aluminum containers yielding reductions in CO_2 emissions in the region of 20–40% (Coelho et al., 2020). An LCA study comparing environmental impacts of reuse and recycling of glass bottles found these reductions were dependent on the number of reuse cycles—two cycles yielded a climate change score of 65% of that for recycled single-use bottles, but at 10 cycles, the impact was 35% (Tua et al., 2020). These reductions level off at around 25 cycles to 30%.

The shift toward globalization has meant that glass containers for food and drink are diverse in size and shape and often originate from remote locations, making container return uneconomical and impractical in terms of logistics. Thus, in many countries, recovery of intact glass bottles has seen a decline, magnifying the need for recycling.

28.6 CLOSED-LOOP RECYCLING

28.6.1 Process

The introduction of waste glass ("cullet") back into the glassmaking furnace will cause it to melt and mix with the other raw materials. This recycling process—sometimes referred to as "remelt"—can be repeated indefinitely without any loss of performance as long as the glass is not contaminated with other material.

The quality of the glass product required defines whether glass recycling can be employed for a given manufacturing process. For instance, the inclusion of cullet in the manufacture of flat glass is not always possible, because of the strict requirements in terms of raw material purity needed to achieve the desired optical properties. However, return of cullet to container manufacture is eminently possible, and can tolerate soda-lime-silica glass from other sources.

For cullet to be recycled back into container manufacture, it is essential that color sorting of glass is conducted. This is because each color of glass can tolerate limited levels of contamination with other colors. As a result of this, quality requirements limit the level of cullet in green containers to 90%. Amber glass and flint glass are more sensitive to contamination, and so technical limits on cullet content are 70% and 60%, respectively. Color sorting can be achieved by collection schemes that require the public to sort their returned bottles. Collection of mixed-color glass from commercial sources has led to the introduction of automated "electronic eye" sorting equipment that is able to identify rogue particles and remove them through air jets, or similar.

Automated color sorting (Chapter 9—Sensor-based sorting) of glass will make up just one stage of several in the cullet processing scheme. These are required because glass from containers is likely to be present alongside other materials. These include materials used for labeling and closure, such as metals, plastics, and paper. There may also be other materials present in the recovery streams, including stone, ceramics, and other types of glass, such as borosilicate materials, which cannot be recycled in soda-lime-silica glass manufacture. Processes include:

- Hand sorting—to remove obvious contamination;
- Crushing;
- Sieving—to also remove some foreign materials such as metals and plastics;
- Magnetic sorting—to remove ferrous metals;
- Eddy current sorting—to remove nonferrous metals;
- Vacuum sorting—where overheard suction is used to pull lighter material, such as paper, from crushed glass.

28.6.2 Environmental Benefits

The closed-loop recycling of soda-lime-silica glass has several environmental benefits. First, it is returning a mineral resource back into the loop—a practice that is an essential aspect of sustainable development. While the minerals used to make glass are not notably scarce, not all these minerals are readily accessible or suitable for glass manufacture. For instance, only a small proportion of available sand deposits are suitable for glass manufacture—only six UK sites produce silica sand suitable for the manufacture of flint container glass (BGS, 2020). As a result, recycling represents a means of reducing pressure on such deposits, and avoiding potential issues of scarcity.

There are more significant environmental benefits. One of these is the reduction of waste. Glass is an inert waste, since it is insoluble and chemically unreactive under normal conditions. In the case of landfill disposal, this is of lesser concern when compared to noninert wastes: glass will not decompose to produce greenhouse gases and is unlikely to leach harmful substances that could threaten groundwater quality. Nonetheless, the disposal of glass in this way represents a wasteful practice that utilizes land that could otherwise be put to more productive use. Landfill space in many areas of the world is at a premium, and reducing disposal is considered an essential part of moving toward sustainable land use.

Disposal through combustion is also an undesirable fate for the material. Glass is clearly noncombustible, and so where combustion is carried out in a waste-to-energy plant with the aim of generating electricity and local heating, no benefit is gained from its presence. Moreover, its high specific heat capacity acts as a sink for energy that would otherwise be put to beneficial use.

The most significant benefit to the environment deriving from the return of cullet to the glassmaking process is the reduction in energy demand and the consequent reductions in pollution.

The processes for extracting and processing raw materials for glass manufacture all require energy. More significantly, melting these materials in the glass furnace requires temperatures between 1400°C and 1600°C. While the temperature requirement is not affected by the inclusion of cullet, energy savings are realized as a result of its behavior at elevated temperatures. To increase the temperature of cullet, energy is required exclusively to raise the temperature and effect melting, while in the case of soda ash, calcium carbonate, and dolomite, additional energy is required to thermally decompose them. The resulting energy saving is shown in Figure 28.2. A study examining the environmental impacts of glass recycling established from industry data that, for each 10% increase in

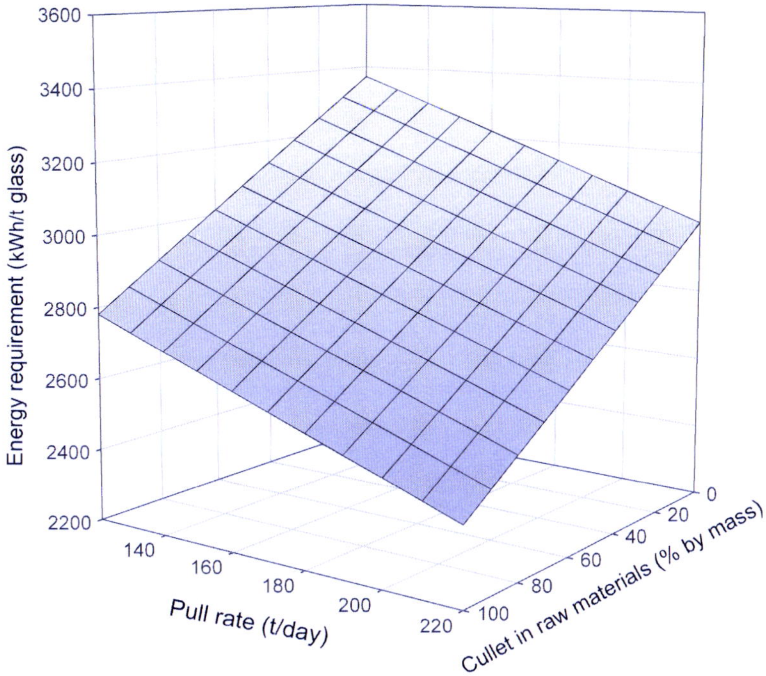

FIGURE 28.2 Diagram illustrating the reductions in energy demand realized through the incorporation of cullet in the raw material feedstock. *Modified from UNIDO (1993).*

cullet use, there was a 2.5–3% saving in energy, and that for every tonne of cullet used there was a 300 kg reduction in CO_2 emissions (Hestin et al., 2016). There are further carbon emission savings through not using conventional raw materials: many of these are carbonate minerals, which evolve CO_2 as they thermally decompose.

28.7 OPEN-LOOP RECYCLING

A number of alternative outlets for glass exist. Some of these also involve remelting glass, but involve forming into very different products. Other applications involve melting glass in the presence of other materials, with different effects depending on the manufacturing conditions. Additionally, there are a number of applications that exploit the properties of granular broken glass. Finally, the possibility of exploiting the chemistry of glass in the manufacture of commercially useful substances has also been explored. All of these outlets are examined in the following text.

While in some cases, the application is an established or growing practice, in other instances the outlet has been explored only from a research and development perspective or at a pilot scale. Nonetheless, such uses have been included on the grounds that feasibility has been demonstrated.

28.7.1 Glass Products

28.7.1.1 Microspheres (Ballotini)

One alternative glass product in which cullet can be recycled by remelting is in the manufacture of *ballotini*—small glass spheres (typically 1–60 μm in diameter) that are highly reflective and used in applications that include road markings and signs, reflective safety clothing, and projection screens. Ballotini are normally required to be colorless, and so only flint cullet can be used in their manufacture. However, novel products including composite panels comprising recycled beads embedded in a polyurethane resin sandwiched between fiberglass sheets for sound insulation have been developed, which do not have color requirements (Aygün and McCann, 2020).

Glass can also be recycled into smaller hollow microspheres, typically having diameters of tens of micrometers. They are manufactured using the flame spheroidization technique, where fine particles of glass coated in gas-forming compounds are sprayed at high speed through a high temperature flame (Dalai et al., 2014). Microspheres potentially have uses in applications including as fillers in polymers (Dalai et al., 2014), for hydrogen storage (Patankar and Kranov, 2010), and as a filler in concrete-like materials, discussed later.

28.7.1.2 Glass Wool

Glass wool insulation consisting of short glass fibers can also be manufactured from cullet. Tolerances for recycled glass are relatively high, and there is no color requirement. One manufacturer is currently using up to 80% cullet. Continuous filament glass fiber cannot currently be made using cullet, due to stricter manufacturing requirements.

Glass wool insulation comes under the broader category of mineral wools. Research has been carried out into a means of converting inorganic byproducts into mineral wools using waste glass as a melting aid and to reduce the viscosity of the melt. Such materials include incinerator bottom ash, sludge from dredging (Scarinci et al., 2000), and wastes from hydrometallurgy (Pelino, 2000).

28.7.1.3 Foamed Glass

Foamed glass is an established commercial product that can tolerate 98% cullet in its manufacture. The manufacture of foamed glass normally involves the introduction of foaming agent particles into glass powder, which is then heated to 700–900°C, where the glass becomes a viscous fluid and bubbles are formed as the foaming agent evolves gas.

Foamed glass can take the form of loose aggregate, pelletized aggregates, and blocks and shapes. Foamed glass blocks and shapes are mostly used for insulation purposes, since they have low thermal conductivities coupled with relatively high compressive strengths. Lightweight aggregates are discussed in a later section.

The foaming agents used include silicon carbide (SiC), manganese oxide (MnO_2), hematite (Fe_2O_3), calcite ($CaCO_3$), and other carbonate minerals (Assefi et al., 2021). The combination of glass with other additives has also been explored, including materials such as fly ash, other glasses (such as aluminosilicate glasses), and mineral wastes, in some cases leading to enhanced strength compared to glass-only foams of the same density (Suvorova et al., 2021).

A slightly modified approach involves the rapid mixing of powdered soda-lime-silica glass (possibly with other inorganic wastes) with an alkali solution and a surfactant. Mixing in the presence of the surfactant leads to the incorporation of air bubbles, while a reaction between the glass and the alkali solution forms calcium silicate hydrates (CSHs) that bind the mixture together. The resulting solidified foam is then sintered at temperatures of 700–1000°C (Ramteke et al., 2021).

28.7.2 Ceramics

The use of glass cullet as an ingredient in ceramic products is also possible. Two approaches can be employed: use as a sintering additive, and in the manufacture of "glass ceramics."

Cullet plays two roles when used as a sintering additive. Firstly, it becomes fluid at lower temperatures than the other constituents, acting to bind the solid particles together, reducing porosity and, therefore, increasing strength. Secondly, mineral particles dissolve into the liquid glass, reducing the required sintering temperature (Asquini et al., 2008).

Research into the use of cullet in conventional clay-based fired ceramics has had much success. Trials of the manufacture of sanitary ware containing 4.5% glass have been shown to reduce the energy requirement compared to the conventional composition by reducing firing temperatures (Hancock, 2011). Glass can be used as a replacement for feldspars in porcelain tiles (Tucci et al., 2004). The optimal level of replacement was 10% of the feldspar. A study examining the use of glass in brickmaking established that optimum performance with respect to compressive strength was achieved at a glass content of 30 wt% (Figure 28.3) (Hasan et al., 2021).

Laminated windscreen glass has been employed in brickmaking, with the polymer layer decomposing during firing to leave porosity, giving the product a higher strength at reduced density (Mörtel and Fuchs, 1997).

Clay need not be the only ceramic employed alongside glass used as a sintering additive. The use of cullet to fuse waste asbestos cement—thermally treated to render it nonhazardous—has been explored, yielding materials with enhanced strength and stain resistance, and reduced firing shrinkage (Ligabue et al., 2020). Other wastes used in this way include lithium mine tailings and copper slags (Adediran et al., 2021).

Glass ceramics are ceramic materials comprising both crystalline and amorphous phases. They are manufactured by holding glassy materials within temperature ranges at which conversion of some of the amorphous material to crystalline phases occurs. Much research has been conducted into the use of cullet in glass ceramics containing other waste products and byproducts, as outlined in Table 28.1.

28.7.3 Abrasive Media

Glass is a hard material, which makes it suitable for use as abrasive media. This normally takes the form of using crushed cullet in a conventional shot-blasting configuration. However,

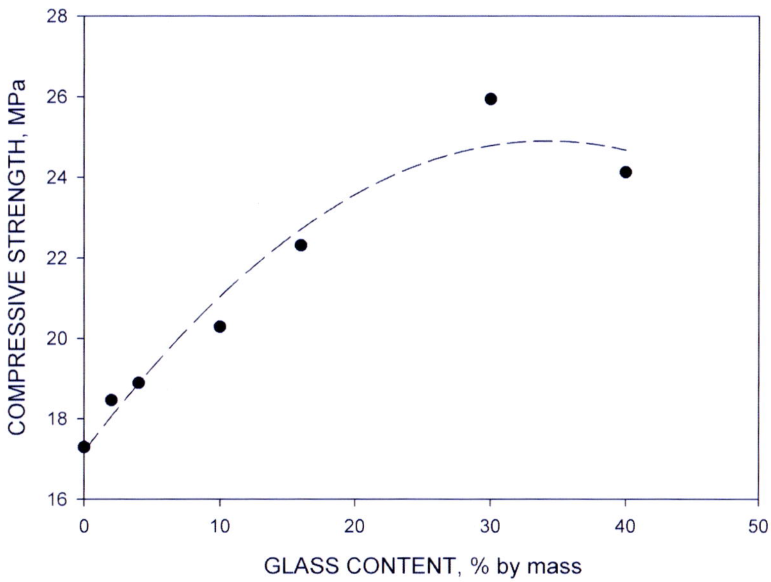

FIGURE 28.3 Influence of the inclusion of glass cullet on the compressive strength of ceramic bricks. *Modified from Hasan et al. (2021).*

TABLE 28.1 Conditions for the production of glass ceramics incorporating glass cullet.

Other inorganic material	Cullet content, wt%	Nucleation temperature and time	Crystallization temperature and time	Reference
Nucleation-crystallization				
Incinerator fly ash/ pollution control residue	35	560°C, 35 min	1000°C, 120 min	Romero et al. (1999)
Incinerator bottom ash	50, 90	1000°C, up to 8 h		Barbieri et al. (2000)
Sinter crystallization				
Dredging sludge	20	940°C, 5 h		Brusatin et al. (2005)
Sunflower husk ash	30, 40, 50	1400°C, 1300°C, 1250°C, no time specified		Quaranta et al. (2011)
Residue from plasma gasification of municipal solid waste	45 (plus 10% kaolin clay)	1000°C, 30 min		Monich et al. (2020)
Eggshell[a]	50 (plus 20% NaCl and 20% $NaHCO_3$)	800°C, 60 min		Hussain et al. (2020)
Petrurgic method				
Coal-derived bottom ash	0–60	Cooling from 1500°C to ambient at 1–10°C/min		Francis et al. (2002)

[a]*Formation of gas from eggshell and other ingredients yields a foamed material used for pollutant adsorption.*

more advanced use of recycled glass in this area has also been initiated, including the use of particles of aluminum oxide embedded in "sponge media" made from recycled glass to clean metal surfaces (Anonymous, 2005). Successful results have also been achieved with media made from molded and sintered glass powder used for vibratory mass finishing in place of polyester resin-bonded ceramic particles (Benjarungroj et al., 2012).

28.7.4 Construction Applications

28.7.4.1 Unbound Aggregate

The strength, stiffness, and hardness of glass make it highly suited for use as an aggregate, once crushed into a granular form. Some of the properties are shown in Table 28.2.

One of the higher value routes to using loose recycled glass aggregate is as granular subbase material in highway construction. Glass satisfies all the requirements for this application, namely high abrasion and impact resistance (Los Angeles test), resistance to freezing and thawing, and nonsusceptibility to "frost heave" (Henry and Morin, 1997).

Often a minimum strength is required, measured using the California bearing ratio (CBR). The CBR minimum requirement will normally be at least 15%, and values for glass (Table 28.2) come relatively close to this. The most effective solution is to combine recycled glass with crushed rock. Typically a mixture containing 15% by mass of glass yields CBR values identical to that of the rock alone (Younus Ali et al., 2011). Some researchers have

TABLE 28.2 Aggregate characteristics of recycled glass for different particle sizes.

	A			B				
	Coarse <19 mm	**Medium <9.5 mm**	**Fine <7.75 mm**	**Coarse 2.50–5.00 mm**	**Medium 1.25–2.50 mm**	**Fine 0.63–1.25 mm**	**Very fine 0.08–0.63 mm**	**Mixed, 0.08–5.00 mm**
Specific gravity	2.48	2.50	2.50	2.49				
Flakiness index, %	94.7	85.4	N/A	–	–	–	–	–
Modified proctor test values								
$\gamma_{d,max}$, kN/m^3	N/A	19.5	17.5	14.22	14.51	14.45	15.91	18.69
w_{opt}, %	N/A	8.8	10.0	–	–	–	–	–
LA abrasion value, %	24.8	25.4	27.7	–	–	–	–	–
California bearing ratio (CBR), %	–	31–32	18–21	–	–	–	–	–
Direct shear test								
σ_n (30–120 kPa)	N/A	52–53°	45–47°	–	–	–	–	–
σ_n (60–240 kPa)	N/A	50–51°	42–43°	–	–	–	–	–
σ_n (120–480 kPa)	N/A	–	40–41°	–	–	–	–	–

A: Data from Disfani et al. (2011, 2012). B: Data from Amlashi et al. (2020).

explored the possibility of utilizing recycled aggregate from demolition as the "rock" constituent in combination with glass, yielding CBR values up to 120% (Chen et al., 2021).

The direct shear strength of recycled glass is also provided in Table 28.2, indicating behavior typical of a well-graded angular sand or gravel, making the material suitable for unbound subbase and similar applications (Disfani et al., 2012).

28.7.4.2 Bituminous Highway Pavements

The use of recycled glass aggregate in bituminous mixtures for highway construction is becoming an increasingly accepted practice. Glass can be used in the base, binder course, and surface course in highway construction. Glass particles are essentially free of porosity, making them susceptible to "stripping"—deterioration of the bituminous mixture due to contact with water. However, the addition of lime rectifies this problem (Su and Chen, 2002).

Benefits of using glass include improved skid resistance, light reflectivity from surface courses, and drainage in wet weather (Su and Chen, 2002). It has also been found that the stiffness modulus and resilient modulus of asphalt increases with glass content, and mixes become less susceptible to fatigue (Tahmoorian et al., 2020). The use of glass in bituminous mixes also has the effect of reducing "rutting," or the formation of depressions in a pavement surface (George et al., 2020).

The possibility of using short glass fibers deriving from scrap fiber-reinforced polymer composites as filler aggregate has been explored

(Yang et al., 2020). The inclusion of the fibers had the effect of enhancing stiffness, resistance to rutting, and creep characteristics, but only significantly with longer fibers (10–12 mm). Resistance to stripping was also improved.

28.7.5 Concrete

The use of glass aggregate in concrete is also possible, although concerns regarding damaging alkali-silica reaction (ASR) have limited this. ASR is an expansive reaction that occurs between aggregates containing reactive silica and hydroxide ions associated with alkali-metal ions from the cement. The best means of avoiding ASR is to include additions in the cement fraction of the concrete that are capable of controlling ASR. These include fly ash, ground granulated blast furnace slag (GGBS), and metakaolin (Hansen et al., 1999; Dhir et al., 2009). Lithium admixtures are effective at controlling the reaction (Topcu et al., 2008). Formulation of concrete with an open structure (Yang et al., 2019) and the use of particles smaller than 1 mm also avoid ASR.

Benefits of using glass as aggregate include improved workability (the fluidity of fresh concrete) (Topcu and Canbaz, 2004), reduced thermal conductivity, increased heat capacity (Alani et al., 2012), and improved fire resistance (Chen et al., 2020a). Glass aggregate has also been used in blocks containing titanium dioxide, and the combination appears to enhance the ability of the units to break down pollutants such as NO_x by photocatalysis (Torres de Rosso and Staub de Melo, 2020). This approach has been taken a step further by "decorating" the surface of glass aggregate particles with nanoparticles of titanium dioxide (Chen et al., 2020b).

28.7.5.1 Light-Weight Aggregate

The foaming of molten glass cullet discussed previously can be employed to produce light-weight aggregate. Such aggregates are now well-established products and have applications in concrete construction as well as in horticulture. Light-weight aggregate can take two forms—granulated aggregate and pelletized aggregate. Granulated aggregate is one possible product of the manufacture of continuous foamed glass sheets, which are broken up into suitable particle sizes.

Pelletization involves forming granules comprising powdered cullet and a foaming agent. These granules are then passed through an inclined rotary furnace during which melting and gas evolution occur. It is common for other materials to be included. Research into such products has included fly ash from coal-fired power generation (Kourti and Cheeseman, 2010) and metallurgical slags (Lim et al., 2020).

Rather than using foamed material, hollow recycled glass microspheres have been used to produce ultra-lightweight concrete, which displays low densities ($<1400\,kg/m^3$) with adequate compressive strength (around 30 MPa) (Wang et al., 2020).

28.7.5.2 Cement Component in Concrete

Soda-lime-silica glass is pozzolanic: in powdered form, it will undergo a reaction with calcium hydroxide to form CSH gel (Dyer and Dhir, 2001), thus allowing it to be used as a partial replacement for Portland cement.

Investigation of borosilicate glass (from lighting tubes) as a filler in concrete has identified the material as being particularly pozzolanic, exceeding the performance of silica fume, which is generally a highly reactive material (Shakhmenko et al., 2012).

There is the possibility of obtaining even more benefit from using powdered glass as a cement by combining with other materials to create "ternary blends." Such combinations have a synergistic effect, with strengths exceeding those of equivalent binary blends often being measured. Blends displaying this synergy have been formulated with fly ash (Saravanakumar et al., 2018), blast furnace slag (Liu et al., 2020), and bagasse ash from sugar production (Joy et al., 2017). More recently, geopolymers have successfully been produced using waste glass in combination with both calcareous fly ash (Xiao et al., 2020), blast furnace slag (Zhang et al., 2020), combinations of siliceous fly ash and blast furnace slag (Huseien et al., 2020), and metakaolin (Si et al., 2020), as well as on its own (Idir et al., 2020).

28.7.6 Filtration Media

The ability of volumes of granular glass to permit rapid percolation of fluid has meant that the material has found application in filtration. Studies have found that it is potentially extremely effective at removing suspended solids from water.

A study examining the effect of particle size on the use of glass filter beds on the tertiary treatment of wastewater identified that a particle size of 0.5–1.45 mm was most appropriate (Horan and Lowe, 2007).

Other researchers have found that the smooth surface of glass limits the extent to which solid particles become attached. This can be rectified through the use of coagulants (Soyer et al., 2010), or the use of dual-media beds containing glass and coarser anthracite (Soyer et al., 2013). Similarly, glass in combination with larger clay balls in a pebble matrix filter has been shown to be suitable for pretreatment of drinking water (Rajapakse and Fenner, 2011).

Another means of potentially enhancing the efficiency of glass as a filtration medium is to use it in the manufacture of foamed particles (essentially granulated lightweight aggregate) whose higher surface area and tortuous porosity make capture of particles more efficient. This approach has been found to be effective in capturing organic pollutants in water (Shishkin et al., 2021).

28.7.7 Zeolites and Other Silicate Products

It has already been seen that solubilization of silicate ions from glass occurs at high pH. This is further enhanced at elevated temperatures. For this reason, research has examined the possibility of producing chemical products through hydrothermal reaction of glasses under high pH conditions. A range of products can be produced, including double-layer silicates and zeolites, which have applications as absorbents, ion exchange media, and desiccants. The conditions used for the synthesis of four such substances from cullet are shown in Table 28.3.

Similar conditions have also been used to produce articles with relatively high tensile strength (Veloza et al., 1999). The process involved using hydrothermal hot pressing to shape mixtures of glass and water at temperatures of about 200°C.

Waste glass can also be used in enhancing the solubility of phosphate (Tumbure et al., 2020).

TABLE 28.3 Conditions used to produce various products from glass cullet using hydrothermal processes.

Product	Temperature	Other reactants	References
Rhodesite, zeolite Na-P1	150°C	NaOH	Grutzeck and Marks (1999)
Tobermorite	100°C	NaOH, CaO	Coleman (2011)
Zeolite A	Extraction at 105°C or 140°C, subsequent reaction temperature with $Na_2O_2Al_2O_3$ not specified, but probably 20–175°C	Extraction of silica using NaOH followed by reaction with $Na_2O \cdot Al_2O_3$	Mavilia and Corigliano (2001)
MCM-41	100°C	NH_4OH and CTAB surfactant	Salam et al. (2020)

This is achieved by sintering apatite-bearing rocks with waste glass at 900°C, causing some of the phosphate ions in the apatite to be substituted by silicate ions in the glass and vice versa. The overall effect is a higher availability of phosphate when the resulting material is used as a fertilizer.

Another high-temperature synthesis application includes the carbothermic reaction of waste glass and its subsequent nitridation to silicon nitride (Topateş, 2020). This involves heating glass in the presence of carbon and nitrogen at a temperature of 1500°C. It has advantages over using conventional SiO_2 sources, in that the lower melting point of glass leads to savings in energy.

28.8 CONCLUSION AND OUTLOOK

The closed-loop recycling of glass into glassmaking is clearly a sustainable practice. The energy savings and resulting reduction in CO_2 emissions mean that it is likely to continue to grow. However, barriers exist to the complete return of all recovered glass to glassmaking, meaning that alternative open-loop outlets for cullet have become established. It can be seen from Figure 18.1 that the extent to which open-loop recycling of glass is conducted is highly variable, reflecting the need to clear surplus cullet, rather than any realization of "better" recycling routes for that material. Thus it is unlikely that the scale of open-loop recycling will expand significantly. However, the diversification of options for open-loop recycling is important, since this is likely to make the process of finding a purpose for surpluses easier, thus smoothing out the market instabilities that unexpected surges in cullet availability might otherwise cause.

The issue of economic viability of alternatives is one that has not been explored in this chapter. In some instances, researchers have included economic analyses. However, a critical influence on viability is the market price for cullet, and this varies considerably around the world and can fluctuate considerably over relatively short periods of time. For this reason, the economics of open-loop recycling of glass are probably best explored on an individual case basis.

The environmental benefits of glass recycling do not stop at energy and CO_2. The world is currently facing a shortage of sand (UNEP, 2019), and while the higher purity silica sand required for glassmaking only makes up a small proportion of the total, glass recycling plays a role in slowing our use of this resource.

It should not be assumed that the future of glass recycling is a straightforward one. For instance, new and growing technologies that employ glass may present future problems to glass recycling. The last few years have seen a rapid expansion of glazing glass products with special functions, including modifying the manner in which light is transmitted through glass, as well as PV glass that acts as means of photovoltaic electricity generation. These products achieve these effects through surface coatings, the handling of which may present future difficulties to glass recyclers, and future challenges to those developing glass recycling technologies.

References

Adediran, A., Lemougna, P.N., Yliniemi, Y., Tanskanen, P., Kinnunen, P., Roning, J., Illikainen, M., 2021. Recycling glass wool as a fluxing agent in the production of clay- and waste-based ceramics. J. Clean. Prod. 289, 125673.

Alani, A., MacMullen, J., Telik, O., Zhang, Z.Y., 2012. Investigation into the thermal performance of recycled glass screed for construction purposes. Constr. Build. Mater. 29, 527–532.

Amlashi, S.M.H., Carter, A., Vaillancourt, M., Bilodeau, J.-P., 2020. Physical and hydraulic properties of recycled glass as granular materials for pavement structure. Can. J. Civ. Eng. 47, 865–874.

Anonymous, 2005. Shipyard takes softer, greener approach to surface prep. J. Prot. Coat. Linings 22, 12–15.

Asquini, L., Furlani, E., Bruckner, S., Maschio, S., 2008. Production and characterization of sintered ceramics from paper mill sludge and glass cullet. Chemosphere 71, 83–89.

Assefi, M., Maroufi, S., Mansuri, I., Sahajwalla, V., 2021. High strength glass foams recycled from LCD waste screens for insulation application. J. Clean. Prod. 280, 124311.

Aygün, H., McCann, F., 2020. Structural and acoustical performance of recycled glass bead panels. Constr. Build. Mater. 258, 119581.

Barbieri, L., Corradi, A., Lancellotti, I., 2000. Bulk and sintered glass-ceramics by recycling municipal incinerator bottom ash. J. Eur. Ceram. Soc. 20, 1637–1643.

Benjarungroj, P., Harrison, P., Vaughan, S., Ren, X.J., Morgan, M., 2012. Investigation of thermally treated recycled glass as a vibratory mass finishing media. Key Eng. Mater. 496, 104–109.

BGS, 2020. Mineral Planning Factsheet: Silica Sand. British Geological Survey. Keyworth, Nottingham. https://www2.bgs.ac.uk/mineralsuk/download/planning_factsheets/mpf_silica_sand.pdf. (Last Accessed 28 January 2022).

Brusatin, G., Bernardo, E., Andreola, F., Barbieri, L., Lancellotti, I., Hreglich, S., 2005. Reutilization of waste inert glass from the disposal of polluted dredging spoils by the obtainment of ceramic products for tiles applications. J. Mater. Sci. 40, 5259–5264.

Chen, B., Zhu, H., Li, B., Sham, M., Li, Z., 2020a. Study on the fire resistance performance of cementitious composites containing recycled glass cullets (RGCs). Constr. Build. Mater. 242, 117992.

Chen, X.-F., Kou, S.-K., Poon, C.S., 2020b. Rheological behaviour, mechanical performance, and NOx removal of photocatalytic mortar with combined clay brick sands-based and recycled glass-based nano-TiO_2 composite photocatalysts. Constr. Build. Mater. 240, 117698.

Chen, Z., Poon, C.S., Li, J.-S., Xue, Q., 2021. Utilization of glass cullet to enhance the performance of recycled aggregate unbound sub-base. J. Clean. Prod. 288, 125083.

Coelho, P.M., Corona, B., Worrell, E., 2020. Reusable vs Single-Use Packaging—A Review of Environmental Impacts. Zero Waste Europe, Brussels.

Coleman, N.J., 2011. 11 Å tobermorite ion exchanger from recycled container glass. Int. J. Environ. Waste Manag. 8, 366–382.

Dalai, S., Vijayalakshmi, S., Shrivastava, P., Sivam, S.P., Sharma, P., 2014. Preparation and characterization of hollow glass microspheres (HGMs) for hydrogen storage using urea as a blowing agent. Microelectron. Eng. 126, 65–70.

Dhir, R.K., Dyer, T.D., Tang, M.C., 2009. Alkali-silica reaction in concrete containing glass. Mater. Struct. 42, 1451–1462.

Disfani, M.M., Arulrajah, A., Bo, M.W., Hankour, R., 2011. Recycled crushed glass in road work applications. Waste Manag. 31, 2341–2351.

Disfani, M.M., Arulrajah, A., Bo, M.W., Sivakugan, N., 2012. Environmental risks of using recycled crushed glass in road work applications. J. Clean. Prod. 20, 170–179.

Dyer, T.D., Dhir, R.K., 2001. Chemical reactions of glass cullet used as a cement component. J. Mater. Civ. Eng. 13, 412–417.

Edgar, R., Holcroft, C., Pudner, M., Hardcastle, G., 2008. UK Glass Manufacture 2008—A Mass Balance Study. British Glass.

Francis, A.A., Rawlings, R.D., Boccaccini, A.R., 2002. Glass-ceramics from mixtures of coal ash and soda lime glass by the petrurgic method. J. Mater. Sci. Lett. 21, 975–980.

George, T.B., Anochie-Boateng, J.K., Jenkins, K.J., 2020. Laboratory performance and modelling behaviour of hot-mix asphalt with recycled crushed glass. J. South Afr. Inst. Civil Eng. 62, 10–22.

Grutzeck, M.W., Marks, J.A., 1999. Synthesis of double-layer silicates from recycled glass cullet: a new type of chemical adsorbent. Environ. Sci. Technol. 33, 312–317.

Hancock, P., 2011. Fluxing Issues—Container Glass in Sanitaryware Applications. Asian Ceramics, pp. 50–53.

Hansen, M.R., Wang, M., Collins, T., Nielsen, M.L., 1999. Recycled glass concrete for pavement patching—alkali-silica reactivity. In: Bank, L.C. (Ed.), 5th ASCE Materials in Engineering. ASCE, pp. 847–851.

Hasan, M.R., Siddika, A., Akanda, M.P.A., Islam, M.R., 2021. Effects of waste glass addition on the physical and mechanical properties of brick. Innov. Infrastruct. Solut. 6, 36. https://link.springer.com/article/10.1007/s41062-020-00401-z.

Henry, K.S., Morin, S.H., 1997. Frost susceptibility of crushed glass used as construction aggregate. J. Cold Reg. Eng. 11, 326–333.

Horan, N.J., Lowe, M., 2007. Full-scale trials of recycled glass as tertiary filter medium for wastewater treatment. Water Res. 41, 253–259.

Hestin, M., de Veron, S., Burgos, S., 2016. Economic Study on Recycling of Building Glass in Europe. Deloitte. https://www.glassforeurope.com/wp-content/uploads/2018/04/Economic-study-on-recycling-of-building-glass-in-Europe-Deloitte.pdf.

Huseien, G.F., Hamzah, H.K., Sam, A.R.M., Khalid, N.H.A., Shah, K.W., Deogrescu, D.P., Mirza, J., 2020. Alkali-activated mortars blended with glass bottle waste nano powder: environmental benefit and sustainability. J. Clean. Prod. 243, 118636.

Hussain, Z., Zada, A., Hussain, K., Naz, M.Y., AbdEl-Salam, N.M., Ibrahim, K.A., 2020. Preparation of activated porous glass adsorbent through thermochemical reforming of ampoules and eggshells for remediation of direct blue dye pollution. Asia-Pac. J. Chem. Eng. e2610.

Idir, R., Cyr, M., Pavoine, A., 2020. Investigations on the durability of alkali-activated recycled glass. Constr. Build. Mater. 236, 117477.

Joy, C.N., Ramakrishnan, K., Snega, M., Ramasundram, S., Venkatasubramanian, C., Muthu, D., 2017. Study on strength and durability characteristics of concrete with ternary blend. IOP Conf. Ser. Earth Environ. Sci. 80, 012022.

Kadam, A.R., Nair, G.B., Dhoble, S.J., 2019. Insights into the extraction of mercury from fluorescent lamps: a review. J. Environ. Chem. Eng. 7, 103279.

Kourti, I., Cheeseman, C.R., 2010. Properties and microstructure of lightweight aggregate produced from lignite coal fly ash and recycled glass. Resour. Conserv. Recycl. 54, 769–775.

Ligabue, M.L., Gualtieri, A.F., Gualtieri, M.L., Malferrari, D., Lusvardi, G., 2020. Recycling of thermally treated cement-asbestos for the production of porcelain stoneware slabs. J. Clean. Prod. 247, 119084.

Lim, Y.C., Shih, Y.-J., Tsai, K.-C., Yang, W.-D., Chen, C.-W., Dong, C.-D., 2020. Recycling dredged harbor sediment to construction materials by sintering with steel slag and waste glass: characteristics, alkali-silica reactivity and metals stability. J. Environ. Manag. 270, 110869.

Liu, T., Wei, H., Zou, D., Zhou, A., Jian, H., 2020. Utilization of waste cathode ray tube funnel glass for ultra-high performance concrete. J. Clean. Prod. 249, 119333.

Mavilia, L., Corigliano, F., 2001. High added value products from off-quality waste glass. In: Dhir, R.K., Limbachiya, M.C., Dyer, T.D. (Eds.), Recycling and Reuse of Glass Cullet., University of Dundee, pp. 63–73.

Monich, P.R., Romero, A.R., Desideri, D., Bernardo, E., 2020. Waste-derived glass-ceramics fired in nitrogen: stabilization and functionalization. Constr. Build. Mater. 232, 117265.

Mörtel, H., Fuchs, F., 1997. Recycling of windshield glasses in fired bricks industry. Key Eng. Mater. 136 (Part 3), 2268–2271.

Osmani M.; Villoria-Sáez P.; Vitale P. (2018). Recovery technologies for construction and demolition waste (MINEA deliverable), COST Action Mining the European Anthroposphere. https://doi.org/10.5281/zenodo.3760465 Accessed 28 January 2022.

Patankar, S.N., Kranov, Y.A., 2010. Hollow glass microsphere HDPE composites for low energy sustainability. Mater. Sci. Eng. A 527, 1361–1366.

Pelino, M., 2000. Recycling of zinc-hydrometallurgy wastes in glass and glass ceramic materials. Waste Manag. 20, 561–568.

Quaranta, N., Unsen, M., López, H., Giansiracusa, C., Roether, J.A., Boccaccini, A.R., 2011. Ash from sunflower husk as raw material for ceramic products. Ceram. Int. 37, 377–385.

Rajapakse, J.P., Fenner, R.A., 2011. Evaluation of alternative media for pebble matrix filtration using clay balls and recycled crushed glass. J. Environ. Eng. 137, 517–524.

Ramteke, D.D., Hujova, M., Kraxner, J., Galusek, D., Romero, A.R., Falcone, R., Bernardo, E., 2021. Up-cycling of 'unrecyclable' glasses in glass-based foams by weak alkali-activation, gel casting and low-temperature sintering. J. Clean. Prod. 278, 123985.

Romero, M., Rawlings, R.D., Rinćon, J.M., 1999. Development of a new glass-ceramic by means of controlled vitrification and crystallisation of inorganic wastes from urban incineration. J. Eur. Ceram. Soc. 19, 2049–2058.

Salam, M.A., AbuKhadra, M.R., Mohamed, A.S., 2020. Effective oxidation of methyl parathion pesticide in water over recycled glass based-MCM-41 decorated by green Co_3O_4 nanoparticles. Environ. Pollut. 259, 113874.

Saravanakumar, P., Prabhakar, S.M., Nishanth, D., Sujith, K. S., 2018. Strength characteristics of ternary blended recycled aggregate concrete. Int. J. Civ. Eng. Technol. 9, 1083–1092.

Scarinci, G., Brusatin, G., Barbieri, L., Corradi, A., Lancellotti, I., Colombo, P., Hreglich, S., Dall'Igna, R., 2000. Vitrification of industrial and natural wastes with production of glass fibres. J. Eur. Ceram. Soc. 20, 2485–2490.

Shakhmenko, G., Korjakins, A., Kara, P., Bumanis, G., 2012. Low-carbon concrete using local industrial by-products. In: Jones, M.R., Newlands, M.D., Halliday, J.E., Csetenyi, L.J., Zheng, L., McCarthy, M.J., Dyer, T.D. (Eds.), Concrete in the Low Carbon Era. University of Dundee, pp. 978–988.

Shishkin, A., Aguedal, H., Goel, G., Peculevica, J., Newport, D., Ozolins, J., 2021. Influence of waste glass in the foaming process of open cell porous ceramic as filtration media for industrial wastewater. J. Clean. Prod. 282, 124546.

Si, R., Dai, Q., Guo, S., Wang, J., 2020. Mechanical property, nanopore structure and drying shrinkage of metakaolin-based geopolymer with waste glass powder. J. Clean. Prod. 242, 118502.

Soyer, E., Akgiray, Ö., Eldem, N.Ö., Saatçi, A.M., 2010. Crushed recycled glass as a filter medium and comparison with silica sand. Clean Soil Air Water 38, 927–935.

Soyer, E., Akgiray, Ö., Eldem, N.Ö., Saatçi, A.M., 2013. On the use of crushed recycled glass instead of silica sand in dual-media filters. Clean Soil Air Water 41, 325–332.

Su, N., Chen, J.S., 2002. Engineering properties of asphalt concrete made with recycled glass. Resour. Conserv. Recycl. 35, 259–274.

Suvorova, O.V., Manakova, N.K., Makarov, D.V., 2021. Use of bulk industrial wastes in the production of glass foam materials. Glas. Ceram. 77, 384–389.

Tahmoorian, F., Yeaman, J., Mirzababaei, M., 2020. Comparisons of the resilient moduli of asphalt mixes containing recycled materials through empirical and experimental methods. J. Mater. Civ. Eng. 32, 04020255.

Topateş, G., 2020. Recycling of waste flat glass in the fabrication of Si_3N_4 powder by the carbo-thermal reduction and nitridation. Int. J. Appl. Ceram. Technol. 17, 555–562.

Topcu, I.B., Canbaz, M., 2004. Properties of concrete containing waste glass. Cem. Concr. Res. 34, 267–274.

Topcu, I.B., Boga, A.R., Bilir, T., 2008. Alkali-silica reactions of mortars produced by using waste glass as fine aggregate and admixtures such as fly ash and Li_2CO_3. Waste Manag. 28, 878–884.

Torres de Rosso, L., Staub de Melo, J.V., 2020. Impact of incorporating recycled glass on the photocatalytic capacity of paving concrete blocks. Constr. Build. Mater. 259, 119778.

Tua, C., Grosso, M., Rigamonti, L., 2020. Reusing glass bottles in Italy: a life cycle assessment evaluation. Procedia CIRP 90, 192–197.

Tucci, A., Esposito, L., Palmonari, C., Rambaldi, E., 2004. Use of soda-lime scrap-glass as a fluxing agent in porcelain stoneware tile mix. J. Eur. Ceram. Soc. 24, 83–92.

Tumbure, A., Bishop, P., Hedley, M.J., Bretherton, M.R., 2020. Increasing phosphorus solubility by sintering igneous Dorowa phosphate rock with recycled glass. J. Therm. Anal. Calorim., 10078-2.

UNEP, 2019. Sand and Sustainability: Finding New Solutions for Environmental Governance of Global Sand Resources. GRID-Geneva, United Nations Environment Programme, Geneva, Switzerland. https://www.unep.org/news-and-stories/press-release/rising-demand-sand-calls-resource-governance. https://wedocs.unep.org/bitstream/handle/20.500.11822/28163/SandSust.pdf?sequence=1&isAllowed=y.

UNIDO, 1993. Output of Seminar on Energy Conservation in Glass Energy. United Nations Industrial Development Organization, Ministry of Internal Trade and Industry. https://open.unido.org/api/documents/4805861/download/HANDY%20MANUAL.%20GLASS%20INDUSTRY.%20OUTPUT%20OF%20A%20SEMINAR%20ON%20ENERGY%20CONSERVATION%20IN%20GLASS%20INDUSTRY%20%2820829.en%29. (Accessed 28 January 2022).

Veloza, Z.M., Yanagizawa, K., Yamasaki, N., 1999. Recycling waste glass by means of the hydrothermal hot pressing method. J. Mater. Sci. Lett. 18, 1811–1813.

Wang, L., Aslani, F., Hajirasouliha, I., Roquino, E., 2020. Ultra-lightweight engineered cementitious composite using waste recycled hollow glass microspheres. J. Clean. Prod. 249, 119331.

Xiao, R., Ma, Y., Jiang, X., Zhang, M., Zhang, Y., Wang, Y., Huang, B., He, Q., 2020. Strength, microstructure, efflorescence behavior and environmental impacts of waste glass geopolymers cured at ambient temperature. J. Clean. Prod. 252, 119610.

Yang, S., Poon, C.S., Ling, T.C., 2019. Distribution of ASR gel in conventional wet-mix glass mortars and mechanically produced dry-mix glass blocks. Constr. Build. Mater. 229, 116916.

Yang, Q., Hong, B., Lin, J., Wang, D., Zhong, J., Oeser, M., 2020. Study on the reinforcement effect and the underlying mechanisms of a bitumen reinforced with recycled glass fiber chips. J. Clean. Prod. 251, 119768.

Younus Ali, M.M., Newman, G., Arulrajah, A., Disfani, M. M., 2011. Application of recycled glass—crushed rock blends in road pavements. Aust. Geomech. 46, 113–121.

Zhang, B., He, P., Poon, C.S., 2020. Optimizing the use of recycled glass materials in alkali activated cement (AAC) based mortars. J. Clean. Prod. 255, 120228.

CHAPTER

29

Lumber

Ambrose Dodoo[1], *Leif Gustavsson*[2], *and Roger Sathre*[3]

[1]Department of Building Technology, Linnaeus University, Växjö, Sweden [2]LG, Silogatan 1, Lund, Sweden [3]Sustainable Built Environment Group, Department of Built Environment and Energy Technology, Linnaeus University, Växjö, Sweden

29.1 INTRODUCTION

Wood from sustainably managed forests can play important roles both as material and as fuel as we transition to a sustainable society. Wood is widely used as an energy source and as a material in diverse applications including furniture and joinery, pulp and paper, packaging, and construction. Postuse wood products may be recovered for cascaded use, including reuse, recycling, and energy recovery, and this can contribute to building a circular bioeconomy. Efficient management of postuse wood products offers significant opportunity to improve resource efficiency and thereby reduce greenhouse gas (GHG) emissions in the built environment (IPCC, 2007).

In contrast to materials such as steel and concrete, which are manufactured through technological processes in factories, wood is produced through natural biological processes that occur in growing trees. The life cycle of wood products begins with forest management activities, e.g., seedling cultivation, tree planting, and forest thinning. This is followed by harvesting and processing of logs into lumber, and the manufacture, use, and end-of-life management of the finished wood products. In addition to the principal flows of round wood and primary wood products, considerable coproducts are generated, such as residues from silviculture, harvesting, primary processing when logs are sawn into lumber, and from secondary processing to make products such as doors and windows (Gustavsson and Sathre, 2011).

By dry weight, wood has an elemental composition of about 50% carbon, 44% oxygen, 6% hydrogen, and trace amounts of several minerals (Pettersen, 1984). These elements return to the environment when a wood product is burned or decayed at the end of its service life. Carbon, oxygen, and hydrogen generally return in the form of CO_2 and H_2O. The elements thus become bioavailable for other trees to use in their growth, leading to continual material cycling.

The use of wood as material or for energy has feedback mechanisms that affect total energy use and GHG emissions. Relatively little energy is needed for the manufacture and processing of wood-based materials compared to nonwood alternatives such as concrete and steel (IPCC, 2007; Gustavsson and Sathre, 2006; Perez-Garcia et al., 2005). Typically, wood-based products use mainly biomass residues for processing energy, and have lower climate impacts than

 https://doi.org/10.1016/B978-0-323-85514-3.00030-0

nonwood alternatives (Gustavsson et al., 2006a). Postuse wood products contain significant amounts of energy stored in chemical bonds that can be recovered and used to substitute for fossil fuels, avoiding fossil emissions (Gustavsson et al., 2006a). Currently, woody biomass provides 9% of the global total primary energy, which is more than the share from all other renewable energy sources and nuclear energy (Food and Agricultural Organization of the United Nations, 2010; International Energy Agency, 2009; International Energy Agency, 2020). Bioenergy accounted for 70% of the global renewable energy in 2017 (World Bioenergy Association, 2019), and short-term projection suggests this trend will continue, with bioenergy constituting 30% of the growth in the global renewable energy supply between 2018 and 2023 (International Energy Agency, 2018). In the European Union, bioenergy is suggested to have an important role in the strategy to achieve a climate-neutral economy by 2050 (European Commission, 2019). Energy recovery from postuse wood will be an increasingly important component of these renewable energy sources.

29.2 WOOD MATERIAL USES

Apparent trends in global production of roundwood, sawnwood, and wood-based panels are summarized in Table 29.1. This is based on data from the Food and Agriculture Organization of the United Nations (FAO) (Food and Agricultural Organization of the United Nations, 2020a) and shows increase in production of these wood materials within the last two decades, with minor drops in roundwood and sawnwood production in 2019 relative to 2018. Roundwood encompasses industrial roundwood as well as fuelwood, while wood-based panels comprise mainly particle board, oriented strand board, fiberboard, and plywood. Industrial roundwood, comprising fence posts, piling, and telephone and electricity poles, accounted for 51% of the produced roundwood in 2019 and fuelwood accounted for the remaining 49%. Sawnwood is processed into value-added wood products for construction works and furniture manufacture and includes beams, boards, boxboards, joists, laths, planks, rafters, scantlings, and sleepers.

TABLE 29.1 Trends in Global Production of Key Wood Materials, Based on FAO's (Food and Agricultural Organization of the United Nations, 2020a) Data.

Material	2019 amount (million m^3)	Change (%) in 2019 compared to:	
		2018	**2000**
Roundwood	3966	−1	+14
Sawnwood	488	−1	+27
Wood-based panels:	357	0	+100
Plywood	*107*	*0*	*+100*
Particle board, OSB and fibreboard	*250*	*0*	*+100*

The level of wood use for building and construction applications differs across countries and regions (Food and Agricultural Organization of the United Nations, 2016; Gustavsson et al., 2006b). For example, while wood is commonly used for construction of both single-family as well as multifamily buildings in North America, it is more common for single-family buildings and less common for multistory apartment building constructions in Nordic countries. In recent years, interest and end markets for wood-based construction products is rising, due mainly to the growing emphasis on sustainability and greener building technologies. For example, window frames made of wood are becoming increasingly preferred and currently constitute 20% of the market share in

Europe, and up to 90% of the market in Nordic countries (United Nations Publications, 2019). The market for innovative wood construction products such as cross-laminated timber (CLT) is growing (Muszynski et al., 2017), especially in Europe where about 70% of the global CLT was produced in 2017 (United Nations Publications, 2019). This product has made it possible to construct tall multistory wood frame structures and wood bridges with long and wide spans.

29.3 POSTUSE WOOD RECOVERY FOR RECYCLING

Postuse wood products and residues produced along the wood value chain are important resources in many countries. In the EU-28, 56.2 Mt of wood waste was generated in 2018, with 3% and 97% being hazardous and nonhazardous waste, respectively (Eurostat, 2020). Postuse wood may be recovered from construction and demolition sites, municipal and industrial waste, furniture and joinery manufacture, packaging, and pallets. Other sources of recovered wood include postuse railroad ties and utility poles, which are often treated with chemical preservative. Falk and McKeever (2004) observed that up to 90% of solid postuse wood may be recovered from deconstruction of wood-frame buildings.

FAO data suggests that a total of 31.6 million tons of postconsumer wood was recovered globally in 2019 (Food and Agricultural Organization of the United Nations, 2020b). This encompassed waste wood from building construction and demolition, engineering works, households, and pallets. The recovered postconsumer wood in 2019 represents an increase of 2.3% and 3.4% over 2017 and 2018, respectively (Food and Agricultural Organization of the United Nations, 2020b). Figure 29.1 shows the recovered postconsumer wood for selected countries in 2019 (Food and Agricultural Organization of the United Nations, 2020b). This is dominated by Germany, followed by France and the United Kingdom. These three countries

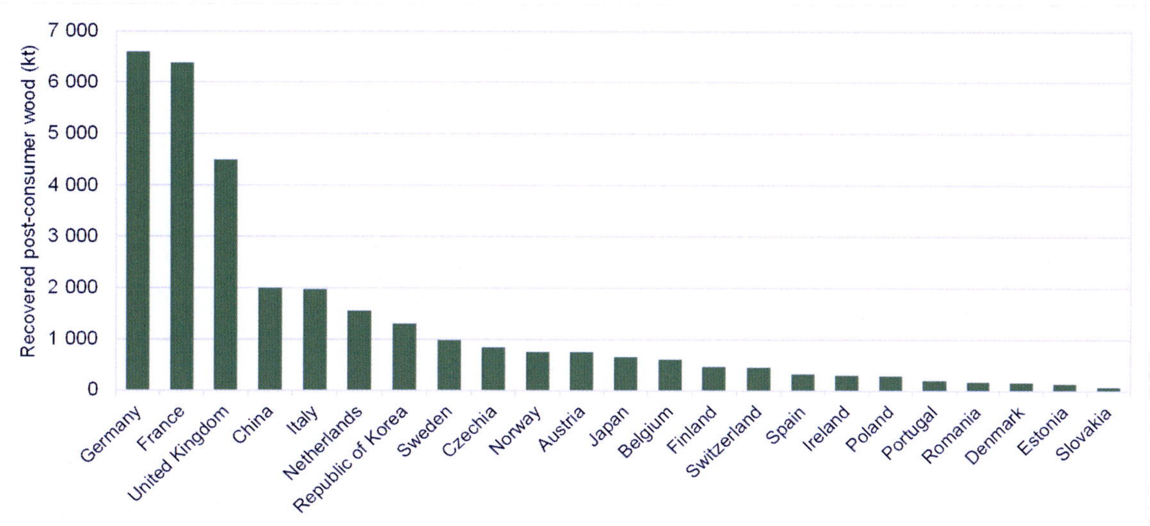

FIGURE 29.1 Recovered wood tonnage in select countries in 2019. *Based on Food and Agricultural Organization of the United Nations (2020b).*

TABLE 29.2 Summary of wood recycling grades in the United Kingdom (PAS (Publicly Available Specification) 111, 2012).

Grade	Wood waste source	Material condition/composition
Grade A: Recycled Wood	Solid wood, wood off-cuts packaging waste, scrap pallets, packing cases, and cable drums	Nails/metal fixings, minor amounts of paint and surface coatings
Grade B: Industrial Feedstock	Up to 60% Grade A material as above, plus building and demolition materials and domestic wood furniture	Nails/metal fixings, some paints, grit, plastics, glass, coatings, binders, and glues. Limited treatment or coating
Grade C: Fuel	All above, and fencing products, flat pack furniture, and high content of panel products such as MDF, chipboard	Nails/metal fixings, paint coatings, glues, paper, plastics, rubber, glass, grit, and coated and treated timber
Grade D: Hazardous Waste	Fencing, transmission poles, railway sleepers, and cooling towers	Chromated copper arsenate (CCA) and creosote treated timber

together accounted for more than half of the reported postconsumer wood recovered. About 87% (27.5 Mt) of the total volume of the reported postconsumer wood came from Europe. Mantau et al. (2010) estimated the amount of recovered postconsumer wood in the EU to be between 58.6 and 67.3 million m^3 by 2030.

In some areas, deposition in landfills is the most common fate for waste and postuse wood (Salazar and Meil, 2009). This, however, is prohibited in the EU and in some states in the United States (Defra, 2012). According to the US Environmental Protection Agency (United States Environmental Protection Agency, 2020a, b), the United States generated 18.1 and 40.8 million tons of wood in municipal solid waste and in construction and demolition waste, respectively, in 2018. About 67% and 73% of the generated wood in municipal solid waste and in construction and demolition waste, respectively, were landfilled (United States Environmental Protection Agency, 2020a, b). About 3.10 Mt (17.1%) and 2.84 Mt (15.7%) of the generated wood in municipal solid waste in the United States were recycled and combusted for energy recovery, respectively, in 2018 (United States Environmental Protection Agency, 2020b).

Typically, postuse wooden materials are transported to material sorting or recycling sites where they are sorted according to size and quality, screened for contaminants, cleaned, and designated for different end-use markets. Sorted clean and large wooden materials are typically used in higher value-added applications, while small wood may be used for low-value purposes (Clean Washington Center, 1997). Waste categorization systems for wood recycling are available in different countries. Table 29.2 summarizes an informative grading system for wood recycling in the United Kingdom (PAS (Publicly Available Specification) 111, 2012). A similar categorization system for Germany is documented by Garcia and Hora (Garcia and Hora, 2017).

29.4 POSTUSE WOOD RECYCLING

Figure 29.2 shows system-wide integrated pathways for efficient material management across the life cycle of a wood product. Recovered

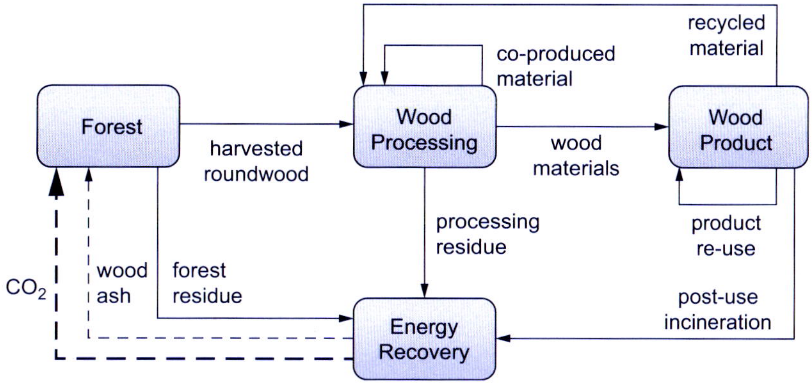

FIGURE 29.2 Schematic diagram of system-wide integrated material flows of wood products.

postuse wood may be reused, recycled into different material applications, recovered as energy, or landfilled with or without the capture of landfill gas (LFG). Reuse of end-of-life products involves the further use of a recovered product in a similar application without reprocessing, while recycling entails reprocessing it to produce a new type of product. For example, large wood building frames may be reused in similar structural applications or be remilled (and recycled) into wood flooring. Landfilling has typically been the default baseline from which recycling benefits may be measured. However, in many European countries the default practice may now be to burn untreated wood in conventional energy plants and treated wood in specific incineration plants. Options for direct combustion of wood for energy purposes include burning in dedicated biomass plants, cofiring in coal-fired power stations, and burning in waste incinerators with municipal solid waste.

29.4.1 Recycling Into Products

The share of recovered wood used as raw material for further reprocessing into products or as bioenergy varies significantly among EU countries. For example, 90% of recovered wood in Sweden is used as bioenergy, while 70% of recovered wood in France is used as raw material for further wood processing (Mantau et al., 2010). Most of the recovered wood waste in Germany, United Kingdom, Italy, and Finland is reported to be used as bioenergy (Garcia and Hora, 2017). In the EU, 9.1 and 9.7 Mt of recovered wood were used as bioenergy and as raw material in 2007, respectively (Mantau et al., 2010). In 2015, recovered postconsumer wood and forest biomass residues contributed at least 49% (222 Mm^3) of the total wood used for energy in the EU (Camia et al., 2021).

Various recycled products can be made from postuse wood, depending on the grade and classification of the waste wood (PAS (Publicly Available Specification) 111, 2012; Garcia and Hora, 2017). Some products that may be made from recovered postuse wood include particleboard, oriented strand board, medium density fiberboard, and animal bedding and mulches. Examples of technologies for recycling wood waste into particleboard or medium density fiberboard are described in Lindvall et al. (2006), Merrild and Christensen (2009), and Kim and Song (2014). Sandberg (2016) and Defra (2006) describe the technical details and various stages of particleboard manufacture. In general, the recycling process entails crushing sorted wood waste into smaller pieces using disk or drum woodchippers. The resulting chips are dried, screened to remove contaminant fractions, and then sorted into different size categories. The contaminant fractions encompass nonwood material present in the chipped wood, including metals, nails, and screws. Resin is mixed with the wood chips, which are then compressed to make panelboards. Other chemical substances may be added to the mixture of the chips and resin depending on the desired performance requirements.

Different published analyses of the energy and environmental implications of particleboard manufacture from recovered wood are available in the literature. Merrild and Christensen (2009) analyzed the energy and global-warming impacts of recycling wood into particleboard or producing particleboard from virgin wood. They found that recycling postuse wood into particleboard results in significant energy and GHG savings compared to the particleboard production from virgin wood, primarily because of the avoided energy for drying virgin wood. Rivela et al. (2006) analyzed the environmental impacts and trade-offs associated with the use of recovered wood for particleboard production or for bioenergy. They found that the use of the recovered wood for particleboard may be a more environmentally favorable option in the studied context. Kim and Song (2014) also found recycling of wood waste into particleboard to give about a factor of 3 climate benefit, compared to using the wood waste for bioenergy, to replace liquefied natural gas.

29.4.2 Wood Cascading

Wood products are well suited for material cascading, which has been suggested as a strategy to increase the efficiency of resource use (Haberl and Geissler, 2000). Cascading is the sequential use of a resource for different purposes as the resource quality degrades over time. The cascade concept includes four dimensions of resource economy: resource quality, utilization time, salvageability, and consumption rate (Sirkin and ten Houten, 1994). In terms of these four characteristics, optimal utilization of wood resources is achieved by matching the resource quality to the task being performed, so as not to use a high-grade resource when a lower-grade one will suffice; increasing the total utility gained from a resource through prolonging the time during which it is used for various purposes; upgrading a resource through salvaging and reprocessing, where appropriate, for additional higher-grade uses; and balancing the usage rate of a resource with the capacity of forest land to regenerate lost resource quality. Effective cascading use of postuse wood materials may further improve resource efficiency (Thonemann and Schumann, 2018), and may also prolong the sequestration of biogenic carbon in wood-based materials. In some cases, particularly when forest resources are limited, it will be beneficial to employ a more complex cascade chain involving multiple material uses before final burning (Sathre and Gustavsson, 2006). The advantages of such wood cascading depend strongly on the relative quantities of wood products entering service and new wood biomass produced by forests. When the use of forest products is limited by the available forest resources, cascading is beneficial by allowing greater usage from the limited forest resources. If the potential production of wood products from primary forest biomass would be larger than the demand of wood products made and used, the benefits of material cascading are questionable. On a global level, the amount of nonwood materials used is much greater than the amount of wood materials used; thus expanding international trade of wood products can provide additional opportunities for cascading. At least two conditions can be imagined in which postuse wood cascading, besides energy recovery, could be beneficial:

(1) total use of woody biomass increases significantly and the primary harvest is limited and

(2) designation of more forest land as protected reserves to increase biodiversity benefits, together with a limited primary harvest.

In the future, if more material and energy services are provided by biomass and fewer are provided by fossil resources, wood cascading will be more important by allowing more intensive use of limited global biomass resources.

Reprocessing of recovered wood in a cascade chain may require altered levels of specific energy use for material processing. There are differences in processing energy required because of the different physical properties of virgin and recovered wood, mainly the lower moisture content of recovered wood. Slightly more energy is needed to saw or chip the dry recovered wood, which is harder than green wood. Substantially less energy is required to kiln-dry the recovered wood than the green wood during production of lumber or particleboard. Drying has the largest single demand for energy in the manufacture of lumber and particleboard made from green wood (Food and Agricultural Organization of the United Nations, 1990). Moisture content, through its effect on heating value, is also important in the comparison of biofuels, e.g., dry recovered wood versus green, freshly harvested biofuel.

29.4.3 Nutrient Cycling

Wood has very small quantities of mineral elements such as Ca, Mg, K, and P, although tree leaves and needles typically have higher concentrations of these elements. To avoid loss of these nutrients from forest ecosystems over the long term, ashes from combusted biomass can be applied to growing forests to ensure that nutrient cycles are closed (Stupak et al., 2007). In the absence of ash recycling, the continued export of nutrients contained in the biomass may lead to nutrient deficiency and reduced forest production. In Sweden, for example, the Swedish Forest Agency has published recommendations regarding the appropriate manner in which ash recycling should be done (Swedish Forest Agency, 2019). The dosage of ash application is calculated in such a way as to balance the removal of nutrients in wood, bark, and foliage with the return of nutrients in ash. Quality standards are specified for ashes, including minimum content of Ca, Mg, K, and P. To avoid the long-term build-up of heavy metals and other contaminants that can be concentrated in the ash, maximum content of trace elements including several heavy metals is also specified. Before wood ashes are applied to the forest, they must be stabilized to slow their dissolution and avoid damage to sensitive flora and fauna. Stabilization can take place both chemically and physically, with the goal that the ashes dissolve slowly over a period of 5–25 years in the field. Ash processing can be done in centralized facilities or can be done with mobile equipment at the locations where the ash is produced. Ashes can be spread in the forest using ground equipment such as converted tractors, or by helicopter.

29.4.4 Preservative Treated Wood Recycling

To prevent the deterioration of wood products while still in service, some wood is treated with chemical wood preservatives that protect the wood from decay organisms. Opportunities for recycling of preservative treated wood are more limited than for untreated wood (Felton and De Groot, 1996). Felton and De Groot (1996) reported a case where pentachlorophenol-treated wooden poles are reprocessed for reuse by a utility company in Alberta. The reprocessing involved shaving sapwood off the preservative-treated wood, followed by regrading and treating of the poles for reuse. Particular concerns in recycling of preservative treated wood include worker exposure to emissions from recycling processes, and interference by preservatives with the bonding of adhesives. Energy recovery from treated wood is possible in specific incineration plants with suitable combustion conditions, flue gas cleaning, and appropriate ash disposal (Townsend et al., 2008).

Two main categories of chemical treatments for wood exist: oil-borne preservatives such as creosote and pentachlorophenol, and waterborne preservatives such as copper- and boron-based solutions (Lebow, 2010). Regulations in many countries define the allowable uses of different types of preservatives, which differ between, e.g., residential and industrial applications. The landscape of chemical wood preservatives has changed significantly in recent decades toward safer materials, and continues to evolve. The use of arsenic in wood preservative solutions, such as the once common chromated copper arsenate (CCA), has been phased out, particularly in residential applications. In the European Union, the Biocidal Products Directive (European Commission, 1998) covers many common wood preservatives including CCA and creosote, leading to increasing restrictions on their use. Nevertheless, significant quantities of chemically treated wood are currently in service and will require postuse management in the coming years.

Several studies have explored the complexities of managing postuse wood that is treated with chemical preservatives. Bolin and Smith (2011a, b) explored the environmental implications of landfilling or energy recovery of preservative treated wood in the US context. The impacts analyzed include energy use, GHG emissions, acidification, eutrophication, smog, and ecological toxicity. The authors found that energy recovery of the preservative-treated wood results in lower impacts for all categories except for eutrophication and water use. They concluded that appropriate combustion of preservative-treated wood for energy recovery should be permitted. In another US study, Jambeck et al. (2007) compared the environmental and economic trade-offs associated with scenarios where treated wood waste is landfilled or combusted for electricity production in a waste-to-energy facility. The ash from the wood combustion was assumed to be landfilled. The economic analysis considered the cost of waste collection, transport, treatment, and disposal, and the revenue generated from the sale of electricity for the combustion for energy scenario. They found the total cost to be nearly twice for the waste-to-energy scenario compared to the landfilling scenario. Nevertheless, more progressive policy instruments and stringent landfill taxes in many parts of the European Union discourage landfilling of postuse materials.

29.4.5 Energy Recovery of Wood

End-of-life management is the single most significant variable affecting the full life cycle energy and carbon profiles of wood products (Gustavsson and Sathre, 2009). Further material usage of recovered wood, such as reusing or reprocessing, can improve the life cycle environmental performance of the material. Furthermore, recovery of energy by burning or gasifying the wood is a resource-efficient final option for postuse wood. Landfilling of wood products carries more uncertainties, and LCA studies have diverged widely in their outcomes depending on decomposition processes and management activities. The life cycle energy balance of the material is improved when recovered wood is used for bioenergy. The carbon balance is mostly affected by avoided fossil carbon emissions if the bioenergy replaces fossil fuels. Due to its dual role as both material and energy resource, wood has relatively more opportunity to improve its life cycle environmental performance.

A meta-analysis of wood product substitution for climate mitigation found that proper end-of-life management resulted in emission reductions averaging 0.7 kg CO_2 per dry kg of wood products (Leskinen et al., 2018). These benefits are primarily due to energy recovery from postuse wood materials used instead of fossil fuels. Benefits are highest when recovered wood is used to substitute for carbon-intensive coal, reaching emission reductions of up to 1.8 kg CO_2 per kg of wood products. The meta-analysis also found that landfilling of end-of-life wood products generally reduced the substitution benefits, due to both the formation of methane in landfills and the reduction in fossil fuel substitution by recovered woody biomass. Studies of landfilled wood products showed high variability, due to the uncertainties regarding biophysical landfill processes and the diversity of assumptions used in the studies, leading to contradictory conclusions of landfill effectiveness. In contrast, studies found that energy recovery from postuse wood provided reliable climate benefits relative to fossil fuel use. These findings are consistent with an earlier meta-analysis (Sathre and O'Connor, 2010), which reported postuse management of wood products to be the single most significant source of variability in the GHG impacts of the wood product life cycle. Studies that analyzed responsible management of end-of-life wood products, in particular using them as bioenergy to substitute for fossil fuels, consistently reported reduced GHG emissions over the product life cycle. Several other studies found climate impact (not benefit) from postuse wood products, resulting from worst-case scenarios of high methane emission from landfilled wood.

Other life cycle studies (e.g., Morris, 2017; Dodoo et al., 2019) have further illuminated the implications of different end-of-life management options for wood products. In the US context, Salazar and Meil (2009) and Carpenter et al. (2013) showed that diverting postuse wood materials from landfill for combustion with energy recovery significantly improved the energy and carbon balances of wood-based building systems. Petersen and Solberg (2002) found that considerably greater GHG and energy benefits are achieved when wood is burned for energy to replace fossil fuels in the Norwegian context. In contrast, they noted that landfilling of wood gives significant atmospheric GHG emissions due to the gradual anaerobic decomposition of the wood, releasing methane. In Swedish studies, Dodoo et al. (2009) and Tettey et al. (2019) showed that replacing fossil fuel with recovered wooden material gives the greatest carbon benefit in the postuse phase of buildings.

Sathre and Gustavsson (2006) analyzed the effects of different postuse management options on the energy and carbon balances of wood lumber. The postuse options included reuse as lumber, reprocessing as particleboard, pulping to form paper products, and burning for energy recovery. They found that several mechanisms affect the energy and carbon balances of recovery wood: direct effects due to different properties and logistics of virgin and recovered materials, substitution effects due to the reduced demand for nonwood materials when wood is reused, and land use effects due to alternative possible land uses when less timber harvest is needed because of wood recovery.

29.5 CASE STUDY SCENARIOS

Here, four scenarios are explored and analyzed to quantify the primary energy use and GHG implications of different end-of-life management options for recovered wood from a building's end-of-life. The case study is based on a four-story wood-frame building with 16 apartments and a total heated floor area of 1190 m^2, located in Växjö, Sweden. Further details of the characteristics of the building are reported in Persson (1998) and in Dodoo et al. (2012). The mass (tons of air-dry) of major materials contained in the building include 59 t of lumber, 18 t of particleboard, 21 t of plywood, 223 t of concrete, 16 t of steel, and 89 t of plasterboard. The impact of production of wood and

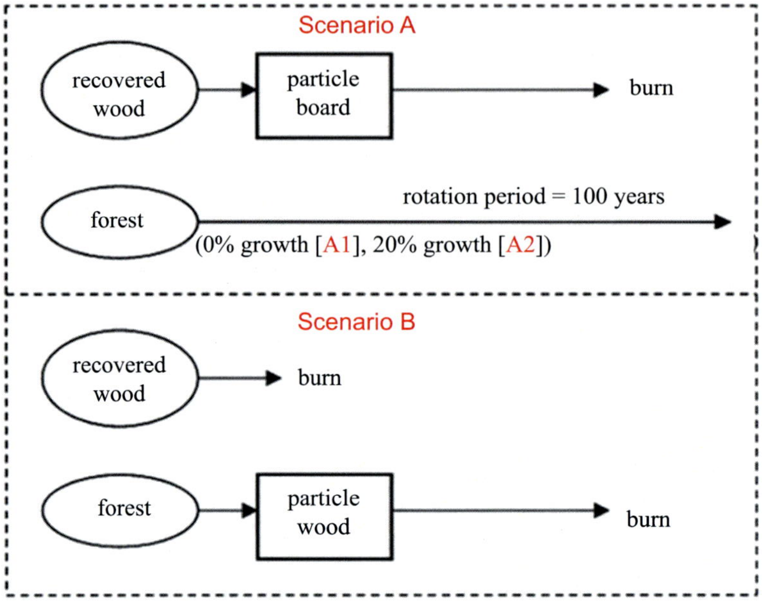

FIGURE 29.3 Comparison of two alternatives for producing particleboard.

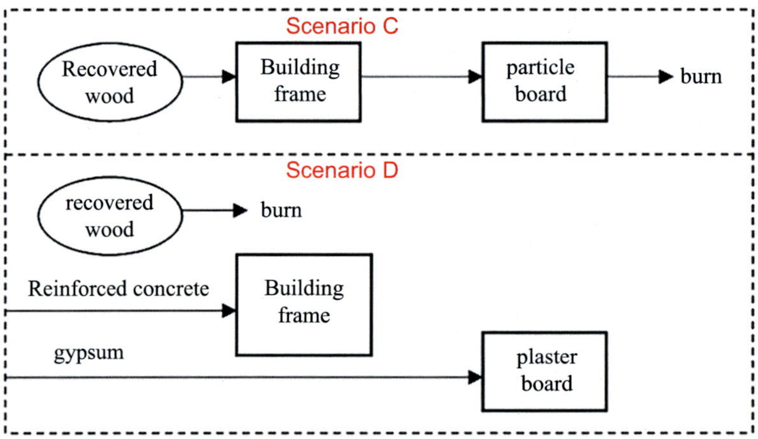

FIGURE 29.4 Comparison of two alternatives for providing building materials.

the impact of replacing wood with other materials is investigated. The four scenarios are (Figures 29.3 and 29.4):

- Scenario A: *Recovered lumber* for particleboard manufacture while forest remains unharvested.
- Scenario B: Virgin wood for particleboard manufacture while *recovered lumber* is recovered of energy (via direct combustion).
- Scenario C: *Recovered lumber* for building materials (frame material initially and thereafter reprocessed into particleboard) and finally recovered of energy.

- Scenario D: Nonwood materials (reinforced concrete and plasterboard) for building materials while *recovered lumber* is recovered of energy.

29.5.1 Primary Energy Use and GHG Balance Analyses

A method developed by Gustavsson et al. (2006a) is used to calculate the primary energy and GHG balances of the scenarios studied. The primary energy used to extract, process, transport, and assemble the materials is calculated, and the lower heating values of the logging and processing residues and of the recovered demolition wood used for energy. The GHG emissions from fossil fuel combustion and cement process reactions are calculated, as well as the potential emissions avoided by replacing fossil fuel with recovered wood, and the carbon stock changes in materials and forests. The recovery and use of LFG from landfilled wood products is considered in scenarios involving landfilling. The LFG emission is estimated based on the default IPCC methodology (IPCC, 2006). Recoverable forest residues at harvest are based on data from Lehtonen et al. (2004) and Sathre and Gustavsson (2006). We assume 70% recovery of available harvest residues (branches and tops) at final forest felling, and 100% of processing and construction residues. The recovery percentage of postuse wood varies with scenario. The assumed lower heating values for the recovered biomass are 4.25 kWh/kg dry biomass for bark and harvest residues, 4.62 kWh/kg dry biomass for processing residues, and 5.17 kWh/kg dry biomass for recovered postuse wood (Gustavsson and Sathre, 2006). The amount of diesel fuel used for biomass recovery, expressed in terms of the heating value of the biomass, is assumed to be 1% for processing residues and postuse wood and 5% for harvest residues (Gustavsson and Sathre, 2006). Specific final energy for building material production is based on Swedish conditions (Sathre and Gustavsson, 2006; Björklund and Tillman, 1997). Electricity is assumed to be produced in coal-fired condensing plants with a conversion efficiency of 40% and high-voltage distribution loss of 2%. Conversion efficiency is assumed to be unchanged when recovered wood replaces coal.

All biogenic carbon flows should be considered linked to the construction of the buildings, starting with the impact on forest and its carbon stock and ending with the end-of-life of wood materials. Such analyses are highly complex to perform and require different type of modeling concepts with a large amount of data that may partly be highly uncertain (Eriksson et al., 2007; Werner et al., 2010; Gustavsson et al., 2017; Xu et al., 2018). However, researchers have developed integrated analysis techniques to clarify connections and potential trade-offs between different forestry strategies, where both biogenic carbon storage in forests and the outcomes of using wood to substitute carbon intensive products in technological systems within society are included. For example, improvements in data availability and modeling sophistication have enabled robust and comprehensive analyses of the climate mitigation effectiveness of different large-scale forestry strategies in Sweden (Gustavsson et al., 2017) and Canada (Xu et al., 2018). In this case study, we use a simplified method assuming that the biogenic carbon storage sequestered or released from wood materials is not included in the analysis, as the wood is to come from sustainably managed forestry, and that the forest activities will not reduce the forest carbon stock.

29.5.2 Particleboard Production From Recovered Lumber or virgin wood

Here we discuss the primary energy and GHG implications of producing particleboard from either recovered lumber or virgin wood.

Figure 29.3 summarizes the scenarios. The analysis considers particleboard in a 100-year life cycle perspective and assumes that the particleboard is combusted for energy at the end of its service life. In *Scenario A*, 90% of the lumber in the case-study building is assumed to be recovered and used as raw material for the particleboard production. In *Scenario B*, particleboard production is from virgin wood, and biomass residues are obtained from forest management and processing activities for the virgin wood. When recovered lumber is reprocessed as particleboard (*Scenario A*), the forest is assumed to remain unharvested and continue to grow by either 0% or 20% over the 100-year life cycle. When the forest is harvested to produce particleboard (*Scenario B*), the recovered lumber is combusted, substituting for coal. An estimated 9% more electricity and 60% less thermal energy are needed for particleboard production with recovered lumber compared to the particleboard made from virgin wood (Sathre and Gustavsson, 2006).

The primary energy balances of all the scenarios are negative, meaning that more energy is available for external use than is used during the product life cycle (Table 29.3). The GHG balances are also negative for all the scenarios. The primary energy and GHG balances are lowest when virgin wood is used for particleboard and the recovered lumber is burned. The difference in process energy between making particleboard from recovered lumber and virgin wood is small in relation to the total energy flow in the production systems.

29.5.3 Cascade Chain for Recovered Wood Products

Here we discuss the primary energy and GHG implications of cascading use of recovered lumber. The recovered lumber is used for building-frame construction and then recycled as particleboard before energy recovery (*Scenario C*). This is compared to a scenario where nonwood alternatives including reinforced concrete frame material and gypsum panelboard are used (*Scenario D*). Here we assume that the forest resources are limited, so the building must be constructed either with recovered lumber or alternate nonwood materials. When nonwood materials are used for construction, the recovered lumber is used in place of coal. A schematic diagram of these scenarios is shown in Figure 29.4.

The primary energy and GHG balances are lowest when recovered lumber is used for the building frame (Table 29.3). This is due mainly to the fossil fuel used to produce concrete and steel, compared to the reuse of the recovered lumber, which is assumed to require no additional processing energy. Significant amounts of biomass residues are recovered at the end of the life cycle of wood products, in contrast to the nonwood alternative materials. The GHG balance is substantially higher for the reinforced concrete materials, owing to the calcination emission of CO_2 during cement production as well as greater fossil fuel use for manufacture of the nonwood materials.

29.6 FUTURE DEVELOPMENTS

Postuse wood products could be managed as part of an integrated flow of material and energy within and between the forestry, manufacturing, construction, energy, and waste management sectors (Figure 29.2). This integration, which valorizes the postuse materials, can bring energetic, economic, and environmental advantages (Gustavsson and Sathre, 2009). Recovery and recycling of wood from demolished buildings is becoming increasingly common. The percentage of end-of-life wood materials that is recoverable is variable and depends on the practical limitations linked to the building design and whether material recovery is facilitated through deconstruction. A high recovery

TABLE 29.3 Primary energy and greenhouse gas emission implications of end-of-life management scenarios for recovered lumber.

	Scenario A1		Scenario A2		Scenario B		Scenario C		Scenario D	
Description	**Primary energy (MWh)**	**Green-house gas (tCO_2 eq.)**	**Primary energy (MWh)**	**Green-house gas (tCO_2 eq.)**	**Primary energy (MWh)**	**Green-house gas (tCO_2 eq.)**	**Primary energy (MWh)**	**Green-house gas (tCO_2 eq.)**	**Primary energy (MWh)**	**Green-house gas (tCO_2 eq.)**
Material recovery & production	60	22	60	22	74	26	58	17	171	68
Heating value of residues	−215		−215		−491		−233		−233	
Substitution for coal by residues		−85		−85		−195		−92		−92
Product carbon stock changes		0		0		0		0		0
Forest carbon stock changes		0		−32		0		0		0
Total	**−155**	**−63**	**−155**	**−95**	**−417**	**−169**	**−175**	**−75**	**−62**	**−24**

In Scenario A, recovered lumber is used as raw material for particleboard, allowing the forest to remain unharvested and continue to grow by 0% (Scenario A1) or by 20% (Scenario A2); in Scenario B, the recovered lumber is burned, and the forest is harvested to produce particleboard; in Scenario C, recovered lumber is used for building materials; and in Scenario D, nonwood products (reinforced concrete and plasterboard) are used for building materials and the recovered lumber is burned.

percentage of demolition wood could be achieved in future as the value of wood as an energy source is more widely recognized, and as more buildings are designed and constructed in ways that facilitate deconstruction to allow greater recycling and reuse of building materials (Kibert, 2003). This may involve the "design for disassembly" of buildings to facilitate the removal of wood products with minimal damage, to maintain their potential for further reuse as a material. Such optimization of end-of-life product recovery and recycling systems may become increasingly important, to gain additional value from the wood as a material before it is ultimately burned to recover its feedstock energy. For example, existing CLT buildings are yet to reach their mid-service life. A future development that might be expected is increased volumes of postuse CLT that may have to be recycled. The size of the CLT panels and the capacity of the current wood crushing equipment might present challenges for recycling such postuse wood into particleboard, for example. Effective recycling logistics for such postuse wood may need to be explored.

29.7 CONCLUDING REMARKS

In this chapter we have explored the implications of end-of-life management options for wood products, and have described several mechanisms through which postuse management of recovered wood materials can affect the life cycle resource efficiency and climate performance of wood products. This analysis shows how efficient management of postuse wood products can contribute to a sustainable, resource-efficient, low-carbon society. Recovering energy from postuse wood material gives significant primary energy and GHG benefits. These benefits of postuse wood materials may be further optimized when postuse wood is reused and recycled for use in a sequence of applications and afterward used to recover the heat content. The benefits of additional wood material use in complex cascade chains, however, depend largely on the additionally available primary forest resources. But additional primary forest resources are limited on a global level and many times on the local level. In such cases complex cascade chains give high benefits.

Life cycle and system perspectives of wood products are needed, so that all the life cycle phases—acquisition of raw material, manufacture, use, and end-of-life—are considered and optimized as a whole. The energy needed to produce the wood products should be considered from natural resources to final services. Primary energy and GHG balances are important metrics when analyzing the resource efficiency and climate mitigation of postuse wood management options. Primary energy use largely determines natural resource efficiency and steers the environmental impacts of material production. More so than with other common materials, appropriate postuse management of wood products is important, because in addition to its use as a physical material, wood can also be used as a sustainable bioenergy source. The increased use of wood products from sustainably managed forests can play an important role in our transition to a low-carbon economy, and postuse management is a critical phase of the wood product life cycle that should be thoughtfully optimized.

References

Björklund, T., Tillman, A.-M., 1997. LCA of Building Frame Structures: Environmental Impact over the Life Cycle of Wooden and Concrete Frames Technical Environmental Planning Report 2. Chalmers University of Technology, Gothenburg, Sweden.

Bolin, C., Smith, S., 2011a. Life cycle assessment of ACQ-treated lumber with comparison to wood plastic composite decking. J. Clean. Prod. 19, 620–629.

Bolin, C.A., Smith, S.T., 2011b. Life cycle assessment of borate-treated lumber with comparison to galvanized steel framing. J. Clean. Prod. 19 (6–7), 630–639.

Camia, A., Giuntoli, J., Jonsson, R., Robert, N., Cazzaniga, N. E., Jasinevičius, G., Avitabile, V., Grassi, G., Barredo, J.I.,

Mubareka, S., 2021. The Use of Woody Biomass for Energy Purposes in the EU. EUR 30548 EN, Publications Office of the European Union, Luxembourg, ISBN: 978-92-76-27867-2, https://doi.org/10.2760/831621. JRC122719.

Carpenter, A., et al., 2013. Life cycle assessment of end-of-life management options for construction and demolition debris. J. Ind. Ecol. 17 (3), 396–406.

Clean Washington Center, 1997. Wood Waste Recovery: Size Reduction Technology Study. Final Report. Accessed at: https://p2infohouse.org/ref/04/03787.htm. 31st January 2021.

Defra, 2006. Secretary of State's Consultation for the A2 Particleboard, Oriented Strand Board and Dry Process Fibreboard Sector. Sector Guidance Note IPPC SGI, Integrated Pollution Prevention and Control (IPPC).

Defra, 2012. Wood Waste Landfill Restrictions in England Call for Evidence. Accessed at: http://www.defra.gov.uk/consult/files/consult-wood-waste-document-20120808.pdf. 31st January 2021.

Dodoo, A., Gustavsson, L., Sathre, R., 2009. Carbon implications of end-of-life management of building materials. Resour. Conserv. Recycl. 53 (5), 276–286.

Dodoo, A., Gustavsson, L., Sathre, R., 2012. Effect of thermal mass on life cycle primary energy balances of a concrete- and a wood-frame building. Appl. Energy 92, 462–472.

Dodoo, A., Gustavsson, L., Tettey, U.Y.A., 2019. Effects of end-of-life management options for materials on primary energy and greenhouse gas balances of building systems. Energy Procedia 158, 4246–4253.

Eriksson, E., et al., 2007. Integrated carbon analysis of forest management practices and wood substitution. Can. J. For. Res. 37 (3), 671–681.

European Commission, 1998. Directive 98/8/EC of the European Parliament and of the Council of 16 February 1998 Concerning the Placing of Biocidal Products on the Market. Accessed at: https://eur-lex.europa.eu/. 31st January 2021.

European Commission, 2019. Facts and Figures on Bioenergy in the EU. Accessed at: https://ec.europa.eu/jrc/en/science-update/facts-and-figures-bioenergy-eu. 1st February 2021.

Eurostat, 2020. Generation of Waste by Waste Category, Hazardousness and NACE Rev. 2 Activity. Accessed at: https://appsso.eurostat.ec.europa.eu/nui/submitViewTableAction.do. 15th January 2021.

Falk, R.H., McKeever, D.B., 2004. Recovering wood for reuse and recycling: a United States perspective. In: European COST E31 Conference: Management of Recovered Wood Recycling Bioenergy and other Options: proceedings, 22–24 April 2004, Thessaloniki. 2004. University Studio Press, Thessaloniki, pp. 29–40.

Felton, C.C., De Groot, R.C., 1996. The recycling potential of preservative-treated wood. For. Prod. J. 46 (7–8), 37–46.

Food and Agricultural Organization of the United Nations, 1990. Energy Conservation in the Mechanical Forest Industries. FAO Forestry Paper 93, Rome,.

Food and Agricultural Organization of the United Nations, 2010. What Wood Fuels Can Do to Mitigate Climate Change. Accessed at: http://www.fao.org/docrep/013/i1756e/i1756e00.htm. 31st January 2021.

Food and Agricultural Organization of the United Nations, 2016. Forestry for a Low-Carbon Future: Integrating Forests and Wood Products in Climate Change Strategies. FAO Forestry Paper 177. Accessed at: http://www.fao.org/3/i5857e/i5857e.pdf. 31st January 2021.

Food and Agricultural Organization of the United Nations, 2020a. Facts and Figures: Global Production and Trade in Forest Products in 2019. Accessed at: http://www.fao.org/forestry/statistics/80938/en/. 20th January 2021.

Food and Agricultural Organization of the United Nations, 2020b. Forestry Production and Trade: Recovered Post-consumer Wood Production Quantity in World. Accessed at: http://www.fao.org/faostat/en/#data/FO/visualize. 14th January 2021.

Garcia, C.A., Hora, G., 2017. State-of-the-art of waste wood supply chain in Germany and selected European countries. Waste Manag. 70, 189–197.

Gustavsson, L., Sathre, R., 2006. Variability in energy and carbon dioxide balances of wood and concrete building materials. Build. Environ. 41 (7), 940–951.

Gustavsson, L., Sathre, R., 2009. A State-of-the-Art Review of Energy and Climate Effects of Wood Product Substitution. School of Technology and Design, Växjö University.

Gustavsson, L., Sathre, R., 2011. Energy and CO_2 analysis of wood substitution in construction. Clim. Chang. 105 (1–2), 129–153.

Gustavsson, L., Pingoud, K., Sathre, R., 2006a. Carbon dioxide balance of wood substitution: comparing concrete- and wood-framed buildings. Mitig. Adapt. Strateg. Glob. Chang. 11 (3), 667–691.

Gustavsson, L., et al., 2006b. The role of wood material for greenhouse gas mitigation. Mitig. Adapt. Strateg. Glob. Chang. 11 (5), 1097–1127.

Gustavsson, L., et al., 2017. Climate change effects of forestry and substitution of carbon-intensive materials and fossil fuels. Renew. Sustain. Energy Rev. 67, 612–624.

Haberl, H., Geissler, S., 2000. Cascade utilization of biomass: strategies for a more efficient use of a scarce resource. Ecol. Eng. 16, 111–121.

International Energy Agency, 2009. Bioenergy—A Sustainable and Reliable Energy Source. IEA, Paris, France.

International Energy Agency, 2018. Market report series: Renewables 2018. Available at: https://webstore.iea.org/market-report-series-renewables-2018.

International Energy Agency, 2020. Key World Energy Statistics 2020. Accessed at: https://webstore.iea.org/download/direct/4093?fileName=Key_World_Energy_Statistics_2020.pdf. 19th January 2021.

IPCC, 2006. Guidelines for National Greenhouse Gas Inventories. Accessed at: http://www.ipcc-nggip.iges.or.jp. 31st January 2021.

IPCC, 2007. Intergovernmental Panel on Climate Change. IPCC Secretariat Geneva.

Jambeck, J., et al., 2007. CCA-treated wood disposed in landfills and life-cycle trade-offs with waste-to-energy and MSW landfill disposal. Waste Manag. 27 (8), S21–S28.

Kibert, C.J., 2003. Deconstruction: the start of a sustainable materials strategy for the built environment. Ind. Environ. 26 (2–3), 84–88.

Kim, M.H., Song, H.B., 2014. Analysis of the global warming potential for wood waste recycling systems. J. Clean. Prod. 69, 199–207.

Lebow, S.T., 2010. Wood preservation. In: Wood Handbook: Wood as an Engineering Material: Chapter 15. Centennial ed. General Technical Report FPL. US Dept. of Agriculture, Forest Service, Forest Products Laboratory, Madison, WI, pp. 15.1–15.28. GTR-190.

Lehtonen, A., et al., 2004. Biomass expansion factors (BEFs) for Scots pine, Norway spruce and birch according to stand age for boreal forests. For. Ecol. Manag. 188 (1), 211–224.

Leskinen, P., et al., 2018. Substitution Effects of Wood-Based Products in Climate Change Mitigation. From Science to Policy 7. European Forest Institute.

Lindvall, H.K., Kirkeby, J., Christensen, T.H., 2006. Indledende miljømæssig vurdering af disponeringen af træaffald i Danmark.

Mantau, U., et al., 2010. Real Potential for Changes in Growth and Use of EU Forests. EUwood, Hamburg. Methodology Report.

Merrild, H., Christensen, T.H., 2009. Recycling of wood for particle board production: accounting of greenhouse gases and global warming contributions. Waste Manag. Res. 27 (8), 781–788.

Morris, J., 2017. Recycle, bury, or burn wood waste biomass?: LCA answer depends on carbon accounting, emissions controls, displaced fuels, and impact costs. J. Ind. Ecol. 21 (4), 844–856.

Muszynski, L., et al., 2017. Insights into the global cross-laminated timber industry. Bioprod. Bus., 77–92.

PAS (Publicly Available Specification) 111, 2012. Specifications for the Requirements and Test Methods for Processing Waste Wood. British Standards Institution.

Perez-Garcia, J., et al., 2005. The environmental performance of renewable building materials in the context of residential construction. Wood Fiber Sci. 37, 3–17.

Persson, S., 1998. Wälludden trähus i fem våningar: Erfarenheter och lärdomar. Tekniska Högskolan i Lund, Avdelningen för Bärande Konstruktioner.

Petersen, A.K., Solberg, B., 2002. Greenhouse gas emissions, life-cycle inventory and cost-efficiency of using laminated wood instead of steel construction: case: beams at Gardermoen airport. Environ. Sci. Pol. 5 (2), 169–182.

Pettersen, R.C., 1984. The chemical composition of wood. In: Rowell, R. (Ed.), The Chemistry of Solid Wood. ACS Publications (Chapter 2).

Rivela, B., et al., 2006. Life cycle assessment of wood wastes: a case study of ephemeral architecture. Sci. Total Environ. 357 (1–3), 1–11.

Salazar, J., Meil, J., 2009. Prospects for carbon-neutral housing: the influence of greater wood use on the carbon footprint of a single-family residence. J. Clean. Prod. 17 (17), 1563–1571.

Sandberg, D., 2016. Additives in wood products—today and future development. In: Environmental Impacts of Traditional and Innovative Forest-Based Bioproducts. Springer, pp. 105–172.

Sathre, R., Gustavsson, L., 2006. Energy and carbon balances of wood cascade chains. Resour. Conserv. Recycl. 47 (4), 332–355.

Sathre, R., O'Connor, J., 2010. Meta-analysis of greenhouse gas displacement factors of wood product substitution. Environ. Sci. Pol. 13 (2), 104–114.

Sirkin, T., ten Houten, M., 1994. The cascade chain: a theory and tool for achieving resource sustainability with applications for product design. Resour. Conserv. Recycl. 10 (3), 213–276.

Stupak, I., et al., 2007. Sustainable utilisation of forest biomass for energy—possibilities and problems: policy, legislation, certification, and recommendations and guidelines in the Nordic, Baltic, and other European countries. Biomass Bioenergy 31 (10), 666–684.

Swedish Forest Agency, 2019. Regler och rekommendationer för skogsbränsleuttag och kompensationsåtgärder (Rules and recommendations for forest fuel extraction and compensation measures). Report 2019/14 (in Swedish).

Tettey, U.Y.A., Dodoo, A., Gustavsson, L., 2019. Carbon balances for a low energy apartment building with different structural frame materials. Energy Procedia 158, 4254–4261.

Thonemann, N., Schumann, M., 2018. Environmental impacts of wood-based products under consideration of cascade utilization: a systematic literature review. J. Clean. Prod. 172, 4181–4188.

Townsend, T., Dubey, B., Solo-Gabriele, H., 2008. Disposal management of preservative-treated wood products. In: Schultz, T., et al. (Eds.), Development of Commercial Wood Preservatives. American Chemical Society, Washington, DC (Chapter 33).

United Nations Publications, 2019. Forest Products Annual Market Review 2017–2018. UN.

United States Environmental Protection Agency, 2020a. Facts and Figures about Materials, Waste and Recycling. Wood: Material-Specific Data. Accessed at: https://www.epa.gov/facts-and-figures-about-materials-waste-and-recycling/wood-material-specific-data. 15th January 2021.

United States Environmental Protection Agency, 2020b. Advancing Sustainable Materials Management: 2018 Fact Sheet. Assessing Trends in Materials Generation and Management in the United States. Accessed at: https://www.epa.gov/sites/production/files/2020-11/documents/2018_ff_fact_sheet.pdf. 15th January 2021.

Werner, F., et al., 2010. National and global greenhouse gas dynamics of different forest management and wood use scenarios: a model-based assessment. Environ. Sci. Pol. 13 (1), 72–85.

World Bioenergy Association, 2019. Global Bioenergy Statistics 2019. Accessed at: https://worldbioenergy.org/uploads/191129%20WBA%20GBS%202019_HQ.pdfWorld. 1st February 2021.

Xu, Z., et al., 2018. Climate change mitigation strategies in the forest sector: biophysical impacts and economic implications in British Columbia, Canada. Mitig. Adapt. Strateg. Glob. Chang. 23 (2), 257–290.

CHAPTER

30

Paper☆

Harald Grossmann[1], Toni Handke[1,], and Tobias Brenner[2]*

[1]Technische Universität Dresden, Dresden, Germany [2]Papiertechnische Stiftung (PTS), Heidenau, Germany

30.1 INTRODUCTION

The principal raw materials used for papermaking are wood and recovered paper or board. Wood must be cut into logs or chips, which can then be converted into chemical or mechanical fiber pulps, depending on which effect dominates the disintegration process. Chemical pulps, which currently represent 33% of all fibers used for papermaking worldwide, contain only long, strong, and flexible cellulose fibers from which all other wood constituents (lignin and hemicellulose) have been reduced. Mechanical pulps (8%) consist of weaker and shorter fibers, which by and large contain all wood components in the same ratio as the wood itself. Both chemical and mechanical pulps are called virgin pulps because they are directly manufactured from wood.

Recovered paper or board is converted into recycled pulp in increasingly complex process chains, the main purpose of which is to remove all unsuitable or potentially detrimental substances from the fibers. All fibers of the resulting recycled (or secondary) pulp therefore have at least once before been part of a paper or board product. They represent 57% of the total fiber consumption (2% are other fiber sources) of the paper industry in the world, making this industry a good example for the circular economy. Still, higher recycling rates are possible as evidenced by recycling rates in selected regions and countries.

Figure 30.1 shows that the annual per capita consumption of paper and paperboard varies significantly between the major regions of the world. The world average in 2019 was close to 52 kg/capita. North America is the region with the highest annual consumption of 196 kg/capita, while this figure is as low as 6 kg/capita for Africa. Consumption levels in Western Europe and North America will remain constant or—even more likely—decline slightly over the long term, while consumption in Asia will increase further. Because of the huge population in this region, it is assumed that the worldwide average paper consumption will also increase. Asia produces nearly half of all the paper in the world (48% in 2019). Note that the

☆ Revised by Ernst Worrell, Copernicus Institute of Sustainable Development, Utrecht University, the Netherlands.

* Currently address: Sappi, Hildesheim, Germany

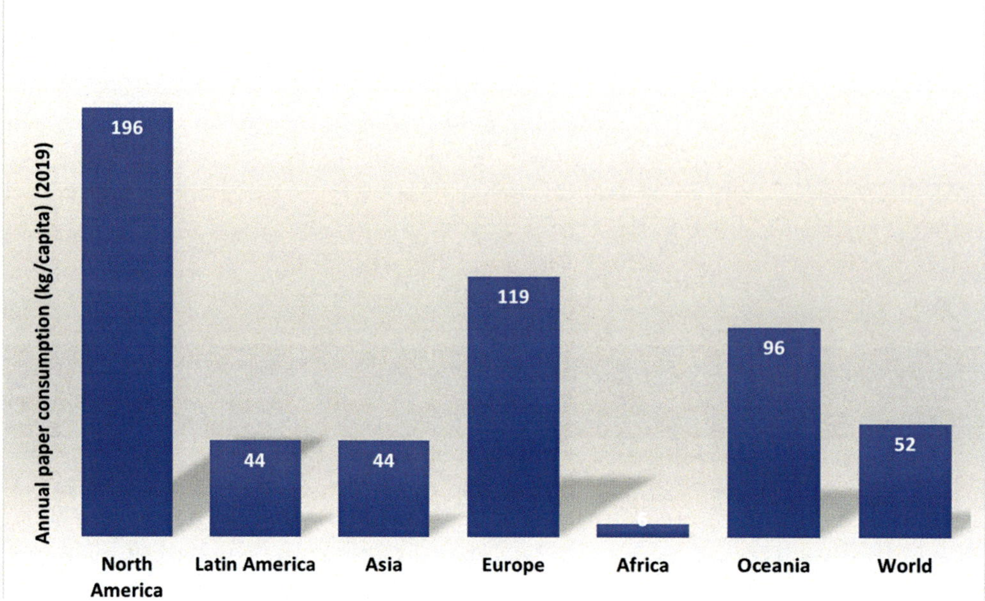

FIGURE 30.1 Annual paper consumption per capita 2019. *Data from FAOstat (2022).*

consumption varies by paper grade. The use of newsprint (for newspapers) declined worldwide [5% in EU by 2020 (CEPI, 2021)], while the consumption of packaging paper increased (41% in the EU by 2020), especially in recent years due to the increase in online shopping. Printing and writing paper (21% in the EU; 2020) and tissue (11% in the EU; 2020) represent the other major paper grades.

During the last few decades, recovered paper and board have become the principal raw materials of the paper industry. The reasons are obvious: recovered paper is significantly cheaper and its conversion into pulp requires far less energy than virgin raw materials (Ervasti et al., 2016). Recycling paper and board, however, is not only economical but also contributes to the energy and resource efficiency of the entire industry. In particular, highly developed countries with limited forest resources like Germany or France, as well as other countries with emerging economies, such as China and India, are highly dependent on recovered paper.

30.2 COLLECTION AND UTILIZATION

The extent to which used paper and board products are recycled is usually characterized by two key indicators: the recovery rate (i.e., the rate of paper recovered for recycling in a country) and the utilization rate (i.e., the amount of recycled paper used by the paper industry in a country). These are applied to a specific geographical region or country.

Recovery rate (%) (Eq. 30.1) is the amount of paper recovered for recycling in a certain region, divided by the amount of paper consumed in this very region, on an annual basis (Göttsching and Pakarinen, 2000):

$$\text{Recovery rate}(\%) = \frac{\text{paper recovered for recycling}(t/y)}{\text{paper consumed}(t/y)} \times 100 \tag{30.1}$$

Utilization rate (%) (Eq. 30.2) is the amount of recovered paper that is used as raw material in the paper industry in a certain region, divided by the paper production of this region, on an annual basis (Göttsching and Pakarinen, 2000):

$$\text{Utilization rate}(\%) = \frac{\text{recovered paper used for paper production } (t/y)}{\text{total paper production } (t/y)} \times 100 \quad (30.2)$$

These definitions, however, although simple and seemingly straightforward, do not allow a meaningful comparison between regions. An obvious—although not the only—reason for this is that they usually do not take into account the amounts of packaging papers that enter together with imported goods, or that leave the region together with exported goods. The packaging paper used for the latter would of course be included in the production statistics, but it would no longer be available for recovery and utilization in this region, and vice versa. The interpretation of the key indicators requires some care; however, they can help to provide a rough idea of paper recycling in a certain region.

Figure 30.2 shows the key indicators. China or India, for example, utilize more recovered paper for their domestic paper production than they collect, i.e., they are net importers of recovered paper. Countries like the United States collect more paper than they use for their own production, which is based primarily on virgin fibers. Japan, France, and Germany recycle almost all available recovered paper and produce most of their papers from this resource.

Products from the paper industry can be subdivided into four classes:

- Packaging paper: case materials, carton board, wrapping papers and other packaging papers,
- Graphic paper: newsprint and other graphic paper,

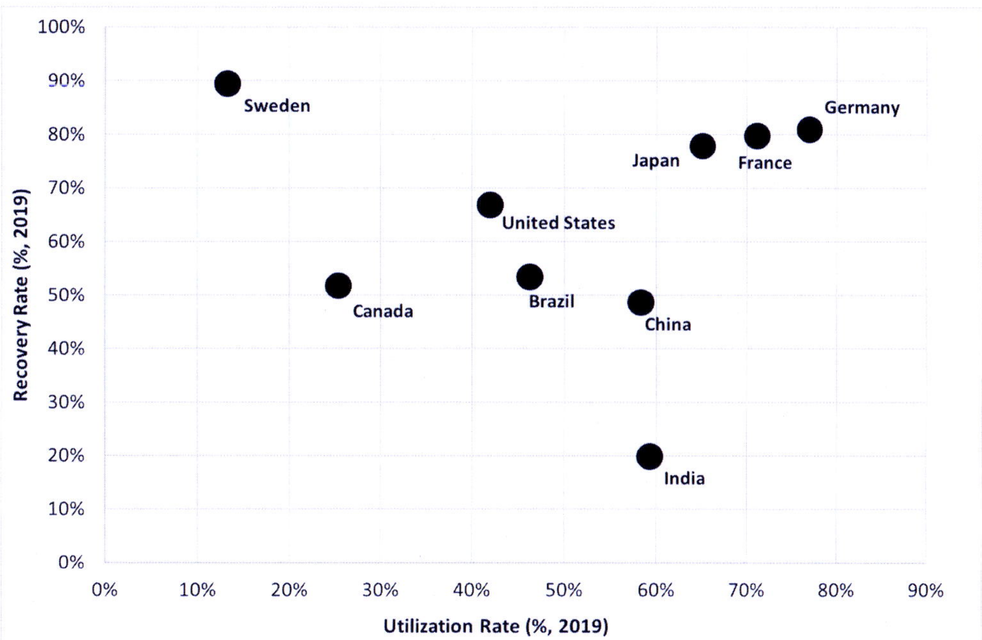

FIGURE 30.2 Recovered paper utilization rate, recycling rate, and collection for selected countries in 2019. *Data from FAOstat (2022).*

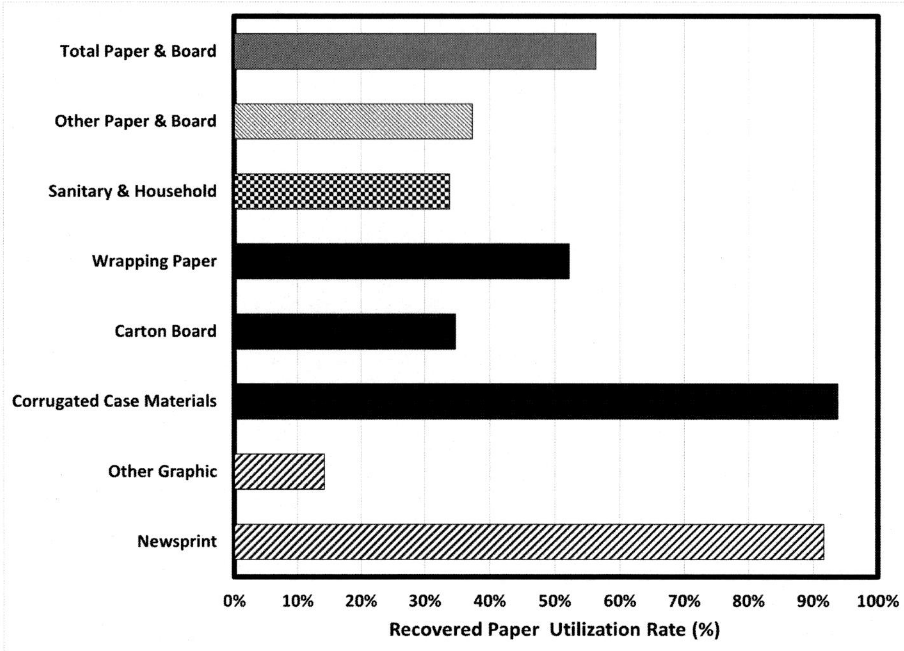

FIGURE 30.3 Recovered paper utilization rates for different paper and board products in Confederation of European Paper Industries (CEPI) countries in 2020. *Data from CEPI (2021).*

- Tissue paper: household and sanitary, and
- Paper for technical and other purposes.

The utilization rates of recovered paper to produce these paper grades vary considerably (i.e., tissue is only minimally recovered and recycled). The highest utilization rates are usually found for corrugated case materials and newsprint (Figure 30.3). On the one hand, these paper grades do not have a sophisticated optical appearance and their product lifetime is rather short. On the other hand, high strength values cannot be achieved with recovered paper and the high ash content prevents it from being used to a greater extent in hygiene paper. Therefore, if high brightness, strength, or softness are required, recovered paper can only be used to a limited degree. The higher the demands placed on the paper; the more energy must be invested in preparing it. Packaging paper requires less energy in stock preparation than a graphic paper that must undergo extensive cleaning.

30.3 COLLECTION AND SORTING SYSTEMS

Supplying paper mills with used fibrous raw material requires the collection and sorting of recovered paper. The collection of recovered paper can be broken down into collection from industry, trade, and administration and private household collections. Collection is organized in various systems (Stawicki and Read, 2010), the most important of which are:

- container collection (drop-off system or pickup system),
- curbside collection (pickup system), and
- underground collection (dropoff system).

In addition to the separate collection of recovered paper, a method known as "co-mingled" collection is practiced in which paper and other waste or metal are collected together. Co-mingled collection is not considered appropriate when high-quality recovered paper is required.

The collected paper can either be prepared in sorting factories according to the subsequent use or delivered directly to the paper mill. The recovered paper that has been collected, and sorted if necessary, is then subdivided into classes based on the composition and content of unusable or contaminating materials. The classification system for recovered paper in the United States is described in the "Guidelines for Paper Stock: PS 2021" (ISRI, 2021). This system defines the allowable shares of prohibited and contaminating materials for individual paper grades and contains terms of the contract such as grade definitions, arbitration, fulfillments by the buyer and seller, and purchase agreement. In Europe, recovered paper is commonly classified according to the European List of Standard Grades of Recovered Paper and Board (EN 643) comprising 67 grades subdivided into five classes. The most important grades, quantitatively speaking, are grades 1.02—mixed papers and boards (sorted), 1.04—supermarket corrugated paper and board, and 1.11 - sorted graphic paper for deinking. In addition, another classification for statistical and commercial purposes has four main groups (Stawicki and Read, 2010):

- Mixed grades: mixed paper and boards,
- OCC: old corrugated containers,
- ONP and OMG: old newspaper and magazines, and
- HG and PS: high-grade deinking and pulp substitutes.

Initially, recovered paper was sorted manually only. The development of efficient sensors and mechanical separators has made it possible to use sufficiently selective semiautomatic or fully automatic systems.

The sorting process for recovered paper consists of the following operations (Blasius et al., 2011):

1. delivery of the recovered paper in loose form or as bales, entry inspection, bale dewiring;
2. mechanical sorting based on differences in physical properties such as shape, size, specific gravity, stiffness;
3. sensor-based sorting and separation by automatic picking, chiefly by blowing;
4. manual sorting.

The recovered paper can be delivered to the paper mill as loose material, which is typical of household collections and paper for deinking. All other recovered paper is traded in bales that are typically $1.2\ m \times 1.0\ m \times 0.8\ m$ and 550 kg.

The fibrous raw material suitable for producing packaging paper includes packaging paper or board as well as used graphic paper. Only used graphic paper or higher grades can be used as fibrous raw material to produce graphic paper or tissue paper (deinking process). This distinction is necessary because of the different preparation processes used for stock preparation in the paper mill.

30.4 STOCK PREPARATION

In addition to the fibrous raw material, recovered paper also contains other substances that are introduced by the paper product itself (printing inks, stickies, mineral particles) or during its use and subsequent collection (e.g., scraps and leftovers, sand, glass). Hence, when producing new paper from recovered paper, the pulp has to be cleaned intensely first, irrespective of the later product. The objective of recovered paper treatment is to produce a homogeneous substance from which contraries (depending on the product) have been eliminated and which provides fibers that enable a stable production process.

For packaging paper, it is particularly important to remove coarse contaminants such as films and foils, adhesives, and consumer residues, and to produce a homogeneous fiber pulp. Figure 30.4 shows the main operations of a stock preparation line to produce "brown" packaging paper. In older mills, the recovered paper is pulped in medium (10%) to high consistency (up to 20%) batch pulpers with a vat design that includes a dumping system for coarse rejects like huge foils or solid matter, as well as a high-consistency cleaner for the removal of heavy impurities such as glass, sand, or metal pieces. State-of the-art technology, however, continues to use systems called drum pulpers. After pulping, the fiber suspension is sometimes stored in a dump chest before it is fed to the first screening section, which is equipped with pressurized screens to remove additional contaminants using perforated or slotted screen baskets.

Packaging papers (e.g., testliner) typically consist of two layers with different strength properties. If the pulps for these layers are produced from the same recovered paper mixture, the entire pulp must be divided into two fractions, one containing significantly longer fibers than the other. The long fiber fraction is then used to produce the high-strength layer.

Graphic paper production from recovered paper is more expensive than packaging paper

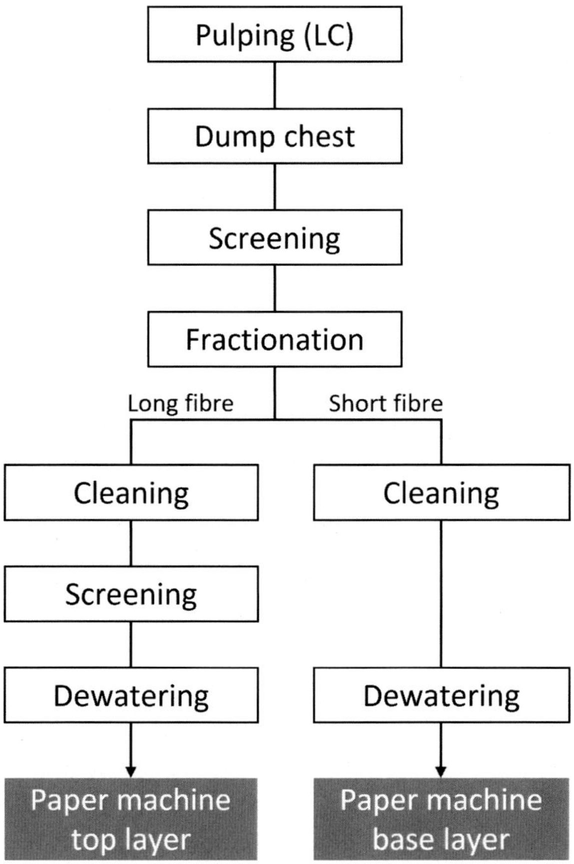

FIGURE 30.4 Concept of stock preparation to produce packaging paper (testliner). LC: low consistency.

because it requires more extensive treatment of the raw material, which consists exclusively of used (or unsold) graphic products. One reason is that the printing ink has to be removed from the pulp by a separate energy-intensive process to achieve high brightness. However, not all printing inks can be removed equally well. While the inks used in "old" printing technologies like coldset offset, heatset offset, and rotogravure present no major problems, virtually no solution has yet been found for water-based flexography or nonimpact printing inks. This might become problematic in the future as these printing technologies rapidly gain importance and are about to replace the older techniques in a number of applications.

Figure 30.5 shows a typical two-loop system for the production of graphic paper. Pulping and sorting are comparable to the stock preparation process used for brown paper, the major difference being the deinking step.

Printing ink particles, once detached from fiber surfaces during pulping, are usually removed by flotation at very low stock consistencies of about 1%. The removal efficiency is a function of the size of the ink particles. The larger they are, the smaller is the probability that they become attached to small air bubbles, which, however, is the prerequisite for being floated.

Ink particles that, due to their size, have not been removed by flotation cause dirt specks in

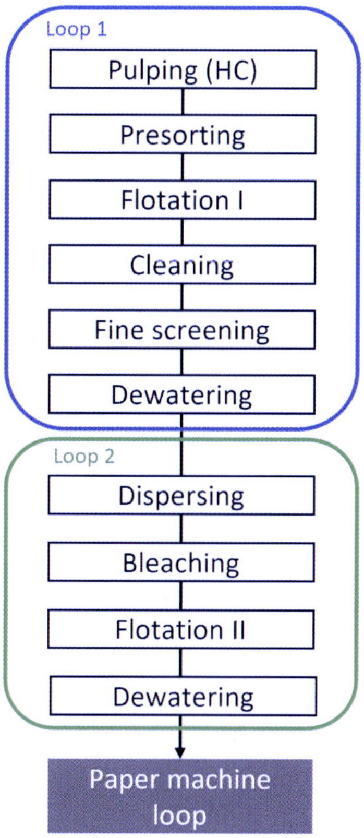

FIGURE 30.5 Concept of stock preparation to produce graphic paper (newsprint). HC: high consistency.

the final paper, which is normally not accepted. This can be avoided by breaking them into smaller pieces no longer visible to the naked eye or floatable in a subsequent flotation step. Comminution of the ink particles is done in a disperser after dewatering the suspension to a consistency of 25–35%. Dewatering also serves to separate the first process water loop, which is highly loaded with impurities, from subsequent loops closer to the paper machine. The high stock consistencies and temperatures during dispersing are beneficial to several additives, in particular bleaching chemicals, which should therefore be added prior to dispersing. The most important process steps will be explained in detail in the following section.

30.4.1 Pulping and Slushing

State-of-the-art paper manufacturing still starts with the formation of a wet web from a highly diluted suspension containing fibers and other components such as minerals, starch, and chemical additives. The first step in recovered paper treatment, therefore, aims at disintegrating the raw material in water and rendering it into a pumpable suspension. This process is called pulping or slushing. In addition, the process can also remove coarse solid contaminants like wooden parts, plastic films, or metal pieces, and mix chemicals into the fiber suspension, especially deinking or bleaching chemicals.

With respect to stock consistency, pulping processes can be divided into three categories:

- low-consistency (LC) pulping: below 6% stock consistency,
- medium-consistency (MC) pulping: 6–12% stock consistency,
- high-consistency (HC) pulping: 12–25% stock consistency.

With respect to the technology used, two different basic processes have become established in the industry. Conventional pulpers in the form of cylindrical vessels with impellers (rotors) at the bottom or the wall and baffles to avoid solid body rotation are suitable for consistencies between 6% and 19%, while drum pulpers, i.e., huge, slowly rotating pipes, are suitable for consistencies of up to 20%.

Drum pulping is always a continuous process. Conventional pulpers can be performed as batch processes as well. In MC and HC pulping, dilution is required to remove the pulp from the pulper. The specific energy demand for pulping is in the range of 15–30 kWh/t paper for conventional pulpers, and 15–20 kWh/t paper for drum pulpers.

Recovered paper always requires—during or after pulping—the separation of coarse impurities that inevitably come with the raw material. Unlike pulpers for virgin fibers, pulpers for recovered paper therefore need a reject handling system.

30.4.2 Pulping in Cylindrical Vessels

Vat-shaped pulpers for the disintegration of recovered paper are equipped with a dumping and reject handling system. For economic reasons, this system includes a primary pulper that creates a pumpable suspension of flakes and single fibers and a secondary pulper or disk screen downstream of the primary pulper that crushes the flakes into single fibers. For different ranges of stock consistency, special rotors are available in the primary pulper—flat rotors for LC and spiral rotors for MC and LC. The coarse contaminants are removed from the primary pulper by a ragging device with a winch. Smaller contaminants are removed as light and heavy rejects in the secondary pulper. The light and heavy rejects are treated in downstream devices to recover the adherent fibers and eventually dewater them to the greatest possible extent.

30.4.3 Drum Pulper

Drum pulpers are primarily used for recovered paper with low wet strength—typically recovered graphic papers. The horizontally aligned system has the shape of a large pipe

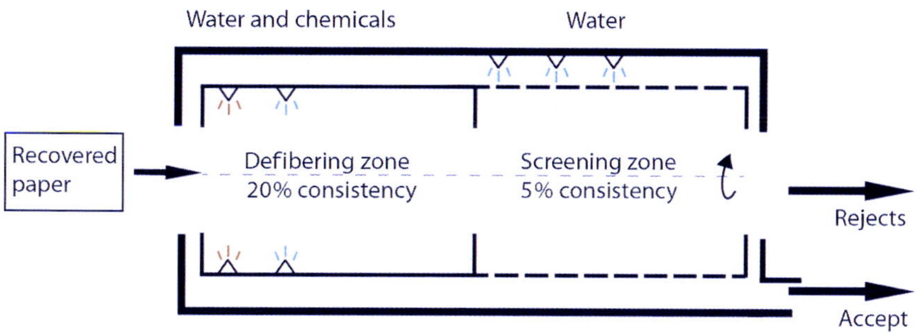

FIGURE 30.6 High-consistency drum pulper.

(up to 4 m in diameter) that rotates at a moderate speed (up to 10 rpm). It consists of two zones (Figure 30.6). The first is a slushing zone where the wet stock (consistency of approximately 20%) is transported upward with the help of baffles aligned perpendicularly to the inner wall of the pipe, to repeatedly fall back again. The stock subsequently enters a pipe with a screening zone where it is diluted to 3–5% and the fibers pass through a screen while all contaminants larger than the screen holes are retained. These coarse contaminants are transported to the end of the screen and removed.

30.4.4 Screening and Cleaning

Two separation techniques have proved valuable for removing contaminants such as shives, stickies, and other solid particles: screening and centrifugal separation. Pressure screens retain particles that are larger or different in shape than the openings of the screens. Centrifugal cleaners accumulate and remove particles that are heavier or lighter than water.

The objective of any separation process is to separate mixed substances into two or more distinct mixtures, at least one of which is enriched with one or more of the mixture's constituents. Recovered paper contains a very wide variety of constituents, many of them in very small quantities. Separation processes generally have three basic mass flows: the "inlet" mass flow, the "accept" mass flow that contains the accepted pulp, and the "reject" with impurities. A separation process is never capable of classifying pulp and impurities with an efficiency of 100%. The smaller the differences between pulp particles and impurities, the poorer will be the separation performance for the two fractions. Therefore some fibers will always end up in the reject mass flow. Individual separation devices can be interconnected to form circuits known as cascade circuits to ultimately reduce the amount of pulp lost into the reject flow.

The specific energy demand for a screening process is 5–20 kWh/t paper, and for a cleaning process 5–15 kWh/t paper.

The contaminants removed by pressurized screens include shives, stickies, waxes, wet strength paper, or plastics. The devices currently used are primarily pressure screens, where the screen has the shape of a basket through which the fiber suspension is pumped. A revolving rotor induces pressure pulses to prevent the screen from clogging. Screening devices are distinguished according to the geometry of their openings. The small holes of perforated screens predominantly retain plane particles, whereas the narrow slots of slotted screens are suitable for removing cubical particles (Hanecker, 2006) (Figure 30.7).

Centrifugal cleaners include a hydrocyclone that creates a spiral flow within the suspension by means of the tangential inlet flow or a rotor. The swirling motion results in a centrifugal field

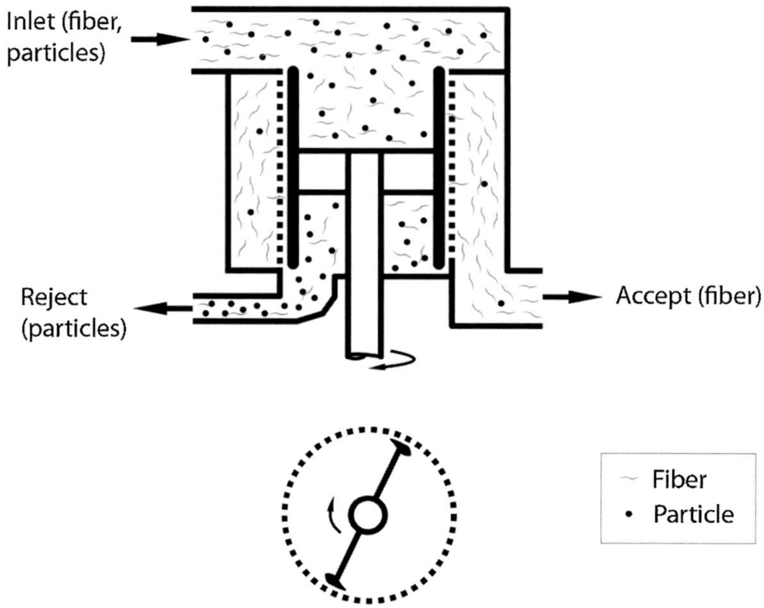

FIGURE 30.7 Pressure screen for low-consistency screening.

whose forces push heavy particles toward the wall and light particles toward the center. Centrifugal cleaning can thus be used to remove heavy impurities such as pieces of glass, metal, or sand as well as lightweight particles such as foam plastics. The removal of different impurities requires specific cleaner designs. Cleaners can be subdivided into heavy particle cleaners, light particle cleaners, and combination cleaners for heavy and light contaminants. Another classification is based on the operating stock consistency. The maximum stock consistency for cleaners is approximately 4.5% (Figure 30.8).

30.4.5 Flotation and Washing

Flotation and washing are processes primarily used to reduce the quantity of minerals in a pulp suspension or, even more frequently, to remove ink particles from the suspension (known as deinking).

Washing is a filtration process for removing from the suspension debris particles smaller than the mesh size of a filter cloth (Holik, 2000). Simultaneously, dissolved and colloidal contaminants are removed with the filtrate. The objective in both cases is to remove substances negatively affecting the papermaking process or finished product quality. Materials removed by washing include fillers and coating particles, fines, micro-stickies, and ink.

Flotation is a separation process for cleaning suspensions. It is based on the use of air bubbles that attach themselves to hydrophobic dirt particles and transport them to the surface of the suspension (see Figure 30.9).

Flotation is performed at very low stock consistencies of 0.8–1.2%. It usually requires the addition of soap or fatty acids that collect the hydrophobic ink particles and eventually agglomerate them. The impact caused by an ink-particle agglomerate hitting an air bubble of matching size is a function of the collision velocity and the mass of the agglomerate—it must be sufficiently high to penetrate the laminar boundary layer of the bubble. As a result,

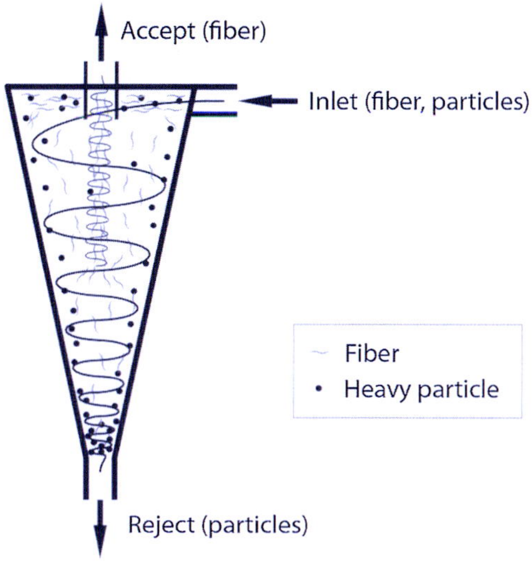

FIGURE 30.8 Hydrocyclone cleaner for the removal of heavy contaminants, showing the separation principle.

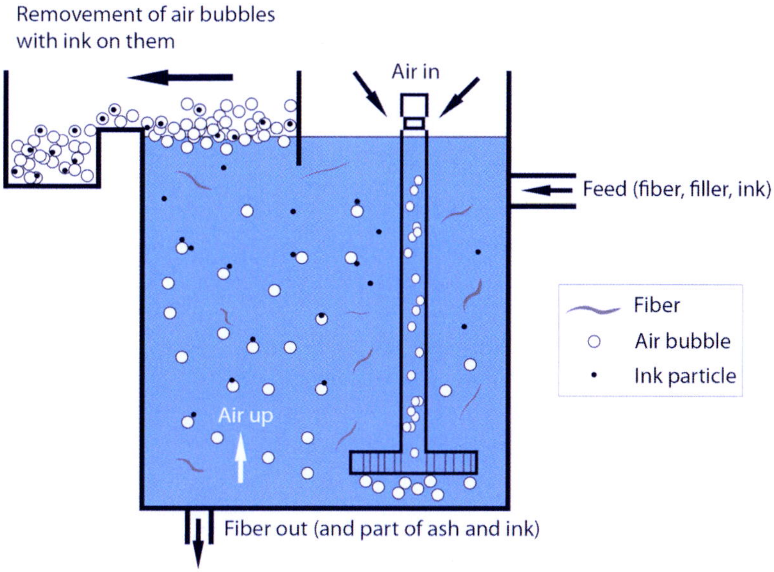

FIGURE 30.9 Flotation cell, showing the separation principle.

the agglomerate adheres to the bubble. If the density of the new ink particle bubble agglomerate is still lower than that of water, it will float, i.e., rise to the foam layer at the top of the flotation cell, where it can be discharged by a rotating foam scraper. The accept, i.e., cleaned pulp, is then fed to several downstream flotation cells, each of which serves to further increase the brightness of the pulp. The foam of all these cells is collected and fed to a secondary flotation line to recover the fibers and minerals unintentionally floated in the primary cells.

The flotation result is controlled primarily by two parameters: the size of the air bubbles and the energy dispersion. With tiny bubbles, rather small particles are discharged, and with larger bubbles, larger particles are discharged. Particles smaller than 10 μm or larger than 100 μm cannot be efficiently removed by flotation. These particles include printing inks, stickies, fillers, coating pigments, and binders. Flotation also removes fillers in some cases. It has a certain de-ashing effect that is often desirable.

The specific energy demand for flotation is 20–50 kWh/t paper, and for washing 20 kWh/t paper (Holik, 2000).

30.4.6 Dewatering

Dewatering increases the stock consistency of the pulp. Compared to the washing process, where a solid-solid separation occurs (e.g., separation of mineral particles from fibers), dewatering separates liquids from solids. Different technologies are suitable for this: disk filters as well as devices with belts or screw presses, all of which are of major importance.

Dewatering is aimed at:

- Ensuring the optimal stock consistency for process steps such as bleaching or dispersing,
- Closing water loops to confine the diffusion of contaminants or additives,
- Saving space in storage vessels, and
- Compacting rejects or sludge for the economic recovery of these materials.

Belt filters and disk filters remove the water up to a discharge consistency of about 10–12% and have an average specific energy demand of 5 kWh/t paper. Belt filter presses and twin wire presses achieve stock consistencies of more than 30% with a specific energy demand of less than 10 kWh/t paper, while screw presses allow a thickening up to 65% consistency with a specific energy demand of 10–15 kWh/t paper. Screw presses (Figure 30.10) require an inlet consistency of at least 10%. These devices are therefore usually combined with belt filters or disk filters.

30.4.7 Dispersing and Bleaching

Dispersing and bleaching are used for products that must meet high optical quality demands, e.g., magazine paper, newspaper, and writing paper. Dispersing and bleaching

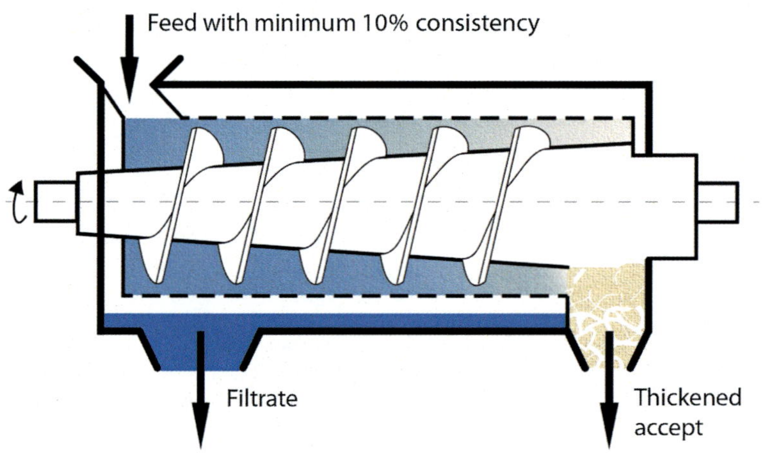

FIGURE 30.10 Screw press.

are not commonly used in the case of brown grades.

The objective of dispersers is to comminute any printing ink particles present in the suspension and separate any printing ink particles still attached to the fibers to eliminate them later by flotation. Comminution and separation serve two aims: producing floatable particles or reducing the size of particles to such an extent that they are below the visibility threshold of the human eye, i.e., <50 μm. This gives the paper product a homogeneous appearance, since the printing ink particles will no longer produce large dirt specks, although the brightness diminishes—the paper turns gray, with typical losses amounting to two to three brightness points. The functional principle of a disperser is quite simple: a rotor rotates against a stator. The rotor/stator plates are equipped with many large cutting surfaces, much like refiners. The energy consumption is high: if dispersing is viewed as a single process step, it requires most of the electric energy used in stock preparation: between 50 and 90 kWh/t paper at a temperature of 60–100°C.

These losses can be compensated in the bleaching step. Bleaching agents are typically added to the pulp just upstream of or in the disperser itself, since the disperser operates at approximately 30% stock consistency and mixes them thoroughly. A typical combination is using hydrogen peroxide in the first step after preflotation and sodium hydroxide in the second step at the end of the process. It is important to couple an oxidative with a reductive bleach to achieve maximum effect.

Problems may occur in hydrogen peroxide bleaching due to high amounts of iron and catalase. Iron ions can be rendered harmless by using sodium silicate, and catalase formation is prevented by high temperatures in the bleaching tower. Once both contaminants are under control, the hydrogen peroxide concentration can be kept at low levels. It is advisable to conduct sodium hydrosulfite bleaching in the absence of oxygen and at high temperatures (70°C) in a bleaching tube. If the reaction were carried out in a batch tower, too much oxygen would greatly inhibit bleaching and the effect would be negligible.

30.5 KEY ISSUES AND FUTURE CHALLENGES

30.5.1 Selectivity

The processes currently used for the treatment of recovered paper are predominantly separation processes designed to eliminate undesired substances such as printing inks, stickies, sand, or plastics, as well as discarded low-quality fibers. In most cases, these substances are used for energy recovery [including production of refuse derived fuel (RDF) pellets] or put in landfills. In selective separation, individual substances can continue to be recovered and utilized. For instance, calcium carbonate or kaolin can be recovered to be reused as coating pigment or internal sizing aid. Separated fines and plastics can be used in biopolymers, wood plastic composites, or paper plastic composites. The selectivity of state-of-the-art separation processes, however, is not yet good enough to achieve this without considerable problems. If their selectivity was increased, however, the products of tomorrow could be manufactured from today's rejects. See Bousios and Worrell (2017) for a discussion on useful applications of recycled paper mill by-products.

30.5.2 Energy Demand

The average energy consumption for the production of 1 t of paper greatly depends on the furnish composition as well as other factors (Laurijssen et al., 2013). Table 30.1 shows some selected products and their specific energy demands, using 2003 data.

TABLE 30.1 Specific energy consumption of various types of paper.

	Heat demand kWh/t	Energy demand kWh/t	Total kWh/t
Sanitary paper	1900	1150	3050
Wood-containing coated	1600	1450	3050
Wood-free coated	1800	1000	2800
Wood-containing uncoated	1250	1370	2620
Wood-free uncoated	1600	770	2370
Newspaper	1000	1200	2200
Packaging paper	1500	500	2000

Data from Bayerisches Landesamt für Umweltschutz (2003).

In Central Europe, packaging paper or newsprint is usually made from 100% recovered paper, thus with a lower energy consumption than paper products containing virgin fibers, e.g., wood-containing coated paper. The difference of about 1000 kWh/t is mainly due to the different processes used for raw material treatment. The use of recovered paper is always less expensive than the use of virgin fibers.

State-of-the-art stock preparation processes must greatly dilute the suspension for deinking to effectively remove the printing ink particles. The large water volumes required for this, however, result in high energy consumption for pumping. To prevent this, the stock consistency must be as high as possible in stock preparation, which cannot be achieved with current flotation technology. Several ongoing research projects are devoted to this theme and are developing and testing alternative slushing and deinking technologies, among others.

The latter include adsorption deinking (Handke et al., 2012) and ultrasound deinking (Großmann et al., 2010), for example. In adsorption deinking, adsorbers take up the printing ink particles at high stock consistencies of 15% and, at the same time, have a dispersing effect (ball mill effect); 90% less water needs to be pumped in the facility owing to the high stock consistency. Ultrasound deinking attempts to separate the printing ink particles from the fibers by cavitation. This is possible when stock consistencies are high.

30.5.3 New Printing Inks

Current flotation technology can remove only hydrophobic printing inks like conventional offset and gravure printing inks. Digital printing inks or water-soluble inks cannot be deinked, or only with great difficulty. Because they offer advantages to printers in some cases, their market share continues to grow. Flexo ink in particular is being used to an ever-greater extent in the graphic sector and causes problems in deinking. The printing ink colloidal dissolves in the water and discolors the entire pulp. This is not a serious problem for brown paper, but is disturbing in the case of graphic paper. The removability of problematic printing inks by conventional technology is a problem that must yet be solved by printing ink manufacturers. Because they lack any incentive to do so, new removal technologies must be developed that can deal with problematic printing inks.

30.5.4 Water Consumption

The water demand of paper mills using recovered paper has been considerably reduced in the last few decades and today ranges between 0 and 10 L/kg paper. In what are termed zero-liquid effluent mills, only the amount of water that is lost through evaporation along the production chain in the dryer section of the paper machine has to be replaced by fresh water.

The specific volume of effluents produced during papermaking declined in Germany from approximately 45 m^3/t paper in the 1970s to 10–11 m^3/t in 2010. The level has stagnated at approximately 10 m^3/t paper since 2000. Some paper mills have completely closed the water circuits in their production facilities, which means that no effluents accumulate and only the water (approximately 1 m^3/t paper) evaporated during production must be replaced by fresh water. A completely closed water circuit, however, gives rise to a variety of problems due to gradually increasing concentrations of dissolved matter and contaminants. Han et al. (2021) provides an overview of wastewater treatment in the recycled paper industry.

30.5.5 Climate Change Mitigation

The internationally agreed upon goals to reduce greenhouse gas emissions in view of the expected climate change can only be realized in the short term in many cases by the increased use of regenerative energy sources, in particular wood. This is reflected by the drastically intensified global competition for this resource noticeable today, and the rise in wood prices.

The lack of forest resources in some emerging countries and the widely accepted necessity to use far more biobased than CO_2-intensive fossil and mineral products (e.g., construction materials) and fuels are bound to increase the competition for wood all over the world. Against this background, it is very likely that the demand for recovered paper and board as well as alternative fibers will increase further. As recycling "frees up" wood for other purposes, recycling contributes to mitigating greenhouse gas emissions (Laurijssen et al., 2010). However, as recycled fibers degrade after each cycle, there will always be a need for virgin pulp. With a growing global paper industry and growing demand of wood for other applications (bioenergy, construction) (Chapter 29—Lumber) this will lead to increased competition for wood.

References

Bayerisches Landesamt für Umweltschutz, 2003. Klimaschutz durch effiziente Energieverwendnung in der Papierindustrie—Nutzung von Niedertemperaturabwärme. sn, Augsburg. ISBN 3-936385-36-x https://www.druckluft-effizient.de/downloads/dokumente/papier.pdf. (Accessed 17 March 2022).

Blasius, K., Escabasse, J.-Y., Farreira, B., 2011. Verbesserte ausbeute. Recycl. Mag. 2.

Bousios, S., Worrell, E., 2017. Towards a multiple input-multiple output paper mill: opportunities for alternative raw materials and sidestream valorisation in the paper and board industry. Resour. Conserv. Recycl. 125, 218–232.

CEPI, 2021. Key Statistics 2020, European Pulp and Paper Industry. CEPI, Confederation of European Paper Industries, Brussels. https://www.cepi.org/wp-content/uploads/2021/07/Key-Stats-2020-FINAL.pdf. (Accessed 17 March 2022).

Ervasti, I., Miranda, R., Kauranen, I., 2016. Paper recycling framework, the "Wheel of Fiber". J. Environ. Manag. 174, 35–44.

FAOstat, 2022. Food and agriculture data. Food and Agriculture Organization of the United Nations, Rome, Italy. www.fao.org/faostat. (Accessed 17 March 2022).

Göttsching, L., Pakarinen, H., 2000. Recycled Fibre and Deinking. s.n. Papermaking Science and Technology, Helsinki. Book 7.

Großmann, H., Fröhlich, H., Wanske, M., 2010. The potential of ultrasound assisted deinking. Norfolk. In: Proceedings of the 2010 TAPPI PEERS Conference.

Han, N., Zhang, J., Hoang, M., Gray, S., Xie, Z., 2021. A review of process and wastewater reuse in the recycled paper industry. Environ. Technol. Innov. 24, 101860.

Handke, T., Schrinner, T., Großmann, H., 2012. Adsorption deinking—a new approach for higher energy efficiencies in paper recycling. In: 2012 TAPPI PEERS Conference: Building a Sustainable Future, pp. 481–492.

Hanecker, E., 2006. Altpapierrohstoff: Rohstoff Altpapier und seine Aufbereitung. In: Strauß, J. (Ed.), Papierherstellung im Überblick, Manuskript EK 697. PTS Papiertechnische Akademie, München.

Holik, H., 2000. Unit operations and equipment in recycled fiber processing. In: Pakarinen, L., Göttsching, H. (Eds.), Papermaking in Science and Technology, Recycled Fiber and Deinking, Part 7. Fapet Oy, Finnland.

ISRI, 2021. ISRI Scrap Specifications Circular. Institute of Scrap Recycling Industries (ISRI), Washington, DC. https://www.isri.org/recycling-commodities-old/scrap-specifications-circular. (Accessed 17 March 2022).

Laurijssen, J., Marsidi, M., Westenbroek, A., Worrell, E., Faaij, A., 2010. Paper and biomass for energy? The impact of paper recycling on biomass availability, energy and CO_2 emissions. Resour. Conserv. Recycl. 12 (54), 1208–1218.

Laurijssen, J., Faaij, A., Worrell, E., 2013. Benchmarking energy use in the paper industry—a benchmarking study on process-unit level. Energy Efficiency 6, 49–63.

Stawicki, B., Read, B., 2010. COST Action E48—The Future of Paper Recycling in Europe: Opportunities and Limitations. Dorset, Great Britain, The Paper Industry Technical Association (PITA).

Further Reading

Blechschmidt, J., 2010. Taschenbuch der Papiertechnik. Carl Hanser, München, ISBN: 978-3-446-41967-4.

Chang, J.C., Beach, R.H., Olivetti, E.A., 2019. Consequential effects of increased use of recycled fiber in the United States pulp and paper industry. J. Clean. Prod. 241, 118133.

Jung, H., et al., 2011. Wasser- und Abwassersituation in der deutschen Papierindustrie. Wochenbl. Pap. 9, 478–481.

CHAPTER

31

Plastic recycling

Li Shen and Ernst Worrell

Copernicus Institute of Sustainable Development, Utrecht University, Utrecht, The Netherlands

31.1 INTRODUCTION

Plastics are synthetic organic polymers, mainly made from petrochemical feedstocks. Since the invention of the first plastic or polymer in the early 1900s (i.e., Bakelite), and the development of polyethylene in the 1930s, the number of plastics has increased dramatically, and so have its uses. As society has steadily increased its use of plastics, plastic waste management has become a growing concern around the world. In 2018, prior to the COVID-19 pandemic, almost 370 Mt plastics were produced annually. The key producers were China (31%), Europe (16%), North America (19%), and the rest of Asia (excluding China; 20%).

Plastics offer many advantages. Plastics are easy to shape, do not corrode or decompose only slowly, and the characteristics can be adapted to the specific needs by using composites or adding specific layers or additives. These characteristics are also the source of some of the problems associated with plastics. From the association of chlorinated plastics with dioxin formation in waste incinerators to the plastic soup in the oceans, the durability and wide use of plastics contributes to significant waste management problems. Recycling of plastics is one of the key strategies to reduce the environmental problems associated with plastic waste. Moreover, petrochemical feedstocks for plastic manufacturing have increased in price over the past decades, are concentrated in a relatively small number of locations, and will be in short supply within the next decades. Hence, plastic recycling will reduce reliance on fossil fuels.

Recycling of plastic wastes, especially postconsumer, has had a slow start. Compared to other commonly used materials such as paper, glass, and metals, recovery and recycling rates are generally low. Even in countries with advanced waste management systems and long experience in recycling, plastic recycling rates are typically much lower than those of other materials. This is partly caused by the huge variety of plastics, additives, and composites used. This variety is one of the key advantages of plastics and reasons for their versatility. Yet, this variation is a problem for recycling.

In this chapter, we first discuss trends in plastics consumption and types of plastics used. This is followed by the current situation with respect to plastics in the waste stage and recycling. We will primarily focus on postconsumer waste streams, as little data is available on preconsumer plastic waste. Then, the typical processes in plastic recycling are discussed, followed by a discussion of the environmental benefits,

Handbook of Recycling. https://doi.org/10.1016/B978-0-323-85514-3.00014-2

using one of the key recycled plastics (PET) as an example. We end with some concluding remarks on the future challenges and opportunities of plastics recycling in a circular economy context.

31.2 USE

A huge variety of plastics exist, and similarly the applications are extremely wide ranging. Plastics are found in virtually all the things we do and use throughout all aspects of life. Still, a few uses dominate current plastics consumption. In the EU member states, packaging is the dominant use of plastic (40%), followed by building and construction (20%), automotive (10%), electrical and electronics (6%), household ware (3%), and a variety of applications (Figure 31.1).

There are many types of plastics. Plastics are subdivided into thermoplasts and thermosets. Thermoplasts do not undergo chemical changes when heated and can be molded again. The main thermoplasts are polyethylene (PE), polypropylene (PP), polystyrene (PS), and polyvinyl chloride (PVC). PE can be subdivided into high-density polyethylene (HDPE), low density polyethylene (LDPE), and linear-low density polyethylene (LLDPE), based on the way the polymer chains are distributed in the plastic. Thermosets can melt and take shape once, i.e., after solidification they stay solid. In the thermosetting process, a chemical reaction occurs that is irreversible. Polyurethane (PUR) is one of the most used thermosets. Recycling of thermosets is more difficult and can only be done in a chemical process.

Figure 31.2 depicts the key plastic types used in Europe in 2019. The global distribution of plastic types is comparable to the distribution found in the European Union. National distribution of uses and plastic types may vary, depending on specific circumstances. Some plastic types are used predominantly in specific applications. For example, the use of PVC in packaging has been reduced in many countries, but PVC, together with PUR and PS, is one of the key plastics used in the building and

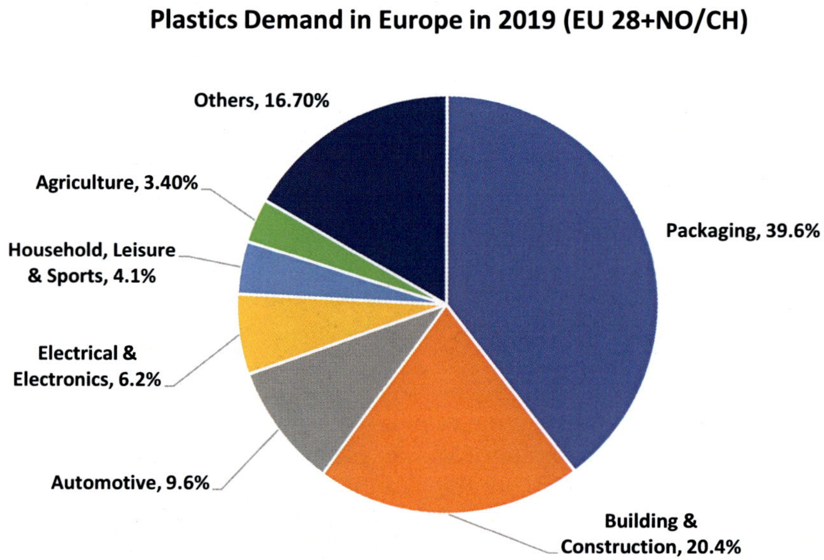

FIGURE 31.1 Plastic demand by application in Europe in 2019 (EU 28, Norway, and Switzerland). *Data from PlasticsEurope (2021).*

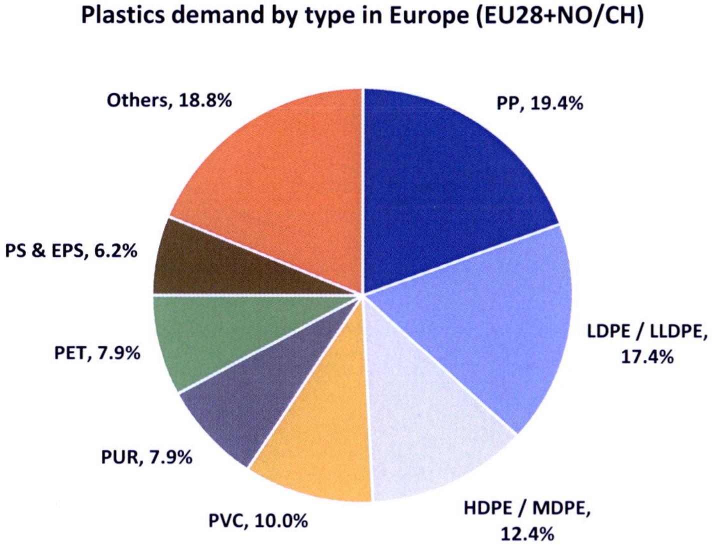

FIGURE 31.2 Plastic demand by type in Europe in 2019 (EU 28, Norway, and Switzerland). *Data from PlasticsEurope (2021).*

construction sector. In the packaging markets, the polyolefins (i.e., PE, PP), PET, and PS dominate. Table 31.1 summarizes some of the key uses of the main plastic types used.

Plastics are an attractive material for many applications, as plastics are easy to shape, and material characteristics may be adapted or even tailored to the specific application. This is often done by adding additives such as plasticizers, filling material, colorants, flame retardants, or by forming composites through adding layers of special materials with, e.g., special barrier properties for oxygen, carbon dioxide, or UV light. While this positively affects the application of the plastic (e.g., reduced material use), this may become a barrier in recycling, effectively limiting or even blocking recycling of the plastic. Typically, additives and fillers are around 20% of the plastic weight and can be even more for some applications. The statistics shown in Figure 31.2 may, hence, contain other materials. Figure 31.2 is based on the main plastic in the product application.

Currently, biobased and/or biodegradable plastics are still less than a minute fraction of the total volume of plastics used. However, in the last decade, the emerging biobased plastics have experienced a rapid growth. The global capacity of the emerging biobased and/or biodegradable plastics has increased from 0.1 Mt in 2003 to 2.1 Mt by 2019. The global production is likely to grow strongly in the next decade. The key bioplastics at this moment are starch plastics and blends (21%), PLA (polylactic acid, 14%), biobased PE (12%), and partially biobased PET (10%).

31.3 RECYCLING

The 2019 plastic use in the European Union was estimated at 51 Mt, based on the consumption by plastic convertors and processors. Post-consumer wastes generated are estimated at 29 Mt. These figures demonstrate that besides the production of industrial wastes, there is also a considerable stock buildup in society. For example, many plastics currently used in construction, representing 20% of total plastic consumption in the EU, end up in buildings with

TABLE 31.1 Typical applications of common plastic types.

Number	Abbreviation and name	Typical applications
1 PETE	PET: polyethylene terephtalate	Bottles and flasks for soft drinks, mineral water, detergents, and pharmaceutical products; blister packs; packaging for ready meals
2 HDPE	HDPE: high density polyethylene	Thick-walled applications such as bottles and flasks, barrels, jerry cans, crates, and jails; films for refuse bags; packaging for carpets and instruments
3 V	PVC: polyvinyl chloride	Blister and press through packs for medication; films for perishables
3 V	PC: polycarbonate	Refillable milk bottles; specific refillable packaging for liquids
4 LDPE	LDPE: low density polyethylene	Foil and film, such as shrink-wraps, tubular film, sacks, and covering wraps for bread, vegetables, fruit, and carrier bags
4 LDPE	LLDPE: linear low-density polyethylene	Ultrathin films: elastic wrap foil or stretch films
5 PP	PP: polypropylene	Buckets, crates, boxes, caps for bottles or flasks, transparent packaging for flowers, plants, confection products; yogurt and dairy product cups; industrial adhesive tapes
6 PS	PS: polystyrene	Food service disposables; boxes and dishes for meat products and vegetables; boxes for ice, for example; boxes for video tapes
6 PS	EPS: expanded polystyrene	Buffer packaging for household devices, electronics and instruments; flasks and pipettes for the medical industry; egg packaging and fast-food packaging
7 OTHER	Other	Other packaging

The numbers refer to the code used to sort plastics for recycling.

long lifetimes. These plastics will become available as waste in the future as these buildings are renovated or demolished.

This stock buildup also explains the considerable differences between the plastic uses and plastic wastes. Figure 31.3 depicts the distribution of use categories in consumption and plastic wastes. Packaging is by far the dominant factor contributing to plastic waste (61%), while it is about 40% of plastic use. Especially, plastic waste from construction, automotive, electronics, and household ware trail behind consumption figures, due to retention of the products in stocks in society.

In 2018, based on figures of PlasticsEurope (PlasticsEurope, 2021), 29.1 Mt of postconsumer plastic waste were collected in the European Union. Of this, more than 75% (22 Mt) was recovered and 25% (7.2 Mt) was disposed of with municipal solid waste. Of the 22 Mt of recovered plastic waste, about 9.4 Mt were sent to recycling, while the remainder was likely used as refuse derived fuel (RDF) or incinerated in MSW incinerators with energy recovery (about 12.4 Mt). While Europe can be considered a leader in plastics recycling, only about 33% of plastic waste is sent to recycling, often "downcycled" and thus not reused in the original application. This is much lower than that of other materials. For example, the recycling rate of paper and board is over 80% in Europe, with a much lower share of downcycled material.

Large differences in recovery and recycling rates can be observed across countries. For example, in 2018, Switzerland, Germany, Austria, Luxembourg, Belgium, Sweden, Denmark, the Netherlands, and Norway had very high recovery rates. These countries had a ban on landfilling of combustible waste, which may have contributed to high recovery rates.

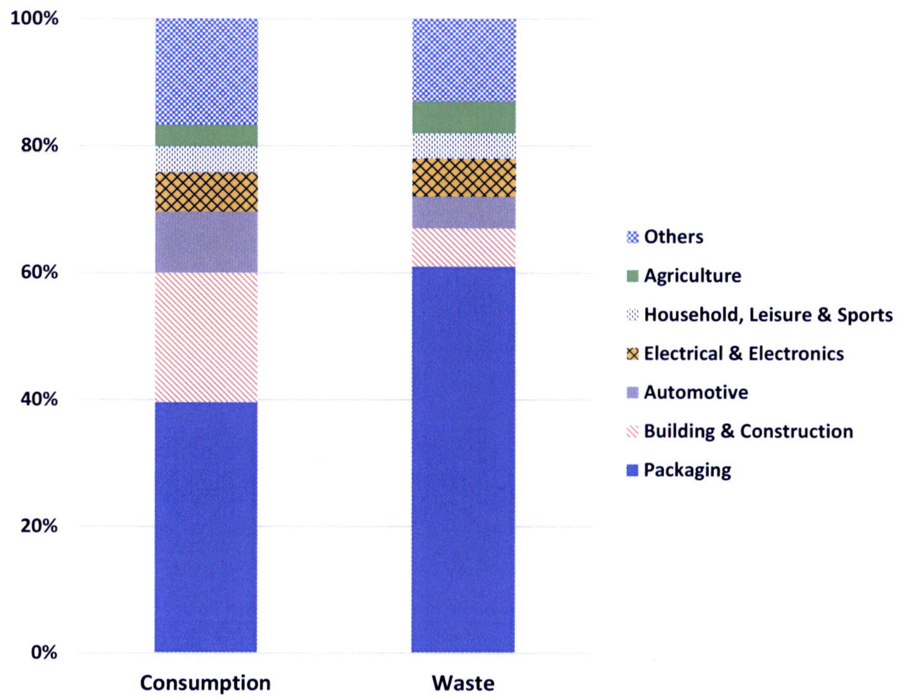

FIGURE 31.3 Plastic consumption and plastic waste generated per sector in EU in 2018. *Data from PlasticsEurope (2021).*

The recovery rates for other member states varied between lows of 20% to about 90%. The recycling rates also varied considerably but are all far below the recovery rate. The highest rate of recycling was seen in Norway (around 45%) and the lowest in Bulgaria at 20%. Some countries do not have energy recovery systems and therefore all recovered wastes were recycled (e.g., Malta), while Germany, Sweden, Belgium, the Netherlands, and Norway had recycling rates of 35–45% compared to very high recovery rates (95–99%), meaning that a large fraction of the recovered plastic is incinerated.

Note that the definition of "recycling rate" is the ratio between collected plastic waste sent to recycling and the total plastic waste generated. The actual recycled plastics are less than the collected plastic waste, due to the loss in sorting and recycling processes.

31.4 MECHANICAL RECYCLING

Mechanical recycling is the main technology used to recycle plastics. Mechanical recycling of plastics in the European Union slowly increased from just below 5 Mt in 2006 to about 9.4 Mt in 2018. Mechanical recycling typically includes four steps. The collected material is first sorted (step 1). The sorted material is shredded (step 2), and then washed and dried (step 3). The material can then be melted and reprocessed to make pellets, which can be used by a manufacturer or convertor, or products directly (step 4). In the following we discuss the key steps in more detail.

Plastics for recycling are collected through various systems. These systems rely mostly on separate collection systems, where the plastics are separated at the source. These can even be focused on specific plastic types (e.g., PET) or products (e.g., bottles). Sometimes, very pure streams of waste plastic are recovered, e.g., through dedicated collection systems for plastic bottles through a refund system. These streams can deliver high-quality recycled material with desired properties, requiring only minimal sorting to remove any impurities (e.g., bottle caps, labels). Alternatively, postcollection separation systems can separate plastics from municipal solid waste or a fraction of combined dry materials (e.g., metals, plastics, paper). Postcollection separation may increase the share of material collected, albeit often at a lower quality due to the increased risk of contamination.

31.4.1 Sorting

The first step in a recycling process is the collection of the waste and transporting it to a sorting plant to sort the plastic mix. The collected mix of plastics usually consists of various types of plastics, especially for postconsumer waste. Preconsumer (production) waste can be (relatively) pure. In the case of postconsumer waste, there are usually still nonplastic impurities in the material, such as labels and little pieces of metal. These are first removed. Separation into various types of plastics is required to improve material quality. The separation of mixed plastics is challenging, and various techniques are applied in varying combinations. The processes and the order in which they are used are defined by the composition of the mixed waste stream. Eddy current separator, sink-float separation, drum separators/screens, induction sorting, X-ray technology, and near infrared (NIR) sensors are the most used separation techniques. Most companies use a combination of different techniques to obtain sufficiently pure streams. The design of a sorting installation may be tailored to the incoming stream of plastic waste, to optimize sorting efficiency. The achievable purity level is a trade-off between (energy) costs and market requirements. This, by definition, will result always in impurities. The maximum achievable purity by separating mixed plastics waste is around 94–95%. High-quality recycled material should have at least a purity of 98% to be used as input into manufacturing processes. This means that in subsequent steps further refining is necessary if high-quality recycled material is to be produced. The key sorting technologies are:

- **Induction sorting**. Material is sent over a conveyor belt with a series of inductive sensors underneath. These sensors locate different types of metal, which are then separated by means of fast air jets.
- **Eddy current separator**. An "eddy current" is an electric current that occurs when changing the magnetic field within a conductor and is used to separate nonferrous metals.
- **Drum separators/screens**. These separate materials based on the particle size. Waste is fed into a large rotating drum, which is perforated with holes of different sizes. Materials smaller than the diameter of the holes drop through, while larger particles remain in the drum.
- **Sink-float separation**. The plastic waste is separated based on the specific weight of the material relative to the fluid. In water, some plastics (e.g., PET, PVC, and PS) will sink, while others will float (e.g., PE, PP, EPS). After the sink-float separation, the fractions still need to be processed in another process to separate the different polymers.
- **X-ray technology**. X-rays can be used to distinguish between different types of material based on density.
- **Near infrared sensors (NIR)**. When plastics are illuminated, they mostly reflect light in the near infrared wavelength spectrum. The NIR sensor can distinguish between different materials based on the way they reflect light. The fractions are then blown with an airjet to separate from the mixed stream. This is currently the preferred method by the industry to accurately identify the many different polymers.

31.4.2 Shredding

The next step is to reduce the size of the scrap, to enable processing larger pieces of plastic waste and to improve the density of the material for more efficient storage and transport. A shredder consists of rotating blades driven by an electric motor, some sort of grid for size grading, and a collection bin. Materials are added to the shredder by a hopper. The product of shredding is a pile of plastic flakes.

31.4.3 Washing

After the plastics are shredded into small flakes they are washed. While most postconsumer waste is washed, this is not true for all input material used for recycling. Some regrind or even agglomerate is processed instantly. Cold or hot water, up to 60°C, may be applied. Cold water use may result in increased use of chemicals (e.g., sodium hydroxide) and mechanical energy. The wastewater from the washing is often treated internally for reuse. The washed plastic flakes are dried until they contain less than 0.1 wt% moisture and are ready for reprocessing.

31.4.4 Reprocessing

There are different techniques for reprocessing, the most common ones being agglomeration and extrusion.

Agglomeration is mainly applied for recycling plastic films. Film is cut into small pieces, heated by friction to allow for agglomeration and subsequently cooled down by injecting water. This is usually carried out in a single machine. The product is referred to as crumbs or agglomerates and is not ideal for further processing. The agglomerates can be mixed with plastic flakes for extrusion. The agglomeration process is very energy intensive, approximately 300–700 kWh/t of plastic. Agglomeration can be avoided for most injection and extrusion grade plastics.

Extrusion is the most applied technique for reprocessing recycled plastic. It is commonly used to manufacture pellets from virgin plastics and used to produce pellets from recycled material. The material is blended and then injected in the extruder from a hopper. It comes into contact with a rotating screw that forces the plastic flakes forward into a heated barrel at the desired melt temperature of the molten plastic, ranging from 200°C to 275°C. The pressure allows the plastic (beads) to mix and melt gradually as they are pushed through the barrel. The melt is degassed to remove oils, waxes, and lubricants. Finally, the molten plastic is pushed through a sieve to remove impurities, cooled, and pelletized.

After the agglomeration or extrusion phase, the agglomerates or pellets are ready for the final processing step, the choice of which is determined by the final product:

- **Injection molding**. The first stage of this manufacturing process is identical to that of extrusion, i.e., the pellets are molten again, but after that the polymer is pressed with high pressure into a split mold. The mixture is pressed into the mold until it is full and, after cooling to allow the plastic to solidify, the mold is opened and the product can be removed.
- **Blow molding**. The spiral screw of the extruder forces the plasticized polymer through a die. A short hollow tube and compressed air is used to expand the tube until it fills the mold and obtains its required shape. This manufacturing technique is used for manufacturing bottles and other containers.
- **Film blowing**. Film blowing is a process used to manufacture items such as plastic bags. It is a technically more complex process and requires high-quality raw materials. The process involves blowing compressed air into a thin tube of polymer to expand it to the point where it becomes a thin film tube.
- **Fiber extrusion**. The melt extruded polymer (polyester) is sent to the spinneret where the spinning of filament takes place. The filaments then pass through a denier setter before they enter the finishing steps where the spun filaments are drawn, dried, cut into staple fiber, and finally baled for sale.

In practice, high-quality recycled material can be made from containers made from HDPE, PP, and PET (especially bottles). Specialized streams can also be recycled to provide high-quality recycled material. For example, in several countries, including the Netherlands, PVC window frames are collected separately, and recycled into material for new window frames. However, other plastic products, especially films and foils, represent a large challenge for the recycling process. Films are hard to sort, and the large use of plasticizers and other additives makes it impossible to guarantee a high purity of the recycled material. Hence, this material may currently be used to replace other materials than plastics (e.g., in construction) or may be incinerated as refuse derived fuel (RDF). The environmental and economic impact of recycling may hence vary considerably from case to case, needing careful analysis.

As shown earlier, the successful examples of high-quality plastic recycling rely on the use of uniform products and materials, requiring the careful separation of different products or plastic types. Plastics offer a wide range of material characteristics (e.g., flexibility, color, barrier, even flame resistance), which can be adapted to the needs of the specific function through the use of additives or layering of multiple materials. The advantage of plastics also represents an important technological barrier for recycling, as the subsequent variety of material compositions limits recyclability. In the transition to a circular economy, the development of materials, composites, or new plastics, including biobased plastics, needs careful considerations to see how these materials will fit in, and affect, current and future collection and recycling infrastructure. Chemical recycling is touted as a way to deal with the large variety of plastics found in today's society. Various chemical recycling technologies are under development (Hann and Connock, 2020).

31.5 CHEMICAL RECYCLING

Typically, plastics are recycled mechanically. However, some plastic cannot be mechanically recycled because of the characteristics of the material (e.g., thermosets) or because of the low purity. The low purity is caused by mixing with other plastics, being composites, being laminated with multiple layers, or containing additives and fillers. Feedstock recycling is then an option. Feedstock recycling can potentially lead to monomers to produce new plastics again by cracking the pyrolysis oil or recovering methanol from the syngas obtained via gasification. These routes are still in their early development stage. The concept is particularly attractive for polyolefins (PP, LDPE, and HDPE) for two reasons. First, unlike polyesters, polyolefins cannot be simply depolymerized. There are few established technological options to obtain the monomers other than feedstock recycling. Second, much of the LDPE and PP are used for film applications that are often heavily contaminated with additives and food residues, making it difficult to mechanically separate different homogeneous monostreams. Presently there are no commercialized operations of feedstock recycling back to plastic monomers, as the pilot plants are now under construction.

In 2008, only 0.07 Mt of plastic was processed by feedstock recycling in the European Union. In Europe, currently about 0.22 Mt of capacity has been installed for feedstock recycling to provide an alternative reducing agent for the iron and steel industry. Low purity, mixed plastics are converted into syngas and used as a reducing agent in blast furnaces to replace coke or injection fuels. While some of the blast furnace gas can be used as a chemical feedstock, most would serve as energy and reductant.

Next to mechanical recycling and feedstock recycling, some plastics, such as PET and PS, can be chemically recycled via depolymerization (Hann and Connock, 2020). In this technology, solvents (hence the name solvolysis) are used to, e.g., depolymerize the plastics into monomers that can be repolymerized to produce virgin polymer. Depolymerization processes often require high purity for plastic feedstock. The state of this technology varies for different plastics. It is currently under development and piloted in a few plants. The chemical recycling of polystyrene is now being implemented in a first plant using the so-called creasolv process (www.creacycle.de), that also allows the removal of bromide-based flame retardants (see, among others, Fraunhofer, 2022). Various chemical recycling process are under development for polyesters like PET and are piloted at various scales. Some cases of chemical recycling of PET are described when impact assessments are discussed.

31.6 IMPACT OF RECYCLING

The environmental impact of recycling depends on many factors, ranging from the energy used for collecting the plastic waste to the type of material and application being replaced by the recycled plastic, which is partly determined by the quality of the recycled material. The markets for recycled plastic are still limited but growing. Applications of the recycled material vary and will affect the overall environmental benefits and economics of recycling.

As a case study, we discuss the results of a life-cycle analysis (LCA) of PET bottle recycling based on Shen et al. (2010a,b). Globally, almost three times as much PET is used for textile production than for packaging. In the European Union, the amount of collected postconsumer PET bottle waste (including the non-PET fractions of caps and labels) has increased from 0.2 Mt in 1998 to 2 Mt in 2018. Roughly about 60% of all used PET bottles in Europe were collected for recycling in 2018 (Eunomia, 2020). If the non-PET fraction of the bottle waste is excluded, the real PET stream that was sorted for recycling was estimated at 52%. In the member states, PET bottle waste is collected with different schemes, of which the deposit return scheme offers very high recycling rate of at least 80%. The PET bottle waste collected via the deposit return systems delivered up to 35% of all collected PET in 2018 in Europe (Eunomia, 2020). It is expected that PET bottle waste collection in Europe will continue to increase, driven by the long-term demand of high viscosity recycled PET.

Figure 31.4 shows the flow sheet of the production of recycled PET flakes. After the baled bottles are opened, loose bottles are sorted by color and material type. Transparent (uncolored) bottles have a higher economic value than blue and green ones. The unwanted color fractions and unwanted materials (e.g., paper and metal) are either sold as by-products, or disposed of in local municipal solid waste (MSW) management facilities. MSW can be incinerated with or without energy recovery, or landfilled, depending on the available local infrastructure. Next, the bottles are sorted. The plant in Europe uses automated sorting, while most Asian producers use manual sorting. Some producers include a step using hot water washing to remove labels before the sorting process. The plastics labels are either sold as by-products (mainly consisting of LDPE and/or PVC) or sent to local MSW management. The bottles are then chopped into flakes, followed by a float separation step to separate PET from other plastics (e.g., HDPE caps) based on density differences. PE obtained from this step is sold as a by-product. The PET flakes are then washed in a cleaning solution, rinsed, and dried. In some production lines, a second chopping step (also called "fine crushing") is required to ensure that the PET flakes meet the quality requirements. Finally, the dried PET flakes are ready to be transported to a pellet plant or a fiber plant.

Mechanical recycling is the physical conversion of flakes into fiber or other products (e.g., bottles or sheets) by melt-extrusion. Currently, there are two ways to produce recycled fiber from mechanical recycling. In flake-to-fiber recycling (1), flakes are off-loaded and they are dried in a column dryer before they are melt-extruded. The extruded polymer is filtered before it passes through the spinneret where filament spinning takes place. After the filaments pass a denier setter, they enter the finishing process where the spun filaments are drawn, dried, cut into staple fiber and baled. The entire process has 1% process solid waste. In many other mechanical recycling plants (2), flakes are first extruded into pellets and then converted into fiber and other products. PET flakes are dried prior to the melt extrusion step. The extruded polymer is further purified through a filtration step. After cooling, the polymer is pelletized and dried. The PET pellets are then delivered to the fiber spinning plant where they are melt-spun into filament fiber. A small amount of ethylene glycol (EG) may be added to meet the final quality requirements.

In chemical recycling, PET polymer is broken down into monomers or oligomers via various depolymerization technologies. Chemical recycling has a higher cost than mechanical recycling. It usually requires a large scale in order to become economically feasible. The important advantage of chemical recycling is that the quality of virgin polyester can be achieved. Current commercially available chemical recycling technologies include glycolysis, methanolysis, and alkaline hydrolysis. The glycolysis of PET yields the oligomer bis-hydroxyl ethylene terephthalate (BHET). The process is usually conducted in a wide range of temperatures 180–250°C with excess ethylene glycol (EG) and in the presence

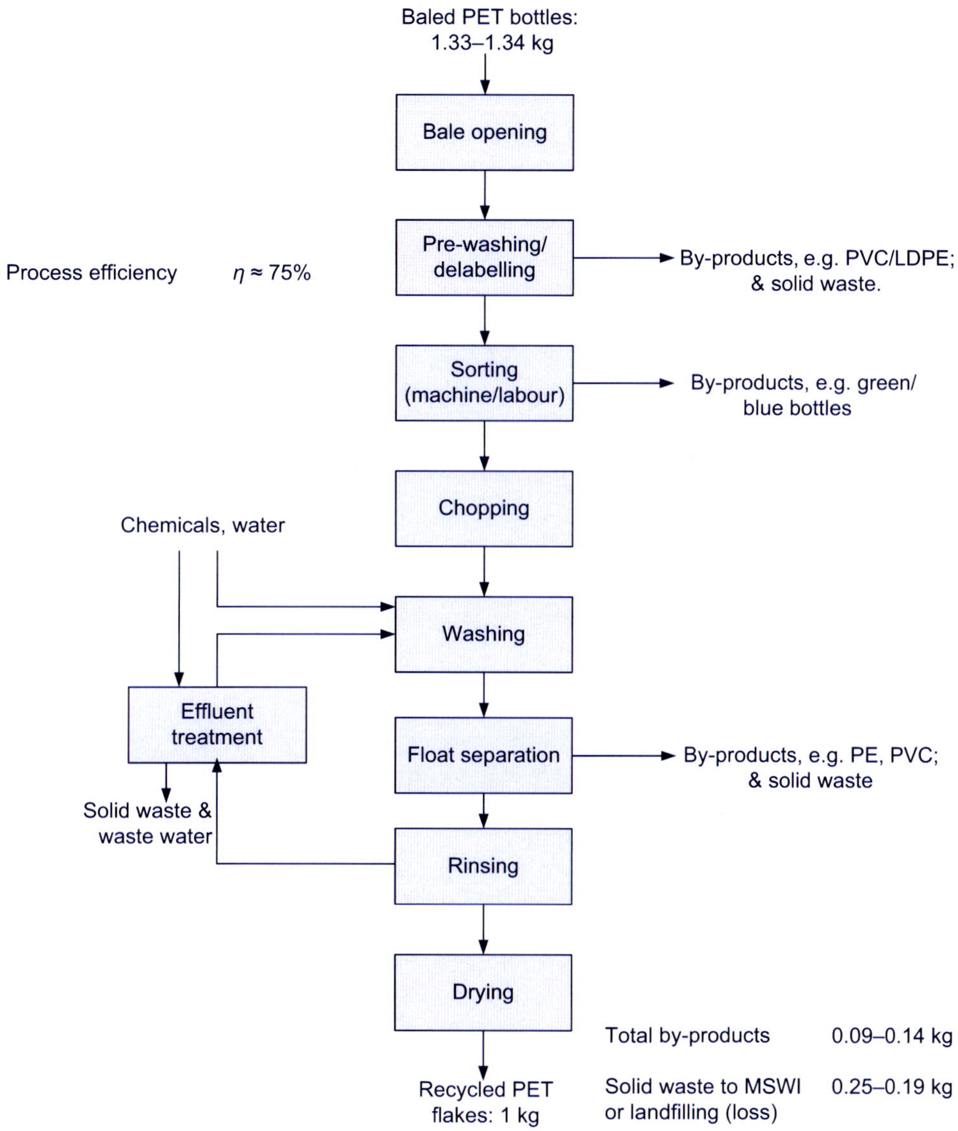

FIGURE 31.4 Schematic presentation of PET bottle recycling (including mass balance). *From Shen et al. (2010b).*

of catalysts. After the glycolysis process, the oligomer passes through a fine filtration step before it is repolymerized into PET. The recycled polymer is then spun into fiber. The entire process creates about 5% process solid waste. In methanolysis, PET is depolymerized with methanol to DMT (dimethyl terephthalate) and EG in the presence of catalysts under the pressure of 2–4 MPa and temperature of 180–280°C. The reaction mix is cooled and DMT is recovered from the mix via precipitation, centrifugation, and crystallization. The recycled polymer is then converted into fiber via spinning and finishing processes. The methanolysis route is used commercially. The repolymerization of the DMT is technically identical with the polymerization process leading to virgin PET. The recycled amorphous PET polymer is sent to the fiber production plant for being upgraded to bottle grade resin for bottle production.

The LCA of the different processes has shown that recycled polyester fibers produced from mechanical recycling have lower environmental impact than virgin polyester. The results (Shen et al., 2010b) show that recycled polyester fibers offer important environmental benefits over virgin polyester. Depending on the allocation of the benefits of open loop recycling, energy savings of 40–85% and reductions in greenhouse gas (GHG) emissions of 25–75% can be achieved.

However, PET fiber used in textiles is a product that cannot be further recycled via mechanical recycling. Chemical recycling is technically possible, but the economic viability of large-scale operation is still to be proven. Recycled fiber produced from chemical recycling offers lower impacts compared to virgin polyester. While mechanical recycling has a better environmental profile than chemical recycling, chemically recycled fibers can be applied in a wider range of applications than mechanically recycled fibers. Another important way of recycling PET bottles is bottle-to-bottle recycling. In this case, a close-loop recycling system is formed. In theory, PET can be recycled multiple times before it is finally converted into fiber. The environmental impact of such "cascading" recycling systems has been studied by Shen et al. (2011). The results show that based on the current global demand of PET bottles and fiber, the recycling system, which includes both bottle-to-fiber recycling and bottle-to-bottle recycling, can offer 20% of impact reduction in both primary energy demand and GHG emissions. Multiple recycling trips can further reduce the environmental impact by maximally six percentage points, due to the lower share of bottle demand (35%) compared to that of the fiber demand (65%).

31.7 CONCLUSIONS AND OUTLOOK

Plastic recycling is still limited compared to most other bulk materials. Recycling rates for plastics are increasing in many countries around the world, while an international market for recycled plastics is developing. To further increase the recycling rate, the industry faces a number of challenges. The key challenge is the quality of the recovered and recycled material. The advantage of plastics (i.e., the tailoring of material characteristics to product needs) is a key barrier to recycling, as the use of fillers, additives, and composites limits the production of high-quality recycled materials if plastic streams are not sufficiently separated in homogeneous streams. Also, contamination with other materials in the collection of plastic limits the ability to come to a cleaner recycled plastic product. Hence plastics recycling needs a systematic approach by integrating collection, recovery, separation, and recycling technology. The solution will consist of an effective combination of these three steps, and may vary for different product streams and waste management system. This is observed in relatively high recycling rates of PET and HDPE, while the recycling rates of films are still very low (typically not exceeding 15%). Dedicated collection and recovery systems will help to increase raw material quality. Furthermore, increasing the recycling rate will also need to include the design stage of products. While common in certain areas, in the key uses of plastics (e.g., packaging), design for recycling is still limited. The use of different plastics for different parts of packaging (e.g., caps, labels, containers), and the use of a variety of additives and fillers are currently barriers for high recycling rates. Improving design practices may reduce this variety and enable better recycling.

New technology will be needed to produce high-purity plastic from recovered material, to improve separation efficiency and effectiveness (e.g., separating the growing number of biobased plastics), and to handle mixed plastic wastes to still produce a high-quality recycled product. Chemical recycling may be part of this and will need further development to decrease costs and energy use. The relatively high energy costs of many chemical recycling technologies do require a systems approach to ensure that plastic recycling results in the best environmental and economic impacts, through smart combinations of product design, collection, separation, mechanical, and chemical recycling technologies.

Finally, currently a large part of the recovered plastic is "downcycled" or used as fuel to generate energy in industrial processes or incinerators. Better monitoring is needed to track actual recycling rates and the fate of the recovered material, and to allow optimization and "quality cascading" of recycled plastic to generate the highest economic and environmental gains. Today's information in most countries does not enable such analysis and optimization. Such a system, and the growing international market, will need an international, widely accepted and used quality certification system, integrated into a transparent monitoring system.

References

Eunomia, 2020. PET Market in Europe: State of Play—Production, Collection and Recycling Data. Bristol, United Kingdom https://www.plasticsrecyclers.eu/plastics-recyclers-publications. (Accessed 17 March 2022).

Fraunhofer, 2022. Recycling Plastics—The Creasolv Process. Fraunhofer Institute for Process Engineering and Packaging IVV, Germany. https://www.ivv.fraunhofer.de/en/recycling-environment/recycling-plastics-creasolv.html. (Accessed 17 March 2022).

Hann, S., Connock, T., 2020. Chemical Recycling—State of Play. Eunomia, Bristol, United Kingdom. https://chemtrust.org/wp-content/uploads/Chemical-Recycling-Eunomia.pdf. https://eunomia.dev.marcom-test.net/reports-tools/final-report-chemical-recycling-state-of-play/. (Accessed 17 March 2022).

PlasticsEurope, 2021. Plastics—The Facts 2021. An Analysis of European Plastics Production, Demand and Waste Data for 2011. PlasticsEurope, Brussels, Belgium. https://plasticseurope.org/knowledge-hub/plastics-the-facts-2021/. (Accessed 17 March 2022).

Shen, L., Worrell, E., Patel, M.K., 2010a. Present and future development in plastics from biomass. Biofuels Bioprod. Biorefin. 4, 25–40.

Shen, L., Worrell, E., Patel, M.K., 2010b. Open-loop recycling: a LCA case study of PET bottle-to-fibre recycling. Resour. Conserv. Recycl. 55, 34–52.

Shen, L., Nieuwlaar, E., Worrell, E., Patel, M.K., 2011. Life cycle energy and GHG emissions of PET recycling: change-oriented effects. Int. J. Life Cycle Anal. 16, 522–536.

Further Reading

Antonopoulos, I., Faraca, G., Tonini, D., 2021. Recycling of post-consumer plastic packaging waste in EU: process efficiencies, material flows, and barriers. Waste Manag. 126, 694–705. https://doi.org/10.1016/j.wasman.2021.04.002.

Brouwer, M., Picuno, C., Thoden van Velzen, E.U., Kuchta, K., De Meester, S., Ragaert, K., 2019. The impact of collection portfolio expansion on key performance indicators of the Dutch recycling system for post-consumer plastic packaging waste, a comparison between 2014 and 2017. Waste Manag. 100, 112–121. https://doi.org/10.1016/J.WASMAN.2019.09.012.

Civancik-uslu, D., Nhu, T.T., Van Gorp, B., Kresovic, U., Larrain, M., Billen, P., Ragaert, K., De Meester, S., Dewulf, J., Huysveld, S., 2021. Moving from linear to circular household plastic packaging in Belgium: prospective life cycle assessment of mechanical and thermochemical recycling. Resour. Conserv. Recycl. 171 (March), 105633. https://doi.org/10.1016/j.resconrec.2021.105633.

European Bioplastics, 2020. Bioplastics: Facts and Figures for 2019. Berlin, Germany https://docs.european-bioplastics.org/publications/EUBP_Facts_and_figures.pdf. (Accessed 17 March 2022).

Gutowski, T., Dahmus, J., Albino, D., Branham, M., 2007. Bayesian material separation model with applications to recycling. In: Proceedings of the 2007 IEEE International Symposium on Electronics and the Environment, pp. 233–238, https://doi.org/10.1109/ISEE.2007.369400.

Joosten, L.A.J., Hekkert, M.P., Worrell, E., 2000. Assessment of the plastic flows in the Netherlands using STREAMS. Resour. Conserv. Recycl. 30, 135–161.

Kuczenski, B., Geyer, R., 2010. Material flow analysis of polyethylene terephthalate in the US, 1996–2007. Resour. Conserv. Recycl. 54, 1161–1169.

Patel, M.K., Jochem, E., Radgen, P., Worrell, E., 1998. Plastic streams in Germany—an analysis of production, consumption and waste generation. Resour. Conserv. Recycl. 24, 191–215.

Patel, M.K., von Thienen, N., Jochem, E., Worrell, E., 2000. Recycling of plastics in Germany. Resour. Conserv. Recycl. 29, 65–90.

Welle, F., 2011. Twenty years of PET bottle to bottle recycling—an overview. Resour. Conserv. Recycl. 55, 865–875.

Welle, F., 2013. Is PET bottle-to-bottle recycling safe? Evaluation of post-consumer recycling processes according to the EFSA guidelines. Resour. Conserv. Recycl. 73, 41–45.

CHAPTER

32

Black rubber products

H.C. van Zyl[1] *and Markus A. Reuter*[2]

[1]Curtin University, Perth, Australia [2]WASM: Minerals, Energy and Chemical Engineering, Faculty of Science and Engineering, Curtin University, Perth, Australia

32.1 INTRODUCTION

Our modern industrial society depends on a wide range of rubber products. Rubber, both natural rubber from trees and synthetic rubber as produced by the chemical industry, is a light opaque brownish color. The pure rubber can be colored by adding mineral colorant, for example red, white, and green, as in household rubber bands.

The largest numbers of rubber products are, however, black and various tones of grey. This is because the rubber is mixed with a substantial volume of carbon black in the process called compounding. The production of carbon black is a global industry and carbon black is one of the most common industrial chemicals produced. Carbon black is sometimes wrongly referred to as soot (as in the German name "Industrieruss") but it is more than just a colorant. It allows the rubber technologist to compound the rubber products for a wide variety of applications.

It is of crucial importance to understand, also in the context of recycling, that a tire contains many different grades of carbon black, which have to comply with rather strict norms to satisfy the quality grades. Carbon black's grades are one of the most important components that determine the main character of the rubber components in the tire. The tread, bead, sidewall, carcass, and belt of the tire each consist of a different rubber compound and often different grades of carbon black. The carbon black grades differ in their primary particle size, aggregate size, and structure, as described in the carbon black manufacturing process review by Okoye et al. (2021).

Carbon black is bonded to the rubber by a physical surface reaction, not a chemical reaction. To meet the rubber compound specification, all these surface specifications of carbon black must be met to ensure that the desired physical bonds between the carbon black and the rubber are established. This imparts the required functionality of each component in the tire and thence whether the required safety can be met.

To achieve the required bonding functionality, the carbon black industry offers the tire manufacturers more than 10 standard grades of carbon black for the rubber industry, as defined by the American Society for Testing and Materials (ASTM), each with its own well-defined specification for particle size, aggregate size, and structure (Himadri, 2021). Additionally, the commercial carbon black must meet very stringent standards for purity and must be free of foreign matter, hard coke particles, and other

Handbook of Recycling. https://doi.org/10.1016/B978-0-323-85514-3.00007-5

contaminants. The difference between carbon black grades with their narrow grade definition and undefined soot from pyrolysis is substantial, and this becomes evident when considering the manufacturing companies' literature on the topic of what carbon black is.

This chapter will solely deal with black rubber products—those rubber products that are a blend of rubber and a substantial amount of carbon black. Grey rubber products are also included in this category, as they contain carbon black and a whitening agent like silica or chalk. This chapter will touch on mechanical rubber products, but the bulk will be devoted to rubber tires and the recycling of these. This is especially relevant as tire production for vehicles is increasing exponentially, exceeding 2.9 billion tires in 2017, and tire waste is nearly proportional to the tire production (Siddika et al., 2019).

While this contribution cannot discuss the art of rubber compounding and the vulcanization process in its entirety, it suffices to say that the modern transport tire, as the key example of this chapter, is a rather complex multimaterial industrial product that must contain different grades of carbon black to perform its overall function.

In this chapter various papers have been cited that lead you, the reader, to a vast body of literature to obtain an overview of this rather complex topic. In the context of the circular economy, it must always be remembered that in the end materials must reach the quality requirements so that the high quality and safety standards of tires are met. Successful recycling of tires is thus determined by if and how the specifications of carbon black can be met. Hopefully, this contribution will provide you with the insight that this may be a rather complex task.

32.2 MECHANICAL RUBBER GOOD (MRGS)

Mechanical rubber goods (MRGs) are sometimes referred to as industrial rubber goods. The term includes all (mainly black) rubber components used in the automotive, energy, mining, and housing industries. MRG is commonly used to describe all the parts and products made from rubber and that require some textile reinforcement (or steel cord) to provide mechanical resistance.

The following list is not complete. Typical MRG applications are tubes and hoses (vehicle, oil, water, and air), (conveyor) belts, extruded profiles like wiper blades and seals, molded products, gaskets, antivibration engine mounts, rubber flooring, roofing, and footwear (Figure 32.1).

As is apparent from the long product listing, the MRG industry operates with a vast number of compound formulations to fulfill the requirements for each specific application. MRGs are thus even more diverse than the tire products. The number of forms, components, and formulations mixed with the rubber varies widely. The rubber itself can be both natural and synthetic. The physical forms—sheets, hoses, profiles—vary widely. The individual component weight is generally less than a commonly used tire; the only exception is conveyor belts, which are large and heavy.

These are the reasons why MRGs are generally rarely recycled and not mentioned when rubber recycling is addressed.

32.3 RUBBER TIRES

Rubber-based tires are universal in their use, and it is barely possible to envision our fast-moving everyday life without them. Rubber tires are often wrongly considered to be nonhigh-technology products. They are in reality very complex products, as they vary in size, application, and components. Tires may simply look like a round piece of rubber, but there are many different components that work together to achieve a common goal of delivering the desired performance (Figure 32.2). Each part of the tire has its own function and rubber

FIGURE 32.1 Industrial rubber products, in this case tubes, hoses, and window wiper rubber used in automotive applications [Evonik].

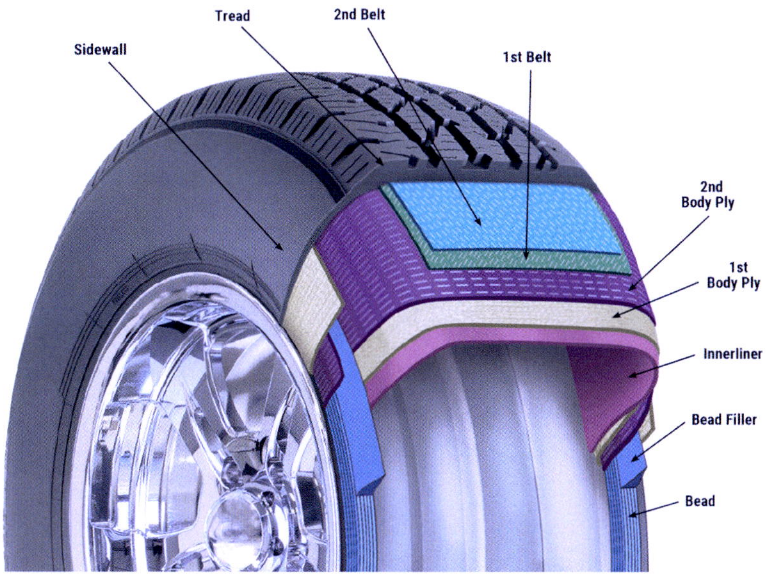

FIGURE 32.2 Structural elements of tires [US Tire Manufacturing Association].

compound. This will be elaborated on further in the section on pyrolysis.

The tread is the thick layer of rubber compound that comes in direct contact with the road surface. This layer is generally wear-resistant and increases the durability of the tire. The rubber compound used is a compromise between long economic life (wear resistance) and safe road holding in wet and dry conditions. Also, it protects other components, including belt and carcass, from failure due to harsh road impacts.

The bead is the part of the tire that mounts onto the rim. Properly designed and constructed, the bead prevents any air leakage from the tire rims. The tire beads are also designed to fit securely around the rim even when the tire has a puncture. It also ensures that the tire does not jump from the rim when it is stressed during cornering.

The sidewall is the part of a tire between the bead and the tread where the manufacturer specifications and warnings are written. It is a rubber composite reinforced with synthetic fiber and/or steel cord that provides strength and flexibility and carries the entire weight of the car.

The carcass is the internal layer of the tire that absorbs vibration and sustains load. The carcass needs to be durable enough to allow stretching throughout the life of a tire. It also helps in sealing the tire properly, and in tubeless tires keeps the air inside the tire.

Lastly, in modern so-called steel-belted tires, the reinforcing belt comprises a metallic cord layer embedded in a rubber compound and is positioned between the tread and the carcass. It provides rigidity and shape to the tire, and at the same time protects the carcass from cuts or punctures of the tread.

Tires are used for different transport systems and purposes, covering the broad categories of tires for passenger cars, trucks, and off-the-road vehicles. They are complex products in their size, variations, composition, and purposes. It is therefore difficult to define a general tire component with lists that summarize the components as used to construct a tire. A generalized summary is given in Table 32.1, while Figure 32.3 compares different tire types.

The rubber in tires contains both natural and synthetic rubber, many different grades of carbon black, reinforcing material, other fillers like clay and silica, and chemicals. The rubber is reinforced with steel and synthetic fibers. Tires are constructed to be used in the transport industries; they are not made with recycling as a main goal and their composition makes them difficult to recycle.

To summarize:

- Truck and off-road tires contain more natural rubber than passenger car tires.

TABLE 32.1 Typical materials used in tire manufacturing in Europe and the United States.

	In United States		In European Union	
Materials	**Passenger tire**	**Truck tire**	**Passenger tire**	**Truck tire**
Natural rubber (%)	14	27	22	30
Synthetic rubber (%)	27	14	23	15
Carbon black (%)	28	28	28	20
Steel (%)	14-15	14-15	13	25
Others[1] (%)	16-17	16-17	14	10
Average weight				
New tire (kg)	11	54	8.5	65
Scrap tire (kg)	9	45	7	56

[1]*Fabric, fillers, accelerators, antiozonants, etc.*

From Sienkiewicz, M., Kucinska-Lipka, J., Janik, H., Balas, A., 2012. Progress in used tyres management in the European Union: a review. Waste Manage., 32(10), 1743.

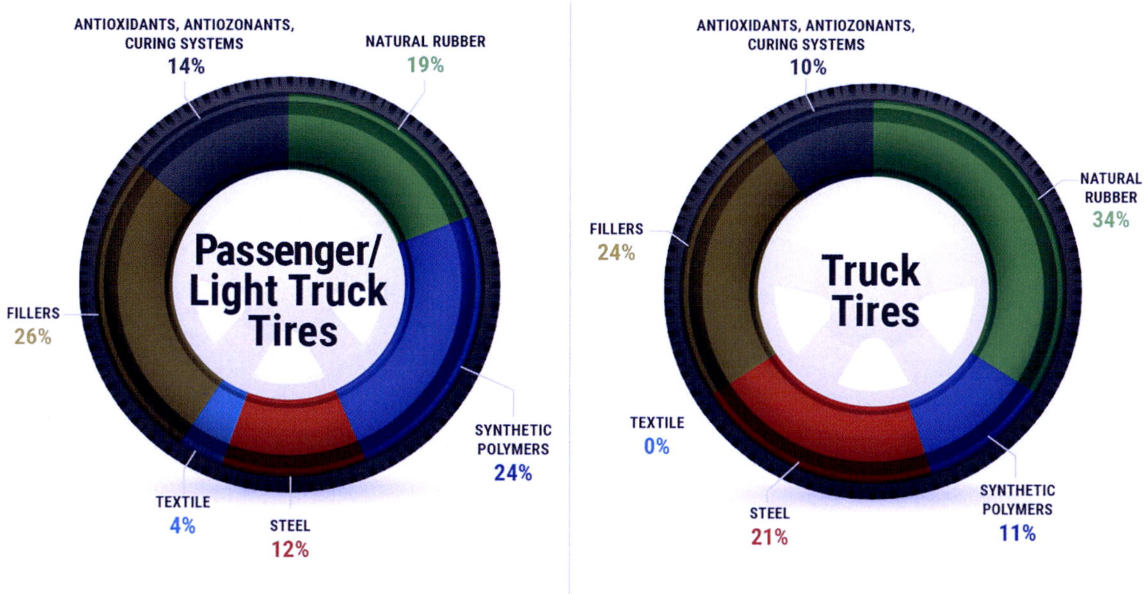

FIGURE 32.3 Tire compositions of passenger/light-duty vehicles (left) and for trucks (right) [US Tire Manufacturing Association].

- Diverse carbon black grades are essential to compound different rubber compounds of the components that make up the tire.
- Silica replaces part of the carbon black in certain types of tires.

When we want to recycle tires and take apart this complex high-tech product and once again obtain high-quality raw materials that can be used to construct new tires, we face a challenging job. The next section will discuss the challenge of recycling old and worn-out tires and the economy of recycling.

32.4 RECYCLING OF TIRES

In general, tires stay in the country where they were sold. The only exception is the export as part of a secondhand car, for instance to Eastern Europe or Africa from Europe. In the case of export to Eastern Europe, an extra set of winter tires is generally included. The recycling routes for tires are summarized in Figure 32.4; the meaning and content of most of the blocks in this diagram are discussed further in the following text.

32.4.1 Landfill and Aggregates—Repurposing

Whole nonshredded tires are not desired in landfills, as they have large void volumes. The tires can trap methane and water in these voids. The tires can then bubble up to the surface, damage the landfill liners, and contaminate the surface water. Additionally, mosquitoes may breed in the trapped water.

Shredded tires are now being used in landfills, replacing other construction materials, for a lightweight back-fill. Shredded tire material may also be used to cover and close landfill sites. Scrap tires as a cover material are cost-effective,

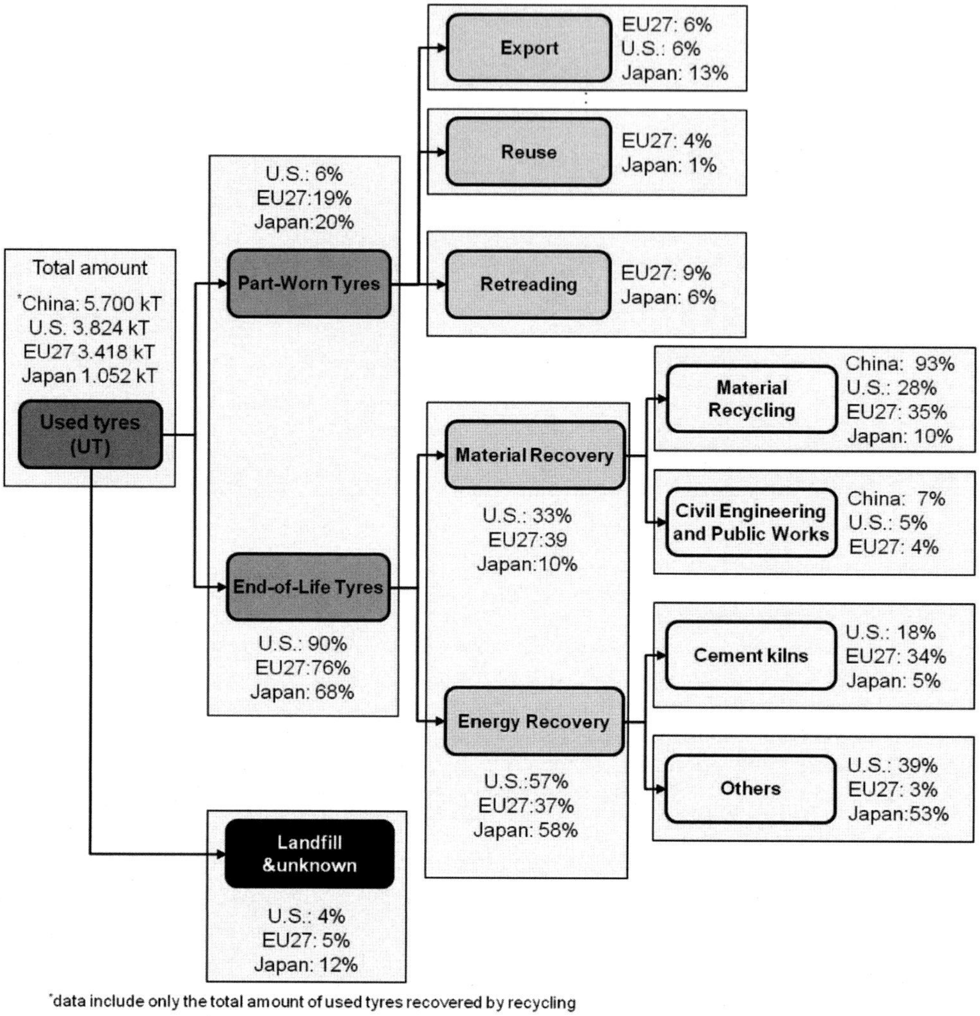

FIGURE 32.4 Waste tire management routes, amounts, and shares based on data from 2009 to 2015. *From Sienkiewicz, M., Janik, H., Borzędowska-Labuda, K., Kucińska-Lipka, J., 2017. Environmentally friendly polymer-rubber composites obtained from waste tyres: a review. J. Clean. Prod., 147, 563.*

since tires can be shredded on-site instead of transporting in other fill materials.

Ground tire materials can also be used as an aggregate substitute in concrete. The shredded rubber tires replace part of the sand and crushed stones in the concrete mass. For a comprehensive review of waste rubber used in concrete products, see Hamdi et al. (2021), Kazmi et al. (2021), da Silva et al. (2015), Strukar et al. (2019), Siddika et al. (2019), Pacheco-Torgal, et al. (2012), Sebola et al. (2018), Picado-Santos et al. (2020), Nanjegowda and Biligiri (2020), Medina et al. (2018), Han et al. (2016), and Li et al. (2019).

Both recycling practices of landfill and aggregate must consider the potential harm to the environment (Gomes et al., 2021). Furthermore, tires can be used in asphalt, as discussed in Presti (2013), for example.

32.4.2 Artificial Reefs—Repurposing

Artificial reefs are built using tires that are bonded together in groups and the tire bundles are sunk into shallow sea or lake water. The tire bundles form hollow caves that then function as safe refuges for small marine creatures. After some years, the tires are crusted with underwater plants and form artificial reefs. The effectiveness of such reefs, as measured by the long-term settlement of marine life, has not yet been finally evaluated. In addition, consult Mohajerani et al. (2020), Diekmann et al. (2019), and Halle et al. (2020) on ecotoxicity and related issues.

32.4.3 Fuel During Cement Manufacture—Secondary Fuel

Old tires can be used as an alternative fuel in the manufacturing of Portland cement in the rotating cement kilns operating between 1000°C and 1300°C. Using tires as fuel has two main drivers:

(i) Tires are a cheap replacement fuel. The main costs of using these as fuels are transport, logistics, and shredding.
(ii) Tires lower the carbon footprint of the process. They are considered a secondary fuel and not coal.

As a result, the carbon footprint of cement production with reference to the use of primary fuels can be lowered, especially as the cement industry contributes significantly to society's carbon footprint through the calcination of limestone. The additional benefit is that the iron in the tire oxidizes and thus is a secondary source of this important component of cement. Please consult, e.g., Pacheco-Torgal et al. (2012), Mohajerani et al. (2020), Hamdi et al. (2021), Siddika et al. (2019), Martínez et al. (2013), Bhatt et al. (2021), and Xiao et al. (2022), who discuss this and the application of created flue dusts in other branches of production.

32.4.4 Reprocessing and Remolding—Reuse

Tires may be recycled to be used in other lower stressed and quality rubber products. In these processes the tires are shredded. During the subsequent physical separation, many of the nonrubber parts, such as steel and synthetic cords, are removed by magnets or screens and thence separated from the fibers. The fibers that may be obtained are a complex mixture of different synthetic fibers. These are not suited to produce new tires and are difficult to recycle into high-grade materials for other applications. The fibers may be used as insulation material. However, attention must be paid to impurities, as these may have undesired effects on the insulation material. The steel cord can be used as scrap steel feedstock in the steel mills. There it is remelted and refined into steel. Rubber crumbs may be reformed to mats and floor covers for sports parks by suitable processing, or transformed and extruded into profiles and then used in park benches and fencing posts. For each of these applications, material properties, quality, toxicity, material integrity, etc. need to be considered. For detailed information, see Martínez et al. (2013), Gomes et al. (2021), Ramarad et al. (2015), and Sathiskumar and Karthikeyan (2019).

This process is generally a downcycling process, since the new products formed from the tire materials, except for the steel cord, have a less stringent material specification. In addition, these materials have a lower economic value by weight than the original tire products.

Looking anew at the design of the tire, material development and use, and manufacturing and recycling processes, from a life cycle perspective is required. This should include economic and environmental considerations. For example, Wemyss et al. (2020) discuss the various material quality issues in detail and review new material compositions and approaches that could be used in tire manufacturing to enhance rubber recycling. Dobrota et al. (2020) investigated the impact of inscribing information about the types of material in the tire on the side of the tire and adapting the tire design. The recycling process would be more efficient, and the obtained materials would be of higher quality, containing fewer impurities. Another example is the BlackCycle project (www.blackcycle-project.eu), a consortium coordinated by Michelin, that aims to create, develop, and optimize a full value chain from end-of-life tire to secondary raw materials.

32.4.5 Retreading—Reuse

Retreading tires is the process to replace the worn tread on used tires with new tread, which helps to extend the life of the tire. The casing is recycled/reused and a new tread is molded onto the intact casing (Figure 32.2). Retreading economically favors larger tires, e.g., mining and transport truck tires. Typically, lower speed applications are favored, as this decreases the possibility of the retread disjoining from the casing during use. Retreading is generally labor intensive and has a considerably lower cost than new tires. Obviously, this lowers the primary carbon footprint of transport per kilometer as only part of the tire is new, i.e., the retread, which carries with it a primary carbon footprint.

32.4.6 Pyrolysis—Chemical Recycling

The first step in the pyrolysis of tires is identical to the reprocessing and remolding process described earlier. The tires are shredded, and the steel and synthetic cords are separated and used as secondary raw materials. The remainder is black rubber crumbs, which is the raw material for the pyrolysis process. Pyrolysis is a thermochemical process where the rubber is reformed to new products by the application of heat. There are three thermochemical processes for tires: gasification, pyrolysis, and combustion, as shown in Figure 32.5 (Arabiourrutia et al., 2020).

The pyrolysis process itself is complex, with many variables like reactor design/type, catalysis, temperature, effect of impurities, and pressure, among others, affecting the principal and by-products produced, as discussed by Arabiourrutia et al. (2020), Martínez et al. (2013), Dwivedi et al. (2020), Lewandowski et al. (2019), Bockstal et al. (2019), Ramarad et al. (2015), Hita et al. (2016), Liu et al. (2020), Xu et al. (2020), Zhang et al. (2021), Mui et al. (2004)).

In essence, in the pyrolysis plant the shredded rubber crumbs from the tires are heated in a reactor vessel without oxygen. The shredded rubber is broken down into smaller molecules, after which the reaction components are separated and exit the reactor as a gas, a solid, and a liquid fraction. The three reaction fractions formed during the pyrolysis can subsequently be used and refined:

- The gas fraction can be burned directly to produce steam for power generation or to supply energy for the pyrolysis process itself.
- The liquid fraction is industrial oil, a valuable distillate that can be used in various applications, also as a fuel oil substitute. However, to date, it cannot be reformed and chemically modified at an industrial scale to monomers and polymers of synthetic rubber, which can then be reused to produce new tires.
- The solids fraction also includes mineral salts of zinc and sulfur, two components used in tire manufacturing. The bulk of solid material from the pyrolysis is the so-called soot (also referred to as char or coke). It is a low-grade

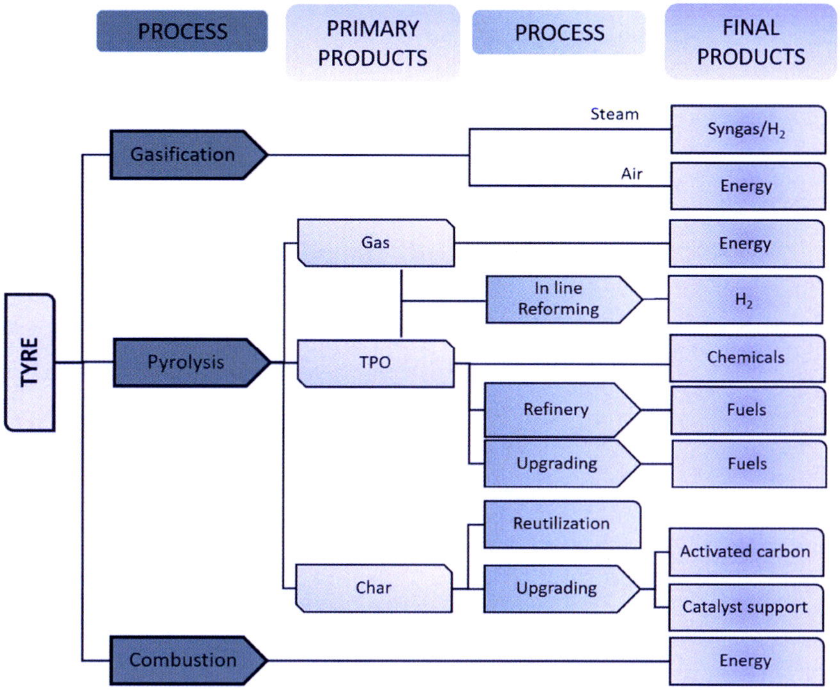

FIGURE 32.5 Summary of thermochemical processes for the valorization of waste tires and the main products obtained. *From Arabiourrutia, M., Lopez, G., Artetxe, M., Alvarez, J., Bilbao, J., Olazar, M., 2020. Waste tyre valorization by catalytic pyrolysis—a review. Renew. Sustain. Energy Rev., 129(2), 109932.*

carbonaceous product that can be used in some black rubber products of lower specification, like mats, sheets, and insulation, but are not used in bulk to produce new tires. The soot gained from the pyrolysis can only partially replace the fresh carbon black, otherwise the mechanical properties of the rubber compound in the tires will not be acceptable (Dwivedi et al., 2020). This is further supported by observing that the advertising campaigns of the tire companies do not advertise their products as containing recycled carbon black from the pyrolysis process. The tire companies stress above all long life, safety, high road mileage, and good road holding, but do not "green wash" their tires by claiming to use recycled carbonous soot from end-of-life tires.

To summarize: The pyrolysis recycling process of tires is at the present time not economical. The by-products produced are mainly used as fuels and have in sum less economic value than the pure replacements like natural gas and raw oil. This may, however, change in future when fossil energy sources become more expensive. This theme continues to be a source of innovation and research; the selected references provide an extensive overview of this topic.

It is recognized that soot (char or coke) from the pyrolysis process is not the same as standard carbon black as specified by ASTM and used by the tire industry. The normalization of the carbon blacks to this standard to meet all the specifications of primary particle size, structure, aggregate size, and purity requires significant further research. However, innovative actions by companies such as Black Bear (blackbearcarbon.com) are important for the true circular economy paradigm.

32.5 CIRCULAR ECONOMY OF TIRES

True circular economy requires the materials in a product, in this case a tire, to be restored into the functional quality that is required to return into a tire.

Due to their complexity and diversity, black rubber products are not the best economic candidate for recycling within the circular economy paradigm, despite a good figure drawn to this effect by Saputra et al. (2021).

In practice it is hard to economically recycle and produce materials of highest quality that can find their way back into tires while meeting all quality criteria. This would be the true definition of the circular economy of tires, i.e., creating grades of material that arise from EoL tires and find their way back into tires. The only actual recycling of tires is the reuse of black rubber components as secondhand spare parts in the automotive industry and in the mining industry. All other reuses are in principle downcycling into products of lesser economic value, as material quality cannot be achieved to truly "close" the loop. Thus full circularity is problematic and therefore speaking of the circular economy of tires is challenging, as in the end too much exergy and therefore cost will be incurred to achieve this.

The words of Saputra et al. (2021) express this well:

> With regard to the ever-increasing feedstock of tires along with WTR (waste tire rubber) generated, a proper valorization technique remains to be one of the most critical challenges that need to be globally addressed. The approach of downsizing WTR to GTR (ground tire rubber), up until this point, is still vastly used due to the easiness of the process. Such GTR can be applied to various products to produce rubberized asphalt, concrete, playground mat, tiles, shock resistance reinforcement in buildings, footwear, conveyor belts, and many more. However, considering that millions of tons of WTRs are generated and wasted annually, it would seem that such an approach is incredibly insufficient. Furthermore, landfilling and open burning have been strictly banned in many countries due to the toxic pollution they generate.

In summary, the main reasons for this are the facts that black rubber exists as many product groups and contains a great number of raw materials, and the raw materials are complexly bound together and functionally linked in both chemical and physical bonds.

- The number of black rubber products is large with different product grades and qualities, making circular economy recycling difficult.
- The list of raw materials in tires is long: from rubber (natural and synthetic), fillers like carbon black (in various precise particle sizes and grades), zinc oxide, and silica as well as sulfur, antioxidants, and oils. These complexly linked materials affect recycled product quality, secondary residues, etc.
- The bonding inside a tire is twofold, i.e., sulfur is chemically bonded to the rubber in the vulcanization process and the fillers are bonded to the rubber with strong physical bonds. To unravel this economically is not so easy and will obviously require enthalpy and entropy/exergy (reflected as an economic value) to achieve, and to achieve once again the material qualities that guarantee safety and integrity of the tire.
- The retreading of tires can be considered a recycling process, but note that in the retreading procedure only part of the tire—at best more than 70% by weight—is recycled or reused. The old run-down treads are skimmed and ground off and the casing of the tire is recapped with a new tread. However, in the end the tire must find its way into other recycling routes after many retreading cycles, when the casing structure's integrity is compromised.

In conclusion, the recycling of black rubber products in general amounts to downcycling, with reuse and repurposing applying to a part of the tire. This includes the reuse of black rubber products as landfill, substitute fillers in concrete, incineration fuel substitute, pyrolysis, shredding, etc. Therefore there is a long way to go before one can speak of the circular economy of tires.

References

A selection of academic journal papers to open the door to a vast body of information

Arabiourrutia, M., Lopez, G., Artetxe, M., Alvarez, J., Bilbao, J., Olazar, M., 2020. Waste tire valorization by catalytic pyrolysis—a review. Renew. Sust. Energ. Rev. 129, 109932.

Bhatt, M., Chakinala, A.G., Joshi, J.B., Sharma, A., Pant, K.K., Shah, K., Sharma, A., 2021. Valorization of solid waste using advanced thermo-chemical process: a review. J. Environ. Chem. Eng. 9(4), 105434.

Bockstal, L., Berchem, T., Schmetz, Q., Richel, A., 2019. Devulcanisation and reclaiming of tires and rubber by physical and chemical processes: a review. J. Clean. Prod. 236, 117574.

da Silva, F.M., Barbosa, L.A.G., Lintz, R.C.C., Jacintho, A.E.P. G.A., 2015. Investigation on the properties of concrete tactile paving blocks made with recycled tire rubber. Constr. Build. Mater. 91, 71–79.

Diekmann, A., Giese, U., Schaumann, A., 2019. Polycyclic aromatic hydrocarbons in consumer goods made from recycled rubber material: a review. Chemosphere 220, 1163–1178.

Dobrota, D., Dobrota, G., Dobrescu, T., 2020. Improvement of waste tire recycling technology based on a new tire markings. J. Clean. Prod. 260, 121141.

Dwivedi, C., Manjare, S., Rajan, S.K., 2020. Recycling of waste tire by pyrolysis to recover carbon black: alternative & environment-friendly reinforcing filler for natural rubber compounds. Compos. Part B 200, 108346.

Gomes, F.O., Rocha, M.R., Alves, A., Ratola, N., 2021. A review of potentially harmful chemicals in crumb rubber used in synthetic football pitches. J. Hazard. Mater. 409, 124998.

Halle, L.L., Palmqvist, A., Kampmann, K., Khan, F.R., 2020. Ecotoxicology of micronized tire rubber: past, present and future considerations. Sci. Total Environ. 706, 135694.

Hamdi, A., Abdelaziz, G., Farhan, K.Z., 2021. Scope of reusing waste shredded tires in concrete and cementitious composite materials: a review. J. Build. Eng. 35, 102014.

Han, L., Zheng, M., Chongtao Wang, C., 2016. Current status and development of terminal blend tire rubber modified asphalt. Constr. Build. Mater. 128, 399–409.

Himadri, 2021. www.himadri.com/carbon_black.

Hita, I., Arabiourrutia, M., Olazar, M., Bilbao, J., Arandes, J.M., Castaño, P., 2016. Opportunities and barriers for producing high quality fuels from the pyrolysis of scrap tires. Renew. Sust. Energ. Rev. 56, 745–759.

Kazmi, S.M.S., Munir, M.J., Wu, Y.-F., 2021. Application of waste tire rubber and recycled aggregates in concrete products: a new compression casting approach. Resour. Conserv. Recycl. 167, 105353.

Lewandowski, W.M., Januszewicz, K., Kosakowsk, W., 2019. Efficiency and proportions of waste tire pyrolysis products depending on the reactor type—a review. J. Anal. Appl. Pyrol. 140, 25–53.

Li, Y., Zhang, S., Wang, R., Dang, F., 2019. Potential use of waste tire rubber as aggregate in cement concrete—a comprehensive review. Constr. Build. Mater. 225, 1183–1201.

Liu, L., Cai, G., Zhang, J., Liu, X., Liu, K., 2020. Evaluation of engineering properties and environmental effect of recycled waste tire-sand/soil in geotechnical engineering: A compressive review. Renew. Sust. Energ. Rev. 126, 109831.

Martínez, J.D., Puy, N., Murillo, R., García, T., Navarro, M.V., Mastral, A.M., 2013. Waste tire pyrolysis—a review. Renew. Sust. Energ. Rev. 23, 179–213.

Medina, N.F., Garcia, R., Hajirasouliha, I., Pilakoutas, K., Guadagnini, M., Raffoul, S., 2018. Composites with recycled rubber aggregates: properties and opportunities in construction. Constr. Build. Mater. 188, 884–897.

Mohajerani, A., Burnett, L., Smith, J.V., Markovski, S., Rodwell, G., Rahman, M.T., Kurmus, H., Mirzababaeib, Arul Arulrajah, M., Horpibulsuk, S., Maghool, F., 2020. Recycling waste rubber tires in construction materials and associated T environmental considerations: a review. Resour. Conserv. Recycl. 155, 104679.

Mui, E.L.K., Ko, D.C.K., McKay, G., 2004. Production of active carbons from waste tires—a review. Carbon 42, 2789–2805.

Nanjegowda, V.H., Biligiri, K.P., 2020. Recyclability of rubber in asphalt roadway systems: a review of applied T research and advancement in technology. Resour. Conserv. Recycl. 155, 104655.

Okoye, C.O., Jones, I., Zhu, M., Zhang, Z., Zhang, D., 2021. Manufacturing of carbon black from spent tire pyrolysis oil—a literature review. J. Clean. Prod. 279, 123336.

Pacheco-Torgal, F., Ding, Y., Jalali, S., 2012. Properties and durability of concrete containing polymeric wastes (tire rubber and polyethylene terephthalate bottles): an overview. Constr. Build. Mater. 30, 714–724.

Picado-Santos, L.G., Capitão, S.D., Neves, J.M.C., 2020. Crumb rubber asphalt mixtures: a literature review. Constr. Build. Mater. 247, 118577.

Presti, D.L., 2013. Recycled tire rubber modified bitumens for road asphalt mixtures: a literature review. Constr. Build. Mater. 49, 863–881.

Ramarad, S., Khalid, M., Ratnam, C.T., Luqman Chuah, A., Rashmi, W., 2015. Waste tire rubber in polymer blends: a review on the evolution, properties and future. Prog. Mater. Sci. 72, 100–140.

Sebola, M.R., Mativeng, P.T., Pretorius, J., 2018. A benchmark survey of waste tire recycling in South Africa to EU practice. CIPR 69, 950–955.

Saputra, R., Walvekar, R., Khalid, M., Mubarak, N.M., Sillanpää, M., 2021. Current progress in waste tire rubber devulcanization. Chemosphere 265, 129033.

Sathiskumar, C., Karthikeyan, S., 2019. Recycling of waste tires and its energy storage application of by-products—a review. Sustain. Mater. Technol. 22, e00125.

Siddika, A., Mamun, M.A.A., Alyousef, R., Amran, Y.H.M., Aslani, F., Alabduljabbar, H., 2019. Properties and utilizations of waste tire rubber in concrete: a review. Constr. Build. Mater. 224, 711–731.

Strukar, K., Šipoš, T.K., Miličević, I., Bušić, R., 2019. Potential use of rubber as aggregate in structural reinforced concrete element—a review. Eng. Struct. 188, 452–468.

Wemyss, A.M., Bowen, C., Plesse, C., Vancaeyzeele, C., Nguyen, G.T.M., Vidal, F., Wan, C., 2020. Dynamic crosslinked rubbers for a green future: a material perspective. Mater. Sci. Eng. R 141, 100561.

Xiao, Z., Pramanik, A., Basak, A.K., Prakash, C., Shankar, S., 2022. Material recovery and recycling of waste tyres—a review. Clean. Mater. 5, 100115.

Xu, J., Yu, J., Xu, J., Sun, C., He, W., Huang, J., Li, G., 2020. High-value utilization of waste tires: a review with focus on modified carbon black from pyrolysis. Sci. Total Environ. 742, 140235.

Zhang, G., Chen, F., Zhang, Y., Zhao, L., Chen, J., Cao, L., Gao, J., Xu, C., 2021. Properties and utilization of waste tire pyrolysis oil: a mini review. Fuel Process. Technol. 211, 106582.

Further Reading

To learn more about carbon black

https://www.thecarycompany.com/media/pdf/specs/orion-what-is-carbon-black.pdf.

http://carbon-black.org/index.php/what-is-carbon-black.

https://www.himadri.com/carbon_black.

www.orioncarbons.com.

CHAPTER

33

Textiles

Jana M. Hawley

College of Merchandising, Hospitality and Tourism, University of North Texas, Denton, TX, United States

33.1 INTRODUCTION

The primary source of textile waste is discarded clothing, although footwear, towels, bedding, and carpet are also part of the mix. The Environmental Protection Agency (EPA) notes that in 2018, American consumers discarded 15.5 million tons of textiles, representing 6.4% of municipal solid waste (US EPA, 2018). Of those 15.5 million tons, 14.7% is recovered as recyclable material and 18.9% incinerated for energy, leaving 66.4% with no alternative but the landfill. These discarded textile products equate to 37 kg of clothing discarded per capita per year in the United States. While the EPA has not released any new data since 2018, we can assume that the situation has only worsened.

Europe is notably ahead of the United States in attention given to sustainability. For example, in 2021, the European Parliament adopted a New Circular Economy Action Plan (European Parliament, 2021). The resolution names textiles as a key product in the comprehensive EU strategy and stresses that it must "prioritize waste prevention and durability, reusability and reparability as well as tackling hazardous and harmful chemicals in line with the waste hierarchy" (Resolution 82–84). The resolution also encourages new business models to address the full range of environmental issues and social impacts. The resolution calls for design and production changes against synthetic microfiber loss known to pollute our oceans (Ocean Clean Wash, 2021). The resolution goes on to underline expressed concerns about textiles' impact on the environment and provides binding targets and overall waste reduction goals.

While the United States has not yet adopted a holistic resolution similar to that in Europe, some municipalities are taking action. For example, New York City law requires businesses to recycle textile materials if they represent more than 10% of their commercial waste (Evans, 2020). This law has inspired entrepreneurs in New York to start new businesses that manage textile waste, particularly preconsumer textile waste. FabScrap is an example of a new company that focuses on the transparent recycling of preconsumer textiles from the garment district and at the same time empowering a community of change-makers who use the textiles to create new products (https://fabscrap.org/).

The fashion industry and western lifestyle are both significant contributors to landfill waste. Fashion itself compounds the problem of textile

Handbook of Recycling. https://doi.org/10.1016/B978-0-323-85514-3.00027-0

waste. Wilson (2003) calls fashion "dress in which the key feature is rapid and continual changing of style. Fashion...is change" (p. 3). Hidden under the glamour of the fashion industry is the alarming environmental impact of the textile industry. For decades, the industry has been subject to severe social and ecological criticism (Bostrom and Micheletti, 2016), including contributions to toxic waste, water shortages, and human rights. While there is a growing number of technological solutions to solve some of the problems, the challenge also requires behavioral changes of consumers, NGOs, governments, and businesses.

For decades, the fashion industry presented four or five new seasonal lines each year. But since the 1960s, the rate of fashion change has increased, resulting in what is commonly known today as "fast fashion." Between 2000 and 2015, the number of garments manufactured doubled from 50 billion to 100 billion and this figure is expected to increase as discretionary incomes rise throughout the world. Furthermore, the fashion industry creates 10% of the world's carbon emissions and 20% of wastewater. Because much of today's fast fashion is made of polyester, it is estimated that fashion contributes about 1.9 million tons of microplastics added to the ocean each year, making textiles the major contributor to global plastic pollution (Okamoto, 2021). Table 33.1 indicates the growth in tons of municipal solid textile waste generated since 1960. This includes textiles waste that is generated, recycled, composted, combusted with energy recovery, and landfilled.

Fast fashion brands such as H&M and Zara have had tremendous company success as they emphasized rapid production of new trends at low costs by optimizing the supply chain, including low-cost labor. Originally, consumers embraced the constant assortments of new merchandise at affordable prices (Segran, 2021). But

TABLE 33.1 1960–2018 Data on textiles in MSW by weight (in thousands of tons).

	1960	1970	1980	1990	2000	2005	2010	2015	2017	2018
Generation	1596	1850	2295	5270	8598	10,440	11,991	14,566	15,319	15,446
Recycled	45	54	145	599	1197	1660	1859	2231	2331	2277
Composted	–	–	–	–	–	–	–	–	–	–
Combustion with energy recovery	–	9	45	798	1705	1914	2059	2775	2875	2921
Landfilled	1551	1787	2104	3873	5696	6866	8072	9560	10,113	10,249
	Source data in thousand US tons									
	1960	**1970**	**1980**	**1990**	**2000**	**2005**	**2010**	**2015**	**2017**	**2018**
Generation	1760	2040	2530	5810	9480	11,510	13,220	16,060	16,890	17,030
Recycled	50	60	160	660	1320	1830	2050	2460	2570	2510
Composted	–	–	–	–	–	–	–	–	–	–
Combustion with energy recovery	–	10	50	880	1880	2110	2270	3060	3170	3220
Landfilled	1710	1970	2320	4270	6280	7570	8900	10,540	11,150	11,300
	1 US ton = 0.907 metric ton									

Means no data available.
Modified from US EPA (2018).

in more recent years, consumers are sounding the alarm as they realize the amount of waste that has been created from "disposable clothing." As a result, H&M has turned its attention to sustainability. As young consumers demand products that are more sustainably produced, company practices have needed to change. But H&M, and many other companies, still face criticism of "green-washing" when marketing messages don't match reality. Not all the blame can be placed on the companies. Without government intervention, mandated transparency, and industry standardization, it is difficult to meet the challenges of a truly sustainable company, even when the conscience of the company wants to make a difference.

33.2 THE RECYCLING PROCESS

The textile recycling industry is one of the oldest and most established recycling industries in the world (Shell, 2020), yet few people understand the industry or how they could participate in the process. Postconsumer municipal waste (PCMW) is produced by the end consumer and includes various items consumers throw away after they are used. Throughout the world, used textile products are salvaged as reclaimed textiles and put to new uses. The textile recycling industry manages more than 1.7B kg of postconsumer textile waste (PCTW) each year. Unfortunately, this is only approximately 15% of the waste, leaving 85% in our landfills (Harmony, 2021). Experts agree that product reuse is favored over recycling, with the least preferable option being landfill disposal. Secondhand clothing is a common reuse strategy, with markets ranging from collectible designer goods to bales of apparel sent to developing countries (Hawley, 2006). Before 2018, Europe's standard method of managing textile waste was incineration. But the 2018 Parliament waste directive, coupled with the recent resolution passed in 2021, outlines important steps to tackle many of the issues that have plagued the textiles industry for centuries (EP, 2021), This has led companies to develop new technologies for recycling textile waste.

Mechanical methods result in the shredding of apparel back to fibers used as raw material for new yarns or products. However, mechanical processing is limited because of the challenge of sorting postconsumer waste into pure categories of color or fiber content. In addition, mechanical processing shortens fibers and, therefore, fiber quality. Some processors are using chemical processing to separate cellulose from polyester or spandex from cotton.

Even though textile recycling begins when consumers decide they no longer want their apparel, much of the discarded clothing and textile waste fails to reach the recycling pipeline. This is often because consumers do not understand the various channels available for used clothing. Consumers often think that clothing is only useful if it remains as clothing. So if the clothing is torn, missing a button, or out of style, it is understood to not be recyclable, so consumers opt to throw it away. Indeed, clothing that is in good enough condition to be distributed as secondhand clothing is best. But clothing can be sorted for many other uses beyond secondhand clothing. Figure 33.1 illustrates the postconsumer waste process.

Generally, there are two categories of textile waste: (1) postindustrial (also referred to as preconsumer) and (2) postconsumer. Postindustrial waste consists of by-product materials from the textile, fiber, and cotton industries remanufactured for the automotive, aeronautical, home building, furniture, mattress, coarse yarn, home furnishings, paper, apparel, and other industries. On the other hand, postconsumer waste is any garment or household article made from manufactured textiles that the owner no longer needs and decides to discard. When these clothing items are donated because they are worn out, damaged, no longer fit, or have gone out of fashion, they often start at a charity such as

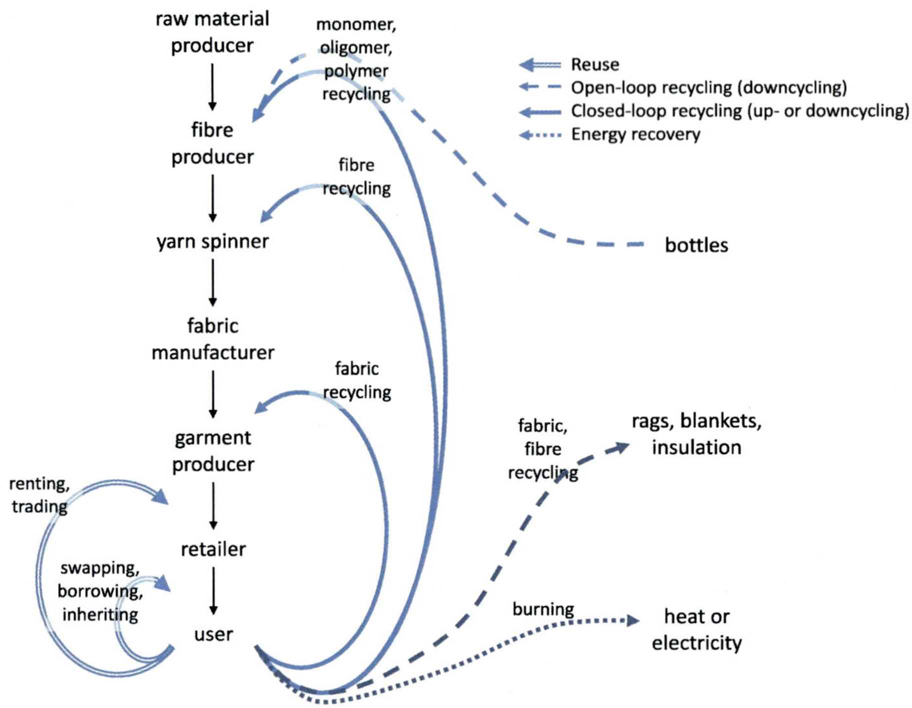

FIGURE 33.1 A classification of textile reuse and recycling routes. *From Sandin and Peters (2018).*

Goodwill, Salvation Army, TRAID, or OXFAM. Charity agencies then sort the clothes and select items for their retail shops, with the surplus sold to textile graders (also known as rag dealers) for pennies on the pound. At regularly scheduled routes, trucks are dispatched to pick up the surplus. The clothes are then taken to recycling warehouses, where the sorting process begins. The primary goal for the rag dealers is to earn profits, but often the business owners are committed to making a positive environmental impact.

Truckloads of used clothing are loaded onto a conveyor belt where the sorting begins. The first step is to perform a crude sort that removes heavy and oversized items such as coats, curtains, and blankets. The second sort separates other readily identifiable items, such as jeans, shirts, and household textiles. As the process proceeds, the categories become more and more refined. For example, once sorters pull all trousers, they are further sorted based on criteria such as women's or men's, woolens or chinos, condition (e.g., tears, missing buttons, stains), or brands and style (Levi's, Dickies, or Wrangler). Workers continue to refine the categories to meet the requirements of specific markets. It is common for a fully integrated rag dealer to sort over 400 grades at any given time. The quality of grading often distinguishes a competitive advantage of one textile recycling company over another. Sorting includes secondhand clothing exported to developing countries, vintage collectibles, and conversion to new products. We also discuss, briefly, the conversion of mattresses and carpet. For the most part, the volume of used clothing is inversely proportional to value (Hawley, 2006).

33.3 SECONDHAND CLOTHING

The largest volume of recycled textiles, approximately 45%, is sorted for secondhand exports, primarily for export to developing countries or disaster relief. Worn apparel is classified under Harmonized System Code #630900. In 2019, the largest exporter of used clothing was the United States, who exported more than $720 M US dollars of worn clothing to places around the world, with much of it sent to developing countries. The other top exporters are United Kingdom (US$486 M), Germany (US$379 M), China (US$372 M), and South Korea (US$312 M). Used clothing ranks 530th among the most traded products, with US$4.46B traded worldwide. Currently, the top destination for secondhand clothing is Ukraine (US$203 M) followed by Pakistan (US$198 M), Ghana (US$168 M), Kenya (US$185 M), and United Arab Emirates (US$151 M) (OEC, 2021).

Once sorted, the goods are compressed into large bales (usually 275–500 kg), wrapped, and warehoused until an order is received. When sorting for used clothing markets, several things are considered: climate, relationships between exporters and importers, and trade laws for used apparel. Used clothing faces several trade barriers, including restrictions based on sanitation, phytosanitation, national security, human/animal health, and deceptive practices (Office of Textiles and Apparel, 2012).

33.4 VINTAGE COLLECTIBLES

An important category of PCMW is the *diamond* category (Hawley, 2006). This category refers to the collectible clothing found on the conveyor belts. Although the volume of collectible clothes is only about 1% of the total volume, the value of these items makes it worth the extra care needed to identify these treasures. For example, some collectible Levi's jeans have sold at auction for more than $10,000 per pair.

Diamonds include vintage collectibles, luxury fibers, couture clothing, event/promotional T-shirts, and highly collectible branded goods such as Harley-Davidson clothing. Many of these mined goods have global markets, and the number of consumers who seek collectibles is on the rise. For example, Japanese consumers seek vintage Americana items such as Ralph Lauren, Levi's, Coach leather goods, or Harley-Davidson, whereas American consumers seek Italian leather and French couture items. Textile graders know the value of this category and have established relationships with vintage shops and online traders for distribution of these collectible items.

Many owners of vintage shops in the United States are members of the National Association of Resale and Thrift Shops. This trade association promotes public education about the used clothing industry and has more than 1000 members that serve thrift shops, resale vendors, and consignment shops.

33.5 CONVERSION TO NEW PRODUCTS

33.5.1 Wipers

Some clothing that has seen the end of its useful life is recycled into wipers for industrial use. T-shirts are a primary source for this category because the cotton fiber content makes an absorbent rag and polishing cloth. Wipers are used in various industries, including automotive, housekeeping, and furniture making.

33.5.2 Fibers

Recycled clothing and textiles that are not wearable or appropriate for wipers are either mechanically or chemically converted to fiber and processed into value-added products. Mechanical operations include cutting, shredding, carding, and processing the fabric. Chemical processes involve enzymatic, thermal, glycolysis, or methanolysis methods (Hawley, 2009). Historically, this was referred to as shoddy and was considered second-rate, at best. Shell (2020) provides a hidden history of shoddy, from the 1800s to the present day. The book also shines light on the possibilities of today's fiber recapture from old clothes.

Once the postconsumer textiles are converted, they are further processed into new products for consumption. These value-added products include stuffing, automotive components, carpet underlays, building materials, and blankets.

In the past decade, more and more companies have been exploring ways to create new products from used textiles and apparel. One such company is GeoHay (www.geohay.com), a South Carolina textile recycling company that makes a versatile line of water management products from recycled textiles. The products have a variety of uses including soil erosion maintenance at construction sites and stormwater management. One of their big success stories was the Deepwater Horizon oil spill in the Gulf of Mexico on April 20, 2010. GeoHay produced oil booms out of recycled textiles that outperformed competitive brands made from new materials and natural hay. Evidence revealed that the oil booms made from recycled fibers were better at removal of particulate and turbidity from the water than booms made from virgin materials (Hillegass, 2010).

Another example is the partnership between Phoenix Fibers and Bonded Logic of Arizona. Phoenix Fibers (www.phxfibers.com) is a closed-loop recycling company that converts 884 ton per month of denim jeans and other cotton textiles into "shoddy" fiber used for insulation in homes, businesses, and automotive. Phoenix Fibers partners with Bonded Logic (www.bondedlogic.com) who turns the shoddy fiber into a variety of products for thermal, acoustical, and construction uses.

Nike continues to demonstrate commitment to the environment with a "move to zero" commitment for zero carbon and zero waste. This commitment has resulted in several programs for recycling used athletic shoes. These include *Reuse-a-Shoe, Nike Trash Talk*, and *Nike Grind*. These programs, launched in the early 1990s, continue to have success. The Reuse-A-Shoe program places recycle bins in more than 150 of US Nike retail stores. Once collected, the shoes are shipped to Memphis, TN, where they are processed into three component parts:

- Nike Grind Rubber, the sole of the shoe, used for outdoor track surfaces,
- Nike Grind Foam, made from the midsole and used for outdoor basketball and tennis courts, and
- Nike Grind Fiber, the shoe's upper fabrics, used to create cushioning pads for athletic arenas (Ecobahn, 2020).

Originally, the Nike Reuse-a-Shoe program was only available through collection bins in retail stores. But the program is now made available online by using the Nike Satchel, an eco-friendly thermoplastic resin container made from sugar cane. The satchel holds two pairs of shoes. It is expected that this program will be expanded to many more countries.

More recently, a plethora of innovative start-ups from all over the world are using the latest in engineering, heavy capital investment, and commitment to close the loop in the fashion industry. Table 33.2 provides a range of innovative companies who are committed to closing the loop on textiles and apparel products.

33.6 CONVERSION OF MATTRESSES

Across the globe, thousands of mattresses are discarded on a regular basis as hotels, universities, and homeowners update their rooms. In the United States alone, companies dispose of between 20 and 40 million mattresses a year, and the UK discards 7 million mattresses

TABLE 33.2 Examples of innovative companies committed to closing the loop on textile waste.

Company	Description	Location	Year launched
Evrnu	An innovative company creating a circular ecosystem to create engineered fibers with extraordinary performance from discarded clothing. A proprietary technology is used to create NuCyl™, made from 100% postconsumer waste Website: https://www.evrnu.com	Seattle, Washington, United States	2019
I-CO ICO Spirit	Large-scale take-back system and logistics network unique to the textile industry. Collects clothing and shoes in more than 60 countries Website: https://www.ico-spirit.com/en/	Ahrensburg, Germany	2018
Green Machine	Developed by the Hong Kong Research Institute of Textiles and Apparel Limited and the H&M Foundation. The world's first technology that can separate and recycle polyester and cotton blend fabrics at scale without quality loss	Global in nature. Operations in Hong Kong and Cambodia	2016
Ambercycle	Uses molecular separations technology to purify and recover new materials from postconsumer polyester clothing Website: https://ambercycle.com/	Los Angeles, California	2015
Re:newcell	Focused on closing the loop of textile waste by dissolving shoddy into a circulose pulp that can then reenter the textile production cycle. Can manage industrial scale production Website: https://www.renewcell.com/	Stockholm, Sweden	2012
Carbios	Using enzymatic process to rethink the end of life for textiles Website: https://www.carbios.com/en/	La Pardieu, France	2011
Circ	Originally a biofuels company, Circ now focuses on hydrothermal processing (water, pressure, chemistry) to manufacture products from textile waste eliminating the need for virgin fibers Website: https://circ.earth/our-purpose/	Danville, Virginia, United States	2011
Looptworks	Upcycles abandoned, preconsumer materials into limited-edition products. The company is expanding into other recycling operations to help close the loop on textile waste Website: https://www.looptworks.com/	Portland, Oregon, United States	2009
Worn Again	Separates blended materials. The focus is on converting polyester, poly/cotton, and PET plastic back into circular raw materials Website: https://wornagain.co.uk	London, England	2005
Martex Fibers	Martex Fibers turns reclaimed textile waste into trademarked labels such as Jimtex™ and Eco-Cotton™ Website: http://www.martexfiber.com/	Pacolet, South Carolina, United States	1970

annually. The numbers have increased recently with the new roll-down mattresses that can be shipped in a box to your doorstep. This means that mattresses are now being replaced at an alarming rate.

Disposing of mattresses poses unique challenges. Not only have landfill fees for mattresses skyrocketed in recent years, but also a single mattress can take up to $0.65\,m^3$ in the landfill, and often the mattresses cause damage to landfill equipment. In addition, mattresses have high transportation costs associated with the recycling process, contain chemical flame retardants, and are labor-intensive to dismantle, making mattress disposal a global environmental nightmare.

At the Hutchinson Correctional Facility (HCF) in Hutchinson, Kansas, inmates dismantle old mattresses from suppliers all over the Midwest. Mattress recycling involves dismantling the mattress into fundamental parts: springs, wool, cotton, foam, and fabric covers. The various components are sold as scrap or repurposed into new products. For example, at HCF, mattress covers and foam padding are made into pet bed cushions, metal springs are sold as metal scrap, and wood from box springs is used to build Adirondack chairs. While often criticized, inmate labor is one of the few cost-effective ways to manage labor-intensive mattress recycling.

The Mattress Recycling Council (www.mattressrecyclingcouncil.org) is a nonprofit organization formed by the US mattress industry to operate recycling programs in states that have passed mattress recycling laws, such as California, Connecticut, and Rhode Island. In a study conducted at UCLA, researchers found that chemically recycled polyurethane mattress foam can be combined with naturally occurring zeolite minerals to produce high-strength composites that are tested to be comparable to ordinary cement and outperformed virgin polyurethane materials. The new discovery reveals that mattress waste composites have more flexibility and crack resistance than traditional materials. This new discovery has the potential of converting thousands of tons of old mattresses to usable construction materials. For example, about 169 ton of recycled mattress foam could be used to build 1 km of highway (Iyer and Srivastava, 2021).

33.7 CONVERSION OF CARPET

Carpet recycling is another important category of textile recycling. Approximately 1.8 million tons of carpet are disposed of each year in the United States. Still, recycling carpet is a complex process that often requires extensive processing to convert it to new products (Wang et al., 2003). Recycling of both residential and commercial carpets has been undertaken on a national scale in the United States, with various end products including high-energy fuel, shingles, flooring, auto parts, and new carpet. The recycled carpet is tested for fiber type (e.g., polyester, nylon, olefin), dismantled into components (backing from face-fiber), and processed into various end uses. Because carpet is heavy and bulky, transportation costs are a significant factor in the cost-efficacy of recycling. Today's carpets are designed to allow for easier capture of components using mechanical and chemical recycling processes. In addition, a supply chain infrastructure is implemented to help reduce transportation costs.

33.8 LANDFILL AND INCINERATION

Although rag sorters work hard to avoid sending PCMW to landfills, postconsumer waste eventually reaches the end of its useful life. As a result, recent research has explored the use of textiles for fuel. Although emission tests of incinerated used fibers are above satisfactory and the British thermal unit (BTU) value is respectable, the process of fueling boiler systems in North American power plants is not feasible.

For decades, textile waste was often incinerated in Europe, but the recent Waste Framework Directive from the European Parliament is changing that. The directive states that all textiles must be collected separately by 2025 (European Parliament, 2021). Basic waste management principles are outlined in the Waste Framework Directive and they state that waste must be managed without endangering human health or harming the environment, including damage to water, air, soil, plants, or animals. In addition, waste cannot cause nuisance through noise or odors and cannot adversely affect the countryside or places of special interest. Textiles are considered a priority material in the directive, with the primary goal to promote repair and reuse. Research will be conducted on textile waste to explore technical, regulatory, economic, and environmental effectiveness of textile recycling. The goal is to find promising new areas for innovative projects.

33.9 CIRCULAR ECONOMY

The circular economy model has recently gained much attention from businesses and the science community. A circular vision is restorative and regenerative by design, with the intention that textiles would never enter the waste stream. The Ellen MacArthur Foundation's report (2017) on the New Textiles Economy considers used textiles a valuable resource with unprecedented business opportunities that will have significant positive impact on the environment and the fashion industry. The present linear process is limited, but the change from linear to circular requires knowledge, technology, awareness, and engagement from all participants in the textile and apparel supply chain. Designing for a circular economy is complex and encompasses an ecosystem: use, repair and reuse, textile collecting and sorting, production of recycled fibers, and production of textiles from recycled fibers. The circular model is a shared system-level vision with inclusion of all aspects of the supply chain and value cycles. In addition, convincing consumers of the value-proposition of circularity can be challenging (Fontell and Heikkilä, 2017). Furthermore, actions to achieve textile waste recovery are complicated, often ambitious, and generally ambiguous. It requires a multidimensional approach that ultimately transforms the way companies and consumers improve processes and behaviors toward an end goal of 100% textile recovery. This circular economy concept aims for business models that reduce environmental footprint by promoting the reduction of raw materials, reuse of goods, and recycling of materials. In 2016, The European Clothing Action Plan set out to encourage industry, scientists, and creatives to reinvent how we produce and design apparel, rethink how we consume, and redefine reuse and recycling (Moorhouse and Moorhouse, 2017). In essence, this was a major push toward circularity of textile and apparel products. More and more companies are embracing the concept of circularity and working hard to implement concepts so that textile waste is diverted from the landfills.

33.10 OUTLOOK

As more people pay attention to environmental issues, including value-added options for recycled textiles, companies must continue to innovate. McDonough and Braungart (2002) point out that the textile industry ranks among the worst industries on environmental impact. It is critical that all of us understand the significant impact of the textiles and fashion industry on climate change. The United Nations Framework Convention on Climate Change (2023) has a goal of zero waste by 2050. The framework lays out a list of strategies, but is 2050 too late?

As we move toward a zero-waste goal, we must give attention to all the options ranging from downcycling to products designed for circularity. Both manufacturers and consumers can reduce the environmental footprint of used textiles by making sure all postindustrial and postconsumer waste enters the recycling pipeline. Along the entire textile supply chain, we must take a holistic approach to the process so that textiles no longer end up in the landfill. Finally, no matter how much attention is given to recycling, upcycling, repurposing, or closing the loop, we must address the most critical issue of mass overconsumption, driven by fast fashion strategies. No matter how many kilograms of clothes are diverted from the landfill, the fast fashion strategy is not sustainable. Until consumer behavior changes, production slows, and commitment to the earth becomes more important than increased sales, we will not be successful. We must embrace both societal and scientific practices across the supply chain to reduce the negative impact of textiles and the fashion industry.

One of the toughest challenges the textile industry faces is how to right-size the industry in a way that supports our environment while at the same time sustains business profitability. Companies are scrambling to find the right balance between consumer, legislative, and goodwill demands to meet environmental goals, and at the same time meeting the demands of shareholders in a capitalist society. As this shift occurs, a new economic model needs to be considered that allows businesses to flourish in an environment in which sustainability becomes more important than sales metrics and stock market performance (Hawley, 2021; Raworth, 2017).

As environmentalism and sustainability continue to emerge as a major public issue, we have to rethink economic models that encourage unlimited growth. The real challenge is to create a culture shift where consumers consume less, companies design for renewal, and the earth thrives. Now is the time.

RECOMMENDED READING

Marcketti S and Karpova E. (2020). The Dangers of Fashion. Bloomsbury Visual Arts.

References

Bostrom, M., Micheletti, M., 2016. Introducing the sustainability challenge of textiles and clothing. J. Consum. Policy 39, 367–375. https://link.springer.com/article/10.1007/s10603-0i6-9336-6.

Ecobahn, 2020. Nike Resue-A-Shoe: Digital Upcycling Shift (cited 2021b December 22) https://theecobahn.com/ideas/nike-reuse-a-shoe-digital-upcycling-shift/.

Ellen MacArthur Foundation, 2017. A New Textiles Economy: Redesigning Fashion's Future (citied 2021c December 20) http://www.ellenmacarthurfoundation.org/publications.

European Parliament, 2021. European Parliament Resolution of 10 February 2021 on the New Circular Economy Action Plan (cited 22 December 2021) https://www.europarl.europa.eu/doceo/document/TA-9-2021-0040_EN.html/.

Evans, L., 2020. The Entrepreneurs Transforming Scraps Into a Fashion Statement. Natural Resources Defense Council (NRDC). (cited 2021d, December 17). Available from: https://www.nrdc.org/stories/entrepreneurs-transforming-scraps-fashion-statement.

Fontell, P., Heikkilä, P., 2017. Model of Circular Business Ecosystem for Textiles. vol. 313 VTT Technical Research Centre of Finland. VTT Technology (cited 2021 June 22) https://cris.vtt.fi/en/publications/model-of-circular-business-ecosystem-for-textiles.

Harmony, 2021. The Facts About Textile Waste (Infographic). (cited 2021, December 20). Available from: https://harmony1.com/textile-waste-infographic/.

Hawley, J.M., 2006. Digging for diamonds: a conceptual framework for understanding reclaimed textile products. Cloth. Text. Res. J. 24 (3), 262–275.

Hawley, J.M., 2009. Understanding and improving textile recycling: a systems perspective. In: Blackburn, R. (Ed.), Sustainable Textiles: Life Cycle and Environmental Impact. Woodhead Publishing, Cambridge, pp. 179–199.

Hawley, J.M., 2021. Right-sized consumption: should doughnut economics inform the textile and apparel industry? J. Text. Sci. Fash. Technol. 9 (2), 2021. https://doi.org/10.33552/JTSFT.2021.09.000707.

Hillegass, R., 2010. Polyester Fibers and GeoHay Support Oil Containment Effort in Gulf of Mexico. (cited 2021f June 23). Available from: https://www.benzinga.com/press-releases/10/05/b299257/polyester-fibers-and-geohay-support-oil-containment-effort-in-gulf-of-m.

Iyer, D., Srivastava, S., 2021. Upcycled Polyurethane Products Based Strong and Robust Inorganic/Organic Composites (Upcycled Mattress Foam). Mattress Recycling Council. (cited 2021 December 17). Available from: https://mattressrecyclingcouncil.org/research-reports/.

McDonough, W., Braungart, M., 2002. Cradle-to-Cradle. North Point Press, New York.

Moorhouse, D., Moorhouse, D., 2017. Sustainable design: circular economy in fashion and textiles. Des. J. 20 (Sup1), S1948–S1959. https://doi.org/10.1080/14606925.2017.1352713.

Ocean Clean Wash, 2021. Clean Laundry, Dirty Ocean. Ocean Clean Wash by Plastic Soup Foundation. (cited 2021, December 13). Available from: https://www.oceancleanwash.org/.

OEC, 2021. Used Clothing. Observatory of Economic Complexity (cited 2023g June 12) https://oec.world/en/profile/hs/used-clothing.

Office of Textiles and Apparel, 2012. Worn (Used) Clothing and Textile Products. Available from: https://otexa.trade.gov/.

Okamoto, K., 2021. Your laundry sheds harmful microfibers. Here's what you can do about it. New York Times (cited 2021 December 17) https://www.nytimes.com/wirecutter/blog/reduce-laundry-microfiber-pollution/.

Raworth, K., 2017. Doughnut Economics: Seven Ways to Think Like a 21st Century Economist. Chelsea Green Publishing. ISBN: 1603586741.

Sandin, G., Peters, G.M., 2018. Environmental impact of textile reuse and recycling—a review. J. Clean. Prod. 184 (2018), 353–365. https://doi.org/10.1016/j.jclepro.2018.02.266.

Segran, E., 2021. H&M is One of Fashion's Biggest Polluters. Now Its Foundation is on a $100 Million Quest to Save the Planet. Fast Company (cited December 21, 2021) https://www.fastcompany.com/90596456/hm-is-one-of-fashions-biggest-polluters-now-its-on-a-100-million-quest-to-save-the-planet.

Shell, H.R., 2020. Shoddy: From Devil's Dust to The Renaissance of Rags. University of Chicago Press.

United Nations Framework Convention on Climate Change, 2023. Race to Zero. (cited June 12, 2023). Available from: https://unfccc.int/climate-action/race-to-zero-campaign.

US EPA, 2018. Advancing Sustainable Materials Management 2018 Fact Sheet. Facts and Figures About Materials, Waste and Recycling. Textiles: Material-Specific Data. US Environmental Protection Agency. 2018 (cited 2021i June 15). Available from: https://www.epa.gov/facts-and-figures-about-materials-waste-and-recycling/textiles-material-specific-data.

Wang, Y., Zhang, Y., Polk, M., Kumar, S., Muzzy, J., 2003. Recycling of carpet and textile fibers. In: Andrady, A.L. (Ed.), Plastics and the Environment. John Wiley & Sons, New York.

Wilson, E., 2003. Adorned in Dreams. University of California Press, Berkeley.

CHAPTER

34

Carbon fibers

Mareen Zöllner, Dorothea Hamann, Thomas Krampitz, Sophia Thüm, and Holger Lieberwirth

Institute for Mineral Processing Machines and Recycling Systems Technology, Technical University Bergakademie Freiberg, Freiberg, Sachsen, Germany

34.1 INTRODUCTION

Technological advances lead to growing and more complex requirements for material properties. The development of composite materials is an answer to those demands. Composites consist of two or more materials attached, coated, in layers, or surrounding one another (as isotropic structures or fiber-reinforced structures). While there are many advantages in applying those materials, their recycling back into the circular economy often poses a challenge. Carbon fiber-reinforced plastic (CFRP) is such a composite material using properties-giving carbon fibers (CFs) and a form-holding matrix. The result is a high-performance compound that is both light and has remarkable strength. This combination creates new possibilities for product development. CFRP can be found in an increasing number of applications, e.g., in aviation, lightweight automotive construction, and in the energy sector (wind turbine blades), to name a few, making CFRP currently the most relevant application area of carbon fibers.

Several factors make recycling considerations for CFRP relevant: the production of the carbon fibers is energy-intensive, and the volumes are growing every year. Laws for protecting natural resources are strict, e.g., in Germany, where only landfilling of materials with less than 5% loss on ignition is allowed. In other countries similar regulations take effect. The recovery of material is preferred to thermal utilization and landfilling if waste prevention and reuse are excluded. In addition, expected future developments with regard to the increasing use of CFRP and an associated growing waste generation make CFRP recycling more and more important.

Many investigations have been performed in this area; nevertheless many questions are still outstanding. Therefore this chapter intends to provide an overview of existing technology and research approaches to the recycling of CFRP.

34.2 MATERIAL USE

Often, the value of a material determines how much technologic and economic effort is spent on recycling. Not only the properties of the material are important, but also the effort

 https://doi.org/10.1016/B978-0-323-85514-3.00043-9

required for production and the expected quantities of waste. In the following, therefore, the market situation is considered in addition to the properties.

34.2.1 Properties

Carbon fibers are synthetic fibers, containing carbon in the range of 95–99% purity and with diameters in the range of 5–10 µm. Structurally they are made of carbon atoms. The arrangement of the C-atoms causes the specific extraordinary properties, so carbon fibers are considered to be one of the most important structural materials in modern industry (Das et al., 2019; Xian and Wang, 2018). Carbon fibers are made of a precursor material, predominantly (≈ 90%) polyacrylonitrile (PAN), but also pitch or rayon (Bledzki et al., 2021; Newcomb, 2016; Prashanth et al., 2017). Carbon fiber production means various heat treatment stages with specific processing conditions depending on different precursors (Huang, 2009). The final fiber is coated with a sizing for resistance against external stress and better fiber-matrix adhesion in the composite. Figure 34.1 shows the range of mechanical property possibilities of carbon fibers (Akpan and Adeosun, 2019).

In comparison to other reinforcing fibers, carbon fibers offer the highest specific modulus and strength (Prashanth et al., 2017). The mechanical parameters depend on manufacturing methods, the fiber type, fiber volume content, fiber orientation, resin types, and the interface between the fiber and resin matrix (Xian and Wang, 2018; Rosato and Rosato, 2005). CFs are highly energy-intensive to produce (Witten and Schuster, 2018); with about 195–595 MJ/kg for carbon fiber production, the energy demand is about 10 times higher than for glass fiber

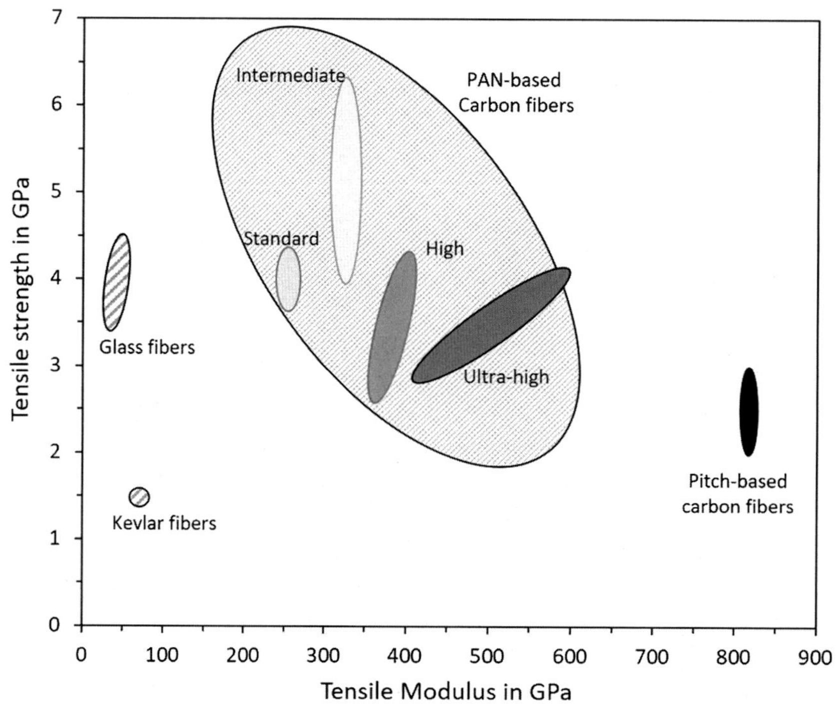

FIGURE 34.1 Range of mechanical properties of carbon and other fibers (Akpan and Adeosun, 2019).

(Bledzki et al., 2021; Zhang et al., 2020). Thus high-energy consumption during production makes carbon fiber an expensive reinforcing material and important regarding the circular economy.

Carbon fiber reinforced composites are mainly characterized by light weight with high strength and stiffness at the same time. Other properties are, for example, high fatigue resistance, high dimensional stability, high electromagnetic interference shielding, and low electrical resistivity (Das et al., 2019; Newcomb, 2016; Prashanth et al., 2017). CFRP is used to make components lighter in order to reduce energy consumption and greenhouse gas emissions by reducing vehicle weight at the same or higher strength compared to other manufacturing materials. Nevertheless, products are often made of CFRP aiming at a "modern" appearance and less at achieving certain mechanical properties.

34.2.2 Markets, Application, and Volumes

Glass and carbon fibers are the most commonly used reinforcement fibers (Das et al., 2019). More than 80% of the annual produced carbon fibers are used to reinforce polymers (Witten and Schuster, 2018). Matrices like metals, ceramics, or concrete can also be used. In the last decade, from 2010 to 2020, global CFRP demand tripled and is still rising (Witten and Schuster, 2018; Bledzki et al., 2021). The global demand for CFRP was 154.7kt in 2018 (Witten and Schuster, 2018). By 2050, it is expected to exceed 190kt (Zhang et al., 2020). Thermosetting polymer matrix systems account for almost two-thirds of the CFRP market. The thermoplastic market share is increasing and accounts for almost one-third of the CFRP share. An insignificant proportion is accounted for by elastomeric or hybrid matrix systems.

The largest application area is aerospace, including defense, with 36% of demand in 2018, followed by the automotive industry with 24% (Bledzki et al., 2021). The demand in the wind energy sector has grown to 13%. CFRP is also increasingly being used in the leisure sector (13%). In civil engineering (5%), the excellent corrosion resistance of carbon fiber reinforcements is often the reason for replacing steel reinforcements.

34.3 RECYCLING

The increasing use of CFRP generates an increasing amount of waste containing carbon fibers. In view of the energy-intensive production of both the fiber and the fiber composite, there are not only legislative reasons to treat and recycle the resulting waste, but also economic and ecologic ones. The recycling of CFRP is complicated because of its specific complex composition. Globally, 10 years ago the largest share of CFRP waste was landfilled. In addition to landfilling, waste incineration is done, aiming at the recovery of the contained energy [about 30MJ/kg (Zhang et al., 2020)]. Since neither incineration nor landfilling lead to a new physical product, they are not classified as recycling. Although there are many approaches to reprocess and recover carbon fibers, the proportion that is recycled to new products is still small. This clearly describes the tremendous challenges that more complex materials, such as CFRP, pose to a real circular economy. As new technologies emerge to create new types of materials, new answers are to be found to recirculate them with high quality into a circular economy.

34.3.1 Volumes of CF-Containing Waste

In each production step, waste or products that do not meet the quality requirements arise. According to Dauguet et al. (2015), this production waste can be subdivided into woven prepreg, nonwoven prepreg, dry carbon fibers (CFs), selvedge waste, and composites. From the pure carbon fiber to the finished

CFRP-composite product, the composition changes at each production step, so the effort and complexity of the required recycling treatment increase. The waste generated during production amounts to 30–50% of the total CFRP production (Zhang et al., 2020; Bledzki et al., 2021; LAGA, 2019). The production waste quantities generated in Europe account for about 10 kt each year (LAGA, 2019). Depending on the type of production waste, as well as quantity and manufacturer, production waste is collected by type and usually processed internally. In some cases, however, the waste is sent abroad or handed over to local disposal companies (LAGA, 2019). Due to improved CFRP processing methods, it can be assumed that the share of production waste will decrease.

On the other hand, the amount of end-of-life (EoL) waste increases. It originates from various applications and is generated after months, years (automotive), or several decades (aerospace) of product use, depending on its lifetime. By 2025 it is predicted that global CFRP waste will increase to 20 kt/year (Zhang et al., 2020). With the highest cumulative growth concentrated in Europe, North America, and Asia, the annual waste to be processed will reach an estimated 34.2 kt in 2050 (Zhang et al., 2020). A characteristic feature of EoL products containing carbon fibers is that the CFRP is often compounded with other materials (metals, plastics, etc.). Additionally, painting or coating may be applied on the surface. Summarizing, Figure 34.2 shows the points of origin and the composition of the CF-containing waste from the manufacturing of a CFRP product onward.

The required operations and the concomitant problems are explored in more detail in the following sections.

34.3.2 Barriers and Limitations

The existing literature on CFRP recycling refers mostly to recycling technologies themselves. Little attention is paid to the analysis of waste streams, the collection of CFRP waste, or identification of CFRP from waste materials. For an effective waste management, waste classification systems are fundamental and prerequisite. Even if all countries and regions adopt similar classification methods, according to Wen et al. (2014), the EU is the only region that has adapted a harmonized approach to managing hazardous and nonhazardous wastes. Yet, there are no material-specific classification numbers for sorted CFRP materials defined; instead the recommended classification numbers overlap with those of other materials such as thermoplastics or glass fiber reinforced plastics (GFRPs) (LAGA, 2019). Consequently, there is no

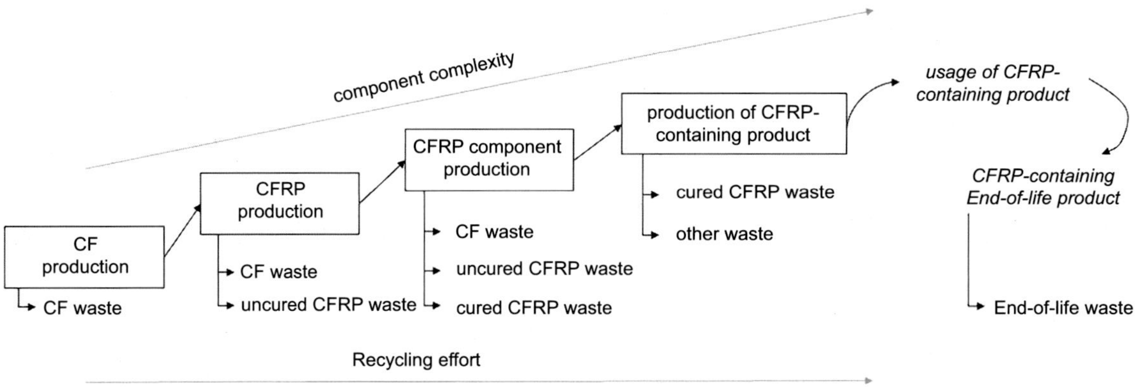

FIGURE 34.2 Waste generation and composition from carbon fiber production to CFRP-containing end-of-life waste.

statistical registration of waste containing CFRP, apart from the certain registrations with some recycling companies that accept CFRP deliveries. As long as there is no completely separate recycling route, small quantities of waste, e.g., consumer articles such as sports gear, are included in mixed municipal or household waste. The resulting fiber dissipation and misallocation to undesirable recycling routes pose hazards.

Even if typically fabricated carbon fibers are not respirable, under certain circumstances (e.g., mechanical or thermal stress) critical fiber dimensions may occur. So fibers may become long-term biopersistent (Kehren et al., 2019; Stockschläder et al., 2018). Upon reaching a length $>5\,\mu m$, a diameter $<3\,\mu m$, and an aspect ratio $>3:1$, fibers are respirable and therefore exposure is relevant, according to WHO specs (World Health Organization, 1997). Such fibers may cause chronic inflammation or increase the risk of developing mesothelioma (Donaldson et al., 2010; Kehren et al., 2019).

Because of this risk and possible damages due to its high electric conductivity in combination with the high thermal resistance of CF, CFRP must be separated from existing waste streams. Araujo-Andrade et al. (2021) reviews different possibilities of sensor-based sorting (SBS) systems for polymers. Differences in thermal conductivity caused by an electromagnetic alternating field are exploited (Warzelhahn, 2019; Oswald-Tranta et al., 2019), for instance. See Chapter 9 Sensor-based sorting for details.

Those new separation methods require an appropriate extensive preconditioning of the waste streams. The separation of metals, which may be necessary in the case of EoL waste, represents the state of the art (Zöllner et al., 2020; Asmatulu et al., 2014). Polymers can be separated via density separation or sensor-based sorting (Zöllner et al., 2020), although further research is required here.

So, the separate collection of CFRP waste and separate recycling routes are to be preferred, with the purpose of avoiding fiber dissipation into other recycling products. If a CFRP waste stream is homogeneous, there are various recycling options, which are considered next.

34.4 RECYCLING TECHNOLOGIES

Before choosing a recycling process, a distinction must first be made regarding thermoplastic or a thermoset matrix material. While CFRP with a thermoplastic matrix is relatively easy to recirculate because it can be melted, in the case of infusible thermoset, fiber and matrix must be separated. Furthermore, the type of waste stream, EoL components or production waste, impacts the choice of the appropriate recycling route. Possible recycling processes are presented in Figure 34.3. Table 34.1 provides an overview of main advantages and disadvantages of the described recycling methods.

34.4.1 Mechanical Recycling

Mechanical recycling of CFRP waste is technologically the most mature recycling method. It includes various process steps for waste comminution: shredding, milling, or grinding. Mainly, the aim of reducing the particle size is to produce granules or liberate the various constituents and to gain short carbon fiber recyclates that can be used as fillers or additives in other applications, such as molding processes (Zhang et al., 2020). In particular, homogeneous CFRP with thermoplastic matrix can be processed into regranules with particle sizes of about $x = 1.0–5.0\,mm$ and directly reused in injection molding processes (Zöllner et al., 2020).

Another special form of mechanical recycling of thermoset CFRP is electrohydraulic fragmentation (EHF) (Orzol et al., 2017). The aim of this method is to separate carbon fibers from the

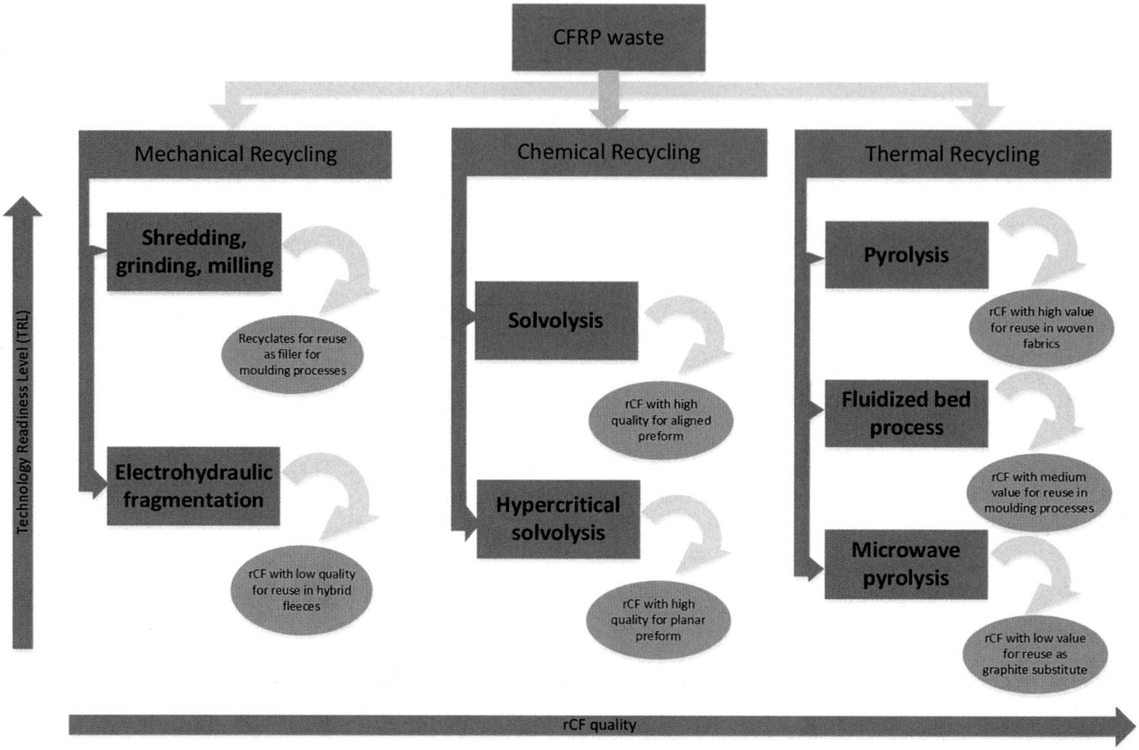

FIGURE 34.3 Overview of CFRP recycling methods and recycling products. *TRL*, technology readiness level; *rCF*, recycled carbon fiber.

matrix by high-voltage impulses that generate high-pressure waves in the surrounding liquid. Generally, the technological readiness level (TRL) of most of the mechanical recycling processes is high, but the value of the recovered products is low. Industry-scale technology for mechanical recycling of thermoset CFRP is not available.

34.4.2 Thermal Recycling

Thermal recycling includes thermal processes in which the polymer matrix is broken down using a temperature increase in a defined atmosphere. As long as the carbon fibers are not mechanically stressed in the process, for instance in a fluidized bed, they may be recovered with little physical damage. Thus recycled carbon fibers (rCFs) from thermal recycling may be of high quality. The quality depends on the respective process and its parameters. Carbon fibers are extremely sensitive to various process conditions, especially oxygen concentration and temperature (Zöllner et al., 2019).

34.4.2.1 *Pyrolysis*

The most noted thermal recycling process for CFRP is pyrolysis. In an inert atmosphere, organic polymer molecules are broken down and the matrix is thus thermally decomposed. The treatment temperature during pyrolysis is approximately 350–600°C, and the dwell time amounts to a few minutes (Rademacker et al., 2018; Limburg et al., 2019). Fibers can be

TABLE 34.1 Advantages and disadvantages of the CFRP recycling treatments.

CFRP recycling treatment	Advantages	Disadvantages
Mechanical recycling	• Simple process management with high TRL • Most mature CFRP recycling method • Low energy demand (except EHF) • Low environmental impact • Possibility to reuse fiber and matrix (esp. thermoplastics) • Independent of matrix material	• Critical fiber dust is created (health and safety hazard) • Low-quality rCF with short lengths • High wear at the processing equipment • Only low value remanufacturing achievable • Arising of nonrecyclable by-products
Thermal recycling	• Possibility of energy recovery from the resin • High fiber quality achievable • High TRL (industry scale) • Possibility of reusing the pyrolysis oil and gas • Low environmental impact • Independent of matrix material	• Removing of the fiber size (coating) • Processing parameters can influence the fiber quality negatively • Deposition of char and matrix residues on the fiber surface possible • Partial high energy input necessary (esp. pyrolysis)
Chemical recycling	• Possibility of reusing the solved matrix material • High fiber quality	• High environmental impact • Homogeneous matrix materials necessary; sorting processes needed • No industry-scale application for CFRP recycling so far (low TRL) • Dependent on matrix material

recovered having good quality. Resulting reaction products can be used as fuels or additives for other petrochemical-based materials (Adams et al., 2014).

In an oxygen-free atmosphere, the polymer matrix does not completely decompose. For this reason, the exposed fibers have to be subjected to another oxidative thermal treatment to completely decompose residual matrix adhesions and char residues (Limburg et al., 2019). Special care is required in this process step, as the fibers react extremely sensitively in the presence of oxygen and can react with pitting and consequently reduced mechanical properties. It should also be noted that the fiber size is no longer present on the fiber surface after pyrolysis. The rCFs are therefore brittle and a new coating before transport and further processing is needed. Otherwise, they can break easily and the length is reduced as an undesirable result.

Regarding the surface properties, with appropriate process parameters the rCF does not show substantial fiber damage and residual matrix or char are also undetectable on the surface (Figure 34.4). At present, the rCFs from pyrolysis processes are those with the highest value.

The pyrolysis process is the state of the art in industry scale for thermal recycling of CFRP. Several companies produce rCF using pyrolysis, for example Gen 2 Carbon (United Kingdom), Carbon Conversions (United States), CFK Valley Stade (Germany), or SGL Automotive Carbon Fibers (United States). Their processing capacity is around 1000–2000 t/a each.

Microwave pyrolysis is a special form of pyrolysis. Analogous to pyrolysis, the material is digested at reduced oxygen atmosphere and high temperatures. The microwaves do not heat the entire furnace chamber, but only the CFRP material. Since carbon fibers are

FIGURE 34.4 SEM image from recycled carbon fibers from the pyrolysis process.

electrically conductive, eddy currents can be induced in the material by means of microwaves, which results in ohmic heating of the material. The material is heated volumetrically, regardless of its thermal conductivity. Process engineering advantages of microwave pyrolysis are a fast regulation of the electromagnetic energy, fast and contactless heating of the material, and a fiber exposure independent of the matrix material (Seiler et al., 2019). Since microwave pyrolysis has so far not been investigated in continuous operation, but only as batch tests, this process is not used on an industrial scale.

34.4.2.2 Fluidized Bed Process

The fluidized bed combustion works in an oxygen-rich atmosphere at elevated temperatures of 450–500°C. The chopped CFRP waste is thermally decomposed into a silica sand bed that is heated and fluidized by hot air. The organic polymer part of the composite is decomposed by high temperature load and then transported, also with the separated fibers, in the air stream with the silica particles. A postprocess treatment including separating and collecting solid particulates (e.g., fibers, fillers, sand particles) is necessary (Asmatulu et al., 2014). Upon the rapid heating and in the oxygen atmosphere, fibers are released by attrition. This process is currently not used on an industrial scale.

34.4.3 Chemical Recycling

Another possibility of separating fibers and matrix is chemical decomposition of the polymer matrix. By using the solvent alcohol or water, the CFRP components are subjected to increased pressure and temperature, and the fibers are separated from the matrix. This process is called solvolysis. The primary aim is to recover the rCF with high quality, but the matrix components can also be recovered and reused in the chemical industry (Shyha and Huo, 2021). The main disadvantage is possible fiber damage due to high thermal stress. The energy consumption is significantly lower compared to thermal processes, but further processing steps are necessary for drying and cleaning.

A special form is the hypercritical solvolysis: fibers are recovered, preferably from a thermoset matrix, by using a supercritical fluid. In the United Kingdom, special studies were conducted to reduce thermal fiber damage and to use environmentally friendly solvents (Jiang et al., 2009; Dauguet et al., 2015). Both processes have not been applied on an industrial scale so far.

34.4.4 Characterization of the rCF

For a high-value reuse of rCF, their properties and quality must first be investigated. The parameters determining that quality are fiber length, fiber damage, fiber purity, and mechanical properties (Zöllner, 2019).

In principle, every recycling process is accompanied by a reduction of the fiber length, e.g., through the recycling process itself or preshredding as preparation for the recycling treatment. This reduction represents a loss of quality, since compounds with shorter reinforcing fibers have lower mechanical properties than those with continuous fibers. The determination of the length of individual long fibers is a challenge, since rCF can be tangled and disordered. Image measurement methods, especially developed for measuring fiber lengths (particles with a high geodetic length or a high aspect ratio), are to be preferred. Crossed-over fibers, a representative selection of the fibers to be measured, the multiscale problem, and fiber agglomeration have to be handled specifically.

Damage to the fiber surfaces can be examined qualitatively by using a scanning electron microscope (SEM). However, due to the high resolutions, only selected areas can be observed here. No quantitative statement of the damage is possible.

The qualitative purity of the rCF can also be determined by using SEM. Residues on the surface of the rCF can be quantified by thermogravimetric or chemical analysis (DIN EN 2564, 1998; Yang et al., 2015; DIN EN 2559, 1997).

The most important mechanical properties of rCF are tensile strength and module. These characteristic values are usually not directly measured on the rCF, but on a compound made of rCF. Single fiber tensile strength tests are difficult with rCF of small length and hardly representative (Pimenta et al., 2010).

34.5 APPLICATIONS OF RCF

The described recycling processes lead to different rCF products. According to the various properties of those products, only specific applications are attainable in each case. Therefore there are many applications and no specific one has been established in industry for all gained CF waste. Major possible applications are listed in the following paragraphs.

To find suitable applications for rCF, it is important to have material streams as clean as possible. Depending on the length of the fibers, various applications can be considered, such as injection molding for fibers under 0.5 mm, or textile processing for fibers over 50 mm (Kreibe et al., 2015). In principle, the longer the fibers contained, the smaller the residual matrix amount, and the more undamaged the sizing, the better are the properties of the recycled product.

Specific applications for rCF are detailed as follows and an overview is given in Figure 34.5. Fischer (2015) distinguishes chopped fibers/milled fibers, secondary chips, compounds, random nonwovens, yarns, and paper as possible rCF semifinished products.

Filler in thermoplastics (Uhlmann and Meier, 2017) analyzed waste from finishing (milling) with fiber lengths of 0.01–10 mm in their studies. The fibers used were pyrolyzed and grounded.

The aim of compounding is to improve the rheological, mechanical, and electrical properties of the thermoplastic (Uhlmann and Meier, 2017). Compounding is possible because polycarbonate and CF are polar materials, so strong interaction occurs between them due to the

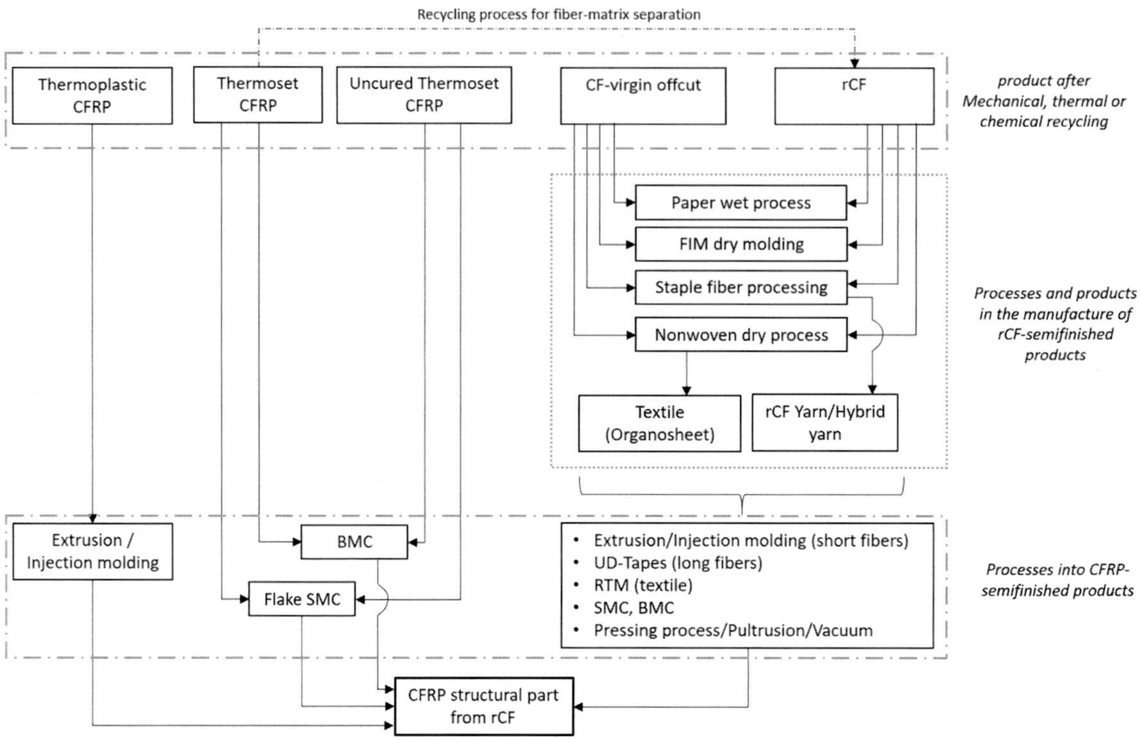

FIGURE 34.5 Overview of CFRP remanufacturing processes.

attraction of polar groups. Therefore the bonding of both materials can occur without an adhesion promoter. The more polar groups present on the surface, the better the bonding effect. At low speed, the completely melted polycarbonate (PC) is bonded to the rCF in the extrusion process (Fette et al., 2015).

Hybrid yarn: Yarn from rCF is made like traditional textile yarn. First, the fibers are aligned and then combed. Fibers that are too short are sorted out. Fleeces are produced, which are spun into slivers and finally into yarns. Carbon staple fibers are added to improve the properties of thermoplastic fibers, but still depend on the length and content of the carbon fibers contained (Hengstermann et al., 2016). Abdkader and Cherif (2019) developed hybrid yarns whose properties reach 80% of the properties (tensile strength) of comparable primary CF yarns: carbon fibers are blended with polymer fibers (PA6) and processed over a special carding machine into a carding belt and then into a draw frame belt to form rCF hybrid yarns.

RCF-UD-tapes, unidirectional tapes from recycled carbon fibers with highly oriented fibers, are another reuse method. Carbon fibers are mixed with other fibers (thermoset), and in carding and stretching processes, first straps and then tapes are formed. These tapes are thermofixed before being stored in rolls or woven. They need to be stored cool in order not to dry out the matrix. In application, tapes are unrolled and placed in the desired pattern (Abdkader and Cherif, 2019).

Nonwovens can be made of pure rCF and as hybrid nonwovens. Parameters influencing the production of hybrid CF yarns and nonwovens are (Pakdel et al., 2020):

- Initial length of rCF and sizing type
- Carding setting and commingling approach
- Production and spinning method

Dry nonwovens consist of pure carbon fibers (rCFs) or mixtures with thermoplastic fibers. Fibers are carded before the layers are cross-lapped and then fixed by a needling or knitting process (Gulich and Hofmann, 2014). These nonwovens have better mechanical properties because of the alignment of the fibers (Wölling et al., 2017). Wet nonwovens are produced in a process evoking paper production: pure, unbowed carbon fibers are dispersed in water to form a paper-like structure on a sieve (Wölling et al., 2017). Several layers are possible without auxiliary agents because fibers become felted. In addition, wet nonwovens are basically used for sheet molding compounds (SMCs). For those, wet nonwovens are impregnated and then pressed to form new components (Rademacker et al., 2018).

The carbon fibers obtained from pyrolysis and solvolysis after crushing are also suitable as reinforcement components of thermoplastic polymers processed in injection molding applications (Meiners and Eversmann, 2014).

More examples are shown in Table 34.2. A great variety of products are technically feasible, but the practical implementation always depends on fiber length, impurities, and product requirements.

TABLE 34.2 Overview of recycling products.

Products	Placement
Extrusion/injection molding	Uhlmann and Meier (2017), Hermann and Witte (2017), Hengstermann (2017), Gude et al. (2018), Fischer (2015), ELG Carbon (2017), and Ansorge (2017)
BMC (bulk molding compound)	Meiners and Eversmann (2014), Hermann and Witte (2017)
SMC (sheet molding compounds)	CFRP-direct recycling: Hermann and Witte (2017) SMC with thermoset matrix, flake SMC (Prepreg from rCF/uncured prepregs): Meiners and Eversmann (2014), Hermann and Witte (2017), and Bartkowski et al. (1990)
Paper—wet process	Gude et al. (2018), Fischer (2015), and Zeisberg (2020)
Nonwoven—dry process	Meiners and Eversmann (2014), Gulich and Hofmann (2014), Hengstermann (2017), Fischer (2015), and Zeisberg (2020)
Yarn	Baz and Gresser (2017), Abdkader and Cherif (2019), staple fiber: Goergen et al. (2016), hybrid: Pakdel et al. (2020), Goergen et al. (2016), rCF-nonwoven made from yarn: Baz and Gresser (2017), reinforcement made from yarn: Zeisberg (2020)
Textile Woven Organic sheets	rCF-woven patches (wovens): Brühl-Saager (2017) Press molded recycled (hybrid nonwoven pressing): Gude et al. (2018), Hengstermann (2017), Andrews (2018), Meiners and Eversmann (2014), hybrid processes (e.g., organic sheets with back-injection of compounding) Hummel et al. (2019)
CF virgin	Hermann and Witte (2017), Pakdel et al. (2020), and Meiners and Eversmann (2014) rCF-short fiber crushed/chopped: ELG Carbon (2017) and Ansorge (2017)
CFRP structural part from rCF	Molded parts from rewetted rCF chips: Christ et al. (2011) Prepreg from rewetted rCF chips in fluidized bed process: Meiners and Eversmann (2014) and Dauner and Huber (2013)

34.6 FUTURE DEVELOPMENTS AND CHALLENGES

Future challenges, technological gaps, and the limits to the recycling of CFRP as discussed in the following text have attracted little attention so far. But they pose an essential part of solving the recycling problem and, hence, overall applicability of this material within a circular economy paradigm. Furthermore, the steadily growing demand for CFRP must be associated with technological developments addressing topics, which are relevant not only for the recycling. Certain solutions can be found in the field of digitization and artificial intelligence.

Future challenges, which are often based on the complexity of the CFRP material, are, in particular:

- separate collection of CFRP
- suitable sorting processes for large-scale use
- in situ identification of the carbon fiber type used in a certain waste material
- large-scale digestion processes for the recovery of clean rCF
- options for assessing the fiber quality on a representative scale
- solving acceptance problems of recyclates in industries with high quality components

Appropriate research in this area will enable progress to be made in the future. The limits of CFRP recycling, such as property reduction through multiple recycling circles, especially reduced fiber lengths, will continue to exist and make high-quality recycling of CFRP difficult in the long term. Moreover, rCFs have a lower quality than respective virgin fibers. Also, fibers from undocumented waste cannot be effectively recycled so far (Limburg and Quicker, 2016). Consequently, CF must be removed from the material cycle and must be disposed of, at least after several cycles. The combustion of CFRP in common waste incineration plants is also technologically impossible. In the hot zones of those plants, the temperatures, oxygen content, and residence times are often insufficient to decompose the thermally rather stable carbon fibers. Research has shown that the fibers break down into small fragments that can be traced with the slags (Stockschläder et al., 2018). Because of their electrical conductivity, carbon fiber fragments can cause electrical short circuits in the electrical dust removal installation and other electrical and gas treatment installations. Fibers were detected in both the slag and the off-gas (Stockschläder et al., 2018). The use of special waste incineration plants is only recommended for those CFs that cannot be recirculated anymore. Current research approaches also pursue the goal of an energetic use of CFRP waste in pyrometallurgical processes, as well as in the furnaces of the calcium carbide or cement industry (Adam and Weimann, 2019; Jiang et al., 2009). The high temperatures there ensure the complete decomposition of carbon fibers. In metallurgy, the contained carbon acts as a reducing agent instead of coke (Schüppel, 2018).

34.6.1 Life Cycle Assessment of Carbon Fibers

Life cycle assessment (LCA) is a method for the combination and assessment of all input and output streams and the potential environmental effects of a product in the course of its life (ISO 14040, 2006).

The environmental impact of CFRP not only depends on the production process, but also on the chosen recycling method. The environmental impacts of individual recycling routes can be represented by means of a life cycle assessment. A "gate-to-gate" approach is necessary to include the production of compound materials made of rCF. Environmental benefits by replacement are primarily achieved by avoiding manufacturing of the energy-intensive virgin carbon fibers. LCA is one of the most sensible methods of mapping these benefits. Meng et al. give an overview of CFRP waste processing routes and compare their credits (Meng et al., 2018). While pyrolysis and solvolysis have the highest energy demands, incineration and pyrolysis have the highest global warming potential. Fluidized bed or mechanical processes have mainly lower environmental impacts.

The urgently needed technical data to assess the recycling process quantitatively are not readily available. Digitalization of the whole life cycle is necessary for the appropriate consideration of all required resources (raw fiber materials, electricity, water, etc.) and emissions throughout all phases of a product's life, including any nonreusable by-products. The LCA method is discussed in more detail in Chapter 47.

34.6.2 Digitalization of the Recycling Process

With the exception of dry fibers and a few selected direct recycling options for CFRP prepregs, the fibers must be recovered from thermoset CFRP (Guo et al., 2017) and processed into semifinished products. The rCF quality depends on the original fiber type (Hengstermann, 2017). Therefore, an important focus is the digital information transfer across all life stages. With such a transfer, the material flow channeling of consumer waste into adaptive processing routes can be optimized, resulting in a potentially high grade rCF with respective high-value application opportunities. Within this process chain, the following technological gaps arise:

- technical identification of CFRP in material streams to track CFRP materials and prevent dissipation of fibers,
- targeted composite disintegration of the highly integral lightweight structures.

While great progress has been made in rCF processing (Hengstermann, 2017; Hengstermann et al., 2016), appropriate applications are still lacking for semifinished rCF products (Rademacker and Heumer, 2014). Obstacles to rCF reuse are

- missing industrial rCF quality assessment methods
- lack of standards for rCF properties
- potential damage to fiber or sizing due to thermal stress or fiber shortening in the only recycling process available on the industrial scale, pyrolysis (Zöllner et al., 2019, 2020)
- still too little knowledge about the mechanical properties of products made with rCF (Gude et al., 2018).

The provision of a consistent quality, guaranteed and standardized by the manufacturer, in processes certified by independent authorities, must be the aim of CFRP developments. The previously mentioned causes are partly interrelated and illustrate the complexity of the CF recycling issues.

Further developments in sorting methods, for example, using tracer-based sorting technologies for the exact detection of the individual materials, will also be investigated in the future.

The digitization of recycling processes, especially recording the energy and mass balances, can simplify future strategies in this area and improve the circularity of the materials. Creating a digital twin of the recycling process can make it easier to derive a preferred process route for each material type aiming at a certain rCF specification. In addition, digitization is essential for a comprehensive environmental assessment using LCA.

34.7 CONCLUSION AND OUTLOOK

The ongoing trend towards development and application of more complex and more functional materials poses new challenges for recirculating them into the circular economy.

This is demonstrated in this chapter exemplarily for carbon fiber applications. The growing use of carbon fibers in CFRP is leading to a steadily increasing amount of waste containing CFRP. Therefore an overview of the market situation, the fields of application of CFRP, and the current waste situation were given as an introduction into this topic. Even though endless recycling is impossible with CF, the idea to recycle and reuse it has its benefits. Quality and length are decisive for the value of the recycled fiber. Various recycling processes for CFRP were presented, compared, and discussed with regard to commercial use. Pyrolysis currently has the highest degree of technical maturity and is the state of the art in CFRP recycling. Previously unresolved challenges and technological gaps were critically examined. For example, there is a need for further research into the preconditions for CF recycling and the avoidance of fiber dissemination into other recycling product streams. The question of how to handle composite EoL waste with low fiber content economically still awaits an answer as well. Despite the acceptance of recycled fibers for reuse being currently low, numerous possibilities for reuse were reviewed and systematized. Only a few of them are industrially mature and allow a recirculation of a minor portion of the materials into the circular economy. Tremendous efforts are still required to develop technologies that allow an increased number of use cycles for high-tech products such as carbon fibers. It has to be acknowledged, however, that many of those products have to be disposed of after a number of cycles, contrary to most products from metals.

References

Abdkader, A., Cherif, C., 2019. Entwicklung einer industriellen Prozesskette von UD-Tapes. Umweltbundesamt; LAGA Ad-hoc-Arbeitskreis; Aufbereitung und Verwertung carbonfaserhaltige Abfälle; TU Dresden; ITM, Dessau-Roßlau.

Adam, C., Weimann, K., 2019. Verwertung Carbonhaltiger Reststoffe in der Pyrometallurgie. Umweltbundesamt; LAGA Ad-hoc-Arbeitskreis; Aufbereitung und Verwertung carbonfaserhaltige Abfälle; BAM, Dessau-Roßlau.

Adams, R.D., Collins, A., Cooper, D., Wingfield-Digby, M., Watts-Farmer, A., Laurence, A., et al., 2014. Recycling of reinforced plastics. Appl. Compos. Mater. 21 (1), 263–284. https://doi.org/10.1007/s10443-013-9380-1.

Akpan, E.I., Adeosun, S.O., 2019. Sustainable Lignin for Carbon Fibers: Principles, Techniques, and Applications. Springer International Publishing, Cham.

Andrews, B., 2018. Standardising Recycled Carbon Backs Its Increasing Acceptance: CCe.V. Carbon Composites Magazin. 01/2018.

Ansorge, F., 2017. CFRTP-Recycling macht den Unterschied. Nürnberg: Carbon Composite e.V. Magazin. 2/2017; Toho Tenax Europe GmbH.

Araujo-Andrade, C., Bugnicourt, E., Philippet, L., Rodriguez-Turienzo, L., Nettleton, D., Hoffmann, L., Schlummer, M., 2021. Review on the photonic techniques suitable for automatic monitoring of the composition of multi-materials wastes in view of their posterior recycling. Waste Manag. Res. 39 (5), 631–651. https://doi.org/10.1177/0734242X21997908.

Asmatulu, E., Twomey, J., Overcash, M., 2014. Recycling of fiber-reinforced composites and direct structural composite recycling concept. J. Compos. Mater. 48 (5), 593–608. https://doi.org/10.1177/0021998313476325.

Bartkowski, K.-D., Wesslein, M., Mischo, H., Kühfusz, R., 1990. Verfahren zur Herstellung einer duroplastischen Pressmasse aus Verbundwerkstoff. Europäische Patentanmeldung.

Baz, S., Gresser, G., 2017. Vorteile von Garnen aus rCF. In: Stuttgart, Denkendorf: Fachtagung Composite Recycling & LCA. Institut für Textil- und Verfahrenstechnik.

Bledzki, A.K., Seidlitz, H., Goracy, K., Urbaniak, M., Rösch, J.J., 2021. Recycling of carbon fiber reinforced composite polymers-review-part 1: volume of production, recycling technologies, legislative aspects. Polymers 13 (2). https://doi.org/10.3390/polym13020300.

Brühl-Saager, M., 2017. Chancen und Grenzen der Direktherstellung von Gewebepatches aus Bauteilen. In: Chemnitz: Carbon Composite e.V.; CCeV-Thementag. GESA.

Christ, M., Öttinger, O., Bode, R., Prefontaine, A., Schwaiger, G., 2011. Verfahren zur Herstellung eines Formteils aus Kohlenstoffwerkstoff unter Verwendung von wiederverwerteten Kohlenstofffasern. Deutsches Patent- und Markenamt, München. Published Patent application; SGL Carbon SE.

Das, T.K., Ghosh, P., Das, N.C., 2019. Preparation, development, outcomes, and application versatility of carbon fiber-based polymer composites: a review. Adv. Compos. Hybrid Mater. 2 (2), 214–233. https://doi.org/10.1007/s42114-018-0072-z.

Dauguet, M., Mantaux, O., Perry, N., Zhao, Y.F., 2015. Recycling of CFRP for high value applications: effect of sizing removal and environmental analysis of the supercritical fluid solvolysis. Procedia CIRP 29, 734–739. https://doi.org/10.1016/j.procir.2015.02.064.

Dauner, W., Huber, S., 2013. Verfahren zur Herstellung eines Halbzeuges zur Verarbeitung zu einem CFK-Bauteil aus C-Faser Abfällen. Declared by Bayerische Motoren Werke AG, 80809 München, DE am 05.11.2013.

DIN EN 2559, 1997. Kohlenstoffaser-Prepregs.

DIN EN 2564, 1998. Kohlenstoffaser-Laminate.

Donaldson, K., Murphy, F.A., Duffin, R., Poland, C.A., 2010. Asbestos, carbon nanotubes and the pleural mesothelium: a review of the hypothesis regarding the role of long fibre retention in the parietal pleura, inflammation and mesothelioma. Part. Fibre Toxicol. 7, 5. https://doi.org/10.1186/1743-8977-7-5.

ELG Carbon, 2017. Recycelte Carbonfasern ebnen den Weg für kostengünstige Leichtbau-Strukturen. Nürnberg: Carbon Composite e.V. Magazin. special edition 2016.

Fette, M., Rademacker, T., Wulfsberg, J., Herrmann, A., Stoess, N., 2015. Resource efficient and sustainable production of secondary structure aircraft components by using recycled carbon fibers for sheet molding compounds. In: Proceedings of the 10th Sampe, pp. 124–129.

Fischer, S., 2015. Technological aspects of CFRP recycling. In: Conference Composite Recycling. SGL Carbon SE, Stuttgart.

Goergen, C., Baz, S., Mitschang, P., 2016. Innovative Organic Sheets Made of Recycled Carbon Fibers. Hanser-Verlag; Kunststoffe international. 5/2016.

Gude, M., Luft, J., Troschitz, J., Krampitz, T., Zöllner, M., Lieberwirth, H., et al., 2018. Abschlussbericht ReLei. Dresden: funding number 02PJ2800-02PJ2808.

Gulich, B., Hofmann, M., 2014. Verarbeitung von rezyklierten Carbonfasern zu Vliesstoffen für die Herstellung von Verbundbauteilen. Springer Fachmedien.

Guo, J., Gao, X., Toma, E., Netzelmann, U., 2017. Anisotropy in carbon fiber reinforced polymer CFRP and its effect on induction thermography. NDT & E Int. 91.

Hengstermann, M., 2017. Standortbestimmung bei Recycling von Carbonfasern. Nürnberg: Carbon Composite e.V.; special edition 2016.

Hengstermann, M., Hasan, M.M., Abdkader, A., Cherif, C., 2016. Influence of fibre length and preparation on mechanical properties of carbon fibre/polyamidea 6 hybrid yarns and composites. Fibres Text. East. Eur 24 (5(119)), 55–62. https://doi.org/10.5604/12303666.1215528.

Hermann, A., Witte, T., 2017. Composite recycling strategy and its first practical implementations at Airbus. In: Stuttgart: Conference Composite Recycling. CTC Stade GmbH.

Huang, X., 2009. Fabrication and properties of carbon fibers. Materials 2 (4), 2369–2403. https://doi.org/10.3390/ma2042369.

Hummel, S., Schemme, M., Karlinger, P., 2019. Variotherme Temperierung für Organobleche. GmbH - Glasfaserhybride mit bester Haftung. In: Plastverarbeiter 2019. 13.08.2019, checked on 30.07.2021.

ISO 14040, 2006. Environmental Management—Life Cycle Assessment—Principles and Framework.

Jiang, G., Pickering, S., Lester, E., Turner, T., Wong, K., Warrior, N., 2009. Characterisation of carbon fibres recycled from carbon fibre/epoxy resin composites using supercritical n-propanol. Compos. Sci. Technol. 69 (2), 192–198. https://doi.org/10.1016/j.compscitech.2008.10.007.

Kehren, D., Simonow, B., Bäger, D., Dziurowitz, N., Wenzlaff, D., Thim, C., et al., 2019. Release of respirable fibrous dust from carbon fibers due to splitting along the fiber axis. Aerosol Air Qual. Res. 19 (10), 2185–2195. https://doi.org/10.4209/aaqr.2019.03.0149.

Kreibe, S., Hartleitner, B., Gottlieb, A., Berkmüller, R., Förster, A., Tronecker, D., 2015. MAI Recycling—Entwicklung ressourceneffizienter CFK-Recyclingverfahren und Prozessketten für die künftige Bereitstellung qualitativ hochwertiger rC-Halbzeuge. Schlussbericht MAI Recycling, Technische Informationsbibliothek (TIB) and bifa Umweltinstitut.

LAGA (2019): Entsorgung faserhaltiger Abfälle. Dessau-Roßlau: Bund/Länder - Arbeitsgemeinschaft Abfall; Umweltbundesamt(Hrsg.); Abschlussbericht.

Limburg, M., Quicker, P., 2016. Disposal of carbon fiber reinforced polymers—problems during recycling and impacts on waste incineration. In: Waste Management Volume 6., Waste-to-Energy, pp. 221–235.

Limburg, M., Stockschläder, J., Quicker, P., 2019. Thermal treatment of carbon fibre reinforced polymers (part 1: recycling). Waste Manag. Res. 37 (1_suppl), 73–82. https://doi.org/10.1177/0734242X18820251.

Meiners, D., Eversmann, B., 2014. Recycling von Carbonfasern. 7 TK-Verlag, Berlin. Recycling und Rohstoffe. 371 ff.

Meng, F., Olivetti, E.A., Zhao, Y., Chang, J.C., Pickering, S.J., McKechnie, J., 2018. Comparing life cycle energy and global warming potential of carbon fiber composite recycling technologies and waste management options. ACS Sustain. Chem. Eng. 6 (8), 9854–9865. https://doi.org/10.1021/acssuschemeng.8b01026.

Newcomb, B.A., 2016. Processing, structure, and properties of carbon fibers. Compos. A: Appl. Sci. Manuf. 91, 262–282. https://doi.org/10.1016/j.compositesa.2016.10.018.

Orzol, C.H.M., Ohlmeyer, M., Krampitz, T., Jäckel, H.-G., Zöllner, M., Lieberwirth, H., 2017. Voruntersuchungen zur elektrohydraulischen Zerkleinerung von CFRP für das Carbonfaser-Recycling. Chem. Ing. Tech. 89 (12), 1773–1781. https://doi.org/10.1002/cite.201700024.

Oswald-Tranta, B., Tuschl, C., Schledjewski, R., 2019. Untersuchungen von CFK Proben mit induktiv angeregter Thermografie. DACH-Tagung, Leoben.

Pakdel, E., Kashi, S., Varley, R., Wang, X., 2020. Recent progress in recycling carbon fibre reinforced composites and dry carbon fibre waste. In: Resources, Conservation and Recycling. Elsevier.

Pimenta, S., Pinho, S.T., Robinson, P., Wong, K.H., Pickering, S.J., 2010. Mechanical analysis and toughening mechanisms of a multiphase recycled CFRP. Compos. Sci. Technol. 70 (12), 1713–1725. https://doi.org/10.1016/j.compscitech.2010.06.017.

Prashanth, S., Subbaya, K.M., Nithin, K., Sachhidananda, S., 2017. Fiber reinforced composites—a review. J. Mater. Sci. Eng. 06 (03). https://doi.org/10.4172/2169-0022.1000341.

Rademacker, T., Heumer, W., 2014. Die Bereitschaft der Industrie war gering, ein Recyclingprodukt einzusetzen, fifty-first ed. vdi Nachrichten.

Rademacker, T., Fette, M., Jüptner, G., 2018. Nachhaltiger Einsatz von Carbonfasern dank CFK_Recycling. Lightweight Des. 11, 12–19. https://doi.org/10.1016/j.compscitech.2010.06.017.

Rosato, D., Rosato, D., 2005. Reinforced Plastics Handbook. Elsevier.

Schüppel, D., 2018. Ergebnisse von Umweltcluster Bayern und MAI Carbon zum Thema Recycling von Carbon. Carbon Compos. Magazin (1), 24–25.

Seiler, E., Hickmann, T., Teipel, U., 2019. Von der Carbonfaser zur Brennstoffzelle - Gewinnung von Sekundärgraphit aus rezyklierten Kohlenstofffasern. In: 5. Symposium Rohstoffinnovationen und Rohstoffeffizienz 2019, pp. 133–142.

Shyha, I., Huo, D., 2021. Advances in Machining of Composite Materials. Springer International Publishing, Cham. checked on 08.07.2021.

Stockschläder, J., Quicker, P., Thiel, C., Beckmann, M., Baumann, W., Wexler, M., et al., 2018. Behandlung von Carbonfaser-haltigen Abfällen in Hausmüllverbrennungsanlagen. Zwischenergebnisse aus praktischen Untersuchungen im industriellen Maßstab. In: Thiel, S., Thomé-Kozmiensky, E., Friedrich, B., Pretz, T., Quicker, P., Senk, D.G., Wotruba, H. (Eds.), Mineralische Nebenprodukte und Abfälle. vol. 5. TK Verlag, pp. 87–105.

Uhlmann, E., Meier, P., 2017. Einsatz von CFK-Fräspartikeln als Füllstoff. Springer-VDI-Verlag GmbH & Co.KG, Düsseldorf, pp. 558–563. wt Werkstatttechnik online; edition 7/8.

Warzelhahn, V., 2019. Recycling, rohstoffliche und thermische Verwertung von CFK-Abfällen. Stuttgart: CCe. V.; Arena 2036.

Wen, X., Luo, Q., Hu, H., Wang, N., Chen, Y., Jin, J., et al., 2014. Comparison research on waste classification between China and the EU, Japan, and the USA. J. Mater. Cycles Waste Manage. 16 (2), 321–334. https://doi.org/10.1007/s10163-013-0190-1.

Witten, E., Schuster, A., 2018. Der Composites-Markt Europa: Marktentwicklungen, Herausforderungen und Chancen. checked on 01.07.2021.

Wölling, J., Schmieg, M., Manis, F., Drechsler, K., 2017. Nonwovens from recycled carbon fibres—comparison of processing technologies. Procedia CIRP 66, 271–276. https://doi.org/10.1016/j.procir.2017.03.281.

World Health Organisation, 1997. Determination of Airborne Fibre Number Concentrations. A Recommended Method, by Phase Contrast Optical Microscopy (Membrane Filter Method). World Health Organization, Geneva. ISBN: 9241544961. https://www.who.int/publications/i/item/9241544961.

Xian, G., Wang, Z., 2018. 2.17 Carbon fiber reinforced plastics—properties. In: Comprehensive Composite Materials II. Elsevier, pp. 342–359.

Yang, J., Liu, W., Wang, J., Tang, T., 2015. Recycling of carbon fibre reinforced epoxy resin composites under various oxygen concentrations in nitrogen–oxygen atmosphere. J. Anal. Appl. Pyrolysis, 253–261.

Zeisberg, 2020. Carbon als Recycling-Werkstoff. Das richtige Verfahren für jede Phase des Produktlebenszyklus. In: DLR Wissenschaftstage 2020. Hannover, p. 29.10.2020.

Zhang, J., Chevali, V.S., Wang, H., Wang, C.-H., 2020. Current status of carbon fibre and carbon fibre composites recycling. Compos. Part B Eng. 193, 108053. https://doi.org/10.1016/j.compositesb.2020.108053.

Zöllner, M., 2019. Untersuchungen zur Analytik freigelegter Kohlenstofffasern aus unterschiedlichen Aufschlussverfahren. Dissertation, TU Bergakademie Freiberg.

Zöllner, M., Lieberwirth, H., Kempkes, P., Fendel, A., 2019. Thermal resistance of carbon fibres/carbon fibre reinforced polymers under stationary atmospheric conditions and varying exposure times. In: Waste Manag. 85, pp. 327–332, https://doi.org/10.1016/j.wasman.2018.12.027.

Zöllner, M., Krampitz, T., Lieberwirth, H., 2020. Aufbereitung kohlenstofffaserverstärkter Thermoplaste zu Granulat für den Wiedereinsatz in Strukturbauteilen. Chem. Ing. Tech. 92 (4), 459–468. https://doi.org/10.1002/cite.201900122.

PART 4

Recycling and the circular economy

CHAPTER

35

From waste management to circular economy—The role of recycling in policy

Ernst Worrell

Copernicus Institute of Sustainable Development, Utrecht University, Utrecht, The Netherlands

35.1 INTRODUCTION

Traditionally, waste management policy has been directed at the sanitary disposal of waste at the lowest direct costs for the stakeholders. As environmental concerns in the past decades increased, waste management has slowly shifted from landfills to sanitary landfills and waste incineration. However, in preindustrial societies, when materials were relatively expensive, earlier waste management strategies built upon formal and informal routes of recycling and these recovered most of the recyclables. This more traditional organization of waste management is still found in developing countries where scavengers or waste-pickers are an important factor in the recycling infrastructure, albeit often part of an informal economy (Chapter 7—The path to inclusive recycling: Developing countries and the informal sector). Also, in times of stress (e.g., economic crisis, war) there is an increased focus on recycling. In fact, during World War II, recycling became patriotic, and countries used massive campaigns to urge more efficient use of resources and recycling. Evidence exists of recycling through remelting of glass and bronze in societies as early as 400 BC, and excavations at Pompeii have shown that recycling was an integral part of the city's economy. In more modern times even money was impossible without recycling, as many banknotes consisted of cotton from recycled clothing. Recycling was then economically more attractive than production from primary resources, also due to the allocation of costs and the negligence of external costs of material production and waste management. Only when low-cost mass production techniques reduced the costs of materials and products, due to the Industrial Revolution, did we see decreasing attention on recycling, as it became economically less attractive.

Since the 1970s, the increased environmental impacts associated with material production and waste disposal, and rising energy costs, have shifted the focus of waste management policy toward an increased role for recycling. The United States adopted the Resource Recovery Act in 1970, introducing the well-known 3 Rs: Reduce, Reuse, Recycle. This started the introduction of various forms of waste hierarchies in most industrialized countries. More recently, increasing costs of raw materials due to rapidly increased demand from a growing and more affluent global population, along with the need to reduce the overall environmental impact of

Handbook of Recycling. **https://doi.org/10.1016/B978-0-323-85514-3.00026-9**

mankind on the world, have raised the need to change the current linear industrial system to a circular (economic) system. In such a world, the stocks of materials in our buildings, equipment, and infrastructure (the "technosphere") represent a large stock of materials, in addition to the stocks found in the earth, and are aptly named the "urban mine." This stock can be managed in a similar way as primary resources, opening ways to change from a linear system of material and waste management toward a circular economy, mimicking natural ecosystems.

This chapter discusses the changing role of recycling as part of (solid) waste management, starting with a brief history, followed by a discussion of recycling as the key to and an integral part of modern waste management. This chapter excludes special or hazardous waste management.

35.2 A BRIEF HISTORY OF WASTE MANAGEMENT

As society industrialized its production and simultaneously saw an increasing share of the population moving into urban areas (as labor became more and more centralized), consumption of materials and resources increased dramatically. As over 90% of the consumed materials and resources end up as waste, this massive increase in materials production and use (see Chapter 1) also generated an increasing waste stream. Due to the increased concentration of consumers, this led to local waste disposal problems, which threatened public health and the local environment. In this era, regulation and policy was hence driven mainly by concerns over pollution impacting human health, with little if any attention paid to environmental concerns or resource efficiency. Waste management was perceived as a technical, end-of-pipe solution to collect and dispose of waste, primarily by landfilling. This was later expanded to include incineration, with the first incinerator build in 1874 in Nottingham, United Kingdom.

In the 1960s, increasing concerns about environmental damage due to our industrial society led to a change toward sanitary landfilling to reduce the impact of landfilling through leaching of hazardous pollutants and other negative impacts (e.g., odor). Simultaneously, as space was becoming more expensive in large urban areas and costs of waste transport and disposal increased, incineration became an economically viable technique to reduce the volume of waste and hence the costs of final disposal. With increasing energy costs in the early 1970s, energy recovery from incineration gained more attention, especially in Europe and Japan. Waste-to-energy facilities were especially developed in those countries with high costs of landfilling (e.g., dense populations in industrialized countries) and energy, and the capacity of this technique is still growing around the world as urbanization and consumption increase.

Yet, this era also saw growing attention to recycling as it reduced the energy costs of producing materials while simultaneously reducing the costs of waste management. The Resource Recovery Act in the United States in 1970 led to changing waste policies around the world. The European Union formalized this in 1975 in a Waste Framework Directive. While the hierarchy has been named differently in various countries, e.g., the "3 Rs" and the "waste ladder," these are essentially similar around the world:

- **Reduce**: preventing the generation of waste through product design
- **Reuse**: reuse of products (e.g., refillable containers)
- **Recycle**: reuse of materials contained in products
- **Recover energy**: incineration in waste-to-energy or industrial (e.g., clinker kilns) facilities
- **Disposal**: sanitary landfilling

This hierarchy forms the basis for modern waste management policy, although in practice the hierarchy is often not observed in the design and implementation of (local) waste management programs. Progress is slow, as today a vast share of all municipal solid waste is still being landfilled. In 2019, of all European municipal solid waste (MSW), 24% was landfilled, 27% incinerated, 30% recycled, and 17% composted. For comparison, in 2010 the recycling rate was 25%. Within the EU, recycling rates still vary widely between member states, from a high in Germany (67% of MSW) to lows in other member states (e.g., 11% in Romania). In the United States, about 50% of municipal solid waste was landfilled, 12% incinerated, 24% recycled, and 9% composted (2018).

More recently, the concerns about the material impact of economic growth have led to (re-)introduction of the term circular economy. This has led to revisiting the waste hierarchy introduced in the 1970s and has now led to up to 10 R-strategies, or other similar frameworks, e.g.:

1.	Rethink/ refuse	Prevent material use
2.	Reduce	Decrease material use (e.g., material efficiency)
3.	Renew	Design products focused on circularity
4.	Reuse	Reuse the *product*
5.	Repair	Repair or maintain product
6.	Refurbish	Revive product
7.	Remanufacture	Make a new product (partially) from used parts
8.	Repurpose	Reuse the product for a (new) service or use
9.	Recycle	Reuse the *materials*
10.	Recover	Recover embodied energy by incinerating

While in line with the original 3 Rs, the various circular economy frameworks have in common that they move the focus from waste management to an integrated approach toward products and materials that helps to retain as much economic value as possible. In practice, this means an increased focus on the products and parts, and less on materials. Yet, recycling remains the foundation of this R-ladder, as all products (and parts) will, after a period as long as possible, reach the end of life and enter the waste management system. For recycling, a distinction should be made between high- and low (er)-quality recycling. It is assumed that high-quality, or closed loop, recycling results in replacing (part of) the primary product or material by reused or recycled materials. In other words, the more of the primary product or material that is replaced, the higher the quality of recycling. Alternatively, the recycled material may replace a different material, which is considered low-quality, or open loop, recycling. In the latter case, the characteristics of the replaced materials should be used to determine the environmental impact of recycling. This type of recycling is generally referred to as downcycling, as the use of the material or product is often of a lower quality and functionality than when the original primary material is replaced. In current statistics no information is available on the nature of the recycling process and the degree to which virgin or primary materials are substituted by the recovered and recycled material, but indications are that the majority of materials are downcycled. Hence, in a circular economy we not only need to move our focus to products (instead of materials) but for materials we need to improve recycling to allow for a better and longer-term use of these materials.

35.3 INTEGRATING RECYCLING IN WASTE MANAGEMENT POLICY DESIGN

Modern integrated waste management involves a wide array of suitable processes and programs including source reduction (prevention), recycling, recovery (e.g., incineration, composting), and final disposal (e.g., landfilling), to meet multiple objectives, including public and environmental health. Increasingly, sustainability, resource efficiency, and circularity are integrated in waste management practices in cities and countries. The European Commission has recognized the importance of resource efficiency and made the transition to a circular economy a cornerstone of the Commission's policy focus to deliver sustainable, smart, and inclusive growth. This focus should help to achieve the European Union's targets on reducing GHG emissions, improving the security of the supply of raw materials, and making the European economy more resilient to price increases of energy and commodities (European Commission, 2020). The strategic goals put an increasing focus on the higher steps of the R-ladder. The Single Use Plastics Directive (Directive 2019/904) (European Commission, 2019) should also be seen in this light, in addition to the need to reduce plastic pollution.

For example, the Netherlands is, as the rest of the world, at the beginning of a transition to a more sustainable production, consumption, and energy system through a circular economy. Waste management is a key element in achieving sustainable resource management. The Netherlands has a long history of (research on) waste-to-energy and saving resources, and has been successful in the past in recovering materials from waste, resulting in high recycling rates and high incineration capacity. The Netherlands bills itself now as one of the leaders in the circular economy (Ministry of Infrastructure and Water Management, 2023). Waste and resources management was also central to the development of the second National Waste Management Plan (Landelijk Afvalbeheer Plan, LAP) in the Netherlands, linking it to a transition to a circular economy. In the plan, next to national goals to reduce the volume of waste, and increasing reuse and recycling, it describes actions for various waste streams that represent large volumes or are (still) problematic within the current waste management system, e.g., mattresses and diapers.

The LAP should also contribute to the reduction of GHG emissions as set out in national policy. It will try to achieve this by focusing on recycling, anaerobic digestion, and incineration. As an example, the model calculations by Corsten et al. (2013) explore various scenarios to evaluate the impact of increased recycling efforts beyond policy goals. The explorative analysis of waste processing showed that waste management in the Netherlands already contributed a lot (in 2010, 33% of municipal solid waste was recycled) to recover materials, saving energy and reducing CO_2 emissions, compared to a situation where waste is not recycled or incinerated. While incineration is important in the Netherlands, the largest contribution in GHG emissions mitigation actually comes from recycling. About 70% of the energy savings were due to current recycling processes and 30% due to incineration with energy recovery. The mitigated CO_2 emissions were solely due to recycling and use of wastes as refuse-derived fuel (RDF). The incineration of waste creates additional CO_2 emissions, despite the avoided emissions in electricity generation, which is a result of the (still) large volume of plastics in the waste.

When recycling of selected materials from the waste stream is increased, this could result in additional emission reductions of 2.3 $MtCO_2$ annually, compared to the reference situation (2008). This is equivalent to a potential improvement of more than 45%. Figure 35.1 depicts the results for the three scenarios. The main contributors to the CO_2 emission reduction potential are found in the optimization of the recycling of plastic (PET, PE/PP), textiles, paper, and organic waste.

The study by Corsten et al. shows that, despite many decades of waste management policy, there was (and is) still a potential for further improvement of the sustainability of waste management in the Netherlands. Especially in a scenario focusing on product and material reuse

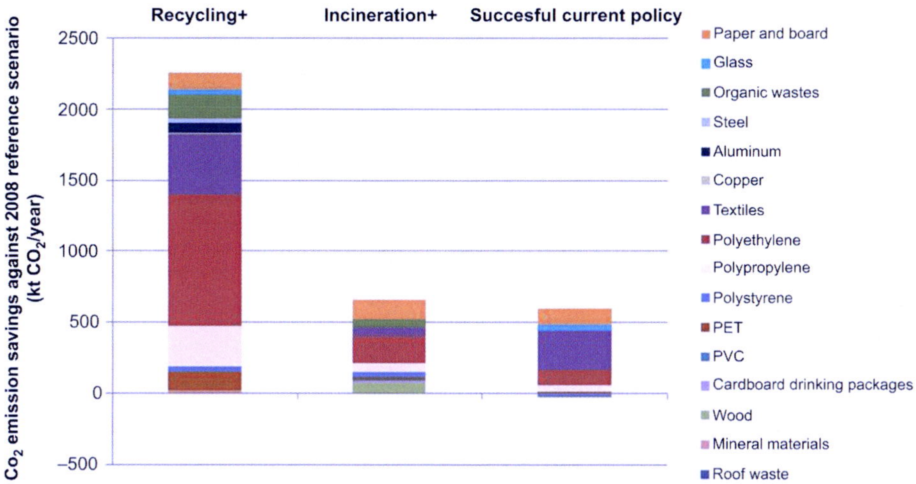

FIGURE 35.1 CO_2 emission reduction potentials of three scenarios for waste management in the Netherlands (based on the 2008 situation), showing that recycling contributes more to emission reduction than other waste management strategies. The Recycling+ scenario assumes increased recycling compared to adopted policy. The Incineration+ scenario assumes similar recycling targets as current policy, while remaining waste is incinerated in high-efficiency waste-to-energy facilities. *From Corsten et al. (2013).*

and high-quality recycling (*Recycling+*), large potential savings are identified. The key materials that will play an important role in achieving the full savings potential are plastics (PET, PE, PP), textiles, paper, and organic waste. A scenario focusing on incineration with higher energy conversion efficiency (*Incineration+*) has the potential of only saving one-third of the CO_2 emission savings achieved in the *Recycling +* scenario. The results confirm that, in terms of energy consumption and CO_2 emission reduction, the waste hierarchy (and with that the newly introduced R-ladders) as used for the basis of European waste management policy is still valid in prioritizing waste management. Analyses and models for other countries show similar results, even though system boundaries (e.g., types of wastes included) may vary in the studies. The transition to a sustainable circular economy and effective waste management system builds on good metrics, which are currently still lacking (Corona et al., 2019). While a wide variety of CE metrics are being developed and applied, none of the current metrics address the CE concept in full (Corona et al., 2019). The challenges of current circularity metrics relate to (1) difficulties in measuring the CE goals in the various sustainability dimensions, (2) evaluating the scarcity of used materials, and (3) underrepresenting the complexities of multiple cycles and the consequences of material downcycling. It is still difficult to represent the benefits of different waste valorization options, even with mature tools such as LCA. A good circularity metric should be sufficiently comprehensive to avoid burden shifting from reduced material consumption to increased environmental, economic, or social impacts (Corona et al., 2019).

The case study of the Netherlands highlighted the need for high-quality recycling, as does the review by Corona et al. (2019). This is currently most often not the case. This is demonstrated by recent reviews on the circularity of the economies around the world. While many of the metrics are incomplete or lack detail, various studies (Circle Economy, 2021; PBL, 2021) show that society, and our waste management system, is still far from reaching any point that could be called circular. This would not only demand a further optimization of recycling, but also of policies beyond the realm of waste management. Hence, a comprehensive and integrated policy is necessary to achieve the full potential of a transition to a circular economy and society, of which waste management is an essential part. This is best achieved with a menu of policy instruments, accounting for the specific characteristics and life cycles of products and services, waste streams, and recycled materials markets.

References

Circle Economy, 2021. The Circularity Gap Report 2021. Circle Economy/PACE, Amsterdam/The Hague, The Netherlands https://www.circle-economy.com/resources/circularity-gap-report-2021. (Accessed 10 March 2022).

Corona, B., Shen, L., Reike, D., Rosales Carreón, J., Worrell, E., 2019. Towards sustainable development through the circular economy – a review and critical assessment on current circularity metrics. Resour. Conserv. Recycl. 151, 104498.

Corsten, M., Worrell, E., van Duin, A., Rouw, M., 2013. The potential contribution of sustainable waste management to greenhouse gas emission reduction in the Netherlands. Resour. Conserv. Recycl. 77, 13–21.

European Commission, 2019. Directive 2019/904 on the reduction of the impact of certain plastic products on the environment. Official Journal of the European Union. 25 June 2019 https://eur-lex.europa.eu/eli/dir/2019/904/oj. (Accessed 10 March 2022).

European Commission, 2020. A New Circular Economy Action Plan for a Cleaner and More Competitive Europe. European Commission, Brussels, Belgium. https://eur-lex.europa.eu/legal-content/EN/TXT/?uri=COM:2020:98:FIN. (Accessed 10 March 2022).

Ministry of Infrastructure and Water Management, 2023. National Circular Economy Programme 2023-2030. Ministry of Infrastructure and Water Management, The Hague, The Netherlands.

PBL (Netherlands Environmental Assessment Agency), 2021. Integrale Circulaire Economie Rapportage 2021. PBL Planbureau voor de Leefomgeving, The Hague, The Netherlands (in Dutch, with English summary) https://www.pbl.nl/publicaties/integrale-circulaire-economie-rapportage-2021. (Accessed 10 March 2022).

CHAPTER

36

Geopolitics of resources and recycling

Benjamin Sprecher[1] *and Ernst Worrell*[2]

[1]Delft University of Technology, Delft, The Netherlands [2]Copernicus Institute of Sustainable Development, Utrecht University, Utrecht, The Netherlands

36.1 INTRODUCTION

Historically, products and materials were reused and recycled in society because materials were too valuable to be wasted. When mass production reduced the costs of materials and products, and rising affluence further reduced the cost of materials as a fraction of income, recycling became less relevant. However, since the 1980s the increased awareness of environmental impacts associated with materials production and waste disposal has revived the attention on recycling.

More recently, increasing costs of raw materials due to rapidly increasing demand from a growing and more affluent global population, and several incidents of severe supply disruptions, have led to concerns for scarcity and political discussions on how to deal with these issues. This has centered around the location of reserves and production of strategic and critical materials. One often quoted example is that of the rare earth elements (REEs), which are increasingly used in modern technology, ranging from efficient electric motors to wind turbines and mobile phones to smart bombs. In 2010, China had a dominating market share of 98% (USGS, 2012). This was largely overlooked outside of specialist circles, right up to the point that China wanted to make a point in a territorial conflict with Japan and blocked exports of rare earths, not only to Japan but to all Western countries. This situation did not last very long. Nevertheless, it still led to massive supply chain disruptions and price increases of up to an order of magnitude for several of the rare earth metals.

Today, around 60% of all rare earth metals are mined and produced in China (USGS, 2022). Nevertheless, the United States is so concerned about this enduring dependency that in 2020 a national emergency was declared when President Trump signed the "Executive Order on Addressing the Threat to the Domestic Supply Chain from Reliance on Critical Minerals from Foreign Adversaries[1]."

Although they are connected, one should differentiate between physical scarcity of materials (Tilton, 2002; Henckens et al., 2014) and criticality of a material (Graedel and Reck, 2015). Scarcity refers to the lack of (future) available

[1] https://trumpwhitehouse.archives.gov/presidential-actions/executive-order-addressing-threat-domestic-supply-chain-reliance-critical-minerals-foreign-adversaries/.

supply of a material due solely to its limited geological availability and is most often a long-term issue spanning several generations. Criticality looks beyond geology to also consider factors such as geopolitics and economics and is often more of a short-term problem. Comparing scarcity and criticality shows that these have many elements in common, but that there are also crucial differences (Table 36.1).

Concerns about scarcity and criticality are not new. The publication of the report "Limits to Growth" by the Club of Rome highlighted resource issues in 1972, and in 1981 the US Environmental Protection Agency commissioned a study on critical materials scarcity for air pollution control equipment (e.g., platinum) (CRA, 1981). Technological progress in catalyst loading, recycling of platinum, and increased mining has helped to secure platinum availability despite strong growth in demand. However, the increasing number of supply chain disruptions due to major events such as the pandemic and conflicts impacting major exporting countries means that criticality is now felt for many materials simultaneously (Habib et al., 2021). This is coming at a time when the world needs to address key challenges like climate change and water availability, while simultaneously supporting unprecedented growth of global consumption of virtually all materials and resources.

TABLE 36.1 The top scarce (based on Henckens et al., 2014) and critical (based on European Commission, 2020) elements in alphabetical order. The European Commission has increased the number of critical materials to 27. A selection of the most important is included in the table.

Potentially scarce elements	Critical elements
Antimony	Antimony
Arsenic	Beryllium
Boron	Boron
Bismuth	Chromium
Cadmium	Cobalt
Chromium	Fluor
Copper	Gallium
Lead	Germanium
Molybdenum	Indium
Nickel	Magnesium
Phosphor	Niobium
Platinum-group (PGM)	Phosphor
Rhenium	Platinum-group (PGM)
Silver, Gold	Rare Earth metals
Tungsten	Silicium
Zinc	Tungsten

This chapter first briefly discusses resource availability and scarcity, followed by a discussion of criticality and the role of supply chain disruptions and a section exploring recycling as a tool in resource management to mitigate the risks of geopolitics around resources. Finally, we give an outlook on other topics that will become increasingly relevant to recycling.

36.2 SCARCITY

Mineral deposits are not distributed evenly. A *mineral deposit* is any accumulation of a mineral or a group of minerals that may be economically recoverable. The value of a deposit depends on availability, access, and economics. Mineral deposits occur where geological processes have concentrated specific minerals in sufficient quantities to be mined (often in so-called basins). A *mineral reserve* is the part of the mineral deposit that has been fully geologically evaluated and is commercially and legally mineable. Reserves may be regarded as "working inventories," which are continually revised as factors such as mining and processing technology change. The *reserve base* includes the mineral reserve plus those parts of the resources that have been quantified and are currently not economical, but could potentially become viable if technology and economics change.

Beyond the reserve base there are speculative estimates of hypothetical resources, referred to as *ultimate reserve*. These are based on assumptions regarding known mineral bodies that may reasonably or statistically be expected to exist under similar geological conditions where mineral discoveries have not yet been made, or in types of deposits that are considered to be uneconomical for the foreseeable future. Uneconomical deposits have been included in the estimates of scarce metals included in Table 36.1, as this list is based on geological availability in the top kilometers of the earth's crust.

The uncertainties associated with resource estimates are large and subject to economics and technological development. Reserve estimates have been constantly revised. Over the past 50 years, the extractive industries sector has succeeded in meeting global demand while the lifetime of reserves and resources (in terms of reserve/production) has continually been extended further into the future. For example, the total amount of copper extracted from the earth is estimated at 320 Mt, compared to the estimated global reserves of 690–3000 Mt. Despite continued mining, reserve estimates of copper have not gone down much over the past 20 years.

Supply of copper has been able to keep up with demand because of technological progress in exploring, mining, and processing mineral ores (coupled with higher prices). Current economically mined copper reserves have a much lower ore grade than historical ore bodies. Moreover, recycling rates of copper are well over 50%. The annual contribution of recycled copper to total copper production is around 35%—a percentage that has remained fairly constant in the last 10 years despite a significant increase in copper mining.

Published reserve figures usually do not reflect the total amount of mineral potentially available and compilation of global reserve figures are not very reliable indicators of long-term availability. These issues are compounded by accounting issues. For a publicly listed mining company to have proven reserves of a metal on its balance sheets means that these reserves must be quantified with a relatively costly standardized method. If a mining company already has 20+ years of proven reserves, it could be seen as a waste of capital to further quantify an ore vein, even though one can be fairly certain that more ore is to be found. This explains, for example, how the *New Scientist* could report in 2007 "that we have, at best, 10 years before we run out of indium. Its impending scarcity could already be reflected in its price: in January 2003 the metal sold for around $60 per kilogram; by August 2006 the price had shot up to over $1000 per kilogram" (Cohen, 2007). Because both demand and price increased drastically, companies started to reevaluate their mineral resources and started reporting vastly more indium reserves than had previously been accounted for (Werner et al., 2017). During 2020–22 the price of indium has fluctuated between US$ 140–270 per kilogram (as retrieved from price.metal.com).

Despite the large uncertainties, mineral reserve estimates are often used to signal future supply problems. However, the key cause for the current debate on availability is not found in the global geological availability of the materials. Rather, problems arise due to the geographical distribution of mineral deposits, access to or control over these deposits, and strategies of developing and emerging economies that aim to exploit their resource base for national development (European Commission, 2020). This debate centers on the concept of *criticality*, which will be discussed next.

36.3 CRITICALITY

Although there is no clear single definition of criticality, in general a material is considered critical if this material is both important to the economy (or for more specific applications, e.g., defense) *and* is susceptible to significant supply disruptions (Graedel and Reck, 2015). Since criticality is heavily dependent on the chance of supply disruptions, researchers have also investigated supply chain resilience as the antithesis of criticality. Resilience is defined as the capacity of a system to tolerate disruptions while retaining its structure and function (Fiksel, 2006). In this perspective, a material can be considered more critical if its supply chain is less resilient (Sprecher et al., 2015).

Criticality has especially come to the forefront as many critical metals play a key role in strategically important technologies ranging from renewable energy technologies to electronics to defense applications. Access to these materials is essential for the future survival of the economy, and hence they have become the object of geopolitical maneuvering.

Critical materials are by definition highly susceptible to supply chain disruptions. There are too many ways in which supply chains can be disrupted to discuss here. But, as illustrated in Figure 36.1, conceptually one can distinguish two axes along which most disruptions can be categorized (Sprecher et al., 2015). One commonly thinks of disruptions as occurring on the supply side, but a sudden change in demand has similar disruptive potential on a supply/demand balance (Tang et al., 2021). These disruptions can occur on a time spectrum from almost instant (e.g., when the Chinese government blocked exports of REEs leading to the 2010 REE crisis) to being a process that plays out over a century or more before suddenly being recognized as a problem (e.g., gradually decreasing ore grades) (Northey et al., 2014).

When analyzing how industries and countries have historically dealt with supply chain disruptions of critical materials, one can distinguish four general strategies:

- Resource efficiency: using materials as efficiently as possible. For example, the amount of platinum required in flue gas catalysts has consistently decreased over time. Another example is the rare earth permanent magnet with high-temperature resistance. These magnets require the relatively scarce metal dysprosium. When dysprosium became extremely

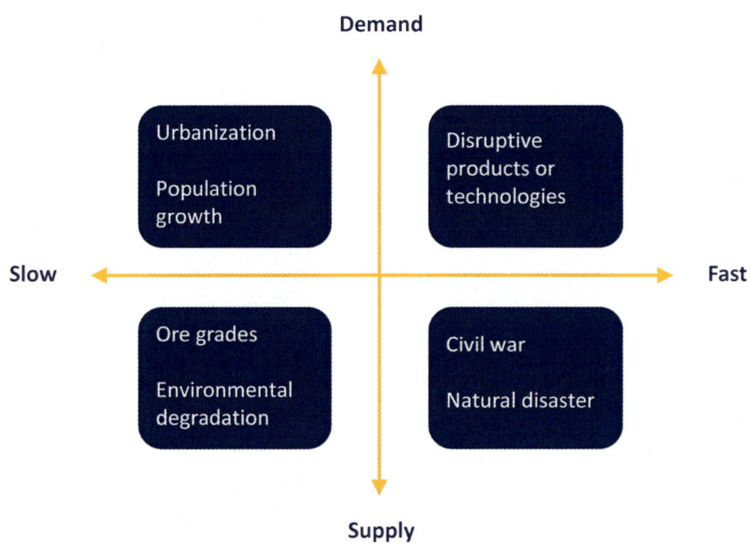

FIGURE 36.1 Conceptual overview of critical materials supply chain disruptions.

expensive during the 2010 REE crisis, Japanese producers of these magnets optimized the physical characteristics of the magnets so that similar temperature resistance could be achieved with lower dysprosium content. One can also redesign systems so that fewer materials are needed. In the same example of the magnets, some customers redesigned their products with better cooling so that a high-temperature resistance was not necessary anymore, thus removing the need for dysprosium altogether.

- Substitution of materials: using alternative materials that can replace the scarce or critical materials. This commonly happens if raw materials become too expensive (e.g., replacing copper with aluminum in power transmission lines), or in special circumstances such as war (e.g., replacement of oil with other fuels). However, physical characteristics must allow for substitution. For some materials, limited or no alternatives may be available. Often critical materials are critical precisely because they are difficult to substitute.
- Recovery and recycling: in the long term this is key to keeping materials in the economy and making them available for future applications. In the past, recycling has mainly been driven by economics and environmental policy, and high recycling rates exist for many metals. However, for many of the critical materials (Table 36.1) recycling rates are low (UNEP, 2011), as no or limited recycling infrastructures exist. Beyond environmental reasons, companies are becoming more interested in recycling to maintain control of certain materials that are crucial for their products to function.
- Stockpiling: this is often seen as the most straightforward manner to prevent massive disruptions but is surprisingly hard to manage correctly. A long-term stockpile essentially acts as an insurance policy. The stockpile ties up capital that could have been deployed elsewhere, and there are costs involved in maintaining the stockpile. When it comes to using the stockpile, one needs to also have the capacity to process the materials that you have stockpiled (Perkins, 1997). A final interesting dynamic in stockpiling behavior is that affected actors often start stockpiling only *after* a disruption hits, thus allowing a limited supply chain hiccup to turn into full blown scarcity, as evidenced by the well-known stockpiling behavior around toilet paper, at the beginning of the COVID pandemic.

Disruptions could largely be avoided if the mining industry would at all times produce sufficient supply to meet demand. So why does this not always happen? Mining developments often respond slowly to changing demand or prices of materials. If opening a new mine takes 10 years, it does not make sense to start building a new mine if there is a price peak, unless the miner is convinced there is a long-term structural supply deficit. Furthermore, similar to the infrastructure of the energy system (Kramer and Haigh, 2009), increasing mining capacity is limited because it often requires a large amount of supporting infrastructure that has limits to how fast it can be scaled up (Sprecher and Kleijn, 2021). Exponentially increasing demand can outstrip the capacity of the mining industry to scale up.

Lack of sufficient production may be the result of the following:

- Environmental and social impacts of mining and production may restrict access or the production of materials, as well as the fact that new supply is often found in very remote locations that lack supporting infrastructure.
- Producing countries may restrict or severely tax exports as part of domestic economic or geopolitical strategies.
- Many critical metals are so-called "companion metals" that are mainly produced as a minor by-product of a major metal such as

zinc or copper. This is determined by naturally occurring metal ratios in minerals (see Fig. 2.2 in Chapter 2). If demand for a companion metal spikes, a mine is not going to massively overproduce its main product to make enough of a minor side-product.

- Small supply chains can have an outsized reliance on a single country or even a single mine. When Glencore announced the closure of the Mutanda mine (Republic of Congo), cobalt prices soared more than 30% (Financial Times, 2019).
- Sufficient investments in new mining technology to address future demand may be limited by current (relatively low) prices of these materials.

The geopolitics of raw materials makes it necessary to reconsider the importance of resource efficiency, as well as the role of recycling. Waste management needs to change from an "end of pipe" approach focused solely on environmental and public health considerations to a policy tool integrated in the life-cycle management of raw materials throughout the supply chain. This means a complete rethinking of our approaches to waste management, which is only slowly taking hold in even the most advanced industrialized countries.

36.4 RECYCLING IN THE GEOPOLITICAL CONTEXT

Material stocks in the manmade world (anthroposphere) are growing rapidly. For example, global in-use steel stocks reached around 12–13 billion tons in 2005, a figure twice that of just 25 years before (Hatayama et al., 2010; Müller et al., 2011). Anthropogenic stocks are the total amount of a material contained within commodities used in human society with a potential for future recovery. These include materials contained in buildings, infrastructure, transport, and appliances (known as in-use stocks), stockpiles by governments or companies held for strategic reasons or supply chain management, as well as waste in tailings (waste rock from mining operations) and landfills, and finally obsolete infrastructure that is not removed, such as underground cables and pipes (known as hibernating stocks). This is also reflected in terms such as "landfill mining," the "urban mine" (e.g., materials stored in the urban infrastructure and buildings), or the "urban forest" (i.e., paper stored in society).

As mineral deposits and related mining operations are becoming beholden to geopolitical maneuvering, managing anthropogenic stocks through recycling becomes increasingly important. In geopolitical terms, recycling can be seen as a strategy to become less vulnerable to supply disruptions and increasing prices. However, per-capita in-use stocks vary between different regions depending on the stage of development. The largest stocks are held in the infrastructure of industrialized countries, with between 11–16 t/capita iron stock, while the most rapid growth of both demand and production takes place in developing countries investing in their infrastructure. Many of these countries—predominantly in Asia and Africa—have an iron stock of less than 1 t/capita (Pauliuk et al., 2013). These countries will absorb any recyclable materials that become available. Thus recycling is not a panacea against imbalances of supply and demand.

There are many challenges when it comes to exploiting the urban mine. Materials only become available when buildings, cars, and appliances reach the end of their useful life, and often it is difficult to (economically) recover critical materials from complex products. Once materials are recovered, they usually need to be reprocessed (refined). Often refining capacity is in the same country as refining of metal ores from mining. Hence, there are three different aspects to recycling in the context of criticality and geopolitics:

- Current or potential future shortages of critical materials, in which recycling represents a key strategy to maintain control of the future supply for a growing need;
- Future imbalance of supply and demand of recoverable and recyclable materials;
- Recycling with the goal of gaining strategic autonomy only works if subsequent refining and processing steps can also be performed domestically.

Scientists have for many years already explored the need for recycling to be able to supply future needs of critical materials. For example, Roelich et al. (2012) models the 2035 demand of neodymium (needed for future wind turbines and premium-efficiency motors) and finds that recycling is essential to meet future demand if the energy system transition is to meet global greenhouse gas emission levels. Similarly, the rapid growth of electric mobility is increasing demand for battery technologies and (critical) materials around the world. For example, if 30% of global passenger transport were electric in 2030, this would require more than the entire global production for several critical metals (Bosch et al., 2019).

Recycling, however, will not be sufficient to meet a rapidly increasing demand over the next decades. This is true for battery metals but also for most of the other critical metals needed for the energy transition. This makes a combined strategy of developing new mines, recovering metals from industrial wastes, recycling, resource efficiency, and searching for alternative technologies necessary. The realization that the resource side of the energy transition will need to be managed has led to regional initiatives and strategies to develop regional supply chains, including raw material supplies (European Commission, 2020).

Less often discussed, but also important on a longer timeframe, is the fact that recycling is essential to keep contamination of steel in check. If scrap is not sorted well enough and recycling rates are high, unwanted elements such as copper can start accumulating in steel, significantly reducing its quality (Chapter 19: Steel) (Daigo et al., 2020).

Overall, only a very small fraction of critical materials is recycled (UNEP, 2011). Hence, there is a strong need for planning and investment in an infrastructure for collection, recovery, and recycling of the materials to ensure sufficient time for developing new resources and increase resource efficiency, even though recycling may not be currently economically attractive. This also requires advances in product design to allow for recovery of these metals (Chapter 5: Material and Product-Centric Recycling). Active policymaking is required to ensure that an effective and efficient recycling infrastructure is developed (Alonso, 2010; Schüler et al., 2011), and that (eco-) design guidelines for recovery of critical materials are implemented at sufficient scale. Policy is especially needed to address the long-term risk of scarcity of elements, as market-based actors do not include these long-term factors in their decision-making. Global policy approaches are essential to safeguard sufficient availability for future generations (Henckens et al., 2016).

Finally, international trade in recyclables does not only lead to countries with low resource self-sufficiency to miss the opportunity to retain these resources, but also leads to high environmental impact in the importing countries. This is especially the case for electronics and plastics, despite international agreements such as the Basel Convention. In the receiving countries, a good recycling and waste management structure is often lacking, leading to environmental pollution and public health concerns. Pollution concerns have led several countries to limit or altogether ban the import of polluted shipments of recyclables. When China, as one of the largest importers of recyclable wastes, banned the import of, e.g., plastic waste, this sent shockwaves around the Western countries that were exporting large volumes of wastes and did not have sufficient domestic capacity to deal with highly polluted recyclables (e.g., United States, Canada, UK, and some European countries). However, this also allowed for a rebalancing of production, consumption, recycling, and waste processing (Wen et al., 2021).

36.5 OUTLOOK

In recent years there is growing concern around the supplies and reserves of strategic and critical materials for economies around the world, especially for those materials for which the global supply is dominated by a limited number of suppliers. Critical materials are those materials for which supply disruptions may damage the economy of a country, while scarcity relates to those materials that may be (in the long term) scarce on a regional or global scale. The increased reliance and economic importance of metals and minerals, both in the current economy as well as for important transitions in our society, has led to a resurgence in geopolitics.

Today, many critical materials show relatively low recycling rates. Recycling can play a key role in securing the future availability of these materials, and can lessen the potential impacts of supply disruptions. However, recycling is not the sole answer to ensure that the services provided by these critical materials can be supplied. Other important means to reduce our reliance on critical materials are product design to extend product lifetime, enable reuse or remanufacturing, and enhance recycling, as well as technological innovation to reduce the need for the raw materials for the product.

Moreover, while this chapter approaches recycling through the geopolitical lens, it is important to stress the crucial ethical aspects of mining (e.g., conflict minerals, environmental inequality) and, hence, our obligation to reduce the need for mining as much as possible through both reducing the overall material intensity of our society and recycling, as well as ensuring that recycling takes place in safe conditions.

References

Alonso, E., 2010. Material Scarcity from the Perspective of Manufacturing Firms: Case Studies of Platinum and Cobalt. Massachusetts Institute of Technology, Cambridge, MA.

Bosch, S., et al., 2019. Metal Demand for Electric Vehicles. Metabolic: https://www.metabolic.nl/publications/metal-demand-for-electric-vehicles.

Charles River Associates (CRA), 1981. Scarcity, Recycling and Substitution of Potentially Critical Materials Used for Vehicular Emissions Control. Prepared for the US Environmental Protection Agency, Ann Arbor, MI.

Cohen, D., 2007. Earth's natural wealth: an audit. New Scientist 2605, 34–41.

Daigo, I., Tajima, K., Hayashi, H., Panasiuk, D., Takeyama, K., Ono, H., et al., 2020. Potential Influences of Impurities on Properties of Recycled Carbon Steel. ISIJ International. ISIJINT-2020.

European Commission, 2020. Critical Raw Materials Resilience: Charting a Path towards greater Security and Sustainability. European Commission, Brussels, Belgium. https://eur-lex.europa.eu/legal-content/EN/TXT/?uri=CELEX:52020DC0474.

Fiksel, J., 2006. Sustainability and resilience: toward a systems approach. Sustain.: Sci. Pract. Policy 2, 14–21.

Financial Times, 2019. Cobalt Rockets as Glencore Plans Closure of Major Mine.

Graedel, T.E., Reck, B.K., 2015. Six years of criticality assessments—what have we learned so far? J. Ind. Ecol. 20, 692–699.

Habib, K., Sprecher, B., Young, S.B., 2021. COVID-19 impacts on metal supply: how does 2020 differ from previous supply chain disruptions? Resour. Conserv. Recycl. 105229–105229.

Hatayama, H., Daigo, I., Matsuno, Y., Adachi, Y., 2010. Outlook of the world steel cycle based on the stock and flow dynamics. Environ. Sci. Technol. 44, 6457–6463.

Henckens, M.L.C.M., Driessen, P.P.J., Worrell, E., 2014. Metal scarcity and sustainability, analyzing the necessity to reduce the extraction of scarce metals. Resour. Conserv. Recycl. 93, 1–8.

Henckens, M.L.C.M., van Ierland, E.C., Driessen, P.P.J., Worrell, E., 2016. Mineral resources: geological scarcity, market price trends, and future generations. Resour. Policy 49, 102–111.

Kramer, G.J., Haigh, M., 2009. No quick switch to low-carbon energy. Nature 462 (7273), 568–569.

Müller, D.B., Wang, T., Duval, B., 2011. Patterns of iron use in societal evolution. Environ. Sci. Technol. 45, 182–188.

Northey, S., Mohr, S., Mudd, G.M., Weng, Z., Giurco, D., 2014. Modelling future copper ore grade decline based on a detailed assessment of copper resources and mining. Resour. Conserv. Recycl. 83, 190–201.

Pauliuk, S., Wang, T., Müller, D.B., 2013. Steel all over the world: estimating in-use stocks of iron for 200 countries. Resour. Conserv. Recycl. 71, 22–30.

Perkins, P.E., 1997. World Metal Markets: The United States Strategic Stockpile and Global Market Influence. Greenwood Publishing Group.

Roelich, K., Kemp Benedict, E., Chadwick, M., Dawkins, E., 2012. Resource Scarcity, Climate Change and the Low Carbon Economy. Stockholm Environmental Institute (SEI), York, UK.

Schüler, D., Buchert, M., Liu, R., Dittrich, S., Merz, C., 2011. Study on Rare Earths and Their Recycling. Öko-Institut, Darmstadt, Germany.

Sprecher, B., Kleijn, R., 2021. Tackling material constraints on the exponential growth of the energy transition. One Earth 4 (3), 335–338.

Sprecher, B., Daigo, I., Murakami, S., Kleijn, R., Vos, M., Kramer, G.J., 2015. Framework for resilience in material supply chains, with a case study from the 2010 rare earth crisis. Environ. Sci. Technol. 49 (11), 6740–6750.

Tang, C., Sprecher, B., Tukker, A., Mogollón, J.M., 2021. The impact of climate policy implementation on lithium, cobalt and nickel demand: the case of the Dutch automotive sector up to 2040. Resour. Policy 74, 102351.

Tilton, J.E., 2002. On Borrowed Time? Assessing the Threat of Mineral Depletion. Resources for the Future, Washington, DC.

U.S. Geological Survey, January 2012. Mineral Commodity Summaries.

U.S. Geological Survey, January 2022. Mineral Commodity Summaries.

UNEP—International Resources Panel, 2011. Recycling Rates of Metals—A Status Report. United Nations Environment Programme, Paris, France.

Wen, Z., Xie, Y., Chen, M., et al., 2021. China's plastic import ban increases prospects of environmental impact mitigation of plastic waste trade flow worldwide. Nat. Commun. 12, 425.

Werner, T.T., Mudd, G.M., Jowitt, S.M., 2017. The world's by-product and critical metal resources. Part III. A global assessment of indium. Ore Geol. Rev. 86, 939–956.

CHAPTER

37

Information and communication

Ernst Worrell

Copernicus Institute of Sustainable Development, Utrecht University, Utrecht, The Netherlands

37.1 INTRODUCTION

Policy instruments are implemented to achieve a set target (e.g., increasing the share of recycling in waste management) and they achieve this by reducing the barriers that inhibit stakeholders from achieving the results, or by stimulating stakeholders to contribute to achieving the target. In practice, many barriers limit the uptake of actions by stakeholders, e.g., households, companies, and waste processors, to increase recovery of recyclable materials and/or products. These barriers may be:

- technical, e.g., technologies to recover materials from a waste stream,
- lack of knowledge, e.g., recognizing recyclable materials, or understand how or where recyclable materials are collected,
- economic, e.g., the costs of separately collecting recyclables, or the low market price of recycled materials,
- organizational, e.g., the infrastructure within an organization or society to collect recyclables,
- a principal agent problem, also known as the renter-landlord problem, e.g., misallocation of investments and benefits,
- lack of interest, e.g., in environmental consciousness, or lack of time or attention due to other needs.

Note that many of these barriers are interrelated, and often multiple barriers may be present and act simultaneously. While policies may be able to reduce the impact or magnitude of the barrier, other barriers may still inhibit the expected outcome. Considering the supply chain of recyclables as a real chain, with the individual links representing stakeholders or actions, it is as strong as its weakest link. Hence, to achieve the desired goal, a good understanding of the barriers throughout the supply chain is needed, as well as an understanding of the drivers of the individual stakeholders. This also implies there is no "silver bullet," or a single policy that will be effective under all circumstances. Hence, generally a portfolio of policy instruments will be needed to achieve a defined (recycling) objective.

Policy instruments can generally be divided into various categories, i.e., communication, economic (Chapter 41—Economic policy instruments), regulatory (Chapter 38—Recycling, law, and the circular economy), or extended producer responsibility schemes (Chapter 39—Extended producer responsibility). Note that

Handbook of Recycling. https://doi.org/10.1016/B978-0-323-85514-3.00033-6

the latter could include a variety of the instruments discussed in the other chapters. Communication or information instruments assume that stakeholders will change the behavior when these are better informed on the benefits and needs of a given activity, e.g., separate collection or recycling, also called proenvironmental behavior. This assumes that there is a knowledge barrier, e.g., that stakeholders are not aware of the benefits to society or to themselves of sorting wastes to allow improved recycling. Virtually any (recycling) policy or program that is implemented (either by governments or organizations) needs some sort of communication to stakeholders, as they need to be informed about the background and benefits of these policies to generate buy-in, and to make implementation effective and efficient in achieving the policy goals. Yet, any communication effort should avoid the risks of greenwashing. Education, publicity, and promotion are essential for the success of any recycling scheme (Evison and Read, 2001). As such, information or communication instruments are rarely implemented in isolation, but are often part of a larger policy portfolio or mix. This also makes it difficult to assess the effectiveness and efficiency of a single policy instrument, including the role of information instruments. In fact, it can be argued, given the importance of affecting the full "chain" of stakeholders and actions, that evaluation of an instrument in isolation is not needed. Rather it is more important to understand the role of the various instruments within the portfolio to improve the overall effectiveness of the portfolio.

The transition to a circular economy really demands revisiting the whole life cycle of a product to find the most effective and sustainable solutions to reduce the impacts of the product. This requires communication and information exchange throughout the life cycle and supply chain of a product, including, e.g., design, material production, product manufacturing, trade, retailers, and citizens/consumers. This chapter focuses on communication around recycling, with a special emphasis on the role of citizens/consumers to improve the effectiveness and efficiency of recovery and recycling programs.

Quality promotion and publicity on a regular basis, will produce better recycling performance figures, whilst poor quality promotion, or none at all, will result in low recycling rates (Evison and Read, 2001). Hence, when planning or changing a recycling service it should include full education and publicity elements. Many studies have identified the factors that contribute to improved recycling behavior (see, e.g., Van der Werff et al., 2019; Wang et al., 2018), and have found that those factors include knowledge (on environment and recycling), moral standards (Halvorsen, 2012), socioeconomic situation, economics (of waste disposal), convenience of recycling (e.g., type, distance, frequency), and behavior of others (e.g., whether the neighbors also recycle). This stresses the role and importance of information to increase the effectiveness, while also supplying the infrastructure and other (e.g., economic) incentives to recycle. In designing an information campaign to support recycling, it is important to understand the target group, the different means of dissemination and communication, and the messaging. In this chapter, based on experiences in various countries and settings, these three elements are discussed.

37.2 TARGET GROUPS

In communicating recycling policy, as in marketing and advertising, a target audience is a specific group of people within the target market at which the message is aimed. A target audience can be formed of people of, e.g., a certain age group, gender, marital status, area (neighborhood), or groups demonstrating common behavior. A combination of factors shapes a common target audience. Determining the appropriate target audience is an important activity in designing an information campaign. As shown earlier, several factors affect recycling behavior. These factors can be used to determine and understand the barriers and subsequent information needs of the target group, and to design the message of the campaign and the best means of reaching the target audience.

It has been found that there is often a need to develop specific targeted information campaigns especially for groups of people in society that are less engaged in recycling. These groups may be found in certain neighborhoods, or in social or age groups. For example, it has been shown that improving adolescents' knowledge of what is recyclable is of considerable importance to campaign designers. However, even high school students with strong proenvironmental attitudes were uncertain about what is recyclable (Prestin and Pearce, 2010). In this case, an effective campaign could begin with messages that increase the knowledge on recyclables and ways to collect, as was also demonstrated in the following example from Poland. In 1995, an integrated program was launched by the aluminum industry to increase recycling of aluminum cans (RECAL). The targeted "School Can Programme" provided technical and educational assistance to schools, municipalities, including educational materials on waste management, containers and bags for aluminum cans, selection magnets, and lists of local collecting depots. The RECAL Programme succeeded in raising public awareness and increasing recovery of aluminum cans in various regions of Poland (Grodzinska-Jurczak and Bartosiewicz, 2001).

Staff within a company is another special target group that can be addressed to increase separation, recovery, and recycling of internally generated production and office waste. Within the company, various audiences may exist (e.g., production sites, offices, production floor, staff, management) and the communication plan should be part of an integrated waste management strategy. Next to providing a recovery and recycling infrastructure, communication is key to the success. Experience has learned that it is important that a clear message and target is communicated from the highest levels of the company, that a common language is developed to ease communication, and that recycling becomes a common aspect of training of (new) staff, and is repeated at least annually. Feedback on the results and success of the recycling program to staff is an important element in corporate programs.

37.3 COMMUNICATION TOOLS

An effective communications program can choose from and combine some or all of the following tools to communicate the intended recycling message to the target audience:

- internet and social media (pages, platforms),
- specific publications and print materials,
- public service broadcasts,
- media relations,
- spokespersons, and
- events (e.g., special days, awards, exhibitions, seminars).

The combination of tools used will depend on the target audience, strategic goals, the advantages and disadvantages of each tool, and available budget. When considering which media to use for promotions, local newspapers are not necessarily the best means of informing the public, while a special (regular) environmental newsletter can effectively put forward policies and strategies. Regular leaflets help to maintain public awareness, and knowledge will decline if frequent reminders are not used (Evison and Read, 2001).

A survey of recycling programs in the UK in 2004 indicated that the publicly preferred option for communication was leaflets (79%), followed by newspapers (34%) and personalized letters (33%) (Mee et al., 2004). With the increasing role of electronic communication, the role of different communication tools may have changed, depending on the specific audience. In the study from 2004, about 66% of the residents were found to have access to the internet at home (66%), but only 15% were prepared to use this as a communication channel (Mee et al., 2004). This likely has changed now, especially for younger stakeholders, where communication through apps or web-based tools may play a key role to engage the public (van der Werff et al., 2019). Mee et al. (2004) claimed that marketing and communications activities had "influenced" some 75% of the residents to recycle more and newsletters (70%) were the most effective communication method. Recycling rates increased from a mere 10% to nearly 50% in 2 years, showing the importance of communication tailored to the stakeholders targeted.

An experiment in the Netherlands engaged a selected target group through a combination of communication methods, including an online platform. The platform added a "gaming" element (through challenges for the participants), as well as a tool for participants to exchange information within the community. It found that an innovative informational strategy consisting of active engagement, e.g., a contract, an online platform, tips on reducing waste, weekly challenge, and comparisons, was effective in promoting waste minimization by households even after the intervention ended (van der Werff et al., 2019). As information increased the extent to which people were aware of the problems caused by their waste and felt that their behavior contributes to solving the problem, households were able to better minimize waste volumes.

A study aimed at understanding the factors to increase compact fluorescent lighting (CFL) recycling in the United States has shown that a lack of knowledge adversely affects the CFL recycling rate. However, the knowledge on recycling was excessively difficult to find (e.g., difficult to locate on websites) and also the collection system was not sufficiently convenient. The analysis did show that information should be provided in a convenient and simple manner when designing or improving collection and recycling from households (Wagner, 2011).

An evaluation of an e-waste program in China demonstrated the need for actionable information to change behavior (Wang et al., 2018). Moreover, increased duration of a program (Van der Werff et al., 2019) or frequency (Wang et al., 2018) are important elements that determine the success rate.

37.4 MESSAGING: INFORMATION AND COMMUNICATION

A recycling program needs a well-designed marketing strategy and campaign to succeed, targeted at the selected audiences. It is important to maintain consistency in the appearance, style and tone, and message in all the communications tools of the campaign. Without this consistency, the message will not be as effective, even if the target audience is exposed to several of the communication products. Worse still, a lack of consistency may confuse or distract the target audience. The message consists of the type of information included, as well as the way it is communicated. As discussed earlier, information regarding what should be recycled, as well as how and where it can be recovered/collected, are key elements. This may also include labeling materials and products for recycling (e.g., the recycling logo, the labeling of plastic types), to allow stakeholders to effectively separate products for recycling. Feedback is another important element of a successful communication strategy. Perrin and Barton (2001) have shown that providing informative feedback to households increased recycling volumes, participation, and weekly set-out levels (Perrin and Barton, 2001).

An analysis of slogans for battery collection and recycling in Switzerland provides an interesting example of the importance of the message itself and the role of culture in shaping an informational program. Hansmann et al. (2009) surveyed the overall impressions of 10 different slogans for the promotion of battery collection and recycling, and subsequently tested two selected slogans in a field experiment at different supermarkets. Slogan effectiveness was positively related with ratings for good and ecological argumentation, creativity, humor, and easy comprehensibility, and negatively related to authoritative wording. The behavioral effectiveness of a humorous slogan urging people to return used batteries, and a factual slogan informing people that batteries were separately collected were investigated in a 9-week field experiment in 21 supermarkets. The informative and easily comprehensible factual slogan achieved an increase of 36% in the weight of returned batteries. In contrast to the initial survey, the prescriptive humorous slogan did not show a positive effect compared to the baseline. The results confirm that factual communication avoids negative reactions, increasing the effectiveness (Hansmann et al., 2009).

Negative or unclear information as well as program changes for specific materials can also negatively affect the behavior of households to separate all types of wastes. In a study (Clarke and Maantay, 2006) of the impact of publicity around program changes in collection of recyclables in New York City, it was shown that in the worst recycling districts of the city, the rate dropped from a high of 35%–21% within a single year. The best 12 recycling districts also suffered a drop in recovery rates, but not to the degree suffered in the low-diversion districts. The strong drop in participation was probably due to the decision to stop collecting plastic, glass, and wax paper containers at an earlier date, and negative publicity on the recycling program prior to the reduction in service. This illustrates that recycling behavior depends to a great degree on attitudes towards the program, and that program changes and negative publicity can be disruptive, perhaps even for longer periods. This example illustrates that communication of program changes should be done carefully to avoid negative impacts on overall program performance (Clarke and Maantay, 2006), and that feedback on the success of a program is essential to maintain long-term behavioral change (Van der Werff et al., 2019).

37.5 CONCLUSION

Information is key to the success of any recycling and recovery program, be it within a country, municipality, or a company. Information should be tailored to the barriers and needs of specific audiences to be effective. Hence, an information campaign needs to start with a careful analysis of the stakeholders targeted, and may need to differentiate groups within the total population to increase the effectiveness of communication. This also affects the means to communicate the message and the formulation of the message itself. Research in various countries has demonstrated that information should be easily accessible, and preferably in the form of dedicated newsletters with a recognizable appearance, style, and message. It has also shown that it is essential to include and repeat simple and basic information on what/how/where to collect the recyclables. Regular information and feedback on program performance and success is essential to maintain the effectiveness of the program over time. An informational strategy may be more likely to succeed when stakeholders are actively engaged, the program is maintained over a longer period of time, and feedback is provided on the success of the program. Information on changes in a program needs to be carefully managed to maintain effectiveness of the program. While messaging is strongly dependent on the culture within a country, region, or municipality, factual information generally provides the best results.

References

Clarke, M.J., Maantay, J.A., 2006. Optimizing recycling in all of New York City's neighborhoods: using GIS to develop the REAP index for improved recycling education, awareness, and participation. Resour. Conserv. Recycl. 46, 128–148.

Evison, T., Read, A.D., 2001. Local authority recycling and waste—awareness publicity/promotion. Resour. Conserv. Recycl. 32, 275–291.

Grodzinska-Jurczak, M., Bartosiewicz, A., 2001. The RECAL Foundation Programme; an example of ecological education in Poland. Resour. Conserv. Recycl. 34, 19–31.

Halvorsen, B., 2012. Effects of norms and policy incentives on household recycling: an international comparison. Resour. Conserv. Recycl. 67, 18–26.

Hansmann, R., Loukopoulos, P., Scholz, R.W., 2009. Characteristics of effective battery recycling slogans: a Swiss field study. Resour. Conserv. Recycl. 53, 218–230.

Mee, N., Clewes, D., Phillips, P.S., Read, A.D., 2004. Effective implementation of a marketing communications strategy for kerbside recycling: a case study from Rushcliffe, UK. Resour. Conserv. Recycl. 42, 1–26.

Perrin, D., Barton, J., 2001. Issues associated with transforming household attitudes and opinions into materials recovery: a review of two kerbside recycling schemes. Resour. Conserv. Recycl. 33, 61–74.

Prestin, A., Pearce, K.E., 2010. We care a lot: formative research for a social marketing campaign to promote school-based recycling. Resour. Conserv. Recycl. 54, 1017–1026.

Wagner, T.P., 2011. Compact fluorescent lights and the impact of convenience and knowledge on household recycling rates. Waste Manag. 31, 1300–1306.

Van der Werff, E., Vrieling, L., van Zuijlen, B., Worrell, E., 2019. Waste minimization by households—a unique informational strategy in The Netherlands. Resour. Conserv. Recycl. 144, 256–266.

Wang, Z., Guo, D., Wang, X., Zhang, B., Wang, B., 2018. How does information publicity influence residents' behaviour intentions around e-waste recycling? Resour. Conserv. Recycl. 133, 1–9.

CHAPTER

38

Recycling, law, and the circular economy

Chris Backes

Utrecht Centre for Water, Oceans and Sustainability Law (UCWOSL), Utrecht University, Utrecht, The Netherlands

38.1 INTRODUCTION

Recycling is not the main or sole issue of legal sources like international conventions, regional legal instruments like EU regulations or directives, or national law. Instead, recycling is traditionally addressed in waste law, hence in legal documents dealing with the management or (transboundary) shipment of waste. During previous decades, waste law has gradually moved its focus from protecting the environment against harmful dumping and disposal practices via promoting recycling to fostering a circular economy in which materials and other sources are efficiently used and material loops are as much closed as possible (Backes, 2020). Legal frameworks for recycling must be considered in this context and development. On the international stage, there is not much "recycling law," except the "Basel Convention on the control of transboundary movements of hazardous wastes and their disposal" (hereafter Basel Convention) and some aspects of international trade law (Van Calster, 2013). The main body of the legal framework for recycling is regional or national. This contribution therefore choses to highlight the legal context for recycling, taking EU law as a case study. The main body of the legal framework for recycling is regional or national. Where appropriate, the Basel Convention and some examples from (some) national law(s) are discussed. The analysis is not meant to be conclusive (Van Calster, 2015).

38.2 REGULATING RECYCLING TO FOSTER A CIRCULAR ECONOMY

Although already in its original version of 1975, article 3 of the EU Waste Framework Directive (hereafter WFD) urged Member States to encourage prevention and recycling of waste, this early legal document aimed at "the protection of human health and the environment against harmful effects caused by the collection, transport, treatment, storage and shipping of waste" (Directive 75/442/EEC, 1975, recital 7 & article 3). However, before that date, the US Resource Recovery Act in the United States in 1970 addressed enhancing recycling as an important aim of waste law but did not substantiate this with very specific legal requirements (Blaschke, 1971). In the renewed WFD of 2008, for the first time, the "waste hierarchy" was laid down in a legally binding provision. A major step at that time was to legally distinguish between recycling and "other recovery" and to define that (reuse and) material recycling should be preferred to energy recovery (Directive 2008/98/EC, article 4). After waste policy had been integrated into the broader approach of the transition to a circular economy (European Commission, 2015, 2020a), the "Circular Economy Package" revision of the WFD in June 2018 comprised the legal implementation of this policy shift and tried to legally "facilitate the transition to more sustainable material management and to a circular economy model" (Directive EU 2018/851, recital 1). The aims of waste law now include, for example, ensuring the security of the supply of certain raw materials (Directive EU 2018/851, recital 36). The EU legal regime for recycling has to be evaluated against this background of the shift of focus of waste law. Legal conditions for recycling are one group of instruments for foster the transition to a more circular economy and should mainly serve this purpose.

38.3 LEGAL INSTRUMENTS FOSTERING RECYCLING

38.3.1 Waste Hierarchy

According to article 4 WFD, the following priority order applies:

(a) prevention
(b) preparing for reuse
(c) recycling
(d) other recovery, e.g., energy recovery
(e) disposal.

The application of the hierarchy should always lead to "options that deliver the best overall environmental outcome." This may require or at least justify deviating from the order in specific cases. Article 4(2) describes such cases as: "where this is justified by life-cycle thinking on the overall impacts of the generation and management of such waste." The political effects of the waste hierarchy are substantial. However, although article 4 is legally binding and Member States have to transpose it into national law (Epiney, 2012), the legal effects of the waste hierarchy are limited. This is partly due to the exceptions referred to in article 4(2), which provide Member States with substantial discretion. The waste hierarchy applies, according to article 4 WFD, to all "legislation and policy." There is, however, no specific legal consequence if the waste hierarchy is not observed. As far as is known, the Court of Justice of the EU (hereafter CJEU) has not yet been asked to decide whether EU legal acts or policy measures comply with the waste hierarchy.

38.3.2 Provisions on Recycling: Definition and Targets

Article 3 sub 17 WFD defines recycling as "any recovery operation by which waste materials are reprocessed into products, materials or substances whether for the original or other purposes." It further clarifies that recycling "includes the reprocessing of organic material but does not include energy recovery and the reprocessing into materials that are to be used as fuels or for backfilling operations." This important clarification was added to the Directive in 2008 (however, article 11(2)(b) WFD, dealt with later, deviates from this and explicitly defines backfilling of construction and demolition waste as a recycling operation). Energy recovery, in the past often falsely called "energy recycling," is explicitly mentioned in article 4 WFD as a form of "other recovery." Also backfilling is qualified as an "other recovery" operation and as such clearly has a lower priority than (material) recycling. Article 10 WFD requires Member States to take all necessary measures to ensure that waste is managed in accordance with the waste hierarchy laid down in article 4.

Whether a certain way of processing waste is to be qualified either as disposal or as a recovery operation has important legal consequences under international and EU law. The international rules on transboundary shipments of waste differ for either disposal or recovery. The same is true for the EU rules on transboundary shipments, but also on the required standards and measure when managing waste. Disposal and (different kinds of) recovery operations are described in Annex IV of the Basel Convention on the Control of Transboundary Movements of Hazardous Wastes and their Disposal, which are copied in EU law. The definition of recovery and disposal in article 3 WFD refers to the Annexis I and II of this Directive. A substantial amount of case law (CJEU; Van Calster, 2015) and a guidance (European Commission, 2012) issued by the European Commission try to further clarify both notions.

An important measure to enhance (the quality of) recycling is the separate collection of certain waste streams. According to article 10(2) and (3) WFD, separate collection is obligatory unless one of the conditions mentioned in article 10(3) applies, for example if separate collection is not technically feasible or is disproportionately costly. These conditions for derogating leave Member States substantial discretion. Article 11(1) requires separate collections at least for paper, metal, plastic, and glass, and by January 1, 2025 also for textiles.

Until the 2018 amendment that ensued from the Circular Economy Package, the Directive did set only one concrete binding target for recovery and recycling. In 2020, 70% (by weight) of the construction and demolition waste (CDW) should be recovered. This obligation has now become article 11(2)(b) WFD, which explicitly makes clear that in this case substitution of other materials in backfilling operations is also accepted as recovering waste. This provision has been controversially discussed in the literature (Backes, 2017). Since the amendment by Directive 2018/851/EU, more targets have been determined. According to article 11(2)(a), in 2020, 50% of household paper, metal, plastic, glass, and comparable streams shall be recovered. In 2025, 55% of the municipal waste as a whole shall be reused or recycled, which figure increases up to 65% in 2035. Recycling and recovery targets can also be found in product(group)-specific directives or regulations, as for example the directive on packaging and packaging waste (Directive 94/92/EC, 1994, article 6). This directive comprises recycling targets for packaging in general, as well as targets for different packaging materials like paper and cardboard, plastics, or aluminum. Other examples include targets for separate collection (article 7) and recycling (article 11) in the directive on waste electrical and electronic equipment (WEEE, Directive 2012/19/EU) as well as collection targets (article 10) and targets concerning recycling efficiency (article 12) in the directive on batteries and accumulators (Directive 2006/66/EC).

However, until now, not all targets have proven to be very effective. For example, while the target for CDW is 70% recovery—including substitution in backfilling operations—the level of recycling and material recovery of CDW varies greatly across the EU, between less than 10% to over 90% (European Commission, 2015). Article 11(3)–(5) WFD offers some possibilities for derogating from these requirements, under specific conditions and timeframes. According to article 11(6), the Commission will consider more targets in 2024, for example for textile waste, as well as stricter targets for recovery of construction and demolition waste, excluding substitution of backfilling materials.

Recycling and recovery targets as discussed here are directed to the Member States. Legally, they have to be distinguished from recycling and recovery targets that apply to producers, and which will be dealt with hereafter in paragraph 37.3.3. Potentially, such targets for Member States are crucial in improving the application of the waste hierarchy and making progress towards a more circular economy, especially in countries that are lagging behind.

It remains to be seen whether most Member States will meet such targets. The advantage of exact figures is that the Commission—if it wants to—can relatively easily take enforcement action if the targets are not met. However, taking enforcement measures against EU Member States because they did not meet the binding targets of article 11 WFD is a capacity of the EU Commission, not an obligation.

38.3.3 Producer Responsibility

Recycling and recovery targets are also addressed within the legal provisions on extended producer responsibility. Article 8 and 8a WFD encourage Member States to impose potentially far-reaching requirements on, among others, producers of goods and those who import or sell materials or goods that are waste or can become waste. These powers serve to "strengthen the re-use and the prevention, recycling and other recovery of waste," thus to implement the waste hierarchy. To that end, the Member States can (or must) impose diverse obligations on producers or importers, among which are targets for separate collection and recycling or recovery, but also, for example, obligations to provide information on the different components and materials used in certain products, as well as the location of dangerous substances and mixtures in these products. This information is to be made available to centers that prepare for reuse and treatment and recycling facilities, for example in article 15 WEEE directive (Directive 2012/19/EU) or in article 8 End-of life vehicles directive (Directive 2000/53/EC), which also require manufacturers to use certain component and material coding standards to facilitate reuse and recycling in the form of manuals or by means of electronic media. Different from the targets mentioned in Chapter 46.3.2, these targets therefore are not only addressed to the Member States, but have to be imposed on the producers and importers who put the respective goods on the market. Legal obligations concerning producer responsibility can, besides in the WFD, also be found in some product-specific directives, namely the directives on waste electrical and electronic equipment (WEEE) (Directive EU 2012/19, articles 4–8), batteries and accumulators (Directive 2006/66/EC, articles 8–12), and end-of-life vehicles (Directive 2000/53/EC, articles 5–8). The proposal for a new regulation on batteries (European Parliament and European Council, 2020) that shall substitute for the current directive on this topic urges the Member States to significantly improve the collection and recycling of portable batteries. The current collection rate of 45% should rise to 65% in 2025 and 70% in 2030 for household batteries and 100% for industrial, automotive, or electric vehicle ones.

38.3.4 Ecodesign

Whether recycling rates are high and recycling is done in a sustainable way is not only decided by legal targets, but also by ensuring that the design of a product fosters its optimal circular use, among which is recycling. Legal basis for ecodesign requirements and binding ecodesign criteria can be found at different places in EU law (Backes, 2017). First of all, this is the case in the so-called "Ecodesign directive" (Directive 2009/125/EC). The full name of the directive better indicates its purpose: "Directive 2009/125/EC ... establishing a framework for the setting of ecodesign requirements for energy related products." Hence, actually, the scope of the directive is limited. The directive covers potentially all products that have an impact on energy consumption during their use (article 2, sub 1). Products falling under the scope of the directive may be placed on the market only if they bear the CE marking sign. This sign may only be affixed to a product if it complies with the requirements imposed on the basis of the

Ecodesign directive. As Annex I makes clear, the criteria that may be taken into account when setting the requirements to gain the CE mark may relate to all phases of the life cycle of the product and should, among other things, take into account the possibilities for reuse, recycling, and recovery of materials. When evaluating the potential for improving the environmental aspects of the goods, parameters should be used like, among others, "use of materials issued from recycling activities" and "ease for reuse and recycling as expressed through: number of materials and components used, use of standard components, time necessary for disassembly, complexity of tools necessary for disassembly, use of component and material coding standards for the identification of components and materials suitable for reuse and recycling, use of easily recyclable materials, easy access to valuable and other recyclable components and materials and more." This far from comprehensive quotation of parts of Annex I of the directive demonstrates that the directive potentially addresses all aspects relevant for fostering circularity of products, among which are recovery and recycling (Schomerus and Spengler, 2012). However, currently this high potential of the legal regime is hardly used. The current regulations for almost all products regulated on the basis of the Ecodesign directive solely concentrate on regulating the energy use and other energy-related topics (Svensson and Dalhammar, 2018). Very recently, the first, however limited, attempts have been made to include some design criteria aiming at fostering circularity in marketing requirements. Examples are the regulations on servers and data storage products, electronic displays, refrigerators, dishwashers, and some other products (Commission Regulation EU 2019/424; Mathieux et al., 2020). Most of these regulations prescribe requirements that ensure reparability, like availability of spare parts. However, some, for example the regulation on displays (Commission Regulation EU 2019/2122, Annex II, sub D), also regulate the ability to be dismantled, hence promote recycling. As the European Commission has announced to more often determine such product requirements that promote recycling and other circular use and to broaden the scope of the Ecodesign directive also to nonenergy-related products (European Commission, 2020b), this legal regime will become more important for fostering recycling in the future.

The legal framework for design requirements fostering recyclability is not limited to the Ecodesign directive, but can sometimes also be found in other places: directives on certain (groups of) products, like the directive on batteries and accumulators or the single-use plastics directive, and provisions in waste law, more especially the provisions on producer responsibility in the WFD, which was already dealt with in the previous paragraph. In order to improve recycling, the directive on batteries and its successor, the proposed Batteries Regulation, for example comprise prohibitions to use certain substances in batteries in order to enhance the profitability of recycling activities and foster investment in new technologies and in additional capacity to recycle the batteries of the future (European Parliament and European Council, 2020). Article 6 of the single-use plastics directive (Directive EU 2019/904) sets certain requirements for containers with caps and lids to facilitate their recycling. Another example is to be found in the directive on end-of-life vehicles (Directive 2000/53/EC, article 4). Product design requirements can also be set on the basis of extended producer responsibility schemes on the basis of the WFD. These may include design requirements supporting multiple use or ensuring that products contain recycled materials or are suitable for reuse and recycling. In other words: Member States may, when concluding EPR regulations, take all measures that contribute to a more circular design and production of goods. Articles 8 and 8a WFD only encourage but make no requirements. Therefore it is important that Member States actively implement EPR regulations in their national legal orders also when EU law does not strictly require this.

Special attention should be given to a new type of regulating of the design of products in order to enhance recyclability, which does not concern the question of how a product should be designed to make it more recyclable, but how much recycled content a new product should contain. This sort of requirement therefore promotes the markets for recycled materials and tries to overcome market failure (Backes, 2017, para. 6.1). First examples in EU law can be found in article 6 (5) Single-use plastics directive (Directive EU 2019/904) on PET bottles or in article 8 of the proposed batteries regulation concerning recycled content in industrial batteries, electric vehicle batteries, and automotive batteries (European Parliament and European Council, 2020).

38.3.5 Standardization and Labeling

Different from, but closely related to, product design requirements are standardization and labeling as a means to foster recyclability. Standardization does not only concern harmonizing product standards but also measurements and research methods. Examples can be found in article 6 (3 and 4) concerning PET bottles and article 8 (9) concerning fishing gear of the single-use plastics directive (Directive EU 2019/904) or article 8 (5) of the WEEE directive (Directive EU 2012/19). Legally, the EU Regulation on standardization is guiding in this field, which mainly is carried out by the European Committee for Standardization (CEN) and the European Committee for Electrotechnical Standardization (CENELEC) (Regulation EU 1025/2012). CEN and CENELEC are involved in the development of "harmonized" product-specific standards set out in the implementing measures under the Ecodesign Directive ("CEN-CENELEC Ecodesign Coordination Group"). The Commission has issued a standardization request for horizontal standards to be developed to support, among others, recyclability (Commission Implementing Decision, M/543, 2015), and has asked European standardization organizations to develop generic standards on the durability, reusability, and recyclability of certain products (European Commission, 2017, p. 5).

Examples for labeling requirements in order to enhance recyclability can be found in article 14 of the proposed batteries regulation or article 4 (2) ELV directive (Directive 2000/53/EC). The Commission has announced to amend the EU Ecolabel Regulation (Regulation EC 66/2010, 2009) to include durability, recyclability, and recycled content more systematically in the EU Ecolabel schemes (European Commission, 2020a).

38.3.6 Taxes and Other Financial Requirements

An important issue when discussing measures to enhance recycling are financial conditions. Recycling is often hampered by higher costs compared to the use of virgin materials, often due to externalization of environmental costs. International and EU law do not provide much in this regard. Taxing and regulating other financial burdens is almost entirely a national competence. There are some exceptions that, however, have a limited scope. The most important one concerns EPR. The core concept of making producers financially responsible for collecting and managing the waste of their product has been implemented in the EPR systems (Chapter 39—Extended producer responsibility). According to article 8a (4) WFD all operating costs of the PROs are covered by the producers and importers participating in the EPR schemes. However, this is something different from all costs the respective products cause, once they have become waste. The responsibility of the producers is limited to realizing the respective targets of an EPR scheme. The financial responsibility corresponds to this and is, therefore, also limited (Chapter 39—Extended producer responsibility). The financial responsibility with regards to products with a long durability has been further limited by a judgment of the CJEU in 2022. The Court decided that producers cannot be forced to pay for the collection and recycling of products that are already on the market (here photovoltaic panels), but only for the (future) costs of the products that will be purchased. The Court therefore declared article 13 (1) WEEE-Directive (Directive 2012/19/EU) invalid, which required producers to also cover the costs of products sold between 2005 and the entry into force of this directive in 2012 (Case C-181/20, 2022, VYSOČINA WIND).

38.4 LEGAL REQUIREMENTS THAT MAY HAMPER RECYCLING

The applicability of waste law and the qualification of a used item as "waste" brings about some consequences that may hamper recycling possibilities and other circular applications (Backes, 2017, para. 5):

- Waste has a negative image. Companies that recover or recycle secondary materials often do not want to be associated with waste processors.
- Handling waste means extra administrative burdens. Companies that deal with waste do need a special permit (or an extra review and extra provisions in their integrated permit), which also means a financial burden.
- If a new idea to recycle something is qualified as a waste recovery operation, a company that does not have a license to treat this sort of waste is not allowed to apply the new circular solution. This may hamper innovative solutions.
- Waste law is an "unsafe area": the interpretation of "waste" and other crucial waste law terms is not clear, despite a huge amount of case law of the European Court of Justice on this issue.

Therefore there have been attempts to limit the scope of waste law without increasing environmental risks. The idea then is that if environmental risks of certain new (recycled) substances and products and of recycling operations can be excluded, application of waste law is no longer needed. In a "zero-waste society," waste no longer exists, but only secondary materials. "Materials law" then ensures the protection of the environment. That, however, would require that the legal rules on substances (chemical law) and products (product law) fully cover and master all risks attached to the end of (the first) life phase of a substance or product. This is not yet the case. In legal terms, this discussion is framed by three issues: the (broad) legal definition of "waste," the exception in this definition for "by-products," and the regulations for ending of the waste status after a recycling process (Backes, 2020).

38.4.1 The (Broad) Definition of "Waste"

The WFD defines "waste" in Article 3(a) as:

> any substance or object which the holder discards or intends or is required to discard.

Although it was devised more than 40 years ago and has not changed substantially for almost 30 years, and there is a constant stream of case law of the CJEU, this definition is not easy to handle. In an opinion from 2019, Advocate General Kokott proposed that the principle of *nulla poena sine lege* (no punishment without law) could probably prohibit criminal sanctioning in this case, as the notion of waste could not be regarded as being sufficiently precise and foreseeable (Opinion Case C-624/17, 2019, Tronex, para. 51). This illustrates the problems that still often occur in practice because of the vagueness of the term "waste." All substances and products can either be goods or waste, depending on the circumstances and especially depending on the subjective intention of the one who holds the substance or product. If a product is used for the purpose it was made for it is not waste. If the holder wants to end this use and wants to get rid of, in legal terms "discard," the product, it is waste. The notion "discard" comprises all possible actions of disposal, but also all actions of recovery and recycling (Case C-419/97, 2000, Arco Chemie, para. 47), as can be deduced from articles 10 and 11 WFD. The Court has again and again stressed that the term "waste" cannot be interpreted restrictively (Case C-419/97, 2000, Arco Chemie, para. 40). The Court has developed a whole list of such indications, such as:

- The way the substance is going to be used concerns "a common method" of disposal or recovery.
- The common perception of what is waste.
- A substance is a residue of the manufacturing process.
- No use other than disposal can be envisaged. It is necessary to take special precautionary measures when the substance is used.
- The risk that the holder will discard the substance in a way likely to harm the environment.

On the other hand, there are also some indications that may, in combination with others, justify the conclusion that a substance is not waste:

- The possibility of selling the substance to another buyer.
- The substance has a significant commercial value.
- The client returned the substance or product with a view to obtaining a refund, pursuant to the sale contract.
- A client accepts a product with the view to remarketing it (Joined cases C-241/12 and C-242/12, 2013, Shell, paras. 41 ff).

38.4.2 By-Products That Are Not Waste

Recycling and other circular applications can be applied without having to respect waste law, if the substance or products qualify as a by-product, which is not deemed to be waste. According to article 5 WFD a substance or object, resulting from a production process, the primary aim of which is not the production of that substance or object, may be regarded as not being waste if:

1. further use of the substance or object is certain;
2. the substance or object can be used directly without any further processing other than normal industrial practice;
3. the substance or object is produced as an integral part of a production process; and
4. further use is lawful, i.e., the substance or object fulfills all relevant product, environmental, and health protection requirements for the specific use and will not lead to overall adverse environmental or human health impacts.

This exception intends to exclude by-products from the definition of waste if there is no need for applying waste law to prevent harm to human health or to the environment. The exception was first developed in case law in 2002 (Case C-9/00, 2002, Palin Granit) and was later codified. Article 5 only applies to substances and objects that are made in a production process. It is not applicable to other substances or objects, for example consumer goods, even if all other criteria are met and there seems to be no danger to human health or the environment.

38.4.3 End of Waste

Article 6 WFD concerns the end of the classification as waste. If waste is recycled or otherwise recovered, at a certain moment it can become a substance or product again, and stops being waste. The holder no longer wants to discard the substance or product, but wants to use or sell it as a secondary substance or product, derived from a recycling or other recovery process. The exact moment when this happens is sometimes difficult to determine. Again, the Court of Justice developed case law on this (C-444/00, 2003, Mayer Parry), which then was codified. According to article 6 WFD, waste that has undergone a recovery operation and complies with certain criteria no longer qualifies as waste. These criteria are:

1. the substance or object is to be used for specific purposes;
2. a market or demand exists for such a substance or object;
3. the substance or object fulfills the technical requirements for the specific purposes and meets the existing legislation and standards applicable to products; and
4. the use of the substance or object will not lead to overall adverse environmental or human health impacts.

According to article 6(2), the Commission can adopt implementing acts to ensure a uniform application of these criteria. For certain waste streams, like glass cullet (Commission Regulation EU 1179/2012), iron, steel, and aluminum scrap (Council Regulation EU 333/2011), and copper scrap (Commission Regulation EU 715/2013), the Commission adopted such implementing acts between 2011 and 2013. Proposals for other streams have, in the end, failed. The Commission planned criteria for paper and compost but did not issue proposals for these streams. In 2014, the Joint Research Centre EC published a proposal for waste plastic (Joint Research Centre EC 2014).

Since the revision of the WFD on the basis of the "circular economy package" in 2018, article 6 (4) WFD empowers member states to "decide on a case-by-case basis whether certain waste has ceased to be waste taking into account the applicable case law." Member States have to notify such decisions to the Commission. The question is whether this new provision brings about any real change in the legal rules. It rather seems to be a political statement, making clear that neither the EU legislator, nor the Commission in the end, has been able to further define comprehensive end-of-waste criteria (Backes, 2017, para. 3.1). Application of this possibility implies the danger that what qualifies as waste and has to be handled according to the conditions of waste law may differ from one Member State to the other, which clearly is not the intention of the WFD.

38.5 CONCLUDING REMARKS

The legal regime governing recycling operations is diverse and sometimes not easy to grasp. There is nothing like a "convention on recycling" or equivalent legal framework on a regional (EU) or national level. Instead, "recycling" law comprises parts of waste law and product law. Currently, recycling law has to be approached as a means to foster the transition to a circular economy. In the EU legal order, which is analyzed in this chapter by way of example, there is a huge variety of legal instruments fostering recycling as one of the means within the hierarchy of circular application of products and substances. These instruments are to be found in general and product-specific waste law and general and product-specific product law. The by and large absence of legal powers to correct market failures with financial means like taxes is striking and can be seen as one of the obstacles for a faster transition to a circular economy.

References

Backes, C.W., 2017. Law for a Circular Economy. Eleven International Publishing, The Hague. ISBN: 978-94-6236-764-7.

Backes, C.W., 2020. The Waste Framework Directive and the circular economy. In: Peeters, M., Eliantonio, M. (Eds.), Research Handbook on EU Environmental Law. Edward Elgar, Northampton, p. 333, https://doi.org/10.4337/9781788970679.00032. Paragraph 3.1.

Blaschke, L.E., 1971. Analysis of the Resource Recovery Act of 1970 and its effect on implementation of solid waste management programs. J. Environ. Health 34 (1 (July/ August, 1971)), 89–92.

Epiney, E., 2012. Umweltrecht der Europäischen Union, third ed. Nomos, p. 553. ISBN 978-3-7089-1843-3.

European Commission, 2012. Guidance on the Interpretation of Key Provisions of Directive 2008/98/EC on Waste, p. 30 ff. https://ec.europa.eu/environment/pdf/waste/framework/guidance_doc.pdf. (last reviewed 3 February 2022).

European Commission, 2015. Closing the Loop – an EU action plan for the Circular Economy, COM(2015), 614 final, 16. https://eur-lex.europa.eu/legal-content/EN/TXT/?uri=CELEX:52015DC0614.

European Commission, 2017. Report from the Commission on the implementation of the Circular Economy Action Plan, COM(2017) 33 final, p. 5 f. https://eur-lex.europa.eu/legal-content/en/ALL/?uri=CELEX:52017DC0033.

European Commission, 2020a. New Circular Economy Action Plan For a cleaner and more competitive Europe, COM(2020) 98 final. https://eur-lex.europa.eu/legal-content/EN/TXT/?uri=COM:2020:98:FIN.

European Commission, 2020b. Inception Impact Assessment 'Sustainable Products Initiative', Ares(2020)4754440. https://ec.europa.eu/info/law/better-regulation/have-your-say/initiatives/12567-Sustainable-products-initiative_en.

European Parliament and European Council, 2020. Proposal for a regulation of the European Parliament and of the Council concerning batteries and waste batteries, COM(2020) 798 final. https://eur-lex.europa.eu/legal-content/EN/TXT/?uri=CELEX:52020PC0798.

Joint Research Centre EC, 2014. End-of-Waste-Criteria for Waste Plastic for Conversion. Technical Proposals. Publications Office of the European Union, Luxembourg. http://publications.jrc.ec.europa.eu/repository/bitstream/JRC91637/2014-jrc91637%20.pdf. (last accessed 3 February 2022).

Mathieux, F., Ardente, F., Bobba, S., 2020. Ten years of scientific support for integrating circular economy requirements in the EU ecodesign directive: overview and lessons learnt. Procedia CIRP 90, 138.

Schomerus, T., Spengler, L., 2012. Integration der Ressourceneffizienz in die Ökodesign Richtlinie – Regelungsrahmen und Regelungsmöglichkeiten – Arbeitspapier. http://www.eup-network.de/fileadmin/user_upload/Produktgruppen/V_526-3_Arbeitspapier_-_Regelungsrahmen_und_Moeglichkeiten__AP2_.pdf. (lastly reviewed 3 February 2022).

Svensson, S., Dalhammar, C., 2018. Regulating recyclability under the ecodesign directive. In: Maitre-Ekern, E., Dalhammar, C., Bugge, H.C. (Eds.), Preventing Environmental Damage From Products. Cambridge University Press, p. 235, https://doi.org/10.1017/9781108500128.009.

Van Calster, G., 2013. China, minerals export, raw and rare earth materials: a perfect storm for WTO dispute settlement. Review of European, Comparative & International Environmental Law 1, 117–122.

Van Calster, G., 2015. EU Waste Law, second ed. Oxford University Press. ISBN 9780198719786.

Directives

Anon., 1975. Directive 442/75/EEC of 15 July 1975 on waste. OJ L194/39 http://data.europa.eu/eli/dir/1975/442/oj.

Anon., 1994. Directive 94/92/EC of 20 December 1994 on packaging and packaging waste. OJ L365/10 http://data.europa.eu/eli/dir/1994/62/oj.

Anon., 2000. Directive 2000/53/EC of 18 September 2000 on end-of life vehicles. OJ L269/34 http://data.europa.eu/eli/dir/2000/53/oj.

Anon., 2006. Directive 2006/66/EC of 6 September 2006 on batteries and accumulators and waste batteries and accumulators and repealing Directive 91/157/EEC. OJ L266/1 http://data.europa.eu/eli/dir/2006/66/oj.

Anon., 2008. Directive 2008/98/EC of 19 November 2008 on waste and repealing certain Directives. OJ L312/3 http://data.europa.eu/eli/dir/2008/98/oj.

Anon., 2009. Directive 2009/125/EC of 21 October 2009 establishing a framework for the setting of ecodesign requirements for energy-related products. OJ L285/10 http://data.europa.eu/eli/dir/2009/125/oj.

Anon., 2012. Directive EU 2012/19 of 4 July 2012 on waste electrical and electronic equipment (WEEE). OJ L197/38 http://data.europa.eu/eli/dir/2012/19/oj.

Anon., 2018. Directive EU 2018/851 of 30 May 2018 amending Directive 2008/98/EC on waste. OJ L150/109 http://data.europa.eu/eli/dir/2018/851/oj.

Anon., 2019. Directive EU 2019/904 of 5 June 2019 on the reduction of the impact of certain plastic products on the environment. OJ L55/1 http://data.europa.eu/eli/dir/2019/904/oj.

Regulations

Anon., 2012. Commission Regulation (EU) 1179/2012 of 10 December 2012 establishing criteria determining when glass cullet ceases to be waste under Directive 2008/98/EC of the European Parliament and of the Council. OJ L337/31 http://data.europa.eu/eli/reg/2012/1179/oj.

Anon., 2013. Commission Regulation (EU) 715/2013 of 25 July 2013 establishing criteria determining when copper scrap ceases to be waste under Directive 2008/98/EC of the European Parliament and of the Council. OJ L201/14 http://data.europa.eu/eli/reg/2013/715/oj.

Anon., 2019. Commission Regulation (EU) 2019/424 of 15 March 2019 laying down ecodesign requirements for servers and data storage products. OJ L 074/46 http://data.europa.eu/eli/reg/2019/424/oj.

Anon., 2019. Commission Regulation (EU) 2019/2122 of 1 October 2019 laying down ecodesign requirements for electronic displays pursuant to Directive 2009/125/EC of the European Parliament and of the Council, amending Commission Regulation (EC) No 1275/2008 and repealing Commission Regulation (EC) 642/2009. OJ L135/41 http://data.europa.eu/eli/reg_del/2019/2122/oj.

Anon., 2011. Council Regulation (EU) 333/2011 of 31 March 2011 establishing criteria determining when certain types of scrap metal cease to be waste under Directive 2008/98/EC of the European Parliament and of the Council. OJ L94/2 http://data.europa.eu/eli/reg/2011/333/oj.

Anon., 2009. Regulation (EC) 66/210 of 25 November 2009 on the EU Ecolabel. OJ L27/1 http://data.europa.eu/eli/reg/2010/66/oj.

Anon., 2012. Regulation (EU) 1025/2012 of 25 October 2012 on European standardisation, amending Council Directives 89/686/EEC and 93/15/EEC and Directives 94/9/EC, 94/25/EC, 95/16/EC, 97/23/EC, 98/34/EC, 2004/22/EC, 2007/23/EC, 2009/23/EC and 2009/105/EC of the European Parliament and of the Council and repealing Council Decision 87/95/EEC and Decision No 1673/2006/EC of the European Parliament and of the Council. OJ L316/12 http://data.europa.eu/eli/reg/2012/1025/oj.

Implementing Decisions

Anon., 2015. M/543 Commission Implementing Decision of 17 December 2015 on a standardization request to the European standardization organizations as regards ecodesign requirements on material efficiency aspects for energy-related products in support of the implementation of Directive 2009/125/EC, COM(2015)9096.

Jurisprudence

Anon., 2000. Case C-419/97, Arco Chemie, ECLI:EU:C:2000:318, 15 June. https://eur-lex.europa.eu/legal-content/EN/TXT/?uri=CELEX:61997CJ0418.

Anon., 2002. Case C-9/00, Palin Granit, ECLI:EU:C:2002:232, 18 April. https://eur-lex.europa.eu/legal-content/EN/TXT/?uri=CELEX:62000CJ0009.

Anon., 2003. Case C-444/00, Mayer Parry, ECLI:EU:C:2003:356, 19 June. https://eur-lex.europa.eu/legal-content/EN/TXT/?uri=CELEX:62000CJ0444.

Anon., 2013. Joined cases C-241/12 and C-242/12, Shell, ECLI:EU:C:2013:821, 12 December. https://www.informea.org/en/content/court-decision/joined-cases-c-24112-and-c-24212-criminal-proceedings-against-shell.

Anon., 2019. Opinion in case C-624/17, Tronex, ECLI:EU:C:2019:150, 28 February. https://eur-lex.europa.eu/legal-content/EN/TXT/?uri=CELEX:62017CC0624.

Anon., 2022. Case C-181/20, VYSOČINA WIND, ECLI:EU:C:2022:51, 25 January. https://curia.europa.eu/juris/document/document.jsf?text=&docid=244211&pageIndex=0&doclang=EN&mode=lst&dir=&occ=first&part=1&cid=216412.

CHAPTER

39

Extended producer responsibility

W.J.V. Vermeulen[1] and K. Campbell-Johnston[2]

[1]Copernicus Institute of Sustainable Development, Utrecht University, Utrecht, The Netherlands
[2]Department of ICT, Strategy and Policy, TNO, The Hague, The Netherlands

39.1 INTRODUCTION

Extended Producer Responsibility (EPR) emerged as a means of operationalizing the "polluter pays principle" to promote cleaner production practices. Building on the notion that those who produce pollution should bear its management costs, it was introduced as a policy instrument in the early 1990s by various northern-western European countries such as Germany, Sweden, France, and the Netherlands (Ayres and Ayres, 2015; Lindhqvist, 2000; Vermeulen and Weterings, 1997). EPR emerged when the landfilling of postconsumer waste was high and where local government bore the costs and responsibilities of waste management. Since then, it has been applied widely, mostly in high-income countries, but also in middle- and low-income countries (OECD, 2014, 2016). In this chapter, we discuss the various perceptions of EPR, as a normative principle and as a policy instrument. After that, we discuss the diverse forms of application and what is known about its effectiveness in practice. Next, we discuss some critical issues around applying EPR and its limitations. Finally, we discuss the role EPR can play in the pursue of a circular economy.

39.2 DEFINING EPR

Extended Producer Responsibility (EPR) was initially defined in a report to the Swedish Ministry of Environment by Lindhqvist as "an environmental protection strategy to reach an environmental objective of a decreased total environmental impact from a product, by making the manufacturer of the product responsible for the entire life-cycle of the product and especially for the take-back, recycling and final disposal of the product" (Lindhqvist, 2000, p. iii).

Alternative definitions (see OECD, 2016) narrow the scope of EPR to the producer's responsibility only being extended to the postconsumer stage of the product life cycle. However, in its original broader interpretation, EPR as a concept contains two interrelated elements: namely, that the producer's "responsibility" should extend to the postconsumer phase of a product's lifecycle, and that increased responsibility would result in product design changes to mitigate pollution and waste management costs. In this way the output side of the value chain and the input side of the value chain are linked to each other. However, even beyond this discussion of broad

Handbook of Recycling. https://doi.org/10.1016/B978-0-323-85514-3.00047-6

and narrow definitions, we see much confusion about the concept, because national governments and supranational bodies apply and define EPR with large diversity. As we will see in Section 39.3, the application practice reflects this diversity even more, due to different choices in the design of EPR policies.

In its original European application, EPR fitted into the shift away from command and control policies to more public-private and collaborative modes of governance (Vermeulen, 2002). Responsibility for solving problems in the commons is shifted toward market actors, with governments setting the ground rules and boundaries in the form of framework regulating, targets, and reporting requirements. The implementation of EPR then may include forms of negotiated agreements, which have been adopted in and adapted to many diverse contexts. The transfer of responsibilities includes different forms or combinations of responsibility. Responsibility can take the form of:

- Liability for (proven) environmental damages caused by the product.
- Financial responsibility for covering the (partial or full) costs for collection, recycling, or final disposal of the product.
- physical responsibility for product management or its effects.
- ownership of the products throughout their life span.
- informative responsibility in providing information on the properties of the product in question (Lindhqvist, 2000; Massarutto, 2014).

In the narrow, output-oriented approach of EPR, this transfer of responsibility applies to the organization of recycling after the user/consumer phase in the product life cycle. Organizing recycling can be done through an *individual* EPR or *collective* EPR: in the first case, companies organize the take back of their own products, while in the second, *producer responsibility organizations* (PROs) are established jointly by producers to organize this. In practice, these are often combined: European regulation and member state regulations ascribe the responsibility to each producer, but they are allowed to organize this collectively, which is the dominant practice in Europe.

Thus these PROs have a central role in EPR practices. As shown in Figure 39.1, this includes (Mayers and Butler, 2013):

- Organizing collection logistics and recycling, e.g., subcontracting waste companies by PROs to carry out day-to-day operations.
- PROs recovering the organizational costs from producers either through fixed fees or variable cost-sharing between producers based on the number of products sold.
- Producers ultimately recovering their take-back costs from their customers, either as part of regular product pricing or visible recycling levies charged on top.

The introduction of collective EPR may be very complex: many thousands of producers and waste collectors need to agree on systematic methods of working together to ensure waste is collected where and when required (Figure 39.1). When established, PROs manage waste services and administer multiple payments to finance these operations. Collection points are organized, when relevant, in collaboration with municipal authorities; collection points in shops and other public places are also created. Possible free riders, companies not participating but benefiting, need to be identified and forced to join. This also includes online sellers within and outside the country (Hilton et al., 2019; Kalimo et al., 2015). How exactly all this is organized and how the roles of public and private actors are defined varies greatly between countries and between waste categories.

EPR is implemented through administrative, economic, and informative instruments. The composition of these instruments determines the precise form of the EPR. Hickle describes

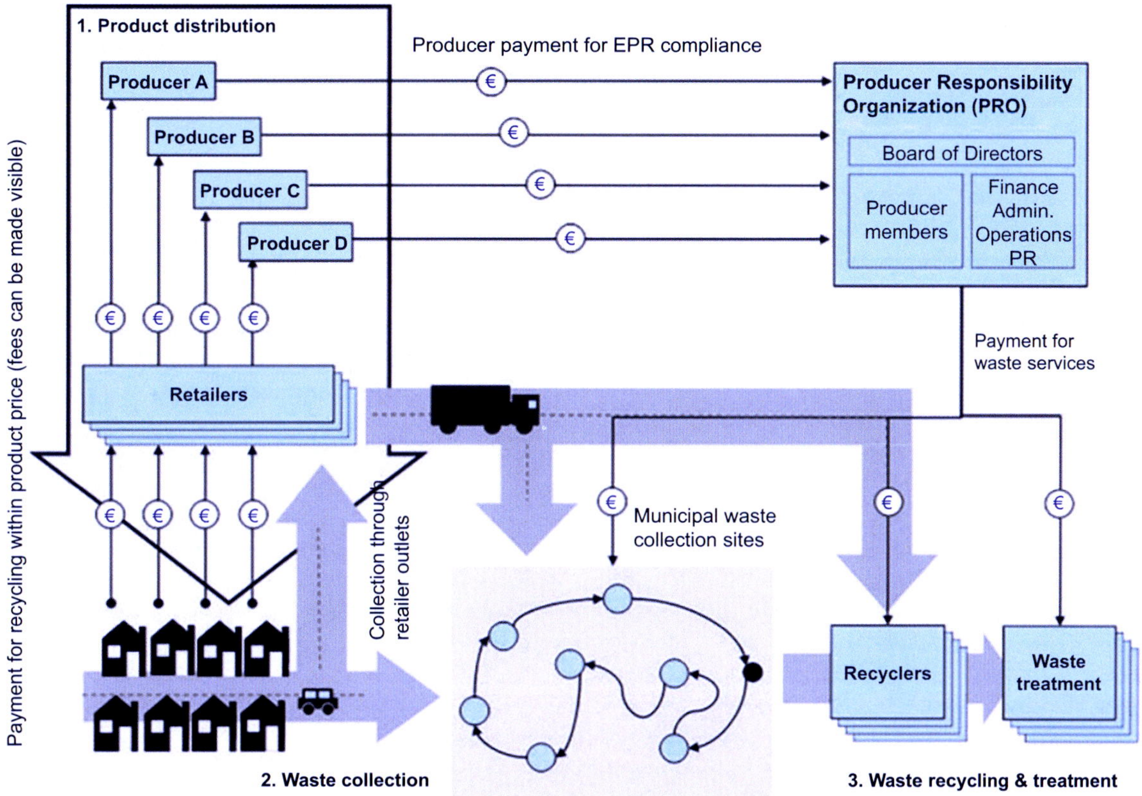

FIGURE 39.1 The role of producer responsibility organizations. *From Mayers and Butler (2013).*

four models with variations in the distribution of roles between producers and public authorities (Hickle, 2014a, p. 63), while the OECD distinguishes four slightly different models (OECD, 2016):

- One single PRO with commercial and/or municipal collection and processing services.
- Multiple PROs with the clearinghouse and commercial and/or municipal collection and processing services.
- Governance structure for tradable credits system.
- Government-run EPR system.

In the broader definition of EPR, as described by the OECD, other recycling-oriented instruments are labeled as EPR. Apart from product take-back requirements, done through collection and recycling targets as described previously, the OECD includes the following in its EPR definition:

- Economic and market-based instruments, e.g., deposit-funds, advanced disposal fees or (virgin) material taxes, landfill taxes.
- Regulation and performance standards, e.g., minimum recycled content in products, landfill bans.
- Information-based instruments, such as reporting requirements or product labeling (OECD, 2016).

This list emphasizes that a government's toolkit for supporting recycling contains more options than the shifting of financial, organizational, and managerial responsibility to producers through *individual* EPR or *collective* EPR. EPR schemes are, in many countries, one of the instruments in a policy mix. Aside from deposit refunds, recycled content requirements, and labeling, other approaches may also include input side-oriented elements, like ecodesign regulations and corporate sustainability programs and reporting requirements. In this chapter, we focus on EPR practices in which *financial and physical responsibility* for collection and recycling of postconsumer waste is established.

39.3 CURRENT PRACTICE OF EPR

Over the last 30 years, EPR-related policies have been steadily adopted by national governments, with the number increasing significantly from the 2000s (Kaffine and O'Reilly, 2013). Globally, most EPR-related policies cover product categories such as electronics, packaging, tires, and batteries, with formal take-back requirements as the most common policy approach (OECD, 2016). EPR for electrical waste has by far received the most attention both from policy and scholarly research related to concerns over health, environment, and resource depletion (Corsini et al., 2017; Ongondo et al., 2011).

39.3.1 EPR Within the European Union

The European Union (EU) has used EPR as a key approach to waste management. The EU has mandated EPR schemes for packaging (94/62/EC; 2018/852), end-of-life vehicles (ELVs) (2000/53/EC), waste electrical and electronic equipment (WEEE) (2002/96/EC; 2012/19/EU), batteries (2006/66/EC) and, most recently, single-use plastic products, e.g., food containers (EU2019/904). These directives broadly require forms of financial and physical responsibility, where producers must finance and organize the postconsumer collection and recovery. They include a broad definition of "producer": one that also contains importers and distributors, not only original equipment manufacturers. They also include collection and recovery targets, for example a minimum of 85% reuse and recovery of the weight of collected end-of-life vehicles, compared to what is placed on the market in the end-of-life vehicles (ELVs) (2000/53/EC).

EU member states have generally followed two organizational approaches to EPR: the clearinghouse model and delegated governance schemes or the single PRO model. In the former, producers in the broader sense register with the government-managed clearinghouse and declare the number of products they placed onto the market. Producers are assigned a collection responsibility equivalent to their market share (Khetriwal et al., 2011). Countries such as Germany, Spain, France, and Italy have adopted this model for the WEEE, corresponding to the second OECD model described earlier. Conversely, in a delegated governance scheme, implementation is charged to PROs, which assume the collection and treatment responsibilities on behalf of all the producers. This corresponds with the first OECD model described before. This mode of organization is prevalent in countries such as the Netherlands, Sweden, Belgium, and Switzerland. In delegated governance schemes, nonprofit organizations are usually set up by producers or trade associations. A final element in the organization of EPR is the producer responsibility providers. These are specialized organizations that can assume collection and treatment responsibilities on behalf of producers or PROs (Khetriwal et al., 2011). Beyond the mandated schemes, member states have deployed EPR for other products, including tires, used oils, textiles, graphic paper, medicines, mobile homes, etc. France is the member state that has used the instrument most frequently, with over 20 schemes (European Commission, 2014).

39.3.2 Global Application

Globally, EPR is predominantly applied in high-income countries. The four most often addressed product categories are electronics, tires, vehicles, and packaging (OECD, 2016), with 90% of the EPR schemes being applied in Europe, North America, Australia, and New Zealand. The uptake in Asia and Africa is far more limited, although examples are available, such as in India (Turaga et al., 2019), Malaysia, Thailand, Vietnam (Akenji et al., 2011), and Nigeria (Odeyingbo et al., 2019) (see also https://weee-forum.org/members/).

In North America, EPR is applied in both the United States and Canada. Canada's Ministry of Environment first outlined working principles on EPR in 2007, with a national plan outlined in 2009. This included broad policy objectives and targets on the number of operational EPR programs and product categories. Although nonbinding, most provinces adopted a similar organizational approach, except Québec, which pursued a program that included progressively increasing collection and recovery targets, coupled with financial penalties for producers or PROs. Following a national framework, key principal differences emerged regarding the scope of products covered, targets, and physical and financial responsibility (Leclerc and Badami, 2020).

Conversely, the United States has followed a much less coordinated but state-driven approach to EPR. Consequently, the adoption of EPR is more fragmented, with states such as Vermont, Maine, and California being early movers and frontrunners (Nash and Bosso, 2013). EPR laws have been introduced in several states for rechargeable batteries, mercury thermostats, auto-switches, paints, and electronics. Although these are not uniform in their requirements, they represent successful applications, including performance incentives and goals (Nash and Bosso, 2013). Since 2010, US states have begun examining the Canadian EPR framework to learn from their successes (Hickle, 2013).

In Asia, EPR for WEEE is also organized in different ways. For example, Taiwan's government initiated EPR in 1998. The design of EPR gave immediate emphasis to modulating the fees paid by producers based both on the collection and recovery costs but also the environmental design of the product. In 2012, producers could reduce their EPR costs by up to 30% if products were certified with ecolabels, including environmental, energy, and water-saving performance (Cheng et al., 2019). To tackle the environmental impacts of WEEE, Japan introduced and enforced the home appliance recycling law from 2001, focusing on washing machines, refrigerators, TVs, and air-conditioning units. This law was amended and broadened in 2008, with additional measures focusing on containers and packaging materials, food waste, and end-of-life vehicles (Ogushi and Kandlikar, 2007). A key difference in this system's organization is the disposal fee, which is paid by consumers themselves at the point of disposal and is not incorporated within the product's price (Ongondo et al., 2011).

In China, the massive demand for electronics and the growth of electronic waste led to calls for EPR (Yu et al., 2008). EPR is a core component of the "Circular Economy Promotion Law," outlined in 2008 (Veenstra et al., 2010). This was developed as a policy approach in 2012, covering five product categories: televisions, washing machines, refrigerators, air conditioners, and computers, adding another nine by 2014 (Zhang et al., 2020).

South Korea's first deposit refund systems, being unsuccessful, were replaced by an EPR system that mandated physical and financial responsibility in the early 2000s (Kim and Mori, 2015; Manomaivibool and Hong, 2014).

Additional studies have focused on applying EPR for WEEE, particularly backyard recycling practices from WEEE shipped (il) legally from, for example, the EU, to developing countries (Nnorom and Osibanjo, 2008). Other global studies have examined the use of EPR for the management of waste tires. For example, research from Ecuador emphasized social sustainability goals included in EPR goals, such as building resilience for vulnerable communities through civil engineering projects (Cecchin et al., 2019). Additionally, research on waste tires in Brazil emphasized the challenges of transferring a policy to the context of "insufficient institutional capacity" resulting in limited waste preventative measures being introduced (Milanez and Bührs, 2009).

39.4 EFFECTIVENESS

Statements about the effectiveness of EPR need to consider the vast diversity of its application around the world (Cahill et al., 2011; Khetriwal et al., 2011; Tasaki et al., 2019) and the double ambition of the instrument: organizing the recycling at the end of the value chain and stimulating the redesign of products towards more sustainable and circular versions. Many studies have analyzed cases of EPR for specific waste streams in particular geographies. For this section, we use some of the meta-evaluative reviews published in the last decade to show the main observations about factors affecting the success of EPR, summarized in Table 39.1.

TABLE 39.1 Strengths and weaknesses of EPR, as mentioned in academic literature.

	Strengths	Weaknesses
Organizing recycling	• EPR schemes do divert waste streams from landfilling or incineration to forms of material recycling, which leads to environmental benefits[3, 4, 7, 8, 10, 11] • National (or in Europe EU) targets are met in frontrunning countries[8, 11] • It uses industry's managerial capacity to organize recycling markets[2, 7]	• Targets and standards are not harmonized and weakly enforced and are not met everywhere[3] • Lack of harmonized definitions[10] • Responsibility for recycling beyond the targeted collection rates is not taken (nonseparated waste, littering, orphans)[5] • Recycling process choices need to be based on better assessments[4, 5, 8, 11] • EPR promotes material recycling over reuse and other R-options[1, 8] • Exports of waste to low-income countries prevail[3, 5]
Efficiency	• Low operation costs (2% to 0.1% of product price), but data incomplete[2, 8, 9, 11] • Higher volume of materials collected in collective EPRs enable more efficient recycling technology[2] • In practice, both competing and single national PROs exist in different countries; views on which is most efficient are contested[3]	• Voluntary PROs face free riding[4, 6] • The level of costs of recycling allocated to producers differs strongly between countries[2] • Data collection and sharing is weak due to cost avoidance[1, 4, 5, 10] • In case recycling is profitable, recycling processors compete with collective systems, cherry-picking the easy gains[3,4]
Stimulating ecodesign	• Being responsible for the end-of-life is assumed to stimulate redesign of products by producers[11, 12]	• Low impact on ecodesign[3, 4, 8, 10] • Weak incentives on ecodesign, fee systems ignore ecodesign efforts[1, 2, 3, 5, 8] • The lack of harmonized legislation hinders impacts on product design[3]

Legend: 1, Lifset et al. (2013); 2, Massarutto (2014); 3, Kunz et al. (2018); 4, Atasu (2019); 5, Kalimo et al. (2015); 6, Hermann et al. (2020); 7, Shan and Yang (2020); 8, Campbell-Johnston et al. (2021); 9, Favot et al. (2018); 10, Pouikli (2020); 11, Vermeulen et al. (2021); 12, Lindhqvist (2000).

A simple answer to the question of effectiveness is not available. All EPR schemes do achieve some level of diversion of waste into recycling, thus limiting landfilling and its environmental impacts (OECD, 2016). Assessing and comparing EPR systems between product categories and countries is difficult, owing to the differences in definitions, reporting and monitoring requirements, and data quality (Ongondo et al., 2011).

39.4.1 Organizing Recycling

Cases of full or almost full collection of post-consumer waste do exist, for example, for cars and tires (OECD, 2016). EPR for tires has resulted in material recycling rates of up to 80–95% in various European countries (Campbell-Johnston et al., 2020; Winternitz et al., 2019). With its high end-of-life material value, worldwide end-of-life vehicle policies in high-income countries, mostly applying EPR, generally achieve 95% or more recovery results (Sakai et al., 2014).

EPR schemes under the EU Directives are meant to apply standardized reporting and monitoring. Despite inevitable data quality issues, we can make some observations as to the effectiveness of these Directives. Table 39.2 shows EPR schemes for glass, packaging, and WEEE for some EU countries, with examples of the highest and the lowest performers.

EPR schemes have proven to contribute to high recycling rates. For example, Germany and Sweden, both early movers in implementing EPR systems, saw recycling rates for glass packaging above 80% of the quantities collected. EU recycling targets for glass were set at 60% of weight in 2008, rising to 70% by 2025, which many countries have already met.

Conversely, recycling targets for plastic packaging were set at 22.5% of weight in 2008 and then rising to 55% by 2025. While the countries illustrated in Table 39.2 have all exceeded the 2008 targets, there is still a ways to go for Hungary, Poland, and Portugal to meet those in 2025.

EU targets for WEEE dictate that from 2016 a 45% collection rate of products sold in the previous 3 years will apply, climbing to 65% in 2019 or 85% of WEEE generated in that year. From 2017, it is clear that, while EPR has led to increased WEEE collection, few countries have met the proposed target. Table 39.2 shows that the best performing EU countries achieved about 60% collection in 2017, while at the lower side 40% was achieved.

WEEE recycling in the United States is organized at state level, with various approaches (Hickle, 2014b; OECD, 2016). The state of Minnesota applies a clearinghouse model, with a financial tax and reward system aiming for 80% recycling of electronics, which claims to be successful, but performance scores like those in Table 39.2 are not available (Alev et al., 2019).

EPR has also been applied for batteries around the world, collection targets being more moderate, aiming at 40–50% collection in the mid-2010s in the EU and Northern America, and 50–80% of that to be recycled (Turner and Nugent, 2016). Japan achieved 50–77% recycling, varying by type of battery (OECD, 2016).

An important element in assessing the success of EPR is the quality of the recycling processes applied. In many cases, lower-level forms of recycling are dominant, including incineration with energy recovery. Considerations of cost-effectiveness may, when the decision making is delegated to market actors, result in cheaper forms of recycling, like with car tires being processed to fillings for artificial sports fields (Campbell-Johnston et al., 2020). A comparison of EPRs for batteries in Europe and North America showed that, in many cases, no requirement on the best technologies to be applied is included in the regulations (Turner and Nugent, 2016). Indeed, post-EPR processing operations have led to downcycling, e.g., materials of lower quality or used in a lower value application. For example, Ortego et al. (2018) showed that the recycling of cars under the EU's EPR scheme in its current form caused

TABLE 39.2 EPR regulated waste streams and performance in some EU countries (2017).

Country	Glass packaging (recycling %)	Plastic packaging (recycling %)	WEEE (collection % 3-year average)
Austria	84.1	33.4	62.4
Belgium	100	45.5	49.4
Germany	84.4	49.7	45.1
Greece	36	41.4	42.4
Hungary	34.2	32	60.6
Poland	63	34.7	45.4
Portugal	49	34.9	53.9
Romania	63	47.6	–
Slovenia	98.5	60.4	39.1
Sweden	93	48.4	56.3
EU28	75.9	41.7	47.6

Data from Eurostat, 2021. https://ec.europa.eu/eurostat/data/database (accessed 12 November 2021).

downcycling and loss of geologically scarce materials. In the United States, Fishbein (2000) showed that for carpets, the quality of the feedstock and choice of recycler within an EPR scheme affect the degree of downcycling. Currently, proposals for more detailed value retention and recycling targets are discussed in Europe. At the same time, a better connection to ecodesign regulation is promoted.

Markets for recycled materials are often poorly developed, and higher prices of recycled materials and quality concerns are barriers to replacing primary raw material with recycled materials. EPR schemes do not play an active role in improving the functioning of markets for secondary materials. It is left to the recycling companies and the original producers to make their choices on economic grounds. Currently, few data are available on the rate of application of recycled materials in EPR-regulated product categories, such as electronic equipment, batteries, plastic packaging, cars, and tires.

Information about the effectiveness of its application in lower- and middle-income countries is much less available. With fewer established waste management systems in place, organizing producer responsibility poses bigger challenges than in the high-income countries (Ferronato and Torretta, 2019). Some countries have introduced it, but it often remains only a concept on paper, or with voluntary uptake, as in Malaysia (Agamuthu and Victor, 2011).

39.4.2 Economic Efficiency

It is argued that delegating responsibility for end-of-life treatment and recycling is efficient (Dubois and Eyckmans, 2015; Winternitz et al., 2019), though limited data on this is available. Collective systems avoid setting up multiple infrastructures and create economies of scale (Monier et al., 2014; Tojo, 2006). In Italy, the costs for organizing recycling of electronic equipment are estimated at 0.4% of the sales prices (Favot et al., 2018), while in the Netherlands, the costs of various EPR schemes are estimated between 2% and 0.2% of the product sales prices (Vermeulen et al., 2021). However, such estimates are very dependent on how EPR is organized and which costs are included in the financial responsibility, making it also hard to compare the efficiency of individual and collective setups of EPR (Kunz et al., 2018).

39.4.3 Stimulating Ecodesign

EPR has been introduced with the ambition to stimulate the redesign of the entire life cycle of products (Van Rossem et al., 2006). However, there is a clear consensus in the scientific community that the application of EPR has so far hardly stimulated producers to widely apply ecodesign or Design for Sustainability (Gottberg et al., 2006; Huisman, 2013; Kautto, 2006; Kemna, 2011; Kunz et al., 2018; Mayers, 2007; Subramanian et al., 2009; Tojo, 2006).

Researchers analyzing motives of producers for applying ecodesign have found no direct connection between ecodesign practices or improved environmental product performance and the participation in the EPR scheme. Design incentives in EPR are not explicit, while only a few of the "producers" addressed in EPR schemes are manufacturers designing original equipment (OEM) and most are thus not in a position to apply ecodesign (Kalimo et al., 2015). Meanwhile, ecodesign and Design for Sustainability approaches address more sustainability aspects than energy use, resource use, or preparations for recycling (Braungart et al., 2007; Crul et al., 2009; Deutz et al., 2013; Ramani et al., 2010), and some approaches are especially designed to address aspects of circularity (Mudgal et al., 2013). While ecodesign approaches have to some extent been adopted in industry, the main drivers for this are related to strategic positioning and market strategies connected to certification and reporting, while research on ecodesign implementation does not show evidence of the influence of EPR schemes (Albino et al., 2009; Gottberg et al., 2006; Richter and Koppejan, 2016; Rossi et al., 2016; Vermeulen et al., 2022).

One of the ways to strengthen the ecodesign incentive in EPR schemes is introducing fee modulation related to the sustainability performance of products (Vermeulen et al., 2021). Examples of this can be found in Canada (Leclerc and Badami, 2020), France, Italy, and Germany, while the EU is moving toward further application of such fee modulations (Sherrington et al., 2020).

39.4.4 General Remarks

We noted in Section 39.2 that EPR was introduced as a form of self-regulation with delegated responsibilities. Such delegation has enabled producers to set up systems in product-, culture-, and country-specific ways, adapting to the specific points of distribution and collection. These are very different for cars, tires, packaging, or batteries. In the same way assessing its success depends on data available about sales volumes, average use times, and stockpiling behavior of consumers, as with batteries or old electronic equipment, as well as secondhand sales practices, which may include cross-border movement, e.g., cars. Such product-specific contextualizing has advantages, but also results in a widely diverse diffusion of EPR practices in time and across the world.

Copying models from collaborative policy cultures (such as in Northwest Europe) may not always work well. Consequently, we see a wide practice of adapting EPR models to national policy cultures and contexts. This also includes the different ways of combining it with other policy measures, e.g., landfill bans and landfill taxation (Niza et al., 2014). Some have suggested solving these issues of confusion and competing or even counterproductive diversity by developing global EPR systems. These have been proposed for plastics (Raubenheimer and Urho, 2020), WEEE (Ongondo et al., 2011), or for specific strategic materials, e.g., platinum, that are often lost in product-oriented EPRs (Wilts et al., 2011).

Such global harmonization, no matter how desired it might be, will most likely not be achieved in the short run. But with successes as described in organizing take-back and recycling of a variety of product categories, the challenge is to take lessons from the successful practices and apply them widely, while also looking for opportunities to strengthen where weaknesses are spotted. Here we see a role for the OECD and UNEP.

39.5 THE FUTURE OF EPR AND CIRCULAR ECONOMY?

EPR will continue to be a crucial policy instrument in promoting recycling. At places where it was implemented successfully, it efficiently organized infrastructures for take-back and recycling.

In recent years many countries around the world have transformed traditional waste management policies into circular economy policies. Crucial in this context are:

(1) a stronger emphasis on extension of the lifetime of products in use;
(2) applying far more options of value retention than the recycling of mixed streams of post-consumer waste; and
(3) a stronger emphasis on producers developing new circular business models (Bocken et al., 2016).

The old 3Rs waste hierarchy (Refuse, Reduce, Recycle) has been replaced by the more detailed 10Rs hierarchy, distinguishing short loops (R0 Refuse, R1 Reduce, R2 Resell/reuse), middle-long loops (R3 Repair, R4 Refurbish, R5 Remanufacture, R6 Repurpose) and long loops (R7 Recycle materials, R8 Recover energy, R9 Remine) (Reike et al., 2018).

EPR systems so far have organized R7 and R8 at low costs and have delegated responsibility to (end-)producers, importers, and retailers selling new products. The new circular policy framing requires recognition of the roles of other (circular) value chain actors. Extending lifetime through repairing and reselling is often in the hands of social enterprises that need to be included in EPR organizations (Bahers and Kim, 2018). Making EPR a better fit in the pursuit of a circular economy requires a more explicit link to rewarding ecodesign improvements of products, rewarding the application of recycled materials, and strengthening the market for recycled materials (Vermeulen et al., 2022). These may very well need the application of a broader and more comprehensive set of policy instruments, such as taxes on virgin materials and the promotion of circular business models.

However, these adjustments still assume continued and unlimited consumption. The R0 Refuse refers to overconsumption and issues that go beyond the interest of producers. An example of this is "planned obsolescence" in products, which benefits both producers and retailers and puts pressure on the consumer to continue consuming. Instruments such as EPR fail to address issues of waste prevention and decreased consumption (Bartl, 2014). EPR systems will not likely be able to address these questions of consumption. It will need to be complemented with other instruments, promoting a more structural transformation of our economies (Calisto Friant et al., 2020).

References

Agamuthu, P., Victor, D., 2011. Policy trends of extended producer responsibility in Malaysia. Waste Manag. Res. 29, 945–953. https://doi.org/10.1177/0734242X11413332.

Akenji, L., Hotta, Y., Bengtsson, M., Hayashi, S., 2011. EPR policies for electronics in developing Asia: an adapted phase-in approach. Waste Manag. Res. 29, 919–930. https://doi.org/10.1177/0734242X11414458.

Albino, V., Balice, A., Dangelico, R.M., 2009. Environmental strategies and green product development: an overview on sustainability-driven companies. Bus. Strateg. Environ. 18, 83–96. https://doi.org/10.1002/bse.638.

Alev, I., Natalie Huang, X., Atasu, A., Toktay, L.B., 2019. A case discussion on market-based extended producer responsibility: the Minnesota electronics recycling act. J. Ind. Ecol. 23, 208–221. https://doi.org/10.1111/jiec.12721.

Atasu, A., 2019. Operational perspectives on extended producer responsibility. J. Ind. Ecol. 23, 744–750. https://doi.org/10.1111/jiec.12816.

Ayres, R., Ayres, L., 2015. A Handbook of Industrial Ecology., https://doi.org/10.4337/9781843765479.

Bahers, J.B., Kim, J., 2018. Regional approach of waste electrical and electronic equipment (WEEE) management in France. Resour. Conserv. Recycl. 129, 45–55. https://doi.org/10.1016/j.resconrec.2017.10.016.

Bartl, A., 2014. Moving from recycling to waste prevention: a review of barriers and enables. Waste Manag. Res. 32, 3–18. https://doi.org/10.1177/0734242X14541986.

Bocken, N.M.P., de Pauw, I., Bakker, C., van der Grinten, B., 2016. Product design and business model strategies for a circular economy. J. Ind. Prod. Eng. 33, 308–320. https://doi.org/10.1080/21681015.2016.1172124.

Braungart, M., McDonough, W., Bollinger, A., 2007. Cradle-to-cradle design: creating healthy emissions—a strategy for eco-effective product and system design. J. Clean. Prod. 15, 1337–1348. https://doi.org/10.1016/j.jclepro.2006.08.003.

Cahill, R., Grimes, S.M., Wilson, D.C., 2011. Review article: extended producer responsibility for packaging wastes and WEEE—a comparison of implementation and the role of local authorities across Europe. Waste Manag. Res. 29, 455–479. https://doi.org/10.1177/0734242X10379455.

Calisto Friant, M., Vermeulen, W.J.V., Salomone, R., 2020. A typology of circular economy discourses: navigating the diverse visions of contested paradigm. Resour. Conserv. Recycl. 161, 104917. https://doi.org/10.1016/j.resconrec.2020.104917.

Campbell-Johnston, K., Friant, M.C., Thapa, K., Lakerveld, D., Vermeulen, W.J.V., Calisto Friant, M., Thapa, K., Lakerveld, D., Vermeulen, W.J.V., 2020. How circular is your tyre: experiences with extended producer responsibility from a circular economy perspective. J. Clean. Prod. 270, 122042. https://doi.org/10.1016/j.jclepro.2020.122042.

Campbell-Johnston, K., de Munck, M., Vermeulen, W.J.V., Backes, C., 2021. Future perspectives on the role of extended producer responsibility within a circular economy: a Delphi study using the case of the Netherlands. Bus. Strateg. Environ. 30 (8), 4054–4067, 1–20 (Submitted).

Cecchin, A., Lamour, M., Joseph Maks Davis, M., Jácome Polit, D., 2019. End-of-life product management as a resilience driver for developing countries: a policy experiment for used tires in Ecuador. J. Ind. Ecol. 23, 1292–1310. https://doi.org/10.1111/jiec.12861.

Cheng, C.-P., Lin, C.-H., Wen, L.-C., Chang, T.-C., 2019. Determining environmental costs: a challenge in a governmental e-waste recycling scheme. Sustain 11. https://doi.org/10.3390/su11195156.

Corsini, F., Rizzi, F., Frey, M., 2017. Extended producer responsibility: the impact of organizational dimensions on WEEE collection from households. Waste Manag. 59, 23–29. https://doi.org/10.1016/j.wasman.2016.10.046.

Crul, M., Diehl, J., Ryan, C., 2009. Design for Sustainability. A Step-By-Step Approach. UNEP, Paris.

Deutz, P., McGuire, M., Neighbour, G., 2013. Eco-design practice in the context of a structured design process: an interdisciplinary empirical study of UK manufacturers. J. Clean. Prod. 39, 117–128. https://doi.org/10.1016/j.jclepro.2012.08.035.

Dubois, M., Eyckmans, J., 2015. Efficient waste management policies and strategic behavior with open Borders. Environ. Resour. Econ. 62, 907–923. https://doi.org/10.1007/s10640-014-9851-3.

European Commission, 2014. Development of Guidance on Extended Producer Responsibility (EPR) Final Report European Commission-DG Environment 2014 in Collaboration With.

Favot, M., Veit, R., Massarutto, A., 2018. The ratio of EPR compliance fees on sales revenues of electrical and electronic equipment in Italy. A circular economy perspective. Resour. Conserv. Recycl. 135, 34–37. https://doi.org/10.1016/j.resconrec.2017.06.012.

Ferronato, N., Torretta, V., 2019. Waste mismanagement in developing countries: a review of global issues. Int. J. Environ. Res. Public Health. https://doi.org/10.3390/ijerph16061060.

Fishbein, B.K., 2000. Carpet take-back: EPR American style. Environ. Qual. Manag. 10, 25–36. https://doi.org/10.1002/1520-6483(200023)10:1<25::AID-TQEM4>3.0.CO;2-4.

Gottberg, A., Morris, J., Pollard, S., Mark-Herbert, C., Cook, M., 2006. Producer responsibility, waste minimisation and the WEEE directive: case studies in eco-design from the European lighting sector. Sci. Total Environ. 359, 38–56. https://doi.org/10.1016/j.scitotenv.2005.07.001.

Hermann, A., Gailhofer, P., Schomerus, T., 2020. Producer Responsibility of Third-Country Producers in e-Commerce. Umweltbundesamt, Dessau-Roßlau.

Hickle, G.T., 2013. Comparative analysis of extended producer responsibility policy in the United States and Canada. J. Ind. Ecol. 17, 249–261. https://doi.org/10.1111/jiec.12020.

Hickle, G.T., 2014a. An examination of governance within extended producer responsibility policy regimes in North America. Resour. Conserv. Recycl. 92, 55–65. https://doi.org/10.1016/j.resconrec.2014.08.007.

Hickle, G.T., 2014b. Moving beyond the "patchwork:" a review of strategies to promote consistency for extended producer responsibility policy in the U.S. J. Clean. Prod. 64, 266–276. https://doi.org/10.1016/j.jclepro.2013.08.013.

Hilton, M., Sherrington, C., Andrew, M., Börkey, P., 2019. Extended Producer Responsibility (EPR) and the Impact of Online Sales.

Huisman, J., 2013. Too big to fail, too academic to function: producer responsibility in the global financial and E-waste crises Huisman too big to fail, too academic to function. J. Ind. Ecol. 17, 172–174. https://doi.org/10.1111/jiec.12012.

Kaffine, D., O'Reilly, P., 2013. What have we learned about extended producer responsibility in the past decade? A survey of the recent EPR economic literature. OECD Pap., 1–39.

Kalimo, H., Lifset, R., Atasu, A., Van Rossem, C., Van Wassenhove, L., 2015. What roles for which stakeholders under extended producer responsibility? Rev. Eur. Comp. Int. Environ. Law 24, 40–57. https://doi.org/10.1111/reel.12087.

Kautto, P., 2006. New instruments—old practices? The implications of environmental management systems and extended producer responsibility for design for the environment. Bus. Strateg. Environ. 15, 377–388. https://doi.org/10.1002/bse.454.

Kemna, R., 2011. MEErP 2011 Methodology Report. COWI Belgium sprl, Brussels/Delft.

Khetriwal, D.S., Widmer, R., Kuehr, R., Huisman, J., 2011. One WEEE, many species: lessons from the European experience. Waste Manag. Res. 29, 954–962. https://doi.org/10.1177/0734242X11413327.

Kim, S., Mori, A., 2015. Revisiting the extended producer responsibility program for metal packaging in South Korea. Waste Manag. 39, 314–320. https://doi.org/10.1016/j.wasman.2015.02.006.

Kunz, N., Mayers, K., Van Wassenhove, L.N., 2018. Stakeholder views on extended producer responsibility and the circular economy. Calif. Manag. Rev. 60, 45–70. https://doi.org/10.1177/0008125617752694.

Leclerc, S.H., Badami, M.G., 2020. Extended producer responsibility for E-waste management: policy drivers and challenges. J. Clean. Prod. 251, 119657. https://doi.org/10.1016/J.JCLEPRO.2019.119657.

Lifset, R., Atasu, A., Tojo, N., 2013. Extended producer responsibility. National, international, and practical perspectives. J. Ind. Ecol. 17, 162–166. https://doi.org/10.1111/jiec.12022.

Lindhqvist, T., 2000. Extended Producer Responsibility in Cleaner Production: Policy Principle to Promote Environmental Improvements of Product Systems. Lund University. http://www.lub.lu.se/luft/diss/tec355.pdf.

Manomaivibool, P., Hong, J.H., 2014. Two decades, three WEEE systems: how far did EPR evolve in Korea's resource circulation policy? Resour. Conserv. Recycl. 83, 202–212. https://doi.org/10.1016/j.resconrec.2013.10.011.

Massarutto, A., 2014. The long and winding road to resource efficiency—an interdisciplinary perspective on extended producer responsibility. Resour. Conserv. Recycl. 85, 11–21. https://doi.org/10.1016/j.resconrec.2013.11.005.

Mayers, C.K., 2007. Strategic, financial, and design implications of extended producer responsibility in Europe: a producer case study. J. Ind. Ecol. https://doi.org/10.1162/jiec.2007.1228.

Mayers, K., Butler, S., 2013. Producer responsibility organizations development and operations. J. Ind. Ecol. 17, 277–289. https://doi.org/10.1111/jiec.12021.

Milanez, B., Bührs, T., 2009. Extended producer responsibility in Brazil: the case of tyre waste. J. Clean. Prod. 17, 608–615. https://doi.org/10.1016/J.JCLEPRO.2008.10.004.

Monier, V., Hestin, M., Cavé, J., Laureysens, I., Watkins, E., Reisinger, H., Porsch, L., 2014. Development of Guidance on Extended Producer Responsibility (EPR)—Final Report. BIO Intelligence Service.

Mudgal, S., Tinetti, B., de Trigo, A.P., Faninge, T., Schischke, K., Proske, M., Geibler, J., von Teubler, J., 2013. Material-Efficiency Ecodesign Report and Module to the Methodology for the Ecodesign of Energy-Related Products (MEErP), Part 2—Enhancing MEErP for Ecodesign. Prepared for: European Commission – DG Enterprise and Industry.

Nash, J., Bosso, C., 2013. Extended producer responsibility in the United States: full speed ahead? Nash and Bosso EPR in the U.S.: full speed ahead? J. Ind. Ecol. 17, 175–185. https://doi.org/10.1111/j.1530-9290.2012.00572.x.

Niza, S., Santos, E., Costa, I., Ribeiro, P., Ferrão, P., 2014. Extended producer responsibility policy in Portugal: a strategy towards improving waste management performance. J. Clean. Prod. 64, 277–287. https://doi.org/10.1016/j.jclepro.2013.07.037.

Nnorom, I.C., Osibanjo, O., 2008. Overview of electronic waste (e-waste) management practices and legislations, and their poor applications in the developing countries. Resour. Conserv. Recycl. 52, 843–858. https://doi.org/10.1016/j.resconrec.2008.01.004.

Odeyingbo, A.O., Nnorom, I.C., Deubzer, O.K., 2019. Used and waste electronics flows into Nigeria: assessment of the quantities, types, sources, and functionality status. Sci. Total Environ. 666, 103–113. https://doi.org/10.1016/j.scitotenv.2019.02.102.

OECD, 2014. The state of play on extended producer responsibility (EPR): opportunities and challenges. In: Global Forum on Environment: Promoting Sustainable Materials Management Through Extended Producer Responsibility (EPR), p. 17.

OECD, 2016. Extended Producer Responsibility: Updated Guidance for Efficient Waste Management. OECD, Paris, https://doi.org/10.1787/9789264256385-en.

Ogushi, Y., Kandlikar, M., 2007. Assessing extended producer responsibility laws in Japan. Environ. Sci. Technol. 41, 4502–4508. https://doi.org/10.1021/es072561x.

Ongondo, F.O., Williams, I.D., Cherrett, T.J., 2011. How are WEEE doing? A global review of the management of electrical and electronic wastes. Waste Manag. 31, 714–730. https://doi.org/10.1016/j.wasman.2010.10.023.

Ortego, A., Valero, A., Valero, A., Iglesias, M., 2018. Downcycling in automobile recycling process: a thermodynamic assessment. Resour. Conserv. Recycl. 136, 24–32. https://doi.org/10.1016/j.resconrec.2018.04.006.

Pouikli, K., 2020. Concretising the role of extended producer responsibility in European Union waste law and policy through the lens of the circular economy. ERA Forum 20, 491–508. https://doi.org/10.1007/s12027-020-00596-9.

Ramani, K., Ramanujan, D., Bernstein, W.Z., Zhao, F., Sutherland, J., Handwerker, C., Choi, J.-K., Kim, H., Thurston, D., 2010. Integrated sustainable life cycle design: a review. J. Mech. Des. 132, 091004. https://doi.org/10.1115/1.4002308.

Raubenheimer, K., Urho, N., 2020. Rethinking global governance of plastics—the role of industry. Mar. Policy 113, 103802. https://doi.org/10.1016/J.MARPOL.2019.103802.

Reike, D., Vermeulen, W.J.V., Witjes, S., 2018. The circular economy: new or refurbished as CE 3.0?—exploring controversies in the conceptualization of the circular economy through a focus on history and resource value retention options. Resour. Conserv. Recycl. 135, 246–264. https://doi.org/10.1016/j.resconrec.2017.08.027.

Richter, J.L., Koppejan, R., 2016. Extended producer responsibility for lamps in Nordic countries: best practices and challenges in closing material loops. J. Clean. Prod. 123, 167–179. https://doi.org/10.1016/j.jclepro.2015.06.131.

Rossi, M., Germani, M., Zamagni, A., 2016. Review of ecodesign methods and tools. Barriers and strategies for an effective implementation in industrial companies. J. Clean. Prod. 129, 361–373. https://doi.org/10.1016/j.jclepro.2016.04.051.

Sakai, S., Yoshida, H., Hiratsuka, J., Vandecasteele, C., Kohlmeyer, R., Rotter, V.S., Passarini, F., Santini, A., Peeler, M., Li, J., Oh, G.-J., Chi, N.K., Bastian, L., Moore, S., Kajiwara, N., Takigami, H., Itai, T., Takahashi, S., Tanabe, S., Tomoda, K., Hirakawa, T., Hirai, Y., Asari, M., Yano, J., 2014. An international comparative study of end-of-life vehicle (ELV) recycling systems. J. Mater. Cycles Waste Manag. 16, 1–20. https://doi.org/10.1007/s10163-013-0173-2.

Shan, H., Yang, J., 2020. Promoting the implementation of extended producer responsibility systems in China: a behavioral game perspective. J. Clean. Prod. 250, 119446. https://doi.org/10.1016/J.JCLEPRO.2019.119446.

Sherrington, C., Papineschi, J., Hilton, M., Massie, A., Jones, P., 2020. Study to Support Preparation of the Commission's Guidance for Extended Producer Responsibility Schemes Final Report. Eunomia Research & Consulting Ltd, Bristol.

Subramanian, R., Gupta, S., Talbot, B., 2009. Product design and supply chain coordination under extended producer responsibility. Prod. Oper. Manag. 18, 259–277. https://doi.org/10.1111/j.1937-5956.2009.01018.x.

Tasaki, T., Tojo, N., Lindhqvist, T., 2019. Differences in perception of extended producer responsibility and product stewardship among stakeholders: an international questionnaire survey and statistical analysis. J. Ind. Ecol. 23, 438–451. https://doi.org/10.1111/jiec.12815.

Tojo, N., 2006. Design change in electrical and electronic equipment. Impacts of extended producer responsibility legislation in Sweden and Japan. In: Smith, T., Sonnenfeld, D., Pellow, D. (Eds.), Challenging the Chip: Labor Rights and Environmental Justice in the Global Electronics Industry, p. 273.

Turaga, R.M.R., Bhaskar, K., Sinha, S., Hinchliffe, D., Hemkhaus, M., Arora, R., Chatterjee, S., Khetriwal, D.S., Radulovic, V., Singhal, P., Sharma, H., 2019. E-waste management in India: issues and strategies. Vikalpa 44, 127–162. https://doi.org/10.1177/0256090919880655.

Turner, J.M., Nugent, L.M., 2016. Charging up battery recycling policies: extended producer responsibility for single-use batteries in the European Union, Canada, and the United States. J. Ind. Ecol. 20, 1148–1158. https://doi.org/10.1111/jiec.12351.

Van Rossem, C., Tojo, N., Lindhqvist, T., 2006. Lost in Transposition? A Study of the Implementation of Individual Producer Responsibility in the WEEE Directive.

Veenstra, A., Wang, C., Fan, W., Ru, Y., 2010. An analysis of E-waste flows in China. Int. J. Adv. Manuf. Technol. 47, 449–459. https://doi.org/10.1007/s00170-009-2356-5.

Vermeulen, W.J.V., 2002. Greening production as co-responsibility. In: Driessen, P.P.J., Glasbergen, P. (Eds.), Greening Society. The Paradigm Shift in Dutch Environmental Politics. Kluwer Academic Publishers, Dordrecht, pp. 67–90.

Vermeulen, W.J.V., Weterings, R.P.M., 1997. Extended producer responsibility: moving from end-of-life management towards public-private commitment in product life cycles. J. Clean Technol. Environ. Toxicol. Occup. Med. 6, 283–298.

Vermeulen, W.J.V., Backes, C.W., de Munck, M.C.J., Campbell-Johnston, K., de Waal, I.M., Carreon, J.R., Boeve, M.N., 2021. White Paper on Pathways for Extended Producer Responsibility on the Road to a Circular Economy. Utrecht University, Utrecht, https://doi.org/10.13140/RG.2.2.11527.93602.

Vermeulen, W., Campbell-Johnston, K., Thapa, K., 2022. Extended producer responsibility and circular economy: three design flaws. Ökologisches Wirtschaften—Fachzeitschrift 37 (1), 21–23 (Oekom Publishers GmbH). https://doi.org/10.14512/oew370121.

Wilts, H., Bringezu, S., Bleischwitz, R., Lucas, R., Wittmer, D., 2011. Challenges of metal recycling and an international covenant as possible instrument of a globally extended producer responsibility. Waste Manag. Res. 29, 902–910. https://doi.org/10.1177/0734242X11415311.

Winternitz, K., Heggie, M., Baird, J., 2019. Extended producer responsibility for waste tyres in the EU: lessons learnt from three case studies—Belgium, Italy and the Netherlands. Waste Manag. 89, 386–396. https://doi.org/10.1016/j.wasman.2019.04.023.

Yu, J., Hills, P., Welford, R., 2008. Extended producer responsibility and eco-design changes: perspectives from China. Corp. Soc. Responsib. Environ. Manag. 15, 111–124. https://doi.org/10.1002/csr.168.

Zhang, D., Cao, Y., Wang, Y., Ding, G., 2020. Operational effectiveness of funding for waste electrical and electronic equipment disposal in China: an analysis based on game theory. Resour. Conserv. Recycl. 152, 104514. https://doi.org/10.1016/j.resconrec.2019.104514.

Further Reading

Driessen, P.P.J.J., Dieperink, C., Laerhoven, F., Runhaar, H. A.C.C., Vermeulen, W.J.V.V., van Laerhoven, F., Runhaar, H.A.C.C., Vermeulen, W.J.V.V., 2012. Towards a conceptual framework for the study of shifts in modes of environmental governance—experiences from the Netherlands. Environ. Policy Gov. 22, 143–160. https://doi.org/10.1002/eet.1580.

European Commission – DG Environment, 2014. Development of Guidance on Extended Producer Responsibility (EPR). European Union, Brussels.

Hobson, K., 2015. Closing the loop or squaring the circle? Locating generative spaces for the circular economy. Prog. Hum. Geogr. 40. https://doi.org/10.1177/0309132514566342. 0309132514566342.

Khetriwal, D.S., Kraeuchi, P., Widmer, R., 2009. Producer responsibility for e-waste management: key issues for consideration—learning from the Swiss experience. J. Environ. Manag. 90, 153–165. https://doi.org/10.1016/j.jenvman.2007.08.019.

OECD, 2006. EPR Policies and Product Design: Economic Theory and Selected Case Studies. OECD, Paris.

Yang, J., Lu, B., Xu, C., 2008. WEEE flow and mitigating measures in China. Waste Manag. 28, 1589–1597. https://doi.org/10.1016/J.WASMAN.2007.08.019.

CHAPTER

40

Exploring the economics of recycling in a dynamic global context

Pieter van Beukering, Onno Kuik, Frans Oosterhuis, and Jan Brusselaers

Institute for Environmental Studies (IVM), Vrije Universiteit Amsterdam, Amsterdam, The Netherlands

40.1 INTRODUCTION

Recycling is generally considered an important strategy for alleviating the pressures of society on the environment. Natural resources are saved, emissions are decreased, and the burden of solid waste is reduced. At the same time, recycling creates employment and attracts investments. In recent years many countries have experienced large increases in recycling. Besides domestic causes, international trade has played an important role in the expansion of the global recycling sector. In the last decades, international trade of recyclable materials has increased significantly. This chapter on the economics of recycling aims to identify the main economic drivers of recycling, address the private and external costs of recycling-related activities, and demonstrate the effectiveness of the most important economic instruments to promote recycling.

40.2 ECONOMIC TRENDS AND DRIVERS

Waste is a by-product of consumption and production activities. Considering the increasing scarcity of natural resources, waste is also seen as a potentially valuable input to production, substituting for virgin resources. Recycling is generally considered as an important strategy for alleviating the pressures of the economy on the environment by saving natural resources, decreasing the emissions of pollutants in the environment, and reducing the burden of solid waste. Recycling also contributes to economic activity, attracts investment, and creates employment. Globally, drivers for recycling vary between high-income countries and low-income countries. In high-income countries, drivers for recycling of waste result from increasing disposal costs, public concern about environment and health, the presence of

Handbook of Recycling. https://doi.org/10.1016/B978-0-323-85514-3.00022-1

valuable resources in waste streams, loss of energy, and the fear of future scarcity of certain natural resources. In low-income countries, recycling is mainly driven by direct economic motives. International trade has played an important role in the expansion of the global recycling sector. As is the case for any commodity, international trade in recyclable waste allows countries to exercise their comparative advantages to increase the efficiency of the allocation of resources of the global recycling industry. But there is also concern that (hazardous) waste is shipped to foreign destinations to avoid costly regulations for disposal and storage.

40.2.1 Generation of Waste

The volume of waste that is generated each year in the world is vast and increasing. Large volumes of waste are generated by mining operations, construction, and agriculture, but it is difficult to put a number on these volumes as they are often not recorded. It has been estimated that the generated waste in the European Union alone amounted to 2.3 billion tonnes in 2018. Several sectors generate this waste, which comes in different forms. The construction and demolition sector accounts for 36% of the EU's waste. Other important sectors are the mining and quarrying sector (27%) and the manufacturing sector (11%). An individual county's economic structure considerably determines the shares of the respective sectors. Some of the EU member states, for example, generate no mining and quarrying waste (EUROSTAT, 2021).

Each of the mentioned sectors generates more waste than the EU's households, accounting for 8% of the EU's waste. This share corresponds to estimates of the importance of municipal solid waste at a global level (OECD, 2020). In the OECD region, the municipal solid waste supply was 675 million tonnes in 2020, which is only a modest increase from 661 million tonnes in 2010 (OECD, 2020). That volume corresponds to 520 kg of waste per year per person living in the OECD region, 20 kg more than in 1990 but 30 kg less than in 2010 (OECD, 2020). At the global level, there is a considerable variance across countries. The World Bank (2021), for example, calculated that the high-income countries, although they only account for 16% of the world population, generate 34% of the world's waste. Lower income countries produce much less waste per capita.

Research has established that the supply of household solid waste grows proportionally with population, grows less than proportionally with household income, and decreases less than proportionally with increasing charges of waste collection services. In economic terms, this means that household demand for waste services has unitary elasticity with respect to population, and is inelastic with respect to income and price (Johnstone and Labonne, 2004). It has been investigated whether the supply of municipal solid waste would eventually start to decline with rising income, according to the so-called Environmental Kuznets Curve (EKC) hypothesis. On the one hand, higher-income families consume more and would therefore produce more waste, but on the other hand it has been suggested that they tend to have a consumption pattern that does not favor waste-intensive goods (Ferrara, 2008). In addition, Zaman (2016) found a positive correlation between per-capita gross income and per-capita resource recovery as well as per-capita waste generation. The evidence on the EKC hypothesis is mixed: some researchers find a "turning point" where household waste supply starts to decline (Abrate and Ferraris, 2010) and others do not (yet) (Mazzanti and Zoboli, 2009; Nicolli et al., 2012).

40.2.2 Recycling

Given this vast amount of waste, there is much interest in reuse or recycling, if not for the potentially valuable materials and energy that can be extracted from the waste, then at least for the increasing costs of disposal and storage. In many countries, an increasing share of industrial and municipal waste is designated for recovery operations, including incineration with energy recovery, composting, and recycling. For example, in the United States, the share of municipal solid waste that is being recycled (excluding composting) has increased from 14% in 1990 to 25% in 2018. Among all OECD countries, Korea has the highest share of recycling of municipal solid waste (64%) (OECD, 2021).

Recycling rates differ significantly per material recycled. According to the industry itself, the global recycling rate of paper is currently 59% (ICFPA, 2021). Europe is the frontrunner with a paper recycling rate of 72% in 2019. Recycling rates for metals vary from very high (gold) to negligible for many specialty metals such as lithium and tellurium. For these metals, the prices are too low and the natural reserves are too large to make recycling profitable (Schmidt et al., 2017). Recycling rates tend to be higher when the metals are used in large quantities in easily recoverable applications, e.g., lead in batteries, steel in automobiles, or when they have a high value. Increasingly, however, small quantities of (rare) metals are used in complex products such as mobile phones (Graedel et al., 2011). In this context, Porter (2002) distinguishes between *economies of scale* in recycling, in which unit costs of recycling go down when the supply of waste material increases, and *diseconomies of scope*, in which unit costs of recycling go up when the number of different recyclable materials and applications increases. Next to these techno-economic factors, key drivers for recycling are the scarcity, costs, and volatility of prices of substitute primary materials, the cost of waste disposal, public pressure, and government regulations.

40.2.3 Circular Economy

During the last decade, the concept of the circular economy has gained momentum as an alternative to more conventional linear economic systems. The circular economy aims to close material loops to (a) decrease the use of virgin materials and (b) reduce waste generation. In this rationale, the achievement of both objectives will reduce the environmental impact of economic activity. For this reason, several countries and cities started to introduce policies to encourage the circular economy. For example, the European Commission adopted its first Circular Economy Action Plan in 2015 and implemented a new action plan in 2020. In the United States, the Environmental Protection Agency called for clear targets for a circular economy, including a 50% national target for recycling. There are many circular strategies to implement but recycling and material recovery is certainly a key strategy in these plans. For that reason, recycling is supported by means of a variety of economic and noneconomic instruments. The economic instruments are discussed in Section 40.4. Part of the noneconomic instruments used to stimulate the circular economy are standards and norms such as a minimum requirement on the recycled content in products, e.g. packaging material, binding minimum recycling rates, extended producer responsibility (Chapter 39—Extended producer responsibility) or ecodesign directives, e.g., facilitating repair or recyclability of products.

Although interest in the circular economy is growing, there is also a lack of consensus on the concept of the circular economy (Kirchherr et al., 2017). Because of this ambiguity, the role of recycling in future circular economies is also the subject of a dynamic debate. Numerous frameworks provide a hierarchy for the different circular economy strategies. For example, the 10R framework is one of the most nuanced and holistic hierarchies in which recycling ranks only on the 8th position, followed by energy recovery and remining. Instead, the hierarchy advocates for waste-avoiding strategies such as repair and repurpose. Reike et al. (2018) sharply observe how recycling is in the bottom of circular economy hierarchies but "at the top when it comes to frequency of use." In this context, empirical research by Ma et al. (2019) even warns for a lock-in effect: recycling efforts can increase resource consumption as they provide few incentives to refrain from the intensive use of raw materials in value chains.

40.2.4 Material Scarcity

Recycling of waste streams is not only interesting from an environmental point of view. Since the last decade, recycling has also gained interest because of strategic and economic reasons, because it can help to secure access to critical raw materials. Those critical raw materials are essential for modern-day economies because they are used in clean technology solutions and modern technology applications. As such, the materials are linked to all industries across all supply chain stages (European Commission, 2021). Nevertheless, many countries and regions struggle to secure reliable and unhindered access to a number of these raw materials. Their access to these materials is threatened by growing international competition for limited natural resources, or by the concentration of production in unstable countries (Buijs et al., 2012). In an attempt to address this challenge, the EU, for example, created a list of critical raw materials that is regularly updated. The number of critical raw materials on this list has steadily grown from 14 in 2011 to 30 in 2020. Recycling is explicitly mentioned as a means to make the EU less dependent on imports and more dependent on domestic raw materials.

40.2.5 International Trade

Recycling has become global business. International trade in recyclable resources is increasing at a fast rate. In 2016, international trade in waste exceeded 200 million tonnes (OECD. Stat, 2018). The most important incentive to international trade is the difference in recycling costs and benefits across countries. Van Beukering (2001) has shown that a number of trade theories predict trade flows of recyclable materials from the high-income countries to the low-income countries. The basic idea is that high-income countries are abundantly endowed with capital and recyclable or secondary resources and low-income countries with labor and primary resources. All other things being equal, this would lead to net export of recyclable resources from high- to low-income countries. This tendency would be enforced by income differences that cause low-income consumers to demand lower quality and therefore cheaper materials and products. Finally, laxer environmental standards of waste handling and disposal in low-income countries may also be an additional factor for trade. A metaanalysis by Kellenberg (2015) confirmed these drivers, adding the presence of organized crime to the issue of little environmental organization. Consequently, developing countries are importing a disproportionately large volume of the world's waste and scrap when viewed in relation to their income. Other empirical observations have supported these predictions. For example, Lyons et al. (2009) showed on the basis of detailed trade statistics that the United States is a large net-exporter of recyclable materials (such as iron and steel, paper, plastics, aluminum, copper, nickel, and zinc) to developing countries, especially Mexico and some Asian countries, including China.

An illuminating way to picture the patterns of international trade in recyclable materials is to make a distinction between the recycle recovery rate and the recycle utilization rate. The recycle recovery rate is the ratio between domestic collection of recyclable material and total consumption of the material (both of recycled and primary origin). The recycle utilization rate is the ratio between the consumption of the recycled material and the total production of the material. For net exporters of recyclable materials, the recovery rate exceeds the utilization rate. For net importers the reverse is true (Grace et al., 1978). Van Beukering (2001) showed that this relationship holds for international trade in recycled paper and that the difference between recovery rate and utilization rate in the North and South also increased between 1970 and 1997. Hence, international trade in waste plays an important role in bridging domestic gaps between demand and supply of recyclable materials. However, there is also some evidence that a part of international trade in waste has less welfare-enhancing consequences when it is in fact a cover-up of dumping waste in places with lax environmental regulations at the expense of human health and the environment. This is called the "waste haven" effect. Using bilateral waste trade data between 92 developed and developing countries, Kellenberg (2012) has indeed found robust statistical evidence of such a "waste haven" effect, even when controlling for relative productivities in recycling industries, effects of the Basel Convention on the Transboundary Movements of Hazardous Wastes and Their Disposal, and other potentially relevant covariates. Hence, while international trade offers a solution to the oversupply of recyclable material in high-income countries and the lack thereof in low-income countries, the present situation is still far away from the ideal of a "closed loop" industrial ecosystem (Lyons et al., 2009). In addition, recent research indicates that the waste haven hypothesis might not be applicable to all situations. E-waste, for example, is not shipped to less regulated countries with cheap labor by default. Instead, e-waste is organized in informal regional hubs, with many of the characteristics of other industrial clusters (Davis et al., 2019).

40.3 ENVIRONMENTAL AND SOCIAL COSTS AND BENEFITS

Recycling is widely assumed to be environmentally beneficial (Craighill and Powell, 1996). It slows down the exhaustion of scarce resources and limits the use of landfill space. Recycling, however, also generates significant environmental impacts through the collection, sorting and processing of materials into new products. Therefore, it is unclear when recycling is to be preferred to the use of virgin goods or when waste recovery should replace landfilling or incineration. Due to differences in environmental effects, studies about the desirability of recycling sometimes lead to opposite conclusions (Leach et al., 1997; Ackerman, 1997; Bartelings et al., 2005; Eshet et al., 2006; Eriksson and Baky, 2010).

40.3.1 Economic Valuation

In economics, environmental and social effects are generally defined as external effects. An external effect, or externality, is said to exist if an economic agent's decision has an influence on another agent's well-being or production possibilities and the former does not (properly) take these effects into account. Because of its unwanted nature, solid waste is often considered an externality. The extent to which solid waste actually is an externality depends on the method by which it is processed. Clearly, if waste is littered or illegally dumped, the externality will be substantially larger than if the waste is recycled or reused in a sustainable manner. For many other waste processing methods, however, this choice between recycling and alternative methods is less straightforward.

Policy makers generally use the level of externalities to determine the preferred ranking and mix of waste management options. Generally, landfilling is considered an environmentally less-favorable option than recycling. However, whether the level of externalities of landfilling always exceed those of recycling or incineration is unclear and requires the relevant external effects to be valued in economic terms. The main reason to express external effects in monetary values is that it allows for the comparison between private costs of various waste management options and the environmental and social costs and benefits related to these options. Economic valuation can express external effects in

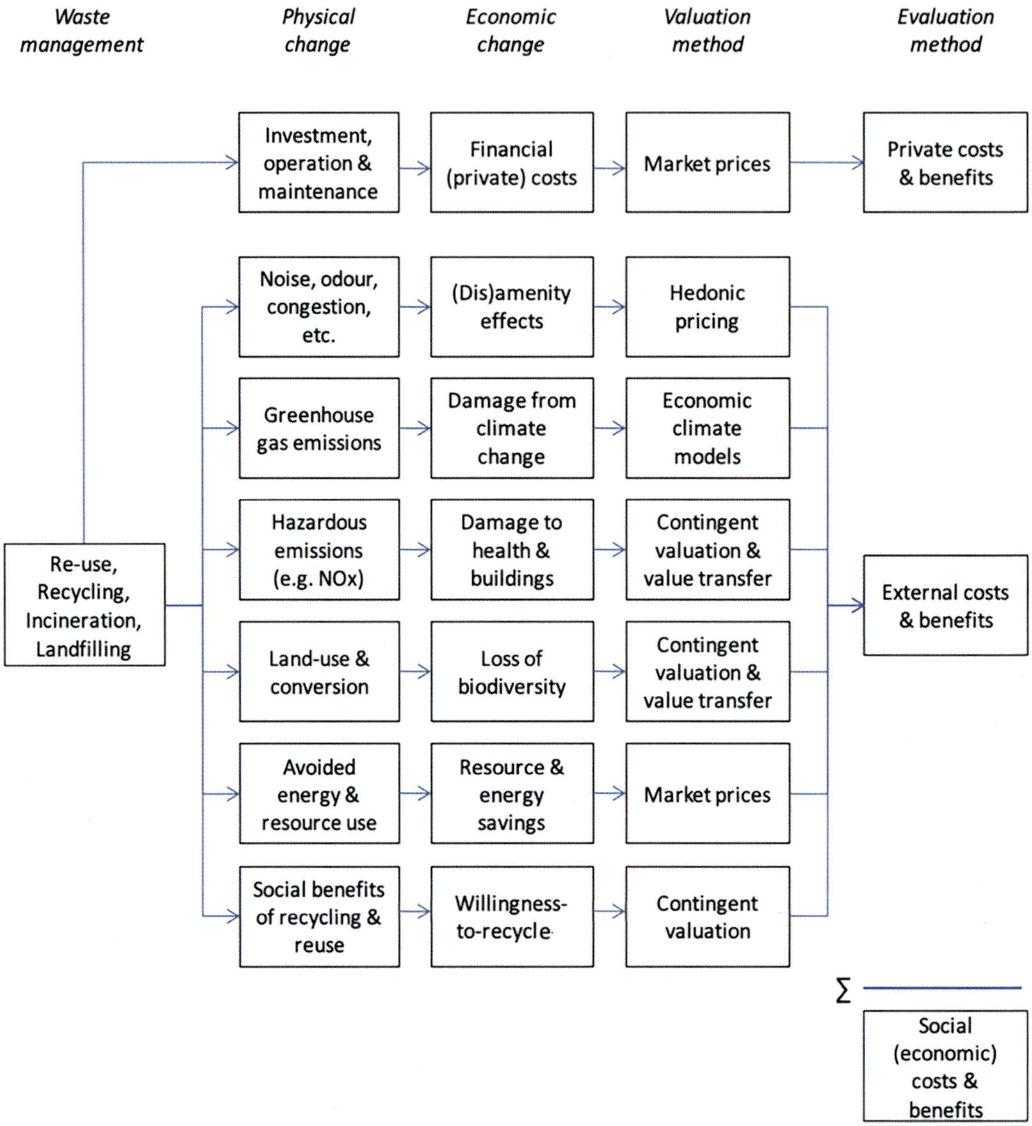

FIGURE 40.1 Economic valuation and the impact pathway approach.

monetary units. This approach facilitates the comparison of multiple environmental impacts and allows for a trade-off between the benefits and costs of many kinds of environmental improvement.

Ideally, economic valuation forms an integral part of the overall environmental assessment of waste management. As shown in Figure 40.1, an example of such an integrated assessment commonly applied in the waste management sector is the "impact pathway approach" (COWI, 2000). This approach proceeds sequentially through the lifecycle or pathway of an economic process, linking impacts to burdens, and subsequently valuing these impacts economically. First, overall emission levels and other external effects are determined in physical terms. Then, the impacts of these effects on economic activities and human wellbeing are assessed. Next, these impacts are translated into monetary values.

An advantage of the impact pathway approach is that it enables the comparison of the benefits of some environmental improvement in the waste sector with the costs to realize such an improvement. Since many studies have applied valuation methods, standard values can be derived for most pollutants or impacts. These allow translating emissions directly into costs bypassing the elaborate impact pathway approach. The valuation of external effects encounters various problems. First, because external effects, by definition, occur outside the market, market values for these effects are generally absent. Therefore special techniques are required for the estimation of external values. Second, external effects of recycling-related processes occur at various locations. When it is not possible to value these effects at each individual location, values estimated at one location need to be transferred to other locations. Therefore a disadvantage of this approach is that the complexity of economic valuation often leads to high degrees of uncertainty on the final results.

40.3.2 External Costs and Benefits

As shown in Figure 40.1, various categories of external effects are associated with waste management. This section illustrates how some of these external effects affect welfare and thus can be economically valued. First, pollution from recycling-related activities can have many impacts on human health, ranging from short periods of coughing to premature death. The valuation of human health impacts remains one of the most controversial aspects of any valuation study. Many reactions to the monetary valuation of these impacts are partly caused by the unfortunate choice of terminology such as the "Value of a Statistical Life" (VSL).

Second, climate change is an important impact category in the valuation of externalities related to waste management. Methane (CH_4) emissions from landfills contribute approximately 11% to the world's total methane emissions (GMI, 2011). Depending on the efficiency of the incinerator and the composition of the burned materials, "waste-to-energy" practices may avoid carbon dioxide (CO_2) emissions through supplying electricity to networks. Recycling is usually less energy intensive than primary production processes and therefore may reduce the impact on global warming. Comprehensive climate models that are linked to economic models calculate the costs of climate change. They include agricultural damage, increased morbidity and mortality, damage caused by sea level rise and by extreme events, and loss of species.

Third, disamenity effects of waste-related processes are likely to make up a significant share of the externalities caused. Landfill sites or incinerators generate substantial social costs for their neighboring population. Disamenity effects may occur in different forms. The trucks that transport the waste to and from the sites may cause noise externalities as well as congestion. The landfill site may emit noxious odors and create visual pollution. Increased health risk, or at least an increased perception of higher health risk, is caused for the people living in the vicinity of an incinerator or landfill.

Fourth, all industrial processes involve the (direct or indirect) use of material resources. These resources may be renewable (e.g., planted forests) or nonrenewable (e.g., fossil fuels, ores). When valuing the use of resources, one may ask whether the actual market price (i.e., the internal costs of one unit of the resource) adequately reflects the real cost to society of using that resource unit. There are two ways in which the social cost of resource use may deviate from its market price. There may be external effects in mining and processing the resource. The environmental impact of (nonrenewable) resource extraction tends to increase as the rate of depletion grows. Ideally, these effects should be incorporated in the analysis and valued accordingly. Moreover, the market price may lead to a higher rate of resource depletion than would be optimal, not taking into account the needs of future generations.

Fifth, recycling may also generate positive externalities as it caters to people's willingness to recycle. Environmental consciousness plays a crucial role in recycling activities (Ackerman, 1997; Jones et al., 2010). Without the voluntary participation of consumers in waste recovery programs, the level of recycling would be significantly less. From the purely private perspective of the consumer, landfilling or incineration of household waste in a distant site or plant might be the optimal way to eliminate waste residues. The increased awareness of pollution, however, can motivate consumers to play their part to alleviate the waste problem. Consumers may be interested in waste management for other reasons than purely money savings on their bills (Sterner and Bartelings, 1999). This "willingness to recycle" can be considered a positive externality.

40.3.3 Private and External Costs

The ultimate purpose of the impact pathway approach is to determine the real costs and benefits of recycling and other waste management options. The external costs of waste management have been the subject of a relatively large number of studies in recent years. Most of these studies focus only on one specific externality. Only a few studies attempt to aggregate a number of the most relevant environmental effects, thereby allowing for a fair comparison, internationally (COWI, 2000; Dijkgraaf and Vollebergh, 2004; Bartelings et al., 2005). By combined published sources, Kinnaman (2006) estimated the marginal cost of landfilling solid waste in Europe between US$5.39 and US$8.76 per ton (Kinnaman, 2006). Isely and Lowen (2007) estimated these costs at US$5.26 per ton in the United States. In a survey reviewing the economics literature on solid waste published since 2000, Kinnaman (2016) concluded that the external marginal cost of landfill disposal is rather small.

Illustrating the composition of the social costs of waste management alternatives, Table 40.1 summarizes the various categories of the private and external costs of landfilling and incineration (Bartelings et al., 2005). On the one hand, private costs are estimated taking into account displacement benefits. On the other hand, externalities are estimated accounting for climate change, other environmental pollution, land use effects, health effects, and disamenity effects.

TABLE 40.1 Private and external costs of incineration and landfilling 1 tonne of waste (€/ton).

Private and external effects	Incineration	Landfill
Gross private costs	*125*	*40*
Displaced costs—energy	*−21*	*−4*
Displaced costs—materials	*−3*	*0*
Net private costs	101	36
Health	*7.09*	*0.70*
Disamenity	*9.09*	*3.50*
Transport-related	*1.67*	*1.25*
Solid waste	*0.11*	*0.00*
Climate change	*0.11*	*4.21*
Other environmental pollution	*0.13*	*0.52*
Displaced effects—energy	*−7.63*	*−1.14*
Total external costs	10.57	9.04
Total social costs	111.57	45.04

Based on Bartelings et al. (2005) and Dijkgraaf and Vollebergh (2004).

Both external and private costs are surrounded by large uncertainties. Kinnaman (2009), for example, reports the marginal cost of operating a municipal recycling program at roughly US$120 per ton for the first ton recycled, yet these costs decrease with economies of scale by an estimated US$2.13 per 1000 tons recycled (Kinnaman, 2009). For external costs the uncertainties are even higher, depending on contextual (e.g., location) and methodological assumptions (e.g., discount rate). The best way to deal with such uncertainties is to conduct an extensive sensitivity analysis to evaluate whether the ranking of preferred waste management options changes with different methodological assumptions (Eriksson and Baky, 2010).

40.4 ECONOMIC INSTRUMENTS

For most products and materials, 100% recycling is neither feasible nor desirable from a social cost-benefit point of view. While there may be disagreement on the optimum level of recycling, many citizens and policymakers feel that current levels are still below that optimum. For certain categories of products and waste, targets are set to increase the recycling rates. Some of these even have a binding character. For example, in the EU minimum recycling rates have been determined for packaging waste, batteries, cars, and electr(*on*)ic equipment. Member States are obliged to achieve these targets by specified dates.

One way of stimulating people and companies to recycle more is by making it financially attractive for them to do so. Policy instruments that aim to provide such a financial incentive are called "economic" (or market-based) instruments. There is no clear distinction between the two terms, and they are often used interchangeably (e.g., in EEA, 2005). Possibly the term "economic instrument" refers more to the financial incentive that the instrument conveys, whereas "market-based" emphasizes the role of the market mechanism in achieving the environmental objective. In that sense, instruments such as ecolabeling could also be called "market based," whereas they do not provide financial incentives and therefore are not "economic instruments."

Economic instruments can be applied in many different ways and at many different points along the value chain from resource extraction to disposal. Basically, we can distinguish two types of economic instruments for recycling:

- those that create a financial disincentive for nonrecycling behavior (such as dumping waste or sending it to landfills and incinerators);
- those that create a financial incentive for recycling.

Taxes and charges (but also penalties on littering) are examples of the first category; subsidies and public facilities belong to the second. In practice, mixtures of incentives and disincentives are also common. Examples include, for instance, tradable landfill permits, differentiated tax rates, and deposit-refund systems. The latter are discussed separately in the next section. We will also address the government's own expenditures (public procurement), which can be used as an incentive for recycling as well.

40.4.1 Taxes and Charges

The distinction between a tax and a charge is related to the destination of the revenue. Tax revenues tend to accrue to the general public budget (although they may sometimes be "earmarked" for specific purposes), while the revenues from charges are spent on purposes that are directly related to the product or activity on which the charge is levied.

"The polluter pays" is a key principle in environmental policy. In the area of waste prevention and recycling, it can be applied in various ways. Taxes at the end of the chain, on landfilling and incineration, improve the competitive position of recycling *vis-à-vis* these disposal options. Most EU Member States nowadays have a tax on landfilling, with rates varying between € 5 and € 100 per tonne. Recent data can be found on https://www.cewep.eu/landfill-taxes-and-bans/. Taxes (or charges) on (potentially) waste-generating products make it more attractive to use less of them, to use them for a longer time, and/or to switch to alternatives that generate less waste. A classic example is the tax on plastic bags in Ireland. After its introduction in 2002, the use of plastic bags dropped by 90% (Convery et al., 2007). Taxes can also be applied at the beginning of the chain, i.e., on natural resources, but such taxes appear to be less effective in terms of promoting recycling (see, e.g., Söderholm, 2006).

Charges for waste collection and processing can also play a role in stimulating recycling. Many households still pay fixed charge rates for municipal waste collection. Making these rates variable, in other words related to the amount of waste offered or "pay-as-you-throw," can incite them to display a less wasteful and more pro-recycling behavior. Differentiating the rates, e.g., by applying a reduced or zero rate to separated types of waste, such as glass, biowaste, and paper, may further enhance this impact. Such differential and variable rate (DVR) based schemes have been shown to be very effective (see, e.g., OECD, 2006; Hogg et al., 2011), especially in the presence of a comprehensive system for the collection of segregated materials for recycling. In some countries, such as South-Korea, DVR schemes are already applied on a nationwide scale.

40.4.2 Subsidies and Public Facilities

Although subsidies in general conflict with the "polluter pays principle," they may still be useful policy instruments in situations where market imperfections (such as transaction and monitoring costs) preclude the use of "first best" tools such as waste disposal fees. For example, Dinan (1993) and Eichner (2005) conclude that the optimum instrument package should include subsidies on reuse or recycling. Likewise, Fullerton and Wu (1998) show that, if market signals cannot be corrected by means of disposal charges, welfare can be improved by subsidies for recycling or recyclability, in combination with product taxes.

In particular, positive incentives for recycling may play a role in cases where there is a high risk of illegal waste disposal practices, for instance because the cost of enforcement would be prohibitively high. An example is ship waste: there are no policemen at sea who might fine for waste dumping. Port reception facilities can make it more attractive for captains to refrain from dumping. Likewise, households can be encouraged to separate their waste by providing a well-organized system of collection facilities, such as glass containers. Clearly, the incentive does not need to be a financial one: recycling behavior can also be made attractive by reducing the time and effort involved.

40.4.3 Deposit-Refund Systems

Deposit-refund schemes (DRS) are basically a combination of two instruments: a tax on the purchase of a certain product, and a subsidy on the separate collection of the same product in its after-use stage. International experiences show that DRS can achieve very high return rates (Ten Brink et al., 2009; Ecorys et al., 2011; Van Beukering et al., 2009) and that they do lead to a reduction in litter (Hogg et al., 2011; Ecorys et al., 2011). On the other hand, the handling and administration costs can be substantial. DRS are widely applied in the area of drinks packaging, mainly on a voluntary basis. Some countries, however, apply mandatory DRS, e.g., Denmark, Germany, and a number of states in the United States. Other product categories to which some countries or states apply DRS include, for instance, lead acid batteries (United States, India).

At the theoretical level, several authors have studied the conditions under which DRS can be considered a useful policy instrument to attain maximum welfare, and how they should be designed. For example, Aalbers and Vollebergh (2008), using a general equilibrium model, found that DRS can provide the optimal incentives for recycling, landfilling, and dumping, taking into account the possibility of waste mixing and the efforts needed to keep waste streams separated. Calcott and Walls (2005) argue that a deposit-refund should be applied to *all* products, combined with a disposal fee (waste tax).

40.4.4 Public Procurement

Government procurement accounts for a substantial proportion of final consumption of raw materials. Within the EU, the aggregated purchases by all governments, for example, account for 27% of final wood demand (Brusselaers et al., 2017). Governments can use their importance as a consumer to entail a demand shock in favor of sustainable value chains. This can ensure both the recycled content in products purchased by governments, as well as the recyclability of the purchased products at the end of their lifetime. In this rationale, public procurement will also foster market dialog and cooperation between procurers and actors in the supply chain. Many sectors and materials are identified as potentially interesting for public, circular procurement, e.g., construction, waste and wastewater management, transportation, food, catering, furniture, and textiles (Alhola et al., 2019).

40.4.5 The Importance of Instrument Mix Design

Economic instruments can play an important role in making recycling a more attractive waste management option and in reducing the attractiveness of other, environmentally less-desirable options. Shaping a recycling policy, however, is not a matter of picking one or more instruments from a preexisting toolbox. In reality, a tailor-made instrument mix will have to be designed that takes into account specific situation characteristics and conditions. These include, among others, the price responsiveness of the producers and consumers involved (in economic terms: the price elasticity of supply and demand), as well as numerous social, cultural, political, and institutional factors. Designing economic instruments for recycling is therefore an art that requires skills and creativity, but which can also benefit from theoretical insights and practical experiences gained by researchers and practitioners.

40.5 CONCLUSION AND OUTLOOK

In the first part of this chapter on the economics of recycling, we explained the main economic drivers and developments of recycling. It shows how recycling takes place in a volatile, dynamic, and globalizing world, which complicates the projections in future developments in the recycling sector and the role of recycling in a "circular economy."

Next, the chapter demonstrates which private and external costs of recycling-related activities need to be taken into account and how these effects can be valued in economic terms. Internalization of externalities leads to better decisions on which waste management policies to pursue. However, a proper valuation study incorporates all externalities across the (international) life cycle. Therefore valuation remains difficult because of spatial and temporal variations and boundary issues.

Finally, the most important economic instruments to promote recycling are identified and explained. Economic instruments may have a larger role to play in waste policies of most countries in the world. Economic instruments are most effective in changing behavior by targeting the subject as directly as possible. Major immediate effects from economic instruments are limited because "low hanging fruit" has already been harvested. "Real world" conditions, e.g., trade effects, long-term contracts, and high costs of recycling, may limit the effectiveness of economic instruments. DVRs (differential and variable rates in waste collection charges) are an essential precondition for the effectiveness of many other economic instruments.

References

Aalbers, R.F.T., Vollebergh, H.R.J., 2008. An economic analysis of mixing wastes. Environ. Resour. Econ. 39 (3), 311–330.

Abrate, G., Ferraris, M., 2010. The Environmental Kuznets Curve in the Municipal Solid Waste Sector (Working Paper No. 1). Hermes, Fondazione Collegio Carlo Alberto, Moncalieri, Italy.

Ackerman, F., 1997. Why Do We Recycle? Markets, Values and Public Policy. Island Press, Washington, DC.

Alhola, K., Ryding, S.O., Salmenperä, H., et al., 2019. Exploiting the potential of public procurement: opportunities for circular economy. J. Ind. Ecol. 23 (1), 96–109.

Bartelings, H., van Beukering, P.J.H., Kuik, O., Linderhof, V., Oosterhuis, F., 2005. Effectiveness of Landfill Taxation. IVM Report R.-05/05, p. 154.

Brusselaers, J., Van Huylenbroeck, G., Buysse, J., 2017. Green public procurement of certified wood: spatial leverage effect and welfare implications. Ecol. Econ. 135, 91–102.

Buijs, B., Sievers, H., Tercero Espinoza, L.A., 2012. Limits to the critical raw materials approach. In: Proceedings of the Institution of Civil Engineers-Waste and Resource Management. ICE Publishing, pp. 201–208.

Calcott, P., Walls, M., 2005. Waste, recycling and "design for environment": roles for markets and policy instruments. Resour. Energy Econ. 27 (4), 287–305.

Convery, F., McDonnell, S., Ferreira, S., 2007. The most popular tax in Europe? Lessons from the Irish plastic bags levy. Environ. Resour. Econ. 38, 1–11.

COWI, 2000. A Study on the Economic Valuation of Environmental Externalities From Landfill Disposal and Incineration of Waste. European Commission DG Environment, Brussels. 43968-4.

Craighill, A.L., Powell, J.C., 1996. Lifecycle assessment and economic evaluation of recycling: a case study. Resour. Conserv. Recycl. 17, 75–96.

Davis, J.-M., Akese, G., Garb, Y., 2019. Beyond the pollution haven hypothesis: where and why do e-waste hubs emerge and what does this mean for policies and interventions? Geoforum 98, 36–45.

Dijkgraaf, E., Vollebergh, H., 2004. Burn or bury? A social cost comparison of final waste disposal methods. Ecol. Econ. 50, 233–247.

Dinan, T.M., 1993. Economic efficiency effects of alternative policies for reducing waste disposal. J. Environ. Econ. Manag. 25 (3), 242–256.

Ecorys, 2011. The role of market-based instruments in achieving a resource efficient economy. In: Report for the European Commission, DG Environment. Ecorys, Cambridge Econometrics and COWI, Rotterdam. October 2011.

EEA, 2005. Market-based instruments for environmental policy in Europe. Technical Report 8/2005. European Environment Agency, Copenhagen.

Eichner, T., 2005. Imperfect competition in the recycling industry. Metroeconomica 56 (1), 1–24.

Eriksson, O., Baky, A., 2010. Identification and testing of potential key parameters in system analysis of municipal solid waste management. Resour. Conserv. Recycl. 54 (12), 1095–1099.

Eshet, T., Ayalon, O., Shechter, M., 2006. Valuation of externalities of selected waste management alternatives: a comparative review and analysis. Resour. Conserv. Recycl. 46 (4), 335–364.

European Commission, 2021. Critical Raw Materials. Available at: https://ec.europa.eu/growth/sectors/raw-materials/specific-interest/critical_en.

EUROSTAT, 2021. In: EUROSTAT (Ed.), Waste Statistics. Luxembourg https://ec.europa.eu/eurostat/statistics-explained/index.php?title=Waste_statistics.

Ferrara, I., 2008. Waste generation and recycling. In: OECD Journal: General Papers. vol. 2008/2. OECD, Paris.

Fullerton, D., Wu, W., 1998. Policies for green design. J. Environ. Econ. Manag. 36 (2), 131–148.

GMI, 2011. Landfill Methane: Reducing Emissions, Advancing Recovery and Use Opportunities. Global Methane Initiative. September 2011.

Grace, R., Turner, K., Walter, I., 1978. Secondary materials and international trade. J. Environ. Econ. Manag. 5 (2), 172–186.

Graedel, T.E., Allwood, J., Birat, J.-P., Reck, B.K., Sibley, S.F., Sonnemann, G., Buchert, M., Hagelüken, C., 2011. Recycling rates of metals—a status report. In: A Report of the Working Group on the Global Metal Flows to the International Resource Panel. UNEP, Nairobi.

Hogg, D., Sherrington, C., Vergunst, T., 2011. A comparative study on economic instruments promoting waste prevention. In: Final Report to Bruxelles Environnement. Eunomia, Bristol. 8 November 2011.

ICFPA, 2021. 2020–2021 Sustainability Progress Report. International Council of Forest & Paper Associations, Brussels.

Isely, P., Lowen, A., 2007. Price and substitution in residential solid waste. Contemp. Econ. Policy 25 (3), 433–443.

Johnstone, N., Labonne, J., 2004. Generation of household solid waste in OECD countries: an empirical analysis using macroeconomic data. Land Econ. 80 (4), 529–538.

Jones, N., Evangelinos, K., Halvadakis, C.P., Losifides, T., Sophoulis, C.M., 2010. Social factors influencing perceptions and willingness to pay for a market-based policy aiming on solid waste management. Resour. Conserv. Recycl. 54 (9), 533–540.

Kellenberg, D., 2012. Trading wastes. J. Environ. Econ. Manag. 64 (1), 68–87.

Kellenberg, D., 2015. The economics of the international trade of waste. Ann. Rev. Resour. Econ. 7 (1), 109–125.

Kinnaman, T.C., 2006. Examining the justification for residential recycling. J. Econ. Perspect. 20 (4), 219–232.

Kinnaman, T.C., 2009. The cost of municipal curbside recycling and waste collection. Working Paper, Department of Economics, Bucknell University.

Kinnaman, T.C., 2016. Understanding the economics of waste: drivers, policies, and external costs. Int. Rev. Environ. Resour. Econ. 8 (3–4), 281–320.

Kirchherr, J., Reike, D., Hekkert, M., 2017. Conceptualizing the circular economy: an analysis of 114 definitions. Resour. Conserv. Recycl. 127, 221–232.

Leach, M.A., Bauen, A., Lucas, D.J., 1997. A system approach to materials flows in sustainable cities: a case study of paper. J. Environ. Plan. Manag. 40 (6), 705–723.

Lyons, D., Rice, M., Wachal, R., 2009. Circuits of scrap: closed loop industrial ecosystems and the geography of US international recyclable material flows 1995-2005. Geogr. J. 175 (4), 286–300.

Ma, B., Li, X., Jiang, Z., et al., 2019. Recycle more, waste more? When recycling efforts increase resource consumption. J. Clean. Prod. 206, 870–877.

Mazzanti, M., Zoboli, R., 2009. Municipal waste Kuznets curves: evidence on socio-economic drivers and policy effectiveness from the EU. Environ. Resour. Econ. 44, 203–230.

Nicolli, F., Mazzanti, M., Lafolla, V., 2012. Waste dynanmics, country heterogeneity and European environmental policy effectiveness. J. Environ. Policy Plan. 14 (4), 371–393.

OECD, 2006. Impacts of unit-based waste collection charges. ENV/EPOC/WGWPR(2005)10/FINAL, 15 May 2006, Organisation for Economic Co-operation and Development, Paris, https://doi.org/10.1787/oecd_papers-v6-art30-en.

OECD, 2020. Environment at a Glance 2020. OECD Publishing, Paris, https://doi.org/10.1787/4ea7d35f-en.

OECD, 2021. In: OECD (Ed.), Municipal Waste, Generation and Treatment. https://stats.oecd.org/Index.aspx?DataSetCode=TRADEENV_IND12.

OECD.Stat, 2018. Trade in Waste and Scrap. Retrieved September 17, 2021 from https://stats.oecd.org/Index.aspx?DataSetCode=TRADEENV_IND12.

Porter, R.C., 2002. The Economics of Waste. Resources for the Future, Washington, DC.

Reike, D., Vermeulen, W.J., Witjes, S., 2018. The circular economy: new or refurbished as CE 3.0?—Exploring controversies in the conceptualization of the circular economy through a focus on history and resource value retention options. Resour. Conserv. Recycl. 135, 246–264.

Schmidt, O., Hawkes, A., Gambhir, A., et al., 2017. The future cost of electrical energy storage based on experience rates. Nat. Energy 2 (8), 1–8.

Söderholm, P., 2006. Environmental taxation in the natural resource extraction sector: is it a good idea? Eur. Environ. 16 (4), 232–245.

Sterner, T., Bartelings, H., 1999. Household waste management in a Swedish municipality: determinants of waste disposal, recycling, and composting. Environ. Resour. Econ. 13, 473–491.

Ten Brink, P., Lutchman, I., Bassi, S., Speck, S., Sheavly, S., Register, K., Woolaway, C., 2009. Guidelines on the Use of Market-Based Instruments to Address the Problem of Marine Litter. Institute for European Environmental Policy (IEEP), Brussels, Belgium.

The World Bank, 2021. Trends in Solid Waste Management. Available at: https://datatopics.worldbank.org/what-a-waste/trends_in_solid_waste_management.html. (Accessed January 9).

Van Beukering, P.J.H., 2001. Recycling, International Trade and the Environment: An Empirical Analysis. Kluwer Academic Publishers, Dordrecht/Boston/London.

Van Beukering, P.J.H., Linderhof, V., Bartelings, H., Oosterhuis, F., 2009. Effectiveness of unit-based pricing of waste in the Netherlands. Waste Manag. 29 (11), 2892–2901.

Zaman, A.U., 2016. A comprehensive study of the environmental and economic benefits of resource recovery from global waste management systems. J. Clean. Prod. 124, 41–50.

CHAPTER

41

Economic policy instruments

Maarten Dubois[1] and Johan Eyckmans[2]

[1]Deloitte Consulting Belgium, Sustainability & Circular Economy, Brussels, Belgium [2]Center for Economics and Corporate Sustainability (CEDON), KU Leuven Brussels, Belgium

41.1 INTRODUCTION

Over the past two decades, the focus in waste management has shifted from solving the environmental impacts of waste disposal to managing material cycles in a sustainable way. Recycling materials from waste streams plays an important role in the new paradigm of sustainable material management, resource efficiency, and circular economy. In particular, it is argued that recycled materials can substitute for the extraction of exhaustible resources, thereby alleviating problems like material scarcity and dependency. In addition, they are believed to reduce environmental impacts related to mining, processing, and production of virgin materials (Damgaard et al., 2009; European Commission, 2011; Acuff and Kaffine, 2013). This chapter will focus on economic drivers and impediments to recycling. We will discuss how recycling incentives of consumers and producers are affected by policy instruments like recycled content standards, taxes on virgin materials, waste disposal fees, recycling subsidy mechanisms, extended producer responsibility (EPR or deposit-refund in the sequel), and recycling certificate schemes.

41.2 CRITERIA TO COMPARE POLICY INSTRUMENTS

Typically, in the environmental economics literature, policy instruments are compared with respect to their achievements on different criteria such as efficiency, effectiveness, equity, or political feasibility (see among others Bohm and Russell, 1985; Callan and Thomas, 2000 or Hanley et al., 2007 for a more elaborate discussion and definition of these criteria). This chapter will focus on three fundamental notions of efficiency. First, the idea of efficiency refers to whether a particular policy instrument is capable of achieving the welfare optimal allocation of resources (notion of *allocative efficiency*). This notion of efficiency includes internalization of all relevant environmental external costs and the allocation over time of nonrenewable resources, i.e., the extraction path of the resource through time. Second, does the policy instrument induce a cost-efficient allocation of efforts if actors are characterized by different costs for sorting and recycling (the idea of *cost* or *static efficiency*)? To achieve a given target at the lowest possible cost, efforts should be differentiated. In particular, actors with low costs should contribute more than their counterparts. Third, does the

Handbook of Recycling. https://doi.org/10.1016/B978-0-323-85514-3.00021-X

instrument foster *dynamic efficiency*, i.e., does it give incentives over time to improve recycling technologies and upstream green product design? This chapter reviews some of the most important contributions to the environmental and resource economics literature, focusing in particular on the interaction of different policy instruments and their effects on recycling behavior.

41.3 BASIC ENVIRONMENTAL POLICY INSTRUMENTS AIMED AT STIMULATING RECYCLING

Pigou (1920) was the first to advocate the use of externality taxes as an effective policy instrument to achieve efficiency in the previously mentioned sense. In the 1980s and 1990s, different authors started to apply this idea to solid waste management and recycling. For example, Bohm (1981) introduced the notion of a deposit-refund scheme, i.e., the combination of a product tax (deposit) and a collection/recycling subsidy (refund). Pearce and Turner (1992) suggested a packaging waste tax to internalize waste disposal externalities and litter costs. In Pearce and Turner (1993) a first systematic comparison between different so-called market-based instruments for solid waste management was presented. This was the starting point of a rich theoretical debate on the merits and drawbacks of different solid waste management policy instruments. Dinan (1993) used a stylized theoretical economic model to compare efficiency of three economic instruments that aim to internalize waste disposal externalities: a *tax on virgin materials*, a *deposit-refund* scheme (combination of a consumption tax and reuse subsidy), and a household *disposal fee*. The household disposal fee is a weight-based fee paid by households to collect waste. Externalities resulting from virgin materials extraction or production are not taken into account. In the model with perfect competition and with two goods that have different constant unit disposal costs (private and external cost), the deposit-refund scheme is shown to be the most efficient. The virgin material tax falls short because it does not allow differentiation between products that give rise to different waste management costs. In a similar way, a household collection fee does not allow differentiation between disposal costs for different products or material streams.

The article by Fullerton and Kinnaman (1995) is generally considered the seminal contribution in the environmental economics literature on the comparison of policy instruments for waste management and their effect on *illegal disposal*. They modeled consumption of a single consumption good that, at its end-of-life stage, can be recycled, disposed of in a controlled way, or disposed of in an illegal way. Illegal disposal is assumed to generate the most adverse environmental impact. The model takes into account the environmental externalities of virgin material extraction but abstracts from import/export flows (closed economy). Consumers are assumed to be identical utility maximizing subjects faced with a limited budget. Firms produce consumption goods using a constant return to scale technology (i.e., unit production costs are constant and independent of production volume) under perfect competition. The model shows that, due to the presence of illegal disposal, the social welfare maximizing outcome cannot be achieved by household *disposal fees* only. If the disposal fee were set at a level to correctly internalize all environmental externalities of disposal, the so-called Pigovian externality tax in environmental economics (see for instance Porter, 2002; Hanley et al., 2007 or Kolstad, 2011), households would face strong incentives to dispose of their waste in an illegal and uncontrolled way. If, instead, households were not charged for disposing their waste, they would not face the full social cost of their actions and tend to overconsume wasteful products. In this case, a *consumption tax* is required to compensate for the implicit consumption subsidy for waste-generating goods. Although an *extraction tax* can internalize the externalities of virgin materials extraction, it should not be used to internalize disposal externalities as long as a consumption tax is feasible. Overall, an important message of this article is that recycling behavior can be affected by several waste management policy instruments that can apply at different stages of materials' life cycle. The net effect is often difficult to predict because of the interacting instruments, externalities, and the possibility of illegal dumping.

Palmer et al. (1997) use a partial equilibrium economic model to assess cost efficiency of three different economic instruments aimed at reducing the amount of disposed waste: a *deposit refund*, an *advanced disposal fee*, and a *recycling subsidy*. The model uses mass balances for material flows used in consumption, recycling, and disposal. Only products with short life cycles are taken into account. To avoid illegal disposal, the household disposal fee is set equal to zero. A numerical simulation shows that the deposit-refund schemes are most desirable because the refund succeeds in stimulating the sorting and recycling in a cost-efficient way, while the consumption tax reduces incentives for consumption of waste-generating products. Since an advance disposal fee works in the same way as a consumption tax or deposit, it effectively reduces consumption. However, if sorting and recycling are not stimulated, the incentive to reduce waste disposal is smaller. Finally, the recycling subsidy in combination with low or even zero household fees for waste collection is shown to be least cost efficient. Although sorting and recycling of waste may be organized in an efficient way, consumption of waste-generating products is implicitly subsidized. In order to reduce total disposal, the subsidies for recycling should be increased to an excessive level.

Using a partial equilibrium model with identical consumers and perfect competition between identical firms, Palmer and Walls (1997) compared efficiency of *deposit-refund systems* with *recycled content standards*, i.e., standards that impose a fraction of recycled material in new products. Both firms and consumers can dispose of waste at constant unit disposal costs (including private and environmental disposal costs). The model uses a mass balance approach for virgin materials going in and disposed materials going out. The model confirms that deposit-refund systems can achieve the social welfare maximizing outcome. In contrast, recycled content standards fail in four aspects. First, standards are typically less efficient than taxes because they fail to internalize externalities of residual emissions. Consequently, if only a recycled content standard is used, overconsumption of wasteful products will prevail. Second, information requirements to impose the efficient level of recycled content are extremely high. It is unrealistic to expect that governments have this detailed information at their disposition. Third, it is well established in the literature (Kolstad, 2011; Tietenberg and Lewis, 2010) that uniform standards are not cost efficient because they do not discriminate between high- and low-cost technologies. Finally, static standards give insufficient incentives for adopting new and more efficient recycling technologies. The intuition for this is that once the standard is reached, there is no incentive for going further.

The review on *virgin material taxes* for aggregates of Söderholm (2011) discusses the theoretical efficiency and effectiveness of implemented taxes in Europe. Although implemented taxes have been effective in reducing the extraction of virgin materials, the author points to important weaknesses. First, virgin material tax rates are typically too low to stimulate recycling from waste products. Other instruments such as disposal taxes or deposit refunds are needed (see Dijkgraaf and Vollebergh (2004) and Dubois (2013) for a discussion of disposal taxes for landfilling and incineration in Western Europe).

Second, the tax can reduce the overall volume extracted but does not give incentives to reduce extraction externalities such as air and water pollution or waste generation. Third, the tax can induce imports or extraction of untaxed resources that also generate environmental effects. The author concludes that a virgin material tax is above all a cost-efficient second-best policy, because domestic extraction is typically easy to monitor.

41.4 EMPIRICAL EVALUATION OF THE EFFECTIVENESS OF POLICY INSTRUMENTS TO STIMULATE RECYCLING

A considerable number of empirical studies have been published in the peer-reviewed scientific literature evaluating ex post the effect on recycling and sorting behavior of introducing waste management policy instruments. Relatively few studies, however, have made use of robust evaluation methodologies that allow for causal inference of the effectiveness of the policies. We focus on the latter type of studies in what follows.

A first strand of literature focuses on municipal solid waste and how different pricing schemes (fixed contribution per year, pay-per-bag, pay-as-you-throw, ...) affect the volume of waste and sorting behavior. Several authors exploited natural experiments (changes in policies and pricing schemes) to detect behavioral responses in aggregated municipal waste collection. The challenge of this type of study is to correctly identify the causal effect of the policy change on the outcome variable because unobserved variables might affect the outcome variable leading to overestimation of the causal effect of the policies. Several approaches have been used to correct for such biases. For example, Kinnaman and Fullerton (2000) and Callan and Thomas (2006) used instrumental variable techniques to control for unobserved variables. But more recently, panel data techniques have become the methodology of choice: see among others, Linderhof et al. (2001), Dijkgraaf and Gradus (2004), Allers and Hoeben (2010), Usui and Takeuchi (2014), De Jaeger and Eyckmans (2015), and Sasao et al. (2021). According to the meta-analysis of Bel and Gradus (2016), a consistent finding of these studies is that weight-based pricing schemes have a stronger impact (i.e., are more price elastic) on waste generation than bag-based or bin-based pricing schemes. In addition, compostable waste seems more price sensitive than other waste fractions, probably because households that have access to a garden can do home composting.

A second strand of literature focuses on specific categories of waste like, for example, single-use plastic bags or beverage containers. Karasik et al. (2020) provide a detailed literature overview for different types of plastic waste policy instruments ranging from product bans to economic incentives like product taxes and deposit refund schemes. For example, Taylor and Villas-Boas (2016) used a difference-in-difference set-up to study the introduction of a single-use plastic bag prohibition combined with a minimum charge for reusable plastic bags or recycled paper bags in the states of Washington and California in the United States. Observational data from different supermarkets were used to compare consumers' carry-out bag use before and after the policy introduction in treated versus nontreated outlets. The researchers highlight that disposable bag policies should be combined with policies regulating the use of reusable bags that stores offer in lieu or alongside disposable bags to avoid negative spillover effects. A similar methodology was used by Berck et al. (2016) to evaluate the introduction of a tax on disposable (mostly lightweight PET) water bottles in the state of Washington in the United States. Their results show that demand for bottled water is rather price inelastic. A tax increase of between 6.5% and 9.5% led to a drop between 2.8% and 5.9% in bottled water consumption. This implies that only imposing a product tax on bottled water to reduce the use of disposable plastic bottles has limited impact on sales and, hence, limited environmental benefits.

41.5 INCENTIVES FOR UPSTREAM GREEN PRODUCT DESIGN

Fullerton and Wu (1998) assessed the efficiency of an extensive list of taxes and subsidies if households are not charged for waste collection. Their article also considers incentives for upstream *green product design*. In their model, three variables should be optimized to achieve a social welfare maximizing outcome: the amount of consumption, the amount of packaging, and the recyclability of packaging. Combinations of taxes on consumption, packaging recyclability, amount of packaging, and recycled inputs were investigated. Identical consumers and producers, perfect competition, and constant returns to scale in the production and waste collection industry were assumed. Firms adapt the attributes of their products (amount of packaging and recyclability) to the demand of consumers, i.e., consumers' willingness to pay for a product with specific packaging attributes has to be taken into account by the firms. Two combinations of instruments were shown to achieve an efficient allocation. First was a tax on consumption combined with a tax on the amount of packaging plus a subsidy for the level of green product design. Second was a tax on consumption—a deposit—combined with a tax on the amount of packaging plus a subsidy for recycled goods—a refund. If the level of green product design cannot be easily observed, the deposit-refund plan is shown to be preferable. Recently, Linderhof et al. (2019) built a numerical application of the Fullerton and Wu (1998) model to simulate the potential impact of introducing mandatory deposit refund schemes for small electric appliances and batteries in the Netherlands. Combining market data on quantities and prices with estimates from the literature for behavioral and technological parameters like demand and supply price elasticities, they were able to quantify the potential impact on recycling rates of different levels of deposit and refund fees.

Also Calcott and Walls (2000) investigated which economic instruments could provide incentives for green product design. They compared a household disposal fee with a deposit-refund plan when heterogeneous products are collected by competitive recycling firms. Transaction costs not only make it impossible to diversify the household collection fee per material stream, but also make subsidies directly to households impossible. These authors used recyclability as an attribute of products and used a mass balance approach so that all selectively collected items were recycled and used as production inputs. Perfect competition applies for producers. If pricing of household waste collection cannot be diversified by the type of waste, the household collection could never give appropriate incentives for green product design. In contrast, a scheme with a consumer tax (deposit) and product-specific subsidy to recycling firms (refund) can reach the second-best level of green product design and efficiency. Calcott and Walls (2005) also developed a more complicated model that introduces *transaction costs* into the recycling market for imperfect sorting by consumers. The level of green product design is defined by one variable that is unobservable to policymakers. They concluded that a second-best outcome could be achieved with a deposit-refund system or a combination of the deposit-refund system and household disposal fee.

An even more general analysis is presented by Bernard (2019) who analyzes a model in which a monopolist durable goods producer has to choose three "design for environment" parameters: energy intensity in the production phase, energy intensity in the consumption phase, and overall durability of the product. The results crucially depend on the cross effects between the different dimensions. When the design dimensions are competitive (in other words, if they are substitutes), targeted emission taxes can lead to unintended shifting of pollution to other phases of the products' life cycle. Cross effects between different design dimensions also play an important role for incentives for green design in EPR regulation in Huang et al. (2019); see also Chapter 39.

41.6 MULTIPRODUCT AND MIXED WASTE STREAMS

Aalbers and Vollebergh (2008) focused on the sorting effort required from households to separate waste into homogeneous streams that could easily be recycled. Their article generalizes the work of Fullerton and Kinnaman (1995) and Choe and Fraser (1999) for *multiproduct streams* that can be mixed in the waste stage by introducing the possibility of a refund or sorting subsidy. Green product design is not incorporated in the model. Consumers purchase portfolios of diversified products. Production, landfilling, and recycling are modeled in a straightforward way with perfect competition and constant returns to scale. The model confirms that deposit-refund schemes work efficiently if they differentiate between waste streams with different external effects. The need for differentiation grows when differences in environmental effects are larger and when the share of mixed disposal is larger. Indeed, a small fraction of hazardous waste can lead to contamination of large quantities of potentially nonhazardous waste. Therefore, policymakers should primarily focus on adequate separation of highly toxic wastes and easily recyclable wastes using differentiated deposit-refund schemes. Maximal extraction of these potentially contaminating waste streams allows for setting up an undifferentiated deposit-refund scheme that limits administrative costs for the remaining waste streams.

Acuff and Kaffine (2013) used the simulation model of Palmer and Walls (1997) but focused on recycling as a way to reduce externalities in production. For example, since the carbon emissions from recycling metals or plastics are significantly lower than the emissions of virgin materials, appropriate waste management policies can contribute to carbon emissions mitigation. Acuff and Kaffine (2013) showed that welfare gains from recycling materials can be significantly bigger than welfare gains from avoiding disposal. Although instruments can simultaneously give incentives to reduce disposal and stimulate recycling, the use of instruments significantly differs if multimaterial goods are considered. Whereas the externality of disposal is typically modeled as a homogeneous externality, the positive externalities of recycling materials are often heterogeneous across materials. These authors therefore conclude that deposit-refund schemes should be diversified per product or waste stream.

41.7 EPR AND RECYCLING CERTIFICATES

In recent decades, EPR has been increasingly implemented, especially in Europe as an instrument to reduce disposal and stimulate recycling. Although implemented EPR schemes resemble the deposit-refund scheme described in the economic literature, Dubois (2012) highlighted the difference. Because implemented EPR schemes typically contain static recycling targets, disposal externalities of nonrecycled waste fractions are not correctly internalized. As a result, incentives for green product design, prevention, and technological innovation are suboptimal. The author concluded that an additional excise tax is required to achieve the welfare maximizing outcome.

To deal with EPR legislation, producers typically establish and finance a collective producer responsibility organization (PRO) that takes the actions needed to achieve the legal collection and recycling targets. For packaging waste, however, the United Kingdom uses *Tradable Recycling Certificates (TRCs)* to implement EPR in its Packaging Recovery Notes scheme. In a TRC scheme, recyclers can obtain recycling certificates if they meet certain quality standards. Producers of consumption goods, on the other hand, are obliged to hand over recycling certificates to the environmental authorities in proportion to their production volume. The idea is that a certificate market will emerge in which producers buy certificates from recyclers. Ino (2011) used a partial equilibrium model inspired by Palmer and Walls (1997) to show that the current targets for waste collection should increase further to achieve the welfare maximizing level.

Matsueda and Nagase (2012) focused on *TRCs in combination with landfill taxes*. They showed that a landfill tax reduces the effectiveness of EPR targets when used in combination with TRCs. The comparison of TRC with other policy instruments highlights that TRCs are not always the most appropriate instrument (Dubois et al., 2013). EPR with a PRO, refunded taxes, and disposal taxes can induce similar effects while avoiding the risks associated with certificate trading, such as, for instance, price volatility. In contrast to disposal taxes, both TRCs and refunded taxes can strengthen international competitiveness of domestic waste treatment companies, which is why the sector often lobbies in favor of these instruments. In order to determine whether EPR is most efficient with a PRO or with TRC, more research is needed.

Recently, the incentives for "design for environment" have regained attention, in particular in the framework of EPR. It has been argued that modulated EPR fee structures, i.e., fees that vary according to more specific product characteristics than weight or volume, can provide much stronger incentives for green design; see Laubinger et al. (2021) for a review of examples and a classification for modulation of EPR fees. As many of the more sophisticated modulated fee schemes are still very recent, it is too early to judge their effectiveness. Several authors also stress the complexity of "design for environment" by means of EPR regulation. For example, Huang et al. (2019) show that there often is a trade-off between design for recycling (i.e., lowering recycling unit costs) and design for durability (i.e., lifetime extension that reduces the volume to be recycled). They also show that regulating by means of collection targets or via recycling targets can have opposing effects on the producers' design choices.

41.8 DURABLE GOODS

The works discussed so far do not consider explicitly the complications that arise with durable goods, i.e., goods that are long lived, such as cars, refrigerators, or washing machines. Runkel (2003) assessed the efficiency of introduction of EPR for *durable goods* under both perfect and imperfect competition. The model considers different vintages (i.e., generations) of products. Under perfect competition, the model confirms that implementation of EPR is desirable if illegal disposal and bounded consumer rationality are an issue. By internalizing the future cost of waste management, EPR not only reduces consumption of waste-generating products but also improves durability (defined as average lifetime of durable goods). However, under imperfect competition with high monopoly rents, the effect is ambiguous because EPR improves incentives for durability but may worsen welfare losses due to noncompetitive behavior by producers. Similarly ambiguous effects were described by Alev et al. (2020). They show that, in contrast to nondurable goods, higher recycling/collection targets in EPR systems for durable goods may lead to adverse environmental effects. In particular, producers may choose to prematurely recycle used products in order to meet EPR targets leading to lower reuse levels, higher production of new goods, and therefore higher negative environmental impacts.

Runkel's model was extended in Eichner and Runkel (2005) to assess the effect of *recyclability* rather than durability of durable consumer goods. The model indicates that the efficiency of EPR may differ in the short and long run. EPR gives incentives for increased recyclability, but at the same time this may lead to more heterogeneity of discarded goods in the short run. The gain in recyclability may therefore be reduced by the increase in heterogeneity. In the long run, however, vintages and recyclability evolve to a steady state, and EPR is once again shown to improve social welfare.

Shinkuma (2008) also focused on the role of EPR for durable goods and on incentives for *reuse* in particular. In a world without transaction costs and illegal disposal, an advance disposal fee is the least efficient instrument because it gives insufficient incentives to *repair* durable goods and the life cycle of products would be shorter than optimal. In a world with illegal disposal, disposal fees become difficult to implement, but the advance disposal fee with subsidies is shown to be a second-best solution.

41.9 IMPERFECT COMPETITION IN PRODUCT, RECYCLING, AND REMANUFACTURING MARKETS

Eichner and Pethig (2001) used a general equilibrium model including production, consumption, and waste management and focused on imperfect markets for industrial recycling and treatment of recycling residues. In contrast with other contributions, they assumed that illegal disposal of households was not an issue. They assessed how a variety of tax-subsidy combinations affect incentives for green design under a wide range of conditions. Fleckinger and Glachant (2010) assessed the interaction of EPR and market power in the production and waste management market. They compared an EPR scheme characterized by individual take-back obligation for producers with collective EPR schemes that allowed for collusion. Under imperfect competition in production and perfect competition for waste management, both individual and collective EPR schemes failed to achieve the welfare maximizing outcome. Under imperfect competition in the waste management sector, a collective EPR scheme leads to a superior outcome because *collusion of producers* increases the negotiation power of producers. This drives down the monopoly rents from the waste management sector and brings consumer welfare closer to social optimum. In the same vein, Bernard (2011) studied the incentives for remanufacturability faced by a duopoly of producers of an original component (i.e., the primary market) for a final good. It is assumed that during the lifetime of the durable good, the original component needs replacement and that all the replacement parts are remanufactured original components. In the secondary market of remanufactured replacement parts, the original component producers are assumed to compete with independent competitive remanufacturers. Consumers are willing to pay a premium for the remanufactured components supplied by the original manufacturers. In such a market setting, the original component producers face little incentive to increase remanufacturability of their original component, as this lowers the cost of the competing independent remanufactures in the aftermarket. But by colluding on remanufacturing quality, Bernard (2011) shows that the original component producers can increase their profits, consumer surplus, and environmental quality. Under some conditions, the regulator could use in this specific model collusion as a substitute for more stringent environmental policy (i.e., remanufacturing minimum standards).

In general, models with imperfectly competitive product or waste and material markets give rise to more ambiguous results, as they are very sensitive to modeling assumptions. An often-recurring finding, however, is that noncompetitive behavior can alleviate environmental externalities because a *monopolist* tends to restrain output to drive up prices (see also Kolstad, 2011; Requate, 2005). The final welfare impact depends on the relative importance of environmental externalities versus the loss of consumer welfare due to excessive pricing.

41.10 POLICY INSTRUMENTS IN AN INTERNATIONAL MARKET FOR WASTE AND MATERIALS

An important evolution in the waste market is the growth of European and *international trade* in waste (EEA, 2012; Kellenberg, 2012). National policymakers, however, still have the authority to determine local waste taxes. The variety of implemented waste policies in Europe is significant as highlighted by the surveys of national waste policies such as Bio IS (2012) and ACR+ (2013). In such a diversified environment, regional policymakers may be tempted to use environmental policy as a strategic instrument to promote domestic interests like protecting local jobs or tax revenues (Kennedy, 1994; Cassing and Kuhn, 2003). If revenues and employment from the recycling sector are important, countries may end up in a "race to the bottom" where countries gradually reduce disposal taxes or "race to the top" in terms of recycling targets (Dubois and Eyckmans, 2015). Anecdotal evidence suggests that the risk for a downward tax spiral is present in Europe. For example, Sweden and Norway first installed an incineration tax in an open border system but abolished it in 2010 due to international waste shipments (Bjorklund and Finnveden, 2007; Dubois, 2013). See also Bernard (2015) who shows that, under certain conditions, increasingly stringent waste regulation in the global North can lead to lower levels of reusability and higher illegal streams of nonreusable waste toward the global South, exacerbating the pollution haven hypothesis.

41.11 RECYCLING AND NONRENEWABLE RESOURCES IN A MACROECONOMIC PERSPECTIVE

The literature we reviewed so far did not consider explicitly the fact that recycling can alleviate the problem of *exhausting nonrenewable resources*. Smith (1972) was one of the first authors to extend the seminal model by Hotelling (1931) to investigate the impact of recycling on the extraction rate of nonrenewable resources, followed by, among others, Lusky (1975, 1976), Hoel (1978), André and Cerdá (2006) and more recently Sørensen (2017). There is also a growing literature that studies the impact of recycling on the growth rate of an economy (see, e.g., Di Vita, 2007; Pittel et al., 2010) in which the substitutability between recycled and virgin material plays a crucial role. Recently, Zhou and Smulders (2021) used a two-sector endogenous growth model to study the impact of closing the loop strategies in the form of a higher refurbishing rate of retired goods. Higher rates of refurbishing lead in their model to lower resource prices and therefore lower incentives to develop less resource-intensive products. In the long run, closing the loops might therefore lead to economic and environmental losses.

Finally, Seyhan et al. (2012) and Hoogmartens et al. (2018) presented numerical applications of the theoretical models for the case of phosphorus and sand extraction, respectively.

Although looming *material scarcity* is an often-used argument for promoting public policy intervention fostering recycling and, more in general, material efficiency (see, among others,

Allwood et al., 2011), many resource economists question this argument. According to the economics perspective, policy interventions should primarily aim at fixing market failures, such as environmental externalities, of which the social costs are reflected inadequately in unregulated market prices. According to Söderholm and Tilton (2012), scarcity is probably less of a concern, even in the long run, than environmental and technological externalities and information asymmetries. They also argued that remediation of these externalities should be done rather with implementation of cleverly designed policy mixes that incentivize firms, consumers, and recyclers rather than picking and supporting particular technology winners.

41.12 CONCLUSION

Recycling can help to reduce the impact of negative environmental externalities related to waste disposal and virgin material extraction and production. Given that environmental externalities are often underpriced in market economies, there remains ample scope for further stimulating recycling. Although the case for public intervention is clear, the remedy is often complicated and involves a clever mix of policy instruments to incentivize the different actors (producers, recyclers, consumers, etc.). A first concern is that charging consumers for disposal of their waste may lead to illegal waste dumping and even greater negative environmental impacts. A second concern relates to imperfect competition in different stages of the life cycle of goods and materials. Many examples are present in the literature in which strategic behavior by firms counteracts the intended environmental benefits of policy instruments. Combinations of taxes on consumption of waste-generating products and subsidies for recycling are often recommended. EPR schemes are particular examples of such approaches. In order to also have beneficial effects on "design for environment," it is recommended to modulate EPR fees in the function of environmental impacts of end-of-life waste. There is a substantial body of empirical literature that demonstrates the economic and environmental effectiveness of certain combinations of economic waste policy instruments, in particular for pricing municipal waste streams, single-use plastic bags, and beverage containers.

Regarding the capability of recycling to overcome material's scarcity, the environmental and resource economics literature is less univocal. Much seems to depend on the substitutability of recycled versus virgin inputs in production. If this substitution is easy, recycling can significantly contribute to a transition toward a more sustainable material management. Resource economists recommend stimulating recycling to reduce environmental externalities related to mining and production of virgin materials. They are, however, more skeptical about the argument that recycling should be stimulated for reasons of alleviating material scarcity and, in the long run, incentives for R&D into less material-intensive products should also be taken into account.

References

Aalbers, R., Vollebergh, H., 2008. An economic analysis of mixing wastes. Environ. Resour. Econ. 39, 311–330.

ACR+, 2013. Plastic Bags: An Inventory of Political Instruments, Brussels. http://www.acrplus.org/index.php/fr/2013-06-11-10-23-55/actualites-de-l-acr/316-plastic-bags-report. (Accessed 09.02.22).

Acuff, K., Kaffine, D.T., 2013. Greenhouse gas emissions, waste and recycling policy. J. Environ. Econ. Manage. 65, 74–86.

Alev, I., Agrawal, V.V., Atasu, A., 2020. Extended producer responsibility for durable products. Manuf. Serv. Oper. Manage. 22 (2), 364–382.

Allers, M.A., Hoeben, C., 2010. Effects of unit-based garbage pricing: a differences-in-differences approach. Environ. Resour. Econ. 45, 405–428.

Allwood, J.M., Ashby, M.F., Gutowski, T.G., Worrell, E., 2011. Material efficiency: a white paper. Resour. Conserv. Recycl. 55 (3), 362–381.

André, J.F., Cerdá, E., 2006. On the dynamics of recycling and natural resources. Environ. Resour. Econ. 33, 199–221.

Bel, G., Gradus, R., 2016. Effects of unit-based pricing on household waste collection demand: a meta-regression analysis. Resour. Energy Econ. 44, 169–182.

Berck, P., Moe-Lange, J., Stevens, A., Villas-Boas, S., 2016. Measuring consumer responses to a bottled water tax policy. Am. J. Agric. Econ. 98 (4), 981–996. https://doi.org/10.1093/ajae/aaw037.

Bernard, S., 2011. Remanufacturing. J. Environ. Econ. Manage. 62, 337–351.

Bernard, S., 2015. North–south trade in reusable goods: green design meets illegal shipments of waste. J. Environ. Econ. Manage. 69, 22–35.

Bernard, S., 2019. Multidimensional green product design. Environ. Resour. Econ. 72, 1183–1202.

Bio IS, 2012. Use of economic instruments and waste management performances. Report for the European Commission DG ENV. Bio Intelligence Service, Paris.

Bjorklund, A., Finnveden, G., 2007. Life cycle assessment of a national policy proposal—the case of a Swedish waste incineration tax. Waste Manage. 27, 1046–1058.

Bohm, P., 1981. Deposit-Refund Systems: Theory and Applications to Environmental, Conservation, and Consumer Policy. Johns Hopkins University Press, Baltimore, p. 175.

Bohm, P., Russell, C.S., 1985. Comparative analysis of alternative policy instruments. In: Kneese, A.V., Sweeney, J.L. (Eds.), Handbook of Natural Resource and Energy Economics. vol. 1. Elsevier, Amsterdam, pp. 395–460.

Calcott, P., Walls, M., 2000. Can downstream waste disposal policies encourage upstream 'design for environment'? Am. Econ. Rev. 90, 233–237.

Calcott, P., Walls, M., 2005. Waste, recycling and 'design for environment': roles for markets and policy instruments. Resour. Energy Econ. 27, 287–305.

Callan, S.J., Thomas, J.M., 2000. Environmental Economics and Management: Theory, Policy and Applications. The Dryden Press, Harcourt College Publishers, p. 708.

Callan, S.J., Thomas, J.M., 2006. Analyzing demand for disposal and recycling services: a systems approach. East. Econ. J. 32, 221–240.

Cassing, J., Kuhn, T., 2003. Strategic environmental policies when waste products are tradable. Rev. Int. Econ. 11 (3), 495–511.

Choe, C., Fraser, I., 1999. An economic analysis of household waste management. J. Environ. Econ. Manage. 38, 234–246.

Damgaard, A., Larsen, A.W., Christensen, T.H., 2009. Recycling of metals: accounting of greenhouse gases and global warming conditions. Waste Manage. Res. 27, 773–780.

De Jaeger, S., Eyckmans, J., 2015. From pay-per-bag to pay-per-kg: the case of Flanders revisited. Waste Manage. Res. 33 (12), 1103–1111.

Di Vita, G., 2007. Exhaustible resources and secondary materials: a macroeconomic analysis. Ecol. Econ. 63, 138–148.

Dijkgraaf, E., Gradus, R.H.J.M., 2004. Cost savings in unit-based pricing of household waste, the case of The Netherlands. Resour. Energy Econ. 26, 353–371.

Dijkgraaf, E., Vollebergh, H., 2004. Burn or bury? A social cost comparison of final waste disposal methods. Ecol. Econ. 50, 233–247.

Dinan, T.M., 1993. Economic efficiency effects of alternative policies for reducing waste disposal. J. Environ. Econ. Manage. 25, 242–256.

Dubois, M., 2012. Extended producer responsibility for consumer waste: the gap between economic theory and implementation. Waste Manage. Res. 30 (9 Suppl), 36–42.

Dubois, M., 2013. Towards a coherent European approach for taxation of combustible waste. Waste Manage. 33, 1776–1783.

Dubois, M., Eyckmans, J., 2015. Efficient waste management policies and strategic behavior with open borders. Environ. Resour. Econ. 62, 907–923.

Dubois, M., Hoogmartens, R., Van Passel, S., Vanderreydt, I., 2013. Opportunities and drawbacks of tradable recycling certificates. In: Proceedings of the 14 International Waste Management and Landfill Symposium, 1–4 October, Sardinia.

EEA, 2012. Movements of Waste Across the EU's Internal and External Borders. European Environment Agency, 2012/7, Copenhagen.

Eichner, T., Pethig, R., 2001. Product design and efficient management of recycling and waste treatment. J. Environ. Econ. Manage. 41, 109–134.

Eichner, T., Runkel, M., 2005. Efficient policies for green design in a vintage durable good model. Environ. Resour. Econ. 30, 259–278.

European Commission, 2011. Roadmap to a Resource Efficient Europe. COM. 571.

Fleckinger, P., Glachant, M., 2010. The organization of extended producer responsibility in waste policy with product differentiation. J. Environ. Econ. Manage. 59, 57–66.

Fullerton, D., Kinnaman, T.C., 1995. Garbage, recycling and illicit burning or dumping. J. Environ. Econ. Manage. 29, 78–91.

Fullerton, D., Wu, W., 1998. Policies for green design. J. Environ. Econ. Manage. 36, 131–148.

Hanley, N., Shogren, J., White, B., 2007. Environmental Economics in Theory and Practice, second ed. Palgrave Macmillan, Basingstoke, Hampshire, UK.

Hoel, M., 1978. Resource extraction and recycling with environmental costs. J. Environ. Econ. Manage. 5, 220–235.

Hoogmartens, R., Eyckmans, J., Van Passel, S., 2018. A Hotelling model for the circular economy including recycling, substitution and waste accumulation. Resour. Conserv. Recycl. 128, 98–109.

Hotelling, H., 1931. The economics of exhaustible resources. J. Polit. Econ. 39 (2), 137–175.

Huang, X. (Natalie), Atasu, A., Toktay, L.B., 2019. Design implications of extended producer responsibility for durable products. Manag. Sci. 65 (6), 2573–2590. https://doi.org/10.1287/mnsc.2018.3072.

Ino, H., 2011. Optimal environmental policy for waste disposal and recycling when firms are not compliant. J. Environ. Econ. Manage. 62, 290–308.

Karasik, R., Vegh, T., Diana, Z., Bering, J., Caldas, J., Pickle, A., Rittschof, D., Virdin, J., 2020. 20 Years of Government Responses to the Global Plastic Pollution Problem: The Plastics Policy Inventory. NI X 20-05. Duke University, Durham, NC.

Kellenberg, D., 2012. Trading wastes. J. Environ. Econ. Manage. 64, 68–87.

Kennedy, P.W., 1994. Equilibrium pollution taxes in open economies with imperfect competition. J. Environ. Econ. Manage. 27, 49–63.

Kinnaman, T.C., Fullerton, D., 2000. Garbage and recycling with endogenous local policy. J. Urban Econ. 48, 419–442.

Kolstad, C.D., 2011. Intermediate Environmental Economics. Oxford University Press, Oxford.

Laubinger, F., Brown, A., Dubois, M., Börkey, P., 2021. Modulated fees for extended producer responsibility schemes (EPR). Environment Working Paper No. 184. OECD, Paris.

Linderhof, V., Kooreman, P., Allers, M., Wiersma, D., 2001. Weight-based pricing in the collection of household waste: the Oostzaan case. Resour. Energy Econ. 23, 359–371.

Linderhof, V., Oosterhuis, F.H., van Beukering, P.J.H., Bartelings, H., 2019. Effectiveness of deposit-refund systems for household waste in the Netherlands: applying a partial equilibrium model. J. Environ. Manage. 232, 842–850.

Lusky, R., 1975. Optimal taxation policies for conservation and recycling. J. Econ. Theory 11, 315–328.

Lusky, R., 1976. A model of recycling and pollution control. Can. J. Econ. 9, 91–101.

Matsueda, N., Nagase, Y., 2012. An economic analysis of the packaging waste recovery system in the UK. Resour. Energy Econ. 34, 669–679.

Palmer, K., Walls, M., 1997. Optimal policies for solid waste disposal: taxes, subsidies and standards. J. Public Econ. 65, 193–205.

Palmer, K., Sigman, H., Walls, M., 1997. The cost of reducing municipal solid waste. J. Environ. Econ. Manage. 33, 128–150.

Pearce, D., Turner, R.K., 1992. Packaging waste and the polluter pays principle – a taxation solution. J. Environ. Plan. Manage. 35, 5–15.

Pearce, D., Turner, R.K., 1993. Market-based approaches to solid waste management. Resour. Conserv. Recycl. 8, 63–90.

Pigou, A.C., 1920. The Economics of Welfare, first ed. MacMillan, London.

Pittel, K., Amigues, J.-P., Kuhn, T., 2010. Recycling under a material balance constraint. Resour. Energy Econ. 32, 379–394.

Porter, R., 2002. The Economics of Waste. Resources for the Future, Washington.

Requate, T., 2005. Environmental Policy under Imperfect Competition—A Survey. Economics working paper, Christian-Albrechts-Universität Kiel, Department of Economics, No. 2005-12 http://econstor.eu/bitstream/10419/22000/1/EWP-2005-12.pdf. (Accessed 09.02.22).

Runkel, M., 2003. Product durability and extended producer responsibility in solid waste management. Environ. Resour. Econ. 24, 161–182.

Sasao, T., De Jaeger, S., De Weerdt, L., 2021. Does weight-based pricing for municipal waste collection contribute to waste reduction? A dynamic panel analysis in Flanders. Waste Manage. 128, 132–141.

Seyhan, D., Weikard, H.-P., van Ierland, E., 2012. An economic model of long-term phosphorus extraction and recycling. Resour. Conserv. Recycl. 61, 103–108.

Shinkuma, T., 2008. Reconsideration of an advance disposal fee policy for end-of-life durable goods. J. Environ. Econ. Manage. 53, 110–121.

Smith, V.L., 1972. Dynamics of waste accumulation: disposal versus recycling. Q. J. Econ. LXXXVI, 600–616.

Söderholm, P., 2011. Taxing virgin natural resources: lessons from aggregates taxation in Europe. Resour. Conserv. Recycl. 55, 911–922.

Söderholm, P., Tilton, J.E., 2012. Material efficiency: an economic perspective. Resour. Conserv. Recycl. 61, 75–82.

Sørensen, P.B., 2017. From the linear economy to the circular economy: a basic model. FinanzArchiv 74, 71–87.

Taylor, R.L., Villas-Boas, S.B., 2016. Bans vs. fees: disposable carryout bag policies and bag usage. Appl. Econ. Perspect. Policy 38 (2), 351–372. https://doi.org/10.1093/aepp/ppv025.

Tietenberg, T., Lewis, L., 2010. Environmental Economics and Policy, sixth ed. Pearson Addison-Wesley, Boston.

Usui, T., Takeuchi, K., 2014. Evaluating unit-based pricing of residential solid waste: a panel data analysis. Environ. Resour. Econ. 58, 245–271.

Zhou, S., Smulders, S., 2021. Closing the loop in a circular economy: saving resources or suffocating innovations? Eur. Econ. Rev. 139, 103857.

CHAPTER

42

Economic aspects of metal recycling

Christian Hagelüken[1] *and Christina Meskers*[2]

[1]Director Government Affairs (Retired), Umicore, Hanau, Germany [2]Department of Industrial Eco-systems, SINTEF Manufacturing, Trondheim, Norway

42.1 INTRODUCTION

Recycling is an essential part of the circular economy. It reduces the consumption of resources and increases resource efficiency by recovering them for further use. Advanced circular economy systems and sophisticated recycling technologies are the backbone needed for this to occur. For effective recycling, seven conditions must be met (Hagelüken, 2012):

- material combinations are technically recyclable
- relevant components are accessible
- economic viability is sufficient
- effective collection mechanisms are in place
- products enter the recycling chain and remain there until final recovery
- technical and organizational setup of the recycling chain is optimal
- sufficient processing capacity is present to recycle the materials available.

This chapter focuses on the microeconomic angle, explaining the economic viablity of recycling using precious metal-containing products as case studies. It complements the chapters in Part 4 of this book, by focusing on the interaction between product, process, and system and business environment perspectives.

Precious metals are valuable by definition, as their prices show (Figure 42.1). Gold (Au) and silver (Ag) have been used since ancient times, while platinum group metals (PGMs) have been found in a wide range of applications for decades. Accordingly, recycling technologies are very mature and effective, and high-yield recycling processes are in place (Chapter 25—Precious and technology metals). The urban mine is much richer than primary resources: motherboards from computers, tablets, etc. contain 100–150 g Au/t, and automotive catalyst ceramic contains 2–3 kg PGMs/t. There is thus a concentration advantage of factor 20 and more. These products' various components are treated in integrated metal refineries, and over 95% process recovery yields are achieved (Hagelüken and Goldmann, 2022; Hagelüken and Meskers, 2010).

One could conclude that the economic and technical prerequisites for successful recycling are excellent. However, the actual end-of-life (EoL) recycling rates over the whole life cycle differ hugely between product types. For jewelry and chemical process catalysts, the EoL recycling rates are well over 90%: that is, over 90% of the precious metals originally used in the products is recovered when the products reach the end of the recycling chain. This results

Handbook of Recycling. https://doi.org/10.1016/B978-0-323-85514-3.00049-X

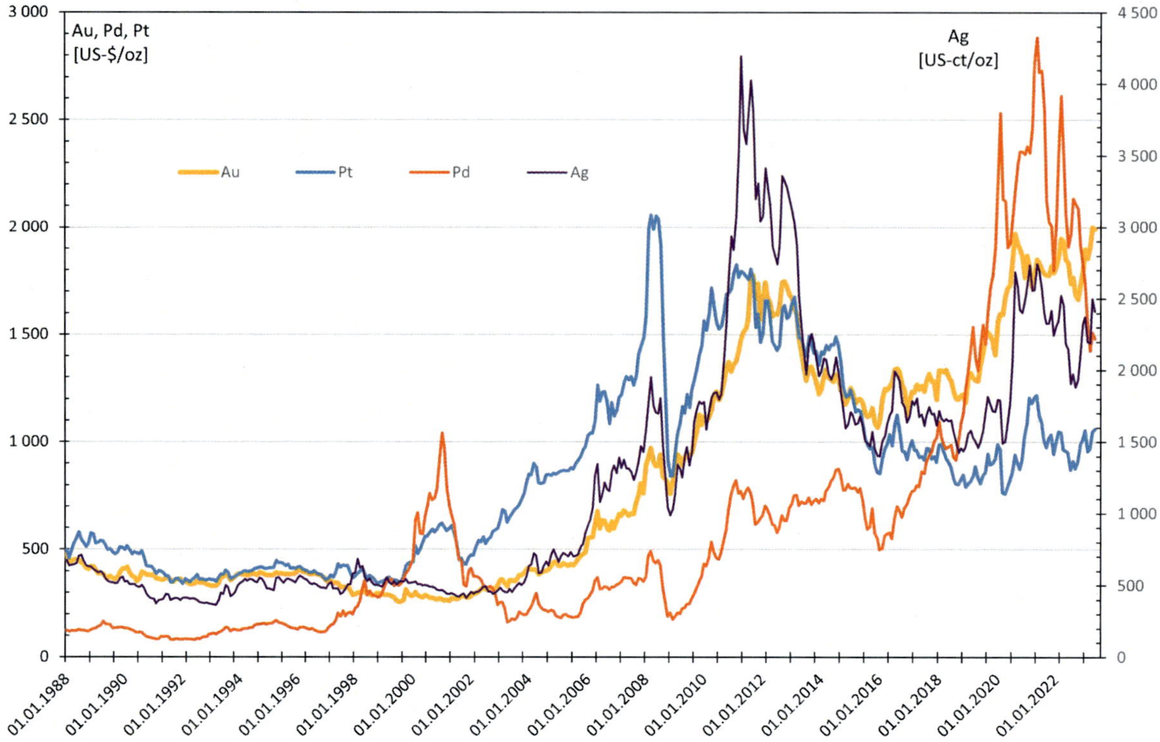

FIGURE 42.1 Evolution of precious metal prices (in USD per troy ounce) over the period 1988–2023. Metal price volatility impacts the recoverable value. *Data from: LBMA (2023).*

in a nearly closed metals loop. In contrast, for automotive catalysts, whose physical characteristics are similar to process catalysts, the EoL recycling rates are only ~60% (global average), and for electronic devices they are only ~25% (Hagelüken and Goldmann, 2022). So, what are the economic drivers or hurdles for recycling and what other factors are relevant?

For starters, recycling is never a single step but *always* a chain of processes (Figure 42.2). It starts with collection, followed by preprocessing, which is a manual (dismantling) or mechanical disintegration of products into certain fractions for further treatment and recovery. In the case of metal recycling, the final step is the chemical-metallurgical end-processing to generate pure metals or metal salts.

The economic actors in each step of the recycling chain make decisions based on factors such as business relationship; metal prices; and input characteristics, including recoverable metal content and requirements linked to policy, product or output fraction, process performance, etc. These interact and impact recycling economics on a company level, and thus affect the recycling chain efficiency and the availability of recycled materials for society. To understand the differences between products/scrap types, we need to look at these. The economic impact of malicious practices on bona fide recyclers and society is also discussed.

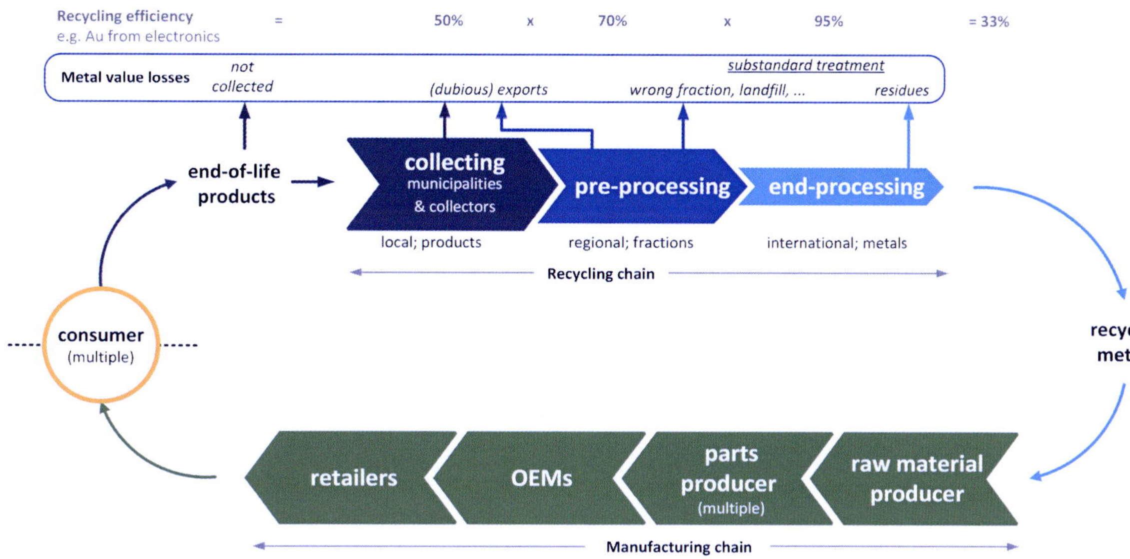

FIGURE 42.2 The recycling chain with causes for metal value losses and the resulting recycling (chain) efficiency, or physical circularity rate (Ressourcenkommission am Umweltbundesamt (KRU), 2023), for gold from electronics (top part of figure). Recycled metals flow back into the manufacturing chain (bottom part of figure). The product user or "consumer" plays an essential role in closing the materials loop. Reuse, repair, and remanufacturing activities are included in "consumer." *Based on: Hagelüken and Meskers (2010).*

42.2 GENERAL PRINCIPLES OF RECYCLING ECONOMICS

42.2.1 Macro- and Microeconomic Perspectives

When considering the economic viability of recycling, we need to distinguish between the microeconomic and the macroeconomic level, as shown in Figure 42.3. To meet the challenges, it is necessary to zoom in and out seamlessly between the levels.

At the **macroeconomic level**, the interactions take place under a systemic and societal perspective. This level includes interactions between actors and stakeholders in different value-chain steps and between value chains. Furthermore, societal benefits and externalized costs caused by inappropriate waste management, policy decisions, and systemic trends appear here. Especially in circular business models, these interactions, the value created, and the impact on the microeconomic level need to be explicitly considered (Brown et al., 2021).

The benefits of state-of-the-art recycling on the macroeconomic level are, among others:

- Increased resource conservation: by complementing the mine supply of raw materials.
- Increased supply security and diversification: accessing metals from local waste sources instead of from countries that may have lower standards for environment, social, and governance (ESG) or bear geopolitical risks.
- Improved responsible sourcing from clean supply chains. Materials can be traced back to compliant recycling plants and don't bear the risk of being derived from substandard mining conditions (unsafe work conditions, child labor, etc.).

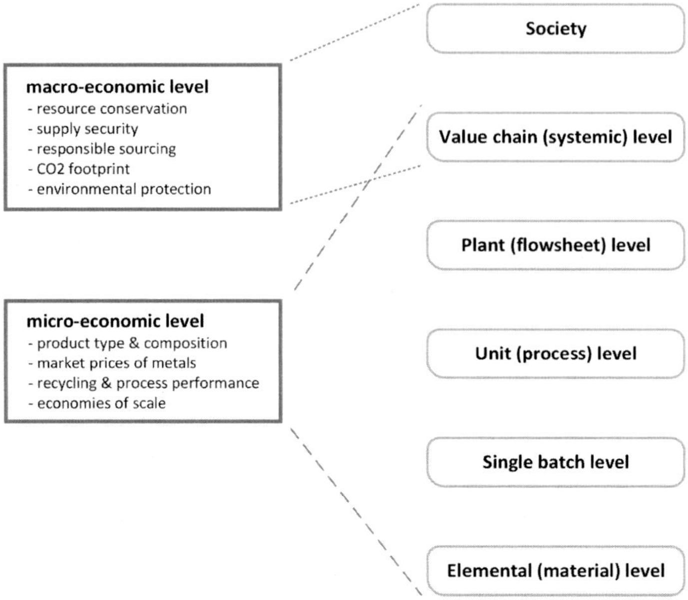

FIGURE 42.3 Factors impacting the economic viability of recycling on microeconomic and macroeconomic levels. *Based on: Hagelüken and Goldmann (2022).*

- Reduced environmental footprint of materials: recycled materials usually have lower environmental impacts (including CO_2 footprint) than primary materials.
- Avoidance of environmental damage and societal costs: landfill or substandard recycling treatment cause pollution and land use, resulting in damage and societal costs.
- Increased employment and local economic activity: jobs requiring different skills at various levels are created in each step of the recycling chain.

These benefits must be included in the economic equation and connected to the microlevel to be realized. Hence an appropriate legal framework is needed, using tools such as (extended) producer obligations, fees, collection and recycling targets, etc. Additionally, external costs should be internalized, e.g., through CO_2 pricing. For details see Chapter 38: Recycling, law and the circular economy; Chapter 39: Extended producer responsibility; Chapter 40: Economic drivers and valuation, and Chapter 41: Economic policy instruments.

The microeconomic level applies to companies. Different levels can be distinguished (Figure 42.3). At all levels the material quality and losses are influenced and determined. When zooming out, the focus changes to (1) the key factors that determine the performance on the level(s) below, and (2) the performance on the plant and enterprise level above, such as finances and long-term planning.

On a microeconomic level the key factors are:

- Product type, composition and lifetime: which recyclable materials are present, in which concentrations, in which components, and how these are connected to each other. The fundamentals are covered in Chapters 2 and 5, and the details in Part 2 of this book. This is the *product perspective* in recycling economics.

- Process performance and resulting plant performance: the material and element yields and losses, throughput time, output quality, energy and water efficiency, environmental and safety performance, waste, residue and emission quantities; for the chosen technologies, and for specific products, scrap fractions, and materials. This is the *process perspective* in recycling economics.
- Market prices of recyclable materials, and of the energy, water, and reagents needed for the recycling processes, as well as labor costs.
- Product availability and access: which EoL products and scrap quantities are available, where and when, to create economies of scale. What are the quality requirements for the outputs, and the market sizes and locations.

These factors are highly distributed in nature and vary over time, and the impact on recycling economics is often underestimated. Each aspect of the microeconomic factors can become a cost driver or a value creator.

42.2.2 Material Value Losses

Both economic perspectives come together when considering the material value losses along the recycling chain. Losses occur at each step (Figure 42.2). The lower the performance of recycling processes (i.e., their quality), the higher are the losses. For an efficient recycling chain, it is thus insufficient to have highly efficient (metallurgical) processes as the final step; efficiency, in the sense of high maintained value, is needed along the entire recycling chain. This total chain efficiency or physical circularity rate is determined by the weakest link in this chain (usually collection) and is derived by multiplying the efficiencies of the individual steps (Ressourcenkommission am Umweltbundesamt (KRU) (2023)). For example, gold recycling from electronics: with a 50% collection efficiency for electronics, a 70% preprocessing efficiency for gold, and a >95% end-processing efficiency for gold, the overall gold recycling efficiency would only be at 33% (Figure 42.2). This means that roughly two-thirds of the gold value is lost. This is close to today's reality, even in Europe.

42.2.3 The Value and Price of Scrap

The material (metal) value losses along the recycling chain impact the waste or scrap price and whether the recycling chain results in a positive or negative bottom line.

To start with the latter, the price for scrap—and principally for any kind of waste—can be determined by calculating the recoverable material value, deducting from this the total recycling chain costs (including recyclers return), and including a recycling fee if needed (Figure 42.4). This calculation can be done for the recycling chain and for individual economic actors. The *recoverable material value* is the sum of the individual material contents, their achievable recycling yields, and their respective market prices. The *recycling chain costs* include:

- the costs of the recycling processes themselves, both operational expenses (OPEX) and capital expenses (CAPEX),
- the costs for environment, health, and safety (EHS) performance,
- costs associated with shipping of the scrap (input) and output materials, and
- administrative costs.

Recycling will not happen if the total costs are higher than the recoverable value for a longer period and when the gap cannot be bridged by a recycling fee or other legislative measures. The recycling fee can be imposed by legislation or can be part of the business model to cover externalized costs.

Three scenarios are possible depending on how the metal losses, recoverable material value, and the recycling costs relate to each other (Figure 42.5). For *positive value recycling* the recovered metal value at the end is

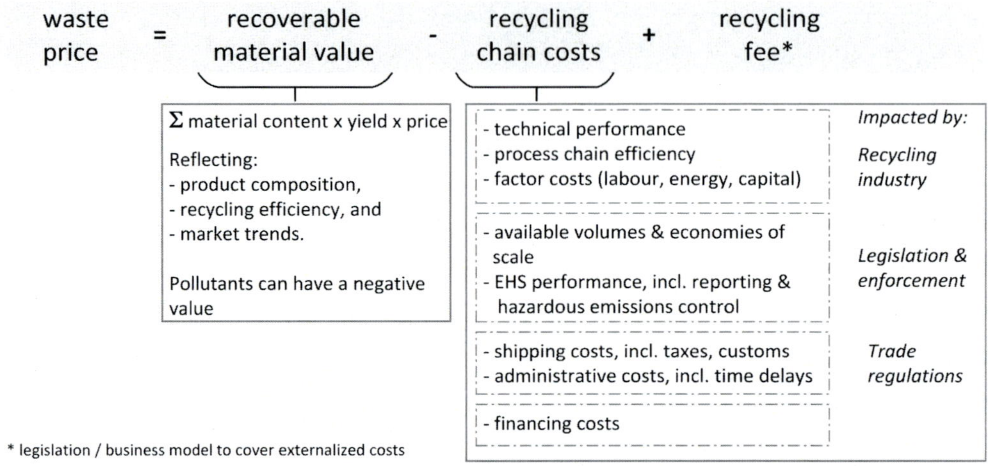

FIGURE 42.4 Principal calculation of waste price, indicating the value and costs to be included. *Adapted from: Hagelüken and Goldmann (2022).*

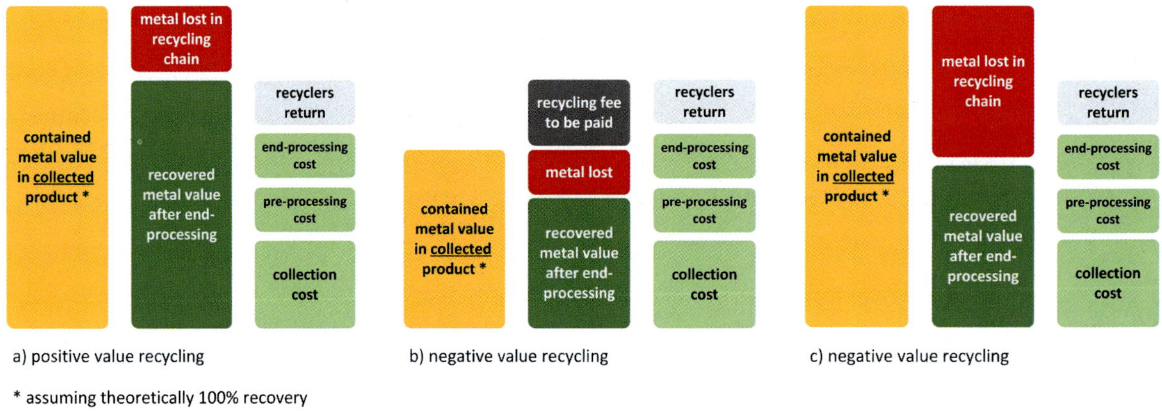

FIGURE 42.5 Three scenarios for recycling value. (A) Positive value recycling: the recovered metal value covers all costs and recyclers return. (B) Negative value recycling: the contained and recovered metal value is insufficient to cover all costs, and a recycling fee is to be paid. (C) Negative value recycling can become positive value recycling if the metal value losses can be reduced. Boxes not to scale.

sufficient to pay for all costs incurred and also covers the recyclers return, i.e., income > costs. *Negative value recycling* occurs when the contained metal value in the product is low, and the recovered value does not cover all recycling costs, i.e., income < costs. A recycling fee (or other business model/policy combination) is necessary to ensure recycling occurs. If the metal losses in the value chain are high, even for products with a high contained material value, the recovered metal value can be insufficient to pay for all costs. Measures that reduce the metal value losses would be necessary to make the economics work. These can include (process) technology-related measures, product design changes, and recycling chain/system level measures, or charging a recycling fee.

42.3 INCOME PERSPECTIVE

For *high grade materials* such as (gold) jewelry, the extremely high metal value density (\$/kg) has driven collection and effective recycling since the beginning of its use. Total gold mining in the history of mankind until 2023 accounted for ~210,000 t. Roughly over 90% is still in use, thanks to the continuously nearly closed recycling loops of coins, jewelry, artifacts, etc. (World Gold Council, 2023). These products are not complex in their composition and recycling processes are relatively easy, highly efficient, and less costly. Therefore collection and recycling work on their own, and legislative incentives are not necessary.

This is also true for high-grade materials, such as jewelry made from Ag, Pt, Pd, and industrial applications like Pt crucibles or Pt-Rh gauze catalysts. Although having a lower value density, high-grade base metal applications, e.g., copper sheet or cables, high-grade aluminum, nickel, lead or steel scraps, etc., usually obtain high EoL recycling rates.

This high-value density is missing in electrical and electronic equipment (EEE). Besides, electronic products are highly complex, containing a wide mix of various substances such as precious metals, base metals, hazardous substances, and other materials. The intricate design of the devices has a high impact on the cost and efficiency of recycling. Recycling of electronic devices, specifically smartphones (global annual sales 1.4–2 billion devices), is used as an example. The specifics of waste electrical and electronic equipment (WEEE) recycling are covered in detail in Chapter 13, Electrical and electronic equipment (WEEE).

42.3.1 Product Perspective

A smartphone is a complex device consisting of, among others (Bookhagen et al., 2020):

- Magnets in the camera and speaker, and in the vibration motor: Dy, Fe, Nd, Pr, Mo, W
- Touchscreen: In, Si (glass), and Sn
- Covering of the circuit board: Al, Sn
- Circuit board, conductor tracks, and SIM contacts: Ag, Au, Cr, Cu, Pd, Si, Ta, Ti
- Battery: Co, Li, Mn, Ni
- Casing: Al, Mg
- Screws: Fe

Furthermore, halogens (Br, F, Cl) and plastics are used in the circuit boards and other parts.

All these components are highly interconnected. The degree of interconnectedness can be impacted by the design choices made. Modular design or design for repairability (a removable battery or replaceable camera, for example) support the value recovery from products.

On a weight basis, the metals make up ~45%, display/glass ~32%, plastics ~17%, and the rest is ~6% in a typical smartphone (put on the market in 2011/2012). The largest contributors to the metal content (expressed in grams) are iron, aluminum, copper, magnesium, silicon, and nickel, while the largest contributors to the metal value are gold (73% of value, 0.034% of weight) and palladium (11% of value, 0.004% of weight). Other metals together make up only 13% of the metal value (Bookhagen et al., 2020). The total metal value in the phone is 1.10 EUR/unit (metal prices Jan–Jun 2020). This is very little compared to the sales price, which easily can exceed 500–1000 EUR. From an economic perspective, it is thus essential to recover *both*:

1. The materials that represent the majority of the weight.
2. The materials that represent the majority of the value, which are present in minute quantities.

The composition of a product, and a scrap stream, is not constant. Within the product "cell phones" there are variations in composition, even for those manufactured in the same year. Differences between models and brands such as the size (weight), the material of the housing (metal vs plastic), and technology choices impact the composition and thus material value of the product. Composition ranges are 116–177 ppm gold, 13–19 ppm palladium, and 33.7–90.7 g/kg copper (Bookhagen et al., 2020), and can be less than 100 ppm Au. The differences in mass and value are considerable. Hence, knowing the spread around the average is just as important as knowing the average itself.

Over time, EEE design and technology have changed considerably. Changes in size, weight, and technology (CRT to LCD/LED) are easily observed in the evolution of TVs (Samsung, 2015). This changed their lifetime, material composition (and value), and the recycling technologies necessary to recover these. Similarly for laptops: the iBook (1999) weighed ~3 kg, while the first MacBook Air (2007) weighed less than 1.5 kg, a 50% reduction. Thus to fill the same recycling plant capacity (in t/y), the number of devices entering the plant needs to double.

In cell phones and other electronics, reducing product weight, miniaturizing components, and thrifting of valuable metals through material efficiency gains all result in a decrease of recoverable metal content (Bangs et al., 2016)—and this is likely to continue. The *average* composition of cell phone handsets arriving at a final metal recovery plant is shown in Figure 42.6 (Bangs et al., 2016; Art, 2021). In 15 years gold decreased 40%, silver about 75%, and palladium 80%. The metal quantity entering a metallurgical recycling plant decreased tremendously. To keep the same Au output (in t/y), ~1.6× more handsets need to be processed.

The metal value of a smartphone fluctuates with the metal prices (Figure 42.1). The high price volatility translates directly into varying recoverable metal values, assuming a similar material composition. When the recoverable metal content decreases at the same time, its impact can be cushioned or exacerbated by the metal prices, as illustrated in Box 42.1.

42.3.2 Process Perspective

Process aspects can impact the income from recycling as the metal content and price are multiplied by the yield, which is determined by the process(es) used. The difficulty in avoiding process losses due to physical, chemical, and thermodynamical limitations, and the other avoidable losses (Figure 42.2) are reflected in the income obtained (Figure 42.5).

During preprocessing, metals can get lost because of insufficient liberation, or even too much liberation and sorting into a fraction from which they cannot be recovered. This includes metals ending up in fractions that go to landfill. For example, very intensive shredding and size reduction of high-value electronics and circuit boards create very small precious metal particles that separate from the printed circuit boards and end up in the dust fraction or stick to other fractions. Improving *presorting* of the recycling input, i.e., creating more homogeneous mixtures of products with similar composition and preprocessing needs; choosing different preprocessing equipment or equipment settings; or using manual/robotic dismantling instead can increase the precious metals yield tremendously and reduce value losses (Meskers and Hagelueken, 2009; Chancerel et al., 2009; Wang et al., 2012).

During end-processing, metal losses occur into slags and other residues. For cell phones and circuit boards these are all the elements that have a high affinity for oxygen, such as Al, Si, Fe, Ta, REE, etc. (Chapter 21, Copper). The tantalum-containing capacitor case in Chapter 25, Precious and technology metals, illustrates that by changing the preprocessing setup to create a separate fraction, additional material value could be obtained. However, the costs were larger than the material value, and separating the capacitors bears a risk of unintended coseparation.

Collaboration along the recycling chain generates value, especially at the interfaces between

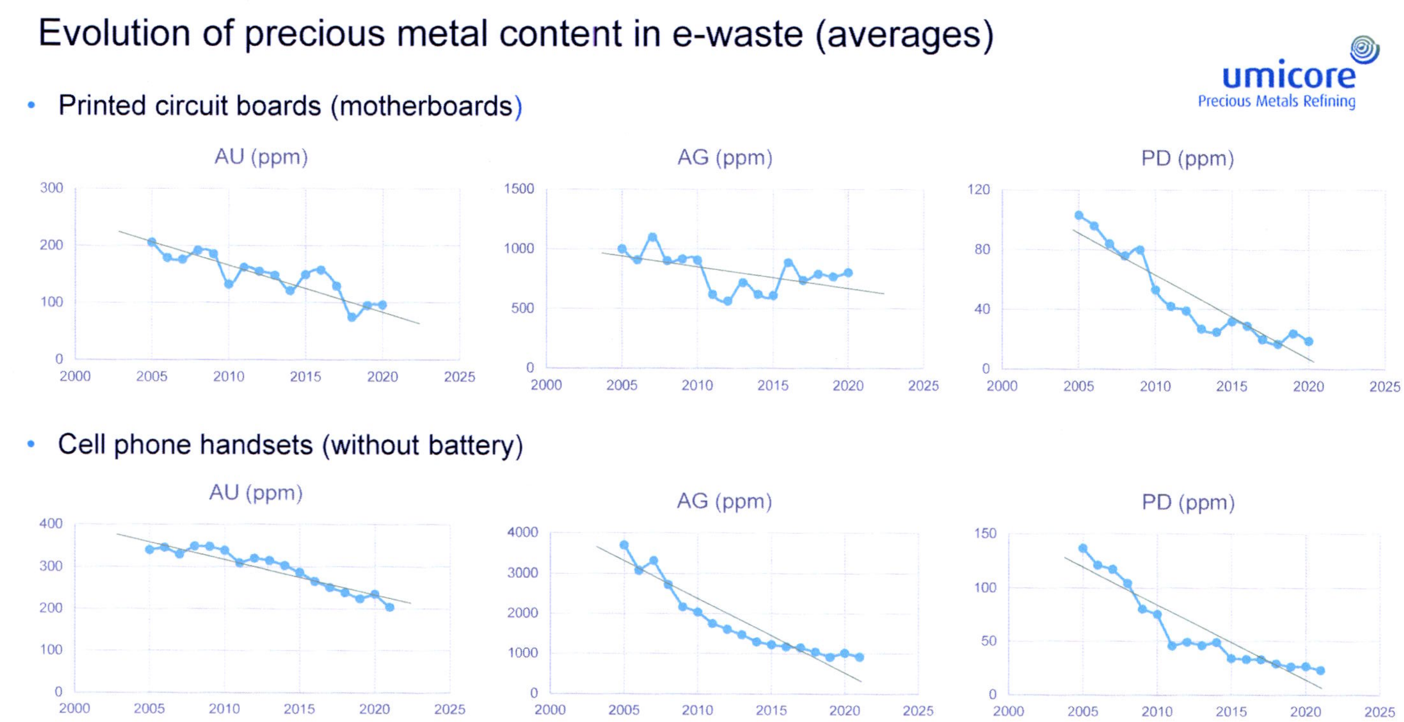

FIGURE 42.6 Evolution of precious metal content in printed circuit boards and cell phone handsets over time (Bangs et al., 2016; Art, 2021). The year indicates the arrival of the cell phones at the final metal recovery plant. *Data from: Umicore.*

BOX 42.1

Impact of metal content and price on the recoverable metal value

The recoverable metal value depends on the metal content, the metal price, and the process yield. The impact of changes in content and price is illustrated using gold from cell phone handsets with data from Figures 42.5 and 42.6 and assuming a constant process yield.

1. **Base case**: Cell phones arrive at the final metal recovery plant in 2005. Their gold content is 330 ppm. The average Au price is 443 USD/oz(=14.24 USD/g). The recoverable Au metal value is 4984 USD/t handsets.
2. **Reduced metal content**: A new shipment of cell phones arrives, containing ~200 ppm Au, 40% less than in the base case. The average Au price is the same. The recoverable Au metal value is 40% lower: 2990 USD/t.
3. **Increased metal price**: The cell phones containing 330 ppm Au arrive at the plant in 2020, when the average Au price is 1771 USD/oz (=56.93 USD/g), ~4× higher than in 2005. The recoverable metal value is now 18,790 USD/t (~ 4× higher).
4. **Increased price and reduced metal content**: A shipment of cell phones arrives in 2020, containing 200 ppm Au, and the average Au price is 1771 USD/oz The recoverable metal value is 11,386 USD/t, 2¼ times higher than the base case.

the steps. Well (pre-)sorted feeds and smart combinations of collected products create economies of scale and support operational excellence. Similarly, optimizing between preprocessing and end-processing creates value (Hagelüken, 2006; Meskers and Hagelueken, 2009; Wang et al., 2012).

To actually close the loop, all the valorizable metals and materials must be of marketable quality and in demand. Otherwise they do not count in the recoverable value calculation. For example, more value may be created from the slags from ferrous and nonferrous metal smelting by using them in cementitious binders, instead of only using them to replace aggregates; see Chapter 27, Cementitious binders incorporating residues.

42.3.3 System and Business Environment Perspective

From an economic perspective the (future) availability of and access to EOL products is as important as the income or revenue generated on the level of an individual product or waste stream (Hagelüken and Meskers, 2010).

Product lifetime is an essential parameter in estimating the current and future availability of products for recycling (Figure 42.7). It varies across products, over time, and likely across geography. For example, compared to EEE put on the market in 2000, the lifespan of EEE put on the market in 2010 declined at ~17% for screens and ~7% for large domestic appliances in the Netherlands (Huisman et al., 2012).

Lifetime extension approaches, via design for recycling approaches and the implementation of reuse, repair, and remanufacturing, delay the moment the entire product becomes available for recycling. Instead, the scrap composition and flow from the product during its life—the broken and replaced parts—and the fractions obtained during preprocessing change. A rigorous implementation of the circular economy fundamentally changes the what, when, where, and how much (i.e., availability and accessibility) of material streams for recycling.

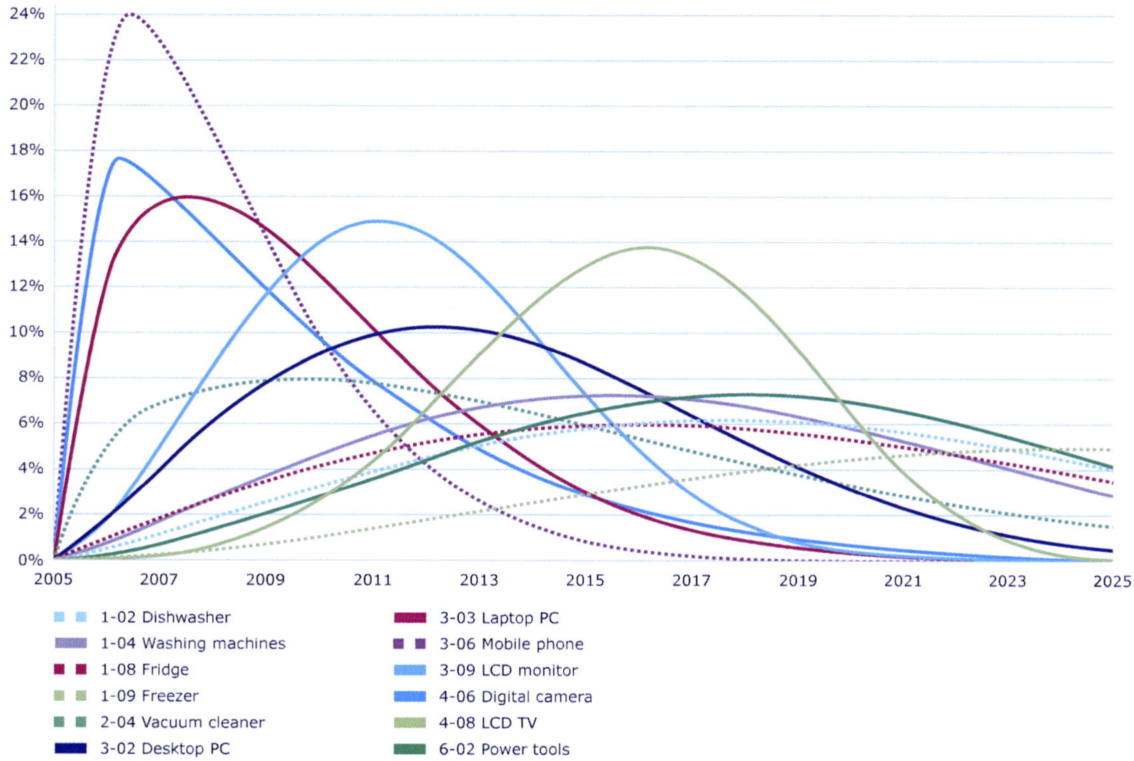

FIGURE 42.7 Residence times of electronic and electrical equipment put on the market in 2005 in the Netherlands, in percent discarded per year, described using Weibull distribution functions. *From: Huisman et al. (2012).*

Access is supported by collection incentives, convenient collection infrastructure, and legislation that enforces comprehensive collection. Considerable value is lost for recycling companies and society when devices are:

- not collected at all (going to landfill),
- collected after a very long time (hibernation),
- leaving the formal recycling chain, e.g., export for (fake) reuse.

The business model and environment have a significant impact on the recycling rates. In a business-to-business (B2B) environment, it is usually easier to close the loop even for lower-value materials, as there are direct relations between the few actors involved, flows are more transparent, and decisions are made based on rational and economic grounds (Hagelüken and Meskers, 2010).

Chemical process catalysts have been recycled in a B2B, closed-loop service model for decades and are used as examples, based on Hagelüken and Meskers (2010), Cimprich et al. (2023), and Hool et al. (2022). These catalysts have a very high metal value and are used in larger quantities at known stationary locations. Typically, a reforming catalyst used in oil refineries (0.4% Pt, 50 t catalyst per reactor) has a metal value of over 100,000 USD/t (metal price 2020). Its performance is regularly monitored, so the moment for replacement is known. The EoL recycling rates are over 90%, as the industrial stakeholders involved strive to minimize losses of the valuable metals.

The service model includes catalyst production and recycling. Usually, catalyst manufacturers also operate in-house catalyst recycling and PGM refining. When the catalyst is sold to the customer, the recycling service is part of the package. The catalyst price itself consists of (1) a manufacturing fee for the catalyst and (2) the value of the contained PGMs. The value fluctuates heavily with the metal market prices. When the spent catalyst is removed from the reactor, it is directly shipped to the catalyst manufacturer, who recycles the PGM and produces the new catalyst. The customer pays the recycling service fee, and the PGMs themselves remain the customer's property (Box 42.2). Only small amounts of additional PGMs need to be purchased to cover the losses during use and recycling (<10%). Hence, the risks from fluctuating metal prices are reduced, as only the initial catalyst (PGM) purchase and the small additional PGMs are subject to market prices.

Similar business approaches exist for, e.g., germanium wafer scrap, Co, Ni, W-containing

BOX 42.2

Commercial procedures in (precious) metal refining

Commercial procedures for (precious) metal-containing products differ from those applied in, e.g., paper or household waste recycling. The latter are usually based on waste volumes/weights and recycling fees per ton. For metals the full material composition is considered.

Metal recycling often uses *toll refining* contracts between supplier (customer) and refiner. The (precious) metals refiner offers the service of recovering the metals for an agreed service charge and credits the recovered metals to the customer, who remains the owner of the metals contained. The recovered metals (*metal return*) can be credited physically (in kilograms) or monetarily (based on metal prices). The service charge and metal return are based on sampling and assaying of each individual customer's lot and the agreed contract terms. This *net value* results from the credited metals, priced at current rates, minus the contracted refining service charges.

The contract typically consists of:

- **metal credit (return)**, which reflects the technical process losses during refining, a percentage of the analytical content of the dry material net weight. The analytical content is determined via sampling and assaying.
- **metals return time**, which reflects the process throughput time to recover and purify the precious metals. It is expressed in days after completion of the sampling process. The metal credit takes place after this.
- **refining price**, which reflects the total recovery costs and the margin of the refiner. It comprises treatment charges (USD/t material received), refining charges (USD/kg of each metal credited), and sampling and assay charges (USD/lot). Additional costs can be charged for handling impurities (*a penalty*).

The accuracy of sampling and assaying is essential. Any mistake has a direct monetary impact for the customer. Also, any undetected metal/material losses prior to the sampling process have a direct impact.

To conduct a toll refining contract, minimum quantities (or lot sizes) are required. For smaller lots the refining charges, especially the fixed costs of sampling and assaying, increase compared to the metal return, reducing the net value.

Based on: Hagelüken and Grehl (2012).

hard metal tools for industrial users, and many more [e.g., Cimprich et al. (2023) and Hool et al. (2022)].

42.4 COST PERSPECTIVE

For complex wastes that contain both valuable and hazardous materials, such as WEEE, considerable costs are incurred for environmentally sound and safe treatment, reporting, and compliance with regulations, including (transboundary) transport and trade. Having all this in place is an absolute necessity for achieving a sustainable circular economy, where all costs are internalized.

Four major cost levels need to be considered:

(1) basic operational costs: all costs related to depollution, processing, proper waste disposal, emission control, waste water management, etc.
(2) operational costs related to quality and services: all costs related to quality, waste characterization, proper reporting and compliance with best available technologies and standards, etc.
(3) office and administrative costs, including overhead.
(4) capital costs: all assets related to processes such as buildings, machines, equipment, patents, etc.

Many of these costs can be directly impacted by the recycling industry itself. Some costs are (partly) beyond its direct control. These are associated with:

- Effort necessary to obtain necessary/available volumes,
- EHS requirements/standards, including reporting and hazardous emissions control, and
- Trade and transport policy and its enforcement.

The impact of time and time delays, linked to, e.g., shipment time, contract return times, and administrative delays, is reflected in the financing cost. The cost perspective is summarized in Figure 42.4.

42.4.1 Product Perspective

The preprocessing costs depend heavily on choices made during product design, both in the selection of the materials and in how and which materials are connected (see Chapters 2 and 5 for details). New technologies and material compositions may require new recycling processes. Early exchange between product development and recyclers can prevent "design for recycling flaws" and support recycling process development. For example, the first LCD TVs and displays were lit by mercury-containing backlights, which required recyclers to manually dismantle the displays to remove the backlights, or to shred the displays in shredders equipped with emission control equipment, so that the mercury could be captured and safely managed. This increased the costs for treatment considerably. The later (LED) displays did not have this problem and could be processed conventionally.

Implementation of design for recycling approaches often also supports other design for recycling strategies, and vice versa. Removable batteries in cell phones support repair as well as the separation of phone and battery, lowering the separation cost and ensuring that both enter the most appropriate end-processing route. Value recovery is maximized.

Finally, the developments in product technology and the spread in product lifespan imply that recycling processes need to be in place that can take care of both the latest product put on the market and products from decade(s) ago.

42.4.2 Process Perspective

In preprocessing and end-processing, costs can increase, especially for heterogeneous feeds and complex compositions, as the number of process steps and complexity of the processes increase. This impacts both CAPEX and OPEX. Hence, mixing very diverse production scraps from the same plant or very diverse EoL consumer products drives costs up and can create unrecyclable mixtures. The additional effort of differentiated collection should be compared to the gains made in processing.

New developments and requirements for the recycling chain and changing requirements on the marketable products (e.g., REACH requirements) are all impacting CAPEX and OPEX.

The costs for compliant WEEE recycling, from the perspective of a preprocessor, are given in Magalini and Huisman (2018). The total compliant treatment cost of flat panel displays is 157 EUR/t. Out of this, 82 EUR (~50%) is for the treatment itself. Compliance (27 EUR), disposal (13 EUR), and depollution of hazardous materials (35 EUR) account for the other half. For cooling and freezing equipment, this increases to almost 60%. A noncompliant recycler—one who does not report on the process, does not have real audits at the plant, and circumvents depollution and proper waste disposal—saves costs by externalizing them to society. The environmental impacts of noncompliance recycling, e.g., CFC emissions from cooling and freezing equipment, are alarming (Magalini and Stillhart, 2019). Clear, mandatory, strictly enforced standards are needed for the treatment of complex metal-containing wastes, especially for WEEE, to combat noncompliant operations.

The spike in energy prices in Europe in early 2022 highlighted the fact that large fluctuations in OPEX can happen. Although recycling generally uses less energy than primary production, energy is still used. For plastic recyclers, energy is the third major cost factor after labor and maintenance, typically accounting for 15–20%. After the 400% energy price increase, it represents up to 70% of OPEX (Packaging Europe, 2022).

42.4.3 System and Business Model Perspective

Assume a company collects about 50,000 cell phones (~5 tons). These create a *net metal value* of up to 50,000 EUR when recycled at a state-of-the-art end-processor. This revenue needs to cover all other costs, including collection and shipment. The biggest impact on collection costs is the first mile, i.e., collection from single households. This can easily eat up the total metal value of a single device. Hence, a (cost) effective collection is decisive for the recycling of lower-grade, complex devices. Providing incentives (e.g., deposits, leasing, etc.), using collection hubs and joint infrastructure, optimizing logistic chains, and, above all, scaling up collection volumes all have a crucial impact on recycling economics from a material quantity and value point of view.

The challenges also highlight that in practice there are two value chains—the recyling and the manufacturing chain—disconnected at the consumer stage (Figure 42.2). This has much to do with the business-to-consumer (B2C) environment (Hagelüken and Meskers, 2010).

Automotive catalysts are used to illustrate this. The value density of automotive catalytic converters or "autocats" is just slightly lower than chemical process catalysts. Despite this, they are handled—like EEE—in a B2C environment.

To accumulate 1 t of autocats requires collection from about 700 ELVs, whose location is difficult to trace. During the 15–20 year lifetime, car ownership is transferred several times. Many vehicles originally registered in OECD countries are exported to developing and transition countries around their EoL, where waste legislation and recycling infrastructure are less established (UNEP, 2020). Because of their metal value, many autocats are either removed prior to export or dismantled from ELVs in these regions. Through opaque chains, they often end up in state-of-the-art recycling plants for PGM recovery. The PGM EoL recycling rate is 50–60%, obtained by multiplying the collection (~70%), transport (~90%), decanning (~95%), and metallurgical refining efficiency (98%) for PGM. Compared to ~30% for gold from cell phones, it is quite high. Besides collection challenges, losses occur during handling and transport, as PGM-containing particles can break off the ceramic and fall out of the autocat.

Collection starts at car workshops, scrap yards, and car dismantlers (Chapter 12, Vehicles (ELV)). The catalyst is cut from the exhaust system and sold to collectors at a price per piece (*telquel price*), which usually follows the PGM prices closely. Car dismantlers and catalyst collectors have quite a good understanding of the metal value of different catalyst types, as this is a highly competitive market. Volumes are consolidated and sold to autocat preprocessors, the *decanners*, where the autocat ceramic, with the PGMs, is removed from the steel can. When several tons of ceramic are accumulated, they are shipped to a PGM refiner. Here, for the first time, the metal content and value is determined by sampling and analysis. PGM recycling is then conducted as a service using toll refining (Box 42.2).

The recycling chain resembles a funnel system with millions of catalyst sources (vehicles) funneled to thousands of collectors and ending up at 10s of decanners (numbers for Germany). While collection and handling take place on a regional level, the final metallurgical recycling is a global business.

42.5 CURRENT AND FUTURE CHALLENGES

42.5.1 Malicious Practices

Unfortunately, the high monetary value of different scraps also triggers malicious and criminal practices. Theft of high-value jewelry and coins has been taking place for centuries. The modern thieves, often international organized crime networks, have extended their portfolio to the "recycling business" (Fox, 2021).

Copper cable theft from public infrastructure and construction sites causes large losses for companies and society. Europol reported an operation across Europe that uncovered over 1000 crimes and seized over 350 t of metal, including about 90 t of copper (Europol, 2018). Cable theft resulted in train delays averaging >1.5 h/day in Belgium (Walker, 2023).

Very high PGM prices have made the theft of catalytic converters a fast-growing crime. The metal value can then quickly exceed 100 EUR/catalyst. The converters are stolen from fully functional, parked cars overnight and find their way via shady buyers and traders into the official recycling chain. Once a catalyst is decanned, it is impossible to trace the true origin of the ceramic material. In London alone, the theft doubled to nearly 3000 cats (Russon, 2019), while in the United States the thefts quadrupled from 2019 to 2020, tripled from 2020 to 2021 to over 50,000 in 2021, with nearly 65,000 thefts in 2022. The cost is large: replacing a converter costs ~3000 USD (Murphy, 2022; Ramirez, 2023).

42.5.2 Scavenging and Substandard Treatment

Malicious practices are also observed in illegal exports of vehicles and electronic scrap, followed by backyard recycling under horrible conditions, with huge human, environmental, and economic impacts (e.g., www.welcome-to-sodom.de and Chapter 7—The path to inclusive recycling: developing countries and the informal sector). Besides substandard and noncompliant recycling, selective *scavenging* is a challenge. EEE products and components and materials with higher economic value or reuse potential are scavenged and do not end up in official takeback systems (products) or with compliant recyclers (components, materials) (Magalini and Huisman, 2018). Cooling and freezing equipment, laptops, and LCD displays are especially underrepresented in formal collection volumes in the EU. Scavenged components are cables, compressors, Cu/Fe coils and motors, batteries, and circuit boards. The total value was estimated at 171 million EUR in 2016 for EU-28. This undermines the economics for recyclers and the extended producer responsibility systems, since part of the income (value) is missing while the costs remain the same (Magalini and Huisman, 2018; Magalini and Stillhart, 2019). The environmental impacts resulting from scavenging are concerning. As long as recycling is driven *mainly* by free-market forces, such malicious practices will continue to happen.

Authorities and official market participants take action through law enforcement (Europol, 2018; Department of Justice, 2022), banning cash payments, enhanced supplier due diligence, and new legislation. The proposed Preventing Auto Recycling Thefts (PART) Act (S. 154 and H. R.621, US Congress) includes better tracking and marking of catalysts and stiffer punishment. The push for and implementation of Digital Product Passports (DPPs), e.g., the EU's Battery Regulation and Sustainable Products Initiative, provides further opportunities for traceability and the verification of actors in the recycling value chain.

42.6 CONCLUSIONS AND OUTLOOK

The *circular economy* already comprises the two basic requirements to close the (metals) loop—the physical and the economic dimensions.

Physically closing a circle means that, by definition, after a product's end-of-life, materials need to find their way back into new products. Comprehensive collection is essential and *must* be combined with a high-quality recycling process chain. Only then can a broad spectrum of materials be recovered with high yields, under sound environmental and working conditions. Thanks to the innovation capacity of the recycling industry, for most complex products there are efficient, high-quality recycling processes in place and/or under development for new products and materials.

The *economic dimension* must be met too. Obtainable recycling revenues (linked to device value) need to cover the total costs of the entire recycling chain (activities and profit margins). Developments in product design, composition, and metal prices are decisive for the economic viability of recycling. Increased product complexity leads to increased technical (process) challenges and recycling costs. Design for recycling approaches can thus enhance economic viability.

Business as ***unusual*** is necessary. Past approaches did not result in a circular economy.

Enhanced transparency and traceability of flows. Digital product passports and other digitalization initiatives can be used so that waste is only treated by authorized, certified recyclers. Time and location stamps can be used for traceability. The access to information by authorized parties, including authorities for transboundary shipment, can further speed up processes. Product lifetime extension measures will influence the future availability and quantity of material streams for recycling. These developments can be anticipated through improved transparency and traceability.

New business models for circularity. This is needed, especially for low-value, high-volume devices, so that critical mass (volume) is created and logistics and processing costs can be reduced. It will require fundamental changes in design, production, distribution, use, repair, remanufacturing, and recycling, as these are all interlinked. All actors are simultaneously supplier and customer. Collaboration and information exchange is thus essential.

Legislation and policy supporting the quality (performance) of recycling and acknowledging the economics associated with that are needed. Extended producer responsibility obligations should reward comprehensive and good recycling and circular business models. Incentives for recovery of critical materials present in lower quantities are currently missing, because of weight-based recycling rates. The difference between reported recycling rates and the physical reality should be overcome.

Appropriate legislation and policy are as important as new business models, more value chain collaboration, and higher transparency on product flows at EoL. The circular economy of electrical vehicle batteries offers an opportunity for *business as* ***unusual***. As a starting point, see Thompson et al. (2020), Neumann et al. (2022), the EU Battery Directive, and the Global Battery Alliance's battery passport (https://www.globalbattery.org/battery-passport/).

References

Art, S., 2021. Electronics recycling: current & future challenges for (end-)processing. In: E-Waste World Conference, Frankfurt, Germany, November 2021.

Bangs, C.E., Meskers, C.E.M., Van Kerckhoven, T., 2016. Trends in electronic products—the canary in the urban mine? In: Electronics Goes Green 2016+, 7–9 September 2016, Berlin, Germany. Fraunhofer Institut, Germany, ISBN: 978-3-00-053763-9.

Bookhagen, B., Bastian, D., Buchholz, P., Faulstich, M., Opper, C., Irrgeher, J., Prohaska, T., Koeberl, C., 2020. Metallic resources in smartphones. Res. Policy 68, 101750. https://doi.org/10.1016/j.resourpol.2020.101750.

Brown, P., Baldassarre, B., Konietzko, J., Bocken, N., Balkenende, R., 2021. A tool for collaborative circular proposition design. J. Clean. Prod. 297, 126354. https://doi.org/10.1016/j.jclepro.2021.126354.

Chancerel, P., Meskers, C.E.M., Hagelueken, C., Rotter, V.S., 2009. Assessment of precious metal flows during preprocessing of waste electrical and electronic equipment. J. Ind. Ecol. 13 (5), 791–810. https://doi.org/10.1111/j.1530-9290.2009.00171.x.

Cimprich, A., Young, S.B., Schrijvers, D., Ku, A.Y., Hagelüken, C., Christmann, P., Eggert, R., Habib, K., Hirohata, A., Hurd, A.J., Lee, M.-H., Peck, D., Petavratzi, E., Tercero Espinoza, L.A., Wäger, P., Hool, A., 2023. The role of industrial actors in the circular economy for critical raw materials: a framework with case studies across a range of industries. Miner. Econ. 36 (2), 301–319. https://doi.org/10.1007/s13563-022-00304-8.

Department of Justice, 2022. Justice Department Announces Takedown of Nationwide Catalytic Converter Theft Ring. The United States Department of Justice. Office of Public Affairs. November 2, 2022 https://www.justice.gov/opa/pr/justice-department-announces-takedown-nationwide-catalytic-converter-theft-ring. Accessed 16 June 2023.

Europol, 2018. Massive Crackdown on Metal Theft: Over 350 Tonnes Seized in 12 EU Member States. Europol. 19 June 2018 https://www.europol.europa.eu/media-press/newsroom/news/massive-crackdown-metal-theft-over-350-tonnes-seized-in-12-eu-member-states. (Accessed 16 June 2023).

Fox, S.J., 2021. The nexus between resources and criminal activities: 'recycling crimes' (metals). Res. Policy 74, 102377. https://doi.org/10.1016/j.resourpol.2021.102377.

Hagelüken, C., 2006. Improving metal returns and eco-efficiency in electronics recycling—a holistic approach for interface optimization between pre-processing and integrated metals smelting and refining. In: Proceedings of the 2006 IEEE International Symposium on Electronics & the Environment, 8-11 May 2006, San Francisco, pp. 218–223.

Hagelüken, C., 2012. Recycling the platinum group metals: a European perspective. Platin. Met. Rev. 56 (1), 29–35. https://doi.org/10.1595/147106712X611733.

Hagelüken, C., Goldmann, D., 2022. Recycling and circular economy—towards a closed loop for metals in emerging clean technologies. Miner. Econ. 35, 539–562. https://doi.org/10.1007/s13563-022-00319-1.

Hagelüken, C., Grehl, M., 2012. Recycling and loop concepts for a sustainable usage. In: Sehrt, U., Grehl, M. (Eds.), Precious Materials Handbook., ISBN: 978-3-8343-3259-2. Umicore AG & Co KG.

Hagelüken, C., Meskers, C.E.M., 2010. Complex life cycles of precious and special metals. In: Graedel, T.E., Van der Voet, E. (Eds.), Linkages of Sustainability. MIT Press, ISBN: 9780262013581, pp. 163–197. https://doi.org/10.7551/mitpress/8455.003.0014.

Hool, A., Schrijvers, D., van Nielen, S., Clifton, A., Ganzeboom, S., Hagelueken, C., Harada, Y., Kim, H., Ku, Y., Meese-Marktscheffel, J., Nemoto, T., 2022. How companies improve critical raw material circularity: 5 use cases. Miner. Econ. 35 (2), 325–335. https://doi.org/10.1007/s13563-022-00315-5.

Huisman, J., van der Maesen, M., Eijsbouts, R.J.J., Wang, F., Baldé, C.P., Wielenga, C.A., 2012. The Dutch WEEE flows. In: United Nations University, ISP—SCYCLE, Bonn, Germany, March 15, 2012. https://collections.unu.edu/eserv/UNU:8735/The-Dutch-WEEE-Flows.pdf. (Accessed 26 June 2023).

LBMA, 2023. LBMA Precious Metal Prices. International Prices for Gold, Silver, Platinum and Palladium. LBMA, London. https://www.lbma.org.uk/prices-and-data/precious-metal-prices#/. (Accessed 23 June 2023).

Magalini, F., Huisman, J., 2018. WEEE recycling economics—the shortcomings of the current business model. In: United Nations University, UNU-VIE SCYCLE, and European Electronics Recyclers Association (EERA). https://greenweee.ro/wp-content/uploads/2019/02/UNU-EERA-brochure-A4-ONLINE.pdf.

Magalini, F., Stillhart, R., 2019. Scavenging of WEEE: Environmental and Economic Consequences for Society. European Electronics Recyclers Association (EERA) and Sofies. www.eera-recyclers.com.

Meskers, C.E.M., Hagelueken, C., 2009. The impact of different pre-processing routes on the metal recovery from PCs. In: Hilty, L., Itoh, H., Hayashi, K., Edelmann, X.D. (Eds.), R'09 Twin World Congress and World Resources Forum, 14-16 September 2009. EMPA Materials Science and Technology, St. Gallen, Switzerland, ISBN: 978-3-905594-54-6.

Murphy, J., 2022. Thefts of this valuable car part have gotten so bad that Congress is trying to fix it. NBC News. 8 July 2022 https://www.nbcnews.com/data-graphics/catalytic-converter-thefts-gotten-bad-congress-trying-fix-rcna35754. (Accessed 16 June 2023).

Neumann, J., Petranikova, M., Meeus, M., Gamarra, J.D., Younesi, R., Winter, M., Nowak, S., 2022. Recycling of lithium-ion batteries—current state of the art, circular economy, and next generation recycling. Adv. Energy Mater. 12 (17), 2102917. https://doi.org/10.1002/aenm.202102917.

Packaging Europe, 2022. Plastic recyclers Europe: high energy prices could drive recycling companies out of business. In: Packaging Europe, 23 September 2022. https://packagingeurope.com/news/plastics-recyclers-europe-high-energy-prices-could-drive-recycling-companies-out-of-business/8759.article. (Accessed 23 June 2023).

Ramirez, I., 2023. Catalytic Converters Are Being Stolen in Record Numbers. Here's Why. VOX Media. 20 May 2023 https://www.vox.com/2023/5/20/23728557/catalytic-converter-theft-scrap-price-replacement-toyota-prius-cars-insurance-protection. (Accessed 16 June 2023).

Ressourcenkommission am Umweltbundesamt (KRU), 2023. Chancen und Grenzen des Recyclings im Kontext der Circular Economy: Rahmenbedingungen, Anforderungen und Handlungsempfehlungen. Umweltbundesamt, Dessau-Roßlau. https://www.umweltbundesamt.de/publikationen/chancen-grenzen-des-recyclings-im-kontext-der.

Russon, M.-A., 2019. Huge rise in catalytic converter thefts. BBC News Services. 20 September 2019 https://www.bbc.com/news/business-49767195. (Accessed 16 June 2023).

Samsung, 2015. History of Samsung things. Samsung Newsroom. January 6 https://news.samsung.com/global/infographic-history-of-samsung-things. (Accessed June 2023).

Thompson, D.L., Hartley, J.M., Lambert, S.M., Shiref, M., Harper, G.D.J., Kendrick, E., Anderson, P., Ryder, K.S., Gaines, L., Abbott, A.P., 2020. The importance of design in lithium ion battery recycling-a critical review. Green Chem. 22 (22), 7585–7603. https://doi.org/10.1039/d0gc02745f.

UNEP, 2020. Used vehicles and the environment—a global overview of used light duty vehicles: flow, scale and regulation. Sustainable Mobility Unit, United Nations Environment Programme. Nairobi https://www.unep.org/resources/report/global-trade-used-vehicles-report. (Accessed 23 June 2023).

Walker, L., 2023. Railway cable thefts tripled in 2022, causing extensive delays and cancellations. The Brussels Times. 6 February 2023 https://www.brusselstimes.com/365047/railway-cable-thefts-tripled-in-2022-causing-extensive-delays-and-cancellations. (Accessed 16 June 2023).

Wang, F., Huisman, J., Meskers, C.E.M., Schluep, M., Stevels, A., Hagelueken, C., 2012. The best-of-two-worlds philosophy: developing local dismantling and global infrastructure network for sustainable e-waste treatment in emerging economies. Waste Management 32 (11), 2134–2146. https://doi.org/10.1016/j.wasman.2012.03.029.

World Gold Council, 2023. Above-Ground Stocks. World Gold Council—Gold Hub. 8 February 2023 https://www.gold.org/goldhub/data/how-much-gold. (Accessed 16 June 2023).

PART 5

Recycling fundamentals

CHAPTER

43

Physical separation fundamentals

Kari Heiskanen

Metso, Espoo, Finland

43.1 PARTICLE SIZE

Any mass of material to be recycled consists of particles. The particles can be characterized by their size. Usually, the shape of particles differs from a sphere, which is the only geometric form with a well-defined unique size and diameter. Depending on the measurement technique, all other geometric and irregular forms have different "sizes."

The most common method for scrap sizing is sieving. The particle size x_A is the characteristic size of an aperture the particle passes. The sieve surfaces are usually woven wire cloths with square apertures. In this case, the size is the side length of the square. The surface can also be made with punched square or round holes.

Another common method to measure particle size is to use settling velocity in liquid or air. Then the size is given as the size of a sphere having the same settling velocity as the particle has. This is called Stokes diameter x_s for small particles settling at laminar velocities (Reynolds number (**Re**) <0.2) or, more generally, drag diameter x_d. Particles can be illuminated by light, and their projected images are measured to obtain the projected area diameter d_p. Particle volume and particle surface area can also be measured, yielding the volume diameter d_v and surface diameter d_s, which are the diameters of a sphere having the same volume or area as the particle.

For fine fractions, laser diffraction is nowadays the most common method. The size is the equivalent diameter of a sphere, having the same optical properties, that produces a similar diffraction pattern as the particle. However, it is more complex than that, as the diffraction pattern measured is the total light energy falling on the sensor array. The particle size distribution is inversely computed using known optical models and diffraction theories. As the particles are suspended in a fluid and pass the measurement cell at some velocity, the measured light energy is an average over some time. Thus the obtained particle size is also a time average of the mass passed through the measurement cell.

There are several average particle sizes, which can be used to estimate recycling system behavior. The most common is the average volume/surface diameter, the Sauter mean diameter (x_{32}). It is the sphere's diameter with the same volume-to-area ratio as all the particles in the whole sample.

Particle sizes and their distributions are most often given in discrete classes due to historical reasons, as sieving naturally gave classes. The class divisions are given usually in geometric series. The international ISO 565/ISO 3310

Handbook of Recycling. **https://doi.org/10.1016/B978-0-323-85514-3.00039-7**

standard is based on a base sieve of 1 mm and has a geometric ratio of R20/3, ≈1.41. Other standards exist as well.

Any set of particles (the sample) will have a particle size distribution. It can be expressed as frequency distribution $q(x)$ or as a cumulative distribution finer than size x, $Q(x)$.

$$Q_r(x) = \int_{x_{\min}}^{x_{\max}} q_r(x)dx \approx \sum_{i=1}^{n} q_{ir}(x) \tag{43.1}$$

Four types of particle size distributions are distinguished by the subscript r in the preceding equation. They are number, length, area, and volume distributions. The most common are the number and the volume (mass) distribution. The convention is that the subscript r is given a value from 0 to 3, subsequently from number to volume distribution. So $Q_3(x)$ means a cumulative volume distribution, and $q_0(x)$ is a number frequency distribution.

Number distributions are important in planning sampling campaigns, as statistical errors and analytical confidence limits depend on the number of particles sampled. Often reaction rates are related to the reactive surface, and the mass balances of such reactions correlate with the mass distributions.

Real distributions can be estimated with simple mathematical distributions. All of them are suitable only for monomodal distributions. The simplest representation is an exponential function called the Gates-Gaudin-Schuhmann (GGS) equation:

$$Q_r(x) = \left(\frac{x}{x_0}\right)^{\alpha} \tag{43.2}$$

where x_0 is the size that all the particles are finer than, and α is the slope (width) of the distribution. A second widely used form of distribution is the Rosin-Rammler-Sperling-Bennett equation:

$$Q_r(x) = 1 - \mathrm{Exp}\left\{-\left(\frac{x}{x_n}\right)^{m}\right\} \tag{43.3}$$

which is capable of describing the ends of the distribution better than GGS. In the equation, x_n is the size that 62.3% of the particles are finer than. Parameter m is the slope (width) of the distribution.

The third equation given here is the log-normal distribution:

$$q_r(x) = \frac{1}{\sigma_{\ln}\sqrt{2\pi}} \mathrm{Exp}\left\{-\frac{1}{2}\left(\frac{\ln(x_r/x_{\mu,r})}{\ln s_g}\right)^2\right\} \tag{43.4}$$

where $x_{\mu,r}$ is the median and $\ln(s_g)$ is $\ln\left(\frac{x_{84,r}}{x_{50,r}}\right)$ in which the number denotes the percentage finer at that size. For multimodal particle distributions, one has to combine two or more distributions. Multimodality is often observed in recycled feeds, as the material properties (e.g., brittleness) in the comminuted product may vary significantly.

43.2 SAMPLING

Sampling is the basis for evaluating a particulate process like recycling. Any property of a set of particles can be treated as a statistical distribution with a mean and a standard deviation. The variability is always a function of particle numbers sampled, not mass.

Careful planning should always be performed before any sampling work. The main aim of planning is to provide reliable statistical information for a defined purpose, such as quality of products, operational efficiency, and operational and feed quality effects.

Gy (2004) has developed a sampling theory that is in general use. The variance caused by the inhomogeneity of the material itself is called the fundamental variance. This error will remain even if the sampling is performed in an ideal way. The fundamental variance is related to the third power of the largest particles present in the sampled material:

$$\mathrm{var}(x_{\mathrm{fundam}}) = \frac{Cx_{95}^3}{n\Delta m} \tag{43.5}$$

where 95% of particles are finer than x_{95}, Δm is increment size, n is the number of increments, and C is a constant depending on property distribution, liberation, particle shape, and the size distribution's width. As can be seen from Eq. (43.5), reducing the size of the largest pieces reduces the fundamental variance rapidly.

There are several sources for error in performing the sampling. The total variance of sampling consists of the fundamental variance and variances of error taking place in assaying and sample selection due to the wrong delimitation of the sample (loss on material from increment) and due to integration errors caused by the discrete sampling of a continuous variability.

$$\text{var}(z) = \text{var}(\text{fundamental}) + \text{var}(\text{assaying}) + \sum \text{var}(\text{sample selection}) \tag{43.6}$$

All particles in the lot to be sampled should have the same probability of ending in the sample (the sample to be representative and the last right-hand term in Eq. (43.6) becomes zero). The sampling dimension can be defined as the spatial directions of a sampler to obtain a representative sample. A one-dimensional example is a material stream falling freely and sampled by a sampler that traverses it. A three-dimensional case is a heap of material, where sampling points should be distributed in three dimensions over the whole volume, a practical impossibility. Three-dimensional sampling situations shall be avoided.

Materials for recycling can be sampled by:

- Random sampling
- Systematic sampling
- Stratified sampling.

These can also be performed as a two-stage process or as a sequential process. True randomness is difficult to obtain, and therefore random sampling is discouraged, especially in three-dimensional cases. A random grab sample will have high variance and will often be prone to errors caused by nonideal sampling. The most accurate sampling method is a systematic sampling from a one-dimensional case, i.e., an automatic sampler sampling a moving stream of material. These rules also apply to all secondary sampling before assaying.

If the material tends to segregate, one can try stratified sampling, where different material strata are sampled separately, and the result is obtained by weighing the results by stratum masses. For very cheap materials, the sampling can be performed by a two-stage process (Eq. 43.7), where the first sampling may be, for example, by randomly selecting the wagons (m) from a train of M wagons to be sampled in more detail:

$$s_{ts}^2 = \left(\frac{M-m}{M-1}\right)\frac{\sigma_b^2}{m} + \frac{\sigma_w^2}{mn} \tag{43.7}$$

where σ_b is the standard deviation between wagons and σ_w within wagons.

A sequential process is often used to estimate if a lot fulfills the required quality standards. A decision can be made if the first sampling set is clearly within or outside defined limits. If the result falls between the specified limits, sampling must continue with a second set. The procedure will be continued until a decision can be made to either accept or reject the lot. Some metrics of interest are computed from several measured variables. For a variable (y) computed from a set of measured variables as $y=f(x_1,x_2,\ldots,x_n)$ the variance is computed as follows:

$$\text{var}(y) = \left(\frac{\partial y}{\partial x_1}\right)^2 \text{var}(x_1) + \left(\frac{\partial y}{\partial x_2}\right)^2 \text{var}(x_2) + \cdots + \left(\frac{\partial y}{\partial x_n}\right)^2 \text{var}(x_n) \tag{43.8}$$

43.3 LIBERATION

The mix of compounds in particles of various origins, from complete devices or parts of them or any by-products like slags, may vary from a single compound to a mix of several compounds. A particle consisting of a single compound is called liberated. A mix of two compounds is called a binary, and, with the same logic, the ternary particles have three compounds.

For optimal processing of compounds, the aim is to maximize the mass of liberated particles in a set of recycled materials. It poses an optimization problem, as breaking the material into too fine particles causes the processes to operate at nonoptimal conditions.

All recycled materials have a specific way of breaking when exposed to a sufficiently large impacting, compressing, or shearing force. By studying the progeny particles, one can form a model of liberation breaking. This kernel function can be determined by either textural modeling or by probabilistic methods (van Schaik et al., 2004; Gay, 2004). The effect of joints in materials to be shredded needs to be added into the liberation models (Castro et al., 2005; Heibeck et al., 2021).

After shredding, the frequency distribution of a property might look like the curve in the following figure (Figure 43.1) taken from a test with aluminum-containing scrap. As observed, there is a large mass of particles that does not possess the property (the range below 10% of Al), and a good mass of particles with a high degree of the defined property.

The distribution gives a possibility of estimating the separation of the material into a stream low in the property (Al) and a stream rich in the property (Al). By calculating the total mass content cumulatively starting from the poorest fraction, we get the *Separability curve* (Figure 43.2) showing how much aluminum is present below a defined limit. The example *original breakage* reveals that 70% of aluminum values are in particles that carry 50% or less nonaluminum matter.

43.4 GRADE-RECOVERY CURVES

There always exists a relation between the grade of a separated product and its recovery (recyclability). There are three fundamental

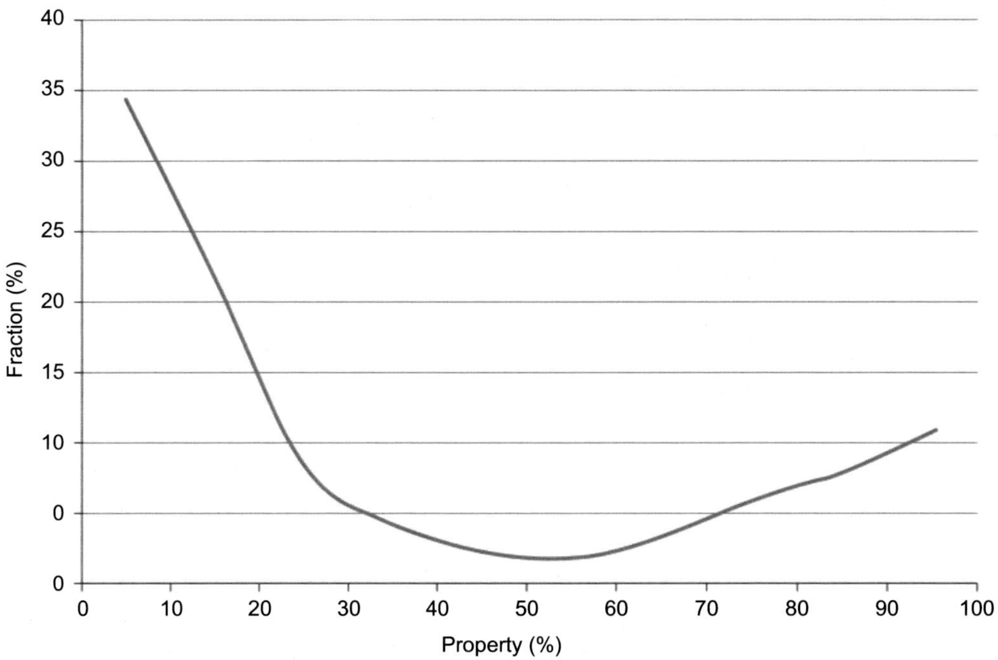

FIGURE 43.1 Frequency distribution of a single property.

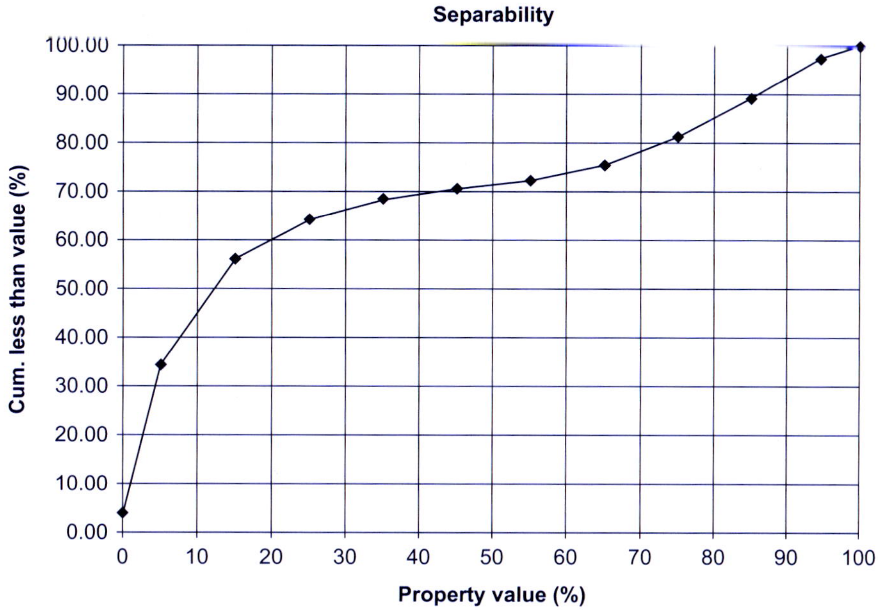

FIGURE 43.2 Separability curve (original breakage).

reasons for that relationship. Firstly, the liberation of particles subjected to separation is not complete. Secondly, the related response to a physical force or chemical potential gradient will cause those different particles to react differently. The third reason is the time of force affecting the particle.

A liberation-based grade-recovery curve (Figure 43.3) can be constructed directly from the separability curve (Figure 43.2) with easy computations. When the entire feed stream is taken as a product, the recovery is 100%, and the product grade is the same as the feed grade, in our aluminum case, 33.7% Al. Rejecting the fractions with less than 10% aluminum reduces the mass by almost 50% but only loses 5% of the aluminum (Figure 43.4) and increases the product quality to 49% aluminum. An entire grade-recovery curve can be constructed by continuing the procedure. As can be seen, in this example material, a 90% Al purity product can only be obtained with a 55% aluminum recovery.

Breaking the material into smaller particles will improve the separability curve (Figure 43.4) compared to the previous separability curve (Figure 43.2). The following experimental value is from the same scrap sample but substantially finer in particle size ("further breakage").

Compared to the original separability curve (Figure 43.2), the central part of the curve is much flatter, indicating that no significant mass of middling particles exists. This leads to the following grade recovery curve (Figure 43.5), where a 90% Al purity product is now obtained with 85% Al recovery: a substantial improvement.

The importance of the liberation-based grade recovery curve depends on the property interactions and the nature of the following processes.

43.5 MASS BALANCES AND PROCESS DYNAMICS

Recycling can also be understood as a material-handling operation, where the material is transported, concentrated, and purified

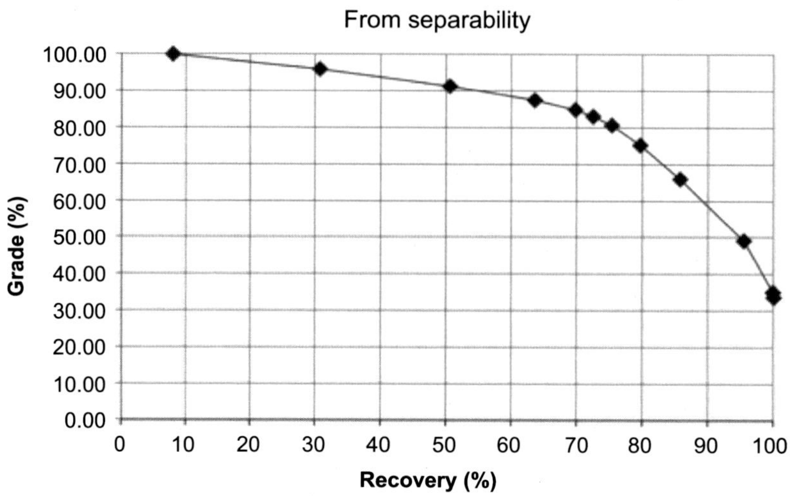

FIGURE 43.3 Grade-recovery curve (original breakage).

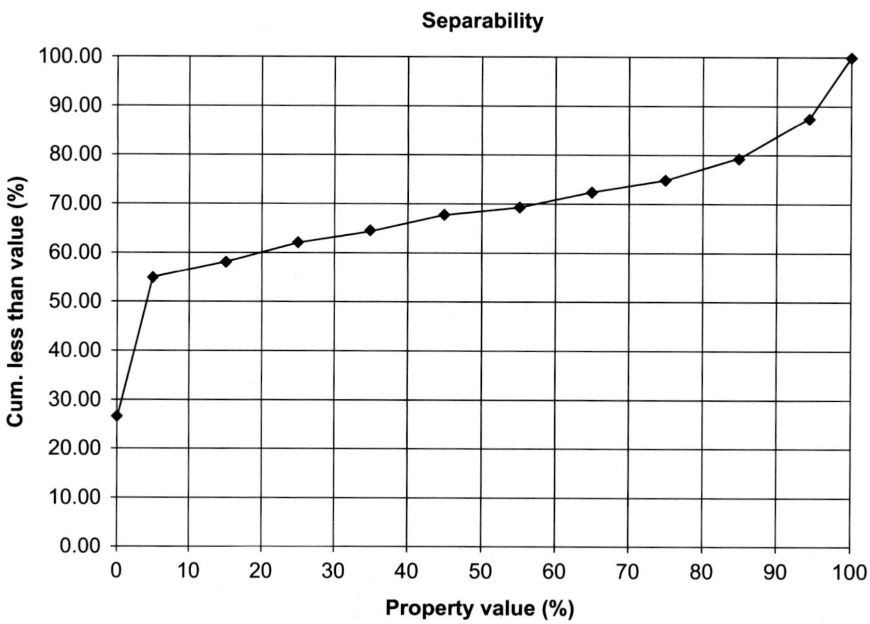

FIGURE 43.4 Separability curve after further breakage.

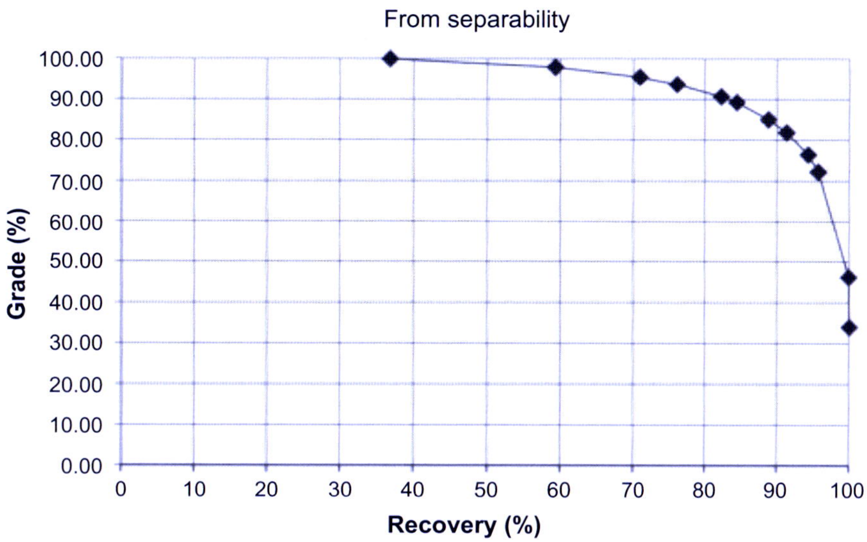

FIGURE 43.5 Grade recovery curve after further breakage.

during the treatment. The process always consists of the units handling the material with temporal hold-ups, units of moving the material, and of storage units.

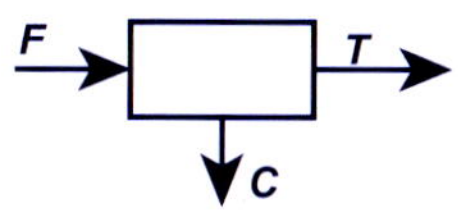

FIGURE 43.6 A simple separator.

43.5.1 Mass Balances

Mass balances can be written either over a single unit or over larger parts of the process. In mechanical recycling, one often starts with steady-state mass balances where the recycled mass and its constituents are assumed to be constant in any flow. For the simple separator (Figure 43.6), we can write

$$\begin{aligned} F &= C + T \\ Fc_{af} &= Cc_{ac} + Tc_{at} \\ Fb_{bf} &= Cc_{bc} + Tc_{bt} \\ &\ldots \\ Fm_{mf} &= Cc_{mc} + Tc_{mt} \end{aligned} \tag{43.9}$$

where capital letters denote total mass flow of feed (F), product (C), and tails (T) and c_{af}, c_{ac}, c_{at}, c_{bf}, c_{bc}, c_{bt}, …, c_{mf}, c_{mc}, c_{mt} are the concentrations of the property of interest (a, b, … m) in streams F, C, and T, respectively.

For a simple separation as in Figure 43.6, recovery (R) of material with the property of interest can be calculated from the mass balance:

$$R_a = \frac{Cc_{ac}}{Fc_{af}} = \frac{c_{af}\left(c_{ac} - c_{at}\right)}{c_{ac}\left(c_{af} - c_{at}\right)} \tag{43.10}$$

Figure 43.7 shows a two-product case (Figure 43.7), where properties a and b are divided into both C and S product streams.

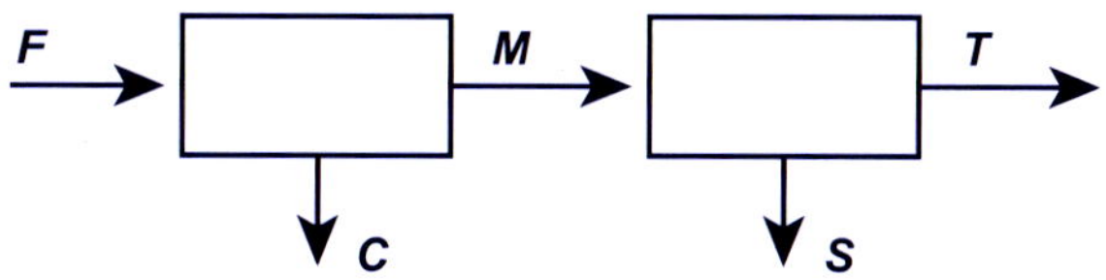

FIGURE 43.7 A two-stream separation case.

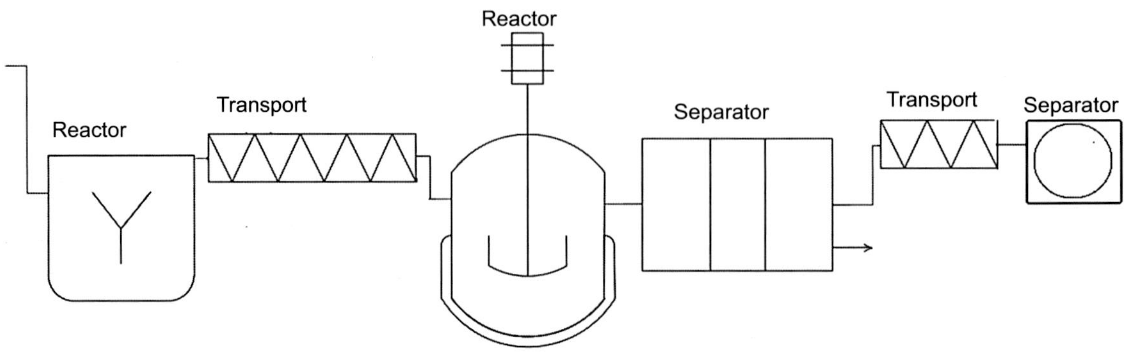

FIGURE 43.8 A recycling flowsheet.

The recoveries become

$$R_a = \frac{c_{ac}\left[\left(c_{af} - c_{as}\right)\left(c_{bs} - c_{bt}\right) - \left(c_{bf} - c_{bs}\right)\left(c_{as} - c_{at}\right)\right]}{c_{af}\left[\left(c_{ac} - c_{as}\right)\left(c_{bs} - c_{bt}\right) - \left(c_{ac} - c_{bs}\right)\left(c_{as} - c_{at}\right)\right]}$$

$$R_b = \frac{c_{bs}\left[\left(c_{ac} - c_{af}\right)\left(c_{af} - c_{bt}\right) - \left(c_{bc} - c_{bf}\right)\left(c_{as} - c_{at}\right)\right]}{c_{bf}\left[\left(c_{ac} - c_{as}\right)\left(c_{bs} - c_{bt}\right) - \left(c_{ac} - c_{bs}\right)\left(c_{as} - c_{at}\right)\right]} \tag{43.11}$$

43.5.2 Process Dynamics

The hold-up or storage variation (stored mass denoted as W) can be expressed as the difference between incoming and outgoing flows in the ideal case:

$$\frac{dW(t)}{dt} = Q_i(t) - Q_o(t) \tag{43.12}$$

The change in the build-up is mathematically an integrating process, but the critical point is that it takes time to change the inventory.

If material is transported a given distance L at a velocity v, it will show a transportation lag of $\tau = L/v$. For a property $p(t)$ entering the transport system at time $t=0$, we can write

$$p_{\text{out}}(t) = p_{\text{in}}(t - \tau) \tag{43.13}$$

In the Laplace domain[1], the transfer function of this is

$$G(s) = e^{-\Delta ts} \tag{43.14}$$

This is a transfer function of a pure time delay (conveyor as an example).

We can use the well-known limiting cases for separators and storage in recycling (Figure 43.8). In the first case, the reactor or storage is not mixed at all (plug flow reactor) (Eq. 43.10) and in the second case, it is instantaneously fully mixed (an ideal (fast) reactor). For the fully mixed case, we have

$$G(s) = \frac{1}{\tau s + 1} \tag{43.15}$$

If such a reactor is "disturbed" with a change in the feed composition, we can multiply the transfer function with the Laplace transform of the disturbance to obtain the response. We get the response in time by performing the inverse Laplace transform back to time-space. If the property entering the fully mixed reactor is a step change with a Laplace transform of $(1/s)$, we get the response

$$C(s) = \frac{1}{s}G(s) = \frac{1}{s}\,\frac{1}{\tau s + 1} => C(t) = 1 - e^{-t/\tau} \tag{43.16}$$

For n reactors in series with the same residence time in all of them:

[1] Laplace transform $F(s) = \int_0^\infty f(t)e^{-st}dt$.

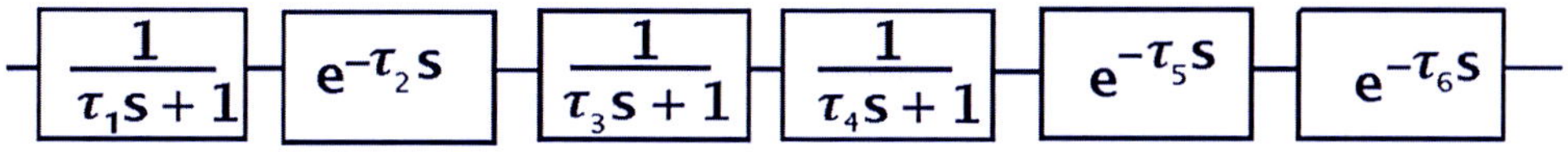

FIGURE 43.9 Flowsheet (Figure 43.8) as a signal flow diagram.

$$G(s) = \frac{1}{(\tau s + 1)^n} \tag{43.17}$$

Using these simple components, we can construct diagrams predicting a flowsheet's dynamic response. Figure 43.8 depicts a simple flowsheet, where the reactors are a crusher and rotating impact mill, respectively.

Figure 43.9 shows the same flowsheet in the Laplace domain showing the residence times and transport lags.

By summing all the transport lags and considering only the two largest residence time constants (the third has only a minor effect on the dynamic response), we get the following signal diagram (Figure 43.10). Using the combined notation of the figure, the variation in the product $p(s)$ related to the required quality setpoint $q(s)$ is

$$p(s) = \frac{G_X(s)G_Y(s)}{1 + G_X(s)G_Y(s)G_M(s)} q(s) \tag{43.18}$$

In designing a control circuit in Figure 43.10, one must ensure that the system is stable and that the control result is adequate. The overall stability of the control circuit can be answered by solving the function. The roots must be negative or have negative real parts. For a constant set-point, the dynamic response of (product/disturbance) is defined by the open-loop transfer function $G_Y(s)G_X(s)G_M(s)$.

43.6 MATERIAL BALANCING

43.6.1 Linear Data Reconciliation

The mass balance equations (Eq. 43.9) always contain random errors, do not close completely, and require data reconciliation, a prerequisite for meaningful recycling computations. It allows obtaining consistent and closed process balances and generates a good understanding of the operation and its trends for further process improvements. Closed balances are also needed for process accounting and performance estimates.

A recycling process can be thought of as a network of connected nodes. A typical flowsheet in recycling can be depicted as in Figure 43.11. Nodes can consist of units combining or separating streams and reactors (chemical and metallurgical reactors and furnaces). The two first kinds of nodes conserve all variables. The reactors only conserve the total mass and the masses of elements but may not conserve other variables.

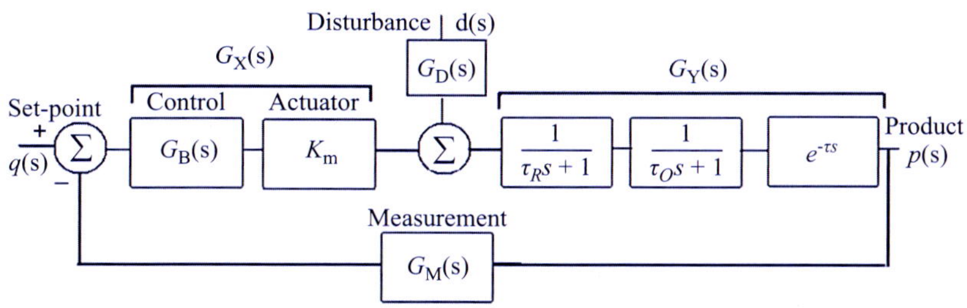

FIGURE 43.10 Signal flow diagram with a feedback control for product quality.

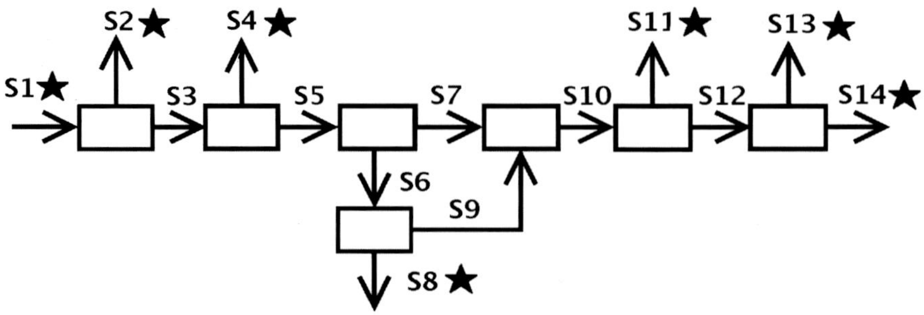

FIGURE 43.11 A typical flowsheet for recycling (star denotes a sampled stream).

A node behavior can be expressed by constraining state equations, which must be fulfilled when performing the data reconciliation. For the balance of the process, we can write in matrix form

$$\mathbf{Cx} = 0 \tag{43.19}$$

where **C** is the constraint equation matrix (with elements *c*), and **x** is the vector of flows connecting the units.

We also need an estimate of the uncertainty of the measured variables (properties). An assumption made in the reconciliation process and formulating the previous equation is that the errors involved are random errors, not gross errors (bias).

Typically, variables are measured only from some of the streams. Thus we have measured and unmeasured variables. Some variables can also be calculated from the information obtained from the other streams. If a variable is not measured but can be computed, it is observable. Of course, if we have no way to get the value of a variable, it is unobservable. When the variable value can be obtained only by the measurement itself, the variable is nonredundant. A redundant variable value can be obtained in at least two independent ways. There are two kinds of redundancy. It is topologically redundant when a variable value is measured and can be back-computed from other measured values. If the size distribution of all the three streams of a size separator (Figure 43.6) is measured, all the measurements are redundant, as we can compute the values of one stream from the two others and have the direct measurement. When one of the measurements is not performed, the remaining stream variables become nonredundant. One of the streams is then nonmeasured but observable. If a second stream measurement is omitted, the two streams not measured become unobservable. A nonredundant variable becomes unobservable if its measurement fails. The second kind of redundancy arises from repeated measurements. This is important, as it gives information on the variable's standard deviation.

Estimability is a slightly broader definition than observability, reserved only for nonmeasured variables. A variable is estimable if it is measured or nonmeasured but observed.

For simple material flow constraints, we get for the matrix **C**:

$$C = \begin{array}{c} \begin{array}{cccccccccccccc} s_1 & s_2 & s_3 & s_4 & s_5 & s_6 & s_7 & s_8 & s_9 & s_{10} & s_{11} & s_{12} & s_{13} & s_{14} \end{array} \\ \begin{bmatrix} 1 & -1 & -1 & & & & & & & & & & & \\ & & 1 & -1 & -1 & & & & & & & & & \\ & & & & 1 & -1 & -1 & & & & & & & \\ & & & & & 1 & & -1 & -1 & & & & & \\ & & & & & & 1 & & 1 & -1 & & & & \\ & & & & & & & & & 1 & -1 & -1 & & \\ & & & & & & & & & & & 1 & -1 & -1 \end{bmatrix} \end{array} \tag{43.20}$$

By arranging the streams as unmeasured and measured and using the basic matrix operators, we can arrange the unmeasured part into observable and unobservable parts by developing the canonical matrix form.

The streams s_3, s_5, s_{12}, and s_{10}, are observable in the example (Figure 43.11). Streams s_6, s_7, and s_9 are unobservable. All the measured streams are nonredundant. A material reconciliation cannot be performed for this sampling scheme. Before sampling campaigns, an analysis of observability must be performed to ensure mass balancing computations. For a detailed treatise, see Rao and Narasimhan (1996).

43.6.2 Nonlinear Data Reconciliation

If the sampling has also given components information, they can be included in data reconciliation. For a single separator, where all streams are measured, we can either minimize the sum of squares of the closure residuals or the sum of squares of the component adjustments. For a steady state, we can write for a unit feed mass (Figure 43.6) for all properties k as follows:

$$c_{fk} - Cc_{ck} - (1 - C)c_{tk} = r_k \tag{43.21}$$

where r_k is closure residual (error). The sum subject to minimization becomes

$$S = \sum_{k=1}^{p} (r_2)^2 \tag{43.22}$$

We get as a result of a least-square computation for the best estimate of C

$$\overline{C} = \frac{\sum_{k=1}^{p} (c_{fk} - c_{tk})(c_{ck} - c_{tk})}{\sum_{k=1}^{p} (c_{ck} - c_{tk})^2} \tag{43.23}$$

To distribute the closure errors, we can write

$$\Delta_{fk} - \overline{C}\Delta_{ck} - (1 - \overline{C})\Delta_{tk} = r_k \tag{43.24}$$

This can be solved by using Lagrangian multipliers (λ). The Lagrangian becomes

$$L = \sum_{k=1}^{p} \left(\Delta_{fk}^2 + \Delta_{ck}^2 + \Delta_{tk}^2\right) + 2\sum_{k=1}^{p} \lambda_k \left(r_k - \Delta_{fk} + \overline{C}\Delta_{ck} + (1 - \overline{C})\Delta_{tk}\right) \tag{43.25}$$

Finding by derivation the minima for the variables (including λ_k) we get after manipulation

$$\begin{aligned} \Delta_{fk} &= \frac{r_k}{1 + \overline{C}^2 + (1 - \overline{C})^2} \\ \Delta_{ck} &= \frac{-\overline{C}r_k}{1 + \overline{C}^2 + (1 - \overline{C})^2} \\ \Delta_{tk} &= \frac{-(1 - \overline{C})r_k}{1 + \overline{C}^2 + (1 - \overline{C})^2} \end{aligned} \tag{43.26}$$

For complex flowsheets, there are several commercially available reconciliation programs (e.g., Metso:Outotec HSC).

43.7 MECHANICAL OPERATIONS

Mechanical separation is based on a balance of forces. The force for separation is chosen according to the properties of the particles to be separated. It can be a body force like gravity, centripetal force, magnetic force, or surface force induced by surface property modifications. This active separation force is directed by the separator design. The trajectory of particles affected by the force becomes different from that of particles not affected by the force. Mass forces can be used to enhance the difference in particle trajectories.

Figure 43.12 depicts a separator flowing from one feed point to two separated streams. All particles start from the same point, but the acting separation force (lower picture) takes the particles to the upper discharge. In contrast, a similar particle without the affecting force will report to the lower discharge.

As argued, separation is a particulate process, where particle size, shape, and density affect the outcome in addition to the active separation force. For large particles, the mass forces are the most important. As the mass decreases to the third power of diminishing particle size and area only to the second power, the surface forces will become dominant at some point. The smaller the particle, the more surface forces like drag and viscosity and even electrostatic and van der Waals forces affect the total force

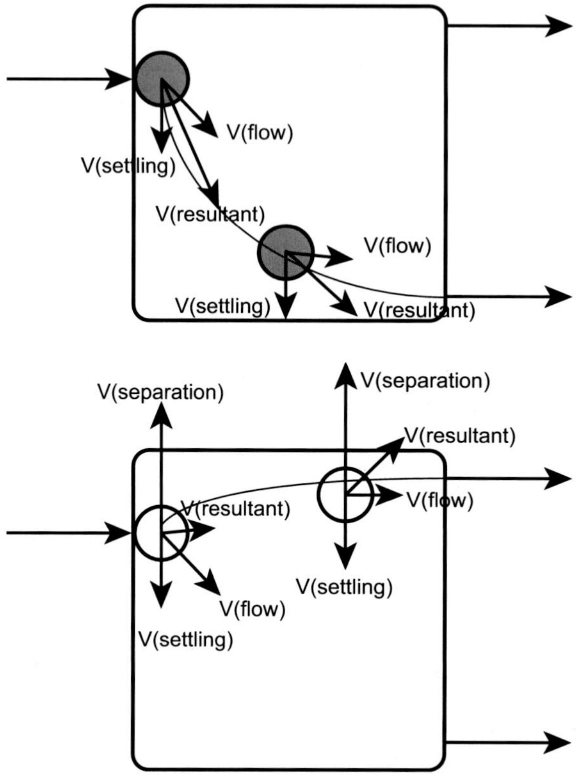

FIGURE 43.12 Particle trajectories in a separator, where a separating force acts on some particle classes (bottom) changing their trajectory, while other particles are unaffected (top).

balance. The task is to optimize particle size for liberation and efficient separation.

There always exists some operational deficiencies for mechanical separations due to property and size distributions, apparent viscosity effects, turbulence, and boundary flows. The subsections to follow discuss such operational variables.

Any particle entering a separator will have a probability to enter one of the product streams. The separation cut-point is the property's value where particles have an equal likelihood of entering any of the two product streams, often denoted as ψ_{50}. In Figure 43.13 the cut-point density is 2705 kg/m^3.

The separation efficiency curve can be calculated from reconciled data as the mass percentage in each property class to report to the chosen product. The shaded areas are misplaced particles. Vertical shading represents light particles reporting in "heavies or sinks" and the horizontal shading heavy particles found in the "light" fraction. The curve's steepness can be used as a measure of separation efficiency, *imperfection*.

$$I = \frac{\psi_{75} - \psi_{25}}{2\psi_{50}} \tag{43.27}$$

A separation efficiency curve can also be computed for secondary effects. In Figure 43.13 the substance Ψ probability curve has a similar slope with an offset. This indicates a slight concentration to the heavy fraction. The reason, in

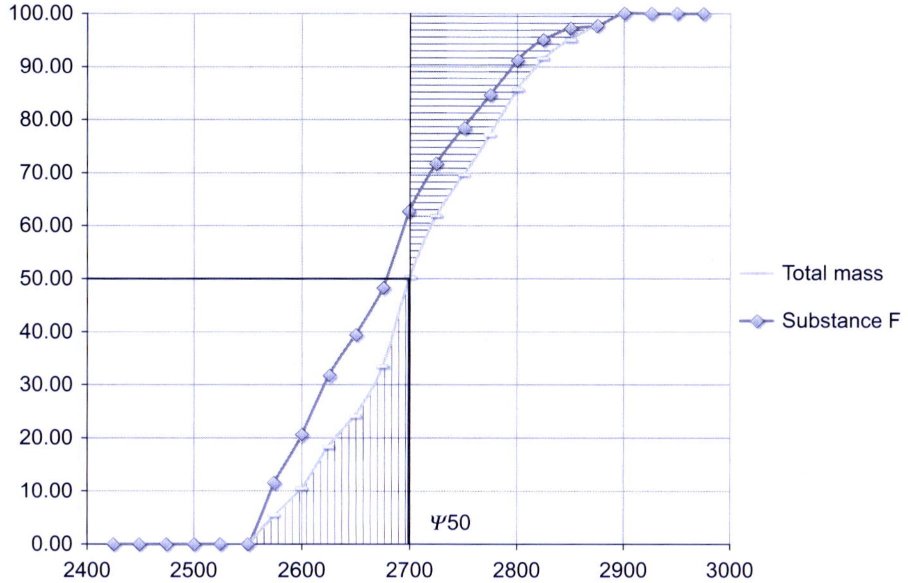

FIGURE 43.13 Separation efficiency curve for density (x-axis) vs probability to sinks (Tromp curve) ($I = 0.023$).

this case, was related to particle shape, but in generic cases requires studying.

43.7.1 The Translational Velocity of Particles

All recycling processes handling discrete entities (particles, droplets, gas bubbles) in continuous media are affected by the translation of such objects. The velocity affects the separation outcome in multiple ways. Velocity differences can be exploited as in a wind sifter. The velocities of fines may be so small that they cannot be separated efficiently as in a dense media separator.

A single particle moves in a fluid medium (liquid or gas), obeying classical mechanics. Two dimensionless numbers help evaluate the behavior of particles settling in a fluid.

Reynolds number **Re** is the ratio of inertial forces to viscous forces. For particles, we can write it to be

$$\mathrm{Re} = \frac{\rho v d}{\mu} \tag{43.28}$$

where d is the characteristic length, i.e., particle diameter, and μ is the dynamic viscosity [Pas]. At low Reynolds numbers, the viscous forces dominate, and the flow around the particle is smooth. A limit is considered typically to be **Re** <0.2. When the Reynolds number increases, inertial forces begin to dominate, and the fluid forms a distinct turbulent wake at the aft of the particle.

The other crucial dimensionless variable is the drag coefficient C_d. When a particle moves through a fluid, it must displace fluid elements from its path. The displacement consumes energy, which can be understood as a force F_d slowing the particle velocity. This drag force has two important components at velocities used in recycling. The first, skin friction, is caused by the fluid viscosity and the second, form drag, by the pressure difference between the fore and aft of the particle:

$$C_d = F_d \frac{2}{\rho_f v^2 A} \tag{43.29}$$

Drag coefficient varies as a function of velocity, particle size and shape, fluid density, and viscosity. The drag coefficient is a function of the Reynolds number **Re**.

The drag force, buoyancy, and gravitational force are the main forces controlling the settling of a particle in a quiescent fluid. The acceleration of a particle is

$$m\frac{dv}{dt} = mg + mg\frac{\rho_f}{\rho_s} + \rho_f v^2 C_d A/2 \tag{43.30}$$

where m is the mass and v is the velocity of the particles, ρ is density (subscript f for fluid and s for solids), C_d is drag coefficient, and A the area of the particle perpendicular to the direction of movement. As the velocity increases, the drag force will increase until the acceleration becomes zero and the particle has obtained its terminal velocity.

For fine particles (typically below 60 μm for solid particles), the drag coefficient can be estimated to be $C_d = 24/\mathrm{Re}$ (for spherical particles). For the terminal velocity, this leads to the well-known Stokes equation:

$$v_{St} = \frac{d^2 g\left(\rho_s - \rho_f\right)}{18\mu} \tag{43.31}$$

For very high Reynolds numbers, the drag coefficient is essentially a constant $C_d \approx 0.44$. This leads to a terminal settling velocity equation for large particles to become (also known as Newton's equation):

$$v_N^2 \approx \frac{3d_p\left(\rho_s - \rho_f\right)}{\rho_f} \tag{43.32}$$

For the intermediate sizes, no closed solutions do exist. Turton and Clark (1987) have presented a useful approximation using dimensionless numbers. For dimensionless velocity v^*, they give as a function of dimensionless size d^*

$$v^* = \left[\left(\frac{18}{d^{*2}}\right)^{0.824} + \left(\frac{0.321}{d^*}\right)^{0.412}\right]^{-1.214} \tag{43.33}$$

The dimensionless size d^* can be obtained from

$$d^* = \left(\frac{3}{4} C_d \,\mathrm{Re}^2\right)^{(1/3)} = d_p\left(\frac{g\left(\rho_s - \rho_f\right)\rho_f}{\eta^2}\right)^{(1/3)} \tag{43.34}$$

and dimensionless velocity v^* from

$$v^* = \left(\frac{4\,\mathrm{Re}}{3C_d}\right)^{(1/3)} = v_S\left(\frac{\rho_f^2}{\eta g\left(\rho_s - \rho_f\right)}\right)^{(1/3)} \tag{43.35}$$

For a known size, calculate first the dimensionless size (Eq. 43.34) and use it to estimate dimensionless velocity (Eq. 43.35) as the last step to solve the actual velocity from Eq. 43.35.

The differences in the settling velocity are large when comparing a piece of steel or a piece of circuit board resin, as can be seen from Figure 43.14. A resin piece having a diameter of 4.3 mm obtains a terminal velocity of 0.1 m/s. A steel particle is only 0.23 mm in diameter to obtain the same terminal velocity.

Nonspherical particles often behave erratically depending on their Reynolds number and the shape itself. In laminar conditions, the particles tend to be oriented to minimize the total drag force. Platy particles tend to wobble and flow erratically. The drag coefficient tends to be a decade higher than for spheres of the same density and mass. Needle-shaped particles translate in a laminar flow with the longest dimension aligned with the flow.

In turbulent flow conditions, the particles tend to become oriented to minimize skin friction. Platy particles are moving, wobbling in a position where the highest surface area is perpendicular to the flow. This causes the drag

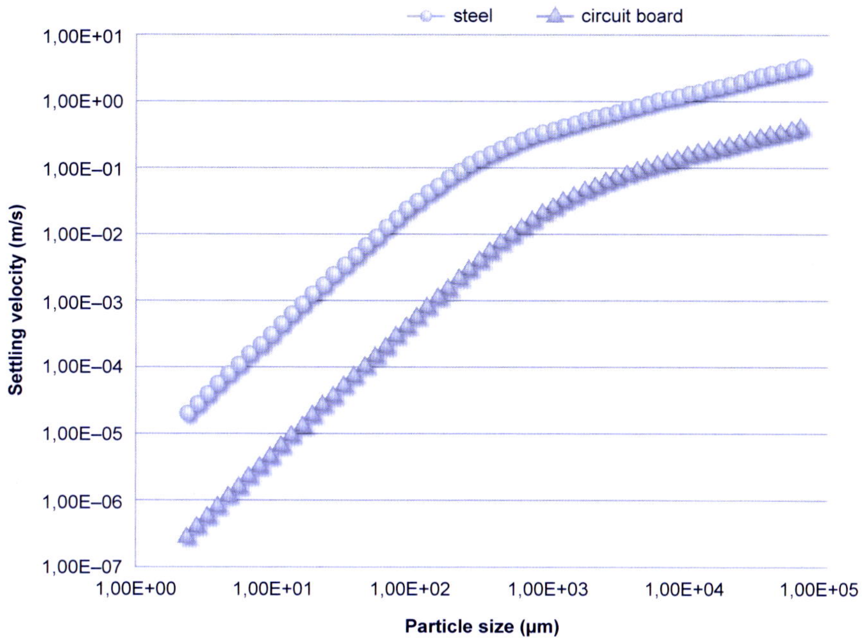

FIGURE 43.14 Particle settling velocity as a function of particle size for two densities (Eqs. 43.34 and 43.35).

coefficient to be up to two decades higher than with a respective sphere. Needle-shaped particles wobble in a turbulent flow.

Generally, the platy particles may be an order or even two orders of magnitude larger (largest dimension) to obtain the same terminal velocity.

43.7.2 Pulp Rheology

Pulp rheology changes affect the flow behavior of all wet separators substantially. Rheological phenomena slow down or even deny particle movement making, for example, heavy media separation impossible if rheology is not controlled.

Particles added to the particle-fluid mixture "pulp" change the flow patterns and how particles move. The most important variable is the volume concentration of solids ϕ in the suspension given by

$$\phi = \frac{\theta_p - 1}{\theta_s - 1} \qquad (43.36)$$

where specific weight $\theta_s = {}^{\rho_s}/_{\rho_{\text{water}}}$.

43.7.2.1 *Apparent Density*

In separators, where solid particles are dispersed in a fluid, the apparent specific density of the dispersion increases as

$$\theta_p = \frac{100}{\left(\frac{P}{\theta_s}\right) + 100 - P} \qquad (43.37)$$

where P is the suspension solid content percentage by mass.

43.7.2.2 *Apparent Viscosity*

Most pure fluids are Newtonian in their behavior. Any minor stress will cause shear, and the fluid moves. The ratio is called viscosity. When the solids content increases in a fluid, the behavior of the fluid resembles increasing viscosity effects. Thomas (1965) proposed the following equation for the viscosity effects of suspended solids:

TABLE 43.1 Parameters for the apparent viscosity equation (Eq. 43.38).

	A	*B*	*C*
Thomas	10.05	0.00273	16.6
Heiskanen and Laapas	14.1	0.0274	−16.6

TABLE 43.2 Parameters for the Richardson and Zaki equation.

Re < 0.2	*b* = 4.65
0.2 < Re < 1.0	*b* = 4.36/Re−0.03
1.0 < Re < 500	*b* = 4.4/Re0.1
Re > 500	*b* = 2.39

$$\frac{\mu_a}{\mu_0} = 1 + 2,5\phi + A\phi^2 + B\exp(C\phi) \quad (43.38)$$

where ϕ is the volume concentration (Eq. 43.38). Heiskanen and Laapas (1979) proposed slightly different parameters for the equation (Table 43.1).

The difference between the predictions is typically below 4% when ϕ is below 30% but increases fast as the Thomas equation predicts substantially higher apparent viscosities at higher volume concentrations.

At high particle volume concentrations, slurries become non-Newtonian, exhibiting critical shear stress. It is a condition when dense media separation starts to show very poor results as particles cannot create sufficient shear to move in the pulp.

43.7.2.3 Hindered Settling

When the solids content of a fluid, notably water, increases, the translational velocity of the particles will decrease. For spherical particles, we get as an experimental equation for the ratio for hindered settling (Richardson and Zaki, 1954):

$$\frac{v_h}{v_{St}} = (1-\phi)^b \quad (43.39)$$

where b is a function of **Re** (Table 43.2).

A small particle having a density of 3000 kg/m^3 in a 30% by weight (12.5% by volume) slurry has only about 55% of the free settling velocity. Increasing the solids fraction by weight to 50% reduces the settling rate further to about 28%. The ratios are 72% and 50% for very large particles, respectively.

To evaluate the effect of particle density in a solid suspension, the following approximate equation can be used for small particles:

$$\frac{x_1}{x_2} \approx \sqrt{\frac{\theta_2 - \theta_p}{\theta_1 - \theta_p}} \quad (43.40)$$

And for large particles:

$$\frac{x_1}{x_2} \approx \frac{\theta_2 - \theta_p}{\theta_1 - \theta_p} \quad (43.41)$$

The size ratio between a steel and an aluminum particle increases by 12% for fine particles and 24% for coarse particles when the specific gravity increases from 1.0 to 1.4.

43.8 CONCLUSION

The metrics and tools described are readily applicable to such recyclable materials, where the interesting component is present in such volume concentrations and structures where liberation is obtainable at processable particle sizes. Applying the tools becomes more difficult if no liberation is easily obtained, as is the case, for example, with coatings. A proxy property correlating with the required property needs to be used in such cases. An example is thin gold coatings having substantial economic value.

References

Castro, M.B., Remmerswaal, J.A.M., Brezet, J.C., van Schaik, A., Reuter, M.A., 2005. A simulation model of the comminution-liberation of recycling streams: relationships between product design and the liberation of materials during recycling. Int. J. Miner. Process. 75 (3–4), 255–281. https://doi.org/10.1016/j.minpro.2004.09.001.

Gay, S., 2004. A liberation model for comminution based on probability theory. Miner. Eng. 17 (4), 525–534. https://doi.org/10.1016/j.mineng.2003.11.012.

Gy, P., 2004. Sampling of discrete materials – a new introduction to the theory of sampling. Chemom. Intell. Lab. Syst. 74 (1), 7–24. https://doi.org/10.1016/j.chemolab.2004.05.012.

Heibeck, M., Rudolph, M., Modler, N., Reuter, M., Filippatos, A., 2021. Characterizing material liberation of multimaterial lightweight structures from shredding experiments and finite element simulations. Miner. Eng. 172, 107142. https://doi.org/10.1016/j.mineng.2021.107142.

Heiskanen, K., Laapas, H., 1979. On the effects of the fluid rheological and flow properties in the wet gravitational classification. In: Proceedings of XIII Int. Mineral Processing Conference, Warsaw, pp. 183–204.

Rao, R.R., Narasimhan, S., 1996. Comparison of techniques for data reconciliation of multicomponent processes. Ind. Eng. Chem. Res. 35 (4), 1362–1368. https://doi.org/10.1021/ie940538b.

Richardson, J.F., Zaki, W.N., 1954. Sedimentation and fluidization: Part I. Trans. Inst. Chem. Eng. 32, 35–53. https://doi.org/10.1016/S0263-8762(97)80006-8.

Thomas, D.G., 1965. Transport characteristics of suspension: VIII. A note on the viscosity of Newtonian suspensions of uniform spherical particles. J. Colloid Sci. 20 (3), 267–277. https://doi.org/10.1016/0095-8522(65)90016-4.

Turton, R., Clark, N., 1987. An explicit relationship to predict spherical particle terminal velocity. Powder Technol. 53, 127–129. https://doi.org/10.1016/0032-5910(87)85007-6.

van Schaik, A., Reuter, M.A., Heiskanen, K., 2004. The influence of particle size reduction and liberation on the recycling rate of end-of-life vehicles. Miner. Eng. 17, 331–347. https://doi.org/10.1016/j.mineng.2003.09.019.

CHAPTER

44

Thermodynamics

Patrick Wollants and B. Blanpain

Department of Materials Engineering, KU Leuven, Leuven, Belgium

44.1 METALS USE AND AVAILABILITY

It is hard to imagine today's life going on without the extensive use of a wide diversity of metals in all kinds of applications—very hard indeed. You could give it a try and tell me how. I do not expect an (easy) answer. The reason is not that living without using (many) metals would not be possible. History teaches us that our ancestors did, but it also teaches that knowledge of producing metals and using them in a variety of applications was inextricably and undeniably linked to power, and to some extent to welfare as well. How willing are you to give up part of your welfare and power...?

At the same time, many reports indicate that metals in ever increasing volumes are the fuel in the transition to renewable energy and a smaller carbon foot print for society. So our present society needs metals. Facing scarcity of energy and metals, we must produce metals at as low an energy cost as possible (optimizing production processes from an energetic point of view is the message) and we must recover, as much as possible, metals from their end-of-life applications to use them again and again and again... (recycling is the message, and also avoiding use of metals in dissipative applications).

Most probably, the concept of closing the metals cycle as part of the hope to be able to create a sustainable world is a beautiful idea of a good mind, but one that can only be realized partially. The real mission is: use less. Use less nonrenewable energy, and use less materials, metals included. Forget the nightmare of growth scenarios, reflect on the essentials and act accordingly. This is the real challenge.

Figure 44.1 gives an overview of the abundance of the chemical elements in the earth's upper continental crust. Metals seldom occur in their native state in nature. Some of the precious metals, such as Au, Ag, Pt, do, but only in limited amounts. Most of the common metals are found in a wide variety of stable minerals, such as oxides, sulfides, fluorides, and halides. Some of them are present in ores, containing concentrations of minerals near the surface of the earth that can be mined economically. The relative abundance of metals in the earth's crust ranges from extremely low (0.001–0.01 ppm for Au, Os, Pt, etc.) to very high (Si 27.7 wt%, Al 8.1 wt%, Fe 5.0 wt%, etc.).

Despite the vast reserves of a number of industrially important metals, it is clear that the growing world population—soon welcoming number 8,000,000,000—cannot hope to be able to consume metals and energy at the rate

Handbook of Recycling. https://doi.org/10.1016/B978-0-323-85514-3.00040-3

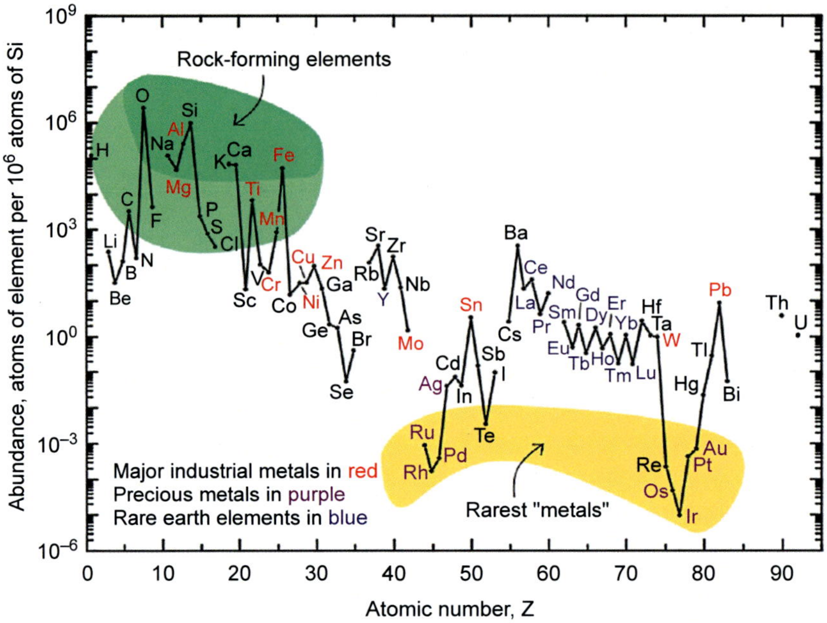

FIGURE 44.1 The abundance (atom fraction) of the chemical elements in the earth's upper continental crust is given here as a function of their atomic number. The rarest elements in the crust (shown in *yellow*) are the densest. The major industrial metals are shown in *red. Based on Haxel et al. (2002). Image from: Wikipedia, the free encyclopedia.*

that has become standard for Western industrialized society. Actually, with the growth of both population and prosperity, especially in developing countries, the prospect of much higher resource consumption levels is "far beyond what is likely sustainable" if realized at all, given finite world resources, as stated in a report by UNEP's International Resource Panel (UNEP 2012).

Some numbers are so hard to imagine that they become overwhelming. Take as an example the production of crude steel. The total crude steel production for 2019 was 1.875 billion tons (World Steel, 2019). With a specific density of around 7.85 tons/m^3, this would represent a volume of $238 \times 10^6 m^3$, a mountain with a surface of $1 km^2$ and a height of 238 m. Another way of looking at these figures is to assume that all this steel would be converted into plate of 1 mm thick. The resulting surface would be $2.38 \times 10^5 km^2$, which corresponds to about a 5.93 km wide strip along the earth's equator. Also, the average CO_2 emission per ton of steel cast equaled 1.83 or in total 3.43 billion tons CO_2 out of 36.44 billion tons CO_2 emissions globally (Ritchie and Roser, 2020), or some 9.4% of the total CO_2 emissions.

In our state of transition, we must realize that, for many reasons, a dramatic change in living conditions and expectations is not likely to occur all of a sudden. In the meantime, we must put all our efforts together to safeguard natural resources as much as possible, to avoid crossing the point of no return, beyond which hope for future generations.

44.2 RECYCLING: AN ENERGY ISSUE

The quantity of available resources is limited. This is a very strong argument in favor of recycling, as much as possible, end-of-life products.

However, recycling does not come without a cost. Depending on the chemical properties and the degree of mixing, more or less energy is required to recover metals in their pure or alloyed state from end-of-life applications. It should be clearly understood that recycling is not creating new resources. Recycling simply postpones the moment at which some metals will no longer be available in the quantities that human mankind requests.

The economic production of metals is not only a question of abundance in the earth's crust. The continued increase in the use of metals over the twentieth century has led to a substantial shift from geological resource-based production, to recovery of metals stored in waste deposits and end-of-life products. It is also an energy issue and, depends on the thermodynamic stability of the primary minerals and on the way they are distributed in the ores, compared with the distribution of metals in end-of-life products and in waste streams.

So we must minimize energy consumption to produce the metals we need and we must minimize the consumption of these metals, certainly in dissipative applications.

44.3 RECYCLING: AN ENTROPY ISSUE

According to the first law of thermodynamics, energy cannot be created or destroyed. Nice. However, the second law of thermodynamics, the "entropy law," makes it clear that the quality of this energy in every process decreases: useful energy is always dissipated as heat in the environment, or, in every real process, entropy is created. In the language of thermodynamics, the entropy of an isolated system (no exchange of heat, work, or matter at its boundaries) can only increase. This introduces the concept of time: an isolated system can never return to a state it was in before. This principle is known as the arrow of time.

From the recycling point of view, this means that a perfect "closing the loop" scenario cannot be written. Recycling requires energy, and in any energetic process energy is dissipated. One could argue that the planet we live on is an open system from the point of view of energy exchange: the sun provides us with "renewable energy." From a materials perspective this is not true. If we want to convert this solar energy into, e.g., electric power, we need suitable infrastructure to realize this. For the construction of this infrastructure, energy and materials are required.

When mixing different metals or compounds (as in waste streams, end-of-life products, etc.), entropy always increases. Therefore to unmix a mixture one must decrease the entropy of that system. This can only be realized by using energy from the environment. Thus recycling is not cost-free: we need energy. We need more energy the more intimately materials are mixed, and the stronger the chemical and physical interactions between these materials are.

So, actually, sustainability does not refer to safeguarding natural resources into eternity, but to extending availability for future generations as long as possible. We play a game we cannot win, but hope exists that it might take a (very) long time to finish the game. From this point of view, the whole concept of recycling is extremely important. Recycling can lead to fewer energy-intensive processes to produce the materials our society seems to need, and extends the horizon of available time left. But the second law cannot be beaten: energy is dissipated all the time, and increased mining and recycling activities, to satisfy growing market demands, only accelerate this irreversible process. So, the real message of importance is: use less of everything, use less energy, and use less materials, whether recycled or not. This change in mindset will most probably prove to be much more important—in the long term—than optimizing recycling processes. But for the moment, we must recycle, also in light of providing enough materials for the transition phase we are in, until we become wise enough to adapt to a more minimalistic mindset. And we must invest in the education of our future generations. They will decide how long we will abuse our planet, and how the gift of life on this beautiful planet can be extended.

44.4 REACTION EQUILIBRIUM

Concerning the primary production of metals and their recovery from end-of-life applications and all types of waste streams, some principles of chemical thermodynamics should be understood.

The simple fact that most metals in nature are not pure but are bound in minerals to oxygen, sulfur, chlorine, etc. reflects that the Gibbs free energy of these minerals is lower than the corresponding Gibbs free energies of the unreacted elements.

Metal ores include mainly oxides, sulfides, carbonates, and halides. As sulfides and carbonates can be easily converted into oxides by roasting or calcination, for example:

$$ZnS + \tfrac{3}{2}O_2 = ZnO + SO_2,$$

extraction of metals is mainly from oxides by chemical or electrolytic reduction.

According to the principles of equilibrium thermodynamics, every system tries to minimize its Gibbs free energy, G, on its way to equilibrium. Consider for example the reaction

$$Si + O_2 = SiO_2.$$

The standard free energy of formation of SiO_2, ΔG^0_{SiO2}, is the difference between the standard Gibbs free energy of the reaction product (1 mol of pure SiO_2) and the standard Gibbs free energy of the reactants (1 mol of pure Si and 1 mol of pure O_2 at a pressure of 1 atm). The Gibbs energy of an element or compound is expressed with respect to a standard state, often the state of the pure element in its stable crystal structure at the given temperature (liquids and solids) and for gases the perfect gas at the given temperature. For any chemical reaction, hence also for the formation of oxides, sulfides, etc., the actual reaction free energy can be expressed, referring to the standard states, as:

$$\Delta G_T = \Delta G^0_T + RT \ln Q,$$

where the reaction quotient Q (take the example of the formation of SiO_2) refers to the actual activities of the reactants and products:

$$Q = \frac{a_{SiO_2}}{a_{Si} . a_{O_2}}.$$

The activities of pure solids and liquids are the activities in the corresponding standard state, and are equal to 1. The activity of the gas at moderate pressures is equal to the actual pressure of the gas:

$$Q = \frac{1}{p_{O_2}}.$$

A reacting system comes to equilibrium when its Gibbs free energy reaches its minimal value, so when the driving force for the reaction becomes zero:

$$\left(\frac{\partial G}{\partial \lambda}\right)_{T,P} = \Delta G_{T,P} = 0.$$

λ is the so-called progress variable. We conclude that at equilibrium

$$\Delta G_T = \Delta G^0_T + RT \ln Q_{eq} = 0,$$

and hence,

$$\Delta G^0_T = -RT \ln Q_{eq} = -RT \ln \left(\frac{1}{p_{O_2}}\right)_{eq}$$
$$= RT \ln p_{O_{2,eq}}.$$

In the case of the formation of SiO_2 the standard reaction free energy predicts the equilibrium pressure of oxygen as a function of temperature. The more negative ΔG^0_T, or the more stable the oxide, the lower will be this equilibrium pressure of oxygen: the oxide hardly decomposes. The negative ΔG^0_T for the formation of a compound indicates directly how much energy is at least required to decompose the compound in its elements: hence to produce the pure metal out of its mineral.

44.5 STABILITY OF COMPOUNDS

The Gibbs free energy is composed of two contributions: enthalpy, H, and entropy, S:

$$G = H - TS.$$

Thus for the formation of the previously mentioned compounds:

$$\Delta G_T^0 = \Delta H_T^0 - T\Delta S_T^0.$$

Ellingham was the first to plot ΔG_T^0 as a function of temperature for oxidation and sulfidation reactions for a series of metals, relevant to the extraction of metals from their ores. Figure 44.2 shows an Ellingham diagram for oxides. In between temperatures at which phase transitions of reactants or products occur, the relation between ΔG_T^0 and T is approximately linear, the slope being equal to the average value of $-\Delta S_T^0$.

From this figure it is clear that the slopes are almost the same for all oxides. This is because the entropy changes in all these cases are similar, being almost entirely due to the condensation of 1 mol of oxygen:

$$-\Delta S_{298}^0 = \left(\frac{2x}{y}\right) S_{298}^0[\mathrm{M}] + S_{298}^0(\mathrm{O_2}) - \left(\frac{2}{y}\right) S_{298}^0\left[\mathrm{M}_x\mathrm{O}_y\right]$$

This Ellingham diagram shows the relative stability of the different oxides and the minimal amount of energy required to produce the metal. The more negative the ΔG_T^0 at a given temperature, the more stable the corresponding oxide and hence the more energy will be required to extract the metal out of it. A metal oxide can be reduced by any metal or element that itself forms a more stable oxide.

In a similar way diagrams can be constructed for the formation of sulfides, halides, nitrides, carbonates, etc. All these diagrams give a good view of the relative stability of the compounds involved.

44.6 THE CARBON TRAGEDY

The unique position of carbon in the Ellingham diagram for oxides appears when we consider the respective reactions:

$$2C + O_2 = 2CO$$

$$C + O_2 = CO_2$$

$$2CO + O_2 = 2CO_2$$

We see that in reaction 3 there is a net loss of 1 mol of gas (slope of ΔG_T^0 versus T curve similar as for most metal oxides), in reaction 2 the number of moles of gas remains the same (entropy change close to zero and hence horizontal line in the Ellingham diagram), and in reaction 1 there is a net production of 1 mol of gas (slope of ΔG_T^0 versus T curve similar in magnitude as for most metal oxides, but opposite in sign). This means that CO becomes more stable than a large number of important metal oxides at sufficiently high temperatures: hence, the key role that carbon plays in the production of metals. CO has a high calorific value; it is further oxidized to CO_2, the heat of combustion being used, for example, to produce electric power, but this at the cost of the unavoidable production of CO_2, which makes the C-based production of steel (and many other metals) in the midterm ecologically unsustainable.

44.7 H_2: AN ALTERNATIVE REDUCER

From the Ellingham diagram we also see that H_2 is great from a reducing point of view, leading to the formation of harmless H_2O as a

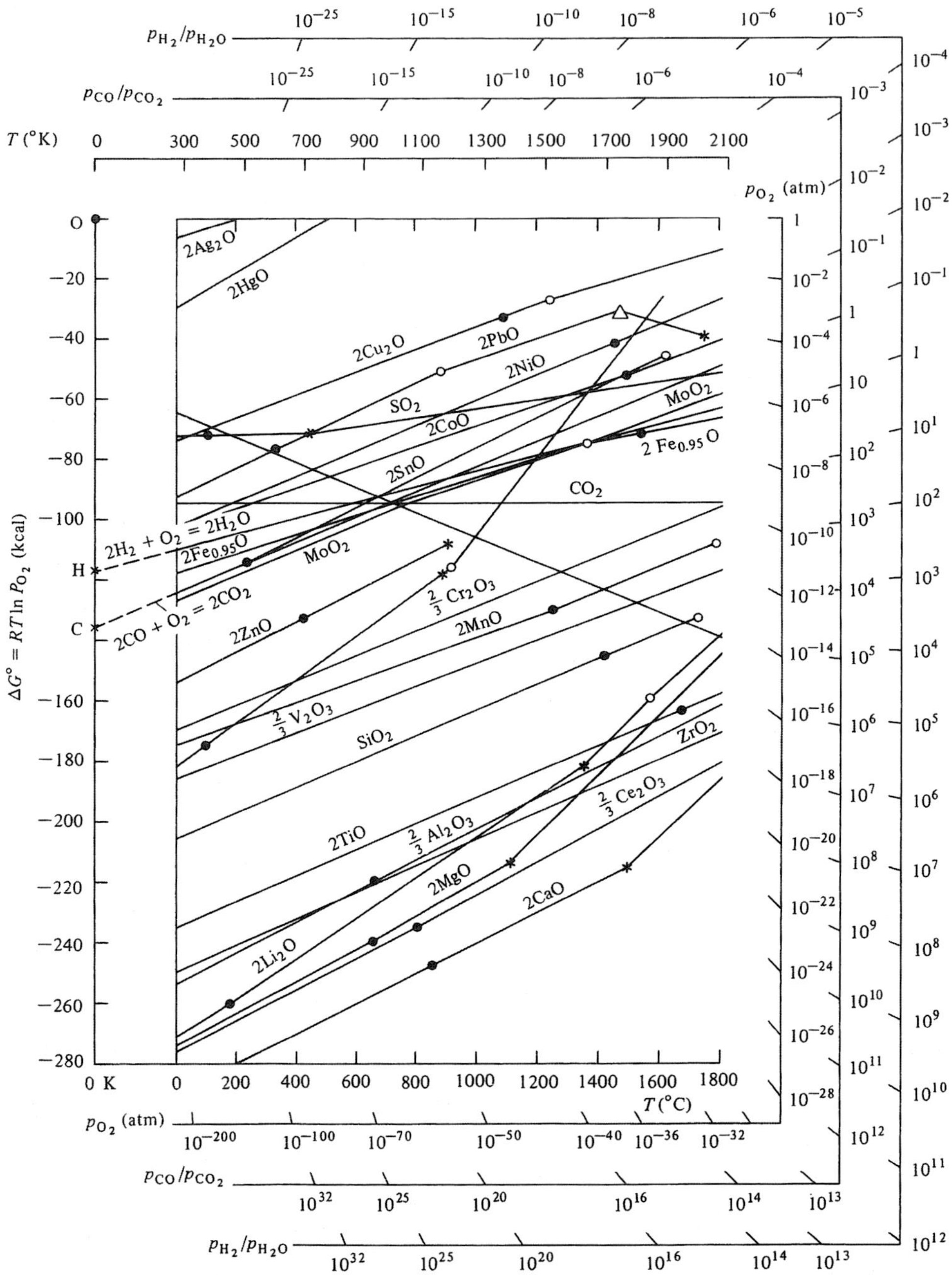

FIGURE 44.2 Standard free energy of formation of some selected oxides, corresponding to the general reaction $(2x/y)$ $M+O_2=(2/y)M_xO_y$. The symbols • and * refer to the melting and boiling points of the metals; the symbols o and △ refer to the melting and boiling points of the oxides. Similar diagrams can be developed based on S, Cl, Br, F, etc. *From Lupis (1983).*

reaction product. From an ecological point of view, it would be great to have access to cheap and vast amounts of hydrogen. There are not that many solutions. Using renewable energy to produce H_2 by the electrolytic dissociation of H_2O is currently being investigated and evaluated.

44.8 VERY STABLE OXIDES

Very low in the diagram we find the lines of very stable oxides: LiO_2, MgO, and Al_2O_3 are among them. They are so stable that C, at acceptable temperatures, cannot be used to reduce them to produce the corresponding metals. To produce Al, for example, the Hall-Héroult process is commonly used. It is the major industrial process for the production of aluminum. It involves dissolving alumina in molten cryolite and electrolyzing the molten salt bath to obtain pure aluminum metal.

The production of aluminum from primary ores requires about 10 times as much energy per kilogram as the production of iron from its primary ores. On the other hand, melting well-sorted scrap is much cheaper in terms of energy consumption. Melting aluminum requires only 5% of the energy required for the production of aluminum starting from primary bauxite. But the issue of purity and composition of the recycled metals and alloys remains important. Iron as an impurity in recycled aluminum, and copper as an impurity in recycled steel, can hardly be removed. For many applications they negatively influence the properties of the recycled metals. It is a major problem. The second law looks around the corner.

44.9 SOLUTIONS AND PURITY LEVELS

44.9.1 Mixing

It is of utmost importance to avoid, as much as possible, mixing of materials (components, elements). Separation is always energy intensive. Dissipative applications of materials should be avoided as well. Dissipated materials cannot be recovered, unless by extremely energy-intensive processes.

The partial molar free energy or chemical potential of an element i, in a liquid or solid solution, is expressed as

$$\mu_i = \mu_i^0 + RT\ln a_i,$$

where in

$$a_i = \frac{p_i}{p_i^0},$$

with p_i being the vapor pressure of component i at the concentration of the given solution at fixed temperature and pressure, and p_i^0 the vapor pressure of component i in its standard state, which in the case of the Raoultian standard state is pure i in its stable crystal structure at the given temperature and pressure. Because partial molar properties are additive, the integral molar free energy of a solution is given by the equation

$$G_m = \sum_i x_i\mu_i^0 + RT\sum_i x_i \ln a_i$$

Hence, the integral molar free energy of mixing is

$$G_m^M = RT\sum_i x_i \ln a_i.$$

As Raoultian activities range from 0 at zero concentration to 1 for the pure component, the integral molar free energy of mixing is always negative: mixing is a spontaneous process, as also illustrated by the fact that the integral molar entropy of mixing is always positive:

$$S_m^M = -\left(\frac{\partial G_m^M}{\partial T}\right)_P = -R\sum_i x_i \ln a_i$$

where we assumed, for the sake of simplicity, that a_i is temperature independent in the temperature range under consideration.

Notice that the value of the chemical potential becomes $-\infty$ at zero concentration of i. This means that the removal of impurities i out of any solution becomes more energy-intensive the lower the residual concentrations are.

44.9.2 Raoultian Activities and Solutions

The activity a_i is proportional to the mole fraction x_i, the proportionality factor γ_i being the activity coefficient:

$$a_i = \gamma_i x_i.$$

For solutions of metals, the activity coefficient γ_i is a measure for the interactions of metal i with the other metals present in the solution. When γ_i equals 1, there is no interaction. Then we have a so-called perfect Raoultian solution. When γ_i is much smaller than 1, there are strong attractive interactions between the metal i and the other metals present in the solution, so separation becomes increasingly difficult (Figure 44.3).

To be able to describe real solution behavior in a quantitative way, one should know how the molar Gibbs free energies of mixing of the phases involved change with composition, or, *how the chemical potentials of the different components depend on the composition*. Thus one should know how the activity coefficients of the different components in the solution depend on temperature, pressure, and composition. This information can be obtained by experiments or by using theoretical models.

Theoretical models for a description of the behavior of different components in, say, a liquid solution, e.g., a liquid metal phase, a liquid metallurgical slag containing a mixture of oxide phases, or a mixture of sulfides, requires insight into the behavior of matter at the atomic level. Therefore, we need strong simplifying assumptions concerning this behavior. For example, in a so-called ideal Raoultian solution, γ_i is equal to 1 over the entire composition range. All components of the solution are then supposed to have identical properties, which in reality is never the case (the components involved wouldn't be different species then). The molar free energy of mixing of an ideal solution is then

$$\Delta G_m^M = RT\sum_i x_i \ln x_i$$

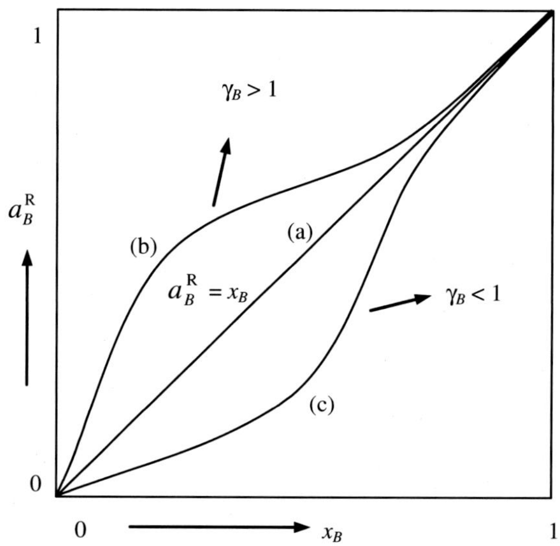

FIGURE 44.3 Schematic representation of the relation between a_B^R and x_B in a binary solution *A-B*: (a) ideal Raoultian behavior, (b) positive deviation from ideal Raoultian behavior, and (c) negative deviation from ideal Raoultian behavior. *From Baelmans and Wollants (2010).*

Then the properties of the solution can be calculated in a straightforward manner, not requiring much experimental input.

Unfortunately, *ideal behavior* never exists, but as a limiting case is approached in certain composition ranges. Yet the concept of ideal behavior is important, as it provides a *reference* for the description of the behavior of mixtures of real gases and of real liquid and solid solutions as well. The difference between real and ideal behavior is the key issue in the thermodynamic description of the composition dependency of the chemical potentials of components in mixtures of real gases and of real liquid and solid solutions.

Many theoretical models have been suggested to describe the composition dependency of the activities and activity coefficients of components in nonideal solutions. These models range from very simple, e.g., the so-called regular solution model, according to which the molar free energy of mixing of a binary solution is given by the equation

$$\Delta G_m^M = RT(x_A \ln x_A + x_B \ln x_B) + \Omega x_A x_B$$

wherein Ω is the regular solution parameter, which, in the basic regular solution model, is a temperature-independent constant. For some solutions, containing components showing very similar properties, the agreement with experimental observations is quite reasonable and good enough to describe the properties of the solution in question with an accuracy that falls within the range of experimental uncertainty. However, there is not a single model by which the properties of all solutions can be described. One can extend the applicability of theoretical models by introducing temperature- and concentration-dependent parameters, and by introducing always more parameters. But finally, at any rate, experimental data are needed to obtain a "perfect" quantitative description of complex nonideal solution behavior.

When a reliable description of the molar Gibbs free energies of mixing of all the possible phases in which the different components of the system, say: CaO, SiO_2, and Al_2O_3, can occur is available, the relative amounts and the composition of the different phases coexisting at equilibrium can be calculated as a function of temperature and pressure. Phase diagrams are the graphical representation of these phase equilibria. They show in a relatively simple way, for example as a function of temperature and concentration, which phases coexist at equilibrium and what is their composition.

Theoretical models are helpful to describe at least in a qualitative way the behavior of solutions, especially when different solutions or phases are contacting each other. The distribution of elements or components between these phases is bound to thermodynamic equilibrium conditions. These conditions show that elements (metals to be recovered) distribute themselves between different phases. In stainless steel making, e.g., Cr, to a certain extent, is picked up by the steelmaking slag. So, Cr is "dissipated." Eventually the slag can be recycled, say as a component for cement production. From the Cr point of view, the remaining Cr in the slag is lost, if no additional recovery process is installed.

Always, the chemical potential of any component *i* decreases strongly at low concentrations of *i*. As the chemical potential is a measure for the amount of energy needed to remove remaining *i* out of the given solution, it becomes clear that the refining of metals can be very expensive in terms of energy and that it is mandatory to avoid mixing of elements during recycling, and, even better, early on in the production or manufacturing process.

44.10 CONCLUSIONS

- The production of metals from primary ores is energy intensive.
- Sometimes the production of metals from recycling is less energy intensive than the production of metals from primary ores.

- Recycling of metals, trying to close the materials cycle, is of the highest priority, both from an energy use and a materials supply perspective.
- Thermodynamics shows that mixing is a spontaneous process by which the entropy increases.
- Based on the principles of equilibrium thermodynamics, the distribution of components between different coexisting phases can be calculated.
- Any activity, also recycling, leads to the dissipation of energy.
- Recycling of metals from very dilute mixtures and solutions is very expensive and should therefore be avoided if possible.
- Careful and selective collection of waste metals is mandatory to avoid unwanted contamination and additional expensive (eventually impossible) refining.
- Dissipative applications should be avoided. This is a top priority.
- C-based reduction processes in the midterm become ecologically unsustainable, due to the massive production of CO_2 and the depletion of fossil fuels. The potential of H_2-based reduction processes based on the use of "renewable" energy in electrolysis as well as electricity-based metal production processes are being explored in detail by many companies in their roadmaps to carbon neutral production.
- The key to a more sustainable use of metals is to use less of them. Much less. Less is beautiful. In the meantime, we need to integrate this absolute truth while providing the materials needed to support the transition of our production and consumer systems. It will be an incredible balancing act as we hopefully are able to steer away from a global apocalypse, as signs of climate change are becoming ever more horrifying.

For further reference, the reader may also consult for instance the textbook by Hayes (2021), in which the reader can obtain more information, not only through the text but also the multitude of references provided.

References

Baelmans, M., Wollants, P., 2010. Inleiding tot de toegepaste thermodynamica. ACCO, Leuven, ISBN: 978 94 6292 661 5.

Haxel, G.B., Hedrick, J.B., Orris, G.J., 2002. Rare Earth Elements—Critical Resources for High Technology. USGS—United States Geological Survey, 2002. https://pubs.usgs.gov/fs/2002/fs087-02/fs087-02.pdf.

Hayes, P., 2021. Process Principles in Minerals and Materials Production—With a Focus on Metal Production and Recycling, fourth ed. Hayes Publishing, Sherwood, Queensland, Australia, ISBN: 978-0-6486989-5-1. eBook edition.

Lupis, C.H.P., 1983. Chemical Thermodynamics of Materials. North Holland, New York, p. 134.

Ritchie, H., Roser, M., 2020. CO_2 and Greenhouse Gas Emissions. Published online at OurWorldInData.org. Retrieved from: https://ourworldindata.org/co2-and-other-greenhouse-gas-emissions. (Online Resource).

UNEP, 2012. Responsible Resource Management for a Sustainable World: Findings From the International Resource Panel. https://wedocs.unep.org/20.500.11822/10580. 978-92-807-3278-8.

World Steel, 2019. World Steel in Figures 2019. World Steel Association, Brussels. https://www.worldsteel.org/steel-by-topic/statistics/annual-production-steel-data/P1_crude_steel_total_pub/CHN/IND. (Accessed 27.12.21).

CHAPTER

45

Exergy—Quantification of resource dissipation

Alejandro Abadías Llamas[1], *Neill J. Bartie*[2], *Christina Meskers*[3], *and Markus A. Reuter*[4]

[1]Nordenham Metall GmbH, Nordenham, Germany [2]Helmholtz Centre Berlin for Materials and Energy (HZB), PVcomB, Berlin, Germany [3]Department of Energy and Process Engineering, Faculty of Engineering Sciences, NTNU—Norwegian University of Science and Technology, Trondheim, Norway [4]WASM: Minerals, Energy and Chemical Engineering, Faculty of Science and Engineering, Curtin University, Perth, Australia

Nomenclature

$\boldsymbol{A}$	incidence matrix
C	velocity (m/s)
e	exergy (kJ/mol or kJ/kg)
E	exergy (kW)
E_{ij}	exergy stream from i to j (kW)
E^*	exergy cost (kW)
$\boldsymbol{E}^*$	exergy cost vector (kW)
F	fuel (kW)
$\boldsymbol{F}$	vector of fuels of the processes (kW)
F^*	fuel exergy cost (kW)
$\boldsymbol{F}^*$	vector of fuel exergy costs (kW)
g	gravitational acceleration (m/s^2)
G_{fi}	Gibbs free energy of the compound i (kJ/mol)
h	enthalpy (kJ/mol or kJ/kg)
I	irreversibility (kW)
k^*	unit exergy cost
$\boldsymbol{K_D}$	diagonal matrix of unit exergy consumptions
$n_{j,\,i}$	moles of the element j per mole of substance i (mol)
P	product (kW)
$\boldsymbol{P}$	vector of products of the processes (kW)
P^*	product exergy cost (kW)
$\boldsymbol{P}^*$	vector of product exergy costs (kW)
q	heat flow (kJ/kg)
R^*	residue cost (kW)
$\boldsymbol{R}^*$	residue cost vector (kW)
s	entropy (kJ/(mol K) or kJ/(kg K))
T	temperature (K)
$\boldsymbol{U_D}$	identity matrix
y	distribution ratios
$\boldsymbol{Y}_e^*$	external assessment vector
z	height (m)
$[\boldsymbol{A}]$	cost matrix
$\langle \boldsymbol{FP} \rangle$	distribution coefficients matrix, y
$\langle \boldsymbol{RP} \rangle$	residue allocation ratios matrix, ψ

Subscripts and Superscripts

−1	inverse matrix
0	environment or system outputs
ch	chemical exergy
e	external resources
F	from fuel
i	unit, component, or chemical compound

https://doi.org/10.1016/B978-0-323-85514-3.00018-X

in input
j unit, component, or chemical compound
k kinetic exergy
out output
P from product
ph physical exergy
q heat exergy
x bifurcation ratios
z potential exergy
$*$ cost

Greek Symbols

α cost matrix component
Δ change
η exergy efficiency
κ unit exergy consumption
ω exergy of the system inputs
ψ residue cost distribution ratio

45.1 INTRODUCTION

More resource efficient metallurgical and recycling processes are key for a sustainable society. Therefore the resource efficiency of these processes should be evaluated thoroughly. Mass and energy balances provide information on the resource use, e.g., raw materials, energy, or water. However, they cannot assess the effect of the Second Law of Thermodynamics (2LT) on the metallurgical or recycling process. Accordingly, the irreversibility of the process cannot be evaluated.

Exergy is based on the 2LT. Therefore, an exergy analysis quantifies the entropy generation or exergy dissipation of a process, which are resources as well. These exergy dissipation indicators give an idea on the effort in terms of additional energy and materials that are needed to process and recycle a material back to usefulness. Accordingly, exergy dissipation should be evaluated in recycling and metallurgical processes, together with other resource consumption indicators such as that obtained from mass and energy balances. This way, pathways for the exergy dissipation minimization along the value chain of metals can be identified.

Moreover, metallurgy and recycling are key players for a circular economy. Unlike mass and energy, exergy is not conserved but dissipated. Accordingly, exergy quantifies the resources dissipated throughout the circular economy, quantifying how circular they are. Therefore exergy is a necessary indicator to evaluate how circular the economy is.

Moreover, exergy allows the integration of different resources such as energy, water, or raw materials in one indicator. These resources have an exergy content measured in energy units such as kilowatts or kilojoules (Szargut, 2005). Accordingly, the resource use in large metallurgical and recycling processing systems can be quantified with one indicator.

45.2 WHAT IS EXERGY?

Exergy expresses the maximum work output that can be extracted from a system when it reaches equilibrium with its environment in a reversible way (Szargut, 2005). It gives an idea of the quality of resources such as energy, water, or raw materials. For example, electricity is an energy source with more quality than a heat flow, even if their energy content is the same. One may also prefer water from mountain springs rather than seawater because of its clarity and the potential to produce hydropower. This quality difference can be quantified by exergy.

The exergy of a system is composed of different types of exergy, such as the kinetic (e_k), potential (e_z), chemical (e_{ch}), and physical (e_{ph}) exergies (Eq. 45.1). These exergies always depend on the reference environment, denoted with the subscript 0.

$$e = e_k + e_z + e_{ch} + e_{ph} \quad [\text{J/kg}] \tag{45.1}$$

The kinetic exergy is calculated through the velocity, C (Eq. 45.2). The potential exergy is determined through the height, z (Eq. 45.3). These two equations represent exergy change through movement.

$$e_k = \frac{1}{2}\left(C^2 - C_0^2\right) \quad [\text{J/kg}] \tag{45.2}$$

$$e_z = g(z - z_0) \quad [\text{J/kg}] \tag{45.3}$$

where g is the gravitational acceleration.

Moreover, the chemical exergy of a substance represents its exergy at environmental temperature and pressure. Accordingly, it represents the maximum work that can be obtained when that substance reaches the state of the reference substances in the environment in a reversible way (Valero and Valero, 2010). Several reference environments have been proposed for the calculation of the chemical exergies of elements and substances. However, Szargut's reference environment is the most used (Szargut, 1989). The chemical exergy of a substance i can be calculated using Eq. (45.4):

$$e_{ch,i} = \Delta G_{f,i} + \sum_i n_{j,i} \cdot e_{ch,j} [\text{J/mol}] \tag{45.4}$$

where $\Delta G_{f,i}$ is the Gibbs free energy of formation of substance i; $n_{j,i}$ is the moles of the element j per mole of substance i; and $e_{ch,j}$ is the standard chemical exergy of the element j in substance i.

The chemical exergy of fuels is difficult to predict because of their chemical heterogeneity and complexity. Nevertheless, their value can be approximated by their heating value (Szargut, 2005).

The physical exergy of a substance is the exergy related with the temperature and pressure variation with respect to the reference environment, without any chemical composition change (Eq. 45.5):

$$e_{ph} = (h - h_0) - T_0(s - s_0) \quad [\text{J/kg}] \tag{45.5}$$

where h and s are the enthalpy and entropy, respectively. The enthalpy and entropy values can be obtained from thermodynamic property tables or other sources. Please see Chapter 44—Thermodynamics for more information on enthalpy and entropy.

Moreover, the exergy content of energy flows depends on the energy type. For example, electrical exergy equals its energy value. However, the exergy of a heat flow q depends on its temperature, and it can be calculated using Eq. (45.6):

$$E_q = Q\left(\frac{T - T_0}{T}\right) \quad [\text{J}] \tag{45.6}$$

Exergy combines the First and Second Laws of Thermodynamics. Accordingly, an exergy balance accounts for the entropy generation or irreversibility of a system; thus unlike mass and energy balances, an exergy balance does not equal to 0. The difference between the exergies of the inputs and outputs of a system equals to the exergy dissipation or irreversibility, I. These inputs and outputs include the exergies of the system (Eq. 45.1) as well as the exergies of the energy flows entering and leaving it, for example electricity and heat:

$$\Delta E = E_{\text{in}} - E_{\text{out}} = I > 0 \quad [\text{J}] \tag{45.7}$$

If a system with several processes is considered, every individual process contributes to the exergy dissipation of the system. Accordingly, the products of that system will have an exergy cost, i.e., the exergy required to produce them (Szargut and Morris, 1987). Thermoeconomics is a methodology that combines thermodynamics and economics, and was developed to calculate the exergy costs of systems (Valero et al., 1986). This methodology will be explained next.

45.3 THERMOECONOMICS

The physical structure of a system is a representation of its process and how they are connected through streams. However, a productive structure represents the purpose of the processes in a system (Torres et al., 1996). Therefore it describes the purpose of each unit operation of the system, known as "Product," and the resources used to achieve these purposes, which are called "Fuel" (Frangopoulos, 1987). The definition of the productive structure of the system is the first step to perform a thermoeconomic analysis.

For example, a steam turbine has one input stream, the steam coming in, and two output streams, the steam coming out and mechanical work, from a physical structure point of view. However, from a productive structure point of view, the product of the turbine is the mechanical work generated, while the difference between the exergy of the steam entering and leaving the turbine would be the resources used, its fuel.

The product, P, of a unit operation, i, of a system with n unit operations can be calculated through Eq. (45.8). Likewise, the resources of the process, or fuel, F, can be calculated through Eq. (45.9) (Torres, 2009).

$$P_i = \sum_{j=0}^{n} E_{ij} \quad (45.8)$$

$$F_i = \sum_{j=0}^{n} E_{ji} \quad (45.9)$$

where E_{ij} is the exergy flow going from the component i to j.

Accordingly, the exergy balance of a unit operation i can be calculated through Eq. (45.10). As a result, the irreversibility, I, of the unit operation i can be calculated.

$$F_i - P_i = I_i \quad (45.10)$$

The exergy efficiency, η, of a unit operation i is defined as the ratio between the product and fuel. Moreover, the inverse of the exergy efficiency, also called the unit exergy consumption, κ, of a unit operation i represents the resources required to produce one unit of product. These efficiency parameters can be calculated through Eq. (45.11).

$$\eta_i = \frac{1}{\kappa_i} = \frac{P_i}{F_i} \quad (45.11)$$

Furthermore, each stream within the system has an exergy cost, which is the exergy used to produce it (Valero et al., 1986). This exergy cost, E^*, includes the exergy of the stream, E, and the irreversibilities generated throughout its production, I. The exergy cost of a stream can be calculated through Eq. (45.12).

$$E^* = E + \sum_{\text{Process}} I \quad (45.12)$$

Additionally, the exergy required to produce one unit of exergy of a stream can be calculated through the ratio of the exergy cost and the exergy of the stream (Eq. 45.13):

$$k^* = \frac{E^*}{E} \quad (45.13)$$

45.3.1 The Exergy Cost Theory

The complexity of industrial systems, where many unit operations, streams, or recirculation of resources are present, makes the process of calculating the exergy cost of all the streams of a system difficult. Therefore the theory of the exergy cost was developed (Lozano and Valero, 1993a). This methodology consists of four rules that allow the calculation of the exergy cost of all the streams of any system. An example of the application of this theory to a simple power plant can be found in the cited article.

- The exergy cost is a conservative property; thus the exergy cost balance equals to zero. Therefore the cost of the input and output streams of a unit operation is the same. One equation per unit operation of the system is obtained through this proposition.
- The exergy cost of the resources entering the system from the environment equals its

exergy if there is no external assessment. Therefore one equation is generated per stream entering the system from the environment.

- If an output stream of a unit operation belongs to its fuel, for example steam leaving a steam turbine, its unit exergy cost equals the one of the input stream it belongs to.
- If the product of a unit operation is composed of several output flows, they have the same unit exergy cost.

The system of equations generated by the application of the theory of the exergy cost can be expressed algebraically, as Eq. (45.14) shows.

$$\begin{bmatrix} A \\ \\ \alpha_e \\ \\ \alpha_x \end{bmatrix} \cdot E^* = \begin{bmatrix} 0 \\ \omega_e \\ 0 \end{bmatrix} \rightarrow [A] \cdot E^* = Y_e^* \rightarrow E^* = [A]^{-1} \cdot Y_e^* \tag{45.14}$$

where $[A]$ is the matrix of costs; E^* is the vector containing the exergy cost of the streams; and Y_e^* is the vector of external assessment containing the external information required to determine the exergy cost of the system. $[A]$ is composed of (Lozano and Valero, 1993a):

- The incidence matrix, A. It can be calculated as the difference between the incidence matrices of fuel, A_F, and product, A_P:

$$A = A_F - A_P \tag{45.15}$$

- The matrix of external assessment, α_e, which defines the inputs of the system. As it is based in the second rule of the theory of the exergy cost, it should satisfy:

$$\alpha_e \cdot E = \alpha_e \cdot E^* = \omega_e \tag{45.16}$$

where ω_e is the vector of external resources containing the exergy values of the inputs of the system.

- The matrix of distribution ratios, α_x. It contains the equations generated in the third and fourth rules of the exergy cost theory.

45.3.2 Symbolic Thermoeconomics

Symbolic thermoeconomics is a methodology with the objective of explaining the cost formation process and its causes (Torres, 2009). It combines the exergy cost theory explained before with symbolic computation to provide a general method for the thermoeconomic analysis of any system as a function of its productive structure and its interaction with the environment. This methodology, which is based in matrix operators, has two approaches: (i) from fuel to product (Fuel-Product representation), and (ii) from product to fuel (Product-Fuel representation). In this section, the Fuel-Product one will be explained. This approach calculates the exergy cost of all the streams of the system from its inputs. It is based on distribution coefficients, y_{ij}, which represent the portion of the product of the component j that is used as resource in i. They can be calculated through Eq. (45.17).

$$y_{ij} = \frac{E_{ji}}{P_j}; \quad \text{where} \sum_{i=0}^{n} y_{ij} = 1 \tag{45.17}$$

Eq. (45.9) can be combined with Eq. (45.17) to provide a relation between the fuel and the product of each component, as Eq. (45.18) shows. It can also be expressed in matrix form (see Eq. 45.19).

$$F_i = E_{0i} + \sum_{j=1}^{n} y_{ij} \cdot P_j \tag{45.18}$$

$$F = F_e + \langle FP \rangle P \tag{45.19}$$

where E_{0i} are the external resources entering the component i; F_e is a vector containing these values for external resources entering each process; F and P are vectors containing the fuel and product values of each process; and $\langle FP \rangle$ is a matrix containing the distribution coefficients y_{ij}.

Eq. (45.11) provides a relation between the fuel and product of a process through the concepts of exergy efficiency and unit exergy consumption. Therefore Eqs. (45.11) and (45.19) can be combined so that the products of each process can be calculated depending on the external resources entering the system, as Eq. (45.20) shows.

$$\boldsymbol{P} = (\boldsymbol{K_D} - \langle \boldsymbol{FP} \rangle)^{-1} \cdot \boldsymbol{F_e} \tag{45.20}$$

where $\boldsymbol{K_D}$ is a diagonal matrix containing the unit exergy consumption of each process.

Furthermore, the exergy cost of the fuel of a component *i* is composed of the cost of the external resources and streams coming from other components entering that process and can be calculated with Eq. (45.21).

$$F_i^* = E_{0i}^* + \sum_{j=1}^{n} E_{ji}^* \tag{45.21}$$

where F_i^* is the exergy cost of the fuel of the process *i*; E_{0i}^* is the exergy cost of the external resources entering the process *i*; and E_{ji}^* is the exergy costs of the stream going from *j* to *i*.

Based on the exergy cost theory, the product streams of a process will have the same unit exergy cost. Accordingly, the distribution coefficients can be also expressed as (Usón et al., 2012a):

$$y_{ij} = \frac{E_{ji}^*}{P_j^*} \tag{45.22}$$

Moreover, the theory of the exergy cost states that the exergy cost of the external resources used in one process equals to its exergy, which leads to:

$$E_{0i}^* = E_{0i} \tag{45.23}$$

Combining Eqs. (45.22) and (45.23) with Eq. (45.21) leads to Eq. (45.24), while Eq. (45.25) represents it in matrix notation.

$$F_i^* = E_{0i}^* + \sum_{j=1}^{n} y_{ij} \cdot P_j^* \tag{45.24}$$

$$\boldsymbol{F}^* = \boldsymbol{F_e} + \langle \boldsymbol{FP} \rangle \cdot \boldsymbol{P}^* \tag{45.25}$$

Considering that the exergy cost is a conservative property, the cost of the product equals the cost of the fuel. Therefore:

$$\boldsymbol{P}^* = \boldsymbol{F}^* \tag{45.26}$$

Then, Eqs. (45.25) and (45.26) can be combined so that the cost of the products of each process can be calculated depending on the external resources entering the system:

$$\boldsymbol{P}^* = (\boldsymbol{U_D} - \langle \boldsymbol{FP} \rangle)^{-1} \cdot \boldsymbol{F_e} \tag{45.27}$$

where $\boldsymbol{U_D}$ is the identity matrix.

45.3.3 Exergy Cost of Residues

The methodology explained before considers that the products of the system are useful. Nevertheless, residues are also produced. The cost of generating and treating these residues should be allocated to the products of the system. There is a methodology for the residue cost allocation (Torres et al., 2008). This methodology considers dissipative processes, which are the processes of the system that treat and dispose of its residues. Accordingly, the output of these processes are residues (R). The residue cost allocation is performed by using the residue cost distribution ratios, ψ, defined by Eq. (45.28).

$$\psi = \frac{R_{rj}^*}{R_{r0}^*} \tag{45.28}$$

where R_{r0}^* is the cost of the dissipative component *r*; and R_{rj}^* is the part of the cost allocated to the component *j*.

The residue cost distribution ratios are arbitrary; thus several residue cost allocation methods have been proposed: for example, an allocation proportional to the exergy of the streams processed in the dissipative process according to the productive structure (Torres et al., 2008). Other researchers proposed an allocation proportionally to the entropy generation throughout the system (Frangopoulos, 2009;

Lozano and Valero, 1993b). An allocation based on the entropy distribution ratios throughout the system was also proposed (Seyyedi et al., 2010). Moreover, this allocation can also be proportional to the contribution of productive components to the cost of the residue (Agudelo et al., 2012).

The residue cost allocated to each process of the system can be expressed in matrix form through Eq. (45.29).

$$\boldsymbol{R}^* = \langle \boldsymbol{RP} \rangle \cdot \boldsymbol{P}^* \tag{45.29}$$

where $\boldsymbol{R}^*$ is a vector containing the residue cost allocated to each process of the system; and $\langle \boldsymbol{RP} \rangle$ is a matrix containing the residue cost allocation ratios, ψ.

When the residue cost is considered, the exergy cost of the products should include it. Accordingly, Eq. (45.26) becomes:

$$\boldsymbol{P}^* = \boldsymbol{F}^* + \boldsymbol{R}^* \tag{45.30}$$

Therefore combining Eqs. (45.25), (45.29), and (45.30) leads to the general equation to obtain the exergy cost of the products of a system depending on its external resources, and considering the residue costs:

$$\boldsymbol{P}^* = \left(\boldsymbol{U}_D - \langle \boldsymbol{FP} \rangle - \langle \boldsymbol{RP} \rangle\right)^{-1} \cdot \boldsymbol{F}_e \tag{45.31}$$

45.4 APPLICATIONS IN INDUSTRIAL PROCESSES

Exergy analyses have been mainly performed in energy systems to evaluate their efficiency and identify areas for improvement: for example, for the assessment of the exergy efficiency and irreversibility of a new combined power and refrigeration cycle (Vidal et al., 2006). Nevertheless, exergy has been used to evaluate other industrial processes. Some researchers performed a review of the exergy analyses applied to the cement industry, e.g., to quantify the exergy efficiency of the unit operations of a cement plant, finding where the largest exergy dissipation occurs (Madlool et al., 2012). Moreover, the industrial symbiosis in Kalundborg, Denmark has also been evaluated through an exergy analysis (Valero et al., 2012).

Exergy analyses have also been applied in metallurgical processes. Ayres and colleagues used exergy to account for resource use and residue generation in the steel, aluminum, copper, lead, and zinc industries (Ayres et al., 2006). The exergy efficiency of the production of nickel from laterites and sulfide ores was also compared (Dominguez et al., 2013). Furthermore, the iron and steel production processes were evaluated to quantify the exergy efficiency of the different unit operations (Hernandez and Cullen, 2016). Recycling processes have also been assessed through exergy.

Exergy analysis has also been used to quantify the exergy losses of the recycling process of several alloys (Amini et al., 2007) or to evaluate the exergy efficiency of a car recycling flowsheet (Ignatenko et al., 2007). Meskers and colleagues used exergy for the evaluation of the recycling process of coated magnesium, ranking the coating components based on exergy dissipation to find the one that leads to the least losses during recycling, so that designers can make an informed decision (Meskers et al., 2007). In order to understand physical particle systems in terms of exergy, statistical exergy has been applied to understand the physical recycling of lithium ion batteries (Velázquez-Martínez et al., 2019a,b). In addition, understanding particulate systems in terms of exergy, recent work developed and analyzed the concepts for crushing and milling, which can also be applied for recycling (Fernandes et al., 2021).

These analyses, mostly done by hand, led to the integration of exergy calculations into HSC 8 in 2013 based on the Szargut standard (www.mogroup.com). This made it possible to easily and efficiently analyze the exergy flow in large flowsheets, as documented by Reuter et al. (2015), and implemented in various examples such as e-waste smelting, copper smelting, as well as pig iron production (Reuter et al., 2015).

In the case of thermoeconomics, this methodology was developed to evaluate and optimize energy systems (Tsatsaronis, 1993). Therefore energy systems have been the main application field for thermoeconomics over the years. Examples of these energy systems are power plants (Usón et al., 2010) or combined heat and power systems (Silveira and Tuna, 2003). Thermoeconomics has also been used for the evaluation of industrial symbiosis cases. For example, the Kalundborg industrial symbiosis, where the unit exergy costs of the products of all the processes within that system have been determined (Usón et al., 2012b). Thermoeconomics has also been applied recently in metallurgical processes, and could be integrated now easily into HSC 10 to perform simulation-based exergy and thermoeconomic analysis of large systems. It was demonstrated for primary copper production using a dedicated tool developed and coupled to the process simulator HSC Sim (Abadías Llamas et al., 2019). The same approach has been applied to the resource efficiency evaluation and optimization in primary zinc production (Abadías Llamas et al., 2020), NdFeB permanent magnet production (Fernandes et al., 2020), and the life cycle of CdTe photovoltaic modules (Bartie et al., 2019). Simulation-based exergy analyses have also been performed for recycling processes. For example, an exergy analysis using HSC Sim and its thermoeconomics calculator for an aluminum recycling process was also performed to quantify the exergy dissipation of the recycling of different aluminum scraps (Hannula et al., 2020). Moreover, Bartie and colleagues have used the same approach to evaluate the resource efficiency of the recycling of photovoltaic modules, finding the room for improvement in each step of the value chain (Bartie et al., 2021).

Please consult Chapter 46—Process simulation—Thermodynamics and process technology to understand recycling systems for more information on some of these examples of simulation-based resource efficiency evaluations of large circular economy systems, which include many of the processes discussed in this book.

45.5 CONCLUSION

Exergy combines the First and Second Laws of Thermodynamics. Accordingly, exergy-based indicators such as exergy dissipation or cost complement classical resource use indicators such as the raw materials, energy, or water use obtained from mass and energy balances. An exergy analysis provides information on how the resources of a system are dissipated because of the Second Law of Thermodynamics. Therefore exergy is an appropriate and complementary indicator for metallurgical and recycling processes, and thus of circular economy systems, since it evaluates how many resources remain in the economy.

Since exergy analysis also can quantify the losses due to the creation of complex mixtures, it is a very useful tool to understand how to reverse the mixing and the exergy cost associated with the reversal process. That is also a proxy for the excess energy required to produce high-quality materials and products once again that can be used in high-tech products.

Recycling processes are basically unmixing the complex product and material mixtures, created by society through its behavior and products used in daily life, into high-quality materials. Quantifying this unmixing and the associated dissipation of resources will help to understand the limits of recycling, of circularity, and of materials usage by society in general.

Obviously, other indicators beside exergy are necessary to perform a comprehensive evaluation of the resource efficiency of metallurgy and recycling. For example, the recovery of specific metals should be evaluated through mass balances. The same happens with the quantification of the environmental impacts of these processes. Exergy cannot evaluate, for example, the carbon footprint or acidification potential of a metallurgical and recycling process, but it does provide the physics basis to understand and mitigate these impacts.

This chapter shows how to calculate the exergy value of the streams of a metallurgical and recycling process, illustrated in Chapter 46—Process simulation—Thermodynamics and process technology to understand recycling systems. Moreover, it explains the basics of the thermoeconomics methodology, which can be used to obtain exergy-based indicators such as the irreversibility of a process and large systems, or the exergy cost of a stream regardless of the complexity of the metallurgical/recycling process or the circular economy systems to be evaluated.

References

Abadías Llamas, A., Valero, A., Valero, A., Torres, C., Hultgren, M., Peltomäki, M., Roine, A., Stelter, M., Reuter, M. A., 2019. Simulation-based exergy, thermo-economic and environmental footprint analysis of primary copper production. Miner. Eng. 131, 51–65. https://doi.org/10.1016/j.mineng.2018.11.007.

Abadías Llamas, A., Bartie, N.J., Heibeck, M., Stelter, M., Reuter, M.A., 2020. Simulation-based exergy analysis of large circular economy systems: zinc production coupled to CdTe photovoltaic module life cycle. J. Sustain. Metall. 6 (1), 34–67. https://doi.org/10.1007/s40831-019-00255-5.

Agudelo, A., Valero, A., Torres, C., 2012. Allocation of waste cost in thermoeconomic analysis. Energy 45 (1), 634–643. https://doi.org/10.1016/j.energy.2012.07.034.

Amini, S.H., Remmerswaal, J.A.M., Castro, M.B.G., Reuter, M.A., 2007. Quantifying the quality loss and resource efficiency of recycling by means of exergy analysis. J. Clean. Prod. 15 (10), 907–913. https://doi.org/10.1016/j.jclepro.2006.01.010.

Ayres, R.U., Ayres, L.W., Masini, A., 2006. An application of exergy accounting to five basic metal industries. In: Sustainable Metals Management: Securing Our Future – Steps Towards a Closed Loop Economy. Springer, pp. 141–194.

Bartie, N.J., Abadías Llamas, A., Heibeck, M., Fröhling, M., Volkova, O., Reuter, M.A., 2019. The simulation-based analysis of the resource efficiency of the circular economy – the enabling role of metallurgical infrastructure. Miner. Process. Extr. Metall. Trans. Inst. Min. Metall. 129 (2), 229–249. https://doi.org/10.1080/25726641.2019.1685243.

Bartie, N.J., Cobos-Becerra, Y.L., Fröhling, M., Schlatmann, R., Reuter, M.A., 2021. The resources, exergetic and environmental footprint of the silicon photovoltaic circular economy: assessment and opportunities. Resour. Conserv. Recycl. 169 (January), 105516. https://doi.org/10.1016/j.resconrec.2021.105516.

Dominguez, A., Valero, A., Valero, A., 2013. Exergy accounting applied to metallurgical systems: the case of nickel processing. Energy 62, 37–45. https://doi.org/10.1016/j.energy.2013.03.089.

Fernandes, I.B., Abadías Llamas, A., Reuter, M.A., 2020. Simulation-based exergetic analysis of NdFeB permanent magnet production to understand large systems. JOM 72 (7), 2754–2769. https://doi.org/10.1007/s11837-020-04185-6.

Fernandes, I.B., Rudolph, M., Hassanzadeh, A., Bachmann, K., Meskers, C., Peuker, U., Reuter, M.A., 2021. The quantification of entropy for multicomponent systems: application to microwave-assisted comminution. Miner. Eng. 170, 107016. https://doi.org/10.1016/j.mineng.2021.107016.

Frangopoulos, C.A., 1987. Thermo-economic functional analysis and optimization. Energy 12 (7), 563–571. https://doi.org/10.1016/0360-5442(87)90097-1.

Frangopoulos, C.A., 2009. Thermoeconomic functional analysis. In: Frangopoulos, C.A. (Ed.), Exergy, Energy System Analysis and Optimization. Encyclopedia of Life Support System (EOLSS) Publishers/UNESCO.

Hannula, J., Godinho, J.R.A., Abadías Llamas, A., Luukkanen, S., Reuter, M.A., 2020. Simulation-based exergy and LCA analysis of aluminum recycling: linking predictive physical separation and re-melting process models with specific alloy production. J. Sustain. Metall. 6 (1), 174–189. https://doi.org/10.1007/s40831-020-00267-6.

Hernandez, A.G., Cullen, J.M., 2016. Unlocking plant-level resource efficiency options: a unified exergy measure. Procedia CIRP 48, 122–127. https://doi.org/10.1016/j.procir.2016.03.111.

Ignatenko, O., van Schaik, A., Reuter, M.A., 2007. Exergy as a tool for evaluation of the resource efficiency of recycling systems. Miner. Eng. 20 (9), 862–874. https://doi.org/10.1016/j.mineng.2007.03.005.

Lozano, M.A., Valero, A., 1993a. Theory of the exergetic cost. Energy 18 (9), 939–960. https://doi.org/10.1016/0360-5442(93)90006-Y.

Lozano, M.A., Valero, A., 1993b. Thermoeconomic analysis of gas turbine cogeneration systems. In: Proceedings of the 1993 ASME Winter Annual Meeting. vol. 30, pp. 311–320.

Madlool, N.A., Saidur, R., Rahim, N.A., Islam, M.R.R., Hossian, M.S.S., 2012. An exergy analysis for cement industries: an overview. Renew. Sust. Energy Rev. 16 (1), 921–932. https://doi.org/10.1016/j.rser.2011.09.013.

Meskers, C.E.M., Xiao, Y., Boom, R., Boin, U., Reuter, M.A., 2007. Evaluation of the recycling of coated magnesium using exergy analysis. Miner. Eng. 20 (9 SPEC. ISS), 913–925. https://doi.org/10.1016/j.mineng.2007.02.006.

Reuter, M.A., van Schaik, A., Gediga, J., 2015. Simulation-based design for resource efficiency of metal production and recycling systems: cases – copper production and recycling, e-waste (LED lamps) and nickel pig iron. Int. J. Life Cycle Assess. 20 (5), 671–693. https://doi.org/10.1007/s11367-015-0860-4.

Seyyedi, S.M., Ajam, H., Farahat, S., 2010. A new criterion for the allocation of residues cost in exergoeconomic analysis of energy systems. Energy 35 (8), 3474–3482. https://doi.org/10.1016/j.energy.2010.04.047.

Silveira, J.L., Tuna, C.E., 2003. Thermoeconomic analysis method for optimization of combined heat and power systems. Part I. Prog. Energy Combust. Sci. 29 (6), 479–485. https://doi.org/10.1016/S0360-1285(03)00041-8.

Szargut, J., 1989. Chemical exergies of the elements. Appl. Energy 32 (4), 269–286.

Szargut, J., 2005. Exergy Method: Technical and Ecological Applications. WITPress. https://books.google.de/books?id=puNSAAAAMAAJ.

Szargut, J., Morris, D.R., 1987. Cumulative exergy consumption and cumulative degree of perfection of chemical processes. Int. J. Energy Res. 11 (2), 245–261. https://doi.org/10.1002/er.4440110207.

Torres, C., 2009. Symbolic thermoeconomic analysis of energy systems. In: Frangopoulos, C.A. (Ed.), Exergy, Energy System Analysis and Optimization. vol. 2, pp. 61–81. EOLSS Publishers/UNESCO.

Torres, C., Serra, L., Valero, A., Lozano, M., 1996. The productive structure and thermoeconomic theories of system optimization. In: Proceedings of the ASME Advanced Energy Systems Division. vol. 36, pp. 429–436.

Torres, C., Valero, A., Rangel, V., Zaleta, A., 2008. On the cost formation process of the residues. Energy 33 (2), 144–152. https://doi.org/10.1016/j.energy.2007.06.007.

Tsatsaronis, G., 1993. Thermoeconomic analysis and optimization of energy systems. Prog. Energy Combust. Sci. 19 (3), 227–257. https://doi.org/10.1016/0360-1285(93)90016-8.

Usón, S., Valero, A., Correas, L., 2010. Energy efficiency assessment and improvement in energy intensive systems through thermoeconomic diagnosis of the operation. Appl. Energy 87 (6), 1989–1995. https://doi.org/10.1016/j.apenergy.2009.12.004.

Usón, S., Valero, A., Agudelo, A., 2012a. Thermoeconomics and Industrial Symbiosis. Effect of by-product integration in cost assessment. Energy 45 (1), 43–51. https://doi.org/10.1016/j.energy.2012.04.016.

Usón, S., Valero, A., Valero, A., Costa, J., 2012b. Thermoeconomic fuel impact approach for assessing resources savings in industrial symbiosis: application to kalundborg ecoindustrial park. In: Proceedings of the 25th International Conference on Efficiency, Cost, Optimization and Simulation of Energy Conversion Systems and Processes, ECOS 2012. vol. 3, pp. 346–356.

Valero, A., Valero, A., 2010. Exergy Analysis of Resources and Processes. Prensas Universitarias de Zaragoza.

Valero, A., Lozano, M.A., Munoz, M., 1986. A general theory of exergy saving I. On the exergetic cost. In: ASME, Advanced Energy Systems Division (Publication). vol. 2, pp. 1–8.

Valero, A., Usón, S., Costa, J., 2012. Exergy analysis of the industrial symbiosis model in Kalundborg. In: Proceedings of the 25th International Conference on Efficiency, Cost, Optimization and Simulation of Energy Conversion Systems and Processes, ECOS 2012. vol. 1, pp. 406–416.

Velázquez-Martínez, O., Porvali, A., van den Boogaart, K.G., Santasalo-Aarnio, A., Lundström, M., Reuter, M.A., Serna-Guerrero, R., 2019a. On the use of statistical entropy analysis as assessment parameter for the comparison of lithium-ion battery recycling processes. Batteries 5 (2). https://doi.org/10.3390/batteries5020041.

Velázquez-Martínez, O., Van Den Boogaart, K.G., Lundström, M., Santasalo-Aarnio, A., Reuter, M.A., Serna-Guerrero, R., 2019b. Statistical entropy analysis as tool for circular economy: proof of concept by optimizing a lithium-ion battery waste sieving system. J. Clean. Prod. 212, 1568–1579. https://doi.org/10.1016/j.jclepro.2018.12.137.

Vidal, A., Best, R., Rivero, R., Cervantes, J., 2006. Analysis of a combined power and refrigeration cycle by the exergy method. Energy 31 (15), 3401–3414. https://doi.org/10.1016/j.energy.2006.03.001.

CHAPTER

46

Process simulation—Thermodynamics and process technology to understand recycling systems

Neill J. Bartie[1], *Alejandro Abadías Llamas*[2], *Christina Meskers*[3], *and Markus A. Reuter*[4]

[1]Helmholtz Centre Berlin for Materials and Energy (HZB), PVcomB, Berlin, Germany [2]Nordenham Metall GmbH, Nordenham, Germany [3]Department of Energy and Process Engineering, Faculty of Engineering Sciences, NTNU—Norwegian University of Science and Technology, Trondheim, Norway [4]WASM: Minerals, Energy and Chemical Engineering, Faculty of Science and Engineering, Curtin University, Perth, Australia

46.1 INTRODUCTION

The complexity that metallurgical processing and recycling systems must cope with is described in Chapter 1—Introduction and Chapter 2—The fundamental limits of circularity quantified by digital twinning of this book and is visualized in the Metal Wheel (Figure 2.2). These systems must be able to direct all materials into appropriate segments of the Metal Wheel to maximally recover elements as compounds, alloys, materials, etc. (Reuter et al., 2019). To really understand the transformation of products and materials during production and recycling, as well as the emissions and losses associated with these processes, one needs to consider the physical, chemical, and thermodynamic properties of the elements and compounds moving through the system. Furthermore, an understanding of production and recycling processes and their operating parameters, including knowledge of reactor technology, mass- and heat transfer, and the kinetics of processes is required. Together, these determine material flows through the complex life cycles that make up circular economy systems and processing infrastructure (Reuter, 2016; Reuter et al., 2019). Keeping track of all these phenomena, how they influence one another, and their impacts on resource efficiency, environment, economics, and society would be virtually impossible without sophisticated tools that are able to keep tabs on the flows and transformations of multiple compounds and mixtures into a multitude of different streams, phases, compounds, and elements.

Handbook of Recycling. https://doi.org/10.1016/B978-0-323-85514-3.00011-7

Product design and the collection of end-of-life products create complex blends of numerous materials, compounds, and elements. As emphasized in Chapter 45—Exergy as well as Chapter 2—the fundamental limits of circularity quantified by digital twinning and Chapter 5—Material and product-centric recycling and design for recycling rules and digital methods, Design for Recycling (DfR) demands the use of tools that can quantify a product's recycling profile and performance reliably. With such tools, DfR hotspots can be identified, appropriate design adjustments can be prioritized, and their impact on resource efficiency can be quantified. Especially with the tremendous increase in the diversity and complexity of products and value chains, getting a clear picture of what is happening in a life cycle system is essential for the transition to a circular economy. This is best achieved by tracking all of the functional materials, products, alloys, etc. and their properties through the system, also in terms of enthalpy and entropy (Chapter 44—Thermodynamics and Chapter 45—Exergy), as they determine what can and cannot be achieved in these processes. Furthermore, to fully understand how to unmix these mixtures of products, wastes, sludges, solutions, dusts, etc., the excess thermodynamic properties of mixing are especially important (Chapter 45—Exergy) as they account for deviations from ideality. The limits of circularity are revealed and can be acted on by policymakers, industry, product designers, and other stakeholders.

Process simulation is a tool that provides a platform for all this detail to be captured and enables the calculation of complex energy and mass balances at high resolution, so also enabling exergy analysis. Simulations are built by developing models of all the unit operations, connected to each other by material or energy streams, in a process. Single processes can be connected to others, creating systems of processes and thence circular economy systems. These simulations can become digital twins of real-world processes, especially when validated with industrial data. Using process simulation to quantify the resource efficiency of processes and systems, and their impacts on sustainability, adds depth to such assessments and offers possibilities beyond that provided by other tools:

- Production and recycling systems can be connected to each other and to energy generation and recovery systems easily to analyze complete systems—streams can be quantified in terms of mass flow (e.g., t/h), energy, entropy, and therefore exergy (e.g., kWh/t), which allows for mass and energy flows to be unified and expressed in energy units (kWh/h).
- The resources and effort required to unmix complex product, scrap, residue or waste mixtures and create high-quality secondary materials, alloys, compounds, plastics, and others for reuse in similar products (i.e., the definition of true circularity) can be quantified more realistically—the thermodynamic phenomena that cause these mixtures to deviate from ideality, which further increases the resources needed to unmix them, can be taken into account via exergy analysis.
- The downward spiral of the circular economy system can be quantified in terms of exergy dissipation, as discussed in Chapter 45—Exergy. The extent to which potential secondary resources are diluted or contaminated in recycling processes, and therefore the resources needed to return them to the original material quality specifications to prevent them from having to be downcycled into lower-quality applications, can be quantified and addressed.
- The quality and realism of environmental impact assessments increase as the use of average, nonspecific, or outdated data from inventory databases can be avoided. Simulation models provide detailed foreground information for current and new processes and technologies, enabling detailed benchmarking and focused optimization and innovation efforts.

- The ability to adapt simulation models fairly quickly for alternative scenarios while maintaining the high detail level strengthens scenario-based decision-making. Processing routes and scenarios can be compared by varying inputs, outputs, operating parameters, and process technologies with the knowledge that the nonlinearities are taken into account via thermodynamics.
- Emerging and noncommercialized or existing technologies, processes, and supply chains can be evaluated and their impact on the circular economy system and its sustainability (environmental, economic, and social) assessed based on detailed physics-based flowsheets and data.
- Stakeholder dialogue, industrial innovation, and cocreation become easier as the level of detail necessary to achieve real-world change is available. Resource needs, production and revenue potential, losses, residue formation, and emissions can be clearly defined at a level of detail that allows for questions about meeting product specifications, complying with environmental, transboundary movement, and REACH regulations, etc. to be answered and communicated clearly.

The prerequisite is that sound chemical, metallurgical, and mechanical processing data; reactor and system information (e.g., reactor technology, flowsheet understanding, economics); and (industrial) operational and engineering know-how (e.g., transfer processes, nonidealities, bottlenecks, and other technical challenges) are used to build and validate simulation models. A simulation will only be as good as the data, know-how, and assumptions used to build it. Data from literature, laboratory experiments, and pilot and industrial processes (where accessible) are all used to obtain the best simulation possible. Considering the broad expertise required to build a simulation, verify and validate the data and assumptions, and analyze the outcomes, this is often done collaboratively in multidisciplinary teams that also have industrial experience.

The simulation flowsheet for a primary Cu production process is shown in Figure 46.1 as an example. Each unit operation contains the physics-based relationships that transform its inputs into products. For instance, the Flash Smelting Furnace unit operation in Figure 46.1 contains all the information needed to convert its input streams (Cu concentrate, dusts, flux, air, and other recycled materials) into its products (Cu matte, slag, dust, offgas, and heat). Connecting the process units using the feed and product streams creates the simulation model for the whole Cu production process. By connecting whole production processes such as that shown in Figure 46.1 with others in the value chain, simulation models for entire life cycles can be created. Additionally, information is generated at the compound and element level for each stream, as shown for the smelting slag in the red square on the righthand side of Figure 46.1.

Clearly, the detail provided by simulation-based approaches goes significantly beyond the material-centric view of steel, Al alloys, and waste streams considered in material flow analysis (MFA) methods (Brunner and Rechberger, 2016) for their analyses. MFA generally cannot provide the full enthalpy and entropy of each stream, also in terms of excess thermodynamic properties, and thus cannot estimate the true efficiency of systems.

In the remainder of this chapter, the basic steps followed to develop a process simulation model are discussed and examples of the information one can extract to assess the efficiency and impacts of a system are highlighted. Finally, the application of process simulation is demonstrated for two cases: Al scrap recycling, and Si solar panel recycling. The Al scrap recycling case illustrates quantitatively the impact of incomplete removal of Cu and Fe impurities on the final alloy quality. The solar panel recycling example shows how simulations can be

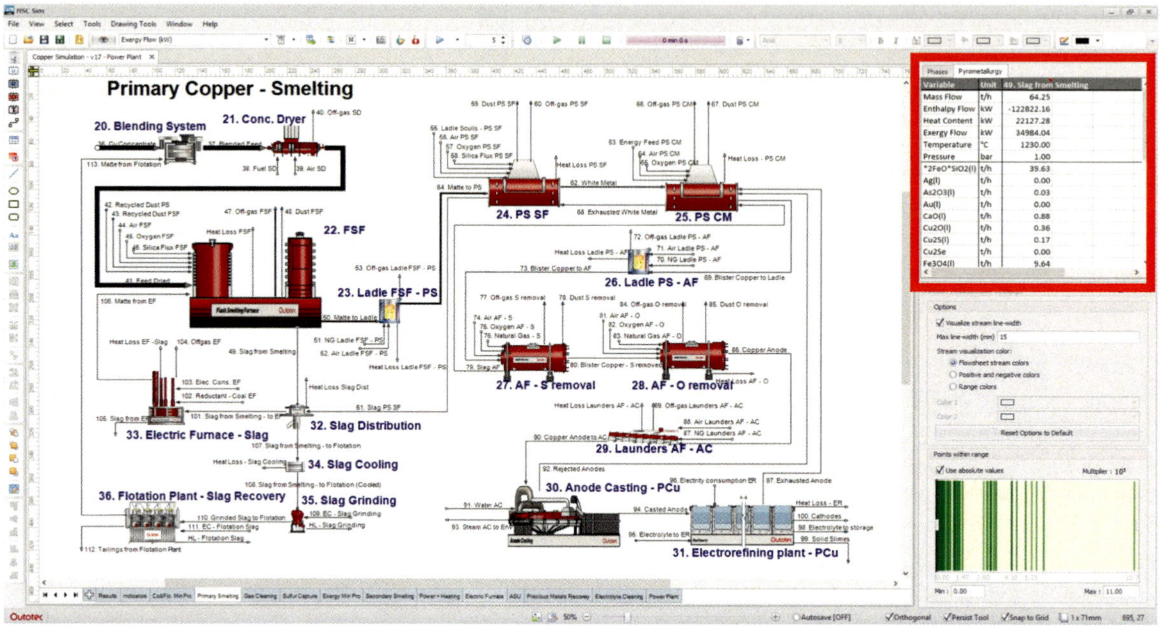

FIGURE 46.1 Flowsheet model of a primary Cu smelting process using the process simulation software HSC Sim (www.mogroup.com), showing the unit operations *(bold blue font)*, material and energy streams, and detail of a stream shown on the right *(red block)*. *From Abadías Llamas et al. (2019).*

used to simultaneously analyze the impacts of process changes and recycling rates on the carbon footprint of a system.

46.2 METHODOLOGY

The key steps for the development of a process simulation model are:

1. **Definition of the system and its boundaries**: everything starts with a clear definition of the system to be studied. The system and its boundaries should be defined in such a way that the objective for which the simulation is being developed is achieved, or that the research questions are answered using the simulation results. The system definition can include, for example, process descriptions, process design criteria, geographical location, feed sources, and any other information needed to clarify what the simulation model will and will not include.
2. **Definition of the flowsheet**: The unit operations that constitute each of the subprocesses in the system are defined. The transformations that occur in each of the unit operations can be:
   a. Chemical: pyro- and hydrometallurgical processes such as oxidative roasting, leaching, carbothermic reduction, fuming, electrolysis, and molten salt processes, among others.
   b. Physical: size reduction, sieving and sorting processes such as shredding, eddy current, magnetic, and electrostatic separations, as well as sorting according to properties like color, shape, etc.
   c. Transport: such as material and energy transport in a process or between industries.

d. Thermal: such as heating (e.g., in induction, electric arc, and plasma furnaces), drying, combustion, pyrolysis, roasting, and cooling in, e.g., heat exchangers and cooling towers.

The material (solids, liquids, and gases) and energy input and output streams, as well as the interactions between the unit operations and subprocesses through internal streams should be identified. Definition of the flowsheet and identification of the input, connecting, and output streams is an important step since it establishes the data that need to be gathered (Step 3) to develop models for the unit operation defined in Step 2.

3. **Collection of data**: The experimental or industrial process calibration data should provide a comprehensive description of the defined unit operations. These are also based on, among others, thermochemical data, fluid flow phenomena, process kinetics and the physical, chemical, or thermal mass and heat transfer phenomena occurring in each unit operation, as discussed by Reuter (2016). The number of parameters required depends on the complexity of the unit operation or reactor model. For example, models of magnetic and eddy current separation processes require particle-level data so that the separation of the scrap into different material fractions can be understood and modeled to reflect different levels of material liberation. A remelting process will require data on the metallurgical process occurring, thermochemical data, operating parameters such as temperature and pressure, the consumption of fluxes and other feed materials, the energy required, mass and heat transfer phenomena, and kinetics as a function of different feed types, among others.

   As the models are intended to represent real-world processes, industrial data are preferred for fine-tuning and validating the process models that lie at the heart of each unit operation. However, industrial data are not always available, and some processes do not yet exist beyond the experimental or pilot scales. In such cases, experimental data, other simulation tools, e.g., FactSage (www.factsage.com) for thermochemical calculations, industrial experience, and literature are other data sources that can be used.
4. **Development of a model of each unit operation and reactor**: The data gathered in the previous step are used to develop the models for every individual unit operation. These can be at varying levels of complexity depending on the availability of data.
5. **Coupling of the individual models of each reactor to create the simulation model of a process, system, circular economy**: When the individual detailed reactor and unit operation models of the circular economy system have been generated, they are coupled through the material and energy streams between them to create meaningful process simulation models. These models represent the flowsheet defined in the second step.
6. **Running the model**: During its development, the process simulation is run repeatedly to iteratively solve the mass, energy, and exergy balances for the unit process models created up to that point, as these are highly nonlinear and become more complex as units are added. System models can have many hundreds of unit operations and many more streams with all the compositional detail contained within them.
7. **Validation and interpretation of the results**: Simulation results are compared and validated with reality and unit operation models are calibrated with industrial data and/or know-how to ensure that the simulation produces realistic results. The validated simulation model can then be used to analyze the system and draw conclusions about its performance to answer the questions defined in Step 1.

While Steps 1 and 2 should happen first, one usually moves between Steps 3 to 7 as required.

46.3 INFORMATION FROM SIMULATION MODELS

The systemic view and the detailed mass and energy balances that process simulation provides enable rigorous resource efficiency and sustainability evaluations for production and recycling processes and the systems they are part of. In this section, the derivation of different resource efficiency and environmental impact indicators from process simulation results are explained.

46.3.1 Resource Efficiency Evaluations

Recycling processes convert linear value chains into circular economy systems. The degree of circularity achieved strongly depends on the effectiveness and efficiency of recycling and, therefore, these need to be measured and maximized. Like production processes, recycling processes consume resources to recover valuable materials from the end-of-life products treated. Their resource efficiencies are determined by the types and quantities of resources consumed, the quantities of wastes and treatable residues generated, the efficiency with which the desired elements, compounds or materials are recovered within that process, and the quality at which they are recovered.

Resource consumption indicators that account for the quantities of materials and energy needed for a process to achieve its intended goal are usually obtained directly as the quantities calculated in the simulated mass and energy balances. Those that account for the generation of entropy in terms of, for example, the exergy dissipation in a process or system, or the exergy cost of a product or residue can be obtained directly from some process simulation tools such as HSC Sim (www.mogroup.com). These are described in more detail in Chapter 45—Exergy.

As mass balances are generated at the element and compound level, the recoveries of valuable secondary resources can also be evaluated at this level. For example, one can calculate the absolute amount of a specific metal or alloying element recovered from the feed materials to a particular recycling process (i.e., its yield) over a certain period of time, or at the material level, the amount of a specific alloy recovered from a given quantity of mixed metal scrap. Recoveries are more commonly expressed in specific terms, such as the quantity of a metal recovered compared to the quantity of that metal in the input to the process, expressed as a fraction or a percentage. An advantage of the detailed stream compositions obtained with process simulation is that recoveries can also be expressed in terms of the actual compounds recovered, such as when a metal is not recovered as an element, but as an oxide, sulfide, or in another form, or at a certain purity in a multicomponent solution. In addition to the quantity recovered, this indicates the purity of the product and whether further processing is needed for it to be transformed into a useful secondary resource. Energy and exergy efficiencies can be calculated using the energy and exergy balances obtained from the simulation model. These indicators are calculated as the ratio of energy or exergy consumed and energy recovered or exergy of the product streams, respectively.

While the static indicators described here are useful on their own, the advantage of process simulation is the relative ease with which input and operating parameters can be changed, simultaneously if necessary, to investigate their effects on the resource consumption and efficiency of processes and whole life-cycle systems. This capability facilitates comprehensive parameter studies that enable optimization for sustainability. The simulation model provides the detail necessary to understand and interpret the values obtained for the different indicators, so that one can gain an understanding of how these indicators change with changes in the system. For instance, the purity of a recovered material usually decreases as the quantity recovered increases. The effects of changes in the system (e.g., different feed material compositions, the use of different technologies, or different product specifications) on this relationship can be evaluated using the indicators described here to find the balance that delivers maximum circularity and sustainability.

46.3.2 Environmental Impact Evaluations

The detailed mass and energy balances can be useful data sources for other methods used to quantify environmental impacts such as Life Cycle Assessment (LCA). LCA is a methodology used for the systematic assessment of the environmental aspects and impacts of product systems, considering their life cycle from raw material acquisition to final disposal, and consists of four phases (see Chapter 47—Life cycle assessment). Process simulation can contribute to the goal and scope definition phase because the system and its objectives already defined in the simulation model can be transferred to the LCA.

Its main contribution would be in the life cycle inventory stage. In LCA, the life cycle inventory analysis is often performed using data from commercial environmental databases. While these contain useful inventory data for many processes, the datasets do not always fully represent the processes being evaluated. For example, some processes and technologies are outdated, some are not sufficiently geography-specific, and datasets for emerging and new technologies are often not available. Process simulation can enhance the inventory analysis stage by generating up-to-date mass and energy balances that can be transferred to the LCA with more detail on the elements and compounds present in each stream of the process. This has been discussed in detail by Reuter et al. (2013, 2015).

46.4 APPLICATIONS

The use of process simulation to evaluate the resource efficiency and environmental impacts of metallurgical and recycling systems is increasingly being reported in the literature. A model developed for the production and recycling of cadmium telluride photovoltaic modules was used to evaluate the performance of the system in terms of material recovery, exergy dissipation, and environmental impacts (Bartie et al., 2019; Heibeck et al., 2020). Reuter and van Schaik (2015) evaluated the recycling of an LED lamp using process simulation. Three recycling options were simulated and compared for end-of-life material recovery efficiency from the modular Fairphone 2 (Reuter et al., 2018). This approach has also been used to evaluate the exergy dissipation throughout the life of an NdFeB permanent magnet (Belo Fernandes et al., 2020; Fernandes et al., 2020).

Two cases are discussed in more detail here to show the value of simulation-based approaches for understanding and assessing recycling systems and the circular economy.

46.4.1 Aluminum Recycling

For the first example, the work of Hannula and colleagues has been selected. In this study, an Al scrap sorting and recycling process is modeled and it shows how simulation can be used to investigate some of the recycling challenges discussed in Chapter 2—The fundamental limits of circularity quantified by digital twinning, Chapter 5—Material and product-centric recycling and design for recycling rules and digital methods, Chapter 8—Physical separation, and Chapter 20—Aluminum. The work is summarized from Hannula et al. (2020) to illustrate how the methodology has been applied.

The objective of the work was to study the effects of particle properties, mechanical separation parameters, and desired alloy type on the resource consumption of an Al recycling process. To this end, the quantity of virgin Al required to dilute impurities and meet predefined alloy specifications was estimated using process simulation (Hannula et al., 2020). Following the steps introduced earlier, the modeling process can be described as follows:

Step 1: The system included the physical separation and metallurgical processing (remelting and alloying) of Al scrap. It excluded the collection and incineration steps, as well as the treatment of the residues generated during remelting.

Step 2: The feed materials consisted of pyrolyzed Al scrap particles, primarily from municipal solid waste, as well as virgin Al and alloying elements. The process included a magnetic separation step, two eddy current separation (ECS) stages that each generate two fractions, and three rotary remelting furnaces (RF)—alloying furnace (AF) combinations. The energy to the furnaces was provided as oxyfuel and electricity. Each RF-AF combination produced a different wrought alloy: 2017A in RF-AF1, 2014A in RF-AF2, and alloy 6024 in RF-AF3 (Figure 46.2). In addition, each RF produced a salt slag with solids.

Step 3: The feed composition was analyzed in detail. It contained 87% Al particles, 4% Cu particles, and 9% steel can particles. The Al particles were from foil (alloy 1200), trays (alloy 8006), cans (alloys 3104, 5182, and 5082), and heavier generic particles (assumed to be made of average cast alloy). The Cu particles were assumed to be pure Cu and the steel cans, tinplate. The data for the ECS model was generated by conducting ECS experiments using industrial scrap samples in which eddy current settings and belt speeds were varied (Hannula et al., 2018). This was followed by characterization of individual particles in each of the collected fractions using X-ray computer microtomography (Hannula et al., 2018). The data for the remelting process was obtained from literature and by using FactSage (www.factsage.com) to obtain the activity coefficients of the elements in the liquid Al and salt slag phases at a temperature of 800°C and a pressure of 1 bar.

Step 4: The models for the individual unit operations were developed and implemented in the HSC Sim simulation tool (www.mogroup.com). The magnetic separator was modeled as a splitter that separates the steel can particles from the nonferrous particles. To account for steel particles possibly trapped within Al particles, the steel recovery during magnetic separation was reduced to 99%. The ECS was represented by a predictive particle-based model that predicts the recovery value of the output streams based on the particle properties and the ECS parameters obtained in step 3. The model, which was developed in MATLAB (www.mathworks.com) and then transferred to the simulator, was described in detail in Hannula et al. (2018).

The metallurgical process was modeled using Gibbs free energy minimization. The operating temperature of both the RF and AF was assumed to be 800°C and a heat loss of 5% of the supplied energy was assumed. All scrap particles entering the RF had a different degree of surface oxidation, depending on their surface-to-mass ratio. To take this into account, the amount of oxide (Al_2O_3) present was estimated as a fixed percentage for each scrap type: 13.6% for foils, 11.5% for trays, 9.4% for cans, and 0.37% for the generic scrap particles. The compounds that formed as a result of reactions between organic contaminants and Al metal were considered to be dissipative losses and were modeled to report to the solids fraction in the slag stream together with the surface oxides. It was assumed that no dross formed in the alloying furnaces (Figure 46.2).

Step 5: The individual models were coupled as shown in Figure 46.2. Each ECS has two outputs—a "high throw" and a "low throw" fraction, which refers to the distance the particles in that fraction travel exiting the ECS. The ECS 1 high throw fraction is transferred to RF 3 while

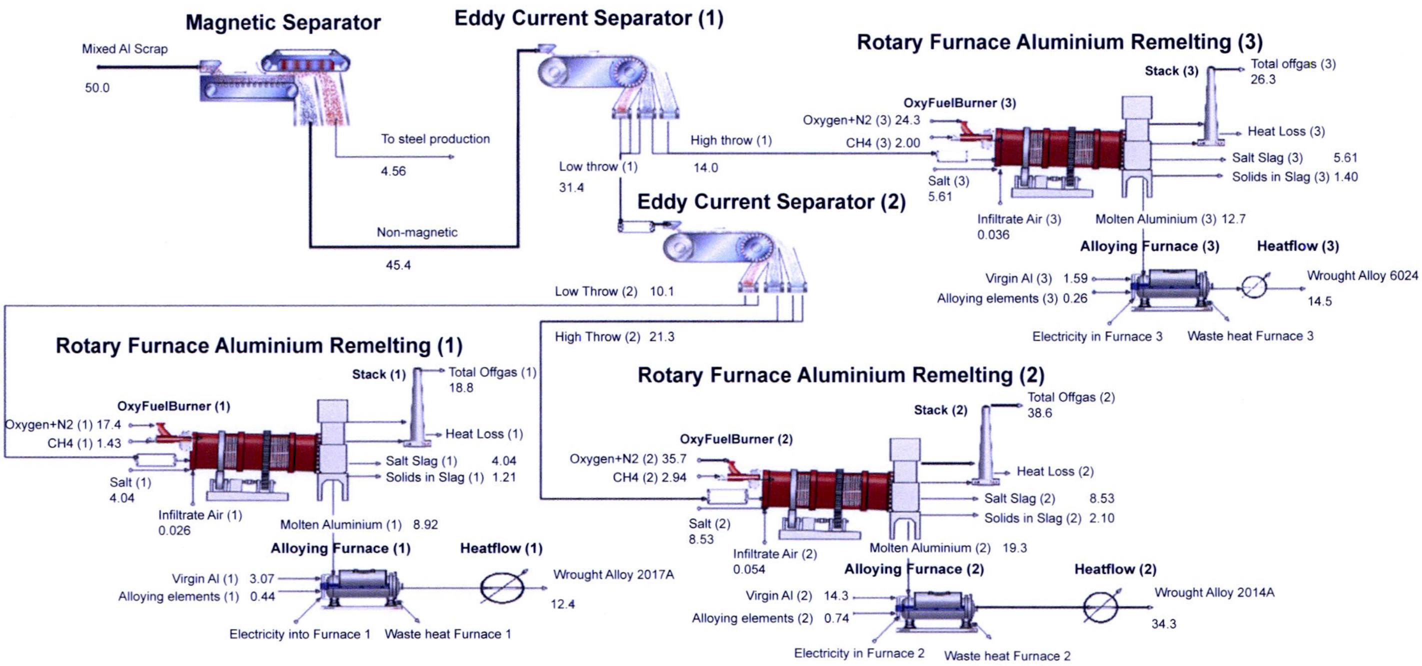

FIGURE 46.2 Flowsheet model used by Hannula et al. The values displayed refer to the total mass flows in Scenario 1. *From Hannula et al. (2020).*

its "low throw" fraction is processed in ECS 2. The ECS 2 high throw fraction is transferred to RF 2, and its low throw fraction to RF 1.

Step 6: The simulation model was run for three different scenarios in which ECS belt speeds and the Cu content of the Al scrap feed material were varied, while the target alloy composition was kept constant. In Scenario 1, different ECS belt speeds were used for ECS 1 and ECS 2, while in Scenario 2 the same belt speed was used throughout. In Scenario 3, the Cu content of the feed material was doubled while using the same belt speeds as in Scenario 1. The mass, energy, and exergy balances were calculated for each of the scenarios to evaluate and compare the effects of the variables on the resource efficiency of the recycling process (Hannula et al., 2020).

Step 7: Using the mass, energy and exergy results, the Resource Index, among other indicators, was determined. This indicator represents the ratio of the exergy content of all the additional inputs, such as virgin Al, fuel, or alloying elements needed to achieve the desired alloy composition and the exergy content of the scrap feed. This index allows one to understand how changes in ECS settings and feed composition affect exergy dissipation (Figure 46.3) and hence resource efficiency. In Scenario 2, the poor separation of Cu and Al in ECS 1 resulted in a much

FIGURE 46.3 Comparison of required additional resources in terms of exergy to process the Al scrap into selected alloys. The Resource Input is compared in three levels: between the different scenarios, between different alloy production inside the scenarios, and between the resources used to produce the alloys. The additional resources are divided into virgin Al, fuel/electricity, alloying elements, and other resources. *From Hannula et al. (2020).*

higher Cu content in the feed stream to RF 3. This Cu needed to be diluted with virgin Al to meet the composition requirements of alloy 6024 (0.9% Cu). In Scenario 3, the doubled Cu content in the scrap feed had the most impact on alloy 2, as most of the extra Cu in the feed entered RF 2, causing the alloy composition to far exceed the 5 wt% Cu limit of alloy 2014A (Hannula et al., 2020).

The interpretation of the simulation outcomes, with respect to the objective "study the effect of particle and mechanical separation parameters and desired alloy type on the resource consumption," was, in the words of Hannula et al. (2020):

> The resource consumption of recycling Al-rich scrap is not only product-specific but very much system-specific. The feed quality (scrap collection and mixing from different sources), the effect of separation stage, and the type of an alloy produced are some of the governing factors in the resource consumption. The results clearly show that an integrated approach linking the whole process chain together is required to fully understand and minimize the resource consumption.

46.4.2 Recycling of Photovoltaic Modules

A process simulation model was developed for the complete life cycle of photovoltaic (PV) modules based on a state-of-the-art Si wafer technology, i.e., a first attempt at digitally twinning the circular economy of PV cells and modules. Apart from creating a simulation model for the whole life cycle system, the aims of the work included investigating the effects of closed-loop Si recycling on the overall carbon footprint, the contribution of different module components to overall exergy dissipation, and the effects of production technology changes on energy consumption and carbon footprint, among others. The overarching aim was to identify hotspots and opportunities to maximize the system's overall resource efficiency and sustainability.

In addition to the Si semiconductor wafers, these modules consist of several layers of materials such as glass, polymer-based encapsulation layers and protective back sheets, antireflection coatings, metal oxide passivation layers, Ag and Al contacts, Sn-coated Cu cell connectors, and other components like junction boxes and cabling. The system selected for simulation included the production processes for metallurgical and solar grade Si, monocrystalline Si ingots and wafers, photovoltaic cells, modules, and the recycling process.

Step 1: For this example, the focus is on the recycling process for which the simulation flowsheet is shown in Figure 46.4. It shows only one of many parts of the simulation (see bottom tabs in Figure 46.4). End-of-life monocrystalline Si modules enter this flowsheet and flow through various process steps with tonnages shown.

Step 2: The process steps inside the system boundary are the dismantling of the modules, separation of the glass layer from the other components, combustion of the polymer layers in the module, recovery of some of the combustion heat to generate power, and the removal of the Sn-coated Cu ribbons used to connect the cells, followed by the etching process that separates Cu, Ag, Al, and Sn from the Si wafers that are reused in the production process. Steps left outside of the system boundary include the treatment of off-gas streams and additional purification of the recovered Ag, Cu, Pb, and Sn. Therefore these steps were not modeled in detail.

Step 3: Models were developed for each of the included process steps using published information on existing PV module recycling processes, the methods and expected efficiencies of promising experimental methods, the reactions that take place in all chemical transformation steps, and the methods and efficiencies of heat recovery and power generation processes. For this step, a significant amount of time is spent on literature research to ensure that the models can be built using detailed and

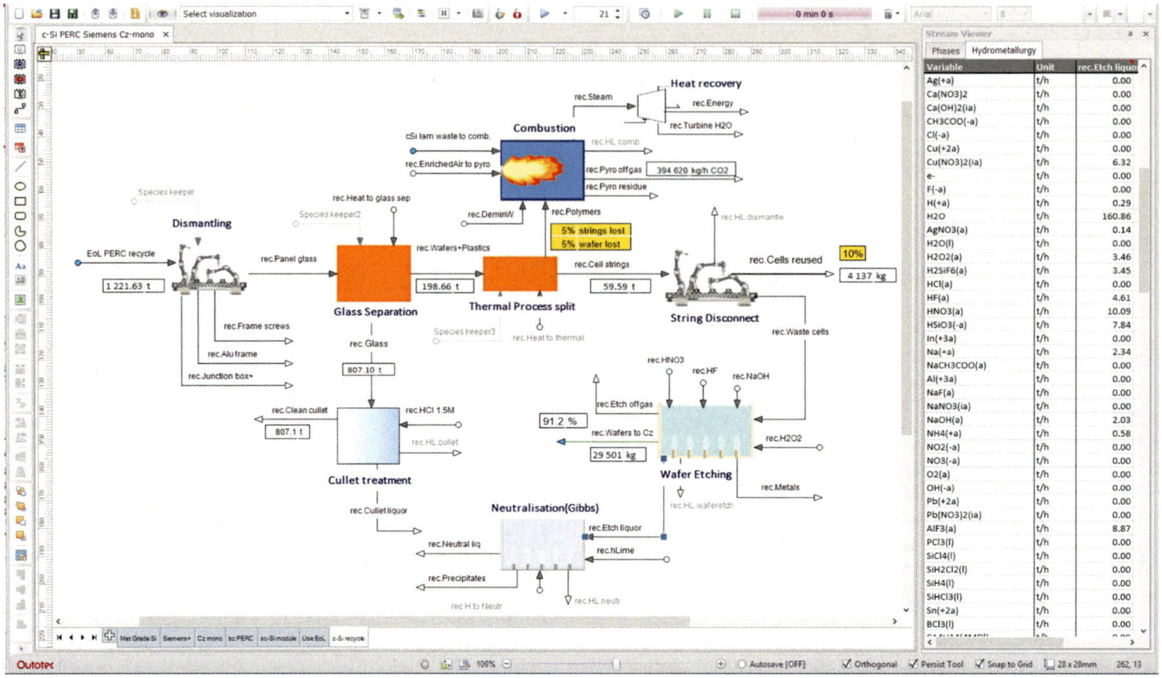

FIGURE 46.4 Si PV module recycling flowsheet.

scientifically sound information. An extract of the information collected and used for this purpose is shown in Box 46.1.

Step 4: Models were developed for the individual processes by implementing the information collected in Step 3. As shown in Box 46.1, some examples include the reactions needed to combust the polymer materials and transform them into the combustion products, the leaching reactions needed to separate Ag, Pb, Cu, Sn, and Al from the Si wafers, and using Gibbs free energy minimization to estimate which species precipitate in the final neutralization step. During the process of setting up these models, further details are sometimes needed. Therefore one can move between Steps 3 and 4 until the models produce acceptable results.

Step 5: As individual models are created in Step 4, they can be connected in smaller subsections to ensure that there are no compatibility issues when the output of one unit is connected as the input to another. More units are progressively connected in this way until all the individual steps shown in Figure 46.4 have been connected and the whole flowsheet converges to a solution.

Step 6: The completed simulation can then be used to evaluate different process parameters and configurations, such as changing the efficiency of the heat recovery process, or not combusting the polymers at all, to investigate their effects on, e.g., the net power consumption of the process or its carbon footprint.

Step 7: For the process configuration and operating parameters selected, simulation results were used to calculate some of the indicators mentioned earlier. Figure 46.5 shows the recycling process recovery efficiencies for Ag, Cu, Pb, Sn, and Si, expressed as a percentage of the amount of that element that enters the

BOX 46.1

Extract of information used to develop the recycling process simulation model

Complete combustion of the polymers (EVA, PET, PVF, and PVB) and a 5% loss of cells and strings are assumed to occur in this step. Polymer combustion is used to estimate direct CO_2 generation (i.e., Scope 1 emission) according to the following chemical reactions:

$$\text{EVA}\quad C_2H_4 + 3O_2 = 2CO_2 + 2H_2O$$
$$2C_4H_6O_2 + 9O_2 = 8CO_2 + 6H_2O$$
$$\text{PVB}\quad C_8H_{14}O + 11O_2 = 8CO_2 + 7H_2O$$
$$\text{PET}\quad C_{10}H_8O_4 + 10O_2 = 10CO_2 + 4H_2O$$
$$\text{PVF}\quad 2C_2H_3F + 5O_2 = 4CO_2 + 2H_2O + 2HF$$

The assumption of complete combustion is conservative in terms of GHG emissions as it likely overestimates the quantity of CO_2 generated. Hydrogen fluoride (HF) formed during the combustion of PVF is absorbed in a scrubber for neutralization. Heat generated during this combustion process is partially recovered to generate electricity.

From string dismantling, it is assumed that 10% of cells are recovered intact for direct reuse. The remaining cell strings proceed to the leaching and etching steps to recover metals and wafer Si for reuse. Firstly, leaching in HNO_3 dissolves only Ag, Pb, and Cu, and precipitates Sn as SnO_2, i.e., the Sn/Pb solder, Sn-coated Cu ribbons, and the Ag contacts. Ag and Cu are recovered in sequential electrowinning (EW) steps with Pb precipitating as hydrated PbO_2 during Cu recovery, while the SnO_2 precipitate is recovered by filtration. The EW steps are not included in the simulation, but their reported metal recoveries are taken into account in overall material efficiency calculations. Next, HF is used to dissolve only the SiN_x and Al-containing layers from the cells. The emitter and BSF are etched away in NaOH. The leaching steps are simulated using the following reactions:

$$3Ag + 4HNO_3 = 3AgNO_3 + 2H_2O + NO$$
$$Pb + 4HNO_3 = Pb(NO_3)_2 + 2NO_2 + H_2O$$
$$Cu + 4HNO_3 = Cu(NO_3)_2 + 2NO_2 + H_2O$$
$$Sn + 4HNO_3 = SnO_2 + 4NO_2 + 2H_2O$$
$$2Al + 6HF = 2AlF_3 + 3H_2$$
$$Si_3N_4 + 18HF + 4H^+ = 3H_2SiF_6 + NH_4^+$$
$$Si + NaOH + 2H_2O = NaSiO_3 + H_2$$

A stream of clean wafers is produced and recycled to the Cz crystallization process as SG-Si feedstock. The remaining acidic liquor is neutralized with hydrated lime ($Ca(OH)_2$) to form calcium fluoride (CaF_2). Gibbs free energy minimization is used in this unit operation to estimate the compositions of the remaining neutral liquor and precipitates.

Adapted from Bartie et al. (2021).

recycling process. This information is useful for estimating the revenue potential from valuable by-products and identifying areas where process innovation would be beneficial, such as improving process efficiency for the recovery of Ag and Cu.

The simulation was also used to investigate the simultaneous effects of recycling process configuration and end-of-life recycling rate on the carbon footprint of the life cycle using direct CO_2 emissions per unit surface area of PV module recycled as the indicator. Two scenarios

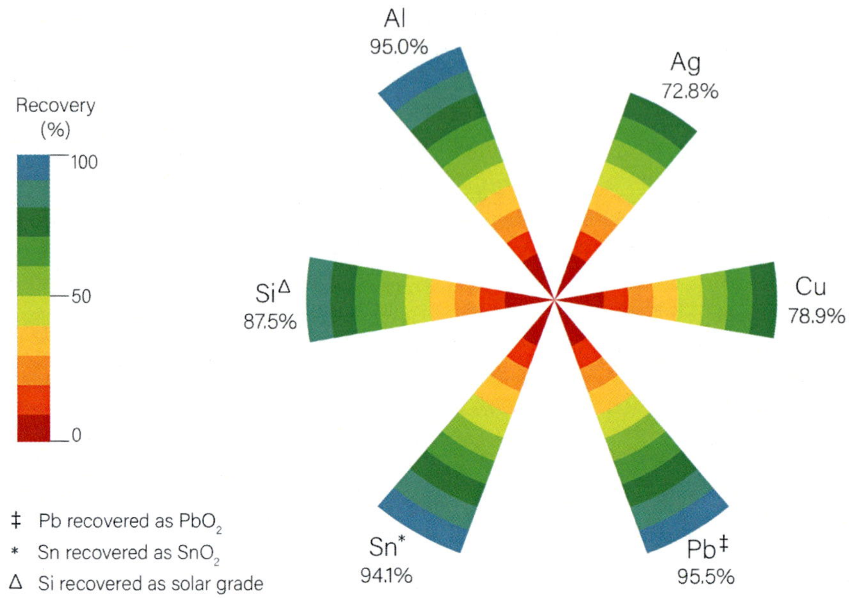

FIGURE 46.5 Resource recovery (%) efficiency of the Si PV recycling process (Bartie et al., 2021).

were evaluated and compared—the base case, in which all polymer layers are combusted during recycling, and a second case in which the rear module support layer is removed for further treatment rather than combusted for heat recovery. Figure 46.6 shows how direct CO_2-equivalent emissions from the whole life cycle changes with end-of-life recycling rate for the two process configurations. Although not discussed in this example, the effects of recycling a Si production residue further upstream in the life cycle is shown on a third axis. Results from the process simulation allow one to quantify and visualize the degree to which process changes could affect the environmental impact. It should be noted that direct process emissions are only part of the total carbon footprint of a process—this is discussed in more detail in Chapter 47—Life cycle assessment. Detailed information about the example presented here can be found in the article written by Bartie et al. (2021).

46.5 CONCLUSION AND OUTLOOK

The predictive nature of simulation models allows for the physics-based estimation of how life-cycle systems respond to changes in, among others, feed material compositions, process configurations, operating conditions, and the technology used. Process simulation thereby enables DfR—the system-wide effects of product design changes on resource consumption and sustainability can be evaluated as early as during the product design phase so as to maximize recyclability. Furthermore, where possible, it allows for residues formed in production and recycling processes to be engineered to meet secondary resource specifications, in the process maximizing resource efficiency. The material and energy consumption, emissions, and residue formation data generated through process simulation can be transferred into various other tools to conduct further analyses. Among these are methods used to assess environmental, economic, and social

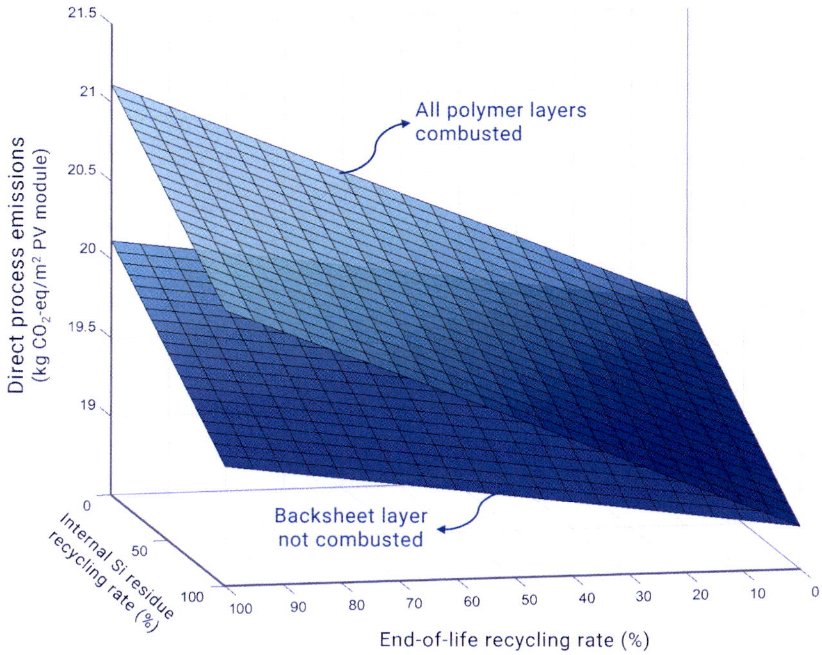

FIGURE 46.6 Direct life cycle CO_2 emissions as a function recycling rate for two polymer recycling process configurations.

impacts. Environmental LCAs, for example, can in some cases be overly reliant on environmental databases in which some datasets are significantly outdated and others do not exist yet because of rapid technology and process development, especially with the increasing focus on more targeted, high-quality recycling. Process simulation can enhance the quality of such assessments by providing up-to-date inventory data. The added flexibility to generate new datasets for alternative scenarios fairly quickly can contribute to identifying process and technology options that would maximize the recovery of potential secondary resources at the highest possible purities within the physics-based, environmental, economic, and societal constraints the system is bound by. The indicators obtained, of which only a few have been mentioned here, can be used to effectively communicate reliable process and sustainability information to industry, consumers, and policymakers to support decision-making and drive progress toward sustainability and circular economy.

The development of process simulation models is not without challenges. A prerequisite for the development and validation of reliable simulation models is the use of scientifically sound data and valid assumptions. Collecting or generating the process and operating data, whether theoretically or from industry, is the most resource-intensive step in the simulation development process. As discussed in this chapter, it requires process know-how and experience. Once a simulation model is in place, however, the value it can add to sustainability and circularity assessments is substantial.

References

Abadías Llamas, A., Valero, A., Valero, A., Torres, C., Hultgren, M., Peltomäki, M., Roine, A., Stelter, M., Reuter, M. A., 2019. Simulation-based exergy, thermo-economic and environmental footprint analysis of primary copper production. Miner. Eng. 131, 51–65. https://doi.org/10.1016/j.mineng.2018.11.007.

Bartie, N.J., Abadías Llamas, A., Heibeck, M., Fröhling, M., Volkova, O., Reuter, M.A., 2019. The simulation-based analysis of the resource efficiency of the circular economy—the enabling role of metallurgical infrastructure. Miner. Process. Extr. Metall. Trans. Inst. Min. Metall. 129 (2), 229–249. https://doi.org/10.1080/25726641.2019.1685243.

Bartie, N.J., Cobos-Becerra, Y.L., Fröhling, M., Schlatmann, R., Reuter, M.A., 2021. The resources, exergetic and environmental footprint of the silicon photovoltaic circular economy: assessment and opportunities. Resour. Conserv. Recycl. 169 (January), 105516. https://doi.org/10.1016/j.resconrec.2021.105516.

Belo Fernandes, I., Abadias Llamas, A., Reuter, M., 2020. Process simulation: NdFeB permanent magnet production. RODARE. https://doi.org/10.14278/rodare.617.

Brunner, P.H., Rechberger, H., 2016. Practical Handbook of Material Flow Analysis. Lewis Publishers, https://doi.org/10.1007/bf02979426.

Fernandes, I.B., Abadías Llamas, A., Reuter, M.A., 2020. Simulation-based exergetic analysis of NdFeB permanent magnet production to understand large systems. JOM 72 (7), 2754–2769. https://doi.org/10.1007/s11837-020-04185-6.

Hannula, J., Kern, M., Luukkanen, S., Roine, A., van den Boogaart, K.G., Reuter, M.A., 2018. Property-based modelling and simulation of mechanical separation processes using dynamic binning and neural networks. Miner. Eng. 126, 52–63. https://doi.org/10.1016/j.mineng.2018.06.017.

Hannula, J., Godinho, J.R.A., Abadías Llamas, A., Luukkanen, S., Reuter, M.A., 2020. Simulation-based exergy and LCA analysis of aluminum recycling: linking predictive physical separation and re-melting process models with specific alloy production. J. Sustain. Metall. 6 (1), 174–189. https://doi.org/10.1007/s40831-020-00267-6.

Heibeck, M., Bartie, N.J., Abadias Llamas, A., Reuter, M.A., 2020. CdTe refining + photovoltaic manufacturing + recycling HSC model. RODARE. https://doi.org/10.14278/rodare.609.

Reuter, M.A., 2016. Digitalizing the circular economy: circular economy engineering defined by the metallurgical internet of things. Metall. Mater. Trans. B Process Metall. Mater. Process. Sci. 47 (6), 3194–3220. https://doi.org/10.1007/s11663-016-0735-5.

Reuter, M.A., van Schaik, A., 2015. Product-centric simulation-based design for recycling: case of LED lamp recycling. J. Sustain. Metall. 1 (1), 4–28. https://doi.org/10.1007/s40831-014-0006-0.

Reuter, M.A., Kojo, I.V., Roine, A., Florin, H., Jåfs, M., 2013. Environmental footprinting of metallurgical processing technology-linking GaBi to HSC sim. Copper 1 (1–4th December), 373–388.

Reuter, M.A., van Schaik, A., Gediga, J., 2015. Simulation-based design for resource efficiency of metal production and recycling systems: cases—copper production and recycling, e-waste (LED lamps) and nickel pig iron. Int. J. Life Cycle Assess. 20 (5), 671–693. https://doi.org/10.1007/s11367-015-0860-4.

Reuter, M.A., van Schaik, A., Ballester, M., 2018. Limits of the circular economy: fairphone modular design pushing the limits. World Metall. – ERZMETALL 71 (2), 68–79.

Reuter, M.A., van Schaik, A., Gutzmer, J., Bartie, N., Abadías Llamas, A., 2019. Challenges of the circular economy: a material, metallurgical, and product design perspective. Annu. Rev. Mater. Res. 49 (1), 253–274. https://doi.org/10.1146/annurev-matsci-070218-010057.

CHAPTER

47

Life cycle assessment (LCA)

Christina Meskers[1], Neill J. Bartie[2], and Markus A. Reuter[3]

[1]Department of Energy and Process Engineering, Faculty of Engineering Sciences, NTNU—Norwegian University of Science and Technology, Trondheim, Norway [2]Helmholtz Centre Berlin for Materials and Energy (HZB), PVcomB, Berlin, Germany [3]WASM: Minerals, Energy and Chemical Engineering, Faculty of Science and Engineering, Curtin University, Perth, Australia

47.1 INTRODUCTION

Environmental life cycle assessment (LCA) quantifies the environmental aspects and *potential* environmental impacts in the areas of ecological health, human health, and resource depletion, throughout a product's or a service's life cycle from raw material extraction through production, use, end-of-life treatment, and disposal (i.e., cradle to grave).

The insights that environmental LCA provides are used to inform the large societal challenges such as climate change, circular economy, energy and mobility transition, land and water usage, sustainable buildings, etc. In contrast, the results of LCA are also used for decision-making on specific processes, products, and services. Hence, the level of detail can range from a single process to entire societal systems that can cover different timescales and geographies. Decision-makers in industry can use it for, e.g., strategic planning, product or process design or redesign, marketing and communication, and benchmarking. Decision-makers in both industry and (non) governmental organizations can use the outcomes from an LCA to inform policymaking, systemic change, or transition scenarios, and to assess trade-offs and risks.

In 2021 the European Commission provided a recommendation on the LCA methods to be used for product environmental footprints (PEFs) and organizational environmental footprints (OEFs), including category and sector specific rules (European Commission, 2021). Their use is already foreseen in the taxonomy regulation (European Union, 2020b) and the sustainable batteries initiative (European Union, 2020a), and use in further regulations is expected.

As Figure 1.4 shows, the environmental LCA is located between the physical resource layer (quantity and quality) and the economic layer. It uses the physical layer as basis for the environmental impact assessment. LCA was initially not developed to address economic or social effects. The methodology has been expanded with Life Cycle Costing (LCC) addressing economic impact (Rebitzer and Seuring, 2003; Rödger et al., 2018) and social LCA (s-LCA) addressing social impact on stakeholders (Poulsen and Jensen, 2005; UNEP, 2020). The stakeholders are, e.g., workers, local community, society, customers, and value chain actors. Impact categories include, among others, human rights, working conditions, health and safety, governance, and cultural heritage. These varieties of LCA all use the same methodological building blocks. The three together form a Life Cycle Sustainability Assessment (LCSA), described in the UNEP/SETAC Life Cycle Initiative (2011).

LCA is a standardized method described in ISO 14040:2006 (and amendments) that can be applied to any product or process, while ISO 14044:2006, ISO 14044:2006/Amd.1:2017, ISO 14044:2006/Amd.2:2020 covers the requirements and guidelines. The framework is described in Figure 47.1.

The main stages (including their interconnections) of an LCA are:

- Goal and scope definition: functional unit, system boundaries, inclusions and exclusions, applied method, intended audience, etc.
- Life-cycle inventory: data collection, data quality check, LCA modeling and inventory generation.
- Life-cycle impact assessment: calculation of the environmental impacts based on the inventory.
- Interpretation of the results and reporting.

A good understanding and execution of each step of the LCA methodology and its iterative cycles are essential to obtain a high-quality, reliable LCA that provides correct insights and conclusions. With each iteration the level of detail and insight increases and the uncertainty of the LCA results reduces.

Some LCA applications are described in their own standards: eco-efficiency assessment of product systems (ISO 14045:2012), water footprint (ISO 14046:2014), carbon footprint of products (ISO 14067:2018), and environmental product declaration (EPD) (ISO 14025:2010), among others. The field of LCA continues to develop further with hybrid input-output life cycle analysis (IO-LCA) and integrated assessment modeling (e.g., Gibon et al., 2015; Arvesen et al., 2018).

It is impossible to cover all aspects and developments in this chapter. Instead, this chapter aims to contribute to life cycle assessment literacy (Viere et al., 2021). It briefly introduces the key concepts of performing an environmental LCA, referring the reader to key literature as a stepping stone for in-depth information, followed by case studies that illustrate the specifics and challenges with LCA in a recycling context. Here, both process and product perspectives will be taken.

47.2 LIFE CYCLE ASSESSMENT—KEY CONCEPTS

To understand the specific attention points/challenges for LCA in a product, materials, process, and recycling (circular economy) context, it is necessary to dive deeper into the details of the LCA methodology and its stages. Figure 47.2 shows the different steps and iterations that take place. When reading through the key concepts, one can notice the similarities between LCA and the seven components for monitoring the physical economy—(Chapter 3) and the steps of the process simulation methodology (Chapter 46—Process simulation).

47.2.1 Goal and Scope Definition

In the goal and scope definition stage, the context in which the assessment is made is defined, as well as the question(s) the LCA will be answering to provide strategic and decision-making support, and the intended target audience(s). Using this, decisions are made about the type of LCA, the system, its boundaries and functional unit, the level of detail necessary, key assumptions, and the environmental effects that are to be reviewed. This stage takes a considerable amount of time. To illustrate the range of

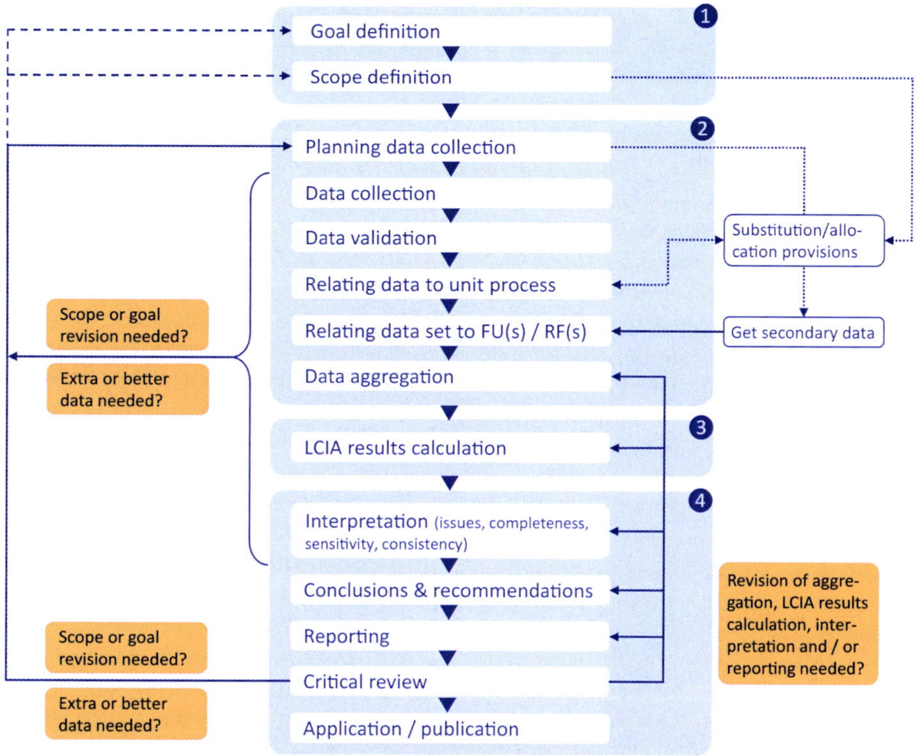

FIGURE 47.1 Framework of life-cycle assessment following ISO 14040:2006/Amd 1:2020 indicating the different phases and the iterative approach in LCA, highlighted for the inventory analysis. ① goal and scope definition ② inventory analysis ③ impact assessment ④ interpretation and reporting. *Based on ISO 14040:2006 and European Commission (2010).*

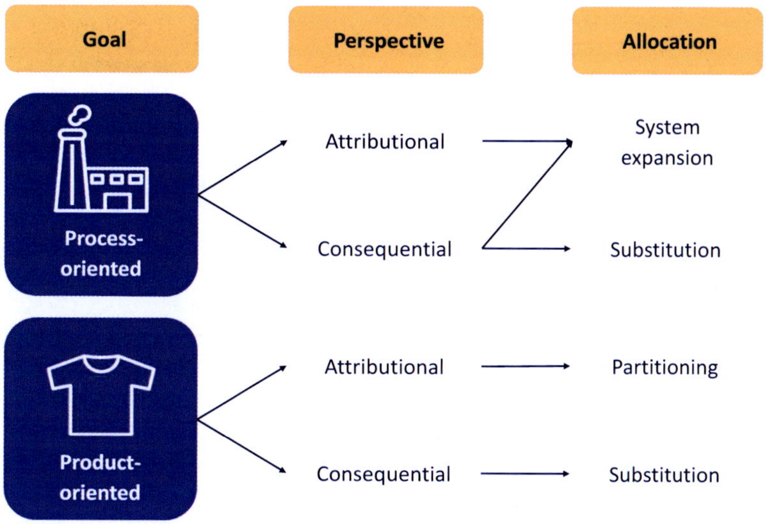

FIGURE 47.2 Framework for consistent recycling and coproduct allocation. *Modified from Schrijvers et al. (2020).*

possibilities, Table 47.1 shows the relationship between research question, functional unit, and type of LCA on the topic of "recycled yttrium" (Schrijvers et al., 2020).

Functional unit needs to be chosen so that they are comparable to each other, and the selection is linked to the goal of study. The functional unit and the variations need to be described in detail. The functional unit must capture the (multi-)functionality and is thus not by default 1 kg material or 1 product, especially when comparing alternatives. The functional unit description answers the questions what, how much, for how long, where, and how well (to which quality). Often not all questions are answered. In Table 47.1, for example, the where and how well questions are not included in the functional unit description.

The system chosen needs to reflect the real-life system adequately and needs to be at a level of detail that serves the purpose of the LCA. To obtain the understanding to represent the real-life system in the LCA involves background literature research, industry, and expert interviews, etc. Understanding the spatial, technological (e.g., feed, process variations), and temporal aspects; the interconnectedness of materials, processes, and value chains; and the distributed nature of properties and attributes within a class are important to translate the real-life system into a model. These aspects are discussed in detail in Part I of this book. Setting the system boundaries means also determining what will be the foreground system, for which primary data will be collected, and what will be the background system, for which generic databases will be used.

Many systems that include waste management, recycling, or coproducts and by-products are multifunctional, as they deliver both the recycling or waste management service and provide material and/or energy flows. This needs to be reflected in the definition of the functional unit and the type/LCA method that is used.

The **types of LCA** are:

- Attributional, focuses on the flows in and out of the life cycle, and the environmental impacts directly linked to the product, process, or value chain under investigation. For background data it uses average background processes.
- Consequential, focuses on the effects of changes in the flows because of decisions made, e.g., product design choices, process choices, etc. It includes both the impacts of the activities taking place in the value chain of the product, and the impacts on other value chains or the global level. The background data it uses are the marginal processes, i.e., the process that can be used to process/produce a quantity right now. It requires a good understanding of and ability to model the dynamics of the economic system, which is a different way of thinking than the engineering perspective of processes and supply chains (Hauschild et al., 2018).

The **orientation** of each type can be process or product oriented. How this relates to the functional unit and research questions is illustrated in Table 47.1. Note that in this example the system boundaries of the attributional product-oriented LCA are the same as the consequential process-oriented LCA (Schrijvers et al., 2020).

Lastly, the **temporal dimension**, or the point in time that is considered instead of taking the situation today, can be:

- Prospective, where a future situation is assessed. It often includes emerging technologies. One challenge in this type of LCA is obtaining sufficiently detailed data to do a good LCA, as emerging technologies and infrastructures are often not included in LCA databases and experimental work has not necessarily reached the operational practice level yet (TRL 7-9). Arvidsson et al. (2018), for example, recommend including

TABLE 47.1 Potential research questions and corresponding (cradle-to-gate) LCA approaches based on the initial subject of "recycled yttrium".

LCA approach	Functional unit	Research question	Modeling specifics
Attributional process-oriented	The treatment of 10.8 kg of phosphorus powder, and the production of 1 kg Y, 2.2 kg glass, 0.2 kg La, 0.1 kg Ce, 0.1 kg Tb, and 0.1 kg Eu	1. What is the accountability for impacts of [*functional unit*] via the recycling process? 2. How can we decrease the accountability for impacts of [*functional unit*] via the recycling process? 3. Does [*functional unit*] via the recycling process have a lower accountability for impacts than the treatment and production of these flows via alternative processes?	Only the LCI of the foreground subsystem is calculated based on attributional background data No allocation is necessary within the foreground subsystem as system expansion is applied
Attributional product-oriented	The consumption of 1 kg of recycled Y	4. What is the accountability for impacts of [*functional unit*]? 5. How can we decrease the accountability for impacts of [*functional unit*]? 6. Does [*functional unit*] have a lower accountability for impacts than the consumption of 1 kg of primary Y?	Partitioning must be applied to identify the cradle-to-gate inventory that is attributed to recycled Y More information is required from the product system that supplies the phosphorus powder, as the recycling process is part of this production system
Consequential process-oriented	The treatment of 10.8 kg of phosphorus powder, and the production of 1 kg Y, 2.2 kg glass, 0.2 kg La, 0.1 kg Ce, 0.1 kg Tb, and 0.1 kg Eu	7. What are the consequences on global impacts of [*functional unit*] via the recycling process? 8. How can we decrease the consequences on global impacts of [*functional unit*] via the recycling process? 9. Does [*functional unit*] via the recycling process have lower consequences on global impacts than the treatment and production of these flows via alternative processes?	Only the LCI of the foreground subsystem is calculated based on consequential background data No allocation is necessary within the foreground subsystem, as system expansion is applied
Consequential process-oriented	The production of 1 kg of recycled Y	10. What are the consequences on global impacts of [*functional unit*]? 11. How can we decrease the consequences on global impacts of [*functional unit*]? 12. Does [*functional unit*] have lower	The LCI of the foreground system is calculated based on consequential background data The ingoing flow of phosphorus powder is modeled by the substitution of its marginal waste treatment process

Continued

TABLE 47.1 Potential research questions and corresponding (cradle-to-gate) LCA approaches based on the initial subject of "recycled yttrium"—cont'd

LCA approach	Functional unit	Research question	Modeling specifics
		consequences on global impacts than the production of 1 kg of primary Y?	The outgoing flows of glass, La, Ce, Tb, and Eu are modeled by the substitution by their marginal users
Consequential product-oriented	The consumption of 1 kg of recycled Y	13. What are the consequences on global impacts of [*functional unit*]? 14. How can we decrease the consequences on global impacts of [*functional unit*]? 15. Does [*functional unit*] have lower consequences on global impacts than the consumption of 1 kg of primary Y?	If the supply of recycled Y is constrained, the LCI represents the substitution of recycled Y by its marginal user If the supply is unconstrained, the LCI of the foreground subsystem is calculated The ingoing flow of phosphorus powder is modeled by the substitution of its marginal waste treatment process The outgoing flows of glass, La, Ce, Tb, and Eu are modeled by the substitution by their marginal users

Y, yttrium; *La*, lanthanum; *Ce*, cerium; *Tb*, terbium; *Eu*, europium. System boundaries for each LCA approach can be found in Schrijvers et al. (2020).
From Schrijvers et al. (2020).

(technology) scenarios and scenario ranges for the foreground and background information.

- Anticipatory, which assesses developments over time, such as changes in geography, markets, raw materials, etc., such as was done, for example, in Ambrose and Kendall (2020) for temporal and spatial changes in lithium production systems over time.

For prospective or anticipatory LCAs, the inventories for foreground and background information need to be adjusted accordingly. Each LCA consists of a combination of the previously mentioned aspects. In other words, a consequential, product-oriented, prospective LCA is conceivable. When in doubt about what to use, make different LCAs and see the impact of the choices on the outcomes.

47.2.2 Inventory Analysis (LCI)

This stage collects the quantitative information about the usage and composition of materials, products, semiproducts, energy and water, and environmental releases (emissions, waste, etc.) in the studied system. It can include changing variables such as geographical differences, variations in product or material composition or technology, etc. All elements, compounds, and types of energy need to be accounted for and the mass (including water) and energy balance needs to be closed (Hauschild et al., 2018). This information is included in a model that represents the real-life system and the relationships in it as well as possible, including respecting the laws of physics and chemistry, nonlinear relationships, and reflecting economic realities (Reuter et al., 2015). The flows are scaled to the functional unit selected in the previous step. The result is the inventory, a list of quantified physical elementary flows for the system associated with the functional unit (Hauschild et al., 2018). Compiling the inventory is an iterative process (Figure 47.1).

Data and its quality forms the foundation of the LCI. Several aspects need to be checked. The data needs to be representative of the actual situation from a geographical perspective, time-related perspective, and a technological perspective.

The data also need to be complete, consistent, and reproducible. For each data source this needs to be assessed. Missing data, or information about its meaning or boundaries, need to be made explicit.

Uncertainty and sensitivity analyses are an essential part of the LCI stage. In the uncertainty analysis, done before the sensitivity analysis, the uncertainty of the result because of uncertainties in each parameter of the LCI is quantified. Tools such as data reconciliation, Monte Carlo simulation, etc. can be used. Hence, for each parameter information on the statistical distribution and statistical parameter values needs to be collected. This can be done via uncertainty factors, probability distributions, or pedigree matrix (Weidema, 1998; Ciroth et al., 2016), for example. The sensitivity analysis's purpose is the systemic identification of the parameters that have the highest influence on the LCA results. In other words, it reflects the uncertainties about the actual system modeled. This can be assessed by developing sensitivity scenarios (Hauschild et al., 2018). As discussed in Part 1 of this book, the composition, properties, etc. of end-of-life products, recyclates, scrap types, etc. are distributed and can vary over time, while sampling errors give uncertainty on the stream composition. These aspects should be addressed when doing the uncertainty and sensitivity analysis. Note that the background inventory can be subjected to sensitivity analysis as well (Kim et al., 2022).

Several tools exist for making the models and inventory, such as Gabi, OpenLCA, Simapro, Umberto, and Brightway2 (Viere et al., 2021). For background data, often LCI databases are used, e.g., ecoinvent, GaBi, etc. The Global LCA Data Access network (GLAD) www.globallcadataacess.org gives an overview.

The utilization of consistent (foreground) input data generated in the physical flows layer, as opposed to disjointed analyses with potentially incompatible system boundaries, facilitates robust assessments. As alluded to in Reuter et al. (2019), digitalization starting with physical quantities and qualities is key to such assessments. Reuter et al. (2013, 2015) developed an approach for linking process simulation (Chapter 46—Process simulation) in HSC Sim with openLCA and GaBi, making it possible to map and export detailed inventory data from the process simulator into the LCA software (Reuter, 2016).

With its roots in process engineering design, where high levels of physical, chemical, and thermodynamic detail and compliance with the laws of conservation and entropy are nonnegotiable, a carefully developed and validated simulation model provides reliable mass, energy, and thermochemical information down to the element and compound level for entire product life cycles and circular economy systems. It captures the complex, nonlinear interactions that fundamentally characterize processing and recycling systems, enabling process optimization and sound, physics-based environmental impact assessments. Using process simulation, one can also generate inventory data for emerging and future technologies.

47.2.3 Impact Assessment (LCIA)

In this step the human and ecological effects are assessed and quantified using the LCI as input. For each item in the inventory, its impact is quantified for each environmental mechanism (or midpoint or impact category), using environmental science knowledge and models. The midpoint categories include human toxicity, carcinogens, climate change (in CO_2-eq.), ozone layer, acidification, eutrophication, ecotoxicity, smog, land use, and resource depletion, which are linked to the three damage categories: human health, ecosystem diversity, and resource scarcity. In this way, the hundreds of flows are converted to several impact categories with clear environmental meaning.

This step is automated in the LCA software: one "only" needs to choose an impact assessment method, such as ReCiPe (Goedkoop et al., 2009), CML, TRACI (USA), ILCD (EU), LIME (Japan), EDIP, IMPACT, etc. But each has its own principles, models, and factors, so the limitations and applicability of the method need to be checked, so that all relevant flows are actually characterized by the chosen method and the method fits the type of LCA done.

For multifunctional or multiproduct systems, often the impacts need to be assigned to the different functions or products. Different allocation procedures are available, depending on the goal and type of the LCA (Schrijvers et al., 2016b; European Commission, 2010) as well as the sustainability concept served (or policy goal pursued), the risk perception, and the resulting eco-efficiency indicators (Frischknecht, 2010; Allacker et al., 2014). This is a highly debated topic in the LCA community.

47.2.4 Interpretation and Communication

Interpretation and communication are an important part of the LCA. The results are interpreted with the original goal, scope, and strategic or decision-making context in mind. Transparency about the assumptions made in the model, inventory, and their impact on the validity of the results is very important. When publishing the results (e.g., in scientific journals) the results and inventories can add to the knowledge base and can serve as improved background information for others. If the target audience is not familiar with LCA, taking time for good communication and visualization of the results, so that the depth and detail of the LCA is reflected, becomes even more important.

47.3 PRODUCT-ORIENTED, ATTRIBUTIONAL LCA

An important part of many LCA studies is the use of metal production inventory data, as background information. The metal trade associations have been performing industry-wide LCAs to obtain a global average LCA for "their" metal; some are further divided in regional average LCAs. Initially a voluntary initiative of the trade association, it is becoming increasingly driven by regulation, such as in the European Union. The LCAs are developed using information from industry actors obtained through surveys and questionnaires. This includes primary metal production data as well as recycled metal production data. Using a standardized methodology allows for comparison over time and comparison between commodities, which can inform decision-making on material selection in the product design phase, for example. Almost all metal trade associations performing LCAs for their metal and metal products show environmental improvement over time. This information is obviously used in the communication to stakeholders like governments and downstream users.

A consistent harmonized methodology between metal commodities is especially important, as there are interconnected carrier metals that are producing similar by-products and products (Chapter 2 in this book). Hence, the choice of the allocation of the environmental impact to the metals needs to be harmonized between the different metal carriers to ensure the results from the LCA studies are comparable across metal commodities and processing routes.

For example, nickel as carrier metal produces copper, cobalt, gold, and platinum group metals as coproducts/by-products, while copper as carrier material produces molybdenum, gold, silver, platinum group metals and zinc as coproducts/by-products. When one would like to know the impacts of the production of copper, nickel, or gold, the impacts of the process need to be divided (allocated) to the different metals that are produced. This allocation can be based on mass or metal price, or another physical property. The difference between the impacts of a metal produced as carrier metal or as by-product metal can be considerable. For example, the kg CO_2-eq./kg metal differs by a factor of 2 for copper from copper ore (4.2 kg CO_2-eq./kg) and copper from nickel processing as by-product (7.2 CO_2-eq./kg) with mass allocation (GaBi 9, 2020). The difference between gold from nickel (689 CO_2-eq./kg Au) and copper carrier metal (12,791 CO_2-eq./kg Au) is nearly a factor of 20 for mass allocation. Compared with gold production from gold as carrier metal (36,793 CO_2-eq./kg Au, World Gold, 2019), the difference is even higher. The impact of price variations on price-based impact allocation for by-product metals is illustrated in this chapter (Section 47.6.2) and in Stamp et al. (2011), for example.

47.4 PROCESS-ORIENTED, ATTRIBUTIONAL LCA

The challenges with allocation, and the use of generic, average (background) data as done in the product-oriented attributional LCAs discussed in the previous section can be overcome using process-oriented attributional LCAs instead. In this way, it is also recognized that materials are not produced as single metals in isolation, but often in processes where multiple metals and other products are produced. Through system expansion, i.e., looking at an entire process or life cycle, allocation can be avoided. A process-oriented approach also allows for comparing the impact of different process routes and feed and process variables and quality on the environmental impacts.

For example, Manjong et al. (2021) assessed the impact of feed and process variations on the extraction and processing chain of minerals for rechargeable batteries. Using a parametric attributional process-oriented model, they investigated the impact of variations and combinations of two variations on the greenhouse warming potential. This includes, among others, the ore grade, mine type, mineral characteristics, concentration of reactants and chemicals, material recovery, energy mix carbon intensity, energy efficiency, and technology variations.

Process-simulation based LCA for metallurgical systems has been implemented by Reuter and coworkers, among others, and has been used for a wide range of cases (Table 47.2).

47.5 END-OF-LIFE TREATMENT IN THE LCA CONTEXT

End-of-life treatment introduces multifunctionality into the LCA. Several allocation procedures can be used, depending on the goal and scope of the LCA (Table 47.1). The goal—scope—allocation possibilities are summarized in Figure 47.2. The choice of method has significant influence on the LCA results (Table 47.3), hence deeply understanding the methods and doing a sensitivity analysis for each method when multiple methods can be applicable is useful.

Furthermore, when discussing end-of-life treatment, other aspects need to be considered, such as the number of life cycles and the type of recycling, which are connected to the quality of the material and the performance of the processes. Closed loop recycling takes place when the (recycled) material can be used in the same product system, while open loop recycling takes place when the recycled material is used in a different product system because its properties have changed. If the properties do not change, it can be considered closed loop recycling. In both cases, downcycling (loss of quality) can take place and needs to be considered.

47.5.1 System Expansion (Process-Oriented LCA)

The LCA guidelines (ISO 14044:2006, ISO 14044:2006/Amd.1:2017, ISO 14044:2006/Amd.2:2020) indicate that allocation should be avoided. If this is not possible **system expansion** is preferred when handling multifunctionality. This means expanding the systems investigated until the same functions are delivered by all systems compared. So instead of a product-oriented LCA, one may need to change to a process-oriented LCA. When using system expansion in a recycling context, the life cycle that produces the recycled material and the life cycle that consumes the recycled material are included in the functional unit and thus are part of the foreground system (e.g., Bartie et al., 2021; Nordelöf et al., 2019). In open loop recycling and for complex products the system can expand quickly, as the papers listed in Table 47.2 show.

In a consequential LCA system, expansion is followed by substitution. As the conventional alternatives are avoided, these are subtracted from the inventory of the functional unit (Schrijvers et al., 2016a).

47.5.2 Partitioning and Substitution (Product-Oriented LCA)

In product-oriented LCAs the LCA calculation rules for recycling need to address whether and how the environmental burden of the primary production of a material is shared between the first user and the subsequent users of that material (Frischknecht, 2010; Allacker et al., 2014; Hauschild et al., 2018; Schrijvers et al., 2016a; European Commission, 2010). The proposed methodologies can roughly be classified into two main approaches that are used in daily LCA practice:

TABLE 47.2 Nonexhaustive overview of process-simulation based LCA studies focusing on metals and metal products, in addition to the papers cited in this chapter.

Authors (year)	Title	Details	DOI
Reuter et al. (2023)	Digital twin for KGHM-Legnica: simulation-based footprinting & exergy allocation of impacts	Processing of copper concentrates and secondary materials into metals, in Poland; industry-verified model	EMC 2023 conference
Rinne, Halli, Aromaa, Lundström (2022)	Alternative method for treating electric arc furnace dust: simulation and LCA	Novel process compared to conventional Waelz process	https://doi.org/10.1007/s40831-022-00531-x
Alexander, Johto, Lindgren, Pesonen and Roine (2021)	Comparison of environmental performance of modern copper smelting technologies	Copper sulfide ores into blister copper, comparing 7 different technologies	https://doi.org/10.1016/j.cesys.2021.100052
Bartie, Cobos-Becerra, Fröhling, Schlatmann, Reuter (2021)	The resources, exergetic, and environmental footprint of the silicon photovoltaic circular economy: assessment and opportunities	From raw materials to modules and their recycling	https://doi.org/10.1016/j.resconrec.2021.105516
Neumann and Reuter (2021)	Simulationsbasierte Berechnung der Umweltwirkung von Zinkdrückguss bei Einsatz von Primär- und Sekundärrohstoffen für die Legierungherstellung	Zinc diecasting process with primary and secondary feed materials, in Western Europe	Erzmetall 74 (6) 325–30
Rinne, Elomaa, Lundström (2021)	LCA and process simulation of prospective battery-grade cobalt sulfate production from Co-Au ores in Finland		https://doi.org/10.1007/s11367-021-01965-3
Rinne, Elomaa, Porvali, Lundström (2021)	Simulation-based LCA of hydrometallurgical recycling of mixed LIB and NiMH waste	Conceptual, experimentally proven hydrometallurgical recycling process	https://doi.org/10.1016/j.resconrec.2021.105586
Rinne, Elomaa, Seisko, Lundström (2021)	Direct cupric chloride leaching of gold from refractory sulfide ore: process simulation and LCA		https://doi.org/10.1080/08827508.2021.1910510
Abadías-Llamas, Bartie, Heibeck, Stelter, Reuter (2020)	Resource efficiency evaluation of pyrometallurgical solutions to minimize iron-rich residues in the roast-leach-electrowinning process	Residue (jarosite) treatment from primary zinc industry	https://doi.org/10.1007/978-3-030-37,070-1_31

Continued

TABLE 47.2 Nonexhaustive overview of process-simulation based LCA studies focusing on metals and metal products, in addition to the papers cited in this chapter—cont'd

Authors (year)	Title	Details	DOI
Abadías-Llamas, Bartie, Heibeck, Stelter, Reuter (2020)	Simulation-based exergy analysis of large circular economy systems: zinc production coupled to CdTe photovoltaic module life cycle	From raw materials to modules and their recycling. Coupling of zinc, lead, and copper flowsheets	https://doi.org/10.1007/s40831-019-00255-5
Belo Fernandes, Abadías-Llamas, Reuter (2020)	Simulation-based exergetic analysis of NdFeB permanent magnet production to understand large systems	From concentrate to magnet and their recycling	https://doi.org/10.1007/s11837-020-04185-6
Elomaa, Rintala, Aromaa, Lundström (2020)	Process simulation based LCA of cyanide-free refractory gold concentrate processing—case study: cupric chloride leaching	Two flowsheet options in development stage, each 3 cases with varying operational parameters	https://doi.org/10.1016/j.mineng.2020.106559
Elomaa, Sinisalo, Rintala, Aromaa, Lundström (2020)	Process simulation and gate-to-gate LCA of hydrometallurgical refractory gold concentrate processing	Two processes for treatment of same concentrate	https://doi.org/10.1007/s11367-019-01723-6
Hannula, Godinho, Abadías-Llamas, Luukanen, Reuter (2020)	Simulation-based exergy and LCA analysis of aluminum recycling: linking predictive physical separation and remelting process models with specific alloy production	Four scrap types, varying physical separation parameters, 3 aluminum target alloys	https://doi.org/10.1007/s40831-020-00267-6
Abadías-Llamas, Valero, Valero, Torres, …, Stelter, Reuter (2019)	Simulation-based exergy, thermo-economic and environmental footprint analysis of primary copper production	From copper ore to refined metal, minor metal production, residue treatment, and scrap processing	https://doi.org/10.1016/j.mineng.2018.11.007
Bartie, Abadías-Llamas, Heibeck, Fröhling, Volkova, Reuter (2019)	The simulation-based analysis of the resource efficiency of the circular economy – the enabling role of metallurgical infrastructure	From raw materials to CdTe PV modules and their recycling. Coupling of zinc, lead, and copper flowsheets	https://doi.org/10.1080/25726641.2019.168543
Pell, Wall, Yan, Li, Zeng (2019)	Mineral processing simulation based environmental LCA for rare earth project development: A case study on the Songwe Hill project	Prefeasibility stage of the project. Several scenarios investigated, in Malawi	https://doi.org/10.1016/j.envman.2019.109353

Ghoddrat, Akbar Rhamdhani, Brooks, Rashidi, Samali (2017)	A thermodynamic-based LCA of precious metal recycling out of waste printed circuit board through secondary copper smelting	Secondary copper recycling with and without printed circuit boards	https://doi.org/10.1016/j.envdev.2017.07.001
Northey, Mudd, Saarivuori, Wess-man-Jääskeläinen, Haque (2016)	Water footprinting and mining: where are the limitations and opportunities?	Mine, mineral processing plant, tailings, and ponds	https://doi.org/10.1016/j.clepro.2016.07.024
Van Schaik and Reuter (2016)	Recycling indices visualizing the performance of the circular economy		Erzmetall 69(4):201–16
Weng, Haque, Mudd, Jowitt (2016)	Assessing the energy requirements and global warming potential of the production of rare earth elements	26 operating and potential REE mining projects. Mining, processing, and refining	https://doi.org/10.1016/j.clepro.2016.08.132
Reuter, Van Schaik, Gediga (2015)	Simulation-based design for resource efficiency of metal production and recycling systems: cases – copper production and recycling, e-waste (LED lamps) and nickel pig iron	Copper production and recycling, LED lamp recycling, nickel pig iron production	https://doi.org/10.1007/s11367-015-0860-4
Northey, Haque, Lovel and Cooksey (2014)	Evaluating the application of water footprint methods to primary metal production systems	Copper, gold, and nickel production: mining, mineral processing, and metal production	https://doi.org/10.1016/j.mineng.2014.07.006
Haque and Norgate (2013)	Estimation of greenhouse gas emissions from ferroalloy production using LCA with particular reference to Australia	From raw materials to ferroalloy: FeMn, SiMn, FeSi, FeNi, FeCr, in Australia	https://doi.org/10.1016/j.clepro.2021.08.010
Reuter, Kojo, Roine, Florin, Jåfs (2013)	Environmental footprinting of metallurgical processing technology—linking GaBi to HSC Sim		Copper 2013 conference
Norgate and Haque (2012)	Using LCA to evaluate some environmental impacts of gold production	Refractory and nonrefractory ores; from mine to metal	https://doi.org/10.1016/j.clepro.2012.01.042
Norgate and Haque (2010)	Energy and greenhouse gas impacts of mining and mineral processing operations	Iron ore, bauxite, and copper concentrate. Surface mining and underground mining, in Australia	https://doi.org/10.1016/j.clepro.2009.09.020

Continued

TABLE 47.2 Nonexhaustive overview of process-simulation based LCA studies focusing on metals and metal products, in addition to the papers cited in this chapter—cont'd

Authors (year)	Title	Details	DOI
Norgate and Jahanshahi (2010)	Low grade ores—smelt, leach or concentrate?	Alternative processing routes for low-grade copper and nickel ores	https://doi.org/10.1016/j.mineng.2009.10.002
Norgate, Jahanshahi, Rankin (2007)	Assessing the environmental impact of metal production processes	Current or potential processes practiced in Australia for Cu, Ni, Al, Pb, Zn, Ti, and stainless steel	https://doi.org/10.1016/j.clepro.2006.06.018
Reuter and Verhoef (2004)	A dynamic model for the assessment of the replacement of lead in solders	Impact of the removal of lead from solders on the coupled flows of the metals Ag, Au, Bi, Cu, Fe, Ni, Pb, Sn, Zn, and PGMs	https://doi.org/10.1007/s11664-004-0100-3

The term life cycle analysis in the titles has been abbreviated to LCA.

- The recycled-content approach (also known as the cut-off or 100:0 approach) based on the principle of the first responsibility, which is a form of **partitioning**.
- The end-of-life recycling approach (also known as substitution, or 0:100, or the avoided or embodied burden approach) based on the principle of the last responsibility, which is used in the context of **substitution**.

The recycled-content approach means that a product has to carry the full environmental burden of the production of its primary material, even if it is subsequently recycled. The system boundary ends after scrap collection to account for the collection rate, and the burden of processing and recycling of waste materials is attributed to the subsequent, scrap consuming product system. When secondary material (e.g., secondary alloy) is used in the manufacturing stage of the life cycle, this will have a burden; it is only the (untreated, not-recycled) scrap that comes into the life cycle that is considered burden-free (Toro et al., 2021; Schrijvers et al., 2016a). This approach provides incentives to use recycled materials due to the lower burden, but provides no incentive to develop recyclable products, because there will be no benefit in terms of environmental impacts for the primary product designed for recycling in the LCA. Atherton and Davies (2006) caution that when the recycled material market is constrained, i.e., there is no excess recycled material available to be used in the process, the increased use of recycled material can lead to decreased use in another life cycle, thus increasing the overall footprint and primary material use. Detailed insight in the product and scrap markets is therefore required to understand the limitations of this method.

Other ways of partitioning are possible, i.e., partitioning the burden of primary material production between different life cycles: for example, the approach that the primary material burden is to be shared equally among all life cycles, cascade recycling approaches, economic partitioning approaches, or approaches that try to use the material quality as partitioning criterion (Schrijvers et al., 2016a). While there is currently no consensus on which of these methods is generally most suitable, the different options are widely acknowledged.

The end-of-life recycling approach gives a benefit to the product if a recyclable material is produced from the end-of-life product; i.e., it gets a credit based on the mass and quality of the recovered secondary material. This approach means that the environmental burden of the primary material production does not remain with the original user of that particular material because the burden is transferred to the subsequent users of the recycled material.

If then these recyclable materials are used as input to the manufacturing stage, then burdens equivalent to the EoL credit should be assigned to the material to avoid double counting. This type of modeling provides incentives for designing recyclable products, but limited incentives to use secondary materials. It may come with the risk that primary materials are used excessively if the recycling potential is overestimated.

The handling of recycling in LCA, especially in product-oriented LCA, is not always straightforward to implement (e.g., Nordelöf et al., 2019 for a critical review on Li-ion battery LCA). This is not only a challenge for metals but also for paper, for example (Hohenthal et al., 2019). Therefore product category rules (PCRs) are published to make comparison within a product category easier (European Commission, 2021). The PCRs describe system boundaries, allocation methods, and functional unit definitions. Another approach is to include the recycling processes, the return of materials, and final waste in the foreground system of the LCA (system expansion) (e.g., Bartie et al., 2021; Nordelöf et al., 2019). In this way a choice between cut-off and end-of-life recycling approaches can be avoided altogether.

47.6 RECYCLING CASES

To illustrate some of the concepts identified in this chapter, several cases are presented. The first case is a product-oriented LCA that illustrates that understanding of the recycled material market dynamics is important for the interpretation, and that a cradle-to-grave footprint is not always lower than a cradle-to-gate footprint. The second case shows the difference between allocation approaches and system expansion, shifting from a product-oriented LCA to a process-oriented LCA. The last case illustrates the nonlinearities in the impacts associated with the recycling of different crystalline silicon scraps in the life cycle.

47.6.1 North American Aluminum Beverage Can Recycling

This study of the Aluminum Association (Toro et al., 2021) assessed the life cycle of 1000 aluminum cans from cradle to grave, using the year 2016 as reference. It included extraction of bauxite, recycling and recovery of the used beverage cans, and all steps in between, excluding the filling of the cans, distribution, and drink consumption. An end-of-life recycling rate of 50.4% and a recycled metal content of 73%, including 50% from postconsumer sources and 23% from preconsumer sources, were used. Internal scrap from rolling mills was excluded. Both a cradle-to-gate footprint, up to and including the manufacturing of the empty can, and a cradle-to-grave footprint, including recycling, were determined. The cradle-to-gate global warming potential is 77.1 kg CO_2 eq., which is lower than the cradle-to-grave global warming potential of 96.8 kg CO_2 eq. This is surprising for beverage cans that are fully recycled at end of life. In this case, the end-of-life recycling rate, which is defined as the percentage of cans going to recycling, is lower than the recycled content of the material used for manufacturing of the can, so the system has a net scrap deficit. Key focus to reduce the cradle-to-grave footprint is increasing the collection rate of beverage cans, so that the percentage of cans going to landfill (49.6%, Toro et al., 2021) is strongly reduced.

47.6.2 CdTe Photovoltaic Module Recycling

Bartie et al. investigated the production and recycling of cadmium-telluride photovoltaic modules (Bartie et al., 2019, 2022; Abadías Llamas et al., 2020). A detailed description and flowsheet of the life cycle itself can be found in Chapter 14—PV and Wind energy. For comparison to the process simulation-based, process-oriented LCA (where closed loop recycling is modeled in the foreground system) and to highlight some of the challenges, the distribution of impacts between products in the CdTe system was calculated using different allocation factors (Table 47.3). The results are generally inconsistent.

TABLE 47.3 Percentage of total impact allocated to the system outputs: copper, zinc, lead, cadmium, tellurium, and PV modules using different allocation parameters.

Allocation parameter	Cu	Zn	Pb	Cd	Te	CdTe PV modules
Mass	41.7	29.7	14.0	0.097	0.011	14.4
Exergy cost	38.0	19.2	6.8	0.033	0.017	35.8
Exergy content	24.0	43.3	4.9	0.071	0.0078	27.7
Economic value[1]	62.9	16.4	6.2	0.055	0.22	14.1

[1] *Based on average commodity prices in 2020 and module average selling price ($0.345/W) in 2019.*
Modified from Bartie et al. (2022).

Subjective or arbitrary decisions would have to be made on the representativeness of reality, which generates an uncertain result that would likely carry low credibility.

The price of Cd is about 3% of that of Te. However, for the CdTe module to function, Cd is equally important to Te. In this case, mass-based allocation would be more appropriate. Looking at the overall value chain system, Cd and Te are by-products of Cu, Zn, and Pb production. The tonnage of Cd and Te produced is mainly driven by the demand for the base metals (Chapters 2 and 46) so economic value-based allocation could make sense, as the metal producer's objective is profit maximization. This especially affects the impact associated with Te. Because it is much more expensive than the other metals (in $/kg), more environmental impact is allocated to it. Even though 11 times more Cd than Te leaves the system as product, Te has 0.22% of the impacts compared to 0.055% for Cd.

47.7 C-SI SCRAP AND PV MODULE RECYCLING

Bartie et al. (2021, 2022) also investigated the production and recycling of crystalline silicon photovoltaic modules, described in Chapters 14 and 46. Within the c-Si PV module life cycle, two main scrap flows are generated: kerf residue during wafer cutting and end-of-life silicon wafers. The scraps reenter the life cycle at different places. Note that the recycling of these scraps is part of the foreground system of the LCA.

The kerf is recycled by returning it to the energy-intensive solar-grade silicon production process (Siemens process), and then proceeds through the Czochralski crystallization, wafering, and module manufacturing steps. In contrast, the silicon wafers obtained from end-of-life recycling are returned to the Czochralski crystallization step directly. The difference in processing route means that recycling 1 kg kerf has a different impact on the power consumption and CO_2 emissions of the entire life cycle compared to recycling 1 kg of silicon wafer from end-of-life modules. Figure 47.3 shows kerf and end-of-life recycling rate variability and the resulting nonlinearity.

47.8 CONCLUSIONS AND OUTLOOK

LCA is gaining even more importance as a tool to benchmark and assess progress on life cycle emission reductions, and as a tool to inform about the impact of decisions regarding the design and adaptation of products, processes, value chains, and systems on the environment.

Environmental LCA considers *all* impact categories, not only water or carbon footprint, avoiding "carbon tunnel vision" (a term coined by Jan Konietzko of Cognizant). The combination with Life Cycle Costing and social LCA to include all dimensions is important in the sustainability transition.

When doing a life cycle assessment that includes treatment of residues, recycling of end-of-life products, or the use of recycled materials in manufacturing, for example, it is essential to be aware of the complexity, multifunctionality, and nonlinearity in the system. Consequently:

- consciously defining the purpose, goal, and scope,
- deeply understanding and carefully making the LCA method selection and decisions,
- pursuing in-depth and physics-based data and system descriptions,
- thoroughly executing uncertainty and sensitivity analysis on model parameters, and
- actively seeking insight from experts to validate results

are crucial in obtaining meaningful insights that lead to actionable results.

LCA that uses a validated process and plant simulation model (Chapter 46) for the LCI offers

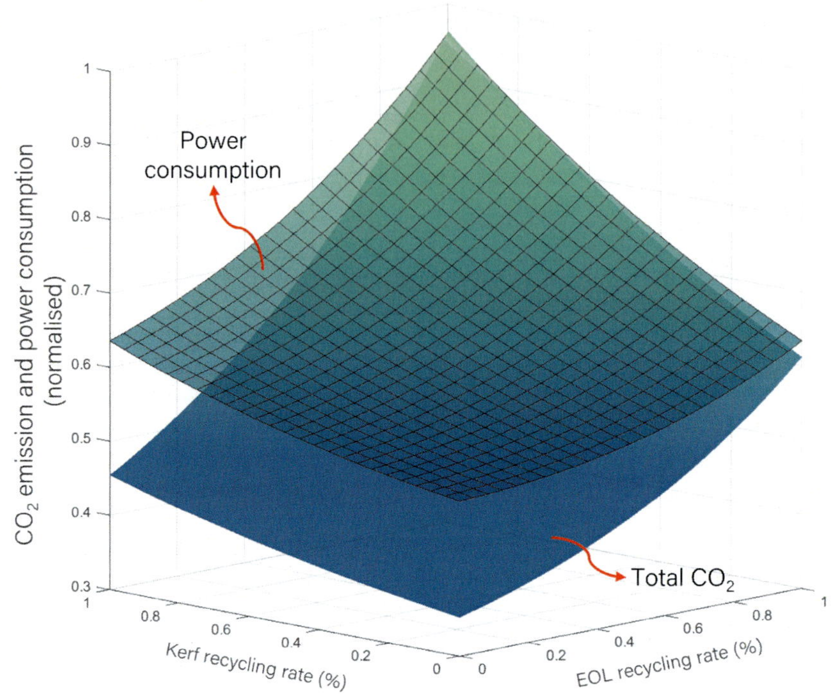

FIGURE 47.3 Variation of power consumption and CO_2 generation (normalized) as a function of recycling rate (Siemens process SG-Si and Cz mono-Si). *From Bartie et al. (2021).*

huge potential. With its roots in process engineering design, where high levels of physical, chemical, and thermodynamic detail, and compliance with the laws of conservation and entropy are nonnegotiable, a carefully developed and validated simulation model provides reliable mass, energy, and thermochemical information down to the element and compound level for entire product life cycles and circular economy systems. It captures the complex, nonlinear interactions that fundamentally characterize processing and recycling systems, enabling process optimization and sound, physics-based environmental impact assessments, see Reuter et al. (2023) for an example.

This combination of the two methods incorporated in software tools like GaBi, OpenLCA, and HSC opens new possibilities, as shown in this chapter. One not yet mentioned is the development of best-practice benchmark libraries for different metals and materials referring to different production routes to show the reduction potential between existing and best practices, while using the same methodology. This is very important for organizations that have committed to the Science Based Target Initiative (https://sciencebasedtargets.org/) for reduction of CO_2 emissions in the supply chain.

LCA literacy (Viere et al., 2021) and knowing one's own LCA literacy level is essential for the future credibility of LCA as a tool. LCA is a knowledge domain in itself that needs to collaborate with subject experts to be rooted in the detailed physical and industrial reality. Otherwise, it cannot provide the necessary, trusted guidance for the transition to circular economy and the SDGs.

References

Abadías Llamas, A., Bartie, N.J., Heibeck, M., Stelter, M., Reuter, M.A., 2020. Simulation-based exergy analysis of large circular economy systems: zinc production coupled to CdTe photovoltaic module life cycle. J. Sustain. Metall. 6 (1), 34–67. https://doi.org/10.1007/s40831-019-00255-5.

Allacker, K., Mathieux, F., Manfredi, S., Pelletier, N., de Camillis, C., Ardente, F., Pant, R., 2014. Allocation solutions for secondary material production and end of life recovery: proposals for product policy initiatives. Resour. Conserv. Recycl. 88, 1–12. https://doi.org/10.1016/j.resconrec.2014.03.016.

Ambrose, H., Kendall, A., 2020. Understanding the future of lithium: part 2, temporally and spatially resolved life-cycle assessment modeling. J. Ind. Ecol. 24 (1), 90–100. https://doi.org/10.1111/jiec.12942.

Arvesen, A., Luderer, G., Pehl, M., Bodirsky, B.L., Hertwich, E.G., 2018. Deriving life cycle assessment coefficients for application in integrated assessment modelling. Environ. Model. Softw. 99, 111–125. https://doi.org/10.1016/j.envsoft.2017.09.010.

Arvidsson, R., Tillman, A.M., Sandén, B.A., Janssen, M., Nordelöf, A., Kushnir, D., Molander, S., 2018. Environmental assessment of emerging technologies: recommendations for prospective LCA. J. Ind. Ecol. 22 (6), 1286–1294. https://doi.org/10.1111/jiec.12690.

Atherton, J., Davies, B., 2006. Materials stewardship—towards the sustainable use of minerals and metals. World Metall ERZMETALL 59 (1), 3–8.

Bartie, N.J., Llamas, A.A., Heibeck, M., Magnus Fröhling, O., Volkova, and M. A. Reuter., 2019. The simulation-based analysis of the resource efficiency of the circular economy—the enabling role of metallurgical infrastructure. Miner. Process. Extr. Metall.: Trans. Inst. Min. Metall. 129 (2), 229–249. https://doi.org/10.1080/25726641.2019.1685243.

Bartie, N.J., Cobos-Becerra, Y.L., Fröhling, M., Schlatmann, R., Reuter, M.A., 2021. The resources, exergetic and environmental footprint of the silicon photovoltaic circular economy: assessment and opportunities. Resour. Conserv. Recycl. 169 (January), 105516. https://doi.org/10.1016/j.resconrec.2021.105516.

Bartie, N.J., Cobos-Becerra, Y.L., Fröhling, M., Schlatmann, R., Reuter, M.A., 2022. Metallurgical infrastructure and technology criticality: the link between photovoltaics, sustainability, and the metals industry. Miner. Econ. 35 (3–4), 503–519. https://doi.org/10.1007/s13563-022-00313-7.

Ciroth, A., Muller, S., Weidema, B., Lesage, P., 2016. Empirically based uncertainty factors for the pedigree matrix in ecoinvent. Int. J. Life Cycle Assess. 21 (9), 1338–1348. https://doi.org/10.1007/s11367-013-0670-5.

European Commission, 2010. International Reference Life Cycle Data System (ILCD) Handbook General Guide for Life Cycle Assessment: Detailed Guidance, first ed. European Commission. Joint Research Centre. Institute for Environment and Sustainability. Publications Office of the European Union. https://eplca.jrc.ec.europa.eu/uploads/ILCD-Handbook-General-guide-for-LCA-DETAILED-GUIDANCE-12March2010-ISBN-fin-v1.0-EN.pdf.

European Commission, 2021. Commission Recommendation of 16.12.2021 on the Use of the Environmental Footprint Methods to Measure and Communicate the Life Cycle Environmental Performance of Products and Organisations. European Commission. https://ec.europa.eu/environment/publications/recommendation-use-environmental-footprint-methods_en.

European Union, 2020a. Proposal for a Regulation of the European Parliament and of the Council Concerning Batteries and Waste Batteries, Repealing Directive 2006/66/EC and Amending Regulation (EU) No 2019/1020, (2020). https://eur-lex.europa.eu/legal-content/EN/TXT/?uri=CELEX:52020PC0798.

European Union, 2020b. Regulation (EU) 2020/852 of the European Parliament and of the Council of 18 June 2020 on the Establishment of a Framework to Facilitate Sustainable Investment and Amending Regulation (EU) 2019/2088 (2020). http://data.europa.eu/eli/reg/2020/852/oj.

Frischknecht, R., 2010. LCI modelling approaches applied on recycling of materials in view of environmental sustainability, risk perception and eco-efficiency. Int. J. Life Cycle Assess. 15 (7), 666–671. https://doi.org/10.1007/s11367-010-0201-6.

GaBi 9., 2020. Software und Datenbank zur Ganzheitlichen Bilanzierung. Sphera Solutions GmbH, Leinfelden-Echterdingen, Germany. https://gabi.sphera.com. Accessed 10 June 2022.

Gibon, T., Wood, R., Arvesen, A., Bergesen, J.D., Suh, S., Hertwich, E.G., 2015. A methodology for integrated, multiregional life cycle assessment scenarios under large-scale technological change. Environ. Sci. Technol. 49 (18), 11218–11226. https://doi.org/10.1021/acs.est.5b01558.

Goedkoop, M., Heijungs, R., Huijbregts, M., De Schryver, A., Struijs, J., Van Zelm, R., 2009. ReCiPe 2008, first ed., Report I: Characterisation. Ministerie van Volkshuisvesting, Ruimtelijke Ordening & Milieubeheer, The Netherlands. https://web.universiteitleiden.nl/cml/ssp/publications/recipe_characterisation.pdf Accessed 27 June 2023.

Hauschild, M.Z., Rosenbaum, R.K., Olsen, S.I., 2018. In: Hauschild, M.Z., Rosenbaum, R.K., Olsen, S.I. (Eds.), Life Cycle Assessment. Springer International Publishing, https://doi.org/10.1007/978-3-319-56475-3.

Hohenthal, C., Leon, J., Dobon, A., Kujanpää, M., Meinl, G., Ringman, J., Hortal, M., Forsström, U., 2019. The ISO 14067 approach to open-loop recycling of paper products: making it operational. J. Clean. Prod. 224, 264–274. https://doi.org/10.1016/j.jclepro.2019.03.179.

ISO 14025:2010, 2010. Environmental Labels and Declarations—Type III Environmental Declarations—Principles and Procedures. ISO. www.iso.org.

ISO 14040:2006, 2006. Environmental Management Life Cycle Assessment Principles and Framework. ISO.

ISO 14040:2006/Amd.1:2020, 2020. Environmental Management-Life Cycle Assessment-Principles and Framework AMENDMENT 1. www.iso.org.

ISO 14044:2006, 2006. Environmental Management Life Cycle Assessment Requirements and Guidelines. ISO. www.iso.org.

ISO 14044:2006/Amd.1:2017, 2017. Environmental Management-Life Cycle Assessment-Requirements and Guidelines AMENDMENT 1. ISO. www.iso.org.

ISO 14044:2006/Amd.2:2020, 2020. Environmental Management-Life Cycle Assessment-Requirements and Guidelines AMENDMENT 2. ISO. www.iso.org.

ISO 14045:2012, 2012. Eco-Efficiency Assessment of Product Systems—Principles, Requirements and Guidelines. ISO. www.iso.org.

ISO 14046:2014, 2014. Environmental Management Water Footprint-Principles, Requirements and Guidelines. www.iso.org.

ISO 14067:2018, 2018. Greenhouse Gases-Carbon Footprint of Products-Requirements and Guidelines for Quantification. ISO. www.iso.org.

Kim, A., Mutel, C.L., Froemelt, A., Hellweg, S., 2022. Global sensitivity analysis of background life cycle inventories. Environ. Sci. Technol. 56 (9), 5874–5885. https://doi.org/10.1021/acs.est.1c07438.

Manjong, N.B., Usai, L., Burheim, O.S., Strømman, A.H., 2021. Life cycle modelling of extraction and processing of battery minerals—a parametric approach. Batteries 7 (3). https://doi.org/10.3390/batteries7030057.

Nordelöf, A., Poulikidou, S., Chordia, M., Bitencourt de Oliveira, F., Tivander, J., Arvidsson, R., 2019. Methodological approaches to end-of-life modelling in life cycle assessments of lithium-ion batteries. Batteries 5 (3), 51. https://doi.org/10.3390/batteries5030051.

Poulsen, P.B., Jensen, A.A., 2005. Working Environment in Life-Cycle Assessment. Society of Environmental Toxicology and Chemistry (SETAC). January 31. ISBN-13: 978-1880611685 https://www.setac.org/page/lca-books.

Rebitzer, G., Seuring, S., 2003. Methodology and application of life cycle costing. Int. J. Life Cycle Assess. 8 (2), 110–111. https://doi.org/10.1007/BF02978436.

Reuter, M.A., 2016. Digitalizing the circular economy: circular economy engineering defined by the metallurgical internet of things. Metall. Mater. Trans. B Process Metall. Mater. Process. Sci. 47 (6), 3194–3220. https://doi.org/10.1007/s11663-016-0735-5.

Reuter, M.A., Kojo, I.V., Roine, A., Florin, H., Jåfs, M., 2013. Environmental footprinting of metallurgical processing technology—linking GaBi to HSC Sim. In: Copper 2013. vol. 1. Chilean Institute of Mining Engineers (IMMCh), Santiago, Chile, pp. 373–388.

Reuter, M.A., van Schaik, A., Gediga, J., 2015. Simulation-based design for resource efficiency of metal production and recycling systems: cases—copper production and recycling, e-waste (LED lamps) and nickel pig Iron. Int. J. Life Cycle Assess. 20 (5), 671–693. https://doi.org/10.1007/s11367-015-0860-4.

Reuter, M.A., van Schaik, A., Gutzmer, J., Bartie, N., Llamas, A.A., 2019. Challenges of the circular economy: a material, metallurgical, and product design perspective. Annu. Rev. Mater. Res. 49 (1), 253–274. https://doi.org/10.1146/annurev-matsci-070218-010057.

Reuter, M.A., et al., 2023. Digital twin for KGHM-Legnica: simulation-based footprinting & exergy allocation of impacts. In: Antrekowitsch, H. (Ed.), Proceedings of EMC 2023, Dusseldorf.

Rödger, J.M., Kjær, L.L., Pagoropoulos, A., 2018. Life cycle costing: an introduction. In: Hauschild, M., Rosenbaum, R., Olsen, S. (Eds.), Life Cycle Assessment. Springer, Cham, https://doi.org/10.1007/978-3-319-56475-3_15.

Schrijvers, D.L., Loubet, P., Sonnemann, G., 2016a. Developing a systematic framework for consistent allocation in LCA. Int. J. Life Cycle Assess. 21 (7), 976–993. https://doi.org/10.1007/s11367-016-1063-3.

Schrijvers, D.L., Loubet, P., Sonnemann, G., 2016b. Critical review of guidelines against a systematic framework with regard to consistency on allocation procedures for recycling in LCA. Int. J. Life Cycle Assess. 21 (7), 994–1008. https://doi.org/10.1007/s11367-016-1069-x.

Schrijvers, D., Loubet, P., Sonnemann, G., 2020. Archetypes of goal and scope definitions for consistent allocation in LCA. Sustainability 12 (14), 5587. https://doi.org/10.3390/su12145587.

Stamp, A., Meskers, C.E.M., Reimer, M., Wäger, P., Althaus, H.J., Scholz, R.W., 2011. Challenges for LCAs of complex systems: the case of a large-scale precious metal refinery plant. In: Finkbeiner, M. (Ed.), Towards Life Cycle Sustainability Management. Springer, Dordrecht, https://doi.org/10.1007/978-94-007-1899-9_24.

Toro, M., Rybl, V., Koffler, C., 2021. Life Cycle Assessment of North American Aluminum Cans. Sphera and The Aluminum Association. https://www.aluminum.org/sites/default/files/2021-10/2021AluminumCanLCAReportFullVersion.pdf.

UNEP, 2020. In: Norris, C.B., Traverso, M., Neugebauer, S., Ekener, E., Schaubroeck, T., Garrido, S.R., Arcese, G. (Eds.), Guidelines for Social Life Cycle Assessment of Products and Organizations 2020. United Nations Environment Programme (UNEP). Retrieved June 10, 2022, from https://www.lifecycleinitiative.org/wp-content/uploads/2021/01/Guidelines-for-Social-Life-Cycle-Assessment-of-Products-and-Organizations-2020-22.1.21sml.pdf.

UNEP/SETAC Life Cycle Initiative, 2011. In: Valdivia, S., Ugaya, C.M.L., Sonnemann, G., Hildenbrand, J. (Eds.), Towards a Life Cycle Sustainability Assessment: Making Informed Choices on Products. https://www.lifecycleinitiative.org/wp-content/uploads/2012/12/2011%20-%20Towards%20LCSA.pdf.

Viere, T., Amor, B., Berger, N., Fanous, R.D., Arduin, R.H., Keller, R., Laurent, A., Loubet, P., Strothmann, P., Weyand, S., Wright, L., Sonnemann, G., 2021. Teaching life cycle assessment in higher education. Int. J. Life Cycle Assess. 26 (3), 511–527. https://doi.org/10.1007/s11367-020-01844-3.

Weidema, B.P., 1998. Multi-user test of the data quality matrix for product life cycle inventory data. Int. J. Life Cycle Assess. 3 (5), 259–265. https://doi.org/10.1007/BF02979832.

World Gold Council, 2019. Gold and Climate Change: Current and Future Impacts. 2019 World Gold Council. https://www.gold.org/goldhub/research/gold-and-climate-change-current-and-future-impacts.

Index

Note: Page numbers followed by *f* indicate figures and *t* indicate tables.

C

Q

R

T

Printed in the United States
by Baker & Taylor Publisher Services